# Instructor's Manual

Randy Gallaher
Kevin Bodden
Lewis and Clark Community College

# Intermediate Algebra

Michael Sullivan, III
Katherine R. Struve

Upper Saddle River, NJ 07458

Editor-in-Chief: Chris Hoag
Executive Editor: Paul Murphy
Senior Marketing Manager: Kate Valentine
Executive Project Manager: Ann Heath
Editorial Assistant: Abigail Rethore
Executive Managing Editor: Kathleen Schiaparelli
Senior Managing Editor: Nicole M. Jackson
Assistant Managing Editor: Karen Bosch Petrov
Production Editor: Ashley M. Booth
Supplement Cover Manager: Paul Gourhan
Supplement Cover Designer: Christopher Kossa
Manufacturing Buyer: Ilene Kahn
Manufacturing Manager: Alexis Heydt-Long

© 2007 Pearson Education, Inc.
Pearson Prentice Hall
Pearson Education, Inc.
Upper Saddle River, NJ 07458

All rights reserved. No part of this book may be reproduced in any form or by any means, without permission in writing from the publisher.

Pearson Prentice Hall™ is a trademark of Pearson Education, Inc.

The author and publisher of this book have used their best efforts in preparing this book. These efforts include the development, research, and testing of the theories and programs to determine their effectiveness. The author and publisher make no warranty of any kind, expressed or implied, with regard to these programs or the documentation contained in this book. The author and publisher shall not be liable in any event for incidental or consequential damages in connection with, or arising out of, the furnishing, performance, or use of these programs.

---

**This work is protected by United States copyright laws and is provided solely for the use of instructors in teaching their courses and assessing student learning. Dissemination or sale of any part of this work (including on the World Wide Web) will destroy the integrity of the work and is not permitted. The work and materials from it should never be made available to students except by instructors using the accompanying text in their classes. All recipients of this work are expected to abide by these restrictions and to honor the intended pedagogical purposes and the needs of other instructors who rely on these materials.**

---

Printed in the United States of America

10 9 8 7 6 5 4 3 2 1

ISBN 0-13-146776-X

Pearson Education Ltd., *London*
Pearson Education Australia Pty. Ltd., *Sydney*
Pearson Education Singapore, Pte. Ltd.
Pearson Education North Asia Ltd., *Hong Kong*
Pearson Education Canada, Inc., *Toronto*
Pearson Educación de Mexico, S.A. de C.V.
Pearson Education—Japan, *Tokyo*
Pearson Education Malaysia, Pte. Ltd.

# Table of Contents – Quick Check Solutions

**Chapter R  Real Numbers and Algebraic Expressions**
R.2   Sets and Classification of Numbers ................................................................................................. 1
R.3   Operations on Signed Numbers; Properties of Real Numbers ......................................................... 2
R.4   Order of Operations ........................................................................................................................... 4
R.5   Algebraic Expressions ....................................................................................................................... 5

**Chapter 1  Linear Equations and Inequalities**
1.1   Linear Equations ................................................................................................................................ 8
1.2   An Introduction to Problem Solving ................................................................................................ 12
1.3   Using Formulas to Solve Problems ................................................................................................. 15
1.4   Linear Inequalities ........................................................................................................................... 17
1.5   Compound Inequalities .................................................................................................................... 20
1.6   Absolute Value Equations and Inequalities .................................................................................... 22

**Chapter 2  Graphs, Relations, and Functions**
2.1   Rectangular Coordinates and Graphs of Equations ........................................................................ 27
2.2   Relations .......................................................................................................................................... 29
2.3   An Introduction to Functions ........................................................................................................... 30
2.4   Functions and Their Graphs ............................................................................................................ 32

**Chapter 3  Linear Functions and Their Graphs**
3.1   Linear Equations and Linear Functions .......................................................................................... 34
3.2   Slope and Equations of Lines .......................................................................................................... 36
3.3   Parallel and Perpendicular Lines ..................................................................................................... 39
3.4   Linear Inequalities in Two Variables .............................................................................................. 41
3.5   Building Linear Models .................................................................................................................. 42

**Chapter 4  Systems of Equations and Inequalities**
4.1   Systems of Linear Equations in Two Variables .............................................................................. 45
4.2   Problem Solving: Systems of Linear Equations ............................................................................. 47
4.3   Systems of Linear Equations in Three Variables ........................................................................... 49
4.4   Using Matrices to Solve Systems .................................................................................................... 51
4.5   Determinants and Cramer's Rule ..................................................................................................... 53
4.6   Systems of Linear Inequalities ........................................................................................................ 54

**Getting Ready for Chapter 5: Polynomials-Integer Exponents** ................................................................ 57

**Chapter 5  Polynomials and Polynomial Functions**
5.1   Adding and Subtracting Polynomials .............................................................................................. 61
5.2   Multiplication of Polynomials ......................................................................................................... 63
5.3   Division of Polynomials; Synthetic Division ................................................................................. 64
5.4   Greatest Common Factor; Factoring by Grouping ......................................................................... 66
5.5   Factoring Trinomials ....................................................................................................................... 67
5.6   Factoring Special Products .............................................................................................................. 70
5.7   Factoring: A General Strategy ........................................................................................................ 71
5.8   Polynomial Equations ...................................................................................................................... 72

**Getting Ready for Chapter 6: Rational Expressions-A Review of Operations on Rational Numbers**........75

**Chapter 6  Rational Expressions and Rational Functions**
6.1   Multiplying and Dividing Rational Expressions..................................................................76
6.2   Adding and Subtracting Rational Expressions ...................................................................77
6.3   Complex Rational Expressions............................................................................................79
6.4   Rational Equations...............................................................................................................80
6.5   Rational Inequalities ............................................................................................................82
6.6   Models Involving Rational Expressions .............................................................................83

**Getting Ready for Chapter 7: Radicals and Rational Exponents-Square Roots and $n$th Roots** .................86

**Chapter 7  Radicals and Rational Exponents**
7.1   $n$th Roots and Rational Exponents......................................................................................87
7.2   Simplify Expressions Using the Laws of Exponents ..........................................................88
7.3   Simplifying Radical Expressions.........................................................................................89
7.4   Adding, Subtracting, and Multiplying Radical Expressions...............................................90
7.5   Rationalizing Radical Expressions ......................................................................................91
7.6   Functions Involving Radicals ..............................................................................................92
7.7   Radical Equations and Their Applications .........................................................................93
7.8   The Complex Number System.............................................................................................96

**Chapter 8  Quadratic Equations and Functions**
8.1   Solving Quadratic Equations by Completing the Square....................................................98
8.2   Solving Quadratic Equations by the Quadratic Formula ....................................................99
8.3   Solving Equations Quadratic in Form ...............................................................................102
8.4   Graphing Quadratic Functions Using Transformations ....................................................105
8.5   Graphing Quadratic Functions Using Properties ..............................................................107
8.6   Quadratic Inequalities.........................................................................................................110

**Chapter 9  Exponential and Logarithmic Functions**
9.1   Composite Functions and Inverse Functions ....................................................................113
9.2   Exponential Functions .......................................................................................................114
9.3   Logarithmic Functions .......................................................................................................116
9.4   Properties of Logarithms ...................................................................................................118
9.5   Exponential and Logarithmic Equations............................................................................119

**Chapter 10  Conics**
10.1  Distance and Midpoint Formulas......................................................................................121
10.2  Circles................................................................................................................................122
10.3  Parabolas............................................................................................................................122
10.4  Ellipses ..............................................................................................................................125
10.5  Hyperbolas.........................................................................................................................126
10.6  Nonlinear Systems of Equations.......................................................................................128

**Chapter 11  Sequences, Series, and The Binomial Theorem**
11.1  Sequences ..........................................................................................................................131
11.2  Arithmetic Sequences........................................................................................................132
11.3  Geometric Sequences and Series ......................................................................................133
11.4  The Binomial Theorem......................................................................................................135

# Table of Contents – Instructor Solutions Manual

**Chapter R  Real Numbers and Algebraic Expressions**
R.1   Success in Mathematics ................................................................................................................. 1
R.2   Sets and Classification of Numbers ............................................................................................... 1
R.3   Operations on Signed Numbers; Properties of Real Numbers ....................................................... 2
R.4   Order of Operations ........................................................................................................................ 6
R.5   Algebraic Expressions .................................................................................................................... 8

**Chapter 1  Linear Equations and Inequalities**
1.1   Linear Equations ........................................................................................................................... 12
1.2   An Introduction to Problem Solving ............................................................................................ 20
1.3   Using Formulas to Solve Problems .............................................................................................. 25
1.4   Linear Inequalities ........................................................................................................................ 28
       *Putting the Concepts Together (Sections 1.1 – 1.4)* ................................................................... 34
1.5   Compound Inequalities ................................................................................................................ 36
1.6   Absolute Value Equations and Inequalities ................................................................................. 42
       *Chapter 1 Review* ....................................................................................................................... 47
       *Chapter 1 Test* ............................................................................................................................. 60
       *Cumulative Review Chapters R – 1* ........................................................................................... 63

**Chapter 2  Graphs, Relations, and Functions**
2.1   Rectangular Coordinates and Graphs of Equations ..................................................................... 66
2.2   Relations ....................................................................................................................................... 72
       *Putting the Concepts Together (Sections 2.1 – 2.2)* ................................................................... 77
2.3   An Introduction to Functions ....................................................................................................... 79
2.4   Functions and Their Graphs ......................................................................................................... 83
       *Chapter 2 Review* ....................................................................................................................... 85
       *Chapter 2 Test* ............................................................................................................................. 96

**Chapter 3  Linear Functions and Their Graphs**
3.1   Linear Equations and Linear Functions ....................................................................................... 99
3.2   Slope and Equations of Lines ..................................................................................................... 107
3.3   Parallel and Perpendicular Lines ................................................................................................ 114
       *Putting the Concepts Together   (Sections 3.1 – 3.3)* ............................................................... 118
3.4   Linear Inequalities in Two Variables ......................................................................................... 121
3.5   Building Linear Models ............................................................................................................. 126
       *Chapter 3 Review* ..................................................................................................................... 130
       *Chapter 3 Test* ........................................................................................................................... 145
       *Cumulative Review Chapters R – 3* ......................................................................................... 149

**Chapter 4  Systems of Equations and Inequalities**
4.1   Systems of Linear Equations in Two Variables ......................................................................... 153
4.2   Problem Solving: Systems of Linear Equations ........................................................................ 160
4.3   Systems of Linear Equations in Three Variables ...................................................................... 165
       *Putting the Concepts Together   (Sections 4.1 – 4.4)* ............................................................... 173
4.4   Using Matrices to Solve Systems ............................................................................................... 175
4.5   Determinants and Cramer's Rule ............................................................................................... 183
4.6   Systems of Linear Inequalities ................................................................................................... 190
       *Chapter 4 Review* ..................................................................................................................... 198
       *Chapter 4 Test* ........................................................................................................................... 222

**Getting Ready for Chapter 5: Polynomials-Integer Exponents** ..................................................................... 228

**Chapter 5  Polynomials and Polynomial Functions**
5.1   Adding and Subtracting Polynomials ........................................................................................... 234
5.2   Multiplication of Polynomials ....................................................................................................... 238
5.3   Division of Polynomials; Synthetic Division .............................................................................. 243
        *Putting the Concepts Together  (Sections 5.1 – 5.3)* ................................................................. 249
5.4   Greatest Common Factor; Factoring by Grouping ..................................................................... 250
5.5   Factoring Trinomials ...................................................................................................................... 253
5.6   Factoring Special Products ............................................................................................................ 263
5.7   Factoring: A General Strategy ....................................................................................................... 267
5.8   Polynomial Equations .................................................................................................................... 270
        *Chapter 5 Review* ........................................................................................................................... 278
        *Chapter 5 Test* ................................................................................................................................. 296
        *Cumulative Review Chapters R – 5* .............................................................................................. 298

**Getting Ready for Chapter 6: Rational Expressions-A Review of Operations on Rational Numbers** ....... 303

**Chapter 6  Rational Expressions and Rational Functions**
6.1   Multiplying and Dividing Rational Expressions ........................................................................ 305
6.2   Adding and Subtracting Rational Expressions ........................................................................... 311
6.3   Complex Rational Expressions ..................................................................................................... 317
        *Putting the Concepts Together  (Sections 6.1 – 6.3)* ................................................................. 325
6.4   Rational Equations .......................................................................................................................... 327
6.5   Rational Inequalities ....................................................................................................................... 335
6.6   Models Involving Rational Expressions ...................................................................................... 341
        *Chapter 6 Review* ........................................................................................................................... 347
        *Chapter 6 Test* ................................................................................................................................. 369

**Getting Ready for Chapter 7: Radicals and Rational Exponents-Square Roots and $n$th Roots** ............. 373

**Chapter 7  Radicals and Rational Exponents**
7.1   $n$th Roots and Rational Exponents ............................................................................................ 375
7.2   Simplify Expressions Using the Laws of Exponents .................................................................. 377
7.3   Simplifying Radical Expressions .................................................................................................. 380
7.4   Adding, Subtracting, and Multiplying Radical Expressions .................................................... 384
7.5   Rationalizing Radical Expressions ............................................................................................... 389
        *Putting the Concepts Together (Sections 7.1 – 7.5)* ................................................................. 394
7.6   Functions Involving Radicals ....................................................................................................... 395
7.7   Radical Equations and Their Applications ................................................................................. 403
7.8   The Complex Number System ...................................................................................................... 412
        *Chapter 7 Review* ........................................................................................................................... 418
        *Chapter 7 Test* ................................................................................................................................. 433
        *Cumulative Review Chapters R – 7* .............................................................................................. 435

**Chapter 8  Quadratic Equations and Functions**
8.1   Solving Quadratic Equations by Completing the Square .......................................................... 440
8.2   Solving Quadratic Equations by the Quadratic Formula .......................................................... 447
8.3   Solving Equations Quadratic in Form ......................................................................................... 459
        *Putting the Concepts Together  (Sections 8.1 – 8.3)* ................................................................. 476
8.4   Graphing Quadratic Functions Using Transformations ........................................................... 479
8.5   Graphing Quadratic Functions Using Properties ...................................................................... 490
8.6   Quadratic Inequalities .................................................................................................................... 506
        *Chapter 8 Review* ........................................................................................................................... 515
        *Chapter 8 Test* ................................................................................................................................. 544

## Chapter 9  Exponential and Logarithmic Functions
- 9.1 Composite Functions and Inverse Functions .................................................................. 550
- 9.2 Exponential Functions ...................................................................................................... 558
- 9.3 Logarithmic Functions ...................................................................................................... 567
  - *Putting the Concepts Together (Sections 9.1 – 9.3)* ..................................................... 574
- 9.4 Properties of Logarithms .................................................................................................. 576
- 9.5 Exponential and Logarithmic Equations .......................................................................... 579
  - *Chapter 9 Review* ........................................................................................................... 587
  - *Chapter 9 Test* ................................................................................................................ 598
  - *Cumulative Review Chapters R – 9* .............................................................................. 601

## Chapter 10  Conics
- 10.1 Distance and Midpoint Formulas ..................................................................................... 606
- 10.2 Circles ................................................................................................................................ 609
- 10.3 Parabolas ........................................................................................................................... 615
- 10.4 Ellipses .............................................................................................................................. 624
- 10.5 Hyperbolas ........................................................................................................................ 634
  - *Putting the Concepts Together (Sections 10.1 – 10.5)* ................................................. 640
- 10.6 Nonlinear Systems of Equations ...................................................................................... 643
  - *Chapter 10 Review* ......................................................................................................... 653
  - *Chapter 10 Test* .............................................................................................................. 672

## Chapter 11  Sequences, Series, and The Binomial Theorem
- 11.1 Sequences .......................................................................................................................... 678
- 11.2 Arithmetic Sequences ....................................................................................................... 682
- 11.3 Geometric Sequences and Series ...................................................................................... 687
  - *Putting the Concepts Together (Sections 11.1 – 11.3)* ................................................. 694
- 11.4 The Binomial Theorem ..................................................................................................... 696
  - *Chapter 11 Review* ......................................................................................................... 699
  - *Chapter 11 Test* .............................................................................................................. 708
  - *Cumulative Review Chapters R – 11* ............................................................................ 710

**Appendix: The Library of Functions** .......................................................................................... 715

# Preface

This Instructor Solutions Manual (ISM) supplements *Intermediate Algebra* by Michael Sullivan, III and Katherine Struve. We hope it will be a valuable resource as you teach your Intermediate Algebra course. This manual consists of two parts. The first portion provides detailed solutions to the Quick Check exercises that are embedded within each section of the textbook. The second portion contains detailed solutions to all other exercises, including the Exercise Sets, Preparing for Sections, Putting the Concepts Together, Chapter Reviews, Chapter Tests, and Cumulative Reviews. For exercises requiring calculators, screenshots from TI-84 Plus graphing calculators have been used. We have made a concerted effort to make this manual easy to read and follow. For each section, problems have been worked using the same methods as the textbook examples. We have also tried to make this manual as error free as possible. If you have suggestions or corrections, please feel free to email us at the addresses below.

We would like to thank Jenny Crawford and Sarah Street for their attention to detail as they accuracy checked the final manuscript for this manual. We would also like to thank Bill Bodden (Kevin's father) and Zeno Forbes (Randy's father-in-law) for checking the early page proofs as the manual evolved.

We offer special thanks Ann Heath at prentice Hall for all her prompt help, often under hectic circumstances. You organization skills are inspiring!

Most importantly, we would like to thank our wives (Angie Bodden and Karen Gallaher) and our children (Shawn, Payton and Logan Bodden, and Ethan, Ben and Annie Gallaher) for their patience and support. Thanks for putting up with us when work had to come first and for enduring the many late hours of typing.

Randy Gallaher and Kevin Bodden
Department of Mathematics
Lewis & Clark Community College
5800 Godfrey Road
Godfrey, IL 62035
rgallahe@lc.edu   kbodden@lc.edu

# Chapter R – Quick Checks

**R.2 Quick Checks**

1. We will let $D$ represent the set of all digits that are less than 5.
   set-builder: $D = \{x \mid x \text{ is a digit less than } 5\}$
   roster: $\{0,1,2,3,4\}$

2. We will let $G$ represent the set of all digits that are greater than or equal to 6.
   set-builder:
   $G = \{x \mid x \text{ is a digit greater than or equal to } 6\}$
   roster: $\{6,7,8,9\}$

3. The statement $B \subseteq A$ is true because all the elements that are in $B$ are also in $A$.

4. The statement $B = C$ is false because there are elements in $B$ ($a$ and $b$) that are not in $C$. There is also an element in $C$, namely $d$, that is not in $B$.

5. The statement $B \subset D$ is false. In order for $B$ to be a proper subset of $D$, it must be the case that all the elements that are in $B$ are also in $D$. In addition, there must be elements in $D$ that are not in $B$. Since $B = D$, this is not the case.

6. The statement $\varnothing \subseteq A$ is true. The empty set is a subset of every set.

7. The statement $5 \in \{0,1,2,3,4,5\}$ is true because 5 is an element of the set.

8. The statement Michigan $\notin$ {Illinois, Indiana, Michigan, Wisconsin} is false because Michigan **is** an element of the set.

9. The statement $\frac{8}{3} \in \{x \mid x = \frac{p}{q} \text{ where } p \text{ and } q \text{ are digits, } q \neq 0\}$ is true because $\frac{8}{3}$ is of the form $\frac{p}{q}$ where $p = 8$ and $q = 3$.

10. 10 and $\frac{12}{4} = 3$ are the only counting numbers.

11. $10, \frac{0}{3} = 0$, and $\frac{12}{4} = 3$ are the whole numbers.

12. $-9, 10, \frac{0}{3} = 0$, and $\frac{12}{4} = 3$ are the integers.

13. $\frac{7}{3}, -9, 10, 4.\overline{56}, \frac{0}{3}, -\frac{4}{7}$, and $\frac{12}{4}$ are the rational numbers.

14. $5.7377377737777\ldots$ and $\pi$ are the irrational numbers. Note that $5.7377377737777\ldots$ is irrational because the decimal portion is non-terminating and non-repeating.

15. All numbers listed except $\sqrt{-4}$ are real numbers. There is no real number whose square is $-4$.

16. a. To truncate, we remove all digits after the third decimal place. So, 5.694392 truncated to three decimal places is 5.694.

    b. To round to three decimal places, we first examine the digit in the fourth decimal place and then truncate. Since the fourth decimal place contains a 3, we do not change the digit in the third decimal place. Thus, 5.694392 rounded to three decimal places is 5.694.

17. a. To truncate, we remove all digits after the second decimal place. So, $-4.9369102$ truncated to two decimal places is $-4.93$.

    b. To round to two decimal places, we first examine the digit in the third decimal place and then truncate. Since the third decimal place contains a 6, we increase the digit in the second decimal place by 1. Thus, $-4.9369102$ rounded to two decimal places is $-4.94$.

18. $\{-2, 0, \frac{1}{2}, 3, 3.5\}$

19. $3 < 6$ because 3 is further to the left on a number line.

20. $-3 < -2$ because $-3$ is further to the left on a number line.

## Chapter R: Real Numbers and Algebraic Expressions

21. $\frac{2}{3} = 0.\overline{6}$ and $\frac{1}{2} = 0.5$.

    $\frac{2}{3} > \frac{1}{2}$ because $\frac{2}{3}$ is further to the right on a number line.

22. $\frac{5}{7} \approx 0.714$.

    $\frac{5}{7} > 0.7$ because $\frac{5}{7}$ is further to the right on a number line.

23. $\frac{2}{3} = 0.\overline{6}$ and $\frac{10}{15} = 0.\overline{6}$.

    $\frac{2}{3} = \frac{10}{15}$ because the numbers are the same.

### R.3 Quick Checks

1. $|6| = 6$ because the distance from 0 to 6 on the real number line is 6 units.

2. $|-10| = 10$ because the distance from 0 to $-10$ on the real number line is 10 units.

3. $18 + (-6)$
   One number is positive, while the other is negative. We determine the absolute value of each number: $|18| = 18$ and $|-6| = 6$. We now subtract the smaller absolute value from the larger and obtain $18 - 6 = 12$. Because the number with the larger absolute value is positive, the sum will be positive. So,
   $18 + (-6) = 12$.

4. $-21 + 10$
   One number is positive, while the other is negative. We determine the absolute value of each number: $|-21| = 21$ and $|10| = 10$. We now subtract the smaller absolute value from the larger and obtain $21 - 10 = 11$. Because the number with the larger absolute value is negative, the sum will be negative. So,
   $-21 + 10 = -11$.

5. $-5.4 + (-1.2)$
   Both numbers are negative, so we first determine their absolute values: $|-5.4| = 5.4$ and $|-1.2| = 1.2$. We now add the absolute values and obtain $5.4 + 1.2 = 6.6$. Because both numbers to be added are negative, the sum is also negative. So, $-5.4 + (-1.2) = -6.6$.

6. $-6.5 + 4.3$
   One number is positive, while the other is negative. We determine the absolute value of each number: $|-6.5| = 6.5$ and $|4.3| = 4.3$. We now subtract the smaller absolute value from the larger and obtain $6.5 - 4.3 = 2.2$. Because the number with the larger absolute value is negative, the sum will be negative. So,
   $-6.5 + 4.3 = -2.2$.

7. $-9 + 9$
   One number is positive, while the other is negative. We determine the absolute value of each number: $|-9| = 9$ and $|9| = 9$. Since the absolute values are the same, the difference will be 0. So, $-9 + 9 = 0$.

8. The additive inverse of 5 is $-5$ because $5 + (-5) = 0$.

9. The additive inverse of $\frac{4}{5}$ is $-\frac{4}{5}$ because $\frac{4}{5} + \left(-\frac{4}{5}\right) = 0$.

10. The additive inverse of $-12$ is $-(-12) = 12$ because $-12 + 12 = 0$.

11. The additive inverse of $-\frac{5}{3}$ is $-\left(-\frac{5}{3}\right) = \frac{5}{3}$ because $-\frac{5}{3} + \frac{5}{3} = 0$.

12. $6 - 2 = 6 + (-2) = 4$

13. $4 - 13 = 4 + (-13) = -9$

14. $-3 - 8 = -3 + (-8) = -11$

15. $12.5 - 3.4 = 12.5 + (-3.4) = 9.1$

**Section R.3 – Quick Checks:** Operations on Signed Numbers; Properties of Real Numbers

16. $-8.5-(-3.4) = -8.5+3.4 = -5.1$

17. $-6.9-9.2 = -6.9+(-9.2) = -16.1$

18. $-6 \cdot (8) = -(6 \cdot 8) = -48$

19. $12 \cdot (-5) = -(12 \cdot 5) = -60$

20. $4 \cdot 14 = 56$

21. $-7 \cdot (-15) = 7 \cdot 15 = 105$

22. $-1.9 \cdot (-2.7) = 1.9 \cdot 2.7 = 5.13$

23. The multiplicative inverse of 10 is $\dfrac{1}{10}$.

24. The multiplicative inverse of $-8$ is $-\dfrac{1}{8}$.

25. The multiplicative inverse of $\dfrac{2}{5}$ is $\dfrac{5}{2}$.

26. The multiplicative inverse of $-\dfrac{1}{5}$ is $-\dfrac{5}{1} = -5$.

27. $\dfrac{2 \cdot 6}{2 \cdot 5} = \dfrac{\cancel{2} \cdot 6}{\cancel{2} \cdot 5} = \dfrac{6}{5}$

28. $\dfrac{-5 \cdot 9}{-2 \cdot 5} = \dfrac{-(5 \cdot 9)}{-(2 \cdot 5)} = \dfrac{5 \cdot 9}{2 \cdot 5} = \dfrac{\cancel{5} \cdot 9}{2 \cdot \cancel{5}} = \dfrac{9}{2}$

29. $\dfrac{25}{15} = \dfrac{5 \cdot 5}{3 \cdot 5} = \dfrac{5 \cdot \cancel{5}}{3 \cdot \cancel{5}} = \dfrac{5}{3}$

30. $\dfrac{-24}{20} = -\dfrac{6 \cdot 4}{5 \cdot 4} = -\dfrac{6 \cdot \cancel{4}}{5 \cdot \cancel{4}} = -\dfrac{6}{5}$

31. $\dfrac{2}{3} \cdot \left(-\dfrac{5}{4}\right) = -\dfrac{2 \cdot 5}{3 \cdot 4}$
$= -\dfrac{2 \cdot 5}{3 \cdot 2 \cdot 2}$
$= -\dfrac{\cancel{2} \cdot 5}{3 \cdot \cancel{2} \cdot 2}$
$= -\dfrac{5}{3 \cdot 2}$
$= -\dfrac{5}{6}$

32. $\dfrac{5}{3} \cdot \dfrac{12}{25} = \dfrac{5 \cdot 12}{3 \cdot 25}$
$= \dfrac{5 \cdot 3 \cdot 4}{3 \cdot 5 \cdot 5}$
$= \dfrac{\cancel{5} \cdot \cancel{3} \cdot 4}{\cancel{3} \cdot \cancel{5} \cdot 5}$
$= \dfrac{4}{5}$

33. $\dfrac{4}{3} \div \dfrac{8}{3} = \dfrac{4}{3} \cdot \dfrac{3}{8}$
$= \dfrac{4 \cdot 3}{3 \cdot 8}$
$= \dfrac{\cancel{4} \cdot \cancel{3}}{\cancel{3} \cdot \cancel{4} \cdot 2}$
$= \dfrac{1}{2}$

34. $\dfrac{\frac{10}{3}}{\frac{5}{12}} = \dfrac{10}{3} \cdot \dfrac{12}{5}$
$= \dfrac{10 \cdot 12}{3 \cdot 5}$
$= \dfrac{2 \cdot \cancel{5} \cdot \cancel{3} \cdot 4}{\cancel{3} \cdot \cancel{5}}$
$= 8$

35. $\dfrac{3}{11} + \dfrac{2}{11} = \dfrac{3+2}{11} = \dfrac{5}{11}$

36. $\dfrac{8}{15} - \dfrac{13}{15} = \dfrac{8-13}{15} = \dfrac{-5}{15} = -\dfrac{\cancel{5}}{3 \cdot \cancel{5}} = -\dfrac{1}{3}$

37. $\dfrac{3}{7} + \dfrac{8}{7} = \dfrac{3+8}{7} = \dfrac{11}{7}$

38. $\dfrac{8}{5} - \dfrac{3}{5} = \dfrac{8-3}{5} = \dfrac{5}{5} = 1$

39. $20 = 2 \cdot 2 \cdot 5$
$15 = 3 \cdot 5$
The common factor is 5. The remaining factors are 2, 2, and 3.
$LCD = 2 \cdot 2 \cdot 3 \cdot 5 = 60$
$\dfrac{3}{20} = \dfrac{3}{20} \cdot \dfrac{3}{3} = \dfrac{3 \cdot 3}{20 \cdot 3} = \dfrac{9}{60}$
$\dfrac{2}{15} = \dfrac{2}{15} \cdot \dfrac{4}{4} = \dfrac{2 \cdot 4}{15 \cdot 4} = \dfrac{8}{60}$

## Chapter R: Real Numbers and Algebraic Expressions

**40.** $18 = 2 \cdot 3 \cdot 3$
$45 = 3 \cdot 3 \cdot 5$
The common factors are 3 and 3. The remaining factors are 2 and 5.
$LCD = 3 \cdot 3 \cdot 2 \cdot 5 = 90$
$\dfrac{5}{18} = \dfrac{5}{18} \cdot \dfrac{5}{5} = \dfrac{5 \cdot 5}{18 \cdot 5} = \dfrac{25}{90}$
$-\dfrac{1}{45} = -\dfrac{1}{45} \cdot \dfrac{2}{2} = -\dfrac{1 \cdot 2}{45 \cdot 2} = -\dfrac{2}{90}$

**41.** $LCD = 2 \cdot 2 \cdot 5 \cdot 3 = 60$
$\dfrac{3}{20} + \dfrac{2}{15} = \dfrac{3}{20} \cdot \dfrac{3}{3} + \dfrac{2}{15} \cdot \dfrac{4}{4}$
$= \dfrac{9}{60} + \dfrac{8}{60}$
$= \dfrac{9+8}{60}$
$= \dfrac{17}{60}$

**42.** $LCD = 2 \cdot 7 \cdot 3 = 42$
$\dfrac{5}{14} - \dfrac{11}{21} = \dfrac{5}{14} + \left(\dfrac{-11}{21}\right)$
$= \dfrac{5}{14} \cdot \dfrac{3}{3} + \left(\dfrac{-11}{21} \cdot \dfrac{2}{2}\right)$
$= \dfrac{15}{42} + \left(\dfrac{-22}{42}\right)$
$= \dfrac{15 + (-22)}{42}$
$= \dfrac{-7}{42} = \dfrac{-1 \cdot 7}{6 \cdot 7}$
$= -\dfrac{1}{6}$

**43.** $LCD = 5 \cdot 5 \cdot 6 = 150$
$\dfrac{-4}{25} - \dfrac{7}{30} = \dfrac{-4}{25} + \left(\dfrac{-7}{30}\right)$
$= \dfrac{-4}{25} \cdot \dfrac{6}{6} + \left(\dfrac{-7}{30} \cdot \dfrac{5}{5}\right)$
$= \dfrac{-24}{150} + \left(\dfrac{-35}{150}\right)$
$= \dfrac{-24 + (-35)}{150}$
$= -\dfrac{59}{150}$

**44.** $LCD = 2 \cdot 3 \cdot 3 \cdot 5 = 90$
$-\dfrac{5}{18} - \dfrac{1}{45} = \dfrac{-5}{18} + \left(\dfrac{-1}{45}\right)$
$= \dfrac{-5}{18} \cdot \dfrac{5}{5} + \left(\dfrac{-1}{45} \cdot \dfrac{2}{2}\right)$
$= \dfrac{-25}{90} + \left(\dfrac{-2}{90}\right)$
$= \dfrac{-25 + (-2)}{90}$
$= \dfrac{-27}{90}$
$= -\dfrac{3 \cdot 9}{10 \cdot 9}$
$= -\dfrac{3}{10}$

**45.** $5(x+3) = 5 \cdot x + 5 \cdot 3$
$= 5x + 15$

**46.** $-6(x+1) = -6 \cdot x + (-6) \cdot 1$
$= -6x - 6$

**47.** $-4(z-8) = -4 \cdot z - (-4) \cdot 8$
$= -4z + 32$

**48.** $\dfrac{1}{3}(6x+9) = \dfrac{1}{3} \cdot 6x + \dfrac{1}{3} \cdot 9$
$= \dfrac{6x}{3} + \dfrac{9}{3}$
$= 2x + 3$

### R.4 Quick Checks

**1.** $4^3 = 4 \cdot 4 \cdot 4 = 64$

**2.** $(-7)^2 = (-7) \cdot (-7) = 49$

**3.** $(-10)^3 = (-10) \cdot (-10) \cdot (-10) = -1000$

**4.** $\left(\dfrac{2}{3}\right)^3 = \dfrac{2}{3} \cdot \dfrac{2}{3} \cdot \dfrac{2}{3} = \dfrac{8}{27}$

**5.** $-8^2 = -(8 \cdot 8) = -64$

**6.** $-(-5)^3 = -\left[(-5) \cdot (-5) \cdot (-5)\right] = -(-125) = 125$

**Section R.5 – Quick Checks:** Algebraic Expressions

7. $5 \cdot 2 + 6 = 10 + 6 = 16$

8. $3 \cdot 2 + 5 \cdot 6 = 6 + 30 = 36$

9. $4 \cdot (5+3) = 4 \cdot 8 = 32$

10. $8 \cdot (9-3) = 8 \cdot 6 = 48$

11. $(12-4) \cdot (18-13) = 8 \cdot 5 = 40$

12. $(4+9) \cdot (6-4) = 13 \cdot 2 = 26$

13. $\dfrac{3+7}{4+9} = \dfrac{10}{13}$

14. $1 - 4 + 8 \cdot 2 + 5 = 1 - 4 + 16 + 5$
    $= 22 - 4$
    $= 18$

15. $25 \cdot [2(8-3)] = 25 \cdot [2(5)]$
    $= 25[10]$
    $= 250$

16. $\dfrac{3+5}{2 \cdot (9-4)} = \dfrac{8}{2 \cdot (5)} = \dfrac{8}{10} = \dfrac{4}{5}$

17. $6 + 5 \cdot 2 = 6 + 10 = 16$

18. $(3+9) \cdot 4 = 12 \cdot 4 = 48$

19. $\dfrac{3+7}{4+5} = \dfrac{10}{9}$

20. $\dfrac{7+5}{4+10} = \dfrac{12}{14} = \dfrac{6}{7}$

21. $4 + [(8-3) \cdot 2] = 4 + [5 \cdot 2]$
    $= 4 + 10$
    $= 14$

22. $[4 \cdot (6-2) - 9] = 4 \cdot 4 - 9$
    $= 16 - 9$
    $= 7$

23. $-8 + 2 \cdot 5^2 = -8 + 2 \cdot 25$
    $= -8 + 50$
    $= 42$

24. $5 \cdot 3 - 3 \cdot 2^3 = 5 \cdot 3 - 3 \cdot 8$
    $= 15 - 24$
    $= -9$

25. $5 \cdot (10-8)^2 = 5 \cdot 2^2$
    $= 5 \cdot 4$
    $= 20$

26. $3 \cdot (-2)^2 + 6 \cdot 3 - 3 \cdot 4^2 = 3 \cdot 4 + 6 \cdot 3 - 3 \cdot 16$
    $= 12 + 18 - 48$
    $= 30 - 48$
    $= -18$

27. $\dfrac{4 + 6^2}{2 \cdot 3 + 2} = \dfrac{4 + 36}{2 \cdot 3 + 2}$
    $= \dfrac{4 + 36}{6 + 2}$
    $= \dfrac{40}{8}$
    $= 5$

28. $-3 \cdot |7^2 - 2 \cdot (8-5)^3| = -3 \cdot |7^2 - 2 \cdot 3^3|$
    $= -3 \cdot |49 - 2 \cdot 27|$
    $= -3 \cdot |49 - 54|$
    $= -3 \cdot |-5|$
    $= -3 \cdot 5$
    $= -15$

**R.5 Quick Checks**

1. The word *sum* indicates addition so we write $3 + 11$.

2. The word *product* indicates multiplication so we write $6 \cdot 7$.

3. The word *quotient* indicates division so we write $\dfrac{y}{4}$.

4. The word *difference* indicates subtraction so we write $3 - z$.

5. *Twice* means to multiply by 2 and *difference* indicates subtraction.
   $2(x-3)$

QC-5

## Chapter R: Real Numbers and Algebraic Expressions

6. The word *plus* indicates addition and the word *ratio* indicates division.
   $5 + \dfrac{z}{2}$

7. $-5x+3$
   $-5(2)+3 = -10+3 = -7$

8. $y^2 - 6y + 1$
   $(-4)^2 - 6(-4) + 1 = 16 + 24 + 1 = 40 + 1 = 41$

9. $\dfrac{w+8}{3w}$
   $\dfrac{(4)+8}{3(4)} = \dfrac{12}{12} = 1$

10. $|4x-5|$
    $\left|4\left(\dfrac{1}{2}\right) - 5\right| = |2-5| = |-3| = 3$

11. $118x$
    $x = 100: \quad 118(100) = 11{,}800 \text{ Yen}$
    $x = 1000: \quad 118(1000) = 118{,}000 \text{ Yen}$
    $x = 10{,}000: \quad 118(10{,}000) = 1{,}180{,}000 \text{ Yen}$

12. $\dfrac{5}{9}(x-32)$
    $x = 32: \quad \dfrac{5}{9}(32-32) = \dfrac{5}{9}(0) = 0° \text{ C}$
    $x = 86: \quad \dfrac{5}{9}(86-32) = \dfrac{5}{9}(54) = 30° \text{ C}$
    $x = 212: \quad \dfrac{5}{9}(212-32) = \dfrac{5}{9}(180) = 100° \text{ C}$

13. $4x - 9x = (4-9)x = -5x$

14. $-2x^2 + 13x^2 = (-2+13)x^2 = 11x^2$

15. $-5x - 3x + 6 - 3 = (-5-3)x + (6-3)$
    $= -8x + 3$

16. $6x - 10x - 4y + 12y = (6-10)x + (-4+12)y$
    $= -4x + 8y$

17. $10y - 3 + 5y + 2 = 10y + 5y - 3 + 2$
    $= (10+5)y + (-3+2)$
    $= 15y - 1$

18. $0.5x^2 + 1.3 + 1.8x^2 - 0.4$
    $= 0.5x^2 + 1.8x^2 + 1.3 - 0.4$
    $= (0.5 + 1.8)x^2 + (1.3 - 0.4)$
    $= 2.3x^2 + 0.9$

19. $4z + 6 - 8z - 3 - 2z$
    $= 4z - 8z - 2z + 6 - 3$
    $= (4 - 8 - 2)z + (6 - 3)$
    $= -6z + 3$

20. $3(x-2) + x = 3 \cdot x - 3 \cdot 2 + x$
    $= 3x - 6 + x$
    $= 3x + x - 6$
    $= 4x - 6$

21. $5(y+3) - 10y - 4 = 5 \cdot y + 5 \cdot 3 - 10y - 4$
    $= 5y + 15 - 10y - 4$
    $= 5y - 10y + 15 - 4$
    $= -5y + 11$

22. $3(z+4) - 2(3z+1)$
    $= 3 \cdot z + 3 \cdot 4 - 2 \cdot 3z - 2 \cdot 1$
    $= 3z + 12 - 6z - 2$
    $= 3z - 6z + 12 - 2$
    $= -3z + 10$

23. $-4(x-2) - (2x+4)$
    $= -4 \cdot x - (-4) \cdot 2 - 2x - 4$
    $= -4x + 8 - 2x - 4$
    $= -4x - 2x + 8 - 4$
    $= -6x + 4$

24. $\dfrac{1}{2}(6x+4) - \dfrac{15x+5}{5} = \dfrac{1}{2} \cdot 6x + \dfrac{1}{2} \cdot 4 - \dfrac{15}{5}x - \dfrac{5}{5}$
    $= 3x + 2 - 3x - 1$
    $= 3x - 3x + 2 - 1$
    $= 1$

## Section R.5 – Quick Checks: Algebraic Expressions

25. $\dfrac{5x-1}{3} + \dfrac{5x+9}{2} = \dfrac{5}{3}x - \dfrac{1}{3} + \dfrac{5}{2}x + \dfrac{9}{2}$

    $= \dfrac{5}{3}x + \dfrac{5}{2}x - \dfrac{1}{3} + \dfrac{9}{2}$

    $= \dfrac{10}{6}x + \dfrac{15}{6}x - \dfrac{2}{6} + \dfrac{27}{6}$

    $= \dfrac{25}{6}x + \dfrac{25}{6}$

    $= \dfrac{25x+25}{6}$ or $\dfrac{25(x+1)}{6}$

26. We need to determine whether the value of the variable causes division by 0. That is, we need to determine if the value makes $x - 4 = 0$.

    a. When $x = 2$, we have that $x - 4 = 2 - 4 = -2$, so 2 is in the domain of the variable.

    b. When $x = 0$, we have that $x - 4 = 0 - 4 = -4$, so 0 is in the domain of the variable.

    c. When $x = 4$, we have that $x - 4 = 4 - 4 = 0$, so 4 is **not** in the domain of the variable.

    d. When $x = -3$, we have that $x - 4 = -3 - 4 = -7$, so $-3$ is in the domain of the variable.

27. We need to determine whether the value of the variable causes division by 0. That is, we need to determine if the value makes $x + 3 = 0$.

    a. When $x = 2$, we have that $x + 3 = 2 + 3 = 5$, so 2 is in the domain of the variable.

    b. When $x = 0$, we have that $x + 3 = 0 + 3 = 3$, so 0 is in the domain of the variable.

    c. When $x = 4$, we have that $x + 3 = 4 + 3 = 7$, so 4 is in the domain of the variable.

    d. When $x = -3$, we have that $x + 3 = -3 + 3 = 0$, so $-3$ is **not** in the domain of the variable.

28. We need to determine whether the value of the variable causes division by 0. That is, we need to determine if the value makes $x^2 + x - 6 = 0$.

    a. When $x = 2$, we have that $x^2 + x - 6 = (2)^2 + (2) - 6 = 4 + 2 - 6 = 0$, so 2 is **not** in the domain of the variable.

    b. When $x = 0$, we have that $x^2 + x - 6 = (0)^2 + (0) - 6 = -6$, so 0 is in the domain of the variable.

    c. When $x = 4$, we have that $x^2 + x - 6 = (4)^2 + (4) - 6 = 14$, so 4 is in the domain of the variable.

    d. When $x = -3$, we have that $x^2 + x - 6 = (-3)^2 + (-3) - 6 = 9 - 3 - 6 = 0$, so $-3$ is **not** in the domain of the variable.

# Chapter 1 – Quick Checks

**1.1 Quick Checks**

1. $-5x+3=-2$
   Let $x=-2$ in the equation.
   $$-5(-2)+3\overset{?}{=}-2$$
   $$10+3\overset{?}{=}-2$$
   $$13\neq -2$$
   $x=-2$ is not a solution to the equation.

   Let $x=1$ in the equation.
   $$-5(1)+3\overset{?}{=}-2$$
   $$-5+3\overset{?}{=}-2$$
   $$-2=-2\ \checkmark$$
   $x=1$ is a solution to the equation.

   Let $x=3$ in the equation.
   $$-5(3)+3\overset{?}{=}-2$$
   $$-15+3\overset{?}{=}-2$$
   $$-12\neq -2$$
   $x=3$ is not a solution to the equation.

2. $3x+2=2x-5$
   Let $x=0$ in the equation.
   $$3(0)+2\overset{?}{=}2(0)-5$$
   $$0+2\overset{?}{=}0-5$$
   $$2\neq -5$$
   $x=0$ is not a solution to the equation.

   Let $x=6$ in the equation.
   $$3(6)+2\overset{?}{=}2(6)-5$$
   $$18+2\overset{?}{=}12-5$$
   $$20\neq 7$$
   $x=6$ is not a solution to the equation.

   Let $x=-7$ in the equation.
   $$3(-7)+2\overset{?}{=}2(-7)-5$$
   $$-21+2\overset{?}{=}-14-5$$
   $$-19=-19\ \checkmark$$
   $x=-7$ is a solution to the equation.

3. $-3(z+2)=4z+1$
   Let $z=-3$ in the equation.
   $$-3((-3)+2)\overset{?}{=}4(-3)+1$$
   $$-3(-1)\overset{?}{=}-12+1$$
   $$3\neq -11$$
   $z=-3$ is not a solution to the equation.

   Let $z=-1$ in the equation.
   $$-3((-1)+2)\overset{?}{=}4(-1)+1$$
   $$-3(1)\overset{?}{=}-4+1$$
   $$-3=-3\ \checkmark$$
   $z=-1$ is a solution to the equation.

   Let $z=2$ in the equation.
   $$-3((2)+2)\overset{?}{=}4(2)+1$$
   $$-3(4)\overset{?}{=}8+1$$
   $$-12\neq 9$$
   $z=2$ is not a solution to the equation.

4. $$3x+8=17$$
   $$3x+8-8=17-8$$
   $$3x=9$$
   $$\frac{3x}{3}=\frac{9}{3}$$
   $$x=3$$
   Check:
   $$3(3)+8\overset{?}{=}17$$
   $$9+8\overset{?}{=}17$$
   $$17=17\ \checkmark$$
   Solution set: $\{3\}$

5. $$-4a-7=1$$
   $$-4a-7+7=1+7$$
   $$-4a=8$$
   $$\frac{-4a}{-4}=\frac{8}{-4}$$
   $$a=-2$$
   Check:
   $$-4(-2)-7\overset{?}{=}1$$
   $$8-7\overset{?}{=}1$$
   $$1=1\ \checkmark$$
   Solution set: $\{-2\}$

**Section 1.1 – Quick Checks:** Linear Equations

6. $5y + 1 = 2$
   $5y + 1 - 1 = 2 - 1$
   $5y = 1$
   $\dfrac{5y}{5} = \dfrac{1}{5}$
   $y = \dfrac{1}{5}$

   Check:
   $5\left(\dfrac{1}{5}\right) + 1 \stackrel{?}{=} 2$
   $1 + 1 \stackrel{?}{=} 2$
   $2 = 2$ ✓

   Solution set: $\left\{\dfrac{1}{5}\right\}$

7. $2x + 3 + 5x + 1 = 4x + 10$
   $7x + 4 = 4x + 10$
   $7x + 4 - 4x = 4x + 10 - 4x$
   $3x + 4 = 10$
   $3x + 4 - 4 = 10 - 4$
   $3x = 6$
   $\dfrac{3x}{3} = \dfrac{6}{3}$
   $x = 2$

   Check:
   $2(2) + 3 + 5(2) + 1 \stackrel{?}{=} 4(2) + 10$
   $4 + 3 + 10 + 1 \stackrel{?}{=} 8 + 10$
   $18 = 18$ ✓

   Solution set: $\{2\}$

8. $4b + 3 - b - 8 - 5b = 2b - 1 - b - 1$
   $-2b - 5 = b - 2$
   $-2b - 5 - b = b - 2 - b$
   $-3b - 5 = -2$
   $-3b - 5 + 5 = -2 + 5$
   $-3b = 3$
   $\dfrac{-3b}{-3} = \dfrac{3}{-3}$
   $b = -1$

   Check:
   $4(-1) + 3 - (-1) - 8 - 5(-1) \stackrel{?}{=} 2(-1) - 1 - (-1) - 1$
   $-4 + 3 + 1 - 8 + 5 \stackrel{?}{=} -2 - 1 + 1 - 1$
   $-3 = -3$ ✓

   Solution set: $\{-1\}$

9. $2w + 8 - 7w + 1 = 3w - 1 + 2w - 5$
   $-5w + 9 = 5w - 6$
   $-5w + 9 - 5w = 5w - 6 - 5w$
   $-10w + 9 = -6$
   $-10w + 9 - 9 = -6 - 9$
   $-10w = -15$
   $\dfrac{-10w}{-10} = \dfrac{-15}{-10}$
   $w = \dfrac{3}{2}$

   Check:
   $2\left(\dfrac{3}{2}\right) + 8 - 7\left(\dfrac{3}{2}\right) + 1 \stackrel{?}{=} 3\left(\dfrac{3}{2}\right) - 1 + 2\left(\dfrac{3}{2}\right) - 5$
   $3 + 8 - \dfrac{21}{2} + 1 \stackrel{?}{=} \dfrac{9}{2} - 1 + 3 - 5$
   $\dfrac{3}{2} = \dfrac{3}{2}$ ✓

   Solution set: $\left\{\dfrac{3}{2}\right\}$

10. $4(x - 1) = 12$
    $4x - 4 = 12$
    $4x - 4 + 4 = 12 + 4$
    $4x = 16$
    $\dfrac{4x}{4} = \dfrac{16}{4}$
    $x = 4$

    Check:
    $4(4 - 1) \stackrel{?}{=} 12$
    $4(3) \stackrel{?}{=} 12$
    $12 = 12$ ✓

    Solution set: $\{4\}$

11. $-2(x - 4) - 6 = 3(x + 6) + 4$
    $-2x + 8 - 6 = 3x + 18 + 4$
    $-2x + 2 = 3x + 22$
    $-2x + 2 - 3x = 3x + 22 - 3x$
    $-5x + 2 = 22$
    $-5x + 2 - 2 = 22 - 2$
    $-5x = 20$
    $\dfrac{-5x}{-5} = \dfrac{20}{-5}$
    $x = -4$

**Chapter 1:** Linear Equations and Inequalities

Check:
$$-2(-4-4)-6 \stackrel{?}{=} 3(-4+6)+4$$
$$-2(-8)-6 \stackrel{?}{=} 3(2)+4$$
$$16-6 \stackrel{?}{=} 6+4$$
$$10 = 10 \checkmark$$
Solution set: $\{-4\}$

12. $4(x+3)-8x = 3(x+2)+x$
$$4x+12-8x = 3x+6+x$$
$$-4x+12 = 4x+6$$
$$-4x+12-4x = 4x+6-4x$$
$$-8x+12 = 6$$
$$-8x+12-12 = 6-12$$
$$-8x = -6$$
$$\frac{-8x}{-8} = \frac{-6}{-8}$$
$$x = \frac{3}{4}$$

Check:
$$4\left(\frac{3}{4}+3\right)-8\left(\frac{3}{4}\right) \stackrel{?}{=} 3\left(\frac{3}{4}+2\right)+\frac{3}{4}$$
$$4\left(\frac{15}{4}\right)-6 \stackrel{?}{=} 3\left(\frac{11}{4}\right)+\frac{3}{4}$$
$$15-6 \stackrel{?}{=} \frac{33}{4}+\frac{3}{4}$$
$$9 = 9 \checkmark$$
Solution set: $\left\{\frac{3}{4}\right\}$

13. $5(x-3)+3(x+3) = 2x-3$
$$5x-15+3x+9 = 2x-3$$
$$8x-6 = 2x-3$$
$$8x-6-2x = 2x-3-2x$$
$$6x-6 = -3$$
$$6x-6+6 = -3+6$$
$$6x = 3$$
$$\frac{6x}{6} = \frac{3}{6}$$
$$x = \frac{1}{2}$$

Check:
$$5\left(\frac{1}{2}-3\right)+3\left(\frac{1}{2}+3\right) \stackrel{?}{=} 2\left(\frac{1}{2}\right)-3$$
$$5\left(-\frac{5}{2}\right)+3\left(\frac{7}{2}\right) \stackrel{?}{=} 1-3$$
$$-\frac{25}{2}+\frac{21}{2} \stackrel{?}{=} -2$$
$$-2 = -2 \checkmark$$
Solution set: $\left\{\frac{1}{2}\right\}$

14. $\frac{3y}{2}+\frac{y}{6} = \frac{10}{3}$
$$6\left(\frac{3y}{2}+\frac{y}{6}\right) = 6\left(\frac{10}{3}\right)$$
$$6\cdot\frac{3y}{2}+6\cdot\frac{y}{6} = 6\cdot\frac{10}{3}$$
$$9y+y = 20$$
$$10y = 20$$
$$\frac{10y}{10} = \frac{20}{10}$$
$$y = 2$$

Check:
$$\frac{3(2)}{2}+\frac{2}{6} \stackrel{?}{=} \frac{10}{3}$$
$$3+\frac{1}{3} \stackrel{?}{=} \frac{10}{3}$$
$$\frac{10}{3} = \frac{10}{3} \checkmark$$
Solution set: $\{2\}$

15. $\frac{3x}{4}-\frac{5}{12} = \frac{5x}{6}$
$$12\left(\frac{3x}{4}-\frac{5}{12}\right) = 12\left(\frac{5x}{6}\right)$$
$$12\cdot\frac{3x}{4}-12\cdot\frac{5}{12} = 12\cdot\frac{5x}{6}$$
$$9x-5 = 10x$$
$$9x-5-9x = 10x-9x$$
$$-5 = x$$

Check:
$$\frac{3(-5)}{4}-\frac{5}{12} \stackrel{?}{=} \frac{5(-5)}{6}$$
$$\frac{-15}{4}-\frac{5}{12} \stackrel{?}{=} -\frac{25}{6}$$
$$\frac{-45}{12}-\frac{5}{12} \stackrel{?}{=} -\frac{25}{6}$$
$$-\frac{50}{12} \stackrel{?}{=} -\frac{25}{6}$$
$$-\frac{25}{6} = -\frac{25}{6} \checkmark$$
Solution set: $\{-5\}$

### Section 1.1 – Quick Checks: Linear Equations

**16.**
$$\frac{x+2}{6} + 2 = \frac{5}{3}$$
$$6\left(\frac{x+2}{6} + 2\right) = 6\left(\frac{5}{3}\right)$$
$$6 \cdot \frac{(x+2)}{6} + 6 \cdot 2 = 6 \cdot \frac{5}{3}$$
$$x + 2 + 12 = 10$$
$$x + 14 = 10$$
$$x + 14 - 14 = 10 - 14$$
$$x = -4$$

Check:
$$\frac{-4+2}{6} + 2 \stackrel{?}{=} \frac{5}{3}$$
$$\frac{-2}{6} + 2 \stackrel{?}{=} \frac{5}{3}$$
$$-\frac{1}{3} + \frac{6}{3} \stackrel{?}{=} \frac{5}{3}$$
$$\frac{5}{3} = \frac{5}{3} \checkmark$$

Solution set: $\{-4\}$

**17.**
$$\frac{4x+3}{9} - \frac{2x+1}{2} = \frac{1}{6}$$
$$18\left(\frac{4x+3}{9} - \frac{2x+1}{2}\right) = 18\left(\frac{1}{6}\right)$$
$$18 \cdot \frac{(4x+3)}{9} - 18 \cdot \frac{(2x+1)}{2} = 18 \cdot \frac{1}{6}$$
$$2(4x+3) - 9(2x+1) = 3$$
$$8x + 6 - 18x - 9 = 3$$
$$-10x - 3 = 3$$
$$-10x - 3 + 3 = 3 + 3$$
$$-10x = 6$$
$$\frac{-10x}{-10} = \frac{6}{-10}$$
$$x = -\frac{3}{5}$$

Check:
$$\frac{4\left(-\frac{3}{5}\right)+3}{9} - \frac{2\left(-\frac{3}{5}\right)+1}{2} \stackrel{?}{=} \frac{1}{6}$$
$$\frac{-\frac{12}{5}+\frac{15}{5}}{9} - \frac{-\frac{6}{5}+\frac{5}{5}}{2} \stackrel{?}{=} \frac{1}{6}$$
$$\frac{\frac{3}{5}}{9} - \frac{\frac{-1}{5}}{2} \stackrel{?}{=} \frac{1}{6}$$
$$\frac{3}{5} \cdot \frac{1}{9} + \frac{1}{5} \cdot \frac{1}{2} \stackrel{?}{=} \frac{1}{6}$$
$$\frac{1}{15} + \frac{1}{10} \stackrel{?}{=} \frac{1}{6}$$
$$\frac{2}{30} + \frac{3}{30} \stackrel{?}{=} \frac{5}{30}$$
$$\frac{5}{30} = \frac{5}{30} \checkmark$$

Solution set: $\left\{-\frac{3}{5}\right\}$

**18.** Our first step is to multiply both sides of the equation by 10 to eliminate the decimals.
$$10(0.2t + 1.4) = 10(0.8)$$
$$2t + 14 = 8$$
$$2t + 14 - 14 = 8 - 14$$
$$2t = -6$$
$$\frac{2t}{2} = \frac{-6}{2}$$
$$t = -3$$

Check:
$$0.2(-3) + 1.4 \stackrel{?}{=} 0.8$$
$$-0.6 + 1.4 \stackrel{?}{=} 0.8$$
$$0.8 = 0.8 \checkmark$$

Solution set: $\{-3\}$

## Chapter 1: Linear Equations and Inequalities

19. Our first step is to multiply both sides of the equation by 100 to eliminate the decimals.
$$100(0.07x - 1.3) = 100(0.05x - 1.1)$$
$$7x - 130 = 5x - 110$$
$$7x - 130 - 5x = 5x - 110 - 5x$$
$$2x - 130 = -110$$
$$2x - 130 + 130 = -110 + 130$$
$$2x = 20$$
$$\frac{2x}{2} = \frac{20}{2}$$
$$x = 10$$
Check:
$$0.07(10) - 1.3 \stackrel{?}{=} 0.05(10) - 1.1$$
$$0.7 - 1.3 \stackrel{?}{=} 0.5 - 1.1$$
$$-0.6 = -0.6 \checkmark$$
Solution set: $\{10\}$

20. Our first step is to multiply both sides of the equation by 10 to eliminate the decimals.
$$10[0.4(y+3)] = 10[0.5(y-4)]$$
$$4(y+3) = 5(y-4)$$
$$4y + 12 = 5y - 20$$
$$4y + 12 - 5y = 5y - 20 - 5y$$
$$-y + 12 = -20$$
$$-y + 12 - 12 = -20 - 12$$
$$-y = -32$$
$$\frac{-y}{-1} = \frac{-32}{-1}$$
$$y = 32$$
Check:
$$0.4(32+3) \stackrel{?}{=} 0.5(32-4)$$
$$0.4(35) \stackrel{?}{=} 0.5(28)$$
$$14 = 14 \checkmark$$
Solution set: $\{32\}$

21. $$4(x+2) = 4x + 2$$
$$4x + 8 = 4x + 2$$
$$4x + 8 - 4x = 4x + 2 - 4x$$
$$8 = 2$$
The last statement is a contradiction. The equation has no solution.
Solution set: $\{\ \}$ or $\varnothing$

22. $$3(x-2) = 2x - 6 + x$$
$$3x - 6 = 3x - 6$$
$$3x - 6 - 3x = 3x - 6 - 3x$$
$$-6 = -6$$
The last statement is an identity. All real numbers are solutions.
Solution set: $\mathbb{R}$

23. $$-4x + 2 + x + 1 = -4(x+2) + 11$$
$$-3x + 3 = -4x - 8 + 11$$
$$-3x + 3 = -4x + 3$$
$$-3x + 3 + 4x = -4x + 3 + 4x$$
$$x + 3 = 3$$
$$x + 3 - 3 = 3 - 3$$
$$x = 0$$
This is a conditional equation.
Solution set: $\{0\}$

24. $$-3(z+1) + 2(z-3) = z + 6 - 2z - 15$$
$$-3z - 3 + 2z - 6 = -z - 9$$
$$-z - 9 = -z - 9$$
$$-z - 9 + z = -z - 9 + z$$
$$-9 = -9$$
The last statement is an identity. All real numbers are solutions.
Solution set: $\mathbb{R}$

### 1.2 Quick Checks

1. $3y = 21$

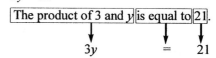

2. $2(3+x) = 5x$

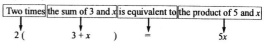

3. $x - 10 = \dfrac{x}{2}$

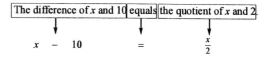

4. $y - 3 = 5y$

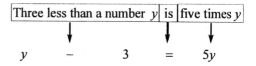

**Section 1.2 – Quick Checks:** *An Introduction to Problem Solving*

5. Let $x$ = the first of the consecutive even integers.
   Second: $x+2$
   Third: $x+4$
   $$x+(x+2)+(x+4)=60$$
   $$x+x+2+x+4=60$$
   $$3x+6=60$$
   $$3x+6-6=60-6$$
   $$3x=54$$
   $$\frac{3x}{3}=\frac{54}{3}$$
   $$x=18$$
   The three consecutive even integers are 18, 20, and 22.
   Check:
   $$18+(18+2)+(18+4)=18+20+22=60$$

6. Let $x$ = the first of the consecutive integers.
   Second: $x+1$
   Third: $x+2$
   $$x+(x+1)+(x+2)=78$$
   $$x+x+1+x+2=78$$
   $$3x+3=78$$
   $$3x+3-3=78-3$$
   $$3x=75$$
   $$\frac{3x}{3}=\frac{75}{3}$$
   $$x=25$$
   The three consecutive integers are 25, 26, and 27.
   Check:
   $$25+(25+1)+(25+2)=25+26+27=78$$

7. Let $w$ = Melody's hourly wage.
   Overtime rate: $1.5w$
   $$40w+6(1.5w)=735$$
   $$40w+9w=735$$
   $$49w=735$$
   $$\frac{49w}{49}=\frac{735}{49}$$
   $$w=15$$
   Melody makes $15 per hour.
   Check:
   $$40(15)+6[1.5(15)]=600+6(22.5)$$
   $$=600+135$$
   $$=735$$

8. Let $w$ = Jim's regular hourly rate.
   Saturday rate: $1.5w$
   Sunday rate: $2w$
   $$30w+6(1.5w)+4(2w)=564$$
   $$30w+9w+8w=564$$
   $$47w=564$$
   $$\frac{47w}{47}=\frac{564}{47}$$
   $$w=12$$
   Jim's regular hourly rate is $12 per hour.
   Check:
   $$30(12)+6[1.5(12)]+4[2(12)]$$
   $$=360+6(18)+4(24)$$
   $$=360+108+96$$
   $$=564$$

9. Let $m$ = the number of miles.
   EZ-Rental cost = Do It Yourself cost
   $$35+0.15m=20+0.25m$$
   $$0.15m=-15+0.25m$$
   $$-0.1m=-15$$
   $$\frac{-0.1m}{-0.1}=\frac{-15}{-0.1}$$
   $$m=150$$
   The rental costs will be the same for 150 miles.
   Check:
   $$35+0.15(150)=35+22.5=\$57.50$$
   $$20+0.25(150)=20+37.5=\$57.50$$

10. Let $m$ = the number of minutes used.
    Company A cost = Company B cost
    $$12+0.1m=0.15m$$
    $$12=0.05m$$
    $$\frac{12}{0.05}=\frac{0.05m}{0.05}$$
    $$240=m$$
    The monthly costs will be the same for 240 minutes.
    Check:
    $$12+0.1(240)=12+24=36$$
    $$0.15(240)=36$$

11. Let $x$ = the desired result.
    $$x=0.4(100)=40$$
    40% of 100 is 40.

*Chapter 1: Linear Equations and Inequalities*

12. Let $x$ = the desired value.
$$8 = 0.05(x)$$
$$\frac{8}{0.05} = \frac{0.05x}{0.05}$$
$$160 = x$$
8 is 5% of 160.

13. Let $p$ = the desired percent (as a decimal).
$$15 = p \cdot 20$$
$$15 = 20p$$
$$\frac{15}{20} = \frac{20p}{20}$$
$$0.75 = p$$
15 is 75% of 20.

14. Let $p$ = the original price.
Discount: $0.3p$
$$p - 0.3p = 21$$
$$0.7p = 21$$
$$\frac{0.7p}{0.7} = \frac{21}{0.7}$$
$$p = 30$$
The original price of the shirt was $30.

15. Let $c$ = Milex's cost for a spark plug.
Mark up: $0.35c$
$$c + 0.35c = 1.62$$
$$1.35c = 1.62$$
$$\frac{1.35c}{1.35} = \frac{1.62}{1.35}$$
$$c = 1.2$$
Milex's cost for each spark plug is $1.20.

16. Note that 1 month is $\frac{1}{12}$ year.
$$I = Prt$$
$$I = (6500)(0.06)\left(\frac{1}{12}\right) = 32.5$$
The interest charge on Dave's car loan after 1 month is $32.50.

17. Note that 6 months is $\frac{1}{2}$ year.
$$I = Prt$$
$$I = (1400)(0.015)\left(\frac{1}{2}\right) = 10.50$$
The interest paid after 6 months would be $10.50.
The balance is the original amount in the account plus the accrued interest.
$1400 + 10.50 = 1410.50$
The account will have a balance of $1410.50.

18. Let $x$ = the amount invested in Aaa-rated bonds.
B-rated bonds: $90,000 - x$
Int. from Aaa-rated + Int. from B-rated = 5400
$$0.05x + 0.09(90,000 - x) = 5400$$
$$0.05x + 8100 - 0.09x = 5400$$
$$8100 - 0.04x = 5400$$
$$-0.04x = -2700$$
$$\frac{-0.04x}{-0.04} = \frac{-2700}{-0.04}$$
$$x = 67,500$$
Sophia should place $67,500 in Aaa-rated bonds and $90,000 - \$67,500 = \$22,500$ in B-rated bonds.

19. Let $x$ = the amount invested in a 5-yr CD.
Corporate bonds: $25,000 - x$
CD interest + bond interest = total interest
$$0.04x + 0.09(25,000 - x) = 0.08(25,000)$$
$$0.04x + 2250 - 0.09x = 2000$$
$$2250 - 0.05x = 2000$$
$$-0.05x = -250$$
$$\frac{-0.05x}{-0.05} = \frac{-250}{-0.05}$$
$$x = 5000$$
Steve should invest $5000 in the CD and $25,000 - \$5000 = \$20,000$ in corporate bonds.

20. Let $x$ = pounds of Tea A.
Tea B pounds: $10 - x$

|  | Tea A | Tea B | Mix |
|---|---|---|---|
| price per pound | 4 | 2.75 | 3.50 |
| pounds | $x$ | $10 - x$ | 10 |
| revenue | $4x$ | $2.75(10-x)$ | $3.5(10)$ |

We want the individual revenues to be the same as the revenue from the mix. Therefore, we get
$$4x + 2.75(10 - x) = 3.5(10)$$
$$4x + 27.5 - 2.75x = 35$$
$$1.25x + 27.5 = 35$$
$$1.25x = 7.5$$
$$\frac{1.25x}{1.25} = \frac{7.5}{1.25}$$
$$x = 6$$
You should blend 6 pounds of Tea A and $10 - 6 = 4$ pounds of Tea B.

### Section 1.3 – Quick Checks: Using Formulas to Solve Problems

**21.** Let $x$ = pounds of cashews.
Cashews: $30 - x$

|  | cashews | peanuts | Mix |
|---|---|---|---|
| price per pound | 6 | 1.5 | 3 |
| pounds | $x$ | $30 - x$ | 30 |
| revenue | $6x$ | $1.5(30-x)$ | $3(30)$ |

We want the individual revenues to be the same as the revenue from the mix. Therefore, we get
$$6x + 1.5(30 - x) = 3(30)$$
$$6x + 45 - 1.5x = 90$$
$$4.5x + 45 = 90$$
$$4.5x = 45$$
$$\frac{4.5x}{4.5} = \frac{45}{4.5}$$
$$x = 10$$
You should blend 10 pounds of cashews and $30 - 10 = 20$ pounds of peanuts.

**22.** We want to know when the distance traveled by both cars will be the same.
Let $t$ = the time of travel by the Cavalier.
Then the time of travel for the BMW will be $t - 2$ since it left two hours later.

Cavalier distance = BMW distance
$$(\text{rate})(\text{time})_{\text{Cavalier}} = (\text{rate})(\text{time})_{\text{BMW}}$$
$$40t = 60(t - 2)$$
$$40t = 60t - 120$$
$$-20t = -120$$
$$t = 6$$
The BMW will catch the Cavalier after the Cavalier has been traveling for 6 hours (4 hours for the BMW). When the BMW catches the Cavalier, both cars will have traveled $40(6) = 240$ miles.

**23.** We want to know when the distance traveled by both the train and the helicopter will be the same.
Let $t$ = the time of travel by the train.
Then the time of travel for the helicopter will be $t - 4$ since it left four hours later.

train distance = helicopter distance
$$(\text{rate})(\text{time})_{\text{train}} = (\text{rate})(\text{time})_{\text{helicopter}}$$
$$50t = 90(t - 4)$$
$$50t = 90t - 360$$
$$-40t = -360$$
$$t = 9$$
The helicopter will catch the train after the train has been traveling for 9 hours (5 hours for the helicopter). When the helicopter catches the train, both the helicopter and the train will have traveled $50(9) = 450$ miles.

### 1.3 Quick Checks

**1.** $A = \pi r^2$

**2.** $V = \pi r^2 h$

**3.** $C = 175x + 7000$

**4.** $s = \frac{1}{2} g t^2$

**5. a.**
$$A = \frac{1}{2} bh$$
$$2 \cdot A = 2 \cdot \frac{1}{2} bh$$
$$2A = bh$$
$$\frac{2A}{b} = \frac{bh}{b}$$
$$\frac{2A}{b} = h \quad \text{or} \quad h = \frac{2A}{b}$$

**b.** $h = \frac{2A}{b} = \frac{2(10)}{4} = 5$

A triangle whose area is 10 square inches and whose base is 4 inches has a height of 5 inches.

**Chapter 1:** Linear Equations and Inequalities

**6. a.**
$$P = 2a + 2b$$
$$P - 2a = 2a + 2b - 2a$$
$$P - 2a = 2b$$
$$\frac{P-2a}{2} = \frac{2b}{2}$$
$$\frac{P-2a}{2} = b \quad \text{or} \quad b = \frac{P-2a}{2}$$

**b.** $b = \frac{P-2a}{2}$
$$b = \frac{60-2(20)}{2} = \frac{60-40}{2} = \frac{20}{2} = 10$$
A parallelogram whose perimeter is 60 cm and whose adjacent length is 20 cm has a length of 10 cm.

**7.**
$$I = P \cdot r \cdot t$$
$$\frac{I}{r \cdot t} = \frac{P \cdot r \cdot t}{r \cdot t}$$
$$\frac{I}{rt} = P \quad \text{or} \quad P = \frac{I}{rt}$$

**8.**
$$Ax + By = C$$
$$Ax + By - Ax = C - Ax$$
$$By = C - Ax$$
$$\frac{By}{B} = \frac{C-Ax}{B}$$
$$y = \frac{C-Ax}{B}$$

**9.**
$$2xh - 4x = 3h - 3$$
$$2xh - 4x - 3h = 3h - 3 - 3h$$
$$2xh - 4x - 3h = -3$$
$$2xh - 4x - 3h + 4x = -3 + 4x$$
$$2xh - 3h = 4x - 3$$
$$h(2x-3) = 4x - 3$$
$$\frac{h(2x-3)}{2x-3} = \frac{4x-3}{2x-3}$$
$$h = \frac{4x-3}{2x-3}$$

**10.**
$$S = na + (n-1)d$$
$$S = na + nd - d$$
$$S = n(a+d) - d$$
$$S + d = n(a+d) - d + d$$
$$S + d = n(a+d)$$
$$\frac{S+d}{a+d} = \frac{n(a+d)}{a+d}$$
$$\frac{S+d}{a+d} = n \quad \text{or} \quad n = \frac{S+d}{a+d}$$

**11.** Let $w$ = the width of the pool in feet. Then the length is $l = w + 10$.
$$P = 2l + 2w$$
$$180 = 2(w+10) + 2w$$
$$180 = 2w + 20 + 2w$$
$$180 = 4w + 20$$
$$160 = 4w$$
$$40 = w$$
The pool is 40 feet wide and $40 + 10 = 50$ feet long.

**12.** Since the bookcase is rectangular, we can consider the height as the length in our formula. Let $w$ = the width of the bookcase in inches. Then the height is $h = w + 32$.
$$P = 2(\text{length}) + 2(\text{width})$$
$$224 = 2(w+32) + 2(w)$$
$$224 = 2w + 64 + 2w$$
$$224 = 4w + 64$$
$$160 = 4w$$
$$40 = w$$
The opening of the bookcase has a width of 40 inches and a height of $40 + 32 = 72$ inches.

**13.**
$$A = 2\pi r^2 + 2\pi rh$$
$$51.8 = 2\pi(1.5)^2 + 2\pi(1.5)h$$
$$51.8 - 2\pi(1.5)^2 = 2\pi(1.5)h$$
$$\frac{51.8 - 2\pi(1.5)^2}{2\pi(1.5)} = h$$

```
(51.8-2π(1.5)²)/
(2π(1.5))
          3.996150701
```

The height of the can is roughly 4.00 inches.

QC - 16

*Section 1.4 – Quick Checks:* Linear Inequalities

**14.**
$$A = 2\pi r^2 + 2\pi rh$$
$$72.7 = 2\pi(1.625)^2 + 2\pi(1.625)h$$
$$72.7 - 2\pi(1.625)^2 = 2\pi(1.625)h$$
$$\frac{72.7 - 2\pi(1.625)^2}{2\pi(1.625)} = h$$

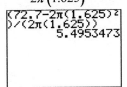

The height of the can is about 5.50 inches.

**1.4 Quick Checks**

1. $-3 \leq x \leq 2$
   Interval: $[-3, 2]$
   Graph:

2. $3 \leq x < 6$
   Interval: $[3, 6)$
   Graph:

3. $x \leq 3$
   Interval: $(-\infty, 3]$
   Graph:

4. $\frac{1}{2} < x < \frac{7}{2}$
   Interval: $\left(\frac{1}{2}, \frac{7}{2}\right)$
   Graph:

5. $(0, 5]$
   Inequality: $0 < x \leq 5$
   Graph:

6. $(-6, 0)$
   Inequality: $-6 < x < 0$
   Graph:

7. $(5, \infty)$
   Inequality: $x > 5$
   Graph:

8. $\left(-\infty, \frac{8}{3}\right]$
   Inequality: $x \leq \frac{8}{3}$
   Graph:

9. $x + 3 > 5$
   $x + 3 - 3 > 5 - 3$
   $x > 2$
   Interval: $(2, \infty)$
   Solution set: $\{x \mid x > 2\}$
   Graph:

10. $\frac{1}{3}x < 2$
    $3 \cdot \frac{1}{3}x \leq 3 \cdot 2$
    $x \leq 6$
    Interval: $(-\infty, 6]$
    Solution set: $\{x \mid x \leq 6\}$
    Graph:

QC – 17

## Chapter 1: Linear Equations and Inequalities

**11.** $4x - 3 < 13$
$4x - 3 + 3 < 13 + 3$
$4x < 16$
$\dfrac{4x}{4} < \dfrac{16}{4}$
$x < 4$
Interval: $(-\infty, 4)$
Solution set: $\{x \mid x < 4\}$
Graph:

**12.** $3x + 1 > x - 5$
$3x + 1 - 1 > x - 5 - 1$
$3x > x - 6$
$3x - x > x - 6 - x$
$2x > -6$
$\dfrac{2x}{2} > \dfrac{-6}{2}$
$x > -3$
Interval: $(-3, \infty)$
Solution set: $\{x \mid x > -3\}$
Graph:

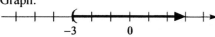

**13.** $-2x + 1 \leq 3x + 11$
$-2x + 1 - 1 \leq 3x + 11 - 1$
$-2x \leq 3x + 10$
$-2x - 3x \leq 3x + 10 - 3x$
$-5x \leq 10$
$\dfrac{-5x}{-5} \geq \dfrac{10}{-5}$
$x \geq -2$
Interval: $[-2, \infty)$
Solution set: $\{x \mid x \geq -2\}$

**14.** $-5x + 12 < x - 3$
$-5x + 12 - 12 < x - 3 - 12$
$-5x < x - 15$
$-5x - x < x - 15 - x$
$-6x < -15$
$\dfrac{-6x}{-6} > \dfrac{-15}{-6}$
$x > \dfrac{5}{2}$
Interval: $\left(\dfrac{5}{2}, \infty\right)$
Solution set: $\left\{x \mid x > \dfrac{5}{2}\right\}$

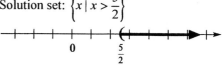

**15.** $4(x - 2) < 3x - 4$
$4x - 8 < 3x - 4$
$4x - 8 + 8 < 3x - 4 + 8$
$4x < 3x + 4$
$4x - 3x < 3x + 4 - 3x$
$x < 4$
Interval: $(-\infty, 4)$
Solution set: $\{x \mid x < 4\}$

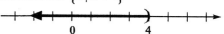

**16.** $-2(x + 1) \geq 4(x + 3)$
$-2x - 2 \geq 4x + 12$
$-2x - 2 + 2 \geq 4x + 12 + 2$
$-2x \geq 4x + 14$
$-2x - 4x \geq 4x + 14 - 4x$
$-6x \geq 14$
$\dfrac{-6x}{-6} \leq \dfrac{14}{-6}$
$x \leq -\dfrac{7}{3}$
Interval: $\left(-\infty, -\dfrac{7}{3}\right]$
Solution set: $\left\{x \mid x \leq -\dfrac{7}{3}\right\}$

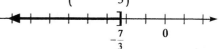

**17.** $7 - 2(x+1) \leq 3(x-5)$
$7 - 2x - 2 \leq 3x - 15$
$-2x + 5 \leq 3x - 15$
$-2x + 5 - 5 \leq 3x - 15 - 5$
$-2x \leq 3x - 20$
$-2x - 3x \leq 3x - 20 - 3x$
$-5x \leq -20$
$\dfrac{-5x}{-5} \geq \dfrac{-20}{-5}$
$x \geq 4$
Interval: $[4, \infty)$
Solution set: $\{x \mid x \geq 4\}$

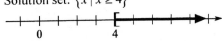

**18.** $\dfrac{3x+1}{5} \geq 2$
$5 \cdot \dfrac{3x+1}{5} \geq 5 \cdot 2$
$3x + 1 \geq 10$
$3x + 1 - 1 \geq 10 - 1$
$3x \geq 9$
$\dfrac{3x}{3} \geq \dfrac{9}{3}$
$x \geq 3$
Interval: $[3, \infty)$
Solution set: $\{x \mid x \geq 3\}$

**19.** $\dfrac{2}{5}x + \dfrac{3}{10} < \dfrac{1}{2}$
$10\left(\dfrac{2}{5}x + \dfrac{3}{10}\right) < 10 \cdot \dfrac{1}{2}$
$4x + 3 < 5$
$4x + 3 - 3 < 5 - 3$
$4x < 2$
$\dfrac{4x}{4} < \dfrac{2}{4}$
$x < \dfrac{1}{2}$
Interval: $\left(-\infty, \dfrac{1}{2}\right)$
Solution set: $\left\{x \mid x < \dfrac{1}{2}\right\}$

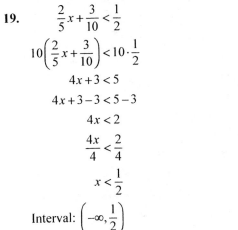

**20.** $\dfrac{1}{2}(x+3) > \dfrac{1}{3}(x-4)$
$6 \cdot \dfrac{1}{2}(x+3) > 6 \cdot \dfrac{1}{3}(x-4)$
$3(x+3) > 2(x-4)$
$3x + 9 > 2x - 8$
$3x + 9 - 9 > 2x - 8 - 9$
$3x > 2x - 17$
$3x - 2x > 2x - 17 - 2x$
$x > -17$
Interval: $(-17, \infty)$
Solution set: $\{x \mid x > -17\}$

**21.** Let $b$ = annual balance in dollars.
The annual cost is the sum of the annual fee and the interest charge. To find the interest charge, we use the simple interest formula $I = P \cdot r \cdot t$ and let $t = 1$ year.

Bank A cost < Bank B cost
$\text{fee}_A + \text{interest}_A < \text{fee}_B + \text{interest}_B$
$25 + 0.099b < 0 + 0.149b$
$25 + 0.099b < 0.149b$
$25 + 0.099b - 0.099b < 0.149b - 0.099b$
$25 < 0.05b$
$\dfrac{25}{0.05} < \dfrac{0.05b}{0.05}$
$500 < b$ or $b > 500$

The card from Bank A will cost less if the annual balance is more than $500.

**22.** Let $x$ = the number of candy boxes.
$R > C$
$12x > 8x + 96$
$12x - 8x > 8x + 96 - 8x$
$4x > 96$
$\dfrac{4x}{4} > \dfrac{96}{4}$
$x > 24$

Revenue will exceed costs when more than 24 boxes of candy are sold.

## Chapter 1: Linear Equations and Inequalities

**1.5 Quick Checks**

1. $A \cap B = \{1, 3, 5\}$

2. $A \cap C = \{2, 4, 6\}$

3. $A \cup B = \{1, 2, 3, 4, 5, 6, 7\}$

4. $A \cup C = \{1, 2, 3, 4, 5, 6, 8\}$

5. $B \cap C = \{\ \}$ or $\emptyset$

6. $B \cup C = \{1, 2, 3, 4, 5, 6, 7, 8\}$

7. $A \cap B$ is the set of all real numbers that are greater than 2 and less than 7.
   
   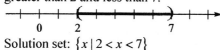
   
   Solution set: $\{x \mid 2 < x < 7\}$
   
   Interval: $(2, 7)$

8. $A \cup C$ is the set of real numbers that are greater than 2 or less than or equal to $-3$.
   
   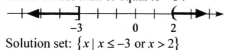
   
   Solution set: $\{x \mid x \leq -3 \text{ or } x > 2\}$
   
   Interval: $(-\infty, -3] \cup (2, \infty)$

9. $2x + 1 \geq 5$ and $-3x + 2 < 5$
   $\quad 2x \geq 4 \qquad\qquad -3x < 3$
   $\quad x \geq 2 \qquad\qquad\ \ x > -1$
   
   We need $x \geq 2$ and $x > -1$.
   
   Solution set: $\{x \mid x \geq 2\}$
   
   Interval: $[2, \infty)$
   
   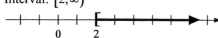

10. $4x - 5 < 7$ and $3x - 1 > -10$
    $\quad 4x < 12 \qquad\qquad 3x > -9$
    $\quad x < 3 \qquad\qquad\ \ x > -3$
    
    We need $x < 3$ and $x > -3$.
    
    Solution set: $\{x \mid -3 < x < 3\}$
    
    Interval: $(-3, 3)$
    
    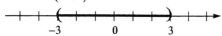

11. $-8x + 3 < -5$ and $\dfrac{2}{3}x + 1 < 3$
    $\quad -8x < -8 \qquad\qquad\qquad \dfrac{2}{3}x < 2$
    $\quad x > 1 \qquad\qquad\qquad\qquad x < 3$
    
    We need $x > 1$ and $x < 3$.
    
    Solution set: $\{x \mid 1 < x < 3\}$
    
    Interval: $(1, 3)$
    
    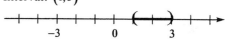

12. $3x - 5 < -8$ and $2x + 1 > 5$
    $\quad 3x < -3 \qquad\qquad 2x > 4$
    $\quad x < -1 \qquad\qquad\ x > 2$
    
    We need $x < -1$ and $x > 2$. Looking at the graphs of the inequalities separately we see that there are no such numbers that satisfy both inequalities. Therefore, the solution set is empty.
    
    Solution set: $\{\ \}$ or $\emptyset$
    
    Interval: none

13. $5x + 1 \leq 6$ and $3x + 2 \geq 5$
    $\quad 5x \leq 5 \qquad\qquad 3x \geq 3$
    $\quad x \leq 1 \qquad\qquad\ x \geq 1$
    
    We need $x \leq 1$ and $x \geq 1$. Looking at the graphs of the inequalities separately we see that the only number that is both less than or equal to 1 and greater than or equal to 1 is the number 1.
    
    Solution set: $\{1\}$
    
    Interval: $[1, 1]$
    
    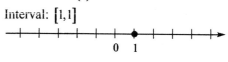

14. $-2 < 3x + 1 < 10$
    $\quad -2 - 1 < 3x + 1 - 1 < 10 - 1$
    $\quad -3 < 3x < 9$
    $\quad \dfrac{-3}{3} < \dfrac{3x}{3} < \dfrac{9}{3}$
    $\quad -1 < x < 3$
    
    Solution set: $\{x \mid -1 < x < 3\}$
    
    Interval: $(-1, 3)$
    
    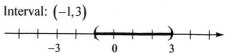

**15.** $0 < 4x - 5 \leq 3$

$0 + 5 < 4x - 5 + 5 \leq 3 + 5$

$5 < 4x \leq 8$

$\dfrac{5}{4} < \dfrac{4x}{4} \leq \dfrac{8}{4}$

$\dfrac{5}{4} < x \leq 2$

Solution set: $\left\{x \mid \dfrac{5}{4} < x \leq 2\right\}$

Interval: $\left(\dfrac{5}{4}, 2\right]$

**16.** $3 \leq -2x - 1 \leq 11$

$3 + 1 \leq -2x - 1 + 1 \leq 11 + 1$

$4 \leq -2x \leq 12$

$\dfrac{4}{-2} \geq \dfrac{-2x}{-2} \geq \dfrac{12}{-2}$

$-2 \geq x \geq -6$ or $-6 \leq x \leq -2$

Solution set: $\{x \mid -6 \leq x \leq -2\}$

Interval: $[-6, -2]$

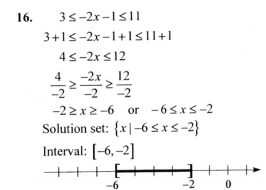

**17.** $x + 3 < 1$ or $x - 2 > 3$

$\quad x < -2 \qquad\quad x > 5$

We need $x < -2$ or $x > 5$.

Solution set: $\{x \mid x < -2 \text{ or } x > 5\}$

Interval: $(-\infty, -2) \cup (5, \infty)$

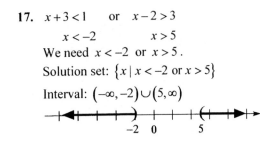

**18.** $3x + 1 \leq 7$ or $2x - 3 > 9$

$\quad 3x \leq 6 \qquad\quad 2x > 12$

$\quad x \leq 2 \qquad\quad\quad x > 6$

We need $x \leq 2$ or $x > 6$.

Solution set: $\{x \mid x \leq 2 \text{ or } x > 6\}$

Interval: $(-\infty, 2] \cup (6, \infty)$

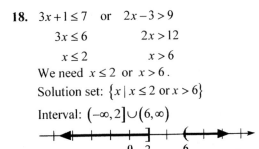

**19.** $2x - 3 \geq 1$ or $6x - 5 \geq 1$

$\quad 2x \geq 4 \qquad\quad 6x \geq 6$

$\quad x \geq 2 \qquad\quad\quad x \geq 1$

We need $x \geq 2$ or $x \geq 1$.

Solution set: $\{x \mid x \geq 1\}$

Interval: $[1, \infty)$

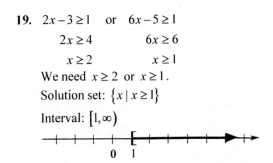

**20.** $\dfrac{3}{4}(x + 4) < 6$ or $\dfrac{3}{2}(x + 1) > 15$

$\quad x + 4 < 8 \qquad\quad x + 1 > 10$

$\quad x < 4 \qquad\quad\quad\quad x > 9$

We need $x < 4$ or $x > 9$.

Solution set: $\{x \mid x < 4 \text{ or } x > 9\}$

Interval: $(-\infty, 4) \cup (9, \infty)$

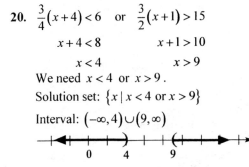

**21.** $3x - 2 > -5$ or $2x - 5 \leq 1$

$\quad 3x > -3 \qquad\quad 2x \leq 6$

$\quad x > -1 \qquad\quad\quad x \leq 3$

If we look at the graph of the inequalities separately, we see that the union of their solution sets is the set of real numbers.

Solution set: $\{x \mid -\infty < x < \infty\}$

Interval: $(-\infty, \infty)$

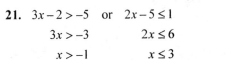

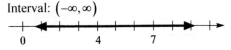

**22.** $-5x - 2 \leq 3$ or $7x - 9 > 5$

$\quad -5x \leq 5 \qquad\quad 7x > 14$

$\quad x \geq -1 \qquad\quad\quad x > 2$

We need $x \geq -1$ or $x > 2$. Since the solution set of the inequality $x > 2$ is a subset of the solution set for the inequality $x \geq -1$, we only need to consider $x \geq -1$.

Solution set: $\{x \mid x \geq -1\}$

Interval: $[-1, \infty)$

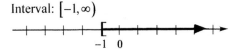

## Chapter 1: Linear Equations and Inequalities

23. Let x = taxable income (in dollars). The federal income tax in the 25% bracket is $4090 plus 25% of the amount over $29,700. In general, the income tax for the 25% bracket is given by
$4090 + 0.25(x - 29,700)$

    Because the federal income tax is between $4,090 and $14,652.50, we have
    $$4090 \le 4090 + 0.25(x - 29,700) \le 14,652.50$$
    $$4090 \le 4090 + 0.25x - 7425 \le 14,652.50$$
    $$4090 \le 0.25x - 3335 \le 14,652.50$$
    $$7425 \le 0.25x \le 17,987.50$$
    $$29,700 \le x \le 71,950$$

    To be in the 25% tax bracket, an individual would have an income between $29,700 and $71,950.

24. Let $x$ = number of minutes. The long distance plan charges $4.95 per month and $0.07 per minute. In general, the charge is given by $4.95 + 0.07x$. Sophia's charges ranged from $13.00 to $22.80, so we get the inequality
    $$13.00 \le 4.95 + 0.07x \le 22.80$$
    $$8.05 \le 0.07x \le 17.85$$
    $$115 \le x \le 255$$

    Over the course of the year, Sophia's monthly minutes were between 115 minutes and 255 minutes.

## 1.6 Quick Checks

1. $|x| = 7$

   $x = 7$ or $x = -7$ because both numbers are 7 units away from 0 on a real number line.

   Solution set: $\{-7, 7\}$

   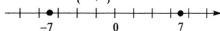

2. $|z| = 1$

   $z = 1$ or $z = -1$ because both numbers are 1 unit away from 0 on a real number line.

   Solution set: $\{-1, 1\}$

   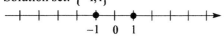

3. $|2x - 3| = 7$

   $2x - 3 = 7$    or    $2x - 3 = -7$
   $2x = 10$             $2x = -4$
   $x = 5$                $x = -2$

   Check:
   Let $x = 5$:            Let $x = -2$:
   $|2(5) - 3| \stackrel{?}{=} 7$      $|2(-2) - 3| \stackrel{?}{=} 7$
   $|10 - 3| \stackrel{?}{=} 7$         $|-4 - 3| \stackrel{?}{=} 7$
   $7 = 7$ ✓            $7 = 7$ ✓

   Solution set: $\{-2, 5\}$

   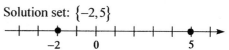

4. $|3x - 2| + 3 = 10$
   $|3x - 2| = 7$
   $3x - 2 = 7$    or    $3x - 2 = -7$
   $3x = 9$              $3x = -5$
   $x = 3$                $x = -\frac{5}{3}$

   Check:
   Let $x = 3$:            Let $x = -\frac{5}{3}$:
   $|3(3) - 2| + 3 \stackrel{?}{=} 10$    $\left|3\left(-\frac{5}{3}\right) - 2\right| + 3 \stackrel{?}{=} 10$
   $|9 - 2| + 3 \stackrel{?}{=} 10$       $|-5 - 2| + 3 \stackrel{?}{=} 10$
   $7 + 3 \stackrel{?}{=} 10$           $7 + 3 \stackrel{?}{=} 10$
   $10 = 10$ ✓         $10 = 10$ ✓

   Solution set: $\left\{-\frac{5}{3}, 3\right\}$

   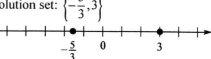

**Section 1.6 – Quick Checks:** Absolute Value Equations and Inequalities

5. $|-5x+2|-2=5$
   $|-5x+2|=7$
   $-5x+2=7$ or $-5x+2=-7$
   $-5x=5 \qquad -5x=-9$
   $x=-1 \qquad x=\dfrac{9}{5}$

   Check:
   Let $x=-1$: $\qquad$ Let $x=\dfrac{9}{5}$
   $|-5(-1)+2|-2\stackrel{?}{=}5 \quad \left|-5\left(\dfrac{9}{5}\right)+2\right|-2\stackrel{?}{=}5$
   $|5+2|-2\stackrel{?}{=}5 \qquad |-9+2|-2\stackrel{?}{=}5$
   $7-2\stackrel{?}{=}5 \qquad\qquad 7-2\stackrel{?}{=}5$
   $5=5 \checkmark \qquad\qquad 5=5 \checkmark$

   Solution set: $\left\{-1,\dfrac{9}{5}\right\}$
   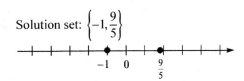

6. $3|x+2|-4=5$
   $3|x+2|=9$
   $|x+2|=3$
   $x+2=3$ or $x+2=-3$
   $x=1 \qquad x=-5$

   Check:
   Let $x=1$: $\qquad$ Let $x=-5$
   $3|1+2|-4\stackrel{?}{=}5 \qquad 3|-5+2|-4\stackrel{?}{=}5$
   $3(3)-4\stackrel{?}{=}5 \qquad 3(3)-4\stackrel{?}{=}5$
   $9-4\stackrel{?}{=}5 \qquad\qquad 9-4\stackrel{?}{=}5$
   $5=5 \checkmark \qquad\qquad 5=5 \checkmark$

   Solution set: $\{-5,1\}$

7. $|5x+3|=-2$
   Since absolute values are never negative, this equation has no solution.
   Solution set: $\{\ \}$ or $\varnothing$

8. $|2x+5|+7=3$
   $|2x+5|=-4$
   Since absolute values are never negative, this equation has no solution.
   Solution set: $\{\ \}$ or $\varnothing$

9. $|x+1|+3=3$
   $|x+1|=0$
   $x+1=0$
   $x=-1$

   Check:
   Let $x=-1$
   $|-1+1|+3\stackrel{?}{=}3$
   $0+3\stackrel{?}{=}3$
   $3=3 \checkmark$

   Solution set: $\{-1\}$

10. $|x-3|=|2x+5|$
    $x-3=2x+5$ or $x-3=-(2x+5)$
    $x=2x+8 \qquad x-3=-2x-5$
    $-x=8 \qquad\qquad x=-2x-2$
    $x=-8 \qquad\qquad 3x=-2$
    $\qquad\qquad\qquad x=-\dfrac{2}{3}$

    Check:
    Let $x=-8$ $\qquad$ Let $x=-\dfrac{2}{3}$
    $|-8-3|\stackrel{?}{=}|2(-8)+5| \quad \left|-\dfrac{2}{3}-3\right|\stackrel{?}{=}\left|2\left(-\dfrac{2}{3}\right)+5\right|$
    $|-11|=|-16+5| \qquad \left|-\dfrac{11}{3}\right|\stackrel{?}{=}\left|-\dfrac{4}{3}+5\right|$
    $11=11 \checkmark \qquad\qquad \dfrac{11}{3}=\dfrac{11}{3} \checkmark$

    Solution set: $\left\{-8,-\dfrac{2}{3}\right\}$

## Chapter 1: Linear Equations and Inequalities

**11.** $|8z+11| = |6z+17|$

$8z+11 = 6z+17$ or $8z+11 = -(6z+17)$
$8z = 6z+6$ $\quad\quad\quad 8z+11 = -6z-17$
$2z = 6$ $\quad\quad\quad\quad\quad 8z = -6z-28$
$z = 3$ $\quad\quad\quad\quad\quad\quad 14z = -28$
$\quad\quad\quad\quad\quad\quad\quad\quad\quad z = -2$

Check:
Let $z = -2$:
$|8(-2)+11| \stackrel{?}{=} |6(-2)+17|$
$|-16+11| \stackrel{?}{=} |-12+17|$
$5 = 5$ ✓

Let $z = 3$:
$|8(3)+11| \stackrel{?}{=} |6(3)+17|$
$|24+11| \stackrel{?}{=} |18+17|$
$35 = 35$ ✓

Solution set: $\{-2, 3\}$

**12.** $|3-2y| = |4y+3|$

$3-2y = 4y+3$ or $3-2y = -(4y+3)$
$-2y = 4y$ $\quad\quad\quad 3-2y = -4y-3$
$-6y = 0$ $\quad\quad\quad\quad -2y = -4y-6$
$y = 0$ $\quad\quad\quad\quad\quad 2y = -6$
$\quad\quad\quad\quad\quad\quad\quad\quad y = -3$

Check:
Let $y = 0$: $\quad\quad$ Let $y = -3$:
$|3-2(0)| \stackrel{?}{=} |4(0)+3|$ $\quad |3-2(-3)| \stackrel{?}{=} |4(-3)+3|$
$|3| \stackrel{?}{=} |3|$ $\quad\quad\quad\quad |3+6| \stackrel{?}{=} |-12+3|$
$3 = 3$ ✓ $\quad\quad\quad\quad\quad 3 = 3$ ✓

Solution set: $\{-3, 0\}$

**13.** $|2x-3| = |5-2x|$

$2x-3 = 5-2x$ or $2x-3 = -(5-2x)$
$2x = 8-2x$ $\quad\quad 2x-3 = -5+2x$
$4x = 8$ $\quad\quad\quad\quad 2x = -2+2x$
$x = 2$ $\quad\quad\quad\quad\quad 0 = -2$ false

The second equation leads to a contradiction. Therefore, the only solution is $x = 2$.

Check:
Let $x = 2$:
$|2(2)-3| \stackrel{?}{=} |5-2(2)|$
$|4-3| \stackrel{?}{=} |5-4|$
$1 = 1$ ✓

Solution set: $\{2\}$

**14.** $|x| \leq 5$

$-5 \leq x \leq 5$

Solution set: $\{x \mid -5 \leq x \leq 5\}$

Interval: $[-5, 5]$

**15.** $|x| < \dfrac{3}{2}$

$-\dfrac{3}{2} < x < \dfrac{3}{2}$

Solution set: $\left\{x \mid -\dfrac{3}{2} < x < \dfrac{3}{2}\right\}$

Interval: $\left(-\dfrac{3}{2}, \dfrac{3}{2}\right)$

**16.** $|x+3| < 5$

$-5 < x+3 < 5$
$-5-3 < x+3-3 < 5-3$
$-8 < x < 2$

Solution set: $\{x \mid -8 < x < 2\}$

Interval: $(-8, 2)$

### Section 1.6 – Quick Checks: Absolute Value Equations and Inequalities

17. $|2x-3| \leq 7$
    $-7 \leq 2x-3 \leq 7$
    $-7+3 \leq 2x-3+3 \leq 7+3$
    $-4 \leq 2x \leq 10$
    $\dfrac{-4}{2} \leq \dfrac{2x}{2} \leq \dfrac{10}{2}$
    $-2 \leq x \leq 5$
    Solution set: $\{x \mid -2 \leq x \leq 5\}$
    Interval: $[-2, 5]$

18. $|7x+2| < -3$
    Since absolute values are never negative, this inequality has no solution.
    Solution set: $\{\ \}$ or $\varnothing$

19. $|x|+4 < 6$
    $|x| < 2$
    $-2 < x < 2$
    Solution set: $\{x \mid -2 < x < 2\}$; Interval: $(-2, 2)$
    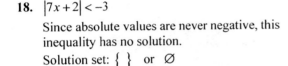

20. $|x-3|+4 \leq 8$
    $|x-3| \leq 4$
    $-4 \leq x-3 \leq 4$
    $-4+3 \leq x-3+3 \leq 4+3$
    $-1 \leq x \leq 7$
    Solution set: $\{x \mid -1 \leq x \leq 7\}$
    Interval: $[-1, 7]$
    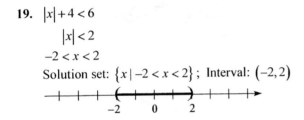

21. $3|2x+1| \leq 9$
    $|2x+1| \leq 3$
    $-3 \leq 2x+1 \leq 3$
    $-3-1 \leq 2x+1-1 \leq 3-1$
    $-4 \leq 2x \leq 2$
    $\dfrac{-4}{2} \leq \dfrac{2x}{2} \leq \dfrac{2}{2}$
    $-2 \leq x \leq 1$
    Solution set: $\{x \mid -2 \leq x \leq 1\}$
    Interval: $[-2, 1]$
    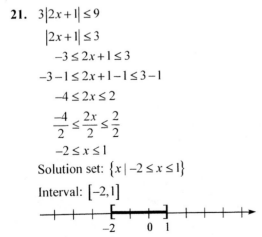

22. $|-3x+1|-5 < 3$
    $|-3x+1| < 8$
    $-8 < -3x+1 < 8$
    $-8-1 < -3x+1-1 < 8-1$
    $-9 < -3x < 7$
    $\dfrac{-9}{-3} > \dfrac{-3x}{-3} > \dfrac{7}{-3}$
    $3 > x > -\dfrac{7}{3}$ or $-\dfrac{7}{3} < x < 3$
    Solution set: $\left\{x \mid -\dfrac{7}{3} < x < 3\right\}$
    Interval: $\left(-\dfrac{7}{3}, 3\right)$
    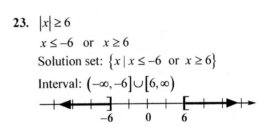

23. $|x| \geq 6$
    $x \leq -6$ or $x \geq 6$
    Solution set: $\{x \mid x \leq -6$ or $x \geq 6\}$
    Interval: $(-\infty, -6] \cup [6, \infty)$
    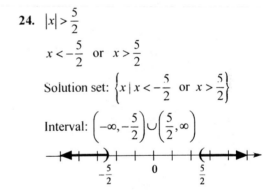

24. $|x| > \dfrac{5}{2}$
    $x < -\dfrac{5}{2}$ or $x > \dfrac{5}{2}$
    Solution set: $\left\{x \mid x < -\dfrac{5}{2}$ or $x > \dfrac{5}{2}\right\}$
    Interval: $\left(-\infty, -\dfrac{5}{2}\right) \cup \left(\dfrac{5}{2}, \infty\right)$

25. $|x+3| > 4$
    $x+3 < -4$ or $x+3 > 4$
    $x < -7$      $x > 1$
    Solution set: $\{x \mid x < -7$ or $x > 1\}$
    Interval: $(-\infty, -7) \cup (1, \infty)$
    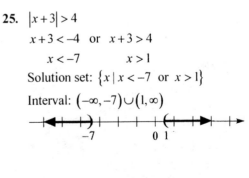

## Chapter 1: Linear Equations and Inequalities

**26.** $|4x-3| \geq 5$

$4x-3 \leq -5$  or  $4x-3 \geq 5$
$4x \leq -2$           $4x \geq 8$
$x \leq -\frac{1}{2}$          $x \geq 2$

Solution set: $\left\{x \mid x \leq -\frac{1}{2} \text{ or } x \geq 2\right\}$

Interval: $\left(-\infty, -\frac{1}{2}\right] \cup [2, \infty)$

**27.** $|-3x+2| > 7$

$-3x+2 < -7$  or  $-3x+2 > 7$
$-3x < -9$              $-3x > 5$
$x > 3$                    $x < -\frac{5}{3}$

Solution set: $\left\{x \mid x < -\frac{5}{3} \text{ or } x > 3\right\}$

Interval: $\left(-\infty, -\frac{5}{3}\right) \cup (3, \infty)$

**28.** $|2x+5| - 2 > -2$

$|2x+5| > 0$
$2x+5 < 0$    or   $2x+5 > 0$
$2x < -5$             $2x > -5$
$x < -\frac{5}{2}$          $x > -\frac{5}{2}$

Solution set: $\left\{x \mid x \neq -\frac{5}{2}\right\}$

Interval: $\left(-\infty, -\frac{5}{2}\right) \cup \left(-\frac{5}{2}, \infty\right)$

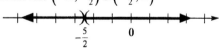

**29.** $|6x-5| \geq 0$

Since absolute values are always nonnegative, all real numbers are solutions to this inequality.
Solution set: $\{x \mid x \text{ is any real number}\}$

Interval: $(-\infty, \infty)$

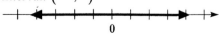

**30.** $|2x+1| > -3$

Since absolute values are always nonnegative, all real numbers are solutions to this inequality.
Solution set: $\{x \mid x \text{ is any real number}\}$

Interval: $(-\infty, \infty)$

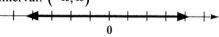

**31.** $|x-4| \leq \frac{1}{32}$

$-\frac{1}{32} \leq x-4 \leq \frac{1}{32}$

$-\frac{1}{32} + 4 \leq x - 4 + 4 \leq \frac{1}{32} + 4$

$-\frac{1}{32} + \frac{128}{32} \leq x \leq \frac{1}{32} + \frac{128}{32}$

$\frac{127}{32} \leq x \leq \frac{129}{32}$

The acceptable belt widths are between $\frac{127}{32}$ inches and $\frac{129}{32}$ inches.

**32.** $|p-9| \leq 1.7$

$-1.7 \leq p - 9 \leq 1.7$
$-1.7 + 9 \leq p - 9 + 9 \leq 1.7 + 9$
$7.3 \leq p \leq 10.7$

The percentage of people that have been shot at is between 7.3 percent and 10.7 percent.

# Chapter 2 – Quick Checks

**2.1 Quick Checks**

1. $A$: quadrant I
   $B$: quadrant IV
   $C$: y-axis
   $D$: quadrant III

   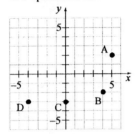

2. $A$: quadrant II
   $B$: x-axis
   $C$: quadrant IV
   $D$: quadrant I

   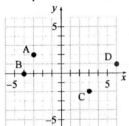

3. a. Let $x = 2$ and $y = -3$.
   $$2(2) - 4(-3) \stackrel{?}{=} 12$$
   $$4 + 12 \stackrel{?}{=} 12$$
   $$16 = 12 \text{ False}$$
   The point $(2, -3)$ is not on the graph.

   b. Let $x = 2$ and $y = -2$.
   $$2(2) - 4(-2) \stackrel{?}{=} 12$$
   $$4 + 8 \stackrel{?}{=} 12$$
   $$12 = 12 \checkmark$$
   The point $(2, -2)$ is on the graph.

   c. Let $x = \frac{3}{2}$ and $y = -\frac{9}{4}$.
   $$2\left(\frac{3}{2}\right) - 4\left(-\frac{9}{4}\right) \stackrel{?}{=} 12$$
   $$3 + 9 \stackrel{?}{=} 12$$
   $$12 = 12 \checkmark$$
   The point $\left(\frac{3}{2}, -\frac{9}{4}\right)$ is on the graph.

4. a. Let $x = 1$ and $y = 4$.
   $$(4) \stackrel{?}{=} (1)^2 + 3$$
   $$4 \stackrel{?}{=} 1 + 3$$
   $$4 = 4 \checkmark$$
   The point $(1, 4)$ is on the graph.

   b. Let $x = -2$ and $y = -1$.
   $$(-1) \stackrel{?}{=} (-2)^2 + 3$$
   $$-1 \stackrel{?}{=} 4 + 3$$
   $$-1 = 7 \text{ False}$$
   The point $(-2, -1)$ is not on the graph.

   c. Let $x = -3$ and $y = 12$.
   $$(12) \stackrel{?}{=} (-3)^2 + 3$$
   $$12 \stackrel{?}{=} 9 + 3$$
   $$12 = 12 \checkmark$$
   The point $(-3, 12)$ is on the graph.

**Chapter 2:** *Graphs, Relations, and Functions*

5. $y = 3x + 1$

| $x$ | $y = 3x + 1$ | $(x, y)$ |
|---|---|---|
| $-2$ | $y = 3(-2) + 1 = -5$ | $(-2, -5)$ |
| $-1$ | $y = 3(-1) + 1 = -2$ | $(-1, -2)$ |
| $0$ | $y = 3(0) + 1 = 1$ | $(0, 1)$ |
| $1$ | $y = 3(1) + 1 = 4$ | $(1, 4)$ |
| $2$ | $y = 3(2) + 1 = 7$ | $(2, 7)$ |

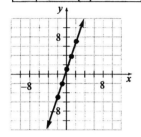

6. $2x + 3y = 8$

$$3y = -2x + 8$$
$$y = -\frac{2}{3}x + \frac{8}{3}$$

| $x$ | $y = -\frac{2}{3}x + \frac{8}{3}$ | $(x, y)$ |
|---|---|---|
| $-5$ | $y = -\frac{2}{3}(-5) + \frac{8}{3} = 6$ | $(-5, 6)$ |
| $-2$ | $y = -\frac{2}{3}(-2) + \frac{8}{3} = 4$ | $(-2, 4)$ |
| $1$ | $y = -\frac{2}{3}(1) + \frac{8}{3} = 2$ | $(1, 2)$ |
| $4$ | $y = -\frac{2}{3}(4) + \frac{8}{3} = 0$ | $(4, 0)$ |
| $7$ | $y = -\frac{2}{3}(7) + \frac{8}{3} = -2$ | $(7, -2)$ |

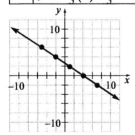

7. $y = x^2 + 3$

| $x$ | $y = x^2 + 3$ | $(x, y)$ |
|---|---|---|
| $-2$ | $y = (-2)^2 + 3 = 7$ | $(-2, 7)$ |
| $-1$ | $y = (-1)^2 + 3 = 4$ | $(-1, 4)$ |
| $0$ | $y = (0)^2 + 3 = 3$ | $(0, 3)$ |
| $1$ | $y = (1)^2 + 3 = 4$ | $(1, 4)$ |
| $2$ | $y = (2)^2 + 3 = 7$ | $(2, 7)$ |

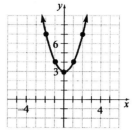

8. $x = y^2 + 2$

| $y$ | $x = y^2 + 2$ | $(x, y)$ |
|---|---|---|
| $-2$ | $x = (-2)^2 + 2 = 6$ | $(6, -2)$ |
| $-1$ | $x = (-1)^2 + 2 = 3$ | $(3, -1)$ |
| $0$ | $x = (0)^2 + 2 = 2$ | $(2, 0)$ |
| $1$ | $x = (1)^2 + 2 = 3$ | $(3, 1)$ |
| $2$ | $x = (2)^2 + 2 = 6$ | $(6, 2)$ |

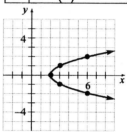

9. $x = (y - 1)^2$

| $y$ | $x = (y - 1)^2$ | $(x, y)$ |
|---|---|---|
| $-1$ | $x = (-1 - 1)^2 = 4$ | $(4, -1)$ |
| $0$ | $x = (0 - 1)^2 = 1$ | $(1, 0)$ |
| $1$ | $x = (1 - 1)^2 = 0$ | $(0, 1)$ |
| $2$ | $x = (2 - 1)^2 = 1$ | $(1, 2)$ |
| $3$ | $x = (3 - 1)^2 = 4$ | $(4, 3)$ |

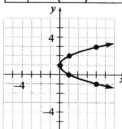

### Section 2.2 – Quick Checks: Relations

10. To find the intercepts, we look for the points where the graph crosses or touches either coordinate axis. From the graph, we see that the intercepts are $(-5, 0)$, $(0, -0.9)$, $(1, 0)$, and $(6.7, 0)$.

    The x-intercepts are the points where the graph crosses or touches the x-axis. From the graph we see that the x-intercepts are $-5$, $1$, and $6.7$.

    The y-intercept is the point where the graph crosses the y-axis. From the graph we see that the y-intercept is $-0.9$.

11. a. Locate the value 250 along the x-axis, go up to the graph, and then read the corresponding value on the y-axis. According to the graph, the cost of refining 250 thousand gallons of gasoline per hour is $200 thousand.

    b. Locate the value 400 along the x-axis, go up to the graph, and then read the corresponding value on the y-axis. According to the graph, the cost of refining 400 thousand gallons of gasoline per hour is $350 thousand.

    c. Since the horizontal axis represents the number of gallons per hour than can be refined, the graph ending at 700 thousand gallons per hour represents the capacity of the refinery per hour.

    d. The intercept is $(0, 100)$. This represents the fixed costs of operating the refinery. That is, refining 0 gallons per hour costs $100 thousand.

### 2.2 Quick Checks

1. The first element of the ordered pair comes from the set 'Friend' and the second element is the corresponding element from the set 'Birthday'.
   {(Max, November 8), (Alesia, January 20), (Trent, March 3), (Yolanda, November 8), (Wanda, July 6), (Elvis, January 8)}

2. The first elements of the ordered pairs make up the domain and the second elements make up the range.

    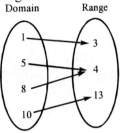

3. The domain is the set of all inputs and the range is the set of all outputs. The inputs are the elements in the set 'Friend' and the outputs are the elements in the set 'Birthday'.
   Domain:
   {Max, Alesia, Trent, Yolanda, Wanda, Elvis}
   Range:
   {January 20, March 3, July 6, November 8, January 8}

4. The domain is the set of all inputs and the range is the set of all outputs. The inputs are the first elements in the ordered pairs and the outputs are the second elements in the ordered pairs.
   Domain:          Range:
   {1, 5, 8, 10}     {3, 4, 13}

5. First we notice that the ordered pairs on the graph are $(-2, 0)$, $(-1, 2)$, $(-1, -2)$, $(2, 3)$, $(3, 0)$, and $(4, -3)$.
   The domain is the set of all x-coordinates and the range is the set of all y-coordinates.
   Domain:                Range:
   $\{-2, -1, 2, 3, 4\}$    $\{-3, -2, 0, 2, 3\}$

6. To find the domain, we first determine the x-values for which the graph exists. The graph exists for all x-values between $-2$ and $4$, inclusive. Thus, the domain is $\{x \mid -2 \leq x \leq 4\}$, or $[-2, 4]$ if we use interval notation.

    To find the range, we first determine the y-values for which the graph exists. The graph exists for all y-values between $-2$ and $2$, inclusive. Thus, the range is $\{y \mid -2 \leq y \leq 2\}$, or $[-2, 2]$ if we use interval notation.

## Chapter 2: Graphs, Relations, and Functions

7. To find the domain, we first determine the x-values for which the graph exists. The graph exists for all x-values on a real number line. Thus, the domain is $\{x \mid x \text{ is any real number}\}$, or $(-\infty, \infty)$ if we use interval notation.

   To find the range, we first determine the y-values for which the graph exists. The graph exists for all y-values on a real number line. Thus, the range is $\{y \mid y \text{ is any real number}\}$, or $(-\infty, \infty)$ if we use interval notation.

8. $y = 3x - 8$

   | $x$ | $y = 3x - 8$ | $(x, y)$ |
   |---|---|---|
   | $-1$ | $y = 3(-1) - 8 = -11$ | $(-1, -11)$ |
   | $0$ | $y = 3(0) - 8 = -8$ | $(0, -8)$ |
   | $1$ | $y = 3(1) - 8 = -5$ | $(1, -5)$ |
   | $2$ | $y = 3(2) - 8 = -2$ | $(2, -2)$ |
   | $3$ | $y = 3(3) - 8 = 1$ | $(3, 1)$ |

   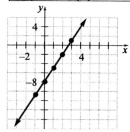

   Domain: $\{x \mid x \text{ is any real number}\}$ or $(-\infty, \infty)$
   Range: $\{y \mid y \text{ is any real number}\}$ or $(-\infty, \infty)$

9. $y = x^2 - 8$

   | $x$ | $y = x^2 - 8$ | $(x, y)$ |
   |---|---|---|
   | $-3$ | $y = (-3)^2 - 8 = 1$ | $(-3, 1)$ |
   | $-2$ | $y = (-2)^2 - 8 = -4$ | $(-2, -4)$ |
   | $0$ | $y = (0)^2 - 8 = -8$ | $(0, -8)$ |
   | $2$ | $y = (2)^2 - 8 = -4$ | $(2, -4)$ |
   | $3$ | $y = (3)^2 - 8 = 1$ | $(3, 1)$ |

   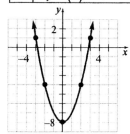

   Domain: $\{x \mid x \text{ is any real number}\}$ or $(-\infty, \infty)$
   Range: $\{y \mid y \geq -8\}$ or $[-8, \infty)$

10. $x = y^2 + 1$

    | $y$ | $x = y^2 + 1$ | $(x, y)$ |
    |---|---|---|
    | $-2$ | $x = (-2)^2 + 1 = 5$ | $(5, -2)$ |
    | $-1$ | $x = (-1)^2 + 1 = 2$ | $(2, -1)$ |
    | $0$ | $x = (0)^2 + 1 = 1$ | $(1, 0)$ |
    | $1$ | $x = (1)^2 + 1 = 2$ | $(2, 1)$ |
    | $2$ | $x = (2)^2 + 1 = 5$ | $(5, 2)$ |

    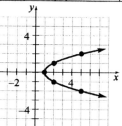

    Domain: $\{x \mid x \geq 1\}$ or $[1, \infty)$
    Range: $\{y \mid y \text{ is any real number}\}$ or $(-\infty, \infty)$

### 2.3 Quick Checks

1. The relation is a function because each element in the domain (Friend) corresponds to exactly one element in the range (Birthday).
   Domain: {Max, Alesia, Trent, Yolanda, Wanda, Elvis}
   Range: {January 20, March 3, July 6, November 8, January 8}

2. The relation is not a function because there is an element in the domain, 210, that corresponds to more than one element in the domain. If 210 is selected from the domain, a single sugar content cannot be determined.

3. The relation is a function because there are no ordered pairs with the same first coordinate, but different second coordinates.
   Domain: $\{-3, -2, -1, 0, 1\}$
   Range: $\{0, 1, 2, 3\}$

4. The relation is not a function because there are two ordered pairs, $(-3, 2)$ and $(-3, 6)$, with the same first coordinate, but different second coordinates.

## Section 2.3 – Quick Checks: An Introduction to Functions

5. $y = -2x + 5$
   The relation is a function since there is only 1 output than can result for each input.

6. $y = \pm 3x$
   The relation is not a function since a single input for $x$ will yield two output values for $y$. For example, if $x = 1$, then $y = \pm 3$.

7. $y = x^2 + 5x$
   The relation is a function since there is only 1 output than can result for each input.

8. $x + y^2 = 9$
   The relation is not a function since a single input for $x$ can yield two output values for $y$. For example, if $x = 0$, then we have
   $$0 + y^2 = 9$$
   $$y^2 = 9$$
   $$y = \pm 3$$

9. The graph is that of a function because every vertical line will cross the graph in at most one point.

10. The graph is not that of a function because a vertical line can cross the graph in more than one point.

11. $f(x) = 3x + 2$
    $f(4) = 3(4) + 2$
    $\phantom{f(4)} = 12 + 2$
    $\phantom{f(4)} = 14$

12. $g(x) = -2x^2 + x - 3$
    $g(-2) = -2(-2)^2 + (-2) - 3$
    $\phantom{g(-2)} = -2(4) - 5$
    $\phantom{g(-2)} = -8 - 5$
    $\phantom{g(-2)} = -13$

13. $f(x) = 3x + 2$
    $f(x-2) = 3(x-2) + 2$
    $\phantom{f(x-2)} = 3x - 6 + 2$
    $\phantom{f(x-2)} = 3x - 4$

14. $f(x) - f(2) = [3x + 2] - [3(2) + 2]$
    $\phantom{f(x) - f(2)} = 3x + 2 - 8$
    $\phantom{f(x) - f(2)} = 3x - 6$

15. $f(x) = -2x + 9$

    | $x$ | $y = f(x) = -2x + 9$ | $(x, y)$ |
    |---|---|---|
    | $-2$ | $f(-2) = -2(-2) + 9 = 13$ | $(-2, 13)$ |
    | $0$ | $f(0) = -2(0) + 9 = 9$ | $(0, 9)$ |
    | $2$ | $f(2) = -2(2) + 9 = 5$ | $(2, 5)$ |
    | $4$ | $f(4) = -2(4) + 9 = 1$ | $(4, 1)$ |
    | $6$ | $f(6) = -2(6) + 9 = -3$ | $(6, -3)$ |

    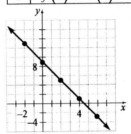

16. $f(x) = x^2 + 2$

    | $x$ | $y = f(x) = x^2 + 2$ | $(x, y)$ |
    |---|---|---|
    | $-3$ | $f(-3) = (-3)^2 + 2 = 11$ | $(-3, 11)$ |
    | $-1$ | $f(-1) = (-1)^2 + 2 = 3$ | $(-1, 3)$ |
    | $0$ | $f(0) = (0)^2 + 2 = 2$ | $(0, 2)$ |
    | $1$ | $f(1) = (1)^2 + 2 = 3$ | $(1, 3)$ |
    | $3$ | $f(3) = (3)^2 + 2 = 11$ | $(3, 11)$ |

    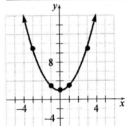

## Chapter 2: Graphs, Relations, and Functions

17. $f(x) = |x - 2|$

| $x$ | $y = f(x) = |x-2|$ | $(x, y)$ |
|---|---|---|
| $-2$ | $f(-2) = |-2-2| = 4$ | $(-2, 4)$ |
| $0$ | $f(0) = |0-2| = 2$ | $(0, 2)$ |
| $2$ | $f(2) = |2-2| = 0$ | $(2, 0)$ |
| $4$ | $f(4) = |4-2| = 2$ | $(4, 2)$ |
| $6$ | $f(6) = |6-2| = 4$ | $(6, 4)$ |

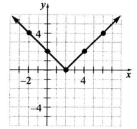

18. a. Independent variable: $t$ (number of days)
    Dependent variable: $A$ (square miles)

    b. $A(t) = 0.25\pi t^2$
    $A(30) = 0.25\pi (30)^2 \approx 706.86$ sq. miles
    After the tanker has been leaking for 30 days, the circular oil slick will cover about 706.86 square miles.

**2.4 Quick Checks**

1. $f(x) = 3x^2 + 2$
   The function tells us to square a number $x$, multiply by 3, and then add 2. Since these operations can be performed on any real number, the domain of $f$ is the set of all real numbers. The domain can be written as $\{x \mid x \text{ is any real number}\}$, or $(-\infty, \infty)$ in interval notation.

2. $h(x) = \dfrac{x+1}{x-3}$
   The function tells us to divide $x+1$ by $x-3$. Since division by 0 is not defined, the denominator $x-3$ can never be 0. Therefore, $x$ can never equal 3. The domain can be written as $\{x \mid x \neq 3\}$, or $(-\infty, 3) \cup (3, \infty)$ in interval notation.

3. $A(r) = \pi r^2$
   Since $r$ represents the radius of the circle, it must take on positive values. Therefore, the domain is $\{r \mid r > 0\}$, or $(0, \infty)$ in interval notation.

4. a. The arrows on the ends of the graph indicate that the graph continues indefinitely. Therefore, the domain is $\{x \mid x \text{ is any real number}\}$, or $(-\infty, \infty)$ in interval notation.
   The function reaches a maximum value of 2, but has no minimum value. Therefore, the range is $\{y \mid y \leq 2\}$, or $(-\infty, 2]$ in interval notation.

   b. The intercepts are $(-2, 0)$, $(0, 2)$, and $(2, 0)$. The x-intercepts are $-2$ and 2, and the y-intercept is 2.

5. a. Since $(-3, -15)$ and $(1, -3)$ are on the graph of $f$, then $f(-3) = -15$ and $f(1) = -3$.

   b. To determine the domain, notice that the graph exists for all real numbers. Thus, the domain is $\{x \mid x \text{ is any real number}\}$, or $(-\infty, \infty)$ in interval notation.

   c. To determine the range, notice that the function can assume any real number. Thus, the range is $\{y \mid y \text{ is any real number}\}$, or $(-\infty, \infty)$ in interval notation.

   d. The intercepts are $(-2, 0)$, $(0, 0)$, and $(2, 0)$. The x-intercepts are $-2$, 0, and 2. The y-intercept is 0.

   e. Since $(3, 15)$ is the only point on the graph where $y = f(x) = 15$, the solution set to $f(x) = 15$ is $\{3\}$.

**Section 2.4 – Quick Checks:** Functions and Their Graphs

**6. a.** When $x = -2$, then
$$f(x) = -3x + 7$$
$$f(-2) = -3(-2) + 7$$
$$= 6 + 7$$
$$= 13$$
Since $f(-2) = 13$, the point $(-2, 13)$ is on the graph. This means the point $(-2, 1)$ is **not** on the graph.

**b.** If $x = 3$, then
$$f(x) = -3x + 7$$
$$f(3) = -3(3) + 7$$
$$= -9 + 7$$
$$= -2$$
The point $(3, -2)$ is on the graph.

**c.** If $f(x) = -8$, then
$$f(x) = -8$$
$$-3x + 7 = -8$$
$$-3x = -15$$
$$x = 5$$
If $f(x) = -8$, then $x = 5$. The point $(5, -8)$ is on the graph.

**7.** Maria's distance from home is a function of time so we put time (in minutes) on the horizontal axis and distance (in blocks) on the vertical axis. Starting at the origin (0, 0), we draw a straight line to the point (5, 5). The ordered pair (5, 5) represents Maria being 5 blocks from home after 5 minutes. From the point (5, 5), we draw a straight line to the point (7, 0) that represents her trip back home. The ordered pair (7, 0) represents Maria being back at home after 7 minutes. Draw a line segment from $(7, 0)$ to $(8, 0)$ to represent the time it takes Maria to find her keys and lock the door. Next, draw a line segment from $(8, 0)$ to $(13, 8)$ that represents her 8 block run in 5 minutes. Then draw a line segment from $(13, 8)$ to $(14, 11)$ that represents her 3 block run in 1 minute. Now draw a horizontal line from $(14, 11)$ to $(16, 11)$ that represents Maria's resting period. Finally, draw a line segment from $(16, 11)$ to $(26, 0)$ that

represents her walk home.

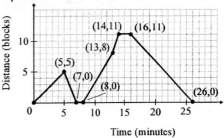

QC – 33

# Chapter 3 – Quick Checks

## 3.1 Quick Checks

1. $y = 2x - 3$

   Let $x = -1, 0, 1,$ and $2$.

   $x = -1$:  $y = 2(-1) - 3$
   $y = -2 - 3$
   $y = -5$

   $x = 0$:  $y = 2(0) - 3$
   $y = 0 - 3$
   $y = -3$

   $x = 1$:  $y = 2(1) - 3$
   $y = 2 - 3$
   $y = -1$

   $x = 2$:  $y = 2(2) - 3$
   $y = 4 - 3$
   $y = 1$

   Thus, the points $(-1, -5)$, $(0, -3)$, $(1, -1)$, and $(2, 1)$ are on the graph.

   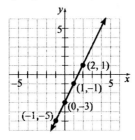

2. $\dfrac{1}{2}x + y = 2$

   Let $x = -2, 0, 2,$ and $4$.

   $x = -2$:  $\dfrac{1}{2}(-2) + y = 2$
   $-1 + y = 2$
   $y = 3$

   $x = 0$:  $\dfrac{1}{2}(0) + y = 2$
   $0 + y = 2$
   $y = 2$

   $x = 2$:  $\dfrac{1}{2}(2) + y = 2$
   $1 + y = 2$
   $y = 1$

   $x = 4$:  $\dfrac{1}{2}(4) + y = 2$
   $2 + y = 2$
   $y = 0$

   Thus, the points $(-2, 3)$, $(0, 2)$, $(2, 1)$, and $(4, 0)$ are on the graph.

   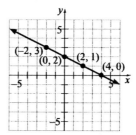

3. $-6x + 3y = 12$

   Let $x = -2, -1, 0,$ and $1$.

   $x = -2$:  $-6(-2) + 3y = 12$
   $12 + 3y = 12$
   $3y = 0$
   $y = 0$

   $x = -1$:  $-6(-1) + 3y = 12$
   $6 + 3y = 12$
   $3y = 6$
   $y = 2$

   $x = 0$:  $-6(0) + 3y = 12$
   $0 + 3y = 12$
   $3y = 12$
   $y = 4$

   $x = 1$:  $-6(1) + 3y = 12$
   $-6 + 3y = 12$
   $3y = 18$
   $y = 6$

   Thus, the points $(-2, 0)$, $(-1, 2)$, $(0, 4)$, and $(1, 6)$ are on the graph.

   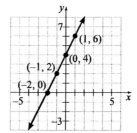

## Section 3.1 – Quick Checks: Linear Equations and Linear Functions

4. $x + y = 4$

   To find the $y$-intercept, let $x = 0$ and solve for $y$.
   
   $x = 0$: $0 + y = 4$
   
   $y = 4$
   
   The $y$-intercept is 4, so the point $(0, 4)$ is on the graph. To find the $x$-intercept, let $y = 0$ and solve for $x$.
   
   $y = 0$: $x + 0 = 4$
   
   $x = 4$
   
   The $x$-intercept is 4, so the point $(4, 0)$ is on the graph.

   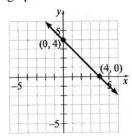

5. $4x - 5y = 20$

   To find the $y$-intercept, let $x = 0$ and solve for $y$.
   
   $x = 0$: $4(0) - 5y = 20$
   
   $-5y = 20$
   
   $y = -4$
   
   The $y$-intercept is $-4$, so the point $(0, -4)$ is on the graph. To find the $x$-intercept, let $y = 0$ and solve for $x$.
   
   $y = 0$: $4x - 5(0) = 20$
   
   $4x = 20$
   
   $x = 5$
   
   The $x$-intercept is 5, so the point $(5, 0)$ is on the graph.

   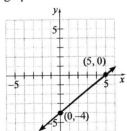

6. $3x - 2y = 0$

   To find the $y$-intercept, let $x = 0$ and solve for $y$.
   
   $x = 0$: $3(0) - 2y = 0$
   
   $-2y = 0$
   
   $y = 0$
   
   The $y$-intercept is 0, so the point $(0,0)$ is on the graph. To find the $x$-intercept, let $y = 0$ and solve for $x$.
   
   $y = 0$: $3x - 2(0) = 0$
   
   $3x = 0$
   
   $x = 0$
   
   The $x$-intercept is 0, so the point $(0,0)$ is on the graph. Since the $x$- and $y$-intercepts both result in the origin, we must find other points on the graph. Let $x = -2$ and 2, and solve for $y$.
   
   $x = -2$: $3(-2) - 2y = 0$
   
   $-6 - 2y = 0$
   
   $-2y = 6$
   
   $y = -3$
   
   $x = 2$: $3(2) - 2y = 0$
   
   $6 - 2y = 0$
   
   $-2y = -6$
   
   $y = 3$
   
   Thus, the points $(-2, -3)$ and $(2, 3)$ are also on the graph.

   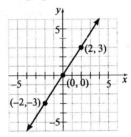

7. The equation $x = 5$ will be a vertical line with $x$-intercept 5.

   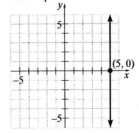

8. The equation $y = -4$ will be a horizontal line with $y$-intercept $-4$.

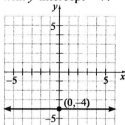

9. a. The independent variable is the number of miles driven, $x$. It does not make sense to drive a negative number of miles, we have that the domain of the function is $\{x \mid x \geq 0\}$ or, using interval notation, $[0, \infty)$.

   b. To determine the $y$-intercept, we find $C(0) = 0.35(0) + 40 = 40$. The $y$-intercept is 40, so the point $(0, 40)$ is on the graph.

   c. $C(80) = 0.35(80) + 40 = 28 + 40 = 68$. If the truck is driven 80 miles, the rental cost will be $68.

   d. We plot the independent variable, *number of miles driven*, on the horizontal axis and the dependent variable, *rental cost*, on the vertical axis. From parts (b) and (c), we have that the points $(0, 40)$ and $(80, 68)$ are on the graph. We find one more point by evaluating the function for $x = 200$: $C(200) = 0.35(200) + 40 = 70 + 40 = 110$. The point $(200, 110)$ is also on the graph.

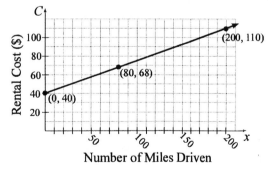

   e. We solve $C(x) = 85.50$:
   $0.35x + 40 = 85.50$
   $0.35x = 45.50$
   $x = 130$
   If the rental cost is $85.50, then the truck was driven 130 miles.

## 3.2 Quick Checks

1. $m = \dfrac{y_2 - y_1}{x_2 - x_1} = \dfrac{12 - 3}{3 - 0} = \dfrac{9}{3} = 3$

   For every 1 unit of increase in $x$, $y$ will increase by 3 units.

2. $m = \dfrac{y_2 - y_1}{x_2 - x_1} = \dfrac{-4 - 3}{3 - (-1)} = \dfrac{-7}{4} = -\dfrac{7}{4}$

   For every 4 units of increase in $x$, $y$ will decrease by 7 units. For every 4 units of decrease in $x$, $y$ will increase by 7 units.

3. $m = \dfrac{y_2 - y_1}{x_2 - x_1} = \dfrac{2 - 2}{-3 - 3} = \dfrac{0}{-6} = 0$

   The line is horizontal.

4. $m = \dfrac{y_2 - y_1}{x_2 - x_1} = \dfrac{-1 - 4}{-2 - (-2)} = \dfrac{-5}{0} =$ undefined

   The line is vertical.

5. $m_1 = \dfrac{4 - 3}{6 - 1} = \dfrac{1}{5}$; $m_2 = \dfrac{8 - 3}{1 - 1} = \dfrac{5}{0} =$ undefined;
   $m_3 = \dfrac{7 - 3}{-3 - 1} = \dfrac{4}{-4} = -1$; $m_4 = \dfrac{3 - 3}{-4 - 1} = \dfrac{0}{-5} = 0$

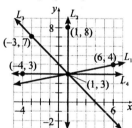

6. a.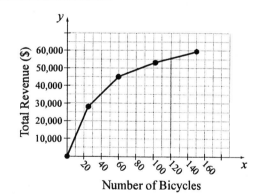

   b. Average rate of change $= \dfrac{28{,}000 - 0}{25 - 0}$
   $= \$1120$ per bicycle

   For each bicycle sold, total revenue increased by $1120 when between 0 and 25 bicycles were sold.

### Section 3.2 – Quick Checks: Slope and Equations of Lines

**c.** Average rate of change $= \dfrac{59,160 - 53,400}{150 - 102}$
$= \$120$ per bicycle

For each bicycle sold, total revenue increased by $120 when between 102 and 150 bicycles were sold.

**d.** No, the revenue does not grow linearly. The average rate of change (slope) is not constant.

**7. a.** Start at $(-1, 3)$. Because $m = \dfrac{1}{3}$, move 3 units to the right and 1 unit up to find point $(2, 4)$.

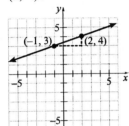

**b.** Start at $(-1, 3)$. Because $m = -4 = -\dfrac{4}{1}$, move 1 unit to the right and 4 units down to find point $(0, -1)$.

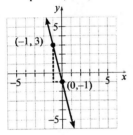

**c.** Because $m = 0$, the line is horizontal passing through the point $(-1, 3)$.

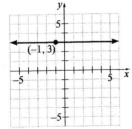

**8.** $y - y_1 = m(x - x_1)$
$y - 5 = 2(x - 3)$

To graph the line, start at $(3, 5)$. Because $m = 2 = \dfrac{2}{1}$, move 1 unit to the right and 2 units up to find point $(4, 7)$.

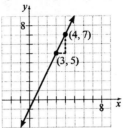

**9.** $y - y_1 = m(x - x_1)$
$y - 3 = -4(x - (-2))$
$y - 3 = -4(x + 2)$

To graph the line, start at $(-2, 3)$. Because $m = -4 = -\dfrac{4}{1}$, move 1 unit to the right and 4 units down to find point $(-1, -1)$.

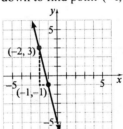

**10.** $y - y_1 = m(x - x_1)$
$y - (-4) = \dfrac{1}{3}(x - 3)$
$y + 4 = \dfrac{1}{3}(x - 3)$

To graph the line, start at $(3, -4)$. Because $m = \dfrac{1}{3}$, move 3 units to the right and 1 unit up to find point $(6, -3)$.

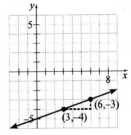

QC – 37

## Chapter 3: Linear Functions and Their Graphs

11. Because the line is horizontal, we know that the slope is $m=0$. Using $(x_1, y_1) = (4,-2)$, we have that
$$y - y_1 = m(x - x_1)$$
$$y - (-2) = 0(x - 4)$$
$$y + 2 = 0$$
$$y = -2$$
To graph the line, we draw a horizontal line that passes through the point

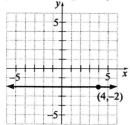

12. $3x - y = 2$
$$-y = -3x + 2$$
$$y = \frac{-3x+2}{-1}$$
$$y = 3x - 2$$
The slope is 3 and the y-intercept is $-2$. To graph the line, begin at $(0,-2)$ and move to the right 1 unit and up 3 units to find the point $(1, 1)$.

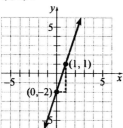

13. $6x + 2y = 8$
$$2y = -6x + 8$$
$$y = \frac{-6x+8}{2}$$
$$y = -3x + 4$$
The slope is $-3$ and the y-intercept is 4. To graph the line, begin at $(0,4)$ and move to the right 1 unit and down 3 units to find the point $(1, 1)$.

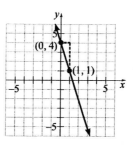

14. $3x - 2y = 7$
$$-2y = -3x + 7$$
$$y = \frac{-3x+7}{-2}$$
$$y = \frac{3}{2}x - \frac{7}{2}$$
The slope is $\frac{3}{2}$ and the y-intercept is $-\frac{7}{2}$. To graph the line, begin at $\left(0,-\frac{7}{2}\right)$ and move to the right 2 units and up 3 units to find the point $\left(2,-\frac{1}{2}\right)$.

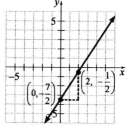

15. $7x + 3y = 0$
$$3y = -7x$$
$$y = -\frac{7}{3}x$$
The slope is $-\frac{7}{3}$ and the y-intercept is 0. To graph the line, begin at $(0,0)$ and move to the right 3 units and down 7 units to find point $(3,-7)$.

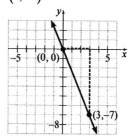

QC – 38

Section 3.3 – Quick Checks: Parallel and Perpendicular Lines

16. $m = \dfrac{y_2 - y_1}{x_2 - x_1} = \dfrac{9-3}{4-1} = \dfrac{6}{3} = 2$

$y - y_1 = m(x - x_1)$
$y - 3 = 2(x - 1)$
$y - 3 = 2x - 2$
$y = 2x + 1$

The slope is 2 and the $y$-intercept is 1.

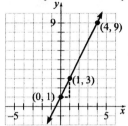

17. $m = \dfrac{y_2 - y_1}{x_2 - x_1} = \dfrac{2-4}{2-(-2)} = \dfrac{-2}{4} = -\dfrac{1}{2}$

$y - y_1 = m(x - x_1)$
$y - 4 = -\dfrac{1}{2}(x - (-2))$
$y - 4 = -\dfrac{1}{2}x - 1$
$y = -\dfrac{1}{2}x + 3$

The slope is $-\dfrac{1}{2}$ and the $y$-intercept is 3.

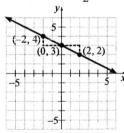

18. $m = \dfrac{y_2 - y_1}{x_2 - x_1} = \dfrac{-4-2}{3-3} = \dfrac{-6}{0} =$ undefined

The slope is undefined, so the line is vertical. The equation of the line is $x = 3$.

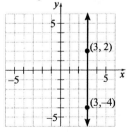

19. a. Let $a$ represent the age of the mother and $W$ represent the weight of the birth weight.

$m = \dfrac{3370 - 3280}{32 - 22} = 9$ grams per year

$W - 3280 = 9(a - 22)$
$W - 3280 = 9a - 198$
$W = 9a + 3082$

Using function notation, we have
$W(a) = 9a + 3082$.

b. $W(30) = 9(30) + 3082 = 3352$

We predict the birth weight would be 3352 grams.

c. The slope indicates that, if the mother's age increases by 1 year, the birth weight will increase by 9 grams.

d. $9a + 3082 = 3310$
$9a = 228$
$a \approx 25.33$

We expect the mother would be about 25 years old.

**3.3 Quick Checks**

1. The slope of $L_1: y = 3x + 1$ is $m_1 = 3$. The slope of $L_2: y = -3x - 3$ is $m_2 = -3$. Since the two slopes are not equal, the lines are not parallel.

2. $\begin{array}{ll} L_1: 6x + 3y = 3 & L_2: -8x - 4y = 12 \\ 3y = -6x + 3 & -4y = 8x + 12 \\ y = \dfrac{-6x+3}{3} & y = \dfrac{8x+12}{-4} \\ y = -2x + 1 & y = -2x - 3 \end{array}$

Since the slopes $m_1 = -2$ and $m_2 = -2$ are equal but the $y$-intercepts are different, the two lines are parallel.

3. $\begin{array}{ll} L_1: -3x + 5y = 10 & L_2: 6x + 10y = 10 \\ 5y = 3x + 10 & 10y = -6x + 10 \\ y = \dfrac{3x+10}{5} & y = \dfrac{-6x+10}{10} \\ y = \dfrac{3}{5}x + 2 & y = -\dfrac{3}{5}x + 1 \end{array}$

Since the slopes $m_1 = \dfrac{3}{5}$ and $m_2 = -\dfrac{3}{5}$ are not equal, the two lines are not parallel.

## Chapter 3: Linear Functions and Their Graphs

4. The slope of the line we seek is $m = 3$, the same as the slope of $y = 3x + 1$. The equation of the line is
$$y - y_1 = m(x - x_1)$$
$$y - 8 = 3(x - 5)$$
$$y - 8 = 3x - 15$$
$$y = 3x - 7$$

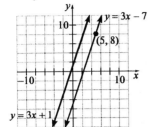

5. The slope of the line we seek is $m = -\dfrac{3}{2}$, the same as the slope of the line
$$3x + 2y = 10$$
$$2y = -3x + 10$$
$$y = \dfrac{-3x + 10}{2}$$
$$y = -\dfrac{3}{2}x + 5$$

Thus, the equation of the line we seek is
$$y - y_1 = m(x - x_1)$$
$$y - 4 = -\dfrac{3}{2}(x - (-2))$$
$$y - 4 = -\dfrac{3}{2}(x + 2)$$
$$y - 4 = -\dfrac{3}{2}x - 3$$
$$y = -\dfrac{3}{2}x + 1$$

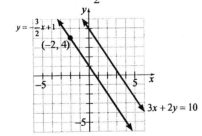

6. The negative reciprocal of $-3$ is $\dfrac{-1}{-3} = \dfrac{1}{3}$. Any line whose slope is $\dfrac{1}{3}$ will be perpendicular to a line whose slope is $-3$.

7. The slope of $L_1 : y = 5x - 3$ is $m_1 = 5$. The slope of $L_2 : y = -\dfrac{1}{5}x - 4$ is $m_2 = -\dfrac{1}{5}$. Since the two slopes are negative reciprocals (i.e., since $5\left(-\dfrac{1}{5}\right) = -1$), the lines are perpendicular.

8. $L_1 : 4x - y = 3 \qquad L_2 : x - 4y = 2$
   $\quad -y = -4x + 3 \qquad -4y = -x + 2$
   $\quad y = \dfrac{-4x + 3}{-1} \qquad y = \dfrac{-x + 2}{-4}$
   $\quad y = 4x - 3 \qquad y = \dfrac{1}{4}x - \dfrac{1}{2}$

Since the slopes $m_1 = 4$ and $m_2 = \dfrac{1}{4}$ are not negative reciprocals (i.e., since $4 \cdot \dfrac{1}{4} = 1 \neq -1$), the two lines are not perpendicular.

9. The slope of the line we seek is $m = -\dfrac{1}{2}$, the negative reciprocal of the slope of the line $y = 2x + 1$. Thus, the equation of the line we seek is
$$y - y_1 = m(x - x_1)$$
$$y - 2 = -\dfrac{1}{2}(x - (-4))$$
$$y - 2 = -\dfrac{1}{2}(x + 4)$$
$$y - 2 = -\dfrac{1}{2}x - 2$$
$$y = -\dfrac{1}{2}x$$

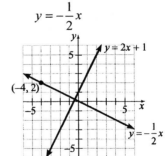

## Section 3.4 – Quick Checks: Linear Inequalities in Two Variables

10. The slope of the line we seek is $m = -\dfrac{4}{3}$, the negative reciprocal of the slope of the line
$3x - 4y = 8$
$-4y = -3x + 8$
$y = \dfrac{-3x + 8}{-4}$
$y = \dfrac{3}{4}x - 2$

Thus, the equation of the line we seek is
$y - y_1 = m(x - x_1)$
$y - (-4) = -\dfrac{4}{3}(x - (-3))$
$y + 4 = -\dfrac{4}{3}x - 4$
$y = -\dfrac{4}{3}x - 8$

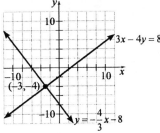

### 3.4 Quick Checks

1. **a.** $x = 4, y = 1$ $\quad -2(4) + 3(1) \overset{?}{\geq} 3$
$-8 + 3 \overset{?}{\geq} 3$
$-5 \overset{?}{\geq} 3$  False

So, $(4, 1)$ is not a solution to $-2x + 3y \geq 3$.

**b.** $x = -1, y = 2$ $\quad -2(-1) + 3(2) \overset{?}{\geq} 3$
$2 + 6 \overset{?}{\geq} 3$
$8 \geq 3$  True

So, $(-1, 2)$ is a solution to $-2x + 3y \geq 3$.

**c.** $x = 2, y = 3$ $\quad -2(2) + 3(3) \overset{?}{\geq} 3$
$-4 + 9 \overset{?}{\geq} 3$
$5 \geq 3$  True

So, $(2, 3)$ is a solution to $-2x + 3y \geq 3$.

**d.** $x = 0, y = 1$ $\quad -2(0) + 3(1) \overset{?}{\geq} 3$
$0 + 3 \overset{?}{\geq} 3$
$3 \geq 3$  True

So, $(0, 1)$ is a solution to $-2x + 3y \geq 3$.

2. Replace the inequality symbol with an equal sign to obtain $y = -2x + 3$. Because the inequality is strict, graph $y = -2x + 3$ using a dashed line.

Test Point: $(0, 0)$: $\quad 0 \overset{?}{<} -2(0) + 3$
$0 \overset{?}{<} 0 + 3$
$0 \overset{?}{<} 3$  True

Therefore, $(0, 0)$ is a solution to $y < -2x + 3$.
Shade the half-plane that contains $(0, 0)$.

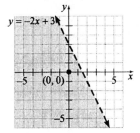

3. Replace the inequality symbol with an equal sign to obtain $6x - 3y = 15$. Because the inequality is non-strict, graph $6x - 3y = 15$ $(y = 2x - 5)$ using a solid line.

Test Point: $(0, 0)$: $\quad 6(0) - 3(0) \overset{?}{\leq} 15$
$0 \leq 15$  True

Therefore, $(0, 0)$ is solution to $6x - 3y \leq 15$.
Shade the half-plane that contains $(0, 0)$.

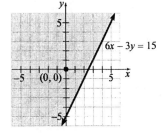

## Chapter 3: Linear Functions and Their Graphs

4. Replace the inequality symbol with an equal sign to obtain $2x + y = 0$. Because the inequality is strict, graph $2x + y = 0$ $(y = -2x)$ using a dashed line.

    Test Point: $(1, 1)$: $2(1) + 1 \overset{?}{<} 0$
    
    $2 + 1 \overset{?}{<} 0$
    
    $3 \overset{?}{<} 0$ False
    
    Therefore, $(1, 1)$ is not a solution to $2x + y < 0$. Shade the half-plane that does not contain $(1, 1)$.

    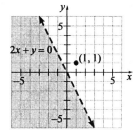

5. **a.** Let $x$ = the number of Chicken Breast filets. Let $y$ = the number of Frosties.
    $430x + 330y \le 800$

    **b.** $x = 1, y = 1$: $430(1) + 330(1) \overset{?}{\le} 800$
    
    $430 + 330 \overset{?}{\le} 800$
    
    $760 \le 800$ True
    
    Yes, Avery can stay within his allotment of Calories by eating one Chicken Breast filet and one Frosty.

    **c.** $x = 2, y = 1$: $430(2) + 330(1) \overset{?}{\le} 800$
    
    $860 + 330 \overset{?}{\le} 800$
    
    $1190 \overset{?}{\le} 800$ False
    
    No, Avery cannot stay within his allotment of Calories by eating two Chicken Breast filets and one Frosty.

### 3.5 Quick Checks

1. **a.** From Example 1, the daily fixed costs were $2000 with a variable cost of $80 per bicycle. The tax of $1 per bicycle changes the variable cost to $81 per bicycle. Thus, the cost function is $C(x) = 81x + 2000$.

    **b.** Label the horizontal axis $x$ and the vertical axis $C$. To graph the function, we plot a few points:
    $C(0) = 81(0) + 2000 = 2000$
    $C(5) = 81(5) + 2000 = 2405$
    $C(10) = 81(10) + 2000 = 2810$
    So the points $(0, 2000)$, $(5, 2405)$, and $(10, 2810)$ are on the graph.

    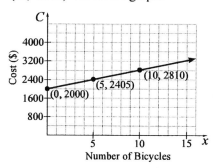

    **c.** $C(18) = 81(18) + 2000 = 3458$
    So, the cost of manufacturing 18 bicycles in a day is $3458.

    **d.** $C(x) = 4025$
    $81x + 2000 = 4025$
    $81x = 2025$
    $x = 25$
    So, 25 bicycles can be manufactured for a cost of $4025.

2. **a.** We let $C(x)$ represent the monthly cost of operating the car after driving $x$ miles, so $C(x) = mx + b$. The monthly cost before the car is driven is $250, so $C(0) = 250$. The $C$-intercept of the linear function is 250. Because the maintenance and gas cost is $0.18 per mile, the slope of the linear function is 0.18. The linear function that relates the monthly cost of operating the car as a of mile driven is $C(x) = 0.18x + 250$.

    **b.** The car cannot be driven a negative distance, the number of miles driven, $x$, must be greater than or equal to zero. In addition, there is no definite maximum number of miles that the car can be driven. Therefore, the implied domain of the function is $\{x \mid x \ge 0\}$, or using interval notation $[0, \infty)$.

## Section 3.5 – Quick Checks: Building Linear Models

**c.** $C(320) = 0.18(320) + 250 = 307.6$
So, the monthly cost of driving 320 miles is $307.60.

**d.** Label the horizontal axis $x$ and the vertical axis $C$. From part (a) we know $C(0) = 250$, and from part (c) we know $C(320) = 307.6$, so $(0, 250)$ and $(320, 307.60)$ are on the graph.

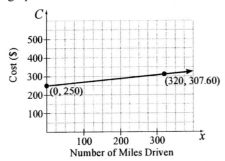

**e.** 
$$C(x) = 282.40$$
$$0.18x + 250 = 282.40$$
$$0.18x = 32.40$$
$$x = 180$$
So, Roberta can drive 180 miles each month for the monthly cost of $282.40.

**3. a.** Because $C$ varies directly with $g$, we know that $C = kg$ for some constant $k$. Because $C = \$22.39$ when $g = 8$ gallons, we obtain
$$C = kg$$
$$22.39 = k \cdot 8$$
$$k = \frac{22.39}{8} = 2.79875$$
So, we have that $C = 2.79875g$, or writing this as a linear function, $C(g) = 2.79875g$.

**b.** $C(5.5) = 2.79875(5.5) = 15.393125$
So, the cost of 5.5 gallons would be approximately $15.39.

**c.** Label the horizontal axis $g$ and the vertical axis $C$. From the problem we know the point $(8, 22.39)$ is on the graph, and from part (b), we know the point $(5.5, 15.39)$ is on the graph.

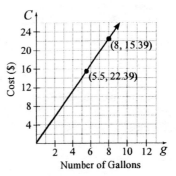

**4. a.**

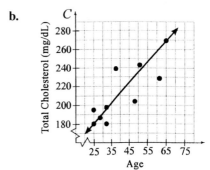

**b.** From the scatter diagram, we can see that as the age increases, the total cholesterol also increases.

**5.** Non-linear

**6.** Linear with a positive slope.

**7. a.** Answers will vary. We will use the points $(25, 180)$ and $(65, 269)$.
$$m = \frac{269 - 180}{65 - 25} = \frac{89}{40} = 2.225$$
$$y - 180 = 2.225(x - 25)$$
$$y - 180 = 2.225x - 55.625$$
$$y = f(x) = 2.225x + 124.375$$

**b.**

QC – 43

**Chapter 3:** *Linear Functions and Their Graphs*

    **c.**   $f(39) = 2.225(39) + 124.375 = 211.15$

         We predict that the total cholesterol of a 39 year old will be approximately 211 mg/dL.

    **d.**   The slope of the linear function is 2.225. This means that, for males, the total cholesterol increases by 2.225 mg/dL for each one-year increase in age. The $y$-intercept, 124.375, would represent the total cholesterol of a male who is 0 years old. Thus, it does not make sense to interpret this $y$-intercept.

# Chapter 4 – Quick Checks

## 4.1 Quick Checks

1. $\begin{cases} 2x+3y=7 & (1) \\ 3x+y=-7 & (2) \end{cases}$

   a. Let $x=3$ and $y=1$ in both equations (1) and (2).

   Equation (1): $2(3)+3(1)\stackrel{?}{=}7$
   $$6+3\stackrel{?}{=}7$$
   $$9 \neq 7$$

   Equation (2): $3(3)+1\stackrel{?}{=}-7$
   $$9+1\stackrel{?}{=}-7$$
   $$10 \neq -7$$

   These values do not satisfy either equation. Therefore, the ordered pair $(3, 1)$ is not a solution.

   b. Let $x=-4$ and $y=5$ in both equations (1) and (2).

   Equation (1): $2(-4)+3(5)\stackrel{?}{=}7$
   $$-8+15\stackrel{?}{=}7$$
   $$7=7$$

   Equation (2): $3(-4)+5 \stackrel{?}{=} -7$
   $$-12+5 \stackrel{?}{=} -7$$
   $$-7 = -7$$

   These values satisfy both equations, so the ordered pair $(-4, 5)$ is a solution.

   c. Let $x=-2$ and $y=-1$ in both equations (1) and (2).

   Equation (1): $2(-2)+3(-1)\stackrel{?}{=}7$
   $$-4-3\stackrel{?}{=}7$$
   $$-7 \neq 7$$

   Equation (2): $3(-2)+(-1) \stackrel{?}{=} -7$
   $$-6-1 \stackrel{?}{=} -7$$
   $$-7 = -7$$

   Although these values satisfy equation (2), they do not satisfy equation (1). Therefore, the ordered pair $(-2,-1)$ is not a solution.

2. $\begin{cases} y=-3x+10 & (1) \\ y=2x-5 & (2) \end{cases}$

   The two equations are in slope-intercept form. Graph each equation and find the point of intersection.

   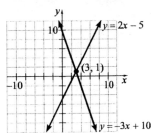

   The solution is the ordered pair $(3, 1)$.

3. $\begin{cases} 2x+y=-1 & (1) \\ -2x+2y=10 & (2) \end{cases}$

   Equation (1) in slope-intercept form is $y=-2x-1$. Equation (2) in slope-intercept form is $y=x+5$. Graph each equation and find the point of intersection.

   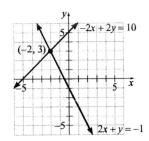

   The solution is the ordered pair $(-2, 3)$.

4. $\begin{cases} y=-3x-5 & (1) \\ 5x+3y=1 & (2) \end{cases}$

   Substituting $-3x-5$ for $y$ in equation (2), we obtain
   $$5x+3(-3x-5)=1$$
   $$5x-9x-15=1$$
   $$-4x-15=1$$
   $$-4x=16$$
   $$x=-4$$

   Substituting $-4$ for $x$ in equation (1), we obtain $y=-3(-4)-5=12-5=7$.

   The solution is the ordered pair $(-4, 7)$.

## Chapter 4: Systems of Linear Equations and Inequalities

**5.** $\begin{cases} 2x + y = -2 & (1) \\ -3x - 2y = -2 & (2) \end{cases}$

Equation (1) solved for $y$ is $y = -2x - 2$.

Substituting $-2x - 2$ for $y$ in equation (2), we obtain
$$-3x - 2(-2x - 2) = -2$$
$$-3x + 4x + 4 = -2$$
$$x + 4 = -2$$
$$x = -6$$

Substituting $-6$ for $y$ in equation (1), we obtain
$$2(-6) + y = -2$$
$$-12 + y = -2$$
$$y = 10$$

The solution is the ordered pair $(-6, 10)$.

**6.** $\begin{cases} -2x + y = 4 & (1) \\ -5x + 3y = 7 & (2) \end{cases}$

Multiply both sides of equation (1) by $-3$ and add the result to equation (2).
$$6x - 3y = -12$$
$$-5x + 3y = 7$$
$$\overline{\phantom{xx}x\phantom{xxxx} = -5}$$

Substituting $-5$ for $x$ in equation (1), we obtain
$$-2(-5) + y = 4$$
$$10 + y = 4$$
$$y = -6$$

The solution is the ordered pair $(-5, -6)$.

**7.** $\begin{cases} -3x + 2y = 3 & (1) \\ 4x - 3y = -6 & (2) \end{cases}$

Multiply both sides of equation (1) by 4, multiply both sides of equation (2) by 3, and add the results.
$$-12x + 8y = 12$$
$$12x - 9y = -18$$
$$\overline{\phantom{xxxx}-y = -6}$$
$$y = 6$$

Substituting 6 for $y$ in equation (1), we obtain
$$-3x + 2(6) = 3$$
$$-3x + 12 = 3$$
$$-3x = -9$$
$$x = 3$$

The solution is the ordered pair $(3, 6)$.

**8.** $\begin{cases} -3x + y = 2 & (1) \\ 6x - 2y = 1 & (2) \end{cases}$

Multiply both sides of equation (1) by 2 and add the result to equation (2).
$$-6x + 2y = 4$$
$$6x - 2y = 1$$
$$\overline{\phantom{xxxx}0 = 5}$$

The equation $0 = 5$ is false, so the system has no solution. The solution set is $\varnothing$ or $\{\ \}$. The system is inconsistent.

The graphs of the equations (shown below) are parallel, which supports the statement that the system has no solution.

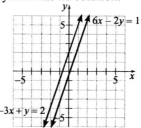

**9.** $\begin{cases} -3x + 2y = 8 & (1) \\ 6x - 4y = -16 & (2) \end{cases}$

Multiply both sides of equation (1) by 2 and add the results.
$$-6x + 4y = 16$$
$$6x - 4y = -16$$
$$\overline{\phantom{xxxx}0 = 0}$$

The equation $0 = 0$ is true, so the system is dependent. The solution is
$\{(x, y) \mid -3x + 2y = 8\}$.

The graphs of the equations (shown below) coincide, which supports the statement that the system is dependent.

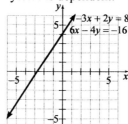

**Section 4.2 – Quick Checks:** *Problem Solving: Systems of Two Equations Containing Two Unknowns*

## 4.2 Quick Checks

1. Let $c$ represent the cost of a cheeseburger and let $s$ represent the cost of a medium shake.
$$\begin{cases} 4c + 2s = 10.10 & (1) \\ 3c + 3s = 10.35 & (2) \end{cases}$$
Multiply both sides of equation (1) by 3, multiply both sides of equation (2) by $-2$, and add the results.
$$12c + 6s = 30.30$$
$$\underline{-6c - 6s = -20.70}$$
$$6c \phantom{-6s} = 9.60$$
$$c = 1.60$$
Substituting 1.60 for $c$ in equation (1), we obtain
$$4(1.60) + 2s = 10.10$$
$$6.40 + 2s = 10.10$$
$$2s = 3.70$$
$$s = 1.85$$
A cheeseburger costs $1.60. A medium shake costs $1.85.

2. Let $x$ and $y$ represent the length and width of the rectangle, respectively.
$$\begin{cases} 2x + 2y = 360 & (1) \\ x = 2y & (2) \end{cases}$$
Substituting $2y$ for $x$ in equation (1), we obtain
$$2(2y) + 2y = 360$$
$$4y + 2y = 360$$
$$6y = 360$$
$$y = 60$$
Substituting 60 for $y$ in equation (2), we obtain $x = 2(60) = 120$.
The length is 120 yards, and the width is 60 yards.

3. Because angles 1 and 3 are supplemental, and because angles 1 and 5 must be equal, we obtain the following system
$$\begin{cases} (x + 3y) + (3x + y) = 180 \\ x + 3y = 5x + y \end{cases}$$
Simplify each equation.
$$\begin{cases} 4x + 4y = 180 & (1) \\ -4x + 2y = 0 & (2) \end{cases}$$
Add the two equations.
$$4x + 4y = 180$$
$$\underline{-4x + 2y = 0}$$
$$6y = 180$$
$$y = 30$$

Substituting 30 for $y$ in equation (2), we obtain
$$-4x + 2(30) = 0$$
$$-4x + 60 = 0$$
$$-4x = -60$$
$$x = 15$$
Thus, $x = 15$ and $y = 30$.

4. Let $a$ represent the amount invested in the Aaa-rated bond, and let $b$ represent the amount invested in the B-rated bond. We organize the information in the table below, using the interest formula: Interest = Principal × rate × time.

|  | Principal $ | Rate % | Time Yr | Interest $ |
|---|---|---|---|---|
| Aaa-rated bond | $a$ | 0.05 | 1 | $0.05a$ |
| B-rated bond | $b$ | 0.10 | 1 | $0.10b$ |
| Total | 120,000 |  |  | 7200 |

Using the columns for Principal and Interest, we obtain the following system.
$$\begin{cases} a + b = 120,000 & (1) \\ 0.05a + 0.10b = 7200 & (2) \end{cases}$$
Solving equation (1) for $b$, we obtain $b = 120,000 - a$. Substituting $120,000 - a$ for $b$ in equation (2), we obtain
$$0.05a + 0.10(120,000 - a) = 7200$$
$$0.05a + 12000 - 0.10a = 7200$$
$$-0.05a + 12000 = 7200$$
$$-0.05a = -4800$$
$$a = 96,000$$
Substituting 96,000 for $a$ into equation (1), we obtain $96,000 + b = 120,000$
$$b = 24,000$$
Maria should invest $96,000 in the Aaa-rated bond and $24,000 in the B-rated bond.

## Chapter 4: Systems of Linear Equations and Inequalities

5. Let $c$ represent the weight of cashews to be used and let $p$ represent the weight of peanuts to be used. We organize the information in the table below, using the formula
Price per pound × Number of pounds = Revenue.

| | Price per pound | Number of pounds | Revenue |
|---|---|---|---|
| Cashews | 7.00 | $c$ | $7c$ |
| Peanuts | 2.50 | $p$ | $2.5p$ |
| Trail mix | 4.00 | 30 | $4(30) = 120$ |

Using the columns for Number of pounds and Revenue, we obtain the following system.
$$\begin{cases} c + p = 30 & (1) \\ 7c + 2.5p = 120 & (2) \end{cases}$$
Solving equation (1) for $c$ yields $c = 30 - p$.
Substituting $30 - p$ for $c$ in equation (2), we obtain
$$7(30 - p) + 2.5p = 120$$
$$210 - 7p + 2.5p = 120$$
$$210 - 4.5p = 120$$
$$-4.5p = -90$$
$$p = 20$$
Substituting 20 for $p$ in equation (1) yields
$$c + 20 = 30$$
$$c = 10$$
The trail mix requires 10 pounds of cashews and 20 pounds of peanuts.

6. Let $a$ represent the airspeed of the plane, and let $w$ represent the impact of wind resistance on the plane. We organize the information in the table below.

| | Distance | Rate | Time |
|---|---|---|---|
| Against Wind (West) | 1200 | $a - w$ | 4 |
| With Wind (East) | 1200 | $a + w$ | 3 |

Using the formula Distance = Rate × Time, we obtain the following system.
$$\begin{cases} 1200 = 4(a - w) \\ 1200 = 3(a + w) \end{cases}$$
Simplify each equation.
$$\begin{cases} 4a - 4w = 1200 & (1) \\ 3a + 3w = 1200 & (2) \end{cases}$$
Multiply both sides of equation (1) by 3, multiply both sides of equation (2) by 4, and add the results
$$12a - 12w = 3600$$
$$12a + 12w = 4800$$
$$\overline{24a \phantom{-12w} = 8400}$$
$$a = 350$$
Substituting 350 for $a$ in equation (2) yields
$$3(350) + 3w = 1200$$
$$1050 + 3w = 1200$$
$$3w = 150$$
$$w = 50$$
The airspeed of the plain is 350 miles per hour. The impact of wind resistance on the plane is 50 miles per hour.

7. a. If one Austrian Pine tree is sold, revenue will be $230. If two trees are sold, revenue will be $230(2) = $460$. If $x$ trees are sold, revenue will be $230x$. The revenue function is $R(x) = 230x$.

   b. The cost function is $C(x) = ax + b$ where $a$ is the variable cost and $b$ is the fixed cost. With $a = 160$ and $b = 2100$, the cost function is $C(x) = 160x + 2100$.

   c.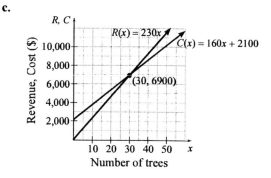

   d. $R(x) = C(x)$
$$230x = 160x + 2100$$
$$70x = 2100$$
$$x = 30$$
$$R(30) = 230(30) = 6900$$
$$C(30) = 160(30) + 2100 = 6900$$
The break-even point is $(30, 6900)$.
If 30 Austrian Pine trees are sold, the cost and revenue will be even at $6900.

QC – 48

**Section 4.3 – Quick Checks:** Systems of Linear Equations in Three Variables

**4.3 Quick Checks**

1. **a.** Substitute $x = 3$, $y = 2$, and $z = -2$ into all three equations.

   Equation (1): $3 + 2 + (-2) \stackrel{?}{=} 3$
   $$3 = 3$$

   Equation (2): $3(3) + 2 - 2(-2) \stackrel{?}{=} -23$
   $$9 + 2 + 4 \stackrel{?}{=} -23$$
   $$15 \neq -23$$

   Equation (3): $-2(3) - 3(2) + 2(-2) \stackrel{?}{=} 17$
   $$-6 - 6 - 4 \stackrel{?}{=} 17$$
   $$-16 \neq 17$$

   Although these values satisfy equation (1), they do not satisfy equations (2) and (3). Therefore, the ordered triple $(3, 2, -2)$ is not a solution of the system.

   **b.** Substitute $x = -4$, $y = 1$, and $z = 6$ into all three equations.

   Equation (1): $-4 + 1 + 6 \stackrel{?}{=} 3$
   $$3 = 3$$

   Equation (2): $3(-4) + 1 - 2(6) \stackrel{?}{=} -23$
   $$-12 + 1 - 12 \stackrel{?}{=} -23$$
   $$-23 = -23$$

   Equation (3): $-2(-4) - 3(1) + 2(6) \stackrel{?}{=} 17$
   $$8 - 3 + 12 \stackrel{?}{=} 17$$
   $$17 = 17$$

   Because these values satisfy all three equations, the ordered pair $(-4, 1, 6)$ is a solution of the system.

2. $\begin{cases} x + y + z = -3 & (1) \\ 2x - 2y - z = -7 & (2) \\ -3x + y + 5z = 5 & (3) \end{cases}$

   Multiply both sides of equation (1) by $-2$ and add the result to equation (2).
   $$-2x - 2y - 2z = 6$$
   $$\underline{\phantom{-}2x - 2y - z = -7}$$
   $$-4y - 3z = -1$$

   Multiply both sides of equation (1) by 3 and add the result to equation (3).

   $$3x + 3y + 3z = -9$$
   $$\underline{-3x + y + 5z = 5}$$
   $$4y + 8z = -4$$

   Rewriting the system, we have
   $\begin{cases} x + y + z = -3 & (1) \\ -4y - 3z = -1 & (2) \\ 4y + 8z = -4 & (3) \end{cases}$

   Add equations (2) and (3).
   $$-4y - 3z = -1$$
   $$\underline{\phantom{-}4y + 8z = -4}$$
   $$5z = -5$$

   Once again rewriting the system, we have
   $\begin{cases} x + y + z = -3 & (1) \\ -4y - 3z = -1 & (2) \\ 5z = -5 & (3) \end{cases}$

   Solving equation (3) for $z$, we obtain
   $$5z = -5$$
   $$z = -1$$

   Back-substituting $-1$ for $z$ in equation (2), we obtain
   $$-4y - 3(-1) = -1$$
   $$-4y + 3 = -1$$
   $$-4y = -4$$
   $$y = 1$$

   Back-substituting 1 for $y$ and $-1$ for $z$ in equation (1), we obtain
   $$x + 1 + (-1) = -3$$
   $$x = -3$$

   The solution is the ordered triple $(-3, 1, -1)$.

3. $\begin{cases} 2x - 4z = -7 & (1) \\ x + 6y = 5 & (2) \\ 2y - z = 2 & (3) \end{cases}$

   Multiply both sides of equation (2) by $-2$ and add the result to equation (1).
   $$-2x - 12y = -10$$
   $$\underline{\phantom{-}2x \phantom{- 12y} - 4z = -7}$$
   $$-12y - 4z = -17$$

   Rewriting the system, we have
   $\begin{cases} 2x - 4z = -7 & (1) \\ -12y - 4z = -17 & (2) \\ 2y - z = 2 & (3) \end{cases}$

   Multiply both sides of equation (3) by 6 and add the result to equation (2).

## Chapter 4: Systems of Linear Equations and Inequalities

$$12y - 6z = 12$$
$$\underline{-12y - 4z = -17}$$
$$-10z = -5$$

Once again rewriting the system, we have
$$\begin{cases} 2x \quad\quad -4z = -7 & (1) \\ -12y - 4z = -17 & (2) \\ \quad\quad -10z = -5 & (3) \end{cases}$$

Solving equation (3) for $z$, we obtain
$$-10z = -5$$
$$z = \frac{-5}{-10} = \frac{1}{2}$$

Back-substituting $\frac{1}{2}$ for $z$ in equation (2), we obtain
$$-12y - 4\left(\frac{1}{2}\right) = -17$$
$$-12y - 2 = -17$$
$$-12y = -15$$
$$y = \frac{-15}{-12} = \frac{5}{4}$$

Back-substituting $\frac{1}{2}$ for $z$ in equation (1), we obtain
$$2x - 4\left(\frac{1}{2}\right) = -7$$
$$2x - 2 = -7$$
$$2x = -5$$
$$x = -\frac{5}{2}$$

The solution is the ordered triple $\left(-\frac{5}{2}, \frac{5}{4}, \frac{1}{2}\right)$.

4. $\begin{cases} x - y + 2z = -7 & (1) \\ -2x + y - 3z = 5 & (2) \\ x - 2y + 3z = 2 & (3) \end{cases}$

Multiply both sides of equation (1) by 2 and add the result to equation (2).
$$2x - 2y + 4z = -14$$
$$\underline{-2x + y - 3z = 5}$$
$$-y + z = -9$$

Multiply both sides of equation (1) by $-1$ and add the result to equation (3).
$$-x + y - 2z = 7$$
$$\underline{x - 2y + 3z = 2}$$
$$-y + z = 9$$

Rewriting the system, we have
$$\begin{cases} x - y + 2z = -7 & (1) \\ -y + z = -9 & (2) \\ -y + z = 9 & (3) \end{cases}$$

Multiply both sides of equation (2) by $-1$ and add the result to equation (3).
$$y - z = 9$$
$$\underline{-y + z = 9}$$
$$0 = 18$$

Once again rewriting the system, we have
$$\begin{cases} x - y + 2z = -7 & (1) \\ -y + z = -9 & (2) \\ 0 = 18 & (3) \text{ False} \end{cases}$$

Equation (3) is a false statement (contradiction). Therefore, the system is inconsistent. The solution set is $\varnothing$ or $\{\ \}$.

5. $\begin{cases} x - y + 3z = 2 & (1) \\ -x + 2y - 5z = -3 & (2) \\ 2x - y + 4z = 3 & (3) \end{cases}$

Add equations (1) and (2).
$$x - y + 3z = 2$$
$$\underline{-x + 2y - 5z = -3}$$
$$y - 2z = -1$$

Multiply both sides of equation (1) by $-2$ and add the result to equation (3).
$$-2x + 2y - 6z = -4$$
$$\underline{2x - y + 4z = 3}$$
$$y - 2z = -1$$

Rewriting the system, we have
$$\begin{cases} x - y + 3z = 2 & (1) \\ y - 2z = -1 & (2) \\ y - 2z = -1 & (3) \end{cases}$$

Multiply both sides of equation (2) by $-1$ and add the result to equation (3).
$$-y + 2z = 1$$
$$\underline{y - 2z = -1}$$
$$0 = 0$$

Once again rewriting the system, we have
$$\begin{cases} x - y + 3z = 2 & (1) \\ y - 2z = -1 & (2) \\ 0 = 0 & (3) \text{ True} \end{cases}$$

Thus, the system is dependent and has an infinite number of solutions.

## Section 4.4 – Quick Checks: Using Matrices to Solve Systems

Solve equation (2) for $y$.
$y = 2z - 1$
Substituting $2z - 1$ for $y$ in equation (1), we obtain
$$x - (2z - 1) + 3z = 2$$
$$x - 2z + 1 + 3z = 2$$
$$x + z + 1 = 2$$
$$x + z = 1$$
$$x = -z + 1$$

The solution to the system is
$\{(x, y, z) | x = -z + 1, y = 2z - 1, z \text{ is any real number}\}$.

6. Let $x$ represent the number of 21-inch mowers, let $y$ represent the number of 24-inch mowers, and let $z$ represent the number of 40-inch riding mowers.
$$\begin{cases} 2x + 3y + 4z = 81 & (1) \\ 3x + 3y + 4z = 95 & (2) \\ x + y + 2z = 35 & (3) \end{cases}$$

Multiply both sides of equation (3) by $-2$ and add the result to equation (1).
$$-2x - 2y - 4z = -70$$
$$\underline{2x + 3y + 4z = 81}$$
$$y \quad\quad\quad\; = 11$$

Multiply both sides of equation (3) by $-3$ and add the result to equation (2).
$$-3x - 3y - 6z = -105$$
$$\underline{3x + 3y + 4z = 95}$$
$$-2z = -10$$

Rewriting the system, we have
$$\begin{cases} 2x + 3y + 4z = 81 & (1) \\ \quad\quad y \quad\quad\; = 11 & (2) \\ \quad\quad\quad -2z = -10 & (3) \end{cases}$$

From equation (2), we know $y = 11$. Solving equation (3) for $z$, we obtain
$$-2z = -10$$
$$z = 5$$

Substituting 11 for $y$ and 5 for $z$ and 11 in equation (1), we obtain
$$2x + 3(11) + 4(5) = 81$$
$$2x + 33 + 20 = 81$$
$$2x + 53 = 81$$
$$2x = 28$$
$$x = 14$$

The company can manufacture 14 twenty-one-inch mowers, 11 twenty-four-inch mowers, and 5 forty-inch rider mowers.

### 4.4 Quick Checks

1. In the augmented matrix, the first column represents the coefficients on the $x$ variable. The second column represents the coefficients on the $y$ variable. The vertical line signifies the equal signs. The third column represents the constants to the right of the equal sign.

$$\begin{bmatrix} 3 & -1 & | & -10 \\ -5 & 2 & | & 0 \end{bmatrix}$$

2. The system
$$\begin{cases} x + 2y - 2z = 11 \\ -x - 2z = 4 \\ 4x - y + z - 3 = 0 \end{cases}$$
gets rearranged as
$$\begin{cases} x + 2y - 2z = 11 \\ -x + 0y - 2z = 4 \\ 4x - y + z = 3 \end{cases}$$
Thus, the augmented matrix is
$$\begin{bmatrix} 1 & 2 & -2 & | & 11 \\ -1 & 0 & -2 & | & 4 \\ 4 & -1 & 1 & | & 3 \end{bmatrix}.$$

3. Since the augmented matrix has two rows, it represents a system of two equations. Because there are two columns to the left of the vertical bar, the system has two variables. If we call the variables $x$ and $y$, the system of equations is
$$\begin{cases} x - 3y = 7 \\ -2x + 5y = -3 \end{cases}$$

4. Since the augmented matrix has three rows, it represents a system of three equations. Because there are three columns to the left of the vertical bar, the system has three variables. If we call the variables $x$, $y$, and $z$, the system of equations is
$$\begin{cases} x - 3y + 2z = 4 \\ 3x \quad\quad\; - z = -1 \\ -x + 4y \quad\quad = 0 \end{cases}$$

5. $\begin{bmatrix} 1 & -2 & | & 5 \\ -4 & 5 & | & -11 \end{bmatrix}$  $(R_2 = 4r_1 + r_2)$

$$= \begin{bmatrix} 1 & -2 & | & 5 \\ 4(1)+(-4) & 4(-2)+5 & | & 4(5)+(-11) \end{bmatrix}$$

$$= \begin{bmatrix} 1 & -2 & | & 5 \\ 0 & -3 & | & 9 \end{bmatrix}$$

## Chapter 4: Systems of Linear Equations and Inequalities

6. We want a 0 in row 1, column 2. We accomplish this by multiplying row 2 by $-5$ and adding the result to row 1. That is, we apply the row operation $R_1 = -5r_2 + r_1$.

$$\begin{bmatrix} 1 & 5 & | & 13 \\ 0 & 1 & | & 2 \end{bmatrix} \quad (R_1 = -5r_2 + r_1)$$

$$= \begin{bmatrix} -5(0)+1 & -5(1)+5 & | & -5(2)+13 \\ 0 & 1 & | & 2 \end{bmatrix}$$

$$= \begin{bmatrix} 1 & 0 & | & 3 \\ 0 & 1 & | & 2 \end{bmatrix}$$

7. Write the augmented matrix of the system and then put it in row echelon form.

$$\begin{bmatrix} 2 & -4 & | & 20 \\ 3 & 1 & | & 16 \end{bmatrix} \quad \left(R_1 = \tfrac{1}{2}r_1\right)$$

$$= \begin{bmatrix} 1 & -2 & | & 10 \\ 3 & 1 & | & 16 \end{bmatrix} \quad (R_2 = -3r_1 + r_2)$$

$$= \begin{bmatrix} 1 & -2 & | & 10 \\ 0 & 7 & | & -14 \end{bmatrix} \quad \left(R_2 = \tfrac{1}{7}r_2\right)$$

$$= \begin{bmatrix} 1 & -2 & | & 10 \\ 0 & 1 & | & -2 \end{bmatrix}$$

From row 2, we have that $y = -2$. Row 1 represents the equation $x - 2y = 10$. Back-substitute $-2$ for $y$ and solve for $x$.

$x - 2(-2) = 10$
$x + 4 = 10$
$x = 6$

The solution is the ordered pair $(6, -2)$.

8. Write the augmented matrix of the system and then put it in row echelon form.

$$\begin{bmatrix} 1 & -1 & 2 & | & 7 \\ 2 & -2 & 1 & | & 11 \\ -3 & 1 & -3 & | & -14 \end{bmatrix} \quad \begin{pmatrix} R_2 = -2r_1 + r_2 \\ R_3 = 3r_1 + r_3 \end{pmatrix}$$

$$= \begin{bmatrix} 1 & -1 & 2 & | & 7 \\ 0 & 0 & -3 & | & -3 \\ 0 & -2 & 3 & | & 7 \end{bmatrix} \quad (\text{Interchange } r_1 \text{ and } r_2)$$

$$= \begin{bmatrix} 1 & -1 & 2 & | & 7 \\ 0 & -2 & 3 & | & 7 \\ 0 & 0 & -3 & | & -3 \end{bmatrix} \quad \begin{pmatrix} R_2 = -\tfrac{1}{2}r_2 \\ R_3 = -\tfrac{1}{3}r_3 \end{pmatrix}$$

$$= \begin{bmatrix} 1 & -1 & 2 & | & 7 \\ 0 & 1 & -\tfrac{3}{2} & | & -\tfrac{7}{2} \\ 0 & 0 & 1 & | & 1 \end{bmatrix}$$

Write the system of equations that corresponds to the row-echelon matrix

$$\begin{cases} x - y + 2z = 7 & (1) \\ y - \tfrac{3}{2}z = -\tfrac{7}{2} & (2) \\ z = 1 & (3) \end{cases}$$

Substituting 1 for $z$ in equation (2), we obtain

$y - \tfrac{3}{2}(1) = -\tfrac{7}{2}$

$y - \tfrac{3}{2} = -\tfrac{7}{2}$

$y = -\tfrac{4}{2} = -2$

Substituting 1 for $z$ and $-2$ for $y$ in equation (1), we obtain

$x - (-2) + 2(1) = 7$
$x + 2 + 2 = 7$
$x + 4 = 7$
$x = 3$

The solution is the ordered triple $(3, -2, 1)$.

9. Write the augmented matrix of the system and then put it in row echelon form.

$$\begin{bmatrix} 1 & 1 & -3 & | & 8 \\ 2 & 3 & -10 & | & 19 \\ -1 & -2 & 7 & | & -11 \end{bmatrix} \quad \begin{pmatrix} R_2 = -2r_1 + r_2 \\ R_3 = r_1 + r_3 \end{pmatrix}$$

$$= \begin{bmatrix} 1 & 1 & -3 & | & 8 \\ 0 & 1 & -4 & | & 3 \\ 0 & -1 & 4 & | & -3 \end{bmatrix} \quad (R_3 = r_2 + r_3)$$

$$= \begin{bmatrix} 1 & 1 & -3 & | & 8 \\ 0 & 1 & -4 & | & 3 \\ 0 & 0 & 0 & | & 0 \end{bmatrix}$$

Write the system of equations that corresponds to the row-echelon matrix.

$$\begin{cases} x + y - 3z = 8 & (1) \\ y - 4z = 3 & (2) \\ 0 = 0 & (3) \end{cases}$$

The statement $0 = 0$ in equation (3) indicates that the system is dependent and has an infinite number of solutions.

Solve equation (2) for $y$.
$y - 4z = 3$
$y = 4z + 3$

### Section 4.5 – Quick Checks: Determinants and Cramer's Rule

Substituting $4z+3$ for $y$ in equation (1), we obtain
$$x+(4z+3)-3z=8$$
$$x+z+3=8$$
$$x=-z+5$$

The solution to the system is
$$\{(x,y,z)|x=-z+5, y=4z+3, z \text{ is any real number}\}.$$

**10.** Write the augmented matrix of the system and then put it in row echelon form.

$$\begin{bmatrix} -1 & 2 & -1 & | & 5 \\ 2 & 1 & 4 & | & 3 \\ 3 & -1 & 5 & | & 0 \end{bmatrix} \quad (R_1 = -1 \cdot r_1)$$

$$= \begin{bmatrix} 1 & -2 & 1 & | & -5 \\ 2 & 1 & 4 & | & 3 \\ 3 & -1 & 5 & | & 0 \end{bmatrix} \quad \begin{pmatrix} R_2 = -2r_1 + r_2 \\ R_3 = -3r_1 + r_3 \end{pmatrix}$$

$$= \begin{bmatrix} 1 & -2 & 1 & | & -5 \\ 0 & 5 & 2 & | & 13 \\ 0 & 5 & 2 & | & 15 \end{bmatrix} \quad \left( R_2 = \frac{1}{5} r_2 \right)$$

$$= \begin{bmatrix} 1 & -2 & 1 & | & -5 \\ 0 & 1 & \frac{2}{5} & | & \frac{13}{5} \\ 0 & 5 & 2 & | & 15 \end{bmatrix} \quad (R_3 = -5r_2 + r_3)$$

$$= \begin{bmatrix} 1 & -2 & 1 & | & -5 \\ 0 & 1 & \frac{2}{5} & | & \frac{13}{5} \\ 0 & 0 & 0 & | & 2 \end{bmatrix}$$

Write the system of equations that corresponds to the row-echelon matrix.
$$\begin{cases} x-2y+z=-5 & (1) \\ y+\frac{2}{5}z=\frac{13}{5} & (2) \\ 0=2 & (3) \end{cases}$$

The statement $0=2$ in equation (3) indicates that the system is inconsistent. The system has no solution. The solution set is $\varnothing$ or $\{\}$.

### 4.5 Quick Checks

**1.** $\begin{vmatrix} 5 & 3 \\ 4 & 6 \end{vmatrix} = 5(6)-4(3)=30-12=18$

**2.** $\begin{vmatrix} -2 & -5 \\ 1 & 7 \end{vmatrix} = -2(7)-1(-5)=-14+5=-9$

**3.** $D = \begin{vmatrix} 3 & 2 \\ -2 & -1 \end{vmatrix} = 3(-1)-(-2)(2) = -3+4 = 1$

$D_x = \begin{vmatrix} 1 & 2 \\ 1 & -1 \end{vmatrix} = 1(-1)-1(2) = -1-2 = -3$

$D_y = \begin{vmatrix} 3 & 1 \\ -2 & 1 \end{vmatrix} = 3(1)-(-2)(1) = 3+2 = 5$

$x = \dfrac{D_x}{D} = \dfrac{-3}{1} = -3; \quad y = \dfrac{D_y}{D} = \dfrac{5}{1} = 5$

Thus, the solution is the ordered pair $(-3, 5)$.

**4.** $D = \begin{vmatrix} 4 & -2 \\ -6 & 3 \end{vmatrix} = 4(3)-(-6)(-2) = 12-12 = 0$

Since $D=0$, Cramer's Rule does not apply.

**5.** $\begin{vmatrix} 2 & -3 & 5 \\ 0 & 4 & -1 \\ 3 & 8 & -7 \end{vmatrix} = 2\begin{vmatrix} 4 & -1 \\ 8 & -7 \end{vmatrix} - (-3)\begin{vmatrix} 0 & -1 \\ 3 & -7 \end{vmatrix} + 5\begin{vmatrix} 0 & 4 \\ 3 & 8 \end{vmatrix}$

$= 2[4(-7)-8(-1)] + 3[0(-7)-3(-1)] + 5[0(8)-3(4)]$
$= 2(-28+8) + 3(0+3) + 5(0-12)$
$= 2(-20) + 3(3) + 5(-12)$
$= -40+9-60$
$= -91$

**6.** $D = \begin{vmatrix} 1 & -1 & 3 \\ 4 & 3 & 1 \\ -2 & 0 & 5 \end{vmatrix}$

$= 1\begin{vmatrix} 3 & 1 \\ 0 & 5 \end{vmatrix} - (-1)\begin{vmatrix} 4 & 1 \\ -2 & 5 \end{vmatrix} + 3\begin{vmatrix} 4 & 3 \\ -2 & 0 \end{vmatrix}$

$= 1[3(5)-0(1)] - (-1)[4(5)-(-2)(1)] + 3[4(0)-(-2)(3)]$
$= 1(15-0) - (-1)(20+2) + 3(0+6)$
$= 1(15) - (-1)(22) + 3(6)$
$= 15+22+18$
$= 55$

$D_x = \begin{vmatrix} -2 & -1 & 3 \\ 9 & 3 & 1 \\ 7 & 0 & 5 \end{vmatrix}$

$= -2\begin{vmatrix} 3 & 1 \\ 0 & 5 \end{vmatrix} - (-1)\begin{vmatrix} 9 & 1 \\ 7 & 5 \end{vmatrix} + 3\begin{vmatrix} 9 & 3 \\ 7 & 0 \end{vmatrix}$

$= -2[3(5)-0(1)] - (-1)[9(5)-7(1)] + 3[9(0)-7(3)]$
$= -2(15-0) - (-1)(45-7) + 3(0-21)$
$= -2(15) - (-1)(38) + 3(-21)$
$= -30+38-63$
$= -55$

## Chapter 4: Systems of Linear Equations and Inequalities

$$D_y = \begin{vmatrix} 1 & -2 & 3 \\ 4 & 9 & 1 \\ -2 & 7 & 5 \end{vmatrix}$$

$$= 1\begin{vmatrix} 9 & 1 \\ 7 & 5 \end{vmatrix} - (-2)\begin{vmatrix} 4 & 1 \\ -2 & 5 \end{vmatrix} + 3\begin{vmatrix} 4 & 9 \\ -2 & 7 \end{vmatrix}$$

$$= 1[9(5) - 7(1)] - (-2)[4(5) - (-2)(1)] +$$
$$3[4(7) - (-2)(9)]$$

$$= 1(45 - 7) - (-2)(20 + 2) + 3(28 + 18)$$

$$= 1(38) - (-2)(22) + 3(46)$$

$$= 38 + 44 + 138$$

$$= 220$$

$$D_z = \begin{vmatrix} 1 & -1 & -2 \\ 4 & 3 & 9 \\ -2 & 0 & 7 \end{vmatrix}$$

$$= 1\begin{vmatrix} 3 & 9 \\ 0 & 7 \end{vmatrix} - (-1)\begin{vmatrix} 4 & 9 \\ -2 & 7 \end{vmatrix} + (-2)\begin{vmatrix} 4 & 3 \\ -2 & 0 \end{vmatrix}$$

$$= 1[3(7) - 0(9)] - (-1)[4(7) - (-2)(9)] +$$
$$(-2)[4(0) - (-2)(3)]$$

$$= 1(21 - 0) - (-1)(28 + 18) + (-2)(0 + 6)$$

$$= 1(21) - (-1)(46) + (-2)(6)$$

$$= 21 + 46 - 12$$

$$= 55$$

$$x = \frac{D_x}{D} = \frac{-55}{55} = -1 \; ; \; y = \frac{D_y}{D} = \frac{220}{55} = 4 \; ;$$

$$z = \frac{D_z}{D} = \frac{55}{55} = 1$$

Thus, the solution is the ordered triple $(-1, 4, 1)$.

### 4.6 Quick Checks

1. $\begin{cases} -4x + y < -5 & (1) \\ 2x - 5y < 10 & (2) \end{cases}$

   **a.** Let $x = 1$ and $y = 2$ in both inequalities (1) and (2).

   Inequality (1): $-4(1) + 2 \overset{?}{<} -5$

   $$-4 + 2 \overset{?}{<} -5$$

   $$-2 \not< -5$$

   Inequality (2): $2(1) - 5(2) \overset{?}{<} 10$

   $$2 - 10 \overset{?}{<} 10$$

   $$-8 < 10$$

   The inequality $-4x + y < -5$ is not true when $x = 1$ and $y = 2$, so $(1, 2)$ is not a solution.

   **b.** Let $x = 3$ and $y = 1$ in both inequalities (1) and (2).

   Inequality (1): $-4(3) + 1 \overset{?}{<} -5$

   $$-12 + 1 \overset{?}{<} -5$$

   $$-11 < -5$$

   Inequality (2): $2(3) - 5(1) \overset{?}{<} 10$

   $$6 - 5 \overset{?}{<} 10$$

   $$1 < 10$$

   These values satisfy both inequalities, so the ordered pair $(3, 1)$ is a solution.

2. $\begin{cases} 2x + y \le 5 \\ -x + y \ge -4 \end{cases}$

   First, graph the inequality $2x + y \le 5$. To do so, replace the inequality symbol with an equal sign to obtain $2x + y = 5$. Because the inequality is non-strict, graph $2x + y = 5$ $(y = -2x + 5)$ using a solid line.

   Test Point: $(0, 0)$: $2(0) + 0 \overset{?}{\le} 5$

   $$0 \le 5$$

   Therefore, the half-plane containing $(0, 0)$ is the solution set of $2x + y \le 5$.

   Second, graph the inequality $-x + y \ge -4$. To do so, replace the inequality symbol with an equal sign to obtain $-x + y = -4$. Because the inequality is non-strict, graph $-x + y = -4$ $(y = x - 4)$ using a solid line.

   Test Point: $(0, 0)$: $-(0) + 0 \overset{?}{\ge} -4$

   $$0 \ge -4$$

   Therefore, the half-plane containing $(0, 0)$ is the solution set of $-x + y \ge -4$.

   The overlapping shaded region (that is, the shaded region in the graph below) is the solution to the system of linear inequalities.

   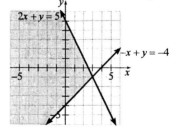

### Section 4.6 – Quick Checks: Systems of Linear Inequalities

3. $\begin{cases} 3x+y>-2 \\ 2x+3y<3 \end{cases}$

   First, graph the inequality $3x+y>-2$. To do so, replace the inequality symbol with an equal sign to obtain $3x+y=-2$. Because the inequality is strict, graph $3x+y=-2$ $(y=-3x-2)$ using a dashed line.

   Test Point: $(0,0)$: $3(0)+0 \overset{?}{>} -2$
   $0>-2$

   Therefore, the half-plane containing $(0,0)$ is the solution set of $3x+y>-2$.

   Second, graph the inequality $2x+3y<3$. To do so, replace the inequality symbol with an equal sign to obtain $2x+3y=3$. Because the inequality is strict, graph $2x+3y=3$ $\left(y=-\dfrac{2}{3}x+1\right)$ using a dashed line.

   Test Point: $(0,0)$: $-2(0)+3(0) \overset{?}{<} 3$
   $0<3$

   Therefore, the half-plane containing $(0,0)$ is the solution set of $2x+3y<3$.

   The overlapping shaded region (that is, the shaded region in the graph below) is the solution to the system of linear inequalities.

   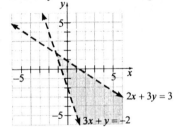

4. $\begin{cases} -2x+3y \geq 9 \\ 6x-9y \geq 9 \end{cases}$

   First, graph the inequality $-2x+3y \geq 9$. To do so, replace the inequality symbol with an equal sign to obtain $-2x+3y=9$. Because the inequality is non-strict, graph $-2x+3y=9$ $\left(y=\dfrac{2}{3}x+3\right)$ using a solid line.

   Test Point: $(0,0)$: $-2(0)+3(0) \overset{?}{\geq} 9$
   $0 \not\geq 9$

   Therefore, the half-plane not containing $(0,0)$ is the solution set of $-2x+3y \geq 9$.

   Second, graph the inequality $6x-9y \geq 9$. To do so, replace the inequality symbol with an equal sign to obtain $6x-9y=9$. Because the inequality is non-strict, graph $6x-9y=9$ $\left(y=\dfrac{2}{3}x-1\right)$ using a solid line.

   Test Point: $(0,0)$: $6(0)-9(0) \overset{?}{\geq} 9$
   $0 \not\geq 9$

   Therefore, the half-plane not containing $(0,0)$ is the solution set of $6x-9y \geq 9$.

   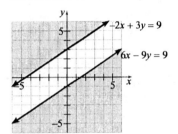

   Because no overlapping region results, there are no points in the Cartesian plane that satisfy both inequalities. The system has no solution, so the solution set is $\varnothing$ or $\{\ \}$.

5. $\begin{cases} x+y \leq 6 \\ 2x+y \leq 10 \\ x \geq 0 \\ y \geq 0 \end{cases}$

   First, graph the inequality $x+y \leq 6$. To do so, replace the inequality symbol with an equal sign to obtain $x+y=6$. Because the inequality is non-strict, graph $x+y=6$ $(y=-x+6)$ using a solid line.

   Test Point: $(0,0)$: $0+0 \overset{?}{\leq} 6$
   $0 \leq 6$

   Therefore, the half-plane containing $(0,0)$ is the solution set of $x+y \leq 6$.

   Next, graph the inequality $2x+y \leq 10$. To do so, replace the inequality symbol with an equal sign to obtain $2x+y=10$. Because the inequality is non-strict, graph $2x+y=10$ $(y=-2x+10)$ using a solid line.

QC – 55

Test Point: $(0,0)$: $2(0)+0\overset{?}{\leq}10$
$0 \leq 10$

Therefore, the half-plane containing $(0,0)$ is the solution set of $2x+y\leq 10$.

The two inequalities $x\geq 0$ and $y\geq 0$ require that the graph be in quadrant I.

The overlapping shaded region (that is, the shaded region in the graph below) is the solution to the system of linear inequalities.

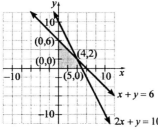

The graph is bounded. The four corner points of the bounded region are $(0, 0)$, $(5, 0)$, $(4, 2)$, and $(0, 6)$.

6. Let $x$ represent the amount to be invested in Treasury notes and let $y$ represent the amount to be invested in corporate bonds. Now, Jack and Mary have at most $25,000 to invest, so $x+y \leq 25,000$. The amount invested in Treasury notes must be at least $10,000, so $x \geq 10,000$. The amount to be invested in corporate bonds must be no more than $15,000, so $y \leq 15,000$. They cannot invest a negative amount in corporate bonds, so we have the non-negativity constraint $y \geq 0$. Thus, we have the following system:

$$\begin{cases} x+y \leq 25,000 \\ x \geq 10,000 \\ y \leq 15,000 \\ y \geq 0 \end{cases}$$

To graph the system, begin with the inequality $x+y \leq 25,000$. To do so, replace the inequality symbol with an equal sign to obtain $x+y = 25,000$. Because the inequality is non-strict, graph $x+y = 25,000$ $(y = -x + 25,000)$ using a solid line.

Test Point: $(0,0)$: $0+0\overset{?}{\leq}25,000$
$0 \leq 25,000$

Thus, the solution set of $x+y \leq 25,000$ is the half-plane containing $(0,0)$.

The inequality $x \geq 10,000$ requires that the graph be to the right of the vertical line $x = 10,000$.

The inequality $y \leq 15,000$ requires that the graph be below the horizontal line $y = 15,000$.

The inequality $y \geq 0$ requires that the graph be above the $x$-axis.

The overlapping shaded region (that is, the shaded region in the graph below) is the solution to the system of linear inequalities.

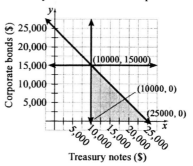

The corner points of the bounded region are $(10000, 15000)$, $(10000, 0)$, and $(25000, 0)$.

# Getting Ready for Chapter 5 – Quick Checks

1. $5^2 \cdot 5 = 5^{2+1} = 5^3 = 125$

2. $(-3)^2 \cdot (-3)^3 = (-3)^{2+3} = (-3)^5 = -243$

3. $y^4 \cdot y^3 = y^{4+3} = y^7$

4. $(5x^2) \cdot (-2x^5) = 5 \cdot (-2) \cdot x^2 \cdot x^5$
   $= -10x^{2+5}$
   $= -10x^7$

5. $(6y^3) \cdot (-y^2) = 6 \cdot (-1) \cdot y^3 \cdot y^2$
   $= -6y^{3+2}$
   $= -6y^5$

6. $\dfrac{5^6}{5^4} = 5^{6-4} = 5^2 = 25$

7. $\dfrac{y^8}{y^6} = y^{8-6} = y^2$

8. $\dfrac{16a^6}{10a^5} = \dfrac{8 \cdot 2}{5 \cdot 2} \cdot \dfrac{a^6}{a^5} = \dfrac{8}{5} \cdot a^{6-5} = \dfrac{8}{5}a$

9. $\dfrac{-24b^5}{16b^3} = \dfrac{-3 \cdot 8}{2 \cdot 8} \cdot \dfrac{b^5}{b^3} = \dfrac{-3}{2} \cdot b^{5-3} = -\dfrac{3}{2}b^2$

10. $5^{-3} = \dfrac{1}{5^3} = \dfrac{1}{125}$

11. $5z^{-7} = 5 \cdot \dfrac{1}{z^7} = \dfrac{5}{z^7}$

12. $\dfrac{1}{x^{-4}} = x^4$

13. $\dfrac{5}{y^{-3}} = 5y^3$

14. $-4^0 = -1 \cdot 4^0 = -1 \cdot 1 = -1$

15. $(-10)^0 = 1$

16. $\left(\dfrac{4}{3}\right)^{-2} = \left(\dfrac{3}{4}\right)^2 = \dfrac{3}{4} \cdot \dfrac{3}{4} = \dfrac{9}{16}$

17. $\left(-\dfrac{1}{4}\right)^{-3} = (-4)^3 = (-4) \cdot (-4) \cdot (-4) = -64$

18. $\left(\dfrac{3}{x}\right)^{-2} = \left(\dfrac{x}{3}\right)^2 = \dfrac{x}{3} \cdot \dfrac{x}{3} = \dfrac{x^{1+1}}{9} = \dfrac{x^2}{9}$

19. $\dfrac{5}{2^{-2}} = 5 \cdot 2^2 = 5 \cdot 4 = 20$

20. $6^3 \cdot 6^{-5} = 6^{3+(-5)} = 6^{-2} = \dfrac{1}{6^2} = \dfrac{1}{36}$

21. $\dfrac{10^{-3}}{10^{-5}} = 10^{-3-(-5)} = 10^{-3+5} = 10^2 = 100$

22. $(4x^2y^3) \cdot (5xy^{-4}) = 4 \cdot 5 \cdot x^2 \cdot x \cdot y^3 \cdot y^{-4}$
    $= 20x^{2+1}y^{3+(-4)}$
    $= 20x^3 y^{-1}$
    $= \dfrac{20x^3}{y}$

23. $\left(\dfrac{3}{4}a^3b\right) \cdot \left(\dfrac{8}{9}a^{-2}b^3\right) = \dfrac{3}{4} \cdot \dfrac{8}{9} \cdot a^3 \cdot a^{-2} \cdot b \cdot b^3$
    $= \dfrac{3 \cdot 4 \cdot 2}{4 \cdot 3 \cdot 3} \cdot a^{3+(-2)}b^{1+3}$
    $= \dfrac{2}{3}a^1b^4$
    $= \dfrac{2}{3}ab^4$

24. $\dfrac{-24b^5}{16b^{-3}} = \dfrac{-24}{16} \cdot \dfrac{b^5}{b^{-3}}$
    $= \dfrac{-3 \cdot 8}{2 \cdot 8} \cdot b^{5-(-3)}$
    $= \dfrac{-3}{2} \cdot b^{5+3}$
    $= -\dfrac{3}{2}b^8$

## Getting Ready for Chapter 5 – Quick Checks

25. $\dfrac{50s^2 t}{15s^5 t^{-4}} = \dfrac{50}{15} \cdot \dfrac{s^2}{s^5} \cdot \dfrac{t}{t^{-4}}$

    $= \dfrac{10 \cdot 5}{3 \cdot 5} \cdot s^{2-5} \cdot t^{1-(-4)}$

    $= \dfrac{10}{3} s^{-3} t^{1+4}$

    $= \dfrac{10 t^5}{3 s^3}$

26. $\left(2^2\right)^3 = 2^{2 \cdot 3} = 2^6 = 64$

27. $\left(5^8\right)^0 = 5^{8 \cdot 0} = 5^0 = 1$

28. $\left[(-4)^3\right]^2 = (-4)^{3 \cdot 2} = (-4)^6 = 4096$

29. $\left(a^3\right)^5 = a^{3 \cdot 5} = a^{15}$

30. $\left(z^3\right)^{-6} = z^{3 \cdot (-6)} = z^{-18} = \dfrac{1}{z^{18}}$

31. $\left(s^{-3}\right)^{-7} = s^{(-3) \cdot (-7)} = s^{21}$

32. $(5y)^3 = 5^3 \cdot y^3 = 125 y^3$

33. $(6y)^0 = 6^0 \cdot y^0 = 1 \cdot 1 = 1$

34. $\left(3x^2\right)^4 = 3^4 \cdot \left(x^2\right)^4$

    $= 81 x^{2 \cdot 4}$

    $= 81 x^8$

35. $\left(4a^3\right)^{-2} = 4^{-2} \cdot \left(a^3\right)^{-2}$

    $= \dfrac{1}{4^2} \cdot a^{3 \cdot (-2)}$

    $= \dfrac{1}{16} a^{-6}$

    $= \dfrac{1}{16 a^6}$

36. $\left(\dfrac{z}{3}\right)^4 = \dfrac{z^4}{3^4} = \dfrac{z^4}{81}$

37. $\left(\dfrac{x}{2}\right)^{-5} = \left(\dfrac{2}{x}\right)^5 = \dfrac{2^5}{x^5} = \dfrac{32}{x^5}$

38. $\left(\dfrac{x^2}{y^3}\right)^4 = \dfrac{\left(x^2\right)^4}{\left(y^3\right)^4} = \dfrac{x^{2 \cdot 4}}{y^{3 \cdot 4}} = \dfrac{x^8}{y^{12}}$

39. $\left(\dfrac{3a^{-2}}{b^4}\right)^3 = \dfrac{3^3 \cdot \left(a^{-2}\right)^3}{\left(b^4\right)^3}$

    $= \dfrac{27 a^{(-2) \cdot 3}}{b^{4 \cdot 3}}$

    $= \dfrac{27 a^{-6}}{b^{12}}$

    $= \dfrac{27}{a^6 b^{12}}$

40. $\dfrac{\left(3x^2 y\right)^2}{12 x y^{-2}} = \dfrac{3^2 \cdot \left(x^2\right)^2 \cdot y^2}{12 x y^{-2}}$

    $= \dfrac{9 x^4 y^2}{12 x y^{-2}}$

    $= \dfrac{3 \cdot 3}{3 \cdot 4} \cdot x^{4-1} \cdot y^{2-(-2)}$

    $= \dfrac{3 x^3 y^4}{4}$

41. $\left(3ab^3\right)^3 \cdot \left(6a^2 b^2\right)^{-2}$

    $= \left(3^3 \cdot a^3 \cdot \left(b^3\right)^3\right) \cdot \left(6^{-2} \cdot \left(a^2\right)^{-2} \cdot \left(b^2\right)^{-2}\right)$

    $= \left(27 a^3 b^9\right) \cdot \left(\dfrac{1}{36} a^{-4} b^{-4}\right)$

    $= \dfrac{27}{36} a^{3+(-4)} b^{9+(-4)}$

    $= \dfrac{3 \cdot 9}{4 \cdot 9} a^{-1} b^5$

    $= \dfrac{3 b^5}{4 a}$

42. $\left(\dfrac{2x^2y^{-1}}{x^{-2}y^2}\right)^2 \cdot \left(\dfrac{4x^3y^2}{xy^{-2}}\right)^{-1}$

$= \left(2x^{2-(-2)}y^{-1-2}\right)^2 \cdot \left(4x^{3-1}y^{2-(-2)}\right)^{-1}$

$= \left(2x^4y^{-3}\right)^2 \cdot \left(4x^2y^4\right)^{-1}$

$= \left(2^2 \cdot \left(x^4\right)^2 \cdot \left(y^{-3}\right)^2\right) \cdot \left(4^{-1} \cdot \left(x^2\right)^{-1} \cdot \left(y^4\right)^{-1}\right)$

$= \left(4x^8y^{-6}\right) \cdot \left(\dfrac{1}{4}x^{-2}y^{-4}\right)$

$= \dfrac{4}{4}x^{8+(-2)}y^{-6+(-4)}$

$= x^6 y^{-10}$

$= \dfrac{x^6}{y^{10}}$

43. Because 532 is greater than 1, we move the decimal point to the left. We move it two places to the left until it is between the 5 and the 3.
$532 = 5.32 \times 10^2$

44. Because the absolute value of $-1,230,000$ is greater than 1, we move the decimal point to the left. We move it six places to the left until it is between the 1 and the 2.
$-1,230,000 = -1.23 \times 10^6$

45. Because the absolute value of 0.034 is less than 1, we move the decimal point to the right. We move it two places to the right until it is between the 3 and the 4.
$0.034 = 3.4 \times 10^{-2}$

46. Because the absolute value of $-0.0000845$ is less than 1, we move the decimal point to the right. We move it five places to the right until it is between the 8 and the 4.
$-0.0000845 = -8.45 \times 10^{-5}$

47. The exponent on 10 is 2 so we move the decimal point 2 places to the right.
$5 \times 10^2 = 5.00 \times 10^2 = 500$

48. The exponent on 10 is 5 so we move the decimal point 5 places to the right.
$9.1 \times 10^5 = 9.10000 \times 10^5 = 910,000$

49. The exponent on 10 is $-4$ so we move the decimal point 4 places to the left.
$1.8 \times 10^{-4} = 00001.8 \times 10^{-4} = 0.00018$

50. The exponent on 10 is $-6$ so we move the decimal point 6 places to the left.
$1 \times 10^{-6} = 0000001 \times 10^{-6} = 0.000001$

51. $(3 \times 10^3) \cdot (2 \times 10^5) = (3 \cdot 2) \times (10^3 \cdot 10^5)$
$= 6 \times 10^8$

52. $(2 \times 10^{-4}) \cdot (4 \times 10^{-7}) = (2 \cdot 4) \times (10^{-4} \cdot 10^{-7})$
$= 8 \times 10^{-11}$

53. $(6 \times 10^{-5}) \cdot (4 \times 10^8) = (6 \cdot 4) \times (10^{-5} \cdot 10^8)$
$= 24 \times 10^3$
$= (2.4 \times 10^1) \times 10^3$
$= 2.4 \times 10^4$

54. $\dfrac{6 \times 10^8}{3 \times 10^6} = \dfrac{6}{3} \times \dfrac{10^8}{10^6} = 2 \times 10^2$

55. $\dfrac{6.8 \times 10^{-8}}{3.4 \times 10^{-5}} = \dfrac{6.8}{3.4} \times \dfrac{10^{-8}}{10^{-5}}$
$= 2 \times 10^{-8-(-5)}$
$= 2 \times 10^{-3}$

56. $\dfrac{4.8 \times 10^7}{9.6 \times 10^3} = \dfrac{4.8}{9.6} \times \dfrac{10^7}{10^3}$
$= 0.5 \times 10^4$
$= (5 \times 10^{-1}) \times 10^4$
$= 5 \times 10^3$

57. $\dfrac{3 \times 10^{-5}}{8 \times 10^7} = \dfrac{3}{8} \times \dfrac{10^{-5}}{10^7}$
$= 0.375 \times 10^{-12}$
$= (3.75 \times 10^{-1}) \times 10^{-12}$
$= 3.75 \times 10^{-13}$

## Getting Ready for Chapter 5 – Quick Checks

**58.** $(8,000,000) \cdot (30,000) = (8 \times 10^6) \cdot (3 \times 10^4)$
$\qquad\qquad\qquad = (8 \cdot 3) \times (10^6 \cdot 10^4)$
$\qquad\qquad\qquad = 24 \times 10^{10}$
$\qquad\qquad\qquad = (2.4 \times 10^1) \times 10^{10}$
$\qquad\qquad\qquad = 2.4 \times 10^{11}$
$\qquad\qquad\qquad = 240,000,000,000$

**59.** $\dfrac{0.000000012}{0.000004} = \dfrac{1.2 \times 10^{-8}}{4 \times 10^{-6}}$
$\qquad\qquad\quad = \dfrac{1.2}{4} \times \dfrac{10^{-8}}{10^{-6}}$
$\qquad\qquad\quad = 0.3 \times 10^{-2}$
$\qquad\qquad\quad = (3 \times 10^{-1}) \times 10^{-2}$
$\qquad\qquad\quad = 3 \times 10^{-3}$
$\qquad\qquad\quad = 0.003$

**60.** $(25,000,000) \cdot (0.00003)$
$\quad = (2.5 \times 10^7) \cdot (3 \times 10^{-5})$
$\quad = (2.5 \cdot 3) \times (10^7 \cdot 10^{-5})$
$\quad = 7.5 \times 10^2$
$\quad = 750$

**61.** $\dfrac{0.000039}{13,000,000} = \dfrac{3.9 \times 10^{-5}}{1.3 \times 10^7}$
$\qquad\qquad\quad = \dfrac{3.9}{1.3} \times \dfrac{10^{-5}}{10^7}$
$\qquad\qquad\quad = 3 \times 10^{-12}$
$\qquad\qquad\quad = 0.000000000003$

# Chapter 5 – Quick Checks

## 5.1 Quick Checks

1. $8x^5$ is a monomial because it is in the form $ax^k$ where $k \geq 0$. The coefficient is $a = 8$ and the degree is $k = 5$.

2. $5x^{-2}$ is not a monomial because the exponent of the variable is $-2$ and $-2$ is not a nonnegative integer.

3. 12 is a monomial because it is in the form $ax^k$ where $k \geq 0$. The coefficient is $a = 12$ and the degree is $k = 0$.

4. $x^{1/3}$ is not a monomial because the exponent of the variable is $\frac{1}{3}$ and $\frac{1}{3}$ is not a nonnegative integer.

5. $3x^5y^2$ is a monomial because the exponents of the variables are nonnegative integers. The coefficient is 3 and the degree is $n = 5 + 2 = 7$.

6. $-2m^3n$ is a monomial because the exponents of the variables are nonnegative integers. The coefficient is $-2$ and the degree is $n = 3 + 1 = 4$.

7. $4ab^{1/2}$ is not a monomial because the exponent ½ is not an integer.

8. $-xy$ is a monomial because the exponents of the variables are nonnegative integers (both are 1 in this case). The coefficient is $-1$ and the degree is $n = 1 + 1 = 2$.

9. $-3x^3 + 7x^2 - x + 5$ is a polynomial with degree $n = 3$.

10. $5z^{-1} + 3$ is not a polynomial because the exponent on the first term, $-1$, is negative.

11. $\dfrac{x-1}{x+1}$ is not a polynomial because it is the quotient of two polynomials and the polynomial in the denominator has degree greater than 0.

12. $\dfrac{3x^2 - 9x + 27}{3} = x^2 - 3x + 9$ is a polynomial. The degree is $n = 2$.

13. $5p^3q - 8pq^2 + pq$ is a polynomial with degree $n = 3 + 1 = 4$.

14. $(2x^2 - 3x + 1) + (4x^2 + 5x - 3)$
$= 2x^2 - 3x + 1 + 4x^2 + 5x - 3$
$= 2x^2 + 4x^2 - 3x + 5x + 1 - 3$
$= (2+4)x^2 + (-3+5)x + (1-3)$
$= 6x^2 + 2x - 2$

15. $(5w^4 - 3w^3 + w - 8) + (-2w^4 + w^3 - 7w^2 + 3)$
$= 5w^4 - 3w^3 + w - 8 - 2w^4 + w^3 - 7w^2 + 3$
$= 5w^4 - 2w^4 - 3w^3 + w^3 - 7w^2 + w - 8 + 3$
$= (5-2)w^4 + (-3+1)w^3 - 7w^2 + w + (-8+3)$
$= 3w^4 - 2w^3 - 7w^2 + w - 5$

16. $(8x^2y + 2x^2y^2 - 7xy^2) + (-3x^2y + 5x^2y^2 + 3xy^2)$
$= 8x^2y + 2x^2y^2 - 7xy^2 - 3x^2y + 5x^2y^2 + 3xy^2$
$= 8x^2y - 3x^2y + 2x^2y^2 + 5x^2y^2 - 7xy^2 + 3xy^2$
$= (8-3)x^2y + (2+5)x^2y^2 + (-7+3)xy^2$
$= 5x^2y + 7x^2y^2 - 4xy^2$

17. $(5x^3 - 6x^2 + x + 9) - (4x^3 + 10x^2 - 6x + 7)$
$= 5x^3 - 6x^2 + x + 9 - 4x^3 - 10x^2 + 6x - 7$
$= 5x^3 - 4x^3 - 6x^2 - 10x^2 + x + 6x + 9 - 7$
$= (5-4)x^3 + (-6-10)x^2 + (1+6)x + (9-7)$
$= x^3 - 16x^2 + 7x + 2$

18. $(8y^3 - 5y^2 + 3y + 1) - (-3y^3 + 6y + 8)$
$= 8y^3 - 5y^2 + 3y + 1 + 3y^3 - 6y - 8$
$= 8y^3 + 3y^3 - 5y^2 + 3y - 6y + 1 - 8$
$= (8+3)y^3 - 5y^2 + (3-6)y + (1-8)$
$= 11y^3 - 5y^2 - 3y - 7$

19. $(8x^2y + 2x^2y^2 - 7xy^2) - (-3x^2y + 5x^2y^2 + 3xy^2)$
$= 8x^2y + 2x^2y^2 - 7xy^2 + 3x^2y - 5x^2y^2 - 3xy^2$
$= 8x^2y + 3x^2y + 2x^2y^2 - 5x^2y^2 - 7xy^2 - 3xy^2$
$= (8+3)x^2y + (2-5)x^2y^2 + (-7-3)xy^2$
$= 11x^2y - 3x^2y^2 - 10xy^2$

## Chapter 5: Polynomials and Polynomial Functions

20. $g(x) = -2x^3 + 7x + 1$

    a. $g(0) = -2(0)^3 + 7(0) + 1$
    $= 0 + 0 + 1$
    $= 1$

    b. $g(2) = -2(2)^3 + 7(2) + 1$
    $= -2(8) + 7(2) + 1$
    $= -16 + 14 + 1$
    $= -1$

    c. $g(-3) = -2(-3)^3 + 7(-3) + 1$
    $= -2(-27) + 7(-3) + 1$
    $= 54 - 21 + 1$
    $= 34$

21. a. For 1994 we have $t = 0$.
    $F(t) = 0.13t^2 - 2.48t + 58.93$
    $F(0) = 0.13(0)^2 - 2.48(0) + 58.93$
    $= 0 - 0 + 58.93$
    $= 58.93$
    In 1994, the fertility rate for women 15-19 years of age was approximately 58.93 births per 1,000 women.

    b. For 2008, we have $t = 14$.
    $F(t) = 0.13t^2 - 2.48t + 58.93$
    $F(14) = 0.13(14)^2 - 2.48(14) + 58.93$
    $= 0.13(196) - 2.48(14) + 58.93$
    $= 25.48 - 34.72 + 58.93$
    $= 49.69$
    In 2008, the fertility rate for women 15-19 years of age is estimated to be 49.69 births per 1,000 women.

22. $f(x) = 3x^2 - x + 1$ and $g(x) = -x^2 + 5x - 6$

    a. $(f+g)(x) = f(x) + g(x)$
    $= (3x^2 - x + 1) + (-x^2 + 5x - 6)$
    $= 3x^2 - x + 1 - x^2 + 5x - 6$
    $= 3x^2 - x^2 - x + 5x + 1 - 6$
    $= 2x^2 + 4x - 5$

    b. $(f-g)(x) = f(x) - g(x)$
    $= (3x^2 - x + 1) - (-x^2 + 5x - 6)$
    $= 3x^2 - x + 1 + x^2 - 5x + 6$
    $= 3x^2 + x^2 - x - 5x + 1 + 6$
    $= 4x^2 - 6x + 7$

    c. Because $(f+g)(x) = 2x^2 + 4x - 5$, we have
    $(f+g)(1) = 2(1)^2 + 4(1) - 5$
    $= 2(1) + 4(1) - 5$
    $= 2 + 4 - 5$
    $= 1$

    d. Because $(f-g)(x) = 4x^2 - 6x + 7$, we have
    $(f-g)(-2) = 4(-2)^2 - 6(-2) + 7$
    $= 4(4) - 6(-2) + 7$
    $= 16 + 12 + 7$
    $= 35$

23. a. The revenue function will remain the same, so we have
    $R(x) = 12x$
    Both the variable and fixed costs will change. Therefore, our new cost function will be
    $C(x) = 8x + 1250$
    The profit function is given by
    $P(x) = R(x) - C(x)$
    $= (12x) - (8x + 1250)$
    $= 12x - 8x - 1250$
    $= 4x - 1250$
    The new profit function will be
    $P(x) = 4x - 1250$.

    b. $P(x) = 4x - 1250$
    $P(800) = 4(800) - 1250$
    $= 3200 - 1250$
    $= 1950$
    If the company manufactures and sells 800 calculators, its profit will be $1950.

## Section 5.2 – Quick Checks: Multiplying Polynomials

**5.2 Quick Checks**

1. $(3x^5) \cdot (2x^2) = (3 \cdot 2)(x^5 \cdot x^2)$
   $= 6x^{5+2}$
   $= 6x^7$

2. $(-7a^3b^2) \cdot (3ab^4) = (-7 \cdot 3)(a^3 \cdot a)(b^2 \cdot b^4)$
   $= -21a^{3+1}b^{2+4}$
   $= -21a^4b^6$

3. $\left(\dfrac{2}{3}x^4\right) \cdot \left(\dfrac{15}{8}x\right) = \left(\dfrac{2}{3} \cdot \dfrac{15}{8}\right)(x^4 \cdot x)$
   $= \dfrac{5}{4}x^{4+1}$
   $= \dfrac{5}{4}x^5$

4. $-3(x+2) = -3 \cdot x + (-3) \cdot 2$
   $= -3x - 6$

5. $5x(x^2 + 3x + 2) = 5x \cdot x^2 + 5x \cdot 3x + 5x \cdot 2$
   $= 5x^3 + 15x^2 + 10x$

6. $2xy(3x^2 - 5xy + 2y^2)$
   $= 2xy \cdot 3x^2 - 2xy \cdot 5xy + 2xy \cdot 2y^2$
   $= 6x^3y - 10x^2y^2 + 4xy^3$

7. $\dfrac{3}{4}y^2\left(\dfrac{4}{3}y^2 + \dfrac{2}{9}y + \dfrac{16}{3}\right)$
   $= \dfrac{3}{4}y^2 \cdot \dfrac{4}{3}y^2 + \dfrac{3}{4}y^2 \cdot \dfrac{2}{9}y + \dfrac{3}{4}y^2 \cdot \dfrac{16}{3}$
   $= \left(\dfrac{3}{4} \cdot \dfrac{4}{3}\right)y^2 \cdot y^2 + \left(\dfrac{3}{4} \cdot \dfrac{2}{9}\right)y^2 \cdot y + \left(\dfrac{3}{4} \cdot \dfrac{16}{3}\right)y^2$
   $= y^4 + \dfrac{1}{6}y^3 + 4y^2$

8. $(x+4)(x+1) = x \cdot x + x \cdot 1 + 4 \cdot x + 4 \cdot 1$
   $= x^2 + x + 4x + 4$
   $= x^2 + 5x + 4$

9. $(3v+5)(2v-3) = 3v \cdot 2v - 3v \cdot 3 + 5 \cdot 2v - 5 \cdot 3$
   $= 6v^2 - 9v + 10v - 15$
   $= 6v^2 + v - 15$

10. $(2a-b)(a+5b) = 2a \cdot a + 2a \cdot 5b - b \cdot a - b \cdot 5b$
    $= 2a^2 + 10ab - ab - 5b^2$
    $= 2a^2 + 9ab - 5b^2$

11. $(2y-3)(y^2 + 4y + 5)$
    $= 2y \cdot y^2 + 2y \cdot 4y + 2y \cdot 5 - 3 \cdot y^2 - 3 \cdot 4y - 3 \cdot 5$
    $= 2y^3 + 8y^2 + 10y - 3y^2 - 12y - 15$
    $= 2y^3 + 5y^2 - 2y - 15$

12. $(z^2 - 3z + 2)(2z^2 + z + 6)$
    $= z^2 \cdot 2z^2 + z^2 \cdot z + z^2 \cdot 6 - 3z \cdot 2z^2 - 3z \cdot z$
    $\quad - 3z \cdot 6 + 2 \cdot 2z^2 + 2 \cdot z + 2 \cdot 6$
    $= 2z^4 + z^3 + 6z^2 - 6z^3 - 3z^2 - 18z$
    $\quad + 4z^2 + 2z + 12$
    $= 2z^4 - 5z^3 + 7z^2 - 16z + 12$

13. $(5y+2)(5y-2) = (5y)^2 - (2)^2$
    $= 25y^2 - 4$

14. $(7y+2z^3)(7y-2z^3) = (7y)^2 - (2z^3)^2$
    $= 49y^2 - 4z^6$

15. $(z-8)^2 = z^2 - 2 \cdot z \cdot 8 + 8^2$
    $= z^2 - 16z + 64$

16. $(6p+5)^2 = (6p)^2 + 2 \cdot 6p \cdot 5 + 5^2$
    $= 36p^2 + 60p + 25$

17. $(4a-3b)^2 = (4a)^2 - 2 \cdot (4a) \cdot (3b) + (3b)^2$
    $= 16a^2 - 24ab + 9b^2$

18. $f(x) = 5x - 3;\ g(x) = x^2 + 3x + 1$

   a. $f(2) = 5(2) - 3 \qquad g(2) = (2)^2 + 3(2) + 1$
      $\quad = 10 - 3 \qquad\qquad\quad = 4 + 6 + 1$
      $\quad = 7 \qquad\qquad\qquad\quad = 11$
      $f(2) \cdot g(2) = 7 \cdot 11 = 77$

   b. $(f \cdot g)(x)$
      $= f(x) \cdot g(x)$
      $= (5x - 3)(x^2 + 3x + 1)$
      $= 5x \cdot x^2 + 5x \cdot 3x + 5x \cdot 1 - 3 \cdot x^2 - 3 \cdot 3x - 3 \cdot 1$
      $= 5x^3 + 15x^2 + 5x - 3x^2 - 9x - 3$
      $= 5x^3 + 12x^2 - 4x - 3$

   c. $(f \cdot g)(2) = 5(2)^3 + 12(2)^2 - 4(2) - 3$
      $= 5(8) + 12(4) - 4(2) - 3$
      $= 40 + 48 - 8 - 3$
      $= 77$

**Chapter 5:** Polynomials and Polynomial Functions

19. $f(x) = x^2 - 2x$

   a. $f(x-3) = (x-3)^2 - 2(x-3)$
   $= x^2 - 2 \cdot x \cdot 3 + 3^2 - 2 \cdot x + 2 \cdot 3$
   $= x^2 - 6x + 9 - 2x + 6$
   $= x^2 - 8x + 15$

   b. $f(x+h) - f(x)$
   $= \left[(x+h)^2 - 2(x+h)\right] - \left[x^2 - 2x\right]$
   $= \left[x^2 + 2xh + h^2 - 2x - 2h\right] - \left[x^2 - 2x\right]$
   $= x^2 + 2xh + h^2 - 2x - 2h - x^2 + 2x$
   $= 2xh + h^2 - 2h$

**5.3 Quick Checks**

1. $\dfrac{9p^4 - 12p^3 + 3p^2}{3p} = \dfrac{9p^4}{3p} - \dfrac{12p^3}{3p} + \dfrac{3p^2}{3p}$
   $= \dfrac{9}{3}p^{4-1} - \dfrac{12}{3}p^{3-1} + \dfrac{3}{3}p^{2-1}$
   $= 3p^3 - 4p^2 + p$

2. $\dfrac{14m^4 - 10m^3 + 2m^2}{2m^2}$
   $= \dfrac{14m^4}{2m^2} - \dfrac{10m^3}{2m^2} + \dfrac{2m^2}{2m^2}$
   $= \dfrac{14}{2}m^{4-2} - \dfrac{10}{2}m^{3-2} + \dfrac{2}{2}m^{2-2}$
   $= 7m^2 - 5m + 1$

3. $\dfrac{x^4 y^4 + 8x^2 y^2 - 4xy}{4x^3 y}$
   $= \dfrac{x^4 y^4}{4x^3 y} + \dfrac{8x^2 y^2}{4x^3 y} - \dfrac{4xy}{4x^3 y}$
   $= \dfrac{1}{4}x^{4-3}y^{4-1} + \dfrac{8}{4}x^{2-3}y^{2-1} - \dfrac{4}{4}x^{1-3}y^{1-1}$
   $= \dfrac{1}{4}xy^3 + 2x^{-1}y - x^{-2}$
   $= \dfrac{xy^3}{4} + \dfrac{2y}{x} - \dfrac{1}{x^2}$

4. $\quad x-4 \overline{\smash{\big)}\, x^3 + 3x^2 - 31x + 21}$ $\quad\quad x^2 + 7x - 3$
   $\underline{-(x^3 - 4x^2)}$
   $\quad\quad 7x^2 - 31x$
   $\quad\quad \underline{-(7x^2 - 28x)}$
   $\quad\quad\quad\quad -3x + 21$
   $\quad\quad\quad\quad \underline{-(-3x + 12)}$
   $\quad\quad\quad\quad\quad\quad 9$

   $\dfrac{x^3 + 3x^2 - 31x + 21}{x-4} = x^2 + 7x - 3 + \dfrac{9}{x-4}$

   Check:
   $(x-4)(x^2 + 7x - 3) + 9$
   $= x^3 + 7x^2 - 3x - 4x^2 - 28x + 12 + 9$
   $= x^3 + 3x^2 - 31x + 21$
   so our answer checks.

5. $\quad 2x-3 \overline{\smash{\big)}\, 2x^3 + 7x^2 - 7x - 12}$ $\quad\quad x^2 + 5x + 4$
   $\underline{-(2x^3 - 3x^2)}$
   $\quad\quad 10x^2 - 7x$
   $\quad\quad \underline{-(10x^2 - 15x)}$
   $\quad\quad\quad\quad 8x - 12$
   $\quad\quad\quad\quad \underline{-(8x - 12)}$
   $\quad\quad\quad\quad\quad\quad 0$

   $\dfrac{2x^3 + 7x^2 - 7x - 12}{2x-3} = x^2 + 5x + 4$

   Check:
   $(2x-3)(x^2 + 5x + 4)$
   $= 2x^3 + 10x^2 + 8x - 3x^2 - 15x - 12$
   $= 2x^3 + 7x^2 - 7x - 12$
   so our answer checks.

**Section 5.3 – Quick Checks:** *Dividing Polynomials; Synthetic Division*

6. 
$$\begin{array}{r} x^3 - 5x^2 + 2 \\ x^2 - 2 \overline{\smash{\big)}\, x^5 - 5x^4 - 2x^3 + 12x^2 + 2} \\ \underline{-(x^5 \phantom{-5x^4} - 2x^3)} \\ -5x^4 \phantom{-2x^3} + 12x^2 \\ \underline{-(-5x^4 \phantom{-2x^3} + 10x^2)} \\ 2x^2 + 2 \\ \underline{-(2x^2 - 4)} \\ 6 \end{array}$$

$$\frac{2 + 12x^2 - 2x^3 - 5x^4 + x^5}{x^2 - 2} = x^3 - 5x^2 + 2 + \frac{6}{x^2 - 2}$$

Check:
$(x^2 - 2)(x^3 - 5x^2 + 2) + 6$
$= x^5 - 5x^4 + 2x^2 - 2x^3 + 10x^2 - 4 + 6$
$= x^5 - 5x^4 - 2x^3 + 12x^2 + 2$
so our answer checks.

7. The divisor is $x - 2$ so $c = 2$.

$$\begin{array}{r|rrrr} 2 & 2 & 1 & -7 & -13 \\ & & 4 & 10 & 6 \\ \hline & 2 & 5 & 3 & -7 \end{array}$$

$$\frac{2x^3 + x^2 - 7x - 13}{x - 2} = 2x^2 + 5x + 3 - \frac{7}{x - 2}$$

8. The divisor is $x + 3$ so $c = -3$.

$$\begin{array}{r|rrrrr} -3 & 1 & 8 & 15 & -2 & -6 \\ & & -3 & -15 & 0 & 6 \\ \hline & 1 & 5 & 0 & -2 & 0 \end{array}$$

$$\frac{x^4 + 8x^3 + 15x^2 - 2x - 6}{x + 3} = x^3 + 5x^2 - 2$$

9. $f(x) = 3x^4 - 4x^3 - 3x^2 + 10x - 5$; $g(x) = x^2 - 2$

a. 
$$\begin{array}{r} 3x^2 - 4x + 3 \\ x^2 - 2 \overline{\smash{\big)}\, 3x^4 - 4x^3 - 3x^2 + 10x - 5} \\ \underline{-(3x^4 \phantom{-4x^3} - 6x^2)} \\ -4x^3 + 3x^2 + 10x \\ \underline{-(-4x^3 \phantom{+3x^2} + 8x)} \\ 3x^2 + 2x - 5 \\ \underline{-(3x^2 \phantom{+2x} - 6)} \\ 2x + 1 \end{array}$$

Thus, $\left(\dfrac{f}{g}\right)(x) = 3x^2 - 4x + 3 + \dfrac{2x+1}{x^2 - 2}$.

b. $\left(\dfrac{f}{g}\right)(3) = 3(3)^2 - 4(3) + 3 + \dfrac{2(3)+1}{(3)^2 - 2}$

$= 3(9) - 4(3) + 3 + \dfrac{6+1}{9-2}$

$= 27 - 12 + 3 + \dfrac{7}{7}$

$= 19$

10. $f(x) = 3x^3 + 10x^2 - 9x - 4$

a. The divisor is $x - 2$ so the Remainder Theorem says that the remainder is $f(2)$.

$f(2) = 3(2)^3 + 10(2)^2 - 9(2) - 4$
$= 3(8) + 10(4) - 9(2) - 4$
$= 24 + 40 - 18 - 4$
$= 42$

When $f(x) = 3x^3 + 10x^2 - 9x - 4$ is divided by $x - 2$, the remainder is 42.

b. The divisor is $x + 4 = x - (-4)$, so the Remainder Theorem says that the remainder is $f(-4)$.

$f(-4) = 3(-4)^3 + 10(-4)^2 - 9(-4) - 4$
$= 3(-64) + 10(16) - 9(-4) - 4$
$= -192 + 160 + 36 - 4$
$= 0$

When $f(x) = 3x^3 + 10x^2 - 9x - 4$ is divided by $x + 4$, the remainder is 0.

## Chapter 5: Polynomials and Polynomial Functions

11. $f(x) = 2x^3 - 9x^2 - 6x + 5$

   a. $f(-2) = 2(-2)^3 - 9(-2)^2 - 6(-2) + 5$
   $= 2(-8) - 9(4) - 6(-2) + 5$
   $= -16 - 36 + 12 + 5$
   $= -35$
   Since $f(c) = f(-2) \neq 0$, we know that $x - c = x + 2$ is not a factor of $f(x)$.

   b. $f(5) = 2(5)^3 - 9(5)^2 - 6(5) + 5$
   $= 2(125) - 9(25) - 6(5) + 5$
   $= 250 - 225 - 30 + 5$
   $= 0$
   Since $f(c) = f(5) = 0$, we know that $x - c = x - 5$ is a factor of $f(x)$.

   $$\begin{array}{r|rrrr} 5) & 2 & -9 & -6 & 5 \\ & & 10 & 5 & -5 \\ \hline & 2 & 1 & -1 & 0 \end{array}$$

   $f(x) = (x-5)(2x^2 + x - 1)$

### 5.4 Quick Checks

1. Look at the coefficients first. The largest number that divides into 5 and 15 evenly is 5, so 5 is part of the GCF. Because 15 does not have a variable factor, the GCF is 5.

2. Look at the coefficients first. The largest number that divides into 4, 10, and 12 evenly is 2, so 2 is part of the GCF. Now look at the variable expressions, $z^3$, $z^2$, and $z$. We choose the variable expression with the smallest exponent. The GCF is $2z$.

3. Look at the coefficients first. The largest number that divides into 3, 9, and 12 evenly is 3, so 3 is part of the GCF. Now look at each variable in the expressions. The smallest exponent on $x$ is 1, so $x$ is part of the GCF. The smallest exponent on $y$ is 3, so $y^3$ is part of the GCF. The GCF is $3xy^3$.

4. First we find that the GCF is $7z$. Next we rewrite each term as the product of the GCF and a remaining factor, then factor out the GCF.
   $7z^2 - 14z = 7z \cdot z - 7z \cdot 2 = 7z(z-2)$

5. First we find that the GCF is $2y$. Next we rewrite each term as the product of the GCF and a remaining factor, then factor out the GCF.
   $6y^3 - 14y^2 + 10y = 2y \cdot 3y^2 - 2y \cdot 7y + 2y \cdot 5$
   $= 2y(3y^2 - 7y + 5)$

6. First we find that the GCF is $2m^2n^2$. Next we rewrite each term as the product of the GCF and a remaining factor, then factor out the GCF.
   $2m^4n^2 + 8m^3n^4 - 6m^2n^5$
   $= 2m^2n^2 \cdot m^2 + 2m^2n^2 \cdot 4mn^2 - 2m^2n^2 \cdot 3n^3$
   $= 2m^2n^2(m^2 + 4mn^2 - 3n^3)$

7. First we find that the GCF is $-5y$. Next we rewrite each term as the product of the GCF and a remaining factor, then factor out the GCF.
   $-5y^2 + 10y = -5y \cdot y + (-5y) \cdot (-2) = -5y(y-2)$

8. First we find that the GCF is $-3a$. Next we rewrite each term as the product of the GCF and a remaining factor, then factor out the GCF.
   $-3a^3 + 6a^2 - 12a$
   $= -3a \cdot a^2 + (-3a) \cdot (-2a) + (-3a) \cdot (4)$
   $= -3a(a^2 - 2a + 4)$

9. First we find that the GCF is $a - 3$. Next we rewrite each term as the product of the GCF and a remaining factor, then factor out the GCF.
   $4a(a-3) + 3(a-3) = (a-3) \cdot 4a + (a-3) \cdot 3$
   $= (a-3)(4a+3)$

10. First we find that the GCF is $w - 5$. Next we rewrite each term as the product of the GCF and a remaining factor, then factor out the GCF.
    $(w+2)(w-5) + (2w+1)(w-5)$
    $= (w-5) \cdot (w+2) + (w-5) \cdot (2w+1)$
    $= (w-5)(w+2+2w+1)$
    $= (w-5)(3w+3)$
    Both terms in the second factor have a common factor of 3. Thus,
    $(w+2)(w-5) + (2w+1)(w-5) = 3(w-5)(w+1)$

## Section 5.5 – Quick Checks: Factoring Trinomials

11. $5x + 5y + bx + by = (5x + 5y) + (bx + by)$
    $= 5(x + y) + b(x + y)$
    $= (x + y)(5 + b)$
    $= (x + y)(b + 5)$

12. $w^3 - 3w^2 + 4w - 12 = (w^3 - 3w^2) + (4w - 12)$
    $= w^2(w - 3) + 4(w - 3)$
    $= (w - 3)(w^2 + 4)$

13. $6z^2 + 2z + 9z + 3 = (6z^2 + 2z) + (9z + 3)$
    $= 2z(3z + 1) + 3(3z + 1)$
    $= (3z + 1)(2z + 3)$

14. $2x^2 + x - 10x - 5 = (2x^2 + x) + (-10x - 5)$
    $= x(2x + 1) + (-5)(2x + 1)$
    $= (2x + 1)(x - 5)$

## 5.5 Quick Checks

1. $y^2 + 9y + 18$
   We are looking for two factors of $c = 18$ whose sum is $b = 9$. Since both $c$ and $b$ are positive, the two factors must both be positive.

   | Factors | 1,18 | 2,9 | 3,6 |
   |---|---|---|---|
   | Sum | 19 | 11 | 9 |

   $y^2 + 9y + 18 = (y + 6)(y + 3)$

2. $p^2 + 14p + 24$
   We are looking for two factors of $c = 24$ whose sum is $b = 14$. Since both $c$ and $b$ are positive, the two factors must both be positive.

   | Factors | 1,24 | 2,12 | 3,8 | 4,6 |
   |---|---|---|---|---|
   | Sum | 25 | 14 | 11 | 10 |

   $p^2 + 14p + 24 = (p + 2)(p + 12)$

3. $q^2 - 6q + 8$
   We are looking for two factors of $c = 8$ whose sum is $b = -6$. Since $c$ is positive the factors have the same sign, and since $b$ is negative both factors are negative.

   | Factors | -1,-8 | -2,-4 |
   |---|---|---|
   | Sum | -9 | -6 |

   $q^2 - 6q + 8 = (q - 4)(q - 2)$

4. $x^2 - 8x + 12$
   We are looking for two factors of $c = 12$ whose sum is $b = -8$. Since $c$ is positive the factors have the same sign, and since $b$ is negative the two factors are negative.

   | Factors | -1,-12 | -2,-6 | -3,-4 |
   |---|---|---|---|
   | Sum | -13 | -8 | -7 |

   $x^2 - 8x + 12 = (x - 2)(x - 6)$

5. $w^2 - 4w - 21$
   We are looking for two factors of $c = -21$ whose sum is $b = -4$. Since $c$ is negative, the two factors will have opposite signs. Since $b$ is negative, the factor with the larger absolute value will be negative.

   | Factors | -21,1 | -7,3 |
   |---|---|---|
   | Sum | -20 | -4 |

   $w^2 - 4w - 21 = (w - 7)(w + 3)$

6. $q^2 - 9q - 36$
   We are looking for two factors of $c = -36$ whose sum is $b = -9$. Since $c$ is negative, the two factors will have opposite signs. Since $b$ is negative, the factor with the larger absolute value will be negative.

   | Factors | -36,1 | -18,2 | -12,3 | -9,4 | -6,6 |
   |---|---|---|---|---|---|
   | Sum | -35 | -16 | -9 | -5 | 0 |

   $q^2 - 9q - 36 = (q - 12)(q + 3)$

7. $t^2 - 5t + 8$
   We are looking for two factors of $c = 8$ whose sum is $b = -5$. Since $c$ is positive the factors have the same sign, and since $b$ is negative the two factors are negative.

   | Factors | -1,-8 | -2,-4 |
   |---|---|---|
   | Sum | -9 | -6 |

   None of the possible factors yield the desire sum, so $t^2 - 5t + 8$ is prime.

8. $2x^3 - 12x^2 - 54x$
   First we factor out the GCF.
   $2x^3 - 12x^2 - 54x = 2x(x^2 - 6x - 27)$
   We are looking for two factors of $c = -27$ whose sum is $b = -6$. Since $c$ is negative, the factors will have opposite signs. Since $b$ is negative, the factor with the larger absolute value will be negative.

   | Factors | −27,1 | −9,3 |
   |---|---|---|
   | Sum | −26 | −6 |

   $2x^3 - 12x^2 - 54x = 2x(x^2 - 6x - 27)$
   $= 2x(x-9)(x+3)$

9. $-3z^2 - 21z - 30$
   Begin by factoring out the GCF.
   $-3z^2 - 21z - 30 = -3(z^2 + 7z + 10)$
   We are looking for two factors of $c = 10$ whose sum is $b = 7$. Since both $c$ and $b$ are positive, the two factors will be positive.

   | Factors | 1,10 | 2,5 |
   |---|---|---|
   | Sum | 11 | 7 |

   $-3z^2 - 21z - 30 = -3(z^2 + 7z + 10)$
   $= -3(z+5)(z+2)$

10. $x^2 + 8xy + 15y^2$
    We are looking for two factors of $c = 15$ whose sum is $b = 8$. Since $c$ and $b$ are both positive, the two factors will be positive.

    | Factors | 1,15 | 3,5 |
    |---|---|---|
    | Sum | 16 | 8 |

    $x^2 + 8xy + 15y^2 = (x+3y)(x+5y)$

11. $m^2 + mn - 20n^2$
    We are looking for two factors of $c = -20$ whose sum is $b = 1$. Since $c$ is negative, the factors will have opposite signs. Since $b$ is positive, the factor with the larger absolute value will be positive.

    | Factors | −1,20 | −2,10 | −4,5 |
    |---|---|---|---|
    | Sum | 19 | 8 | 1 |

    $m^2 + mn - 20n^2 = (m-4n)(m+5n)$

12. $6x^2 + 11x + 3$
    There are no common factors, so $a = 6, b = 11, c = 3$. We find that $a \cdot c = 6 \cdot 3 = 18$. Therefore, we are looking for two factors of 18 whose sum is $b = 11$. Since the product and the sum are both positive, both factors must be positive and neither can be greater than 11.

    | factor 1 | factor 2 | sum | |
    |---|---|---|---|
    | 3 | 6 | 9 | too small |
    | 2 | 9 | 11 | ← okay |

    $6x^2 + 11x + 3 = 6x^2 + 2x + 9x + 3$
    $= (6x^2 + 2x) + (9x + 3)$
    $= 2x(3x+1) + 3(3x+1)$
    $= (3x+1)(2x+3)$

13. $2b^2 + 7b - 15$
    There are no common factors, so $a = 2$, $b = 7$, and $c = -15$. Thus, $a \cdot c = 2 \cdot (-15) = -30$. We are looking for two factors of $-30$ whose sum is $b = 7$. Since the product is negative, the two factors will have opposite signs. Since the sum is positive, the factor with the larger absolute value will be positive.

    | factor 1 | factor 2 | sum | |
    |---|---|---|---|
    | −2 | 15 | 13 | too big |
    | −5 | 6 | 1 | too small |
    | −3 | 10 | 7 | ← okay |

    $2b^2 + 7b - 15 = 2b^2 - 3b + 10b - 15$
    $= (2b^2 - 3b) + (10b - 15)$
    $= b(2b-3) + 5(2b-3)$
    $= (2b-3)(b+5)$

14. $8x^2 + 14x + 5$
    First note that there are no common factors and that $a = 8$, $b = 14$, and $c = 5$.
    Since $c$ is positive the signs in our factors will be the same. Since $b$ is positive, the two signs will be positive. We will consider factorizations with this form:
    $(\_x + \_)(\_x + \_)$.
    Since $a = 8$ can be factored as $1 \cdot 8$ and $2 \cdot 4$, we have the following forms:
    $(x + \_)(8x + \_)$
    $(2x + \_)(4x + \_)$
    $|c| = |5| = 5$ can be factored as $1 \cdot 5$. This gives us

### Section 5.5 – Quick Checks: Factoring Trinomials

the following possibilities:
$(x+1)(8x+5) \to 8x^2 +13x+5$
$(2x+1)(4x+5) \to 8x^2 +14x+5$
$(x+5)(8x+1) \to 8x^2 +41x+5$
$(2x+5)(4x+1) \to 8x^2 +22x+5$
The correct factorization is
$8x^2 +14x+5 = (2x+1)(4x+5)$.

15. $12y^2 +32y-35$

    First note that there are no common factors and that $a=12$, $b=32$, and $c=-35$.
    Since $c$ is negative, the signs in our factors will be opposites. We will consider factorizations with this form:
    $(\_y+\_)(\_y-\_)$.
    If our choice results in a middle term with the wrong sign, we simply switch the signs of the factors.
    Since $a=12$ can be factored as $1\cdot 12$, $2\cdot 6$, and $3\cdot 4$, we have the following forms:
    $(y+\_)(12y-\_)$
    $(2y+\_)(6y-\_)$
    $(3y+\_)(4y-\_)$
    $|c|=|-35|=35$ can be factored as $1\cdot 35$ and $5\cdot 7$. This gives us the following possibilities:
    $(y+1)(12y-35) \to 12y^2 -23y-35$
    $(2y+1)(6y-35) \to 12y^2 -64y-35$
    $(3y+1)(4y-35) \to 12y^2 -101y-35$
    $(y+5)(12y-7) \to 12y^2 +53y-35$
    $(2y+5)(6y-7) \to 12y^2 +16y-35$
    $(3y+5)(4y-7) \to 12y^2 -y-35$
    $(y+35)(12y-1) \to 12y^2 +419y-35$
    $(2y+35)(6y-1) \to 12y^2 +208y-35$
    $(3y+35)(4y-1) \to 12y^2 +137y-35$
    $(y+7)(12y-5) \to 12y^2 +79y-35$
    $(2y+7)(6y-5) \to 12y^2 +32y-35$
    $(3y+7)(4y-5) \to 12y^2 +13y-35$
    The correct factorization is
    $12y^2 +32y-35 = (2y+7)(6y-5)$.

16. $30x^2 +7xy-2y^2$

    First note that there are no common factors and that $a=30$, $b=7$, and $c=-2$.
    Since $c$ is negative, the signs in our factors will be opposites. We will consider factorizations of the form:
    $(\_x+\_y)(\_x-\_y)$.
    If our choice results in a middle term with the wrong sign, we simply switch the signs of the factors.
    Since $a=30$ can be factored as $1\cdot 30$, $2\cdot 15$, $3\cdot 10$, and $5\cdot 6$, we have the following forms:
    $(x+\_y)(30x-\_y)$
    $(2x+\_y)(15x-\_y)$
    $(3x+\_y)(10x-\_y)$
    $(5x+\_y)(6x-\_y)$
    $|c|=|-2|=2$ can be factored as $1\cdot 2$. Since the original expression had no common factors, the binomials we select cannot have a common factor. This gives us the following possibilities:
    $(x+2y)(30x-y) \to 30x^2 +59xy-2y^2$
    $(2x+y)(15x-2y) \to 30x^2 +11xy-2y^2$
    $(3x+2y)(10x-y) \to 30x^2 +17xy-2y^2$
    $(5x+2y)(6x-y) \to 30x^2 +7xy-2y^2$
    The correct factorization is
    $30x^2 +7xy-2y^2 = (5x+2y)(6x-y)$.

17. $-6y^2 +23y+4$

    Begin by factoring out $-1$.
    $-6y^2 +23y+4 = -(6y^2 -23y-4)$
    Focusing on the reduced trinomial, we have $a=6$, $b=-23$, and $c=-4$. Thus, $a\cdot c = 6\cdot(-4) = -24$. We are looking for two factors of $-24$ whose sum is $b=-23$. Since the product is negative, the two factors have opposite signs. Since the sum is negative, the factor with the larger absolute value will be negative.

*Chapter 5:* Polynomials and Polynomial Functions

| factor 1 | factor 2 | sum |           |
|----------|----------|-----|-----------|
| −8       | 3        | −5  | too big   |
| −12      | 2        | −10 | too big   |
| −24      | 1        | −23 | ← okay    |

$$-6y^2 + 23y + 4 = -(6y^2 - 23y - 4)$$
$$= -(6y^2 - 24y + y - 4)$$
$$= -\left[(6y^2 - 24y) + (y - 4)\right]$$
$$= -\left[6y(y - 4) + 1(y - 4)\right]$$
$$= -(y - 4)(6y + 1)$$

18. $-9x^2 - 21xy - 10y^2$

Begin by factoring out $-1$.
$$-9x^2 - 21xy - 10y^2 = -(9x^2 + 21xy + 10y^2)$$

Focusing on the reduced trinomial, we have $a = 9$, $b = 21$, and $c = 10$. Thus, $a \cdot c = 9 \cdot 10 = 90$. We are looking for two factors of 90 whose sum is $b = 21$. Since the product is positive, the two factors have the same sign. Since the sum is also positive, the factors will both be positive.

| factor 1 | factor 2 | sum |            |
|----------|----------|-----|------------|
| 3        | 30       | 33  | too big    |
| 9        | 10       | 19  | too small  |
| 6        | 15       | 21  | ← okay     |

$$-9x^2 - 21xy - 10y^2$$
$$= -(9x^2 + 21xy + 10y^2)$$
$$= -(9x^2 + 6xy + 15xy + 10y^2)$$
$$= -\left[(9x^2 + 6xy) + (15xy + 10y^2)\right]$$
$$= -\left[3x(3x + 2y) + 5y(3x + 2y)\right]$$
$$= -(3x + 2y)(3x + 5y)$$

19. $y^4 - 2y^2 - 24$

Let $u = y^2$. Then $u^2 = (y^2)^2 = y^4$ and we get
$$y^4 - 2y^2 - 24 = (y^2)^2 - 2(y^2) - 24$$
$$= u^2 - 2u - 24$$
$$= (u - 6)(u + 4)$$
$$= (y^2 - 6)(y^2 + 4)$$

20. $4(x - 3)^2 + 5(x - 3) - 6$

Let $u = x - 3$. Then $u^2 = (x - 3)^2$ and we get
$$4(x - 3)^2 + 5(x - 3) - 6$$
$$= 4u^2 + 5u - 6$$
$$= (4u - 3)(u + 2)$$
$$= (4(x - 3) - 3)((x - 3) + 2)$$
$$= (4x - 12 - 3)(x - 3 + 2)$$
$$= (4x - 15)(x - 1)$$

**5.6 Quick Checks**

1. $x^2 - 18x + 81 = x^2 - 2 \cdot 9 \cdot x + 9^2$
$$= (x - 9)^2$$

2. $4x^2 + 20xy + 25y^2 = (2x)^2 + 2 \cdot 2x \cdot 5y + (5y)^2$
$$= (2x + 5y)^2$$

3. $18p^4 - 84p^2 + 98 = 2(9p^4 - 42p^2 + 49)$
$$= 2\left((3p^2)^2 - 2 \cdot (3p^2) \cdot 7 + 7^2\right)$$
$$= 2(3p^2 - 7)^2$$

4. $z^2 - 16 = z^2 - 4^2$
$$= (z - 4)(z + 4)$$

5. $16m^2 - 81n^2 = (4m)^2 - (9n)$
$$= (4m - 9n)(4m + 9n)$$

6. $4a^2 - 9b^4 = (2a)^2 - (3b^2)^2$
$$= (2a - 3b^2)(2a + 3b^2)$$

7. $3b^4 - 48 = 3(b^4 - 16)$
$$= 3\left((b^2)^2 - 4^2\right)$$
$$= 3(b^2 - 4)(b^2 + 4)$$
$$= 3(b^2 - 2^2)(b^2 + 4)$$
$$= 3(b - 2)(b + 2)(b^2 + 4)$$

**Section 5.7 – Quick Checks:** *Factoring: A General Strategy*

8. $p^2 - 8p + 16 - q^2 = (p^2 - 8p + 16) - q^2$
$= (p^2 - 2 \cdot 4 \cdot p + 4^2) - q^2$
$= (p-4)^2 - q^2$
$= (p-4-q)(p-4+q)$

9. $z^3 + 64 = z^3 + 4^4$
$= (z+4)(z^2 - 4 \cdot z + 4^2)$
$= (z+4)(z^2 - 4z + 16)$

10. $125p^3 - 216q^6$
$= (5p)^3 - (6q^2)^3$
$= (5p - 6q^2)((5p)^2 + 5p \cdot 6q^2 + (6q^2)^2)$
$= (5p - 6q^2)(25p^2 + 30pq^2 + 36q^4)$

11. $32m^3 + 500n^6$
$= 4(8m^3 + 125n^6)$
$= 4((2m)^3 + (5n^2)^3)$
$= 4(2m + 5n^2)((2m)^2 - 2m \cdot 5n^2 + (5n^2)^2)$
$= 4(2m + 5n^2)(4m^2 - 10mn^2 + 25n^4)$

12. $(x+1)^3 - 27x^3$
$= (x+1)^3 - (3x)^3$
$= (x+1-3x)((x+1)^2 + (x+1) \cdot 3x + (3x)^2)$
$= (-2x+1)(x^2 + 2x + 1 + 3x^2 + 3x + 9x^2)$
$= (-2x+1)(13x^2 + 5x + 1)$

**5.7 Quick Checks**

1. $2p^2q - 8pq^2 - 90q^3 = 2q(p^2 - 4pq - 45q^2)$
$= 2q(p - 9q)(p + 5q)$

2. $-45x^2y + 66xy + 27y = -3y(15x^2 - 22x - 9)$
$= -3y(3x+1)(5x-9)$

3. $81x^2 - 100y^2 = (9x)^2 - (10y)^2$
$= (9x - 10y)(9x + 10y)$

4. $-3m^2n + 147n = -3n(m^2 - 49)$
$= -3n(m^2 - 7^2)$
$= -3n(m-7)(m+7)$

5. $p^2 - 16pq + 64q^2 = p^2 - 2 \cdot p \cdot 8q + (8q)^2$
$= (p - 8q)^2$

6. $20x^2 + 60x + 45 = 5(4x^2 + 12x + 9)$
$= 5((2x)^2 + 2 \cdot 2x \cdot 3 + 3^2)$
$= 5(2x + 3)^2$

7. $64y^3 - 125 = (4y)^3 - 5^3$
$= (4y - 5)((4y)^2 + 4y \cdot 5 + 5^2)$
$= (4y - 5)(16y^2 + 20y + 25)$

8. $-16m^3 - 2n^3 = -2(8m^3 + n^3)$
$= -2((2m)^3 + n^3)$
$= -2(2m + n)((2m)^2 - 2m \cdot n + n^2)$
$= -2(2m + n)(4m^2 - 2mn + n^2)$

9. $10z^2 - 15z + 35 = 5(2z^2 - 3z + 7)$

Using $a = 2$, $b = -3$, and $c = 7$, we get $a \cdot c = 2 \cdot 7 = 14$. We are looking for two factors of 14 whose sum is $b = -3$. There are no such factors, so the above factorization is complete.

10. $6xy^2 + 81x^3 = 3x(2y^2 + 27x^2)$

The expression in parentheses cannot be factored any further.

11. $2x^3 + 5x^2 + 4x + 10 = (2x^3 + 5x^2) + (4x + 10)$
$= x^2(2x + 5) + 2(2x + 5)$
$= (2x + 5)(x^2 + 2)$
$= (x^2 + 2)(2x + 5)$

**Chapter 5:** *Polynomials and Polynomial Functions*

12. $9x^3 + 3x^2 - 9x - 3 = 3(3x^3 + x^2 - 3x - 1)$
$= 3((3x^3 + x^2) + (-3x - 1))$
$= 3(x^2(3x+1) - 1(3x+1))$
$= 3(3x+1)(x^2 - 1)$
$= 3(3x+1)(x-1)(x+1)$

13. $4x^2 + 4xy + y^2 - 81 = (4x^2 + 4xy + y^2) - 81$
$= ((2x)^2 + 2 \cdot 2x \cdot y + y^2) - 81$
$= (2x+y)^2 - 9^2$
$= (2x+y-9)(2x+y+9)$

14. $16 - m^2 - 8mn - 16n^2$
$= 16 - (m^2 + 8mn + 16n^2)$
$= 16 - (m^2 + 2 \cdot m \cdot 4n + (4n)^2)$
$= 4^2 - (m+4n)^2$
$= (4 - (m+4n))(4 + (m+4n))$
$= (4 - m - 4n)(4 + m + 4n)$

**5.8 Quick Checks**

1. $x(x+7) = 0$
$x = 0$ or $x + 7 = 0$
$\phantom{x = 0 \text{ or }} x = -7$
Check: $(-7)(-7+7) \stackrel{?}{=} 0$
$(-7)(0) \stackrel{?}{=} 0$
$0 = 0$ ✓
The solution set is $\{-7, 0\}$.

2. $(x-3)(4x+3) = 0$
$x - 3 = 0$ or $4x + 3 = 0$
$x = 3 \phantom{aaa} 4x = -3$
$\phantom{x = 3 \text{ or } 4x} x = -\dfrac{3}{4}$

Check:
$(3-3)(4(3)+3) \stackrel{?}{=} 0$
$0(15) \stackrel{?}{=} 0$
$0 = 0$ ✓
$\left(-\dfrac{3}{4} - 3\right)\left(4\left(-\dfrac{3}{4}\right) + 3\right) \stackrel{?}{=} 0$
$\left(-\dfrac{15}{4}\right)(0) \stackrel{?}{=} 0$
$0 = 0$ ✓
The solution set is $\left\{-\dfrac{3}{4}, 3\right\}$.

3. $p^2 - 5p + 6 = 0$
$(p-3)(p-2) = 0$
$p - 3 = 0$ or $p - 2 = 0$
$p = 3 \phantom{aaaa} p = 2$
Check:
$(2)^2 - 5(2) + 6 \stackrel{?}{=} 0$
$4 - 10 + 6 \stackrel{?}{=} 0$
$0 = 0$ ✓
$(3)^2 - 5(3) + 6 \stackrel{?}{=} 0$
$9 - 15 + 6 \stackrel{?}{=} 0$
$0 = 0$ ✓
The solution set is $\{2, 3\}$.

4. $3t^2 - 14t = 5$
$3t^2 - 14t - 5 = 0$
$(3t+1)(t-5) = 0$
$3t + 1 = 0$ or $t - 5 = 0$
$3t = -1 \phantom{aaa} t = 5$
$t = -\dfrac{1}{3}$
Check:
$3\left(-\dfrac{1}{3}\right)^2 - 14\left(-\dfrac{1}{3}\right) \stackrel{?}{=} 5$
$\dfrac{1}{3} + \dfrac{14}{3} \stackrel{?}{=} 5$
$5 = 5$ ✓
$3(5)^2 - 14(5) \stackrel{?}{=} 5$
$75 - 70 \stackrel{?}{=} 5$
$5 = 5$ ✓
The solution set is $\left\{-\dfrac{1}{3}, 5\right\}$.

**Section 5.8 – Quick Checks:** *Polynomial Equations*

5.  $4y^2 + 8y + 3 = y^2 - 1$
    $3y^2 + 8y + 4 = 0$
    $(3y+2)(y+2) = 0$
    $3y+2 = 0$ or $y+2 = 0$
    $3y = -2$ $\quad\quad y = -2$
    $y = -\dfrac{2}{3}$

    Check:
    $4(-2)^2 + 8(-2) + 3 \stackrel{?}{=} (-2)^2 - 1$
    $16 - 16 + 3 \stackrel{?}{=} 4 - 1$
    $3 = 3$ ✓
    $4\left(-\dfrac{2}{3}\right)^2 + 8\left(-\dfrac{2}{3}\right) + 3 \stackrel{?}{=} \left(-\dfrac{2}{3}\right)^2 - 1$
    $\dfrac{16}{9} - \dfrac{16}{3} + 3 \stackrel{?}{=} \dfrac{4}{9} - 1$
    $-\dfrac{5}{9} = -\dfrac{5}{9}$ ✓

    The solution set is $\left\{-2, -\dfrac{2}{3}\right\}$.

6.  $x(x+3) = -2$
    $x^2 + 3x = -2$
    $x^2 + 3x + 2 = 0$
    $(x+1)(x+2) = 0$
    $x+1 = 0$ or $x+2 = 0$
    $x = -1$ $\quad\quad x = -2$

    Check:
    $(-2)(-2+3) \stackrel{?}{=} -2$
    $-2(1) \stackrel{?}{=} -2$
    $-2 = -2$ ✓
    $(-1)(-1+3) \stackrel{?}{=} -2$
    $(-1)(2) \stackrel{?}{=} -2$
    $-2 = -2$ ✓

    The solution set is $\{-2, -1\}$.

7.  $(x-3)(x+5) = 9$
    $x^2 - 3x + 5x - 15 = 9$
    $x^2 + 2x - 24 = 0$
    $(x+6)(x-4) = 0$
    $x+6 = 0$ or $x-4 = 0$
    $x = -6$ $\quad\quad x = 4$

    Check:
    $(-6-3)(-6+5) \stackrel{?}{=} 9$
    $(-9)(-1) \stackrel{?}{=} 9$
    $9 = 9$ ✓
    $(4-3)(4+5) \stackrel{?}{=} 9$
    $(1)(9) \stackrel{?}{=} 9$
    $9 = 9$ ✓

    The solution set is $\{-6, 4\}$.

8.  $y^3 - y^2 - 9y + 9 = 0$
    $(y^3 - y^2) + (-9y + 9) = 0$
    $y^2(y-1) - 9(y-1) = 0$
    $(y-1)(y^2 - 9) = 0$
    $(y-1)(y-3)(y+3) = 0$
    $y-1 = 0$ or $y-3 = 0$ or $y+3 = 0$
    $y = 1$ $\quad\quad y = 3$ $\quad\quad y = -3$

    Check:
    $(-3)^3 - (-3)^2 - 9(-3) + 9 \stackrel{?}{=} 0$
    $-27 - 9 + 27 + 9 \stackrel{?}{=} 0$
    $0 = 0$ ✓
    $(1)^3 - (1)^2 - 9(1) + 9 \stackrel{?}{=} 0$
    $1 - 1 - 9 + 9 \stackrel{?}{=} 0$
    $0 = 0$ ✓
    $(3)^3 - (3)^2 - 9(3) + 9 \stackrel{?}{=} 0$
    $27 - 9 - 27 + 9 \stackrel{?}{=} 0$
    $0 = 0$ ✓

    The solution set is $\{-3, 1, 3\}$.

9.  $g(x) = x^2 - 8x + 3$

    a.  $g(x) = 12$
        $x^2 - 8x + 3 = 12$
        $x^2 - 8x - 9 = 0$
        $(x-9)(x+1) = 0$
        $x-9 = 0$ or $x+1 = 0$
        $x = 9$ $\quad\quad x = -1$

        The solution set is $\{-1, 9\}$. The points $(-1, 12)$ and $(9, 12)$ are on the graph of $g$.

*Chapter 5:* Polynomials and Polynomial Functions

**b.** 
$$g(x) = -4$$
$$x^2 - 8x + 3 = -4$$
$$x^2 - 8x + 7 = 0$$
$$(x-7)(x-1) = 0$$
$$x - 7 = 0 \quad \text{or} \quad x - 1 = 0$$
$$x = 7 \qquad\qquad x = 1$$
The solution set is $\{1, 7\}$. The points $(1, -4)$ and $(7, -4)$ are on the graph of $g$.

**10.**
$$h(x) = 2x^2 + 3x - 20$$
$$h(x) = 0$$
$$2x^2 + 3x - 20 = 0$$
$$(2x - 5)(x + 4) = 0$$
$$2x - 5 = 0 \quad \text{or} \quad x + 4 = 0$$
$$2x = 5 \qquad\qquad x = -4$$
$$x = \tfrac{5}{2}$$
The zeros are $-4$ and $\tfrac{5}{2}$; the x-intercepts are also $-4$ and $\tfrac{5}{2}$.

**11.** Let $w$ = the width of the plot. Then the length is given by $w + 6$.
$$A = \text{length} \cdot \text{width}$$
$$= (w + 6) \cdot w$$
$$135 = w(w + 6)$$
$$135 = w^2 + 6w$$
$$0 = w^2 + 6w - 135$$
$$0 = (w + 15)(w - 9)$$
$$w + 15 = 0 \quad \text{or} \quad w - 9 = 0$$
$$w = -15 \qquad\qquad w = 9$$
Since the width must be positive, we discard the negative solution. The width is 9 miles and the length is $9 + 6 = 15$ miles.

**12.** Let $x$ = the number of boxes ordered. The price per box is then given by $100 - (x - 30) = 130 - x$.
$$\text{Cost} = \text{price} \cdot \text{quantity}$$
$$4200 = (130 - x)x$$
$$4200 = 130x - x^2$$
$$x^2 - 130x + 4200 = 0$$
$$(x - 60)(x - 70) = 0$$
$$x - 60 = 0 \quad \text{or} \quad x - 70 = 0$$
$$x = 60 \qquad\qquad x = 70$$
Since the number of boxes ordered must be between 30 and 65, we discard the second solution. The customer ordered 60 boxes of CDs.

**13.** $s(t) = -16t^2 + 160t$

**a.**
$$s(t) = 384$$
$$-16t^2 + 160t = 384$$
$$16t^2 - 160t + 384 = 0$$
$$t^2 - 10t + 24 = 0$$
$$(t - 4)(t - 6) = 0$$
$$t - 4 = 0 \quad \text{or} \quad t - 6 = 0$$
$$t = 4 \qquad\qquad t = 6$$
The rocket will be 384 feet above the ground after 4 seconds and after 6 seconds.

**b.**
$$s(t) = 0$$
$$-16t^2 + 160t = 0$$
$$t^2 - 10t = 0$$
$$t(t - 10) = 0$$
$$t = 0 \quad \text{or} \quad t - 10 = 0$$
$$t = 10$$
The rocket will strike the ground after 10 seconds.

# Getting Ready for Chapter 6 – Quick Checks

1. $\dfrac{13 \cdot 5}{13 \cdot 6} = \dfrac{\cancel{13}^1 \cdot 5}{\cancel{13}_1 \cdot 6} = \dfrac{5}{6}$

2. $\dfrac{80}{12} = \dfrac{20 \cdot 4}{3 \cdot 4} = \dfrac{20 \cdot \cancel{4}^1}{3 \cdot \cancel{4}_1} = \dfrac{20}{3}$

3. $\dfrac{5}{7} \cdot \left(-\dfrac{21}{10}\right) = -\dfrac{5}{7} \cdot \dfrac{3 \cdot 7}{2 \cdot 5} = -\dfrac{\cancel{5}^1}{\cancel{7}_1} \cdot \dfrac{3 \cdot \cancel{7}^1}{2 \cdot \cancel{5}_1} = -\dfrac{3}{2}$

4. $\dfrac{35}{15} \cdot \dfrac{3}{14} = \dfrac{5 \cdot 7}{3 \cdot 5} \cdot \dfrac{3}{2 \cdot 7} = \dfrac{\cancel{5}^1 \cdot \cancel{7}^1}{\cancel{3}_1 \cdot \cancel{5}_1} \cdot \dfrac{\cancel{3}^1}{2 \cdot \cancel{7}_1} = \dfrac{1}{2}$

5. $\dfrac{\frac{4}{5}}{\frac{12}{25}} = \dfrac{4}{5} \cdot \dfrac{25}{12} = \dfrac{\cancel{4}^1}{\cancel{5}_1} \cdot \dfrac{\cancel{5}^1 \cdot 5}{\cancel{4}_1 \cdot 3} = \dfrac{5}{3}$

6. $\dfrac{24}{35} \div \left(-\dfrac{8}{7}\right) = -\dfrac{24}{35} \cdot \dfrac{7}{8}$
$= -\dfrac{8 \cdot 3}{5 \cdot 7} \cdot \dfrac{7}{8}$
$= -\dfrac{\cancel{8}^1 \cdot 3}{5 \cdot \cancel{7}_1} \cdot \dfrac{\cancel{7}^1}{\cancel{8}_1}$
$= -\dfrac{3}{5}$

7. $\dfrac{11}{12} + \dfrac{5}{12} = \dfrac{11+5}{12} = \dfrac{16}{12} = \dfrac{4 \cdot 4}{4 \cdot 3} = \dfrac{4}{3}$

8. $\dfrac{3}{18} - \dfrac{13}{18} = \dfrac{3-13}{18} = \dfrac{-10}{18} = -\dfrac{2 \cdot 5}{2 \cdot 9} = -\dfrac{5}{9}$

9. $25 = 5 \cdot 5$
$15 = 3 \cdot 5$
LCD $= 3 \cdot 5 \cdot 5 = 75$
$\dfrac{3}{25} = \dfrac{3}{25} \cdot \dfrac{3}{3} = \dfrac{9}{75}$
$\dfrac{2}{15} = \dfrac{2}{15} \cdot \dfrac{5}{5} = \dfrac{10}{75}$

10. $18 = 2 \cdot 3 \cdot 3$
$63 = 3 \cdot 3 \cdot 7$
LCD $= 2 \cdot 3 \cdot 3 \cdot 7 = 126$
$\dfrac{5}{18} = \dfrac{5}{18} \cdot \dfrac{7}{7} = \dfrac{35}{126}$
$-\dfrac{1}{63} = -\dfrac{1}{63} \cdot \dfrac{2}{2} = -\dfrac{2}{126}$

11. $4 = 2 \cdot 2$
5 is prime
LCD $= 2 \cdot 2 \cdot 5 = 20$
$\dfrac{3}{4} + \dfrac{1}{5} = \dfrac{3}{4} \cdot \dfrac{5}{5} + \dfrac{1}{5} \cdot \dfrac{4}{4}$
$= \dfrac{15}{20} + \dfrac{4}{20} = \dfrac{15+4}{20}$
$= \dfrac{19}{20}$

12. $20 = 2 \cdot 2 \cdot 5$
$15 = 3 \cdot 5$
LCD $= 2 \cdot 2 \cdot 3 \cdot 5 = 60$
$\dfrac{3}{20} + \dfrac{2}{15} = \dfrac{3}{20} \cdot \dfrac{3}{3} + \dfrac{2}{15} \cdot \dfrac{4}{4}$
$= \dfrac{9}{60} + \dfrac{8}{60} = \dfrac{9+8}{60}$
$= \dfrac{17}{60}$

13. $14 = 2 \cdot 7$
$21 = 3 \cdot 7$
LCD $= 2 \cdot 3 \cdot 7 = 42$
$\dfrac{5}{14} - \dfrac{11}{21} = \dfrac{5}{14} \cdot \dfrac{3}{3} - \dfrac{11}{21} \cdot \dfrac{2}{2}$
$= \dfrac{15}{42} - \dfrac{22}{42} = \dfrac{15-22}{42}$
$= -\dfrac{7}{42} = -\dfrac{1 \cdot 7}{6 \cdot 7}$
$= -\dfrac{1}{6}$

# Chapter 6 – Quick Checks

## 6.1 Quick Checks

1. We need to find all values of $x$ that cause the denominator $x+6$ to equal 0.
   $x+6=0$
   $x=-6$
   Thus, the domain of $\dfrac{x-4}{x+6}$ is $\{x|x \neq -6\}$.

2. We need to find all values of $x$ that cause the denominator $z^2+3z-28$ to equal 0.
   $z^2+3z-28=0$
   $(z+7)(x-4)=0$
   $z+7=0$ or $z-4=0$
   $z=-7$ or $z=4$
   Thus, the domain of $\dfrac{z^2-9}{z^2+3z-28}$ is $\{z|z \neq -4, z \neq 7\}$.

3. $\dfrac{x^2-7x+12}{x^2+4x-21} = \dfrac{(x-3)(x-4)}{(x-3)(x+7)}$
   $= \dfrac{\cancel{(x-3)}(x-4)}{\cancel{(x-3)}(x+7)} = \dfrac{x-4}{x+7}$

4. $\dfrac{z^3-64}{2z^2-3z-20} = \dfrac{(z-4)(z^2+4z+16)}{(2z+5)(z-4)}$
   $= \dfrac{\cancel{(z-4)}(z^2+4z+16)}{(2z+5)\cancel{(z-4)}}$
   $= \dfrac{z^2+4z+16}{2z+5}$

5. $\dfrac{3w^2+13w-10}{2-3w} = \dfrac{(3w-2)(w+5)}{-1(3w-2)}$
   $= \dfrac{\cancel{(3w-2)}(w+5)}{-1\cancel{(3w-2)}}$
   $= \dfrac{w+5}{-1} = -(w+5)$

6. $\dfrac{p^2-9}{p^2+5p+6} \cdot \dfrac{3p^2-p-2}{2p-6}$
   $= \dfrac{(p-3)(p+3)}{(p+2)(p+3)} \cdot \dfrac{(3p+2)(p-1)}{2(p-3)}$
   $= \dfrac{\cancel{(p-3)}\cancel{(p+3)}}{(p+2)\cancel{(p+3)}} \cdot \dfrac{(3p+2)(p-1)}{2\cancel{(p-3)}}$
   $= \dfrac{(3p+2)(p-1)}{2(p+2)}$

7. $\dfrac{2x+8}{2x^2+11x+12} \cdot \dfrac{2x^2-3x-9}{6-2x}$
   $= \dfrac{2(x+4)}{(2x+3)(x+4)} \cdot \dfrac{(2x+3)(x-3)}{-1 \cdot 2(x-3)}$
   $= \dfrac{\cancel{2}\cancel{(x+4)}}{\cancel{(2x+3)}\cancel{(x+4)}} \cdot \dfrac{\cancel{(2x+3)}\cancel{(x-3)}}{-1 \cdot \cancel{2}\cancel{(x-3)}}$
   $= \dfrac{1}{-1} = -1$

8. $\dfrac{m^2+2mn+n^2}{2m^2+3mn+n^2} \cdot \dfrac{2m^2-5mn-3n^2}{3n-m}$
   $= \dfrac{(m+n)(m+n)}{(2m+n)(m+n)} \cdot \dfrac{(2m+n)(m-3n)}{-1(m-3n)}$
   $= \dfrac{\cancel{(m+n)}(m+n)}{\cancel{(2m+n)}\cancel{(m+n)}} \cdot \dfrac{\cancel{(2m+n)}\cancel{(m-3n)}}{-1\cancel{(m-3n)}}$
   $= \dfrac{m+n}{-1} = -(m+n)$

9. $\dfrac{\dfrac{12a^4}{5b^2}}{\dfrac{4a^2}{15b^5}} = \dfrac{12a^4}{5b^2} \cdot \dfrac{15b^5}{4a^2}$
   $= \dfrac{\cancel{12}^3 \cdot \cancel{a^4} \cdot a^2}{\cancel{5} \cdot \cancel{b^2}} \cdot \dfrac{\cancel{15}^3 \cdot \cancel{b^5} \cdot b^3}{4\cancel{a^2}} = 9a^2b^3$

**Section 6.2 – Quick Checks:** Adding and Subtracting Rational Expressions

10. $\dfrac{\dfrac{m^2-5m}{m-7}}{\dfrac{2m}{m^2-6m-7}} = \dfrac{m^2-5m}{m-7} \cdot \dfrac{m^2-6m-7}{2m}$

    $= \dfrac{m(m-5)}{m-7} \cdot \dfrac{(m-7)(m+1)}{2\cdot m}$

    $= \dfrac{\cancel{m}(m-5)}{\cancel{m-7}} \cdot \dfrac{\cancel{(m-7)}(m+1)}{2\cdot \cancel{m}}$

    $= \dfrac{(m-5)(m+1)}{2}$

11. We need to find all values of $x$ that cause the denominator $x^2+x-30$ to equal 0.

    $x^2+x-30=0$
    $(x+6)(x-5)=0$
    $x+6=0$ or $x-5=0$
    $x=-6$ or $x=5$

    Thus, the domain of $R(x)=\dfrac{2x}{x^2+x-30}$ is $\{x\,|\,x\neq -6, x\neq 5\}$.

12. **a.** $R(x)=f(x)\cdot g(x)$

    $= \dfrac{x^2-4x-5}{3x-5} \cdot \dfrac{3x^2+4x-15}{x^2-2x-15}$

    $= \dfrac{(x-5)(x+1)}{3x-5} \cdot \dfrac{(3x-5)(x+3)}{(x-5)(x+3)}$

    $= \dfrac{\cancel{(x-5)}(x+1)}{\cancel{3x-5}} \cdot \dfrac{\cancel{(3x-5)}\cancel{(x+3)}}{\cancel{(x-5)}\cancel{(x+3)}}$

    $= x+1$

    The domain of $f(x)$ is $\left\{x\,\middle|\,x\neq \dfrac{5}{3}\right\}$.

    The domain of $g(x)$ is $\{x\,|\,x\neq -3, x\neq 5\}$.

    Therefore, the domain of $R(x)$ is $\left\{x\,\middle|\,x\neq -3, x\neq \dfrac{5}{3}, x\neq 5\right\}$.

    **b.** $H(x)=\dfrac{f(x)}{h(x)}=\dfrac{\dfrac{x^2-4x-5}{3x-5}}{\dfrac{4x^2+7x+3}{9x^2-15x}}$

    $= \dfrac{x^2-4x-5}{3x-5} \cdot \dfrac{9x^2-15x}{4x^2+7x+3}$

    $= \dfrac{(x-5)(x+1)}{3x-5} \cdot \dfrac{3x(3x-5)}{(4x+3)(x+1)}$

    $= \dfrac{(x-5)\cancel{(x+1)}}{\cancel{3x-5}} \cdot \dfrac{3x\cancel{(3x-5)}}{(4x+3)\cancel{(x+1)}}$

    $= \dfrac{3x(x-5)}{4x+3}$

    The domain of $f(x)$ is $\left\{x\,\middle|\,x\neq \dfrac{5}{3}\right\}$.

    The domain of $h(x)$ is $\left\{x\,\middle|\,x\neq 0, x\neq \dfrac{5}{3}\right\}$.

    Because the denominator of $H(x)$ cannot equal 0, we must exclude those values of $x$ such that $4x^2+7x+3=0$. These values are $x=-\dfrac{3}{4}$ and $x=-1$. Therefore, the domain of $H(x)$ is $\left\{x\,\middle|\,x\neq -1, x\neq -\dfrac{3}{4}, x\neq 0, x\neq \dfrac{5}{3}\right\}$.

**6.2 Quick Checks**

1. $\dfrac{x^2-3x-1}{x-2}+\dfrac{x^2-2x+3}{x-2} = \dfrac{x^2-3x-1+(x^2-2x+3)}{x-2}$

   $= \dfrac{2x^2-5x+2}{x-2}$

   $= \dfrac{(2x-1)(x-2)}{x-2}$

   $= \dfrac{(2x-1)\cancel{(x-2)}}{\cancel{x-2}}$

   $= 2x-1$

2. $\dfrac{4x+3}{x+5}-\dfrac{x-6}{x+5} = \dfrac{4x+3-(x-6)}{x+5}$

   $= \dfrac{4x+3-x+6}{x+5}$

   $= \dfrac{3x+9}{x+5}$ or $\dfrac{3(x+3)}{x+5}$

3. $\dfrac{4x}{x-5}+\dfrac{3}{5-x} = \dfrac{4x}{x-5}+\dfrac{3}{-1(x-5)}$

   $= \dfrac{4x}{x-5}+\dfrac{-3}{x-5} = \dfrac{4x-3}{x-5}$

4. $8x^2y = 2^3\cdot x^2\cdot y$
   $12xy^3 = 2^2\cdot 3\cdot x\cdot y^3$
   $LCD = 2^3\cdot 3\cdot x^2\cdot y^3 = 24x^2y^3$

*Chapter 6:* Rational Expressions and Rational Functions

5. $x^2 - 5x - 14 = (x+2)(x-7)$
   $x^2 + 4x + 4 = (x+2)^2$
   LCD $= (x+2)^2 (x-7)$

6. $10a = 2 \cdot 5 \cdot a$
   $15a^2 = 3 \cdot 5 \cdot a^2$
   LCD $= 2 \cdot 3 \cdot 5 \cdot a^2 = 30a^2$
   $\dfrac{3}{10a} + \dfrac{4}{15a^2} = \dfrac{3}{10a} \cdot \dfrac{3a}{3a} + \dfrac{4}{15a^2} \cdot \dfrac{2}{2}$
   $= \dfrac{9a}{30a^2} + \dfrac{8}{30a^2} = \dfrac{9a+8}{30a^2}$

7. $x-1$
   $x+2$
   LCD $= (x-1)(x+2)$
   $\dfrac{3x}{x-1} + \dfrac{x+5}{x+2} = \dfrac{3x}{x-1} \cdot \dfrac{x+2}{x+2} + \dfrac{x+5}{x+2} \cdot \dfrac{x-1}{x-1}$
   $= \dfrac{3x^2 + 6x}{(x-1)(x+2)} + \dfrac{x^2 + 4x - 5}{(x-1)(x+2)}$
   $= \dfrac{3x^2 + 6x + (x^2 + 4x - 5)}{(x-1)(x+2)}$
   $= \dfrac{4x^2 + 10x - 5}{(x-1)(x+2)}$

8. $2x^2 + 7x + 6 = (2x+3)(x+2)$
   $x^2 + 6x + 8 = (x+2)(x+4)$
   LCD $= (2x+3)(x+2)(x+4)$
   $\dfrac{x-1}{2x^2 + 7x + 6} + \dfrac{x-1}{x^2 + 6x + 8}$
   $= \dfrac{x-1}{(2x+3)(x+2)} + \dfrac{x-1}{(x+2)(x+4)}$
   $= \dfrac{x-1}{(2x+3)(x+2)} \cdot \dfrac{x+4}{x+4} + \dfrac{x-1}{(x+2)(x+4)} \cdot \dfrac{2x+3}{2x+3}$
   $= \dfrac{x^2 + 3x - 4}{(2x+3)(x+2)(x+4)} + \dfrac{2x^2 + x - 3}{(2x+3)(x+2)(x+4)}$
   $= \dfrac{x^2 + 3x - 4 + (2x^2 + x - 3)}{(2x+3)(x+2)(x+4)}$
   $= \dfrac{3x^2 + 4x - 7}{(2x+3)(x+2)(x+4)}$
   $= \dfrac{(3x+7)(x-1)}{(2x+3)(x+2)(x+4)}$

9. $2x^2 + x - 6 = (2x-3)(x+2)$
   $x^2 + 4x + 4 = (x+2)^2$
   LCD $= (2x-3)(x+2)^2$
   $\dfrac{3x+4}{2x^2 + x - 6} - \dfrac{x-1}{x^2 + 4x + 4}$
   $= \dfrac{3x+4}{(2x-3)(x+2)} - \dfrac{x-1}{(x+2)^2}$
   $= \dfrac{3x+4}{(2x-3)(x+2)} \cdot \dfrac{x+2}{x+2} - \dfrac{x-1}{(x+2)^2} \cdot \dfrac{2x-3}{2x-3}$
   $= \dfrac{3x^2 + 10x + 8}{(2x-3)(x+2)^2} - \dfrac{2x^2 - 5x + 3}{(2x-3)(x+2)^2}$
   $= \dfrac{3x^2 + 10x + 8 - (2x^2 - 5x + 3)}{(2x-3)(x+2)^2}$
   $= \dfrac{3x^2 + 10x + 8 - 2x^2 + 5x - 3}{(2x-3)(x+2)^2}$
   $= \dfrac{x^2 + 15x + 5}{(2x-3)(x+2)^2}$

10. $x^2 - 4 = (x-2)(x+2)$
    $x-2$
    $x+2$
    LCD $= (x-2)(x+2)$
    $\dfrac{4}{x^2 - 4} - \dfrac{x+3}{x-2} + \dfrac{x+3}{x+2}$
    $= \dfrac{4}{(x-2)(x+2)} - \dfrac{x+3}{x-2} + \dfrac{x+3}{x+2}$
    $= \dfrac{4}{(x-2)(x+2)} - \dfrac{x+3}{x-2} \cdot \dfrac{x+2}{x+2} + \dfrac{x+3}{x+2} \cdot \dfrac{x-2}{x-2}$
    $= \dfrac{4}{(x-2)(x+2)} - \dfrac{x^2 + 5x + 6}{(x-2)(x+2)} + \dfrac{x^2 + x - 6}{(x-2)(x+2)}$
    $= \dfrac{4 - (x^2 + 5x + 6) + x^2 + x - 6}{(x-2)(x+2)}$
    $= \dfrac{4 - x^2 - 5x - 6 + x^2 + x - 6}{(x-2)(x+2)}$
    $= \dfrac{-4x - 8}{(x-2)(x+2)}$
    $= \dfrac{-4(x+2)}{(x-2)(x+2)}$
    $= \dfrac{-4}{x-2}$

### Section 6.3 – Quick Checks: Complex Rational Expressions

**6.3 Quick Checks**

1. Write the numerator of the complex rational expression as a single rational expression. Note that the LCD of the rational expressions in the numerator is $4z$.

$$\frac{z}{4} - \frac{4}{z} = \frac{z}{4} \cdot \frac{z}{z} - \frac{4}{z} \cdot \frac{4}{4} = \frac{z^2}{4z} - \frac{16}{4z} = \frac{z^2 - 16}{4z}$$

The denominator of the complex rational expression is already a single rational expression: $\frac{z+4}{16}$.

Rewrite the complex rational expression using the single rational expressions and simplify.

$$\frac{\frac{z}{4} - \frac{4}{z}}{\frac{z+4}{16}} = \frac{\frac{z^2 - 16}{4z}}{\frac{z+4}{16}} = \frac{z^2 - 16}{4z} \cdot \frac{16}{z+4}$$

$$= \frac{(z-4)(z+4)}{4z} \cdot \frac{16}{z+4}$$

$$= \frac{(z-4)\cancel{(z+4)}}{\cancel{4}z} \cdot \frac{\cancel{16}^{4}}{\cancel{z+4}} = \frac{4(z-4)}{z}$$

2. Write the numerator of the complex rational expression as a single rational expression. Note that the LCD of the rational expressions in the numerator is $(x+1)(x+2)$.

$$\frac{2x}{x+1} - \frac{x^2 - 3}{x^2 + 3x + 2} = \frac{2x}{x+1} - \frac{x^2 - 3}{(x+1)(x+2)}$$

$$= \frac{2x}{x+1} \cdot \frac{x+2}{x+2} - \frac{x^2 - 3}{(x+1)(x+2)}$$

$$= \frac{2x^2 + 4x}{(x+1)(x+2)} - \frac{x^2 - 3}{(x+1)(x+2)}$$

$$= \frac{2x^2 + 4x - (x^2 - 3)}{(x+1)(x+2)}$$

$$= \frac{2x^2 + 4x - x^2 + 3}{(x+1)(x+2)}$$

$$= \frac{x^2 + 4x + 3}{(x+1)(x+2)}$$

Write the denominator of the complex rational expression as a single rational expression. Note that the LCD of the rational expressions in the denominator is $x+2$.

$$4 + \frac{4}{x+2} = 4 \cdot \frac{x+2}{x+2} + \frac{4}{x+2}$$

$$= \frac{4x+8}{x+2} + \frac{4}{x+2} = \frac{4x+12}{x+2} = \frac{4(x+3)}{x+2}$$

Rewrite the complex rational expression using the single rational expressions and simplify.

$$\frac{\frac{2x}{x+1} - \frac{x^2 - 3}{x^2 + 3x + 2}}{4 + \frac{4}{x+2}} = \frac{\frac{x^2 + 4x + 3}{(x+1)(x+2)}}{\frac{4(x+3)}{x+2}}$$

$$= \frac{x^2 + 4x + 3}{(x+1)(x+2)} \cdot \frac{x+2}{4(x+3)}$$

$$= \frac{(x+1)(x+3)}{(x+1)(x+2)} \cdot \frac{x+2}{4(x+3)}$$

$$= \frac{\cancel{(x+1)}\cancel{(x+3)}}{\cancel{(x+1)}\cancel{(x+2)}} \cdot \frac{\cancel{x+2}}{4\cancel{(x+3)}}$$

$$= \frac{1}{4}$$

3. The LCD of all the denominators in the complex rational expression is $16z$.

$$\frac{\frac{z}{4} - \frac{4}{z}}{\frac{z+4}{16}} = \frac{\frac{z}{4} - \frac{4}{z}}{\frac{z+4}{16}} \cdot \frac{16z}{16z} = \frac{\frac{z}{4} \cdot 16z - \frac{4}{z} \cdot 16z}{\frac{z+4}{16} \cdot 16z}$$

$$= \frac{\frac{z}{\cancel{4}} \cdot \cancel{16}^{4} z - \frac{4}{\cancel{z}} \cdot 16\cancel{z}}{\frac{z+4}{\cancel{16}} \cdot \cancel{16}z}$$

$$= \frac{4z^2 - 64}{z(z+4)} = \frac{4(z-4)(z+4)}{z(z+4)}$$

$$= \frac{4(z-4)\cancel{(z+4)}}{z\cancel{(z+4)}} = \frac{4(z-4)}{z}$$

## Chapter 6: Rational Expressions and Rational Functions

4. The LCD of all the denominators in the complex rational expression is $(x+5)(x+1)$.

$$\frac{\frac{x+2}{x+5} - \frac{x+2}{x+1}}{\frac{2x+1}{x+1} - 1} = \frac{\frac{x+2}{x+5} - \frac{x+2}{x+1}}{\frac{2x+1}{x+1} - 1} \cdot \frac{(x+5)(x+1)}{(x+5)(x+1)}$$

$$= \frac{\frac{x+2}{x+5} \cdot (x+5)(x+1) - \frac{x+2}{x+1} \cdot (x+5)(x+1)}{\frac{2x+1}{x+1} \cdot (x+5)(x+1) - 1 \cdot (x+5)(x+1)}$$

$$= \frac{\frac{x+2}{\cancel{x+5}} \cdot \cancel{(x+5)}(x+1) + \frac{x+2}{\cancel{x+1}} \cdot (x+5)\cancel{(x+1)}}{\frac{2x+1}{\cancel{x+1}} \cdot (x+5)\cancel{(x+1)} - 1 \cdot (x+5)(x+1)}$$

$$= \frac{(x+2)(x+1) - (x+2)(x+5)}{(2x+1)(x+5) - (x+5)(x+1)}$$

$$= \frac{(x+2)[x+1-(x+5)]}{(x+5)[2x+1-(x+1)]}$$

$$= \frac{(x+2)[x+1-x-5]}{(x+5)[2x+1-x-1]}$$

$$= \frac{(x+2)(-4)}{(x+5)x} = \frac{-4(x+2)}{x(x+5)}$$

5. Rewrite the expression so that it does not contain any negative exponents.

$$\frac{3a^{-1} + b^{-1}}{9a^{-2} - b^{-2}} = \frac{\frac{3}{a} + \frac{1}{b}}{\frac{9}{a^2} - \frac{1}{b^2}}$$

Using Method 1, begin by writing the numerator of the complex rational expression as a single rational expression. Note that the LCD of the rational expressions in the numerator is $ab$.

$$\frac{3}{a} + \frac{1}{b} = \frac{3}{a} \cdot \frac{b}{b} + \frac{1}{b} \cdot \frac{a}{a} = \frac{3b}{ab} + \frac{a}{ab} = \frac{3b+a}{ab}$$

Next, write the denominator of the complex rational expression as a single rational expression. Note that the LCD of the rational expressions in the denominator is $a^2b^2$.

$$\frac{9}{a^2} - \frac{1}{b^2} = \frac{9}{a^2} \cdot \frac{b^2}{b^2} - \frac{1}{b^2} \cdot \frac{a^2}{a^2}$$

$$= \frac{9b^2}{a^2b^2} - \frac{a^2}{a^2b^2} = \frac{9b^2 - a^2}{a^2b^2}$$

Rewrite the complex rational expression using the single rational expressions and simplify.

$$\frac{\frac{3}{a} + \frac{1}{b}}{\frac{9}{a^2} - \frac{1}{b^2}} = \frac{\frac{3b+a}{ab}}{\frac{9b^2 - a^2}{a^2b^2}} = \frac{3b+a}{ab} \cdot \frac{a^2b^2}{9b^2 - a^2}$$

$$= \frac{3b+a}{ab} \cdot \frac{a^2b^2}{(3b-a)(3b+a)}$$

$$= \frac{\cancel{3b+a}}{\cancel{ab}} \cdot \frac{\cancel{a^2b^2}^{ab}}{(3b-a)\cancel{(3b+a)}}$$

$$= \frac{ab}{3b-a}$$

Using Method 2, the LCD of all the denominators in the complex rational expression is $a^2b^2$.

$$\frac{3a^{-1} + b^{-1}}{9a^{-2} - b^{-2}} = \frac{\frac{3}{a} + \frac{1}{b}}{\frac{9}{a^2} - \frac{1}{b^2}} = \frac{\frac{3}{a} + \frac{1}{b}}{\frac{9}{a^2} - \frac{1}{b^2}} \cdot \frac{a^2b^2}{a^2b^2}$$

$$= \frac{\frac{3}{a} \cdot a^2b^2 + \frac{1}{b} \cdot a^2b^2}{\frac{9}{a^2} \cdot a^2b^2 - \frac{1}{b^2} \cdot a^2b^2}$$

$$= \frac{3ab^2 + a^2b}{9b^2 - a^2}$$

$$= \frac{ab(3b+a)}{(3b-a)(3b+a)}$$

$$= \frac{ab\cancel{(3b+a)}}{(3b-a)\cancel{(3b+a)}}$$

$$= \frac{ab}{3b-a}$$

### 6.4 Quick Checks

1. 
$$\frac{x-4}{x^2+4} = \frac{3}{3x+2}$$

$$(x^2+4)(3x+2)\left(\frac{x-4}{x^2+4}\right) = (x^2+4)(3x+2)\left(\frac{3}{3x+2}\right)$$

$$(3x+2)(x-4) = 3(x^2+4)$$

$$3x^2 - 10x - 8 = 3x^2 + 12$$

$$-10x - 8 = 12$$

$$-10x = 20$$

$$x = -2$$

**Section 6.4 – Quick Checks:** *Rational Equations*

Check:
$$\frac{-2-4}{(-2)^2+4} \stackrel{?}{=} \frac{3}{3(-2)+2}$$
$$\frac{-2-4}{4+4} \stackrel{?}{=} \frac{3}{-6+2}$$
$$\frac{-6}{8} \stackrel{?}{=} \frac{3}{-4}$$
$$-\frac{3}{4} = -\frac{3}{4} \leftarrow \text{True}$$

The solution checks, so the solution set is $\{-2\}$.

2. $$\frac{5}{x} + \frac{1}{4} = \frac{3}{2x} - \frac{3}{2}$$
$$4x\left(\frac{5}{x}+\frac{1}{4}\right) = 4x\left(\frac{3}{2x}-\frac{3}{2}\right)$$
$$4x \cdot \frac{5}{x} + 4x \cdot \frac{1}{4} = 4x \cdot \frac{3}{2x} - 4x \cdot \frac{3}{2}$$
$$20 + x = 6 - 6x$$
$$20 + 7x = 6$$
$$7x = -14$$
$$x = -2$$

Check:
$$\frac{5}{-2} + \frac{1}{4} \stackrel{?}{=} \frac{3}{2(-2)} - \frac{3}{2}$$
$$-\frac{5}{2} + \frac{1}{4} \stackrel{?}{=} \frac{3}{-4} - \frac{3}{2}$$
$$-\frac{10}{4} + \frac{1}{4} \stackrel{?}{=} -\frac{3}{4} - \frac{6}{4}$$
$$-\frac{9}{4} = -\frac{9}{4} \leftarrow \text{True}$$

Both solutions check, so the solution set is $\{-2\}$.

3. First, we find the domain of the variable, $x$.
$x^2 + 5x + 4 = (x+1)(x+4)$, so $x \neq -1, x \neq -4$ in the first term.
$x^2 - 3x - 4 = (x+1)(x-4)$, so $x \neq -1, x \neq 4$ in the second term.
$x^2 - 16 = (x+4)(x-4)$, so $x \neq -4, x \neq 4$ in the third term.
$$\frac{3}{x^2+5x+4} + \frac{2}{x^2-3x-4} = \frac{4}{x^2-16}$$
$$\frac{3}{(x+1)(x+4)} + \frac{2}{(x+1)(x-4)} = \frac{4}{(x+4)(x-4)}$$
$$(x+1)(x+4)(x-4)\left(\frac{3}{(x+1)(x+4)}+\frac{2}{(x+1)(x-4)}\right) =$$
$$(x+1)(x+4)(x-4)\left(\frac{4}{(x+4)(x-4)}\right)$$

$$3(x-4) + 2(x+4) = 4(x+1)$$
$$3x - 12 + 2x + 8 = 4x + 4$$
$$5x - 4 = 4x + 4$$
$$x - 4 = 4$$
$$x = 8$$

Check:
$$\frac{3}{8^2+5(8)+4} + \frac{2}{8^2-3(8)-4} \stackrel{?}{=} \frac{4}{8^2-16}$$
$$\frac{3}{108} + \frac{2}{36} \stackrel{?}{=} \frac{4}{48}$$
$$\frac{3}{108} + \frac{6}{108} \stackrel{?}{=} \frac{1}{12}$$
$$\frac{9}{108} \stackrel{?}{=} \frac{1}{12}$$
$$\frac{1}{12} = \frac{1}{12} \leftarrow \text{True}$$

The solution set is $\{8\}$.

4. First, we find the domain of the variable, $z$.
$z^2 + 2z - 3 = (z-1)(z+3)$, so $z \neq -3, z \neq 1$ in the first term.
$z^2 + z - 2 = (z-1)(z+2)$, so $z \neq -2, z \neq 1$ in the second term.
$z^2 + 5z + 6 = (z+2)(z+3)$, so $z \neq -3, z \neq -2$ in the third term.
$$\frac{5}{z^2+2z-3} - \frac{3}{z^2+z-2} = \frac{1}{z^2+5z+6}$$
$$\frac{5}{(z-1)(z+3)} - \frac{3}{(z-1)(z+2)} = \frac{1}{(z+2)(z+3)}$$
$$(z-1)(z+2)(z+3)\left(\frac{5}{(z-1)(z+3)}-\frac{3}{(z-1)(z+2)}\right) =$$
$$(z-1)(z+2)(z+3)\left(\frac{1}{(z+2)(z+3)}\right)$$
$$5(z+2) - 3(z+3) = z - 1$$
$$5z + 10 - 3z - 9 = z - 1$$
$$2z + 1 = z - 1$$
$$z + 1 = -1$$
$$z = -2$$

Notice that $z = -2$ is not in the domain of the variable $z$. Therefore, there is no solution to the equation, and the solution set is $\varnothing$ or $\{\ \}$.

QC – 81

5. First, we find the domain of the variable, $z$.
   $z+4$, so $z \neq -4$ in the first term.
   $z-3$, so $z \neq 3$ in the second term.
   $z^2 + z - 12 = (z+4)(z-3)$, so $z \neq -4, z \neq 3$ in the third term.

   $$\frac{z+1}{z+4} + \frac{z+1}{z-3} = \frac{z^2+z+16}{z^2+z-12}$$

   $$\frac{z+1}{z+4} + \frac{z+1}{z-3} = \frac{z^2+z+16}{(z+4)(z-3)}$$

   $$(z+4)(z-3)\left(\frac{z+1}{z+4} + \frac{z+1}{z-3}\right)$$

   $$= (z+4)(z-3)\left(\frac{z^2+z+16}{(z+4)(z-3)}\right)$$

   $(z-3)(z+1) + (z+4)(z+1) = z^2 + z + 16$
   $z^2 - 2z - 3 + z^2 + 5z + 4 = z^2 + z + 16$
   $2z^2 + 3z + 1 = z^2 + z + 16$
   $z^2 + 2z - 15 = 0$
   $(z+5)(z-3) = 0$
   $z + 5 = 0$ or $z - 3 = 0$
   $z = -5$ or $z = 3$

   Since $z = 3$ is not in the domain of the variable, it is an extraneous solution.

   Check $z = -5$:

   $$\frac{-5+1}{-5+4} + \frac{-5+1}{-5-3} \stackrel{?}{=} \frac{(-5)^2 + (-5) + 16}{(-5)^2 + (-5) - 12}$$

   $$\frac{-4}{-1} + \frac{-4}{-8} \stackrel{?}{=} \frac{25 + (-5) + 16}{25 + (-5) - 12}$$

   $$4 + \frac{1}{2} \stackrel{?}{=} \frac{36}{8}$$

   $$\frac{9}{2} = \frac{9}{2} \quad \leftarrow \text{True}$$

   The solution checks, so the solution set is $\{-5\}$.

6. $f(x) = 1$
   $$2x - \frac{3}{x} = 1$$
   $$x\left(2x - \frac{3}{x}\right) = x \cdot 1$$
   $2x^2 - 3 = x$
   $2x^2 - x - 3 = 0$
   $(2x-3)(x+1) = 0$

   $2x - 3 = 0$ or $x + 1 = 0$
   $x = \frac{3}{2}$ or $x = -1$

   The solution set is $\left\{-1, \frac{3}{2}\right\}$. Thus, $f(-1) = 1$ and $f\left(\frac{3}{2}\right) = 1$, so the points $(-1, 1)$ and $\left(\frac{3}{2}, 1\right)$ are on the graph of $f$.

7. $C(t) = 4$
   $$\frac{50t}{t^2 + 6} = 4$$
   $$(t^2 + 6)\left(\frac{50t}{t^2 + 6}\right) = (t^2 + 6) \cdot 4$$
   $50t = 4t^2 + 24$
   $-4t^2 + 50t - 24 = 0$
   $-2(2t - 1)(t - 12) = 0$
   $2t - 1 = 0$ or $t - 12 = 0$
   $t = \frac{1}{2}$ or $t = 12$

   The concentration of the drug will be 4 milligrams per liter after $\frac{1}{2}$ hour and after 12 hours.

### 6.5 Quick Checks

1. $\dfrac{x-7}{x+3} \geq 0$

   The rational expression will equal 0 when $x = 7$. It is undefined when $x = -3$. Thus, we separate the real number line into the intervals $(-\infty, -3)$, $(-3, 7)$, and $(7, \infty)$. Determine where the numerator and denominator are positive and negative and where the quotient is positive and negative.

   | Interval | $(-\infty, -3)$ | $-3$ | $(-3, 7)$ | $7$ | $(7, \infty)$ |
   |---|---|---|---|---|---|
   | $x - 7$ | Neg | Neg | Neg | 0 | Pos |
   | $x + 3$ | Neg | 0 | Pos | Pos | Pos |
   | $\dfrac{x-7}{x+3}$ | Pos | Undef | Neg | 0 | Pos |

## Section 6.6 – Quick Checks: Models Involving Rational Expressions

The rational function is undefined at $x = -3$, so $-3$ is not part of the solution. The inequality is non-strict, so 7 is part of the solution. Now, $\dfrac{x-7}{x+3}$ is greater than zero where the quotient is positive. The solution is $\{x \mid x < -3 \text{ or } x \geq 7\}$ in set-builder notation; the solution is $(-\infty, -3) \cup [7, \infty)$ in interval notation.

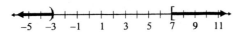

2. 
$$\dfrac{4x+5}{x+2} < 3$$
$$\dfrac{4x+5}{x+2} - 3 < 0$$
$$\dfrac{4x+5}{x+2} - \dfrac{3(x+2)}{x+2} < 0$$
$$\dfrac{4x+5-3x-6}{x+2} < 0$$
$$\dfrac{x-1}{x+2} < 0$$

The rational expression will equal 0 when $x = 1$. It is undefined when $x = -2$. Thus, we separate the real number line into the intervals $(-\infty, -2)$, $(-2, 1)$, and $(1, \infty)$. Determine where the numerator and denominator are positive and negative and where the quotient is positive and negative.

| Interval | $(-\infty, -2)$ | $-2$ | $(-2, 1)$ | $1$ | $(1, \infty)$ |
|---|---|---|---|---|---|
| $x - 1$ | Neg | Neg | Neg | 0 | Pos |
| $x + 2$ | Neg | 0 | Pos | Pos | Pos |
| $\dfrac{x-1}{x+2}$ | Pos | Undef | Neg | 0 | Pos |

The rational function is undefined at $x = -2$, so $-2$ is not part of the solution. The inequality is strict, so 1 is not part of the solution. Now, $\dfrac{x-1}{x+2}$ is less than zero where the quotient is negative. The solution is $\{x \mid -2 < x < 1\}$ in set-builder notation; the solution is $(-2, 1)$ in interval notation.

### 6.6 Quick Checks

1. a. 
$$Y = \dfrac{G}{1-b}$$
$$(1-b)Y = (1-b)\left(\dfrac{G}{1-b}\right)$$
$$Y - Yb = G$$
$$Y = G + Yb$$
$$Y - G = Yb$$
$$\dfrac{Y-G}{Y} = b \quad \text{or} \quad b = \dfrac{Y-G}{Y}$$

b. Substitute $G = \$100$ billion and $Y = \$1,000$ billion into $b = \dfrac{Y-G}{Y}$.
$$b = \dfrac{1,000 \text{ billion} - 100 \text{ billion}}{1000 \text{ billion}}$$
$$= \dfrac{900 \text{ billion}}{1000 \text{ billion}} = 0.9$$

2. 
$$\dfrac{AB}{AC} = \dfrac{DE}{DF}$$
$$\dfrac{3x}{8} = \dfrac{6}{x}$$
$$8x\left(\dfrac{3x}{8}\right) = 8x\left(\dfrac{6}{x}\right)$$
$$3x^2 = 48$$
$$3x^2 - 48 = 0$$
$$3(x+4)(x-4) = 0$$
$$x + 4 = 0 \quad \text{or} \quad x - 4 = 0$$
$$x = -4 \quad \text{or} \quad x = 4$$

Since the length of DF cannot be negative, we discard $x = -4$ and keep $x = 4$. The length of $AB$ is $3x = 3 \cdot 4 = 12$.

In summary, $AB = 12$ and $DF = 4$.

3. Let $p$ represent the population of the United States in 2005.
$$\dfrac{20}{100,000} = \dfrac{59,850}{x}$$
$$100,000x\left(\dfrac{20}{100,000}\right) = 100,000x\left(\dfrac{59,850}{x}\right)$$
$$20x = 5,985,000,000$$
$$x = \dfrac{5,985,000,000}{20}$$
$$x = 299,250,000$$

In 2005, the population of the United States was 299,250,000 people.

## Chapter 6: Rational Expressions and Rational Functions

**4.** Let $t$ represent the time required to fill the pool using the two hoses together.

$$\begin{pmatrix}\text{Part done} \\ \text{by Juan and} \\ \text{Maria's hose} \\ \text{in 1 hour}\end{pmatrix} + \begin{pmatrix}\text{Part done by} \\ \text{the neighbor's} \\ \text{hose in 1 hour}\end{pmatrix} = \begin{pmatrix}\text{Part done} \\ \text{together} \\ \text{in 1 hour}\end{pmatrix}$$

$$\frac{1}{30} + \frac{1}{24} = \frac{1}{t}$$

$$120t\left(\frac{1}{30}+\frac{1}{24}\right)=120t\left(\frac{1}{t}\right)$$

$$4t+5t=120$$
$$9t=120$$
$$t=\frac{120}{9}$$
$$t=\frac{40}{3}=13\frac{1}{3}$$

Using both hoses, it will take $\frac{40}{3}$ hours, or 13 hours and 20 minutes, to fill the pool.

**5.** Let $t$ represent the time required to fill the pool.

$$\begin{pmatrix}\text{Portion of} \\ \text{pool filled in} \\ \text{1 minute by} \\ \text{the air pump}\end{pmatrix} - \begin{pmatrix}\text{Portion of} \\ \text{air released} \\ \text{from the pool} \\ \text{in 1 minute}\end{pmatrix} = \begin{pmatrix}\text{Portion of} \\ \text{pool filled in} \\ \text{1 minute with} \\ \text{valve open}\end{pmatrix}$$

$$\frac{1}{20} - \frac{1}{50} = \frac{1}{t}$$

$$100t\left(\frac{1}{20}-\frac{1}{50}\right)=100t\left(\frac{1}{t}\right)$$

$$5t-2t=100$$
$$3t=100$$
$$t=\frac{100}{3}\approx 33.3$$

To fill the pool with the valve open, it will take $\frac{100}{3}=33\frac{1}{3}$ minutes or 33 minutes, 20 seconds.

**6.** Let $x$ represent the speed of the canoe in still water.

|  | Distance (miles) | Rate (mph) | Time (hours) |
|---|---|---|---|
| Upstream | 12 | $x-2$ | $\dfrac{12}{x-2}$ |
| Downstream | 12 | $x+2$ | $\dfrac{12}{x+2}$ |

Trip up + Trip down = 8

$$\frac{12}{x-2}+\frac{12}{x+2}=8$$

$$(x-2)(x+2)\left(\frac{12}{x-2}+\frac{12}{x+2}\right)=(x-2)(x+2)\cdot 8$$

$$12(x+2)+12(x-2)=8(x^2-4)$$
$$12x+24+12x-24=8(x^2-4)$$
$$24x=8(x^2-4)$$
$$3x=x^2-4$$
$$0=x^2-3x-4$$
$$0=(x+1)(x-4)$$
$$x+1=0 \quad \text{or} \quad x-4=0$$
$$x=-1 \quad \text{or} \quad x=4$$

We disregard $-1$ because the speed of the canoe must be positive. Thus, the speed of the canoe in still water is 4 miles per hour.

**7. a.** $V=\dfrac{k}{l}$

$$500=\frac{k}{30}$$
$$k=500(30)=15{,}000$$

Thus, the function that relates the rate of vibration to the length of the string is

$$V(l)=\frac{15{,}000}{l}.$$

**b.** $V(50)=\dfrac{15{,}000}{50}=300$

The rate of vibration is 300 oscillations per second.

**8.** Let $K$ represent the kinetic energy, $m$ represent the mass, and $v$ represent the velocity. Then,

$$K=kmv^2$$
$$4455=k\cdot 110\cdot 9^2$$
$$4455=8910k$$
$$\frac{1}{2}=k$$

Thus, $K=\dfrac{1}{2}mv^2$.

For $m=140$ and $v=5$, $K=\dfrac{1}{2}(140)(5)^2=1750$.

The kinetic energy of a 140 kg lineman running at 5 meters per second is 1750 joules.

**Section 6.6 – Quick Checks:** Models Involving Rational Expressions

9. $R = \dfrac{kl}{d^2}$

   $1.24 = \dfrac{k \cdot 432}{4^2}$

   $1.24 = \dfrac{432k}{16}$

   $1.24 = 27k$

   $k = \dfrac{1.24}{27} \approx 0.0459$

   Thus, $R = \dfrac{0.0459l}{d^2}$.

   For $l = 282$ and $d = 3$, $R = \dfrac{0.0459(282)}{3^2} \approx 1.44$.

   The resistance is approximately 1.44 ohms when the length of the wire is 282 feet and the diameter is 3 millimeters.

# Getting Ready for Chapter 7 – Quick Checks

1. $\sqrt{81} = 9$ because $9^2 = 81$.

2. $\sqrt{900} = 30$ because $30^2 = 900$.

3. $\sqrt{\dfrac{1}{4}} = \dfrac{1}{2}$ because $\left(\dfrac{1}{2}\right)^2 = \dfrac{1}{4}$.

4. $\sqrt{0.16} = 0.4$ because $(0.4)^2 = 0.16$

5. $\left(\sqrt{13}\right)^2 = 13$ because $\left(\sqrt{c}\right)^2 = c$ if $c \geq 0$.

6. $5\sqrt{9} = 5\sqrt{3^2} = 5 \cdot 3 = 15$

7. $\sqrt{36 + 64} = \sqrt{100} = \sqrt{10^2} = 10$

8. $\sqrt{36} + \sqrt{64} = \sqrt{6^2} + \sqrt{8^2} = 6 + 8 = 14$

9. $\sqrt{25 - 4 \cdot 3 \cdot (-2)} = \sqrt{25 + 24} = \sqrt{49} = \sqrt{7^2} = 7$

10. $\sqrt{400}$ is rational because $20^2 = 400$. Thus, $\sqrt{400} = 20$.

11. $\sqrt{40}$ is irrational because 40 is not a perfect square. There is no rational number whose square is 40. Using a calculator, we find $\sqrt{40} \approx 6.32$.

12. $\sqrt{-25}$ is not a real number. There is no real number whose square is $-25$.

13. $-\sqrt{196}$ is a rational number because $14^2 = 196$. Thus, $-\sqrt{196} = -14$.

14. $\sqrt{(-14)^2} = |-14| = 14$

15. $\sqrt{z^2} = |z|$

16. $\sqrt{(2x+3)^2} = |2x+3|$

17. $\sqrt{p^2 - 12p + 36} = \sqrt{(p-6)^2} = |p-6|$

# Chapter 7 – Quick Checks

## 7.1 Quick Checks

1. $\sqrt[3]{125} = 5$ since $5^3 = 125$.

2. $\sqrt[4]{81} = 3$ since $3^4 = 81$.

3. $\sqrt[3]{-216} = -6$ since $(-6)^3 = -216$.

4. $\sqrt[4]{-32}$ is not a real number.

5. $\sqrt[5]{\dfrac{1}{32}} = \dfrac{1}{2}$ since $\left(\dfrac{1}{2}\right)^5 = \dfrac{1}{32}$.

6. $\sqrt[3]{50} \approx 3.68$
   ```
   ³√(50)
        3.684031499
   ```

7. $\sqrt[4]{80} \approx 2.99$
   ```
   4*√(80)
        2.990697562
   ```

8. $\sqrt[5]{40} \approx 2.09$
   ```
   5*√(40)
        2.091279105
   ```

9. Because the index is even, we have
   $\sqrt[4]{5^4} = |5| = 5$.

10. Because the index is even, we have $\sqrt[6]{z^6} = |z|$.

11. Because the index is odd, we have
    $\sqrt[7]{(3x-2)^7} = 3x - 2$.

12. Because the index is even, we have
    $\sqrt[8]{(-2)^8} = |-2| = 2$.

13. Because the index is odd, we have
    $\sqrt[5]{\dfrac{-32}{243}} = \sqrt[5]{\left(\dfrac{-2}{3}\right)^5} = -\dfrac{2}{3}$.

14. $25^{1/2} = \sqrt{25} = \sqrt{5^2} = 5$

15. $(-27)^{1/3} = \sqrt[3]{-27} = \sqrt[3]{(-3)^3} = -3$

16. $-64^{1/2} = -\sqrt{64} = -\sqrt{8^2} = -8$

17. $(-64)^{1/2} = \sqrt{-64}$ is not a real number.

18. $b^{1/2} = \sqrt{b}$

19. $\sqrt[5]{8b} = (8b)^{1/5}$

20. $\sqrt[8]{\dfrac{mn^5}{3}} = \left(\dfrac{mn^5}{3}\right)^{1/8}$

21. $16^{3/2} = \left(\sqrt{16}\right)^3 = 4^3 = 64$

22. $27^{2/3} = \left(\sqrt[3]{27}\right)^2 = 3^2 = 9$

23. $-16^{3/4} = -\left(\sqrt[4]{16}\right)^3 = -2^3 = -8$

24. $(-64)^{2/3} = \left(\sqrt[3]{-64}\right)^2 = (-4)^2 = 16$

25. $(-25)^{5/2} = \left(\sqrt{-25}\right)^5$ is not a real number.

26. $50^{2/3} \approx 13.57$
    ```
    50^(2/3)
         13.57208808
    ```

27. $40^{0.15} \approx 1.74$
    ```
    40^0.15
         1.739037707
    ```

## Chapter 7: Radicals and Rational Exponents

**28.** $\sqrt[8]{a^3} = \left(a^3\right)^{1/8} = a^{3/8}$

**29.** $\left(\sqrt[4]{12ab^3}\right)^9 = \left(\left(12ab^3\right)^{1/4}\right)^9 = \left(12ab^3\right)^{9/4}$

**30.** $81^{-1/2} = \dfrac{1}{81^{1/2}} = \dfrac{1}{\sqrt{81}} = \dfrac{1}{9}$

**31.** $\dfrac{1}{8^{-2/3}} = 8^{2/3} = \left(\sqrt[3]{8}\right)^2 = 2^2 = 4$

**32.** $(13x)^{-3/2} = \dfrac{1}{(13x)^{3/2}}$

### 7.2 Quick Checks

**1.** $5^{3/4} \cdot 5^{1/6} = 5^{\frac{3}{4}+\frac{1}{6}} = 5^{\frac{9}{12}+\frac{2}{12}} = 5^{\frac{11}{12}}$

**2.** $\dfrac{32^{6/5}}{32^{3/5}} = 32^{\frac{6}{5}-\frac{3}{5}} = 32^{3/5} = \left(\sqrt[5]{32}\right)^3 = 2^3 = 8$

**3.** $\left(100^{3/8}\right)^{4/3} = 100^{\frac{3}{8} \cdot \frac{4}{3}} = 100^{1/2} = \sqrt{100} = 10$

**4.** $\left(a^{3/2} \cdot b^{5/4}\right)^{2/3} = a^{\frac{3}{2} \cdot \frac{2}{3}} \cdot b^{\frac{5}{4} \cdot \frac{2}{3}} = ab^{5/6}$

**5.** $\left(8x^{3/4}y^{-1}\right)^{2/3} = 8^{2/3}\left(x^{3/4}\right)^{2/3}\left(y^{-1}\right)^{2/3}$
$= 4x^{\frac{3}{4} \cdot \frac{2}{3}} y^{-1 \cdot \frac{2}{3}}$
$= 4x^{1/2} y^{-2/3}$
$= \dfrac{4x^{1/2}}{y^{2/3}}$

**6.** $\left(\dfrac{25x^{1/2}y^{3/4}}{x^{-3/4}y}\right)^{1/2} = \left(25x^{\frac{1}{2}-\left(-\frac{3}{4}\right)}y^{\frac{3}{4}-1}\right)^{1/2}$
$= \left(25x^{5/4}y^{-1/4}\right)^{1/2}$
$= 25^{1/2}\left(x^{5/4}\right)^{1/2}\left(y^{-1/4}\right)^{1/2}$
$= 5x^{\frac{5}{4} \cdot \frac{1}{2}} y^{-\frac{1}{4} \cdot \frac{1}{2}}$
$= 5x^{5/8} y^{-1/8}$
$= \dfrac{5x^{5/8}}{y^{1/8}}$

**7.** $8\left(125a^{3/4}b^{-1}\right)^{2/3} = 8 \cdot 125^{2/3}\left(a^{3/4}\right)^{2/3}\left(b^{-1}\right)^{2/3}$
$= 8 \cdot 25 a^{\frac{3}{4} \cdot \frac{2}{3}} b^{-1 \cdot \frac{2}{3}}$
$= 200 a^{1/2} b^{-2/3}$
$= \dfrac{200a^{1/2}}{b^{2/3}}$

**8.** $\sqrt[10]{36^5} = \left(36^5\right)^{1/10} = 36^{5 \cdot \frac{1}{10}} = 36^{1/2} = \sqrt{36} = 6$

**9.** $\sqrt[4]{16a^8b^{12}} = \left(16a^8b^{12}\right)^{1/4}$
$= 16^{1/4}\left(a^8\right)^{1/4}\left(b^{12}\right)^{1/4}$
$= 2a^{8 \cdot \frac{1}{4}} b^{12 \cdot \frac{1}{4}}$
$= 2a^2 b^3$

**10.** $\dfrac{\sqrt[3]{x^2}}{\sqrt[4]{x}} = \dfrac{x^{2/3}}{x^{1/4}} = x^{\frac{2}{3}-\frac{1}{4}} = x^{5/12} = \sqrt[12]{x^5}$

**11.** $\sqrt[4]{\sqrt[3]{a^2}} = \sqrt[4]{a^{2/3}} = \left(a^{2/3}\right)^{1/4} = a^{\frac{2}{3} \cdot \frac{1}{4}} = a^{1/6} = \sqrt[6]{a}$

**12.** $8x^{3/2} + 3x^{1/2}(4x+3) = 8x^{\frac{2}{2}+\frac{1}{2}} + 3x^{1/2}(4x+3)$
$= 8x \cdot x^{1/2} + 3x^{1/2}(4x+3)$
$= x^{1/2}\left(8x + 3(4x+3)\right)$
$= x^{1/2}(8x + 12x + 9)$
$= x^{1/2}(20x + 9)$

**13.** $9x^{1/3} + x^{-2/3}(3x+1) = 9x^{\frac{3}{3}-\frac{2}{3}} + x^{-2/3}(3x+1)$
$= 9x \cdot x^{-2/3} + x^{-2/3}(3x+1)$
$= x^{-2/3}\left(9x + (3x+1)\right)$
$= x^{-2/3}(12x + 1)$
$= \dfrac{12x+1}{x^{2/3}}$

## Section 7.3 – Quick Checks: Simplifying Radical Expressions

**7.3 Quick Checks**

1. $\sqrt{11}\cdot\sqrt{7} = \sqrt{11\cdot 7} = \sqrt{77}$

2. $\sqrt[4]{6}\cdot\sqrt[4]{7} = \sqrt[4]{6\cdot 7} = \sqrt[4]{42}$

3. $\sqrt{x-5}\cdot\sqrt{x+5} = \sqrt{(x-5)(x+5)} = \sqrt{x^2-25}$

4. $\sqrt[7]{5p}\cdot\sqrt[7]{4p^3} = \sqrt[7]{5p\cdot 4p^3} = \sqrt[7]{20p^4}$

5. $\sqrt{48} = \sqrt{16\cdot 3} = \sqrt{16}\cdot\sqrt{3} = 4\sqrt{3}$

6. $4\sqrt[3]{54} = 4\sqrt[3]{27\cdot 2} = 4\sqrt[3]{27}\cdot\sqrt[3]{2} = 4\cdot 3\sqrt[3]{2} = 12\sqrt[3]{2}$

7. $\sqrt{200a^2} = \sqrt{200}\cdot\sqrt{a^2}$
$= \sqrt{100\cdot 2}\cdot\sqrt{a^2}$
$= \sqrt{100}\cdot\sqrt{2}\cdot\sqrt{a^2}$
$= 10\sqrt{2}|a|$ or $10|a|\sqrt{2}$

8. $\sqrt[4]{40}$ cannot be simplified further.

9. $\dfrac{6+\sqrt{45}}{3} = \dfrac{6+\sqrt{9\cdot 5}}{3} = \dfrac{6+\sqrt{9}\cdot\sqrt{5}}{3}$
$= \dfrac{6+3\sqrt{5}}{3} = \dfrac{3(2+\sqrt{5})}{3}$
$= 2+\sqrt{5}$

10. $\dfrac{-2+\sqrt{32}}{4} = \dfrac{-2+\sqrt{16\cdot 2}}{4} = \dfrac{-2+\sqrt{16}\cdot\sqrt{2}}{4}$
$= \dfrac{-2+4\sqrt{2}}{4} = \dfrac{2(-1+2\sqrt{2})}{4}$
$= \dfrac{-1+2\sqrt{2}}{2}$

11. $\sqrt{75a^6} = \sqrt{25\cdot 3\cdot a^6}$
$= \sqrt{25}\cdot\sqrt{3}\cdot\sqrt{a^6}$
$= \sqrt{25}\cdot\sqrt{3}\cdot\sqrt{(a^3)^2}$
$= 5\sqrt{3}\cdot a^3$
$= 5a^3\sqrt{3}$

12. $\sqrt{18a^5} = \sqrt{9\cdot 2\cdot a^4\cdot a}$
$= \sqrt{9}\cdot\sqrt{2}\cdot\sqrt{a^4}\cdot\sqrt{a}$
$= \sqrt{9}\cdot\sqrt{2}\cdot\sqrt{(a^2)^2}\cdot\sqrt{a}$
$= 3\sqrt{2}\cdot a^2\sqrt{a}$
$= 3a^2\sqrt{2a}$

13. $\sqrt[3]{128x^6y^{10}} = \sqrt[3]{64\cdot 2\cdot x^6\cdot y^9\cdot y}$
$= \sqrt[3]{64}\cdot\sqrt[3]{2}\cdot\sqrt[3]{x^6}\cdot\sqrt[3]{y^9}\cdot\sqrt[3]{y}$
$= \sqrt[3]{64}\cdot\sqrt[3]{2}\cdot\sqrt[3]{(x^2)^3}\cdot\sqrt[3]{(y^3)^3}\cdot\sqrt[3]{y}$
$= 4x^2y^3\sqrt[3]{2y}$

14. $\sqrt[4]{16a^5b^{11}} = \sqrt[4]{16\cdot a^4\cdot a\cdot b^8\cdot b^3}$
$= \sqrt[4]{16}\cdot\sqrt[4]{a^4}\cdot\sqrt[4]{a}\cdot\sqrt[4]{b^8}\cdot\sqrt[4]{b^3}$
$= \sqrt[4]{16}\cdot\sqrt[4]{a^4}\cdot\sqrt[4]{a}\cdot\sqrt[4]{(b^2)^4}\cdot\sqrt[4]{b^3}$
$= 2ab^2\sqrt[4]{ab^3}$

15. $\sqrt{6}\cdot\sqrt{8} = \sqrt{6\cdot 8}$
$= \sqrt{48}$
$= \sqrt{16\cdot 3}$
$= \sqrt{16}\cdot\sqrt{3}$
$= 4\sqrt{3}$

16. $\sqrt[3]{12a^2}\cdot\sqrt[3]{10a^4} = \sqrt[3]{12a^2\cdot 10a^4}$
$= \sqrt[3]{120a^6}$
$= \sqrt[3]{8\cdot 15\cdot a^6}$
$= \sqrt[3]{8}\cdot\sqrt[3]{a^6}\cdot\sqrt[3]{15}$
$= 2a^2\sqrt[3]{15}$

17. $4\sqrt[3]{8a^2b^5}\cdot\sqrt[3]{6a^2b^4} = 4\sqrt[3]{8a^2b^5\cdot 6a^2b^4}$
$= 4\sqrt[3]{48a^4b^9}$
$= 4\sqrt[3]{8\cdot 6\cdot a^3\cdot a\cdot b^9}$
$= 4\sqrt[3]{8}\cdot\sqrt[3]{6}\cdot\sqrt[3]{a^3}\cdot\sqrt[3]{a}\cdot\sqrt[3]{b^9}$
$= 4\cdot 2ab^3\cdot\sqrt[3]{6a}$
$= 8ab^3\sqrt[3]{6a}$

## Chapter 7: Radicals and Rational Exponents

18. $\sqrt{\dfrac{13}{49}} = \dfrac{\sqrt{13}}{\sqrt{49}} = \dfrac{\sqrt{13}}{7}$

19. $\sqrt[3]{\dfrac{27p^3}{8}} = \dfrac{\sqrt[3]{27p^3}}{\sqrt[3]{8}} = \dfrac{3p}{2}$

20. $\sqrt[4]{\dfrac{3q^4}{16}} = \dfrac{\sqrt[4]{3q^4}}{\sqrt[4]{16}} = \dfrac{q\sqrt[4]{3}}{2}$

21. $\dfrac{\sqrt{12a^5}}{\sqrt{3a}} = \sqrt{\dfrac{12a^5}{3a}} = \sqrt{4a^4} = 2a^2$

22. $\dfrac{\sqrt[3]{-24x^2}}{\sqrt[3]{3x^{-1}}} = \sqrt[3]{\dfrac{-24x^2}{3x^{-1}}} = \sqrt[3]{-8x^3} = -2x$

23. $\dfrac{\sqrt[3]{250a^5b^{-2}}}{\sqrt[3]{2ab}} = \sqrt[3]{\dfrac{250a^5b^{-2}}{2ab}}$
    $= \sqrt[3]{125a^4b^{-3}}$
    $= \sqrt[3]{\dfrac{125a^4}{b^3}}$
    $= \dfrac{5a\sqrt[3]{a}}{b}$

24. $\sqrt[4]{5} \cdot \sqrt[3]{3} = 5^{1/4} \cdot 3^{1/3}$
    $= 5^{3/12} \cdot 3^{4/12}$
    $= \left(5^3\right)^{1/12} \cdot \left(3^4\right)^{1/12}$
    $= \left[\left(5^3\right)\left(3^4\right)\right]^{1/12}$
    $= (10{,}125)^{1/12}$
    $= \sqrt[12]{10{,}125}$

25. $\sqrt{10} \cdot \sqrt[3]{12} = 10^{1/2} \cdot 12^{1/3}$
    $= 10^{3/6} \cdot 12^{2/6}$
    $= \left(10^3\right)^{1/6} \cdot \left(12^2\right)^{1/6}$
    $= \left[\left(10^3\right)\left(12^2\right)\right]^{1/6}$
    $= \left(2^3 \cdot 5^3 \cdot 4^2 \cdot 3^2\right)^{1/6}$
    $= \left(2^3 \cdot 5^3 \cdot 2^4 \cdot 3^2\right)^{1/6}$
    $= \left(2^7 \cdot 5^3 \cdot 3^2\right)^{1/6}$
    $= 2\left(2 \cdot 5^3 \cdot 3^2\right)^{1/6}$
    $= 2(2250)^{1/6}$
    $= 2\sqrt[6]{2250}$

### 7.4 Quick Checks

1. $9\sqrt{13y} + 4\sqrt{13y} = (9+4)\sqrt{13y} = 13\sqrt{13y}$

2. $\sqrt[4]{5} + 9\sqrt[4]{5} - 3\sqrt[4]{5} = (1+9-3)\sqrt[4]{5} = 7\sqrt[4]{5}$

3. $4\sqrt{18} - 3\sqrt{8} = 4 \cdot \sqrt{9} \cdot \sqrt{2} - 3 \cdot \sqrt{4} \cdot \sqrt{2}$
    $= 4 \cdot 3\sqrt{2} - 3 \cdot 2\sqrt{2}$
    $= 12\sqrt{2} - 6\sqrt{2}$
    $= (12-6)\sqrt{2}$
    $= 6\sqrt{2}$

4. $-5x\sqrt[3]{54x} + 7\sqrt[3]{2x^4}$
    $= -5x \cdot \sqrt[3]{27} \cdot \sqrt[3]{2x} + 7 \cdot \sqrt[3]{x^3} \cdot \sqrt[3]{2x}$
    $= -5x \cdot 3\sqrt[3]{2x} + 7x \cdot \sqrt[3]{2x}$
    $= -15x\sqrt[3]{2x} + 7x\sqrt[3]{2x}$
    $= (-15+7)x\sqrt[3]{2x}$
    $= -8x\sqrt[3]{2x}$

5. $7\sqrt{10} - 6\sqrt{3}$ cannot be simplified further.

**Section 7.5 – Quick Checks:** *Rationalizing Radical Expressions*

6. $\sqrt[3]{8z^4} - 2z\sqrt[3]{-27z} + \sqrt[3]{125z}$
   $= \sqrt[3]{8z^3} \cdot \sqrt[3]{z} - 2z \cdot \sqrt[3]{-27} \cdot \sqrt[3]{z} + \sqrt[3]{125} \cdot \sqrt[3]{z}$
   $= 2z\sqrt[3]{z} + -2z(-3)\sqrt[3]{z} + 5\sqrt[3]{z}$
   $= 2z\sqrt[3]{z} + 6z\sqrt[3]{z} + 5\sqrt[3]{z}$
   $= (2z + 6z + 5)\sqrt[3]{z}$
   $= (8z + 5)\sqrt[3]{z}$

7. $\sqrt{25m} - 3\sqrt[4]{m^2} = \sqrt{25m} - 3m^{2/4}$
   $= \sqrt{25m} - 3m^{1/2}$
   $= \sqrt{25m} - 3\sqrt{m}$
   $= \sqrt{25} \cdot \sqrt{m} - 3\sqrt{m}$
   $= 5\sqrt{m} - 3\sqrt{m}$
   $= (5-3)\sqrt{m}$
   $= 2\sqrt{m}$

8. $\sqrt{6}(3 - 5\sqrt{6}) = \sqrt{6} \cdot 3 - 5\sqrt{6} \cdot \sqrt{6}$
   $= 3\sqrt{6} - 5\sqrt{36}$
   $= 3\sqrt{6} - 5 \cdot 6$
   $= 3\sqrt{6} - 30$
   $= 3(\sqrt{6} - 10)$

9. $\sqrt[3]{12}(3 - \sqrt[3]{2}) = \sqrt[3]{12} \cdot 3 - \sqrt[3]{12} \cdot \sqrt[3]{2}$
   $= 3\sqrt[3]{12} - \sqrt[3]{24}$
   $= 3\sqrt[3]{12} - 2\sqrt[3]{3}$

10. $(2 - 7\sqrt{3})(5 + 4\sqrt{3})$
    $= 2 \cdot 5 + 2 \cdot 4\sqrt{3} - 7\sqrt{3} \cdot 5 - 7\sqrt{3} \cdot 4\sqrt{3}$
    $= 10 + 8\sqrt{3} - 35\sqrt{3} - 28\sqrt{9}$
    $= 10 - 27\sqrt{3} - 28 \cdot 3$
    $= 10 - 27\sqrt{3} - 84$
    $= -74 - 27\sqrt{3}$

11. $(5\sqrt{2} + \sqrt{3})^2 = (5\sqrt{2} + \sqrt{3})(5\sqrt{2} + \sqrt{3})$
    $= (5\sqrt{2})^2 + 2(5\sqrt{2})(\sqrt{3}) + (\sqrt{3})^2$
    $= 25\sqrt{4} + 10\sqrt{6} + \sqrt{9}$
    $= 25 \cdot 2 + 10\sqrt{6} + 3$
    $= 50 + 10\sqrt{6} + 3$
    $= 53 + 10\sqrt{6}$

12. $(\sqrt{7} - 3\sqrt{2})^2 = (\sqrt{7} - 3\sqrt{2})(\sqrt{7} - 3\sqrt{2})$
    $= (\sqrt{7})^2 - 2(\sqrt{7})(3\sqrt{2}) + (3\sqrt{2})^2$
    $= \sqrt{49} - 6\sqrt{14} + 9\sqrt{4}$
    $= 7 - 6\sqrt{14} + 9 \cdot 2$
    $= 7 - 6\sqrt{14} + 18$
    $= 25 - 6\sqrt{14}$

13. $(\sqrt{3} + \sqrt{2})(\sqrt{3} - \sqrt{2}) = (\sqrt{3})^2 - (\sqrt{2})^2$
    $= \sqrt{9} - \sqrt{4}$
    $= 3 - 2$
    $= 1$

**7.5 Quick Checks**

1. $\dfrac{1}{\sqrt{3}} = \dfrac{1}{\sqrt{3}} \cdot \dfrac{\sqrt{3}}{\sqrt{3}} = \dfrac{\sqrt{3}}{\sqrt{9}} = \dfrac{\sqrt{3}}{3}$

2. $\dfrac{\sqrt{5}}{\sqrt{8}} = \dfrac{\sqrt{5}}{2\sqrt{2}} = \dfrac{\sqrt{5}}{2\sqrt{2}} \cdot \dfrac{\sqrt{2}}{\sqrt{2}} = \dfrac{\sqrt{10}}{2\sqrt{4}} = \dfrac{\sqrt{10}}{2 \cdot 2} = \dfrac{\sqrt{10}}{4}$

3. $\dfrac{5}{\sqrt{10x}} = \dfrac{5}{\sqrt{10x}} \cdot \dfrac{\sqrt{10x}}{\sqrt{10x}}$
   $= \dfrac{5\sqrt{10x}}{\sqrt{100x^2}}$
   $= \dfrac{5\sqrt{10x}}{10x}$
   $= \dfrac{\sqrt{10x}}{2x}$

4. $\dfrac{4}{\sqrt[3]{3}} = \dfrac{4}{\sqrt[3]{3}} \cdot \dfrac{\sqrt[3]{3^2}}{\sqrt[3]{3^2}} = \dfrac{4\sqrt[3]{9}}{\sqrt[3]{27}} = \dfrac{4\sqrt[3]{9}}{3}$

## Chapter 7: Radicals and Rational Exponents

5. $\sqrt[3]{\dfrac{3}{20}} = \dfrac{\sqrt[3]{3}}{\sqrt[3]{20}} = \dfrac{\sqrt[3]{3}}{\sqrt[3]{2^2 \cdot 5}} \cdot \dfrac{\sqrt[3]{2 \cdot 5^2}}{\sqrt[3]{2 \cdot 5^2}}$

$= \dfrac{\sqrt[3]{150}}{\sqrt[3]{1000}}$

$= \dfrac{\sqrt[3]{150}}{10}$

6. $\dfrac{3}{\sqrt[4]{p}} = \dfrac{3}{\sqrt[4]{p}} \cdot \dfrac{\sqrt[4]{p^3}}{\sqrt[4]{p^3}} = \dfrac{3\sqrt[4]{p^3}}{p}$

7. $\dfrac{4}{\sqrt{3}+1} = \dfrac{4}{\sqrt{3}+1} \cdot \dfrac{\sqrt{3}-1}{\sqrt{3}-1}$

$= \dfrac{4(\sqrt{3}-1)}{(\sqrt{3}+1)(\sqrt{3}-1)}$

$= \dfrac{4(\sqrt{3}-1)}{(\sqrt{3})^2 - 1^2}$

$= \dfrac{4(\sqrt{3}-1)}{3-1}$

$= \dfrac{4(\sqrt{3}-1)}{2}$

$= 2(\sqrt{3}-1)$

8. $\dfrac{\sqrt{2}}{\sqrt{6}-\sqrt{2}} = \dfrac{\sqrt{2}}{\sqrt{6}-\sqrt{2}} \cdot \dfrac{\sqrt{6}+\sqrt{2}}{\sqrt{6}+\sqrt{2}}$

$= \dfrac{\sqrt{2}(\sqrt{6}+\sqrt{2})}{(\sqrt{6}-\sqrt{2})(\sqrt{6}+\sqrt{2})}$

$= \dfrac{\sqrt{2}(\sqrt{6}+\sqrt{2})}{(\sqrt{6})^2 - (\sqrt{2})^2}$

$= \dfrac{\sqrt{12}+\sqrt{4}}{6-2}$

$= \dfrac{2\sqrt{3}+2}{4}$

$= \dfrac{2(\sqrt{3}+1)}{4}$

$= \dfrac{\sqrt{3}+1}{2}$

9. $\dfrac{\sqrt{5}+4}{\sqrt{5}-\sqrt{2}} = \dfrac{\sqrt{5}+4}{\sqrt{5}-\sqrt{2}} \cdot \dfrac{\sqrt{5}+\sqrt{2}}{\sqrt{5}+\sqrt{2}}$

$= \dfrac{(\sqrt{5}+4)(\sqrt{5}+\sqrt{2})}{(\sqrt{5}-\sqrt{2})(\sqrt{5}+\sqrt{2})}$

$= \dfrac{\sqrt{25}+\sqrt{10}+4\sqrt{5}+4\sqrt{2}}{(\sqrt{5})^2 - (\sqrt{2})^2}$

$= \dfrac{5+\sqrt{10}+4\sqrt{5}+4\sqrt{2}}{5-2}$

$= \dfrac{5+\sqrt{10}+4\sqrt{5}+4\sqrt{2}}{3}$

### 7.6 Quick Checks

1. $f(x) = \sqrt{3x+7}$

   a. $f(3) = \sqrt{3(3)+7} = \sqrt{9+7} = \sqrt{16} = 4$

   b. $f(7) = \sqrt{3(7)+7} = \sqrt{21+7} = \sqrt{28} = 2\sqrt{7}$

2. $g(x) = \sqrt[3]{2x+7}$

   a. $g(-4) = \sqrt[3]{2(-4)+7} = \sqrt[3]{-8+7} = \sqrt[3]{-1} = -1$

   b. $g(10) = \sqrt[3]{2(10)+7} = \sqrt[3]{20+7} = \sqrt[3]{27} = 3$

3. $H(x) = \sqrt{x+6}$

   The function tells us to take the square root of $x+6$. We can only take the square root of numbers that are greater than or equal to 0. Thus, we need
   $x+6 \geq 0$
   $x \geq -6$
   The domain of $H$ is $\{x \mid x \geq -6\}$ or the interval $[-6, \infty)$.

4. $g(t) = \sqrt[5]{3t-1}$

   The function tells us to take the fifth root of $3t-1$. Since we can take the fifth root of any real number, the domain of $g$ is all real numbers $\{t \mid t \text{ is any real number}\}$ or the interval $(-\infty, \infty)$.

## Section 7.7 – Quick Checks: Radical Equations and Their Applications

5. $F(m) = \sqrt[4]{6-3m}$

    The function tells us to take the fourth root of $6-3m$. We can only take the fourth root of numbers that are greater than or equal to 0. Thus, we need
    $$6 - 3m \geq 0$$
    $$-3m \geq -6$$
    $$m \leq 2$$
    The domain of $F$ is $\{m \mid m \leq 2\}$ or the interval $(-\infty, 2]$.

6. $f(x) = \sqrt{x+3}$

    a. The function tells us to take the square root of $x+3$. We can only take the square root of numbers that are greater than or equal to 0. Thus, we need
    $$x + 3 \geq 0$$
    $$x \geq -3$$
    The domain of $f$ is $\{x \mid x \geq -3\}$ or the interval $[-3, \infty)$.

    b.
    | $x$ | $f(x) = \sqrt{x+3}$ | $(x, y)$ |
    |---|---|---|
    | $-3$ | $f(-3)\sqrt{-3+3} = 0$ | $(-3, 0)$ |
    | $-2$ | $f(-2) = \sqrt{-2+3} = 1$ | $(-2, 1)$ |
    | $1$ | $f(1) = \sqrt{1+3} = 2$ | $(1, 2)$ |
    | $6$ | $f(6) = \sqrt{6+3} = 3$ | $(6, 3)$ |

    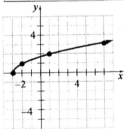

    c. Based on the graph, the range is $\{y \mid y \geq 0\}$ or the interval $[0, \infty)$.

7. $G(x) = \sqrt[3]{x} - 1$

    a. The function tells us to take the cube root of $x$ and then subtract 1. We can take the cube root of any real number so the domain is all real numbers, $\{x \mid x \text{ is any real number}\}$ or the interval $(-\infty, \infty)$.

    b.
    | $x$ | $G(x) = \sqrt[3]{x} - 1$ | $(x, y)$ |
    |---|---|---|
    | $-8$ | $G(-8) = \sqrt[3]{-8} - 1 = -3$ | $(-8, -3)$ |
    | $-1$ | $G(-1) = \sqrt[3]{-1} - 1 = -2$ | $(-1, -2)$ |
    | $0$ | $G(0) = \sqrt[3]{0} - 1 = -1$ | $(0, -1)$ |
    | $1$ | $G(1) = \sqrt[3]{1} - 1 = 0$ | $(1, 0)$ |
    | $8$ | $G(8) = \sqrt[3]{8} - 1 = 1$ | $(8, 1)$ |

    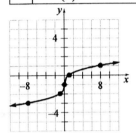

    c. Based on the graph, the range is all real numbers, $\{y \mid y \text{ is any real number}\}$ or the interval $(-\infty, \infty)$.

### 7.7 Quick Checks

1. $\sqrt{3x+1} - 4 = 0$
$$\sqrt{3x+1} = 4$$
$$\left(\sqrt{3x+1}\right)^2 = 4^2$$
$$3x + 1 = 16$$
$$3x + 1 - 1 = 16 - 1$$
$$3x = 15$$
$$\frac{3x}{3} = \frac{15}{3}$$
$$x = 5$$

Check:
$$\sqrt{3(5)+1} \stackrel{?}{=} 4$$
$$\sqrt{15+1} \stackrel{?}{=} 4$$
$$\sqrt{16} \stackrel{?}{=} 4$$
$$4 = 4 \checkmark$$

The solution set is $\{5\}$.

## Chapter 7: Radicals and Rational Exponents

2. $\sqrt{6x+4} - 5 = 3$
$\sqrt{6x+4} - 5 + 5 = 3 + 5$
$\sqrt{6x+4} = 8$
$\left(\sqrt{6x+4}\right)^2 = 8^2$
$6x + 4 = 64$
$6x + 4 - 4 = 64 - 4$
$6x = 60$
$\dfrac{6x}{6} = \dfrac{60}{6}$
$x = 10$

Check:
$\sqrt{6(10)+4} - 5 \stackrel{?}{=} 3$
$\sqrt{64} - 5 \stackrel{?}{=} 3$
$8 - 5 \stackrel{?}{=} 3$
$3 = 3 \checkmark$

The solution set is $\{10\}$.

3. $\sqrt{2x+3} + 8 = 6$
$\sqrt{2x+3} + 8 - 8 = 6 - 8$
$\sqrt{2x+3} = -2$
$\left(\sqrt{2x+3}\right)^2 = (-2)^2$
$2x + 3 = 4$
$2x + 3 - 3 = 4 - 3$
$2x = 1$
$\dfrac{2x}{2} = \dfrac{1}{2}$
$x = \dfrac{1}{2}$

Check:
$\sqrt{2\left(\dfrac{1}{2}\right)+3} + 8 \stackrel{?}{=} 6$
$\sqrt{4} + 8 \stackrel{?}{=} 6$
$2 + 8 \stackrel{?}{=} 6$
$10 \neq 6$

The solution does not check. Since there are no other possible solutions, the equation has no real solution.

4. $\sqrt{2x+1} = x - 1$
$\left(\sqrt{2x+1}\right)^2 = (x-1)^2$
$2x + 1 = x^2 - 2x + 1$
$x^2 - 4x = 0$
$x(x-4) = 0$
$x = 0$  or  $x - 4 = 0$
$\phantom{x = 0 \text{ or }}x = 4$

Check:
$\sqrt{2(0)+1} \stackrel{?}{=} 0 - 1 \qquad \sqrt{2(4)+1} \stackrel{?}{=} 4 - 1$
$\sqrt{1} \stackrel{?}{=} -1 \qquad\qquad \sqrt{9} \stackrel{?}{=} 3$
$1 \neq -1 \qquad\qquad\quad 3 = 3 \checkmark$

The solution set is $\{4\}$.

5. $\sqrt[3]{6x+3} - 4 = -1$
$\sqrt[3]{6x+3} - 4 + 4 = -1 + 4$
$\sqrt[3]{6x+3} = 3$
$\left(\sqrt[3]{6x+3}\right)^3 = 3^3$
$6x + 3 = 27$
$6x + 3 - 3 = 27 - 3$
$6x = 24$
$\dfrac{6x}{6} = \dfrac{24}{6}$
$x = 4$

Check:
$\sqrt[3]{6(4)+3} - 4 \stackrel{?}{=} -1$
$\sqrt[3]{27} - 4 \stackrel{?}{=} -1$
$3 - 4 \stackrel{?}{=} -1$
$-1 = -1 \checkmark$

The solution set is $\{4\}$.

**Section 7.7 – Quick Checks:** Radical Equations and Their Applications

6. $(2x-3)^{1/3} - 7 = -4$
$(2x-3)^{1/3} - 7 + 7 = -4 + 7$
$(2x-3)^{1/3} = 3$
$\left[(2x-3)^{1/3}\right]^3 = 3^3$
$2x - 3 = 27$
$2x - 3 + 3 = 27 + 3$
$2x = 30$
$\dfrac{2x}{2} = \dfrac{30}{2}$
$x = 15$

Check:
$(2(15)-3)^{1/3} - 7 \stackrel{?}{=} -4$
$(27)^{1/3} - 7 \stackrel{?}{=} -4$
$3 - 7 \stackrel{?}{=} -4$
$-4 = -4$ ✓
The solution set is $\{15\}$.

7. $\sqrt[3]{m^2 + 4m + 4} = \sqrt[3]{2m + 7}$
$\left(\sqrt[3]{m^2 + 4m + 4}\right)^3 = \left(\sqrt[3]{2m+7}\right)^3$
$m^2 + 4m + 4 = 2m + 7$
$m^2 + 4m + 4 - 2m - 7 = 2m + 7 - 2m - 7$
$m^2 + 2m - 3 = 0$
$(m+3)(m-1) = 0$
$m + 3 = 0$ or $m - 1 = 0$
$m = -3$ $\qquad$ $m = 1$

Check:
$\sqrt[3]{(-3)^2 + 4(-3) + 4} \stackrel{?}{=} \sqrt[3]{2(-3) + 7}$
$\sqrt[3]{9 - 12 + 4} \stackrel{?}{=} \sqrt[3]{-6 + 7}$
$\sqrt[3]{1} \stackrel{?}{=} \sqrt[3]{1}$
$1 = 1$ ✓

$\sqrt[3]{(1)^2 + 4(1) + 4} \stackrel{?}{=} \sqrt[3]{2(1) + 7}$
$\sqrt[3]{1 + 4 + 4} \stackrel{?}{=} \sqrt[3]{2 + 7}$
$\sqrt[3]{9} = \sqrt[3]{9}$ ✓

The solution set is $\{-3, 1\}$

8. $\sqrt{2x+1} - \sqrt{x+4} = 1$
$\sqrt{2x+1} = 1 + \sqrt{x+4}$
$\left(\sqrt{2x+1}\right)^2 = \left(1 + \sqrt{x+4}\right)^2$
$2x + 1 = 1 + 2\sqrt{x+4} + \left(\sqrt{x+4}\right)^2$
$2x + 1 = 1 + 2\sqrt{x+4} + x + 4$
$2x + 1 = 5 + x + 2\sqrt{x+4}$
$x - 4 = 2\sqrt{x+4}$
$(x-4)^2 = \left(2\sqrt{x+4}\right)^2$
$x^2 - 8x + 16 = 4(x+4)$
$x^2 - 8x + 16 = 4x + 16$
$x^2 - 12x = 0$
$x(x - 12) = 0$
$x = 0$ or $x - 12 = 0$
$\qquad\qquad x = 12$

Check:
$\sqrt{2(0)+1} - \sqrt{0+4} \stackrel{?}{=} 1 \qquad \sqrt{2(12)+1} - \sqrt{(12)+4} \stackrel{?}{=} 1$
$\sqrt{1} - \sqrt{4} \stackrel{?}{=} 1 \qquad\qquad \sqrt{25} - \sqrt{16} \stackrel{?}{=} 1$
$1 - 2 \stackrel{?}{=} 1 \qquad\qquad\qquad 5 - 4 \stackrel{?}{=} 1$
$-1 \neq 1 \qquad\qquad\qquad\qquad 1 = 1$ ✓

The solution set is $\{12\}$.

9. a. $T = 2\pi \sqrt{\dfrac{L}{32}}$
$\dfrac{T}{2\pi} = \sqrt{\dfrac{L}{32}}$
$\left(\dfrac{T}{2\pi}\right)^2 = \left(\sqrt{\dfrac{L}{32}}\right)^2$
$\dfrac{T^2}{4\pi^2} = \dfrac{L}{32}$
$L = \dfrac{32T^2}{4\pi^2}$
$L = \dfrac{8T^2}{\pi^2}$

b. $L = \dfrac{8T^2}{\pi^2} = \dfrac{8(2\pi)^2}{\pi^2} = \dfrac{8 \cdot 4\pi^2}{\pi^2} = 32$ feet.

## Chapter 7: Radicals and Rational Exponents

### 7.8 Quick Checks

1. $\sqrt{-36} = \sqrt{36 \cdot (-1)} = \sqrt{36} \cdot \sqrt{-1} = 6i$

2. $\sqrt{-5} = \sqrt{5 \cdot (-1)} = \sqrt{5} \cdot \sqrt{-1} = \sqrt{5}\,i$

3. $\sqrt{-12} = \sqrt{12 \cdot (-1)} = \sqrt{12} \cdot \sqrt{-1} = 2\sqrt{3}\,i$

4. $4 + \sqrt{-100} = 4 + \sqrt{100 \cdot (-1)}$
   $= 4 + \sqrt{100} \cdot \sqrt{-1}$
   $= 4 + 10i$

5. $-2 - \sqrt{-8} = -2 - \sqrt{8 \cdot (-1)}$
   $= -2 - \sqrt{8} \cdot \sqrt{-1}$
   $= -2 - 2\sqrt{2}\,i$

6. $\dfrac{6 - \sqrt{-72}}{3} = \dfrac{6 - \sqrt{72 \cdot (-1)}}{3}$
   $= \dfrac{6 - \sqrt{72} \cdot \sqrt{-1}}{3}$
   $= \dfrac{6 - 6\sqrt{2}\,i}{3}$
   $= \dfrac{3(2 - 2\sqrt{2}\,i)}{3}$
   $= 2 - 2\sqrt{2}\,i$

7. $(4 + 6i) + (-3 + 5i) = 4 + 6i - 3 + 5i$
   $= (4 - 3) + (6 + 5)i$
   $= 1 + 11i$

8. $(4 - 2i) - (-2 + 7i) = 4 - 2i + 2 - 7i$
   $= (4 + 2) + (-2 - 7)i$
   $= 6 - 9i$

9. $\left(4 - \sqrt{-4}\right) + \left(-7 + \sqrt{-9}\right) = (4 - 2i) + (-7 + 3i)$
   $= 4 - 2i - 7 + 3i$
   $= (4 - 7) + (-2 + 3)i$
   $= -3 + i$

10. $3i(5 - 4i) = 3i \cdot 5 - 3i \cdot 4i$
    $= 15i - 12i^2$
    $= 15i - 12(-1)$
    $= 12 + 15i$

11. $(-2 + 5i)(4 - 2i) = -2 \cdot 4 - 2 \cdot (-2i) + 5i \cdot 4 - 5i \cdot 2i$
    $= -8 + 4i + 20i - 10i^2$
    $= -8 + 24i - 10(-1)$
    $= 2 + 24i$

12. $\sqrt{-9} \cdot \sqrt{-36} = 3i \cdot 6i = 18i^2 = -18$

13. $\left(2 + \sqrt{-36}\right)\left(4 - \sqrt{-25}\right)$
    $= (2 + 6i)(4 - 5i)$
    $= 2 \cdot 4 - 2 \cdot 5i + 6i \cdot 4 - 6i \cdot 5i$
    $= 8 - 10i + 24i - 30i^2$
    $= 8 + 14i - 30(-1)$
    $= 38 + 14i$

14. $(3 - 8i)(3 + 8i) = 3^2 - (8i)^2$
    $= 9 - 64i^2$
    $= 9 - 64(-1)$
    $= 73$

15. $(-2 + 5i)(-2 - 5i) = (-2)^2 - (5i)^2$
    $= 4 - 25i^2$
    $= 4 - 25(-1)$
    $= 29$

16. $\dfrac{-4 + i}{3i} = \dfrac{-4 + i}{3i} \cdot \dfrac{3i}{3i} = \dfrac{-12i + 3i^2}{9i^2}$
    $= \dfrac{-12i - 3}{-9} = \dfrac{-3}{-9} + \dfrac{-12}{-9}i$
    $= \dfrac{1}{3} + \dfrac{4}{3}i$

**Section 7.8 – Quick Checks:** The Complex Number System

17. $\dfrac{4+3i}{1-3i} = \dfrac{4+3i}{1-3i} \cdot \dfrac{1+3i}{1+3i}$

$= \dfrac{(4+3i)(1+3i)}{(1-3i)(1+3i)}$

$= \dfrac{4+12i+3i+9i^2}{1-9i^2}$

$= \dfrac{4+15i-9}{1+9}$

$= \dfrac{-5+15i}{10}$

$= -\dfrac{5}{10} + \dfrac{15}{10}i$

$= -\dfrac{1}{2} + \dfrac{3}{2}i$

18. $i^{43} = i^{40} \cdot i^3$

$= \left(i^4\right)^{10} \cdot i^3$

$= (1)^{10} \cdot (-i)$

$= -i$

19. $i^{98} = i^{96} \cdot i^2$

$= \left(i^4\right)^{24} \cdot i^2$

$= (1)^{24} \cdot (-1)$

$= -1$

# Chapter 8 – Quick Checks

**8.1 Quick Checks**

1.  $p^2 = 48$
    $p = \pm\sqrt{48}$
    $p = \pm 4\sqrt{3}$
    The solution set is $\{-4\sqrt{3},\ 4\sqrt{3}\}$.

2.  $3b^2 = 75$
    $b^2 = 25$
    $b = \pm\sqrt{25}$
    $b = \pm 5$
    The solution set is $\{-5,\ 5\}$.

3.  $s^2 - 81 = 0$
    $s^2 = 81$
    $s = \pm\sqrt{81}$
    $s = \pm 9$
    The solution set is $\{-9,\ 9\}$.

4.  $d^2 = -72$
    $d = \pm\sqrt{-72}$
    $d = \pm 6\sqrt{2}\,i$
    The solution set is $\{-6\sqrt{2}\,i,\ 6\sqrt{2}\,i\}$.

5.  $3q^2 + 27 = 0$
    $3q^2 = -27$
    $q^2 = -9$
    $q = \pm\sqrt{-9}$
    $q = \pm 3i$
    The solution set is $\{-3i,\ 3i\}$.

6.  $(y+3)^2 = 100$
    $y+3 = \pm\sqrt{100}$
    $y+3 = \pm 10$
    $y = -3 \pm 10$
    $y = -3-10$ or $y = -3+10$
    $y = -13$ or $y = 7$
    The solution set is $\{-13,\ 7\}$.

7.  $(q-5)^2 + 20 = 4$
    $(q-5)^2 = -16$
    $q-5 = \pm\sqrt{-16}$
    $q-5 = \pm 4i$
    $q = 5 \pm 4i$
    The solution set is $\{5-4i,\ 5+4i\}$.

8.  Start: $p^2 + 14p$
    Add: $\left(\frac{1}{2} \cdot 14\right)^2 = 49$
    Result: $p^2 + 14p + 49$
    Factored Form: $(p+7)^2$

9.  Start: $w^2 + 3w$
    Add: $\left(\frac{1}{2} \cdot 3\right)^2 = \frac{9}{4}$
    Result: $w^2 + 3w + \frac{9}{4}$
    Factored Form: $\left(w + \frac{3}{2}\right)^2$

10. $b^2 + 2b - 8 = 0$
    $b^2 + 2b = 8$
    $b^2 + 2b + \left(\frac{1}{2} \cdot 2\right)^2 = 8 + \left(\frac{1}{2} \cdot 2\right)^2$
    $b^2 + 2b + 1 = 8 + 1$
    $(b+1)^2 = 9$
    $b+1 = \pm\sqrt{9}$
    $b+1 = \pm 3$
    $b = -1 \pm 3$
    $b = -4$ or $b = 2$
    The solution set is $\{-4,\ 2\}$.

**Section 8.2 – Quick Checks:** Solving Quadratic Equations by the Quadratic Formula

11.
$$2q^2 + 6q - 1 = 0$$
$$\frac{2q^2 + 6q - 1}{2} = \frac{0}{2}$$
$$q^2 + 3q - \frac{1}{2} = 0$$
$$q^2 + 3q = \frac{1}{2}$$
$$q^2 + 3q + \left[\frac{1}{2} \cdot 3\right]^2 = \frac{1}{2} + \left[\frac{1}{2} \cdot 3\right]^2$$
$$q^2 + 3q + \frac{9}{4} = \frac{1}{2} + \frac{9}{4}$$
$$\left(q + \frac{3}{2}\right)^2 = \frac{11}{4}$$
$$q + \frac{3}{2} = \pm\sqrt{\frac{11}{4}}$$
$$q + \frac{3}{2} = \pm\frac{\sqrt{11}}{2}$$
$$q = -\frac{3}{2} \pm \frac{\sqrt{11}}{2}$$

The solution set is $\left\{\frac{-3}{2} - \frac{\sqrt{11}}{2}, \frac{-3}{2} + \frac{\sqrt{11}}{2}\right\}$, or $\left\{\frac{-3 - \sqrt{11}}{2}, \frac{-3 + \sqrt{11}}{2}\right\}$.

12.
$$c^2 = a^2 + b^2$$
$$c^2 = 3^2 + 4^2$$
$$= 9 + 16$$
$$= 25$$
$$c = \sqrt{25}$$
$$= 5$$
The hypotenuse is 5 units long.

13.
$$c^2 = a^2 + b^2$$
$$c^2 = 6^2 + 6^2$$
$$= 36 + 36$$
$$= 72$$
$$c = \sqrt{72}$$
$$= 6\sqrt{2}$$
The hypotenuse is $6\sqrt{2}$ units long.

14. Let $d$ represent the unknown distance.

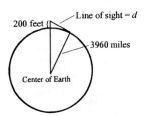

Since the line of sight and the two lines drawn from the center of Earth form a right triangle, we know Hypotenuse$^2$ = Leg$^2$ + Leg$^2$. We also know 200 feet = $\frac{200}{5280}$ mile, so the hypotenuse is $\left(3960 + \frac{200}{5280}\right)$ miles. Since one leg is the line of sight, $d$, and the other leg is 3960 miles, we have that

$$d^2 + 3960^2 = \left(3960 + \frac{200}{5280}\right)^2$$
$$d^2 = \left(3960 + \frac{200}{5280}\right)^2 - 3960^2$$
$$d^2 \approx 300.001435$$
$$d \approx \sqrt{300.001435}$$
$$\approx 17.32$$

The sailor can see approximately 17.32 miles.

**8.2 Quick Checks**

1. $2x^2 - 3x - 9 = 0$
For this equation, $a = 2$, $b = -3$, and $c = -9$.
$$x = \frac{-(-3) \pm \sqrt{(-3)^2 - 4(2)(-9)}}{2(2)}$$
$$= \frac{3 \pm \sqrt{9 + 72}}{4}$$
$$= \frac{3 \pm \sqrt{81}}{4}$$
$$= \frac{3 \pm 9}{4}$$
$$x = \frac{3 - 9}{4} \text{ or } x = \frac{3 + 9}{4}$$
$$= \frac{-6}{4} \text{ or } = \frac{12}{4}$$
$$= -\frac{3}{2} \text{ or } = 3$$

The solution set is $\left\{-\frac{3}{2}, 3\right\}$.

## Chapter 8: Quadratic Equations and Functions

2. $2x^2 + 7x = 4$
   $2x^2 + 7x - 4 = 0$
   For this equation, $a = 2$, $b = 7$, and $c = -4$.
   $$x = \frac{-7 \pm \sqrt{7^2 - 4(2)(-4)}}{2(2)}$$
   $$= \frac{-7 \pm \sqrt{49 + 32}}{4}$$
   $$= \frac{-7 \pm \sqrt{81}}{4}$$
   $$= \frac{-7 \pm 9}{4}$$
   $x = \frac{-7-9}{4}$ or $x = \frac{-7+9}{4}$
   $= \frac{-16}{4}$ or $= \frac{2}{4}$
   $= -4$ or $= \frac{1}{2}$
   The solution set is $\left\{-4, \frac{1}{2}\right\}$.

3. $4z^2 + 1 = 8z$
   $4z^2 - 8z + 1 = 0$
   For this equation, $a = 4$, $b = -8$, and $c = 1$.
   $$z = \frac{-(-8) \pm \sqrt{(-8)^2 - 4(4)(1)}}{2(4)}$$
   $$= \frac{8 \pm \sqrt{64 - 16}}{8}$$
   $$= \frac{8 \pm \sqrt{48}}{8}$$
   $$= \frac{8 \pm 4\sqrt{3}}{8}$$
   $$= \frac{8}{8} \pm \frac{4\sqrt{3}}{8}$$
   $$= 1 \pm \frac{\sqrt{3}}{2}$$
   The solution set is $\left\{1 - \frac{\sqrt{3}}{2}, 1 + \frac{\sqrt{3}}{2}\right\}$.

4. $4w + \frac{25}{w} = 20$
   $w\left(4w + \frac{25}{w}\right) = w(20)$
   $4w^2 + 25 = 20w$
   $4w^2 - 20w + 25 = 0$

For this equation, $a = 4$, $b = -20$, and $c = 25$.
$$w = \frac{-(-20) \pm \sqrt{(-20)^2 - 4(4)(25)}}{2(4)}$$
$$= \frac{20 \pm \sqrt{400 - 400}}{8}$$
$$= \frac{20 \pm \sqrt{0}}{8}$$
$$= \frac{20}{8}$$
$$= \frac{5}{2}$$
The solution set is $\left\{\frac{5}{2}\right\}$.

5. $z^2 + 2z + 26 = 0$
   For this equation, $a = 1$, $b = 2$, and $c = 26$.
   $$z = \frac{-2 \pm \sqrt{2^2 - 4(1)(26)}}{2(1)}$$
   $$= \frac{-2 \pm \sqrt{4 - 104}}{2}$$
   $$= \frac{-2 \pm \sqrt{-100}}{2}$$
   $$= \frac{-2 \pm 10i}{2}$$
   $$= \frac{-2}{2} \pm \frac{10i}{2}$$
   $$= -1 \pm 5i$$
   The solution set is $\{-1 - 5i, -1 + 5i\}$.

6. $2z^2 + 5z + 4 = 0$
   For this equation, $a = 2$, $b = 5$, and $c = 4$.
   $b^2 - 4ac = 5^2 - 4(2)(4) = 25 - 32 = -7$
   Because $b^2 - 4ac = -7$ is negative, the quadratic equation will have two complex solutions that are not real. The solutions will be complex conjugates of each other.

7. $4y^2 + 12y = -9$
   $4y^2 + 12y + 9 = 0$
   For this equation, $a = 4$, $b = 12$, and $c = 9$.
   $b^2 - 4ac = 12^2 - 4(4)(9) = 144 - 144 = 0$
   Because $b^2 - 4ac = 0$, the quadratic equation will have one repeated real solution.

**Section 8.2 – Quick Checks:** Solving Quadratic Equations by the Quadratic Formula

8. $2x^2 - 4x + 1 = 0$
   For this equation, $a = 2$, $b = -4$, and $c = 1$.
   $b^2 - 4ac = (-4)^2 - 4(2)(1) = 16 - 8 = 8$
   Because $b^2 - 4ac = 8$ is positive, but not a perfect square, the quadratic equation will have two irrational solutions.

9. $5n^2 - 45 = 0$
   Because this equation has no linear term, solve by using the square root method.
   $5n^2 = 45$
   $n^2 = 9$
   $n = \pm\sqrt{9}$
   $= \pm 3$
   The solution set is $\{-3, 3\}$.

10. $-2y^2 + 5y - 6 = 0$
    $2y^2 - 5y + 6 = 0$
    Because this equation does not easily factor, solve by using the quadratic formula. For this equation, $a = 2$, $b = -5$, and $c = 6$.
    $y = \dfrac{-(-5) \pm \sqrt{(-5)^2 - 4(2)(6)}}{2(2)}$
    $= \dfrac{5 \pm \sqrt{25 - 48}}{4}$
    $= \dfrac{5 \pm \sqrt{-23}}{4}$
    $= \dfrac{5 \pm \sqrt{23}\, i}{4}$
    The solution set is $\left\{\dfrac{5 - \sqrt{23}\,i}{4},\ \dfrac{5 + \sqrt{23}\,i}{4}\right\}$ or
    $\left\{\dfrac{5}{4} - \dfrac{\sqrt{23}}{4}i,\ \dfrac{5}{4} + \dfrac{\sqrt{23}}{4}i\right\}$.

11. $3w^2 + 2w = 5$
    $3w^2 + 2w - 5 = 0$
    Because this equation factors easily, solve by factoring.
    $(3w + 5)(w - 1) = 0$
    $3w + 5 = 0$   or   $w - 1 = 0$
    $3w = -5$   or   $w = 1$
    $w = -\dfrac{5}{3}$
    The solution set is $\left\{-\dfrac{5}{3}, 1\right\}$.

12. a. $R(x) = 600$
    $-0.005x^2 + 4x = 600$
    $-0.005x^2 + 4x - 600 = 0$
    For this equation, $a = -0.005$, $b = 4$, and $c = -600$.
    $x = \dfrac{-4 \pm \sqrt{4^2 - 4(-0.005)(-600)}}{2(-0.005)}$
    $= \dfrac{-4 \pm \sqrt{4}}{-0.01}$
    $= \dfrac{-4 \pm 2}{-0.01}$
    $x = \dfrac{-6}{-0.01}$   or   $x = \dfrac{-2}{-0.01}$
    $= 600$   or   $= 200$
    If revenue is to be $600 per day, then either 200 or 600 DVDs must be rented.

    b. $R(x) = 800$
    $-0.005x^2 + 4x = 800$
    $-0.005x^2 + 4x - 800 = 0$
    For this equation, $a = -0.005$, $b = 4$, and $c = -800$.
    $x = \dfrac{-4 \pm \sqrt{4^2 - 4(-0.005)(-800)}}{2(-0.005)}$
    $= \dfrac{-4 \pm \sqrt{0}}{-0.01}$
    $= \dfrac{-4 \pm 0}{-0.01}$
    $x = \dfrac{-4}{-0.01} = 400$
    If revenue is to be $800 per day, then 400 DVDs must be rented.

13. Let $w$ represent the width of the rectangle. Then $w + 14$ will represent the length.
    $w^2 + (w + 14)^2 = 34^2$
    $w^2 + w^2 + 28w + 196 = 1156$
    $2w^2 + 28w - 960 = 0$
    $w^2 + 14w - 480 = 0$
    $(w - 16)(w + 30) = 0$
    $w - 16 = 0$   or   $w + 30 = 0$
    $w = 16$   or   $w = -30$
    We disregard $w = -30$ because $w$ represents the width of the rectangle, which must be positive. Thus, $w = 16$ is the only viable answer. Now, $w + 14 = 16 + 14 = 30$. Thus, the dimensions of the rectangle are 16 meters by 30 meters.

## Chapter 8: Quadratic Equations and Functions

### 8.3 Quick Checks

1. $x^4 - 13x^2 + 36 = 0$
$(x^2)^2 - 13(x^2) + 36 = 0$
Let $u = x^2$.
$u^2 - 13u + 36 = 0$
$(u-4)(u-9) = 0$
$u - 4 = 0$ or $u - 9 = 0$
$u = 4$ or $u = 9$
$x^2 = 4$ or $x^2 = 9$
$x = \pm\sqrt{4}$ or $x = \pm\sqrt{9}$
$x = \pm 2$ or $x = \pm 3$

Check:
$x = -2$: $(-2)^4 - 13(-2)^2 + 36 \stackrel{?}{=} 0$
$16 - 13 \cdot 4 + 36 \stackrel{?}{=} 0$
$16 - 52 + 36 \stackrel{?}{=} 0$
$0 = 0$ ✓

$x = 2$: $2^4 - 13(2)^2 + 36 \stackrel{?}{=} 0$
$16 - 13 \cdot 4 + 36 \stackrel{?}{=} 0$
$16 - 52 + 36 \stackrel{?}{=} 0$
$0 = 0$ ✓

$x = -3$: $(-3)^4 - 13(-3)^2 + 36 \stackrel{?}{=} 0$
$81 - 13 \cdot 9 + 36 \stackrel{?}{=} 0$
$81 - 117 + 36 \stackrel{?}{=} 0$
$0 = 0$ ✓

$x = 3$: $3^4 - 13(3)^2 + 36 \stackrel{?}{=} 0$
$81 - 13 \cdot 9 + 36 \stackrel{?}{=} 0$
$81 - 117 + 36 \stackrel{?}{=} 0$
$0 = 0$ ✓

All check; the solution set is $\{-3, -2, 2, 3\}$.

2. $p^4 - 7p^2 = 18$
$p^4 - 7p^2 - 18 = 0$
$(p^2)^2 - 7(p^2) - 18 = 0$
Let $u = p^2$.
$u^2 - 7u - 18 = 0$
$(u+2)(u-9) = 0$
$u + 2 = 0$ or $u - 9 = 0$
$u = -2$ or $u = 9$
$p^2 = -2$ or $p^2 = 9$
$p = \pm\sqrt{-2}$ or $p = \pm\sqrt{9}$
$p = \pm\sqrt{2}\,i$ or $p = \pm 3$

Check:
$p = -\sqrt{2}\,i$: $(\sqrt{2}\,i)^4 - 7(\sqrt{2}\,i)^2 \stackrel{?}{=} 18$
$4i^4 - 7 \cdot 2i^2 \stackrel{?}{=} 18$
$4(1) - 7 \cdot 2(-1) \stackrel{?}{=} 18$
$4 - 7 \cdot 2(-1) \stackrel{?}{=} 18$
$4 + 14 \stackrel{?}{=} 18$
$18 = 18$ ✓

$p = \sqrt{2}\,i$: $(-\sqrt{2}\,i)^4 - 7(-\sqrt{2}\,i)^2 \stackrel{?}{=} 18$
$4i^4 - 7 \cdot 2i^2 \stackrel{?}{=} 18$
$4(1) - 7 \cdot 2(-1) \stackrel{?}{=} 18$
$4 - 7 \cdot 2(-1) \stackrel{?}{=} 18$
$4 + 14 \stackrel{?}{=} 18$
$18 = 18$ ✓

$p = -3$: $(-3)^4 - 7(-3)^2 \stackrel{?}{=} 18$
$81 - 7 \cdot 9 \stackrel{?}{=} 18$
$81 - 63 \stackrel{?}{=} 18$
$18 = 18$ ✓

$p = 3$: $3^4 - 7(3)^2 \stackrel{?}{=} 18$
$81 - 7 \cdot 9 \stackrel{?}{=} 18$
$81 - 63 \stackrel{?}{=} 18$
$18 = 18$ ✓

All check; the solution set is $\{-\sqrt{2}\,i,\ \sqrt{2}\,i,\ -3,\ 3\}$.

**Section 8.3 – Quick Checks:** *Solving Equations Quadratic in Form*

3. $(p^2-2)^2 - 9(p^2-2) + 14 = 0$

   Let $u = p^2 - 2$.
   $$u^2 - 9u + 14 = 0$$
   $$(u-2)(u-7) = 0$$
   $u - 2 = 0$    or    $u - 7 = 0$
   $u = 2$    or    $u = 7$
   $p^2 - 2 = 2$    or    $p^2 - 2 = 7$
   $p^2 = 4$    or    $p^2 = 9$
   $p = \pm\sqrt{4}$    or    $p = \pm\sqrt{9}$
   $p = \pm 2$    or    $p = \pm 3$

   Check:
   $p = -2$: $((-2)^2 - 2)^2 - 9((-2)^2 - 2) + 14 \stackrel{?}{=} 0$
   $(4-2)^2 - 9(4-2) + 14 \stackrel{?}{=} 0$
   $2^2 - 9 \cdot 2 + 14 \stackrel{?}{=} 0$
   $4 - 18 + 14 \stackrel{?}{=} 0$
   $0 = 0$ ✓

   $p = 2$: $(2^2 - 2)^2 - 9(2^2 - 2) + 14 \stackrel{?}{=} 0$
   $(4-2)^2 - 9(4-2) + 14 \stackrel{?}{=} 0$
   $2^2 - 9 \cdot 2 + 14 \stackrel{?}{=} 0$
   $4 - 18 + 14 \stackrel{?}{=} 0$
   $0 = 0$ ✓

   $p = -3$: $((-3)^2 - 2)^2 - 9((-3)^2 - 2) + 14 \stackrel{?}{=} 0$
   $(9-2)^2 - 9(9-2) + 14 \stackrel{?}{=} 0$
   $7^2 - 9 \cdot 7 + 14 \stackrel{?}{=} 0$
   $49 - 63 + 14 \stackrel{?}{=} 0$
   $0 = 0$ ✓

   $p = 3$: $(3^2 - 2)^2 - 9(3^2 - 2) + 14 \stackrel{?}{=} 0$
   $(9-2)^2 - 9(9-2) + 14 \stackrel{?}{=} 0$
   $7^2 - 9 \cdot 7 + 14 \stackrel{?}{=} 0$
   $49 - 63 + 14 \stackrel{?}{=} 0$
   $0 = 0$ ✓

   All check; the solution set is $\{-3, -2, 2, 3\}$.

4. $2(2z^2-1)^2 + 5(2z^2-1) - 3 = 0$

   Let $u = 2z^2 - 1$.
   $$2u^2 + 5u - 3 = 0$$
   $$(2u-1)(u+3) = 0$$
   $2u - 1 = 0$    or    $u + 3 = 0$
   $u = \dfrac{1}{2}$    or    $u = -3$
   $2z^2 - 1 = \dfrac{1}{2}$    or    $2z^2 - 1 = -3$
   $2z^2 = \dfrac{3}{2}$    or    $2z^2 = -2$
   $z^2 = \dfrac{3}{4}$    or    $z^2 = -1$
   $z = \pm\sqrt{\dfrac{3}{4}}$    or    $z = \pm\sqrt{-1}$
   $z = \pm\dfrac{\sqrt{3}}{2}$    or    $z = \pm i$

   Check:
   $z = -\dfrac{\sqrt{3}}{2}$: $2\left(2\left(-\dfrac{\sqrt{3}}{2}\right)^2 - 1\right)^2 + 5\left(2\left(-\dfrac{\sqrt{3}}{2}\right)^2 - 1\right) - 3 \stackrel{?}{=} 0$
   $2\left(2\left(\dfrac{3}{4}\right) - 1\right)^2 + 5\left(2\left(\dfrac{3}{4}\right) - 1\right) - 3 \stackrel{?}{=} 0$
   $2\left(\dfrac{3}{2} - 1\right)^2 + 5\left(\dfrac{3}{2} - 1\right) - 3 \stackrel{?}{=} 0$
   $2\left(\dfrac{1}{2}\right)^2 + 5\left(\dfrac{1}{2}\right) - 3 \stackrel{?}{=} 0$
   $2\left(\dfrac{1}{4}\right) + 5\left(\dfrac{1}{2}\right) - 3 \stackrel{?}{=} 0$
   $\dfrac{1}{2} + \dfrac{5}{2} - 3 \stackrel{?}{=} 0$
   $0 = 0$ ✓

   $z = \dfrac{\sqrt{3}}{2}$: $2\left(2\left(\dfrac{\sqrt{3}}{2}\right)^2 - 1\right)^2 + 5\left(2\left(\dfrac{\sqrt{3}}{2}\right)^2 - 1\right) - 3 \stackrel{?}{=} 0$
   $2\left(2\left(\dfrac{3}{4}\right) - 1\right)^2 + 5\left(2\left(\dfrac{3}{4}\right) - 1\right) - 3 \stackrel{?}{=} 0$
   $2\left(\dfrac{3}{2} - 1\right)^2 + 5\left(\dfrac{3}{2} - 1\right) - 3 \stackrel{?}{=} 0$
   $2\left(\dfrac{1}{2}\right)^2 + 5\left(\dfrac{1}{2}\right) - 3 \stackrel{?}{=} 0$
   $2\left(\dfrac{1}{4}\right) + 5\left(\dfrac{1}{2}\right) - 3 \stackrel{?}{=} 0$
   $\dfrac{1}{2} + \dfrac{5}{2} - 3 \stackrel{?}{=} 0$
   $0 = 0$ ✓

## Chapter 8: Quadratic Equations and Functions

$z = -i$: $2\left(2(-i)^2 - 1\right)^2 + 5\left(2(-i)^2 - 1\right) - 3 \stackrel{?}{=} 0$

$2\left(2i^2 - 1\right)^2 + 5\left(2i^2 - 1\right) - 3 \stackrel{?}{=} 0$

$2\left(2(-1) - 1\right)^2 + 5\left(2(-1) - 1\right) - 3 \stackrel{?}{=} 0$

$2(-2 - 1)^2 + 5(-2 - 1) - 3 \stackrel{?}{=} 0$

$2(-3)^2 + 5(-3) - 3 \stackrel{?}{=} 0$

$2(9) + 5(-3) - 3 \stackrel{?}{=} 0$

$18 - 15 - 3 \stackrel{?}{=} 0$

$0 = 0$ ✓

$z = i$: $2\left(2i^2 - 1\right)^2 + 5\left(2i^2 - 1\right) - 3 \stackrel{?}{=} 0$

$2\left(2(-1) - 1\right)^2 + 5\left(2(-1) - 1\right) - 3 \stackrel{?}{=} 0$

$2(-2 - 1)^2 + 5(-2 - 1) - 3 \stackrel{?}{=} 0$

$2(-3)^2 + 5(-3) - 3 \stackrel{?}{=} 0$

$2(9) + 5(-3) - 3 \stackrel{?}{=} 0$

$18 - 15 - 3 \stackrel{?}{=} 0$

$0 = 0$ ✓

All check; the solution set is $\left\{-\dfrac{\sqrt{3}}{2}, \dfrac{\sqrt{3}}{2}, -i, i\right\}$.

5. $\quad 3w - 14\sqrt{w} + 8 = 0$

$3\left(\sqrt{w}\right)^2 - 14\left(\sqrt{w}\right) + 8 = 0$

Let $u = \sqrt{w}$.

$3u^2 - 14u + 8 = 0$

$(3u - 2)(u - 4) = 0$

$3u - 2 = 0 \quad$ or $\quad u - 4 = 0$

$u = \dfrac{2}{3} \quad$ or $\quad u = 4$

$\sqrt{w} = \dfrac{2}{3} \quad$ or $\quad \sqrt{w} = 4$

$w = \left(\dfrac{2}{3}\right)^2 \quad$ or $\quad w = 4^2$

$w = \dfrac{4}{9} \quad$ or $\quad w = 16$

Check:

$w = \dfrac{4}{9}$: $3\left(\dfrac{4}{9}\right) - 14\sqrt{\dfrac{4}{9}} + 8 \stackrel{?}{=} 0$

$\dfrac{4}{3} - 14 \cdot \dfrac{2}{3} + 8 \stackrel{?}{=} 0$

$\dfrac{4}{3} - \dfrac{28}{3} + 8 \stackrel{?}{=} 0$

$0 = 0$ ✓

$w = 16$: $3(16) - 14\sqrt{16} + 8 \stackrel{?}{=} 0$

$48 - 14 \cdot 4 + 8 \stackrel{?}{=} 0$

$48 - 56 + 8 \stackrel{?}{=} 0$

$0 = 0$ ✓

Both check; the solution set is $\left\{\dfrac{4}{9}, 16\right\}$.

6. $\quad 2q - 9\sqrt{q} - 5 = 0$

$2\left(\sqrt{q}\right)^2 - 9\left(\sqrt{q}\right) - 5 = 0$

Let $u = \sqrt{q}$.

$2u^2 - 9u - 5 = 0$

$(2u + 1)(u - 5) = 0$

$2u + 1 = 0 \quad$ or $\quad u - 5 = 0$

$u = -\dfrac{1}{2} \quad$ or $\quad u = 5$

$\sqrt{q} = -\dfrac{1}{2} \quad$ or $\quad \sqrt{q} = 5$

$q = \left(-\dfrac{1}{2}\right)^2 \quad$ or $\quad q = 5^2$

$q = \dfrac{1}{4} \quad$ or $\quad q = 25$

Check:

$q = \dfrac{1}{4}$: $2\left(\dfrac{1}{4}\right) - 9\sqrt{\dfrac{1}{4}} - 5 \stackrel{?}{=} 0$

$\dfrac{1}{2} - 9 \cdot \dfrac{1}{2} - 5 \stackrel{?}{=} 0$

$\dfrac{1}{2} - \dfrac{9}{2} - 5 \stackrel{?}{=} 0$

$-9 \neq 0$ ✗

$q = 25$: $2(25) - 9\sqrt{25} - 5 \stackrel{?}{=} 0$

$50 - 9 \cdot 5 - 5 \stackrel{?}{=} 0$

$50 - 45 - 5 \stackrel{?}{=} 0$

$0 = 0$ ✓

$q = \dfrac{1}{4}$ does not check; the solution set is $\{25\}$.

## Section 8.4 – Quick Checks: Graphing Quadratic Functions Using Transformations

**7.** $5x^{-2} + 12x^{-1} + 4 = 0$
$5(x^{-1})^2 + 12(x^{-1}) + 4 = 0$
Let $u = x^{-1}$.
$5u^2 + 12u + 4 = 0$
$(5u+2)(u+2) = 0$
$5u + 2 = 0 \quad \text{or} \quad u + 2 = 0$
$u = -\dfrac{2}{5} \quad \text{or} \quad u = -2$
$x^{-1} = -\dfrac{2}{5} \quad \text{or} \quad x^{-1} = -2$
$\dfrac{1}{x} = -\dfrac{2}{5} \quad \text{or} \quad \dfrac{1}{x} = -2$
$x = -\dfrac{5}{2} \quad \text{or} \quad x = -\dfrac{1}{2}$

Check:
$x = -\dfrac{5}{2}: \; 5\left(-\dfrac{5}{2}\right)^{-2} + 12\left(-\dfrac{5}{2}\right)^{-1} + 4 \stackrel{?}{=} 0$
$5\left(-\dfrac{2}{5}\right)^{2} + 12\left(-\dfrac{2}{5}\right) + 4 \stackrel{?}{=} 0$
$5\left(\dfrac{4}{25}\right) + 12\left(-\dfrac{2}{5}\right) + 4 \stackrel{?}{=} 0$
$\dfrac{4}{5} - \dfrac{24}{5} + 4 \stackrel{?}{=} 0$
$0 = 0 \; \checkmark$

$x = -\dfrac{1}{2}: \; 5\left(-\dfrac{1}{2}\right)^{-2} + 12\left(-\dfrac{1}{2}\right)^{-1} + 4 \stackrel{?}{=} 0$
$5(-2)^2 + 12(-2) + 4 \stackrel{?}{=} 0$
$5(4) + 12(-2) + 4 \stackrel{?}{=} 0$
$20 - 24 + 4 \stackrel{?}{=} 0$
$0 = 0 \; \checkmark$

Both check; the solution set is $\left\{-\dfrac{5}{2}, -\dfrac{1}{2}\right\}$.

**8.** $p^{2/3} - 4p^{1/3} - 5 = 0$
$\left(p^{1/3}\right)^2 - 4\left(p^{1/3}\right) - 5 = 0$
Let $u = p^{1/3}$.
$u^2 - 4u - 5 = 0$
$(u+1)(u-5) = 0$
$u + 1 = 0 \quad \text{or} \quad u - 5 = 0$
$u = -1 \quad \text{or} \quad u = 5$

$p^{1/3} = -1 \quad \text{or} \quad p^{1/3} = 5$
$\left(p^{1/3}\right)^3 = (-1)^3 \quad \text{or} \quad \left(p^{1/3}\right)^3 = 5^3$
$p = -1 \quad \text{or} \quad p = 125$

Check:
$p = -1: \; (-1)^{2/3} - 4 \cdot (-1)^{1/3} - 5 \stackrel{?}{=} 0$
$\left(\sqrt[3]{-1}\right)^2 - 4\left(\sqrt[3]{-1}\right) - 5 \stackrel{?}{=} 0$
$(-1)^2 - 4(-1) - 5 \stackrel{?}{=} 0$
$1 + 4 - 5 \stackrel{?}{=} 0$
$0 = 0 \; \checkmark$

$p = 125: \; (125)^{2/3} - 4 \cdot (125)^{1/3} - 5 \stackrel{?}{=} 0$
$\left(\sqrt[3]{125}\right)^2 - 4\left(\sqrt[3]{125}\right) - 5 \stackrel{?}{=} 0$
$(5)^2 - 4(5) - 5 \stackrel{?}{=} 0$
$25 - 20 - 5 \stackrel{?}{=} 0$
$0 = 0 \; \checkmark$

Both check; the solution set is $\{-1, 125\}$.

### 8.4 Quick Checks

**1.** Begin with the graph of $y = x^2$, then shift the graph up 5 unit to obtain the graph of $f(x) = x^2 + 5$.

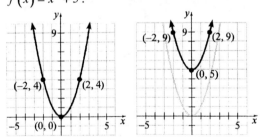

**2.** Begin with the graph of $y = x^2$, then shift the graph down 2 units to obtain the graph of $f(x) = x^2 - 2$.

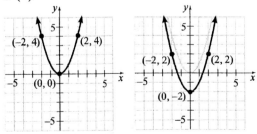

## Chapter 8: Quadratic Equations and Functions

3. Begin with the graph of $y = x^2$, then shift the graph to the left 5 units to obtain the graph of $f(x) = (x+5)^2$.

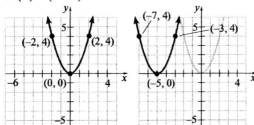

4. Begin with the graph of $y = x^2$, then shift the graph to the right 1 unit to obtain the graph of $f(x) = (x-1)^2$.

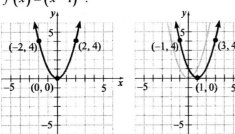

5. Begin with the graph of $y = x^2$, then shift the graph 3 units to the right to obtain the graph of $y = (x-3)^2$. Shift this graph up 2 units to obtain the graph of $f(x) = (x-3)^2 + 2$.

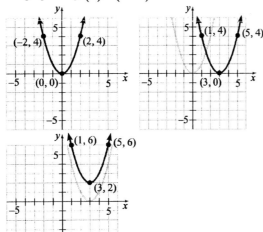

6. Begin with the graph of $y = x^2$, then shift the graph 1 unit to the left to obtain the graph of $y = (x+1)^2$. Shift this graph down 4 units to obtain the graph of $f(x) = (x+1)^2 - 4$.

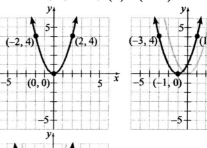

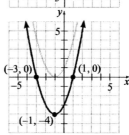

7. Begin with the graph of $y = x^2$, then vertically stretch the graph by a factor of 3 (multiply each $y$-coordinate by 3) to obtain the graph of $f(x) = 3x^2$.

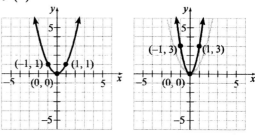

8. Begin with the graph of $y = x^2$, then multiply the $y$-coordinates by $-\dfrac{1}{4}$ to obtain the graph of $f(x) = -\dfrac{1}{4}x^2$.

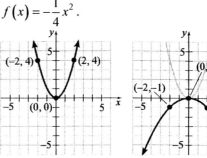

## Section 8.5 – Quick Checks: Graphing Quadratic Functions Using Properties

9. Begin with the graph of $y = x^2$, then shift the graph 2 units to the left to obtain the graph of $y = (x+2)^2$. Multiply the y-coordinates by $-3$ to obtain the graph of $y = -3(x+2)^2$. Lastly, shift the graph up 1 unit to obtain the graph of $f(x) = -3(x+2)^2 + 1$.

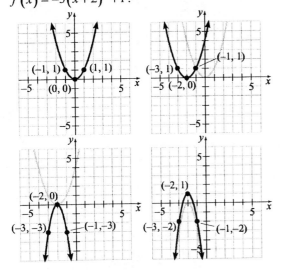

The domain is $\{x \mid x \text{ is and real number}\}$ or, using interval notation, $(-\infty, \infty)$. The range is $\{y \mid y \leq 1\}$ or using interval notation $(-\infty, 1]$.

10. Use completing the square to write the function in the form $y = a(x-h)^2 + k$.

$$f(x) = 2x^2 - 8x + 5$$
$$= (2x^2 - 8x) + 5$$
$$= 2(x^2 - 4x) + 5$$
$$= 2(x^2 - 4x + 4) + 5 - 8$$
$$= 2(x-2)^2 - 3$$

Begin with the graph of $y = x^2$, then shift the graph right 2 units to obtain the graph of $y = (x-2)^2$. Vertically stretch this graph by a factor of 2 (multiply the y-coordinates by 2) to obtain the graph of $y = 2(x-2)^2$. Lastly, shift the graph down 3 units to obtain the graph of $f(x) = 2(x-2)^2 - 3$.

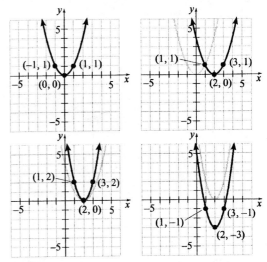

The domain is $\{x \mid x \text{ is and real number}\}$ or, using interval notation, $(-\infty, \infty)$. The range is $\{y \mid y \geq -3\}$ or, using interval notation, $[-3, \infty)$.

11. Consider the form $f(x) = a(x-h)^2 + k$. From the graph, we know that the vertex is $(-1, 2)$. So $h = -1$ and $k = 2$, and we have that

$$f(x) = a(x-(-1))^2 + 2$$
$$f(x) = a(x+1)^2 + 2$$

The graph also passes through the point $(0, 1)$ which means that $f(0) = 1$. Substituting these values into the functon, we can solve for $a$:

$$f(x) = a(x+1)^2 + 2$$
$$1 = a(0+1)^2 + 2$$
$$1 = a(1)^2 + 2$$
$$1 = a + 2$$
$$-1 = a$$

The quadratic function is $f(x) = -(x+1)^2 + 2$.

### 8.5 Quick Checks

1. For the function $f(x) = x^2 - 4x - 12$, we see that $a = 1$, $b = -4$, and $c = -12$. The parabola opens up because $a = 1 > 0$. The x-coordinate of the vertex is $x = -\dfrac{b}{2a} = -\dfrac{(-4)}{2(1)} = 2$. The y-coordinate of the vertex is

$$f\left(-\frac{b}{2a}\right) = f(2)$$
$$= (2)^2 - 4(2) - 12$$
$$= 4 - 8 - 12$$
$$= -16$$

Thus, the vertex is $(2, -16)$ and the axis of symmetry is the line $x = 2$.
The $y$-intercept is
$$f(0) = (0)^2 - 4(0) - 12 = -12.$$
Now, $b^2 - 4ac = (-4)^2 - 4(1)(-12) = 64 > 0$.
The parabola will have two distinct $x$-intercepts. We find these by solving
$$f(x) = 0$$
$$x^2 - 4x - 12 = 0$$
$$(x-6)(x+2) = 0$$
$$x - 6 = 0 \text{ or } x + 2 = 0$$
$$x = 6 \text{ or } x = -2$$
Finally, the $y$-intercept point, $(0, -12)$, is two units to the left of the axis of symmetry. Therefore, if we move two units to the right of the axis of symmetry, we obtain the point $(4, -12)$ which must also be on the graph.

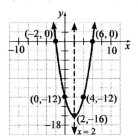

2. For the function $f(x) = -3x^2 + 12x - 7$, we see that $a = -3$, $b = 12$, and $c = -7$. The parabola opens down because $a = -3 < 0$. The $x$-coordinate of the vertex is
$$x = -\frac{b}{2a} = -\frac{12}{2(-3)} = 2.$$ The $y$-coordinate of the vertex is
$$f\left(-\frac{b}{2a}\right) = f(2)$$
$$= -3(2)^2 + 12(2) - 7$$
$$= -12 + 24 - 7$$
$$= 5$$
Thus, the vertex is $(2, 5)$ and the axis of symmetry is the line $x = 2$. The $y$-intercept is
$$f(0) = -3(0)^2 + 12(0) - 7 = -7.$$
Now, $b^2 - 4ac = 12^2 - 4(-3)(-7) = 60 > 0$.
The parabola will have two distinct $x$-intercepts. We find these by solving
$$f(x) = 0$$
$$-3x^2 + 12x - 7 = 0$$
$$x = \frac{-12 \pm \sqrt{60}}{2(-3)}$$
$$= \frac{-12 \pm 2\sqrt{15}}{-6}$$
$$= \frac{6 \pm \sqrt{15}}{3}$$
$$x \approx 0.71 \text{ or } x \approx 3.29$$
Finally, the $y$-intercept point, $(0, -7)$, is two units to the left of the axis of symmetry. Therefore, if we move two units to the right of the axis of symmetry, we obtain the point $(4, -7)$ which must also be on the graph.

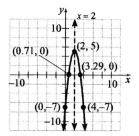

3. For the function $f(x) = x^2 + 6x + 9$, we see that $a = 1$, $b = 6$, and $c = 9$. The parabola opens up because $a = 1 > 0$. The $x$-coordinate of the vertex is $x = -\frac{b}{2a} = -\frac{6}{2(1)} = -3$. The $y$-coordinate of the vertex is
$$f\left(-\frac{b}{2a}\right) = f(-3)$$
$$= (-3)^2 + 6(-3) + 9$$
$$= 9 - 18 + 9$$
$$= 0$$
Thus, the vertex is $(-3, 0)$ and the axis of symmetry is the line $x = -3$. The $y$-intercept is $f(0) = 0^2 + 6(0) + 9 = 9$. Now, $b^2 - 4ac = 6^2 - 4(1)(9) = 36 - 36 = 0$. Since the discriminant is 0, the $x$-coordinate of the

## Section 8.5 – Quick Checks: Graphing Quadratic Functions Using Properties

vertex is the only x-intercept, $x = -3$. Finally, the y-intercept point, $(0, 9)$, is three units to the right of the axis of symmetry. Therefore, if we move three units to the left of the axis of symmetry, we obtain the point $(-6, 9)$ which must also be on the graph.

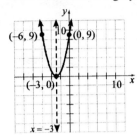

4. For the function $G(x) = -3x^2 + 9x - 8$, we see that $a = -3$, $b = 9$, and $c = -8$. The parabola opens down because $a = -3 < 0$. The x-coordinate of the vertex is
$$x = -\frac{b}{2a} = -\frac{9}{2(-3)} = \frac{9}{6} = \frac{3}{2}.$$
The y-coordinate of the vertex is
$$G\left(-\frac{b}{2a}\right) = G\left(\frac{3}{2}\right)$$
$$= -3\left(\frac{3}{2}\right)^2 + 9\left(\frac{3}{2}\right) - 8$$
$$= -\frac{27}{4} + \frac{27}{2} - 8$$
$$= -\frac{5}{4}$$
Thus, the vertex is $\left(\frac{3}{2}, -\frac{5}{4}\right)$ and the axis of symmetry is the line $x = \frac{3}{2}$. The y-intercept is $G(0) = -3(0)^2 + 9(0) - 8 = -8$. Now, $b^2 - 4ac = 9^2 - 4(-3)(-8) = 81 - 96 = -15$. Since the discriminant is negative, there are no x-intercepts. Finally, the y-intercept point, $(0, -8)$, is three-halves units to the left of the axis of symmetry. Therefore, if we move three-halves units to the right of the axis of symmetry, we obtain the point $(3, -8)$ which must also be on the graph.

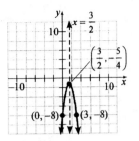

5. If we compare $f(x) = 2x^2 - 8x + 1$ to $f(x) = ax^2 + bx + c$, we find that $a = 2$, $b = -8$, and $c = 1$. Because $a > 0$, we know the graph will open up, so the function will have a minimum value. The minimum value occurs at
$$x = -\frac{b}{2a} = -\frac{(-8)}{2(2)} = 2.$$
The minimum value of the function is
$$f\left(-\frac{b}{2a}\right) = f(2) = 2(2)^2 - 8(2) + 1 = -7.$$ So, the minimum value is $-7$ and it occurs when $x = 2$.

6. If we compare $G(x) = -x^2 + 10x + 8$ to $G(x) = ax^2 + bx + c$, we find that $a = -1$, $b = 10$, and $c = 8$. Because $a < 0$, we know the graph will open down, so the function will have a maximum value. The maximum value occurs at $x = -\frac{b}{2a} = -\frac{10}{2(-1)} = 5$.
The maximum value of the function is
$$G\left(-\frac{b}{2a}\right) = G(5) = -5^2 + 10(5) + 8 = 33.$$ So, the maximum value is 33 and it occurs when $x = 5$.

7. **a.** We first recognize that the revenue function is a quadratic function whose graph opens down since $a = -0.5 < 0$. This means that the function indeed has a maximum value. The maximum value occurs when
$$p = -\frac{b}{2a} = -\frac{75}{2(-0.5)} = 75.$$
The revenue will be maximized when the calculators are sold at a price of $75.

**b.** The maximum revenue is obtained by evaluating the revenue function at the price found in part (a).
$$R(75) = -0.5(75)^2 + 75(75) = 2812.5$$
The maximum daily revenue is $2812.50.

## Chapter 8: Quadratic Equations and Functions

8. Let $l$ = length and $w$ = width.
   The area of a rectangle is given by $A = l \cdot w$.
   Before we can work on maximizing the area, we need to get the function in terms of one independent variable.
   The 1000 yards of fence will form the perimeter of the rectangle. That is, we have
   $2l + 2w = 1000$.
   We can solve this equation for $l$ and substitute the result in the area equation.
   $2l + 2w = 1000$
   $l + w = 500$
   $l = 500 - w$
   Thus, the area equation becomes
   $A = l \cdot w$
   $= (500 - w) \cdot w$
   $= -w^2 + 500w$

   The area function is a quadratic function whose graph opens down since $a = -1 < 0$. This means that the function indeed has a maximum. This occurs when $w = -\dfrac{b}{2a} = -\dfrac{500}{2(-1)} = 250$.

   The maximum area can be found by substituting this value for $w$ in the area function.
   $A = -250^2 + 500(250) = 62{,}500$ square yards

   Since $l = 500 - w$ and $w = 250$, the length will be $l = 500 - 250 = 250$ yards.

   The rectangular field will have a maximum area of 62,500 square yards when the field measures 250 yards by 250 yards.

9. Let $x$ represent the number of boxes in excess of 30. Revenue is price times quantity. If 30 boxes of CDs are sold, the revenue will be $100(30)$. If 31 boxes of CDs are sold, the revenue will be $99(31)$. If 32 boxes of CDs are sold, the revenue will be $98(32)$. In general, if $x$ boxes in excess of 30 are sold, then the number of boxes will be $30 + x$, and the price per box will be $100 - x$. Thus, revenue will be
   $R(x) = (100 - x)(30 + x)$
   $= 3000 + 100x - 30x - x^2$
   $= -x^2 + 70x + 3000$

   Now, this revenue function is a quadratic function whose graph opens down since $a = -1 < 0$. This means that the function indeed has a maximum value. The maximum value occurs when $x = -\dfrac{b}{2a} = -\dfrac{70}{2(-1)} = 35$.

   The maximum revenue will be
   $R(35) = -35^2 + 70(35) + 3000 = 4225$.

   Now recall that $x$ is the number of boxes in excess of 30. Therefore, $30 + 30 = 65$ boxes of CDs should be sold in order to maximize revenue. The maximum revenue will be $4,225.

### 8.6 Quick Checks

1. We graph $f(x) = x^2 + 3x - 10$. We see that $a = 1$, $b = 3$, and $c = -10$. The parabola opens up because $a = 1 > 0$. The $x$-coordinate of the vertex is $x = -\dfrac{b}{2a} = -\dfrac{3}{2(1)} = -\dfrac{3}{2}$. The $y$-coordinate of the vertex is

   $f\left(-\dfrac{b}{2a}\right) = f\left(-\dfrac{3}{2}\right)$
   $= \left(-\dfrac{3}{2}\right)^2 + 3\left(-\dfrac{3}{2}\right) - 10$
   $= \dfrac{9}{4} - \dfrac{9}{2} - 10$
   $= -\dfrac{49}{4}$

   Thus, the vertex is $\left(-\dfrac{3}{2}, -\dfrac{49}{4}\right)$ and the axis of symmetry is the line $x = -\dfrac{3}{2}$.

   The $y$-intercept is $f(0) = 0^2 + 3(0) - 10 = -10$.
   Now, $b^2 - 4ac = 3^2 - 4(1)(-10) = 49 > 0$. The parabola will have two distinct $x$-intercepts. We find these by solving
   $f(x) = 0$
   $x^2 + 3x - 10 = 0$
   $(x + 5)(x - 2) = 0$
   $x + 5 = 0 \quad \text{or} \quad x - 2 = 0$
   $x = -5 \quad \text{or} \quad x = 2$

   Finally, the $y$-intercept point, $(0, -10)$, is three-haves units to the right of the axis of symmetry. Therefore, if we move three-halves units to the left of the axis of symmetry, we obtain the point $(-3, -10)$ which must also be on the graph.

# Section 8.6 – Quick Checks: Quadratic Inequalities

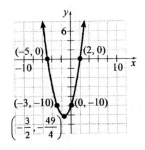

From the graph, we can see that $f(x) = x^2 + 3x - 10$ is greater than 0 for $x < -5$ or $x > 2$. Because the inequality is non-strict, we include the x-intercepts in the solution. So, the solution is $\{x \mid x \leq -5 \text{ or } x \geq 2\}$ using set-builder notation; the solution is $(-\infty, -5] \cup [2, \infty)$ using interval notation.

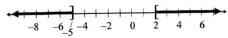

2. $x^2 + 3x - 10 \geq 0$

   Solve: $x^2 + 3x - 10 = 0$
   $(x-2)(x+5) = 0$
   $x - 2 = 0$ or $x + 5 = 0$
   $x = 2$ or $x = -5$

   Determine where each factor is positive and negative and where the product of these factors is positive and negative.

   | Interval | $(-\infty, -5)$ | $-5$ | $(-5, 2)$ | $2$ | $(2, \infty)$ |
   |---|---|---|---|---|---|
   | $x - 2$ | Neg | Neg | Neg | 0 | Pos |
   | $x + 5$ | Neg | 0 | Pos | Pos | Pos |
   | $(x-2)(x+5)$ | Pos | 0 | Neg | 0 | Pos |

   The inequality is non-strict, so –5 and 2 are part of the solution. Now, $(x-2)(x+5)$ is greater than zero where the product is positive. The solution is $\{x \mid x \leq -5 \text{ or } x \geq 2\}$ in set-builder notation; the solution is $(-\infty, -5] \cup [2, \infty)$ in interval notation.

3. $-x^2 > 2x - 24$
   $0 > x^2 + 2x - 24$
   $x^2 + 2x - 24 < 0$

   Solve: $x^2 + 2x - 24 = 0$
   $(x-4)(x+6) = 0$
   $x - 4 = 0$ or $x + 6 = 0$
   $x = 4$ or $x = -6$

   Determine where each factor is positive and negative and where the product of these factors is positive and negative.

   | Interval | $(-\infty, -6)$ | $-6$ | $(-6, 4)$ | $4$ | $(4, \infty)$ |
   |---|---|---|---|---|---|
   | $x - 4$ | Neg | Neg | Neg | 0 | Pos |
   | $x + 6$ | Neg | 0 | Pos | Pos | Pos |
   | $(x-4)(x+6)$ | Pos | 0 | Neg | 0 | Pos |

   The inequality is strict, so –6 and 4 are not part of the solution. Now, $(x+6)(x-4)$ is less than zero where the product is negative. The solution is $\{x \mid -6 < x < 4\}$ in set-builder notation; the solution is $(-6, 4)$ in interval notation.

4. $3x^2 > -x + 5$
   $3x^2 + x - 5 > 0$

   **Graphical Method:**
   To graph $f(x) = 3x^2 + x - 5$, we notice that $a = 3$, $b = 1$, and $c = -5$. The parabola opens up because $a = 3 > 0$. The x-coordinate of the vertex is $x = -\dfrac{b}{2a} = -\dfrac{1}{2(3)} = -\dfrac{1}{6}$. The y-coordinate of the vertex is
   $$f\left(-\dfrac{b}{2a}\right) = f\left(-\dfrac{1}{6}\right)$$
   $$= 3\left(-\dfrac{1}{6}\right)^2 + \left(-\dfrac{1}{6}\right) - 5$$
   $$= -\dfrac{61}{12}$$

   Thus, the vertex is $\left(-\dfrac{1}{6}, -\dfrac{61}{12}\right)$ and the axis of symmetry is the line $x = -\dfrac{1}{6}$. The y-intercept is $f(0) = 3(0)^2 + 0 - 5 = -5$.

Now, $b^2 - 4ac = 1^2 - 4(3)(-5) = 61 > 0$. The parabola will have two distinct $x$-intercepts. We find these by solving
$$f(x) = 0$$
$$3x^2 + x - 5 = 0$$
$$x = \frac{-1 \pm \sqrt{61}}{2(3)}$$
$$= \frac{-1 \pm \sqrt{61}}{6}$$
$$x \approx -1.47 \text{ or } x \approx 1.14$$

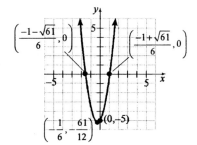

From the graph, we can see that $f(x) = 3x^2 + x - 5$ is greater than 0 for $x < \frac{-1 - \sqrt{61}}{6}$ or $x > \frac{-1 + \sqrt{61}}{6}$. Because the inequality is strict, we do not include the $x$-intercepts in the solution. So, the solution is $\left\{ x \,\middle|\, x < \frac{-1 - \sqrt{61}}{6} \text{ or } x > \frac{-1 + \sqrt{61}}{6} \right\}$ using set-builder notation; the solution is $\left( -\infty, \frac{-1 - \sqrt{61}}{6} \right) \cup \left( \frac{-1 + \sqrt{61}}{6}, \infty \right)$ using interval notation.

**Algebraic Method:**

Solve: $3x^2 + x - 5 = 0$
$$x = \frac{-1 \pm \sqrt{61}}{2(3)}$$
$$= \frac{-1 + \sqrt{61}}{6}$$
$$x \approx -1.47 \text{ or } x \approx 1.14$$

Determine where each factor is positive and negative and where the product of these factors is positive and negative.

| Interval | $(-\infty, -1.47)$ | $(-1.47, 1.14)$ | $(1.14, \infty)$ |
|---|---|---|---|
| | | $\frac{-1-\sqrt{61}}{6}$ | $\frac{-1+\sqrt{61}}{6}$ |
| $x - \left( \frac{-1-\sqrt{61}}{6} \right)$ | Neg  0 | Pos  Pos | Pos |
| $x - \left( \frac{-1+\sqrt{61}}{6} \right)$ | Neg  Neg | Neg  0 | Pos |
| $\left[ x - \left( \frac{-1-\sqrt{61}}{6} \right) \right]\left[ x - \left( \frac{-1+\sqrt{61}}{6} \right) \right]$ | Pos  0 | Neg  0 | Pos |

The inequality is strict, so $\frac{-1-\sqrt{61}}{6}$ and $\frac{-1+\sqrt{61}}{6}$ are not part of the solution. Now, $\left[ x - \left( \frac{-1-\sqrt{61}}{6} \right) \right]\left[ x - \left( \frac{-1+\sqrt{61}}{6} \right) \right]$ is greater than zero where the product is positive. So, the solution is $\left\{ x \,\middle|\, x < \frac{-1-\sqrt{61}}{6} \text{ or } x > \frac{-1+\sqrt{61}}{6} \right\}$ using set-builder notation; the solution is $\left( -\infty, \frac{-1-\sqrt{61}}{6} \right) \cup \left( \frac{-1+\sqrt{61}}{6}, \infty \right)$ using interval notation.

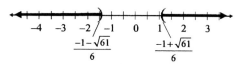

# Chapter 9 – Quick Checks

## 9.1 Quick Checks

1. $f(x) = 4x - 3$; $g(x) = x^2 + 1$

   a. $g(2) = (2)^2 + 1 = 4 + 1 = 5$
   $f(5) = 4(5) - 3 = 20 - 3 = 17$
   $(f \circ g)(2) = f(g(2)) = f(5) = 17$

   b. $f(2) = 4(2) - 3 = 8 - 3 = 5$
   $g(5) = (5)^2 + 1 = 25 + 1 = 26$
   $(g \circ f)(2) = g(f(2)) = g(5) = 26$

   c. $f(-3) = 4(-3) - 3 = -12 - 3 = -15$
   $f(-15) = 4(-15) - 3 = -60 - 3 = -63$
   $(f \circ f)(-3) = f(f(-3)) = f(-15) = -63$

2. $f(x) = x^2 - 3x + 1$; $g(x) = 3x + 2$

   a. $(f \circ g)(x) = f(g(x))$
   $= (3x + 2)^2 - 3(3x + 2) + 1$
   $= 9x^2 + 12x + 4 - 9x - 6 + 1$
   $= 9x^2 + 3x - 1$

   b. $(g \circ f)(x) = g(f(x))$
   $= 3(x^2 - 3x + 1) + 2$
   $= 3x^2 - 9x + 3 + 2$
   $= 3x^2 - 9x + 5$

   c. $(f \circ g)(-2) = 9(-2)^2 + 3(-2) - 1$
   $= 9(4) + 3(-2) - 1$
   $= 36 - 6 - 1$
   $= 29$

3. Since the two friends (inputs) Max and Yolanda share the same birthday (output) of November 8, this function is not one-to-one.

4. The function is one-to-one because there are no two distinct inputs that correspond to the same output.

5. The graph fails the horizontal line test. For example, the line $y = -1$ will intersect the graph of $f$ in four places. Therefore, the function is not one-to-one.

6. The graph passes the horizontal line test because every horizontal line will intersect the graph of $f$ exactly once. Thus, the function is one-to-one.

7. The inverse of the one-to-one function is:

   | Right Tibia | Right Humerus |
   |---|---|
   | 36.05 | → 24.80 |
   | 35.57 | → 24.59 |
   | 34.58 | → 24.29 |
   | 34.20 | → 23.81 |
   | 34.73 | → 24.87 |

   The domain of the inverse function is $\{36.05, 35.57, 34.58, 34.20, 34.73\}$. The range of the inverse function is $\{24.80, 24.59, 24.29, 23.81, 24.87\}$.

8. To obtain the inverse, we switch the $x$- and $y$-coordinates: $\{(3, -3), (2, -2), (1, -1), (0, 0), (-1, 1)\}$
   The domain of the inverse function is $\{3, 2, 1, 0, -1\}$. The range of the inverse function is $\{-3, -2, -1, 0, 1\}$.

9. To plot the inverse, switch the $x$- and $y$-coordinates of each point and connect the corresponding points. The graph of the function (shaded) and the line $y = x$ (dashed) are included for reference.

   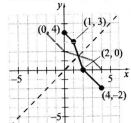

10. $g(x) = 5x - 1$
    $y = 5x - 1$
    $x = 5y - 1$
    $x + 1 = 5y$
    $\dfrac{x + 1}{5} = y$
    $g^{-1}(x) = \dfrac{x + 1}{5}$

    Check: $g(g^{-1}(x)) = 5\left(\dfrac{x+1}{5}\right) - 1 = x + 1 - 1 = x$

    $g^{-1}(g(x)) = \dfrac{(5x - 1) + 1}{5} = \dfrac{5x}{5} = x$

## Chapter 9: Exponential and Logarithmic Functions

11. $f(x) = x^5 + 3$

    $y = x^5 + 3$

    $x = y^5 + 3$

    $x - 3 = y^5$

    $\sqrt[5]{x-3} = y$

    $f^{-1}(x) = \sqrt[5]{x-3}$

    Check:

    $f(f^{-1}(x)) = (\sqrt[5]{x-3})^5 + 3 = x - 3 + 3 = x$

    $f^{-1}(f(x)) = \sqrt[5]{x^5 + 3 - 3} = \sqrt[5]{x^5} = x$

### 9.2 Quick Checks

1. a. $2^{1.7} \approx 3.249009585$

   b. $2^{1.73} \approx 3.317278183$

   c. $2^{1.732} \approx 3.321880096$

   d. $2^{1.7321} \approx 3.32211036$

   e. $2^{\sqrt{3}} \approx 3.321997085$

2. Locate some points on the graph of $f(x) = 4^x$.

| $x$ | $f(x) = 4^x$ | $(x, f(x))$ |
|---|---|---|
| $-2$ | $f(-2) = 4^{-2} = \dfrac{1}{4^2} = \dfrac{1}{16}$ | $\left(-2, \dfrac{1}{16}\right)$ |
| $-1$ | $f(-1) = 4^{-1} = \dfrac{1}{4^1} = \dfrac{1}{4}$ | $\left(-1, \dfrac{1}{4}\right)$ |
| 0 | $f(0) = 4^0 = 1$ | $(0, 1)$ |
| 1 | $f(1) = 4^1 = 4$ | $(1, 4)$ |

Plot the points and connect them with a smooth curve.

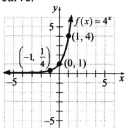

The domain of $f$ is all real numbers or, using interval notation, $(-\infty, \infty)$. The range of $f$ is $\{y \mid y > 0\}$ or, using interval notation, $(0, \infty)$.

3. Locate some points on the graph of $f(x) = \left(\dfrac{1}{4}\right)^x$.

| $x$ | $f(x) = \left(\dfrac{1}{4}\right)^x$ | $(x, f(x))$ |
|---|---|---|
| $-1$ | $f(-1) = \left(\dfrac{1}{4}\right)^{-1} = 4^1 = 4$ | $(-1, 4)$ |
| 0 | $f(0) = \left(\dfrac{1}{4}\right)^0 = 1$ | $(0, 1)$ |
| 1 | $f(1) = \left(\dfrac{1}{4}\right)^1 = \dfrac{1}{4}$ | $\left(1, \dfrac{1}{4}\right)$ |
| 2 | $f(2) = \left(\dfrac{1}{4}\right)^2 = \dfrac{1}{16}$ | $\left(2, \dfrac{1}{16}\right)$ |

Plot the points and connect them with a smooth curve.

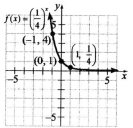

The domain of $f$ is all real numbers or, using interval notation, $(-\infty, \infty)$. The range of $f$ is $\{y \mid y > 0\}$ or, using interval notation, $(0, \infty)$.

4. Locate some points on the graph of $f(x) = 2^{x-1}$.

| $x$ | $f(x) = 2^{x-1}$ | $(x, f(x))$ |
|---|---|---|
| $-1$ | $f(-1) = 2^{-1-1} = 2^{-2} = \dfrac{1}{2^2} = \dfrac{1}{4}$ | $\left(-1, \dfrac{1}{4}\right)$ |
| 0 | $f(0) = 2^{0-1} = 2^{-1} = \dfrac{1}{2^1} = \dfrac{1}{2}$ | $\left(0, \dfrac{1}{2}\right)$ |
| 1 | $f(1) = 2^{1-1} = 2^0 = 1$ | $(1, 1)$ |
| 2 | $f(2) = 2^{2-1} = 2^1 = 2$ | $(2, 2)$ |
| 3 | $f(3) = 2^{3-1} = 2^2 = 4$ | $(3, 4)$ |
| 4 | $f(4) = 2^{4-1} = 2^3 = 8$ | $(4, 8)$ |

Plot the points and connect them with a smooth curve.

### Section 9.2 – Quick Checks: Exponential Functions

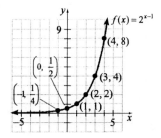

The domain of $f$ is all real numbers or, using interval notation, $(-\infty, \infty)$. The range of $f$ is $\{y \mid y > 0\}$ or, using interval notation, $(0, \infty)$.

5. Locate some points on the graph of $f(x) = 3^x + 1$.

| $x$ | $f(x) = 3^x + 1$ | $(x, f(x))$ |
|---|---|---|
| $-2$ | $f(-2) = 3^{-2} + 1 = \frac{1}{9} + 1 = \frac{10}{9}$ | $\left(-2, \frac{10}{9}\right)$ |
| $-1$ | $f(-1) = 3^{-1} + 1 = \frac{1}{3} + 1 = \frac{4}{3}$ | $\left(-1, \frac{4}{3}\right)$ |
| $0$ | $f(0) = 3^0 + 1 = 1 + 1 = 2$ | $(0, 2)$ |
| $1$ | $f(1) = 3^1 + 1 = 3 + 1 = 4$ | $(1, 4)$ |
| $2$ | $f(2) = 3^2 + 1 = 9 + 1 = 10$ | $(2, 10)$ |

Plot the points and connect them with a smooth curve.

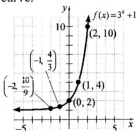

The domain of $f$ is all real numbers or, using interval notation, $(-\infty, \infty)$. The range of $f$ is $\{y \mid y > 1\}$ or, using interval notation, $(1, \infty)$.

6. $5^{x-4} = 5^{-1}$
$x - 4 = -1$
$x = 3$
The solution set is $\{3\}$.

7. $3^{x+2} = 81$
$3^{x+2} = 3^4$
$x + 2 = 4$
$x = 2$
The solution set is $\{2\}$.

8. $e^{x^2} = e^x \cdot e^{4x}$
$e^{x^2} = e^{5x}$
$x^2 = 5x$
$x^2 - 5x = 0$
$x(x - 5) = 0$
$x = 0 \quad \text{or} \quad x - 5 = 0$
$\qquad\qquad\qquad x = 5$
The solution set is $\{0, 5\}$.

9. $\dfrac{2^{x^2}}{8} = 2^{2x}$

$\dfrac{2^{x^2}}{2^3} = 2^{2x}$

$2^{x^2 - 3} = 2^{2x}$
$x^2 - 3 = 2x$
$x^2 - 2x - 3 = 0$
$(x+1)(x-3) = 0$
$x + 1 = 0 \quad \text{or} \quad x - 3 = 0$
$x = -1 \quad \text{or} \quad x = 3$
The solution set is $\{-1, 3\}$.

10. a. $F(10) = 1 - e^{-0.25(10)} \approx 0.918$.
The likelihood that a person will arrive within 10 minutes of 3:00 P.M. is 0.918, or 91.8%.

   b. $F(25) = 1 - e^{-0.25(25)} \approx 0.998$.
The likelihood that a person will arrive within 25 minutes of 3:00 P.M. is 0.998, or 99.8%.

11. a. $A(10) = 10\left(\dfrac{1}{2}\right)^{10/18.72} \approx 6.91$.
After 10 days, approximately 6.91 grams of thorium-227 will be left in the sample.

   b. $A(18.72) = 10\left(\dfrac{1}{2}\right)^{18.72/18.72} = 5$.
After 18.72 days, 5 grams of thorium-227 will be left in the sample.

   c. $A(74.88) = 10\left(\dfrac{1}{2}\right)^{74.88/18.72} = 0.625$.
After 74.88 days, 0.625 gram of thorium-227 will be left in the sample.

   d. $A(100) = 10\left(\dfrac{1}{2}\right)^{100/18.72} \approx 0.247$.
After 100 days, approximately 0.247 gram of thorium-227 will be left in the sample.

## Chapter 9: Exponential and Logarithmic Functions

**12.** We use the compound interest formula with $P = \$2000$, $r = 0.12$, and $n = 12$, so that

$$A = 2000\left(1 + \frac{0.12}{12}\right)^{12t}$$
$$= 2000(1 + 0.01)^{12t}$$
$$= 2000(1.01)^{12t}$$

**a.** The value of the account after $t = 1$ years is
$$A = 2000(1.01)^{12(1)}$$
$$= 2000(1.01)^{12} \approx \$2,253.65$$

**b.** The value of the account after $t = 15$ years is
$$A = 2000(1.01)^{12(15)}$$
$$= 2000(1.01)^{180} \approx \$11,991.60$$

**c.** The value of the account after $t = 30$ years is
$$A = 2000(1.01)^{12(30)}$$
$$= 2000(1.01)^{360} \approx \$71,899.28$$

### 9.3 Quick Checks

**1.** If $4^3 = w$, then $3 = \log_4 w$.

**2.** If $p^{-2} = 8$, then $-2 = \log_p 8$.

**3.** If $5^b = 125$, then $b = \log_5 125$.

**4.** If $y = \log_2 16$, then $2^y = 16$.

**5.** If $5 = \log_a 20$, then $a^5 = 20$.

**6.** If $-3 = \log_5 z$, then $5^{-3} = z$.

**7.** Let $y = \log_5 25$. Then,
$$5^y = 25$$
$$5^y = 5^2$$
$$y = 2$$
Thus, $\log_5 25 = 2$.

**8.** Let $y = \log_2 \frac{1}{8}$. Then,
$$2^y = \frac{1}{8}$$
$$2^y = \frac{1}{2^3} = 2^{-3}$$
$$y = -3$$
Thus, $\log_2 \frac{1}{8} = -3$.

**9.** $g(25)$ means to evaluate $\log_5 x$ at $x = 25$. So, we want to know the value of $\log_5 25$.
Let $y = \log_5 25$. Then,
$$5^y = 25$$
$$5^y = 5^2$$
$$y = 2$$
Thus, $g(25) = 2$.

**10.** $g\left(\frac{1}{5}\right)$ means to evaluate $\log_5 x$ at $x = \frac{1}{5}$. So, we want to know the value of $\log_5 \left(\frac{1}{5}\right)$.
Let $y = \log_5 \left(\frac{1}{5}\right)$. Then,
$$5^y = \frac{1}{5}$$
$$5^y = 5^{-1}$$
$$y = -1$$
Thus, $g\left(\frac{1}{5}\right) = -1$.

**11.** The domain of $g(x) = \log_8(x+3)$ is the set of all real numbers $x$ such that
$$x + 3 > 0$$
$$x > -3$$
Thus, the domain of $g(x) = \log_8(x+3)$ is $\{x | x > -3\}$ or, using interval notation, $(-3, \infty)$.

**12.** The domain of $F(x) = \log_2(5 - 2x)$ is the set of all real numbers $x$ such that
$$5 - 2x > 0$$
$$-2x > -5$$
$$x < \frac{-5}{-2}$$
$$x < \frac{5}{2}$$
Thus, the domain of $F(x) = \log_2(5 - 2x)$ is $\left\{x | x < \frac{5}{2}\right\}$ or, using interval notation, $\left(-\infty, \frac{5}{2}\right)$.

13. Rewrite $y = f(x) = \log_4 x$ as $x = 4^y$. Locate some points on the graph of $x = 4^y$.

| $y$ | $x = 4^y$ | $(x, y)$ |
|---|---|---|
| $-2$ | $x = 4^{-2} = \dfrac{1}{4^2} = \dfrac{1}{16}$ | $\left(\dfrac{1}{16}, -2\right)$ |
| $-1$ | $x = 4^{-1} = \dfrac{1}{4^1} = \dfrac{1}{4}$ | $\left(\dfrac{1}{4}, -1\right)$ |
| $0$ | $x = 4^0 = 1$ | $(1, 0)$ |
| $1$ | $x = 4^1 = 4$ | $(4, 1)$ |

Plot the points and connect them with a smooth curve.

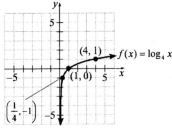

The domain of $f$ is $\{x \mid x > 0\}$ or, using interval notation, $(0, \infty)$. The range of $f$ is all real numbers or, using interval notation, $(-\infty, \infty)$.

14. Rewrite $y = f(x) = \log_{1/4} x$ as $x = \left(\dfrac{1}{4}\right)^y$. Locate some points on the graph of $x = \left(\dfrac{1}{4}\right)^y$.

| $y$ | $x = \left(\dfrac{1}{4}\right)^y$ | $(x, y)$ |
|---|---|---|
| $-1$ | $x = \left(\dfrac{1}{4}\right)^{-1} = 4^1 = 4$ | $(4, -1)$ |
| $0$ | $x = \left(\dfrac{1}{4}\right)^0 = 1$ | $(1, 0)$ |
| $1$ | $x = \left(\dfrac{1}{4}\right)^1 = \dfrac{1}{4}$ | $\left(\dfrac{1}{4}, 1\right)$ |
| $2$ | $x = \left(\dfrac{1}{4}\right)^2 = \dfrac{1}{16}$ | $\left(\dfrac{1}{16}, 2\right)$ |

Plot the points and connect them with a smooth curve.

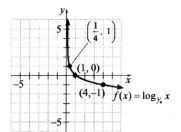

The domain of $f$ is $\{x \mid x > 0\}$ or, using interval notation, $(0, \infty)$. The range of $f$ is all real numbers or, using interval notation, $(-\infty, \infty)$.

15. $\log 1400 \approx 3.146$

16. $\ln 4.8 \approx 1.569$

17. $\log 0.3 \approx -0.523$

18. $\log_3(5x+1) = 4$
    $5x+1 = 3^4$
    $5x+1 = 81$
    $5x = 80$
    $x = 16$

   Check: $\log_3(5 \cdot 16 + 1) \overset{?}{=} 4$
   $\log_3(80+1) \overset{?}{=} 4$
   $\log_3(81) \overset{?}{=} 4$
   $4 = 4$ ✓

   The solution set is $\{16\}$.

19. $\log_x 16 = 2$
    $x^2 = 16$
    $x = \pm\sqrt{16}$
    $x = \pm 4$

   Since the base of a logarithm must always be positive, we know that $a = -4$ is extraneous. We check the potential solution $x = 4$.

   Check: $\log_4 16 \overset{?}{=} 2$
   $2 = 2$ ✓

   The solution set is $\{4\}$.

## Chapter 9: Exponential and Logarithmic Functions

20. $\ln x = -2$
    $x = e^{-2}$
    
    Check: $\ln e^{-2} \stackrel{?}{=} -2$
    $-2 = -2$ ✓
    
    The solution set is $\{e^{-2}\}$.

21. $\log(x - 20) = 4$
    $x - 20 = 10^4$
    $x - 20 = 10,000$
    $x = 10,020$
    
    Check: $\log(10,020 - 20) \stackrel{?}{=} 4$
    $\log(10,000) \stackrel{?}{=} 4$
    $4 = 4$ ✓
    
    The solution set is $\{10,020\}$.

22. We evaluate $L(x) = 10 \log \dfrac{x}{10^{-12}}$ at $x = 10^{-2}$.
    
    $L(10^{-2}) = 10 \log \dfrac{10^{-2}}{10^{-12}}$
    $= 10 \log 10^{-2-(-12)}$
    $= 10 \log 10^{10}$
    $= 10(10)$
    $= 100$
    
    The loudness of an MP3 player on "full blast" is 100 decibels.

### 9.4 Quick Checks

1. $\log_5 1 = 0$

2. $\ln 1 = 0$

3. $\log_4 4 = 1$

4. $\log 10 = 1$

5. $12^{\log_{12} \sqrt{2}} = \sqrt{2}$

6. $10^{\log 0.2} = 0.2$

7. $\log_8 8^{1.2} = 1.2$

8. $\log 10^{-4} = -4$

9. $\log_4(9 \cdot 5) = \log_4 9 + \log_4 5$

10. $\log(5w) = \log 5 + \log w$

11. $\log_7 \left(\dfrac{9}{5}\right) = \log_7 9 - \log_7 5$

12. $\ln\left(\dfrac{p}{3}\right) = \ln p - \ln 3$

13. $\log_2 \left(\dfrac{3m}{n}\right) = \log_2(3m) - \log_2 n$
    $= \log_2 3 + \log_2 m - \log_2 n$

14. $\ln\left(\dfrac{q}{3p}\right) = \ln q - \ln(3p)$
    $= \ln q - (\ln 3 + \ln p)$
    $= \ln q - \ln 3 - \ln p$

15. $\log_2 5^{1.6} = 1.6 \log_2 5$

16. $\log b^5 = 5 \log b$

17. $\log_4(a^2 b) = \log_4 a^2 + \log_4 b$
    $= 2 \log_4 a + \log_4 b$

18. $\log_3 \left(\dfrac{9m^4}{\sqrt[3]{n}}\right) = \log_3(9m^4) - \log_3 \sqrt[3]{n}$
    $= \log_3(3^2 m^4) - \log_3 n^{1/3}$
    $= \log_3 3^2 + \log_3 m^4 - \log_3 n^{1/3}$
    $= 2 + 4 \log_3 m - \dfrac{1}{3} \log_3 n$

19. $\log_8 4 + \log_8 16 = \log_8(4 \cdot 16)$
    $= \log_8 64$
    $= 2$

20. $\log_3(x+4) - \log_3(x-1) = \log_3 \left(\dfrac{x+4}{x-1}\right)$

21. $\log_5 x - 3 \log_5 2 = \log_5 x - \log_5 2^3$
    $= \log_5 x - \log_5 8$
    $= \log_5 \dfrac{x}{8}$

QC – 118

Section 9.5 –Quick Checks: Exponential and Logarithmic Equations

22. $\log_2(x+1) + \log_2(x+2) - 2\log_2 x$
$= \log_2[(x+1)(x+2)] - \log_2 x^2$
$= \log_2(x^2 + 3x + 2) - \log_2 x^2$
$= \log_2\left(\dfrac{x^2 + 3x + 2}{x^2}\right)$

23. Using common logarithms:
$\log_3 32 = \dfrac{\log 32}{\log 3} \approx 3.155$

24. Using natural logarithms:
$\log_{\sqrt{2}} \sqrt{7} = \dfrac{\ln \sqrt{7}}{\ln \sqrt{2}} \approx 2.807$

**9.5 Quick Checks**

1. $2\log_4 x = \log_4 9$
$\log_4 x^2 = \log_4 9$
$x^2 = 9$
$x = \pm\sqrt{9}$
$x = \pm 3$
The apparent solution $x = -3$ is extraneous because the argument of a logarithm must be positive. The solution set is $\{3\}$.

2. $\log_4(x-6) + \log_4 x = 2$
$\log_4[x(x-6)] = 2$
$\log_4(x^2 - 6x) = 2$
$x^2 - 6x = 4^2$
$x^2 - 6x = 16$
$x^2 - 6x - 16 = 0$
$(x+2)(x-8) = 0$
$x + 2 = 0$ or $x - 8 = 0$
$x = -2$ or $x = 8$
The apparent solution $x = -2$ is extraneous because it causes the argument of a logarithm to be negative. The solution set is $\{8\}$.

3. $2^x = 11$
$\log 2^x = \log 11$
$x \log 2 = \log 11$
$x = \dfrac{\log 11}{\log 2} \approx 3.459$

The solution set is $\left\{\dfrac{\log 11}{\log 2}\right\} \approx \{3.459\}$. If we had taken the natural logarithm of both sides, the solution set would be $\left\{\dfrac{\ln 11}{\ln 2}\right\} \approx \{3.459\}$.

4. $5^{2x} = 3$
$\log 5^{2x} = \log 3$
$2x \log 5 = \log 3$
$x = \dfrac{\log 3}{2\log 5} \approx 0.341$

The solution set is $\left\{\dfrac{\log 3}{2\log 5}\right\} \approx \{0.341\}$. If we had taken the natural logarithm of both sides, the solution set would be $\left\{\dfrac{\ln 3}{2\ln 5}\right\} \approx \{0.341\}$.

5. $e^{2x} = 5$
$\ln e^{2x} = \ln 5$
$2x = \ln 5$
$x = \dfrac{\ln 5}{2} \approx 0.805$

The solution set is $\left\{\dfrac{\ln 5}{2}\right\} \approx \{0.805\}$.

6. $3e^{-4x} = 20$
$e^{-4x} = \dfrac{20}{3}$
$\ln e^{-4x} = \ln\left(\dfrac{20}{3}\right)$
$-4x = \ln\left(\dfrac{20}{3}\right)$
$x = \dfrac{\ln\left(\dfrac{20}{3}\right)}{-4} \approx -0.474$

The solution set is $\left\{\dfrac{\ln(20/3)}{-4}\right\} \approx \{-0.474\}$.

## Chapter 9: Exponential and Logarithmic Functions

**7. a.** We need to determine the time until $A = 9$ grams. So we solve the equation

$$9 = 10\left(\frac{1}{2}\right)^{t/18.72}$$

$$0.9 = \left(\frac{1}{2}\right)^{t/18.72}$$

$$\log 0.9 = \log\left(\frac{1}{2}\right)^{t/18.72}$$

$$\log 0.9 = \frac{t}{18.72}\log\left(\frac{1}{2}\right)$$

$$\frac{18.72 \cdot \log 0.9}{\log\left(\frac{1}{2}\right)} = t$$

So, $t = \dfrac{18.72 \cdot \log 0.9}{\log\left(\frac{1}{2}\right)} \approx 2.85$ days.

Thus, 9 grams of thorium-227 will be left after approximately 2.85 days.

**b.** We need to determine the time until $A = 3$ grams. So we solve the equation

$$3 = 10\left(\frac{1}{2}\right)^{t/18.72}$$

$$0.3 = \left(\frac{1}{2}\right)^{t/18.72}$$

$$\log 0.3 = \log\left(\frac{1}{2}\right)^{t/18.72}$$

$$\log 0.3 = \frac{t}{18.72}\log\left(\frac{1}{2}\right)$$

$$\frac{18.72 \cdot \log 0.3}{\log\left(\frac{1}{2}\right)} = t$$

So, $t = \dfrac{18.72 \cdot \log 0.3}{\log\left(\frac{1}{2}\right)} \approx 32.52$ days.

Thus, 3 grams of thorium-227 will be left after approximately 32.52 days.

**8.** We first write the model with the parameters $P = 2000$, $r = 0.06$, and $n = 12$ to obtain

$$A = 2000\left(1 + \frac{0.06}{12}\right)^{12t} \text{ or } A = 2000(1.005)^{12t}.$$

**a.** We need to determine the time until $A = \$3000$, so we solve the equation

$$3000 = 2000(1.005)^{12t}$$

$$1.5 = (1.005)^{12t}$$

$$\log 1.5 = \log(1.005)^{12t}$$

$$\log 1.5 = 12t \log 1.005$$

$$\frac{\log 1.5}{12 \log 1.005} = t$$

So, $t = \dfrac{\log 1.5}{12 \log 1.005} \approx 6.77$. Thus, after approximately 6.77 years (6 years, 9 months), the account will be worth $3000.

**b.** We need to determine the time until $A = \$4000$, so we solve the equation

$$4000 = 2000(1.005)^{12t}$$

$$2 = (1.005)^{12t}$$

$$\log 2 = \log(1.005)^{12t}$$

$$\log 2 = 12t \log 1.005$$

$$\frac{\log 2}{12 \log 1.005} = t$$

So, $t = \dfrac{\log 2}{12 \log 1.005} \approx 11.58$. Thus, after approximately 11.58 years (11 years, 7 months), the account will be worth $4000.

# Chapter 10 – Quick Checks

## 10.1 Quick Checks

1. Using the distance formula with $P_1 = (3, 8)$ and $P_2 = (0, 4)$, the length is
$$d(P_1, P_2) = \sqrt{(x_2 - x_1)^2 + (y_2 - y_1)^2}$$
$$= \sqrt{(0-3)^2 + (4-8)^2}$$
$$= \sqrt{(-3)^2 + (-4)^2}$$
$$= \sqrt{9+16}$$
$$= \sqrt{25}$$
$$= 5$$

2. Using the distance formula with $P_1 = (-2, -5)$ and $P_2 = (4, 7)$, the length is
$$d(P_1, P_2) = \sqrt{(x_2 - x_1)^2 + (y_2 - y_1)^2}$$
$$= \sqrt{(4-(-2))^2 + (7-(-5))^2}$$
$$= \sqrt{6^2 + 12^2}$$
$$= \sqrt{36+144}$$
$$= \sqrt{180}$$
$$= 6\sqrt{5} \approx 13.42$$

3. a.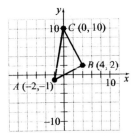

   b. $d(A, B) = \sqrt{(4-(-2))^2 + (2-(-1))^2}$
   $= \sqrt{6^2 + 3^2}$
   $= \sqrt{36+9}$
   $= \sqrt{45}$
   $= 3\sqrt{5}$

   $d(B, C) = \sqrt{(0-4)^2 + (10-2)^2}$
   $= \sqrt{(-4)^2 + 8^2}$
   $= \sqrt{16+64}$
   $= \sqrt{80}$
   $= 4\sqrt{5}$

   $d(A, C) = \sqrt{(0-(-2))^2 + (10-(-1))^2}$
   $= \sqrt{2^2 + 11^2}$
   $= \sqrt{4+121}$
   $= \sqrt{125}$
   $= 5\sqrt{5}$

   c. To determine if triangle $ABC$ is a right triangle, we check to see if
   $$[d(A,B)]^2 + [d(B,C)]^2 \stackrel{?}{=} [d(A,C)]^2$$
   $$(3\sqrt{5})^2 + (4\sqrt{5})^2 \stackrel{?}{=} (5\sqrt{5})^2$$
   $$9 \cdot 5 + 16 \cdot 5 \stackrel{?}{=} 25 \cdot 5$$
   $$45 + 80 \stackrel{?}{=} 125$$
   $$125 = 125 \leftarrow \text{True}$$
   Therefore, triangle $ABC$ is a right triangle.

   d. The length of the "base" of the triangle is $d(A, B) = 3\sqrt{5}$ and the length of the "height" of the triangle is $d(B, C) = 4\sqrt{5}$. Thus, the area of triangle $ABC$ is
   $$\text{Area} = \frac{1}{2} \cdot \text{base} \cdot \text{height} = \frac{1}{2} \cdot 3\sqrt{5} \cdot 4\sqrt{5} = 30$$
   square units.

4. $M = \left(\dfrac{x_1 + x_2}{2}, \dfrac{y_1 + y_2}{2}\right)$
   $= \left(\dfrac{3+0}{2}, \dfrac{8+4}{2}\right)$
   $= \left(\dfrac{3}{2}, \dfrac{12}{2}\right)$
   $= \left(\dfrac{3}{2}, 6\right)$

5. $M = \left(\dfrac{x_1 + x_2}{2}, \dfrac{y_1 + y_2}{2}\right)$
   $= \left(\dfrac{-2+4}{2}, \dfrac{-5+10}{2}\right)$
   $= \left(\dfrac{2}{2}, \dfrac{5}{2}\right)$
   $= \left(1, \dfrac{5}{2}\right)$

## Chapter 10: Conics

**10.2 Quick Checks**

1. We are given that $h = 2$, $k = 4$, and $r = 5$.
   Thus, the equation of the circle is
   $$(x-h)^2 + (y-k)^2 = r^2$$
   $$(x-2)^2 + (y-4)^2 = 5^2$$
   $$(x-2)^2 + (y-4)^2 = 25$$

2. We are given that $h = -2$, $k = 0$, and $r = \sqrt{2}$.
   Thus, the equation of the circle is
   $$(x-h)^2 + (y-k)^2 = r^2$$
   $$(x-(-2))^2 + (y-0)^2 = (\sqrt{2})^2$$
   $$(x+2)^2 + y^2 = 2$$

3. $(x-3)^2 + (y-1)^2 = 4$
   $(x-3)^2 + (y-1)^2 = 2^2$
   The center is $(h,k) = (3, 1)$; the radius is $r = 2$.

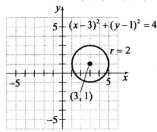

4. $(x+5)^2 + y^2 = 16$
   $(x-(-5))^2 + (y-0)^2 = 4^2$
   The center is $(h, k) = (-5, 0)$; the radius is $r = 4$.

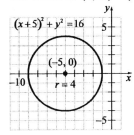

5. $x^2 + y^2 - 6x - 4y + 4 = 0$
   $(x^2 - 6x) + (y^2 - 4y) = -4$
   $(x^2 - 6x + 9) + (y^2 - 4y + 4) = -4 + 9 + 4$
   $(x-3)^2 + (y-2)^2 = 9$
   $(x-3)^2 + (y-2)^2 = 3^2$
   The center is $(h, k) = (3, 2)$; the radius is $r = 3$.

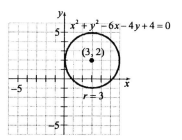

6. $2x^2 + 2y^2 - 16x + 4y - 38 = 0$
   $$\frac{2x^2 + 2y^2 - 16x + 4y - 38}{2} = \frac{0}{2}$$
   $x^2 + y^2 - 8x + 2y - 19 = 0$
   $(x^2 - 8x) + (y^2 + 2y) = 19$
   $(x^2 - 8x + 16) + (y^2 + 2y + 1) = 19 + 16 + 1$
   $(x-4)^2 + (y+1)^2 = 36$
   $(x-4)^2 + (y-(-1))^2 = 6^2$
   The center is $(h, k) = (4, -1)$; the radius is $r = 6$.

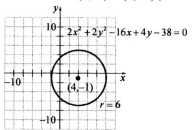

**10.3 Quick Checks**

1. Notice that $y^2 = 8x$ is of the form $y^2 = 4ax$, where $4a = 8$, so that $a = 2$. Now, the graph of an equation of the form $y^2 = 4ax$ will be a parabola that opens to the right with the vertex at the origin, focus at $(a, 0)$, and directrix of $x = -a$. Thus, the graph of $y^2 = 8x$ is a parabola that opens to the right with vertex $(0,0)$, focus $(2, 0)$, and directrix $x = -2$. To help graph the parabola, we plot the two points on the graph above and below the focus. Let $x = 2$:
   $$y^2 = 8(2)$$
   $$y^2 = 16$$
   $$y = \pm 4$$
   The points $(2, -4)$ and $(2, 4)$ are on the graph.

### Section 10.3 – Quick Checks: Parabolas

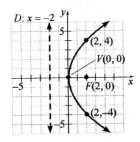
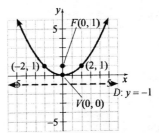

2. Notice that $y^2 = -20x$ is of the form $y^2 = -4ax$, where $-4a = -20$, so that $a = 5$. Now, the graph of an equation of the form $y^2 = -4ax$ will be a parabola that opens to the left with the vertex at the origin, focus at $(-a, 0)$, and directrix of $x = a$. Thus, the graph of $y^2 = -20x$ is a parabola that opens to the left with vertex $(0, 0)$, focus $(-5, 0)$, and directrix $x = 5$. To help graph the parabola, we plot the two points on the graph above and below the focus. Let $x = 5$:
$$y^2 = -20(-5)$$
$$y^2 = 100$$
$$y = \pm 10$$
The points $(-5, -10)$ and $(-5, 10)$ are on the graph.

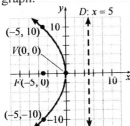

3. Notice that $x^2 = 4y$ is of the form $x^2 = 4ay$, where $4a = 4$, so that $a = 1$. Now, the graph of an equation of the form $x^2 = 4ay$ will be a parabola that opens up with the vertex at the origin, focus at $(0, a)$, and directrix of $y = -a$. Thus, the graph of $x^2 = 4y$ is a parabola that opens up with vertex $(0, 0)$, focus $(0, 1)$, and directrix $y = -1$. To help graph the parabola, we plot the two points on the graph to the left and to the right of the focus. Let $y = 1$:
$$x^2 = 4(1)$$
$$x^2 = 4$$
$$x = \pm 2$$
The points $(-2, 1)$ and $(2, 1)$ are on the graph.

4. Notice that $x^2 = -12y$ is of the form $x^2 = -4ay$, where $-4a = -12$, so that $a = 3$. Now, the graph of an equation of the form $x^2 = -4ay$ will be a parabola that opens down with the vertex at the origin, focus at $(0, -a)$, and directrix of $y = a$. Thus, the graph of $x^2 = -12y$ is a parabola that opens down with vertex $(0, 0)$, focus $(0, -3)$, and directrix $y = 3$. To help graph the parabola, we plot the two points on the graph to the left and to the right of the focus. Let $y = -3$:
$$x^2 = -12(-3)$$
$$x^2 = 36$$
$$x = \pm 6$$
The points $(-6, -3)$ and $(6, -3)$ are on the graph.

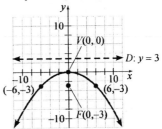

5. The distance from the vertex $(0, 0)$ to the focus $(0, -8)$ is $a = 8$. Because the focus lies on the negative $y$-axis, we know that the parabola will open down and the axis of symmetry is the $y$-axis. This means the equation of the parabola is of the form $x^2 = -4ay$ with $a = 8$:
$$x^2 = -4(8)y$$
$$x^2 = -32y$$
The directrix is the line $y = 8$. To help graph the parabola, we plot the two points on the graph to the left and right of the focus. Let $y = -8$:
$$x^2 = -32(-8) = 256$$
$$x = \pm 16$$
The points $(-16, -8)$ and $(16, -8)$ are on the graph.

QC – 123

## Chapter 10: Conics

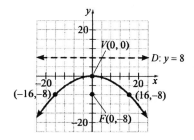

6. The vertex is at the origin and the axis of symmetry is the x-axis, so the parabola either opens left or right. Because the graph contains the point $(3, 2)$, which is in quadrant I, the parabola must open right. Therefore, the equation of the parabola is of the form $y^2 = 4ax$. Now $y = 2$ when $x = 3$, so
$$y^2 = 4ax$$
$$2^2 = 4a(3)$$
$$4 = 12a$$
$$a = \frac{4}{12} = \frac{1}{3}$$
The equation of the parabola is
$$y^2 = 4\left(\frac{1}{3}\right)x$$
$$y^2 = \frac{4}{3}x$$
With $a = \frac{1}{3}$, we know that the focus is $\left(\frac{1}{3}, 0\right)$ and the directrix is the line $x = -\frac{1}{3}$. To help graph the parabola, we plot the two points on the graph to the left and right of the focus. Let $x = \frac{1}{3}$:
$$y^2 = \frac{4}{3}\left(\frac{1}{3}\right)$$
$$y^2 = \frac{4}{9}$$
$$y = \pm\frac{2}{3}$$
The points $\left(\frac{1}{3}, -\frac{2}{3}\right)$ and $\left(\frac{1}{3}, \frac{2}{3}\right)$ are on the graph.

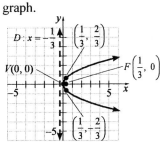

7. We complete the square in $y$ to write the equation in standard form:
$$y^2 - 4y - 12x - 32 = 0$$
$$y^2 - 4y = 12x + 32$$
$$y^2 - 4y + 4 = 12x + 32 + 4$$
$$y^2 - 4y + 4 = 12x + 36$$
$$(y-2)^2 = 12(x+3)$$
Notice that the equation is of the form $(y-k)^2 = 4a(x-h)$. The graph is a parabola that opens right with vertex $(h, k) = (-3, 2)$. Since $4a = 12$, we have that $a = 3$. The focus is $(h+a, k) = (-3+3, 2) = (0, 2)$, and the directrix is $x = h - a = -3 - 3 = -6$. To help graph the parabola, we plot the two points on the graph above and below the focus. Let $x = 0$:
$$(y-2)^2 = 12(0+3)$$
$$(y-2)^2 = 12(3)$$
$$(y-2)^2 = 36$$
$$y - 2 = \pm 6$$
$$y = 2 \pm 6$$
$$y = -4 \text{ or } y = 8$$
The points $(0, -4)$ and $(0, 8)$ are on the graph.

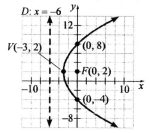

8. The receiver should be located at the focus of the satellite dish, so we need to find where the focus of the satellite dish is. To solve this problem, we draw a parabola on a Cartesian plane so that the vertex is the origin and the focus is on the positive y-axis. The width of the parabola is 4 feet, and the depth is 6 inches $= 0.5$ feet. Therefore, we know two points on the graph of the parabola: $(-2, 0.5)$ and $(2, 0.5)$.

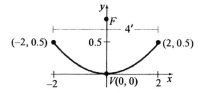

QC – 124

The equation of the parabola has the form $x^2 = 4ay$. Since $(2, 0.5)$ is a point on the graph, we have
$$2^2 = 4a(0.5)$$
$$4 = 2a$$
$$2 = a$$
The receiver should be located 2 feet from the base of the dish along its axis of symmetry.

## 10.4 Quick Checks

1. $\dfrac{x^2}{9} + \dfrac{y^2}{4} = 1$

   The larger number, 9, is in the denominator of the $x^2$-term. This means that the major axis is the $x$-axis and that the equation of the ellipse is of the form $\dfrac{x^2}{a^2} + \dfrac{y^2}{b^2} = 1$, so that $a^2 = 9$ and $b^2 = 4$. The center of the ellipse is $(0,0)$. Because $b^2 = a^2 - c^2$, or $c^2 = a^2 - b^2$, we have that $c^2 = 9 - 4 = 5$, so that $c = \pm\sqrt{5}$. Since the major axis is the $x$-axis, the foci are $\left(-\sqrt{5}, 0\right)$ and $\left(\sqrt{5}, 0\right)$. To find the $x$-intercepts (vertices), let $y = 0$; to find the $y$-intercepts, let $x = 0$:

   $x$-intercepts:
   $$\dfrac{x^2}{9} + \dfrac{0^2}{4} = 1$$
   $$\dfrac{x^2}{9} = 1$$
   $$x^2 = 9$$
   $$x = \pm 3$$

   $y$-intercepts:
   $$\dfrac{0^2}{9} + \dfrac{y^2}{4} = 1$$
   $$\dfrac{y^2}{4} = 1$$
   $$y^2 = 4$$
   $$y = \pm 2$$

   The intercepts are $(-3, 0)$, $(3, 0)$, $(0, -2)$, and $(0, 2)$.

   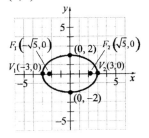

2. $\dfrac{x^2}{16} + \dfrac{y^2}{36} = 1$

   The larger number, 36, is in the denominator of the $y^2$-term. This means that the major axis is the $y$-axis and that the equation of the ellipse is of the form $\dfrac{x^2}{b^2} + \dfrac{y^2}{a^2} = 1$, so that $a^2 = 36$ and $b^2 = 16$. The center of the ellipse is $(0,0)$. Because $b^2 = a^2 - c^2$, or $c^2 = a^2 - b^2$, we have that $c^2 = 36 - 16 = 20$, so that $c = \pm\sqrt{20} = \pm 2\sqrt{5}$. Since the major axis is the $y$-axis, the foci are $\left(0, -2\sqrt{5}\right)$ and $\left(0, 2\sqrt{5}\right)$. To find the $x$-intercepts, let $y = 0$; to find the $y$-intercepts (vertices), let $x = 0$:

   $x$-intercepts:
   $$\dfrac{x^2}{16} + \dfrac{0^2}{36} = 1$$
   $$\dfrac{x^2}{16} = 1$$
   $$x^2 = 16$$
   $$x = \pm 4$$

   $y$-intercepts:
   $$\dfrac{0^2}{16} + \dfrac{y^2}{36} = 1$$
   $$\dfrac{y^2}{36} = 1$$
   $$y^2 = 36$$
   $$y = \pm 6$$

   The intercepts are $(-4, 0)$, $(4, 0)$, $(0, -6)$, and $(0, 6)$.

   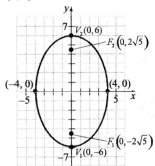

3. The given focus $(0, 3)$ and the given vertex $(0, 7)$ lie on the $y$-axis. Thus, the major axis is the $y$-axis, and the equation of the ellipse is of the form $\dfrac{x^2}{b^2} + \dfrac{y^2}{a^2} = 1$. The distance from the center of the ellipse to the vertex is $a = 7$ units. The distance from the center of the ellipse to the focus is $c = 3$ units. Because $b^2 = a^2 - c^2$, we have that $b^2 = 7^2 - 3^2 = 49 - 9 = 40$. So, the equation of the ellipse is $\dfrac{x^2}{40} + \dfrac{y^2}{49} = 1$.

## Chapter 10: Conics

To help graph the ellipse, find the x-intercepts:

Let $y = 0$: $\dfrac{x^2}{40} + \dfrac{0^2}{49} = 1$

$\dfrac{x^2}{40} = 1$

$x^2 = 40$

$x = \pm\sqrt{40} = \pm 2\sqrt{10}$

The x-intercepts are $\left(-2\sqrt{10}, 0\right)$ and $\left(2\sqrt{10}, 0\right)$.

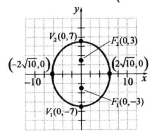

4. $9x^2 + y^2 + 54x - 2y + 73 = 0$

$9x^2 + y^2 + 54x - 2y + 73 = 0$

$9x^2 + 54x + y^2 - 2y = -73$

$9\left(x^2 + 6x\right) + \left(y^2 - 2y\right) = -73$

$9\left(x^2 + 6x + 9\right) + \left(y^2 - 2y + 1\right) = -73 + 9(9) + 1$

$9(x+3)^2 + (y-1)^2 = 9$

$\dfrac{9(x+3)^2 + (y-1)^2}{9} = \dfrac{9}{9}$

$\dfrac{(x+3)^2}{1} + \dfrac{(y-1)^2}{9} = 1$

The center of the ellipse is $(h, k) = (-3, 1)$. Because the larger number, 9, is the denominator of the $y^2$-term, the major axis is parallel to the y-axis. Because $a^2 = 9$ and $b^2 = 1$, we have that $c^2 = a^2 - b^2 = 9 - 1 = 8$. The vertices are $a = 3$ units below and above the center at $V_1(-3, -2)$ and $V_2(-3, 4)$. The foci are $c = \sqrt{8} = 2\sqrt{2}$ units below and above the center at $F_1\left(-3, 1 - 2\sqrt{2}\right)$ and $F_2\left(-3, 1 + 2\sqrt{2}\right)$. We plot the points $b = 1$ unit to the left and right of the center point at $(-4, 1)$ and $(-2, 1)$.

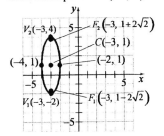

5. To solve the problem, we draw the ellipse on a Cartesian plane so that the center of the ellipse is at the origin and the major axis is along the x-axis. The equation of the ellipse is of the form $\dfrac{x^2}{a^2} + \dfrac{y^2}{b^2} = 1$. Since the length of the hall is 100 feet, the distance from the center of the room to each vertex is $a = \dfrac{100}{2} = 50$ feet. The distance from the center of the room to each focus is $c = 30$ feet.

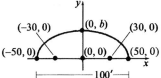

Now, because $b^2 = a^2 - c^2$, we have that $b^2 = 50^2 - 30^2 = 2500 - 900 = 1600$. Thus, the equation that describes the room is

$\dfrac{x^2}{2500} + \dfrac{y^2}{1600} = 1$.

The height of the room at its center is $b = \sqrt{1600} = 40$ feet.

### 10.5 Quick Checks

1. $\dfrac{x^2}{36} - \dfrac{y^2}{64} = 1$

Notice the equation is of the form $\dfrac{x^2}{a^2} - \dfrac{y^2}{b^2} = 1$.

Because the $x^2$-term is first, the transverse axis is the x-axis and the hyperbola opens left and right. The center of the hyperbola is the origin. We have that $a^2 = 36$ and $b^2 = 64$. Because $c^2 = a^2 + b^2$, we have that $c^2 = 36 + 64 = 100$, so that $c = \sqrt{100} = 10$. The vertices are $(\pm a, 0) = (\pm 6, 0)$, and the foci are $(\pm c, 0) = (\pm 10, 0)$. To help graph the hyperbola, we plot the points on the graph above and below the foci. Let $x = \pm 10$:

$\dfrac{(\pm 10)^2}{36} - \dfrac{y^2}{64} = 1$

$\dfrac{100}{36} - \dfrac{y^2}{64} = 1$

$\dfrac{25}{9} - \dfrac{y^2}{64} = 1$

$$-\frac{y^2}{64} = -\frac{16}{9}$$
$$y^2 = \frac{1024}{9}$$
$$y = \pm\frac{32}{3}$$

The points above and below the foci are $\left(-10, -\frac{32}{3}\right)$, $\left(-10, \frac{32}{3}\right)$, $\left(10, -\frac{32}{3}\right)$, and $\left(10, \frac{32}{3}\right)$.

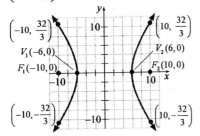

2. $\dfrac{y^2}{9} - \dfrac{x^2}{16} = 1$

Notice the equation is of the form $\dfrac{y^2}{a^2} - \dfrac{x^2}{b^2} = 1$.

Because the $y^2$-term is first, the transverse axis is the $y$-axis and the hyperbola opens up and down. The center of the hyperbola is the origin. We have that $a^2 = 9$ and $b^2 = 16$. Because $c^2 = a^2 + b^2$, we have that $c^2 = 9 + 16 = 25$, so that $c = \sqrt{25} = 5$. The vertices are $(0, \pm a) = (0, \pm 3)$, and the foci are $(0, \pm c) = (0, \pm 5)$. To help graph the hyperbola, we plot the points on the graph to the left and right of the foci. Let $y = \pm 5$:

$$\frac{(\pm 5)^2}{9} - \frac{x^2}{16} = 1$$
$$\frac{25}{9} - \frac{x^2}{16} = 1$$
$$-\frac{x^2}{16} = -\frac{16}{9}$$
$$x^2 = \frac{256}{9}$$
$$x = \pm\frac{16}{3}$$

The points to the left and right of the foci are $\left(-\frac{16}{3}, -5\right)$, $\left(-\frac{16}{3}, 5\right)$, $\left(\frac{16}{3}, -5\right)$, and $\left(\frac{16}{3}, 5\right)$.

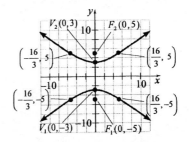

3. The given vertices $(\pm 4, 0)$ and focus $(6, 0)$ all lie on the $x$-axis. Thus, the transverse axis is the $x$-axis and the hyperbola opens left and right. The center of the hyperbola is the midpoint between the two vertices. Therefore, the center is $(0, 0)$ and the equation of the hyperbola is of the form $\dfrac{x^2}{a^2} - \dfrac{y^2}{b^2} = 1$. Now, the distance from the center $(0, 0)$ to each vertex is $a = 4$ units. Likewise, the distance from the center to the given focus $(6, 0)$ is $c = 6$ units. Also, $b^2 = c^2 - a^2$, so $b^2 = 6^2 - 4^2 = 36 - 16 = 20$. Thus, the equation of the hyperbola is $\dfrac{x^2}{16} - \dfrac{y^2}{20} = 1$.

To help graph the hyperbola, we first find the focus that was not given. Since the center is at $(0, 0)$ and since one focus is at $(6, 0)$, the other focus must be at $(-6, 0)$. Next, plot the points on the graph to the left and right of the foci. Let $x = \pm 6$:

$$\frac{(\pm 6)^2}{16} - \frac{y^2}{20} = 1$$
$$\frac{36}{16} - \frac{y^2}{20} = 1$$
$$\frac{9}{4} - \frac{y^2}{20} = 1$$
$$-\frac{y^2}{20} = -\frac{5}{4}$$
$$y^2 = 25$$
$$y = \pm 5$$

The points to the left and right of the foci are $(-6, -5)$, $(-6, 5)$, $(6, -5)$, and $(6, 5)$.

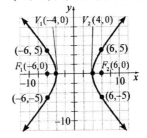

## Chapter 10: Conics

**4.** $x^2 - 9y^2 = 9$

$$\frac{x^2 - 9y^2}{9} = \frac{9}{9}$$

$$\frac{x^2}{9} - \frac{y^2}{1} = 1$$

Notice that the equation is of the form $\frac{x^2}{a^2} - \frac{y^2}{b^2} = 1$. The center of the hyperbola is $(0,0)$. Because the $x^2$-term is first, the hyperbola opens left and right. The transverse axis is along the $x$-axis. We have that $a^2 = 9$ and $b^2 = 1$. Because $c^2 = a^2 + b^2$, we have that $c^2 = 9 + 1 = 10$, so that $c = \sqrt{10}$. The vertices are $(\pm a, 0) = (\pm 3, 0)$, and the foci are $(\pm c, 0) = (\pm\sqrt{10}, 0)$. The asymptotes are of the form $y = -\frac{b}{a}x$ and $y = \frac{b}{a}x$. Since $a = 3$ and $b = 1$, the equations of the asymptotes are $y = -\frac{1}{3}x$ and $y = \frac{1}{3}x$. To help graph the hyperbola, we form the rectangle using the points $(\pm a, 0) = (\pm 3, 0)$ and $(0, \pm b) = (0, \pm 1)$. The diagonals are the asymptotes.

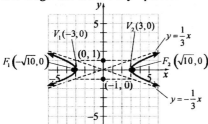

**5.** $\frac{y^2}{16} - \frac{x^2}{9} = 1$

Notice the equation is of the form $\frac{y^2}{a^2} - \frac{x^2}{b^2} = 1$. Because the $y^2$-term is first, the transverse is along the $y$-axis and the hyperbola opens up and down. The center of the hyperbola is $(0,0)$. We have that $a^2 = 16$ and $b^2 = 9$. Because $c^2 = a^2 + b^2$, we have that $c^2 = 16 + 9 = 25$, so that $c = \sqrt{25} = 5$. The vertices are $(0, \pm a) = (0, \pm 4)$, and the foci are $(0, \pm c) = (0, \pm 5)$. The asymptotes are of the form $y = -\frac{a}{b}x$ and $y = \frac{a}{b}x$. Since $a = 4$ and $b = 3$, the equations of the asymptotes are $y = -\frac{4}{3}x$ and $y = \frac{4}{3}x$. To help graph the hyperbola, we form the rectangle using the points $(0, \pm a) = (0, \pm 4)$ and $(\pm b, 0) = (\pm 3, 0)$. The diagonals are the asymptotes.

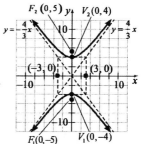

### 10.6 Quick Checks

**1.** $\begin{cases} 2x + y = -1 \\ x^2 - y = 4 \end{cases}$

First, graph each equation in the system.

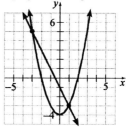

The system apparently has two solutions. Now solve the first equation for $y$: $y = -2x - 1$. Substitute the result for $y$ into the second equation:

$x^2 - (-2x - 1) = 4$

$x^2 + 2x + 1 = 4$

$x^2 + 2x - 3 = 0$

$(x - 1)(x + 3) = 0$

$x = 1$ or $x = -3$

Substitute these $x$-values into the first equation to find the corresponding $y$-values:

$x = 1$: $2(1) + y = -1$

$2 + y = -1$

$y = -3$

$x = -3$: $2(-3) + y = -1$

$-6 + y = -1$

$y = 5$

Both pairs check, so the solutions are $(-3, 5)$ and $(1, -3)$.

QC – 128

Section 10.6 – Quick Checks: Nonlinear Systems of Equations

2. $\begin{cases} 2x + y = 0 \\ (x-4)^2 + (y+2)^2 = 9 \end{cases}$

First, graph each equation in the system.

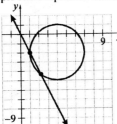

The system apparently has two solutions. Now solve the first equation for $y$: $y = -2x$.

Substitute the result for $y$ into the second equation:
$$(x-4)^2 + (-2x+2)^2 = 9$$
$$x^2 - 8x + 16 + 4x^2 - 8x + 4 = 9$$
$$5x^2 - 16x + 11 = 0$$
$$(5x - 11)(x - 1) = 0$$
$$x = \frac{11}{5} \quad \text{or} \quad x = 1$$

Substitute these $x$-values into the first equation to find the corresponding $y$-values:

$x = \frac{11}{5}$: $2\left(\frac{11}{5}\right) + y = 0$
$$\frac{22}{5} + y = 0$$
$$y = -\frac{22}{5}$$

$x = 1$: $2(1) + y = 0$
$$2 + y = 0$$
$$y = -2$$

Both pairs check, so the solutions are $(1, -2)$ and $\left(\frac{11}{5}, -\frac{22}{5}\right)$.

3. $\begin{cases} x^2 + y^2 = 16 \\ x^2 - 2y = 8 \end{cases}$

First, graph each equation in the system.

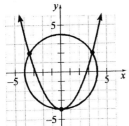

The system apparently has three solutions. Now multiply the second equation by $-1$ and add the result to the first equation:
$$\begin{array}{r} x^2 + y^2 = 16 \\ -x^2 \phantom{+y^2} + 2y = -8 \\ \hline y^2 + 2y = 8 \end{array}$$
$$y^2 + 2y - 8 = 0$$
$$(y+4)(y-2) = 0$$
$$y = -4 \quad \text{or} \quad y = 2$$

Substitute these $y$-values into the first equation to find the corresponding $x$-values:

$y = -4$: $x^2 + (-4)^2 = 16$
$$x^2 + 16 = 16$$
$$x^2 = 0$$
$$x = 0$$

$y = 2$: $x^2 + 2^2 = 16$
$$x^2 + 4 = 16$$
$$x^2 = 12$$
$$x = \pm\sqrt{12} = \pm 2\sqrt{3}$$

All three pairs check, so the solutions are $(0, -4)$, $(-2\sqrt{3}, 2)$ and $(2\sqrt{3}, 2)$.

4. $\begin{cases} x^2 - y = -4 \\ x^2 + y^2 = 9 \end{cases}$

First, graph each equation in the system.

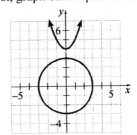

The system apparently has no solution. Now multiply the first equation by $-1$ and add the result to the second equation:
$$\begin{array}{r} -x^2 \phantom{+y^2} + y = 4 \\ x^2 + y^2 \phantom{+y} = 9 \\ \hline y^2 + y = 13 \end{array}$$
$$y^2 + y - 13 = 0$$
$$y = \frac{-1 \pm \sqrt{1^2 - 4(1)(-13)}}{2(1)}$$
$$= \frac{-1 \pm \sqrt{53}}{2}$$

Substitute these $y$-values into the first equation to find the corresponding $x$-values:

## Chapter 10: Conics

$y = \dfrac{-1-\sqrt{53}}{2}: \quad x^2 - \left(\dfrac{-1-\sqrt{53}}{2}\right) = -4$

$\qquad\qquad\qquad x^2 = -4 + \left(\dfrac{-1-\sqrt{53}}{2}\right)$

$\qquad\qquad\qquad x^2 = \dfrac{-9-\sqrt{53}}{2}$

$\qquad\qquad\qquad x = \pm\sqrt{\dfrac{-9-\sqrt{53}}{2}} \quad$ (not real)

$y = \dfrac{-1+\sqrt{53}}{2}: \quad x^2 - \left(\dfrac{-1+\sqrt{53}}{2}\right) = -4$

$\qquad\qquad\qquad x^2 = -4 + \left(\dfrac{-1+\sqrt{53}}{2}\right)$

$\qquad\qquad\qquad x^2 = \dfrac{-9+\sqrt{53}}{2}$

$\qquad\qquad\qquad x = \pm\sqrt{\dfrac{-9+\sqrt{53}}{2}} \quad$ (not real)

Since both *y*-values result in non-real *x*-values (because the values under each of the radicals are negative), the system of equations is inconsistent. The solution set is ∅.

# Chapter 11 – Quick Checks

**11.1 Quick Checks**

1. $a_n = 2n - 3$
   $a_1 = 2(1) - 3 = 2 - 3 = -1$
   $a_2 = 2(2) - 3 = 4 - 3 = 1$
   $a_3 = 2(3) - 3 = 6 - 3 = 3$
   $a_4 = 2(4) - 3 = 8 - 3 = 5$
   $a_5 = 2(5) - 3 = 10 - 3 = 7$
   The first five terms of the sequence are $-1$, 1, 3, 5, and 7.

2. $b_n = (-1)^n \cdot 4n$
   $b_1 = (-1)^1 \cdot 4(1) = -4$
   $b_2 = (-1)^2 \cdot 4(2) = 8$
   $b_3 = (-1)^3 \cdot 4(3) = -12$
   $b_4 = (-1)^4 \cdot 4(4) = 16$
   $b_5 = (-1)^5 \cdot 4(5) = -20$
   The first five terms of the sequence are $-4$, 8, $-12$, 16, and $-20$.

3. $5, 7, 9, 11, \ldots$
   The terms are consecutive odd numbers with the first term being 5. We can write the terms as follows:
   $5 = 2(1) + 3$
   $7 = 2(2) + 3$
   $9 = 2(3) + 3$
   $11 = 2(4) + 3$
   Notice that each term is 3 more than twice the term number. A formula for the $n$th term is given by $a_n = 2n + 3$.

4. $\dfrac{1}{2}, -\dfrac{1}{3}, \dfrac{1}{4}, -\dfrac{1}{5}, \ldots$
   The terms alternate sign with the first term being positive. So, $(-1)^{n+1}$ must be part of the formula. Each term is a fraction with a numerator of 1 and a denominator that is 1 more than the term number. A formula for the $n$th term is given by
   $b_n = (-1)^{n+1} \cdot \dfrac{1}{n+1}$.

5. $\displaystyle\sum_{i=1}^{3}(4i - 1) = (4 \cdot 1 - 1) + (4 \cdot 2 - 1) + (4 \cdot 3 - 1)$
   $= 4 - 1 + 8 - 1 + 12 - 1$
   $= 21$

6. $\displaystyle\sum_{i=1}^{5}(i^3 + 1) = (1^3 + 1) + (2^3 + 1) + (3^3 + 1)$
   $\qquad + (4^3 + 1) + (5^3 + 1)$
   $= (1 + 1) + (8 + 1) + (27 + 1)$
   $\qquad + (64 + 1) + (125 + 1)$
   $= 2 + 9 + 28 + 65 + 126$
   $= 230$

7. $1 + 4 + 9 + \ldots + 144$
   Notice that each term is a perfect square. We can rewrite the sum as $1^2 + 2^2 + 3^2 + \ldots + 12^2$. Thus, the sum has 12 terms, each of the form $i^2$.
   $1 + 4 + 9 + \ldots + 144 = \displaystyle\sum_{i=1}^{12} i^2$

8. Begin by writing the first term as $\dfrac{1}{1}$. We notice that each term is a fraction with a numerator of 1 and a denominator that is a power of 2. We can write the sum as
   $\dfrac{1}{2^0} + \dfrac{1}{2^1} + \dfrac{1}{2^2} + \ldots + \dfrac{1}{2^5}$
   The exponent on 2 in the denominator is always 1 less than the term number. Thus, the sum has 6 terms, each of the form $\dfrac{1}{2^{i-1}}$.
   $1 + \dfrac{1}{2} + \dfrac{1}{4} + \ldots + \dfrac{1}{32} = \displaystyle\sum_{i=1}^{6}\left(\dfrac{1}{2^{i-1}}\right)$

## Chapter 11: Sequences, Series, and the Binomial Theorem

### 11.2 Quick Checks

1. $-1-(-3) = -1+3 = 2$
   $1-(-1) = 1+1 = 2$
   $3-1 = 2$
   $5-3 = 2$
   The sequence $-3, -1, 1, 3, 5, \ldots$ is arithmetic because the difference between consecutive terms is constant. The first term is $a = -3$ and the common difference is $d = 2$.

2. $9-3 = 6$, $27-9 = 18$, $81-27 = 54$
   The sequence is not arithmetic because the difference between consecutive terms is not constant.

3. $a_n - a_{n-1} = [3n-8] - [3(n-1) - 8]$
   $= [3n-8] - [3n-11]$
   $= 3n - 8 - 3n + 11$
   $= 3$
   The sequence is arithmetic because the difference between consecutive terms is constant. The first term is $a = a_1 = 3(1) - 8 = -5$ and the common difference is $d = 3$.

4. $b_n - b_{n-1} = [n^2 - 1] - [(n-1)^2 - 1]$
   $= [n^2 - 1] - [n^2 - 2n + 1 - 1]$
   $= [n^2 - 1] - [n^2 - 2n]$
   $= n^2 - 1 - n^2 + 2n$
   $= 2n - 1$
   The sequence is not arithmetic because the difference between consecutive terms is not constant.

5. $c_n - c_{n-1} = [5 - 2n] - [5 - 2(n-1)]$
   $= [5 - 2n] - [5 - 2n + 2]$
   $= [5 - 2n] - [7 - 2n]$
   $= 5 - 2n - 7 + 2n$
   $= -2$
   The sequence is arithmetic because the difference between consecutive terms is constant. The first term is $a = c_1 = 5 - 2(1) = 3$ and the common difference is $d = -2$.

6. **a.** We have that $a_5 = 25$ and $d = 6$.
   $a_n = a + (n-1)d$
   $a_5 = a + (5-1)d$
   $25 = a + 4(6)$
   $25 = a + 24$
   $1 = a$
   Therefore, the $n$th term of the sequence is given by
   $a_n = 1 + (n-1)(6) = 1 + 6n - 6$
   $a_n = 6n - 5$

   **b.** $a_{14} = 6(14) - 5 = 84 - 5 = 79$

7. **a.** We are given $a_5 = 7$ and $a_{13} = 31$. The $n$th term of an arithmetic sequence is given by $a_n = a + (n-1)d$. Therefore, we have
   $\begin{cases} a_5 = a + (5-1)d \\ a_{13} = a + (13-1)d \end{cases}$ or
   $\begin{cases} 7 = a + 4d \quad (1) \\ 31 = a + 12d \quad (2) \end{cases}$
   This is a system of linear equations in $a$ and $d$. We can solve the system by using elimination. Subtract equation (2) from equation (1) to obtain
   $-24 = -8d$
   $3 = d$
   Let $d = 3$ in equation (1) and solve for $a$.
   $7 = a + 4(3)$
   $7 = a + 12$
   $-5 = a$
   The first term is $a = -5$ and the common difference is $d = 3$.

   **b.** A formula for the $n$th term is
   $a_n = a + (n-1)d = -5 + (n-1)(3)$
   $= -5 + 3n - 3$
   $a_n = 3n - 8$

8. We have $a = 5$ and $d = 2$, and wish to find $S_{100}$.
   $S_n = \dfrac{n}{2}[2a + (n-1)d]$
   $S_{100} = \dfrac{100}{2}[2(5) + (100-1)(2)] = 50[10 + 198]$
   $= 50(208) = 10,400$

**Section 11.3 – Quick Checks:** Geometric Sequences and Series

9. The first term is $a=1$ and the common difference is $d=5-1=4$. We wish to find $S_{70}$.

   $S_n = \dfrac{n}{2}[2a+(n-1)d]$
   
   $S_{70} = \dfrac{70}{2}[2(1)+(70-1)(4)]$
   
   $= 35[2+276]$
   
   $= 35(278)$
   
   $= 9,730$

10. We have $a=4$ and $a_{50}=298$, and wish to find the sum of the first 50 terms, $S_{50}$.

    $S_n = \dfrac{n}{2}[a+a_n]$
    
    $S_{50} = \dfrac{50}{2}[a+a_{50}]$
    
    $= \dfrac{50}{2}[4+298]$
    
    $= 25(302)$
    
    $= 7,550$

11. $a_n = -3n+100$
    
    $a = a_1 = -3(1)+100 = 97$
    
    $a_{75} = -3(75)+100 = -125$
    
    $S_n = \dfrac{n}{2}[a+a_n]$
    
    $S_{75} = \dfrac{75}{2}[a+a_{75}]$
    
    $S_{75} = \dfrac{75}{2}[97+(-125)]$
    
    $= \dfrac{75}{2}(-28)$
    
    $= -1,050$

12. If we let $a_n$ = the number of seats in the $n$th row, then we have an arithmetic sequence with first term $a=20$ and common difference $d=2$. Thus, the total number of seats in the 30 rows is given by $S_{30}$.

    $S_n = \dfrac{n}{2}[2a+(n-1)d]$
    
    $S_{30} = \dfrac{30}{2}[2a+(30-1)d] = 15[2(20)+29(2)]$
    
    $= 15(98) = 1,470$
    
    There are 1,470 seats in the 30 rows.

**11.3 Quick Checks**

1. $\dfrac{8}{4}=2,\ \dfrac{16}{8}=2,\ \dfrac{32}{16}=2,\ \dfrac{64}{32}=2$

   The sequence is geometric because the ratio of consecutive terms is constant. The first term is $a=4$ and the common ratio is $r=2$.

2. $\dfrac{10}{5}=2,\ \dfrac{16}{10}=\dfrac{8}{5}$

   The sequence is not geometric because the ratio of consecutive terms is not constant.

3. $\dfrac{3}{9}=\dfrac{1}{3},\ \dfrac{1}{3},\ \dfrac{\frac{1}{3}}{1}=\dfrac{1}{3},\ \dfrac{\frac{1}{9}}{\frac{1}{3}}=\dfrac{1}{9}\cdot 3=\dfrac{1}{3}$

   The sequence is geometric because the ratio of consecutive terms is constant. The first term is $a=9$ and the common ratio is $r=\dfrac{3}{9}=\dfrac{1}{3}$.

4. $\dfrac{a_n}{a_{n-1}} = \dfrac{5^n}{5^{n-1}} = \dfrac{5\cdot 5^{n-1}}{5^{n-1}} = 5$

   The sequence is geometric because the ratio of consecutive terms is constant. The first term is $a=a_1=5^1=5$ and the common ratio is $r=5$.

5. $\dfrac{b_n}{b_{n-1}} = \dfrac{n^2}{(n-1)^2} = \dfrac{n^2}{n^2-2n+1}$

   The sequence is not geometric because the ratio of consecutive terms is not constant.

6. $\dfrac{c_n}{c_{n-1}} = \dfrac{5\left(\frac{2}{3}\right)^n}{5\left(\frac{2}{3}\right)^{n-1}} = \dfrac{5\cdot \frac{2}{3}\cdot \left(\frac{2}{3}\right)^{n-1}}{5\left(\frac{2}{3}\right)^{n-1}} = \dfrac{2}{3}$

   The sequence is geometric because the ratio of consecutive terms is constant. The first term is $a=a_1=5\left(\dfrac{2}{3}\right)^1=\dfrac{10}{3}$ and the common ratio is $r=\dfrac{2}{3}$.

7. The $n$th term of a geometric sequence is given by $a_n = a\cdot r^{n-1}$. We are given $a=5$ and $r=2$. Therefore, a formula for the $n$th term of the sequence is $a_n = 5\cdot 2^{n-1}$.
   
   $a_9 = 5\cdot 2^{9-1} = 5\cdot 2^8 = 5(256) = 1,280$

QC – 133

## Chapter 11: Sequences, Series, and the Binomial Theorem

**8.** The $n$th term of a geometric sequence is given by $a_n = a \cdot r^{n-1}$. We are given that the first term is $a = 50$ and the common ratio is $r = \dfrac{25}{50} = \dfrac{1}{2}$.
Therefore, a formula for the $n$th term of the sequence is $a_n = 50 \cdot \left(\dfrac{1}{2}\right)^{n-1}$.

$a_9 = 50 \cdot \left(\dfrac{1}{2}\right)^{9-1} = 50 \cdot \left(\dfrac{1}{2}\right)^8 = \dfrac{50}{256}$

$= \dfrac{25}{128} = 0.1953125$

**9.** We wish to find the sum of the first 13 terms of a geometric sequence with first term $a = 3$ and common ratio $r = \dfrac{6}{3} = 2$.

$S_n = a \cdot \dfrac{1 - r^n}{1 - r}$

$S_{13} = 3 \cdot \dfrac{1 - 2^{13}}{1 - 2} = 3(8191) = 24{,}573$

**10.** We wish to find the sum of the first 10 terms of a geometric sequence with first term $a = 8\left(\dfrac{1}{2}\right)^1 = 4$ and common ratio $r = \dfrac{1}{2}$.

$S_n = a \cdot \dfrac{1 - r^n}{1 - r}$

$S_{10} = 4 \cdot \dfrac{1 - \left(\dfrac{1}{2}\right)^{10}}{1 - \dfrac{1}{2}} = 4\left(\dfrac{\tfrac{1023}{1024}}{\tfrac{1}{2}}\right) = 4\left(\dfrac{1023}{512}\right)$

$= \dfrac{1023}{128} = 7.9921875$

**11.** This is an infinite geometric series with $a = 10$ and $r = \dfrac{\tfrac{5}{2}}{10} = \dfrac{5}{2} \cdot \dfrac{1}{10} = \dfrac{1}{4}$. Since the common ratio is between $-1$ and 1, we can use the formula for the sum of an infinite geometric series to find

$S_\infty = \dfrac{a}{1 - r}$

$S_\infty = \dfrac{10}{1 - \dfrac{1}{4}} = \dfrac{10}{\tfrac{3}{4}} = 10 \cdot \dfrac{4}{3} = \dfrac{40}{3}$

**12.** This is an infinite geometric series with $a = a_1 = \left(\dfrac{1}{3}\right)^1 = \dfrac{1}{3}$ and $r = \dfrac{1}{3}$. Since the common ratio is between $-1$ and 1, we can use the formula for the sum of an infinite geometric series to find

$S_\infty = \dfrac{a}{1 - r}$

$S_\infty = \dfrac{\tfrac{1}{3}}{1 - \tfrac{1}{3}} = \dfrac{\tfrac{1}{3}}{\tfrac{2}{3}} = \dfrac{1}{3} \cdot \dfrac{3}{2} = \dfrac{1}{2}$

**13.** The line over the 2 indicates that the 2 repeats indefinitely. That is, we can write
$0.\overline{2} = 0.2 + 0.02 + 0.002 + 0.0002 + \ldots$
This is an infinite geometric series with $a = 0.2$ and common ratio $r = 0.1$. Since the common ratio is between $-1$ and 1, we can use the formula for the sum of a geometric series to find that
$0.\overline{2} = 0.2 + 0.02 + 0.002 + 0.0002 + \ldots$

$= \dfrac{0.2}{1 - 0.1}$

$= \dfrac{0.2}{0.9}$

$= \dfrac{2}{9}$

**14.** The total impact of the $500 tax rebate on the U.S. economy is

$\$500 + \$500(0.95) + \$500(0.95)^2 + \$500(0.95)^3 + \ldots$

This is an infinite geometric series with first term $a = 500$ and common ratio $r = 0.95$. The sum of this series is

$\$500 + \$500(0.95) + \$500(0.95)^2 + \ldots$

$= \dfrac{\$500}{1 - 0.95}$

$= \dfrac{\$500}{0.05}$

$= \$10{,}000$

The U.S. economy will grow by $10,000 because of the child tax credit to Roberta.

### Section 11.4 – Quick Checks: The Binomial Theorem

15. This is an ordinary annuity with $n = 4 \cdot 30 = 120$ payments with deposits of $P = \$500$. The interest rate per payment period is $i = \dfrac{0.08}{4} = 0.02$. This gives

$$A = 500 \left[ \dfrac{(1+0.02)^{120} - 1}{0.02} \right]$$
$$= 500(488.258152)$$
$$= 244,129.08$$

After 30 years, the IRA will be worth $244,129.08.

### 11.4 Quick Checks

1. $9! = 9 \cdot 8 \cdot 7 \cdot 6 \cdot 5 \cdot 4 \cdot 3 \cdot 2 \cdot 1 = 362,880$

2. $\dfrac{7!}{3!} = \dfrac{7 \cdot 6 \cdot 5 \cdot 4 \cdot 3!}{3!} = 7 \cdot 6 \cdot 5 \cdot 4 = 840$

3. $\binom{7}{1} = \dfrac{7!}{1! \cdot (7-1)!} = \dfrac{7!}{6!} = \dfrac{7 \cdot 6!}{6!} = 7$

4. $\binom{6}{3} = \dfrac{6!}{3! \cdot (6-3)!} = \dfrac{6 \cdot 5 \cdot 4 \cdot 3!}{3! \cdot 3!} = \dfrac{6 \cdot 5 \cdot 4}{3 \cdot 2 \cdot 1} = 20$

5. $(x+2)^4 = \binom{4}{0}x^4 + \binom{4}{1}2^1 \cdot x^{4-1} + \binom{4}{2}2^2 \cdot x^{4-2} + \binom{4}{3}2^3 \cdot x^{4-3} + \binom{4}{4}2^4$
$= 1 \cdot x^4 + 4 \cdot 2 \cdot x^3 + 6 \cdot 4 \cdot x^2 + 4 \cdot 8 \cdot x + 1 \cdot 16$
$= x^4 + 8x^3 + 24x^2 + 32x + 16$

6. $(2p - 1)^5$
$= (2p + (-1))^5$
$= \binom{5}{0}(2p)^5 + \binom{5}{1}(-1)^1 \cdot (2p)^{5-1} + \binom{5}{2}(-1)^2 \cdot (2p)^{5-2} + \binom{5}{3}(-1)^3 \cdot (2p)^{5-3} + \binom{5}{4}(-1)^4 \cdot (2p)^{5-4} + \binom{5}{5}(-1)^5$
$= 1 \cdot (2p)^5 + 5 \cdot (-1) \cdot (2p)^4 + 10 \cdot (-1)^2 \cdot (2p)^3 + 10 \cdot (-1)^3 \cdot (2p)^2 + 5 \cdot (-1)^4 \cdot (2p) + 1 \cdot (-1)^5$
$= 32p^5 + 5(-1)(16p^4) + 10(1)(8p^3) + 10(-1)(4p^2) + 5(1)(2p) + 1(-1)$
$= 32p^5 - 80p^4 + 80p^3 - 40p^2 + 10p - 1$

# Chapter R

**R.1 Exercises**

Answers will vary

**R.2 Exercises**

2. subset; $\subseteq$

4. irrational numbers

6. False. A set with no elements is called the empty set, but it is denoted by $\{\ \}$ or $\varnothing$.

8. False. Decimals that terminate or repeat represent rational numbers, but decimals that do not repeat and do not terminate represent irrational numbers.

10. True. The set of rational numbers and the set of irrational numbers are disjoint.

12. Suppose we have $a + b = c$ where $a$ is a rational number and $b$ is an irrational number. If we assume that $c$ is a rational number, then $c + (-a) = b$ must be rational since the sum of two rational numbers is rational. This contradicts our initial statement that $b$ was irrational. Therefore, $c$ must be irrational.

14. Since there are an infinite number of rational numbers, it would not be possible to list all the elements of the set of rational numbers.

16. $0.45$ is a terminating decimal while $0.\overline{45}$ is a repeating decimal. Both numbers are rational.

18. There is no positive real number that is "closest" to 0. Since every continuous segment on the real number line contains an infinite number of numbers, we can always find a smaller positive number that is closer to 0 than the one we selected.
    For example,
    $0.1, 0.01, 0.001, 0.0001, 0.00001, 0.000001, \ldots$

20. $\{1, 2, 3\}$

22. $\{-3, -2, -1, 0, 1, 2, 3, 4, 5\}$

24. $\varnothing$ or $\{\ \}$

26. The statement $A \subseteq C$ is true since all the elements of $A$ are also elements of $C$.

28. The statement $A \subset C$ is true because all the elements in set $A$ are also in set $C$ and there are elements in set $C$ that are not in set $A$.

30. The statement $D \subset B$ is false. In order for set $D$ to be a proper subset of set $B$, all the elements of set $D$ must be in the set $B$ and there must be elements in set $B$ that are not in set $D$. Since $B = D$, the statement is false.

32. The statement $\varnothing \subset B$ is true because the empty set is a subset of every set, and the set $B$ is not empty.

34. $4.\overline{5} \in \{x \mid x \text{ is a rational number}\}$

36. $0 \notin \{x \mid x \text{ is a counting number}\}$

38. a. 13

    b. 0, 13

    c. $-4.5656\ldots, 0, 2.43, 13$

    d. $\sqrt{2}$

    e. $-4.5656\ldots, 0, \sqrt{2}, 2.43, 13$

40. a. 15

    b. $-\dfrac{6}{1}, 15$

    c. $-\dfrac{6}{1}, 7.3, 15$

    d. $\sqrt{2} + \pi$

    e. $-\dfrac{6}{1}, \sqrt{2} + \pi, 7.3, 15$

42. a. To truncate, we remove all digits after the second decimal place. So, $-93.432101$ truncated to two decimal places is $-93.43$.

    b. To round to two decimal places, we first examine the digit in the third decimal place and then truncate. Since the third decimal place contains a 2, we do not change the digit in the second decimal place. Thus,

## Chapter R: Real Numbers and Algebraic Expressions

-93.432101 rounded to two decimal places is $-93.43$.

**44. a.** To truncate, we remove all digits after the second decimal place. So, 9.9999 truncated to two decimal places is $9.99$.

  **b.** To round to two decimal places, we first examine the digit in the third decimal place and then truncate. Since the third decimal place contains a 9, we increase the digit in the second decimal place by 1. Because the digit in the second decimal place is a 9, we will need to carry. So, 9.9999 rounded to two decimal places is 10.00.

**46.**

**48.** $4 > 2$ since 4 is further to the right on a real number line.

**50.** $\frac{2}{3} > \frac{2}{5}$ since $\frac{2}{3}$ is further to the right on a real number line.

**52.** $-\frac{8}{3} < -\frac{8}{5}$ since $-\frac{8}{3}$ is further to the left on a real number line.

### R.3 Exercises

**2.** distributive property

**4.** False. $|0| = 0$ which is not a positive number.

**6.** True

**8.** Suppose 0 had a multiplicative inverse, $a$; then, $a \cdot 0 = 1$ but $a \cdot 0 = 0$. Therefore, 0 cannot have a multiplicative inverse.

**10.** 2 is not a factor of both the numerator and denominator, so it cannot be divided out.

**12.** No, the grouping of subtractions is important. For example:
$$10 - (5-2) \stackrel{?}{=} (10-5) - 2$$
$$10 - 3 \stackrel{?}{=} 5 - 2$$
$$7 \neq 3$$

**54.** $0 - 1349 = -1349$
Since sea level is an elevation of 0 ft, the Dead Sea has an elevation of $-1349$ feet.

**56.** $-\$0.08$
The value of Sun Microsystems common stock fell \$0.08 during the first quarter of 2004.

**58.** $57 - 80 = -23$
The departure from the normal high temperature was $-23°F$.

**60.** Answers will vary. One possible starting point:
http://www-gap.dcs.st-and.ac.uk/~history/HistTopics/Zero.html

**62.** Answers will vary. One possible starting point:
http://www-gap.dcs.st-and.ac.uk/~history/Mathematicians/Euler.html

**64.** Rounded: $\frac{19}{7} \approx 2.7143$

Truncated: $\frac{19}{7} \approx 2.7142$

```
19/7
      2.714285714
```

**14.** No, the grouping of divisions is important. For example:
$$50 \div (10 \div 5) \stackrel{?}{=} (50 \div 10) \div 5$$
$$50 \div 2 \stackrel{?}{=} 5 \div 5$$
$$25 \neq 1$$

**16. a.** $-\left(\frac{16}{5}\right) = -\frac{16}{5}$

The additive inverse of $\frac{16}{5}$ is $-\frac{16}{5}$.

**b.** $\frac{1}{16/5} = \frac{5}{16}$

The multiplicative inverse of $\frac{16}{5}$ is $\frac{5}{16}$.

**c.** $\frac{1}{16/5} = \frac{5}{16}$

The reciprocal of $\frac{16}{5}$ is $\frac{5}{16}$.

**Section R.3** *Operations on Signed Numbers; Properties of Real Numbers*

18. a. $-(-73) = 73$
    The additive inverse of $-73$ is $73$.

    b. $\dfrac{1}{-73} = -\dfrac{1}{73}$
    The multiplicative inverse of $-73$ is $-\dfrac{1}{73}$.

    c. $\dfrac{1}{-73} = -\dfrac{1}{73}$
    The reciprocal of $-73$ is $-\dfrac{1}{73}$.

20. a. $-\left(-\dfrac{1}{5}\right) = \dfrac{1}{5}$
    The additive inverse of $-\dfrac{1}{5}$ is $\dfrac{1}{5}$.

    b. $\dfrac{1}{(-1/5)} = -\dfrac{5}{1} = -5$
    The multiplicative inverse of $-\dfrac{1}{5}$ is $-5$.

    c. $\dfrac{1}{(-1/5)} = -\dfrac{5}{1} = -5$
    The reciprocal of $-\dfrac{1}{5}$ is $-5$.

22. a. $-(10) = -10$
    The additive inverse of $10$ is $-10$.

    b. $\dfrac{1}{(10)} = \dfrac{1}{10}$
    The multiplicative inverse of $10$ is $\dfrac{1}{10}$.

    c. $\dfrac{1}{(10)} = \dfrac{1}{10}$
    The reciprocal of $10$ is $\dfrac{1}{10}$.

24. a. $-(-6) = 6$
    The additive inverse of $-6$ is $6$.

    b. $\dfrac{1}{(-6)} = -\dfrac{1}{6}$
    The multiplicative inverse of $-6$ is $-\dfrac{1}{6}$.

    c. $\dfrac{1}{(-6)} = -\dfrac{1}{6}$
    The reciprocal of $-6$ is $-\dfrac{1}{6}$.

26. a. $-\left(-\dfrac{5}{4}\right) = \dfrac{5}{4}$
    The additive inverse of $-\dfrac{5}{4}$ is $\dfrac{5}{4}$.

    b. $\dfrac{1}{(-5/4)} = -\dfrac{1}{(5/4)} = -\dfrac{4}{5}$
    The multiplicative inverse of $-\dfrac{5}{4}$ is $-\dfrac{4}{5}$.

    c. $\dfrac{1}{(-5/4)} = -\dfrac{1}{(5/4)} = -\dfrac{4}{5}$
    The reciprocal of $-\dfrac{5}{4}$ is $-\dfrac{4}{5}$.

28. $3(y-5) = 3 \cdot y - 3 \cdot 5 = 3y - 15$

30. $5(x+4) = 5 \cdot x + 5 \cdot 4 = 5x + 20$

32. $(3x+y) \cdot 2 = 3x \cdot 2 + y \cdot 2 = 6x + 2y$

34. $-\dfrac{2}{3}(3x+15) = -\dfrac{2}{3} \cdot 3x + \left(-\dfrac{2}{3}\right) \cdot 15$
    $= -\dfrac{6}{3}x - \dfrac{30}{3}$
    $= -2x - 10$

36. $\dfrac{7 \cdot y}{35} = \dfrac{7 \cdot y}{7 \cdot 5} = \dfrac{\cancel{7} \cdot y}{\cancel{7} \cdot 5} = \dfrac{y}{5}$

38. $\dfrac{40}{16} = \dfrac{8 \cdot 5}{8 \cdot 2} = \dfrac{\cancel{8} \cdot 5}{\cancel{8} \cdot 2} = \dfrac{5}{2}$

40. $-6 + 10 = 4$

42. $9 + (-3) = 6$

44. $|-4| + 12 = 4 + 12 = 16$

46. $-8.2 - 4.5 = -8.2 + (-4.5) = -12.7$

48. $-5 \cdot (-15) = 5 \cdot 15 = 75$

ISM – 3

Chapter R: *Real Numbers and Algebraic Expressions*

**50.** $7 \cdot (-15) = -(7 \cdot 15) = -105$

**52.** $\dfrac{2}{5} \cdot \dfrac{15}{8} = \dfrac{2 \cdot 15}{5 \cdot 8} = \dfrac{2 \cdot 3 \cdot 5}{5 \cdot 2 \cdot 4} = \dfrac{\cancel{2} \cdot 3 \cdot \cancel{5}}{\cancel{5} \cdot \cancel{2} \cdot 4} = \dfrac{3}{4}$

**54.** $-\dfrac{7}{3} \cdot \left(-\dfrac{12}{35}\right) = \dfrac{7}{3} \cdot \dfrac{12}{35} = \dfrac{7 \cdot 3 \cdot 4}{3 \cdot 5 \cdot 7} = \dfrac{\cancel{7} \cdot \cancel{3} \cdot 4}{\cancel{3} \cdot 5 \cdot \cancel{7}} = \dfrac{4}{5}$

**56.** $-\dfrac{10}{3} \div \dfrac{15}{21} = -\dfrac{10}{3} \cdot \dfrac{21}{15}$
$= -\dfrac{2 \cdot 5 \cdot 3 \cdot 7}{3 \cdot 3 \cdot 5}$
$= -\dfrac{2 \cdot \cancel{5} \cdot \cancel{3} \cdot 7}{3 \cdot \cancel{3} \cdot \cancel{5}}$
$= -\dfrac{14}{3}$

**58.** $\dfrac{\frac{18}{7}}{\frac{3}{14}} = \dfrac{18}{7} \cdot \dfrac{14}{3} = \dfrac{18 \cdot 14}{7 \cdot 3}$
$= \dfrac{3 \cdot 6 \cdot 7 \cdot 2}{7 \cdot 3}$
$= \dfrac{\cancel{3} \cdot 6 \cdot \cancel{7} \cdot 2}{\cancel{7} \cdot \cancel{3}}$
$= 12$

**60.** $\dfrac{8}{5} - \dfrac{18}{5} = \dfrac{8-18}{5} = -\dfrac{10}{5} = -\dfrac{5 \cdot 2}{5} = -\dfrac{\cancel{5} \cdot 2}{\cancel{5}} = -2$

**62.** $LCD = 2 \cdot 2 \cdot 2 \cdot 3 \cdot 3 = 72$
$\dfrac{3}{8} - \dfrac{7}{18} = \dfrac{3}{8} \cdot \dfrac{9}{9} - \dfrac{7}{18} \cdot \dfrac{4}{4}$
$= \dfrac{27}{72} - \dfrac{28}{72}$
$= \dfrac{27-28}{72}$
$= -\dfrac{1}{72}$

**64.** $LCD = 2 \cdot 2 \cdot 2 \cdot 3 \cdot 3 \cdot 5 = 360$
$-\dfrac{17}{45} - \dfrac{23}{24} = -\dfrac{17}{45} \cdot \dfrac{8}{8} - \dfrac{23}{24} \cdot \dfrac{15}{15}$
$= -\dfrac{136}{360} - \dfrac{345}{360}$
$= \dfrac{-136 + (-345)}{360}$
$= -\dfrac{481}{360}$

**66.** $\dfrac{2}{3} \div 8 = \dfrac{2}{3} \cdot \dfrac{1}{8} = \dfrac{2 \cdot 1}{3 \cdot 2 \cdot 4} = \dfrac{\cancel{2} \cdot 1}{3 \cdot \cancel{2} \cdot 4} = \dfrac{1}{12}$

**68.** $|-5.4 + 10.5| = |5.1| = 5.1$

**70.** $-|-5 \cdot 9| = -|-45| = -(45) = -45$

**72.** $\left|\dfrac{4}{3} - \dfrac{8}{7}\right| = \left|\dfrac{4}{3} \cdot \dfrac{7}{7} - \dfrac{8}{7} \cdot \dfrac{3}{3}\right|$
$= \left|\dfrac{28}{21} - \dfrac{24}{21}\right|$
$= \left|\dfrac{4}{21}\right|$
$= \dfrac{4}{21}$

**74.** $\left|\dfrac{4}{15} - \dfrac{1}{6}\right| = \left|\dfrac{4}{15} \cdot \dfrac{2}{2} - \dfrac{1}{6} \cdot \dfrac{5}{5}\right|$
$= \left|\dfrac{8}{30} - \dfrac{5}{30}\right|$
$= \left|\dfrac{8-5}{30}\right|$
$= \left|\dfrac{3}{30}\right|$
$= \dfrac{3}{30} = \dfrac{3}{3 \cdot 10} = \dfrac{\cancel{3}}{\cancel{3} \cdot 10} = \dfrac{1}{10}$

**76.** $-\dfrac{10}{21} + 6 = -\dfrac{10}{21} + \dfrac{6}{1} \cdot \dfrac{21}{21} = -\dfrac{10}{21} + \dfrac{126}{21}$
$= \dfrac{-10 + 126}{21} = \dfrac{116}{21}$

**78.** $\dfrac{20}{\frac{5}{4}} = \dfrac{20}{1} \cdot \dfrac{4}{5} = \dfrac{5 \cdot 4 \cdot 4}{5} = \dfrac{\cancel{5} \cdot 4 \cdot 4}{\cancel{5}} = \dfrac{16}{1} = 16$

**80.** $9 + (-9) = 0$
Additive Inverse Property

**82.** $\dfrac{a}{a} = 1, \ a \neq 0$
Division Property

**84.** $3(x - 4) = 3x - 12$
Distributive Property

**Section R.3** Operations on Signed Numbers; Properties of Real Numbers

86. $\dfrac{0}{6} = 0$

    Division Property

88. $78 - 49 = 78 + (-49) = 29$

    The difference in life expectancy from 1950 to 1998 is 29 years.

90. $400 - 20 - 45 - 60 - 105 + 150$
    $= (400 + 150) - (20 + 45 + 60 + 105)$
    $= 550 - 230$
    $= 550 + (-230)$
    $= 320$

    Paul's balance at the end of the month was $320.

92. $535 - (-8) = 535 + 8 = 543$

    The difference between the highest and lowest elevations in Louisiana is 543 feet.

94. Multiplying two numbers, say $a$ and $b$, means that we add the number $a$ a total of $b$ times. Since subtraction is the same as adding a negative, we will subtract a total of $b$ times if $a$ is negative. On a number line, we would start at 0 and move to the left $a$ units (since $a$ is negative) a total of $b$ times. Since we keep moving to the left, the end result will be negative. For example, $-5 \cdot 3$ means we would subtract the number 5 three times. That is, $-5 \cdot 3 = -5 - 5 - 5 = -15$
    On a number line we would get:

    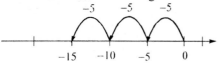

96. $d(P,Q) = |10 - (-4)| = |10 + 4| = |14| = 14$

97. $d(P,Q) = |7.2 - (-3.2)| = |7.2 + 3.2| = |10.4| = 10.4$

100. $d(P,Q) = \left|\dfrac{6}{5} - \left(-\dfrac{10}{3}\right)\right| = \left|\dfrac{6}{5} + \dfrac{10}{3}\right|$
    $= \left|\dfrac{6}{5} \cdot \dfrac{3}{3} + \dfrac{10}{3} \cdot \dfrac{5}{5}\right| = \left|\dfrac{18}{15} + \dfrac{50}{15}\right|$
    $= \left|\dfrac{18 + 50}{15}\right| = \left|\dfrac{68}{15}\right|$
    $= \dfrac{68}{15}$

102. a. $a \cdot 0 = 0$

    b. $a \cdot 0 = 0$
    $a \cdot (b + (-b)) = 0$

    c. $a \cdot (b + (-b)) = 0$
    $ab + a(-b) = 0$

    d. If $a < 0$ and $b > 0$, then $ab < 0$ since the product of a negative and a positive is negative. In addition, $-b < 0$. Now, if $ab < 0$ then $a(-b)$ must be positive so that the sum $ab + a(-b)$ is 0. Since $a < 0$ and $-b < 0$, we have that the product of two negative numbers is a positive number.

104. $\left\{\pm\dfrac{9}{4}, \pm\dfrac{9}{2}, \pm 9, \pm\dfrac{3}{4}, \pm\dfrac{3}{2}, \pm 3, \pm\dfrac{1}{4}, \pm\dfrac{1}{2}, \pm 1\right\}$

106. ```
-2.9+(-6.3)
           -9.2
```

108. ```
-5.4(-4.8)
         25.92
```

110. ```
-(3/8)+(1/10)▶Fr
ac
           -11/40
```
```

Chapter R: Real Numbers and Algebraic Expressions

112. abs(-3.65)*abs(5
.4)
          19.71

114. -(3/4)/(9/20)▸Fr
ac
          -5/3

### R.4 Exercises

2. parentheses (grouping symbols); exponents; multiplication; division; addition; subtraction

4. True

6. While numerically equivalent, $-4^3$ and $(-4)^3$ are different in the order of operations. To evaluate $-4^3$, we cube 4 and then multiply the result by $-1$. To evaluate $(-4)^3$, we multiply 4 by $-1$ and then cube the result.

8. Answers will vary.

10. $(-3)^2 + (-2)^2 = 9 + 4 = 13$

12. $-5 \cdot 3 + 12 = -15 + 12 = -3$

14. $2 + 5 \cdot (8 - 5) = 2 + 5 \cdot 3 = 2 + 15 = 17$

16. $3[15 - (7 - 3)] = 3[15 - 4] = 3(11) = 33$

18. $\dfrac{12-4}{-2} = \dfrac{8}{-2} = \dfrac{4 \cdot 2}{-1 \cdot 2} = \dfrac{4 \cdot \cancel{2}}{-1 \cdot \cancel{2}} = -4$

20. $|6 \cdot 2 - 5 \cdot 3| = |12 - 15| = |-3| = 3$

22. $15 \cdot \dfrac{3}{5} + 4 \cdot 3 = \dfrac{45}{5} + 12 = 9 + 12 = 21$

24. $2[25 - 2(10 - 4)] = 2[25 - 2(6)]$
$= 2[25 - 12]$
$= 2(13)$
$= 26$

26. $-2(5-2) - (-5)^2 = -2(3) - (-5)^2$
$= -2(3) - (25)$
$= -6 - 25$
$= -31$

28. $-2(4 + |2 \cdot 3 - 5^2|) = -2(4 + |2 \cdot 3 - 25|)$
$= -2(4 + |6 - 25|)$
$= -2(4 + |-19|)$
$= -2(4 + 19)$
$= -2(23)$
$= -46$

30. $|6 \cdot (3 \cdot 2 - 10)| = |6 \cdot (6 - 10)|$
$= |6(-4)|$
$= |-24|$
$= 24$

32. $\dfrac{2 \cdot 3^2}{4^2 - 4} = \dfrac{2 \cdot 9}{16 - 4} = \dfrac{18}{16 - 4} = \dfrac{18}{12} = \dfrac{\cancel{6} \cdot 3}{\cancel{6} \cdot 2} = \dfrac{3}{2}$

34. $\dfrac{3 \cdot (5 + 2^2)}{2 \cdot 3^3} = \dfrac{3(5+4)}{2 \cdot 27} = \dfrac{3 \cdot 9}{2 \cdot 3 \cdot 9} = \dfrac{\cancel{3} \cdot \cancel{9}}{2 \cdot \cancel{3} \cdot \cancel{9}} = \dfrac{1}{2}$

36. $\dfrac{4[3 + 2(8 - 6)]}{5[14 - 2(2 + 3)]} = \dfrac{4[3 + 2(2)]}{5[14 - 2(5)]}$
$= \dfrac{4[3 + 4]}{5[14 - 10]}$
$= \dfrac{4(7)}{5(4)} = \dfrac{\cancel{4} \cdot 7}{5 \cdot \cancel{4}}$
$= \dfrac{7}{5}$

38. $\left(\dfrac{3^2}{29 - 3 \cdot 2^3}\right) \cdot \dfrac{5}{4 + 5} = \left(\dfrac{9}{29 - 3 \cdot 8}\right) \cdot \dfrac{5}{9}$
$= \left(\dfrac{9}{29 - 24}\right) \cdot \dfrac{5}{9}$
$= \dfrac{9}{5} \cdot \dfrac{5}{9}$
$= 1$

**Section R.4** Order of Operations

40. $2^5 + 4(-5) + 4^2 = 32 + 4(-5) + 16$
$= 32 - 20 + 16$
$= 12 + 16$
$= 28$

42. $\dfrac{5 \cdot (37 - 6^2)}{6 \cdot 2 - 3^2} + \dfrac{7 \cdot 2 - 4^2}{5 + 4} = \dfrac{5(37 - 36)}{6 \cdot 2 - 9} + \dfrac{7 \cdot 2 - 16}{5 + 4}$
$= \dfrac{5(1)}{12 - 9} + \dfrac{14 - 16}{5 + 4}$
$= \dfrac{5}{3} + \dfrac{-2}{9}$
$= \dfrac{5}{3} \cdot \dfrac{3}{3} - \dfrac{2}{9}$
$= \dfrac{15}{9} - \dfrac{2}{9}$
$= \dfrac{13}{9}$

44. $\dfrac{\dfrac{4}{4^2 - 1} - \dfrac{3}{5 \cdot (7 - 5)}}{\dfrac{-(-2)^2}{4 \cdot 7 + 2}} = \dfrac{\dfrac{4}{16 - 1} - \dfrac{3}{5(2)}}{\dfrac{-(4)}{28 + 2}} = \dfrac{\dfrac{4}{15} - \dfrac{3}{10}}{\dfrac{-4}{30}}$
$= \dfrac{\dfrac{4}{15} \cdot \dfrac{2}{2} - \dfrac{3}{10} \cdot \dfrac{3}{3}}{\dfrac{-4}{30}} = \dfrac{\dfrac{8}{30} - \dfrac{9}{30}}{\dfrac{-4}{30}}$
$= \dfrac{\dfrac{-1}{30}}{\dfrac{-4}{30}} = \left(-\dfrac{1}{30}\right) \cdot \left(-\dfrac{30}{4}\right)$
$= \dfrac{1}{4}$

46. $-2 \cdot (3 - 5) = 4$

48. $(3 + 5) \cdot (6 - 3) = 24$

50. $4 \cdot 3.1416 \cdot 3^2 = 4 \cdot 3.1416 \cdot 9$
$= 12.5664 \cdot 9$
$= 113.0976$
$\approx 113.10$
The surface area is about 113.10 square centimeters.

52. $\dfrac{10^2 \cdot 8}{2.5} = \dfrac{100 \cdot 8}{2.5} = \dfrac{800}{2.5} = 320$
The engine has a rating of 320 horsepower.

54. Answers will vary.

56. $\dfrac{105 + 80 + 115 + 95 + 105}{5} = \dfrac{500}{5} = \dfrac{\cancel{5} \cdot 100}{\cancel{5}} = 100$

58. $\dfrac{65 - 50}{10} = \dfrac{15}{10} = \dfrac{3 \cdot \cancel{5}}{2 \cdot \cancel{5}} = \dfrac{3}{2}$

60. $3 - \left(\dfrac{6}{5}\right)^3 = 3 - \dfrac{216}{125} = \dfrac{375}{125} - \dfrac{216}{125} = \dfrac{159}{125}$

62. $\dfrac{3^2 - 2^3}{1 + 3 \cdot 2} = \dfrac{9 - 8}{1 + 6} = \dfrac{1}{7}$

64. $2.3^4 \cdot \dfrac{4}{11} - (3.7)^2 \cdot \dfrac{8}{3} \approx -26.33$

66. $6.3^2 + 4.2^2 = 39.69 + 17.64 = 57.33$

Chapter R: Real Numbers and Algebraic Expressions

**R.5 Exercises**

2. constant

4. False. Like terms contain the same variables with the same corresponding exponents. The terms $4x^2$ and $-10y^2$ do not have the same variables.

6. Answers will vary. A *variable* is a letter used to represent any number from a given set of numbers. A *constant* is a letter used to represent a fixed (though possibly unknown) value.

8. Answers will vary. Like terms are terms that have the same variable(s) and the same corresponding exponents on the variables. The Distributive Property is used to first remove parentheses and then later 'in reverse' to combine the coefficients of like terms.

10. $-5x+1$
    $-5(3)+1 = -15+1 = -14$

12. $y^2 - 4y + 5$
    $(3)^2 - 4(3) + 5 = 9 - 12 + 5 = -3 + 5 = 2$

14. $-2z^2 + z + 3$
    $-2(-4)^2 + (-4) + 3 = -2(16) - 4 + 3$
    $= -32 - 1$
    $= -33$

16. $\dfrac{4z+3}{z^2-4}$
    $\dfrac{4(3)+3}{(3)^2-4} = \dfrac{12+3}{9-4} = \dfrac{15}{5} = \dfrac{\cancel{5} \cdot 3}{\cancel{5}} = 3$

18. $\dfrac{2x^2+5x+2}{x^2+5x+6}$
    $\dfrac{2(3)^2+5(3)+2}{(3)^2+5(3)+6} = \dfrac{2(9)+15+2}{9+15+6}$
    $= \dfrac{18+17}{30}$
    $= \dfrac{35}{30} = \dfrac{\cancel{5} \cdot 7}{\cancel{5} \cdot 6}$
    $= \dfrac{7}{6}$

20. $|x^2 - 6x + 1|$
    $|(2)^2 - 6(2) + 1| = |4 - 12 + 1| = |-7| = 7$

22. $|4z - 1|$
    $\left|4\left(-\dfrac{5}{2}\right) - 1\right| = \left|-\dfrac{20}{2} - 1\right| = \left|-\dfrac{\cancel{2} \cdot 10}{\cancel{2}} - 1\right|$
    $= |-10 - 1|$
    $= |-11|$
    $= 11$

24. $\dfrac{|3-5z|}{(z-4)^2}$
    $\dfrac{|3-5(4)|}{((4)-4)^2} = \dfrac{|3-20|}{(4-4)^2} = \dfrac{|-17|}{0}$
    The expression is undefined when $z = 4$ since it yields division by 0 which is undefined.

26. $5y + 2y = (5+2)y = 7y$

28. $8x - 9x + 1 = (8-9)x + 1 = -x + 1$

30. $-10x + 6 + 4x - x + 1 = -10x + 4x - x + 6 + 1$
    $= (-10 + 4 - 1)x + (6+1)$
    $= -7x + 7$

32. $\dfrac{3}{10}y + \dfrac{4}{15}y = \left(\dfrac{3}{10} + \dfrac{4}{15}\right)y$
    $= \left(\dfrac{9}{30} + \dfrac{8}{30}\right)y$
    $= \dfrac{17}{30}y$

34. $-x - 3x^2 + 4x - x^2 = -3x^2 - x^2 - x + 4x$
    $= (-3-1)x^2 + (-1+4)x$
    $= -4x^2 + 3x$

36. $2.5y - 1.8 - 1.4y + 0.4 = 2.5y - 1.4y - 1.8 + 0.4$
    $= (2.5 - 1.4)y + (-1.8 + 0.4)$
    $= 1.1y - 1.4$

38. $10y + 3 - 2y + 6 + y = 10y - 2y + y + 3 + 6$
    $= (10 - 2 + 1)y + (3+6)$
    $= 9y + 9$

**Section R.5** Algebraic Expressions

40. $3(2y+5)-6(y+2) = 6y+15-6y-12$
$\phantom{3(2y+5)-6(y+2)} = 6y-6y+15-12$
$\phantom{3(2y+5)-6(y+2)} = (6-6)y+(15-12)$
$\phantom{3(2y+5)-6(y+2)} = 0y+3$
$\phantom{3(2y+5)-6(y+2)} = 3$

42. $\dfrac{1}{2}(20x-14)+\dfrac{1}{3}(6x+9)$
$= \dfrac{1}{2}\cdot 20x - \dfrac{1}{2}\cdot 14 + \dfrac{1}{3}\cdot 6x + \dfrac{1}{3}\cdot 9$
$= 10x - 7 + 2x + 3$
$= 10x + 2x - 7 + 3$
$= (10+2)x + (-7+3)$
$= 12x - 4$

44. $4(w+2)+3(4w+3) = 4w+8+12w+9$
$\phantom{4(w+2)+3(4w+3)} = 4w+12w+8+9$
$\phantom{4(w+2)+3(4w+3)} = (4+12)w+(8+9)$
$\phantom{4(w+2)+3(4w+3)} = 16w+17$

46. $-4(w-3)-(2w+1) = -4(w-3)-1\cdot(2w+1)$
$\phantom{-4(w-3)-(2w+1)} = -4w+12-2w-1$
$\phantom{-4(w-3)-(2w+1)} = -4w-2w+12-1$
$\phantom{-4(w-3)-(2w+1)} = (-4-2)w+(12-1)$
$\phantom{-4(w-3)-(2w+1)} = -6w+11$

48. $\dfrac{4}{3}(5y+1)-\dfrac{2}{5}(3y-4) = \dfrac{20}{3}y+\dfrac{4}{3}-\dfrac{6}{5}y+\dfrac{8}{5}$
$= \dfrac{20}{3}y-\dfrac{6}{5}y+\dfrac{4}{3}+\dfrac{8}{5}$
$= \left(\dfrac{20}{3}-\dfrac{6}{5}\right)y+\left(\dfrac{4}{3}+\dfrac{8}{5}\right)$
$= \left(\dfrac{100}{15}-\dfrac{18}{15}\right)y+\left(\dfrac{20}{15}+\dfrac{24}{15}\right)$
$= \dfrac{82}{15}y+\dfrac{44}{15}$

50. $\dfrac{1}{4}\left(\dfrac{2}{3}x-\dfrac{1}{2}\right)+\dfrac{1}{10}\left(\dfrac{5}{2}x-\dfrac{15}{4}\right)$
$= \dfrac{2}{12}x-\dfrac{1}{8}+\dfrac{5}{20}x-\dfrac{15}{40}$
$= \dfrac{1}{6}x-\dfrac{1}{8}+\dfrac{1}{4}x-\dfrac{3}{8}$
$= \dfrac{1}{6}x+\dfrac{1}{4}x-\dfrac{1}{8}-\dfrac{3}{8}$
$= \left(\dfrac{1}{6}+\dfrac{1}{4}\right)x+\left(-\dfrac{1}{8}-\dfrac{3}{8}\right)$
$= \left(\dfrac{2}{12}+\dfrac{3}{12}\right)x-\dfrac{4}{8}$
$= \dfrac{5}{12}x-\dfrac{1}{2}$

52. $0.4(2.9x-1.6)-2.7(0.3x+6.2)$
$= 1.16x-0.64-0.81x-16.74$
$= 1.16x-0.81x-0.64-16.74$
$= (1.16-0.81)x+(-0.64-16.74)$
$= 0.35x-17.38$

54. $9.3(0.2x-0.8)+3.8(1.3x+6.3)$
$= 1.86x-7.44+4.94x+23.94$
$= 1.86x+4.94x-7.44+23.94$
$= (1.86+4.94)x+(-7.44+23.94)$
$= 6.8x+16.5$

56. $10-y$

58. $\dfrac{x}{5}$

60. $z+30$

62. $\dfrac{x+5}{10}$

64. $3z+\dfrac{z}{8}$

66. $2x-\dfrac{y}{3}$

68. We need to determine whether the value of the variable causes division by 0. That is, we need to determine if the value makes $x+1=0$.

   a. When $x=5$, we have $x+1=5+1=6$, so 5 is in the domain of the variable.

ISM – 9

Chapter R: Real Numbers and Algebraic Expressions

b. When $x = -1$, we have $x + 1 = -1 + 1 = 0$, so $-1$ is **not** in the domain of the variable.

c. When $x = -4$, we have $x + 1 = -4 + 1 = -3$, so $-4$ is in the domain of the variable.

d. When $x = 0$, we have $x + 1 = 0 + 1 = 1$, so $0$ is in the domain of the variable.

70. We need to determine whether the value of the variable causes division by 0. That is, we need to determine if the value makes $x - 1 = 0$.

    a. When $x = 5$, we have $x - 1 = 5 - 1 = 4$, so 5 is in the domain of the variable.

    b. When $x = -1$, we have $x - 1 = -1 - 1 = -2$, so $-1$ is in the domain of the variable.

    c. When $x = -4$, we have $x - 1 = -4 - 1 = -5$, so $-4$ is in the domain of the variable.

    d. When $x = 0$, we have $x - 1 = 0 - 1 = -1$, so 0 is in the domain of the variable.

72. We need to determine whether the value of the variable causes division by 0. That is, we need to determine if the value makes $x^2 + 5x + 4 = 0$.

    a. When $x = 5$, we have
    $x^2 + 5x + 4 = (5)^2 + 5(5) + 4$,
    $= 25 + 25 + 4$
    $= 54$
    so 5 is in the domain of the variable.

    b. When $x = -1$, we have
    $x^2 + 5x + 4 = (-1)^2 + 5(-1) + 4$,
    $= 1 - 5 + 4$
    $= 0$
    so $-1$ is **not** in the domain of the variable.

    c. When $x = -4$, we have
    $x^2 + 5x + 4 = (-4)^2 + 5(-4) + 4$,
    $= 16 - 20 + 4$
    $= 0$
    so $-4$ is **not** in the domain of the variable.

    d. When $x = 0$, we have
    $x^2 + 5x + 4 = (0)^2 + 5(0) + 4$,
    $= 0 + 0 + 4$
    $= 4$
    so 0 is in the domain of the variable.

74. For $h = 2$ cm,
$\frac{1}{2}(h+2)h = \frac{1}{2}(2+2)(2) = 4$ sq cm

For $h = 5$ cm,
$\frac{1}{2}(h+2)h = \frac{1}{2}(5+2)(5) = \frac{35}{2} = 17.5$ sq cm

For $h = 10$ cm,
$\frac{1}{2}(h+2)h = \frac{1}{2}(10+2)(10) = 60$ sq cm

76. For $x = 20$ watches,
$30x + 1000 = 30(20) + 1000$
$= 600 + 1000$
$= 1600$
The cost of manufacturing 20 watches is \$1600.

For $x = 30$ watches,
$30x + 1000 = 30(30) + 1000$
$= 900 + 1000$
$= 1900$
The cost of manufacturing 30 watches is \$1900.

For $x = 40$ watches,
$30x + 1000 = 30(40) + 1000$
$= 1200 + 1000$
$= 2200$
The cost of manufacturing 40 watches is \$2200.

78. Let $p$ = regular price of a computer in dollars.
"\$50 off regular price" is $p - 50$.
When $p = 890$, $p - 50 = 890 - 50 = 840$.
When the original price is \$890, the discount price is \$840.

80. Let $x$ = Marissa's age in years.
Marissa's mother is then $2(x+3)$ years old.
When $x = 18$, we have
$2(x+3) = 2(18+3)$
$= 2(21)$
$= 42$
When Marissa is 18 years of age, her mother is 42 years of age.

**Section R.5** Algebraic Expressions

82. $Z = \dfrac{X - \mu}{\sigma}$

    When $X = 40$, $\mu = 50$, and $\sigma = 10$, we have

    $Z = \dfrac{40 - 50}{10} = \dfrac{-10}{10} = \dfrac{-1 \cdot \cancel{10}}{1 \cdot \cancel{10}} = -1$

84. Answers may vary. One possible answer is: "Three more than 5 times a number x"

86. Answers may vary. One possible answer is: "5 times the sum of a number x and 3"

88. Answers may vary. One possible answer is: "The quotient of a number t and 3 decreased by the product of 2 and the number t"

90. a. $-5x + 9$ when $x = 4$.

    ```
    4→X: -5X+9
                -11
    ```

    b. $-5x + 9$ when $x = -3$.

    ```
    -3→X: -5X+9
                24
    ```

92. a. $-9x^2 + x - 5$ when $x = 3$.

    ```
    3→X: -9X²+X-5
                -83
    ```

    b. $-9x^2 + x - 5$ when $x = -4$.

    ```
    -4→X: -9X²+X-5
                -153
    ```

94. a. $\dfrac{2y^2 + 5}{3y - 1}$ when $y = 3$.

    ```
    3→Y: (2Y²+5)/(3Y-1)▶Frac
                23/8
    ```

    b. $\dfrac{2y^2 + 5}{3y - 1}$ when $y = -8$.

    ```
    -8→Y: (2Y²+5)/(3Y-1)▶Frac
                -133/25
    ```

96. a. $|-3x^2 + 5x - 2|$ when $x = 6$.

    ```
    6→X: abs(-3X²+5X-2)
                80
    ```

    b. $|-3x^2 + 5x - 2|$ when $x = -4$.

    ```
    -4→X: abs(-3X²+5X-2)
                70
    ```

# Chapter 1

## 1.1 Preparing for Linear Equations

1. The additive inverse of 5 is $-5$ because $5+(-5)=0$.

2. The multiplicative inverse of $-3$ is $-\dfrac{1}{3}$ because $(-3)\cdot\left(-\dfrac{1}{3}\right)=1$.

3. $\dfrac{1}{5}\cdot 5x = \dfrac{5\cdot x}{5} = \dfrac{\cancel{5}\cdot x}{\cancel{5}} = x$

4. $8 = 2\cdot 2\cdot 2$
   $12 = 2\cdot 2\cdot 3$
   The common factors are 2 and 2. The remaining factors are 2 and 3.
   $LCD = 2\cdot 2\cdot 2\cdot 3 = 24$

5. $6(z-2) = 6\cdot z - 6\cdot 2 = 6z - 12$

6. The coefficient of $-4x$ is $-4$.

7. $4(y-2)-y+5 = 4y-8-y+5$
   $\phantom{4(y-2)-y+5}= 4y-y-8+5$
   $\phantom{4(y-2)-y+5}= 3y-3$

8. $-5(x+3)-8$
   $-5((-2)+3)-8 = -5(1)-8$
   $\phantom{-5((-2)+3)-8}= -5-8$
   $\phantom{-5((-2)+3)-8}= -13$

9. Since division by 0 is undefined, we need to determine if the value for the variable makes $x+3=0$.
   When $x=3$, we have $x+3=(3)+3=6$, so 3 is in the domain of the variable.
   When $x=-3$, we have $x+3=(-3)+3=0$, so $-3$ is **not** in the domain of the variable.

## 1.1 Exercises

2. solutions; satisfy

4. True

6. True

8. "solve" is used with equations and means to find all solutions to an equation. That is, find the values that make the statement true. "simplify" is used with expressions and means to write the expression in a simpler form.

10. $-4x-3=-15$
    Let $x=-2$ in the equation.
    $-4(-2)-3\stackrel{?}{=}-15$
    $8-3\stackrel{?}{=}-15$
    $5\ne -15$
    $x=-2$ is **not** a solution to the equation.
    Let $x=1$ in the equation.
    $-4(1)-3\stackrel{?}{=}-15$
    $-4-3\stackrel{?}{=}-15$
    $-7\ne -15$
    $x=1$ is **not** a solution to the equation.
    Let $x=3$ in the equation.
    $-4(3)-3\stackrel{?}{=}-15$
    $-12-3\stackrel{?}{=}-15$
    $-15 = -15$ T
    $x=3$ is a solution to the equation.

12. $6x+1=-2x+9$
    Let $x=-2$ in the equation.
    $6(-2)+1\stackrel{?}{=}-2(-2)+9$
    $-12+1\stackrel{?}{=}4+9$
    $-11\ne 13$
    $x=-2$ is **not** a solution to the equation.
    Let $x=1$ in the equation.

## Section 1.1 Linear Equations

$6(1)+1\stackrel{?}{=}-2(1)+9$

$6+1\stackrel{?}{=}-2+9$

$7=7$ T

$x=1$ is a solution to the equation.
Let $x=4$ in the equation.

$6(4)+1\stackrel{?}{=}-2(4)+9$

$24+1\stackrel{?}{=}-8+9$

$25\neq 1$

$x=4$ is **not** a solution to the equation.

14. $3(t+1)-t=4t+9$

Let $t=-3$ in the equation.

$3((-3)+1)-(-3)\stackrel{?}{=}4(-3)+9$

$3(-2)+3\stackrel{?}{=}-12+9$

$-6+3\stackrel{?}{=}-3$

$-3=-3$ T

$t=-3$ is a solution to the equation.
Let $t=-1$ in the equation.

$3((-1)+1)-(-1)\stackrel{?}{=}4(-1)+9$

$3(0)+1\stackrel{?}{=}-4+9$

$1\neq 5$

$t=-1$ is **not** a solution to the equation.
Let $t=2$ in the equation.

$3((2)+1)-(2)\stackrel{?}{=}4(2)+9$

$3(3)-2\stackrel{?}{=}8+9$

$9-2\stackrel{?}{=}17$

$7\neq 17$

$t=2$ is **not** a solution to the equation.

16. $8x-6=18$

$8x-6+6=18+6$

$8x=24$

$\dfrac{8x}{8}=\dfrac{24}{8}$

$x=3$

Check:

$8(3)-6\stackrel{?}{=}18$

$24-6\stackrel{?}{=}18$

$18=18$ T

This is a conditional equation.
Solution set: $\{3\}$

18. $-6x-5=13$

$-6x-5+5=13+5$

$-6x=18$

$\dfrac{-6x}{-6}=\dfrac{18}{-6}$

$x=-3$

Check:

$-6(-3)-5\stackrel{?}{=}13$

$18-5\stackrel{?}{=}13$

$13=13$ T

This is a conditional equation.
Solution set: $\{-3\}$

20. $8y+3=5$

$8y+3-3=5-3$

$8y=2$

$\dfrac{8y}{8}=\dfrac{2}{8}$

$y=\dfrac{1}{4}$

Check:

$8\left(\dfrac{1}{4}\right)+3\stackrel{?}{=}5$

$2+3\stackrel{?}{=}5$

$5=5$ T

This is a conditional equation.
Solution set: $\left\{\dfrac{1}{4}\right\}$

## Chapter 1: Linear Equations and Inequalities

**22.**
$$-7t - 3 + 5t = 11$$
$$-2t - 3 = 11$$
$$-2t - 3 + 3 = 11 + 3$$
$$-2t = 14$$
$$\frac{-2t}{-2} = \frac{14}{-2}$$
$$t = -7$$

Check:
$$-7(-7) - 3 + 5(-7) \stackrel{?}{=} 11$$
$$49 - 3 - 35 \stackrel{?}{=} 11$$
$$11 = 11 \text{ T}$$

This is a conditional equation.
Solution set: $\{-7\}$

**24.**
$$-5z + 3 = -3z + 1$$
$$-5z + 3 - 3 = -3z + 1 - 3$$
$$-5z = -3z - 2$$
$$-5z + 3z = -3z - 2 + 3z$$
$$-2z = -2$$
$$\frac{-2z}{-2} = \frac{-2}{-2}$$
$$z = 1$$

Check:
$$-5(1) + 3 \stackrel{?}{=} -3(1) + 1$$
$$-5 + 3 \stackrel{?}{=} -3 + 1$$
$$-2 = -2 \text{ T}$$

This is a conditional equation.
Solution set: $\{1\}$

**26.**
$$-6x + 2 + 2x + 9 + x = 5x + 10 - 6x + 11$$
$$-3x + 11 = -x + 21$$
$$-3x + 11 - 11 = -x + 21 - 11$$
$$-3x = -x + 10$$
$$-3x + x = -x + 10 + x$$
$$-2x = 10$$
$$\frac{-2x}{-2} = \frac{10}{-2}$$
$$x = -5$$

Check:
$$-6(-5) + 2 + 2(-5) + 9 + (-5) \stackrel{?}{=} 5(-5) + 10 - 6(-5) + 11$$
$$30 + 2 - 10 + 9 - 5 \stackrel{?}{=} -25 + 10 + 30 + 11$$
$$41 - 15 \stackrel{?}{=} 51 - 25$$
$$26 = 26 \text{ T}$$

This is a conditional equation.
Solution set: $\{-5\}$

**28.**
$$4(z - 2) = 12$$
$$\frac{4(z - 2)}{4} = \frac{12}{4}$$
$$z - 2 = 3$$
$$z - 2 + 2 = 3 + 2$$
$$z = 5$$

Check:
$$4((5) - 2) \stackrel{?}{=} 12$$
$$4(3) \stackrel{?}{=} 12$$
$$12 = 12 \text{ T}$$

This is a conditional equation.
Solution set: $\{5\}$

**30.**
$$5(s + 3) = 3s + 2s$$
$$5s + 15 = 5s$$
$$5s + 15 - 5s = 5s - 5s$$
$$15 = 0$$

This statement is a contradiction. The equation has no solution.
Solution set: $\{\ \}$ or $\varnothing$

**32.**
$$10(x - 1) - 4x = 2x - 1 + 4(x + 1)$$
$$10x - 10 - 4x = 2x - 1 + 4x + 4$$
$$6x - 10 = 6x + 3$$
$$6x - 10 - 6x = 6x + 3 - 6x$$
$$-10 = 3$$

This statement is a contradiction. The equation has no solution.
Solution set: $\{\ \}$ or $\varnothing$

**34.** $8(w+2) - 3w = 7(w+2) + 2(1-w)$
$8w + 16 - 3w = 7w + 14 + 2 - 2w$
$5w + 16 = 5w + 16$
$5w + 16 - 5w = 5w + 16 - 5w$
$16 = 16$

This statement is an identity. All real numbers are solutions.

Solution set: $\{w \mid w \text{ is a real number}\}$ or $\mathbb{R}$

**36.** $\dfrac{3x}{2} + \dfrac{x}{6} = -\dfrac{5}{3}$

$6\left(\dfrac{3x}{2} + \dfrac{x}{6}\right) = 6\left(-\dfrac{5}{3}\right)$

$6 \cdot \dfrac{3x}{2} + 6 \cdot \dfrac{x}{6} = 6\left(-\dfrac{5}{3}\right)$

$9x + x = -10$

$10x = -10$

$\dfrac{10x}{10} = \dfrac{-10}{10}$

$x = -1$

Check:
$\dfrac{3(-1)}{2} + \dfrac{(-1)}{6} \stackrel{?}{=} -\dfrac{5}{3}$

$-\dfrac{3}{2} - \dfrac{1}{6} \stackrel{?}{=} -\dfrac{5}{3}$

$-\dfrac{5}{3} = -\dfrac{5}{3}$ T

This is a conditional equation.
Solution set: $\{-1\}$

**38.** $\dfrac{z-2}{4} + \dfrac{2z-3}{6} = 7$

$12\left(\dfrac{z-2}{4} + \dfrac{2z-3}{6}\right) = 12(7)$

$3(z-2) + 2(2z-3) = 84$

$3z - 6 + 4z - 6 = 84$

$7z - 12 = 84$

$7z - 12 + 12 = 84 + 12$

$7z = 96$

$\dfrac{7z}{7} = \dfrac{96}{7}$

$z = \dfrac{96}{7}$

Check:
$\dfrac{\frac{96}{7} - 2}{4} + \dfrac{2\left(\frac{96}{7}\right) - 3}{6} \stackrel{?}{=} 7$

$\dfrac{\frac{96}{7} - \frac{14}{7}}{4} + \dfrac{\frac{192}{7} - \frac{21}{7}}{6} \stackrel{?}{=} 7$

$\dfrac{\frac{82}{7}}{4} + \dfrac{\frac{171}{7}}{6} \stackrel{?}{=} 7$

$\dfrac{82}{28} + \dfrac{171}{42} \stackrel{?}{=} 7$

$\dfrac{41}{14} + \dfrac{57}{14} \stackrel{?}{=} 7$

$7 = 7$ T

This is a conditional equation.

Solution set: $\left\{\dfrac{96}{7}\right\}$

**40.** $\dfrac{r}{2} + 2(r-1) = \dfrac{5r}{2} + 4$

$2\left(\dfrac{r}{2} + 2(r-1)\right) = 2\left(\dfrac{5r}{2} + 4\right)$

$r + 4(r-1) = 5r + 8$

$r + 4r - 4 = 5r + 8$

$5r - 4 = 5r + 8$

$5r - 4 - 5r = 5r + 8 - 5r$

$-4 = 8$

This statement is a contradiction. The equation has no solution. Solution set: $\{\ \}$ or $\varnothing$

**42.** $\dfrac{2x+1}{3} - \dfrac{6x-1}{4} = -\dfrac{5}{12}$

$12\left(\dfrac{2x+1}{3} - \dfrac{6x-1}{4}\right) = 12\left(-\dfrac{5}{12}\right)$

$4(2x+1) - 3(6x-1) = -5$

$8x + 4 - 18x + 3 = -5$

$-10x + 7 = -5$

$-10x + 7 - 7 = -5 - 7$

$-10x = -12$

$\dfrac{-10x}{-10} = \dfrac{-12}{-10}$

$x = \dfrac{6}{5}$

## Chapter 1: Linear Equations and Inequalities

Check:
$$\frac{2\left(\frac{6}{5}\right)+1}{3} - \frac{6\left(\frac{6}{5}\right)-1}{4} \stackrel{?}{=} -\frac{5}{12}$$
$$\frac{\frac{12}{5}+\frac{5}{5}}{3} - \frac{\frac{36}{5}-\frac{5}{5}}{4} \stackrel{?}{=} -\frac{5}{12}$$
$$\frac{\frac{17}{5}}{3} - \frac{\frac{31}{5}}{4} \stackrel{?}{=} -\frac{5}{12}$$
$$\frac{17}{15} - \frac{31}{20} \stackrel{?}{=} -\frac{5}{12}$$
$$\frac{68}{60} - \frac{93}{60} \stackrel{?}{=} -\frac{5}{12}$$
$$-\frac{25}{60} \stackrel{?}{=} -\frac{5}{12}$$
$$-\frac{5}{12} \stackrel{?}{=} -\frac{5}{12} \, T$$

This is a conditional equation.
Solution set: $\left\{\frac{6}{5}\right\}$

**44.**
$$\frac{3x+1}{4} - \frac{7x-4}{2} = \frac{26}{3}$$
$$12\left(\frac{3x+1}{4} - \frac{7x-4}{2}\right) = 12\left(\frac{26}{3}\right)$$
$$3(3x+1) - 6(7x-4) = 4(26)$$
$$9x + 3 - 42x + 24 = 104$$
$$-33x + 27 = 104$$
$$-33x + 27 - 27 = 104 - 27$$
$$-33x = 77$$
$$\frac{-33x}{-33} = \frac{77}{-33}$$
$$x = -\frac{7}{3}$$

Check:
$$\frac{3\left(-\frac{7}{3}\right)+1}{4} - \frac{7\left(-\frac{7}{3}\right)-4}{2} \stackrel{?}{=} \frac{26}{3}$$
$$\frac{-7+1}{4} - \frac{-\frac{49}{3}-4}{2} \stackrel{?}{=} \frac{26}{3}$$
$$-\frac{3}{2} + \frac{\frac{61}{3}}{2} \stackrel{?}{=} \frac{26}{3}$$
$$-\frac{9}{6} + \frac{61}{6} \stackrel{?}{=} \frac{26}{3}$$
$$\frac{52}{6} \stackrel{?}{=} \frac{26}{3}$$
$$\frac{26}{3} = \frac{26}{3} \, T$$

This is a conditional equation.
Solution set: $\left\{-\frac{7}{3}\right\}$

**46.**
$$0.3z + 0.8 = -0.1$$
$$10(0.3z + 0.8) = 10(-0.1)$$
$$3z + 8 = -1$$
$$3z + 8 - 8 = -1 - 8$$
$$3z = -9$$
$$\frac{3z}{3} = \frac{-9}{3}$$
$$z = -3$$

Check:
$$0.3(-3) + 0.8 \stackrel{?}{=} -0.1$$
$$-0.9 + 0.8 \stackrel{?}{=} -0.1$$
$$-0.1 = -0.1 \, T$$

This is a conditional equation.
Solution set: $\{-3\}$

**48.**
$$0.12y - 5.26 = 0.05y + 1.25$$
$$100(0.12y - 5.26) = 100(0.05y + 1.25)$$
$$12y - 526 = 5y + 125$$
$$12y - 526 + 526 = 5y + 125 + 526$$
$$12y = 5y + 651$$
$$12y - 5y = 5y + 651 - 5y$$
$$7y = 651$$
$$\frac{7y}{7} = \frac{651}{7}$$
$$y = 93$$

Check: $0.12(93) - 5.26 \stackrel{?}{=} 0.05(93) + 1.25$
$$11.16 - 5.26 \stackrel{?}{=} 4.65 + 1.25$$
$$5.9 = 5.9 \, T$$

This is a conditional equation.
Solution set: $\{93\}$

**50.** $0.9(z-3) - 0.2(z-5) = 0.4(z+1) + 0.3z - 2.1$
$0.9z - 2.7 - 0.2z + 1.0 = 0.4z + 0.4 + 0.3z - 2.1$
$0.7z - 1.7 = 0.7z - 1.7$
$0.7z - 1.7 - 0.7z = 0.7z - 1.7 - 0.7z$
$-1.7 = -1.7$

This statement is an identity. The solution set is all real numbers.

Solution set: $\{z \mid z \text{ is any real number}\}$ or $\mathbb{R}$

**52.** $\frac{4}{5}(y-4) + 3 = \frac{2}{3}(y+1) + \frac{4}{15}$

$15\left[\frac{4}{5}(y-4) + 3\right] = 15\left[\frac{2}{3}(y+1) + \frac{4}{15}\right]$

$12(y-4) + 45 = 10(y+1) + 4$
$12y - 48 + 45 = 10y + 10 + 4$
$12y - 3 = 10y + 14$
$12y - 3 + 3 = 10y + 14 + 3$
$12y = 10y + 17$
$12y - 10y = 10y + 17 - 10y$
$2y = 17$
$y = \frac{17}{2}$

Check:
$\frac{4}{5}\left(\frac{17}{2} - 4\right) + 3 \stackrel{?}{=} \frac{2}{3}\left(\frac{17}{2} + 1\right) + \frac{4}{15}$

$\frac{4}{5}\left(\frac{9}{2}\right) + 3 \stackrel{?}{=} \frac{2}{3}\left(\frac{19}{2}\right) + \frac{4}{15}$

$\frac{18}{5} + \frac{15}{5} \stackrel{?}{=} \frac{19}{3} + \frac{4}{15}$

$\frac{33}{5} \stackrel{?}{=} \frac{95}{15} + \frac{4}{15}$

$\frac{33}{5} \stackrel{?}{=} \frac{99}{15}$

$\frac{33}{5} = \frac{33}{5}$ T

This is a conditional equation.

Solution set: $\left\{\frac{17}{2}\right\}$

**54.** $4y + 5 = 7$
$4y + 5 - 5 = 7 - 5$
$4y = 2$
$\frac{4y}{4} = \frac{2}{4}$
$y = \frac{1}{2}$

Check:
$4\left(\frac{1}{2}\right) + 5 \stackrel{?}{=} 7$
$2 + 5 \stackrel{?}{=} 7$
$7 = 7$ T

This is a conditional equation.

Solution set: $\left\{\frac{1}{2}\right\}$

**56.** $-5x + 5 + 3x + 7 = 5x - 6 + x + 12$
$-2x + 12 = 6x + 6$
$-2x + 12 - 12 = 6x + 6 - 12$
$-2x = 6x - 6$
$-2x - 6x = 6x - 6 - 6x$
$-8x = -6$
$\frac{-8x}{-8} = \frac{-6}{-8}$
$x = \frac{3}{4}$

Check:
$-5\left(\frac{3}{4}\right) + 5 + 3\left(\frac{3}{4}\right) + 7 \stackrel{?}{=} 5\left(\frac{3}{4}\right) - 6 + \frac{3}{4} + 12$

$-\frac{15}{4} + 12 + \frac{9}{4} \stackrel{?}{=} \frac{15}{4} + 6 + \frac{3}{4}$

$-\frac{15}{4} + \frac{48}{4} + \frac{9}{4} \stackrel{?}{=} \frac{15}{4} + \frac{24}{4} + \frac{3}{4}$

$\frac{42}{4} = \frac{42}{4}$ T

This is a conditional equation.

Solution set: $\left\{\frac{3}{4}\right\}$

**58.**
$$7(x+2) = 5(x-2) + 2(x+12)$$
$$7x + 14 = 5x - 10 + 2x + 24$$
$$7x + 14 = 7x + 14$$
$$7x + 14 - 7x = 7x + 14 - 7x$$
$$14 = 14$$

This statement is an identity. The solution set is all real numbers.

Solution set: $\{x \mid x \text{ is any real number}\}$ or $\mathbb{R}$

**60.**
$$13z - 8(z+1) = 2(z-3) + 3z$$
$$13z - 8z - 8 = 2z - 6 + 3z$$
$$5z - 8 = 5z - 6$$
$$5z - 8 - 5z = 5z - 6 - 5z$$
$$-8 = -6$$

This statement is a contradiction. The equation has no solution.

Solution set: $\{\ \}$ or $\varnothing$

**62.**
$$\frac{z-4}{6} - \frac{2z+1}{9} = \frac{1}{3}$$
$$18\left(\frac{z-4}{6} - \frac{2z+1}{9}\right) = 18\left(\frac{1}{3}\right)$$
$$3(z-4) - 2(2z+1) = 6$$
$$3z - 12 - 4z - 2 = 6$$
$$-z - 14 = 6$$
$$-z - 14 + 14 = 6 + 14$$
$$-z = 20$$
$$\frac{-z}{-1} = \frac{20}{-1}$$
$$z = -20$$

Check:
$$\frac{-20-4}{6} - \frac{2(-20)+1}{9} \stackrel{?}{=} \frac{1}{3}$$
$$\frac{-24}{6} - \frac{-39}{9} \stackrel{?}{=} \frac{1}{3}$$
$$-\frac{12}{3} + \frac{13}{3} \stackrel{?}{=} \frac{1}{3}$$
$$\frac{1}{3} = \frac{1}{3} \text{ T}$$

This is a conditional equation.
Solution set: $\{-20\}$

**64.**
$$-0.8y + 0.3 = 0.2y - 3.7$$
$$-0.8y + 0.3 - 0.3 = 0.2y - 3.7 - 0.3$$
$$-0.8y = 0.2y - 4.0$$
$$-0.8y - 0.2y = 0.2y - 4.0 - 0.2y$$
$$-1.0y = -4.0$$
$$\frac{-1.0y}{-1.0} = \frac{-4.0}{-1.0}$$
$$y = 4$$

Check:
$$-0.8(4) + 0.3 \stackrel{?}{=} 0.2(4) - 3.7$$
$$-3.2 + 0.3 \stackrel{?}{=} 0.8 - 3.7$$
$$-2.9 = -2.9 \text{ T}$$

This is a conditional equation.
Solution set: $\{4\}$

(Note: Alternatively, we could have started by multiplying both sides of the equation by 10 to clear the decimals.)

**66.**
$$0.5(x+3) = 0.2(x-6)$$
$$0.5x + 1.5 = 0.2x - 1.2$$
$$0.5x + 1.5 - 1.5 = 0.2x - 1.2 - 1.5$$
$$0.5x = 0.2x - 2.7$$
$$0.5x - 0.2x = 0.2x - 2.7 - 0.2x$$
$$0.3x = -2.7$$
$$\frac{0.3x}{0.3} = \frac{-2.7}{0.3}$$
$$x = -9$$

Check:
$$0.5(-9+3) \stackrel{?}{=} 0.2(-9-6)$$
$$0.5(-6) \stackrel{?}{=} 0.2(-15)$$
$$-3 = -3 \text{ T}$$

This is a conditional equation.
Solution set: $\{-9\}$

(Note: Alternatively, we could have started by multiplying both sides of the equation by 10 to clear the decimals.)

Section 1.1 Linear Equations

68. $\frac{1}{5}(2a-5)-4=\frac{1}{2}(a+4)-\frac{7}{10}$

$10\left[\frac{1}{5}(2a-5)-4\right]=10\left[\frac{1}{2}(a+4)-\frac{7}{10}\right]$

$2(2a-5)-40=5(a+4)-7$

$4a-10-40=5a+20-7$

$4a-50=5a+13$

$4a-50+50=5a+13+50$

$4a=5a+63$

$4a-5a=5a+63-5a$

$-a=63$

$a=-63$

Check:

$\frac{1}{5}(2(-63)-5)-4\overset{?}{=}\frac{1}{2}(-63+4)-\frac{7}{10}$

$\frac{1}{5}(-131)-4\overset{?}{=}\frac{1}{2}(-59)-\frac{7}{10}$

$-\frac{131}{5}-\frac{20}{5}\overset{?}{=}\frac{5}{10}(-59)-\frac{7}{10}$

$-\frac{151}{5}\overset{?}{=}-\frac{295}{10}-\frac{7}{10}$

$-\frac{151}{5}\overset{?}{=}-\frac{302}{10}$

$-\frac{151}{5}=-\frac{151}{5}$ T

This is a conditional equation.
Solution set: $\{-63\}$

70. $ax+6=20$

Let $x=7$.

$a(7)+6=20$

$7a+6-6=20-6$

$7a=14$

$\frac{7a}{7}=\frac{14}{7}$

$a=2$

Check:
$2x+6=20$

$2x+6-6=20-6$

$2x=14$

$\frac{2x}{2}=\frac{14}{2}$

$x=7$

72. $ax+3=2x+3(x+1)$

$ax+3=2x+3x+3$

$ax+3=5x+3$

To have the solution set be all real numbers, we need both sides of the linear equation to be identical. This can be achieved by letting $a=5$.

74. We need to set the denominator equal to 0 and solve the resulting equation.

$5x+8=0$

$5x+8-8=0-8$

$5x=-8$

$\frac{5x}{5}=\frac{-8}{5}$

$x=-\frac{8}{5}$

Check:

$5\left(-\frac{8}{5}\right)+8\overset{?}{=}0$

$-8+8\overset{?}{=}0$

$0=0$ T

We must exclude $-\frac{8}{5}$ from the domain.

76. We need to set the denominator equal to 0 and solve the resulting equation.

$3x+1=0$

$3x+1-1=0-1$

$3x=-1$

$\frac{3x}{3}=\frac{-1}{3}$

$x=-\frac{1}{3}$

Check:

$3\left(-\frac{1}{3}\right)+1\overset{?}{=}0$

$-1+1\overset{?}{=}0$

$0=0$ T

We must exclude $x=-\frac{1}{3}$ from the domain.

## Chapter 1: Linear Equations and Inequalities

**78.** We need to set the denominator equal to 0 and solve the resulting equation.
$$4(x-3)+2=0$$
$$4x-12+2=0$$
$$4x-10=0$$
$$4x-10+10=0+10$$
$$4x=10$$
$$\frac{4x}{4}=\frac{10}{4}$$
$$x=\frac{5}{2}$$

Check:
$$4\left(\frac{5}{2}-3\right)+2\stackrel{?}{=}0$$
$$4\left(-\frac{1}{2}\right)+2\stackrel{?}{=}0$$
$$-2+2\stackrel{?}{=}0$$
$$0=0 \;\text{T}$$

We must exclude $x=\frac{5}{2}$ from the domain.

**80.** 
$$26x+9x=539$$
$$35x=539$$
$$\frac{35x}{35}=\frac{539}{35}$$
$$x=15.4$$
Your regular hourly rate is $15.40 per hour.

**82.** 
$$6020=0.15(x-14,600)+1460$$
$$6020=0.15x-2190+1460$$
$$6020=0.15x-730$$
$$6020+730=0.15x-730+730$$
$$6750=0.15x$$
$$\frac{6750}{0.15}=\frac{0.15x}{0.15}$$
$$45,000=x$$
You and your spouse earned $45,000 in 2005.

### 1.2 Preparing for Problem Solving

**1.** $12+z$
The keyword 'sum' indicates addition.

**2.** $4x$
The keyword 'product' indicates multiplication.

**3.** $y-87$
The keywords 'decreased by' indicate subtraction. Since order matters here, we subtract the quantity after the keywords.

**4.** $\dfrac{z}{12}$
The keyword 'quotient' indicates division. The first quantity is the numerator and the second quantity is the denominator.

**5.** $4(x+7)$
The keyword 'times' indicates multiplication and the keyword 'sum' indicates addition. Because the statement says 'times the sum' we multiply 4 by the entire sum, not just the $x$.

**6.** $4x+7$
The keyword 'times' indicates multiplication and the keyword 'sum' indicates addition. After the keywords 'sum of', we look for the two quantities to add. These are separated by the keyword 'and'. Thus, the first addend is 'four times a number $x$' while the second addend is the number '7'.

### 1.2 Exercises

**2.** uniform motion

**4.** True

**6.** Answers will vary. Assumptions are typically made to simplify a problem or to make it more manageable.

**8.** Answers will vary.

**10.** Let $x$ = desired result.
$x=1.5(70)=105$
150% of 70 is 105.

**12.** Let $x$ = original number.
$$0.9x=50$$
$$\frac{0.9x}{0.9}=\frac{50}{0.9}$$
$$x=\frac{500}{9}$$
50 is 90% of $\dfrac{500}{9}$.

**Section 1.2** An Introduction to Problem Solving

14. Let $x$ = percent as a decimal.
$$120x = 90$$
$$\frac{120x}{120} = \frac{90}{120}$$
$$x = \frac{9}{12} = 0.75$$
90 is 75% of 120.

16. $$10 - z = 6$$
$$10 - z - 10 = 6 - 10$$
$$-z = -4$$
$$\frac{-z}{-1} = \frac{-4}{-1}$$
$$z = 4$$

18. $$2y + 3 = 16$$
$$2y + 3 - 3 = 16 - 3$$
$$2y = 13$$
$$\frac{2y}{2} = \frac{13}{2}$$
$$y = \frac{13}{2}$$

20. $$x + 4 = 2x$$
$$x + 4 - x = 2x - x$$
$$4 = x \text{ or } x = 4$$

22. $$5x = 3x - 10$$
$$5x - 3x = 3x - 10 - 3x$$
$$2x = -10$$
$$\frac{2x}{2} = \frac{-10}{2}$$
$$x = -5$$

24. $$0.4x = x - 10$$
$$0.4x - x = x - 10 - x$$
$$-0.6x = -10$$
$$\frac{-0.6x}{-0.6} = \frac{-10}{-0.6}$$
$$x = \frac{50}{3}$$

26. First number: $x$
Second number: $x + 8$
$$x + (x + 8) = 56$$
$$2x + 8 = 56$$
$$2x + 8 - 8 = 56 - 8$$
$$2x = 48$$
$$\frac{2x}{2} = \frac{48}{2}$$
$$x = 24$$
Check:
$$24 + (24 + 8) = 24 + 32 = 56$$
The two numbers are 24 and 32.

28. First of the consecutive odd integers: $x$
Second consecutive odd integer: $x + 2$
Third consecutive odd integer: $x + 4$
Fourth consecutive odd integer: $x + 6$
$$x + (x + 2) + (x + 4) + (x + 6) = 104$$
$$4x + 12 = 104$$
$$4x + 12 - 12 = 104 - 12$$
$$4x = 92$$
$$\frac{4x}{4} = \frac{92}{4}$$
$$x = 23$$
Check:
$$23 + (23 + 2) + (23 + 4) + (23 + 6)$$
$$= 23 + 25 + 27 + 29$$
$$= 104$$
The integers are 23, 25, 27, and 29.

30. Let $x$ = score on Mark's final exam.
$$\frac{3x + 65 + 79 + 83 + 68}{7} = 70$$
$$\frac{3x + 295}{7} = 70$$
$$7 \cdot \frac{3x + 295}{7} = 7 \cdot 70$$
$$3x + 295 = 490$$
$$3x + 295 - 295 = 490 - 295$$
$$3x = 195$$
$$\frac{3x}{3} = \frac{195}{3}$$
$$x = 65$$
Check:
$$\frac{3(65) + 65 + 79 + 83 + 68}{7} = \frac{195 + 295}{7} = \frac{490}{7} = 70$$

*Chapter 1: Linear Equations and Inequalities*

Mark needs a 65 on his final exam to have an average of 70.

32. Let $x$ = Maria's monthly sales.
 First offer: $2500 + 0.03x$
 Second offer: $1500 + 0.035x$
 $$2500 + 0.03x = 1500 + 0.035x$$
 $$2500 + 0.03x - 2500 = 1500 + 0.035x - 2500$$
 $$0.03x = 0.035x - 1000$$
 $$0.03x - 0.035x = 0.035x - 100 - 0.035x$$
 $$-0.005x = -1000$$
 $$\frac{-0.005x}{-0.005} = \frac{-1000}{-0.005}$$
 $$x = 200,000$$

 Check:
 $$2500 + 0.03(200,000) \stackrel{?}{=} 1500 + 0.035(200,000)$$
 $$2500 + 6000 \stackrel{?}{=} 1500 + 7000$$
 $$8500 = 8500 \text{ T}$$
 Maria would need $200,000 in monthly sales for the two salaries to be equivalent.

34. Let $x$ = amount Linda should pay.
 Amount Judy should pay: $\frac{3}{4}x$
 $$x + \frac{3}{4}x = 21$$
 $$\frac{4}{4}x + \frac{3}{4}x = 21$$
 $$\frac{7}{4}x = 21$$
 $$\frac{4}{7} \cdot \frac{7}{4}x = \frac{4}{7} \cdot 21$$
 $$x = 12$$

 Check:
 $$12 + \frac{3}{4}(12) \stackrel{?}{=} 21$$
 $$12 + 9 \stackrel{?}{=} 21$$
 $$21 = 21 \text{ T}$$
 Linda should pay $12 and Judy should pay $9.

36. Let $c$ = total cost including tax.
 $$c = 39.83 + 0.0625(39.83)$$
 $$= 39.83 + 2.49$$
 $$= 42.32$$
 The final bill, including tax, would be $42.32.

38. Let $c$ = cost of text to bookstore.
 $$c + 0.3c = 95$$
 $$1.3c = 95$$
 $$\frac{1.3c}{1.3} = \frac{95}{1.3}$$
 $$c = 73.08$$

 Check:
 $$73.08 + 0.3(73.08) = 73.08 + 21.92 = 95$$
 The cost of the textbook to the bookstore is $73.08.

40. Let $x$ = original price of shirt.
 Discount price: $x - 0.3x = 0.7x$
 $$0.7x = 28$$
 $$\frac{0.7x}{0.7} = \frac{28}{0.7}$$
 $$x = 40$$

 Check:
 $$0.7(40) \stackrel{?}{=} 28$$
 $$28 = 28 \text{ T}$$
 The shirts originally cost $40.

42. Let $x$ = stopping distance of Nissan Altima.
 Mazda 6s: $x + 18$
 Honda Accord EX: $x + 26$
 $$x + (x + 18) + (x + 26) = 734$$
 $$3x + 44 = 734$$
 $$3x + 44 - 44 = 734 - 44$$
 $$3x = 690$$
 $$\frac{3x}{3} = \frac{690}{3}$$
 $$x = 230$$

 Check: $230 + (230 + 18) + (230 + 26)$
 $= 230 + 248 + 256 = 734$
 The Nissan Altima requires 230 feet to stop, the Mazda 6s requires 248 feet to stop, and the Honda Accord EX requires 256 feet to stop.

## Section 1.2 An Introduction to Problem Solving

**44.** Let $x$ = Michelson's score.
Love: $x - 22$
$$x + (x - 22) = 570$$
$$2x - 22 = 570$$
$$2x - 22 + 22 = 570 + 22$$
$$2x = 592$$
$$\frac{2x}{2} = \frac{592}{2}$$
$$x = 296$$

Check:
$296 + (296 - 22) = 296 + 274 = 570$
Phil Michelson required 296 strokes.

**46.** Let $b$ = amount invested in bonds.
Stocks: $b + 6,000$
$$b + (b + 6,000) = 40,000$$
$$2b + 6,000 = 40,000$$
$$2b + 6,000 - 6,000 = 40,000 - 6,000$$
$$2b = 34,000$$
$$\frac{2b}{2} = \frac{34,000}{2}$$
$$b = 17,000$$

Check:
$17,000 + (17,000 + 6,000)$
$= 17,000 + 23,000$
$= 40,000$
$17,000 should be invested in bonds and $23,000 should be invested in stocks.

**48.** Let $x$ = amount invested in stocks.
Bonds: $\frac{2}{3}x$
$$x + \frac{2}{3}x = 60,000$$
$$\frac{5}{3}x = 60,000$$
$$\frac{3}{5} \cdot \frac{5}{3}x = \frac{3}{5} \cdot 60,000$$
$$x = 36,000$$

Check:
$36,000 + \frac{2}{3}(36,000) = 36,000 + 24,000 = 60,000$
$36,000 should be invested in stocks and $24,000 should be invested in bonds.

**50.** Let $I$ = interest charge.
1 month = $\frac{1}{12}$ year
$$I = P \cdot r \cdot t$$
$$= (70,000)(0.06)\left(\frac{1}{12}\right)$$
$$= 350$$
The interest charge on Faye's loan after 1 month is $350.

**52.** Let $x$ = amount loaned at 6%
Unsecured loan: $2,000,000 - x$
$$0.06x + 0.14(2,000,000 - x) = 0.12(2,000,000)$$
$$0.06x + 280,000 - 0.14x = 240,000$$
$$-0.08x + 280,000 = 240,000$$
$$-0.08x = -40,000$$
$$\frac{-0.08x}{-0.08} = \frac{-40,000}{-0.08}$$
$$x = 500,000$$
Patrick should lend out $500,000 at 6% and $1,500,000 at 14%.

**54.** Let $x$ = amount invested in savings account.
Amount in stocks: $10,000 - x$
Here we need the simple interest formula $I = P \cdot r \cdot t$. We want the sum of the interest earned from the two sources to equal the interest that would be obtained by investing all the money in a single source at 7%.
$$0.02x + 0.10(10,000 - x) = 0.07(10,000)$$
$$0.02x + 1000 - 0.10x = 700$$
$$1000 - 0.08x = 700$$
$$1000 - 0.08x - 1000 = 700 - 1000$$
$$-0.08x = -300$$
$$\frac{-0.08x}{-0.08} = \frac{-300}{-0.08}$$
$$x = 3750$$
Johnny should invest $3750 in the savings account and $10,000 - 3,750 = \$6,250$ in the stock fund.

Check:
$$0.02(3750) + 0.1(6250) \stackrel{?}{=} 0.07(10,000)$$
$$75 + 625 \stackrel{?}{=} 700$$
$$700 = 700 \text{ T}$$

## Chapter 1: Linear Equations and Inequalities

**56.** Let $x =$ pounds of almonds.
Peanuts: $50 - x$
$$6.50x + 4.00(50 - x) = 5.00(50)$$
$$6.5x + 200 - 4x = 250$$
$$2.5x + 200 = 250$$
$$2.5x = 50$$
$$\frac{2.5x}{2.5} = \frac{50}{2.5}$$
$$x = 20$$

Check:
$$6.5(20) + 4(50 - 20) = 130 + 4(30)$$
$$= 130 + 120$$
$$= 250$$
$$= 5(50)$$

The store should mix 20 pounds of almonds with 30 pounds of peanuts to form the mixture.

**58.** Let $n =$ number of nickels.
Dimes: $48 - n$
$$0.05n + 0.10(48 - n) = 4.50$$
$$0.05n + 4.8 - 0.10n = 4.50$$
$$-0.05n + 4.8 = 4.50$$
$$-0.05n = -0.30$$
$$\frac{-0.05n}{-0.05} = \frac{-0.30}{-0.05}$$
$$n = 6$$

Check:
$$0.05(6) + 0.10(48 - 6) = 0.30 + 0.10(42)$$
$$= 0.30 + 4.20$$
$$= 4.50$$

Diana has 6 nickels and 42 dimes in her bank.

**60.** Let $x =$ number of liters drained and replaced
Gallons left after draining: $15 - x$
antifreeze$_{\text{after drain}}$ + antifreeze$_{\text{add}}$ = antifreeze$_{\text{final}}$
$$0.4(15 - x) + 1.0(x) = 0.5(15)$$
$$6 - 0.4x + x = 7.5$$
$$0.6x + 6 = 7.5$$
$$0.6x = 1.5$$
$$\frac{0.6x}{0.6} = \frac{1.5}{0.6}$$
$$x = 2.5$$

Check:
$$0.4(15 - 2.5) + 1(2.5) = 0.4(12.5) + 2.5$$
$$= 5 + 2.5$$
$$= 7.5$$
$$= 0.5(15)$$

2.5 liters of the mixture should be drained and replaced with pure antifreeze.

**62.** Let $x =$ time spent running (hours).
Time swimming: $2.5 - x$
For this problem, we will need the distance traveled formula: $d = r \cdot t$
tot. dist. = dist. run + dist. swam
tot. dist. = (run rate)(time) + (swim rate)(time)
$$15 = 7(x) + 2(2.5 - x)$$
$$15 = 7x + 5 - 2x$$
$$15 = 5x + 5$$
$$10 = 5x$$
$$\frac{10}{5} = \frac{5x}{5}$$
$$2 = x$$

Check:
$$7(2) + 2(2.5 - 2) = 14 + 2(0.5) = 14 + 1 = 15$$

We now know that the time spent running is 2 hours and the time spent swimming is 0.5 hours. However, the question asked for distances. We can use the distance traveled formula here.

Run: distance $= 7(2) = 14$ miles

Swim: distance $= 2(0.5) = 1$ mile

The race consists of running for 14 miles and swimming for 1 mile.

**64.** Let $v =$ speed of slow plane.
Speed of fast plane: $v + 40$
distance$_{\text{total}}$ = distance$_{\text{slow plane}}$ + distance$_{\text{fast plane}}$
$$720 = 0.75v + 0.75(v + 40)$$
$$720 = 0.75v + 0.75v + 30$$
$$720 = 1.5v + 30$$
$$690 = 1.5v$$
$$\frac{690}{1.5} = \frac{1.5v}{1.5}$$
$$460 = v$$

Check:
$0.75(460) + 0.75(460 + 40)$
$= 345 + 0.75(500)$
$= 345 + 375 = 720$
The slow plane is traveling at a rate of 460 miles per hour and the fast plane is traveling at a rate of 500 miles per hour.

66. Let $v$ = speed of slow cyclist.
    Speed of fast cyclist: $v + 3$
    distance$_{total}$ = distance$_{slow\ bike}$ + distance$_{fast\ bike}$
    $$162 = 6v + 6(v+3)$$
    $$162 = 6v + 6v + 18$$
    $$162 = 12v + 18$$
    $$144 = 12v$$
    $$\frac{144}{12} = \frac{12v}{12}$$
    $$12 = v$$

    Check:
    $6(12) + 6(12 + 3) = 72 + 6(15) = 72 + 90 = 162$
    The slow cyclist is traveling at a rate of 12 miles per hour and the fast cyclist is traveling at a rate of 15 miles per hour.

68. Let $v$ = speed of slow train.
    Speed of fast train: $v + 10$
    distance$_{total}$ = distance$_{slow\ train}$ + distance$_{fast\ train}$
    $$715 = 5.5v + 5.5(v+10)$$
    $$715 = 5.5v + 5.5v + 55$$
    $$715 = 11v + 55$$
    $$660 = 11v$$
    $$\frac{660}{11} = \frac{11v}{11}$$
    $$60 = v$$

    Check:
    $5.5(60) + 5.5(60 + 10) = 330 + 5.5(70)$
    $\phantom{5.5(60) + 5.5(60 + 10)} = 330 + 385$
    $\phantom{5.5(60) + 5.5(60 + 10)} = 715$
    The slow train is traveling at a rate of 60 miles per hour and the fast train is traveling at a rate of 70 miles per hour.

70. Let $p$ = original price.
    Sale price: $0.7p$ (which is 30% off of the original price)

    profit$_{per\ shirt}$ = revenue$_{per\ shirt}$ − cost$_{per\ shirt}$
    $$6 = 0.7p - 15$$
    $$21 = 0.7p$$
    $$\frac{21}{0.7} = \frac{0.7p}{0.7}$$
    $$30 = p$$
    The shirts should originally be priced at $30.

72. Answers will vary. One possibility:
    *"Kiersten has a calling card plan that charges a flat monthly rate of $10, plus $0.14 per minute used. If her bill for one month was $50, how many minutes did she use that month?"*

## 1.3 Preparing for Formulas

1. a. To round to three decimal places, we first examine the digit in the fourth decimal place and then truncate. Since the fourth decimal place contains a 4, we do not change the digit in the third decimal place. Thus, 3.00343 rounded to three decimal places is 3.003.

   b. To truncate, we remove all digits after the third decimal place. So, 3.00343 truncated to three decimal places is 3.003.

2. a. To round to two decimal places, we first examine the digit in the third decimal place and then truncate. Since the third decimal place contains a 7, we increase the digit in the second decimal place by 1. Thus, 14.957 rounded to two decimal places is 14.96.

   b. To truncate, we remove all digits after the second decimal place. So, 14.957 truncated to two decimal places is 14.95.

## 1.3 Exercises

2. $A = \pi r^2$

4. False; the formula for the perimeter of a rectangle is $P = 2l + 2w$ or $P = 2(l+w)$, where $w$ is the width and $l$ is the length.

6. No, the area would increase by a factor of 4. Let $r$ = original radius and $R = 2r$ be the new radius.
   $$\pi R^2 = \pi(2r)^2 = \pi \cdot 4r^2 = 4 \cdot \pi r^2$$

## Chapter 1: Linear Equations and Inequalities

If the length of a side of a cube is doubled, the volume increases by a factor of 8.
Let $s$ = original length of a side and $S = 2s$ be the new length.
$V = S^3 = (2s)^3 = 8s^3$.

**8.** $A = \dfrac{1}{2} b \cdot h$

**10.** $R = \$800 \cdot x$

**12.** $y = kx$

$\dfrac{y}{x} = \dfrac{kx}{x}$

$\dfrac{y}{x} = k$

**14.** $y = mx + b$

$y - b = mx + b - b$

$y - b = mx$

$\dfrac{y-b}{x} = \dfrac{mx}{x}$

$\dfrac{y-b}{x} = m$

**16.** $E = \dfrac{Z \cdot \sigma}{\sqrt{n}}$

$\sqrt{n} \cdot E = \sqrt{n} \cdot \dfrac{Z \cdot \sigma}{\sqrt{n}}$

$\sqrt{n} \cdot E = Z \cdot \sigma$

$\dfrac{\sqrt{n} \cdot E}{E} = \dfrac{Z \cdot \sigma}{E}$

$\sqrt{n} = \dfrac{Z \cdot \sigma}{E}$

**18.** $S - rS = a - ar^5$

$S(1-r) = a - ar^5$

$\dfrac{S(1-r)}{(1-r)} = \dfrac{a - ar^5}{1-r}$

$S = \dfrac{a - ar^5}{1-r}$

$S = \dfrac{a(1-r^5)}{1-r}$

**20.** $p + \dfrac{1}{2}\rho v^2 + \rho gy = a$

$p + \dfrac{1}{2}\rho v^2 + \rho gy - p = a - p$

$\dfrac{1}{2}\rho v^2 + \rho gy = a - p$

$\rho\left(\dfrac{1}{2}v^2 + gy\right) = a - p$

$\dfrac{\rho\left(\dfrac{1}{2}v^2 + gy\right)}{\left(\dfrac{1}{2}v^2 + gy\right)} = \dfrac{a-p}{\left(\dfrac{1}{2}v^2 + gy\right)}$

$\rho = \dfrac{a-p}{\dfrac{1}{2}v^2 + gy}$

**22.** $A = \dfrac{1}{2}h(B+b)$

$2 \cdot A = 2 \cdot \dfrac{1}{2}h(B+b)$

$2A = h(B+b)$

$\dfrac{2A}{h} = \dfrac{h(B+b)}{h}$

$\dfrac{2A}{h} = B + b$

$\dfrac{2A}{h} - B = B + b - B$

$\dfrac{2A}{h} - B = b \quad \text{or} \quad b = \dfrac{2A - Bh}{h}$

**24.** $-4x + y = 12$

$-4x + y + 4x = 12 + 4x$

$y = 12 + 4x \quad \text{or} \quad y = 4x + 12$

**26.** $4x + 2y = 20$

$4x + 2y - 4x = 20 - 4x$

$2y = 20 - 4x$

$\dfrac{2y}{2} = \dfrac{20 - 4x}{2}$

$y = 10 - 2x \quad \text{or} \quad y = -2x + 10$

**Section 1.3** Using Formulas to Solve Problems

28. $5x - 6y = 18$
$5x - 6y - 5x = 18 - 5x$
$-6y = 18 - 5x$
$\dfrac{-6y}{-6} = \dfrac{18 - 5x}{-6}$
$y = \dfrac{18}{-6} - \dfrac{5x}{-6}$
$y = -3 + \dfrac{5x}{6}$ or $y = \dfrac{5}{6}x - 3$

30. $\dfrac{2}{3}x - \dfrac{5}{2}y = 5$
$\dfrac{2}{3}x - \dfrac{5}{2}y - \dfrac{2}{3}x = 5 - \dfrac{2}{3}x$
$-\dfrac{5}{2}y = 5 - \dfrac{2}{3}x$
$-\dfrac{2}{5} \cdot \left(-\dfrac{5}{2}y\right) = -\dfrac{2}{5} \cdot \left(5 - \dfrac{2}{3}x\right)$
$y = -2 + \dfrac{4}{15}x$ or $y = \dfrac{4}{15}x - 2$

32. a. $A = 2\pi r h + 2\pi r^2$
$A - 2\pi r^2 = 2\pi r h + 2\pi r^2 - 2\pi r^2$
$A - 2\pi r^2 = 2\pi r h$
$\dfrac{A - 2\pi r^2}{2\pi r} = \dfrac{2\pi r h}{2\pi r}$
$\dfrac{A - 2\pi r^2}{2\pi r} = h$

b. $h = \dfrac{A - 2\pi r^2}{2\pi r} = \dfrac{72\pi - 2\pi(4)^2}{2\pi(4)}$
$= \dfrac{72\pi - 32\pi}{8\pi} = 5$
The height of the cylinder is 5 centimeters.

34. a. $M = -0.85A + 217$
$M - 217 = -0.85A + 217 - 217$
$M - 217 = -0.85A$
$\dfrac{M - 217}{-0.85} = \dfrac{-0.85A}{-0.85}$
$\dfrac{217 - M}{0.85} = A$

b. $A = \dfrac{217 - M}{0.85} = \dfrac{217 - 160}{0.85} \approx 67.06$
Someone with a maximum heart rate of 160 should be about 67 years old.

36. a. $P = D - 0.03(I - 142{,}700)$
$P - D = D - 0.03(I - 142{,}700) - D$
$P - D = -0.03(I - 142{,}700)$
$\dfrac{P - D}{-0.03} = \dfrac{-0.03(I - 142{,}700)}{-0.03}$
$\dfrac{D - P}{0.03} = I - 142{,}700$
$\dfrac{D - P}{0.03} + 142{,}700 = I - 142{,}700 + 142{,}700$
$\dfrac{D - P}{0.03} + 142{,}700 = I$

b. $I = \dfrac{D - P}{0.03} + 142{,}700$
$= \dfrac{18{,}000 - 16{,}941}{0.03} + 142{,}700$
$= 35{,}300 + 142{,}700$
$= 178{,}000$
The adjusted gross income for the couple was $178,000.

38. Let $x$ = measure of smaller angle (in degrees).
Larger angle: $2x$
$x + 2x = 180$
$3x = 180$
$\dfrac{3x}{3} = \dfrac{180}{3}$
$x = 60$
The smaller angle measures 60° and its supplement measures 120°.

40. Let $x$ = measure of one angle (in degrees).
Complementary angle: $2x - 30$
$x + (2x - 30) = 90$
$x + 2x - 30 = 90$
$3x - 30 = 90$
$3x - 30 + 30 = 90 + 30$
$3x = 120$
$\dfrac{3x}{3} = \dfrac{120}{3}$
$x = \dfrac{120}{3} = 40$
The smaller angle measures 40° and its complement measures 50°.

42. Let $w$ = width of the window (in inches).
Length: $2w$

ISM – 27

## Chapter 1: Linear Equations and Inequalities

$$P = 2(l+w)$$
$$120 = 2(2w+w)$$
$$120 = 2(3w)$$
$$120 = 6w$$
$$\frac{120}{6} = \frac{6w}{6}$$
$$20 = w$$

The width of the window is 20 inches and the length is 40 inches.

**44.** Let $w$ = width of rectangle. Length: $2w$
Radius of circles: $\frac{w}{2}$
$$P = 2(l+w)$$
$$36 = 2(2w+w)$$
$$36 = 2(3w)$$
$$36 = 6w$$
$$\frac{36}{6} = \frac{6w}{6}$$
$$6 = w$$

The width of the rectangle is 6 centimeters, so the radius of each circle is 3 centimeters.
$$A = \pi r^2 = \pi(3)^2$$
$$= 9\pi \approx 28.27 \text{ sq. cm.}$$

Each circular window has an area of about 28.27 square centimeters.

**46.** Let $a$ = measure of first angle.
Second angle: $3a$
Third angle: $a + 20$
$$a + (3a) + (a+20) = 180$$
$$5a + 20 = 180$$
$$5a + 20 - 20 = 180 - 20$$
$$5a = 160$$
$$\frac{5a}{5} = \frac{160}{5}$$
$$a = 32$$

The interior angles measure $32°$, $96°$, and $52°$.

**48. a.** The diameter of the base is 12 feet so the radius is 6 feet (half the radius).
$$A = \pi r^2 = \pi(6)^2 = 36\pi \approx 113.1 \text{ sq. ft.}$$
The area of the base is about 113.1 square feet.

**b.** Since 4 inches is $\frac{1}{3}$ foot, we get
$$V = A \cdot h = 36\pi \cdot \frac{1}{3} = 12\pi \approx 37.7 \text{ cu. ft.}$$
You would need to purchase about 37.7 cubic feet of cement.

**50. a.** Since the width of the window is 3 feet, the radius of the semicircle is 1.5 feet.
$$A_{\text{total}} = A_{\text{rectangle}} + A_{\text{semicircle}}$$
$$A = 8 \cdot 3 + \frac{1}{2} \cdot \pi(1.5)^2 \approx 27.53 \text{ sq. ft.}$$
The window has an area of about 27.53 square feet.

**b.** The perimeter is half the circumference of a circle with radius $r = 1.5$ plus the width and twice the length of the rectangle.
$$P = \frac{1}{2} \cdot 2\pi r + 2l + w$$
$$= \frac{1}{2} \cdot 2\pi(1.5) + 2(8) + 3 = 1.5\pi + 19 \approx 23.71$$
The perimeter of the window is about 23.71 feet.

**c.** Cost = (price per unit area)(area)
$$C = (8.25)(27.53)$$
$$= 227.12$$
Glass for the window would cost $227.12.

### 1.4 Preparing for Linear Inequalities

**1.** $3 < 6$ because 6 is further to the right on a real number line.

**2.** $-3 > -6$ because $-3$ is further to the right on a real number line.

**3.** $\frac{1}{2} = 0.5$ because both numbers are at the same point on a real number line.

**4.** $\frac{2}{3} = \frac{10}{15} > \frac{9}{15} = \frac{3}{5}$ because $\frac{2}{3}$ is further to the right on a real number line.

**5.** False; strict inequalities are either $<$ or $>$.

**6.** $\{x \mid x \text{ is a digit that is divisible by 3}\}$

## 1.4 Exercises

2. multiplication property

4. True. The addition property says that adding or subtracting the same amount from both sides of the inequality does not change the direction of the inequality.

6. False. The multiplication property says that we must switch the direction of the inequality if we multiply or divide both sides by a negative number. Therefore, if $a < b$ and $c < 0$, we must have $\dfrac{a}{c} > \dfrac{b}{c}$.

8. The notation $(a, b)$ represents the interval $a < x < b$. Writing $(4, -\infty)$ would mean $4 < x < -\infty$. This implies that $4 < -\infty$ which is incorrect. When writing interval notation, we write the numbers from smaller to larger.

10. $x + 6 < 9$
$x + 6 - 6 < 9 - 6$
$x < 3$
$\{x \mid x < 3\}\ ;\ (-\infty, 3)$

12. $4x \geq 20$
$\dfrac{4x}{4} \geq \dfrac{20}{4}$
$x \geq 5$
$\{x \mid x \geq 5\}\ ;\ [5, \infty)$

14. $-8x > 32$
$\dfrac{-8x}{-8} < \dfrac{32}{-8}$
$x < -4$
$\{x \mid x < -4\}\ ;\ (-\infty, -4)$

16. $\dfrac{3}{8}x < \dfrac{9}{16}$
$\dfrac{8}{3} \cdot \dfrac{3}{8}x < \dfrac{8}{3} \cdot \dfrac{9}{16}$
$x < \dfrac{3}{2}$
$\left\{x \mid x < \dfrac{3}{2}\right\}\ ;\ \left(-\infty, \dfrac{3}{2}\right)$

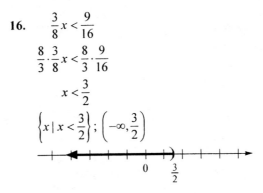

18. $5x - 4 \leq 16$
$5x - 4 + 4 \leq 16 + 4$
$5x \leq 20$
$\dfrac{5x}{5} \leq \dfrac{20}{5}$
$x \leq 4$
$\{x \mid x \leq 4\}\ ;\ (-\infty, 4]$

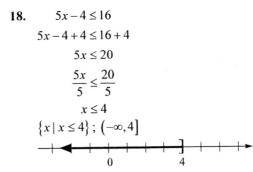

19. $-3x + 1 > 13$
$-3x + 1 - 1 > 13 - 1$
$-3x > 12$
$\dfrac{-3x}{-3} < \dfrac{12}{-3}$
$x < -4$
$\{x \mid x < -4\}\ ;\ (-\infty, -4)$

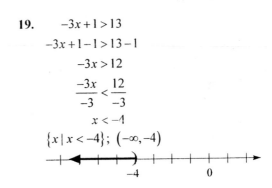

20. $-6x - 5 < 13$
$-6x - 5 + 5 < 13 + 5$
$-6x < 18$
$\dfrac{-6x}{-6} > \dfrac{18}{-6}$
$x > -3$
$\{x \mid x > -3\}\ ;\ (-3, \infty)$

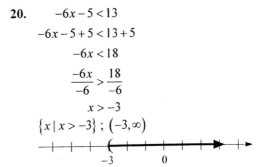

## Chapter 1: Linear Equations and Inequalities

**22.**
$$8x+3 \geq 5x-9$$
$$8x+3-3 \geq 5x-9-3$$
$$8x \geq 5x-12$$
$$8x-5x \geq 5x-12-5x$$
$$3x \geq -12$$
$$\frac{3x}{3} \geq \frac{-12}{3}$$
$$x \geq -4$$
$\{x \mid x \geq -4\}$ ; $[-4,\infty)$

**24.**
$$3x+4 \geq 5x-8$$
$$3x+4-4 \geq 5x-8-4$$
$$3x \geq 5x-12$$
$$3x-5x \geq 5x-12-5x$$
$$-2x \geq -12$$
$$\frac{-2x}{-2} \leq \frac{-12}{-2}$$
$$x \leq 6$$
$\{x \mid x \leq 6\}$ ; $(-\infty, 6]$

**26.** 
$$3(x-2)+5 > 4(x+1)+x$$
$$3x-6+5 > 4x+4+x$$
$$3x-1 > 5x+4$$
$$3x-1+1 > 5x+4+1$$
$$3x > 5x+5$$
$$3x-5x > 5x+5-5x$$
$$-2x > 5$$
$$\frac{-2x}{-2} < \frac{5}{-2}$$
$$x < -\frac{5}{2}$$
$\left\{x \mid x < -\frac{5}{2}\right\}$ ; $\left(-\infty, -\frac{5}{2}\right)$

**28.**
$$-3(x+4)+5x < 4(x+3)-14$$
$$-3x-12+5x < 4x+12-14$$
$$2x-12 < 4x-2$$
$$2x-12+12 < 4x-2+12$$
$$2x < 4x+10$$
$$2x-4x < 4x+10-4x$$
$$-2x < 10$$
$$\frac{-2x}{-2} > \frac{10}{-2}$$
$$x > -5$$
$\{x \mid x > -5\}$ ; $(-5, \infty)$

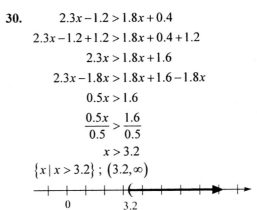

**30.**
$$2.3x-1.2 > 1.8x+0.4$$
$$2.3x-1.2+1.2 > 1.8x+0.4+1.2$$
$$2.3x > 1.8x+1.6$$
$$2.3x-1.8x > 1.8x+1.6-1.8x$$
$$0.5x > 1.6$$
$$\frac{0.5x}{0.5} > \frac{1.6}{0.5}$$
$$x > 3.2$$
$\{x \mid x > 3.2\}$ ; $(3.2, \infty)$

**32.**
$$\frac{2x-3}{3} > \frac{4}{3}$$
$$3 \cdot \frac{2x-3}{3} > 3 \cdot \frac{4}{3}$$
$$2x-3 > 4$$
$$2x-3+3 > 4+3$$
$$2x > 7$$
$$\frac{2x}{2} > \frac{7}{2}$$
$$x > \frac{7}{2}$$
$\left\{x \mid x > \frac{7}{2}\right\}$ ; $\left(\frac{7}{2}, \infty\right)$

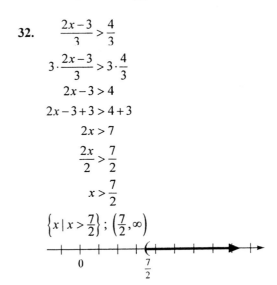

**34.** 
$$\frac{1}{3}(3x+5) < \frac{1}{6}(x+4)$$
$$6 \cdot \frac{1}{3}(3x+5) < 6 \cdot \frac{1}{6}(x+4)$$
$$2(3x+5) < x+4$$
$$6x+10 < x+4$$
$$6x+10-10 < x+4-10$$
$$6x < x-6$$
$$6x-x < x-6-x$$
$$5x < -6$$
$$\frac{5x}{5} < \frac{-6}{5}$$
$$x < -\frac{6}{5}$$
$$\left\{x \mid x < -\frac{6}{5}\right\}; \left(-\infty, -\frac{6}{5}\right)$$

**36.** 
$$\frac{2}{3} - \frac{5}{6}x > 2$$
$$\frac{2}{3} - \frac{5}{6}x - \frac{2}{3} > 2 - \frac{2}{3}$$
$$-\frac{5}{6}x > \frac{6}{3} - \frac{2}{3}$$
$$-\frac{5}{6}x > \frac{4}{3}$$
$$\left(-\frac{6}{5}\right)\cdot\left(-\frac{5}{6}x\right) < -\frac{6}{5} \cdot \frac{4}{3}$$
$$x < -\frac{8}{5}$$
$$\left\{x \mid x < -\frac{8}{5}\right\}; \left(-\infty, -\frac{8}{5}\right)$$

**38.** 
$$-3(2x+1) \leq 2[3x-2(x-5)]$$
$$-6x-3 \leq 2[3x-2x+10]$$
$$-6x-3 \leq 2(x+10)$$
$$-6x-3 \leq 2x+20$$
$$-6x-3+3 \leq 2x+20+3$$
$$-6x \leq 2x+23$$
$$-6x-2x \leq 2x+23-2x$$
$$-8x \leq 23$$
$$x \geq -\frac{23}{8}$$
$$\left\{x \mid x \geq -\frac{23}{8}\right\}; \left[-\frac{23}{8}, \infty\right)$$

**40.** 
$$7(x+2) - 4(2x+3) < -2[5x-2(x+3)] + 7x$$
$$7x+14-8x-12 < -2[5x-2x-6] + 7x$$
$$-x+2 < -2(3x-6) + 7x$$
$$-x+2 < -6x+12+7x$$
$$-x+2 < x+12$$
$$-x+2-2 < x+12-2$$
$$-x < x+10$$
$$-x-x < x+10-x$$
$$-2x < 10$$
$$\frac{-2x}{-2} > \frac{10}{-2}$$
$$x > -5$$
$$\{x \mid x > -5\}; (-5, \infty)$$

## Chapter 1: Linear Equations and Inequalities

**42.**
$$\frac{5}{6}(3x-2) - \frac{2}{3}(4x-1) < -\frac{2}{9}(2x+5)$$
$$18 \cdot \frac{5}{6}(3x-2) - 18 \cdot \frac{2}{3}(4x-1) < -18 \cdot \frac{2}{9}(2x+5)$$
$$15(3x-2) - 12(4x-1) < -4(2x+5)$$
$$45x - 30 - 48x + 12 < -8x - 20$$
$$-3x - 18 < -8x - 20$$
$$-3x - 18 + 18 < -8x - 20 + 18$$
$$-3x < -8x - 2$$
$$-3x + 8x < -8x - 2 + 8x$$
$$5x < -2$$
$$\frac{5x}{5} < \frac{-2}{5}$$
$$x < -\frac{2}{5}$$
$$\left\{x \mid x < -\frac{2}{5}\right\}; \left(-\infty, -\frac{2}{5}\right)$$

**44.**
$$\frac{2}{5}x + \frac{3}{10} < \frac{1}{2}$$
$$10 \cdot \frac{2}{5}x + 10 \cdot \frac{3}{10} < 10 \cdot \frac{1}{2}$$
$$4x + 3 < 5$$
$$4x + 3 - 3 < 5 - 3$$
$$4x < 2$$
$$\frac{4x}{4} < \frac{2}{4}$$
$$x < \frac{1}{2}$$
$$\left\{x \mid x < \frac{1}{2}\right\}; \left(-\infty, \frac{1}{2}\right)$$

**46.**
$$\frac{x}{12} \geq \frac{x}{2} - \frac{2x+1}{4}$$
$$12 \cdot \frac{x}{12} \geq 12 \cdot \frac{x}{2} - 12 \cdot \frac{2x+1}{4}$$
$$x \geq 6x - 3(2x+1)$$
$$x \geq 6x - 6x - 3$$
$$x \geq -3$$
$$\{x \mid x \geq -3\}; [-3, \infty)$$

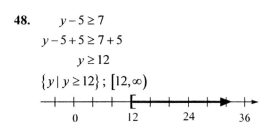

**48.**
$$y - 5 \geq 7$$
$$y - 5 + 5 \geq 7 + 5$$
$$y \geq 12$$
$$\{y \mid y \geq 12\}; [12, \infty)$$

**50.**
$$-5x < 30$$
$$\frac{-5x}{-5} > \frac{30}{-5}$$
$$x > -6$$
$$\{x \mid x > -6\}; (-6, \infty)$$

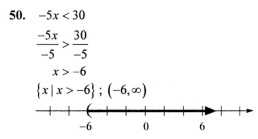

**52.**
$$4x + 3 \geq -6x - 2$$
$$4x + 3 - 3 \geq -6x - 2 - 3$$
$$4x \geq -6x - 5$$
$$4x + 6x \geq -6x - 5 + 6x$$
$$10x \geq -5$$
$$\frac{10x}{10} \geq \frac{-5}{10}$$
$$x \geq -\frac{1}{2}$$
$$\left\{x \mid x \geq -\frac{1}{2}\right\}; \left[-\frac{1}{2}, \infty\right)$$

**54.**
$$5(y+7) < 6(y+4)$$
$$5y + 35 < 6y + 24$$
$$5y + 35 - 35 < 6y + 24 - 35$$
$$5y < 6y - 11$$
$$5y - 6y < 6y - 11 - 6y$$
$$-y < -11$$
$$\frac{-y}{-1} > \frac{-11}{-1}$$
$$y > 11$$
$$\{y \mid y > 11\} \;;\; (11, \infty)$$

**56.**
$$2(5-x) - 3 \le 4 - 5x$$
$$10 - 2x - 3 \le 4 - 5x$$
$$-2x + 7 \le 4 - 5x$$
$$-2x + 7 - 7 \le 4 - 5x - 7$$
$$-2x \le -5x - 3$$
$$-2x + 5x \le -5x - 3 + 5x$$
$$3x \le -3$$
$$\frac{3x}{3} \le \frac{-3}{3}$$
$$x \le -1$$
$$\{x \mid x \le -1\} \;;\; (-\infty, -1]$$

**58.**
$$3[1 + 2(x-4)] \ge 3x + 3$$
$$3[1 + 2x - 8] \ge 3x + 3$$
$$3(2x - 7) \ge 3x + 3 \text{ wrong}$$
$$6x - 21 \ge 3x + 3$$
$$6x - 21 + 21 \ge 3x + 3 + 21$$
$$6x \ge 3x + 24$$
$$6x - 3x \ge 3x + 24 - 3x$$
$$3x \ge 24$$
$$\frac{3x}{3} \ge \frac{24}{3}$$
$$x \ge 8$$
$$\{x \mid x \ge 8\} \;;\; [8, \infty)$$

**60.**
$$\frac{b}{3} + \frac{5}{6} < \frac{11}{12}$$
$$\frac{b}{3} + \frac{5}{6} - \frac{5}{6} < \frac{11}{12} - \frac{5}{6}$$
$$\frac{b}{3} < \frac{11}{12} - \frac{10}{12}$$
$$\frac{b}{3} < \frac{1}{12}$$
$$3 \cdot \frac{b}{3} < 3 \cdot \frac{1}{12}$$
$$b < \frac{1}{4}$$
$$\left\{b \mid b < \frac{1}{4}\right\} \;;\; \left(-\infty, \frac{1}{4}\right)$$

**62.**
$$3x - 2 < 7$$
$$3x - 2 + 2 < 7 + 2$$
$$3x < 9$$
$$\frac{3x}{3} < \frac{9}{3}$$
$$x < 3$$
$$\{x \mid x < 3\} \;;\; (-\infty, 3)$$

**64.**
$$2y + 3 > 13$$
$$2y + 3 - 3 > 13 - 3$$
$$2y > 10$$
$$\frac{2y}{2} > \frac{10}{2}$$
$$y > 5$$
$$\{y \mid y > 5\} \;;\; (5, \infty)$$

## Chapter 1: Linear Equations and Inequalities

66. Let $x$ = score on final exam.
$$\frac{94+83+88+92+2x}{6} \geq 90$$
$$\frac{357+2x}{6} \geq 90$$
$$6 \cdot \frac{357+2x}{6} \geq 6 \cdot 90$$
$$357+2x \geq 540$$
$$357+2x-357 \geq 540-357$$
$$2x \geq 183$$
$$\frac{2x}{2} \geq \frac{183}{2}$$
$$x \geq 91.5$$
Mark's final exam score must be at least 91.5 in order for him to earn an $A$.

68. Let $x$ = number of cheeseburgers.
$$21x+16+8 \leq 87$$
$$21x+24 \leq 87$$
$$21x+24-24 \leq 87-24$$
$$21x \leq 63$$
$$\frac{21x}{21} \leq \frac{63}{21}$$
$$x \leq 3$$
You can order up to three cheeseburgers to keep the total fat content to no more than 87 grams.

70. Let $x$ = number of miles.
$$39.95+0.65x \leq 125.75$$
$$39.95+0.65x-39.95 \leq 125.75-39.95$$
$$0.65x \leq 85.8$$
$$\frac{0.65x}{0.65} \leq \frac{85.8}{0.65}$$
$$x \leq 132$$
You can drive up to 132 miles and be within budget.

72. Let $t$ = number of years since 1990.
$$26t+411 > 1000$$
$$26t+411-411 > 1000-411$$
$$26t > 589$$
$$\frac{26t}{26} > \frac{589}{26}$$
$$t > \frac{589}{26} \approx 22.65$$
Total private health expenditures will exceed $1 trillion in 2013.

74. Let $x$ = annual sales.
$$24,300+0.03x > 60,000$$
$$24,300+0.03x-24,300 > 60,000-24,300$$
$$0.03x > 35,700$$
$$\frac{0.03x}{0.03} > \frac{35,700}{0.03}$$
$$x > 1,190,000$$
Al must have annual sales of more than $1,190,000 to meet his salary goal.
$$\frac{1,190,000}{15,000} \approx 79.33$$
Al would need to sell at least 80 cars to meet his salary goal.

76. 
$$D > S$$
$$1800-12p > -2800+13p$$
$$1800-12p-1800 > -2800+13p-1800$$
$$-12p > 13p-4600$$
$$-12p-13p > 13p-4600-13p$$
$$-25p > -4600$$
$$\frac{-25p}{-25} < \frac{-4600}{-25}$$
$$p < 184$$
If the price of the digital camera is less than $184, demand will exceed supply.

78. $-3(x-2)+7x > 2(2x+5)$
$$-3x+6+7x > 4x+10$$
$$4x+6 > 4x+10$$
$$4x+6-4x > 4x+10-4x$$
$$6 > 10$$
This statement is a contradiction. There are no solutions to the inequality.

**Putting the Concepts Together (Sections 1.1-1.4)**

1. a. Let $x = -3$ in the equation.
$$5(2(-3)-3)+1 \stackrel{?}{=} 2(-3)-6$$
$$5(-6-3)+1 \stackrel{?}{=} -6-6$$
$$5(-9)+1 \stackrel{?}{=} -12$$
$$-45+1 \stackrel{?}{=} -12$$
$$-44 \neq -12$$
$x = -3$ is **not** a solution to the equation.

**Putting the Concepts Together (Sections 1.1 – 1.4)**

b. Let $x = 1$ in the equation.
$$5(2(1)-3)+1 \stackrel{?}{=} 2(1)-6$$
$$5(2-3)+1 \stackrel{?}{=} 2-6$$
$$5(-1)+1 \stackrel{?}{=} -4$$
$$-5+1 \stackrel{?}{=} -4$$
$$-4 = -4$$
$x = 1$ is a solution to the equation.

2. $3(2x-1)+6 = 5x-2$
$6x-3+6 = 5x-2$
$6x+3 = 5x-2$
$6x+3-5x = 5x-2-5x$
$x+3 = -2$
$x+3-3 = -2-3$
$x = -5$
Solution set: $\{-5\}$

3. $\dfrac{7}{3}x + \dfrac{4}{5} = \dfrac{5x+12}{15}$
$15\left(\dfrac{7}{3}x + \dfrac{4}{5}\right) = 15\left(\dfrac{5x+12}{15}\right)$
$35x + 12 = 5x + 12$
$35x + 12 - 12 = 5x + 12 - 12$
$35x = 5x$
$35x - 5x = 5x - 5x$
$30x = 0$
$\dfrac{30x}{30} = \dfrac{0}{30}$
$x = 0$
Solution set: $\{0\}$

4. $5 - 2(x+1) + 4x = 6(x+1) - (3+4x)$
$5 - 2x - 2 + 4x = 6x + 6 - 3 - 4x$
$3 + 2x = 2x + 3$
These expressions are equivalent, so the statement is an identity.

5. Let $x$ represent the number. $x - 3 = \dfrac{1}{2}x + 2$

6. Let $x$ represent the number.
$\dfrac{x}{2} < x + 5$

7. Let $x$ = liters of 20% solution.
 Amount of 40% solution: $16 - x$ liters
 tot. acid = acid from 20% + acid from 40%
 $(\%)(\text{liters}) = (\%)(\text{liters}) + (\%)(\text{liters})$
 $0.35(16) = 0.2(x) + 0.4(16-x)$
 $5.6 = 0.2x + 6.4 - 0.4x$
 $5.6 = 6.4 - 0.2x$
 $5.6 - 6.4 = 6.4 - 0.2x - 6.4$
 $-0.8 = -0.2x$
 $\dfrac{-0.8}{-0.2} = \dfrac{-0.2x}{-0.2}$
 $4 = x$
The chemist needs to mix 4 liters of the 20% solution with 12 liters of the 40% solution.

8. Let $t$ = time of travel in hours.
 $d = r \cdot t$
 total distance = dist. for car 1 + dist. for car 2
 $255 = 30(t) + 45(t)$
 $255 = 75t$
 $\dfrac{255}{75} = \dfrac{75t}{75}$
 $3.4 = t$
After 3.4 hours, the two cars will be 255 miles apart.

9. $3x - 2y = 4$
$3x - 2y - 3x = 4 - 3x$
$-2y = -3x + 4$
$\dfrac{-2y}{-2} = \dfrac{-3x+4}{-2}$
$y = \dfrac{-3x}{-2} + \dfrac{4}{-2}$
$y = \dfrac{3}{2}x - 2$

10. $A = P + Prt$
$A - P = P + Prt - P$
$A - P = Prt$
$\dfrac{A-P}{Pt} = \dfrac{Prt}{Pt}$
$\dfrac{A-P}{Pt} = r$ or $r = \dfrac{A-P}{Pt}$

## Chapter 1: Linear Equations and Inequalities

**11.**  **a.**  $V = \pi r^2 h$

$\dfrac{V}{\pi r^2} = \dfrac{\pi r^2 h}{\pi r^2}$

$\dfrac{V}{\pi r^2} = h$ or $h = \dfrac{V}{\pi r^2}$

**b.**  $h = \dfrac{V}{\pi r^2} = \dfrac{294\pi}{\pi(7)^2} = \dfrac{294}{49} = 6$ inches.

**12.**  **a.**  Interval: $(-3, \infty)$
Graph:
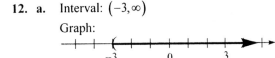

**b.**  Interval: $(2, 5]$
Graph:
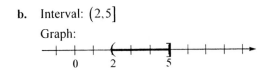

**13.**  **a.**  Inequality: $x \leq -1.5$
Graph:
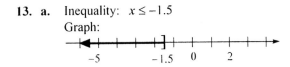

**b.**  Inequality: $-3 < x \leq 1$
Graph:
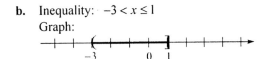

**14.**  $2x + 3 \leq 4x - 9$
$2x + 3 - 3 \leq 4x - 9 - 3$
$2x \leq 4x - 12$
$2x - 4x \leq 4x - 12 - 4x$
$-2x \leq -12$
$\dfrac{-2x}{-2} \geq \dfrac{-12}{-2}$
$x \geq 6$
Interval: $[6, \infty)$
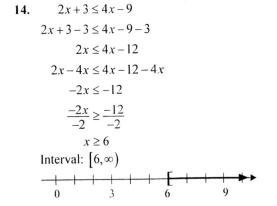

**15.**  $-3 > 3x - (x + 5)$
$-3 > 3x - x - 5$
$-3 > 2x - 5$
$-3 + 5 > 2x - 5 + 5$
$2 > 2x$
$\dfrac{2}{2} > \dfrac{2x}{2}$
$1 > x$ or $x < 1$
Interval: $(-\infty, 1)$
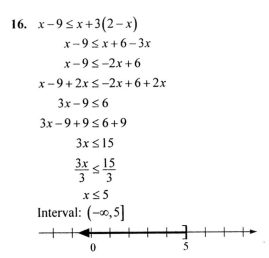

**16.**  $x - 9 \leq x + 3(2 - x)$
$x - 9 \leq x + 6 - 3x$
$x - 9 \leq -2x + 6$
$x - 9 + 2x \leq -2x + 6 + 2x$
$3x - 9 \leq 6$
$3x - 9 + 9 \leq 6 + 9$
$3x \leq 15$
$\dfrac{3x}{3} \leq \dfrac{15}{3}$
$x \leq 5$
Interval: $(-\infty, 5]$
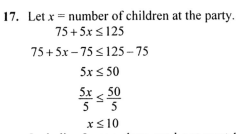

**17.**  Let $x$ = number of children at the party.
$75 + 5x \leq 125$
$75 + 5x - 75 \leq 125 - 75$
$5x \leq 50$
$\dfrac{5x}{5} \leq \dfrac{50}{5}$
$x \leq 10$
Including Logan, there can be at most 10 children at the party. Therefore, Logan can invite at most 9 children to the party.

### 1.5 Preparing for Compound Inequalities

**1.**  $\{x \mid -2 \leq x \leq 4,\ x \text{ is an integer}\}$
$\{-2, -1, 0, 1, 2, 3, 4\}$

**2.**  $\{x \mid 0 < x < 3,\ x \text{ is an integer}\}$
$\{1, 2\}$

## Section 1.5 Compound Inequalities

### 1.5 Exercises

**2.** and; or

**4.** True. If the two sets have no elements in common, the intersection will be the empty set.

**6.** True

**8.** Answers may vary. When writing compound inequalities compactly, we get the variable by itself in the 'middle', but the direction of the inequalities must be the same.

**10.** To be a solution, $x = -1$ must satisfy one or both of the individual inequalities.

$$5x - 2 \stackrel{?}{\leq} 13 \qquad 2x - 5 \stackrel{?}{>} 3$$
$$5(-1) - 2 \stackrel{?}{\leq} 13 \qquad 2(-1) - 5 \stackrel{?}{>} 3$$
$$-5 - 2 \stackrel{?}{\leq} 13 \qquad -2 - 5 \stackrel{?}{>} 3$$
$$-7 \leq 13 \; T \qquad -7 \not> 3$$

Because $x = -1$ makes the first inequality true, it is a solution. It is not necessary to be a solution to both.

**12.** $A \cup C = \{2, 3, 4, 5, 6, 7, 8, 9\}$

**14.** $A \cap C = \{4, 6\}$

**16.** $B \cup C = \{1, 2, 3, 4, 5, 6, 7, 9\}$

**18. a.** $A \cap B = \{\;\}$ or $\varnothing$

**b.** $A \cup B = \{x \mid x < 1 \text{ or } x \geq 4\}$
Interval: $(-\infty, 1) \cup [4, \infty)$

**20. a.** $E \cap F = \{x \mid -2 \leq x \leq 2\}$
Interval: $[-2, 2]$

**b.** $E \cup F = \{x \mid x \text{ is a real number}\}$
Interval: $(-\infty, \infty)$

**22.** $x \leq 5$ and $x > 0$
Solution set: $\{x \mid 0 < x \leq 5\}$
Interval: $(0, 5]$

**24.** $x < 0$ or $x \geq 6$
Solution set: $\{x \mid x < 0 \text{ or } x \geq 6\}$
Interval: $(-\infty, 0) \cup [6, \infty)$

**26.** $6x - 2 \leq 10$ and $10x > -20$
$\quad 6x \leq 12 \qquad\qquad x > -2$
$\quad x \leq 2$
We need $x \leq 2$ and $x > -2$.
Solution set: $\{x \mid -2 < x \leq 2\}$
Interval: $(-2, 2]$

**28.** $x - 3 \leq 2$ and $6x + 5 \geq -1$
$\quad x \leq 5 \qquad\qquad 6x \geq -6$
$\qquad\qquad\qquad\qquad x \geq -1$
We need $x \leq 5$ and $x \geq -1$.
Solution set: $\{x \mid -1 \leq x \leq 5\}$
Interval: $[-1, 5]$

**30.** $7x + 2 \geq 9$ and $4x + 3 \leq 7$
$\quad 7x \geq 7 \qquad\qquad 4x \leq 4$
$\quad x \geq 1 \qquad\qquad x \leq 1$
We need $x \geq 1$ and $x \leq 1$.
Solution set: $\{x \mid x = 1\}$

**32.** $x + 3 \leq 5$ or $x - 2 \geq 3$
$\quad x \leq 2 \qquad\qquad x \geq 5$
We need $x \leq 2$ or $x \geq 5$.
Solution set: $\{x \mid x \leq 2 \text{ or } x \geq 5\}$
Interval: $(-\infty, 2] \cup [5, \infty)$

## Chapter 1: Linear Equations and Inequalities

**34.** $4x+3 > -5$ or $8x-5 < 3$
$\qquad 4x > -8 \qquad\qquad 8x < 8$
$\qquad x > -2 \qquad\qquad x < 1$
We need $x > -2$ or $x < 1$.
Solution set: $\{x \mid x \text{ is a real number}\}$
Interval: $(-\infty, \infty)$

**36.** $3x \geq 7x+8$ or $x < 4x-9$
$\qquad -4x \geq 8 \qquad\qquad -3x < -9$
$\qquad x \leq -2 \qquad\qquad x > 3$
We need $x \leq -2$ or $x > 3$.
Solution set: $\{x \mid x \leq -2 \text{ or } x > 3\}$
Interval: $(-\infty, -2] \cup (3, \infty)$

**38.** $-10 < 6x+8 \leq -4$
$-10-8 < 6x+8-8 \leq -4-8$
$-18 < 6x \leq -12$
$\dfrac{-18}{6} < \dfrac{6x}{6} \leq \dfrac{-12}{6}$
$-3 < x \leq -2$
Solution set: $\{x \mid -3 < x \leq -2\}$
Interval: $(-3, -2]$

**40.** $-12 < 7x+2 \leq 6$
$-12-2 < 7x+2-2 \leq 6-2$
$-14 < 7x \leq 4$
$\dfrac{-14}{7} < \dfrac{7x}{7} \leq \dfrac{4}{7}$
$-2 < x \leq \dfrac{4}{7}$
Solution set: $\left\{x \mid -2 < x \leq \dfrac{4}{7}\right\}$
Interval: $\left(-2, \dfrac{4}{7}\right]$

**42.** $-\dfrac{4}{5}x-5 > 3$ or $7x-3 > 4$
$\qquad -\dfrac{4}{5}x > 8 \qquad\qquad 7x > 7$
$\qquad -4x > 40 \qquad\qquad x > 1$
$\qquad x < -10$
We need $x < -10$ or $x > 1$.
Solution set: $\{x \mid x < -10 \text{ or } x > 1\}$
Interval: $(-\infty, -10) \cup (1, \infty)$

**44.** $-6 < -3x+6 \leq 4$
$-6-6 < -3x+6-6 \leq 4-6$
$-12 < -3x \leq -2$
$\dfrac{-12}{-3} > \dfrac{-3x}{-3} \geq \dfrac{-2}{-3}$
$4 > x \geq \dfrac{2}{3}$ or $\dfrac{2}{3} \leq x < 4$
Solution set: $\left\{x \mid \dfrac{2}{3} \leq x < 4\right\}$
Interval: $\left[\dfrac{2}{3}, 4\right)$

**46.** $0 < \dfrac{3}{2}x-3 \leq 3$
$0+3 < \dfrac{3}{2}x-3+3 \leq 3+3$
$3 < \dfrac{3}{2}x \leq 6$
$2 \cdot 3 < 2 \cdot \dfrac{3}{2}x \leq 2 \cdot 6$
$6 < 3x \leq 12$
$\dfrac{6}{3} < \dfrac{3x}{3} \leq \dfrac{12}{3}$
$2 < x \leq 4$
Solution set: $\{x \mid 2 < x \leq 4\}$
Interval: $(2, 4]$

ISM − 38

**Section 1.5** Compound Inequalities

48. $-3 < -4x + 1 < 17$
$-3 - 1 < -4x + 1 - 1 < 17 - 1$
$-4 < -4x < 16$
$\dfrac{-4}{-4} > \dfrac{-4x}{-4} > \dfrac{16}{-4}$
$1 > x > -4$ or $-4 < x < 1$
Solution set: $\{x \mid -4 < x < 1\}$
Interval: $(-4, 1)$

50. $\dfrac{2}{3}x + 2 \le 4$ or $\dfrac{5x-3}{3} \ge 4$
$\dfrac{2}{3}x \le 2$   $5x - 3 \ge 12$
$x \le 3$   $5x \ge 15$
   $x \ge 3$
We need $x \le 3$ or $x \ge 3$.
Solution set: $\{x \mid x \text{ is a real number}\}$
Interval: $(-\infty, \infty)$

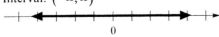

52. $x - \dfrac{3}{2} \le \dfrac{5}{4}$   and   $-\dfrac{2}{3}x - \dfrac{2}{9} < \dfrac{8}{9}$
$4\left(x - \dfrac{3}{2}\right) \le 4\left(\dfrac{5}{4}\right)$   $9\left(-\dfrac{2}{3}x - \dfrac{2}{9}\right) < 9\left(\dfrac{8}{9}\right)$
$4x - 6 \le 5$   $-6x - 2 < 8$
$4x \le 11$   $-6x < 10$
$x \le \dfrac{11}{4}$   $\dfrac{-6x}{-6} > \dfrac{10}{-6}$
   $x > -\dfrac{5}{3}$
We need $x > -\dfrac{5}{3}$ and $x \le \dfrac{11}{4}$.
Solution set: $\left\{x \mid -\dfrac{5}{3} < x \le \dfrac{11}{4}\right\}$
Interval: $\left(-\dfrac{5}{3}, \dfrac{11}{4}\right]$

54. $-4 \le \dfrac{4x-3}{3} < 3$
$3(-4) \le 3\left(\dfrac{4x-3}{3}\right) < 3(3)$
$-12 \le 4x - 3 < 9$
$-12 + 3 \le 4x - 3 + 3 < 9 + 3$
$-9 \le 4x < 12$
$\dfrac{-9}{4} \le \dfrac{4x}{4} < \dfrac{12}{4}$
$-\dfrac{9}{4} \le x < 3$
Solution set: $\left\{x \mid -\dfrac{9}{4} \le x < 3\right\}$
Interval: $\left[-\dfrac{9}{4}, 3\right)$

56. $-6 < -3(x - 2) < 15$
$-6 < -3x + 6 < 15$
$-6 - 6 < -3x + 6 - 6 < 15 - 6$
$-12 < -3x < 9$
$\dfrac{-12}{-3} > \dfrac{-3x}{-3} > \dfrac{9}{-3}$
$4 > x > -3$   or   $-3 < x < 4$
Solution set: $\{x \mid -3 < x < 4\}$
Interval: $(-3, 4)$

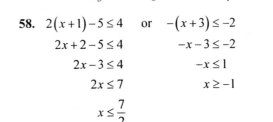

58. $2(x+1) - 5 \le 4$   or   $-(x+3) \le -2$
$2x + 2 - 5 \le 4$   $-x - 3 \le -2$
$2x - 3 \le 4$   $-x \le 1$
$2x \le 7$   $x \ge -1$
$x \le \dfrac{7}{2}$
We need $x \ge -1$ or $x \le \dfrac{7}{2}$.
Solution set: $\{x \mid x \text{ is a real number}\}$
Interval: $(-\infty, \infty)$

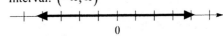

## Chapter 1: Linear Equations and Inequalities

**60.**  
$5x - 1 < 9$ and $5x > -20$  
$5x - 1 + 1 < 9 + 1$   $\dfrac{5x}{5} > \dfrac{-20}{5}$  
$5x < 10$   $x > -4$  
$\dfrac{5x}{5} < \dfrac{10}{5}$  
$x < 2$  

We need $x < 2$ and $x > -4$.  
Solution set: $\{x \mid -4 < x < 2\}$  
Interval: $(-4, 2)$  

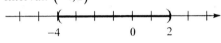

**62.** $3(x+7) < 24$ or $6(x-4) > -30$  
$\dfrac{3(x+7)}{3} < \dfrac{24}{3}$   $\dfrac{6(x-4)}{6} > \dfrac{-30}{6}$  
$x + 7 < 8$   $x - 4 > -5$  
$x + 7 - 7 < 8 - 7$   $x - 4 + 4 > -5 + 4$  
$x < 1$   $x > -1$  

We need $x < 1$ or $x > -1$.  
Solution set: $\{x \mid x \text{ is a real number}\}$  
Interval: $(-\infty, \infty)$  

**64.** $-8 \leq 5x - 3 \leq 4$  
$-8 + 3 \leq 5x - 3 + 3 \leq 4 + 3$  
$-5 \leq 5x \leq 7$  
$\dfrac{-5}{5} \leq \dfrac{5x}{5} \leq \dfrac{7}{5}$  
$-1 \leq x \leq \dfrac{7}{5}$  

Solution set: $\left\{x \mid -1 \leq x \leq \dfrac{7}{5}\right\}$  
Interval: $\left[-1, \dfrac{7}{5}\right]$  

**66.** $3x - 8 < -14$ or $4x - 5 > 7$  
$3x - 8 + 8 < -14 + 8$   $4x - 5 + 5 > 7 + 5$  
$3x < -6$   $4x > 12$  
$\dfrac{3x}{3} < \dfrac{-6}{3}$   $\dfrac{4x}{4} > \dfrac{12}{4}$  
$x < -2$   $x > 3$  

We need $x < -2$ or $x > 3$.

Solution set: $\{x \mid x < -2 \text{ or } x > 3\}$  
Interval: $(-\infty, -2) \cup (3, \infty)$  

**68.** $-5 < 2x + 7 \leq 5$  
$-5 - 7 < 2x + 7 - 7 \leq 5 - 7$  
$-12 < 2x \leq -2$  
$\dfrac{-12}{2} < \dfrac{2x}{2} \leq \dfrac{-2}{2}$  
$-6 < x \leq -1$  

Solution set: $\{x \mid -6 < x \leq -1\}$  
Interval: $(-6, -1]$  

**70.** $\dfrac{x}{2} \leq -4$ or $\dfrac{2x-1}{3} \geq 2$  
$2 \cdot \dfrac{x}{2} \leq 2 \cdot (-4)$   $3 \cdot \dfrac{2x-1}{3} \geq 3 \cdot 2$  
$x \leq -8$   $2x - 1 \geq 6$  
   $2x - 1 + 1 \geq 6 + 1$  
   $2x \geq 7$  
   $\dfrac{2x}{2} \geq \dfrac{7}{2}$  
   $x \geq \dfrac{7}{2}$  

We need $x \leq -8$ or $x \geq \dfrac{7}{2}$.  
Solution set: $\left\{x \mid x \leq -8 \text{ or } x \geq \dfrac{7}{2}\right\}$  
Interval: $(-\infty, -8] \cup \left[\dfrac{7}{2}, \infty\right)$  

**72.** $-15 < -3(x+2) \leq 1$  
$-15 < -3x - 6 \leq 1$  
$-15 + 6 < -3x - 6 + 6 \leq 1 + 6$  
$-9 < -3x \leq 7$  
$\dfrac{-9}{-3} > \dfrac{-3x}{-3} \geq \dfrac{7}{-3}$  
$3 > x \geq -\dfrac{7}{3}$ or $-\dfrac{7}{3} \leq x < 3$  

Solution set: $\left\{x \mid -\dfrac{7}{3} \leq x < 3\right\}$

Section 1.5 Compound Inequalities

Interval: $\left[-\frac{7}{3}, 3\right)$

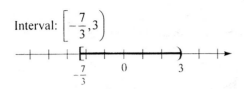

74. $-2 < x < 3$
$-2 - 3 < x - 3 < 3 - 3$
$-5 < x - 3 < 0$
We need $a = -5$ and $b = 0$.

76. $2 < x < 12$
$\frac{2}{2} < \frac{x}{2} < \frac{12}{2}$
$1 < \frac{x}{2} < 6$
We need $a = 1$ and $b = 6$.

78. $-4 < x < 3$
$2(-4) < 2x < 2(3)$
$-8 < 2x < 6$
$-8 - 7 < 2x - 7 < 6 - 7$
$-15 < 2x - 7 < -1$
We need $a = -15$ and $b = -1$.

80. Let $x$ = diastolic blood pressure.
$60 < x < 90$

82. Let $x$ = final exam score.
$70 \le \frac{67 + 72 + 81 + 75 + 3x}{7} \le 79$
$70 \le \frac{295 + 3x}{7} \le 79$
$7(70) \le 7\left(\frac{295 + 3x}{7}\right) \le 7(79)$
$490 \le 295 + 3x \le 553$
$490 - 295 \le 295 + 3x - 295 \le 553 - 295$
$195 \le 3x \le 258$
$\frac{195}{3} \le \frac{3x}{3} \le \frac{258}{3}$
$65 \le x \le 86$
Jack needs to score between 65 and 86 (inclusive) on the final exam to earn a C.

84. Let $x$ = weekly wages.
$800 \le x \le 900$
$800 - 517 \le x - 517 \le 900 - 517$
$283 \le x - 517 \le 383$
$0.28(283) \le 0.28(x - 517) \le 0.28(383)$
$79.24 \le 0.28(x - 517) \le 107.24$
$79.24 + 69.6 \le 0.28(x - 517) + 69.6 \le 107.24 + 69.6$
$148.84 \le 0.28(x - 517) + 69.6 \le 176.84$
The amount withheld ranges between $148.84

86. Let $x$ = number of kwh.
$39.11 \le 0.075895(x - 350) + 31.52 \le 69.47$
$7.59 \le 0.075895(x - 350) \le 37.95$
$\frac{7.59}{0.075895} \le x - 350 \le \frac{37.95}{0.075895}$
$\frac{7.59}{0.075895} + 350 \le x \le \frac{37.95}{0.075895} + 350$
$450 \le x \le 850$ (approx.)
The electricity usage ranged from 450 to 850 kwh.

88. a. Here we have $a = 3$, $b = 4$, and $c = 5$.
$b - a < c < b + a$
$4 - 3 < 5 < 4 + 3$
$1 < 5 < 7$ T
These sides could form a triangle.

b. Here we have $a = 4$, $b = 7$, and $c = 12$.
$b - a < c < b + a$
$7 - 4 < 12 < 7 + 4$
$3 < 12 < 11$ false
These sides could not form a triangle.

c. Here we have $a = 3$, $b = 3$, and $c = 5$.
$b - a < c < b + a$
$3 - 3 < 5 < 3 + 3$
$0 < 5 < 6$ T
These sides could form a triangle.

d. Here we have $a = 1$, $b = 9$, and $c = 10$.
$b - a < c < b + a$
$9 - 1 < 10 < 9 + 1$
$8 < 10 < 10$ false
These sides could not form a triangle.

## Chapter 1: Linear Equations and Inequalities

**90.**
$$x - 3 \leq 3x + 1 \leq x + 11$$
$$x - 3 - x \leq 3x + 1 - x \leq x + 11 - x$$
$$-3 \leq 2x + 1 \leq 11$$
$$-3 - 1 \leq 2x + 1 - 1 \leq 11 - 1$$
$$-4 \leq 2x \leq 10$$
$$\frac{-4}{2} \leq \frac{2x}{2} \leq \frac{10}{2}$$
$$-2 \leq x \leq 5$$
Solution set: $\{x \mid -2 \leq x \leq 5\}$
Interval: $[-2, 5]$

**92.**
$$4x + 1 > 2(2x - 1)$$
$$4x + 1 > 4x - 2$$
$$4x + 1 - 4x > 4x - 2 - 4x$$
$$1 > -2$$
This is an identity. All real numbers are solutions. If during simplification, the variable terms all cancel out and an identity results, then the solution set consists of all real numbers.

### 1.6 Preparing for Absolute Value Equations and Inequalities

**1.** $|3| = 3$ because the distance from 0 to 3 on a real number line is 3 units.

**2.** $|-4| = 4$ because the distance from 0 to $-4$ on a real number line is 4 units.

**3.** $|-1.6| = 1.6$ because the distance from 0 to $-1.6$ on a real number line is 1.6 units.

**4.** $|0| = 0$ because the distance from 0 to 0 on a real number line is 0 units.

**5.** The distance between 0 and 5 on a real number line can be expressed as $|5|$.

**6.** The distance between 0 and $-8$ on a real number line can be expressed as $|-8|$.

### 1.6 Exercises

**2.** $-a < u < a$

**4.** True; the absolute value of a number represents the distance of the number from 0 on a real number line. Since distance is never negative, absolute value is never negative.

**6.** False; $|u| > a$ is equivalent to $u < -a$ or $u > a$.

**8.** $|z| = 9$
$z = -9$ or $z = 9$
Solution set: $\{-9, 9\}$
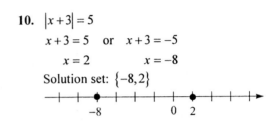

**10.** $|x + 3| = 5$
$x + 3 = 5$ or $x + 3 = -5$
$x = 2$ \qquad $x = -8$
Solution set: $\{-8, 2\}$
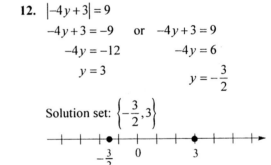

**12.** $|-4y + 3| = 9$
$-4y + 3 = -9$ or $-4y + 3 = 9$
$-4y = -12$ \qquad $-4y = 6$
$y = 3$ \qquad $y = -\frac{3}{2}$
Solution set: $\left\{-\frac{3}{2}, 3\right\}$
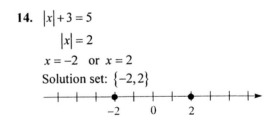

**14.** $|x| + 3 = 5$
$|x| = 2$
$x = -2$ or $x = 2$
Solution set: $\{-2, 2\}$

**16.** $|3y + 1| - 5 = -3$
$|3y + 1| = 2$
$3y + 1 = -2$ or $3y + 1 = 2$
$3y = -3$ \qquad $3y = 1$
$y = -1$ \qquad $y = \frac{1}{3}$

Solution set: $\left\{-1, \frac{1}{3}\right\}$

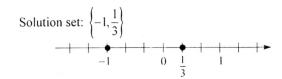

18. $3|y-4|+4=16$
    $3|y-4|=12$
    $|y-4|=4$
    $y-4=-4$ or $y-4=4$
    $y=0$ $\qquad$ $y=8$
    Solution set: $\{0, 8\}$

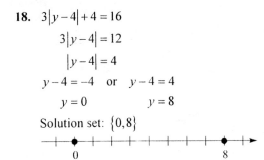

20. $|-2x|+9=9$
    $|-2x|=0$
    $-2x=0$
    $x=0$
    Solution set: $\{0\}$

22. $\left|\dfrac{2x-3}{5}\right|=2$
    $\dfrac{2x-3}{5}=2$ or $\dfrac{2x-3}{5}=-2$
    $2x-3=10$ $\qquad$ $2x-3=-10$
    $2x=13$ $\qquad$ $2x=-7$
    $x=\dfrac{13}{2}$ $\qquad$ $x=-\dfrac{7}{2}$
    Solution set: $\left\{-\dfrac{7}{2}, \dfrac{13}{2}\right\}$

24. $|5y-2|=|4y+7|$
    $5y-2=4y+7$ or $5y-2=-(4y+7)$
    $5y=4y+9$ $\qquad$ $5y-2=-4y-7$
    $y=9$ $\qquad$ $5y=-4y-5$
    $\qquad$ $9y=-5$
    $\qquad$ $y=-\dfrac{5}{9}$

Solution set: $\left\{-\dfrac{5}{9}, 9\right\}$

26. $|5x+3|=|12-4x|$
    $5x+3=12-4x$ or $5x+3=-(12-4x)$
    $5x=9-4x$ $\qquad$ $5x+3=-12+4x$
    $9x=9$ $\qquad$ $5x=-15+4x$
    $x=1$ $\qquad$ $x=-15$
    Solution set: $\{-15, 1\}$

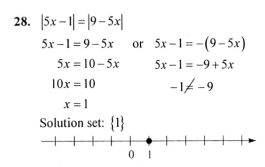

28. $|5x-1|=|9-5x|$
    $5x-1=9-5x$ or $5x-1=-(9-5x)$
    $5x=10-5x$ $\qquad$ $5x-1=-9+5x$
    $10x=10$ $\qquad$ $-1\ne -9$
    $x=1$
    Solution set: $\{1\}$

30. $|x|\le \dfrac{5}{4}$
    $-\dfrac{5}{4}\le x \le \dfrac{5}{4}$
    Solution set: $\left\{x \mid -\dfrac{5}{4}\le x \le \dfrac{5}{4}\right\}$
    Interval: $\left[-\dfrac{5}{4}, \dfrac{5}{4}\right]$

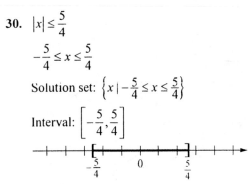

32. $|y+4|<6$
    $-6<y+4<6$
    $-10<y<2$
    Solution set: $\{y \mid -10 < y < 2\}$
    Interval: $(-10, 2)$

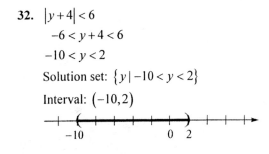

## Chapter 1: Linear Equations and Inequalities

**34.** $|4x-3| \leq 9$

$-9 \leq 4x - 3 \leq 9$

$-6 \leq 4x \leq 12$

$-\dfrac{3}{2} \leq x \leq 3$

Solution set: $\left\{x \mid -\dfrac{3}{2} \leq x \leq 3\right\}$

Interval: $\left[-\dfrac{3}{2}, 3\right]$

**36.** $|4x+3| \leq 0$

$4x + 3 = 0$

$4x = -3$

$x = -\dfrac{3}{4}$

Solution set: $\left\{-\dfrac{3}{4}\right\}$

Interval: $\left[-\dfrac{3}{4}, -\dfrac{3}{4}\right]$

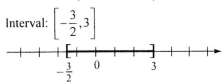

**38.** $|x+4| \geq 7$

$x + 4 \leq -7$ or $x + 4 \geq 7$

$x \leq -11 \qquad\qquad x \geq 3$

Solution set: $\{x \mid x \leq -11 \text{ or } x \geq 3\}$

Interval: $(-\infty, -11] \cup [3, \infty)$

**40.** $|-5y+3| > 7$

$-5y + 3 < -7$ or $-5y + 3 > 7$

$-5y < -10 \qquad -5y > 4$

$y > 2 \qquad\qquad y < -\dfrac{4}{5}$

Solution set: $\left\{y \mid y < -\dfrac{4}{5} \text{ or } y > 2\right\}$

Interval: $\left(-\infty, -\dfrac{4}{5}\right) \cup (2, \infty)$

**42.** $3|z| + 8 > 2$

$3|z| > -6$

$|z| > -2$

Since $|z| \geq 0 > -2$ for all $z$, all real numbers are solutions.

Solution set: $\{z \mid z \text{ is a real number}\}$

Interval: $(-\infty, \infty)$

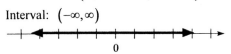

**44.** $3|y+2| - 2 > 7$

$3|y+2| > 9$

$|y+2| > 3$

$y + 2 < -3$ or $y + 2 > 3$

$y < -5 \qquad\qquad y > 1$

Solution set: $\{y \mid y < -5 \text{ or } y > 1\}$

Interval: $(-\infty, -5) \cup (1, \infty)$

**46.** $|-9x+2| \geq -1$

Since absolute value is always $\geq 0$, this inequality is true for all real numbers.

Solution set: $\{x \mid x \text{ is a real number}\}$

Interval: $(-\infty, \infty)$

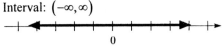

**48.** $|-3x+2| - 7 \leq -2$

$|-3x+2| \leq 5$

$-5 \leq -3x + 2 \leq 5$

$-7 \leq -3x \leq 3$

$\dfrac{7}{3} \geq x \geq -1$ or $-1 \leq x \leq \dfrac{7}{3}$

Solution set: $\left\{x \mid -1 \leq x \leq \dfrac{7}{3}\right\}$

Interval: $\left[-1, \dfrac{7}{3}\right]$

### Section 1.6 Absolute Value Equations and Inequalities

50. $|8x+3| \geq 3$

    $8x+3 \leq -3$ or $8x+3 \geq 3$

    $8x \leq -6 \qquad\qquad 8x \geq 0$

    $x \leq -\dfrac{3}{4} \qquad\qquad x \geq 0$

    Solution set: $\left\{ x \mid x \leq -\dfrac{3}{4} \text{ or } x \geq 0 \right\}$

    Interval: $\left(-\infty, -\dfrac{3}{4}\right] \cup [0, \infty)$

52. $|3-5x| < |-7|$

    $|3-5x| < 7$

    $-7 < 3-5x < 7$

    $-10 < -5x < 4$

    $2 > x > -\dfrac{4}{5}$ or $-\dfrac{4}{5} < x < 2$

    Solution set: $\left\{ x \mid -\dfrac{4}{5} < x < 2 \right\}$

    Interval: $\left(-\dfrac{4}{5}, 2\right)$

    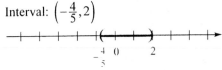

54. $|(3x+2)-8| < 0.01$

    $|3x+2-8| < 0.01$

    $|3x-6| < 0.01$

    $-0.01 < 3x-6 < 0.01$

    $5.99 < 3x < 6.01$

    $1.997 < x < 2.003$ (approx.)

    Solution set: $\{x \mid 1.997 < x < 2.003\}$

    Interval: $(1.997, 2.003)$

56. $|x| > \dfrac{8}{3}$

    $x < -\dfrac{8}{3}$ or $x > \dfrac{8}{3}$

    Solution set: $\left\{ x \mid x < -\dfrac{8}{3} \text{ or } x > \dfrac{8}{3} \right\}$

    Interval: $\left(-\infty, -\dfrac{8}{3}\right) \cup \left(\dfrac{8}{3}, \infty\right)$

58. $|4x+3| = 1$

    $4x+3 = -1$ or $4x+3 = 1$

    $4x = -4 \qquad\qquad 4x = -2$

    $x = -1 \qquad\qquad x = -\dfrac{1}{2}$

    Solution set: $\left\{-1, -\dfrac{1}{2}\right\}$

60. $8|y| = 32$

    $|y| = 4$

    $y = -4$ or $y = 4$

    Solution set: $\{-4, 4\}$

62. $|7y-3| < 11$

    $-11 < 7y-3 < 11$

    $-8 < 7y < 14$

    $-\dfrac{8}{7} < y < 2$

    Solution set: $\left\{ y \mid -\dfrac{8}{7} < y < 2 \right\}$

    Interval: $\left(-\dfrac{8}{7}, 2\right)$

64. $|3x-4| = -9$

    No solution. Absolute value is never negative.

    Solution set: $\varnothing$ or $\{\ \}$

66. $|5y+3| > 2$

    $5y+3 < -2$ or $5y+3 > 2$

    $5y < -5 \qquad\qquad 5y > -1$

    $y < -1 \qquad\qquad y > -\dfrac{1}{5}$

    Solution set: $\left\{ y \mid y < -1 \text{ or } y > -\dfrac{1}{5} \right\}$

Interval: $(-\infty, -1) \cup \left(-\frac{1}{5}, \infty\right)$

**68.** $|4y+3| - 8 \geq -3$

$|4y+3| \geq 5$

$4y+3 \leq -5$ or $4y+3 \geq 5$

$4y \leq -8 \qquad 4y \geq 2$

$y \leq -2 \qquad y \geq \frac{1}{2}$

Solution set: $\left\{y \mid y \leq -2 \text{ or } y \geq \frac{1}{2}\right\}$

Interval: $(-\infty, -2] \cup \left[\frac{1}{2}, \infty\right)$

**70.** $|3z-2| = |z+6|$

$3z-2 = z+6$ or $3z-2 = -(z+6)$

$3z = z+8 \qquad 3z-2 = -z-6$

$2z = 8 \qquad 3z = -z-4$

$z = 4 \qquad 4z = -4$

$\qquad\qquad z = -1$

Solution set: $\{-1, 4\}$

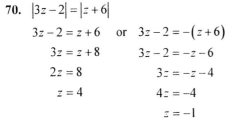

**72.** $|4x+1| > 0$

Since absolute value is always nonnegative, all real numbers are solutions except where
$4x+1 = 0$

$4x = -1$

$x = -\frac{1}{4}$

Thus, all real numbers are solutions except $x = -\frac{1}{4}$.

Solution set: $\left\{x \mid x \neq -\frac{1}{4}\right\}$

Interval: $\left(-\infty, -\frac{1}{4}\right) \cup \left(-\frac{1}{4}, \infty\right)$

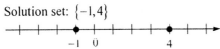

**74.** $\left|\frac{1}{2}x - 3\right| = \left|\frac{2}{3}x + 1\right|$

$\frac{1}{2}x - 3 = \frac{2}{3}x + 1$ or $\frac{1}{2}x - 3 = -\frac{2}{3}x - 1$

$\frac{1}{2}x = \frac{2}{3}x + 4 \qquad \frac{1}{2}x = -\frac{2}{3}x + 2$

$6\left(\frac{1}{2}x\right) = 6\left(\frac{2}{3}x + 4\right) \qquad 6\left(\frac{1}{2}x\right) = 6\left(-\frac{2}{3}x + 2\right)$

$3x = 4x + 24 \qquad 3x = -4x + 12$

$-x = 24 \qquad 7x = 12$

$x = -24 \qquad x = \frac{12}{7}$

Solution set: $\left\{-24, \frac{12}{7}\right\}$

**76.** $|x - (-4)| < 2$

$|x+4| < 2$

$-2 < x+4 < 2$

$-2-4 < x+4-4 < 2-4$

$-6 < x < -2$

Solution set: $\{x \mid -6 < x < -2\}$

Interval: $(-6, -2)$

**78.** $|2x-7| > 3$

$2x-7 < -3$ or $2x-7 > 3$

$2x < 4 \qquad 2x > 10$

$x < 2 \qquad x > 5$

Solution set: $\{x \mid x < 2 \text{ or } x > 5\}$

Interval: $(-\infty, 2) \cup (5, \infty)$

**80.** $|x - 6.125| \leq 0.0005$

$-0.0005 \leq x - 6.125 \leq 0.0005$

$6.1245 \leq x \leq 6.1255$

The acceptable rod lengths are between 6.1245 inches and 6.1255 inches, inclusive.

**82.** $\left|\frac{x-266}{16}\right| > 1.96$

$\frac{x-266}{16} < -1.96$ or $\frac{x-266}{16} > 1.96$

$x - 266 < -31.36 \qquad x - 266 > 31.36$

$x < 234.64 \qquad x > 297.36$

### Section 1.6 Absolute Value Equations and Inequalities

Gestation periods less than 234.64 days or more than 297.36 days would be considered unusual.

84. $|5x-3| > -5$ has a solution set containing all real numbers because $|5x-3| \geq 0 > -5$ for any $x$.

86. $|x-5| = |5-x|$

$$x - 5 = 5 - x \quad \text{or} \quad x - 5 = -(5-x)$$
$$x = 10 - x \qquad\qquad x - 5 = x - 5$$
$$2x = 10 \qquad\qquad\qquad 0 = 0$$
$$x = 5$$

Solution set: $\{x \mid x \text{ is a real number}\}$

The result is reasonable because the expressions inside the absolute values are opposites of each other. The have the same magnitude, but different signs, so their absolute values will be the same. This type of equation is called an identity.

88. $|y| + y = 3$

$$|y| = 3 - y$$
$$y = 3 - y \quad \text{or} \quad y = -(3-y)$$
$$2y = 3 \qquad\qquad y = y - 3$$
$$y = \frac{3}{2} \qquad\qquad 0 \neq -3$$

We also need $3 - y \geq 0$ so the absolute value is not negative. This means we need $y \leq 3$ so the solution above works.

Solution set: $\left\{\frac{3}{2}\right\}$

90. $y - |-y| = 12$

$$|-y| = y - 12$$
$$-y = y - 12 \quad \text{or} \quad -y = -(y-12)$$
$$-2y = -12 \qquad\qquad -y = -y + 12$$
$$y = 6 \qquad\qquad\qquad 0 \neq 12$$

We also need $y - 12 \geq 0$ so that $|-y|$ is not negative. This means we need $y \geq 12$. Thus, there are no solutions to the equation.

Solution set: $\{\ \}$ or $\varnothing$

91. $|4x+1| = x - 2$

$$4x + 1 = x - 2 \quad \text{or} \quad 4x + 1 = -(x-2)$$
$$4x = x - 3 \qquad\qquad 4x + 1 = -x + 2$$
$$3x = -3 \qquad\qquad\quad 4x = -x + 1$$
$$x = -1 \qquad\text{or}\qquad 5x = 1$$
$$x = -1 \quad \text{or} \quad x = \frac{1}{5}$$

We also need $x - 2 \geq 0$ so that the absolute value is not negative. This means we need $x \geq 2$ so neither solution above works.

Check:

$$|4(-1)+1| \stackrel{?}{=} (-1) - 2 \qquad \left|4\left(\frac{1}{5}\right)+1\right| \stackrel{?}{=} \frac{1}{5} - 2$$
$$|-4+1| \stackrel{?}{=} -3 \qquad\qquad \left|\frac{4}{5}+1\right| \stackrel{?}{=} -\frac{9}{5}$$
$$|-3| \stackrel{?}{=} -3 \qquad\qquad\quad \left|\frac{9}{5}\right| \stackrel{?}{=} -\frac{9}{5}$$
$$3 \neq -3 \qquad\qquad\qquad \frac{9}{5} \neq -\frac{9}{5}$$

Solution set: $\varnothing$ or $\{\ \}$

93. $|x+5| = -(x+5)$

Since we have $|u| = -u$, we need $u \leq 0$ so the absolute value will not be negative. Thus, we can say that $x + 5 \leq 0$ or $x \leq -5$.

Solution set: $\{x \mid x \leq -5\}$

Interval: $(-\infty, -5]$

### Chapter 1 Review

1. $3x - 4 = 6 + x$

Let $x = 5$ in the equation.

$$3(5) - 4 \stackrel{?}{=} 6 + (5)$$
$$15 - 4 \stackrel{?}{=} 11$$
$$11 = 11 \quad \text{T}$$

$x = 5$ is a solution to the equation.

Let $x = 6$ in the equation.

$$3(6) - 4 \stackrel{?}{=} 6 + (6)$$
$$18 - 4 \stackrel{?}{=} 12$$
$$14 \neq 12$$

$x = 6$ is **not** a solution to the equation.

## Chapter 1: Linear Equations and Inequalities

2. $-1-4x = 2(3-2x)-7$

    Let $x = -2$ in the equation.
    $$-1-4(-2) \stackrel{?}{=} 2(3-2(-2))-7$$
    $$-1+8 \stackrel{?}{=} 2(3+4)-7$$
    $$7 \stackrel{?}{=} 2(7)-7$$
    $$7 \stackrel{?}{=} 14-7$$
    $$7 = 7 \text{ T}$$
    $x = -2$ is a solution to the equation.

    Let $x = -1$ in the equation.
    $$-1-4(-1) \stackrel{?}{=} 2(3-2(-1))-7$$
    $$-1+4 \stackrel{?}{=} 2(3+2)-7$$
    $$3 \stackrel{?}{=} 2(5)-7$$
    $$3 \stackrel{?}{=} 10-7$$
    $$3 = 3 \text{ T}$$
    $x = -1$ is a solution to the equation.

3. $4y-(1-y)+5 = -6-2(3y-5)-2y$

    Let $y = -2$ in the equation.
    $$4(-2)-(1-(-2))+5 \stackrel{?}{=} -6-2(3(-2)-5)-2(-2)$$
    $$-8-(3)+5 \stackrel{?}{=} -6-2(-6-5)+4$$
    $$-11+5 \stackrel{?}{=} -2-2(-11)$$
    $$-6 \neq 20$$
    $y = -2$ is **not** a solution to the equation.

    Let $y = 0$ in the equation.
    $$4(0)-(1-0)+5 \stackrel{?}{=} -6-2(3(0)-5)-2(0)$$
    $$0-1+5 \stackrel{?}{=} -6-2(-5)-0$$
    $$4 \stackrel{?}{=} -6+10$$
    $$4 = 4 \text{ T}$$
    $y = 0$ is a solution to the equation.

4. $\dfrac{w-7}{3} - \dfrac{w}{4} = -\dfrac{7}{6}$

    Let $w = -14$ in the equation.
    $$\frac{-14-7}{3} - \frac{-14}{4} \stackrel{?}{=} -\frac{7}{6}$$
    $$\frac{-21}{3} + \frac{14}{4} \stackrel{?}{=} -\frac{7}{6}$$
    $$-7 + \frac{7}{2} \stackrel{?}{=} -\frac{7}{6}$$
    $$-\frac{7}{2} \neq -\frac{7}{6}$$
    $w = -14$ is **not** a solution to the equation.

    Let $w = 7$ in the equation.
    $$\frac{7-7}{3} - \frac{7}{4} \stackrel{?}{=} -\frac{7}{6}$$
    $$0 - \frac{7}{4} \stackrel{?}{=} -\frac{7}{6}$$
    $$-\frac{7}{4} \neq -\frac{7}{6}$$
    $w = 7$ is **not** a solution to the equation.

5. $2w+9 = 15$
    $$2w+9-9 = 15-9$$
    $$2w = 6$$
    $$\frac{2w}{2} = \frac{6}{2}$$
    $$w = 3$$
    Check: $2(3)+9 \stackrel{?}{=} 15$
    $$6+9 \stackrel{?}{=} 15$$
    $$15 = 15 \text{ T}$$
    This is a conditional equation. Solution set: $\{3\}$

6. $-4 = 8-3y$
    $$-4-8 = 8-3y-8$$
    $$-12 = -3y$$
    $$\frac{-12}{-3} = \frac{-3y}{-3}$$
    $$4 = y$$
    Check: $-4 \stackrel{?}{=} 8-3(4)$
    $$-4 \stackrel{?}{=} 8-12$$
    $$-4 = -4 \text{ T}$$
    This is a conditional equation. Solution set: $\{4\}$

## Chapter 1 Review

7. $2x + 5x - 1 = 20$
$7x - 1 = 20$
$7x - 1 + 1 = 20 + 1$
$7x = 21$
$x = 3$
Check: $2(3) + 5(3) - 1 \stackrel{?}{=} 20$
$6 + 15 - 1 \stackrel{?}{=} 20$
$20 = 20$ T
This is a conditional equation. Solution set: $\{3\}$

8. $7x + 5 - 8x = 13$
$-x + 5 = 13$
$-x + 5 - 5 = 13 - 5$
$-x = 8$
$x = -8$
Check: $7(-8) + 5 - 8(-8) \stackrel{?}{=} 13$
$-56 + 5 + 64 \stackrel{?}{=} 13$
$13 = 13$ T
This is a conditional equation. Solution set $\{-8\}$

9. $-2(x - 4) = 8 - 2x$
$-2x + 8 = 8 - 2x$
$-2x + 8 - 8 = 8 - 2x - 8$
$-2x = -2x$
$-2x + 2x = -2x + 2x$
$0 = 0$
This statement is an identity. The equation is true for all real numbers. Solution set:
$\{x \mid x \text{ is a real number}\}$, or $\mathbb{R}$

10. $3(2r + 1) - 5 = 9(r - 1) - 3r$
$6r + 3 - 5 = 9r - 9 - 3r$
$6r - 2 = 6r - 9$
$6r - 2 + 2 = 6r - 9 + 2$
$6r = 6r - 7$
$6r - 6r = 6r - 7 - 6r$
$0 = -7$
This statement is a contradiction. The equation has no solution. Solution set: $\{\ \}$ or $\emptyset$

11. $\dfrac{2y + 3}{4} - \dfrac{y}{2} = 5$
$4\left(\dfrac{2y+3}{4}\right) - 4\left(\dfrac{y}{2}\right) = 4(5)$
$2y + 3 - 2y = 20$
$3 = 20$
This statement is a contradiction. The equation has no solution. Solution set: $\{\ \}$ or $\emptyset$

12. $\dfrac{x}{3} + \dfrac{2x}{5} = \dfrac{x - 20}{15}$
$15\left(\dfrac{x}{3}\right) + 15\left(\dfrac{2x}{5}\right) = 15\left(\dfrac{x-20}{15}\right)$
$5x + 6x = x - 20$
$11x = x - 20$
$11x - x = x - 20 - x$
$10x = -20$
$\dfrac{10x}{10} = \dfrac{-20}{10}$
$x = -2$
Check:
$\dfrac{-2}{3} + \dfrac{2(-2)}{5} \stackrel{?}{=} \dfrac{(-2) - 20}{15}$
$-\dfrac{2}{3} - \dfrac{4}{5} \stackrel{?}{=} -\dfrac{22}{15}$
$-\dfrac{10}{15} - \dfrac{12}{15} \stackrel{?}{=} -\dfrac{22}{15}$
$-\dfrac{22}{15} = -\dfrac{22}{15}$ T
This is a conditional equation.
Solution set: $\{-2\}$

13. $0.2(x - 6) + 1.75 = 4.25 + 0.1(3x + 10)$
$0.2x - 1.2 + 1.75 = 4.25 + 0.3x + 1$
$0.2x + 0.55 = 5.25 + 0.3x$
$0.2x - 0.3x + 0.55 = 5.25 + 0.3x - 0.3x$
$-0.1x + 0.55 = 5.25$
$-0.1x + 0.55 - 0.55 = 5.25 - 0.55$
$-0.1x = 4.7$
$\dfrac{-0.1x}{-0.1} = \dfrac{4.7}{-0.1}$
$x = -47$
Check:

## Chapter 1: Linear Equations and Inequalities

$0.2(-47-6)+1.75 \stackrel{?}{=} 4.25+0.1(3(-47)+10)$

$0.2(-53)+1.75 \stackrel{?}{=} 4.25+0.1(-141+10)$

$-10.6+1.75 \stackrel{?}{=} 4.25+0.1(-131)$

$-8.85 \stackrel{?}{=} 4.25-13.1$

$-8.85 = -8.85$ T

This is a conditional equation.
Solution set: $\{-47\}$

(Note: Alternatively, we could have started by multiplying both sides of the equation by 100 to clear the decimals.)

14. $2.1w-3(2.4-0.2w)=0.9(3w-5)-2.7$

    $2.1w-7.2+0.6w=2.7w-4.5-2.7$

    $2.7w-7.2=2.7w-7.2$

    $0=0$

    This statement is an identity. The equation is true for all real numbers. Solution set:
    $\{w \mid w \text{ is a real number}\}$, or $\mathbb{R}$

    (Note: Alternatively, we could have started by multiplying both sides of the equation by 10 to clear the decimals.)

15. We need to set the denominator equal to 0 and solve the resulting equation.

    $2x+3=0$

    $2x+3-3=0-3$

    $2x=-3$

    $\dfrac{2x}{2}=\dfrac{-3}{2}$

    $x=-\dfrac{3}{2}$

    Check:

    $2\left(-\dfrac{3}{2}\right)+3 \stackrel{?}{=} 0$

    $-3+3 \stackrel{?}{=} 0$

    $0=0$ T

    Solution set: $\left\{-\dfrac{3}{2}\right\}$. Thus, $x=-\dfrac{3}{2}$ must be excluded from the domain.

16. We need to set the denominator equal to 0 and solve the resulting equation.

    $6(x-1)+3=0$

    $6x-6+3=0$

    $6x-3=0$

    $6x-3+3=0+3$

    $6x=3$

    $\dfrac{6x}{6}=\dfrac{3}{6}$

    $x=\dfrac{1}{2}$

    Check:

    $6\left(\dfrac{1}{2}-1\right)+3 \stackrel{?}{=} 0$

    $6\left(-\dfrac{1}{2}\right)+3 \stackrel{?}{=} 0$

    $-3+3 \stackrel{?}{=} 0$

    $0=0$ T

    Solution set: $\left\{\dfrac{1}{2}\right\}$. Thus, $x=\dfrac{1}{2}$ must be excluded from the domain.

17. $2370=0.06(x-9000)+315$

    $2370=0.06x-540+315$

    $2370=0.06x-225$

    $2370+225=0.06x-225+225$

    $2595=0.06x$

    $\dfrac{2595}{0.06}=\dfrac{0.06x}{0.06}$

    $43,250=x$

    Her Missouri taxable income was $43,250.

18. $x+4(x-10)=69.75$

    $x+4x-40=69.75$

    $5x-40=69.75$

    $5x-40+40=69.75+40$

    $5x=109.75$

    $\dfrac{5x}{5}=\dfrac{109.75}{5}$

    $x=21.95$

    The regular club price for a DVD is $21.95.

19. Let $x$ = the number.
    $3x+7=22$

20. Let $x$ = the number.
    $x-3=\dfrac{x}{2}$

ISM − 50

## Chapter 1 Review

**21.** Let $x$ = the number.
$0.2x = x - 12$

**22.** Let $x$ = the number.
$6x = 2x - 4$

**23.** Let $x$ = Payton's age.
Shawn's age: $x + 8$
$$x + (x+8) = 18$$
$$2x + 8 = 18$$
$$2x + 8 - 8 = 18 - 8$$
$$2x = 10$$
$$\frac{2x}{2} = \frac{10}{2}$$
$$x = 5$$
Payton is 5 years old and Shawn is 13 years old.

**24.** Let $x$ = the first odd integer.
Second odd integer: $x + 2$
Third odd integer: $x + 4$
Fourth odd integer: $x + 6$
Fifth odd integer: $x + 8$
$$x + (x+2) + (x+4) + (x+6) + (x+8) = 125$$
$$5x + 20 = 125$$
$$5x + 20 - 20 = 125 - 20$$
$$5x = 105$$
$$\frac{5x}{5} = \frac{105}{5}$$
$$x = 21$$
The five odd integers are 21, 23, 25, 27, and 29.

**25.** Let $x$ = the final exam score.
$$\frac{85 + 81 + 84 + 77 + 2x}{6} = 80$$
$$\frac{327 + 2x}{6} = 80$$
$$6 \cdot \frac{327 + 2x}{6} = 6 \cdot 80$$
$$327 + 2x = 480$$
$$327 + 2x - 327 = 480 - 327$$
$$2x = 153$$
$$\frac{2x}{2} = \frac{153}{2}$$
$$x = 76.5$$
Logan needs to get a score of 76.5 on the final exam to have an average of 80.

**26.** Here we can use the simple interest formula $I = P \cdot r \cdot t$. Remember that $t$ is in years and that the interest rate should be written as a decimal.

$$I = 3200(0.0425)\left(\frac{1}{12}\right)$$
$$I = \frac{34}{3} \approx 11.33$$
After 1 month, Cherie will accrue about $11.33 in interest.

**27.** Let $x$ = the original price.
Discount: $0.3x$
sale price = original price − discount
$$94.50 = x - 0.3x$$
$$94.50 = 0.7x$$
$$\frac{94.50}{0.7} = \frac{0.7x}{0.7}$$
$$135 = x$$
The original price of the sleeping bag was $135.00.

**28.** Let $x$ = the federal minimum wage.
Increase: $0.65x$
$$8.50 = x + 0.65x$$
$$8.50 = 1.65x$$
$$\frac{8.50}{1.65} = \frac{1.65x}{1.65}$$
$$5.15 \approx x$$
The federal minimum wage was $5.15 (per hour).

**29.** Let $x$ = pounds of blueberries.
Strawberries: $12 - x$
tot. rev. = rev. from b.b. + rev. from s.b.
$$(\text{price})(\text{lbs}) = (\text{price})(\text{lbs}) + (\text{price})(\text{lbs})$$
$$(12.95)(12) = (10.95)(x) + (13.95)(12 - x)$$
$$155.4 = 10.95x + 167.4 - 13.95x$$
$$155.4 = 167.4 - 3x$$
$$155.4 - 167.4 = 167.4 - 3x - 167.4$$
$$-12 = -3x$$
$$\frac{-12}{-3} = \frac{-3x}{-3}$$
$$4 = x$$
The store should mix 4 pounds of chocolate covered blueberries with 8 pounds of chocolate covered strawberries.

**30.** Let $x$ = pounds of baseball gum.
Soccer gum: $10 - x$

## Chapter 1: Linear Equations and Inequalities

tot. rev. = rev. from b.b. + rev. from s.b.
$$(\text{price})(\text{lbs}) = (\text{price})(\text{lbs}) + (\text{price})(\text{lbs})$$
$$(3.75)(10) = (3.50)(x) + (4.50)(10-x)$$
$$37.5 = 3.5x + 45 - 4.5x$$
$$37.5 = 45 - x$$
$$37.5 - 45 = 45 - x - 45$$
$$-7.5 = -x$$
$$7.5 = x$$
The store should mix 7.5 pounds of baseball gumballs with 2.5 pounds of soccer gumballs.

**31.** Let $x$ = amount invested at 8%.
Amount at 18%: $8000 - x$
tot. return = ret. from 8% + ret. from 18%
$$(\%)(\$) = (\%)(\$) + (\%)(\$)$$
$$(0.12)(8000) = (0.08)(x) + (0.18)(8000-x)$$
$$960 = 0.08x + 1440 - 0.18x$$
$$960 = 1440 - 0.1x$$
$$960 - 1440 = 1440 - 0.1x - 1440$$
$$-480 = -0.1x$$
$$\frac{-480}{-0.1} = \frac{-0.1x}{-0.1}$$
$$4800 = x$$
Angie should invest $4800 at 8% and $3200 at 18% to achieve a 12% return.

**32.** Let $x$ = liters drained = liters added.
We can write the following equation for the amount of antifreeze:
final amt = amt now − amt. drain + amt add
$$(\%)(L) = (\%)(L) - (\%)(L) + (\%)(L)$$
$$(0.5)(10.1) = (0.3)(10.1) - (0.3)(x) + (1.0)(x)$$
$$5.05 = 3.03 - 0.3x + x$$
$$5.05 = 3.03 + 0.7x$$
$$5.05 - 3.03 = 3.03 + 0.7x - 3.03$$
$$2.02 = 0.7x$$
$$\frac{2.02}{0.7} = \frac{0.7x}{0.7}$$
$$2.89 \approx x$$
About 2.89 liters would need to be drained and replaced with pure antifreeze.

**33.** For this problem, we will need the distance traveled formula: $d = r \cdot t$.
Let $x$ = miles driven at 60 mph.
Miles at 70 mph: $300 - x$

tot. time = time at 60 mph + time at 70 mph
$$4.5 = \frac{x}{60} + \frac{300-x}{70}$$
$$420(4.5) = 420\left(\frac{x}{60} + \frac{300-x}{70}\right)$$
$$1890 = 7x + 1800 - 6x$$
$$1890 = x + 1800$$
$$1890 - 1800 = x + 1800 - 1800$$
$$90 = x$$
Josh drove 90 miles at 60 miles per hour and 210 miles at 70 miles per hour.

**34.** For this problem, we will need the distance traveled formula: $d = r \cdot t$
We also need to note that 50 minutes is $\frac{5}{6}$ of an hour.
Let $x$ = speed of the F14.
Speed of F15: $x + 200$.
tot. dist. = dist. of F14 + dist. of F15
$$2200 = \frac{5}{6}x + \frac{5}{6}(x+200)$$
$$2200 = \frac{5}{6}x + \frac{5}{6}x + \frac{500}{3}$$
$$2200 = \frac{5}{3}x + \frac{500}{3}$$
$$3(2200) = 3\left(\frac{5}{3}x + \frac{500}{3}\right)$$
$$6600 = 5x + 500$$
$$6600 - 500 = 5x + 500 - 500$$
$$6100 = 5x$$
$$\frac{6100}{5} = \frac{5x}{5}$$
$$1220 = x$$
The F14 is traveling at a speed of 1220 miles per hour and the F15 is traveling at a speed of 1420 miles per hour.

**35.** $y = \dfrac{k}{x}$
$$y \cdot x = \frac{k}{x} \cdot x$$
$$xy = k$$
$$\frac{xy}{y} = \frac{k}{y}$$
$$x = \frac{k}{y}$$

## Chapter 1 Review

36.  $F = \dfrac{9}{5}C + 32$

$F - 32 = \dfrac{9}{5}C + 32 - 32$

$F - 32 = \dfrac{9}{5}C$

$\dfrac{5}{9}(F - 32) = \dfrac{5}{9}\left(\dfrac{9}{5}C\right)$

$\dfrac{5}{9}(F - 32) = C$

37.  $P = 2L + 2W$

$P - 2L = 2L + 2W - 2L$

$P - 2L = 2W$

$\dfrac{P - 2L}{2} = \dfrac{2W}{2}$

$\dfrac{P - 2L}{2} = W$

38.  $p = m_1 v_1 + m_2 v_2$

$p - m_1 v_1 = m_1 v_1 + m_2 v_2 - m_1 v_1$

$p - m_1 v_1 = m_2 v_2$

$\dfrac{p - m_1 v_1}{v_2} = \dfrac{m_2 v_2}{v_2}$

$\dfrac{p - m_1 v_1}{v_2} = m_2$

39.  $PV = nRT$

$\dfrac{PV}{nR} = \dfrac{nRT}{nR}$

$\dfrac{PV}{nR} = T$

40.  $S = 2LW + 2LH + 2WH$

$S - 2LH = 2LW + 2LH + 2WH - 2LH$

$S - 2LH = 2LW + 2WH$

$S - 2LH = W(2L + 2H)$

$\dfrac{S - 2LH}{2L + 2H} = \dfrac{W(2L + 2H)}{2L + 2H}$

$\dfrac{S - 2LH}{2L + 2H} = W$

41.  $3x + 4y = 2$

$3x + 4y - 3x = 2 - 3x$

$4y = -3x + 2$

$\dfrac{4y}{4} = \dfrac{-3x + 2}{4}$

$y = -\dfrac{3}{4}x + \dfrac{1}{2}$

42.  $-5x + 4y = 10$

$-5x + 4y + 5x = 10 + 5x$

$4y = 5x + 10$

$\dfrac{4y}{4} = \dfrac{5x + 10}{4}$

$y = \dfrac{5}{4}x + \dfrac{5}{2}$

43.  $4.8x - 1.2y = 6$

$4.8x - 1.2y - 4.8x = 6 - 4.8x$

$-1.2y = -4.8x + 6$

$\dfrac{-1.2y}{-1.2} = \dfrac{-4.8x + 6}{-1.2}$

$y = 4x - 5$

44.  $\dfrac{2}{5}x + \dfrac{1}{3}y = 8$

$\dfrac{2}{5}x + \dfrac{1}{3}y - \dfrac{2}{5}x = 8 - \dfrac{2}{5}x$

$\dfrac{1}{3}y = -\dfrac{2}{5}x + 8$

$3\left(\dfrac{1}{3}y\right) = 3\left(-\dfrac{2}{5}x + 8\right)$

$y = -\dfrac{6}{5}x + 24$

45.  $C = \dfrac{5}{9}(3221.6 - 32)$

$C = \dfrac{5}{9}(3189.6)$

$C = 1772$

The melting point of platinum is $1772°C$.

46.  Let $x$ = the measure of the congruent angle.
Third angle: $x - 30$

$x + x + (x - 30) = 180$

$3x - 30 = 180$

$3x - 30 + 30 = 180 + 30$

$3x = 210$

$\dfrac{3x}{3} = \dfrac{210}{3}$

$x = 70$

The angles measure $70°, 70°$, and $40°$.

47.  Let $w$ = width of the window in feet.
Length: $l = w + 8$

## Chapter 1: Linear Equations and Inequalities

$$2l + 2w = P$$
$$2(w+8) + 2w = 76$$
$$2w + 16 + 2w = 76$$
$$4w + 16 = 76$$
$$4w + 16 - 16 = 76 - 16$$
$$4w = 60$$
$$\frac{4w}{4} = \frac{60}{4}$$
$$w = 15$$
$$l = w + 8 = 15 + 8 = 23$$

The window measures 15 feet by 23 feet.

**48. a.**
$$C = 2.95 + 0.04x$$
$$C - 2.95 = 2.95 + 0.04x - 2.95$$
$$C - 2.95 = 0.04x$$
$$\frac{C - 2.95}{0.04} = \frac{0.04x}{0.04}$$
$$25C - 73.75 = x \quad \text{or} \quad x = 25C - 73.75$$

**b.**
$$x = 25(20) - 73.75$$
$$x = 500 - 73.75$$
$$x = 426.25$$

Debbie can talk for 426 minutes in one month and not spend more than $20 on long distance.

**49.** Let $x$ = thickness in feet.
$$V = l \cdot w \cdot h$$
$$80 = 12 \cdot 18 \cdot x$$
$$80 = 216x$$
$$\frac{80}{216} = \frac{216x}{216}$$
$$\frac{10}{27} = x \quad \text{or} \quad x \approx 0.37$$

The patio will be $\frac{10}{27}$ of a foot thick (i.e. about 4.44 inches).

**50. a.**
$$A = \pi s(R + r)$$
$$\frac{A}{\pi s} = \frac{\pi s(R + r)}{\pi s}$$
$$\frac{A}{\pi s} = R + r$$
$$\frac{A}{\pi s} - R = r \quad \text{or} \quad r = \frac{A - \pi R s}{\pi s}$$

**b.**
$$r = \frac{10\pi - \pi(3)(2)}{\pi(2)}$$
$$r = \frac{10 - 6}{2}$$
$$r = \frac{4}{2} = 2$$

The radius of the top of the frustum is 2 feet.

**51. a.**
$$C = 23.121 + 0.05947(x - 300)$$
$$C = 23.121 + 0.05947x - 17.841$$
$$C = 5.28 + 0.05947x$$
$$C - 5.28 = 5.28 + 0.05947x - 5.28$$
$$C - 5.28 = 0.05947x$$
$$\frac{C - 5.28}{0.05947} = \frac{0.05947x}{0.05947}$$
$$x = \frac{C - 5.28}{0.05947}$$

**b.**
$$x = \frac{115.30 - 5.28}{0.05947}$$
$$x = \frac{110.02}{0.05947}$$
$$x \approx 1850$$

Approximately 1850 kwh were used.

**52.** Let $x$ = the measure of the angle.
Complement: $90 - x$
Supplement: $180 - x$
$$(90 - x) + (180 - x) = 150$$
$$90 - x + 180 - x = 150$$
$$270 - 2x = 150$$
$$270 - 2x - 270 = 150 - 270$$
$$-2x = -120$$
$$\frac{-2x}{-2} = \frac{-120}{-2}$$
$$x = 60$$

The angle measures $60°$.

**53.** $2 < x \leq 7$
Interval: $(2, 7]$

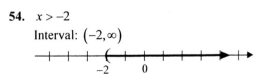

**54.** $x > -2$
Interval: $(-2, \infty)$

ISM − 54

**55.** $(-\infty, 4]$
Inequality: $x \leq 4$
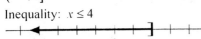

**56.** $[-1, 3)$
Inequality: $-1 \leq x < 3$
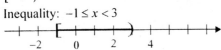

**57.** $5 \leq x \leq 9$
$2 \cdot 5 \leq 2 \cdot x \leq 2 \cdot 9$
$10 \leq 2x \leq 18$
$10 - 3 \leq 2x - 3 \leq 18 - 3$
$7 \leq 2x - 3 \leq 15$
Therefore, $a = 7$ and $b = 15$.

**58.** $-2 < x < 0$
$3(-2) < 3(x) < 3(0)$
$-6 < 3x < 0$
$-6 + 5 < 3x + 5 < 0 + 5$
$-1 < 3x + 5 < 5$
Therefore, $a = -1$ and $b = 5$.

**59.** $3x + 12 \leq 0$
$3x + 12 - 12 \leq 0 - 12$
$3x \leq -12$
$\dfrac{3x}{3} \leq \dfrac{-12}{3}$
$x \leq -4$
Solution set: $\{x \mid x \leq -4\}$
Interval: $(-\infty, -4]$
Graph:

**60.** $2 < 1 - 3x$
$2 - 1 < 1 - 3x - 1$
$1 < -3x$
$\dfrac{1}{-3} > \dfrac{-3x}{-3}$
$-\dfrac{1}{3} > x$ or $x < -\dfrac{1}{3}$
Solution set: $\{x \mid x < -\tfrac{1}{3}\}$
Interval: $(-\infty, -\tfrac{1}{3})$

Graph:
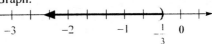

**61.** $-7 \leq 3(h+1) - 8$
$-7 \leq 3h + 3 - 8$
$-7 \leq 3h - 5$
$-7 + 5 \leq 3h - 5 + 5$
$-2 \leq 3h$
$\dfrac{-2}{3} \leq \dfrac{3h}{3}$
$-\dfrac{2}{3} \leq h$ or $h \geq -\dfrac{2}{3}$
Solution set: $\{h \mid h \geq -\tfrac{2}{3}\}$
Interval: $[-\tfrac{2}{3}, \infty)$
Graph:
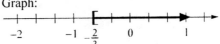

**62.** $-7x - 8 < -22$
$-7x - 8 + 8 < -22 + 8$
$-7x < -14$
$\dfrac{-7x}{-7} > \dfrac{-14}{-7}$
$x > 2$
Solution set: $\{x \mid x > 2\}$
Interval: $(2, \infty)$
Graph:
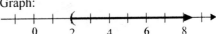

**63.** $3(p-2) + (5-p) > 2 - (p-3)$
$3p - 6 + 5 - p > 2 - p + 3$
$2p - 1 > 5 - p$
$2p + p - 1 > 5 - p + p$
$3p - 1 > 5$
$3p - 1 + 1 > 5 + 1$
$3p > 6$
$\dfrac{3p}{3} > \dfrac{6}{3}$
$p > 2$
Solution set: $\{p \mid p > 2\}$
Interval: $(2, \infty)$

Graph:

64. $2(x+1)+1 > 2(x-2)$
$2x+2+1 > 2x-4$
$2x+3 > 2x-4$
$2x+3-2x > 2x-4-2x$
$3 > -4$
This statement is true for all real numbers.
Solution set: $\{x \mid \text{any real number}\}$
Interval: $(-\infty, \infty)$
Graph:
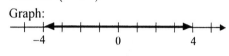

65. $5(x-1)-7x > 2(2-x)$
$5x-5-7x > 4-2x$
$-2x-5 > 4-2x$
$-2x-5+2x > 4-2x+2x$
$-5 > 4$ false
This statement is false for all real numbers. The inequality has no solutions.
Solution set: $\{\ \}$ or $\emptyset$

66. $0.03x+0.10 > 0.52-0.07x$
$0.03x+0.10-0.10 > 0.52-0.07x-0.10$
$0.03x > 0.42-0.07x$
$0.03x+0.07x > 0.42-0.07x+0.07x$
$0.1x > 0.42$
$\dfrac{0.1x}{0.1} > \dfrac{0.42}{0.1}$
$x > 4.2$
Solution set: $\{x \mid x > 4.2\}$
Interval: $(4.2, \infty)$
Graph:

67. $-\dfrac{4}{9}w + \dfrac{7}{12} < \dfrac{5}{36}$
$36\left(-\dfrac{4}{9}w + \dfrac{7}{12}\right) < 36\left(\dfrac{5}{36}\right)$
$-16w+21 < 5$
$-16w+21-21 < 5-21$
$-16w < -16$
$\dfrac{-16w}{-16} > \dfrac{-16}{-16}$
$w > 1$
Solution set: $\{w \mid w > 1\}$
Interval: $(1, \infty)$
Graph:
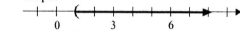

68. $\dfrac{2}{5}y - 20 > \dfrac{2}{3}y + 12$
$15\left(\dfrac{2}{5}y - 20\right) > 15\left(\dfrac{2}{3}y + 12\right)$
$6y - 300 > 10y + 180$
$6y - 300 - 10y > 10y + 180 - 10y$
$-4y - 300 > 180$
$-4y - 300 + 300 > 180 + 300$
$-4y > 480$
$\dfrac{-4y}{-4} < \dfrac{480}{-4}$
$y < -120$
Solution set: $\{y \mid y < -120\}$
Interval: $(-\infty, -120)$
Graph:
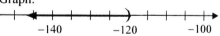

69. Let $x$ = number of people attending.
$7.5x + 150 \leq 600$
$7.5x + 150 - 150 \leq 600 - 150$
$7.5x \leq 450$
$\dfrac{7.5x}{7.5} \leq \dfrac{450}{7.5}$
$x \leq 60$
To stay within budget, no more than 60 people can attend the banquet.

**70.** Let $x$ = number of miles driven daily (average)
Miles charged: $x - 150$
$$43.46 + 0.25(x - 150) \leq 60$$
$$43.46 + 0.25x - 37.5 \leq 60$$
$$0.25x + 5.96 \leq 60$$
$$0.25x + 5.96 - 5.96 \leq 60 - 5.96$$
$$0.25x \leq 54.04$$
$$\frac{0.25x}{0.25} \leq \frac{54.04}{0.25}$$
$$x \leq 216.16$$
To stay within budget, you can drive an average of 216 miles per day.

**71.** Let $x$ = number of candy bars sold.
$$1.0x > 50 + 0.6x$$
$$1.0x - 0.6x > 50 + 0.6x - 0.6x$$
$$0.4x > 50$$
$$\frac{0.4x}{0.4} > \frac{50}{0.4}$$
$$x > 125$$
The band must sell more than 125 candy bars to make a profit.

**72.** Let $x$ = number of DVDs purchased.
$$24.95 + 9.95(x - 1) \leq 72$$
$$24.95 + 9.95x - 9.95 \leq 72$$
$$15 + 9.95x \leq 72$$
$$15 + 9.95x - 15 \leq 72 - 15$$
$$9.95x \leq 57$$
$$\frac{9.95x}{9.95} \leq \frac{57}{9.95}$$
$$x \leq \frac{57}{9.95} \approx 5.73$$
You can purchase up to 5 DVDs and still be within budget.

**73.** $A \cup B = \{-1, 0, 1, 2, 3, 4, 6, 8\}$

**74.** $A \cap C = \{2, 4\}$

**75.** $B \cap C = \{1, 2, 3, 4\}$

**76.** $A \cup C = \{1, 2, 3, 4, 6, 8\}$

**77. a.** $A \cap B = \{x \mid 2 < x \leq 4\}$
Interval: $(2, 4]$
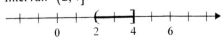

**b.** $A \cup B = \{x \mid x \text{ is a real number}\}$
Interval: $(-\infty, \infty)$
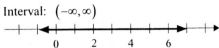

**78. a.** $E \cap F = \{\ \} \text{ or } \emptyset$

**b.** $E \cup F = \{x \mid x < -2 \text{ or } x \geq 3\}$
Interval: $(-\infty, -2) \cup [3, \infty)$
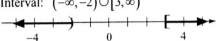

**79.** $x < 4$ and $x + 3 > 2$
$x > -1$
We need $x < 4$ and $x > -1$.
Solution set: $\{x \mid -1 < x < 4\}$
Interval: $(-1, 4)$
Graph:
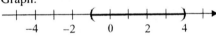

**80.** $3 < 2 - x < 7$
$$3 - 2 < 2 - x - 2 < 7 - 2$$
$$1 < -x < 5$$
$$-1(1) > -1(-x) > -1(5)$$
$$-1 > x > -5$$
$$-5 < x < -1$$
Solution set: $\{x \mid -5 < x < -1\}$
Interval: $(-5, -1)$
Graph:
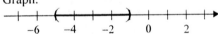

**81.** $x + 3 < 1$ or $x > 2$
$x < -2$
We need $x < -2$ or $x > 2$.
Solution set: $\{x \mid x < -2 \text{ or } x > 2\}$
Interval: $(-\infty, -2) \cup (2, \infty)$
Graph:
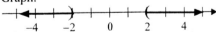

## Chapter 1: Linear Equations and Inequalities

**82.** $x+6 \geq 10$ or $x \leq 0$
$x \geq 4$
We need $x \geq 4$ or $x \leq 0$.
Solution set: $\{x \mid x \leq 0 \text{ or } x \geq 4\}$
Interval: $(-\infty, 0] \cup [4, \infty)$
Graph:
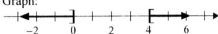

**83.** $3x+2 \leq 5$ and $-4x+2 \leq -10$
$3x \leq 3 \qquad -4x \leq -12$
$x \leq 1 \qquad x \geq 3$
We need $x \leq 1$ and $x \geq 3$.
Solution set: $\{\ \}$ or $\varnothing$

**84.** $1 \leq 2x+5 < 13$
$1-5 \leq 2x+5-5 < 13-5$
$-4 \leq 2x < 8$
$\dfrac{-4}{2} \leq \dfrac{2x}{2} < \dfrac{8}{2}$
$-2 \leq x < 4$
Solution set: $\{x \mid -2 \leq x < 4\}$
Interval: $[-2, 4)$
Graph:
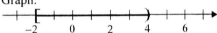

**85.** $x-3 \leq -5$ or $2x+1 > 7$
$x \leq -2 \qquad 2x > 6$
$\qquad\qquad x > 3$
We need $x \leq -2$ or $x > 3$.
Solution set: $\{x \mid x \leq -2 \text{ or } x > 3\}$
Interval: $(-\infty, -2] \cup (3, \infty)$
Graph:
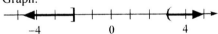

**86.** $3x+4 > -2$ or $4-2x \geq -6$
$3x > -6 \qquad -2x \geq -10$
$x > -2 \qquad x \leq 5$
We need $x > -2$ or $x \leq 5$.
Solution set: $\{x \mid -\infty < x < \infty\}$
Interval: $(-\infty, \infty)$
Graph:
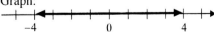

**87.** $\dfrac{1}{3}x > 2$ or $\dfrac{2}{5}x < -4$
$x > 6 \qquad x < -10$
We need $x < -10$ or $x > 6$.
Solution set: $\{x \mid x < -10 \text{ or } x > 6\}$
Interval: $(-\infty, -10) \cup (6, \infty)$
Graph:
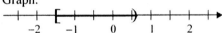

**88.** $x + \dfrac{3}{2} \geq 0$ and $-2x + \dfrac{3}{2} > \dfrac{1}{4}$
$x \geq -\dfrac{3}{2} \qquad -2x > -\dfrac{5}{4}$
$\qquad\qquad x < \dfrac{5}{8}$
We need $x \geq -\dfrac{3}{2}$ and $x < \dfrac{5}{8}$.
Solution set: $\{x \mid -\dfrac{3}{2} \leq x < \dfrac{5}{8}\}$
Interval: $\left[-\dfrac{3}{2}, \dfrac{5}{8}\right)$
Graph:

**89.** $70 \leq x \leq 75$

**90.** Let $x$ = number of kilowatt-hours. Then, the number of kilowatt-hours *above* 300 is given by the expression $x - 300$. Therefore, we need to solve the following inequality:
$50.28 \leq 23.12 + 0.05947(x - 300) \leq 121.43$
$27.16 \leq 0.05947(x - 300) \leq 98.31$
$456.7 \leq x - 300 \leq 1653.1$
$756.7 \leq x \leq 1953.1$
The electric usage varied from roughly 756.7 kilowatt-hours up to roughly 1953.1 kilowatt-hours.

**91.** $|x| = 4$
$x = 4$ or $x = -4$
Solution set: $\{-4, 4\}$
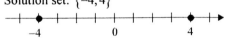

ISM − 58

92. $|3x-5|=4$

$3x-5=4$ or $3x-5=-4$

$3x=9 \qquad 3x=1$

$x=3 \qquad x=\dfrac{1}{3}$

Solution set: $\left\{\dfrac{1}{3},3\right\}$

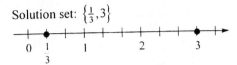

93. $|-y+4|=9$

$-y+4=9$ or $-y+4=-9$

$-y=5 \qquad -y=-13$

$y=-5 \qquad y=13$

Solution set: $\{-5,13\}$

94. $-3|x+2|-5=-8$

$-3|x+2|=-3$

$|x+2|=1$

$x+2=1$ or $x+2=-1$

$x=-1 \qquad x=-3$

Solution set: $\{-3,-1\}$

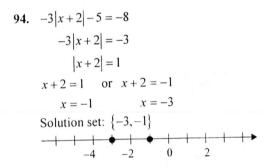

95. $|2w-7|=-3$

This equation has no solution since an absolute value can never yield a negative result.

Solution set: $\{\ \}$ or $\varnothing$

96. $|x+3|=|3x-1|$

$x+3=3x-1$ or $x+3=-(3x-1)$

$-2x=-4 \qquad x+3=-3x+1$

$x=2 \qquad 4x=-2$

$\qquad\qquad x=-\dfrac{1}{2}$

Solution set: $\left\{-\dfrac{1}{2},2\right\}$

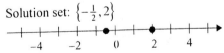

97. $|x|<2$

$-2<x<2$

Solution set: $\{x\,|\,-2<x<2\}$

Interval: $(-2,2)$

98. $|x|\geq \dfrac{7}{2}$

$x\leq -\dfrac{7}{2}$ or $x\geq \dfrac{7}{2}$

Solution set: $\left\{x\,\Big|\,x\leq -\dfrac{7}{2}\text{ or }x\geq \dfrac{7}{2}\right\}$

Interval: $\left(-\infty,-\dfrac{7}{2}\right]\cup\left[\dfrac{7}{2},\infty\right)$

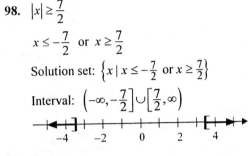

99. $|x+2|\leq 3$

$-3\leq x+2\leq 3$

$-3-2\leq x+2-2\leq 3-2$

$-5\leq x\leq 1$

Solution set: $\{x\,|\,-5\leq x\leq 1\}$

Interval: $[-5,1]$

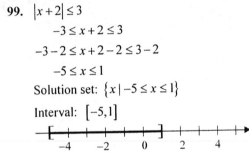

100. $|4x-3|\geq 1$

$4x-3\leq -1$ or $4x-3\geq 1$

$4x\leq 2 \qquad 4x\geq 4$

$x\leq \dfrac{1}{2} \qquad x\geq 1$

Solution set: $\left\{x\,\Big|\,x\leq \dfrac{1}{2}\text{ or }x\geq 1\right\}$

Interval: $\left(-\infty,\dfrac{1}{2}\right]\cup[1,\infty)$

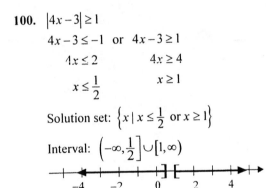

101. $3|x|+6\geq 1$

$3|x|\geq -5$

$|x|\geq -\dfrac{5}{3}$

Since the result of an absolute value is always nonnegative, any real number is a solution to this inequality.

Solution set: $\{x\,|\,x\text{ is a real number}\}$

Interval: $(-\infty,\infty)$

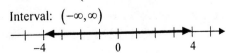

**102.** $|7x+5|+4<3$

$|7x+5|<-1$

Since the result of an absolute value is never negative, this inequality has no solutions.
Solution set: $\{\ \}$ or $\varnothing$

**103.** $|(x-3)-2|\leq 0.01$

$|x-3-2|\leq 0.01$

$|x-5|\leq 0.01$

$-0.01\leq x-5\leq 0.01$

$-0.01+5\leq x-5+5\leq 0.01+5$

$4.99\leq x\leq 5.01$

Solution set: $\{x\mid 4.99\leq x\leq 5.01\}$

Interval: $[4.99, 5.01]$

**104.** $\left|\dfrac{2x-3}{4}\right|>1$

$\dfrac{2x-3}{4}<-1$ or $\dfrac{2x-3}{4}>1$

$2x-3<-4 \qquad 2x-3>4$

$2x<-1 \qquad 2x>7$

$x<-\dfrac{1}{2} \qquad x>\dfrac{7}{2}$

Solution set: $\left\{x\mid x<-\dfrac{1}{2}\text{ or }x>\dfrac{7}{2}\right\}$

Interval: $\left(-\infty,-\dfrac{1}{2}\right)\cup\left(\dfrac{7}{2},\infty\right)$

**105.** $|x-0.503|\leq 0.001$

$-0.001\leq x-0.503\leq 0.001$

$0.502\leq x\leq 0.504$

The acceptable diameters of the bearing are between 0.502 inches and 0.504 inches, inclusive.

**106.** $\left|\dfrac{x-40}{2}\right|>1.96$

$\dfrac{x-40}{2}<-1.96$ or $\dfrac{x-40}{2}>1.96$

$x-40<-3.92 \qquad x-40>3.92$

$x<36.08 \qquad x>43.92$

Tensile strengths below 36.08 lb/in² or above 43.92 lb/in² would be considered unusual.

**Chapter 1 Test**

**1. a.** Let $x=6$ in the equation.

$3(6-7)+5\stackrel{?}{=}6-4$

$3(-1)+5\stackrel{?}{=}2$

$-3+5\stackrel{?}{=}2$

$2=2$ true

$x=6$ is a solution to the equation.

**b.** Let $x=-2$ in the equation.

$3(-2-7)+5\stackrel{?}{=}-2-4$

$3(-9)+5\stackrel{?}{=}-6$

$-27+5\stackrel{?}{=}-6$

$-22\neq -6$

$x=-2$ is **not** a solution to the equation.

**2. a.** Interval: $(-4,\infty)$

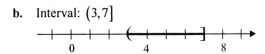

**b.** Interval: $(3,7]$

**3.** Let $x$ represent the number.

$3x-8=x+4$

**4.** Let $x$ represent the number.

$\dfrac{2}{3}x+2(x-5)>7$

**5.** $5x-(x-2)=6+2x$

$5x-x+2=6+2x$

$4x+2=6+2x$

$4x+2-2=6+2x-2$

$4x=4+2x$

$4x-2x=4+2x-2x$

$2x=4$

$\dfrac{2x}{2}=\dfrac{4}{2}$

$x=2$

Solution set: $\{2\}$

This is a conditional equation.

6. $|2x+5|-3=0$
   $|2x+5|=3$
   $2x+5=-3$ or $2x+5=3$
   $2x=-8 \qquad 2x=-2$
   $x=-4 \qquad x=-1$
   Solution set: $\{-4,-1\}$
   This is a conditional equation.

7. $7+(x-3)=3(x+1)-2x$
   $7+x-3=3x+3-2x$
   $4+x=x+3$
   $4+x-x=x+3-x$
   $\qquad 4=3$ false
   This statement is a contradiction. The equation has no solution. Solution set: $\{\ \}$ or $\varnothing$

8. $\quad x+2 \le 3x-4$
   $x+2-3x \le 3x-4-3x$
   $-2x+2 \le -4$
   $-2x+2-2 \le -4-2$
   $-2x \le -6$
   $\dfrac{-2x}{-2} \ge \dfrac{-6}{-2}$
   $x \ge 3$
   Solution set: $\{x \mid x \ge 3\}$
   Interval: $[3,\infty)$

9. $4x+7 > 2x-3(x-2)$
   $4x+7 > 2x-3x+6$
   $4x+7 > -x+6$
   $4x+7+x > -x+6+x$
   $5x+7 > 6$
   $5x+7-7 > 6-7$
   $5x > -1$
   $\dfrac{5x}{5} > \dfrac{-1}{5}$
   $x > -\dfrac{1}{5}$
   Solution set: $\{x \mid x > -\tfrac{1}{5}\}$
   Interval: $\left(-\tfrac{1}{5},\infty\right)$
   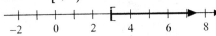

10. $\quad -x+4 \le x+3$
    $-x+4-x \le x+3-x$
    $-2x+4 \le 3$
    $-2x+4-4 \le 3-4$
    $-2x \le -1$
    $\dfrac{-2x}{-2} \ge \dfrac{-1}{-2}$
    $x \ge \dfrac{1}{2}$
    Solution set: $\{x \mid x \ge \tfrac{1}{2}\}$
    Interval: $\left[\tfrac{1}{2},\infty\right)$
    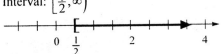

11. $x+2<8$ and $2x+5 \ge 1$
    $\quad x<6 \qquad\quad 2x \ge -4$
    $\qquad\qquad\qquad\quad x \ge -2$
    We need $x \ge -2$ and $x < 6$.
    Solution set: $\{x \mid -2 \le x < 6\}$
    Interval: $[-2, 6)$

12. $x > 4$ or $2(x-1)+3 < -2$
    $\qquad\qquad 2x-2+3 < -2$
    $\qquad\qquad 2x+1 < -2$
    $\qquad\qquad 2x+1-1 < -2-1$
    $\qquad\qquad 2x < -3$
    $\qquad\qquad \dfrac{2x}{2} < \dfrac{-3}{2}$
    $\qquad\qquad x < -\dfrac{3}{2}$
    We need $x > 4$ or $x < -\dfrac{3}{2}$.
    Solution set: $\{x \mid x < -\tfrac{3}{2} \text{ or } x > 4\}$
    Interval: $\left(-\infty, -\tfrac{3}{2}\right) \cup (4, \infty)$
    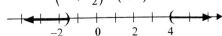

## Chapter 1: Linear Equations and Inequalities

13.  $7x + 4y = 3$
$7x + 4y - 7x = 3 - 7x$
$4y = -7x + 3$
$\dfrac{4y}{4} = \dfrac{-7x + 3}{4}$
$y = -\dfrac{7}{4}x + \dfrac{3}{4}$

14. **a.** $A \cup B = \{1, 3, 5, 6, 7, 9, 12\}$

**b.** $A \cap B = \{3, 9\}$

15. Let $x$ = Glen's weekly sales.
$400 + 0.08x \geq 750$
$400 + 0.08x - 400 \geq 750 - 400$
$0.08x \geq 350$
$\dfrac{0.08x}{0.08} \geq \dfrac{350}{0.08}$
$x \geq 4375$
Glen's weekly sales must be at least $4375 for him to earn at least $750.

16. Let $x$ = width in inches.
$|x - 60| \leq 1$
$-1 \leq x - 60 \leq 1$
$-1 + 60 \leq x - 60 + 60 \leq 1 + 60$
$59 \leq x \leq 61$
The Crescent Rod can fit openings with widths between 59 inches and 61 inches.

17. Let $x$ = number of children at the party.
$75 + 5x = 145$
$75 + 5x - 75 = 145 - 75$
$5x = 70$
$\dfrac{5x}{5} = \dfrac{70}{5}$
$x = 14$
There were 14 children at Payton's party.

18. Let $x$ = the width of the sandbox.
length: $x + 2$
$P = 2L + 2W$
$20 = 2(x + 2) + 2(x)$
$20 = 2x + 4 + 2x$
$20 = 4x + 4$
$20 - 4 = 4x + 4 - 4$
$16 = 4x$
$\dfrac{16}{4} = \dfrac{4x}{4}$
$4 = x$
The sandbox has a width of 4 feet and a length of 6 feet.

19. Let $x$ = liters of 10% solution.
Amount of 40% solution: $12 - x$ liters
tot. acid = acid from 10% + acid from 40%
$(\%)(\text{liters}) = (\%)(\text{liters}) + (\%)(\text{liters})$
$0.2(12) = 0.1(x) + 0.4(12 - x)$
$2.4 = 0.1x + 4.8 - 0.4x$
$2.4 = 4.8 - 0.3x$
$2.4 - 4.8 = 4.8 - 0.3x - 4.8$
$-2.4 = -0.3x$
$\dfrac{-2.4}{-0.3} = \dfrac{-0.3x}{-0.3}$
$8 = x$
The chemist needs to mix 8 liters of the 10% solution with 4 liters of the 40% solution.

20. Let $t$ = time to catch up in hours.
Running time for A: $t + 0.5$
Running time for B: $t$
The total distance traveled by both runners will be the same when they meet. Therefore, we have
distance A = distance B
$(\text{rate})(\text{time}) = (\text{rate})(\text{time})$
$8(t + 0.5) = 10(t)$
$8t + 4 = 10t$
$8t + 4 - 8t = 10t - 8t$
$4 = 2t$
$\dfrac{4}{2} = \dfrac{2t}{2}$
$2 = t$
It will take contestant B two hours to catch up to contestant A.

## Cumulative Review Chapters R–1

1. a. (i) 27.235
      (ii) 27.236
   b. (i) 1.0
      (ii) 1.1

2.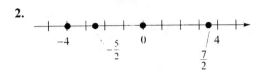

3. $-|-14| = -(14) = -14$

4. $(-3) + 4 - 7 = 1 - 7 = -6$

5. $\dfrac{-3(12)}{-6} = \dfrac{-36}{-6} = 6$

6. $(-3)^4 = (-3) \cdot (-3) \cdot (-3) \cdot (-3) = 81$

7. $5 - 2(1-4)^3 + 5 \cdot 3 = 5 - 2(-3)^3 + 5 \cdot 3$
   $= 5 + 54 + 15 = 74$

8. $\dfrac{2}{3} + \dfrac{1}{2} - \dfrac{1}{4} = \dfrac{8}{12} + \dfrac{6}{12} - \dfrac{3}{12} = \dfrac{8+6-3}{12} = \dfrac{11}{12}$

9. $3(2)^2 + 2(2) - 7 = 3(4) + 4 - 7 = 12 - 3 = 9$

10. $4a^2 - 6a + a^2 - 12 + 2a - 1$
    $= 4a^2 + a^2 - 6a + 2a - 12 - 1$
    $= 5a^2 - 4a - 13$

11. a. Evaluate the denominator when $x = -2$.
    $(-2)^2 + (-2) - 2 = 4 - 2 - 2 = 0$
    Since $x = -2$ makes the denominator 0, it is not in the domain.

    b. Evaluate the denominator when $x = 0$.
    $(0)^2 + (0) - 2 = -2$
    Since $x = 0$ does not make the denominator equal 0, it is in the domain.

12. $3(x+2) - 4(2x-1) + 8 = 3x + 6 - 8x + 4 + 8$
    $= 3x - 8x + 6 + 4 + 8$
    $= -5x + 18$

13. Let $x = 3$ in the equation.
    $3 - (2(3) + 3) \stackrel{?}{=} 5(3) - 1$
    $3 - (6+3) \stackrel{?}{=} 15 - 1$
    $3 - 9 \stackrel{?}{=} 14$
    $-6 \ne 14$
    $x = 3$ is **not** a solution to the equation.

14. $4x - 3 = 2(3x - 2) - 7$
    $4x - 3 = 6x - 4 - 7$
    $4x - 3 = 6x - 11$
    $4x - 3 - 6x = 6x - 11 - 6x$
    $-2x - 3 = -11$
    $-2x - 3 + 3 = -11 + 3$
    $-2x = -8$
    $\dfrac{-2x}{-2} = \dfrac{-8}{-2}$
    $x = 4$
    Solution set: $\{4\}$

15. $\dfrac{x+1}{3} = x - 4$
    $3\left(\dfrac{x+1}{3}\right) = 3(x-4)$
    $x + 1 = 3x - 12$
    $x + 1 - 3x = 3x - 12 - 3x$
    $-2x + 1 = -12$
    $-2x + 1 - 1 = -12 - 1$
    $-2x = -13$
    $\dfrac{-2x}{-2} = \dfrac{-13}{-2}$
    $x = \dfrac{13}{2}$
    Solution set: $\left\{\dfrac{13}{2}\right\}$

16. $2|x-1| + 2 = 9$
    $2|x-1| = 7$
    $|x-1| = \dfrac{7}{2}$
    $x - 1 = -\dfrac{7}{2}$ or $x - 1 = \dfrac{7}{2}$
    $x = -\dfrac{5}{2}$     $x = \dfrac{9}{2}$
    Solution set: $\left\{-\dfrac{5}{2}, \dfrac{9}{2}\right\}$

## Chapter 1: Linear Equations and Inequalities

17. $2x - 5y = 6$
$2x - 5y - 2x = 6 - 2x$
$-5y = -2x + 6$
$\dfrac{-5y}{-5} = \dfrac{-2x + 6}{-5}$
$y = \dfrac{2}{5}x - \dfrac{6}{5}$

18. $\dfrac{x+3}{2} \leq \dfrac{3x-1}{4}$
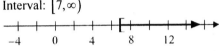
$2x + 6 \leq 3x - 1$
$2x + 6 - 3x \leq 3x - 1 - 3x$
$-x + 6 \leq -1$
$-x + 6 - 6 \leq -1 - 6$
$-x \leq -7$
$x \geq 7$
Solution set: $\{x \mid x \geq 7\}$
Interval: $[7, \infty)$

```
—+—+—+—+—+—[—+—+—+—+—>
 -4    0    4    8    12
```

19. $|x - 4| \leq 3$
$-3 \leq x - 4 \leq 3$
$-3 + 4 \leq x - 4 + 4 \leq 3 + 4$
$1 \leq x \leq 7$
Solution set: $\{x \mid 1 \leq x \leq 7\}$
Interval: $[1, 7]$

```
—+—+—[—+—+—+—]—+—+—+—>
 -4    0    4    8    12
```

20. $|x + 7| > 2$
$x + 7 < -2$ or $x + 7 > 2$
$x < -9$      $x > -5$
Solution set: $\{x \mid x < -9 \text{ or } x > -5\}$
Interval: $(-\infty, -9) \cup (-5, \infty)$

```
<—+—+—)—+—+(—+—+—+—+—>
 -12  -8   -4   0    4
```

21. Let $x$ = score on final exam.
$93 \leq \dfrac{94 + 95 + 90 + 97 + 2x}{6} \leq 100$
$93 \leq \dfrac{376 + 2x}{6} \leq 100$
$6(93) \leq 6\left(\dfrac{376 + 2x}{6}\right) \leq 6(100)$
$558 \leq 376 + 2x \leq 600$
$558 - 376 \leq 376 + 2x - 376 \leq 600 - 376$
$182 \leq 2x \leq 224$
$\dfrac{182}{2} \leq \dfrac{2x}{2} \leq \dfrac{224}{2}$
$91 \leq x \leq 112$
Shawn needs to score at least 91 on the final exam to earn an A (assuming the maximum score on the exam is 100).

22. Let $x$ = weight in pounds.
$0.2x - 2 \geq 30$
$0.2x - 2 + 2 \geq 30 + 2$
$0.2x \geq 32$
$\dfrac{0.2x}{0.2} \geq \dfrac{32}{0.2}$
$x \geq 160$
A person 62 inches tall would be considered obese if they weighed 160 pounds or more.

23. Let $x$ = measure of the smaller angle.
Larger angle: $15 + 2x$
$x + (15 + 2x) = 180$
$3x + 15 = 180$
$3x + 15 - 15 = 180 - 15$
$3x = 165$
$\dfrac{3x}{3} = \dfrac{165}{3}$
$x = 55$
The angles measure $55°$ and $125°$.

24. Let $h$ = height of cylinder in inches.
$S = 2\pi r^2 + 2\pi r h$
$100 = 2\pi(2)^2 + 2\pi(2)h$
$100 = 8\pi + 4\pi h$
$100 - 8\pi = 4\pi h$
$h = \dfrac{100 - 8\pi}{4\pi} \approx 5.96$
The cylinder should be about 5.96 inches tall.

**25.** Let $x$ = first even integer.
Second even integer: $x+2$
Third even integer: $x+4$
$$x+(x+2) = 22+(x+4)$$
$$2x+2 = x+26$$
$$2x+2-x = x+26-x$$
$$x+2 = 26$$
$$x+2-2 = 26-2$$
$$x = 24$$
The three consecutive even integers are 24, 26, and 28.

# Chapter 2

## 2.1 Preparing for Rectangular Coordinates and Graphs of Equations

1.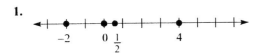

2.  $3x - 5(x+2) = 4$

    a.  Let $x = 0$ in the equation.
    $$3(0) - 5(0+2) \stackrel{?}{=} 4$$
    $$0 - 5(2) \stackrel{?}{=} 4$$
    $$-10 = 4 \text{ False}$$
    $x = 0$ is not a solution to the equation.

    b.  Let $x = -3$ in the equation.
    $$3(-3) - 5(-3+2) \stackrel{?}{=} 4$$
    $$-9 - 5(-1) \stackrel{?}{=} 4$$
    $$-9 + 5 \stackrel{?}{=} 4$$
    $$-4 = 4 \text{ False}$$
    $x = -3$ is not a solution to the equation.

    c.  Let $x = -7$ in the equation.
    $$3(-7) - 5(-7+2) \stackrel{?}{=} 4$$
    $$-21 - 5(-5) \stackrel{?}{=} 4$$
    $$-21 + 25 \stackrel{?}{=} 4$$
    $$4 = 4 \text{ True}$$
    $x = -7$ is a solution to the equation.

3.  a.  Let $x = 0$:
    $$2x^2 - 3x + 1 = 2(0)^2 - 3(0) + 1$$
    $$= 0 - 0 + 1$$
    $$= 1$$

    b.  Let $x = 2$:
    $$2x^2 - 3x + 1 = 2(2)^2 - 3(2) + 1$$
    $$= 2(4) - 6 + 1$$
    $$= 8 - 6 + 1$$
    $$= 3$$

    c.  Let $x = -3$:
    $$2x^2 - 3x + 1 = 2(-3)^2 - 3(-3) + 1$$
    $$= 2(9) + 9 + 1$$
    $$= 18 + 9 + 1$$
    $$= 28$$

4.  $$3x + 2y = 8$$
    $$3x + 2y - 3x = 8 - 3x$$
    $$2y = 8 - 3x$$
    $$\frac{2y}{2} = \frac{8 - 3x}{2}$$
    $$y = \frac{8}{2} - \frac{3x}{2}$$
    $$y = 4 - \frac{3}{2}x \text{ or } y = -\frac{3}{2}x + 4$$

5.  $|-4| = 4$ because $-4$ is 4 units away from 0 on a real number line.

### 2.1 Exercises

2.  $x < 0$; $y > 0$

4.  True

6.  True

8.  The graph of an equation is a visual representation of the solution set for the equation.

10. The y-coordinate of an x-intercept is always 0 since x-intercepts lie on the x-axis and all points on the x-axis have a y-coordinate of 0.

    The x-coordinate of a y-intercept is always 0 since y-intercepts lie on the y-axis and all points on the y-axis have an x-coordinate of 0.

12. $A: (1, 4)$; in quadrant I

    $B: (-5, 6)$; in quadrant II

    $C: (-5, 0)$; on the x-axis

    $D: (-3, -4)$; in quadrant III

    $E: (0, -3)$; on the y-axis

    $F: (4, -2)$; in quadrant IV

Section 2.1  *Rectangular Coordinates and Graphs of Equations*

14. *A*: quadrant II; *B*: x-axis; *C*: quadrant IV; *D*: quadrant III; *E*: quadrant I; *F*: y-axis

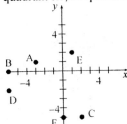

16. **a.** Let $x = 1$ and $y = 7$.
$$-4(1) + 3(7) \stackrel{?}{=} 18$$
$$-4 + 21 \stackrel{?}{=} 18$$
$$17 = 18 \text{ False}$$
The point $(1, 7)$ is not on the graph.

**b.** Let $x = 0$ and $y = 6$.
$$-4(0) + 3(6) \stackrel{?}{=} 18$$
$$0 + 18 \stackrel{?}{=} 18$$
$$18 = 18 \text{ True}$$
The point $(0, 6)$ is on the graph.

**c.** Let $x = -3$ and $y = 10$.
$$-4(-3) + 3(10) \stackrel{?}{=} 18$$
$$12 + 30 \stackrel{?}{=} 18$$
$$42 = 18 \text{ False}$$
The point $(-3, 10)$ is not on the graph.

**d.** Let $x = \frac{3}{2}$ and $y = 4$.
$$-4\left(\frac{3}{2}\right) + 3(4) \stackrel{?}{=} 18$$
$$-6 + 12 \stackrel{?}{=} 18$$
$$6 = 18 \text{ False}$$
The point $\left(\frac{3}{2}, 4\right)$ is not on the graph.

18. **a.** Let $x = 2$ and $y = 2$.
$$2 \stackrel{?}{=} (2)^3 - 3(2)$$
$$2 \stackrel{?}{=} 8 - 6$$
$$2 = 2 \text{ True}$$
The point $(2, 2)$ is on the graph.

**b.** Let $x = 3$ and $y = 8$.
$$8 \stackrel{?}{=} (3)^3 - 3(3)$$
$$8 \stackrel{?}{=} 27 - 9$$
$$8 = 18 \text{ False}$$
The point $(3, 8)$ is not on the graph.

**c.** Let $x = -3$ and $y = -18$.
$$-18 \stackrel{?}{=} (-3)^3 - 3(-3)$$
$$-18 \stackrel{?}{=} -27 + 9$$
$$-18 = -18 \text{ True}$$
The point $(-3, -18)$ is on the graph.

**d.** Let $x = 0$ and $y = 0$.
$$0 \stackrel{?}{=} (0)^3 - 3(0)$$
$$0 = 0 \text{ True}$$
The point $(0, 0)$ is on the graph.

20. **a.** Let $x = 0$ and $y = 1$.
$$(0)^2 + (1)^2 \stackrel{?}{=} 1$$
$$0 + 1 \stackrel{?}{=} 1$$
$$1 = 1 \text{ True}$$
The point $(0, 1)$ is on the graph.

**b.** Let $x = 1$ and $y = 1$.
$$(1)^2 + (1)^2 \stackrel{?}{=} 1$$
$$1 + 1 \stackrel{?}{=} 1$$
$$2 = 1 \text{ False}$$
The point $(1, 1)$ is not on the graph.

**c.** Let $x = \frac{1}{2}$ and $y = \frac{1}{2}$.
$$\left(\frac{1}{2}\right)^2 + \left(\frac{1}{2}\right)^2 \stackrel{?}{=} 1$$
$$\frac{1}{4} + \frac{1}{4} \stackrel{?}{=} 1$$
$$\frac{1}{2} = 1 \text{ False}$$
The point $\left(\frac{1}{2}, \frac{1}{2}\right)$ is not on the graph.

## Chapter 2: Graphs, Relations, and Functions

**d.** Let $x = \dfrac{\sqrt{3}}{2}$ and $y = \dfrac{1}{2}$.

$$\left(\dfrac{\sqrt{3}}{2}\right)^2 + \left(\dfrac{1}{2}\right)^2 \stackrel{?}{=} 1$$

$$\dfrac{3}{4} + \dfrac{1}{4} \stackrel{?}{=} 1$$

$$1 = 1 \text{ True}$$

The point $\left(\dfrac{\sqrt{3}}{2}, \dfrac{1}{2}\right)$ is on the graph.

**22.** The intercepts are $(3,0)$ and $(0,2)$.

**24.** The intercepts are $(-2,0)$, $(3,0)$, and $(0,9)$.

**26.** $y = 2x$

| $x$ | $y = 2x$ | $(x,y)$ |
|---|---|---|
| $-2$ | $y = 2(-2) = -4$ | $(-2,-4)$ |
| $-1$ | $y = 2(-1) = -2$ | $(-1,-2)$ |
| $0$ | $y = 2(0) = 0$ | $(0,0)$ |
| $1$ | $y = 2(1) = 2$ | $(1,2)$ |
| $2$ | $y = 2(2) = 4$ | $(2,4)$ |

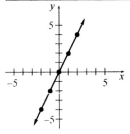

**28.** $y = -\dfrac{1}{3}x$

| $x$ | $y = -\dfrac{1}{3}x$ | $(x,y)$ |
|---|---|---|
| $-6$ | $y = -\dfrac{1}{3}(-6) = 2$ | $(-6,2)$ |
| $-3$ | $y = -\dfrac{1}{3}(-3) = 1$ | $(-3,1)$ |
| $0$ | $y = -\dfrac{1}{3}(0) = 0$ | $(0,0)$ |
| $3$ | $y = -\dfrac{1}{3}(3) = -1$ | $(3,-1)$ |
| $6$ | $y = -\dfrac{1}{3}(6) = -2$ | $(6,-2)$ |

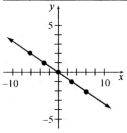

**30.** $y = x - 2$

| $x$ | $y = x - 2$ | $(x,y)$ |
|---|---|---|
| $-3$ | $y = (-3) - 2 = -5$ | $(-3,-5)$ |
| $-1$ | $y = (-1) - 2 = -3$ | $(-1,-3)$ |
| $0$ | $y = (0) - 2 = -2$ | $(0,-2)$ |
| $1$ | $y = (1) - 2 = -1$ | $(1,-1)$ |
| $3$ | $y = (3) - 2 = 1$ | $(3,1)$ |

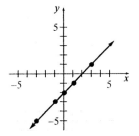

**Section 2.1** Rectangular Coordinates and Graphs of Equations

32. $y = -4x + 2$

| $x$ | $y = -4x + 2$ | $(x, y)$ |
|---|---|---|
| $-2$ | $y = -4(-2) + 2 = 10$ | $(-2, 10)$ |
| $-1$ | $y = -4(-1) + 2 = 6$ | $(-1, 6)$ |
| $0$ | $y = -4(0) + 2 = 2$ | $(0, 2)$ |
| $1$ | $y = -4(1) + 2 = -2$ | $(1, -2)$ |
| $2$ | $y = -4(2) + 2 = -6$ | $(2, -6)$ |

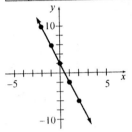

36. $3x + y = 9$
    $y = -3x + 9$

| $x$ | $y = -3x + 9$ | $(x, y)$ |
|---|---|---|
| $-1$ | $y = -3(-1) + 9 = 12$ | $(-1, 12)$ |
| $0$ | $y = -3(0) + 9 = 9$ | $(0, 9)$ |
| $1$ | $y = -3(1) + 9 = 6$ | $(1, 6)$ |
| $2$ | $y = -3(2) + 9 = 3$ | $(2, 3)$ |
| $3$ | $y = -3(3) + 9 = 0$ | $(3, 0)$ |

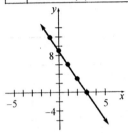

34. $y = -\dfrac{1}{2}x + 2$

| $x$ | $y = -\dfrac{1}{2}x + 2$ | $(x, y)$ |
|---|---|---|
| $-4$ | $y = -\dfrac{1}{2}(-4) + 2 = 4$ | $(-4, 4)$ |
| $-2$ | $y = -\dfrac{1}{2}(-2) + 2 = 3$ | $(-2, 3)$ |
| $0$ | $y = -\dfrac{1}{2}(0) + 2 = 2$ | $(0, 2)$ |
| $2$ | $y = -\dfrac{1}{2}(2) + 2 = 1$ | $(2, 1)$ |
| $4$ | $y = -\dfrac{1}{2}(4) + 2 = 0$ | $(4, 0)$ |

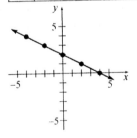

38. $y = x^2 - 2$

| $x$ | $y = x^2 - 2$ | $(x, y)$ |
|---|---|---|
| $-3$ | $y = (-3)^2 - 2 = 7$ | $(-3, 7)$ |
| $-2$ | $y = (-2)^2 - 2 = 2$ | $(-2, 2)$ |
| $0$ | $y = (0)^2 - 2 = -2$ | $(0, -2)$ |
| $2$ | $y = (2)^2 - 2 = 2$ | $(2, 2)$ |
| $3$ | $y = (3)^2 - 2 = 7$ | $(3, 7)$ |

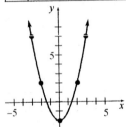

## Chapter 2: Graphs, Relations, and Functions

**40.** $y = -2x^2 + 8$

| $x$ | $y = -2x^2 + 8$ | $(x, y)$ |
|---|---|---|
| $-2$ | $y = -2(-2)^2 + 8 = 0$ | $(-2, 0)$ |
| $-1$ | $y = -2(-1)^2 + 8 = 6$ | $(-1, 6)$ |
| $0$ | $y = -2(0)^2 + 8 = 8$ | $(0, 8)$ |
| $1$ | $y = -2(1)^2 + 8 = 6$ | $(1, 6)$ |
| $2$ | $y = -2(2)^2 + 8 = 0$ | $(2, 0)$ |

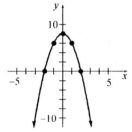

**42.** $y = |x| - 2$

| $x$ | $y = |x| - 2$ | $(x, y)$ |
|---|---|---|
| $-4$ | $y = |-4| - 2 = 2$ | $(-4, 2)$ |
| $-2$ | $y = |-2| - 2 = 0$ | $(-2, 0)$ |
| $0$ | $y = |0| - 2 = -2$ | $(0, -2)$ |
| $2$ | $y = |2| - 2 = 0$ | $(2, 0)$ |
| $4$ | $y = |4| - 2 = 2$ | $(4, 2)$ |

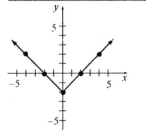

**44.** $y = -|x|$

| $x$ | $y = -|x|$ | $(x, y)$ |
|---|---|---|
| $-4$ | $y = -|-4| = -4$ | $(-4, -4)$ |
| $-2$ | $y = -|-2| = -2$ | $(-2, -2)$ |
| $0$ | $y = -|0| = 0$ | $(0, 0)$ |
| $2$ | $y = -|2| = -2$ | $(2, -2)$ |
| $4$ | $y = -|4| = -4$ | $(4, -4)$ |

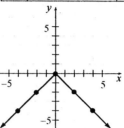

**46.** $y = -x^3$

| $x$ | $y = -x^3$ | $(x, y)$ |
|---|---|---|
| $-3$ | $y = -(-3)^3 = 27$ | $(-3, 27)$ |
| $-2$ | $y = -(-2)^3 = 8$ | $(-2, 8)$ |
| $-1$ | $y = -(-1)^3 = 1$ | $(-1, 1)$ |
| $0$ | $y = -(0)^3 = 0$ | $(0, 0)$ |
| $1$ | $y = -(1)^3 = -1$ | $(1, -1)$ |
| $2$ | $y = -(2)^3 = -8$ | $(2, -8)$ |
| $3$ | $y = -(3)^3 = -27$ | $(3, -27)$ |

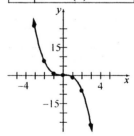

## Section 2.1 Rectangular Coordinates and Graphs of Equations

**48.** $y = x^3 - 2$

| $x$ | $y = x^3 - 2$ | $(x, y)$ |
|---|---|---|
| $-3$ | $y = (-3)^3 - 2 = -29$ | $(-3, -29)$ |
| $-2$ | $y = (-2)^3 - 2 = -10$ | $(-2, -10)$ |
| $-1$ | $y = (-1)^3 - 2 = -3$ | $(-1, -3)$ |
| $0$ | $y = (0)^3 - 2 = -2$ | $(0, -2)$ |
| $1$ | $y = (1)^3 - 2 = -1$ | $(1, -1)$ |
| $2$ | $y = (2)^3 - 2 = 6$ | $(2, 6)$ |
| $3$ | $y = (3)^3 - 2 = 25$ | $(3, 25)$ |

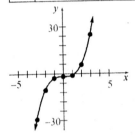

**50.** $x^2 + y = 5$

$y = -x^2 + 5$

| $x$ | $y = -x^2 + 5$ | $(x, y)$ |
|---|---|---|
| $-3$ | $y = -(-3)^2 + 5 = -4$ | $(-3, -4)$ |
| $-2$ | $y = -(-2)^2 + 5 = 1$ | $(-2, 1)$ |
| $0$ | $y = -(0)^2 + 5 = 5$ | $(0, 5)$ |
| $2$ | $y = -(2)^2 + 5 = 1$ | $(2, 1)$ |
| $3$ | $y = -(3)^2 + 5 = -4$ | $(3, -4)$ |

**52.** $x = y^2 + 2$

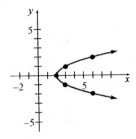

| $y$ | $x = y^2 + 2$ | $(x, y)$ |
|---|---|---|
| $-2$ | $x = (-2)^2 + 2 = 6$ | $(6, -2)$ |
| $-1$ | $x = (-1)^2 + 2 = 3$ | $(3, -1)$ |
| $0$ | $x = (0)^2 + 2 = 2$ | $(2, 0)$ |
| $1$ | $x = (1)^2 + 2 = 3$ | $(3, 1)$ |
| $2$ | $x = (2)^2 + 2 = 6$ | $(6, 2)$ |

**54.** If $(a, -2)$ is a point on the graph, then we have

$-2 = -3(a) + 5$

$-2 = -3a + 5$

$-7 = -3a$

$\dfrac{7}{3} = a$

We need $a = \dfrac{7}{3}$.

**56.** If $(-2, b)$ is a point on the graph, then we have

$b = -2(-2)^2 + 3(-2) + 1$

$b = -2(4) + 3(-2) + 1$

$b = -8 - 6 + 1$

$b = -13$

We need $b = -13$.

**58.**
   a. According to the graph, the object had a height of about 175 feet after 1.5 seconds.

   b. According to the graph, the object reached a maximum height of 196 feet after 2.5 seconds.

   c. The t-intercept is $t = 6$. This represents the time needed for the object to hit the ground.

   The y-intercept is $y = 96$. This represents the initial height of the object when it was thrown.

**60.**
   a. According to the graph, a temperature of 10°C will have a wind chill of 6°C when the wind is blowing 4 meters per second.

   b. According to the graph, a temperature of 10°C will have a wind chill of −3.7°C when the wind is blowing 20 meters per second.

## Chapter 2: Graphs, Relations, and Functions

c. The intercept on the horizontal axis is $(10, 0)$. This represents the wind speed (in meters per second) required for the wind chill to be $0°C$ (the freezing point of water).

The intercept on the vertical axis is $(0, 10)$. This represents the apparent temperature (in $°C$) when there is no wind blowing.

62. The set of all points of the form $(x, 2)$ form a horizontal line with a y-intercept of 2.

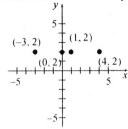

64. Answers will vary. One possible graph is below.

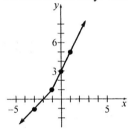

66. Answers will vary. One possibility:
All three points have a y-coordinate of 3. Therefore, they all lie on the horizontal line with equation $y = 3$.

68. $y = -5x + 8$

70. $y = 2x^2 - 4$

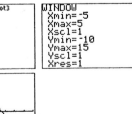

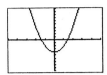

72. $y - x^2 = -15$

$y = x^2 - 15$

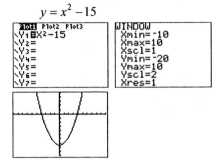

74. $y = -x^3 + 3x$

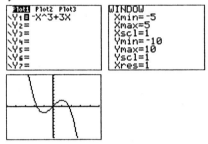

### 2.2 Preparing for Relations

1. Inequality: $-4 \leq x \leq 4$
   Interval: $[-4, 4]$
   We use square brackets in interval notation because the inequalities are not strict.

2. Interval: $[2, \infty)$
   Inequality: $2 \leq x < \infty$ or $x \geq 2$
   The square bracket indicates a non-strict inequality while the parenthesis indicates a strict inequality.

### 2.2 Exercises

2. domain; range

4. False; the range may be the set of all real numbers, but not necessarily.

6. The four methods for describing a relation are mapping, set of ordered pairs, graphing, and an

equation. Ordered pairs are appropriate if there are a finite number of values in the domain. If there are an infinite (or very large) number of elements in the domain, a graph is more appropriate.

8. $\{(30, \$7.09), (35, \$7.09), (40, \$8.40), (45, \$11.29)\}$

   **Domain:** $\{30, 35, 40, 45\}$

   **Range:** $\{\$7.09, \$8.40, \$11.29\}$

10. $\{(\text{Northeast}, \$41984), (\text{Midwest}, \$42679), (\text{South}, \$37442), (\text{West}, \$42720)\}$

    **Domain:** {Northeast, Midwest, South, West}
    **Range:** {$41984, $42679, $37442, $42720}

12.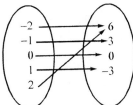

    Domain: $\{-2, -1, 0, 1, 2\}$
    Range: $\{-3, 0, 3, 6\}$

14.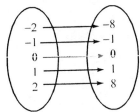

    Domain: $\{-2, -1, 0, 1, 2\}$
    Range: $\{-8, -1, 0, 1, 8\}$

16.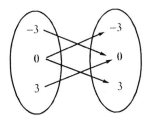

    Domain: $\{-3, 0, 3\}$
    Range: $\{-3, 0, 3\}$

18. Domain: $\{-3, -2, -1, 1, 3\}$
    Range: $\{-3, -1, 0, 1, 3\}$

20. Domain: $\{x \mid -3 \leq x \leq 3\}$ or $[-3, 3]$
    Range: $\{y \mid -2 \leq y \leq 4\}$ or $[-2, 4]$

22. Domain: $\{x \mid -5 \leq x \leq 3\}$ or $[-5, 3]$
    Range: $\{y \mid -1 \leq y \leq 3\}$ or $[-1, 3]$

24. Domain: $\{x \mid x \geq -2\}$ or $[-2, \infty)$
    Range: $\{y \mid y \geq -1\}$ or $[-1, \infty)$

26. $y = 2x$

    | $x$ | $y = 2x$ | $(x, y)$ |
    |---|---|---|
    | $-2$ | $y = 2(-2) = -4$ | $(-2, -4)$ |
    | $-1$ | $y = 2(-1) = -2$ | $(-1, -2)$ |
    | $0$ | $y = 2(0) = 0$ | $(0, 0)$ |
    | $1$ | $y = 2(1) = 2$ | $(1, 2)$ |
    | $2$ | $y = 2(2) = 4$ | $(2, 4)$ |

    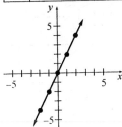

    Domain: $\{x \mid x \text{ is a real number}\}$ or $(-\infty, \infty)$
    Range: $\{y \mid y \text{ is a real number}\}$ or $(-\infty, \infty)$

28. $y = -\dfrac{1}{3}x$

    | $x$ | $y = -\dfrac{1}{3}x$ | $(x, y)$ |
    |---|---|---|
    | $-6$ | $y = -\dfrac{1}{3}(-6) = 2$ | $(-6, 2)$ |
    | $-3$ | $y = -\dfrac{1}{3}(-3) = 1$ | $(-3, 1)$ |
    | $0$ | $y = -\dfrac{1}{3}(0) = 0$ | $(0, 0)$ |
    | $3$ | $y = -\dfrac{1}{3}(3) = -1$ | $(3, -1)$ |
    | $6$ | $y = -\dfrac{1}{3}(6) = -2$ | $(6, -2)$ |

## Chapter 2: Graphs, Relations, and Functions

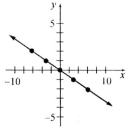

Domain: $\{x \mid x \text{ is a real number}\}$ or $(-\infty, \infty)$
Range: $\{y \mid y \text{ is a real number}\}$ or $(-\infty, \infty)$

30. $y = x - 2$

| $x$ | $y = x - 2$ | $(x, y)$ |
|---|---|---|
| $-3$ | $y = (-3) - 2 = -5$ | $(-3, -5)$ |
| $-1$ | $y = (-1) - 2 = -3$ | $(-1, -3)$ |
| $0$ | $y = (0) - 2 = -2$ | $(0, -2)$ |
| $1$ | $y = (1) - 2 = -1$ | $(1, -1)$ |
| $3$ | $y = (3) - 2 = 1$ | $(3, 1)$ |

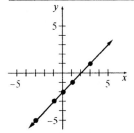

Domain: $\{x \mid x \text{ is a real number}\}$ or $(-\infty, \infty)$
Range: $\{y \mid y \text{ is a real number}\}$ or $(-\infty, \infty)$

32. $y = -4x + 2$

| $x$ | $y = -4x + 2$ | $(x, y)$ |
|---|---|---|
| $-2$ | $y = -4(-2) + 2 = 10$ | $(-2, 10)$ |
| $-1$ | $y = -4(-1) + 2 = 6$ | $(-1, 6)$ |
| $0$ | $y = -4(0) + 2 = 2$ | $(0, 2)$ |
| $1$ | $y = -4(1) + 2 = -2$ | $(1, -2)$ |
| $2$ | $y = -4(2) + 2 = -6$ | $(2, -6)$ |

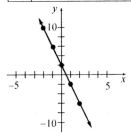

Domain: $\{x \mid x \text{ is a real number}\}$ or $(-\infty, \infty)$
Range: $\{y \mid y \text{ is a real number}\}$ or $(-\infty, \infty)$

34. $y = -\dfrac{1}{2}x + 2$

| $x$ | $y = -\dfrac{1}{2}x + 2$ | $(x, y)$ |
|---|---|---|
| $-4$ | $y = -\dfrac{1}{2}(-4) + 2 = 4$ | $(-4, 4)$ |
| $-2$ | $y = -\dfrac{1}{2}(-2) + 2 = 3$ | $(-2, 3)$ |
| $0$ | $y = -\dfrac{1}{2}(0) + 2 = 2$ | $(0, 2)$ |
| $2$ | $y = -\dfrac{1}{2}(2) + 2 = 1$ | $(2, 1)$ |
| $4$ | $y = -\dfrac{1}{2}(4) + 2 = 0$ | $(4, 0)$ |

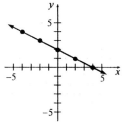

Domain: $\{x \mid x \text{ is a real number}\}$ or $(-\infty, \infty)$
Range: $\{y \mid y \text{ is a real number}\}$ or $(-\infty, \infty)$

36. $3x + y = 9$
$y = -3x + 9$

| $x$ | $y = -3x + 9$ | $(x, y)$ |
|---|---|---|
| $-1$ | $y = -3(-1) + 9 = 12$ | $(-1, 12)$ |
| $0$ | $y = -3(0) + 9 = 9$ | $(0, 9)$ |
| $1$ | $y = -3(1) + 9 = 6$ | $(1, 6)$ |
| $2$ | $y = -3(2) + 9 = 3$ | $(2, 3)$ |
| $3$ | $y = -3(3) + 9 = 0$ | $(3, 0)$ |

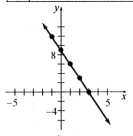

Domain: $\{x \mid x \text{ is a real number}\}$ or $(-\infty, \infty)$
Range: $\{y \mid y \text{ is a real number}\}$ or $(-\infty, \infty)$

38. $y = x^2 - 2$

| $x$ | $y = x^2 - 2$ | $(x, y)$ |
|---|---|---|
| $-3$ | $y = (-3)^2 - 2 = 7$ | $(-3, 7)$ |
| $-2$ | $y = (-2)^2 - 2 = 2$ | $(-2, 2)$ |
| $0$ | $y = (0)^2 - 2 = -2$ | $(0, -2)$ |
| $2$ | $y = (2)^2 - 2 = 2$ | $(2, 2)$ |
| $3$ | $y = (3)^2 - 2 = 7$ | $(3, 7)$ |

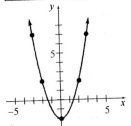

Domain: $\{x \mid x \text{ is a real number}\}$ or $(-\infty, \infty)$

Range: $\{y \mid y \geq -2\}$ or $[-2, \infty)$

40. $y = -2x^2 + 8$

| $x$ | $y = -2x^2 + 8$ | $(x, y)$ |
|---|---|---|
| $-2$ | $y = -2(-2)^2 + 8 = 0$ | $(-2, 0)$ |
| $-1$ | $y = -2(-1)^2 + 8 = 6$ | $(-1, 6)$ |
| $0$ | $y = -2(0)^2 + 8 = 8$ | $(0, 8)$ |
| $1$ | $y = -2(1)^2 + 8 = 6$ | $(1, 6)$ |
| $2$ | $y = -2(2)^2 + 8 = 0$ | $(2, 0)$ |

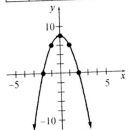

Domain: $\{x \mid x \text{ is a real number}\}$ or $(-\infty, \infty)$

Range: $\{y \mid y \leq 8\}$ or $(-\infty, 8]$

42. $y = |x| - 2$

| $x$ | $y = |x| - 2$ | $(x, y)$ |
|---|---|---|
| $-4$ | $y = |-4| - 2 = 2$ | $(-4, 2)$ |
| $-2$ | $y = |-2| - 2 = 0$ | $(-2, 0)$ |
| $0$ | $y = |0| - 2 = -2$ | $(0, -2)$ |
| $2$ | $y = |2| - 2 = 0$ | $(2, 0)$ |
| $4$ | $y = |4| - 2 = 2$ | $(4, 2)$ |

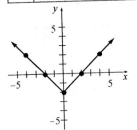

Domain: $\{x \mid x \text{ is a real number}\}$ or $(-\infty, \infty)$

Range: $\{y \mid y \geq -2\}$ or $[-2, \infty)$

44. $y = -|x|$

| $x$ | $y = -|x|$ | $(x, y)$ |
|---|---|---|
| $-4$ | $y = -|-4| = -4$ | $(-4, -4)$ |
| $-2$ | $y = -|-2| = -2$ | $(-2, -2)$ |
| $0$ | $y = -|0| = 0$ | $(0, 0)$ |
| $2$ | $y = -|2| = -2$ | $(2, -2)$ |
| $4$ | $y = -|4| = -4$ | $(4, -4)$ |

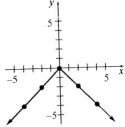

Domain: $\{x \mid x \text{ is a real number}\}$ or $(-\infty, \infty)$

Range: $\{y \mid y \leq 0\}$ or $(-\infty, 0]$

## Chapter 2: Graphs, Relations, and Functions

**46.** $y = -x^3$

| $x$ | $y = -x^3$ | $(x, y)$ |
|---|---|---|
| $-3$ | $y = -(-3)^3 = 27$ | $(-3, 27)$ |
| $-2$ | $y = -(-2)^3 = 8$ | $(-2, 8)$ |
| $-1$ | $y = -(-1)^3 = 1$ | $(-1, 1)$ |
| $0$ | $y = -(0)^3 = 0$ | $(0, 0)$ |
| $1$ | $y = -(1)^3 = -1$ | $(1, -1)$ |
| $2$ | $y = -(2)^3 = -8$ | $(2, -8)$ |
| $3$ | $y = -(3)^3 = -27$ | $(3, -27)$ |

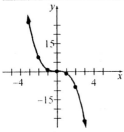

Domain: $\{x \mid x \text{ is a real number}\}$ or $(-\infty, \infty)$

Range: $\{y \mid y \text{ is a real number}\}$ or $(-\infty, \infty)$

**48.** $y = x^3 - 2$

| $x$ | $y = x^3 - 2$ | $(x, y)$ |
|---|---|---|
| $-3$ | $y = (-3)^3 - 2 = -29$ | $(-3, -29)$ |
| $-2$ | $y = (-2)^3 - 2 = -10$ | $(-2, -10)$ |
| $-1$ | $y = (-1)^3 - 2 = -3$ | $(-1, -3)$ |
| $0$ | $y = (0)^3 - 2 = -2$ | $(0, -2)$ |
| $1$ | $y = (1)^3 - 2 = -1$ | $(1, -1)$ |
| $2$ | $y = (2)^3 - 2 = 6$ | $(2, 6)$ |
| $3$ | $y = (3)^3 - 2 = 25$ | $(3, 25)$ |

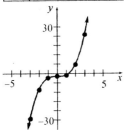

Domain: $\{x \mid x \text{ is a real number}\}$ or $(-\infty, \infty)$

Range: $\{y \mid y \text{ is a real number}\}$ or $(-\infty, \infty)$

**50.** $x^2 + y = 5$

$y = -x^2 + 5$

| $x$ | $y = -x^2 + 5$ | $(x, y)$ |
|---|---|---|
| $-3$ | $y = -(-3)^2 + 5 = -4$ | $(-3, -4)$ |
| $-2$ | $y = -(-2)^2 + 5 = 1$ | $(-2, 1)$ |
| $0$ | $y = -(0)^2 + 5 = 5$ | $(0, 5)$ |
| $2$ | $y = -(2)^2 + 5 = 1$ | $(2, 1)$ |
| $3$ | $y = -(3)^2 + 5 = -4$ | $(3, -4)$ |

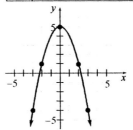

Domain: $\{x \mid x \text{ is a real number}\}$ or $(-\infty, \infty)$

Range: $\{y \mid y \leq 5\}$ or $(-\infty, 5]$

**52.** $x = y^2 + 2$

| $y$ | $x = y^2 + 2$ | $(x, y)$ |
|---|---|---|
| $-2$ | $x = (-2)^2 + 2 = 6$ | $(6, -2)$ |
| $-1$ | $x = (-1)^2 + 2 = 3$ | $(3, -1)$ |
| $0$ | $x = (0)^2 + 2 = 2$ | $(2, 0)$ |
| $1$ | $x = (1)^2 + 2 = 3$ | $(3, 1)$ |
| $2$ | $x = (2)^2 + 2 = 6$ | $(6, 2)$ |

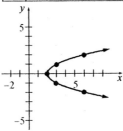

Domain: $\{x \mid x \geq 2\}$ or $[2, \infty)$

Range: $\{y \mid y \text{ is a real number}\}$ or $(-\infty, \infty)$

**54.** According to the graph:

Domain: $\{x \mid 0 \leq x \leq 6\}$ or $[0, 6]$

Range: $\{y \mid 0 \leq y \leq 196\}$ or $[0, 196]$

The ball will be in the air for 6 seconds. Starting at an initial height of 96 feet, the ball will reach a maximum height of 196 feet before hitting the ground.

56. According to the graph:
Domain: $\{x \mid x \geq 0\}$
Range: $\{y \mid y \leq 10\}$
The wind speed will be at least 0 meters per second and the wind chill will be less than or equal to 10°C (the outside temperature).

58. Actual graphs will vary but all should be vertical lines.

## Chapter 2 Putting the Concepts Together (Sections 2.1-2.2)

1. $A$: x-axis; $B$: quadrant II; $C$: quadrant I; $D$: y-axis; $E$: quadrant III; $F$: quadrant IV.

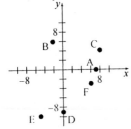

2. a. Let $x = 1$ and $y = \frac{5}{2}$.
$$\frac{5}{2} \stackrel{?}{=} 4(1) - \frac{3}{2}$$
$$\frac{5}{2} \stackrel{?}{=} 4 - \frac{3}{2}$$
$$\frac{5}{2} = \frac{5}{2} \text{ True}$$
The point $\left(1, \frac{5}{2}\right)$ is on the graph.

b. Let $x = \frac{1}{2}$ and $y = \frac{1}{2}$.
$$\frac{1}{2} \stackrel{?}{=} 4\left(\frac{1}{2}\right) - \frac{3}{2}$$
$$\frac{1}{2} \stackrel{?}{=} 2 - \frac{3}{2}$$
$$\frac{1}{2} = \frac{1}{2} \text{ True}$$
The point $\left(\frac{1}{2}, \frac{1}{2}\right)$ is on the graph.

c. Let $x = \frac{1}{4}$ and $y = \frac{1}{4}$.
$$\frac{1}{4} \stackrel{?}{=} 4\left(\frac{1}{4}\right) - \frac{3}{2}$$
$$\frac{1}{4} \stackrel{?}{=} 1 - \frac{3}{2}$$
$$\frac{1}{4} \neq -\frac{1}{2}$$
The point $\left(\frac{1}{4}, \frac{1}{4}\right)$ is not on the graph.

3. $y = |x| + 3$

| $x$ | $y = |x| + 3$ | $(x, y)$ |
|---|---|---|
| $-3$ | $y = |-3| + 3 = 6$ | $(-3, 6)$ |
| $-1$ | $y = |-1| + 3 = 4$ | $(-1, 4)$ |
| $0$ | $y = |0| + 3 = 3$ | $(0, 3)$ |
| $1$ | $y = |1| + 3 = 4$ | $(1, 4)$ |
| $3$ | $y = |3| + 3 = 6$ | $(3, 6)$ |

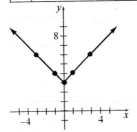

4. $y = \frac{1}{2}x^2 - 1$

| $x$ | $y = \frac{1}{2}x^2 - 1$ | $(x, y)$ |
|---|---|---|
| $-4$ | $y = \frac{1}{2}(-4)^2 - 1 = 7$ | $(-4, 7)$ |
| $-2$ | $y = \frac{1}{2}(-2)^2 - 1 = 1$ | $(-2, 1)$ |
| $0$ | $y = \frac{1}{2}(0)^2 - 1 = -1$ | $(0, -1)$ |
| $2$ | $y = \frac{1}{2}(2)^2 - 1 = 1$ | $(2, 1)$ |
| $4$ | $y = \frac{1}{2}(4)^2 - 1 = 7$ | $(4, 7)$ |

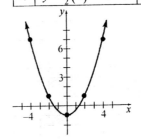

## Chapter 2: Graphs, Relations, and Functions

5. The intercepts are $(0,-5)$ and $(4,0)$. The x-intercept is 4 and the y-intercept is $-5$.

6. a. The average selling price of a new home in January of 2003 was approximately $230,000.

   b. The average selling price was highest in July of 2003 when it was approximately $248,000.

   c. The average selling price increased the most between September 2002 and October 2002. The average selling price increased approximately $16,000 during that period.

7. $\{(-2,-1),(-1,0),(0,1),(1,2),(2,3)\}$

8. a. Domain: $\{x \mid -4 \le x \le 5\}$ or $[-4,5]$
      Range: $\{y \mid -6 \le y \le -1\}$ or $[-6,-1]$

   b. Domain: $\{-4,-1,0,3,6\}$
      Range: $\{-3,-2,2,6\}$

9. a. $y = |x-2| - 3$

   | $x$ | $y = |x-2| - 3$ | $(x,y)$ |
   |---|---|---|
   | $-2$ | $y = |-2-2| - 3 = 1$ | $(-2,1)$ |
   | $0$ | $y = |-0-2| - 3 = -1$ | $(0,-1)$ |
   | $2$ | $y = |2-2| - 3 = -3$ | $(2,-3)$ |
   | $4$ | $y = |4-2| - 3 = -1$ | $(4,-1)$ |
   | $6$ | $y = |6-2| - 3 = 1$ | $(6,1)$ |

   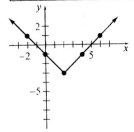

   Domain: $\{x \mid x \text{ is a real number}\}$ or $(-\infty,\infty)$
   Range: $\{y \mid y \ge -3\}$ or $[-3,\infty)$

   b. $y = \frac{1}{2}x^2 + 1$

   | $x$ | $y = \frac{1}{2}x^2 + 1$ | $(x,y)$ |
   |---|---|---|
   | $-4$ | $y = \frac{1}{2}(-4)^2 + 1 = 9$ | $(-4,9)$ |
   | $-2$ | $y = \frac{1}{2}(-2)^2 + 1 = 3$ | $(-2,3)$ |
   | $0$ | $y = \frac{1}{2}(0)^2 + 1 = 1$ | $(0,1)$ |
   | $2$ | $y = \frac{1}{2}(2)^2 + 1 = 3$ | $(2,3)$ |
   | $4$ | $y = \frac{1}{2}(4)^2 + 1 = 9$ | $(4,9)$ |

   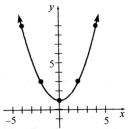

   Domain: $\{x \mid x \text{ is a real number}\}$ or $(-\infty,\infty)$
   Range: $\{y \mid y \ge 1\}$ or $[1,\infty)$

10. From the graph, we see that the object will be in the air from $t = 0$ seconds to $t = 3.8$ seconds (when it hits the ground). The ball reaches a maximum height of 105 feet and a minimum height of 0 feet (when it is on the ground). Therefore, we have
    Domain: $\{t \mid 0 \le t \le 3.8\}$
    Range: $\{h \mid 0 \le h \le 105\}$

## Section 2.3  An Introduction to Functions

### 2.3 Preparing for an Introduction to Functions

1. a. Let $x = 1$:
   $$2x^2 - 5x = 2(1)^2 - 5(1) = 2 - 5 = -3$$

   b. Let $x = 4$:
   $$2x^2 - 5x = 2(4)^2 - 5(4)$$
   $$= 2(16) - 20$$
   $$= 32 - 20$$
   $$= 12$$

   c. Let $x = -3$:
   $$2x^2 - 5x = 2(-3)^2 - 5(-3)$$
   $$= 2(9) + 15$$
   $$= 18 + 15$$
   $$= 33$$

2. Inequality: $x \leq 5$
   Interval: $(-\infty, 5]$

3. Interval: $(2, \infty)$
   Set notation: $\{x \mid x > 2\}$
   The inequality is strict since the parenthesis was used instead of a square bracket.

### 2.3 Exercises

2. dependent; independent

4. False; every function is a relation, but not every relation is a function.

6. Answers will vary. A vertical line is defined by a single $x$ value (e.g. $x = 3$). If the vertical line crosses the graph of the relation more than once, this indicates that a single $x$-value corresponds to more than one $y$-value and, therefore, the graph is not of a function.

8. Independent implies 'free' while dependent implies a dependence on something else. We are 'free' to pick a value for the independent variable (as long as the value is in the domain). Once we do so, the value of the dependent variable is already determined. The value of $y$ depends on what we choose to input for $x$.

10. Function. Each animal in the domain corresponds to exactly one gestation period in the range.
    Domain: $\{$Cat, Dog, Goat, Pig, Rabbit$\}$
    Range: $\{31, 63, 115, 151\}$

12. Not a function. The domain element $A$ for the exam grade corresponds to two different study times in the range.
    Domain: $\{A, B, C, D\}$
    Range: $\{1, 3.5, 4, 5, 6\}$

14. Function. There are no ordered pairs that have the same first coordinate, but different second coordinates.
    Domain: $\{-1, 0, 1, 2\}$
    Range: $\{-2, -5, 1, 4\}$

16. Not a function. Each ordered pair has the same first coordinate, but different second coordinates.
    Domain: $\{-2\}$
    Range: $\{-3, 1, 3, 9\}$

18. Function. There are no ordered pairs that have the same first coordinate, but different second coordinates.
    Domain: $\{-5, -2, 5, 7\}$
    Range: $\{-3, 1, 3\}$

20. $y = -6x + 3$
    Since there is only one output $y$ that can result from any given input $x$, this relation is a function.

22. $6x - 3y = 12$
    $$-3y = -6x + 12$$
    $$y = \frac{-6x + 12}{-3}$$
    $$y = 2x - 4$$
    Since there is only one output $y$ that can result from any given input $x$, this relation is a function.

24. $y = \pm 2x^2$
    Since a given input $x$ can result in more than one output $y$, this relation is not a function.

*Chapter 2: Graphs, Relations, and Functions*

26. $y = x^3 - 3$

   Since there is only one output $y$ that can result from any given input $x$, this relation is a function.

28. $y^2 = x$

   Since a given input $x$ can result in more than one output $y$, this relation is not a function. For example, if $x = 1$ then $y^2 = 1$ which means that $y = 1$ or $y = -1$.

30. Not a function. The graph fails the vertical line test so it is not the graph of a function.

32. Not a function. The graph fails the vertical line test so it is not the graph of a function.

34. Function. The graph passes the vertical line test so it is the graph of a function.

36. Not a function. The graph fails the vertical line test so it is not the graph of a function.

38. a. $f(0) = 3(0) + 1 = 0 + 1 = 1$

   b. $f(3) = 3(3) + 1 = 9 + 1 = 10$

   c. $f(-2) = 3(-2) + 1 = -6 + 1 = -5$

   d. $f(-x) = 3(-x) + 1 = -3x + 1$

   e. $-f(x) = -(3x + 1) = -3x - 1$

   f. $f(x+2) = 3(x+2) + 1$
   $= 3x + 6 + 1$
   $= 3x + 7$

   g. $f(2x) = 3(2x) + 1 = 6x + 1$

   h. $f(x+h) = 3(x+h) + 1 = 3x + 3h + 1$

40. a. $f(0) = -2(0) - 3 = 0 - 3 = -3$

   b. $f(3) = -2(3) - 3 = -6 - 3 = -9$

   c. $f(-2) = -2(-2) - 3 = 4 - 3 = 1$

   d. $f(-x) = -2(-x) - 3 = 2x - 3$

   e. $-f(x) = -(-2x - 3) = 2x + 3$

   f. $f(x+2) = -2(x+2) - 3$
   $= -2x - 4 - 3$
   $= -2x - 7$

   g. $f(2x) = -2(2x) - 3 = -4x - 3$

   h. $f(x+h) = -2(x+h) - 3 = -2x - 2h - 3$

42. $f(x) = -2x^2 + x + 1$
   $f(-3) = -2(-3)^2 + (-3) + 1$
   $= -2(9) - 3 + 1$
   $= -20$

44. $g(h) = -h^2 + 5h - 1$
   $g(4) = -(4)^2 + 5(4) - 1$
   $= -16 + 20 - 1$
   $= 3$

46. $G(z) = 2|z + 5|$
   $G(-6) = 2|-6 + 5| = 2|-1| = 2 \cdot 1 = 2$

48. $h(q) = \dfrac{3q^2}{q+2}$
   $h(2) = \dfrac{3(2)^2}{2+2} = \dfrac{3(4)}{4} = 3$

**Section 2.3** An Introduction to Functions

**50.** $g(x) = -3x + 5$

| $x$ | $y = g(x) = -3x + 5$ | $(x, y)$ |
|---|---|---|
| $-2$ | $g(-2) = -3(-2) + 5 = 11$ | $(-2, 11)$ |
| $-1$ | $g(-1) = -3(-1) + 5 = 8$ | $(-1, 8)$ |
| $0$ | $g(0) = -3(0) + 5 = 5$ | $(0, 5)$ |
| $1$ | $g(1) = -3(1) + 5 = 2$ | $(1, 2)$ |
| $2$ | $g(2) = -3(2) + 5 = -1$ | $(2, -1)$ |

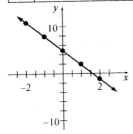

**52.** $F(x) = x^2 + 1$

| $x$ | $y = F(x) = x^2 + 1$ | $(x, y)$ |
|---|---|---|
| $-2$ | $F(-2) = (-2)^2 + 1 = 5$ | $(-2, 5)$ |
| $-1$ | $F(-1) = (-1)^2 + 1 = 2$ | $(-1, 2)$ |
| $0$ | $F(0) = 0^2 + 1 = 1$ | $(0, 1)$ |
| $1$ | $F(1) = 1^2 + 1 = 2$ | $(1, 2)$ |
| $2$ | $F(2) = 2^2 + 1 = 5$ | $(2, 5)$ |

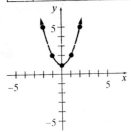

**54.** $H(x) = |x + 1|$

| $x$ | $y = H(x) = |x + 1|$ | $(x, y)$ |
|---|---|---|
| $-5$ | $H(-5) = |-5 + 1| = 4$ | $(-5, 4)$ |
| $-3$ | $H(-3) = |-3 + 1| = 2$ | $(-3, 2)$ |
| $-1$ | $H(-1) = |-1 + 1| = 0$ | $(-1, 0)$ |
| $1$ | $H(1) = |1 + 1| = 2$ | $(1, 2)$ |
| $3$ | $H(3) = |3 + 1| = 4$ | $(3, 4)$ |

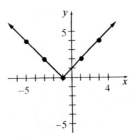

**56.** $h(x) = x^3 - 3$

| $x$ | $y = h(x) = x^3 - 3$ | $(x, y)$ |
|---|---|---|
| $-2$ | $h(-2) = (-2)^3 - 3 = -11$ | $(-2, -11)$ |
| $-1$ | $h(-1) = (-1)^3 - 3 = -4$ | $(-1, -4)$ |
| $0$ | $h(0) = 0^3 - 3 = -3$ | $(0, -3)$ |
| $1$ | $h(1) = 1^3 - 3 = -2$ | $(1, -2)$ |
| $2$ | $h(2) = 2^3 - 3 = 5$ | $(2, 5)$ |

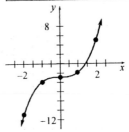

**58.** $f(x) = -2x^2 + 5x + C;\ f(-2) = -15$

$$-15 = -2(-2)^2 + 5(-2) + C$$
$$-15 = -2(4) - 10 + C$$
$$-15 = -8 - 10 + C$$
$$-15 = -18 + C$$
$$3 = C$$

**60.** $f(x) = \dfrac{-x + B}{x - 5};\ f(3) = -1$

$$-1 = \dfrac{-3 + B}{3 - 5}$$
$$-1 = \dfrac{-3 + B}{-2}$$
$$2 = -3 + B$$
$$5 = B$$

## Chapter 2: Graphs, Relations, and Functions

**62.** $A = \dfrac{1}{2}bh$

If $b = 8$ cm, we have $A(h) = \dfrac{1}{2}(8)h = 4h$.

$A(5) = 4(5) = 20$ square centimeters.

**64.** Let $p$ = price of items sold, and
$G$ = gross weekly salary.
$G(p) = 250 + 0.15p$
$G(10,000) = 250 + 0.15(10,000) = 1750$
Roberta's gross weekly salary is $1750.

**66. a.** The dependent variable is the number of housing units, $N$, and the independent variable is the number of rooms, $r$.

**b.** $N(3) = -1.33(3)^2 + 14.68(3) - 17.09$
$= -11.97 + 44.04 - 17.09$
$= 14.98$
In 2005, there were 14.98 million housing units with 3 rooms.

**c.** $N(0)$ would be the number of housing units with 0 rooms. It is impossible to have a housing unit with no rooms.

**68. a.** The dependent variable is the trip length, $T$, and the independent variable is the number of years since 1969, $x$.

**b.** $T(35) = 0.01(35)^2 - 0.12(35) + 8.89$
$= 12.25 - 4.2 + 8.89$
$= 16.94$
In 2004 (35 years after 1969), the average vehicle trip length was 16.94 miles.

**c.** $T(0) = 0.01(0)^2 - 0.12(0) + 8.89$
$= 8.89$
In 1969, the average vehicle trip length was 8.89 miles.

**70.** Every function is a relation, but not every relation is a function. To be a function, each element in the domain of the relation must correspond to exactly one element in the range of the relation.

**72.** $f(x) = -2x^2 + x + 1$

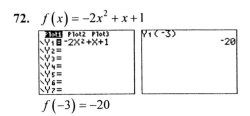

$f(-3) = -20$

**74.** $g(h) = \sqrt{2h+1}$

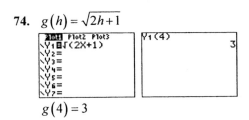

$g(4) = 3$

**76.** $G(z) = 2|z+5|$

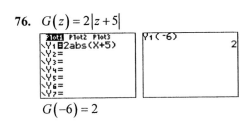

$G(-6) = 2$

**78.** $h(q) = \dfrac{3q^2}{q+2}$

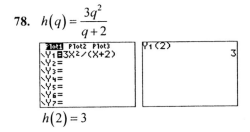

$h(2) = 3$

### 2.4 Preparing for Functions and Their Graphs

**1.** We need to determine whether the value of the variable causes division by 0. That is, we need to determine if the value makes $x - 5 = 0$.

**a.** When $x = -1$, we have $x - 5 = -1 - 5 = -6$, so $-1$ is in the domain of the variable.

**b.** When $x = 5$, we have $x - 5 = 5 - 5 = 0$, so 5 is not in the domain of the variable.

**c.** When $x = 0$, we have $x - 5 = 0 - 5 = -5$, so 0 is in the domain of the variable.

2.  $3x - 12 = 0$
    $3x - 12 + 12 = 0 + 12$
    $3x = 12$
    $\dfrac{3x}{3} = \dfrac{12}{3}$
    $x = 4$
    The solution set is $\{4\}$.

## 2.4 Exercises

2. $(-2, 4)$

4. False; if a graph has more than one y-intercept, it would fail the vertical line test and so would not be the graph of a function.

6. If a graph had more than one y-intercept, then a single input, $x = 0$, would yield more than one output (the y-intercepts). Thus, the graph could not be of a function.

8. Answers will vary. The range is the set of all possible output values.

10. $G(x) = -8x + 3$
    Since each operation in the function can be performed for any real number, the domain of the function is all real numbers.
    Domain: $\{x \mid x \text{ is a real number}\}$ or $(-\infty, \infty)$

12. $H(x) = \dfrac{x+5}{2x+1}$
    The function tells us to divide $x + 5$ by $2x + 1$. Since division by 0 is not defined, the denominator can never equal 0.
    $2x + 1 = 0$
    $2x = -1$
    $x = -\dfrac{1}{2}$
    Domain: $\left\{x \mid x \neq -\dfrac{1}{2}\right\}$ or $\left(-\infty, -\dfrac{1}{2}\right) \cup \left(-\dfrac{1}{2}, \infty\right)$

14. $s(t) = 2t^2 - 5t + 1$
    Since each operation in the function can be performed for any real number, the domain of the function is all real numbers.
    Domain: $\{t \mid t \text{ is a real number}\}$ or $(-\infty, \infty)$

16. $H(q) = \dfrac{1}{6q+5}$
    The function tells us to divide 1 by $6q + 5$. Since division by 0 is not defined, the denominator can never equal 0.
    $6q + 5 = 0$
    $6q = -5$
    $q = -\dfrac{5}{6}$
    Domain: $\left\{q \mid q \neq -\dfrac{5}{6}\right\}$ or $\left(-\infty, -\dfrac{5}{6}\right) \cup \left(-\dfrac{5}{6}, \infty\right)$

18. $f(x) = \dfrac{4x-9}{7}$
    Since each operation in the function can be performed for any real number, the domain of the function is all real numbers.
    Domain: $\{x \mid x \text{ is a real number}\}$ or $(-\infty, \infty)$

20. a. Domain: $\{x \mid x \text{ is a real number}\}$ or $(-\infty, \infty)$
    Range: $\{y \mid y \text{ is a real number}\}$ or $(-\infty, \infty)$

    b. The intercepts are $(0, -1)$ and $(3, 0)$. The x-intercept is 3 and the y-intercept is $-1$.

22. a. Domain: $\{x \mid x \text{ is a real number}\}$ or $(-\infty, \infty)$
    Range: $\{y \mid y \leq 4\}$ or $(-\infty, 4]$

    b. The intercepts are $(-1, 0)$, $(3, 0)$, and $(0, 3)$. The x-intercepts are $-1$ and 3, and the y-intercept is 3.

24. a. Domain: $\{x \mid x \text{ is a real number}\}$ or $(-\infty, \infty)$
    Range: $\{y \mid y \text{ is a real number}\}$ or $(-\infty, \infty)$

    b. The intercepts are $(-2, 0)$, $(1, 0)$, $(4, 0)$, and $(0, 2)$. The x-intercepts are $-2$, 1, and 4, and the y-intercept is 2.

26. a. Domain: $\{x \mid x \text{ is a real number}\}$ or $(-\infty, \infty)$
    Range: $\{y \mid y \geq 0\}$ or $[0, \infty)$

## Chapter 2: Graphs, Relations, and Functions

b. The intercepts are $(-1,0)$, $(2,0)$, and $(0,4)$. The x-intercepts are $-1$ and $2$, and the y-intercept is $4$.

28. a. Domain: $\{x \mid x \leq 2\}$ or $(-\infty, 2]$
    Range: $\{y \mid y \leq 3\}$ or $(-\infty, 3]$

    b. The intercepts are $(-2,0)$, $(2,0)$, and $(0,3)$. The x-intercepts are $-2$ and $2$, and the y-intercept is $3$.

30. a. $g(-3) = -2$

    b. $g(5) = 2$

    c. $g(6) = 3$

    d. $g(-5)$ is positive since the graph is above the x-axis when $x = -5$.

    e. $g(x) = 0$ for $x = \{-4, 3\}$.

    f. Domain: $\{x \mid -6 \leq x \leq 6\}$ or $[-6, 6]$

    g. Range: $\{y \mid -3 \leq y \leq 4\}$ or $[-3, 4]$

    h. The x-intercepts are $-4$ and $3$.

    i. The y-intercept is $-3$.

    j. $g(x) = -2$ for $x = \{-3, 2\}$.

    k. $g(x) = 3$ for $x = \{-5, 6\}$

32. a. From the table, when $x = 3$ the value of the function is $8$. Therefore, $G(3) = 8$

    b. From the table, when $x = 7$ the value of the function is $5$. Therefore, $G(7) = 5$

    c. From the table, $G(x) = 5$ when $x = 0$ and when $x = 7$.

    d. The x-intercept is the value of $x$ that makes the function equal $0$. From the table,

$G(x) = 0$ when $x = -4$. Therefore, the x-intercept is $x = -4$.

e. The y-intercept is the value of the function when $x = 0$. From the table, when $x = 0$ the value of the function is $5$. Therefore, the y-intercept is $y = 5$.

34. a. $f(-2) = 3(-2) + 5 = -6 + 5 = -1$
    Sine $f(-2) = -1$, the point $(-2, 1)$ is not on the graph of the function.

    b. $f(4) = 3(4) + 5 = 12 + 5 = 17$
    The point $(4, 17)$ is on the graph.

    c. $3x + 5 = -4$
    $3x = -9$
    $x = -3$
    The point $(-3, -4)$ is on the graph.

36. a. $H(3) = \frac{2}{3}(3) - 4 = 2 - 4 = -2$
    Since $H(3) = -2$, the point $(3, -2)$ is on the graph of the function.

    b. $H(6) = \frac{2}{3}(6) - 4 = 4 - 4 = 0$
    The point $(6, 0)$ is on the graph.

    c. $\frac{2}{3}x - 4 = -4$
    $\frac{2}{3}x = 0$
    $x = 0$
    The point $(0, -4)$ is on the graph.

38. $A(h) = \frac{5}{2}h$
    Since the height cannot have a negative length, the domain is all nonnegative real numbers.
    Domain: $\{h \mid h \geq 0\}$ or $[0, \infty)$

40. $G(p) = 350 + 0.12p$
    Since price will not be negative and there is no necessary upper limit, the domain is all non-negative real numbers, or $\{p \mid 0 \leq p \leq \infty\}$.

42. Answers may vary. The graph of $R$ is a parabola that is open down. The graph crosses the p-axis (i.e. the x-axis) when $p = 0$ and $p = 200$. For

values of *p* that are greater than $200, the revenue function will be negative. Since revenue is $\geq 0$, values greater than $200 are not in the domain.

44. a. Graph (II). Temperatures generally fluctuate during the year from very cold in the winter to very hot in the summer. Thus, we would expect to see a graph that oscillates.

   b. Graph (I). The number of bacteria will grow more quickly over time as the bacteria divide. We would expect to see a graph that increases slowly at first but becomes steeper as time goes on.

   c. Graph (V). Since the person is riding at a constant speed, we expect the distance to increase at a constant rate. The graph should be linear with a positive slope.

   d. Graph (III). We would expect the pizza to cool off quickly when it is first removed from the oven. The rate of cooling should slow as time goes on as the pizza temperature approaches the room temperature.

   e. Graph (IV). The value of a car decreases rapidly at first and then more slowly as time goes on. The value should approach 0 as time goes on (ignoring antique autos).

46.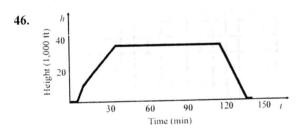

48. Answers will vary. One possibility:

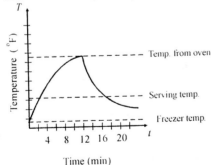

50. Answers will vary. One possibility: For the first 100 days, the depth of the lake is fairly constant. Then there is a increase in depth, possibly due to spring rains, followed by a large decrease, possibly due to a hot summer. Towards the end of the year the depth increases back to its original level, possibly due to snow and ice accumulation.

52. Answers will vary. One possibility:

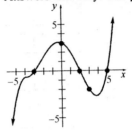

**Chapter 2 Review**

1. A: quadrant IV; B: quadrant III; C: y-axis; D: quadrant II; E: x-axis; F: quadrant I.

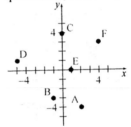

2. A: x-axis; B: quadrant I; C: quadrant III; D: quadrant II; E: quadrant IV, F: y-axis.

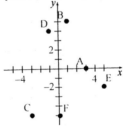

3. a. Let $x = 3$ and $y = 1$.
   $$3(3) - 2(1) \stackrel{?}{=} 7$$
   $$9 - 2 \stackrel{?}{=} 7$$
   $$7 = 7 \text{ True}$$
   The point $(3, 1)$ is on the graph.

b. Let $x = 2$ and $y = -1$.

$$3(2) - 2(-1) \stackrel{?}{=} 7$$
$$6 + 2 \stackrel{?}{=} 7$$
$$8 = 7 \text{ False}$$

The point $(2, -1)$ is not on the graph.

c. Let $x = 4$ and $y = 0$.

$$3(4) - 2(0) \stackrel{?}{=} 7$$
$$12 - 0 \stackrel{?}{=} 7$$
$$12 = 7 \text{ False}$$

The point $(4, 0)$ is not on the graph.

d. Let $x = \frac{1}{3}$ and $y = -3$.

$$3\left(\frac{1}{3}\right) - 2(-3) \stackrel{?}{=} 7$$
$$1 + 6 \stackrel{?}{=} 7$$
$$7 = 7 \text{ True}$$

The point $\left(\frac{1}{3}, -3\right)$ is on the graph.

4. a. Let $x = -1$ and $y = 3$.

$$3 \stackrel{?}{=} 2(-1)^2 - 3(-1) + 2$$
$$3 \stackrel{?}{=} 2(1) + 3 + 2$$
$$3 \stackrel{?}{=} 2 + 3 + 2$$
$$3 = 7 \text{ False}$$

The point $(-1, 3)$ is not on the graph.

b. Let $x = 1$ and $y = 1$.

$$1 \stackrel{?}{=} 2(1)^2 - 3(1) + 2$$
$$1 \stackrel{?}{=} 2(1) - 3 + 2$$
$$1 \stackrel{?}{=} 2 - 3 + 2$$
$$1 = 1 \text{ True}$$

The point $(1, 1)$ is on the graph.

c. Let $x = -2$ and $y = 16$.

$$16 \stackrel{?}{=} 2(-2)^2 - 3(-2) + 2$$
$$16 \stackrel{?}{=} 2(4) + 6 + 2$$
$$16 \stackrel{?}{=} 8 + 6 + 2$$
$$16 = 16 \text{ True}$$

The point $(-2, 16)$ is on the graph.

d. Let $x = \frac{1}{2}$ and $y = \frac{3}{2}$.

$$\frac{3}{2} \stackrel{?}{=} 2\left(\frac{1}{2}\right)^2 - 3\left(\frac{1}{2}\right) + 2$$
$$\frac{3}{2} \stackrel{?}{=} 2\left(\frac{1}{4}\right) - \frac{3}{2} + 2$$
$$\frac{3}{2} \stackrel{?}{=} \frac{1}{2} - \frac{3}{2} + 2$$
$$\frac{3}{2} = 1 \text{ False}$$

The point $\left(\frac{1}{2}, \frac{3}{2}\right)$ is not on the graph.

5. $y = x + 2$

| $x$ | $y = x + 2$ | $(x, y)$ |
|---|---|---|
| $-3$ | $y = (-3) + 2 = -1$ | $(-3, -1)$ |
| $-1$ | $y = (-1) + 2 = 1$ | $(-1, 1)$ |
| $0$ | $y = (0) + 2 = 2$ | $(0, 2)$ |
| $1$ | $y = (1) + 2 = 3$ | $(1, 3)$ |
| $3$ | $y = (3) + 2 = 5$ | $(3, 5)$ |

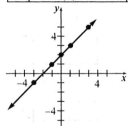

6. $2x + y = 3$
   $y = -2x + 3$

| $x$ | $y = -2x + 3$ | $(x, y)$ |
|---|---|---|
| $-3$ | $y = -2(-3) + 3 = 9$ | $(-3, 9)$ |
| $-1$ | $y = -2(-1) + 3 = 5$ | $(-1, 5)$ |
| $0$ | $y = -2(0) + 3 = 3$ | $(0, 3)$ |
| $1$ | $y = -2(1) + 3 = 1$ | $(1, 1)$ |
| $3$ | $y = -2(3) + 3 = -3$ | $(3, -3)$ |

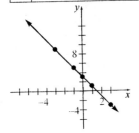

7. $y = 2x^2 - 3$

| $x$ | $y = 2x^2 - 3$ | $(x, y)$ |
|---|---|---|
| $-2$ | $y = 2(-2)^2 - 3 = 5$ | $(-2, 5)$ |
| $-1$ | $y = 2(-1)^2 - 3 = -1$ | $(-1, -1)$ |
| $0$ | $y = 2(0)^2 - 3 = -3$ | $(0, -3)$ |
| $1$ | $y = 2(1)^2 - 3 = -1$ | $(1, -1)$ |
| $2$ | $y = 2(2)^2 - 3 = 5$ | $(2, 5)$ |

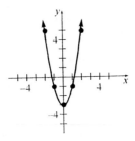

8. $y = -x^2 + 4$

| $x$ | $y = -x^2 + 4$ | $(x, y)$ |
|---|---|---|
| $-3$ | $y = -(-3)^2 + 4 = -5$ | $(-3, -5)$ |
| $-1$ | $y = -(-1)^2 + 4 = 3$ | $(-1, 3)$ |
| $0$ | $y = -(0)^2 + 4 = 4$ | $(0, 4)$ |
| $1$ | $y = -(1)^2 + 4 = 3$ | $(1, 3)$ |
| $3$ | $y = -(3)^2 + 4 = -5$ | $(3, -5)$ |

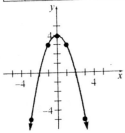

9. $y = -|x| - 2$

| $x$ | $y = -|x| - 2$ | $(x, y)$ |
|---|---|---|
| $-3$ | $y = -|-3| - 2 = -5$ | $(-3, -5)$ |
| $-1$ | $y = -|-1| - 2 = -3$ | $(-1, -3)$ |
| $0$ | $y = -|0| - 2 = -2$ | $(0, -2)$ |
| $1$ | $y = -|1| - 2 = -3$ | $(1, -3)$ |
| $3$ | $y = -|3| - 2 = -5$ | $(3, -5)$ |

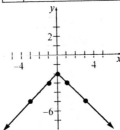

10. $y = |x + 2| - 1$

| $x$ | $y = |x + 2| - 1$ | $(x, y)$ |
|---|---|---|
| $-5$ | $y = |-5 + 2| - 1 = 2$ | $(-5, 2)$ |
| $-3$ | $y = |-3 + 2| - 1 = 0$ | $(-3, 0)$ |
| $-1$ | $y = |-1 + 2| - 1 = 0$ | $(-1, 0)$ |
| $0$ | $y = |0 + 2| - 1 = 1$ | $(0, 1)$ |
| $1$ | $y = |1 + 2| - 1 = 2$ | $(1, 2)$ |

## Chapter 2: Graphs, Relations, and Functions

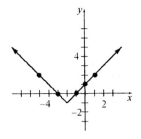

**11.** $y = x^3 + 2$

| $x$ | $y = x^3 + 2$ | $(x, y)$ |
|---|---|---|
| $-2$ | $y = (-2)^3 + 2 = -6$ | $(-2, -6)$ |
| $-1$ | $y = (-1)^3 + 2 = 1$ | $(-1, 1)$ |
| $0$ | $y = (0)^3 + 2 = 2$ | $(0, 2)$ |
| $1$ | $y = (1)^3 + 2 = 3$ | $(1, 3)$ |
| $2$ | $y = (2)^3 + 2 = 10$ | $(2, 10)$ |

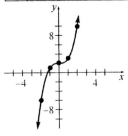

**12.** $y = -x^3 + 1$

| $x$ | $y = -x^3 + 1$ | $(x, y)$ |
|---|---|---|
| $-2$ | $y = -(-2)^3 + 1 = 9$ | $(-2, 9)$ |
| $-1$ | $y = -(-1)^3 + 1 = 2$ | $(-1, 2)$ |
| $0$ | $y = -(0)^3 + 1 = 1$ | $(0, 1)$ |
| $1$ | $y = -(1)^3 + 1 = 0$ | $(1, 0)$ |
| $2$ | $y = -(2)^3 + 1 = -7$ | $(2, -7)$ |

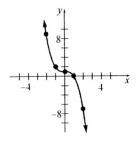

**13.** $x = y^2 + 1$

| $y$ | $x = y^2 + 1$ | $(x, y)$ |
|---|---|---|
| $-2$ | $x = (-2)^2 + 1 = 5$ | $(5, -2)$ |
| $-1$ | $x = (-1)^2 + 1 = 2$ | $(2, -1)$ |
| $0$ | $x = (0)^2 + 1 = 1$ | $(1, 0)$ |
| $1$ | $x = (1)^2 + 1 = 2$ | $(2, 1)$ |
| $2$ | $x = (2)^2 + 1 = 5$ | $(5, 2)$ |

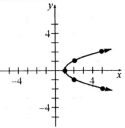

**14.** $y = \dfrac{1}{x-2}$

| $x$ | $y = \dfrac{1}{x-2}$ | $(x, y)$ |
|---|---|---|
| $-2$ | $y = \dfrac{1}{-2-2} = -\dfrac{1}{4}$ | $\left(-2, -\tfrac{1}{4}\right)$ |
| $0$ | $y = \dfrac{1}{0-2} = -\dfrac{1}{2}$ | $\left(0, -\tfrac{1}{2}\right)$ |
| $1$ | $y = \dfrac{1}{1-2} = -1$ | $(1, -1)$ |
| $3$ | $y = \dfrac{1}{3-2} = 1$ | $(3, 1)$ |
| $4$ | $y = \dfrac{1}{4-2} = \dfrac{1}{2}$ | $\left(4, \tfrac{1}{2}\right)$ |
| $6$ | $y = \dfrac{1}{6-2} = \dfrac{1}{4}$ | $\left(6, \tfrac{1}{4}\right)$ |

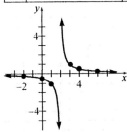

**15.** The intercepts are $(-3, 0)$, $(0, -1)$, and $(0, 3)$.

**16.** The intercepts are $(-2, 0)$, $(0, 0)$, and $(2, 0)$.

**17. a.** Since 2250 is less than 3000, the monthly bill will still be $40. Using the graph, we find that when $x = 2.25$ (2250 minutes) the y-value is 40, or $40.

b. For 12,000 minutes, we have $x = 12$. According to the graph, the monthly bill for 12,000 minutes will be about $500.

18. a. According to the graph, the winning time in the 1999 Kentucky Derby was 2:03, or 123 seconds.

b. To find the fastest time, we look for the lowest point on the graph. The lowest point on the graph occurs for the year 2001. Therefore, the fastest winning time for the Kentucky Derby between 1995 and 2004 occurred in 2001. The winning time was roughly 2:00, or 120 seconds.

19. {(Cent, 2.500), (Nickel, 5.000), (Dime, 2.268), (Quarter, 5.670), (Half Dollar, 11.340), (Dollar, 8.100)}
    **Domain:**
    {Cent, Nickel, Dime, Quarter, Half Dollar, Dollar}
    **Range:**
    {2.268, 2.500, 5.000, 5.670, 8.100, 11.340}

20. {(70, $6.99), (90, $9.99), (120, $9.99), (128, $12.99), (446, $49.99)}
    **Domain:**
    {70, 90, 120, 128, 446}
    **Range:**
    {$6.99, $9.99, $12.99, $49.99}

21. Domain: $\{-4,-2,2,3,6\}$
    Range: $\{-9,-1,5,7,8\}$
    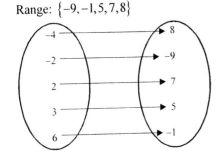

22. Domain: $\{-2,1,3,5\}$
    Range: $\{1,4,7,8\}$
    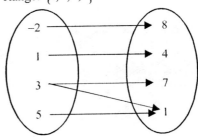

23. Domain: $\{x \mid x \text{ is a real number}\}$ or $(-\infty, \infty)$
    Range: $\{y \mid y \text{ is a real number}\}$ or $(-\infty, \infty)$

24. Domain: $\{x \mid -6 \leq x \leq 4\}$ or $[-6, 4]$
    Range: $\{y \mid -4 \leq y \leq 6\}$ or $[-4, 6]$

25. Domain: $\{2\}$
    Range: $\{y \mid y \text{ is a real number}\}$ or $(-\infty, \infty)$

26. Domain: $\{x \mid x \geq -1\}$ or $[-1, \infty)$
    Range: $\{y \mid y \geq -2\}$ or $[-2, \infty)$

27. $y = x + 2$

| $x$ | $y = x + 2$ | $(x, y)$ |
|---|---|---|
| $-3$ | $y = (-3) + 2 = -1$ | $(-3, -1)$ |
| $-1$ | $y = (-1) + 2 = 1$ | $(-1, 1)$ |
| $0$ | $y = (0) + 2 = 2$ | $(0, 2)$ |
| $1$ | $y = (1) + 2 = 3$ | $(1, 3)$ |
| $3$ | $y = (3) + 2 = 5$ | $(3, 5)$ |

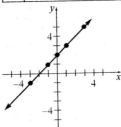

Domain: $\{x \mid x \text{ is a real number}\}$ or $(-\infty, \infty)$
Range: $\{y \mid y \text{ is a real number}\}$ or $(-\infty, \infty)$

28. $2x + y = 3$
$y = -2x + 3$

| $x$ | $y = -2x + 3$ | $(x, y)$ |
|---|---|---|
| $-3$ | $y = -2(-3) + 3 = 9$ | $(-3, 9)$ |
| $-1$ | $y = -2(-1) + 3 = 5$ | $(-1, 5)$ |
| $0$ | $y = -2(0) + 3 = 3$ | $(0, 3)$ |
| $1$ | $y = -2(1) + 3 = 1$ | $(1, 1)$ |
| $3$ | $y = -2(3) + 3 = -3$ | $(3, -3)$ |

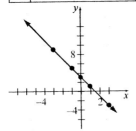

Domain: $\{x \mid x \text{ is a real number}\}$ or $(-\infty, \infty)$
Range: $\{y \mid y \text{ is a real number}\}$ or $(-\infty, \infty)$

29. $y = 2x^2 - 3$

| $x$ | $y = 2x^2 - 3$ | $(x, y)$ |
|---|---|---|
| $-2$ | $y = 2(-2)^2 - 3 = 5$ | $(-2, 5)$ |
| $-1$ | $y = 2(-1)^2 - 3 = -1$ | $(-1, -1)$ |
| $0$ | $y = 2(0)^2 - 3 = -3$ | $(0, -3)$ |
| $1$ | $y = 2(1)^2 - 3 = -1$ | $(1, -1)$ |
| $2$ | $y = 2(2)^2 - 3 = 5$ | $(2, 5)$ |

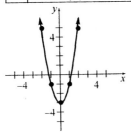

Domain: $\{x \mid x \text{ is a real number}\}$ or $(-\infty, \infty)$
Range: $\{y \mid y \geq -3\}$ or $[-3, \infty)$

30. $y = -x^2 + 4$

| $x$ | $y = -x^2 + 4$ | $(x, y)$ |
|---|---|---|
| $-3$ | $y = -(-3)^2 + 4 = -5$ | $(-3, -5)$ |
| $-1$ | $y = -(-1)^2 + 4 = 3$ | $(-1, 3)$ |
| $0$ | $y = -(0)^2 + 4 = 4$ | $(0, 4)$ |
| $1$ | $y = -(1)^2 + 4 = 3$ | $(1, 3)$ |
| $3$ | $y = -(3)^2 + 4 = -5$ | $(3, -5)$ |

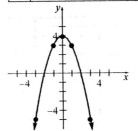

Domain: $\{x \mid x \text{ is a real number}\}$ or $(-\infty, \infty)$
Range: $\{y \mid y \leq 4\}$ or $(-\infty, 4]$

31. $y = -|x| - 2$

| $x$ | $y = -|x| - 2$ | $(x, y)$ |
|---|---|---|
| $-3$ | $y = -|-3| - 2 = -5$ | $(-3, -5)$ |
| $-1$ | $y = -|-1| - 2 = -3$ | $(-1, -3)$ |
| $0$ | $y = -|0| - 2 = -2$ | $(0, -2)$ |
| $1$ | $y = -|1| - 2 = -3$ | $(1, -3)$ |
| $3$ | $y = -|3| - 2 = -5$ | $(3, -5)$ |

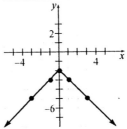

Domain: $\{x \mid x \text{ is a real number}\}$ or $(-\infty, \infty)$
Range: $\{y \mid y \leq -2\}$ or $(-\infty, -2]$

**32.** $y = |x+2| - 1$

| $x$ | $y = |x+2| - 1$ | $(x, y)$ |
|---|---|---|
| $-5$ | $y = |-5+2| - 1 = 2$ | $(-5, 2)$ |
| $-3$ | $y = |-3+2| - 1 = 0$ | $(-3, 0)$ |
| $-1$ | $y = |-1+2| - 1 = 0$ | $(-1, 0)$ |
| $0$ | $y = |0+2| - 1 = 1$ | $(0, 1)$ |
| $1$ | $y = |1+2| - 1 = 2$ | $(1, 2)$ |

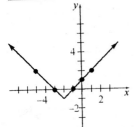

Domain: $\{x \mid x \text{ is a real number}\}$ or $(-\infty, \infty)$

Range: $\{y \mid y \geq -1\}$ or $[-1, \infty)$

**33.** $y = x^3 + 2$

| $x$ | $y = x^3 + 2$ | $(x, y)$ |
|---|---|---|
| $-2$ | $y = (-2)^3 + 2 = -6$ | $(-2, -6)$ |
| $-1$ | $y = (-1)^3 + 2 = 1$ | $(-1, 1)$ |
| $0$ | $y = (0)^3 + 2 = 2$ | $(0, 2)$ |
| $1$ | $y = (1)^3 + 2 = 3$ | $(1, 3)$ |
| $2$ | $y = (2)^3 + 2 = 10$ | $(2, 10)$ |

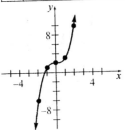

Domain: $\{x \mid x \text{ is a real number}\}$ or $(-\infty, \infty)$

Range: $\{y \mid y \text{ is a real number}\}$ or $(-\infty, \infty)$

**34.** $y = -x^3 + 1$

| $x$ | $y = -x^3 + 1$ | $(x, y)$ |
|---|---|---|
| $-2$ | $y = -(-2)^3 + 1 = 9$ | $(-2, 9)$ |
| $-1$ | $y = -(-1)^3 + 1 = 2$ | $(-1, 2)$ |
| $0$ | $y = -(0)^3 + 1 = 1$ | $(0, 1)$ |
| $1$ | $y = -(1)^3 + 1 = 0$ | $(1, 0)$ |
| $2$ | $y = -(2)^3 + 1 = -7$ | $(2, -7)$ |

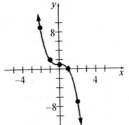

Domain: $\{x \mid x \text{ is a real number}\}$ or $(-\infty, \infty)$

Range: $\{y \mid y \text{ is a real number}\}$ or $(-\infty, \infty)$

**35.** $x = y^2 + 1$

| $y$ | $x = y^2 + 1$ | $(x, y)$ |
|---|---|---|
| $-2$ | $x = (-2)^2 + 1 = 5$ | $(5, -2)$ |
| $-1$ | $x = (-1)^2 + 1 = 2$ | $(2, -1)$ |
| $0$ | $x = (0)^2 + 1 = 1$ | $(1, 0)$ |
| $1$ | $x = (1)^2 + 1 = 2$ | $(2, 1)$ |
| $2$ | $x = (2)^2 + 1 = 5$ | $(5, 2)$ |

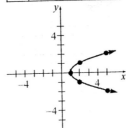

Domain: $\{x \mid x \geq 1\}$ or $[1, \infty)$

Range: $\{y \mid y \text{ is a real number}\}$ or $(-\infty, \infty)$

36. $y = \dfrac{1}{x-2}$

| $x$ | $y = \dfrac{1}{x-2}$ | $(x,y)$ |
|---|---|---|
| $-2$ | $y = \dfrac{1}{-2-2} = -\dfrac{1}{4}$ | $\left(-2,-\dfrac{1}{4}\right)$ |
| $0$ | $y = \dfrac{1}{0-2} = -\dfrac{1}{2}$ | $\left(0,-\dfrac{1}{2}\right)$ |
| $1$ | $y = \dfrac{1}{1-2} = -1$ | $(1,-1)$ |
| $3$ | $y = \dfrac{1}{3-2} = 1$ | $(3,1)$ |
| $4$ | $y = \dfrac{1}{4-2} = \dfrac{1}{2}$ | $\left(4,\dfrac{1}{2}\right)$ |
| $6$ | $y = \dfrac{1}{6-2} = \dfrac{1}{4}$ | $\left(6,\dfrac{1}{4}\right)$ |

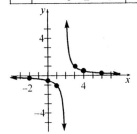

Domain: $\{x \mid x \neq 2\}$ or $(-\infty, 2) \cup (2, \infty)$

Range: $\{y \mid y \neq 0\}$ or $(-\infty, 0) \cup (0, \infty)$

37. a. Domain: $\{x \mid 0 \le x \le 44.64\}$ or $[0, 44.64]$

Range: $\{y \mid 40 \le y \le 2122\}$ or $[40, 2122]$

The monthly cost will be at least $40 but no more than $2122. The number of minutes used must be between 0 and 44,640.

b. Answers may vary. There are at most 31 days in a month. Since 31 days is equivalent to 44,640 minutes, this must be the largest value in the domain. It is not possible to talk for a negative number of minutes, so the domain should begin at 0.

38. Domain: $\{t \mid 0 \le t \le 4\}$ or $[0, 4]$

Range: $\{y \mid 0 \le y \le 121\}$ or $[0, 121]$

The ball will be in the air from 0 to 4 seconds and will reach heights from 0 feet up to a maximum of 121 feet.

39. a. Not a function. The domain element $-1$ corresponds to two different values in the range.

Domain: $\{-1, 5, 7, 9\}$

Range: $\{-2, 0, 2, 3, 4\}$

b. Function. Each animal corresponds to exactly one typical lifespan.
Domain: {Camel, Macaw, Deer, Fox, Tiger, Crocodile}
Range: $\{14, 22, 35, 45, 50\}$

40. a. Function. There are no ordered pairs that have the same first coordinate, but different second coordinates.

Domain: $\{-3, -2, 2, 4, 5\}$

Range: $\{-1, 3, 4, 7\}$

b. Not a function. The domain element 'Blue' corresponds to two different types of cars in the range.
Domain: {Red, Blue, Green, Black}
Range: {Camry, Taurus, Windstar, Durango}

41. $3x - 5y = 18$
$-5y = -3x + 18$
$y = \dfrac{-3x + 18}{-5}$
$y = \dfrac{3}{5}x - \dfrac{18}{5}$

Since there is only one output $y$ that can result from any given input $x$, this relation is a function.

42. $x^2 + y^2 = 81$
$y^2 = 81 - x^2$

Since a given input $x$ can result in more than one output $y$, this relation is not a function. For example, if $x = 0$ then $y = 9$ or $y = -9$.

43. $y = \pm 10x$

Since a given input $x$ can result in more than one output $y$, this relation is not a function.

44. $y = x^2 - 14$

Since there is only one output $y$ that can result from any given input $x$, this relation is a function.

45. Not a function. The graph fails the vertical line test so it is not the graph of a function.

46. Function. The graph passes the vertical line test so it is the graph of a function.

47. Function. The graph passes the vertical line test so it is the graph of a function.

48. Not a function. The graph fails the vertical line test so it is not the graph of a function.

49. a. $f(-2) = (-2)^2 + 2(-2) - 5$
    $= 4 - 4 - 5$
    $= -5$

    b. $f(3) = (3)^2 + 2(3) - 5$
    $= 9 + 6 - 5$
    $= 10$

50. a. $g(0) = \dfrac{2(0)+1}{(0)-3}$
    $= \dfrac{0+1}{-3}$
    $= -\dfrac{1}{3}$

    b. $g(2) = \dfrac{2(2)+1}{(2)-3}$
    $= \dfrac{4+1}{-1}$
    $= -5$

51. a. $F(5) = -2(5) + 7$
    $= -10 + 7$
    $= -3$

    b. $F(-x) = -2(-x) + 7$
    $= 2x + 7$

52. a. $G(7) = 2(7) + 1$
    $= 14 + 1$
    $= 15$

    b. $G(x+h) = 2(x+h) + 1$
    $= 2x + 2h + 1$

53. $f(x) = 2x - 5$

| $x$ | $y = f(x) = 2x - 5$ | $(x, y)$ |
|---|---|---|
| $-1$ | $f(-1) = 2(-1) - 5 = -7$ | $(-1, -7)$ |
| $0$ | $f(0) = 2(0) - 5 = -5$ | $(0, -5)$ |
| $1$ | $f(1) = 2(1) - 5 = -3$ | $(1, -3)$ |
| $2$ | $f(2) = 2(2) - 5 = -1$ | $(2, -1)$ |
| $3$ | $f(3) = 2(3) - 5 = 1$ | $(3, 1)$ |

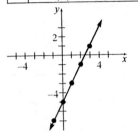

54. $g(x) = x^2 - 3x + 2$

| $x$ | $y = g(x) = x^2 - 3x + 2$ | $(x, y)$ |
|---|---|---|
| $-1$ | $g(-1) = (-1)^2 - 3(-1) + 2 = 6$ | $(-1, 6)$ |
| $0$ | $g(0) = (0)^2 - 3(0) + 2 = 2$ | $(0, 2)$ |
| $1$ | $g(1) = (1)^2 - 3(1) + 2 = 0$ | $(1, 0)$ |
| $2$ | $g(2) = (2)^2 - 3(2) + 2 = 0$ | $(2, 0)$ |
| $3$ | $g(3) = (3)^2 - 3(3) + 2 = 2$ | $(3, 2)$ |
| $4$ | $g(4) = (4)^2 - 3(4) + 2 = 6$ | $(4, 6)$ |

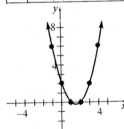

## Chapter 2: Graphs, Relations, and Functions

55. $h(x) = (x-1)^3 - 3$

| $x$ | $y = h(x) = (x-1)^3 - 3$ | $(x, y)$ |
|---|---|---|
| $-1$ | $h(-1) = (-1-1)^3 - 3 = -11$ | $(-1, -11)$ |
| $0$ | $h(0) = (0-1)^3 - 3 = -4$ | $(0, -4)$ |
| $1$ | $h(1) = (1-1)^3 - 3 = -3$ | $(1, -3)$ |
| $2$ | $h(2) = (2-1)^3 - 3 = -2$ | $(2, -2)$ |
| $3$ | $h(3) = (3-1)^3 - 3 = 5$ | $(3, 5)$ |

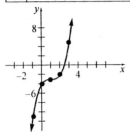

56. $f(x) = |x+1| - 4$

| $x$ | $y = f(x) = |x+1| - 4$ | $(x, y)$ |
|---|---|---|
| $-5$ | $f(-5) = |-5+1| - 4 = 0$ | $(-5, 0)$ |
| $-3$ | $f(-3) = |-3+1| - 4 = -2$ | $(-3, -2)$ |
| $-1$ | $f(-1) = |-1+1| - 4 = -4$ | $(-1, -4)$ |
| $1$ | $f(1) = |1+1| - 4 = -2$ | $(1, -2)$ |
| $3$ | $f(3) = |3+1| - 4 = 0$ | $(3, 0)$ |

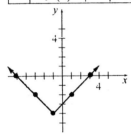

57. **a.** The dependent variable is the population, $P$, and the independent variable is the number of years after 1900, $t$.

   **b.** $P(110) = 0.144(110)^2 - 6.613(110) + 104.448$
   $= 1119.418$
   According to the model, the population of Orange County will be roughly 1,119,418 in 2010.

   **c.** $P(-70) = 0.144(-70)^2 - 6.613(-70) + 104.448$
   $= 1272.958$
   According to the model, the population of Orange County was roughly 1,272,958 in 1830. This is not reasonable. (The population of the entire Florida territory was roughly 35,000 in 1830.)

58. **a.** The dependent variable is the annual wage, $W$, and the independent variable is age, $a$.

   **b.** $W(30) = -0.058(30)^2 + 5.410(30) - 73.839$
   $= 36.261$
   According to the model, a 30-year old Wyoming resident working in the mining industry in 2000 made about $36,261 annually on average.

   **c.** $W(16) = -0.058(16)^2 + 5.410(16) - 73.839$
   $= -2.127$
   $W(16)$ represents the average annual salary of a 16-year old Wyoming resident working in the mining industry in 2000. This result is unreasonable since annual salaries cannot be negative.

59. $f(x) = -\frac{3}{2}x + 5$
   Since each operation in the function can be performed for any real number, the domain of the function is all real numbers.
   Domain: $\{x \mid x \text{ is a real number}\}$ or $(-\infty, \infty)$

60. $g(w) = \frac{w-9}{2w+5}$
   The function tells us to divide $w-9$ by $2w+5$. Since division by 0 is not defined, the denominator can never equal 0.
   $2w + 5 = 0$
   $2w = -5$
   $w = -\frac{5}{2}$
   Thus, the domain is all real numbers except $-\frac{5}{2}$.
   Domain: $\{w \mid w \neq -\frac{5}{2}\}$ or $\left(-\infty, -\frac{5}{2}\right) \cup \left(-\frac{5}{2}, \infty\right)$

61. $h(t) = \frac{t+2}{t-5}$
   The function tells us to divide $t+2$ by $t-5$. Since division by 0 is not defined, the denominator can never equal 0.
   $t - 5 = 0$
   $t = 5$
   Thus, the domain of the function is all real numbers except 5.
   Domain: $\{t \mid t \neq 5\}$ or $(-\infty, 5) \cup (5, \infty)$

62. $F(x) = \dfrac{3}{x-2}$

    The function tells us to divide 3 by $x-2$. We can't divide by 0, so $x \neq 2$.
    Domain: $\{x \mid x \neq 2\}$ or $(-\infty, 2) \cup (2, \infty)$

63. $G(t) = 3t^2 + 4t - 9$

    Since each operation in the function can be performed for any real number, the domain of the function is all real numbers.
    Domain: $\{t \mid t \text{ is a real number}\}$ or $(-\infty, \infty)$

64. $H(x) = x^4 - 2x^3 + 7$

    Since each operation in the function can be performed for any real number, the domain of the function is all real numbers.
    Domain: $\{x \mid x \text{ is a real number}\}$ or $(-\infty, \infty)$

65. a. Domain: $\{x \mid x \text{ is a real number}\}$ or $(-\infty, \infty)$
       Range: $\{y \mid y \text{ is a real number}\}$ or $(-\infty, \infty)$

    b. The intercepts are $(0,2)$ and $(4,0)$. The x-intercept is 4 and the y-intercept is 2.

66. a. Domain: $\{x \mid x \text{ is a real number}\}$ or $(-\infty, \infty)$
       Range: $\{y \mid y \geq -3\}$ or $[-3, \infty)$

    b. The intercepts are $(-2,0)$, $(2,0)$, and $(0,-3)$. The x-intercepts are $-2$ and 2, and the y-intercept is $-3$.

67. a. Domain: $\{x \mid x \text{ is a real number}\}$ or $(-\infty, \infty)$
       Range: $\{y \mid y \text{ is a real number}\}$ or $(-\infty, \infty)$

    b. The intercepts are $(0,0)$ and $(2,0)$. The x-intercepts are 0 and 2; the y-intercept is 0.

68. a. Domain: $\{x \mid x \geq -3\}$ or $[-3, \infty)$
       Range: $\{y \mid y \geq 1\}$ or $[1, \infty)$

    b. The only intercept is $(0,3)$. There are no x-intercepts, but there is a y-intercept of 3.

69. a. Domain: $\{x \mid x \text{ is a real number}\}$ or $(-\infty, \infty)$
       Range: $\{y \mid y \geq -4\}$ or $[-4, \infty)$

    b. The intercepts are $(-1,0)$, $(3,0)$, and $(0,-2)$. The x-intercepts are $-1$ and 3, and the y-intercept is $-2$.

70. a. Domain: $\{x \mid x \text{ is a real number}\}$ or $(-\infty, \infty)$
       Range: $\{y \mid y \leq 0\}$ or $(-\infty, 0]$

    b. The intercepts are $(-2,0)$, $(2,0)$, and $(0,-4)$. The x-intercepts are $-2$ and 2, and the y-intercept is $-4$.

71. a. $h(3) = 2(3) - 7 = 6 - 7 = -1$
       Since $h(3) = -1$, the point $(3,-1)$ is on the graph of the function.

    b. $h(-2) = 2(-2) - 7 = -4 - 7 = -11$
       The point $(-2,-11)$ is on the graph of the function.

    c. $h(x) = 4$
       $2x - 7 = 4$
       $2x = 11$
       $x = \dfrac{11}{2}$
       The point $\left(\dfrac{11}{2}, 4\right)$ is on the graph of $h$.

72. a. $g(-5) = \dfrac{3}{5}(-5) + 4 = -3 + 4 = 1$
       Since $g(-5) = 1$, the point $(-5,2)$ is not on the graph of the function.

    b. $g(3) = \dfrac{3}{5}(3) + 4 = \dfrac{9}{5} + 4 = \dfrac{29}{5}$
       The point $\left(3, \dfrac{29}{5}\right)$ is on the graph of the function.

    c. $g(x) = -2$
       $\dfrac{3}{5}x + 4 = -2$
       $\dfrac{3}{5}x = -6$
       $x = -10$
       The point $(-10,-2)$ is on the graph of $g$.

## Chapter 2: Graphs, Relations, and Functions

73.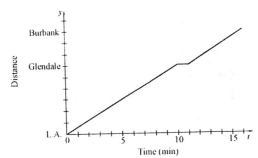

74.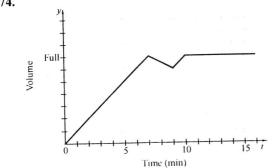

### Chapter 2 Test

1. $A$: quadrant IV; $B$: y-axis; $C$: x-axis; $D$: quadrant I; $E$: quadrant III; $F$: quadrant II.

   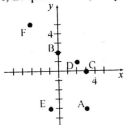

2. a. Let $x = -2$ and $y = 4$.

   $4 \stackrel{?}{=} 3(-2)^2 + (-2) - 5$

   $4 \stackrel{?}{=} 3(4) - 7$

   $4 \stackrel{?}{=} 12 - 7$

   $4 = 5$ False

   The point $(-2, 4)$ is not on the graph.

   b. Let $x = -1$ and $y = -3$.

   $-3 \stackrel{?}{=} 3(-1)^2 + (-1) - 5$

   $-3 \stackrel{?}{=} 3(1) - 6$

   $-3 \stackrel{?}{=} 3 - 6$

   $-3 = -3$ True

   The point $(-1, -3)$ is on the graph.

   c. Let $x = 2$ and $y = 9$.

   $9 \stackrel{?}{=} 3(2)^2 + (2) - 5$

   $9 \stackrel{?}{=} 3(4) - 3$

   $9 \stackrel{?}{=} 12 - 3$

   $9 = 9$ True

   The point $(2, 9)$ is on the graph.

3. $y = 4x - 1$

   | $x$ | $y = 4x - 1$ | $(x, y)$ |
   |---|---|---|
   | $-1$ | $y = 4(-1) - 1 = -5$ | $(-1, -5)$ |
   | $0$ | $y = 4(0) - 1 = -1$ | $(0, -1)$ |
   | $1$ | $y = 4(1) - 1 = 3$ | $(1, 3)$ |

   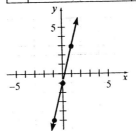

4. $y = 4x^2$

| $x$ | $y = 4x^2$ | $(x, y)$ |
|---|---|---|
| $-1$ | $y = 4(-1)^2 = 4$ | $(-1, 4)$ |
| $-\frac{1}{2}$ | $y = 4\left(-\frac{1}{2}\right)^2 = 1$ | $\left(-\frac{1}{2}, 1\right)$ |
| $0$ | $y = 4(0)^2 = 0$ | $(0, 0)$ |
| $\frac{1}{2}$ | $y = 4\left(\frac{1}{2}\right)^2 = 1$ | $\left(\frac{1}{2}, 1\right)$ |
| $1$ | $y = 4(1)^2 = 4$ | $(1, 4)$ |

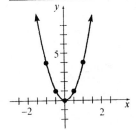

5. The intercepts are $(-3, 0)$, $(0, 1)$, and $(0, 3)$. The x-intercept is $-3$ and the y-intercepts are 1 and 3.

6. a. The unemployment rate in May was about 4.9%.

   b. The unemployment rate was highest in January when it was approximately 5.9%.

   c. The unemployment rate was lowest in November when it was approximately 4.4%.

   d. The unemployment rate gradually declined over the year except for a jump during the summer months. This might be due to recent graduates who are looking for work.

7. Domain: $\{-4, 2, 5, 7\}$
   Range: $\{-7, -2, -1, 3, 8, 12\}$

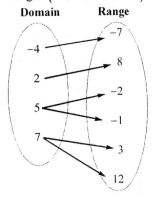

8. Domain: $\left\{x \mid -\frac{5\pi}{2} \le x \le \frac{5\pi}{2}\right\}$ or $\left[-\frac{5\pi}{2}, \frac{5\pi}{2}\right]$
   Range: $\{y \mid 1 \le y \le 5\}$ or $[1, 5]$

9. $y = x^2 - 3$

| $x$ | $y = x^2 - 3$ | $(x, y)$ |
|---|---|---|
| $-2$ | $y = (-2)^2 - 3 = 1$ | $(-2, 1)$ |
| $-1$ | $y = (-1)^2 - 3 = -2$ | $(-1, -2)$ |
| $0$ | $y = (0)^2 - 3 = -3$ | $(0, -3)$ |
| $1$ | $y = (1)^2 - 3 = -2$ | $(1, -2)$ |
| $2$ | $y = (2)^2 - 3 = 1$ | $(2, 1)$ |

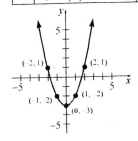

Domain: $\{x \mid x \text{ is a real number}\}$ or $(-\infty, \infty)$
Range: $\{y \mid y \ge -3\}$ or $[-3, \infty)$

10. Function. Each element in the domain corresponds to exactly one element in the range.
    Domain: $\{-5, -3, 0, 2\}$
    Range: $\{3, 7\}$

11. Not a function. The graph fails the vertical line test so it is not the graph of a function.
    Domain: $\{x \mid x \le 3\}$ or $(-\infty, 3]$
    Range: $\{y \mid y \text{ is a real number}\}$ or $(-\infty, \infty)$

12. No, $y = \pm 5x$ is not a function because a single input, $x$, can yield two different outputs. For example, if $x = 1$ then $y = -5$ or $y = 5$.

13. $f(x + h) = -3(x + h) + 11 = -3x - 3h + 11$

14. a. $g(-2) = 2(-2)^2 + (-2) - 1 = 2(4) - 3$
       $= 8 - 3 = 5$

## Chapter 2: Graphs, Relations, and Functions

b. $g(0) = 2(0)^2 + (0) - 1 = 0 + 0 - 1 = -1$

c. $g(3) = 2(3)^2 + (3) - 1 = 2(9) + 2$
$= 18 + 2 = 20$

15. $f(x) = x^2 + 3$

| $x$ | $y = f(x) = x^2 + 3$ | $(x, y)$ |
|---|---|---|
| $-2$ | $f(-2) = (-2)^2 + 3 = 7$ | $(-2, 7)$ |
| $-1$ | $f(-1) = (-1)^2 + 3 = 4$ | $(-1, 4)$ |
| $0$ | $f(0) = (0)^2 + 3 = 3$ | $(0, 0)$ |
| $1$ | $f(1) = (1)^2 + 3 = 4$ | $(1, 4)$ |
| $2$ | $f(2) = (2)^2 + 3 = 7$ | $(2, 7)$ |

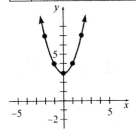

16. a. The dependent variable is the ticket price, $P$, and the independent variable is the number of years after 1989, $x$.

b. $P(15) = 0.13(15) + 3.76 = 5.71$
According to the model, the average ticket price in 2004 ($x = 15$) was $5.71.

17. a. The dependent variable is the number of registered climbers, $N$, and the independent variable is the number of years after 1960, $x$.

b. $N(43) = 271.4(43) + 836.83 \approx 12,507$
According to the model, there were about 12,507 registered climbers on Mt. Rainier in 2003 ($x = 43$).

c. Answers may vary. The value obtained in part (b) is much higher than the actual number. This illustrates the dangers of predicting outside the scope of given data. Beginning in 2001, Mt. Rainier began seeing a decline in the number of registered climbers which continued at least through 2003.

18. The function tells us to divide $-15$ by $x + 2$. Since we can't divide by zero, we need $x \neq -2$. That is, we need $x < -2$ or $x > -2$.
Domain: $\{x \mid x \neq -2\}$ or the interval $(-\infty, -2) \cup (-2, \infty)$.

19. a. $h(2) = -5(2) + 12$
$= -10 + 12$
$= 2$
Since $h(2) = 2$, the point $(2, 2)$ is on the graph of the function.

b. $h(3) = -5(3) + 12$
$= -15 + 12$
$= -3$
Since $h(3) = -3$, the point $(3, -3)$ is on the graph of the function.

c. $h(x) = 0$
$-5x + 12 = 0$
$-5x = -12$
$x = \dfrac{12}{5}$
The point $\left(\dfrac{12}{5}, 0\right)$, or $(2.4, 0)$, is on the graph of $h$.

20. a. The car stops accelerating when the speed stops increasing. Thus, the car stops accelerating after 6 seconds.

b. The car has a constant speed when the graph is horizontal. Thus, the car maintains a constant speed for 18 seconds.

# Chapter 3

## 3.1 Preparing for Linear Equations and Linear Functions

1.  $3x + 12 = 0$
    $3x + 12 - 12 = 0 - 12$
    $3x = -12$
    $\dfrac{3x}{3} = \dfrac{-12}{3}$
    $x = -4$
    The solution set is $\{-4\}$.

2.  $-2x + y = -4$
    Let $x = 0, 1, 2,$ and $3$.
    $x = 0:\ -2(0) + y = -4$
    $\qquad\qquad 0 + y = -4$
    $\qquad\qquad y = -4$
    $x = 1:\ -2(1) + y = -4$
    $\qquad\qquad -2 + y = -4$
    $\qquad\qquad y = -2$
    $x = 2:\ -2(2) + y = -4$
    $\qquad\qquad -4 + y = -4$
    $\qquad\qquad y = 0$
    $x = 3:\ -2(3) + y = -4$
    $\qquad\qquad -6 + y = -4$
    $\qquad\qquad y = 2$
    Thus, the points $(0,-4)$, $(1,-2)$, $(2,0)$, and $(3,2)$ are on the graph.

    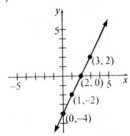

3.  Yes, the equation $-2x + y = -4$ is a function. The graph passes the vertical line test.

4.  $f(3) = 2(3)^2 - 5 = 2(9) - 5 = 18 - 5 = 13$

5.  The domain is $\{x \mid -5 \leq x \leq 5\}$ or, using interval notation, $[-5, 5]$. The range is $\{y \mid -2 \leq y \leq 5\}$ or, using interval notation, $[-2, 5]$.

6.  $f(3) = 5$ means that $y = 5$ when $x = 3$. Thus, the point $(3, 5)$ is on the graph.

## 3.1 Exercises

2. line

4. FALSE. Linear equations of the form $x = a$ (i.e., vertical lines) are not functions.

6. FALSE. A horizontal line is an equation of the form $y = b$ where $b$ is the $y$-intercept of the graph of the equation.

8. No. Vertical lines are the only lines that can have an $x$-intercept, but no $y$-intercept. Vertical lines are not functions.

10. $x + y = -3$
    Let $x = -2, -1, 0,$ and $1$.
    $x = -2:\ -2 + y = -3 \qquad x = -1:\ -1 + y = -3$
    $\qquad\qquad y = -1 \qquad\qquad\qquad\qquad y = -2$
    $x = 0:\ 0 + y = -3 \qquad\ x = 1:\ 1 + y = -3$
    $\qquad\qquad y = -3 \qquad\qquad\qquad\qquad y = -4$
    Thus, the points $(-2,-1)$, $(-1,-2)$, $(0,-3)$, and $(1,-4)$ are on the graph.

    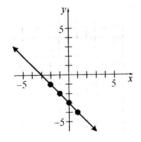

## Chapter 3: Linear Functions and Their Graphs

**12.** $2x - y = -8$

Let $x = -2, -1, 0,$ and $1$.

$x = -2$: $2(-2) - y = -8$
$-4 - y = -8$
$-y = -4$
$y = 4$

$x = -1$: $2(-1) - y = -8$
$-2 - y = -8$
$-y = -6$
$y = 6$

$x = 0$: $2(0) - y = -8$
$-y = -8$
$y = 8$

$x = 1$: $2(1) - y = -8$
$2 - y = -8$
$-y = -10$
$y = 10$

Thus, the points $(-2, 4)$, $(-1, 6)$, $(0, 8)$, and $(1, 10)$ are on the graph.

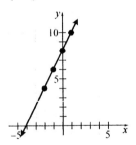

**14.** $-5x + y = 10$

Let $x = -2, -1, 0,$ and $1$.

$x = -3$: $-5(-3) + y = 10$
$15 + y = 10$
$y = -5$

$x = -2$: $-5(-2) + y = 10$
$10 + y = 10$
$y = 0$

$x = -1$: $-5(-1) + y = 10$
$5 + y = 10$
$y = 5$

$x = 0$: $-5(0) + y = 10$
$y = 10$

Thus, the points $(-3, -5)$, $(-2, 0)$, $(-1, 5)$, and $(0, 10)$ are on the graph.

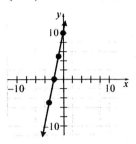

**16.** $2x - \dfrac{3}{2}y = 10$

Let $x = -1, 2, 5,$ and $8$.

$x = -1$: $2(-1) - \dfrac{3}{2}y = 10$
$-2 - \dfrac{3}{2}y = 10$
$-\dfrac{3}{2}y = 12$
$y = -8$

$x = 2$: $2(2) - \dfrac{3}{2}y = 10$
$4 - \dfrac{3}{2}y = 10$
$-\dfrac{3}{2}y = 6$
$y = -4$

$x = 5$: $2(5) - \dfrac{3}{2}y = 10$
$10 - \dfrac{3}{2}y = 10$
$-\dfrac{3}{2}y = 0$
$y = 0$

$x = 8$: $2(8) - \dfrac{3}{2}y = 10$
$16 - \dfrac{3}{2}y = 10$
$-\dfrac{3}{2}y = -6$
$y = 4$

Thus, the points $(-1, -8)$, $(2, -4)$, $(5, 0)$, and $(8, 4)$ are on the graph.

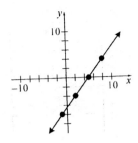

**18.** $-7x + 3y = 9$

Let $x = -3, 0, 3,$ and $6$.

$x = -6$: $-7(-6) + 3y = 9$
$42 + 3y = 9$
$3y = -33$
$y = -11$

$x = -3$: $-7(-3) + 3y = 9$
$21 + 3y = 9$
$3y = -12$
$y = -4$

$x = 0$: $-7(0) + 3y = 9$
$3y = 9$
$y = 3$

$x = 3$: $-7(3) + 3y = 9$
$-21 + 3y = 9$
$3y = 30$
$y = 10$

Thus, the points $(-6,-11)$, $(-3,-4)$, $(0,3)$, and $(3,10)$ are on the graph.

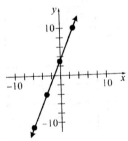

**20.** $x + y = -2$

To find the $y$-intercept, let $x = 0$ and solve for $y$.
$x = 0$: $0 + y = -2$
$y = -2$

The $y$-intercept is $-2$, so the point $(0,-2)$ is on the graph. To find the $x$-intercept, let $y = 0$ and solve for $x$.
$y = 0$: $x + 0 = -2$
$x = -2$

The $x$-intercept is $-2$, so the point $(-2,0)$ is on the graph.

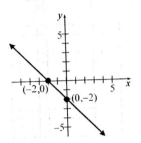

**22.** $-2x + y = 4$

To find the $y$-intercept, let $x = 0$ and solve for $y$.
$x = 0$: $-2(0) + y = 4$
$0 + y = 4$
$y = 4$

The $y$-intercept is 4, so the point $(0,4)$ is on the graph. To find the $x$-intercept, let $y = 0$ and solve for $x$.
$y = 0$: $-2x + 0 = 4$
$-2x = 4$
$x = -2$

The $x$-intercept is $-2$, so the point $(-2,0)$ is on the graph.

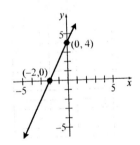

**24.** $-4x + 3y = 24$

To find the $y$-intercept, let $x = 0$ and solve for $y$.
$x = 0$: $-4(0) + 3y = 24$
$3y = 24$
$y = 8$

The $y$-intercept is 8, so the point $(0,8)$ is on the graph. To find the $x$-intercept, let $y = 0$ and solve for $x$.

$y=0$: $-4x+3(0)=24$
$-4x=24$
$x=-6$

The $x$-intercept is $-6$, so the point $(-6,0)$ is on the graph.

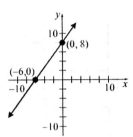

26. $7x-2y=10$

To find the $y$-intercept, let $x=0$ and solve for $y$.
$x=0$: $7(0)-2y=10$
$-2y=10$
$y=-5$

The $y$-intercept is $-5$, so the point $(0,-5)$ is on the graph. To find the $x$-intercept, let $y=0$ and solve for $x$.
$y=0$: $7x-2(0)=10$
$7x=10$
$x=\dfrac{10}{7}$

The $x$-intercept is $\dfrac{10}{7}$, so the point $\left(\dfrac{10}{7},0\right)$ is on the graph.

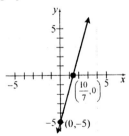

28. $\dfrac{1}{4}x+\dfrac{1}{5}y=2$

To find the $y$-intercept, let $x=0$ and solve for $y$.
$x=0$: $\dfrac{1}{4}(0)+\dfrac{1}{5}y=2$
$\dfrac{1}{5}y=2$
$y=10$

The $y$-intercept is 10, so the point $(0,10)$ is on the graph. To find the $x$-intercept, let $y=0$ and solve for $x$.
$y=0$: $\dfrac{1}{4}x+\dfrac{1}{5}(0)=2$
$\dfrac{1}{4}x=2$
$x=8$

The $x$-intercept is 8, so the point $(8,0)$ is on the graph.

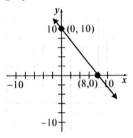

30. $4x+3y=0$

To find the $y$-intercept, let $x=0$ and solve for $y$.
$x=0$: $4(0)+3y=0$
$3y=0$
$y=0$

The $y$-intercept is 0, so the point $(0,0)$ is on the graph. To find the $x$-intercept, let $y=0$ and solve for $x$.
$y=0$: $4x+3(0)=0$
$4x=0$
$x=0$

The $x$-intercept is 0, so the point $(0,0)$ is on the graph. Since the $x$- and $y$-intercepts both result in the origin, we must find other points on the graph. Let $x=-3$ and 3, and solve for $y$.
$x=-3$: $4(-3)+3y=0$
$-12+3y=0$
$3y=12$
$y=4$
$x=3$: $4(3)+3y=0$
$12+3y=0$
$3y=-12$
$y=-4$

Thus, the points $(-3,4)$ and $(3,-4)$ are also on the graph.

### Section 3.1 Linear Equations and Linear Functions

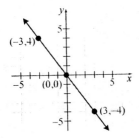

**32.** $5x - 3y = 0$

To find the y-intercept, let $x = 0$ and solve for y.

$x = 0$: $5(0) - 3y = 0$
$-3y = 0$
$y = 0$

The y-intercept is 0, so the point $(0,0)$ is on the graph. To find the x-intercept, let $y = 0$ and solve for x.

$y = 0$: $5x - 3(0) = 0$
$5x = 0$
$x = 0$

The x-intercept is 0, so the point $(0,0)$ is on the graph. Since the x- and y-intercepts both result in the origin, we must find other points on the graph. Let $x = -3$ and 3, and solve for y.

$x = -3$: $5(-3) - 3y = 0$
$-15 - 3y = 0$
$-3y = 15$
$y = -5$

$x = 3$: $5(3) - 3y = 0$
$15 - 3y = 0$
$-3y = -15$
$y = 5$

Thus, the points $(-3,-5)$ and $(3,5)$ are also on the graph.

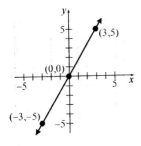

**34.** $-\dfrac{3}{2}x + \dfrac{3}{4}y = 0$

To find the y-intercept, let $x = 0$ and solve for y.

$x = 0$: $-\dfrac{3}{2}(0) + \dfrac{3}{4}y = 0$
$\dfrac{3}{4}y = 0$
$y = 0$

The y-intercept is 0, so the point $(0,0)$ is on the graph. To find the x-intercept, let $y = 0$ and solve for x.

$y = 0$: $-\dfrac{3}{2}x + \dfrac{3}{4}(0) = 0$
$-\dfrac{3}{2}x = 0$
$x = 0$

The x-intercept is 0, so the point $(0,0)$ is on the graph. Since the x- and y-intercepts both result in the origin, we must find other points on the graph. Let $x = -2$ and 2, and solve for y.

$x = -2$: $-\dfrac{3}{2}(-2) + \dfrac{3}{4}y = 0$
$3 + \dfrac{3}{4}y = 0$
$\dfrac{3}{4}y = -3$
$y = -4$

$x = 2$: $-\dfrac{3}{2}(2) + \dfrac{3}{4}y = 0$
$-3 + \dfrac{3}{4}y = 0$
$\dfrac{3}{4}y = 3$
$y = 4$

Thus, the points $(-2,-4)$ and $(2,4)$ are also on the graph.

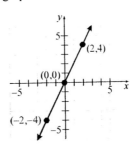

36. The equation $x = 5$ will be a vertical line with $x$-intercept 5.

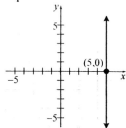

38. The equation $y = 6$ will be a horizontal line with $y$-intercept 6.

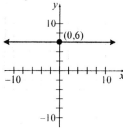

40. The equation $x = -8$ will be a vertical line with $x$-intercept $-8$.

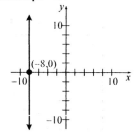

42. $3y + 20 = -10$
$3y = -30$
$y = -10$

The equation $y = -10$ will be a horizontal line with $y$-intercept $-10$.

44. **a.** The independent variable is total sales, $s$. It would not make sense for total sales to be negative. Thus, the domain of $I$ is $\{s \mid s \geq 0\}$ or, using interval notation, $[0, \infty)$.

**b.** $I(0) = 0.01(0) + 20{,}000 = 20{,}000$

If Tanya's total sales for the year are $0, her income will be $20,000. In other words, her base salary is $20,000.

**c.** Evaluate $I$ at $s = 500{,}000$.
$I(500{,}000) = 0.01(500{,}000) + 20{,}000$
$= 25{,}000$
If Tanya sells $500,000 in books for the year, her salary will be $25,000.

**d.** Evaluate $I$ at $m = 0$, 500000, and 1000000.
$I(0) = 0.01(0) + 20{,}000 = 20{,}000$
$I(500{,}000) = 0.01(500{,}000) + 20{,}000$
$= 25{,}000$
$I(1{,}000{,}000) = 0.01(1{,}000{,}000) + 20{,}000$
$= 30{,}000$
Thus, the points $(0, 20000)$, $(500000, 25000)$, and $(1000000, 30000)$ are on the graph.

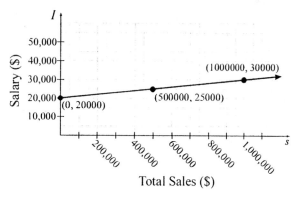

**e.** We must solve $I(s) = 45{,}000$.
$0.01s + 20{,}000 = 45{,}000$
$0.01s = 25{,}000$
$s = 2{,}500{,}000$
For Tanya's income to be $45,000, her total sales would have to be $2,500,000.

46. **a.** The independent variable is payroll, $p$. The payroll tax only applies if the payroll is $128 million or more. Thus, the domain of $T$ is $\{p \mid p \geq 128\}$ or, using interval notation, $[128, \infty)$.

**b.** Evaluate $T$ at $p = 160$ million.
$T(160) = 0.225(160 - 128) = 7.2$
The luxury tax whose payroll is $160 million would be $7.2 million.

**c.** Evaluate $T$ at $p = 128$ million, 200 million, and 300 million.
$T(128) = 0.225(128 - 128) = 0$
$T(200) = 0.225(200 - 128) = 16.2$ million

$T(300) = 0.225(300-128) = 38.7$ million

Thus, the points $(128 \text{ million}, 0)$, $(200 \text{ million}, 16.2 \text{ million})$, and $(300 \text{ million}, 38.7 \text{ million})$ are on the graph.

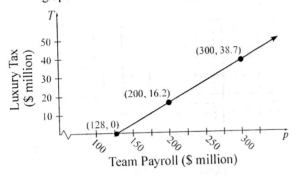
Team Payroll ($ million)

d. We must solve $T(p) = 11,700,000$.
$$0.225(p - 128,000,000) = 11,700,000$$
$$0.225p - 28,800,000 = 11,700,000$$
$$0.225p = 40,500,000$$
$$p = 180,000,000$$
For the luxury tax to be $11.7 million, the payroll of the team would be $180 million.

48. a. The independent variable is age, $a$. The dependent variable is the birth rate, $B$.

b. We are told in the problem that $a$ is restricted from 15 to 44, inclusive. Thus, the domain of $B$ is $\{a \mid 15 \leq a \leq 44\}$ or, using interval notation, $[15, 44]$.

c. Evaluate $B$ at $a = 22$.
$$B(22) = 1.73(22) - 14.56$$
$$= 23.5$$
The multiple birth rate of 22 year-old women is 23.5 births per 1000 women.

d. Evaluate $H$ at $a = 15$, 30, and 44.
$$B(15) = 1.73(22) - 14.56$$
$$= 11.39$$
$$B(30) = 1.73(30) - 14.56$$
$$= 37.34$$
$$B(44) = 1.73(44) - 14.56$$
$$= 61.56$$
Thus, the points $(15, 11.39)$, $(30, 37.34)$, and $(44, 61.56)$ are on the graph.

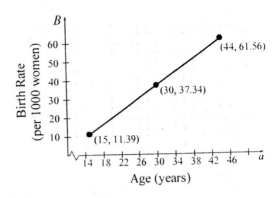

e. We must solve $B(a) = 49.45$.
$$1.73a - 14.56 = 49.45$$
$$1.73a = 64.01$$
$$a = 37$$
The age of women with a multiple birth rate of 49.45 is 37 years.

50. a. The point $(2, 1)$ is on the graph of $g$, so $g(2) = 1$. Thus, the solution of $g(x) = 1$ is $x = 2$.

b. The point $(6, -1)$ is on the graph of $g$, so $g(6) = -1$. Thus, the solution of $g(x) = -1$ is $x = 6$.

c. The point $(-4, 4)$ is on the graph of $g$, so $g(-4) = 4$. Thus, the solution of $g(x) = 4$ is $x = -4$.

d. The intercepts of $y = g(x)$ are $(0, 2)$ and $(4, 0)$. The $y$-intercept is 2 and the $x$-intercept is 4.

e. Recall that the $x$-intercept of a function is also called the zero of the function. Thus, the zero of $y = g(x)$ is 4.

## Chapter 3: Linear Functions and Their Graphs

**52.** The graphs of the three lines are shown below.

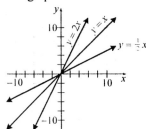

Answers will vary. All three lines all pass through the origin and slant upward from left to right. The graph of $y = x$ slants more steeply than the graph of $y = \frac{1}{2}x$. The graph of $y = 2x$ slants even more steeply than that of $y = x$.

The graph of $y = ax$, with $a > 0$, will be a line that passes through the origin and slants upward from left to right.

**54.** Fifteen percent of $C$ is equal to $0.15C$. Then,
$T = C + 0.15C$
$\phantom{T} = 1.15C$
Now, from Problem 45, $C = 1.5m + 2$. Thus,
$T = 1.15C$
$\phantom{T} = 1.15(1.5m + 2)$
$\phantom{T} = 1.725m + 2.3$
Using function notation, the linear function relating total cost, $T$, to the number of miles, $m$, is $T(m) = 1.725m + 2.3$.

**56.** $x + y = -2$
$y = -x - 2$

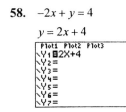

**58.** $-2x + y = 4$
$y = 2x + 4$

**60.** $-4x + 3y = 24$
$3y = 4x + 24$
$y = \dfrac{4x + 24}{3}$
$y = \dfrac{4}{3}x + 8$

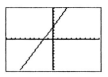

**62.** $7x - 2y = 10$
$-2y = -7x + 10$
$y = \dfrac{-7x + 10}{-2}$
$y = \dfrac{7}{2}x - 5$

**64.** $\dfrac{1}{4}x + \dfrac{1}{5}y = 2$
$\dfrac{1}{5}y = -\dfrac{1}{4}x + 2$
$y = 5\left(-\dfrac{1}{4}x + 2\right)$
$y = -\dfrac{5}{4}x + 10$

Section 3.1 Linear Equations and Linear Functions

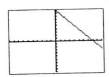

66. $4x + 3y = 0$

$3y = -4x$

$y = -\dfrac{4}{3}x$

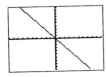

**3.2 Preparing for Slope and Equations of Lines**

1. $\dfrac{5-2}{-2-4} = \dfrac{3}{-6} = -\dfrac{1}{2}$

2. $-2(x+3) = -2 \cdot x + (-2) \cdot 3 = -2x - 6$

3. $\quad 4x + 3y = 15$
$-4x + 4x + 3y = -4x + 15$
$\quad 3y = -4x + 15$
$\quad \dfrac{3y}{3} = \dfrac{-4x + 15}{3}$
$\quad y = -\dfrac{4}{3}x + 5$

4. For $F(z) = 4z - 3$, the independent variable is $z$, and the dependent variable is $F$.

**3.2 Exercises**

2. $\dfrac{4}{10} = \dfrac{2}{5}$

4. FALSE. The slope of line $L$ would be
$m = \dfrac{y_2 - y_1}{x_2 - x_1}$, $x_1 \ne x_2$.

6. TRUE

8. Standard form: $Ax + By = C$
Slope-intercept form: $y = mx + b$
Point-slope form: $y - y_1 = m(x - x_1)$
Horizontal line: $y = b$
Vertical line: $x = a$

10. A horizontal line has one $y$-intercept, but no $x$-intercept.

12. No. Answers may vary. One possibility follows: Every line must have at least one intercept. Vertical lines have one $x$-intercept, but no $y$-intercept. Horizontal lines have one $y$-intercept, but no $x$-intercept. Lines that are neither vertical nor horizontal will have one $x$-intercept and one $y$-intercept.

14. a. $m = \dfrac{y_2 - y_1}{x_2 - x_1} = \dfrac{-2-0}{3-0} = \dfrac{-2}{3} = -\dfrac{2}{3}$

   b. For every 3 units of increase in $x$, $y$ will decrease by 2 units. For every 3 units of decrease in $x$, $y$ will increase by 2 units.

16. a. $m = \dfrac{y_2 - y_1}{x_2 - x_1} = \dfrac{2-(-3)}{4-(-2)} = \dfrac{5}{6}$

   b. For every 6 units of increase in $x$, $y$ will increase by 5 units.

18. $m = \dfrac{y_2 - y_1}{x_2 - x_1} = \dfrac{5-0}{-2-0} = \dfrac{5}{-2} = -\dfrac{5}{2}$

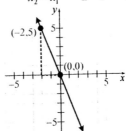

ISM − 107

## Chapter 3: Linear Functions and Their Graphs

20. $m = \dfrac{y_2 - y_1}{x_2 - x_1} = \dfrac{10-1}{7-4} = \dfrac{9}{3} = 3$

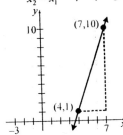

22. $m = \dfrac{y_2 - y_1}{x_2 - x_1} = \dfrac{11-(-1)}{-2-3} = \dfrac{12}{-5} = -\dfrac{12}{5}$

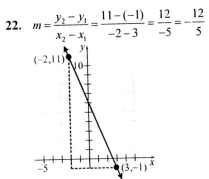

24. $m = \dfrac{y_2 - y_1}{x_2 - x_1} = \dfrac{3-(-4)}{-1-1} = \dfrac{7}{-2} = -\dfrac{7}{2}$

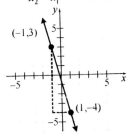

26. $m = \dfrac{y_2 - y_1}{x_2 - x_1} = \dfrac{1-1}{2-(-3)} = \dfrac{0}{5} = 0$

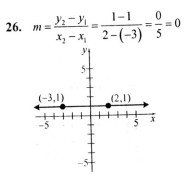

28. $m = \dfrac{y_2 - y_1}{x_2 - x_1} = \dfrac{-3-1}{4-4} = \dfrac{-4}{0} =$ undefined

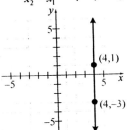

30. $m = \dfrac{y_2 - y_1}{x_2 - x_1} = \dfrac{\frac{13}{4} - \frac{5}{2}}{\frac{13}{9} - \frac{7}{3}} = \dfrac{\frac{13}{4} - \frac{10}{4}}{\frac{13}{9} - \frac{21}{9}} = \dfrac{\frac{3}{4}}{-\frac{8}{9}} = \dfrac{3}{4}\left(-\dfrac{9}{8}\right) = -\dfrac{27}{32}$

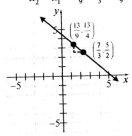

32. $m = \dfrac{\frac{3}{7} - \left(-\frac{15}{4}\right)}{-\frac{12}{5} - \frac{1}{3}} = \dfrac{\frac{12}{28} + \frac{105}{28}}{-\frac{36}{15} - \frac{5}{15}} = \dfrac{\frac{117}{28}}{-\frac{41}{15}} = \dfrac{117}{28}\left(-\dfrac{15}{41}\right) = -\dfrac{1755}{1148}$

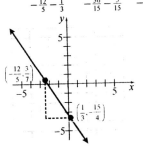

34. Start at $(-1, 4)$. Because $m = 2 = \dfrac{2}{1}$, move 1 unit to the right and 2 units up to find point $(0, 6)$.

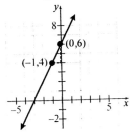

36. Start at $(-1, 5)$. Because $m = -4 = -\dfrac{4}{1}$, move 1 unit to the right and 4 units down to find

## Section 3.2 Slope and Equations of Lines

point $(0,1)$.

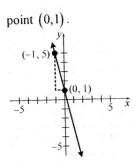

38. Start at $(-2,-5)$. Because $m = \dfrac{4}{3}$, move 3 units to the right and 4 units up to find point $(1,-1)$.

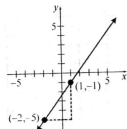

40. Start at $(3,3)$. Because $m = -\dfrac{1}{2}$, move 2 unit to the right and 1 unit down to find point $(5,2)$. We can also move 2 units to the left and 1 unit up to find point $(1,4)$.

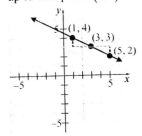

42. Because $m$ is undefined, the line is vertical passing through the point $(-5,2)$.

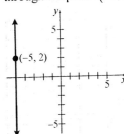

44. Answers will vary. One possible answer follows. We know $m = -2 = -\dfrac{2}{1}$. Beginning at the point

$(1,4)$, and using a change in $x$ of 1 and a change of $y$ of $-2$, $1+1=2$ and $4+(-2)=2$, so $(2,2)$ is on the line. Next, beginning at $(2,2)$, $2+1=3$ and $2+(-2)=0$, so $(3,0)$ is on the line. Now, beginning at $(3,0)$, $3+1=4$ and $0+(-2)=-2$, so $(4,-2)$ is on the line. In summary, the points $(2,2)$, $(3,0)$, and $(4,-2)$ will all be on the line.

46. Answers will vary. One possible answer follows. We know $m = -\dfrac{2}{3}$. Beginning at the point $(1,-3)$, and using a change in $x$ of $-3$ and a change of $y$ of 2, $1+(-3)=-2$ and $-3+2=-1$, so $(-2,-1)$ is on the line. Next, beginning at $(-2,-1)$, $-2+(-3)=-5$ and $-1+2=1$, so $(-5,1)$ is on the line. Now, beginning at $(-5,1)$, $-5+(-3)=-8$ and $1+2=3$, so $(-8,3)$ is on the line. In summary, the points $(-2,-1)$, $(-5,1)$, and $(-8,3)$ will all be on the line.

48. $m = \dfrac{y_2 - y_1}{x_2 - x_1} = \dfrac{-1-1}{5-(-5)} = \dfrac{-2}{10} = -\dfrac{1}{5}$

$y - y_1 = m(x - x_1)$

$y - 1 = -\dfrac{1}{5}(x - (-5))$

$y - 1 = -\dfrac{1}{5}(x + 5)$

$y - 1 = -\dfrac{1}{5}x - 1$

$y = -\dfrac{1}{5}x$

50. $m = \dfrac{y_2 - y_1}{x_2 - x_1} = \dfrac{4-0}{4-(-2)} = \dfrac{4}{6} = \dfrac{2}{3}$

$y - y_1 = m(x - x_1)$

$y - 0 = \dfrac{2}{3}(x - (-2))$

$y = \dfrac{2}{3}(x + 2)$

$y = \dfrac{2}{3}x + \dfrac{4}{3}$

**Chapter 3:** *Linear Functions and Their Graphs*

52. $m = \dfrac{y_2 - y_1}{x_2 - x_1} = \dfrac{4-(-2)}{2-2} = \dfrac{6}{0} =$ undefined

    So, the line is vertical. The equation of the line is $x = 2$.

54. $y - y_1 = m(x - x_1)$
    $y - 0 = -1(x - 0)$
    $y = -x$

56. $y - y_1 = m(x - x_1)$
    $y - (-1) = 4(x - 2)$
    $y + 1 = 4x - 8$
    $y = 4x - 9$

58. $y - y_1 = m(x - x_1)$
    $y - 1 = \dfrac{1}{2}(x - 2)$
    $y - 1 = \dfrac{1}{2}x - 1$
    $y = \dfrac{1}{2}x$

60. $y - y_1 = m(x - x_1)$
    $y - (-3) = -\dfrac{4}{3}(x - 1)$
    $y + 3 = -\dfrac{4}{3}x + \dfrac{4}{3}$
    $y = -\dfrac{4}{3}x - \dfrac{5}{3}$

62. $y - y_1 = m(x - x_1)$
    $y - (-2) = 0(x - 3)$
    $y + 2 = 0$
    $y = -2$

64. $m = \dfrac{y_2 - y_1}{x_2 - x_1} = \dfrac{-3 - 0}{4 - 0} = -\dfrac{3}{4}$
    $y - y_1 = m(x - x_1)$
    $y - 0 = -\dfrac{3}{4}(x - 0)$
    $y = -\dfrac{3}{4}x$

66. $m = \dfrac{y_2 - y_1}{x_2 - x_1} = \dfrac{7 - 3}{3 - 1} = \dfrac{4}{2} = 2$
    $y - y_1 = m(x - x_1)$
    $y - 3 = 2(x - 1)$
    $y - 3 = 2x - 2$
    $y = 2x + 1$

68. $m = \dfrac{y_2 - y_1}{x_2 - x_1} = \dfrac{6 - 1}{1 - (-3)} = \dfrac{5}{4}$
    $y - y_1 = m(x - x_1)$
    $y - 1 = \dfrac{5}{4}(x - (-3))$
    $y - 1 = \dfrac{5}{4}x + \dfrac{15}{4}$
    $y = \dfrac{5}{4}x + \dfrac{19}{4}$

70. $m = \dfrac{y_2 - y_1}{x_2 - x_1} = \dfrac{-4 - (-4)}{1 - (-3)} = \dfrac{0}{4} = 0$
    $y - y_1 = m(x - x_1)$
    $y - (-4) = 0(x - (-3))$
    $y + 4 = 0$
    $y = -4$

72. $m = \dfrac{y_2 - y_1}{x_2 - x_1} = \dfrac{-1 - 1}{1 - (-5)} = \dfrac{-2}{6} = -\dfrac{1}{3}$
    $y - y_1 = m(x - x_1)$
    $y - 1 = -\dfrac{1}{3}(x - (-5))$
    $y - 1 = -\dfrac{1}{3}(x + 5)$
    $y - 1 = -\dfrac{1}{3}x - \dfrac{5}{3}$
    $y = -\dfrac{1}{3}x - \dfrac{2}{3}$

73. $m = \dfrac{y_2 - y_1}{x_2 - x_1} = \dfrac{4 - 4}{-4 - 2} = \dfrac{0}{-6} = 0$
    $y - y_1 = m(x - x_1)$
    $y - 4 = 0(x - 2)$
    $y - 4 = 0$
    $y = 4$

74. $m = \dfrac{y_2 - y_1}{x_2 - x_1} = \dfrac{-4-1}{3-3} = \dfrac{-5}{0} =$ undefined

So, the line is vertical. The equation of the line is $x = 3$.

76. $m = \dfrac{y_2 - y_1}{x_2 - x_1} = \dfrac{\frac{11}{4} - \frac{9}{2}}{-\frac{4}{9} - \frac{2}{3}} = \dfrac{-\frac{7}{4}}{-\frac{10}{9}} = \left(-\dfrac{7}{4}\right)\left(-\dfrac{9}{10}\right) = \dfrac{63}{40}$

$y - y_1 = m(x - x_1)$

$y - \dfrac{9}{2} = \dfrac{63}{40}\left(x - \dfrac{2}{3}\right)$

$y - \dfrac{9}{2} = \dfrac{63}{40}x - \dfrac{21}{20}$

$y = \dfrac{63}{40}x + \dfrac{69}{20}$

78. For $y = 3x + 2$, the slope is 3 and the $y$-intercept is 2. Begin at $(0,2)$ and move to the right 1 unit and up 3 units to find point $(1,5)$.

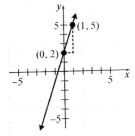

80. For $y = -7x$, the slope is $-7$ and the $y$-intercept is 0. Begin at $(0,0)$ and move to the right 1 unit and down 7 units to find point $(1,-7)$. We can also move 1 unit to the left and 7 units up to find point $(-1, 7)$.

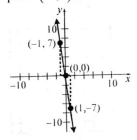

82. $-3x + y = 1$

$y = 3x + 1$

The slope is 3 and the $y$-intercept is 1. Begin at $(0,1)$ and move to the right 1 unit and up 3 units to find the point $(1,4)$.

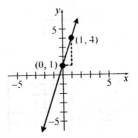

84. $3x + 6y = 12$

$6y = -3x + 12$

$y = \dfrac{-3x + 12}{6}$

$y = -\dfrac{1}{2}x + 2$

The slope is $-\dfrac{1}{2}$ and the $y$-intercept is 2. Begin at $(0,2)$ and move to the right 2 units and down 1 unit to find the point $(2,1)$. We can also move 2 units to the left and 1 unit up to find the point $(-2, 3)$.

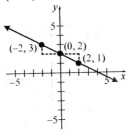

86. $2x - 5y - 10 = 0$

$-5y = -2x + 10$

$y = \dfrac{-2x + 10}{-5}$

$y = \dfrac{2}{5}x - 2$

The slope is $\dfrac{2}{5}$ and the $y$-intercept is $-2$. Begin at $(0,-2)$ and move to the right 5 units and up 2 units to find the point $(5,0)$.

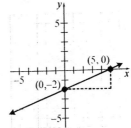

**88.** For $y = -4$, the slope is 0 and the $y$-intercept is $-4$. It is a horizontal line.

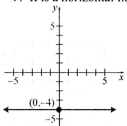

**90.** For $g(x) = -6x + 2$, the slope is $-6$ and the $y$-intercept is 2. Begin at $(0,2)$ and move to the right 1 unit and down 6 units to find the point $(1,-4)$. We can also move 1 unit to the left and 3 units up to find the point $(-1,8)$.

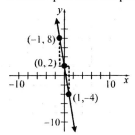

**92.** For $G(x) = \frac{1}{2}x - 5$, the slope is $\frac{1}{2}$ and the $y$-intercept is $-5$. Begin at $(0,-5)$ and move to the right 2 units and up 1 unit to find the point $(2,-4)$.

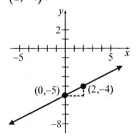

**94.** The $y$-axis is a vertical line passing through the origin. Thus, the slope is undefined and the $x$-intercept is 0. So, the equation of the $y$-axis is $x = 0$.

**96.** Since $g(1) = 5$ and $g(5) = 17$, the points $(1, 5)$ and $(5, 17)$ are on the graph of $g$. Thus,
$$m = \frac{y_2 - y_1}{x_2 - x_1} = \frac{17 - 5}{5 - 1} = \frac{12}{4} = 3.$$

$$y - y_1 = m(x - x_1)$$
$$y - 5 = 3(x - 1)$$
$$y - 5 = 3x - 3$$
$$y = 3x + 2 \text{ or } g(x) = 3x + 2$$

Finally, $g(-3) = 3(-3) + 2 = -9 + 2 = -7$.

**98.** Since $F(2) = 5$ and $F(-3) = 9$, the points $(2, 5)$ and $(-3, 9)$ are on the graph of $F$. Thus,
$$m = \frac{y_2 - y_1}{x_2 - x_1} = \frac{9 - 5}{-3 - 2} = \frac{4}{-5} = -\frac{4}{5}.$$
$$y - y_1 = m(x - x_1)$$
$$y - 5 = -\frac{4}{5}(x - 2)$$
$$y - 5 = -\frac{4}{5}x + \frac{8}{5}$$
$$y = -\frac{4}{5}x + \frac{33}{5} \text{ or } F(x) = -\frac{4}{5}x + \frac{33}{5}$$

Finally, $F\left(-\frac{3}{2}\right) = -\frac{4}{5}\left(-\frac{3}{2}\right) + \frac{33}{5} = \frac{6}{5} + \frac{33}{5} = \frac{39}{5}$.

**100. a.**

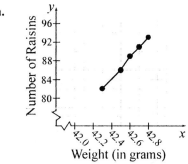

**b.** Average rate of change $= \frac{86 - 82}{42.5 - 42.3}$
= 20 raisins per gram
Between weights of 42.3 and 42.5 grams, the number of raisins is increasing at a rate of 20 raisins per gram.

**c.** Average rate of change $= \frac{93 - 91}{42.8 - 42.7}$
= 20 raisins per gram
Between weights of 42.7 and 42.8 grams, the number of raisins is increasing at a rate of 20 raisins per gram.

**d.** No, the number of raisins is not *perfectly* linearly related to weight. The average rate of change (slope) is not constant. However, it is close.

**102. a.**

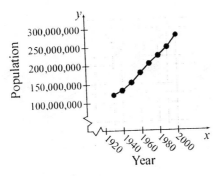

**b.** Average rate of change
$$= \frac{132,164,569 - 123,202,624}{1940 - 1930}$$
$$= 896,194.5 \text{ people per year}$$
Between 1930 and 1940, the population increased a rate of 896,194.5 people per year.

**c.** Average rate of change
$$= \frac{281,421,906 - 248,709,873}{2000 - 1990}$$
$$= 3,271,203.3 \text{ people per year}$$
Between 1990 and 2000, the population increased a rate of 3,271,203.3 people per year.

**d.** No. The population is not linearly related to year. The average rate of change (slope) is not constant.

**104. a.** Let $x$ represent the number pounds of fish caught (in millions). Let $P$ represent the price per pound (in cents).
$$m = \frac{39.1 - 33.2}{9,069 - 10,467}$$
$$\approx -0.00422 \text{ cents per million pounds}$$
$$P - 33.2 = -0.00422(x - 10,467)$$
$$P - 33.2 = -0.00422x + 44.171$$
$$P = -0.00422x + 77.371$$
Using function notation,
$P(x) = -0.00422x + 77.371$.

**b.** $P(9,830) = -0.00422(9,830) + 77.371$
$$\approx 35.9$$
If 9,830 millions pounds of fish are caught, the price per pound of fish caught will be approximately 35.9 cents.

**c.** The slope ($-0.00422$) indicates that the price per pound decreases at a rate of 0.00422 cents per pound per million pounds caught.

**d.** $-0.00422x + 77.371 = 38.4$
$$-0.00422x = -38.971$$
$$x = \frac{-38.971}{-0.00422} \approx 9235$$
If the price per pound is 38.4 cents, then approximately 9,235 million pounds of fish were caught.

**106. a.** Let $x$ represent the area of the North Chicago apartment and $R$ represent the rent.
$$m = \frac{1660 - 1507}{970 - 820} = \$1.02 \text{ per square foot}$$
$$R - 1507 = 1.02(x - 820)$$
$$R - 1507 = 1.02x - 836.4$$
$$R = 1.02x + 670.6$$
Using function notation, $R(x) = 1.02x + 670.6$.

**b.** $R(900) = 1.02(900) + 670.6$
$$= 1588.6$$
The rent for a 900 square foot apartment in North Chicago would be $1588.60.

**c.** The slope indicates that the rent of a North Chicago apartment increases at a rate of $1.02 per square foot.

**d.** $1.02x + 670.6 = 1300$
$$1.02x = 629.4$$
$$x = \frac{629.4}{1.02} \approx 617$$
If the rent is $1300, then the area of the apartment would be approximately 617 square feet.

**108.** Let $F$ represent the Fahrenheit temperature and $C$ represent the Celsius temperature.
$$m = \frac{100 - 0}{212 - 32} = \frac{100}{180} = \frac{5}{9}$$
$$C - 0 = \frac{5}{9}(F - 32)$$
$$C = \frac{5}{9}(F - 32)$$
Finally, $C(60) = \frac{5}{9}(60 - 32) \approx 15.6$

Thus, $60°F$ is equivalent to $15.6°C$.

## Chapter 3: Linear Functions and Their Graphs

110. The graph shown has both a negative slope and a negative *y*-intercept. Two of the given equations meet both of these conditions: those in parts (c) and (d), $y = -\frac{2}{3}x - 3$ and $4x + 3y = -5$.

112. Graph the three equations $-3x + y = 1$, $0x + y = 1$, and $3x + y = 1$ on the same Cartesian plane.

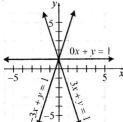

    Answers may vary. The graphs of the members of the family $Cx + y = 1$ will all have a *y*-intercept of 1.

114. For every 100 units of horizontal change in the road, the vertical change must be 15 units or less.

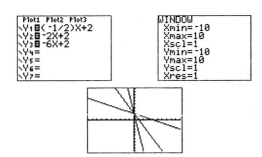

    Answers may vary. For $m < 0$, the graph will slant downward from left to right. The slant will be steeper as $|m|$ becomes larger.

116. The settings and graphs are shown below.

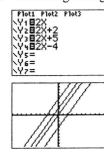

Answers may vary. For each graph of $y = 2x + b$, the slope is be 2 and the *y*-intercept is *b*. Thus, the graphs are parallel.

### 3.3 Preparing for Parallel and Perpendicular Lines

1. The reciprocal of 3 is $\frac{1}{3}$ because $3 \cdot \frac{1}{3} = 1$.

2. The reciprocal of $-\frac{3}{5}$ is $-\frac{5}{3}$ because $\left(-\frac{3}{5}\right)\left(-\frac{5}{3}\right) = 1$.

### 3.3 Exercises

2. $-1$

4. TRUE. The two lines have the same slope (3) and different *y*-intercepts ($-1$ and 1).

6. The slope of horizontal line is 0, which has no reciprocal. Likewise, the slope of a vertical line is undefined, so it has no reciprocal.

8. a. Slope of parallel line: $m = -8/5$.
   b. Slope of perpendicular line: $m = 5/8$.

10. a. Slope of parallel line: $m = 0$.
    b. Slope of perpendicular line: $m$ is undefined.

12. Perpendicular. The two slopes, $m_1 = 3$ and $m_2 = -1/3$, are negative reciprocals.

14.  $-3x - y = 3$         $6x + 2y = 9$
     $-y = 3x + 3$          $2y = -6x + 9$
     $y = \frac{3x+3}{-1}$   $y = \frac{-6x+9}{2}$
     $y = -3x - 3$           $y = -3x + \frac{9}{2}$

    Parallel. The two lines have equal slopes, $m = -3$, but different *y*-intercepts.

ISM –114

16. 
$$10x - 3y = 5$$
$$-3y = -10x + 5$$
$$y = \frac{-10x+5}{-3}$$
$$y = \frac{10}{3}x - \frac{5}{3}$$

$$5x + 6y = 3$$
$$6y = -5x + 3$$
$$y = \frac{-5x+3}{6}$$
$$y = -\frac{5}{6}x + \frac{1}{2}$$

Neither. The two slopes, $m_1 = -10/3$ and $m_2 = -5/6$, are not equal and are not negative reciprocals.

18.
$$\frac{1}{2}x - \frac{3}{2}y = 3$$
$$-\frac{3}{2}y = -\frac{1}{2}x + 3$$
$$-\frac{2}{3}\left(-\frac{3}{2}y\right) = -\frac{2}{3}\left(-\frac{1}{2}x+3\right)$$
$$y = \frac{1}{3}x - 2$$

$$2x + \frac{2}{3}y = 1$$
$$\frac{2}{3}y = -2x + 1$$
$$\frac{3}{2}\left(\frac{2}{3}y\right) = \frac{3}{2}(-2x+1)$$
$$y = -3x + \frac{3}{2}$$

Perpendicular. The two slopes, $m_1 = 1/3$ and $m_2 = -3$, are negative reciprocals.

20. $m_1 = \dfrac{3-1}{4-1} = \dfrac{2}{3}$; $m_2 = \dfrac{-3-3}{3-(-1)} = \dfrac{-6}{4} = -\dfrac{3}{2}$

Perpendicular. The two slopes are negative reciprocals.

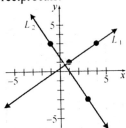

22. $m_1 = \dfrac{2-0}{0-(-3)} = \dfrac{2}{3}$; $m_2 = \dfrac{5-8}{2-4} = \dfrac{-3}{-2} = \dfrac{3}{2}$

Neither. The two slopes are not equal and are not negative reciprocals.

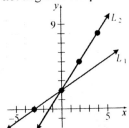

24. $m_1 = \dfrac{-4-(-3)}{5-1} = \dfrac{-1}{4} = -\dfrac{1}{4}$; $m_2 = \dfrac{2-4}{8-0} = \dfrac{-2}{8} = -\dfrac{1}{4}$

Parallel. The two lines have equal slopes, but different y-intercepts.

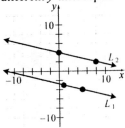

26. Since line $L$ is parallel to $y = -\dfrac{1}{2}x + 1$, the slope of $L$ is $m = -1/2$. From the graph, the y-intercept of $L$ is $-1$. Thus, the equation of line $L$ is $y = -\dfrac{1}{2}x - 1$.

28. Since line $L$ is perpendicular to $y = \dfrac{2}{3}x + 1$, the slope of $L$ is $m = -3/2$, the negative reciprocal of $2/3$. From the graph the y-intercept is $-2$. Thus, the equation of line $L$ is $y = -\dfrac{3}{2}x - 2$.

30. Since line $L$ is perpendicular to $y = 3$ and since $y = 3$ is a horizontal line, then $L$ must be a vertical line. From the graph, the x-intercept of $L$ is 1. Thus, the equation of line $L$ is $x = 1$.

32. The slope of the line we seek is $m = -3$, the same as the slope of $y = -3x + 1$. The equation of the line is
$$y - y_1 = m(x - x_1)$$
$$y - 5 = -3(x - 2)$$
$$y - 5 = -3x + 6$$
$$y = -3x + 11$$

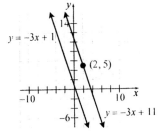

34. The slope of the line we seek is $m = -1/4$, the negative reciprocal of the slope of $y = 4x + 3$. The equation of the line is
$$y - y_1 = m(x - x_1)$$
$$y - 1 = -\frac{1}{4}(x - 4)$$
$$y - 1 = -\frac{1}{4}x + 1$$
$$y = -\frac{1}{4}x + 2$$

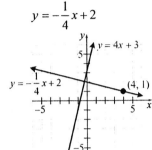

36. The line we seek is vertical, parallel to the vertical line $x = -2$. Now a vertical line passing through the point $(2, 5)$ must also pass through the point $(2, 0)$. That is, the $x$-intercept is 2. Thus, the equation of the line is $x = 2$.

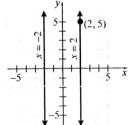

38. The line we seek is vertical, perpendicular to the horizontal line $y = 8$. Now a vertical line passing through the point $(2, -4)$ must also pass through the point $(2, 0)$. That is, the $x$-intercept is 2. Thus, the equation of the line is $x = 2$.

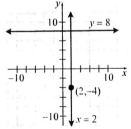

40. The slope of the line we seek is $m = -2$, the same as the slope of the line
$$2x + y = 5$$
$$y = -2x + 5$$
Thus, the equation of the line we seek is
$$y - y_1 = m(x - x_1)$$
$$y - 3 = -2(x - (-4))$$
$$y - 3 = -2(x + 4)$$
$$y - 3 = -2x - 8$$
$$y = -2x - 5$$

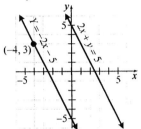

42. The slope of the line we seek is $m = -5/2$, the negative reciprocal of the slope of the line
$$-2x + 5y - 3 = 0$$
$$5y = 2x + 3$$
$$y = \frac{2x + 3}{5}$$
$$y = \frac{2}{5}x + \frac{3}{5}$$
Thus, the equation of the line we seek is
$$y - y_1 = m(x - x_1)$$
$$y - (-3) = -\frac{5}{2}(x - 2)$$
$$y + 3 = -\frac{5}{2}x + 5$$
$$y = -\frac{5}{2}x + 2$$

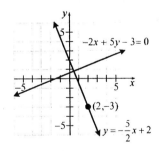

44. The slope of the line we seek is $m = 1/3$, the same as the slope of the line

$3x + y = 1$
$y = -3x + 1$

Thus, the equation of the line we seek is
$y - y_1 = m(x - x_1)$
$y - (-1) = \frac{1}{3}(x - 3)$
$y + 1 = \frac{1}{3}x - 1$
$y = \frac{1}{3}x - 2$

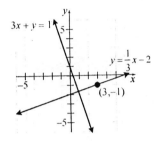

46. a. $A = (-2, -2)$, $B = (3, 1)$, $C = (-5, 3)$.

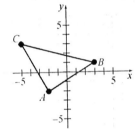

b. Slope of $\overline{AB} = \frac{1-(-2)}{3-(-2)} = \frac{3}{5}$

Slope of $\overline{AC} = \frac{3-(-2)}{-5-(-2)} = \frac{5}{-3} = -\frac{5}{3}$

Because the slopes are negative reciprocals, segments $\overline{AB}$ and $\overline{AC}$ are perpendicular. Thus, triangle $ABC$ is a right triangle.

48. a. $A = (-2, -1)$, $B = (4, 1)$, $C = (5, 5)$, $D = (-1, 3)$.

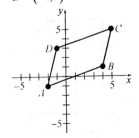

b. Slope of $\overline{AB} = \frac{1-(-1)}{4-(-2)} = \frac{2}{6} = \frac{1}{3}$

Slope of $\overline{BC} = \frac{5-1}{5-4} = \frac{4}{1} = 4$

Slope of $\overline{CD} = \frac{3-5}{-1-5} = \frac{-2}{-6} = \frac{1}{3}$

Slope of $\overline{DA} = \frac{-1-3}{-2-(-1)} = \frac{-4}{-1} = 4$

Because the slopes of $\overline{AB}$ and $\overline{CD}$ are equal, $\overline{AB}$ and $\overline{CD}$ are parallel. Because the slopes of $\overline{BC}$ and $\overline{DA}$ are equal, $\overline{BC}$ and $\overline{DA}$ are parallel. Thus, quadrilateral $ABCD$ is a parallelogram.

50. The line
$2x - 3y = 8$
$-3y = -2x + 8$
$y = \frac{2}{3}x - \frac{8}{3}$

has a slope of $2/3$. The line
$-6x + By = 3$
$By = 6x + 3$
$y = \frac{6x + 3}{B}$
$y = \frac{6}{B}x + \frac{3}{B}$

has a slope of $6/B$. For the two lines to be perpendicular, the product of their slopes must be $-1$. Use this fact to determine the value of $B$:
$\frac{2}{3} \cdot \left(\frac{6}{B}\right) = -1$
$\frac{4}{B} = -1$
$4 = -B$
$B = -4$

52. In the graph, the line with the positive slope has a positive $y$-intercept, and the line with the negative slope has a negative $y$-intercept. The only pair of equations that meets these criteria is the pair in part (a). (Note: For the equations in part (b), the equation with the negative slope has a positive $y$-intercept. For the equations in part (c), the equation with the positive slope has a negative $y$-intercept. For the equations in parts (d) and (e), the equations with the negative slopes have positive $y$-intercepts, and the

## Chapter 3: Linear Functions and Their Graphs

equations with the positive slopes have negative *y*-intercepts.)

**Putting the Concepts Together (Sections 3.1 – 3.3)**

1. Find the *y*-intercept by letting $x = 0$ and solving for *y*.
   $x = 0$: $y = -0 + 6$
   $y = 6$
   
   The *y*-intercept is 6, so the point (0, 6) is on the graph. Find the *x*-intercept by letting $y = 0$ and solving for *x*.
   $y = 0$: $0 = -x + 6$
   $x = 6$
   
   The *x*-intercept is 6, so the point (6, 0) is on the graph.

   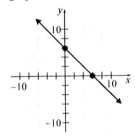

2. Let $x = -4, 0, 4,$ and $8$.
   $x = -4$: $3(-4) - 4y = 0$
   $-12 - 4y = 0$
   $-4y = 12$
   $y = -3$
   
   $x = 0$: $3(0) - 4y = 0$
   $y = 0$
   
   $x = 4$: $3(4) - 4y = 0$
   $12 - 4y = 0$
   $-4y = -12$
   $y = 3$
   
   $x = 8$: $3(8) - 4y = 0$
   $24 - 4y = 0$
   $-4y = -24$
   $y = 6$
   
   Thus, the points (-4, -3), (0, 0), (4, 3), and (8, 6) are on the graph.

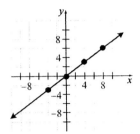

3. Find the *y*-intercept by letting $x = 0$ and solving for *y*.
   $x = 0$: $2(0) - 5y = 20$
   $-5y = 20$
   $y = -4$
   
   The *y*-intercept is -4, so the point (0, -4) is on the graph. Find the *x*-intercept by letting $y = 0$ and solving for *x*.
   $y = 0$: $2x - 5(0) = 20$
   $2x = 20$
   $x = 10$
   
   The *x*-intercept is 10, so the point (10, 0) is on the graph.

   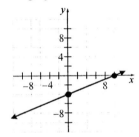

4. Let $x = -2, 2, 6,$ and $10$.
   $x = -2$: $\frac{1}{2}(-2) - \frac{2}{3}y = 1$
   $-1 - \frac{2}{3}y = 1$
   $-\frac{2}{3}y = 2$
   $-\frac{3}{2}\left(-\frac{2}{3}y\right) = -\frac{3}{2}(2)$
   $y = -3$

ISM –118

# Putting the Concepts Together (Sections 3.1 – 3.3)

$x = 6$: $\quad \frac{1}{2}(6) - \frac{2}{3}y = 1$

$\quad\quad\quad\quad 3 - \frac{2}{3}y = 1$

$\quad\quad\quad\quad -\frac{2}{3}y = -2$

$\quad\quad\quad\quad -\frac{3}{2}\left(-\frac{2}{3}y\right) = -\frac{3}{2}(-2)$

$\quad\quad\quad\quad y = 3$

$x = 10$: $\quad \frac{1}{2}(10) - \frac{2}{3}y = 1$

$\quad\quad\quad\quad 5 - \frac{2}{3}y = 1$

$\quad\quad\quad\quad -\frac{2}{3}y = -4$

$\quad\quad\quad\quad -\frac{3}{2}\left(-\frac{2}{3}y\right) = -\frac{3}{2}(-4)$

$\quad\quad\quad\quad y = 6$

Thus, the points (-2, -3), (2, 0), (6, 3), and (10, 6) are on the graph.

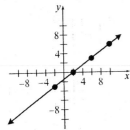

5. The equation $y = 6$ will be a horizontal line with $y$-intercept 6.

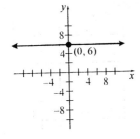

6. For $g(x) = -\frac{3}{4}x + 1$, the slope is $-\frac{3}{4}$ and the $y$-intercept is 1. Begin at (0, 1) and move to the right 4 units and down 3 units to find the point (4, -2). We can also move 4 units to the left and 3 units up to find the point (-4, 4).

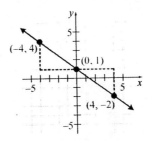

7. $m = \frac{y_2 - y_1}{x_2 - x_1} = \frac{7 - 5}{3 - (-1)} = \frac{2}{4} = \frac{1}{2}$;

For every 2 units of increase in $x$, $y$ will increase by 1 unit.

8. We convert the equation to the form $y = mx + b$.

$4x - 5y = 20$

$-5y = -4x + 20$

$y = \frac{-4x + 20}{-5}$

$y = \frac{4}{5}x - 4$

The slope is $\frac{4}{5}$, and the $y$-intercept is $-4$.

9. Start at (-2, -4). Because $m = \frac{3}{4}$, move 4 units to the right and 3 units up to find point (2, -1).

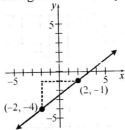

10. $\quad 12x - 4y = 1 \quad\quad\quad x - 3y = -12$

$\quad\quad -4y = -12x + 1 \quad\quad -3y = -x - 12$

$\quad\quad y = \frac{-12x + 1}{-4} \quad\quad y = \frac{-x - 12}{-3}$

$\quad\quad y = 3x - \frac{1}{4} \quad\quad\quad y = \frac{1}{3}x + 4$

Neither. The slopes of the two lines, 3 and 1/3, are not equal and are not negative reciprocals.

11. $y - y_1 = m(x - x_1)$
$y - (-4) = \frac{3}{2}(x - 6)$
$y + 4 = \frac{3}{2}x - 9$
$y = \frac{3}{2}x - 13$

12. $m = \frac{y_2 - y_1}{x_2 - x_1} = \frac{-3 - 7}{4 - (-1)} = \frac{-10}{5} = -2$
$y - y_1 = m(x - x_1)$
$y - 7 = -2(x - (-1))$
$y - 7 = -2x - 2$
$y = -2x + 5$

13. The slope of the line we seek is $m = \frac{2}{3}$, the same as the slope of the line
$2x - 3y = 12$
$-3y = -2x + 12$
$y = \frac{-2x + 12}{-3}$
$y = \frac{2}{3}x - 4$
Thus, the equation of the line we seek is:
$y - 4 = \frac{2}{3}(x - (-3))$
$y - 4 = \frac{2}{3}x + 2$
$y = \frac{2}{3}x + 6$

14. The slope of the line we seek is $m = -\frac{3}{4}$, the negative reciprocal of the line
$4x - 3y = 3$
$-3y = -4x + 3$
$y = \frac{-4x + 3}{-3}$
$y = \frac{4}{3}x - 1$

Thus, the equation of the line we seek is:
$y - (-1) = -\frac{3}{4}(x - 4)$
$y + 1 = -\frac{3}{4}x + 3$
$y = -\frac{3}{4}x + 2$

15. a. The total weight will be 40 pounds times the number of television sets $x$, plus the 5700 pounds from the weight of the van. Thus, the function is $W(x) = 40x + 5700$.

b. The independent variable is the number of television sets, $x$, which cannot be negative. Thus, the domain of $W$ is $\{x \mid x \geq 0\}$ or, using interval notation, $[0, \infty)$.

c. $W(62) = 40(62) + 5700 = 8180$
If 62 televisions sets are loaded on the van, the total weight will be 8180 pounds.

d. Evaluate $W$ at $x = 0, 50,$ and $100$.
$W(0) = 40(0) + 5700 = 5700$
$W(50) = 40(50) + 5700 = 7700$
$W(100) = 40(100) + 5700 = 9700$
Thus, the points (0, 5700), (50, 7700), and (100, 9700) are on the graph.

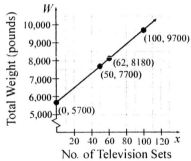

e. We must solve $W(x) = 8580$.
$40x + 5700 = 8580$
$40x = 2880$
$x = 72$
The total weight is 8580 pounds when 72 television sets are loaded on the van.

# Putting the Concepts Together (Sections 3.1 – 3.3)

16. a.

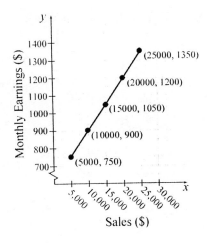

b. Average rate of change
$$= \frac{900-750}{10,000-5,000}$$
$= \$0.03$ per dollar of sales

Between monthly sales of \$5,000 and \$10,000, the earnings increase at a rate of \$0.03 for each additional dollar of sales.

c. Average rate of change
$$= \frac{1350-1200}{25,000-20,000}$$
$= \$0.03$ per dollar of sales

Between monthly sales of \$20,000 and \$25,000, the earnings increase at a rate of \$0.03 for each additional dollar of sales.

d. Yes. The monthly earnings are linearly related to sales. The average rate of change (slope) is constant.

## 3.4 Preparing for Linear Inequalities in Two Variables

1. $3(4)+1 \overset{?}{\geq} 7$
$12+1 \overset{?}{\geq} 7$
$13 \geq 7 \leftarrow$ True
Yes, $x=4$ does satisfy the inequality $3x+1 \geq 7$.

2. $-4x-3 > 9$
$-4x-3+3 > 9+3$
$-4x > 12$
$\dfrac{-4x}{-4} < \dfrac{12}{-4}$
$x < -3$

The solution set is $\{x | x < -3\}$ or, using interval notation, $(-\infty, -3)$.

## 3.4 Exercises

2. solid

4. TRUE.

6. Answers may vary. If we use a test point that lies on the line separating the two half-planes, then we would not be able to tell which half-plane to shade. We must use a test point not on the line separating the two half-planes, so that we can tell which half-plane to shade.

8. a. $x=2, y=-1$ $\quad 2(2)+(-1) \overset{?}{>} -3$
$4+(-1) \overset{?}{>} -3$
$3 \overset{?}{>} -3$ True
So, $(2,-1)$ is a solution to $2x+y>-3$.

b. $x=1, y=3$ $\quad 2(1)+(-3) \overset{?}{>} -3$
$2+(-3) \overset{?}{>} -3$
$-1 \overset{?}{>} -3$ True
So, $(1,-3)$ is a solution to $2x+y>-3$.

c. $x=-5, y=4$ $\quad 2(-5)+4 \overset{?}{>} -3$
$-6 \overset{?}{>} -3$ False
So, $(-5,4)$ is not a solution to $2x+y>-3$.

10. a. $x=1, y=-1$ $\quad 2(1)-5(-1) \overset{?}{\leq} 2$
$2+5 \overset{?}{\leq} 2$
$7 \leq 2$ False
So, $(1,-1)$ is not a solution to $2x-5y \leq 2$.

ISM –121

## Chapter 3: Linear Functions and Their Graphs

**b.** $x = 3, y = 0$   $2(3) - 5(0) \stackrel{?}{\leq} 2$

$6 + 0 \stackrel{?}{\leq} 2$

$6 \stackrel{?}{\leq} 2$   False

So, $(3, 0)$ is not a solution to $2x - 5y \leq 2$.

**c.** $x = -3, y = -2$   $2(-3) - 5(-2) \stackrel{?}{\leq} 2$

$-6 + 10 \stackrel{?}{\leq} 2$

$4 \stackrel{?}{\leq} 2$   False

So, $(-3, -2)$ is a not solution to $2x - 5y \leq 2$.

**12.** Replace the inequality symbol with an equal sign to obtain $y = -2$. Because the inequality is strict, graph $y = -2$ using a dashed line.

Test Point: $(0, 0)$:   $0 \stackrel{?}{<} -2$   False

Therefore, $(0, 0)$ is not a solution to $y < -2$. Shade the half-plane that does not contain $(0, 0)$.

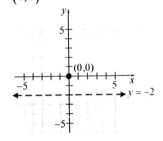

**14.** Replace the inequality symbol with an equal sign to obtain $x = 7$. Because the inequality is strict, graph $x = 7$ using a dashed line.

Test Point: $(0, 0)$:   $0 \stackrel{?}{<} 7$   True

Therefore, $(0, 0)$ is a solution to $x < 7$. Shade the half-plane that contains $(0, 0)$.

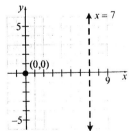

**16.** Replace the inequality symbol with an equal sign to obtain $y = \frac{2}{3}x$. Because the inequality is non-strict, graph $y = \frac{2}{3}x$ using a solid line.

Test Point: $(0, 1)$:   $1 \stackrel{?}{\geq} \frac{2}{3}(0)$

$1 \stackrel{?}{\geq} 0$   True

Therefore, $(0, 1)$ is a solution to $y \geq \frac{2}{3}x$.

Shade the half-plane that contains $(0, 1)$.

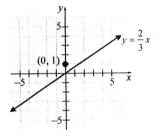

**18.** Replace the inequality symbol with an equal sign to obtain $y = -3x + 1$. Because the inequality is strict, graph $y = -3x + 1$ using a dashed line.

Test Point: $(0, 0)$:   $0 \stackrel{?}{<} -3(0) + 1$

$0 \stackrel{?}{<} 1$   True

Therefore, $(0, 0)$ is a solution to $y < -3x + 1$.

Shade the half-plane that contains $(0, 0)$.

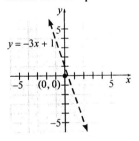

**20.** Replace the inequality symbol with an equal sign to obtain $y = -\frac{4}{3}x + 5$. Because the inequality is non-strict, graph $y = -\frac{4}{3}x + 5$ using a solid line.

Test Point: $(0, 0)$:   $0 \stackrel{?}{\geq} -\frac{4}{3}(0) + 5$

$0 \stackrel{?}{\geq} 5$   False

Therefore, $(0,0)$ is not a solution to $y \geq -\dfrac{4}{3}x+5$. Shade the half-plane that does not contain $(0,0)$.

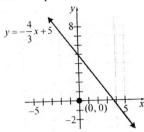

22. Replace the inequality symbol with an equal sign to obtain $-4x+y=-5$. Because the inequality is non-strict, graph $-4x+y=-5$ $(y=4x-5)$ using a solid line.

    Test Point: $(0,0)$: $-4(0)+0 \overset{?}{\geq} -5$
    
    $0 \overset{?}{\geq} -5$   True

    Therefore, $(0,0)$ is solution to $-4x+y \geq -5$. Shade the half-plane that contains $(0,0)$.

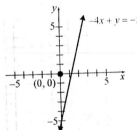

24. Replace the inequality symbol with an equal sign to obtain $3x+4y=12$. Because the inequality is non-strict, graph $3x+4y=12$ $\left(y=-\dfrac{3}{4}x+3\right)$ using a solid line.

    Test Point: $(0,0)$: $3(0)+4(0) \overset{?}{\geq} 12$
    
    $0 \overset{?}{\geq} 12$   False

    Therefore, $(0,0)$ is not a solution to $3x+4y \geq 12$. Shade the half-plane that does not contain $(0,0)$.

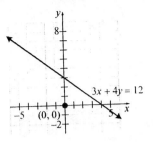

26. Replace the inequality symbol with an equal sign to obtain $-5x+3y=30$. Because the inequality is strict, graph $-5x+3y=30$ $\left(y=\dfrac{5}{3}x+10\right)$ using a dashed line.

    Test Point: $(0,0)$: $-5(0)+3(0) \overset{?}{<} 30$
    
    $0 < 30$   True

    Therefore, $(0,0)$ is a solution to $-5x+3y<30$. Shade the half-plane that contains $(0,0)$.

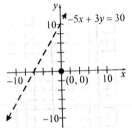

28. Replace the inequality symbol with an equal sign to obtain $\dfrac{x}{3}-\dfrac{y}{4}=1$. Because the inequality is non-strict, graph $\dfrac{x}{3}-\dfrac{y}{4}=1$ $\left(y=\dfrac{4}{3}x-4\right)$ using a solid line.

    Test Point: $(0,0)$: $\dfrac{0}{3}-\dfrac{0}{4} \overset{?}{\leq} 1$
    
    $0 \overset{?}{\leq} 1$   True

    Therefore, $(0,0)$ is a solution to $\dfrac{x}{3}-\dfrac{y}{4} \leq 1$. Shade the half-plane that contains $(0,0)$.

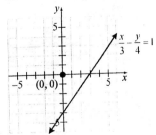

## Chapter 3: Linear Functions and Their Graphs

30. Replace the inequality symbol with an equal sign to obtain $\frac{5}{4}x - \frac{3}{5}y = 2$. Because the inequality is non-strict, graph $\frac{5}{4}x - \frac{3}{5}y = 2$ $\left(y = \frac{25}{12}x - \frac{10}{3}\right)$ using a solid line.

    Test Point $(0,0)$: $\frac{5}{4}(0) - \frac{3}{5}(0) \overset{?}{\leq} 2$
    
    $0 \overset{?}{\leq} 2$  True
    
    Therefore, $(0,0)$ is a solution to $\frac{5}{4}x - \frac{3}{5}y \leq 2$.
    Shade the half-plane that contains $(0,0)$.

    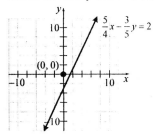

32. **a.** Let $x$ = the number of Model A computers.
    Let $y$ = the number of Model B computers.
    $45x + 65y \geq 4000$

    **b.** $x = 50$, $y = 28$: $45(50) + 65(28) \overset{?}{\geq} 4000$
    
    $4070 \geq 4000$  True
    
    Juanita will make her sales goal by selling 50 Model A and 28 Model B computers.

    **c.** $x = 41$, $y = 33$: $45(41) + 65(33) \overset{?}{\geq} 4000$
    
    $3990 \overset{?}{\geq} 4000$  False
    
    Juanita will not make her sales goal by selling 44 Model A and 33 Model B computers.

34. **a.** Let $x$ = the number Gummy Bears.
    Let $y$ = the number suckers.
    $0.10x + 0.25y \leq 3.00$

    **b.** $x = 18$, $y = 5$: $0.10(18) + 0.25(5) \overset{?}{\leq} 3.00$
    
    $3.05 \overset{?}{\leq} 3.00$  False
    
    Johnny cannot buy 18 Gummy Bears and 5 suckers.

    **c.** $x = 19$, $y = 4$: $0.10(19) + 0.25(4) \overset{?}{\leq} 3.00$
    
    $2.90 \overset{?}{\leq} 3.00$  True

    Johnny can buy 19 Gummy Bears and 4 suckers.

36. The graph contains a line passing through the points $(0,1)$ and $(2,-3)$. The slope of the line is $m = \frac{-3-1}{2-0} = \frac{-4}{2} = -2$. The y-intercept is 1.
    Thus, the equation of the line is $y = -2x + 1$.
    Because the half-plane that is shaded is below the line, our inequality will contain a "less than." Because the line is solid, this means that the inequality is non-strict. That is, our inequality will consist of "less than or equal to." Thus, the linear inequality is $y \leq -2x + 1$.

38. The graph contains a line passing through the points $(-1,-3)$ and $(2,1)$. The slope of the line is $m = \frac{1-(-3)}{2-(-1)} = \frac{4}{3}$.
    
    The equation of the line is
    $y - (-3) = \frac{4}{3}(x - (-1))$
    $y + 3 = \frac{4}{3}x + \frac{4}{3}$
    $y = \frac{4}{3}x - \frac{5}{3}$
    
    Because the half-plane that is shaded is above the line, our inequality will contain a "greater than." Because the line is dashed, this means that the inequality is strict. That is, our inequality will consist only of "greater than."
    Thus, the linear inequality is $y > \frac{4}{3}x - \frac{5}{3}$.

40. $y < -2$

ISM –124

42. $y \geq \dfrac{2}{3}x$

44. $y < -3x + 1$

46. $y \geq -\dfrac{4}{3}x + 5$

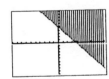

48. $-4x + y \geq -5$
    $y \geq 4x - 5$

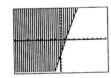

49. $2x + 5y \leq -10$
    $5y \leq -2x - 10$
    $y \leq \dfrac{-2x - 10}{5}$
    $y \leq -\dfrac{2}{5}x - 2$

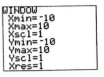

50. $3x + 4y \geq 12$
    $4y \geq -3x + 12$
    $y \geq \dfrac{-3x + 12}{4}$
    $y \geq -\dfrac{3}{4}x + 3$

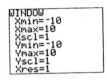

### 3.5 Preparing for Linear Inequalities in Two Variables

1. Step 1: Identify what you are looking for.
   Step 2: Give names to the unknowns.
   Step 3: Translate the problem into the language of mathematics
   Step 4: Solve the equation(s) found in Step 3.
   Step 5: Check the reasonableness to your answer
   Step 6: Answer the question.

## Chapter 3: Linear Functions and Their Graphs

2.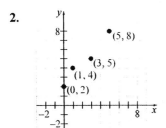

### 3.5 Exercises

2. Variation

4. TRUE

6. Answers will vary. It would mean that the student's average depends on the amount of time he or she spent studying. Though it is not stated specifically, the teacher also means that constant of proportionality will be positive. That is, the more the student studies, the higher his or her average will be, and the less the student studies, the lower his or her average will be.

8. a. $y = kx$
   $15 = k \cdot 3$
   $k = \dfrac{15}{3} = 5$

   b. $y = 5x$

   c. $y = 5(5) = 25$

10. a. $y = kx$
    $4 = k \cdot 20$
    $k = \dfrac{4}{20} = \dfrac{1}{5}$

    b. $y = \dfrac{1}{5}x$

    c. $y = \dfrac{1}{5}(35) = 7$

12. a. $y = kx$
    $8 = k \cdot 12$
    $k = \dfrac{8}{12} = \dfrac{2}{3}$

    b. $y = \dfrac{2}{3}x$

    c. $y = \dfrac{2}{3}(20) = \dfrac{40}{3}$

14. Linear with negative slope

16. Nonlinear

    b. Answers will vary. We will use the points $(4, 1.8)$ and $(9, 2.6)$.

18. a.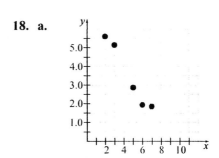

    b. Answers will vary. We will use the points $(2, 5.7)$ and $(7, 1.8)$.
    $m = \dfrac{1.8 - 5.7}{7 - 2} = \dfrac{-3.9}{5} = -0.78$
    $y - 5.7 = -0.78(x - 2)$
    $y - 5.7 = -0.78x + 1.56$
    $y = -0.78x + 7.26$

    c.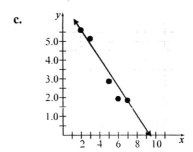

20. a.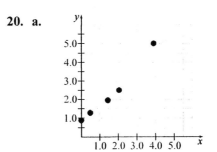

    b. Answers will vary. We will use the points $(0, 0.8)$ and $(3.9, 5.0)$.

$$m = \frac{5.0 - 0.8}{3.9 - 0} = \frac{4.2}{3.9} \approx 1.08$$
$$y - 0.8 = 1.08(x - 0)$$
$$y - 0.8 = 1.08x$$
$$y = 1.08x + 0.8$$

c.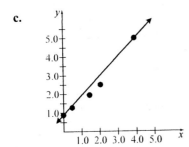

22. a. $R(m) = 0.15m + 129.50$

    b. The number of miles, $m$, is the independent variable. The rental cost, $R$, is the dependent variable.

    c. Because the number miles cannot be negative, the number of miles must be greater than or equal to zero. That is, the implied domain is $\{m \mid m \geq 0\}$, or using interval notation $[0, \infty)$.

    d. $R(860) = 0.15(860) + 129.50 = 258.50$
    If 860 miles are driven, the rental cost will be $258.50.

    e. $0.15m + 129.50 = 213.80$
    $0.15m = 84.30$
    $m = 562$
    If the rental cost is $213.80, then 562 miles were driven.

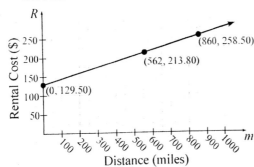

24. a. The machine will depreciate by $\dfrac{\$1,200,000}{20} = \$60,000$ per year. Thus, the slope is $-60,000$. The $y$-intercept will be $1,200,000, the initial value of the machine. The linear function that represents book value, $V$, of the machine after $x$ years is $V(x) = -60,000x + 1,200,000$.

    b. Because the machine cannot have a negative age, the age, $x$, must be greater than or equal to 0. After 20 years, the book value will be $V(20) = -60,000(20) + 1,200,000 = 0$, and the book value cannot be negative. Therefore the implied domain of function is $\{x \mid 0 \leq x \leq 20\}$, or using interval notation $[0, 20]$.

    c. $V(3) = -60,000(3) + 1,200,000 = 1,020,000$
    After three years, the book value of the machine will be $1,020,000.

    d. The intercepts are $(0, 1200000)$ and $(20, 0)$. The $y$-intercept is 1,200,000 and the $x$-intercept is 20.

    e. $-60,000x + 1,200,000 = 480,000$
    $-60,000x = -720,000$
    $x = 12$
    The book value of the machine will be $480,000 after 12 years.

    f.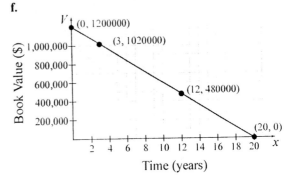

26. a. Because $p$ varies directly with $b$, we know that $p = kb$ for some constant $k$. We know that $p = \$980.50$ when $b = \$120,000$, so, $\begin{aligned} 980.50 &= k \cdot 120000 \\ k &= 0.0081708 \end{aligned}$
    Therefore, we have $p = 0.0081708b$ or,

## Chapter 3: Linear Functions and Their Graphs

using function notation,
$p(b) = 0.0081708b$.

**b.** $p(150,000) = 0.0081708(150,000)$
$\approx 1225.62$

The monthly payment would be approximately \$1225.62 if \$150,000 is borrowed.

**c.**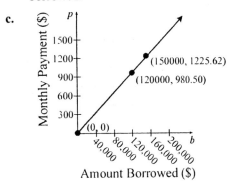

**28. a.** Because $E$ varies directly with $d$, we know that $E = kd$ for some constant $k$. We know that $E = 520$ when $d = 600$ lbs, so,
$520 = k \cdot 600$
$k = 0.86666667$
Therefore, we have $E = 0.86666667d$ or, using function notation,
$E(d) = 0.86666667d$.

**b.** $E(700) = 0.86666667(700) = 606.67$
\$700 will convert 606.67 euros.

**c.**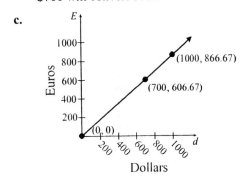

**30.** Because $C$ is directly proportional to $r$, we know that $C = kr$ for some constant $k$. We know that $C = 10\pi$ inches when $r = 5$ inches, so
$10\pi = k \cdot 5$
$k = 2\pi$
Thus, $C = 2\pi r$ or, using function notation, $C(r) = 2\pi r$. Finally, $C(8) = 2\pi(8) = 16\pi$.
Therefore, if the radius of a circle is 8 inches, then the circumference of the circle is $16\pi$ inches.

**32. a.**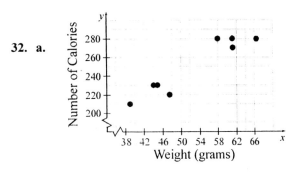

**b.** Linear.

**c.** Answers will vary. We will use the points $(39.52, 210)$ and $(66.45, 280)$.
$m = \dfrac{280 - 210}{66.45 - 39.52} = \dfrac{70}{26.93} \approx 2.599$
$y - 210 = 2.599(x - 39.52)$
$y - 210 = 2.599x - 102.712$
$y = 2.599x + 107.288$

**d.**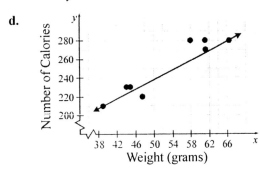

**e.** $x = 62.3$: $y = 2.599(62.3) + 107.288$
$\approx 269$
We predict that a candy bar weighing 62.3 grams will contain 269 calories.

**f.** The slope of the line found is 2.599 calories per gram. This means that if the weight of a candy bar is increased by 1 gram, then the number of calories will increase by 2.599.

**34. a.** No, the relation does not represent a function. The $h$-coordinate 26.75 is paired with the two different $C$-coordinates 17.3 and 17.5.

ISM –128

**b.**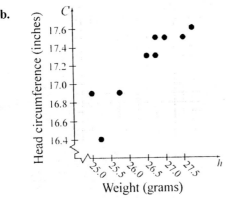

**c.** Answers will vary. We will use the points $(25, 16.9)$ and $(27.75, 17.6)$.

$$m = \frac{17.6 - 16.9}{27.75 - 25} = \frac{0.7}{2.75} = 0.255$$

$$C - 16.9 = 0.255(h - 25)$$
$$C - 16.9 = 0.255h - 6.375$$
$$C = 0.255h + 10.525$$

**d.**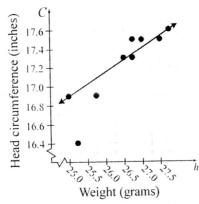

**e.** Let $C$ represent the head circumference (in inches), and let $h$ represent the height (in inches).
$$C(h) = 0.255h + 10.525$$

**f.** $C(26.5) = 0.255(26.5) + 10.525 \approx 17.28$

We predict that the head circumference will be 17.28 inches if the height is 26.5 inches.

**g.** The slope of the line found is 0.255. This means that if the height increases by one inch, then the head circumference increases by 0.255 inch.

**36.** Since $y$ is directly proportional to $x^2$, we know that $y = kx^2$ for some constant $k$. If $x$ is doubled, then $y = k(2x)^2 = 4kx^2$, which means that $y$ is

**38. a.** The scatter diagram and window settings are shown below.

**b.** As shown below, the line of best fit is approximately $y = 2.884x + 97.587$.

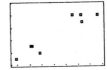

**40. a.** The scatter diagram and window settings are shown below.

**b.** As shown below, the line of best fit is approximately $y = 0.373x + 7.327$.

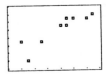

## Chapter 3 Review

1. Let $x = -2, -1, 0,$ and $1$.

   $x = -2:\quad -2 + y = 7$
   $\qquad\qquad\ \ y = 9$

   $x = -1:\quad -1 + y = 7$
   $\qquad\qquad\ \ y = 8$

   $x = 0:\quad 0 + y = 7$
   $\qquad\qquad y = 7$

   $x = 1:\quad 1 + y = 7$
   $\qquad\qquad y = 6$

   Thus, the points $(-2, 9)$, $(-1, 8)$, $(0, 7)$, and $(1, 6)$ are on the graph.

   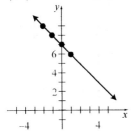

2. Let $x = -2, -1, 0,$ and $1$.

   $x = -2:\quad -2 - y = -4$
   $\qquad\qquad\ \ -y = -2$
   $\qquad\qquad\ \ \ \ y = 2$

   $x = -1:\quad -1 - y = -4$
   $\qquad\qquad\ \ -y = -3$
   $\qquad\qquad\ \ \ \ y = 3$

   $x = 0:\quad 0 - y = -4$
   $\qquad\qquad -y = -4$
   $\qquad\qquad\ \ y = 4$

   $x = 1:\quad 1 - y = -4$
   $\qquad\qquad -y = -5$
   $\qquad\qquad\ \ y = 5$

   Thus, the points $(-2, 2)$, $(-1, 3)$, $(0, 4)$, and $(1, 5)$ are on the graph.

   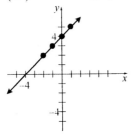

3. Let $x = -2, 0, 2,$ and $4$.

   $x = -2:\quad 5(-2) - 2y = 6$
   $\qquad\qquad\ \ -10 - 2y = 6$
   $\qquad\qquad\qquad -2y = 16$
   $\qquad\qquad\qquad\ \ \ \ y = -8$

   $x = 0:\quad 5(0) - 2y = 6$
   $\qquad\qquad\ \ -2y = 6$
   $\qquad\qquad\ \ \ \ y = -3$

   $x = 2:\quad 5(2) - 2y = 6$
   $\qquad\qquad\ 10 - 2y = 6$
   $\qquad\qquad\qquad -2y = -4$
   $\qquad\qquad\qquad\ \ \ \ y = 2$

   $x = 4:\quad 5(4) - 2y = 6$
   $\qquad\qquad\ 20 - 2y = 6$
   $\qquad\qquad\qquad -2y = -14$
   $\qquad\qquad\qquad\ \ \ \ y = 7$

   Thus, the points $(-2, -8)$, $(0, -3)$, $(2, 2)$, and $(4, 7)$ are on the graph.

   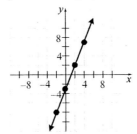

4. Let $x = -4, -2, 0,$ and $2$.

   $x = -4:\quad -3(-4) + 2y = 8$
   $\qquad\qquad\ \ 12 + 2y = 8$
   $\qquad\qquad\qquad\ 2y = -4$
   $\qquad\qquad\qquad\ \ \ y = -2$

   $x = -2:\quad -3(-2) + 2y = 8$
   $\qquad\qquad\ \ 6 + 2y = 8$
   $\qquad\qquad\qquad 2y = 2$
   $\qquad\qquad\qquad\ \ y = 1$

   $x = 0:\quad -3(0) + 2y = 8$
   $\qquad\qquad\ \ 2y = 8$
   $\qquad\qquad\ \ \ y = 4$

   $x = 2:\quad -3(2) + 2y = 8$
   $\qquad\qquad\ \ -6 + 2y = 8$
   $\qquad\qquad\qquad 2y = 14$
   $\qquad\qquad\qquad\ \ y = 7$

   Thus, the points $(-4, -2)$, $(-2, 1)$, $(0, 4)$, and $(2, 7)$ are on the graph.

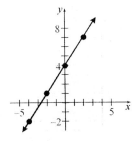

5. To find the *y*-intercept, let $x = 0$ and solve for *y*.
$$x = 0: \quad 5(0) + 3y = 30$$
$$3y = 30$$
$$y = 10$$
The *y*-intercept is 10, so $(0,10)$ is on the graph.
To find the *x*-intercept, let $y = 0$ and solve for *x*.
$$y = 0: \quad 5x + 3(0) = 30$$
$$5x = 30$$
$$x = 6$$
The *x*-intercept is 6, so $(6,0)$ is on the graph.

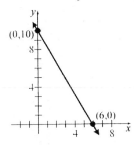

6. To find the *y*-intercept, let $x = 0$ and solve for *y*.
$$x = 0: \quad 4(0) + 3y = 0$$
$$4y = 0$$
$$y = 0$$
The *y*-intercept is 0, so the point $(0,0)$ is on the graph. To find the *x*-intercept, let $y = 0$ and solve for *x*.
$$y = 0: \quad 4x + 3(0) = 0$$
$$4x = 0$$
$$x = 0$$
The *x*-intercept is 0, so the point $(0,0)$ is on the graph. Since the *x*- and *y*-intercepts both result in the origin, we must find other points on the graph. Let $x = -3$ and 3 and solve for *y*.
$$x = -3: \quad 4(-3) + 3y = 0$$
$$-12 + 3y = 0$$
$$3y = 12$$
$$y = 4$$

$$x = 3: \quad 4(3) + 3y = 0$$
$$12 + 3y = 0$$
$$3y = -12$$
$$y = -4$$
Thus, the points $(-3,4)$ and $(3,-4)$ are also on the graph.

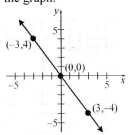

7. To find the *y*-intercept, let $x = 0$ and solve for *y*.
$$x = 0: \quad \frac{3}{4}(0) - \frac{1}{2}y = 1$$
$$-\frac{1}{2}y = 1$$
$$y = -2$$
The *y*-intercept is –2, so the point $(0,-2)$ is on the graph. To find the *x*-intercept, let $y = 0$ and solve for *x*.
$$y = 0: \quad \frac{3}{4}x - \frac{1}{2}(0) = 1$$
$$\frac{3}{4}x = 1$$
$$x = \frac{4}{3}$$
The *x*-intercept is $\frac{4}{3}$, so the point $\left(\frac{4}{3},0\right)$ is on the graph.

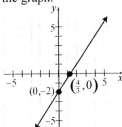

8. To find the *y*-intercept, let $x = 0$ and solve for *y*.
$$x = 0: \quad 4(0) + y = 8$$
$$y = 8$$
The *y*-intercept is 8, so the point $(0,8)$ is on the graph. To find the *x*-intercept, let $y = 0$ and solve for *x*.

## Chapter 3: Linear Functions and Their Graphs

$y = 0$: $4x + 0 = 8$
$4x = 8$
$x = 2$

The $x$-intercept is 2, so the point $(2,0)$ is on the graph.

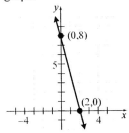

9. The equation $x = 4$ will be a vertical line with $x$-intercept 4.

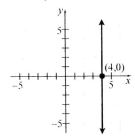

10. The equation $y = -8$ will be a horizontal line with $y$-intercept -8.

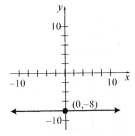

11. The equation $x = -2$ will be a vertical line with $x$-intercept -2.

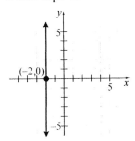

12. The equation $y = 5$ will be a horizontal line with $y$-intercept 5.

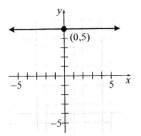

13. **a.** The independent variable is the number of long-distance minutes used, $x$. It would not make sense to talk for a negative number of minutes. Thus, the domain of $C$ is $\{x \mid x \geq 0\}$ or, using interval notation, $[0, \infty)$.

  **b.** $C(235) = 0.07(235) + 5 = 21.45$
  The cost for 235 minutes of long-distance calls during one month is $21.45.

  **c.** Evaluate $C$ at $x = 0$, 100, and 500.
  $C(0) = 0.07(0) + 5 = 5$
  $C(100) = 0.07(100) + 5 = 12$
  $C(500) = 0.07(500) + 5 = 40$
  Thus, the points $(0,5)$, $(100,12)$, and $(500,40)$ are on the graph.

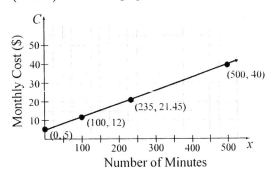

  **d.** We must solve $C(x) = 75$.
  $0.07x + 5 = 75$
  $0.07x = 70$
  $x = 1000$
  In one month, a person can purchase 1000 minutes of long distance for $75.

14. **a.** The independent variable is the number of years after purchase, $x$. The dependent variable is the value of the computer, $V$.

  **b.** We are told that the value function $V$ is for 0 to 5 years, inclusive. Thus, the domain is $\{x \mid 0 \leq x \leq 5\}$ or, using interval notation, $[0, 5]$.

c. The initial value of the computer will be the value at $x = 0$ years.
$V(0) = 1800 - 360(0) = 1800$
The initial value of the computer is $1800.

d. $V(2) = 1800 - 360(2) = 1080$
After 2 years, the value of the computer is $1080.

e. Evaluate $V$ at $x = 3, 4,$ and $5$.
$V(3) = 1800 - 360(3) = 720$
$V(4) = 1800 - 360(4) = 360$
$V(5) = 1800 - 360(5) = 0$
Thus, the points $(3, 720)$, $(4, 360)$, and $(5, 0)$ are on the graph.

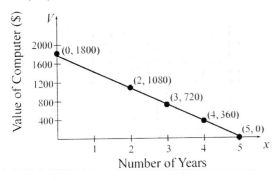

f. We must solve $V(x) = 0$.
$1800 - 360x = 0$
$-360x = -1800$
$x = 5$
After 5 years, the computer's value will be $0.

15. a. $m = \dfrac{y_2 - y_1}{x_2 - x_1} = \dfrac{4 - (-1)}{2 - (-2)} = \dfrac{5}{4}$

b. For every 4 units of increase in $x$, $y$ will increase by 5 units.

16. a. $m = \dfrac{y_2 - y_1}{x_2 - x_1} = \dfrac{-1 - 1}{5 - (-3)} = \dfrac{-2}{8} = -\dfrac{1}{4}$

b. For every 4 units of increase in $x$, $y$ will decrease by 1 unit. For every 4 units of decrease in $x$, $y$ will increase by 1 unit.

17. $m = \dfrac{y_2 - y_1}{x_2 - x_1} = \dfrac{-1 - 5}{2 - (-1)} = \dfrac{-6}{3} = -2$

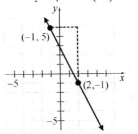

18. $m = \dfrac{y_2 - y_1}{x_2 - x_1} = \dfrac{-1 - 5}{0 - 4} = \dfrac{-6}{-4} = \dfrac{3}{2}$

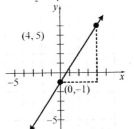

19. a.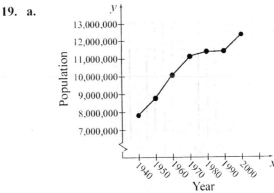

b. Average rate of change
$= \dfrac{8{,}712{,}176 - 7{,}897{,}241}{1950 - 1940}$
$= 81{,}493.5$ people per year
Between 1940 and 1950, the population of Illinois increased at a rate of 81,493.5 people per year.

c. Average rate of change
$= \dfrac{11{,}430{,}602 - 11{,}427{,}409}{1990 - 1980}$
$= 319.3$ people per year
Between 1980 and 1990, the population of Illinois increased at a rate of 319.3 people per year.

## Chapter 3: Linear Functions and Their Graphs

**d.** Average rate of change
$$= \frac{12{,}419{,}293 - 11{,}430{,}602}{2000 - 1990}$$
$= 98{,}869.1$ people per year

Between 1990 and 2000, the population of Illinois increased at a rate of 98,869.1 people per year.

**e.** No. The population of Illinois is not linearly related to the year. The average rate of change (slope) is not constant.

**20. a.**

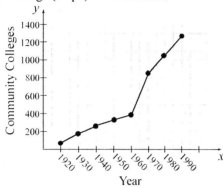

**b.** Average rate of change
$$= \frac{178 - 70}{1930 - 1920}$$
$= 10.8$ community colleges per year

Between 1920 and 1930, the number of public community colleges in the U.S. increased at a rate of 10.8 colleges per year.

**c.** Average rate of change
$$= \frac{847 - 390}{1970 - 1960}$$
$= 45.7$ community colleges per year

Between 1960 and 1970, the number of public community colleges in the U.S. increased at a rate of 45.7 colleges per year.

**d.** Average rate of change
$$= \frac{1282 - 1049}{1990 - 1980}$$
$= 23.3$ community colleges per year

Between 1980 and 1990, the number of public community colleges in the U.S. increased at a rate of 23.3 colleges per year.

**e.** No. The number of public community colleges in the U.S. is not linearly related to the year. The average rate of change (slope) is not constant.

**21.** Start at $(-1,-5)$. Because $m = 4 = \dfrac{4}{1}$, move 1 unit to the right and 4 units up to find point $(0,-1)$.

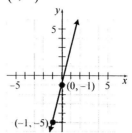

**22.** Start at $(3,2)$. Because $m = -\dfrac{2}{3}$, move 3 unit to the right and 2 units down to find point $(6,0)$. We can also move 3 units to the left and 2 units up to find point $(0,4)$.

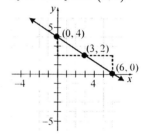

**23.** Because $m$ is undefined, the line is vertical passing through the point $(2,-4)$.

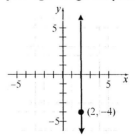

**24.** Because $m = 0$, the line is horizontal passing through the point $(-3,1)$.

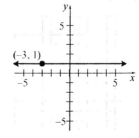

25. $m = \dfrac{y_2 - y_1}{x_2 - x_1} = \dfrac{-1-3}{6-(-2)} = \dfrac{-4}{8} = -\dfrac{1}{2}$

$y - y_1 = m(x - x_1)$

$y - (-1) = -\dfrac{1}{2}(x - 6)$

$y + 1 = -\dfrac{1}{2}x + 3$

$y = -\dfrac{1}{2}x + 2 \text{ or } x + 2y = 4$

26. $m = \dfrac{y_2 - y_1}{x_2 - x_1} = \dfrac{6-(-3)}{2-(-1)} = \dfrac{9}{3} = 3$

$y - y_1 = m(x - x_1)$

$y - 6 = 3(x - 2)$

$y - 6 = 3x - 6$

$y = 3x \text{ or } 3x - y = 0$

27. $y - y_1 = m(x - x_1)$

$y - 2 = -1(x - 3)$

$y - 2 = -x + 3$

$y = -x + 5 \text{ or } x + y = 5$

28. $y - y_1 = m(x - x_1)$

$y - (-4) = \dfrac{3}{5}(x - (-10))$

$y + 4 = \dfrac{3}{5}(x + 10)$

$y + 4 = \dfrac{3}{5}x + 6$

$y = \dfrac{3}{5}x + 2 \text{ or } 3x - 5y = -10$

29. $m = \dfrac{y_2 - y_1}{x_2 - x_1} = \dfrac{5-2}{-3-6} = \dfrac{3}{-9} = -\dfrac{1}{3}$

$y - y_1 = m(x - x_1)$

$y - 2 = -\dfrac{1}{3}(x - 6)$

$y - 2 = -\dfrac{1}{3}x + 2$

$y = -\dfrac{1}{3}x + 4 \text{ or } x + 3y = 12$

30. $m = \dfrac{y_2 - y_1}{x_2 - x_1} = \dfrac{3-3}{4-(-2)} = \dfrac{0}{6} = 0$

$y - y_1 = m(x - x_1)$

$y - 3 = 0(x - 4)$

$y - 3 = 0$

$y = 3$

31. $m = \dfrac{y_2 - y_1}{x_2 - x_1} = \dfrac{-7-(-1)}{1-4} = \dfrac{-6}{-3} = 2$

$y - y_1 = m(x - x_1)$

$y - (-7) = 2(x - 1)$

$y + 7 = 2x - 2$

$y = 2x - 9 \text{ or } 2x - y = 9$

32. $m = \dfrac{y_2 - y_1}{x_2 - x_1} = \dfrac{-1-2}{8-(-1)} = \dfrac{-3}{9} = -\dfrac{1}{3}$

$y - y_1 = m(x - x_1)$

$y - (-1) = -\dfrac{1}{3}(x - 8)$

$y + 1 = -\dfrac{1}{3}x + \dfrac{8}{3}$

$y = -\dfrac{1}{3}x + \dfrac{5}{3} \text{ or } x + 3y = 5$

33. For $y = 4x - 6$, the slope is 4 and the y-intercept is -6. Begin at $(0, -6)$ and move to the right 1 unit and up 4 units to find point $(1, -2)$.

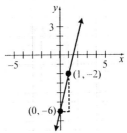

34. $2x + 3y = 12$

$3y = -2x + 12$

$y = \dfrac{-2x + 12}{3}$

$y = -\dfrac{2}{3}x + 4$

The slope is $-\dfrac{2}{3}$ and the y-intercept is 4. Begin at $(0, 4)$ and move to the right 3 units and down

2 units to find the point $(3,2)$. We can also move 3 units to the left and 2 units up to find the point $(-3,6)$.

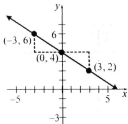

35. $x - y = 4$
    $-y = -x + 4$
    $y = x - 4$

The slope is 1 and the y-intercept is -4. Begin at $(0,-4)$ and move to the right 1 unit and up 1 unit to find the point $(1,-3)$.

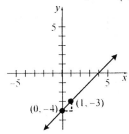

36. $2x + 6y = 3$
    $6y = -2x + 3$
    $y = \dfrac{-2x + 3}{6}$
    $y = -\dfrac{1}{3}x + \dfrac{1}{2}$

The slope is $-\dfrac{1}{3}$ and the y-intercept is $\dfrac{1}{2}$.

Begin at $\left(0, \dfrac{1}{2}\right)$ and move to the right 3 units and down 1 unit to find the point $\left(3, -\dfrac{1}{2}\right)$. We can also move to the left 3 units and up 1 unit to find the point $\left(-3, \dfrac{3}{2}\right)$.

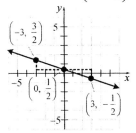

37. To find the y-intercept, find $g(0)$.
    $g(0) = 2(0) - 6 = -6$
    The y-intercept is -6, so the point $(0,-6)$ is on the graph. To find the x-intercept, solve the equation $g(x) = 0$.
    $2x - 6 = 0$
    $2x = 6$
    $x = 3$
    The x-intercept is 3, so the point $(3,0)$ is on the graph.

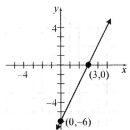

38. To find the y-intercept, find $H(0)$.
    $H(0) = -\dfrac{4}{3}(0) + 5 = 5$
    The y-intercept is 5, so the point $(0,5)$ is on the graph. To find the x-intercept, solve the equation $H(x) = 0$.
    $-\dfrac{4}{3}x + 5 = 0$
    $-\dfrac{4}{3}x = -5$
    $x = \dfrac{15}{4} = 3.75$
    The x-intercept is $\dfrac{15}{4}$, so the point $\left(\dfrac{15}{4}, 0\right)$ is on the graph.

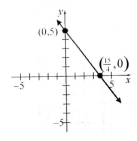

**39.** To find the y-intercept, find $F(0)$.

$F(0) = -(0) - 3 = -3$

The y-intercept is -3, so the point $(0, -3)$ is on the graph. To find the x-intercept, solve $F(x) = 0$.

$-x - 3 = 0$
$-x = 3$
$x = -3$

The x-intercept is -3, so the point $(-3, 0)$ is on the graph.

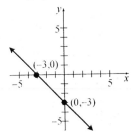

**40.** To find the y-intercept, find $f(0)$.

$f(0) = \frac{3}{4}(0) - 3 = -3$

The y-intercept is -3, so the point $(0, -3)$ is on the graph. To find the x-intercept, solve $f(x) = 0$.

$\frac{3}{4}x - 3 = 0$
$\frac{3}{4}x = 3$
$x = 4$

The x-intercept is 4, so the point $(4, 0)$ is on the graph.

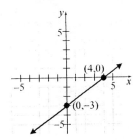

**41. a.** Let $x$ represent the number of years since 1996. Then $x = 0$ represents 1996 and $x = 5$ represents 2001. Let $E$ represent the percentage of electronically-filed returns.

$m = \frac{30.7 - 12.6}{5 - 0} = 3.62$ percent

$E - 12.6 = 3.62(x - 0)$
$E - 12.6 = 3.62x$
$E = 3.62x + 12.6$

In function notation, $E(x) = 3.62x + 12.6$.

**b.** The year 2004 corresponds to $x = 8$.
$E(8) = 3.62(8) + 12.6 = 41.56$

If the linear trend continues, 41.56% of U.S. Federal Tax Returns will be filed electronically in 2004.

**c.** The slope (3.62) indicates that electronically-filed tax returns are increasing at a rate of 3.62% per year.

**d.** $3.62x + 12.6 = 48.8$
$3.62x = 36.2$
$x = 10$

Now, $x = 10$ correspond to the year 2006. Thus, 48.8% of U.S. Federal Tax Returns will be filed electronically in the year 2006.

**42. a.** Let $x$ represent the age of men and $H$ represent the maximum recommended heart rate for men under stress.

$m = \frac{160 - 200}{60 - 20} = -1$ beats per minute per year

$H - 200 = -1(x - 20)$
$H - 200 = -x + 20$
$H = -x + 220$

In function notation, $H(x) = -x + 220$.

**b.** $H(45) = -(45) + 220 = 175$

The maximum recommended heart rate for a 45 year old man under stress is 175 beats per minute.

c. The slope (-1) indicates that the maximum recommended heart rate for men under stress decreases at a rate of 1 beat per minute per year.

d. $-x + 220 = 168$
$-x = -52$
$x = 52$
The maximum recommended heart rate under stress is 168 beats per minute for 52 year old men.

43. A line parallel to $L$ will have a slope equal to the slope of $L$. Thus, the slope of the line parallel to $L$ is $-\frac{3}{8}$.

44. A line perpendicular to $L$ will have a slope that is the negative reciprocal to the slope of $L$. Thus, the slope of the line perpendicular to $L$ is
$$\frac{-1}{-\frac{3}{8}} = -1 \cdot -\frac{8}{3} = \frac{8}{3}.$$

45. $x - 3y = 9$      $9x + 3y = -3$
$-3y = -x + 9$    $3y = -9x - 3$
$y = \frac{-x+9}{-3}$   $y = \frac{-9x-3}{3}$
$y = \frac{1}{3}x - 3$   $y = -3x - 1$

Perpendicular. The two slopes, $m_1 = \frac{1}{3}$ and $m_2 = -3$, are negative reciprocals.

46. $6x - 8y = 16$    $3x + 4y = 28$
$-8y = -6x + 16$   $4y = -3x + 28$
$y = \frac{-6x+16}{-8}$   $y = \frac{-3x+28}{4}$
$y = \frac{3}{4}x - 2$   $y = -\frac{3}{4}x + 7$

Neither. The two slopes are $m_1 = \frac{3}{4}$ and $m_2 = -\frac{3}{4}$. The two slopes are not equal and are not negative reciprocals.

47. $2x - y = 3$       $-6x + 3y = 0$
$-y = -2x + 3$     $3y = 6x$
$y = \frac{-2x+3}{-1}$   $y = \frac{6x}{3}$
$y = 2x - 3$       $y = 2x$

Parallel. The two lines have equal slopes, $m = 2$, but different $y$-intercepts.

48. Perpendicular. The line $x = 2$ is vertical, and the line $y = 2$ is horizontal. Vertical lines are perpendicular to horizontal lines.

49. The slope of the line we seek is $m = -2$, the same as the slope of $y = -2x - 5$. The equation of the line is:
$y - 2 = -2(x - 1)$
$y - 2 = -2x + 2$
$y = -2x + 4$

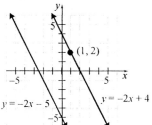

50. The slope of the line we seek is $m = \frac{5}{2}$, the same as the slope of the line
$5x - 2y = 8$
$-2y = -5x + 8$
$y = \frac{-5x+8}{-2}$
$y = \frac{5}{2}x - 4$

Thus, the equation of the line we seek is:
$y - 3 = \frac{5}{2}(x - 4)$
$y - 3 = \frac{5}{2}x - 10$
$y = \frac{5}{2}x - 7$

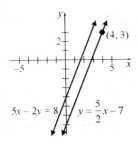

**51.** The line we seek is vertical, parallel to the vertical line $x = -3$. Now a vertical line passing through the point $(1, -4)$ must also pass through the point $(1, 0)$. That is, the x-intercept is 1. Thus, the equation of the line is $x = 1$.

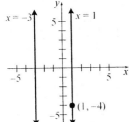

**52.** The slope of the line we seek is $m = -\dfrac{1}{3}$, the negative reciprocal of the slope of $y = 3x + 7$. The equation of the line is:

$$y - 2 = -\dfrac{1}{3}(x - 6)$$
$$y - 2 = -\dfrac{1}{3}x + 2$$
$$y = -\dfrac{1}{3}x + 4$$

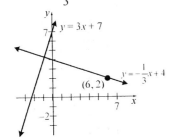

**53.** The slope of the line we seek is $m = \dfrac{4}{3}$, the negative reciprocal of the line

$3x + 4y = 6$
$4y = -3x + 6$
$y = \dfrac{-3x + 6}{4}$
$y = -\dfrac{3}{4}x + \dfrac{3}{2}$

Thus, the equation of the line we seek is:

$$y - (-2) = \dfrac{4}{3}(x - (-3))$$
$$y + 2 = \dfrac{4}{3}x + 4$$
$$y = \dfrac{4}{3}x + 2$$

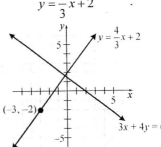

**54.** The line we seek is horizontal, perpendicular to the vertical line $x = 2$. Now a horizontal line passing through the point $(5, -4)$ must also pass through the point $(0, -4)$. That is, the y-intercept is -4. Thus, the equation of the line is $y = -4$.

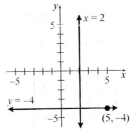

**55. a.** $x = 4, y = -2 \quad 5(4) + 3(-2) \stackrel{?}{\leq} 15$

$$20 - 6 \stackrel{?}{\leq} 15$$
$$14 \stackrel{?}{\leq} 15 \quad \text{True}$$

So, $(4, -2)$ is a solution to $5x + 3y \leq 15$.

**b.** $x = -6, y = 15 \quad 5(-6) + 3(15) \stackrel{?}{\leq} 15$

$$-30 + 45 \stackrel{?}{\leq} 15$$
$$15 \leq 15 \quad \text{True}$$

So, $(-6, 15)$ is a solution to $5x + 3y \leq 15$.

c. $x = 5, y = -1$  $5(5) + 3(-1) \stackrel{?}{\leq} 15$
$25 - 3 \stackrel{?}{\leq} 15$
$22 \stackrel{?}{\leq} 15$  False

So, $(5, -1)$ is not a solution to $5x + 3y \leq 15$.

56. a. $x = 2, y = 3$  $2 - 2(3) \stackrel{?}{>} -4$
$2 - 6 \stackrel{?}{>} -4$
$-4 \stackrel{?}{>} -4$  False

So, $(2, 3)$ is not a solution to $x - 2y > -4$.

b. $x = 5, y = -2$  $5 - 2(-2) \stackrel{?}{>} -4$
$5 + 4 \stackrel{?}{>} -4$
$9 \stackrel{?}{>} -4$  True

So, $(5, -2)$ is a solution to $x - 2y > -4$.

c. $x = -1, y = 3$  $(-1) - 2(3) \stackrel{?}{>} -4$
$-1 - 6 \stackrel{?}{>} -4$
$-7 \stackrel{?}{>} -4$  False

So, $(-1, 3)$ is not a solution to $x - 2y > -4$.

57. Replace the inequality symbol with an equal sign to obtain $y = 3x - 2$. Because the inequality is strict, graph $y = 3x - 2$ using a dashed line.

Test Point: $(0, 0)$:  $0 \stackrel{?}{<} 3(0) - 2$
$0 \stackrel{?}{<} -2$  False

Therefore, $(0, 0)$ is not a solution to $y < 3x - 2$.
Shade the half-plane that does not contain $(0, 0)$.

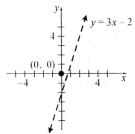

58. Replace the inequality symbol with an equal sign to obtain $2x - 4y = 12$. Because the inequality is non-strict, graph $2x - 4y = 12$ $\left(y = \frac{1}{2}x - 3\right)$ using a solid line.

Test Point: $(0, 0)$:  $2(0) - 4(0) \stackrel{?}{\leq} 12$
$0 \stackrel{?}{\leq} 12$  True

Therefore, $(0, 0)$ is a solution to $2x - 4y \leq 12$.
Shade the half-plane that contains $(0, 0)$.

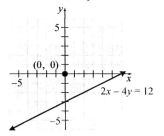

59. Replace the inequality symbol with an equal sign to obtain $3x + 4y = 20$. Because the inequality is strict, graph $3x + 4y > 20$ $\left(y = -\frac{3}{4}x + 5\right)$ using a dashed line.

Test Point: $(0, 0)$:  $3(0) + 4(0) \stackrel{?}{>} 20$
$0 \stackrel{?}{>} 20$  False

Therefore, $(0, 0)$ is not a solution to $3x + 4y > 20$. Shade the half-plane that does not contain $(0, 0)$.

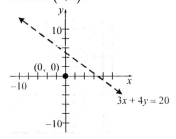

60. Replace the inequality symbol with an equal sign to obtain $y = 5$. Because the inequality is non-strict, graph $y = 5$ using a solid line.

Test Point: $(0, 0)$:  $0 \stackrel{?}{>} 5$  False

Therefore, $(0, 0)$ is not a solution to $y > 5$.

Shade the half-plane that does not contain $(0,0)$.

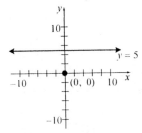

61. Replace the inequality symbol with an equal sign to obtain $2x+3y=0$. Because the inequality is strict, graph $2x+3y=0$ $\left(y=-\dfrac{2}{3}x\right)$ using a dashed line.

    Test Point: $(1,1)$: $2(1)+3(1)\stackrel{?}{<}0$

    $2+3\stackrel{?}{<}0$

    $5<0$ False

    Therefore, $(1,1)$ is not a solution to $2x+3x<0$. Shade the half-plane that does not contain $(1,1)$.

    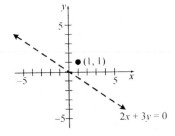

62. Replace the inequality symbol with an equal sign to obtain $x=-8$. Because the inequality is non-strict, graph $x=-8$ using a solid line.

    Test Point: $(0,0)$: $0\stackrel{?}{>}-8$ True

    Therefore, $(0,0)$ is a solution to $x>-8$. Shade the half-plane that contains $(0,0)$.

    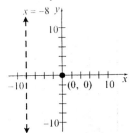

63. a. Let $x$ = the number of movie tickets.
    Let $y$ = the number of music CDs.
    $7.50x+15y\leq 60$

    b. $x=5, y=2$: $7.50(5)+15(2)\stackrel{?}{\leq}60$

    $37.50+30\stackrel{?}{\leq}60$

    $67.50\leq 60$ False

    No, Ethan cannot buy 5 movie tickets and 2 music tickets on his $60 budget.

    c. $x=2, y=2$: $7.50(2)+15(2)\stackrel{?}{\leq}60$

    $15+30\stackrel{?}{\leq}60$

    $45\leq 60$ True

    Yes, Ethan can buy 2 movie tickets and 2 music tickets on his $60 budget.

64. a. Let $x$ = the number of candy bars.
    Let $y$ = the number of candles.
    $0.50x+2y\geq 1000$

    b. $x=500, y=350$:

    $0.50(500)+2(350)\stackrel{?}{\geq}1000$

    $250+700\stackrel{?}{\geq}1000$

    $950\geq 1000$ False

    No, the math club will not earn enough money for the field trip by selling 500 candy bars and 350 candles.

    c. $x=600, y=400$:

    $0.50(600)+2(400)\stackrel{?}{\geq}1000$

    $300+800\stackrel{?}{\geq}1000$

    $1100\geq 1000$ True

    Yes, the math club will earn enough money for the field trip by selling 600 candy bars and 400 candles.

65. a. $C(m)=0.12m+35$

    b. The number of miles driven, $m$, is the independent variable. The rental cost, $C$, is the dependent variable.

    c. Because the number of miles cannot be negative, the it must be greater than or equal to zero. That is, the implied domain is $\{m\mid m\geq 0\}$ or, using interval notation, $[0,\infty)$.

## Chapter 3: Linear Functions and Their Graphs

**d.** $C(124) = 0.12(124) + 35 = 49.88$

If 124 miles are driven during a one-day rental, the charge will be $49.88.

**e.** $0.12m + 35 = 67.16$
$0.12m = 32.16$
$m = 268$

If the charge for a one-day rental is $67.16, then 268 miles were driven.

**f.**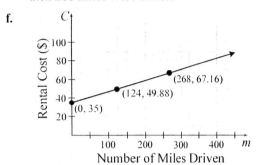

**66. a.** $B(x) = 3.50x + 33.99$

**b.** The number of pay-per-view movies watched, $x$, is the independent variable. The monthly bill, $B$, is the dependent variable.

**c.** Because the number pay-per-view movies watched cannot be negative, it must be greater than or equal to zero. That is, the implied domain is $\{x \mid x \geq 0\}$ or, using interval notation, $[0, \infty)$.

**d.** $B(5) = 3.50(5) + 33.99 = 51.49$

If 5 pay-per-view movies are watched one month, the bill will be $51.49.

**e.** $3.50x + 33.99 = 58.49$
$3.50x = 24.50$
$x = 7$

If the bill one month is $58.49, then 7 pay-per-view movies were watched.

**f.**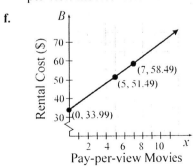

**67. a.** The tractor will depreciate by $\dfrac{\$84,600}{12} = \$7050$ per year. Thus, the slope is $-7050$. The y-intercept will be $84,600, the initial value of the tractor. The linear function that represents book value, $V$, of the tractor after $x$ years is
$V(x) = -7050x + 84,600$.

**b.** Because the tractor cannot have a negative age, the age, $x$, must be greater than or equal to 0. After 12 years, the book value will be $V(12) = -7050(12) + 84,600 = 0$, and the book value cannot be negative. Therefore the implied domain of function is $\{x \mid 0 \leq x \leq 12\}$ or, using interval notation, $[0, 12]$.

**c.** $V(5) = -7050(5) + 84,600 = 49,350$

After five years, the book value of the tractor will be $49,350.

**d.** The intercepts are $(0, 84600)$ and $(12, 0)$. The y-intercept is 84,600 and the x-intercept is 12.

**e.** $-7050x + 84,600 = 28,200$
$-7050x = -56,400$
$x = 8$

After eight years, the book value of the tractor will be $28,200.

**f.**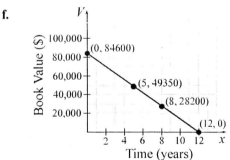

**68. a.** $y = kx$
$108 = k \cdot 9$
$k = \dfrac{108}{9} = 12$

**b.** $y = 12x$

**c.** $y = 12(4) = 48$

# Chapter 3 Review

**69. a.** $y = kx$
$11 = k \cdot 6$
$k = \dfrac{11}{6}$

**b.** $y = \dfrac{11}{6}x$

**c.** $y = \dfrac{11}{6}(24) = 44$

**70.** Because recommended dosage $d$ is directly proportional to weight $w$, we know that $d = kw$ for some constant $k$. We know that $d = 3024$ milligrams when $w = 168$ pounds, so
$3024 = k \cdot 168$
$k = 18$
Thus, $d = 18w$ or, using function notation, $d(w) = 18w$. Finally, $d(146) = 18(146) = 2628$. Therefore, the recommended dosage for a 146 pound person is 2628 milligrams.

**71.** Because perimeter $P$ varies directly with side length $s$, we know that $P = ks$ for some constant $k$. We know that $P = 32$ feet when $s = 8$ feet, so
$32 = k \cdot 8$
$k = 4$
Thus, $P = 4s$ or, using function notation, $P(s) = 4s$. Finally, $P(6) = 4(6) = 24$. Therefore, the perimeter of a square with a side length of 6 feet is 24 feet.

**72 a.**

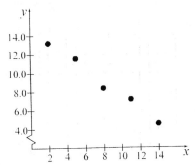

**b.** Answers will vary. We will use the points $(2, 13.3)$ and $(14, 4.6)$.
$m = \dfrac{4.6 - 13.3}{14 - 2} = \dfrac{-8.7}{12} = -0.725$
$y - 13.3 = -0.725(x - 2)$
$y - 13.3 = -0.725x + 1.45$
$y = -0.725x + 14.75$

**c.**

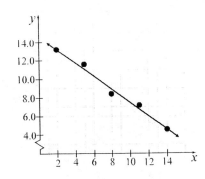

**73. a.**

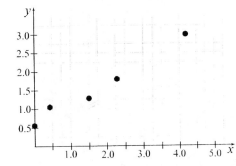

**b.** Answers will vary. We will use the points $(0, 0.6)$ and $(4.2, 3.0)$.
$m = \dfrac{3.0 - 0.6}{4.2 - 0} = \dfrac{2.4}{4.2} = \dfrac{4}{7}$
$y - 0.6 = \dfrac{4}{7}(x - 0)$
$y - 0.6 = \dfrac{4}{7}x$
$y = \dfrac{4}{7}x + 0.6$

**c.**

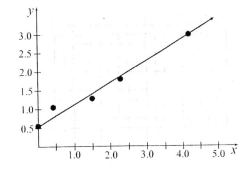

## Chapter 3: Linear Functions and Their Graphs

**74. a.**

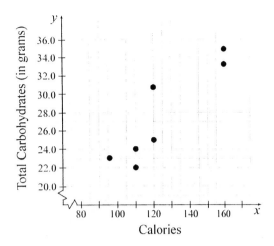

**b.** Approximately linear.

**c.** Answers will vary. We will use the points $(96, 23.2)$ and $(160, 33.3)$.

$$m = \frac{33.3 - 23.2}{160 - 96} = \frac{10.1}{64} \approx 0.158$$

$$y - 23.2 = 0.158(x - 96)$$
$$y - 23.2 = 0.158x - 15.168$$
$$y = 0.158x + 8.032$$

**d.**

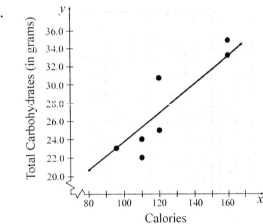

**e.** $x = 140$: $y = 0.158(140) + 8.032$
$= 30.152$

We predict that a one-cup serving of cereal having 140 calories will have approximately 30.2 grams of total carbohydrates.

**f.** The slope of the line found is 0.158. This means that, in a one-cup serving of cereal, total carbohydrates will increase by 0.158 grams for each one-calorie increase.

**75. a.**

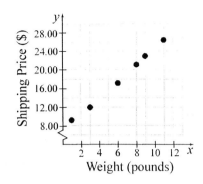

**b.** Approximately linear.

**c.** Answers will vary. We will use the points $(1, 9.25)$ and $(11, 26.50)$.

$$m = \frac{26.50 - 9.25}{11 - 1} = \frac{17.25}{10} = 1.725$$

$$y - 9.25 = 1.725(x - 1)$$
$$y - 9.25 = 1.725x - 1.725$$
$$y = 1.725x + 7.525$$

**d.**

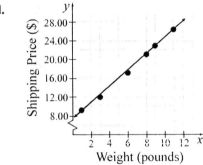

**e.** $x = 5$: $y = 1.725(5) + 7.525$
$= 16.15$

We predict that the FedEx 2Day® price for shipping a 5-pound package will be $16.15.

**f.** The slope of the line found is 1.725. This means that the FedEx 2Day shipping price increases by $1.725 for each one-pound increase in the weight of a package.

# Chapter 3 Test

1. Find the $y$-intercept by letting $x = 0$ and solving for $y$.
$$x = 0: \quad 0 - y = 8$$
$$-y = 8$$
$$y = -8$$
The $y$-intercept is -8, so the point (0, -8) is on the graph. Find the $x$-intercept by letting $y = 0$ and solving for $x$.
$$y = 0: \quad x - 0 = 8$$
$$x = 8$$
The $x$-intercept is 8, so the point (8, 0) is on the graph.

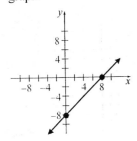

2. Let $x = -5, 0, 5,$ and $10$.
$$x = -5: \quad 3(-5) + 5y = 0$$
$$-15 + 5y = 0$$
$$5y = 15$$
$$y = 3$$

$$x = 0: \quad 3(0) + 5y = 0$$
$$0 + 5y = 0$$
$$y = 0$$

$$x = 5: \quad 3(5) + 5y = 0$$
$$15 + 5y = 0$$
$$5y = -15$$
$$y = -3$$

$$x = 10: \quad 3(10) + 5y = 0$$
$$30 + 5y = 0$$
$$5y = -30$$
$$y = -6$$

Thus, the points (-5, 3), (0, 0), (5, -3), and (10, -6) are on the graph.

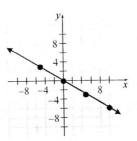

3. Find the $y$-intercept by letting $x = 0$ and solving for $y$.
$$x = 0: \quad 3(0) + 2y = 12$$
$$2y = 12$$
$$y = 6$$
The $y$-intercept is 6, so the point (0, 6) is on the graph. Find the $x$-intercept by letting $y = 0$ and solving for $x$.
$$y = 0: \quad 3x + 2(0) = 12$$
$$3x = 12$$
$$x = 4$$
The $x$-intercept is 4, so the point (4, 0) is on the graph.

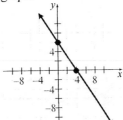

4. Let $x = -1, 0, 1,$ and $2$.
$$x = -1: \quad \frac{3}{2}(-1) - \frac{1}{4}y = 1$$
$$-\frac{3}{2} - \frac{1}{4}y = 1$$
$$-4\left(-\frac{3}{2} - \frac{1}{4}y\right) = -4(1)$$
$$6 + y = -4$$
$$y = -10$$

$$x = 0: \quad \frac{3}{2}(0) - \frac{1}{4}y = 1$$
$$-\frac{1}{4}y = 1$$
$$-4\left(-\frac{1}{4}y\right) = -4(1)$$
$$y = -4$$

$x = 1$: $\quad \dfrac{3}{2}(1) - \dfrac{1}{4}y = 1$

$\dfrac{3}{2} - \dfrac{1}{4}y = 1$

$-4\left(\dfrac{3}{2} - \dfrac{1}{4}y\right) = -4(1)$

$-6 + y = -4$

$y = 2$

$x = 2$: $\quad \dfrac{3}{2}(2) - \dfrac{1}{4}y = 1$

$3 - \dfrac{1}{4}y = 1$

$-\dfrac{1}{4}y = -2$

$-4\left(-\dfrac{1}{4}y\right) = -4(-2)$

$y = 8$

Thus, the points (-1, -10), (0, -4), (1, 2), and (2, 8) are on the graph.

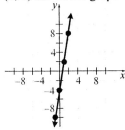

5. The equation $x = -7$ will be a vertical line with $x$-intercept -7.

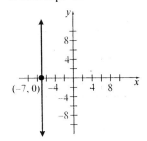

6. For $f(x) = -\dfrac{2}{3}x + 6$, the slope is $-\dfrac{2}{3}$ and the $y$-intercept is 6. Begin at (0, 6) and move to the right 3 units and down 2 units to find the point (3, 4). We can also move 3 units to the left and 2 units up to find the point (-3, 8).

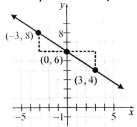

7. $m = \dfrac{y_2 - y_1}{x_2 - x_1} = \dfrac{6 - (-2)}{-1 - 5} = \dfrac{8}{-6} = -\dfrac{4}{3}$;

For every 3 units of increase in $x$, $y$ will decrease by 4 units. For every 3 units of decrease in $x$, $y$ will increase by 4 units.

8. Start at (2, -4). Because $m = -\dfrac{3}{5}$, move 5 units to the right and 3 units down to find point (7, -7). We can also move 5 units to the left and 3 units up to find point (-3, -1).

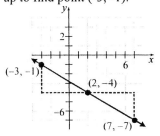

9. $8x - 2y = 1 \qquad\qquad x + 4y = -2$

$-2y = -8x + 1 \qquad\quad 4y = -x - 2$

$y = \dfrac{-8x + 1}{-2} \qquad\quad y = \dfrac{-x - 2}{4}$

$y = 4x - \dfrac{1}{2} \qquad\qquad y = -\dfrac{1}{4}x - \dfrac{1}{2}$

Perpendicular. The two lines have slopes that are negative reciprocals.

10. $y - y_1 = m(x - x_1)$

$y - 1 = 4(x - (-3))$

$y - 1 = 4(x + 3)$

$y - 1 = 4x + 12$

$y = 4x + 13$ or $4x - y = -13$

11. $m = \dfrac{y_2 - y_1}{x_2 - x_1} = \dfrac{7-1}{-3-6} = \dfrac{6}{-9} = -\dfrac{2}{3}$

$y - y_1 = m(x - x_1)$

$y - 1 = -\dfrac{2}{3}(x - 6)$

$y - 1 = -\dfrac{2}{3}x + 4$

$y = -\dfrac{2}{3}x + 5$ or $2x + 3y = 15$

12. The slope of the line we seek is $m = \dfrac{1}{5}$, the same as the slope of the line
$x - 5y = 15$
$-5y = -x + 15$
$y = \dfrac{-x + 15}{-5}$
$y = \dfrac{1}{5}x - 3$

Thus, the equation of the line we seek is:

$y - (-1) = \dfrac{1}{5}(x - 10)$

$y + 1 = \dfrac{1}{5}x - 2$

$y = \dfrac{1}{5}x - 3$ or $x - 5y = 15$

13. The slope of the line we seek is $m = -\dfrac{1}{3}$, the negative reciprocal of the slope of the line
$3x - y = 4$
$-y = -3x + 4$
$y = 3x - 4$

Thus, the equation of the line we seek is:

$y - 2 = -\dfrac{1}{3}(x - 6)$

$y - 2 = -\dfrac{1}{3}x + 2$

$y = -\dfrac{1}{3}x + 4$ or $x + 3y = 12$

14. a. $x = 3, y = -1$  $3(3) - (-1) \stackrel{?}{>} 10$

$9 + 1 \stackrel{?}{>} 10$

$10 \stackrel{?}{>} 10$  False

So, $(3, -1)$ is not a solution to $3x - y > 10$.

b. $x = 4, y = 5$  $3(4) - 5 \stackrel{?}{>} 10$

$12 - 5 \stackrel{?}{>} 10$

$7 \stackrel{?}{>} 10$  False

So, $(4, 5)$ is not a solution to $3x - y > 10$.

c. $x = 5, y = 3$  $3(5) - 3 \stackrel{?}{>} 10$

$15 - 3 \stackrel{?}{>} 10$

$12 \stackrel{?}{>} 10$  True

So, $(5, 3)$ is a solution to $3x - y > 10$.

15. Replace the inequality symbol with an equal sign to obtain $y = -2x + 1$. Because the inequality is non-strict, graph $y = -2x + 1$ using a solid line.

Test Point: $(0, 0)$: $0 \stackrel{?}{\leq} -2(0) + 1$

$0 \stackrel{?}{\leq} 1$  True

Therefore, $(0, 0)$ is a solution to $y \leq -2x + 1$.

Shade the half-plane that contains $(0, 0)$.

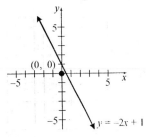

16. Replace the inequality symbol with an equal sign to obtain $5x - 2y = 0$. Because the inequality is strict, graph $5x - 2y = 0$ $\left(y = \dfrac{5}{2}x\right)$ using a dashed line.

Test Point: $(1, 1)$: $5(1) - 2(1) \stackrel{?}{<} 0$

$5 - 2 \stackrel{?}{<} 0$

$3 \stackrel{?}{<} 0$  False

Therefore, $(1, 1)$ is not a solution to $5x - 2y < 0$.

Shade the half-plane that does not contain $(1, 1)$.

## Chapter 3: Linear Functions and Their Graphs

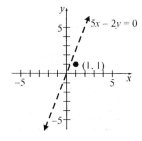

17. a. The profit is $30 - $12 = $18 times the number of shelves sold $x$, minus the $100 for renting the display. Thus, the function is $P(x) = 18x - 100$.

   b. The independent variable is the number of shelves sold, $x$. Henry could not sell a negative number of shelves. Thus, the domain of $P$ is $\{x \mid x \geq 0\}$ or, using interval notation, $[0, \infty)$.

   c. $P(34) = 18(34) - 100$
   $= 512$
   If Henry sells 34 shelves, his profit will be $512.

   d. Evaluate $P$ at $x = 0, 20$, and $50$.
   $P(0) = 18(0) - 100$
   $= -100$
   $P(20) = 18(20) - 100$
   $= 260$
   $P(50) = 18(50) - 100$
   $= 800$
   Thus, the points $(0, -100)$, $(20, 260)$, and $(50, 800)$ are on the graph.

   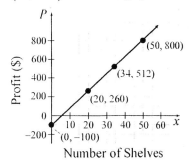

   e. We must solve $P(x) = 764$.
   $18x - 100 = 764$
   $18x = 864$
   $x = 48$

   If Henry sells 48 shelves, his profit will be $764.

18. a.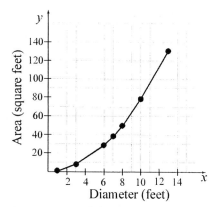

   b. Average rate of change
   $= \dfrac{7.07 - 0.79}{3 - 1}$
   $= 3.14$ square feet per foot
   Between diameters of 1 foot and 3 feet, the area of the circle increases at a rate of 3.14 square feet per foot.

   c. Average rate of change
   $= \dfrac{132.73 - 78.54}{13 - 10}$
   $\approx 18.06$ square feet per foot
   Between diameters of 10 feet and 13 feet, the area of the circle increases at a rate of approximately 18.06 square feet per foot.

   d. No. The area of circles is not linearly related to the diameter. The average rate of change (slope) is not constant.

19. Because recommended dosage $E$ is directly proportional to time worked $t$, we know that $E = kt$ for some constant $k$. We know that $E = \$296.40$ when $t = 24$ hours, so
   $296.40 = k \cdot 24$
   $k = 12.35$
   Thus, $E = 12.35t$ or, using function notation, $E(t) = 12.35t$. Finally,
   $E(10) = 12.35(10) = 123.5$. Therefore, Benjamin will earn $123.50 for 10 hours of work.

**20. a.**

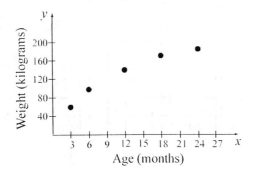

**b.** Approximately linear.

**c.** Answers will vary. We will use the points $(6, 95)$ and $(18, 170)$.

$$m = \frac{170 - 95}{18 - 6} = \frac{75}{12} = 6.25$$
$$y - 95 = 6.25(x - 6)$$
$$y - 95 = 6.25x - 37.5$$
$$y = 6.25x + 57.5$$

**d.**

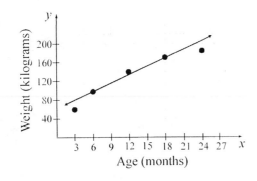

**e.** $x = 9$: $y = 6.25(9) + 57.5$
$= 113.75$

We predict that a nine month old Shetland pony will weigh 113.75 kilograms.

**f.** The slope of the line found is 6.25. This means that a Shetland pony's weight will increase by 6.25 kilograms for each one-month increase in age.

## Cumulative Review Chapters R – 3

**1.** $200 \div 25 \cdot (-2) = 8 \cdot (-2) = -16$

**2.** $\dfrac{3}{4} + \dfrac{1}{6} - \dfrac{2}{3} = \dfrac{9}{12} + \dfrac{2}{12} - \dfrac{8}{12}$
$= \dfrac{9 + 2 - 8}{12}$
$= \dfrac{3}{12}$
$= \dfrac{1}{4}$

**3.** $\dfrac{8 - 3(5 - 3^2)}{7 - 2 \cdot 6} = \dfrac{8 - 3(5 - 9)}{7 - 12}$
$= \dfrac{8 - 3(-4)}{-5}$
$= \dfrac{8 + 12}{-5}$
$= \dfrac{20}{-5}$
$= -4$

**4.** $(-3)^3 + 3(-3)^2 - 5(-3) - 7$
$= -27 + 3(9) - 5(-3) - 7$
$= -27 + 27 + 15 - 7$
$= 0 + 15 - 7$
$= 15 - 7$
$= 8$

**5.** $8m - 5m^2 - 3 + 9m^2 - 3m - 6$
$= -5m^2 + 9m^2 + 8m - 3m - 3 - 6$
$= 4m^2 + 5m - 9$

**6.** $8(n + 2) - 7 = 6n - 5$
$8n + 16 - 7 = 6n - 5$
$8n + 9 = 6n - 5$
$8n = 6n - 14$
$2n = -14$
$n = -7$

The solution set is {-7}.

## Chapter 3: Linear Functions and Their Graphs

7. $$\frac{2}{5}x + \frac{1}{6} = -\frac{2}{3}$$
$$30\left(\frac{2}{5}x + \frac{1}{6}\right) = 30\left(-\frac{2}{3}\right)$$
$$12x + 5 = -20$$
$$12x = -25$$
$$x = \frac{-25}{12}$$
The solution set is $\left\{-\frac{25}{12}\right\}$.

8. $$\frac{|2x-5|}{3} + 1 = 4$$
$$\frac{|2x-5|}{3} = 3$$
$$|2x-5| = 9$$
$2x - 5 = -9$ or $2x - 5 = 9$
$2x = -4$     $2x = 14$
$x = -2$     $x = 7$
The solution set is {-2, 7}.

9. $$A = \frac{1}{2}h(b+B)$$
$$2(A) = 2\left(\frac{1}{2}h(b+B)\right)$$
$$2A = h(b+B)$$
$$2A = hb + hB$$
$$2A - hb = hB$$
$$\frac{2A - hb}{h} = B$$
$$B = \frac{2A - hb}{h} \text{ or } B = \frac{2A}{h} - b$$

10. $6x - 7 > -31$
$6x > -24$
$x > -4$
$\{x \mid x > -4\}$ or $(-4, \infty)$

    ![number line]

11. $5(x-3) \geq 7(x-4) + 3$
$5x - 15 \geq 7x - 28 + 3$
$5x - 15 \geq 7x - 25$
$5x \geq 7x - 10$
$-2x \geq -10$
$x \leq 5$
$\{x \mid x \leq 5\}$ or $(-\infty, 5]$

    ![number line]

12. $|2x+5| < 9$
$-9 < 2x + 5 < 9$
$-9 - 5 < 2x + 5 - 5 < 9 - 5$
$-14 < 2x < 4$
$$\frac{-14}{2} < \frac{2x}{2} < \frac{4}{2}$$
$-7 < x < 2$
$\{x \mid -7 < x < 2\}$ or $(-7, 2)$

    ![number line]

13.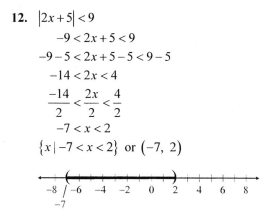

14. Because the x-coordinate -1 is paired with two different y-coordinates, 3 and 6, the relation is not a function.
Domain = {-1, 0, 1, 2}
Range = {-5, -2, 3, 4, 6}

15. The relation fails the vertical line test. That is, a vertical line will cross the graph in two places. Thus, the relation is not a function.
Domain = $\{x \mid 1 \leq x \leq 5\}$ = $[1, 5]$
Range = $\{y \mid -2 \leq y \leq 5\}$ = $[-2, 5]$

16. The relation passes the vertical line test. That is, a vertical line will only ever cross the graph in one place. Thus, the relation is a function.

Domain = {x|x is any real number} = (−∞, ∞)
Range = {y|y is any real number} = (−∞, ∞)

17. a. $H(1) = \dfrac{3(1)+5}{1-2} = \dfrac{3+5}{1-2} = \dfrac{8}{-1} = -8$

   b. $H\left(\dfrac{1}{2}\right) = \dfrac{3\left(\dfrac{1}{2}\right)+5}{\dfrac{1}{2}-2} = \dfrac{\dfrac{3}{2}+5}{\dfrac{1}{2}-2}$
   
   $= \dfrac{2\left(\dfrac{3}{2}+5\right)}{2\left(\dfrac{1}{2}-2\right)} = \dfrac{3+10}{1-4} = \dfrac{13}{-3} = -\dfrac{13}{3}$

18. a. $g(-4) = 2(-4) - 3 = -8 - 3 = -11$

    b. $g(x+h) = 2(x+h) - 3 = 2x + 2h - 3$

19. For $y = -\dfrac{1}{2}x + 4$, the slope is $-\dfrac{1}{2}$ and the y-intercept is 4. Begin at (0, 4) and move to the right 2 units and down 1 units to find the point (2, 3). We can also move 2 units to the left and 1 units up to find the point (-2, 5).

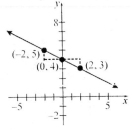

20. $4x - 5y = 15$
    $-5y = -4x + 15$
    $y = \dfrac{-4x+15}{-5}$
    $y = \dfrac{4}{5}x - 3$
    
    The slope is $\dfrac{4}{5}$ and the y-intercept is -3.
    
    Begin at the point (0, -3) and move to the right 5 units and up 4 units to find the point (5, 1).

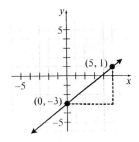

21. $m = \dfrac{y_2 - y_1}{x_2 - x_1} = \dfrac{10-(-2)}{-6-3} = \dfrac{12}{-9} = -\dfrac{4}{3}$
    $y - y_1 = m(x - x_1)$
    $y - 10 = -\dfrac{4}{3}(x-(-6))$
    $y - 10 = -\dfrac{4}{3}(x+6)$
    $y - 10 = -\dfrac{4}{3}x - 8$
    $y = -\dfrac{4}{3}x + 2$ or $4x + 3y = 6$

22. The slope of the line we seek is $m = -3$, the same as the slope of the line $y = -3x + 10$
    Thus, the equation of the line we seek is:
    $y - 7 = -3(x-(-5))$
    $y - 7 = -3(x+5)$
    $y - 7 = -3x - 15$
    $y = -3x - 8$ or $3x + y = -8$

23. Replace the inequality symbol with an equal sign to obtain $x - 3y = 12$. Because the inequality is strict, graph $x - 3y = 12$ $\left(y = \dfrac{1}{3}x - 4\right)$ using a dashed line.
    
    Test Point: $(0,0)$: $0 - 3(0) \overset{?}{>} 12$
    $0 \overset{?}{>} 12$   False
    
    Therefore, $(0,0)$ is a not a solution to $x - 3y > 12$. Shade the half-plane that does not contain $(0,0)$.

## Chapter 3: Linear Functions and Their Graphs

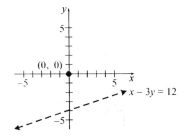

24. **a.** The independent variable is $t$, the number of years after 2000. The dependent variable is $C$, the projected number of cell phones that will be in use that year.

   **b.** $C(5) = 14.27(5) + 106.39 = 177.74$;
   $t = 5$ represents 5 years after 2000 which is the year 2005. $C(5)$ represents the projected number of cell phones that will be in use during the year 2005. That is, the function projects that 177.74 million cell phones will be in use in 2005.

   **c.** $C(-8) = 14.27(-8) + 106.39 = -7.77$;
   $t = -8$ represents 8 years before 2000 which is 1992. $C(-8)$ would represent the projected number of cell phones that was in use during the year 1992. That is, the function projects that -7.77 million cell phones were in use in 1992. This result is not reasonable. The number of cell phones in use could not be negative.

25. Because the amount of total carbohydrates $C$ is directly proportional to volume $V$, we know that $C = kV$ for some constant $k$. We know that $C = 21$ grams when $V = \dfrac{3}{4}$ cups, so

$$21 = k \cdot \dfrac{3}{4}$$
$$k = \dfrac{4}{3} \cdot 21$$
$$k = 28$$

Thus, $C = 28V$ or, using function notation, $C(V) = 28V$. Finally, $C(2) = 28(2) = 56$.
Therefore, 2 cups of Peanut Butter Crunch ® will contain 56 grams of total carbohydrates.

# Chapter 4

## 4.1 Preparing for Systems of Linear Equations in Two Variables

1. Substitute $x = 5$ and $y = 4$, and simplify:
   $$2x - 3y = 2(5) - 3(4)$$
   $$= 10 - 12$$
   $$= -2$$

2. $2(4) - 3(-1) \stackrel{?}{=} 11$
   $8 + 3 \stackrel{?}{=} 11$
   $11 = 11 \quad \leftarrow \text{True}$
   Yes, $(4, -1)$ is on the graph of $2x - 3y = 11$.

3. $y = 3x - 7$
   Let $x = 0, 1,$ and $2$.
   $x = 0:\ y = 3(0) - 7$
   $\phantom{x = 0:\ }y = 0 - 7$
   $\phantom{x = 0:\ }y = -7$
   $x = 1:\ y = 3(1) - 7$
   $\phantom{x = 1:\ }y = 3 - 7$
   $\phantom{x = 1:\ }y = -4$
   $x = 2:\ y = 3(2) - 7$
   $\phantom{x = 2:\ }y = 6 - 7$
   $\phantom{x = 2:\ }y = -1$
   Thus, the points $(0, -7)$, $(1, -4)$, and $(2, -1)$ are on the graph.

   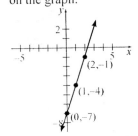

4. The slope of the line we seek is $m = -3$, the same as the slope of $y = -3x + 1$. The equation of the line is
   $$y - y_1 = m(x - x_1)$$
   $$y - 3 = -3(x - 2)$$
   $$y - 3 = -3x + 6$$
   $$y = -3x + 9$$

5. $4x - 3y = 15$
   $-3y = -4x + 15$
   $y = \dfrac{-4x + 15}{-3}$
   $y = \dfrac{4}{3}x - 5$
   The slope is $\dfrac{4}{3}$ and the $y$-intercept is $-5$.

6. The additive inverse of 4 is $-4$ because $4 + (-4) = 0$.

7. $2x - 3(-3x + 1) = -36$
   $2x + 9x - 3 = -36$
   $11x - 3 = -36$
   $11x = -33$
   $x = \dfrac{-33}{11} = -3$
   The solution set is $\{-3\}$.

## 4.1 Exercises

2. inconsistent

4. FALSE. If the graphs of the equations are parallel lines, then the system will have no solution.

6. TRUE

8. The lines may intersect at one point (consistent and independent), the lines may be parallel (inconsistent), or the lines may coincide (consistent and dependent).

10. Answers may vary. One possibility follows: When using either the method of substitution or the method of elimination, if the variables drop out leaving a true statement, then the system is consistent, but dependent. If the variables drop out leaving a false statement, then the system is inconsistent.

12. a. Let $x = -5, y = 3$ in both equations (1) and (2).

## Chapter 4: Systems of Linear Equations and Inequalities

$$\begin{cases} -5 - 2(3) = -5 - 6 = -11 \\ 3(-5) + 2(3) = -15 + 6 = -9 \neq -1 \end{cases}$$

Although these values satisfy equation (1), they do not satisfy equation (2). Therefore, the ordered pair $(-5, 3)$ is not a solution of the system.

b. Let $x = -3$, $y = 4$ in both equations (1) and (2).

$$\begin{cases} -3 - 2(4) = -3 - 8 = -11 \\ 3(-3) + 2(4) = -9 + 8 = -1 \end{cases}$$

Because these values satisfy both equations (1) and (2), the ordered pair $(-3, 4)$ is a solution of the system.

14. a. Let $x = -2$, $y = -1$ in both equations (1) and (2).

$$\begin{cases} -3(-2) + (-1) = 6 - 1 = 5 \\ 6(-2) - 2(-1) = -12 + 2 = -10 \neq 6 \end{cases}$$

Although these values satisfy equation (1), they do not satisfy equation (2). Therefore, the ordered pair $(-2, -1)$ is not a solution of the system.

b. Let $x = 2$, $y = 0$ in both equations (1) and (2).

$$\begin{cases} -3(2) + 0 = -6 + 0 = -6 \neq 5 \\ 6(2) - 2(0) = 12 - 0 = 12 \neq 6 \end{cases}$$

Because these values do not satisfy the equations, the ordered pair $(2, 0)$ is not a solution of the system.

16. The graphs of the linear equations intersect at the point $(-1, 2)$. Thus, the solution is $(-1, 2)$.

18. The graphs of the linear equations are coincident. The system is dependent. The solution is $\{(x, y) \mid 3x + y = 1\}$.

20. $\begin{cases} y = -2x + 4 & (1) \\ y = 2x - 4 & (2) \end{cases}$

The two equations are in slope-intercept form. Graph each equation and find the point of intersection.

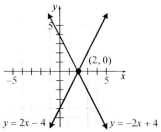

The solution is the ordered pair $(2, 0)$.

22. $\begin{cases} y = -\dfrac{2}{3}x + 3 & (1) \\ 2x + 3y = 9 & (2) \end{cases}$

Equation (2) in slope-intercept form is $y = -\dfrac{2}{3}x + 3$, which is the same as equation (1).

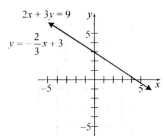

The two lines are coincident. The system is dependent. The solution is $\left\{(x, y) \mid y = -\dfrac{2}{3}x + 3\right\}$.

24. $\begin{cases} -x + 2y = -9 & (1) \\ 2x + y = -2 & (2) \end{cases}$

Equation (1) in slope-intercept form is $y = \dfrac{1}{2}x - \dfrac{9}{2}$. Equation (2) in slope-intercept form is $y = -2x - 2$. Graph each equation and find the point of intersection.

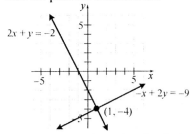

The solution is the ordered pair $(1, -4)$.

ISM – 154

**Section 4.1** *Systems of Linear Equations in Two Variables*

26. $\begin{cases} y = -3x - 4 & (1) \\ y = 4x + 17 & (2) \end{cases}$

    Substituting $-3x - 4$ for $y$ in equation (2), we obtain
    $-3x - 4 = 4x + 17$
    $-7x - 4 = 17$
    $-7x = 21$
    $x = -3$
    Substituting $-3$ for $x$ in equation (2), we obtain
    $y = 4(-3) + 17 = -12 + 17 = 5$.
    The solution is the ordered pair $(-3, 5)$.

28. $\begin{cases} y = \dfrac{1}{2}x & (1) \\ x - 4y = -4 & (2) \end{cases}$

    Substituting $\dfrac{1}{2}x$ for $y$ in equation (2), we obtain
    $x - 4\left(\dfrac{1}{2}x\right) = -4$
    $x - 2x = -4$
    $-x = -4$
    $x = 4$
    Substituting 4 for $x$ in equation (1), we obtain
    $y = \dfrac{1}{2}(4) = 2$.
    The solution is the ordered pair $(4, 2)$.

30. $\begin{cases} -2x + 4y = 9 & (1) \\ x - 2y = -3 & (2) \end{cases}$

    Equation (2) solved for $x$ is $x = 2y - 3$.
    Substituting $2y - 3$ for $x$ in equation (1), we obtain
    $-2(2y - 3) + 4y = 9$
    $-4y + 6 + 4y = 9$
    $6 = 9$
    The system has no solution. The solution set is $\varnothing$ or $\{\ \}$. The system is inconsistent.

32. $\begin{cases} 3x + 2y = 0 & (1) \\ 6x + 2y = 5 & (2) \end{cases}$

    Equation (1) solved for $y$ is $y = -\dfrac{3}{2}x$.
    Substituting $-\dfrac{3}{2}x$ for $y$ in equation (2), we obtain

    $6x + 2\left(-\dfrac{3}{2}x\right) = 5$
    $6x - 3x = 5$
    $3x = 5$
    $x = \dfrac{5}{3}$
    Substituting $\dfrac{5}{3}$ for $x$ in equation (1), we obtain
    $3\left(\dfrac{5}{3}\right) + 2y = 0$
    $5 + 2y = 0$
    $2y = -5$
    $y = -\dfrac{5}{2}$
    The solution is the ordered pair $\left(\dfrac{5}{3}, -\dfrac{5}{2}\right)$.

34. $\begin{cases} -4x + y = 8 & (1) \\ x - \dfrac{y}{4} = -2 & (2) \end{cases}$

    Equation (1) solved for $y$ is $y = 4x + 8$.
    Substituting $4x + 8$ for $y$ in equation (2), we obtain
    $x - \dfrac{4x + 8}{4} = -2$
    $x - x - 2 = -2$
    $-2 = -2$
    The system is dependent. The solution is $\{(x, y) \mid -4x + y = 8\}$.

36. $\begin{cases} x + y = 5000 & (1) \\ 0.04x + 0.08y = 340 & (2) \end{cases}$

    Equation (1) solved for $x$ is $x = 5000 - y$.
    Substituting $5000 - y$ for $x$ in equation (2), we obtain
    $0.04(5000 - y) + 0.08y = 340$
    $200 - 0.04y + 0.08y = 340$
    $0.04y + 200 = 340$
    $0.04y = 140$
    $y = 3500$
    Substituting 3500 for $y$ in equation (1), we obtain
    $x + 3500 = 5000$
    $x = 1500$
    The solution is the ordered pair $(1500, 3500)$.

## Chapter 4: Systems of Linear Equations and Inequalities

**38.** $\begin{cases} x+y=-6 & (1) \\ -2x-y=0 & (2) \end{cases}$

Add equations (1) and (2).

$\begin{cases} x+y=-6 \\ -2x-y=0 \end{cases}$
$\overline{-x\phantom{+y}=-6}$
$x=6$

Substituting 6 for $x$ in equation (1), we obtain
$6+y=-6$
$y=-12$

The solution is the ordered pair $(6,-12)$.

**40.** $\begin{cases} -3x+2y=-5 & (1) \\ 2x-\phantom{2}y=10 & (2) \end{cases}$

Multiply both sides of equation (2) by 2, and add the result to equation (1).

$\begin{cases} -3x+2y=-5 \\ 4x-2y=20 \end{cases}$
$\overline{\phantom{-3}x\phantom{+2y}=15}$

Substituting 15 for $x$ in equation (2), we obtain
$2(15)-y=10$
$30-y=10$
$-y=-20$
$y=20$

The solution is the ordered pair $(15,\ 20)$.

**42.** $\begin{cases} 6x-4y=6 & (1) \\ -3x+2y=3 & (2) \end{cases}$

Multiply both sides of equation (2) by 2 and add the result to equation (1).

$\begin{cases} 6x-4y=6 \\ -6x+4y=6 \end{cases}$
$\overline{\phantom{6x-4y}0=12}$

The system has no solution. The solution set is $\varnothing$ or $\{\ \}$. The system is inconsistent.

**44.** $\begin{cases} x+2y=-\dfrac{8}{3} & (1) \\ 3x-3y=5 & (2) \end{cases}$

Multiply both sides of equation (1) by $-3$ and add the result to equation (2).

$\begin{cases} -3x-6y=8 & (1) \\ 3x-3y=5 & (2) \end{cases}$
$\overline{\phantom{-3x}-9y=13}$
$y=-\dfrac{13}{9}$

Substituting $-\dfrac{13}{9}$ for $y$ in equation (2), we obtain

$3x-3\left(-\dfrac{13}{9}\right)=5$

$3x+\dfrac{13}{3}=5$

$3\left(3x+\dfrac{13}{3}\right)=3(5)$

$9x+13=15$
$9x=2$
$x=\dfrac{2}{9}$

The solution is the ordered pair $\left(\dfrac{2}{9},-\dfrac{13}{9}\right)$.

**46.** $\begin{cases} \dfrac{5}{4}x-\dfrac{1}{2}y=6 & (1) \\ -\dfrac{5}{3}x+\dfrac{2}{3}y=-8 & (2) \end{cases}$

Multiply both sides of equation (1) by 4, multiply both sides of equation (2) by 3, and add the results.

$\begin{cases} 5x-2y=24 \\ -5x+2y=-24 \end{cases}$
$\overline{\phantom{5x-2y}0=0}$

The system is dependent. The solution is

$\left\{(x,y)\ \Big|\ \dfrac{5}{4}x-\dfrac{1}{2}y=6\right\}$.

**48.** $\begin{cases} 0.04x+0.06y=2.1 & (1) \\ 0.06x-0.03y=0.15 & (2) \end{cases}$

Multiply both sides of equation (2) by 2, and add the result to equation (1).

$\begin{cases} 0.04x+0.06y=2.1 \\ 0.12x-0.06y=0.3 \end{cases}$
$\overline{0.16x\phantom{+0.06y}=2.4}$
$x=15$

Substituting 15 for $x$ in equation (1), we obtain

## Section 4.1 Systems of Linear Equations in Two Variables

$0.04(15) + 0.06y = 2.1$
$0.6 + 0.06y = 2.1$
$0.06y = 1.5$
$y = 25$

The solution is the ordered pair $(15, 25)$.

**50.** $\begin{cases} 2x + y = -1 & (1) \\ -3x - 2y = 7 & (2) \end{cases}$

Because the coefficient of $y$ in equation (1) is 1, we use substitution to solve the system.
Equation (1) solved for $y$ is $y = -2x - 1$.
Substituting $-2x - 1$ for $y$ in equation (2), we obtain
$-3x - 2(-2x - 1) = 7$
$-3x + 4x + 2 = 7$
$x + 2 = 7$
$x = 5$
Substituting 5 for $x$ in equation (1), we obtain
$2(5) + y = -1$
$10 + y = -1$
$y = -11$

The solution is the ordered pair $(5, -11)$.

**52.** $\begin{cases} y = \dfrac{1}{2}x + 2 & (1) \\ x - 2y = -4 & (2) \end{cases}$

Because equation (1) is already solved for $y$, we use substitution to solve the system.
Substituting $\dfrac{1}{2}x + 2$ for $y$ in equation (2), we obtain
$x - 2\left(\dfrac{1}{2}x + 2\right) = -4$
$x - x - 4 = -4$
$-4 = -4$
The system is dependent. The solution is
$\left\{(x, y) \,\middle|\, y = \dfrac{1}{2}x + 2\right\}$.

**54.** $\begin{cases} 12x + 45y = 0 & (1) \\ 8x + 6y = 24 & (2) \end{cases}$

Because none of variables have a coefficient of 1, we use elimination to solve the system.
Multiply both sides of equation (1) by 2, multiply both sides of equation (2) by $-3$, and add the results.

$\begin{cases} 24x + 90y = 0 \\ -24x - 18y = -72 \end{cases}$
$\overline{\phantom{xxxxx}72y = -72}$
$y = -1$

Substituting $-1$ for $y$ in equation (2), we obtain
$8x + 6(-1) = 24$
$8x - 6 = 24$
$8x = 30$
$x = \dfrac{30}{8} = \dfrac{15}{4}$

The solution is the ordered pair $\left(\dfrac{15}{4}, -1\right)$.

**56.** $\begin{cases} \dfrac{1}{3}x - \dfrac{1}{2}y = -5 & (1) \\ -\dfrac{4}{5}x + \dfrac{6}{5}y = 1 & (2) \end{cases}$

Because none of the variables have a coefficient of 1, we use elimination to solve the system.
Multiply both sides of equation (1) by 12, multiply both sides of equation (2) by 5, and add the results.

$\begin{cases} 4x - 6y = -60 \\ -4x + 6y = 5 \end{cases}$
$\overline{\phantom{xxxxx}0 = -55}$

The system has no solution. The solution set is $\varnothing$ or $\{\ \}$. The system is inconsistent.

**58.** $\begin{cases} 4x - 2y = 8 & (1) \\ -10x + 5y = 5 & (2) \end{cases}$

Write each equation in slope-intercept form.

$4x - 2y = 8 \qquad\qquad -10x + 5y = 5$
$-2y = -4x + 8 \qquad\qquad 5y = 10x + 5$
$y = \dfrac{-4x + 8}{-2} \qquad\qquad y = \dfrac{10x + 5}{5}$
$y = 2x - 4 \qquad\qquad y = 2x + 1$

Since the equations have the same slope but different $y$-intercepts, the lines are parallel.

Chapter 4: Systems of Linear Equations and Inequalities

Thus, the system in inconsistent and has no solution.

60. $\begin{cases} 2x - y = -5 & (1) \\ -4x + 3y = 9 & (2) \end{cases}$

Write each equation in slope-intercept form.

$2x - y = -5 \qquad -4x + 3y = 9$
$-y = -2x - 5 \qquad 3y = 4x + 9$
$y = 2x + 5 \qquad y = \dfrac{4}{3}x + 3$

Since the equations have different slopes, the lines are neither parallel nor coincident. Thus, the system must have exactly one solution.

62. a. Equation of line through $(-1, 3)$ and $(3, 1)$:

$m = \dfrac{1 - 3}{3 - (-1)} = \dfrac{-2}{4} = -\dfrac{1}{2}$

$y - 1 = -\dfrac{1}{2}(x - 3)$

$y - 1 = -\dfrac{1}{2}x + \dfrac{3}{2}$

$y = -\dfrac{1}{2}x + \dfrac{5}{2}$

Equation of line through $(-1, -2)$ and $(3, 6)$:

$m = \dfrac{y_2 - y_1}{x_2 - x_1} = \dfrac{6 - (-2)}{3 - (-1)} = \dfrac{8}{4} = 2$

$y - 6 = 2(x - 3)$
$y - 6 = 2x - 6$
$y = 2x$

b. Solve the system formed by the equations found in part a.

$\begin{cases} y = 2x & (1) \\ y = -\dfrac{1}{2}x + \dfrac{5}{2} & (2) \end{cases}$

Substituting $2x$ for $y$ in equation (2), we obtain

$2x = -\dfrac{1}{2}x + \dfrac{5}{2}$

$2(2x) = 2\left(-\dfrac{1}{2}x + \dfrac{5}{2}\right)$

$4x = -x + 5$
$5x = 5$
$x = 1$

Substituting 1 for $x$ in equation (1), we obtain
$y = 2(1) = 2$.
The solution is the ordered pair $(1, 2)$.

c. The slopes are 2 and $-\dfrac{1}{2}$, which are negative reciprocals. The diagonals of the rhombus are perpendicular.

64. One of the lines in the graph has a positive slope and a positive y-intercept. The other line in the graph has a negative slope and a positive y-intercept. Now, the first equation in part (b) has a negative slope and positive y-intercept, and the second equation has a positive slope and a positive y-intercept, so the system in part (b) could have the graph shown. The equations in (a) both have negative slope, so the system cannot have the graph shown. The first equation in part (c) has a negative y-intercept, so the system cannot have the graph shown.

66. Answers will vary. One possibility follows:

$\begin{cases} x + y = 8 \\ x - y = -2 \end{cases}$

68. We can find the centroid of the triangle by finding the intersection point of any two of the triangle's medians. First, we must find the equations of two of the medians.

Equation of line through $(3, 1)$ and $(4, 6)$:

$m = \dfrac{6 - 1}{4 - 3} = \dfrac{5}{1} = 5$

$y - 1 = 5(x - 3)$
$y - 1 = 5x - 15$
$y = 5x - 14$

Equation of line through $(0, 0)$ and $(5, 4)$:

$m = \dfrac{y_2 - y_1}{x_2 - x_1} = \dfrac{4 - 0}{5 - 0} = \dfrac{4}{5}$

$y - 0 = \dfrac{4}{5}(x - 0)$

$y = \dfrac{4}{5}x$

Solve the system $\begin{cases} y = 5x - 14 & (1) \\ y = \dfrac{4}{5}x & (2) \end{cases}$

Substitute $\frac{4}{5}x$ for $y$ into equation (1).

$$\frac{4}{5}x = 5x - 14$$
$$5\left(\frac{4}{5}x\right) = 5(5x - 14)$$
$$4x = 25x - 70$$
$$-21x = -70$$
$$x = \frac{-70}{-21} = \frac{10}{3}$$

Substitute $\frac{10}{3}$ for $x$ in equation (2).

$$y = \frac{4}{5}\left(\frac{10}{3}\right) = \frac{8}{3}.$$

The centroid of the triangle is $\left(\frac{10}{3}, \frac{8}{3}\right)$.

70. $\begin{cases} y = \frac{3}{2}x - 4 & (1) \\ y = -\frac{1}{4}x + 3 & (2) \end{cases}$

The solution is the ordered pair $(4, 2)$.

72. $\begin{cases} -6x - 2y = 4 & (1) \\ 5x + 3y = -2 & (2) \end{cases}$

Writing each equation in slope-intercept form, we obtain $y = -3x - 2$ and $y = -\frac{5}{3}x - \frac{2}{3}$.

 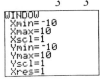

The solution is the ordered pair $(-1, 1)$.

73. $\begin{cases} 4x - 3y = 1 & (1) \\ -8x + 6y = -2 & (2) \end{cases}$

Writing each equation in slope-intercept form, we obtain the same equation for both, $y = \frac{4}{3}x - \frac{1}{3}$.

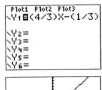 

The two lines are coincident. The system is dependent. The solution is $\{(x, y) \mid 4x - 3y = 1\}$.

74. $\begin{cases} -2x + 5y = -2 & (1) \\ 4x - 10y = 1 & (2) \end{cases}$

Writing each equation in slope-intercept form, we obtain $y = \frac{2}{5}x - \frac{2}{5}$ and $y = \frac{2}{5}x - \frac{1}{10}$.

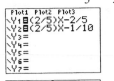 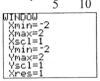

The system has no solution. The solution set is $\varnothing$ or $\{\ \}$. The system is inconsistent.

76. $\begin{cases} x - 3y = 21 & (1) \\ x + 6y = -2 & (2) \end{cases}$

Writing each equation in slope-intercept form, we obtain $y = \frac{1}{3}x - 7$ and $y = -\frac{1}{6}x - \frac{1}{3}$.

 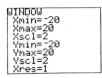

## Chapter 4: Systems of Linear Equations and Inequalities

The solution is approximately the ordered pair $(13.33, -2.56)$.

### 4.2 Preparing for Problem Solving

1. Step 1: Identify what you are looking for.
   Step 2: Give names to the unknowns.
   Step 3: Translate the problem into the language of mathematics.
   Step 4: Solve the equation(s) found in Step 3.
   Step 5: Check the reasonableness of your answer.
   Step 6: Answer the question.

2. Since the total to be invested is $25,000 and the amount to be invested in stocks is $s$, the amount to be invested in bonds is $25,000 - s$.

3. The total interest for the year would be $0.125(\$3500) = \$437.50$. Since there are 12 months in a year, the interest that would be paid after one month would be $\dfrac{\$437.50}{12} \approx \$36.46$.

4. A linear cost function will be of the form $C(x) = mx + b$. For this problem, the variable cost is $m = \$15$ and the fixed cost is $b = \$500$. Thus, the cost function is $C(x) = 15x + 500$.

### 4.2 Exercises

2. Answers will vary.

4. Let $x$ and $y$ represent the two numbers.
   $\begin{cases} x + y = 25 & (1) \\ x - y = 3 & (2) \end{cases}$
   Adding equations (1) and (2) result in
   $2x = 28$
   $x = 14$
   Substituting 14 for $x$ in equation (1), we obtain
   $14 + y = 25$
   $y = 11$
   The two numbers are 14 and 11.

6. Let $x$ represent the first number and let $y$ represent the second number.
   $\begin{cases} 3x - y = 118 & (1) \\ 2x + y = 147 & (2) \end{cases}$
   Adding equations (1) and (2) results in
   $5x = 265$
   $x = 53$
   Substituting 53 for $x$ in equation (2), we obtain
   $2(53) + y = 147$
   $106 + y = 147$
   $y = 41$
   The 1st number is 53 and the 2nd number is 41.

8. Let $x$ represent the first number and let $y$ represent the second number.
   $\begin{cases} 4x + y = 68 & (1) \\ x - 2y = -1 & (2) \end{cases}$
   Multiply both sides of equation (1) by 2 and add the result to equation (1).
   $\begin{cases} 8x + 2y = 136 \\ x - 2y = -1 \end{cases}$
   $9x \quad\quad = 135$
   $x = 15$
   Substituting 15 for $x$ in equation (1), we obtain
   $4(15) + y = 68$
   $60 + y = 68$
   $y = 8$
   The first number is 15 and the second number is 8.

10. a. $\begin{cases} R(x) = 16x \\ C(x) = 7x + 3645 \end{cases}$

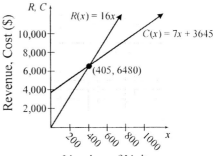

**Section 4.2** Problem Solving: Systems of Two Linear Equations with Two Unknowns

b. $R(x) = C(x)$
$16x = 7x + 3645$
$9x = 3645$
$x = 405$

$R(405) = 16(405) = 6480$
$C(405) = 7(405) + 3645 = 6480$

The break-even point is $(405, 6480)$.

12. Let $d$ represent the number of dimes and let $q$ represent the number of quarters.
$\begin{cases} 0.10d + 0.25q = 6.75 & (1) \\ d = q + 8 & (2) \end{cases}$
Substituting $q + 8$ for $d$ in equation (1), we get
$0.10(q + 8) + 0.25q = 6.75$
$0.10q + 0.80 + 0.25q = 6.75$
$0.35q + 0.80 = 6.75$
$0.35q = 5.95$
$q = 17$
Substituting 17 for $q$ in equation (2), we obtain
$d = 17 + 8 = 25$.
Johnny has 17 quarters and 25 dimes.

14. Let $b$ represent the number of grams of carbohydrates a McDonald's sausage biscuit and let $j$ represent the number of grams of carbohydrates in an orange juice.
$\begin{cases} 2b + j = 98 & (1) \\ 3b + 2j = 168 & (2) \end{cases}$
Multiply both sides of equation (1) by $-2$ and add the result to equation (1).
$\begin{cases} -4b - 2j = -196 \\ 3b + 2j = 168 \end{cases}$
$-b = -28$
$b = 28$
Substituting 28 for $b$ in equation (1), we obtain
$2(28) + j = 98$
$56 + j = 98$
$j = 42$
A McDonald's sausage biscuit contains 28 grams of carbohydrates. An orange juice contains 42 grams of carbohydrates.

16. Let $x$ and $y$ represent the length and width of the rectangle, respectively.
$\begin{cases} 2x + 2y = 260 & (1) \\ y = x - 15 & (2) \end{cases}$

Substituting $x - 15$ for $y$ in equation (1), we obtain
$2x + 2(x - 15) = 260$
$2x + 2x - 30 = 260$
$4x - 30 = 260$
$4x = 290$
$x = 72.5$
Substituting 72.5 for $x$ in equation (2), we obtain
$y = 72.5 - 15 = 57.5$.
The dimensions of the rectangle are 72.5 cm by 57.5 cm.

18. Let $x$ represent the length of one of the remaining sides and let $y$ represent the length of the other side.
$\begin{cases} x + y + 2(12) = 100 & (1) \\ y = 2x - 14 & (2) \end{cases}$
Substituting $2x - 14$ for $y$ in equation (1), we obtain
$x + (2x - 14) + 2(12) = 100$
$x + 2x - 14 + 24 = 100$
$3x + 10 = 100$
$3x = 90$
$x = 30$
Substituting 30 for $x$ in equation (2), we obtain
$y = 2(30) - 14 = 46$.
The lengths of the remaining two sides are 30 cm and 46 cm.

20. Because angles 1 and 2 are supplemental, and because angles 1 and 3 must be equal, we obtain the following system.
$\begin{cases} (2x + 3y) + 4x = 180 & (1) \\ 2x + 3y = 5x + y + 1 & (2) \end{cases}$
Simplify each equation.
$\begin{cases} 6x + 3y = 180 & (1) \\ -3x + 2y = 1 & (2) \end{cases}$
Multiply both sides of equation (2) by 2 and add the result to equation (1).
$\begin{cases} 6x + 3y = 180 \\ -6x + 4y = 2 \end{cases}$
$7y = 182$
$y = 26$

Substituting 26 for $y$ in equation (1), we obtain
$$6x + 3(26) = 180$$
$$6x + 78 = 180$$
$$6x = 102$$
$$x = 17$$
Thus, $x = 17$ and $y = 26$.

22. Because angles 1 and 3 are equal, and because angles 1 and 2 are supplemental, and because, we obtain the following system.
$$\begin{cases} 5x + 7y = 15x - 9y & (1) \\ (5x + 7y) + (10x + 5y) = 180 & (2) \end{cases}$$
Simplify each equation.
$$\begin{cases} -10x + 16y = 0 & (1) \\ 15x + 12y = 180 & (2) \end{cases}$$
Multiply both sides of equation (1) by 2, multiply both sides of equation (2) by 3, and add the results.
$$\begin{cases} -30x + 48y = 0 \\ 30x + 24y = 360 \end{cases}$$
$$72y = 360$$
$$y = 5$$
Substituting 5 for $y$ in equation (2), we obtain
$$15x + 12(5) = 180$$
$$15x + 60 = 180$$
$$15x = 120$$
$$x = 8$$
Thus, $x = 8$ and $y = 5$.

24. Let $s$ represent the amount to be invested in stocks and let $b$ represent the amount to be invested in bonds.
$$\begin{cases} s + b = 80,000 & (1) \\ 3b = 2s & (2) \end{cases}$$
Write equation (2) in standard form.
$$\begin{cases} s + b = 80,000 & (1) \\ -2s + 3b = 0 & (2) \end{cases}$$
Multiply both sides of equation (1) by 2 and add the result to equation (2).
$$\begin{cases} 2s + 2b = 160,000 \\ -2s + 3b = 0 \end{cases}$$
$$5b = 160,000$$
$$b = 32,000$$
Substituting 32,000 for $b$ in equation (1), we obtain

$$s + 32,000 = 80,000$$
$$s = 48,000$$
Invest $48,000 in stocks and $32,000 in bonds.

26. Let $x$ represent the amount invested in the 1sr stock that earned 13% and let $y$ represent the amount invested in the $2^{nd}$ stock that declined by 5%. We organize the information in the table below, using the interest formula:
Interest = Principal × rate × time or $I = Prt$.

|  | $P$ | $r$ | $t$ | $I$ |
|---|---|---|---|---|
| $1^{st}$ stock | $x$ | 0.13 | 1 | $0.13x$ |
| $2^{nd}$ stock | $y$ | −0.05 | 1 | $-0.05y$ |
| Total | 70,000 |  | 1 | 2800 |

Using the columns for Principal and Interest, we obtain the following system.
$$\begin{cases} x + y = 70,000 & (1) \\ 0.13x - 0.05y = 2800 & (2) \end{cases}$$
Solving equation (1) for $y$, we obtain $y = 70,000 - x$. Substituting $70,000 - x$ for $y$ in equation (2), we obtain
$$0.13x - 0.05(70,000 - x) = 2800$$
$$0.13x - 3500 + 0.05x = 2800$$
$$0.18x - 3500 = 2800$$
$$0.18x = 6300$$
$$x = 35,000$$
Substituting 35,000 for $x$ in equation (1) yields
$$35,000 + y = 70,000$$
$$y = 35,000$$
Thus, $35,000 was invested in each of the two stocks.

28. Let $a$ represent the weight of chocolate-covered almonds to be used and let $p$ represent the weight of chocolate-covered peanuts to be used. We organize the information in the table below, using the formula
Price per pound × number of pounds = Revenue
.

|  | Price per pound | Number of pounds | Revenue |
|---|---|---|---|
| Almonds | 6.50 | $a$ | $6.5a$ |
| Peanuts | 4.00 | $p$ | $4p$ |
| Bridge Mix | 6.00 | 50 | $6(50) = 300$ |

Using the columns for Number of pounds and Revenue, we obtain the following system.

### Section 4.2 Problem Solving: Systems of Two Linear Equations with Two Unknowns

$\begin{cases} a + p = 50 & (1) \\ 6.5a + 4p = 300 & (2) \end{cases}$

Solving equation (1) for $a$ yields $a = 50 - p$.
Substituting $50 - p$ for $a$ in equation (2), we obtain $6.5(50 - p) + 4p = 300$
$$325 - 6.5p + 4p = 300$$
$$325 - 2.5p = 300$$
$$-2.5p = -25$$
$$p = 10$$

Substituting 10 for $p$ in equation (1) yields
$$a + 10 = 50$$
$$a = 40$$

The bridge mix requires 10 pounds of chocolate-covered peanuts and 40 pounds of chocolate-covered almonds.

30. Let $x$ represent the number of units of the first powder and let $y$ represent the number of units of the second powder to be used. Then the amount of vitamin $B_{12}$ in the two powders is $0.20x + 0.50y$. The amount of vitamin E in the two powders is $0.40x + 0.30y$. This results in the following system:

$\begin{cases} 0.20x + 0.50y = 20 & (1) \\ 0.40x + 0.30y = 12 & (2) \end{cases}$

Multiply both sides of equation (1) by $-2$, and add the result to equation (2).

$\begin{cases} -0.40x - 1.00y = -40 \\ \underline{0.40x + 0.30y = 12} \end{cases}$
$$-0.70y = -28$$
$$y = 40$$

Substituting 40 for $x$ in equation (1), we obtain
$$0.20x + 0.50(40) = 20$$
$$0.20x + 20 = 20$$
$$0.20x = 0$$
$$x = 0$$

To fill the prescription, the pharmacist should use 0 units of the first powder and 40 units of the second powder. No mixture is really needed after all.

32. Let $p$ represent the airspeed of the Piper Arrow and let $w$ represent the impact of the wind. The rate at which the plane travels with the wind is $p + w$, and the rate against the wind is $p - w$.

Using the formula rate $\times$ time = distance, we obtain the following system:

$\begin{cases} (p + w) \cdot 3 = 510 & (1) \\ (p - w) \cdot 3 = 390 & (2) \end{cases}$

Divide both sides of equation (1) by 3, divide both sides of equation (2) by 3, and add the results.

$\begin{cases} p + w = 170 \\ \underline{p - w = 130} \end{cases}$
$$2p \phantom{+w} = 300$$
$$p = 150$$

Substituting 150 for $p$ in equation (1) yields
$$(150 + w) \cdot 3 = 510$$
$$150 + w = 170$$
$$w = 20$$

The airspeed of the Piper Arrow is 150 miles per hour. The impact of the wind is 20 miles per hour.

34. Let $t$ represent the time for Enrique's brother to run half the distance Enrique has run. Let $d$ represent the distance traveled by Enrique's brother. Then the time for Enrique is $t + \dfrac{1}{2}$, and the distance traveled by the Enrique $2d$. We obtain the following system by utilizing the formula rate $\times$ time = distance:

$\begin{cases} 8t = d & (1) \\ 6\left(t + \dfrac{1}{2}\right) = 2d & (2) \end{cases}$

Substituting $8t$ for $d$ in equation (2) yields
$$6\left(t + \dfrac{1}{2}\right) = 2(8t)$$
$$6t + 3 = 16t$$
$$3 = 10t$$
$$t = \dfrac{3}{10}$$

It will take Enrique's brother $\dfrac{3}{10}$ hour, or 18 minutes, to run half the distance that Enrique has run.

36. a. $\begin{cases} S(p) = 8p - 30 \\ D(p) = -35p + 995 \end{cases}$

ISM – 163

## Chapter 4: Systems of Linear Equations and Inequalities

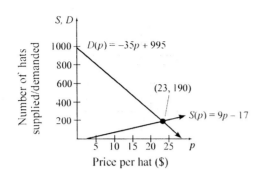

b. $S(p) = D(p)$
$9p - 17 = -35p + 995$
$44p - 17 = 995$
$44p = 1012$
$p = 23$

$S(23) = 9(23) - 17 = 207 - 17 = 190$
$D(23) = -35(23) + 995 = -805 + 995 = 190$

The equilibrium price is $23. The equilibrium quantity is 190 baseball hats.

**38. a.** $\begin{cases} M(t) = 8.92t + 272.11 \\ F(t) = 9.44t + 244.92 \end{cases}$

b. $M(t) = F(t)$
$8.92t + 272.11 = 9.44t + 244.92$
$-0.52t = -27.19$
$t \approx 52.3$

The average weekly earning of 16 – 24 year old males will equal the average weekly earnings of 16 – 24 year old females approximately 52 years after 1990; that is in approximately 2042.

$M(30) = 0.327(30) + 24.24 = 34.05$
$F(30) = 0.532(30) + 18.12 = 34.08$

The proportion of will be approximately 34%.

**40. a.** Let $x$ represent the amount of sales, let $A$ represent the pay from option $A$, and let $B$ represent the pay from option $B$.
$\begin{cases} A(x) = 15,000 + 0.01x \\ B(x) = 25,000 + 0.0075x \end{cases}$

b.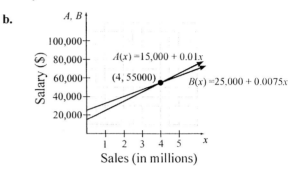

c. $A(x) = B(x)$
$15,000 + 0.01x = 25,000 + 0.0075x$
$0.0025x = 10,000$
$x = 4,000,000$

$4,000,000 of sales is required for the two options to result in the same salary is.

$A(4,000,000) = 15,000 + 0.01(4,000,000)$
$= 55,000$
$B(4,000,000) = 25,000 + 0.0075(4,000,000)$
$= 55,000$

The annual salary would be $55,000.

**42. a.** $R(x) = 0.30x$

b. $C(x) = 40 + 0.03x + 0.07x$
$= 40 + 0.10x$

c.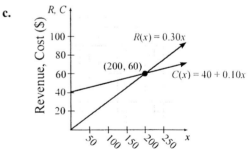

d. $R(x) = C(x)$
$0.30x = 40 + 0.10x$
$0.20x = 40$
$x = 200$

To break even, Audra must sell 200 cups if lemonade.

**Section 4.2** *Problem Solving: Systems of Two Linear Equations with Two Unknowns*

$R(200) = 0.30(200) = 60$
$C(200) = 40 + 0.10(200) = 60$

The cost and revenue for making and selling 200 cups of lemonade will both be $60.

## 4.3 Preparing for Systems of Linear Equations in Three Variables

1. Substitute $x = 1$, $y = -2$, and $z = 3$:
$3x - 2y + 4z = 3(1) - 2(-2) + 4(3)$
$= 3 + 4 + 12$
$= 19$

## 4.3 Exercises

2. consistent; dependent

4. FALSE. For example, if the graphs of the equations are parallel planes, then the system will have no solution.

6. TRUE

8. Answers will vary. One possibility follows: We must eliminate the same variable in the first step so that we will obtain a system of two equations containing two variables, which we can then solve using the methods of Section 4.2. Once we solve for the two variables, we can then back-substitute in order to find the third variable.

10. a. Substitute $x = 3$, $y = 2$, $z = 1$ into all three equations.
$$\begin{cases} 2(3) + 2 - 2(1) = 6 + 2 - 2 = 6 \\ -2(3) + 2 + 5(1) = -6 + 2 + 5 = 1 \\ 2(3) + 3(2) + 1 = 6 + 6 + 1 = 13 \end{cases}$$
Because these values satisfy all three equations, the ordered triple $(3, 2, 1)$ is a solution of the system.

    b. Substitute $x = -3$, $y = 5$, $z = 2$ into all thre equations).
$$\begin{cases} 2(10) + (-4) - 2(5) = 20 - 4 - 10 = 6 \\ -2(10) + (-4) + 5(5) = -20 - 4 + 25 = 1 \\ 2(10) + 3(-4) + 5 = 20 - 12 + 5 = 13 \end{cases}$$
Because these values satisfy all three equations, the ordered triple $(10, -4, 5)$ is a solution of the system.

12. $\begin{cases} x + 2y - z = 4 & (1) \\ 2x - y + 3z = 8 & (2) \\ -2x + 3y - 2z = 10 & (3) \end{cases}$

Multiply both sides of equation (1) by $-2$ and add the result to equation (2).
$-2x - 4y + 2z = -8$
$\underline{2x - y + 3z = 8}$
$-5y + 5z = 0 \quad (4)$

Add equations (2) and (3).
$2x - y + 3z = 8$
$\underline{-2x + 3y - 2z = 10}$
$2y + z = 18 \quad (5)$

Multiply both sides of equation (5) by $-5$ and add the result to equation (4).
$-5y + 5z = 0$
$\underline{-10y - 5z = -90}$
$-15y \quad\quad = -90$
$y = 6$

Substituting 4 for $y$ in equation (4), we obtain
$-5(6) + 5z = 0$
$-30 + 5z = 0$
$5z = 30$
$z = 6$

Substituting 6 for $y$ and 6 for $z$ in equation (1), we obtain
$x + 2(6) - (6) = 4$
$x + 12 - 6 = 4$
$x + 6 = 4$
$x = -2$

The solution is the ordered triple $(-2, 6, 6)$.

14. $\begin{cases} x + 2y - 3z = -19 & (1) \\ 3x + 2y - z = 13 & (2) \\ -2x - y + 3z = 25 & (3) \end{cases}$

Multiply equation (1) by $-3$ and add the result to equation (2).
$-3x - 6y + 9z = 57$
$\underline{3x + 2y - z = -9}$
$-4y + 8z = 48 \quad (4)$

Multiply both sides of equation (1) by 2 and add the result to equation (2).
$2x + 4y - 6z = -38$
$\underline{-2x - y + 3z = 26}$
$3y - 3z = -12 \quad (5)$

**Chapter 4:** *Systems of Linear Equations and Inequalities*

Divide equation (4) by 4, divide equation (5) by 3, and add the results.
$$-y+2z=12$$
$$\underline{y-\ z=-4}$$
$$z=8$$
Substituting 8 for $z$ in equation (4) and solving for $y$, we obtain
$$-4y+8(8)=48$$
$$-4y+64=48$$
$$-4y=-16$$
$$y=4$$
Substituting 4 for $y$ and 8 for $z$ in equation (1), we obtain
$$x+2(4)-3(8)=-19$$
$$x+8-24=-19$$
$$x-16=-19$$
$$x=-3$$
The solution is the ordered triple $(-3,4,8)$.

16. $\begin{cases} x-y+3z=2 & (1) \\ -2x+3y-8z=-1 & (2) \\ 2x-2y+4z=7 & (3) \end{cases}$

Multiply both sides of equation (1) by 2 and add the result to equation (2).
$$2x-2y+6z=4$$
$$\underline{-2x+3y-8z=-1}$$
$$y-2z=3 \quad (4)$$
Add equations (2) and (3).
$$-2x+3y-8z=-1$$
$$\underline{2x-2y+4z=7}$$
$$y-4z=6 \quad (5)$$
Multiply equation (4) by $-1$ and add the result to equation to equation (5).
$$-y+2z=-3$$
$$\underline{y-4z=6}$$
$$-2z=3$$
$$z=-\frac{3}{2}$$
Substituting $-\frac{3}{2}$ for $z$ in equation (4), we obtain
$$y-2\left(-\frac{3}{2}\right)=3$$
$$y+3=3$$
$$y=0$$

Substituting 0 for $y$ and $-\frac{3}{2}$ for $z$ in equation (3), we obtain
$$2x-2(0)+4\left(-\frac{3}{2}\right)=7$$
$$2x-0-6=7$$
$$2x-6=7$$
$$2x=13$$
$$x=\frac{13}{2}$$
The solution is the ordered triple $\left(\frac{13}{2},0,-\frac{3}{2}\right)$.

18. $\begin{cases} 2x+2y-z=-7 & (1) \\ x+2y-3z=-8 & (2) \\ 4x-2y+z=-11 & (3) \end{cases}$

Multiply both sides of equation (2) by $-2$ and add the result to equation (1).
$$2x+2y-z=-7$$
$$\underline{-2x-4y+6z=16}$$
$$-2y+5z=9 \quad (4)$$
Multiply both sides of equation (1) by $-2$ and add the result to equation (3).
$$-4x-4y+2z=14$$
$$\underline{4x-2y+z=-11}$$
$$-6y+3z=3 \quad (5)$$
Multiply both sides of equation (4) by $-3$ and add the result to equation (5).
$$6y-15z=-27$$
$$\underline{-6y+3z=3}$$
$$-12z=-24$$
$$z=2$$
Substituting 2 for $z$ in equation (4), we obtain
$$-2y+5(2)=9$$
$$-2y+10=9$$
$$-2y=-1$$
$$y=\frac{-1}{-2}=\frac{1}{2}$$
Substituting $\frac{1}{2}$ for $y$ and 2 for $z$ in equation (1), we obtain

### Section 4.3 Systems of Linear Equations with Three Unknowns

$$2x + 2\left(\frac{1}{2}\right) - 2 = -7$$
$$2x + 1 - 2 = -7$$
$$2x - 1 = -7$$
$$2x = -6$$
$$x = -3$$

The solution is the ordered triple $\left(-3, \frac{1}{2}, 2\right)$.

20. $\begin{cases} 2x \phantom{+3y} + z = -7 & (1) \\ \phantom{2x+} 3y - 2z = 17 & (2) \\ -4x - y \phantom{+2z} = 7 & (3) \end{cases}$

Multiply both sides of equation (1) by 2 and add the result to equation (3).
$$4x \phantom{-y} + 2z = -14$$
$$-4x - y \phantom{+2z} = 7$$
$$\overline{\phantom{-4x} -y + 2z = -7} \quad (4)$$

Add equation (4) to equation (2).
$$3y - 2z = 17$$
$$-y + 2z = -7$$
$$\overline{2y \phantom{+2z} = 10}$$
$$y = 5$$

Substituting 5 for $y$ in equation (2), we obtain
$$3(5) - 2z = 17$$
$$15 - 2z = 17$$
$$-2z = 2$$
$$z = 1$$

Substituting 5 for $y$ in equation (3), we obtain
$$-4x - 5 = 7$$
$$-4x = 12$$
$$x = -3$$

The solution is the ordered triple $(-3, 5, -1)$.

22. $\begin{cases} x \phantom{+3y} - 3z = -3 & (1) \\ \phantom{x+} 3y + 4z = -5 & (2) \\ 3x - 2y \phantom{+4z} = 6 & (3) \end{cases}$

Multiply both sides of equation (1) by $-3$ and add the result to equation (3).
$$-3x \phantom{-2y} + 9z = 9$$
$$3x - 2y \phantom{+9z} = 6$$
$$\overline{\phantom{-3x} -2y + 9z = 15} \quad (4)$$

Multiply both sides of equation (2) by 2, multiply both sides of equation (4) by 3, and add

the results.
$$6y + 8z = -10$$
$$-6y + 27z = 45$$
$$\overline{35z = 35}$$
$$z = 1$$

Substituting 1 for $z$ in equation (1), we obtain
$$x - 3(1) = -3$$
$$x - 3 = 3$$
$$x = 0$$

Substituting 0 for $x$ in equation (3), we obtain
$$3(0) - 2y = 6$$
$$-2y = 6$$
$$y = -3$$

The solution is the ordered triple $(0, -3, 1)$.

24. $\begin{cases} x - y + 2z = 3 & (1) \\ 2x + y - 2z = 1 & (2) \\ 4x - y + 2z = 0 & (3) \end{cases}$

Multiply both sides of equation (1) by $-2$ and add the result to equation (2).
$$-2x + 2y - 4z = -6$$
$$2x + y - 2z = 1$$
$$\overline{\phantom{-2x} 3y - 6z = -5} \quad (4)$$

Multiply both sides of equation (1) by $-4$ and add the result to equation (3).
$$-4x + 4y - 8z = -12$$
$$4x - y + 2z = 0$$
$$\overline{\phantom{-4x} 3y - 6z = -12} \quad (5)$$

Multiply both sides of equation (4) by $-1$ and add the result to equation (5).
$$-3y + 6z = 5$$
$$3y - 6z = -12$$
$$\overline{0 = -7 \quad \text{False}}$$

The system has no solution. The solution set is $\emptyset$ or $\{\ \}$. The system is inconsistent.

26. $\begin{cases} x + 2y - z = -4 & (1) \\ -2x + 4y - z = 6 & (2) \\ 2x + 2y + 3z = 5 & (3) \end{cases}$

Multiply both sides of equation (1) by 2 and add the result to equation (2).

## Chapter 4: Systems of Linear Equations and Inequalities

$$2x + 4y - 2z = -8$$
$$\underline{-2x + 4y - \phantom{2}z = 6}$$
$$8y - 3z = -2 \qquad (4)$$

Add equations (2) and (3).
$$-2x + 4y - z = 6$$
$$\underline{2x + 2y + 3z = 1}$$
$$6y + 2z = 7 \qquad (5)$$

Multiply both sides of equation (4) by $-3$, multiply both sides of equation (5) by 4, and add the results.
$$-24y + 9z = 6$$
$$\underline{24y + 8z = 28}$$
$$17z = 34$$
$$z = 2$$

Substituting 2 for $z$ in equation (4), we obtain
$$8y - 3(2) = -2$$
$$8y - 6 = -2$$
$$8y = 4$$
$$y = \frac{4}{8} = \frac{1}{2}$$

Substituting $\frac{1}{2}$ for $y$ and 2 for $z$ in equation (1), we obtain
$$x + 2\left(\frac{1}{2}\right) - 2 = -4$$
$$x + 1 - 2 = -4$$
$$x - 1 = -4$$
$$x = -3$$

The solution is the ordered triple $\left(-3, \frac{1}{2}, 2\right)$.

28. $\begin{cases} x + y - 2z = 3 & (1) \\ -2x - 3y + z = -7 & (2) \\ x + 2y + z = 4 & (3) \end{cases}$

Multiply both sides of equation (1) by 2 and add the result to equation (2).
$$2x + 2y - 4z = 6$$
$$\underline{-2x - 3y + \phantom{2}z = -7}$$
$$-y - 3z = -1 \qquad (4)$$

Multiply both sides of equation (3) by 2 and add the result to equation (2).

$$-2x - 3y + \phantom{2}z = -7$$
$$\underline{2x + 4y + 2z = 8}$$
$$y + 3z = 1 \qquad (5)$$

Add equations (4) and (5).
$$-y - 3z = -1$$
$$\underline{y + 3z = 1}$$
$$0 = 0 \qquad \text{True}$$

Thus, the system is dependent and has an infinite number of solutions.
Solve equation (5) for $y$.
$$y = -3z + 1$$

Substituting $-3z + 1$ for $y$ in equation (1), we obtain
$$x + (-3z + 1) - 2z = 3$$
$$x - 5z + 1 = 3$$
$$x = 5z + 2$$

The solution to the system is $\{(x, y, z) | x = 5z + 2,$ $y = -3z + 1, z \text{ is any real number}\}$.

30. $\begin{cases} x + y + z = 4 & (1) \\ 2x + 3y - z = 8 & (2) \\ x + y - z = 3 & (3) \end{cases}$

Multiply equation (1) by $-2$ and add the result to equation (2).
$$-2x - 2y - 2z = -8$$
$$\underline{2x + 3y - \phantom{2}z = 8}$$
$$y - 3z = 0 \qquad (4)$$

Multiply both sides of equation (1) by $-1$ and add the result to equation (3).
$$x + y + z = 4$$
$$\underline{-x - y + z = -3}$$
$$2z = 1$$
$$z = \frac{1}{2}$$

Substituting $\frac{1}{2}$ for $z$ in equation (4), we obtain
$$y - 3\left(\frac{1}{2}\right) = 0$$
$$y - \frac{3}{2} = 0$$
$$y = \frac{3}{2}$$

Substituting $\frac{3}{2}$ for $y$ and $\frac{1}{2}$ for $z$ in equation (1),

Section 4.3 Systems of Linear Equations with Three Unknowns

we obtain
$$x + \frac{3}{2} + \frac{1}{2} = 4$$
$$x + 2 = 4$$
$$x = 2$$

The solution is the ordered triple $\left(2, \frac{3}{2}, \frac{1}{2}\right)$.

32. $\begin{cases} x + \frac{1}{2}y + \frac{1}{2}z = \frac{3}{2} & (1) \\ -x + 2y + 3z = 1 & (2) \\ 3x + 4y + 5z = 7 & (3) \end{cases}$

Multiply equation (1) by 2, multiply equation (2) by 2, and add the results.
$$2x + y + z = 3$$
$$-2x + 4y + 6z = 2$$
$$\overline{\phantom{xx}5y + 7z = 5} \quad (4)$$

Multiply both sides of equation (1) by 2, multiply both sides of equation (2) by 2, and add the results.
$$-3x + 6y + 9z = 3$$
$$3x + 4y + 5z = 7$$
$$\overline{\phantom{xxx}10y + 14z = 10} \quad (5)$$

Multiply both sides of equation (4) by $-2$ and add the result to equation (5).
$$-10y - 14z = -10$$
$$10y + 14z = 10$$
$$\overline{\phantom{xxx}0 = 0} \quad \text{True}$$

Thus, the system is dependent and has an infinite number of solutions.
Solve equation (5) for $y$.
$$5y + 7z = 5$$
$$5y = -7z + 5$$
$$y = \frac{-7z + 5}{5}$$
$$y = -\frac{7}{5}z + 1$$

Substituting $-\frac{7}{5}z + 1$ for $y$ in equation (1), we obtain
$$3x + 4\left(-\frac{7}{5}z + 1\right) + 5z = 7$$
$$3x - \frac{28}{5}z + 4 + 5z = 7$$
$$5\left(3x - \frac{28}{5}z + 4 + 5z\right) = 5(7)$$
$$15x - 28z + 20 + 25z = 35$$
$$15x - 3z + 20 = 35$$
$$15x = 3z + 15$$
$$x = \frac{3z + 15}{15} = \frac{1}{5}z + 1$$

The solution to the system is $\left\{(x, y, z) \,\middle|\, x = \frac{1}{5}z + 1, \right.$
$\left. y = -\frac{7}{5}z + 1, z \text{ is any real number}\right\}$.

34. Answers will vary. One possibility follows.
For the ordered triple $(-4, 1, -3)$, evaluate the expressions $x + y + z$, $x - y + z$, and $x + y - z$:
$$x + y + z = -4 + 1 + (-3) = -6$$
$$x - y + z = -4 - 1 + (-3) = -8$$
$$x + y - z = -4 + 1 - (-3) = 0$$
So, the system below has the solution $(-4, 1, -3)$.
$$\begin{cases} x + y + z = -6 \\ x - y + z = -8 \\ x + y - z = 0 \end{cases}$$

36. a. If $f(1) = 2$, then $a(1)^2 + b(1) + c = 2$ or
   $a + b + c = 2$.
   If $f(2) = 9$, then $a(2)^2 + b(2) + c = 9$ or
   $4a + 2b + c = 9$.

   b. To find $a$, $b$, and $c$, we must solve the following system:
   $\begin{cases} a - b + c = 6 & (1) \\ a + b + c = 2 & (2) \\ 4a + 2b + c = 9 & (3) \end{cases}$
   Add equations (1) and (2).
   $$a - b + c = 6$$
   $$a + b + c = 2$$
   $$\overline{\phantom{xx}2a + 2c = 8} \quad (4)$$
   Multiply both sides of equation (1) by 2 and add the result to equation (3).

$2a - 2b + 2c = 12$
$4a + 2b + c = 9$
───────────────
$6a \phantom{+2b} + 3c = 21$ \quad (5)

Divide both sides of equation (4) by 2, divide equation (5) by $-3$, and add the results.
$a + c = 4$
$-2a - c = -7$
───────────────
$-a \phantom{+c} = -3$
$a = 3$

Substituting 3 for $a$ in equation (4), we obtain
$2(3) + 2c = 8$
$6 + 2c = 8$
$2c = 2$
$c = 1$

Substituting 3 for $a$ and 1 for $c$ in equation (2), we obtain
$3 + b + 1 = 2$
$b + 4 = 2$
$b = -2$

Thus, the quadratic equation that contains $(-1, 6)$, $(1, 2)$, and $(2, 9)$ is
$f(x) = 3x^2 - 2x + 1$.

**38.** Rewrite the system with each equation in standard form.
$$\begin{cases} i_1 - i_2 + i_3 = 0 & (1) \\ -5i_1 \phantom{+8i_2} + 8i_3 = 8 & (2) \\ \phantom{-5i_1 +} 6i_2 + 8i_3 = 48 & (3) \end{cases}$$

Multiply both sides of equation (1) by 5 and add the result to equation (2).
$5i_1 - 5i_2 + 5i_3 = 0$
$-5i_1 \phantom{-5i_2} + 8i_3 = 8$
───────────────
$-5i_2 + 13i_3 = 8$ \quad (4)

Multiply both sides of equation (3) by 5, multiply both sides of equation (4) by 6, and add the results.
$30i_2 + 40i_3 = 240$
$-30i_2 + 78i_3 = 48$
───────────────
$118i_3 = 288$
$i_3 = \dfrac{288}{118} = \dfrac{144}{59}$

Substituting $\dfrac{144}{59}$ for $i_3$ in equation (2), we obtain
$-5i_1 + 8\left(\dfrac{144}{59}\right) = 8$
$-5i_1 + \dfrac{1152}{59} = 8$
$59\left(-5i_1 + \dfrac{1152}{59}\right) = 59(8)$
$-295i_1 + 1152 = 472$
$-295i_1 = -680$
$i_1 = \dfrac{-680}{-295} = \dfrac{136}{59}$

Substituting $\dfrac{144}{59}$ for $i_3$ and $\dfrac{136}{59}$ for $i_1$ in equation (1), we obtain
$\dfrac{136}{59} - i_2 + \left(\dfrac{144}{59}\right) = 0$
$-i_2 + \dfrac{280}{59} = 0$
$\dfrac{280}{59} = i_2$

The currents are $i_1 = \dfrac{136}{59}$, $i_2 = \dfrac{280}{59}$, and $i_3 = \dfrac{144}{59}$.

**40.** Let $r$ represent the number of orchestra seats, let $m$ represent the number of main floor seats, and let $b$ represent the number of balcony seats.
$$\begin{cases} r + m + b = 600 & (1) \\ 80r + 60m + 25b = 33{,}500 & (2) \\ 80r + \dfrac{3}{5}(60m) + \dfrac{4}{5}(25b) = 24{,}640 & (3) \end{cases}$$

Simplify equation (3) and rewrite the system.
$80r + \dfrac{3}{5}(60m) + \dfrac{4}{5}(25b) = 24{,}640$
$80r + 36m + 20b = 24{,}640$

Rewriting the system, we have
$$\begin{cases} r + m + b = 600 & (1) \\ 80r + 60m + 25b = 33{,}500 & (2) \\ 80r + 36m + 20b = 24{,}640 & (3) \end{cases}$$

Multiply both sides of equation (1) by $-80$ and add the result to equation (2).
$-80r - 80m - 80b = -48{,}000$
$80r + 60m + 25b = 33{,}500$
───────────────
$-20m - 55b = -14{,}500$ \quad (4)

**Section 4.3** Systems of Linear Equations with Three Unknowns

Multiply both sides of equation (3) by $-1$ and add the result to equation (2).
$$80r + 60m + 25b = 33,500$$
$$\underline{-80r - 36m - 20b = -24,640}$$
$$24m + 5b = 8860 \quad (5)$$

Multiply both side of equation (5) by 11 and add the result to equation (4).
$$-20m - 55b = -14,500$$
$$\underline{264m + 55b = 97,460}$$
$$244m \phantom{+ 55b} = 82,960$$
$$m = 340$$

Substituting 340 for $m$ in equation (5), we obtain
$$24(340) + 5b = 8860$$
$$8160 + 5b = 8860$$
$$5b = 700$$
$$b = 140$$

Substituting 340 for $m$ and 140 for $b$ in equation (1), we obtain
$$r + 340 + 140 = 600$$
$$r + 480 = 600$$
$$r = 120$$

There are 120 orchestra seats, 340 main flor seats, and 140 balcony seats in the stadium.

42. Let $p$ represent the number of Broccoli and Cheese Baked Potatoes, let $s$ represent the number of Chicken BLT Salads, and let $c$ represent the number medium Cokes.
$$\begin{cases} 480p + 310s + 140c = 1325 & \text{(Calories)} \quad (1) \\ 80p + 10s + 37c = 172 & \text{(carbs)} \quad (2) \\ 9p + 33s \phantom{+ 37c} = 63 & \text{(protein)} \quad (3) \end{cases}$$

Multiply both sides of equation (2) by $-6$ and add the result to equation (1).
$$480p + 310s + 140c = 1325$$
$$\underline{-480p - 60s - 222c = -1032}$$
$$250s - 82c = 293 \quad (4)$$

Multiply both sides of equation (1) by $-9$, multiply both sides of equation (4) by 80, and add the results.
$$-720p - 90s - 333c = -1548$$
$$\underline{720p + 2640s \phantom{- 333c} = 5040}$$
$$2550s - 333c = 3492 \quad (5)$$

Multiply both sides of equation (4) by 333, multiply both sides of equation (5) by -82, and add the results.
$$83,250s - 27,306c = 97,569$$
$$\underline{-209,100s + 27,306c = -286,344}$$
$$-125,850s \phantom{+ 27,306c} = -188,755$$
$$s = 1.5$$

Substituting 1.5 for $s$ in equation (4), we obtain
$$250(1.5) - 82c = 293$$
$$375 - 82c = 293$$
$$-82c = -82$$
$$c = 1$$

Substituting 1.5 for $s$ in equation (3), we obtain
$$9p + 33(1.5) = 63$$
$$9p + 49.5 = 63$$
$$9p = 13.5$$
$$p = 1.5$$

Antonio can have 1.5 Broccoli and Cheese Baked Potatoes, 1.5 Chicken BLT Salads, and 1 medium Coke.

44. Let $s$ represent the amount to be invested in the savings account, let $t$ represent the amount to be invested in treasury bonds, let $m$ represent the amount to be invested in the mutual fund.
$$\begin{cases} s + t + m = 15,000 & (1) \\ 0.02s + 0.05t + 0.10m = 720 & (2) \\ s = 2m & (3) \end{cases}$$

Rewrite the system in standard form.
$$\begin{cases} s + t + m = 15,000 & (1) \\ 0.02s + 0.05t + 0.10m = 720 & (2) \\ s - 2m = 0 & (3) \end{cases}$$

Multiply both sides of equation (1) by $-0.05$ and add the result to equation (2).
$$-0.05s - 0.05t - 0.05m = -750$$
$$\underline{0.02s + 0.05t + 0.10m = 720}$$
$$-0.03s \phantom{+ 0.05t} + 0.05m = -30 \quad (4)$$

Multiply both sides of equation (3) by 0.03 and add the result to equation (4).
$$0.03s - 0.06m = 0$$
$$\underline{-0.03s + 0.05m = -30}$$
$$-0.01m = -30$$
$$m = 3000$$

## Chapter 4: Systems of Linear Equations and Inequalities

Substituting 3000 for $m$ in equation (3), we obtain
$$s - 2(3000) = 0$$
$$s - 6000 = 0$$
$$x = 6000$$

Substituting 3000 for $m$ and 6000 for $s$ in equation (1), we obtain
$$6000 + t + 3000 = 15,000$$
$$t + 9000 = 15,000$$
$$t = 6000$$

Delu should invest $6000 in the savings account, $6000 in treasury bills, and $3000 in the mutual fund.

**46.** $\begin{cases} \frac{1}{4}x + \frac{1}{4}y + \frac{1}{2}z = 6 & (1) \\ -\frac{1}{8}x + \frac{1}{2}y - \frac{1}{5}z = -5 & (2) \\ \frac{1}{2}x + \frac{1}{2}y - \frac{1}{2}z = -3 & (3) \end{cases}$

Simplify each equation and rewrite the system.

(1) $4\left(\frac{1}{4}x + \frac{1}{4}y + \frac{1}{2}z\right) = 4(6)$
$$x + y + 2z = 24$$

(2) $40\left(-\frac{1}{8}x + \frac{1}{2}y - \frac{1}{5}z\right) = 40(-5)$
$$-5x + 20y - 8z = -200$$

(3) $2\left(\frac{1}{2}x + \frac{1}{2}y - \frac{1}{2}z\right) = 2(-3)$
$$x + y - z = -6$$

Rewriting the system, we have
$\begin{cases} x + y + 2z = 24 & (1) \\ -5x + 20y - 8z = -200 & (2) \\ x + y - z = -6 & (3) \end{cases}$

Multiply equation (1) by 5 and add the result to equation (2).
$$5x + 5y + 10z = 120$$
$$\underline{-5x + 20y - 8z = -200}$$
$$25y + 2z = -80 \quad (4)$$

Multiply equation (2) by $-1$ and add the result to equation (3).
$$-x - y - 2z = -24$$
$$\underline{x + y - z = -6}$$
$$-3z = -30$$
$$z = 10$$

Substituting 10 for $z$ in equation (4), we obtain
$$25y + 2(10) = -80$$
$$25y + 20 = -80$$
$$25y = -100$$
$$y = -4$$

Substituting $-4$ for $y$ and 10 for $z$ in equation (1), we obtain
$$x + (-4) + 2(10) = 24$$
$$x - 4 + 20 = 24$$
$$x + 16 = 24$$
$$x = 8$$

The solution is the ordered triple $(8, -4, 10)$.

**48.** $\begin{cases} x + y + z + w = 0 & (1) \\ 2x - 3y - z + w = -17 & (2) \\ 3x + y + 2z - w = 8 & (3) \\ -x + 2y - 3z + 2w = -7 & (4) \end{cases}$

Add equations (1) and (4).
$$x + y + z + w = 0$$
$$\underline{-x + 2y - 3z + 2w = -7}$$
$$3y - 2z + 3w = -7 \quad (5)$$

Multiply both sides of equation (4) by 2 and add the result to equation (2).
$$-2x + 4y - 6z + 4w = -14$$
$$\underline{2x - 3y - z + w = -17}$$
$$y - 7z + 5w = -31 \quad (6)$$

Multiply both sides of equation (4) by 3 and add the result to equation (2).
$$-3x + 6y - 9z + 6w = -21$$
$$\underline{3x + y + 2z - w = 8}$$
$$7y - 7z + 5w = -13 \quad (7)$$

Equations (5), (6), and (7) form a system of three equations with three unknowns.

Multiply both sides of equation (6) by $-3$ and add the result to equation (5).
$$-3y + 21z - 15w = 93$$
$$\underline{3y - 2z + 3w = -7}$$
$$19z - 12w = 86 \quad (9)$$

Multiply both sides of equation (6) by $-7$ and add the result to equation (7).
$$-7y + 49z - 35w = 217$$
$$\underline{7y - 7z + 5w = -13}$$
$$42z - 30w = 204 \quad (10)$$

## Section 4.3 Systems of Linear Equations with Three Unknowns

Multiply both sides of equation (9) by 5, multiply both sides of equation (10) by $-2$, and add the results.

$$\begin{aligned} 95z - 60w &= 430 \\ -84z + 60w &= -408 \\ \hline 11z &= 22 \\ z &= 2 \end{aligned}$$

Substituting 2 for $z$ in equation (9), we obtain
$$\begin{aligned} 19(2) - 12w &= 86 \\ 38 - 12w &= 86 \\ -12w &= 48 \\ w &= -4 \end{aligned}$$

Substituting 2 for $z$ and $-4$ for $w$ in equation (6), we obtain
$$\begin{aligned} y - 7(2) + 5(-4) &= -31 \\ y - 14 - 20 &= -31 \\ y - 34 &= -31 \\ y &= 3 \end{aligned}$$

Substituting 2 for $z$, $-4$ for $w$, and 3 for $y$ in equation (1), we obtain
$$\begin{aligned} x + 3 + 2 + (-4) &= 0 \\ x + 1 &= 0 \\ x &= -1 \end{aligned}$$

The solution is $(-1, 3, 2, -4)$.

### Putting the Concepts Together (Sections 4.1 – 4.3)

1. The graphs of the linear equations intersect at the point $(-3, 5)$. Thus, the solution is $(-3, 5)$.

2. $\begin{cases} 3x + y = 7 & (1) \\ -2x + 3y = -12 & (2) \end{cases}$

   Equation (1) in slope-intercept form is $y = -3x + 7$. Equation (2) in slope-intercept form is $y = \frac{2}{3}x - 4$. Graph each equation and find the point of intersection.

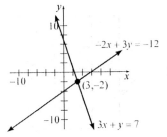

The solution is the ordered pair $(3, -2)$.

3. $\begin{cases} x = 2 - 3y & (1) \\ 3x + 10y = 5 & (2) \end{cases}$

   Because equation (1) is solved for $x$, we use substitution to solve the system.
   Substituting $2 - 3y$ for $x$ in equation (2), we obtain
   $$\begin{aligned} 3(2 - 3y) + 10y &= 5 \\ 6 - 9y + 10y &= 5 \\ 6 + y &= 5 \\ y &= -1 \end{aligned}$$
   Substituting $-1$ for $y$ in equation (1), we obtain
   $x = 2 - 3(-1) = 2 + 3 = 5$.
   The solution is the ordered pair $(5, -1)$.

4. $\begin{cases} 4x + 3y = -1 & (1) \\ 2x - y = 3 & (2) \end{cases}$

   Because the coefficient of $y$ in equation (2) is $-1$, we use substitution to solve the system.
   Equation (2) solved for $y$ is $y = 2x - 3$.
   Substituting $2x - 3$ for $y$ in equation (1), we obtain
   $$\begin{aligned} 4x + 3(2x - 3) &= -1 \\ 4x + 6x - 9 &= -1 \\ 10x - 9 &= -1 \\ 10x &= 8 \\ x &= \frac{8}{10} = \frac{4}{5} \end{aligned}$$
   Substituting $\frac{4}{5}$ for $x$ in equation (2), we obtain
   $$\begin{aligned} 2\left(\frac{4}{5}\right) - y &= 3 \\ \frac{8}{5} - y &= \frac{15}{5} \\ -y &= \frac{7}{5} \\ y &= -\frac{7}{5} \end{aligned}$$
   The solution is the ordered pair $\left(\frac{4}{5}, -\frac{7}{5}\right)$.

5. $\begin{cases} x + 3y = 8 & (1) \\ \frac{1}{5}x + \frac{1}{2}y = 1 & (2) \end{cases}$

   Because equation (2) contains fractions, we choose to use the elimination method to solve

ISM – 173

## Chapter 4: Systems of Linear Equations and Inequalities

the system.
Multiply both sides of equation (1) by 2, multiply both side of equation (2) by -10, and add the results.
$$2x + 6y = 16$$
$$-2x - 5y = -10$$
$$\overline{\qquad\qquad y = 6}$$
Substituting 6 for $y$ in equation (1), we obtain
$$x + 3(6) = 8$$
$$x + 18 = 8$$
$$x = -10$$
The solution is the ordered pair (-10, 6).

6. $\begin{cases} 8x - 4y = 12 & (1) \\ -10x + 5y = -15 & (2) \end{cases}$

Because none of variables have a coefficient of 1, we use elimination to solve the system.
Multiply both sides of equation (1) by 5, multiply both sides of equation (2) by 4, and add the results.
$$40x - 20y = 60$$
$$-40x + 20y = -60$$
$$\overline{\qquad 0 = 0 \quad \text{True}}$$
The system is dependent. The solution is
$\{(x, y) \mid 8x - 4y = 12\}$.

7. $\begin{cases} 2x + y + 3z = 10 & (1) \\ x - 2y + z = 10 & (2) \\ -4x + 3y + 2z = 5 & (3) \end{cases}$

Multiply both sides of equation (2) by -2 and add the result to equation (1).
$$-2x + 4y - 2z = -20$$
$$2x + y + 3z = 10$$
$$\overline{\qquad 5y + z = -10 \quad (4)}$$

Multiply both sides of equation (1) by 2 and add the result to equation (3).
$$4x + 2y + 6z = 20$$
$$-4x + 3y + 2z = 5$$
$$\overline{\qquad 5y + 8z = 25 \quad (5)}$$

Multiply both sides of equation (4) by -1 and add the result to equation (5).
$$-5y - z = 10$$
$$5y + 8z = 25$$
$$\overline{\qquad 7z = 35}$$
$$z = 5$$

Substituting 5 for $z$ in equation (4), we obtain
$$5y + 5 = -10$$
$$5y = -15$$
$$y = -3$$

Substituting 5 for $z$ and -3 for $y$ in equation (1), we obtain
$$2x + (-3) + 3(5) = 10$$
$$2x - 3 + 15 = 10$$
$$2x + 12 = 10$$
$$2x = -2$$
$$x = -1$$
The solution is the ordered triple (-1, -3, 5).

8. $\begin{cases} x + 2y - 2z = 3 & (1) \\ x + 3y - 4z = 6 & (2) \\ 4x + 5y - 2z = 6 & (3) \end{cases}$

Multiply both sides of equation (1) by -1 and add the result to equation (2).
$$-x - 2y + 2z = -3$$
$$x + 3y - 4z = 6$$
$$\overline{\qquad y - 2z = 3 \quad (4)}$$

Multiply both sides of equation (1) by -4 and add the result to equation (3).
$$-4x - 8y + 8z = -12$$
$$4x + 5y - 2z = 6$$
$$\overline{\qquad -3y + 6z = -6 \quad (5)}$$

Multiply both sides of equation (4) by 3 and add the result to equation (5).
$$3y - 6z = 9$$
$$-3y + 6z = -6$$
$$\overline{\qquad 0 = 3 \quad \text{False}}$$
The system has no solution. The solution set is $\varnothing$ or $\{\ \}$. The system is inconsistent.

## Putting the Concepts Together (Sections 4.1 – 4.3)

Substituting 5 for $z$ and -3 for $y$ in equation (1), we obtain
$$2x + (-3) + 3(5) = 10$$
$$2x - 3 + 15 = 10$$
$$2x + 12 = 10$$
$$2x = -2$$
$$x = -1$$
The solution is the ordered triple (-1, -3, 5).

8. $\begin{cases} x + 2y - 2z = 3 & (1) \\ x + 3y - 4z = 6 & (2) \\ 4x + 5y - 2z = 6 & (3) \end{cases}$

Multiply both sides of equation (1) by -1 and add the result to equation (2).
$$-x - 2y + 2z = -3$$
$$\underline{x + 3y - 4z = 6}$$
$$y - 2z = 3 \quad (4)$$

Multiply both sides of equation (1) by -4 and add the result to equation (3).
$$-4x - 8y + 8z = -12$$
$$\underline{4x + 5y - 2z = 6}$$
$$-3y + 6z = -6 \quad (5)$$

Multiply both sides of equation (4) by 3 and add the result to equation (5).
$$3y - 6z = 9$$
$$\underline{-3y + 6z = -6}$$
$$0 = 3 \quad \text{False}$$

The system has no solution. The solution set is $\emptyset$ or $\{\ \}$. The system is inconsistent.

9. Let $a$ represent the number of adult tickets and let $y$ represent the number of youth tickets sold.
$\begin{cases} 9a + 5y = 5925 & (1) \\ a + y = 825 & (2) \end{cases}$

Multiply both sides of equation (2) by $-5$ and add the result to equation (1).
$$-5a - 5y = -4125$$
$$\underline{9a + 5y = 5925}$$
$$4a = 1800$$
$$a = 450$$

Substituting 450 for $a$ in equation (2), we obtain
$$450 + y = 825$$
$$y = 375$$

The museum sold 450 adult tickets and 375 youth tickets.

10. Let $r$ represent the number of orchestra seats, let $m$ represent the number of mezzanine seats, and let $b$ represent the number of balcony seats.
$\begin{cases} r + m + b = 1200 & (1) \\ 65r + 48m + 35b = 55,640 & (2) \\ b = m + 150 & (3) \end{cases}$

Writing the equations in the system in standard form, we obtain
$\begin{cases} r + m + b = 1200 & (1) \\ 65r + 48m + 35b = 55,640 & (2) \\ m - b = -150 & (3) \end{cases}$

Multiply equation (1) by -65 and add the result to equation (2).
$$-65r - 65m - 65b = -78,000$$
$$\underline{65r + 48m + 35b = 55,640}$$
$$-17m - 30b = -22,360 \quad (4)$$

Multiply equation (3) by 17 and add the result to equation (4).
$$17m - 17b = -2,550$$
$$\underline{-17m - 30b = -22,360}$$
$$-47b = -24,910$$
$$b = 530$$

Substituting 530 for $b$ in equation (3), we obtain
$$m - 530 = -150$$
$$m = 380$$

Substituting 530 for $b$ and 380 for $m$ in equation (1), we obtain
$$r + 380 + 530 = 1200$$
$$r + 910 = 1200$$
$$r = 290$$

The theatre contains 290 orchestra seats, 380 mezzanine seats, and 530 balcony seats.

### 4.4 Preparing for Using Matrices to Solve Systems

1. For the expression $4x - 2y + z$, the coefficients are 4, $-2$, and 1.

2. $$x - 4y = 3$$
$$x - 4y + 4y = 3 + 4y$$
$$x = 4y + 3$$

3. Substitute $x = 1$, $y = -3$, and $z = 2$:
$$3x - 2y + z = 3(1) - 2(-3) + 2$$
$$= 3 + 6 + 2$$
$$= 11$$

### 4.4 Exercises

2. augmented

4. FALSE. The augmented matrix of a system of two equations containing two unknowns has 2 rows and 3 columns.

6. TRUE

8. If in any row all the entries to the left of the vertical rule are zeros and the entry to the right of the vertical line is not zero, then the system is inconsistent.

10. Answers may vary. One possibility follows: Multiply each entry in row 2 by $\frac{1}{5}$ (or divide each entry of row 2 by 5). That is, use the row operation $R_2 = \frac{1}{5} r_2$. This operation places a 1 in the second row, second column entry and moves the matrix one step closer to being in row-echelon form.

12. $\begin{bmatrix} -1 & 1 & | & 6 \\ 5 & -1 & | & -3 \end{bmatrix}$

14. $\begin{bmatrix} 1 & 1 & -1 & | & 2 \\ -2 & 1 & -4 & | & 13 \\ 3 & -1 & -2 & | & -4 \end{bmatrix}$

16. Write each equation in the system in standard form.
$$\begin{cases} 6x + 4y = -2 \\ -x - y = -1 \end{cases}$$
Thus, the augmented matrix is
$$\begin{bmatrix} 6 & 4 & | & -2 \\ -1 & -1 & | & -1 \end{bmatrix}$$

18. Write each equation in the system in standard form.
$$\begin{cases} 2x + 7y = 1 \\ -x - 6z = 5 \\ 5x + 2y - 4z = -1 \end{cases}$$
Thus, the augmented matrix is
$$\begin{bmatrix} 2 & 7 & 0 & | & 1 \\ -1 & 0 & -6 & | & 5 \\ 5 & 2 & -4 & | & -1 \end{bmatrix}$$

20. a. $\begin{bmatrix} 1 & 5 & | & 7 \\ 3 & 11 & | & 13 \end{bmatrix}$ $(R_2 = -3r_1 + r_2)$
$$= \begin{bmatrix} 1 & 5 & | & 7 \\ -3(1)+3 & -3(5)+11 & | & -3(7)+13 \end{bmatrix}$$
$$= \begin{bmatrix} 1 & 5 & | & 7 \\ 0 & -4 & | & -8 \end{bmatrix}$$

b. $\begin{bmatrix} 1 & 5 & | & 7 \\ 0 & -4 & | & -8 \end{bmatrix}$ $\left(R_2 = -\frac{1}{4} r_2\right)$
$$= \begin{bmatrix} 1 & 5 & | & 7 \\ -\frac{1}{4}(0) & -\frac{1}{4}(-4) & | & -\frac{1}{4}(-8) \end{bmatrix}$$
$$= \begin{bmatrix} 1 & 5 & | & 7 \\ 0 & 1 & | & 2 \end{bmatrix}$$

22. a. $\begin{bmatrix} 1 & -1 & 1 & | & 6 \\ -2 & 1 & -3 & | & 3 \\ 3 & 2 & -2 & | & -5 \end{bmatrix}$ $(R_2 = 2r_1 + r_2)$
$$= \begin{bmatrix} 1 & -1 & 1 & | & 6 \\ 2(1)+(-2) & 2(-1)+1 & 2(1)+(-3) & | & 2(6)+3 \\ 3 & 2 & -2 & | & -5 \end{bmatrix}$$
$$= \begin{bmatrix} 1 & -1 & 1 & | & 6 \\ 0 & -1 & -1 & | & 15 \\ 3 & 2 & -2 & | & -5 \end{bmatrix}$$

b. $\begin{bmatrix} 1 & -1 & 1 & | & 6 \\ 0 & -1 & -1 & | & 15 \\ 3 & 2 & -2 & | & -5 \end{bmatrix}$ $(R_3 = -3r_1 + r_3)$
$$= \begin{bmatrix} 1 & -1 & 1 & | & 6 \\ 0 & -1 & -1 & | & 15 \\ -3(1)+3 & -3(-1)+2 & -3(1)+(-2) & | & -3(6)+(-5) \end{bmatrix}$$
$$= \begin{bmatrix} 1 & -1 & 1 & | & 6 \\ 0 & -1 & -1 & | & 15 \\ 0 & 5 & -5 & | & -23 \end{bmatrix}$$

**24.a.** $\begin{bmatrix} 1 & -3 & 4 & | & 11 \\ 0 & 3 & 6 & | & -12 \\ 0 & -2 & -3 & | & 8 \end{bmatrix}$ $\left(R_2 = \frac{1}{3}r_2\right)$

$= \begin{bmatrix} 1 & -3 & 4 & | & 11 \\ \frac{1}{3}(0) & \frac{1}{3}(3) & \frac{1}{3}(6) & | & \frac{1}{3}(-12) \\ 0 & -2 & -3 & | & 8 \end{bmatrix}$

$= \begin{bmatrix} 1 & -3 & 4 & | & 11 \\ 0 & 1 & 2 & | & -4 \\ 0 & -2 & -3 & | & 8 \end{bmatrix}$

**b.** $\begin{bmatrix} 1 & -3 & 4 & | & 11 \\ 0 & 1 & 2 & | & -4 \\ 0 & -2 & -3 & | & 8 \end{bmatrix}$ $(R_3 = 2r_2 + r_3)$

$= \begin{bmatrix} 1 & -3 & 4 & | & 11 \\ 0 & 1 & 2 & | & -4 \\ 2(0)+0 & 2(1)+(-2) & 2(2)+(-3) & | & 2(-4)+8 \end{bmatrix}$

$= \begin{bmatrix} 1 & -3 & 4 & | & 11 \\ 0 & 1 & 2 & | & -4 \\ 0 & 0 & 1 & | & 0 \end{bmatrix}$

**26.** $\begin{cases} x - 2y = 3 & (1) \\ y = -5 & (2) \end{cases}$

This system is consistent and independent.
Substituting $-5$ for $y$ in equation (1), we obtain
$x - 2(-5) = 3$
$x + 10 = 3$
$x = -7$
The solution is the ordered pair $(-7, -5)$.

**28.** $\begin{cases} x - 2y - 4z = 6 & (1) \\ y - 5z = -3 & (2) \\ 0 = 0 & (3) \end{cases}$

This system is consistent and dependent.

**30.** $\begin{cases} x + 2y - z = -7 & (1) \\ y + 2z = -4 & (2) \\ z = -3 & (3) \end{cases}$

This system is consistent and independent.
Substituting $-3$ for $z$ in equation (2), we obtain
$y + 2(-3) = -4$
$y - 6 = -4$
$y = 2$
Substituting $-3$ for $z$ and 2 for $y$ in equation (1),

we obtain
$x + 2(2) - (-3) = -7$
$x + 4 + 3 = -7$
$x + 7 = -7$
$x = -14$
The solution is the ordered triple $(-14, 2, -3)$.

**32.** Write the augmented matrix of the system and then put it in row echelon form.

$\begin{bmatrix} 1 & 5 & | & 2 \\ -2 & 3 & | & 9 \end{bmatrix}$ $(R_2 = 2r_1 + r_2)$

$= \begin{bmatrix} 1 & 5 & | & 2 \\ 0 & 13 & | & 13 \end{bmatrix}$ $\left(R_2 = \frac{1}{13}r_2\right)$

$= \begin{bmatrix} 1 & 5 & | & 2 \\ 0 & 1 & | & 1 \end{bmatrix}$

Write the system of equations that corresponds to the row-echelon matrix

$\begin{cases} x + 5y = 2 & (1) \\ y = 1 & (2) \end{cases}$

This system is consistent and independent.
Substituting 1 for $y$ in equation (1), we obtain
$x + 5(1) = 2$
$x + 5 = 2$
$x = -3$
The solution is the ordered pair $(-3, 1)$.

**34.** Write the augmented matrix of the system and then put it in row echelon form.

$\begin{bmatrix} 5 & -2 & | & 3 \\ -15 & 6 & | & -9 \end{bmatrix}$ $(R_2 = 3r_1 + r_2)$

$= \begin{bmatrix} 5 & -2 & | & 3 \\ 0 & 0 & | & 0 \end{bmatrix}$

The system is dependent. The solution to the system is $\{(x, y) \mid 5x - 2y = 3\}$.

**36.** Write the augmented matrix of the system and then put it in row echelon form.

$\begin{bmatrix} 3 & 3 & | & -1 \\ 2 & 1 & | & 1 \end{bmatrix}$ $\left(R_1 = \frac{1}{3}r_1\right)$

$= \begin{bmatrix} 1 & 1 & | & -\frac{1}{3} \\ 2 & 1 & | & 1 \end{bmatrix}$ $(R_2 = 2r_1 - r_2)$

$= \begin{bmatrix} 1 & 1 & | & -\frac{1}{3} \\ 0 & 1 & | & -\frac{5}{3} \end{bmatrix}$

## Chapter 4: Systems of Linear Equations and Inequalities

Write the system of equations that corresponds to the row-echelon matrix

$$\begin{cases} x + y = -\dfrac{1}{3} & (1) \\ \phantom{x+}y = -\dfrac{5}{3} & (2) \end{cases}$$

This system is consistent and independent.

Substituting $-\dfrac{5}{3}$ for $y$ in equation (1), we obtain

$$x + \left(-\dfrac{5}{3}\right) = -\dfrac{1}{3}$$

$$x = \dfrac{4}{3}$$

The solution is the ordered pair $\left(\dfrac{4}{3}, -\dfrac{5}{3}\right)$.

**38.** Write the augmented matrix of the system and then put it in row echelon form.

$$\begin{bmatrix} 5 & -2 & | & 10 \\ 2 & -\frac{4}{5} & | & 4 \end{bmatrix} \quad \left(R_1 = \dfrac{1}{5}r_1\right)$$

$$= \begin{bmatrix} 1 & -\frac{2}{5} & | & 2 \\ 2 & -\frac{4}{5} & | & 4 \end{bmatrix} \quad (R_2 = -2r_1 + r_2)$$

$$= \begin{bmatrix} 1 & -\frac{2}{5} & | & 2 \\ 0 & 0 & | & 0 \end{bmatrix}$$

The system is dependent. The solution to the system is $\{(x,y) \mid 5x - 2y = 10\}$.

**40.** Write the augmented matrix of the system and then put it in row echelon form.

$$\begin{bmatrix} 1 & 1 & 1 & | & 5 \\ 2 & -1 & 3 & | & -3 \\ -1 & 2 & -1 & | & 10 \end{bmatrix} \quad \left(\begin{array}{l} R_2 = -2r_1 + r_2 \\ R_3 = r_1 + r_3 \end{array}\right)$$

$$= \begin{bmatrix} 1 & 1 & 1 & | & 5 \\ 0 & -3 & 1 & | & -13 \\ 0 & 3 & 0 & | & 15 \end{bmatrix} \quad (\text{Interchange } r_2 \text{ and } r_3)$$

$$= \begin{bmatrix} 1 & 1 & 1 & | & 5 \\ 0 & 3 & 0 & | & 15 \\ 0 & -3 & 1 & | & -13 \end{bmatrix} \quad \left(R_2 = \dfrac{1}{3}r_2\right)$$

$$= \begin{bmatrix} 1 & 1 & 1 & | & 5 \\ 0 & 1 & 0 & | & 5 \\ 0 & -3 & 1 & | & -13 \end{bmatrix} \quad (R_3 = 3r_2 + r_3)$$

$$= \begin{bmatrix} 1 & 1 & 1 & | & 5 \\ 0 & 1 & 0 & | & 5 \\ 0 & 0 & 1 & | & 2 \end{bmatrix}$$

Write the system of equations that corresponds to the row-echelon matrix

$$\begin{cases} x + y + z = 5 & (1) \\ \phantom{x+}y\phantom{+z} = 5 & (2) \\ \phantom{x+y+}z = 2 & (3) \end{cases}$$

This system is consistent and independent.

We have $y = 5$ and $z = 2$. Substituting 5 for $y$ and 2 for $z$ in equation (1), we obtain

$$x + 5 + 2 = 5$$

$$x + 7 = 5$$

$$x = -2$$

The solution is the ordered triple $(-2, 5, 2)$.

**42.** Write the augmented matrix of the system and then put it in row echelon form.

$$\begin{bmatrix} 2 & -1 & 2 & | & 13 \\ -1 & 2 & -1 & | & -14 \\ 3 & 1 & -2 & | & -13 \end{bmatrix} \quad (\text{Interchange } r_1 \text{ and } r_2)$$

$$= \begin{bmatrix} -1 & 2 & -1 & | & -14 \\ 2 & -1 & 2 & | & 13 \\ 3 & 1 & -2 & | & -13 \end{bmatrix} \quad (R_1 = -1 \cdot r_1)$$

$$= \begin{bmatrix} 1 & -2 & 1 & | & 14 \\ 2 & -1 & 2 & | & 13 \\ 3 & 1 & -2 & | & -13 \end{bmatrix} \quad \left(\begin{array}{l} R_2 = -2r_1 + r_2 \\ R_3 = -3r_1 + r_3 \end{array}\right)$$

$$= \begin{bmatrix} 1 & -2 & 1 & | & 14 \\ 0 & 3 & 0 & | & -15 \\ 0 & 7 & -5 & | & -55 \end{bmatrix} \quad \left(R_2 = \dfrac{1}{3}r_2\right)$$

$$= \begin{bmatrix} 1 & -2 & 1 & | & 14 \\ 0 & 1 & 0 & | & -5 \\ 0 & 7 & -5 & | & -55 \end{bmatrix} \quad (R_3 = -7r_2 + r_3)$$

$$= \begin{bmatrix} 1 & -2 & 1 & | & 14 \\ 0 & 1 & 0 & | & -5 \\ 0 & 0 & -5 & | & -20 \end{bmatrix} \quad \left(R_3 = -\dfrac{1}{5}r_3\right)$$

$$= \begin{bmatrix} 1 & -2 & 1 & | & 14 \\ 0 & 1 & 0 & | & -5 \\ 0 & 0 & 1 & | & 4 \end{bmatrix}$$

Write the system of equations that corresponds to the row-echelon matrix

$$\begin{cases} x - 2y + z = 14 & (1) \\ \phantom{x-2}y\phantom{+z} = -5 & (2) \\ \phantom{x-2y+}z = 4 & (3) \end{cases}$$

We have $y = -5$ and $z = 4$. Substituting $-5$ for $y$ and 4 for $y$ in equation (1), we obtain

ISM –178

$x - 2(-5) + 4 = 14$

$x + 10 + 4 = 14$

$x + 14 = 14$

$x = 0$

The solution is the ordered triple $(0, -5, 4)$.

**44.** Write the augmented matrix of the system and then put it in row echelon form.

$$\begin{bmatrix} -1 & 2 & 1 & | & 1 \\ 2 & -1 & 3 & | & -3 \\ -1 & 5 & 6 & | & 2 \end{bmatrix} \quad (R_1 = -1 \cdot r_1)$$

$$= \begin{bmatrix} 1 & -2 & -1 & | & -1 \\ 2 & -1 & 3 & | & -3 \\ -1 & 5 & 6 & | & 2 \end{bmatrix} \quad \begin{pmatrix} R_2 = -2r_1 + r_2 \\ R_3 = r_1 + r_3 \end{pmatrix}$$

$$= \begin{bmatrix} 1 & -2 & -1 & | & -1 \\ 0 & 3 & 5 & | & -1 \\ 0 & 3 & 5 & | & 1 \end{bmatrix} \quad (R_3 = -1 \cdot r_2 + r_3)$$

$$= \begin{bmatrix} 1 & -2 & -1 & | & -1 \\ 0 & 3 & 5 & | & -1 \\ 0 & 0 & 0 & | & 2 \end{bmatrix}$$

The system is inconsistent. The system has no solution. The solution set is $\varnothing$ or $\{\ \}$.

**46.** Write the augmented matrix of the system and then put it in row echelon form.

$$\begin{bmatrix} -1 & 4 & -3 & | & 1 \\ 3 & 1 & -1 & | & -3 \\ 1 & 9 & -7 & | & -1 \end{bmatrix} \quad (R_1 = -1 \cdot r_1)$$

$$= \begin{bmatrix} 1 & -4 & 3 & | & -1 \\ 3 & 1 & -1 & | & -3 \\ 1 & 9 & -7 & | & -1 \end{bmatrix} \quad \begin{pmatrix} R_2 = -3r_1 + r_2 \\ R_3 = -1r_1 + r_3 \end{pmatrix}$$

$$= \begin{bmatrix} 1 & -4 & 3 & | & -1 \\ 0 & 13 & -10 & | & 0 \\ 0 & 13 & -10 & | & 0 \end{bmatrix} \quad (R_3 = -1r_2 + r_3)$$

$$= \begin{bmatrix} 1 & -4 & 3 & | & -1 \\ 0 & 13 & -10 & | & 0 \\ 0 & 0 & 0 & | & 0 \end{bmatrix}$$

The system is dependent and has an infinite number of solutions.

Write the system of equations that corresponds to the row-echelon matrix

$$\begin{cases} x - 4y + 3z = -1 & (1) \\ 13y - 10z = 0 & (2) \\ 0 = 0 & (3) \end{cases}$$

Solve equation (2) for $y$.

$13y - 10z = 0$

$13y = 10z$

$y = \dfrac{10}{13} z$

Substituting $\dfrac{10}{13} z$ for $y$ in equation (1), we obtain

$x - 4\left(\dfrac{10}{13} z\right) + 3z = -1$

$x - \dfrac{40}{13} z + 3z = -1$

$x - \dfrac{1}{13} z = -1$

$x = \dfrac{1}{13} z - 1$

The solution to the system is $\left\{ (x, y, z) \,\middle|\, x = \dfrac{1}{13} z - 1, \right.$ $\left. y = \dfrac{10}{13} z, z \text{ is any real number} \right\}$.

**48.** Write the augmented matrix of the system and then put it in row echelon form.

$$\begin{bmatrix} 1 & 1 & 1 & | & 8 \\ 2 & 3 & 1 & | & 19 \\ 2 & 2 & 4 & | & 21 \end{bmatrix} \quad \begin{pmatrix} R_2 = -2r_1 + r_2 \\ R_3 = -2r_1 + r_3 \end{pmatrix}$$

$$= \begin{bmatrix} 1 & 1 & 1 & | & 8 \\ 0 & 1 & -1 & | & 3 \\ 0 & 0 & 2 & | & 5 \end{bmatrix} \quad \left( R_3 = \dfrac{1}{2} r_3 \right)$$

$$= \begin{bmatrix} 1 & 1 & 1 & | & 8 \\ 0 & 1 & -1 & | & 3 \\ 0 & 0 & 1 & | & \frac{5}{2} \end{bmatrix}$$

Write the system of equations that corresponds to the row-echelon matrix

$$\begin{cases} x + y + z = 8 & (1) \\ y - z = 3 & (2) \\ z = \dfrac{5}{2} & (3) \end{cases}$$

This system is consistent and independent.

ISM –179

Substituting $\dfrac{5}{2}$ for $z$ in equation (2), we obtain

$$y - \dfrac{5}{2} = 3$$
$$y = \dfrac{11}{2}$$

Substituting $\dfrac{5}{2}$ for $z$ and $\dfrac{11}{2}$ for $y$ in equation (1), we obtain

$$x + \dfrac{5}{2} + \dfrac{11}{2} = 8$$
$$x + 8 = 8$$
$$x = 0$$

The solution is the ordered triple $\left(0, \dfrac{11}{2}, \dfrac{5}{2}\right)$.

50. Write the augmented matrix of the system and then put it in row echelon form.

$$\begin{bmatrix} 2 & 1 & 3 & | & 3 \\ 2 & -3 & 0 & | & 7 \\ 0 & 4 & 6 & | & -2 \end{bmatrix} \quad \left(R_1 = \dfrac{1}{2}r_1\right)$$

$$= \begin{bmatrix} 1 & \tfrac{1}{2} & \tfrac{3}{2} & | & \tfrac{3}{2} \\ 2 & -3 & 0 & | & 7 \\ 0 & 4 & 6 & | & -2 \end{bmatrix} \quad (R_2 = -2r_1 + r_2)$$

$$= \begin{bmatrix} 1 & \tfrac{1}{2} & \tfrac{3}{2} & | & \tfrac{3}{2} \\ 0 & -4 & -3 & | & 4 \\ 0 & 4 & 6 & | & -2 \end{bmatrix} \quad (R_3 = r_2 + r_3)$$

$$= \begin{bmatrix} 1 & \tfrac{1}{2} & \tfrac{3}{2} & | & \tfrac{3}{2} \\ 0 & -4 & -3 & | & 4 \\ 0 & 0 & 3 & | & 2 \end{bmatrix} \quad \left(\begin{array}{l} R_2 = -\tfrac{1}{4}r_2 \\ R_3 = \tfrac{1}{3}r_3 \end{array}\right)$$

$$= \begin{bmatrix} 1 & \tfrac{1}{2} & \tfrac{3}{2} & | & \tfrac{3}{2} \\ 0 & 1 & \tfrac{3}{4} & | & -1 \\ 0 & 0 & 1 & | & \tfrac{2}{3} \end{bmatrix}$$

Write the system of equations that corresponds to the row-echelon matrix

$$\begin{cases} x + \dfrac{1}{2}y + \dfrac{3}{2}z = \dfrac{3}{2} & (1) \\ y + \dfrac{3}{4}z = -1 & (2) \\ z = \dfrac{2}{3} & (3) \end{cases}$$

This system is consistent and independent.

Substituting $\dfrac{2}{3}$ for $z$ in equation (2), we obtain

$$y + \dfrac{3}{4}\left(\dfrac{2}{3}\right) = -1$$
$$y + \dfrac{1}{2} = -1$$
$$y = -\dfrac{3}{2}$$

Substituting $-\dfrac{3}{2}$ for $y$ and $\dfrac{2}{3}$ for $z$ in equation (1), we obtain

$$x + \dfrac{1}{2}\left(-\dfrac{3}{2}\right) + \dfrac{3}{2}\left(\dfrac{2}{3}\right) = \dfrac{3}{2}$$
$$x - \dfrac{3}{4} + 1 = \dfrac{3}{2}$$
$$x = \dfrac{5}{4}$$

The solution is the ordered triple $\left(\dfrac{5}{4}, -\dfrac{3}{2}, \dfrac{2}{3}\right)$.

52. Write the augmented matrix of the system and then put it in row echelon form.

$$\begin{bmatrix} 1 & 2 & -1 & | & 3 \\ 0 & 1 & 1 & | & 1 \\ 1 & 0 & -3 & | & 2 \end{bmatrix} \quad (R_3 = -1r_1 + r_3)$$

$$= \begin{bmatrix} 1 & 2 & -1 & | & 3 \\ 0 & 1 & 1 & | & 1 \\ 0 & -2 & -2 & | & -1 \end{bmatrix} \quad (R_3 = 2r_2 + r_3)$$

$$= \begin{bmatrix} 1 & 2 & -1 & | & 3 \\ 0 & 1 & 1 & | & 1 \\ 0 & 0 & 0 & | & 1 \end{bmatrix}$$

The system is inconsistent. The system has no solution. The solution set is $\varnothing$ or $\{\ \}$.

54. **a.** If $f(1) = 0$, then $a(1)^2 + b(1) + c = 0$ or $a + b + c = 0$.

If $f(2) = -3$, then $a(2)^2 + b(2) + c = -3$ or $4a + 2b + c = -3$.

**b.** To find $a$, $b$, and $c$, we must solve the following system:

$$\begin{cases} a - b + c = -6 & (1) \\ a + b + c = 0 & (2) \\ 4a + 2b + c = -3 & (3) \end{cases}$$

Write the augmented matrix of the system and then put it in row echelon form.

$$\begin{bmatrix} 1 & -1 & 1 & | & -6 \\ 1 & 1 & 1 & | & 0 \\ 4 & 2 & 1 & | & -3 \end{bmatrix} \quad \begin{pmatrix} R_2 = -1r_1 + r_2 \\ R_3 = -4r_1 + r_3 \end{pmatrix}$$

$$= \begin{bmatrix} 1 & -1 & 1 & | & -6 \\ 0 & 2 & 0 & | & 6 \\ 0 & 6 & -3 & | & 21 \end{bmatrix} \quad \left( R_2 = \frac{1}{2}r_2 \right)$$

$$= \begin{bmatrix} 1 & -1 & 1 & | & -6 \\ 0 & 1 & 0 & | & 3 \\ 0 & 6 & -3 & | & 21 \end{bmatrix} \quad (R_3 = -6r_2 + r_3)$$

$$= \begin{bmatrix} 1 & -1 & 1 & | & -6 \\ 0 & 1 & 0 & | & 3 \\ 0 & 0 & -3 & | & 3 \end{bmatrix} \quad \left( R_3 = -\frac{1}{3}r_3 \right)$$

$$= \begin{bmatrix} 1 & -1 & 1 & | & -6 \\ 0 & 1 & 0 & | & 3 \\ 0 & 0 & 1 & | & -1 \end{bmatrix}$$

Write the system of equations that corresponds to the row-echelon matrix

$$\begin{cases} a - b + c = -6 & (1) \\ b = 3 & (2) \\ c = -1 & (3) \end{cases}$$

This system is consistent and independent. We have $b = 3$ and $c = -1$. Substituting $-3$ for $y$ and $1$ for $z$ in equation (1), we obtain

$$a - 3 + (-1) = -6$$
$$a - 4 = -6$$
$$a = -2$$

Thus, the quadratic equation that contains $(-1, -6)$, $(1, 0)$, and $(2, -3)$ is

$$f(x) = -2x^2 + 3x - 1.$$

56. Let $s$ represent the amount to be invested in the savings account, let $t$ represent the amount to be invested in Treasury bonds, and let $m$ represent the amount to be invested in the municipal fund.

$$\begin{cases} s + t + m = 12,000 & (1) \\ 0.02s + 0.04t + 0.09m = 440 & (2) \\ s - t = 4000 & (3) \end{cases}$$

Write the augmented matrix of the system and then put it in row echelon form.

$$\begin{bmatrix} 1 & 1 & 1 & | & 12,000 \\ 0.02 & 0.04 & 0.09 & | & 440 \\ 1 & -1 & 0 & | & 4,000 \end{bmatrix} \quad \begin{pmatrix} R_2 = -0.02r_1 + r_2 \\ R_3 = -1r_3 + r_1 \end{pmatrix}$$

$$= \begin{bmatrix} 1 & 1 & 1 & | & 12,000 \\ 0 & 0.02 & 0.07 & | & 200 \\ 0 & 2 & 1 & | & 8,000 \end{bmatrix} \quad (R_2 = 50r_2)$$

$$= \begin{bmatrix} 1 & 1 & 1 & | & 12,000 \\ 0 & 1 & 3.5 & | & 10,000 \\ 0 & 2 & 1 & | & 8,000 \end{bmatrix} \quad (R_3 = -2r_2 + r_3)$$

$$= \begin{bmatrix} 1 & 1 & 1 & | & 12,000 \\ 0 & 1 & 3.5 & | & 10,000 \\ 0 & 0 & -6 & | & -12,000 \end{bmatrix} \quad \left( R_3 = -\frac{1}{6}r_3 \right)$$

$$= \begin{bmatrix} 1 & 1 & 1 & | & 12,000 \\ 0 & 1 & 3.5 & | & 10,000 \\ 0 & 0 & 1 & | & 2,000 \end{bmatrix}$$

Write the system of equations that corresponds to the row-echelon matrix

$$\begin{cases} s + t + m = 12,000 & (1) \\ t + 3.5m = 10,000 & (2) \\ m = 2,000 & (3) \end{cases}$$

Substituting 2,000 for $m$ in equation (2), we obtain

$$t + 3.5(2000) = 10,000$$
$$t + 7000 = 10,000$$
$$t = 3,000$$

Substituting 3,000 for $t$ and 2,000 for $m$ in equation (1), we obtain

$$s + 3,000 + 2,000 = 12,000$$
$$s + 5,000 = 12,000$$
$$s = 7,000$$

Therefore, Marlon should invest $7,000 in the savings account, $3,000 in treasury bonds, and $2,000 in the mutual fund.

58. Write the augmented matrix of the system and then put it in reduced row echelon form.

$$\begin{bmatrix} 2 & 1 & | & -1 \\ -3 & -2 & | & -3 \end{bmatrix} \quad (R_1 = 2r_1 + r_2)$$

$$= \begin{bmatrix} 1 & 0 & | & -5 \\ -3 & -2 & | & -3 \end{bmatrix} \quad (R_2 = 3r_1 + r_2)$$

$$= \begin{bmatrix} 1 & 0 & | & -5 \\ 0 & -2 & | & -18 \end{bmatrix} \quad \left( R_2 = -\frac{1}{2}r_2 \right)$$

$$= \begin{bmatrix} 1 & 0 & | & -5 \\ 0 & 1 & | & 9 \end{bmatrix}$$

## Chapter 4: Systems of Linear Equations and Inequalities

Write the system of equations that corresponds to the reduced row echelon matrix
$$\begin{cases} x = -5 \\ y = 9 \end{cases}$$
The solution is the ordered pair $(-5, 9)$.

**60.** Write the augmented matrix of the system and then put it in reduced row echelon form.

$$\begin{bmatrix} 1 & 1 & 1 & | & 3 \\ 3 & -2 & 2 & | & 38 \\ -2 & 0 & -3 & | & -19 \end{bmatrix} \quad \begin{pmatrix} R_2 = -3r_1 + r_2 \\ R_3 = 2r_1 + r_3 \end{pmatrix}$$

$$= \begin{bmatrix} 1 & 1 & 1 & | & 3 \\ 0 & -5 & -1 & | & 29 \\ 0 & 2 & -1 & | & -13 \end{bmatrix} \quad \begin{pmatrix} R_1 = 5r_1 + r_2 \\ R_3 = 2r_2 + 5r_3 \end{pmatrix}$$

$$= \begin{bmatrix} 5 & 0 & 4 & | & 44 \\ 0 & -5 & -1 & | & 29 \\ 0 & 0 & -7 & | & -7 \end{bmatrix} \quad \left( R_3 = -\frac{1}{7}r_3 \right)$$

$$= \begin{bmatrix} 5 & 0 & 4 & | & 44 \\ 0 & -5 & -1 & | & 29 \\ 0 & 0 & 1 & | & 1 \end{bmatrix} \quad \begin{pmatrix} R_1 = -4r_3 + r_1 \\ R_2 = r_3 + r_2 \end{pmatrix}$$

$$= \begin{bmatrix} 5 & 0 & 0 & | & 40 \\ 0 & -5 & 0 & | & 30 \\ 0 & 0 & 1 & | & 1 \end{bmatrix} \quad \begin{pmatrix} R_1 = \frac{1}{5}r_1 \\ R_2 = -\frac{1}{5}r_2 \end{pmatrix}$$

$$= \begin{bmatrix} 1 & 0 & 0 & | & 8 \\ 0 & 1 & 0 & | & -6 \\ 0 & 0 & 1 & | & 1 \end{bmatrix}$$

Write the system of equations that corresponds to the reduced row echelon matrix
$$\begin{cases} x = 8 \\ y = -6 \\ z = 1 \end{cases}$$
The solution is the ordered triple $(8, -6, 1)$.

**62.** Write the augmented matrix of the system.
$$\begin{bmatrix} 3 & 2 & | & 4 \\ -5 & -3 & | & -4 \end{bmatrix}$$

Enter the system into a 2 by 3 matrix, [A]. Then, use the **ref(** command along with the ▶**frac** command to write the matrix in row echelon form with the entries in fractional form.

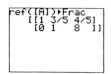

Thus, the row echelon matrix is
$$\begin{bmatrix} 1 & \frac{3}{5} & | & \frac{4}{5} \\ 0 & 1 & | & 8 \end{bmatrix}$$
Write the system of equations that corresponds to the row echelon matrix.
$$\begin{cases} x + \frac{3}{5}y = \frac{4}{5} & (1) \\ y = 8 & (2) \end{cases}$$
Substituting 8 for $y$ in equation (1), we obtain
$$x + \frac{3}{5}(8) = \frac{4}{5}$$
$$x + \frac{24}{5} = \frac{4}{5}$$
$$x = -\frac{20}{5} = -4$$
The solution is the ordered pair $(-4, 8)$.

**64.** Write the augmented matrix of the system.
$$\begin{bmatrix} 2 & -3 & -4 & | & 16 \\ -3 & 1 & 2 & | & -23 \\ 4 & 3 & -1 & | & 13 \end{bmatrix}$$

Enter the system into a 3 by 4 matrix, [A]. Then, use the **ref(** command along with the ▶**frac** command to write the matrix in row echelon form with the entries in fractional form. Since the entire matrix does not fit on the screen, we need to scroll right to see the rest of it.

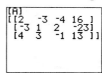

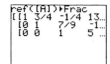 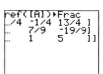

$$\begin{cases} x + \frac{3}{4}y - \frac{1}{4}z = \frac{13}{4} & (1) \\ y + \frac{7}{9}z = -\frac{19}{9} & (2) \\ z = 5 & (3) \end{cases}$$

Substituting 5 for $z$ in equation (2), we obtain
Thus, the row echelon matrix is

Write the system of equations that corresponds to the row-echelon matrix

$$y + \frac{7}{9}(5) = -\frac{19}{9}$$
$$y + \frac{35}{9} = -\frac{19}{9}$$
$$y = -\frac{54}{9} = -6$$

Substituting 5 for $z$ and $-6$ for $y$ in equation (1), we obtain

$$x + \frac{3}{4}(-6) - \frac{1}{4}(5) = \frac{13}{4}$$
$$x - \frac{9}{2} - \frac{5}{4} = \frac{13}{4}$$
$$x - \frac{23}{4} = \frac{13}{4}$$
$$x = \frac{36}{4} = 9$$

The solution is the ordered triple $(9, -6, 5)$.

## 4.5 Preparing for Determinants and Cramer's Rule

1. $4 \cdot 2 - 3 \cdot (-3) = 8 + 9 = 17$

2. $\frac{18}{6} = 3$

## 4.5 Exercises

2. $ad - bc$

4. FALSE. For example, $\begin{vmatrix} 1 & 1 & 1 \\ 1 & 1 & 1 \\ 1 & 1 & 1 \end{vmatrix} = 0$.

6. If the coefficient matrix is zero, then Cramer's Rule cannot be used because the denominator of a fraction cannot be zero.

   If the coefficient matrix is zero, then the system is either inconsistent and has no solution, or the system is dependent and has infinitely many solutions.

8. The system is inconsistent and the solution set is $\varnothing$ or $\{\ \}$.

10. $\begin{vmatrix} 5 & 3 \\ 2 & 4 \end{vmatrix} = 5(4) - 2(3) = 20 - 6 = 14$

12. $\begin{vmatrix} -8 & 5 \\ -4 & 3 \end{vmatrix} = -8(3) - (-4)(5) = -24 + 20 = -4$

14. $\begin{vmatrix} -2 & 1 & 6 \\ -3 & 2 & 5 \\ 1 & 0 & -2 \end{vmatrix} = -2\begin{vmatrix} 2 & 5 \\ 0 & -2 \end{vmatrix} - 1\begin{vmatrix} -3 & 5 \\ 1 & -2 \end{vmatrix} + 6\begin{vmatrix} -3 & 2 \\ 1 & 0 \end{vmatrix}$

$= -2[2(-2) - 0(5)] - 1[-3(-2) - 1(5)]$
$\qquad + 6[-3(0) - 1(2)]$
$= -2(-4 - 0) - 1(6 - 5) + 6(0 - 2)$
$= -2(-4) - 1(1) + 6(-2)$
$= 8 - 1 - 12$
$= -5$

16. $\begin{vmatrix} 8 & 4 & -1 \\ 2 & -7 & 1 \\ 0 & 5 & -3 \end{vmatrix} = 8\begin{vmatrix} -7 & 1 \\ 5 & -3 \end{vmatrix} - 4\begin{vmatrix} 2 & 1 \\ 0 & -3 \end{vmatrix} + (-1)\begin{vmatrix} 2 & -7 \\ 0 & 5 \end{vmatrix}$

$= 8[(-7)(-3) - 5(1)] - 4[2(-3) - 0(1)]$
$\qquad + (-1)[2(5) - 0(-7)]$
$= 8(21 - 5) - 4(-6 - 0) + (-1)(10 - 0)$
$= 8(16) - 4(-6) + (-1)(10)$
$= 128 + 24 - 10$
$= 142$

## Chapter 4: Systems of Linear Equations and Inequalities

18. $\begin{vmatrix} -3 & 4 & -2 \\ 1 & -2 & 0 \\ 0 & 6 & 6 \end{vmatrix}$

$= -3\begin{vmatrix} -2 & 0 \\ 6 & 6 \end{vmatrix} - 4\begin{vmatrix} 1 & 0 \\ 0 & 6 \end{vmatrix} + (-2)\begin{vmatrix} 1 & -2 \\ 0 & 6 \end{vmatrix}$

$= -3[-2(6) - 6(0)] - 4[1(6) - 0(0)]$
$\qquad + (-2)[1(6) - 0(-6)]$

$= -3(-12 - 0) - 4(6 - 0) + (-2)(6 - 0)$

$= -3(-12) - 4(6) + (-2)(6)$

$= 36 - 24 - 12$

$= 0$

20. $D = \begin{vmatrix} 1 & 1 \\ 1 & -1 \end{vmatrix} = 1(-1) - 1(1) = -1 - 1 = -2$

$D_x = \begin{vmatrix} 6 & 1 \\ 4 & -1 \end{vmatrix} = 6(-1) - 4(1) = -6 - 4 = -10$

$D_y = \begin{vmatrix} 1 & 6 \\ 1 & 4 \end{vmatrix} = 1(4) - 1(6) = 4 - 6 = -2$

$x = \dfrac{D_x}{D} = \dfrac{-10}{-2} = 5$; $y = \dfrac{D_y}{D} = \dfrac{-2}{-2} = 1$

Thus, the solution is the ordered pair $(5, 1)$.

22. $D = \begin{vmatrix} 2 & 4 \\ 3 & 2 \end{vmatrix} = 2(2) - 3(4) = 4 - 12 = -8$

$D_x = \begin{vmatrix} -6 & 4 \\ 7 & 2 \end{vmatrix} = -6(2) - 7(4) = -12 - 28 = -40$

$D_y = \begin{vmatrix} 2 & -6 \\ 3 & 7 \end{vmatrix} = 2(7) - 3(-6) = 14 + 18 = 32$

$x = \dfrac{D_x}{D} = \dfrac{-40}{-8} = 5$; $y = \dfrac{D_y}{D} = \dfrac{32}{-8} = -4$

Thus, the solution is the ordered pair $(5, -4)$.

24. $D = \begin{vmatrix} 2 & 4 \\ 3 & 6 \end{vmatrix} = 2(6) - 3(4) = 12 - 12 = 0$

Since $D = 0$, we know that Cramer's Rule does not apply and that this is not a consistent and independent system. Next, we will find $D_x$, $D_y$, and $D_z$ in order to investigate the system further.

$D_x = \begin{vmatrix} 6 & 4 \\ 1 & 6 \end{vmatrix} = 6(6) - 1(4) = 36 - 4 = 32$

Because $D = 0$ and because the determinant $D_x$ does not equal 0, the system is inconsistent and has no solution. The solution set is $\emptyset$ or $\{\ \}$.

26. First, we must write the system in standard form:
$\begin{cases} 3x - 6y = 2 \\ x + 2y = 4 \end{cases}$

$D = \begin{vmatrix} 3 & -6 \\ 1 & 2 \end{vmatrix} = 3(2) - 1(-6) = 6 + 6 = 12$

$D_x = \begin{vmatrix} 2 & -6 \\ 4 & 2 \end{vmatrix} = 2(2) - 4(-6) = 4 + 24 = 28$

$D_y = \begin{vmatrix} 3 & 2 \\ 1 & 4 \end{vmatrix} = 3(4) - 1(2) = 12 - 2 = 10$

$x = \dfrac{D_x}{D} = \dfrac{28}{12} = \dfrac{7}{3}$; $y = \dfrac{D_y}{D} = \dfrac{10}{12} = \dfrac{5}{6}$

Thus, the solution is the ordered pair $\left(\dfrac{7}{3}, \dfrac{5}{6}\right)$.

28. $D = \begin{vmatrix} 1 & 1 & -1 \\ 1 & 2 & 1 \\ -1 & -1 & 2 \end{vmatrix} = 1\begin{vmatrix} 2 & 1 \\ -1 & 2 \end{vmatrix} - 1\begin{vmatrix} 1 & 1 \\ -1 & 2 \end{vmatrix} + (-1)\begin{vmatrix} 1 & 2 \\ -1 & -1 \end{vmatrix}$

$= 1[2(2) - (-1)(1)] - 1[1(2) - (-1)(1)] +$
$\qquad (-1)[1(-1) - (-1)(2)]$

$= 1(4 + 1) - 1(2 + 1) + (-1)(-1 + 2)$

$= 1(5) - 1(3) + (-1)(1)$

$= 5 - 3 - 1$

$= 1$

$D_x = \begin{vmatrix} 6 & 1 & -1 \\ 6 & 2 & 1 \\ -7 & -1 & 2 \end{vmatrix} = 6\begin{vmatrix} 2 & 1 \\ -1 & 2 \end{vmatrix} - 1\begin{vmatrix} 6 & 1 \\ -7 & 2 \end{vmatrix} + (-1)\begin{vmatrix} 6 & 2 \\ -7 & -1 \end{vmatrix}$

$= 6[2(2) - (-1)(1)] - 1[6(2) - (-7)(1)]$
$\qquad + (-1)[6(-1) - (-7)(2)]$

$= 6(4 + 1) - 1(12 + 7) + (-1)(-6 + 14)$

$= 6(5) - 1(19) + (-1)(8)$

$= 30 - 19 - 8$

$= 3$

$D_y = \begin{vmatrix} 1 & 6 & -1 \\ 1 & 6 & 1 \\ -1 & -7 & 2 \end{vmatrix} = 1\begin{vmatrix} 6 & 1 \\ -7 & 2 \end{vmatrix} - 6\begin{vmatrix} 1 & 1 \\ -1 & 2 \end{vmatrix} + (-1)\begin{vmatrix} 1 & 6 \\ -1 & -7 \end{vmatrix}$

$= 1[6(2) - (-1)(1)] - 6[1(2) - (-1)(1)] +$
$\qquad (-1)[1(-7) - (-1)(6)]$

$= 1(12 + 1) - 6(2 + 1) + (-1)(-7 + 6)$

$= 1(13) - 6(3) + (-1)(-1)$

$= 13 - 18 + 1$

$= -4$

ISM − 184

### Section 4.5 Determinants and Cramer's Rule

$D_z = \begin{vmatrix} 1 & 1 & 6 \\ 1 & 2 & 6 \\ -1 & -1 & -7 \end{vmatrix} = 1\begin{vmatrix} 2 & 6 \\ -1 & -7 \end{vmatrix} - 1\begin{vmatrix} 1 & 6 \\ -1 & -7 \end{vmatrix} + 6\begin{vmatrix} 1 & 2 \\ -1 & -1 \end{vmatrix}$

$= 1\big[2(-7)-(-1)(6)\big] - 1\big[1(-7)-(-1)(6)\big] +$
$\qquad\qquad 6\big[1(-1)-(-1)(2)\big]$

$= 1(-14+6) - 1(-7+6) + 6(-1+2)$

$= 1(-8) - 1(-1) + 6(1)$

$= -8 + 1 + 6$

$= -1$

$x = \dfrac{D_x}{D} = \dfrac{3}{1} = 3$; $\; y = \dfrac{D_y}{D} = \dfrac{2}{1} = 2$; $\; z = \dfrac{D_z}{D} = \dfrac{-1}{1} = -1$

Thus, the solution is the ordered triple $(3, 2, -1)$.

**30.** $D = \begin{vmatrix} 1 & 1 & 1 \\ -2 & -3 & -1 \\ 2 & -1 & -3 \end{vmatrix} = 1\begin{vmatrix} -3 & -1 \\ -1 & -3 \end{vmatrix} - 1\begin{vmatrix} -2 & -1 \\ 2 & -3 \end{vmatrix} + 1\begin{vmatrix} -2 & -3 \\ 2 & -1 \end{vmatrix}$

$= 1\big[-3(-3)-(-1)(-1)\big] - 1\big[-2(-3)-2(-1)\big]$
$\qquad\qquad + 1\big[-2(-1)-2(-3)\big]$

$= 1(9-1) - 1(6+2) + 1(2+6)$

$= 1(8) - 1(8) + 1(8)$

$= 8 - 8 + 8$

$= 8$

$D_x = \begin{vmatrix} -3 & 1 & 1 \\ 1 & -3 & -1 \\ -5 & -1 & -3 \end{vmatrix} = -3\begin{vmatrix} -3 & -1 \\ -1 & -3 \end{vmatrix} - 1\begin{vmatrix} 1 & -1 \\ -5 & -3 \end{vmatrix} + 1\begin{vmatrix} 1 & -3 \\ -5 & -1 \end{vmatrix}$

$= -3\big[-3(-3)-(-1)(-1)\big] - 1\big[1(-3)-(-5)(-1)\big]$
$\qquad\qquad + 1\big[1(-1)-(-5)(-3)\big]$

$= -3(9-1) - 1(-3-5) + 1(-1-15)$

$= -3(8) - 1(-8) + 1(-16)$

$= -24 + 8 - 16$

$= -32$

$D_y = \begin{vmatrix} 1 & -3 & 1 \\ -2 & 1 & -1 \\ 2 & -5 & -3 \end{vmatrix} = 1\begin{vmatrix} 1 & -1 \\ -5 & -3 \end{vmatrix} - (-3)\begin{vmatrix} -2 & -1 \\ 2 & -3 \end{vmatrix} + 1\begin{vmatrix} -2 & 1 \\ 2 & -5 \end{vmatrix}$

$= 1\big[1(-3)-(-5)(-1)\big] - (-3)\big[-2(-3)-2(-1)\big]$
$\qquad\qquad + 1\big[-2(-5)-2(1)\big]$

$= 1(-3-5) - (-3)(6+2) + 1(10-2)$

$= 1(-8) - (-3)(8) + 1(8)$

$= -8 + 24 + 8$

$= 24$

$D_z = \begin{vmatrix} 1 & 1 & -3 \\ -2 & -3 & 1 \\ 2 & -1 & -5 \end{vmatrix} = 1\begin{vmatrix} -3 & 1 \\ -1 & -5 \end{vmatrix} - 1\begin{vmatrix} -2 & 1 \\ 2 & -5 \end{vmatrix} + (-3)\begin{vmatrix} -2 & -3 \\ 2 & -1 \end{vmatrix}$

$= 1\big[-3(-5)-(-1)(1)\big] - 1\big[-2(-5)-2(1)\big]$
$\qquad\qquad + (-3)\big[-2(-1)-2(-3)\big]$

$= 1(15+1) - 1(10-2) + (-3)(2+6)$

$= 1(16) - 1(8) + (-3)(8)$

$= 16 - 8 - 24$

$= -16$

$x = \dfrac{D_x}{D} = \dfrac{-32}{8} = -4$; $\; y = \dfrac{D_y}{D} = \dfrac{24}{8} = 3$;

$z = \dfrac{D_z}{D} = \dfrac{-16}{8} = -2$

Thus, the solution is the ordered triple $(-4, 3, -2)$.

**32.** $D = \begin{vmatrix} -1 & 2 & -1 \\ 2 & 1 & 2 \\ -1 & 7 & -1 \end{vmatrix} = -1\begin{vmatrix} 1 & 2 \\ 7 & -1 \end{vmatrix} - 2\begin{vmatrix} 2 & 2 \\ -1 & -1 \end{vmatrix} + (-1)\begin{vmatrix} 2 & 1 \\ -1 & 7 \end{vmatrix}$

$= -1\big[1(-1)-7(2)\big] - 2\big[2(-1)-(-1)(2)\big]$
$\qquad\qquad + (-1)\big[2(7)-(-1)(1)\big]$

$= -1(-1-14) - 2(-2+2) + (-1)(14+1)$

$= -1(-15) - 2(0) + (-1)(15) = 15 - 0 - 15 = 0$

Since $D = 0$, we know that Cramer's Rule does not apply and that this is not a consistent and independent system. Next, we will find $D_x$, $D_y$, and $D_z$ in order to investigate the system further.

$D_x = \begin{vmatrix} 2 & 2 & -1 \\ -6 & 1 & 2 \\ 0 & 7 & -1 \end{vmatrix} = 2\begin{vmatrix} 1 & 2 \\ 7 & -1 \end{vmatrix} - 2\begin{vmatrix} -6 & 2 \\ 0 & -1 \end{vmatrix} + (-1)\begin{vmatrix} -6 & 1 \\ 0 & 7 \end{vmatrix}$

$= 2\big[1(-1)-7(2)\big] - 2\big[(-6)(-1)-0(2)\big]$
$\qquad\qquad + (-1)\big[-6(7)-0(1)\big]$

$= 2(-1-14) - 2(6-0) + (-1)(-42-0)$

$= 2(-15) - 2(6) + (-1)(-42) = -30 - 12 + 42 = 0$

$D_y = \begin{vmatrix} -1 & 2 & -1 \\ 2 & -6 & 2 \\ -1 & 0 & -1 \end{vmatrix}$

$= -1\begin{vmatrix} -6 & 2 \\ 0 & -1 \end{vmatrix} - 2\begin{vmatrix} 2 & 2 \\ -1 & -1 \end{vmatrix} + (-1)\begin{vmatrix} 2 & -6 \\ -1 & 0 \end{vmatrix}$

$= -1\big[-6(-1)-0(2)\big] - 2\big[2(-1)-(-1)(2)\big]$
$\qquad\qquad + (-1)\big[2(0)-(-1)(-6)\big]$

$= -1(6-0) - 2(-2+2) + (-1)(0-6)$

$= -1(6) - 2(0) + (-1)(-6) = -6 - 0 + 6 = 0$

## Chapter 4: Systems of Linear Equations and Inequalities

$D_z = \begin{vmatrix} -1 & 2 & 2 \\ 2 & 1 & -6 \\ -1 & 7 & 0 \end{vmatrix} = -1\begin{vmatrix} 1 & -6 \\ 7 & 0 \end{vmatrix} - 2\begin{vmatrix} 2 & -6 \\ -1 & 0 \end{vmatrix} + 2\begin{vmatrix} 2 & 1 \\ -1 & 7 \end{vmatrix}$

$= -1[1(0) - 7(-6)] - 2[2(0) - (-1)(-6)]$
$\qquad + 2[2(7) - (-1)(1)]$
$= -1(0 + 42) - 2(0 - 6) + 2(14 + 1)$
$= -1(42) - 2(-6) + 2(15) = -42 + 12 + 30 = 0$

Because $D = 0$ and because all the determinants $D_x$, $D_y$, and $D_z$ equal 0, the system is consistent and dependent. We know that there are infinitely many solutions, but we cannot yet tell what the solution set is. Next we use the augmented matrix of the system in order to determine the solution set.

$\begin{bmatrix} -1 & 2 & -1 & | & 2 \\ 2 & 1 & 2 & | & -6 \\ -1 & 7 & -1 & | & 0 \end{bmatrix} \quad (R_1 = -1 \cdot r_1)$

$= \begin{bmatrix} 1 & -2 & 1 & | & -2 \\ 2 & 1 & 2 & | & -6 \\ -1 & 7 & -1 & | & 0 \end{bmatrix} \quad \begin{pmatrix} R_2 = -2r_1 + r_2 \\ R_3 = r_1 + r_3 \end{pmatrix}$

$= \begin{bmatrix} 1 & -2 & 1 & | & -2 \\ 0 & 5 & 0 & | & -2 \\ 0 & 5 & 0 & | & -2 \end{bmatrix} \quad (R_3 = -1r_2 + r_3)$

$= \begin{bmatrix} 1 & -2 & 1 & | & -2 \\ 0 & 5 & 0 & | & -2 \\ 0 & 0 & 0 & | & 0 \end{bmatrix}$

We have confirmed that this system is dependent and has an infinite number of solutions.

Write the system of equations that corresponds to the row-echelon matrix
$\begin{cases} x - 2y + z = -2 & (1) \\ 5y = -2 & (2) \\ 0 = 0 & (3) \end{cases}$

Solve equation (2) for $y$.
$5y = -2$
$y = -\frac{2}{5}$

Substituting $-\frac{2}{5}$ for $y$ in equation (1), we obtain
$x - 2\left(-\frac{2}{5}\right) + z = -2$
$x + \frac{4}{5} + z = -2$
$x + z = -\frac{14}{5}$
$x = -z - \frac{14}{5}$

The solution to the system is $\{(x, y, z) | x = -z - \frac{14}{5},$ $y = -\frac{2}{5}, z$ is any real number$\}$

34. $D = \begin{vmatrix} 1 & -2 & -1 \\ 2 & 2 & 1 \\ 1 & -4 & 3 \end{vmatrix} = 1\begin{vmatrix} 2 & 1 \\ -4 & 3 \end{vmatrix} - (-2)\begin{vmatrix} 2 & 1 \\ 1 & 3 \end{vmatrix} + (-1)\begin{vmatrix} 2 & 2 \\ 1 & -4 \end{vmatrix}$

$= 1[2(3) - (-4)(1)] - (-2)[2(3) - 1(1)]$
$\qquad + (-1)[2(-4) - 1(2)]$
$= 1(6 + 4) - (-2)(6 - 1) + (-1)(-8 - 2)$
$= 1(10) - (-2)(5) + (-1)(-10)$
$= 10 + 10 + 10$
$= 30$

$D_x = \begin{vmatrix} 1 & -2 & -1 \\ 3 & 2 & 1 \\ 14 & -4 & 3 \end{vmatrix}$

$= 1\begin{vmatrix} 2 & 1 \\ -4 & 3 \end{vmatrix} - (-2)\begin{vmatrix} 3 & 1 \\ 14 & 3 \end{vmatrix} + (-1)\begin{vmatrix} 3 & 2 \\ 14 & -4 \end{vmatrix}$

$= 1[2(3) - (-4)(1)] - (-2)[3(3) - 14(1)]$
$\qquad + (-1)[3(-4) - 14(2)]$
$= 1(6 + 4) - (-2)(9 - 14) + (-1)(-12 - 28)$
$= 1(10) - (-2)(-5) + (-1)(-40)$
$= 10 - 10 + 40$
$= 40$

$D_y = \begin{vmatrix} 1 & 1 & -1 \\ 2 & 3 & 1 \\ 1 & 14 & 3 \end{vmatrix} = 1\begin{vmatrix} 3 & 1 \\ 14 & 3 \end{vmatrix} - 1\begin{vmatrix} 2 & 1 \\ 1 & 3 \end{vmatrix} + (-1)\begin{vmatrix} 2 & 3 \\ 1 & 14 \end{vmatrix}$

$= 1[3(3) - 14(1)] - 1[2(3) - 1(1)]$
$\qquad + (-1)[2(14) - 1(3)]$
$= 1(9 - 14) - 1(6 - 1) + (-1)(28 - 3)$
$= 1(-5) - 1(5) + (-1)(25)$
$= -5 - 5 - 25$
$= -35$

$D_z = \begin{vmatrix} 1 & -2 & 1 \\ 2 & 2 & 3 \\ 1 & -4 & 14 \end{vmatrix} = 1\begin{vmatrix} 2 & 3 \\ -4 & 14 \end{vmatrix} - (-2)\begin{vmatrix} 2 & 3 \\ 1 & 14 \end{vmatrix} + 1\begin{vmatrix} 2 & 2 \\ 1 & -4 \end{vmatrix}$

$= 1[2(14) - (-4)(3)] - (-2)[2(14) - 1(3)]$
$\qquad + 1[2(-4) - 1(2)]$
$= 1(28 + 12) - (-2)(28 - 3) + 1(-8 - 2)$
$= 1(40) - (-2)(25) + 1(-10)$
$= 40 + 50 - 10$
$= 80$

## Section 4.5 Determinants and Cramer's Rule

$x = \dfrac{D_x}{D} = \dfrac{40}{30} = \dfrac{4}{3}$ ; $y = \dfrac{D_y}{D} = \dfrac{-35}{30} = -\dfrac{7}{6}$ ;

$z = \dfrac{D_z}{D} = \dfrac{80}{30} = \dfrac{8}{3}$

The solution is the ordered triple $\left(\dfrac{4}{3}, -\dfrac{7}{6}, \dfrac{8}{3}\right)$.

**36.** $D = \begin{vmatrix} 1 & 2 & 1 \\ -1 & -3 & -2 \\ 2 & 0 & -3 \end{vmatrix} = 1\begin{vmatrix} -3 & -2 \\ 0 & -3 \end{vmatrix} - 2\begin{vmatrix} -1 & -2 \\ 2 & -3 \end{vmatrix} + 1\begin{vmatrix} -1 & -3 \\ 2 & 0 \end{vmatrix}$

$= 1[-3(-3) - 0(-2)] - 2[-1(-3) - 2(-2)]$
$\qquad + 1[-1(0) - 2(-3)]$

$= 1(9+0) - 2(3+4) + 1(0+6)$

$= 1(9) - 2(7) + 1(6)$

$= 9 - 14 + 6$

$= 1$

$D_x = \begin{vmatrix} 0 & 2 & 1 \\ 3 & -3 & -2 \\ 7 & 0 & -3 \end{vmatrix} = 0\begin{vmatrix} -3 & -2 \\ 0 & -3 \end{vmatrix} - 2\begin{vmatrix} 3 & -2 \\ 7 & -3 \end{vmatrix} + 1\begin{vmatrix} 3 & -3 \\ 7 & 0 \end{vmatrix}$

$= 0[-3(-3) - 0(-2)] - 2[3(-3) - 7(-2)]$
$\qquad + 1[3(0) - 7(-3)]$

$= 0(9+0) - 2(-9+14) + 1(0+21)$

$= 0(9) - 2(5) + 1(21)$

$= 0 - 10 + 21$

$= 11$

$D_y = \begin{vmatrix} 1 & 0 & 1 \\ -1 & 3 & -2 \\ 2 & 7 & -3 \end{vmatrix} = 1\begin{vmatrix} 3 & -2 \\ 7 & -3 \end{vmatrix} - 0\begin{vmatrix} -1 & -2 \\ 2 & -3 \end{vmatrix} + 1\begin{vmatrix} -1 & 3 \\ 2 & 7 \end{vmatrix}$

$= 1[3(-3) - 7(-2)] - 0[-1(-3) - 2(-2)]$
$\qquad + 1[-1(7) - 2(3)]$

$= 1(-9+14) - 0(3+4) + 1(-7-6)$

$= 1(5) - 0(7) + 1(-13)$

$= 5 - 0 - 13$

$= -8$

$D_z = \begin{vmatrix} 1 & 2 & 0 \\ -1 & -3 & 3 \\ 2 & 0 & 7 \end{vmatrix} = 1\begin{vmatrix} -3 & 3 \\ 0 & 7 \end{vmatrix} - 2\begin{vmatrix} -1 & 3 \\ 2 & 7 \end{vmatrix} + 0\begin{vmatrix} -1 & -3 \\ 2 & 0 \end{vmatrix}$

$= 1[-3(7) - 0(3)] - 2[-1(7) - 2(3)]$
$\qquad + 0[-1(0) - 2(-3)]$

$= 1(-21 - 0) - 2(-7 - 6) + 0(0 + 6)$

$= 1(-21) - 2(-13) + 0(6)$

$= -21 + 26 + 0$

$= 5$

$x = \dfrac{D_x}{D} = \dfrac{11}{1} = 11$ ; $y = \dfrac{D_y}{D} = \dfrac{-8}{1} = -8$ ; $z = \dfrac{D_z}{D} = \dfrac{5}{1} = 5$

The solution is the ordered triple $(11, -8, 5)$.

**38.** $D = \begin{vmatrix} 2 & 4 & 0 \\ -2 & 0 & 1 \\ 0 & -4 & -3 \end{vmatrix} = 2\begin{vmatrix} 0 & 1 \\ -4 & -3 \end{vmatrix} - 4\begin{vmatrix} -2 & 1 \\ 0 & -3 \end{vmatrix} + 0\begin{vmatrix} -2 & 0 \\ 0 & -4 \end{vmatrix}$

$= 2[0(-3) - (-4)(1)] - 4[-2(-3) - 0(1)]$
$\qquad + 0[-2(-4) - 0(0)]$

$= 2(0+4) - 4(6-0) + 0(8-0)$

$= 2(4) - 4(6) + 0(8)$

$= 8 - 24 + 0$

$= -16$

$D_x = \begin{vmatrix} 0 & 4 & 0 \\ -5 & 0 & 1 \\ 1 & -4 & -3 \end{vmatrix} = 0\begin{vmatrix} 0 & 1 \\ -4 & -3 \end{vmatrix} - 4\begin{vmatrix} -5 & 1 \\ 1 & -3 \end{vmatrix} + 0\begin{vmatrix} -5 & 0 \\ 1 & -4 \end{vmatrix}$

$= 0[0(-3) - (-4)(1)] - 4[-5(-3) - 1(1)]$
$\qquad + 0[-5(-4) - 1(0)]$

$= 0(0+4) - 4(15-1) + 0(20-0)$

$= 0(4) - 4(14) + 0(20)$

$= 0 - 56 + 0$

$= -56$

$D_y = \begin{vmatrix} 2 & 0 & 0 \\ -2 & -5 & 1 \\ 0 & 1 & -3 \end{vmatrix} = 2\begin{vmatrix} -5 & 1 \\ 1 & -3 \end{vmatrix} - 0\begin{vmatrix} -2 & 1 \\ 0 & -3 \end{vmatrix} + 0\begin{vmatrix} -2 & -5 \\ 0 & 1 \end{vmatrix}$

$= 2[-5(-3) - 1(1)] - 0[-2(-3) - 0(1)]$
$\qquad + 0[-2(1) - 0(-5)]$

$= 2(15-1) - 0(6-0) + 0(-2+0)$

$= 2(14) - 0(6) + 0(-2)$

$= 28 - 0 + 0$

$= 28$

$D_z = \begin{vmatrix} 2 & 4 & 0 \\ -2 & 0 & -5 \\ 0 & -4 & 1 \end{vmatrix} = 2\begin{vmatrix} 0 & -5 \\ -4 & 1 \end{vmatrix} - 4\begin{vmatrix} -2 & -5 \\ 0 & 1 \end{vmatrix} + 0\begin{vmatrix} -2 & 0 \\ 0 & -4 \end{vmatrix}$

$= 2[0(1) - (-4)(-5)] - 4[-2(1) - 0(-5)]$
$\qquad + 0[-2(-4) - 0(0)]$

$= 2(0-20) - 4(-2+0) + 0(8-0)$

$= 2(-20) - 4(-2) + 0(8)$

$= -40 + 8 + 0$

$= -32$

## Chapter 4: Systems of Linear Equations and Inequalities

$x = \dfrac{D_x}{D} = \dfrac{-56}{-16} = \dfrac{7}{2}$; $y = \dfrac{D_y}{D} = \dfrac{28}{-16} = -\dfrac{7}{4}$;

$z = \dfrac{D_z}{D} = \dfrac{-32}{-16} = 2$

The solution is the ordered triple $\left(\dfrac{7}{2}, -\dfrac{7}{4}, 2\right)$.

**40.** $\begin{vmatrix} -2 & x \\ 3 & 4 \end{vmatrix} = 1$

$-2(4) - 3(x) = 1$

$-8 - 3x = 1$

$-3x = 9$

$x = -3$

**42.** $\begin{vmatrix} 2 & x & -1 \\ 3 & 5 & 0 \\ -4 & 1 & 2 \end{vmatrix} = 0$

$2\begin{vmatrix} 5 & 0 \\ 1 & 2 \end{vmatrix} - x\begin{vmatrix} 3 & 0 \\ -4 & 2 \end{vmatrix} + (-1)\begin{vmatrix} 3 & 5 \\ -4 & 1 \end{vmatrix} = 0$

$2[5(2) - 1(0)] - x[3(2) - (-4)(0)]$
$\qquad + (-1)[3(1) - (-4)(5)] = 0$

$2(10 - 0) - x(6 - 0) + (-1)(3 + 20) = 0$

$2(10) - x(6) + (-1)(23) = 0$

$20 - 6x - 23 = 0$

$-6x - 3 = 0$

$-6x = 3$

$x = \dfrac{3}{-6} = -\dfrac{1}{2}$

**44. a.** Triangle $ABC$ is graphed below.

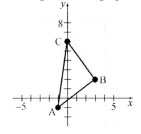

**b.** $D = \dfrac{1}{2}\begin{vmatrix} -1 & 3 & 0 \\ -1 & 2 & 6 \\ 1 & 1 & 1 \end{vmatrix}$

$= \dfrac{1}{2}\left(-1\begin{vmatrix} 2 & 6 \\ 1 & 1 \end{vmatrix} - 3\begin{vmatrix} -1 & 6 \\ 1 & 1 \end{vmatrix} + 0\begin{vmatrix} -1 & 2 \\ 1 & 1 \end{vmatrix}\right)$

$= \dfrac{1}{2}\left[-1(2 \cdot 1 - 1 \cdot 6) - 3(-1 \cdot 1 - 1 \cdot 6) + 0\right]$

$= \dfrac{1}{2}\left[-1(-4) - 3(-7) + 0\right]$

$= \dfrac{1}{2}(4 + 21 + 0)$

$= \dfrac{1}{2}(25)$

$= 12.5$

Thus, the area of triangle $ABC$ is $|12.5| = 12.5$.

**46. a.** Parallelogram $ABCD$ is graphed below.

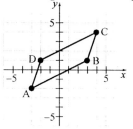

**b.** Triangle $ABC$ is formed by the points $(-3, -2)$, $(3, 1)$, and $(4, 4)$.

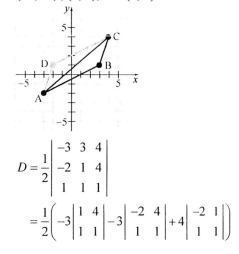

$D = \dfrac{1}{2}\begin{vmatrix} -3 & 3 & 4 \\ -2 & 1 & 4 \\ 1 & 1 & 1 \end{vmatrix}$

$= \dfrac{1}{2}\left(-3\begin{vmatrix} 1 & 4 \\ 1 & 1 \end{vmatrix} - 3\begin{vmatrix} -2 & 4 \\ 1 & 1 \end{vmatrix} + 4\begin{vmatrix} -2 & 1 \\ 1 & 1 \end{vmatrix}\right)$

ISM −188

## Section 4.5 Determinants and Cramer's Rule

$$= \frac{1}{2}\left[-3(1\cdot 1-1\cdot 4)-3(-2\cdot 1-1\cdot 4)\right.$$
$$\left.+4(-2\cdot 1-1\cdot 1)\right]$$
$$= \frac{1}{2}\left[-3(-3)-3(-6)+4(-3)\right]$$
$$= \frac{1}{2}(9+18-12)$$
$$= \frac{1}{2}(15)$$
$$= 7.5$$

Thus, the area of triangle $ABC$ is $|7.5| = 7.5$.

**c.** Triangle $ADC$ is formed by the points $(-3,-2)$, $(-2, 1)$, and $(4, 4)$.

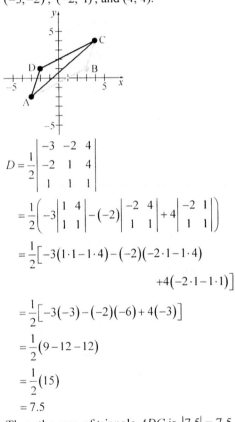

$$D = \frac{1}{2}\begin{vmatrix} -3 & -2 & 4 \\ -2 & 1 & 4 \\ 1 & 1 & 1 \end{vmatrix}$$

$$= \frac{1}{2}\left(-3\begin{vmatrix} 1 & 4 \\ 1 & 1 \end{vmatrix}-(-2)\begin{vmatrix} -2 & 4 \\ 1 & 1 \end{vmatrix}+4\begin{vmatrix} -2 & 1 \\ 1 & 1 \end{vmatrix}\right)$$

$$= \frac{1}{2}\left[-3(1\cdot 1-1\cdot 4)-(-2)(-2\cdot 1-1\cdot 4)\right.$$
$$\left.+4(-2\cdot 1-1\cdot 1)\right]$$
$$= \frac{1}{2}\left[-3(-3)-(-2)(-6)+4(-3)\right]$$
$$= \frac{1}{2}(9-12-12)$$
$$= \frac{1}{2}(15)$$
$$= 7.5$$

Thus, the area of triangle $ADC$ is $|7.5| = 7.5$.

**d.** The areas of triangle $ABC$ and triangle $ADC$ are equal. (That is, the diagonal of the parallelogram forms two triangles of equal area.) Thus, the area of parallelogram $ABCD$ is $7.5 + 7.5 = 15$.

**48. a.**

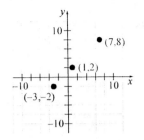

From the graph, the three points appear to be collinear.

**b.** $$\begin{vmatrix} -3 & -2 & 1 \\ 1 & 2 & 1 \\ 7 & 8 & 1 \end{vmatrix} = -3\begin{vmatrix} 2 & 1 \\ 8 & 1 \end{vmatrix} - (-2)\begin{vmatrix} 1 & 1 \\ 7 & 1 \end{vmatrix} + 1\begin{vmatrix} 1 & 2 \\ 7 & 8 \end{vmatrix}$$
$$= -3\left[2(1)-8(1)\right]-(-2)\left[1(1)-7(1)\right]$$
$$+1\left[1(8)-7(2)\right]$$
$$= -3(2-8)-(-2)(1-7)+1(8-14)$$
$$= -3(-6)-(-2)(-6)+1(-6)$$
$$= 18-12-6$$
$$= 0$$

Since the determinant is 0, the three points are collinear.

**c.** The slope between $(-3, -2)$ and $(1, 2)$ is
$$m = \frac{y_2 - y_1}{x_2 - x_1} = \frac{2-(-2)}{1-(-3)} = \frac{4}{4} = 1$$

The slope between $(1, 2)$ and $(7, 8)$ is
$$m = \frac{y_2 - y_1}{x_2 - x_1} = \frac{8-2}{7-1} = \frac{6}{6} = 1$$

The slope between $(-3, -2)$ and $(7, 8)$ is
$$m = \frac{y_2 - y_1}{x_2 - x_1} = \frac{8-(-2)}{7-(-3)} = \frac{10}{10} = 1$$

Since the slopes between each pair of points are equal, the three points must be collinear.

**50.** $$\begin{vmatrix} -3 & 1 \\ 6 & 5 \end{vmatrix} = -3(5)-6(1)$$
$$= -15-6$$
$$= -21$$

Multiplying the entries in column 2 by 3 and recomputing the determinant, we obtain
$$\begin{vmatrix} -3 & 3 \\ 6 & 15 \end{vmatrix} = -3(15)-6(3)$$
$$= -45-18$$
$$= -63$$

The value of the second determinant is three times the value of the first determinant.

ISM –189

Answers may vary. In general, it is true that, if the entries of a column of a determinant are multiplied by a constant, then the value of the new determinant will be that constant times the original value of the determinant.

52. Using $A = D$, $B = D_x$, and $C = D_y$.

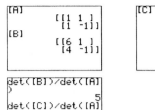

The solution is the ordered pair $(5, 1)$.

54. Using $A = D$, $B = D_x$, and $C = D_y$.

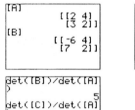

The solution is the ordered pair $(5, -4)$.

56. Using $A = D$, $B = D_x$, $C = D_y$, and $D = D_z$.

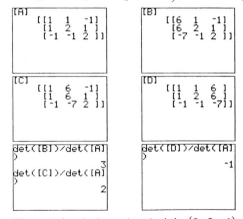

The solution is the ordered triple $(3, 2, -1)$.

## 4.6 Preparing for Systems of Linear Inequalities

1. $-3(-2) + 2 \overset{?}{\geq} 7$
   $6 + 2 \overset{?}{\geq} 7$
   $8 \geq 7 \leftarrow$ True
   Yes, $x = -2$ does satisfy the inequality $-3x + 2 \geq 7$.

2. $-2x + 1 > 7$
   $-2x + 1 - 1 > 7 - 1$
   $-2x > 6$
   $\dfrac{-2x}{-2} < \dfrac{6}{-2}$
   $x < -3$

   The solution set is $\{x | x < -3\}$ or, using interval notation, $(-\infty, -3)$.

3. Replace the inequality symbol with an equal sign to obtain $3x + 2y = -6$. Because the inequality is strict, graph $3x + 2y = -6$ $\left(y = -\dfrac{3}{2}x - 3\right)$ using a dashed line.

   Test Point: $(0, 0)$: $3(0) + 2(0) \overset{?}{>} -6$
   $0 + 0 \overset{?}{>} -6$
   $0 \overset{?}{>} -6$ True

   Therefore, $(0, 0)$ is solution to $3x + 2y > -6$. Shade the half-plane that contains $(0, 0)$.

   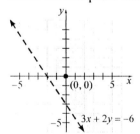

## 4.6 Exercises

2. corner points

4. FALSE. For example, the following system of linear inequalities has no solution:
   $\begin{cases} x + y < 1 \\ x + y > 4 \end{cases}$

## Section 4.6 Systems of Linear Inequalities

6. No, the corner point would not be a solution to the system. Solutions must satisfy all of the inequalities in the system. However, the corner point would only satisfy the non-strict inequality. Since it would not satisfy the strict inequality, it would not be a solution to the system.

8. Yes, it is possible for a system to have a straight line as the solution set. Examples may vary. One possibility follows: The solution to the following system of linear inequalities is the line $x+y=1$:
$$\begin{cases} x+y \le 1 \\ x+y \ge 1 \end{cases}$$

10. a. Let $x=-3$ and $y=6$ in each inequality.
$$\begin{cases} x+y=-3+6=3 \ge 2 \\ -3x+y=-3(-3)+6=15 \not\le 10 \end{cases}$$
The inequality $-3x+y \le 10$ is not true when $x=-3$ and $y=6$, so $(-3, 6)$ is not a solution.

    b. Let $x=4$ and $y=1$ in each inequality.
$$\begin{cases} x+y=4+1=5 \ge 2 \\ -3x+y=-3(4)+1=-11 \le 10 \end{cases}$$
Both inequalities are true when $x=4$ and $y=1$, so $(4, 1)$ is a solution.

12. a. Let $x=1$ and $y=3$ in each inequality.
$$\begin{cases} 5x+2y=5(1)+2(3)=11 \not< 10 \\ 4x-3y=4(1)-3(3)=-5 < 24 \end{cases}$$
The inequality $5x+2y<10$ is not true when $x=1$ and $y=3$, so $(1, 3)$ is not a solution.

    b. Let $x=1$ and $y=1$ in each inequality.
$$\begin{cases} 5x+2y=5(1)+2(1)=7<10 \\ 4x-3y=4(1)-3(1)=1<24 \end{cases}$$
Both inequalities are true when $x=1$ and $y=1$, so $(1, 1)$ is a solution.

14. a. Let $x=4$ and $y=2$ in each inequality.
$$\begin{cases} x+y=4+2=6 \ge 6 \\ 2x+y=2(4)+2=10 \ge 10 \\ x=4 \ge 0 \\ y=2 \ge 0 \end{cases}$$
All four inequalities are true when $x=4$ and $y=4$, so $(4, 2)$ is a solution.

    b. Let $x=2$ and $y=5$ in each inequality.
$$\begin{cases} x+y=2+5=7 \ge 6 \\ 2x+y=2(2)+6=9 \not\ge 10 \\ x=2 \ge 0 \\ y=5 \ge 0 \end{cases}$$
The inequality $2x+y \ge 10$ is not true when $x=2$ and $y=5$, so $(2, 5)$ is not a solution.

16. $\begin{cases} x+y \ge 2 \\ -3x+y \le 10 \end{cases}$

    First, graph the inequality $x+y \ge 2$. To do so, replace the inequality symbol with an equal sign to obtain $x+y=2$. Because the inequality is non-strict, graph $x+y=2$ $(y=-x+2)$ using a solid line.

    Test Point: $(0,0)$: $0+0=0 \not\ge 2$

    Therefore, the half-plane not containing $(0,0)$ is the solution set of $x+y \ge 2$.

    Second, graph the inequality $-3x+y \le 10$. To do so, replace the inequality symbol with an equal sign to obtain $-3x+y=10$. Because the inequality is non-strict, graph $-3x+y=10$ $(y=3x+10)$ using a solid line.

    Test Point: $(0,0)$: $-3(0)+0=0 \le 10$

    Therefore, the half-plane containing $(0,0)$ is the solution set of $-3x+y \le 10$.

    The overlapping shaded region (that is, the shaded region in the graph below) is the solution to the system of linear inequalities.

    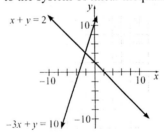

18. $\begin{cases} 5x+2y<10 \\ 4x-3y<24 \end{cases}$

    First, graph the inequality $5x+2y<10$. To do so, replace the inequality symbol with an equal sign to obtain $5x+2y=10$. Because the inequality is strict, graph $5x+2y=10$

ISM –191

$\left(y = -\dfrac{5}{2}x + 2\right)$ using a dashed line.

Test Point: $(0,0)$: $5(0)+2(0)=0<10$

Therefore, the half-plane containing $(0,0)$ is the solution set of $5x+2y<10$.

Second, graph the inequality $4x-3y<24$. To do so, replace the inequality symbol with an equal sign to obtain $4x-3y=24$. Because the inequality is strict, graph $4x-3y=24$ $\left(y=\dfrac{4}{3}x-8\right)$ using a dashed line.

Test Point: $(0,0)$: $4(0)-3(0)=0<24$

Therefore, the half-plane containing $(0,0)$ is the solution set of $4x-3y<24$.

The overlapping shaded region (that is, the shaded region in the graph below) is the solution to the system of linear inequalities.

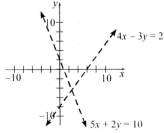

20. $\begin{cases} -x+\dfrac{1}{3}y<3 \\ \dfrac{4}{3}x+y\geq 4 \end{cases}$

First, graph the inequality $-x+\dfrac{1}{3}y<3$. To do so, replace the inequality symbol with an equal sign to obtain $-x+\dfrac{1}{3}y=3$. Because the inequality is strict, graph $-x+\dfrac{1}{3}y=3$ $(y=3x+9)$ using a dashed line.

Test Point: $(0,0)$: $-0+\dfrac{1}{3}(0)=0<3$

Therefore, the half-plane containing $(0,0)$ is the solution set of $-x+\dfrac{1}{3}y<3$.

Second, graph the inequality $\dfrac{4}{3}x+y\geq 4$. To do so, replace the inequality symbol with an equal sign to obtain $\dfrac{4}{3}x+y=4$. Because the inequality is non-strict, graph $\dfrac{4}{3}x+y=4$ $\left(y=-\dfrac{4}{3}x+4\right)$ using a solid line.

Test Point: $(0,0)$: $\dfrac{4}{3}(0)+0=0\not\geq 4$

Therefore, the half-plane not containing $(0,0)$ is the solution set of $\dfrac{4}{3}x+y\geq 4$.

The overlapping shaded region (that is, the shaded region in the graph below) is the solution to the system of linear inequalities.

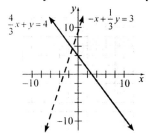

22. $\begin{cases} 3x-2y<-6 \\ -6x+4y>12 \end{cases}$

First, graph the inequality $3x-2y<-6$. To do so, replace the inequality symbol with an equal sign to obtain $3x-2y=-6$. Because the inequality is strict, graph $3x-2y=-6$ $\left(y=\dfrac{3}{2}x-2\right)$ using a dashed line.

Test Point: $(0,0)$: $3(0)-2(0)=0\not<-6$

Therefore, the half-plane not containing $(0,0)$ is the solution set of $3x-2y<-6$.

Second, graph the inequality $-6x+4y>12$. To do so, replace the inequality symbol with an equal sign to obtain $-6x+4y=12$. Because the inequality is strict, graph $-6x+4y=12$ $\left(y=\dfrac{3}{2}x-2\right)$ using a dashed line.

Test Point: $(0,0)$: $-6(0)+4(0)=0\not>12$

Therefore, the half-plane not containing $(0,0)$ is the solution set of $-6x+4y>12$.

The overlapping shaded region (that is, the shaded region in the graph below) is the solution to the system of linear inequalities. In this case, notice that the two inequalities that make up the system both have the exact same solution set. Therefore, the overlapping shaded region is the solution set of both individual inequalities.

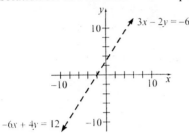

24. $\begin{cases} 4x+3y>-9 \\ -8x-6y>12 \end{cases}$

First, graph the inequality $4x+3y>-9$. To do so, replace the inequality symbol with an equal sign to obtain $4x+3y=-9$. Because the inequality is strict, graph $4x+3y=-9$ $\left(y=-\dfrac{4}{3}x-3\right)$ using a dashed line.

Test Point: $(0,0)$: $4(0)+3(0)=0>-9$

Therefore, the half-plane containing $(0,0)$ is the solution set of $4x+3y>-9$.

Second, graph the inequality $-8x-6y>12$. To do so, replace the inequality symbol with an equal sign to obtain $-8x-6y=12$. Because the inequality is strict, graph $-8x-6y=12$ $\left(y=-\dfrac{4}{3}x-2\right)$ using a dashed line.

Test Point: $(0,0)$: $-8(0)-6(0)=0\not>12$

Therefore, the half-plane not containing $(0,0)$ is the solution set of $-8x-6y>12$.

The overlapping shaded region (that is, the shaded region in the graph below) is the solution to the system of linear inequalities.

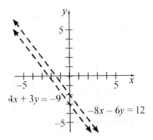

26. $\begin{cases} y\le 4 \\ x\ge -1 \end{cases}$

First, graph the inequality $y\le 4$. To do so, replace the inequality symbol with an equal sign to obtain $y=4$. Because the inequality is non-strict, graph $y=4$ using a solid line.

Test Point: $(0,0)$: $0\le 4$

Therefore, the half-plane containing $(0,0)$ is the solution set of $y\le 4$.

Second, graph the inequality $x\ge -1$. To do so, replace the inequality symbol with an equal sign to obtain $x=-1$. Because the inequality is non-strict, graph $x=-1$ using a solid line.

Test Point: $(0,0)$: $0\ge -1$

Therefore, the half-plane containing $(0,0)$ is the solution set of $x\ge -1$.

The overlapping shaded region (that is, the shaded region in the graph below) is the solution to the system of linear inequalities.

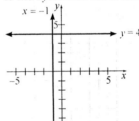

28. $\begin{cases} x+y\le 9 \\ 3x+2y\le 24 \\ x\ge 0 \\ y\ge 0 \end{cases}$

First, graph the inequality $x+y\le 9$. To do so, replace the inequality symbol with an equal sign to obtain $x+y=9$. Because the inequality is non-strict, graph $x+y=9$ $(y=-x+9)$ using a solid line.

Test Point: $(0,0)$: $0+0=0 \leq 9$

Therefore, the half-plane containing $(0,0)$ is the solution set of $x+y \leq 9$.

Next, graph the inequality $3x+2y \leq 24$. To do so, replace the inequality symbol with an equal sign to obtain $3x+2y=24$. Because the inequality is non-strict, graph $3x+2y=24$ $\left(y=-\dfrac{3}{2}x+12\right)$ using a solid line.

Test Point: $(0,0)$: $-3(0)+2(0)=0 \leq 24$

Therefore, the half-plane containing $(0,0)$ is the solution set of $3x+2y \leq 24$.

The two inequalities $x \geq 0$ and $y \geq 0$ require that the graph be in quadrant I.

The overlapping shaded region (that is, the shaded region in the graph below) is the solution to the system of linear inequalities.

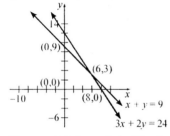

The graph is bounded. The four corner points of the bounded region are $(0, 0)$, $(8, 0)$, $(6, 3)$, and $(0, 9)$.

30. $\begin{cases} x+y \geq 8 \\ x+3y \geq 12 \\ x \geq 0 \\ y \geq 0 \end{cases}$

First, graph the inequality $x+y \geq 8$. To do so, replace the inequality symbol with an equal sign to obtain $x+y=8$. Because the inequality is non-strict, graph $x+y=8$ $(y=-x+8)$ using a solid line.

Test Point: $(0,0)$: $0+0=0 \not\geq 8$

Therefore, the half-plane not containing $(0,0)$ is the solution set of $x+y \geq 8$.

Next, graph the inequality $x+3y \geq 12$. To do so, replace the inequality symbol with an equal sign to obtain $x+3y=12$. Because the inequality is non-strict, graph $x+3y=12$ $\left(y=-\dfrac{1}{3}x+4\right)$ using a solid line.

Test Point: $(0,0)$: $0+3(0)=0 \not\geq 12$

Therefore, the half-plane not containing $(0,0)$ is the solution set of $x+3y \geq 12$.

The two inequalities $x \geq 0$ and $y \geq 0$ require that the graph be in quadrant I.

The overlapping shaded region (that is, the shaded region in the graph below) is the solution to the system of linear inequalities.

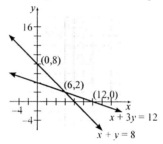

The graph is unbounded. The three corner points of the unbounded region are $(0, 8)$, $(6, 2)$, and $(12, 0)$.

32. $\begin{cases} 2x+3y \leq 36 \\ x+y \leq 14 \\ 3x+y \leq 30 \\ x \geq 0 \\ y \geq 0 \end{cases}$

First, graph the inequality $2x+3y \leq 36$. To do so, replace the inequality symbol with an equal sign to obtain $2x+3y=36$. Because the inequality is non-strict, graph $2x+3y=36$ $\left(y=-\dfrac{2}{3}x+12\right)$ using a solid line.

Test Point: $(0,0)$: $2(0)+3(0)=0 \leq 36$

Therefore, the half-plane containing $(0,0)$ is the solution set of $2x+3y \leq 36$.

Second, graph the inequality $x+y \leq 14$. To do so, replace the inequality symbol with an equal sign to obtain $x+y=14$. Because the inequality is non-strict, graph $x+y=14$ $(y=-x+14)$ using a solid line.

Test Point: $(0,0)$: $0+0=0 \leq 14$

Therefore, the half-plane containing $(0,0)$ is the solution set of $x+y \leq 14$.

Third, graph the inequality $3x+y \leq 30$. To do so, replace the inequality symbol with an equal sign to obtain $3x+y=30$. Because the inequality is non-strict, graph $3x+y=30$ $(y=-3x+30)$ using a solid line.

Test Point: $(0,0)$: $3(0)+0=0 \leq 30$

Therefore, the half-plane containing $(0,0)$ is the solution set of $3x+y \leq 30$.

The two inequalities $x \geq 0$ and $y \geq 0$ require that the graph be in quadrant I.

The overlapping shaded region (that is, the shaded region in the graph below) is the solution to the system of linear inequalities.

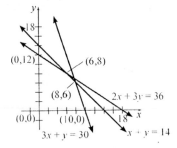

The graph is bounded. The five corner points of the bounded region are $(0, 0)$, $(10, 0)$, $(8, 6)$, $(6, 8)$, and $(0, 12)$.

34. $\begin{cases} 7x+3y \geq 45 \\ 5x+3y \geq 39 \\ x \geq 0 \\ y \geq 3 \end{cases}$

First, graph the inequality $7x+3y \geq 45$. To do so, replace the inequality symbol with an equal sign to obtain $7x+3y=45$. Because the inequality is non-strict, graph $7x+3y=45$ $\left(y=-\dfrac{7}{3}x+15\right)$ using a solid line.

Test Point: $(0,0)$: $7(0)+3(0)=0 \not\geq 45$

Therefore, the half-plane not containing $(0,0)$ is the solution set of $7x+3y \geq 45$.

Second, graph the inequality $5x+3y \geq 39$. To do so, replace the inequality symbol with an

equal sign to obtain $5x+3y=39$. Because the inequality is non-strict, graph $5x+3y=39$ $\left(y=-\dfrac{5}{3}x+13\right)$ using a solid line.

Test Point: $(0,0)$: $5(0)+3(0)=0 \not\geq 39$

Therefore, the half-plane not containing $(0,0)$ is the solution set of $5x+3y \geq 39$.

Third, the inequality $x \geq 0$ requires that the graph be to the right of the $y$-axis.

Fourth, graph the inequality $y \geq 3$. To do so, replace the inequality symbol with an equal sign to obtain $y=3$. Because the inequality is non-strict, graph $y=3$ using a solid line.

Test Point: $(0,0)$: $0 \not\geq 3$

Therefore, the half-plane not containing $(0,0)$ is the solution set of $y \geq 3$.

The overlapping shaded region (that is, the shaded region in the graph below) is the solution to the system of linear inequalities.

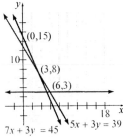

The graph is unbounded. The three corner points of the unbounded region are $(0, 15)$, $(3, 8)$, and $(6, 3)$.

36. a. Let $x$ represent the number of color ink jet printers and let $y$ represent the number of black-and-white laser printers. Now, there are at most 80 hours available for molding, so $2x+3y \leq 80$. Also, there are at most 120 hours available for assembly, so $3x+4y \leq 120$. Finally, The number of printers cannot be negative, so we have the non-negativity constraints $x \geq 0$ and $y \geq 0$.

Thus, we have the following system:
$\begin{cases} 2x+3y \leq 80 \\ 3x+4y \leq 120 \\ x \geq 0 \\ y \geq 0 \end{cases}$

## Chapter 4: Systems of Linear Equations and Inequalities

b. First, graph the inequality $2x+3y \leq 80$. To do so, replace the inequality symbol with an equal sign to obtain $2x+3y=80$. Because the inequality is non-strict, graph $2x+3y=80$ $\left(y=-\dfrac{2}{3}x+\dfrac{80}{3}\right)$ using a solid line.

Test Point: $(0,0)$: $2(0)+3(0)=0 \leq 80$
Thus, the solution set of $2x+3y \leq 80$ is the half-plane containing $(0,0)$.

Second, graph the inequality $3x+4y \leq 120$. To do so, replace the inequality symbol with an equal sign to obtain $3x+4y=120$. Because the inequality is non-strict, graph $3x+4y=120$ $\left(y=-\dfrac{1}{3}x+\dfrac{7}{3}\right)$ using a solid line.

Test Point: $(0,0)$: $3(0)+4(0)=0 \leq 120$
Thus, the solution set of $3x+4y \leq 120$ is the half-plane containing $(0,0)$.

The two inequalities $x \geq 0$ and $y \geq 0$ require that the graph be in quadrant I.

The overlapping shaded region (that is, the shaded region in the graph below) is the solution to the system of linear inequalities.

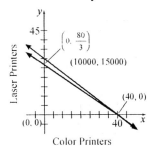

The three corner points of the bounded region are $(0,0)$, $(40,0)$ and $\left(0, \dfrac{80}{3}\right)$.

38. a. Let $x$ represent the number of premium mixes and let $y$ represent the number of standard mixes. Now, there are at most 1600 ounces of cashews that can be used, so $12x+6y \leq 1600$. Also, there are at most 1920 ounces of peanuts that can be used, so $4x+6y \leq 1920$. Now, the numbers of mixes cannot be negative, so we have the non-negativity constraints $x \geq 0$ and $y \geq 0$.

Thus, we have the following system:
$$\begin{cases} 12x+6y \leq 1600 \\ 4x+6y \leq 1920 \\ x \geq 0 \\ y \geq 0 \end{cases}$$

b. First, graph the inequality $12x+6y \leq 1600$. To do so, replace the inequality symbol with an equal sign to obtain $12x+6y=1600$. Because the inequality is non-strict, graph $12x+6y=1600$ $\left(y=-2x+\dfrac{800}{3}\right)$ using a solid line.

Test Point: $(0,0)$: $12(0)+6(0)=0 \leq 1600$
Thus, the solution set of $12x+6y \leq 1600$ is the half-plane containing $(0,0)$.

Second, graph the inequality $4x+6y \leq 1920$. To do so, replace the inequality symbol with an equal sign to obtain $4x+6y=1920$. Because the inequality is non-strict, graph $4x+6y=1920$ $\left(y=-\dfrac{2}{3}x+320\right)$ using a solid line.

Test Point: $(0,0)$: $4(0)+6(0)=0 \leq 1920$
Thus, the solution set of $4x+6y \leq 1920$ is the half-plane containing $(0,0)$.

The two inequalities $x \geq 0$ and $y \geq 0$ require that the graph be in quadrant I.

The overlapping shaded region (that is, the shaded region in the graph below) is the solution to the system of linear inequalities.

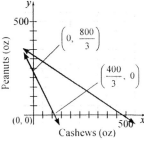

The four corner points of the bounded region are $(0,0)$, $\left(\dfrac{400}{3}, 0\right)$, $(20000, 5000)$, and $\left(0, \dfrac{800}{3}\right)$.

**40.** The graph contains a line passing through the points $(0,-1)$ and $(3,1)$. The slope of the line is $m = \dfrac{1-(-1)}{3-0} = \dfrac{2}{3}$. The y-intercept is $-1$.

Thus, the equation of the line is $y = \dfrac{2}{3}x - 1$.

Because the shaded region above this line, our inequality will contain a "greater than." Because the line is solid, this means that the inequality is non-strict. That is, our inequality will consist of "greater than or equal to." Thus, the linear inequality is $y \geq \dfrac{2}{3}x - 1$.

The second line in the graph passes through the points $(0,-8)$ and $(3,1)$. The slope of the line is $m = \dfrac{1-(-8)}{3-0} = \dfrac{9}{3} = 3$. The y-intercept is $-8$.

Thus, the equation of the line is $y = 3x - 8$.

Because the shaded region is above this line, our inequality will contain a "greater than." Because the line is solid, this means that the inequality is non-strict. That is, our inequality will consist of "greater than or equal to." Thus, the linear inequality is $y \geq 3x - 8$.

Thus, the system of linear inequalities is

$$\begin{cases} y \geq \dfrac{2}{3}x - 1 \\ y \geq 3x - 8 \end{cases}$$

**42.** The graph contains a line passing through the points $(5,7)$ and $(9,3)$. The slope of the line is $m = \dfrac{3-7}{9-4} = \dfrac{-4}{5} = -\dfrac{4}{5}$. Using the point-slope form of the equation, we obtain

$$y - 3 = -\dfrac{4}{5}(x-9)$$
$$y - 3 = -\dfrac{4}{5}x + \dfrac{36}{5}$$
$$y = -\dfrac{4}{5}x + \dfrac{51}{5}$$

Thus, the equation of the line is $y = -\dfrac{4}{5}x + \dfrac{51}{5}$.

Because the shaded region is below this line, our inequality will contain a "less than." Because the line is solid, this means that the inequality is non-strict. That is, our inequality will consist of "less than or equal to." Thus, the linear inequality is $y \leq -\dfrac{4}{5}x + \dfrac{51}{5}$ or $4x + 5y \leq 51$.

The second line in the vertical line with a x-intercept 5. The equation of this line is $x = 5$. Because the shaded region is to the right of this line, our inequality will contain a "greater than." Because the line is solid, this means that the inequality is non-strict. That is, our inequality will consist of "greater than or equal to." Thus, the linear inequality is $x \geq 5$.

The third line in the horizontal line with a y-intercept 3. The equation of this line is $y = 3$. Because the shaded region is above this line, our inequality will contain a "greater than." Because the line is solid, this means that the inequality is non-strict. That is, our inequality will consist of "greater than or equal to." Thus, the linear inequality is $y \geq 3$.

Thus, the system of linear inequalities is

$$\begin{cases} 4x + 5y \leq 51 \\ x \geq 5 \\ y \geq 3 \end{cases}$$

**44.** $\begin{cases} x + y \geq 2 & (y \geq -x + 2) \\ -3x + y \leq 10 & (y \leq 3x + 10) \end{cases}$

**46.** $\begin{cases} 5x + 2y < 10 & \left(y < -\dfrac{5}{2}x + 5\right) \\ 4x - 3y < 24 & \left(y > \dfrac{4}{3}x - 8\right) \end{cases}$

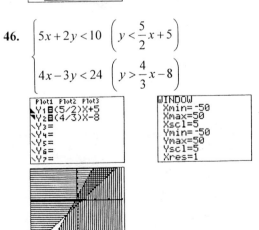

48. $\begin{cases} x+y \leq 9 & (y \leq -x+9) \\ 3x+2y \leq 24 & \left(y \leq -\dfrac{3}{2}x+12\right) \\ x \geq 0 \\ y \geq 0 \end{cases}$

The inequalities $x \geq 0$ and $y \geq 0$ restrict the graph to the first quadrant.

   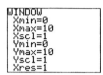

## Chapter 4 Review

1. a. Let $x = 3$, $y = -1$ in both equations (1) and (2).

   $\begin{cases} 3+3(-1) = 3-3 = 0 \neq -2 \\ 2(3)-(-1) = 6+1 = 7 \neq 10 \end{cases}$

   These values do not satisfy either equation. Therefore, the ordered pair (3, -1) is not a solution of the system.

   b. Let $x = 4$, $y = -2$ in both equations (1) and (2).

   $\begin{cases} 4+3(-2) = 4-6 = -2 \\ 2(4)-(-2) = 8+2 = 10 \end{cases}$

   Because these values satisfy both equations, the ordered pair (4, -2) is a solution of the system.

2. a. Let $x = \dfrac{1}{2}$, $y = 1$ in both equations (1) and (2).

   $\begin{cases} -2\left(\dfrac{1}{2}\right)+5(1) = -1+5 = 4 \\ 4\left(\dfrac{1}{2}\right)-5(1) = 2-5 = -3 \end{cases}$

   Because these values satisfy both equations, the ordered pair $\left(\dfrac{1}{2}, 1\right)$ is a solution of the system.

   b. Let $x = -1$, $y = \dfrac{1}{3}$ in both equations (1) and (2).

   $\begin{cases} -2(-1)+5\left(\dfrac{1}{3}\right) = 2+\dfrac{5}{3} = \dfrac{11}{3} \neq 4 \\ 4(-1)-5\left(\dfrac{1}{3}\right) = -4-\dfrac{5}{3} = -\dfrac{17}{3} \neq -3 \end{cases}$

   These values do not satisfy either equation. Therefore, the ordered pair $\left(-1, \dfrac{1}{3}\right)$ is not a solution of the system.

3. a. Let $x = 3$, $y = 6$ in both equation (1) and (2).

   $\begin{cases} 6(3)-5(6) = 18-30 = -12 \\ 3-6 = -3 \end{cases}$

   These values satisfy both equations. Therefore, the ordered pair (3, 6) is a solution of the system.

   b. Let $x = 1$, $y = 4$ in both equations (1) and (2).

   $\begin{cases} 6(1)-5(4) = 6-20 = -14 \neq -12 \\ 1-4 = -3 \end{cases}$

   Although these values satisfy equation (2), they do not satisfy the equation (1). Therefore, the ordered pair (1, 4) is not a solution of the system.

4. a. Let $x = 3$, $y = 0$ in both equations (1) and (2).

   $\begin{cases} 3(3)-0 = 9-0 = 9 \\ 8(3)+3(0) = 24+0 = 24 \neq 7 \end{cases}$

   Although these values satisfy equation (1), they do not satisfy the equation (2). Therefore, the ordered pair (3, 0) is not a solution of the system.

   b. Let $x = 2$, $y = -3$ in both equations (1) and (2).

   $\begin{cases} 3(2)-(-3) = 6+3 = 9 \\ 8(2)+3(-3) = 16-9 = 7 \end{cases}$

   Because these values satisfy both equations, the ordered pair (2, -3) is a solution of the system.

5. The graphs of the linear equations intersect at the point (4, 2). Thus, the solution is (4, 2).

6. The graphs of the linear equations intersect at the point (2, -1). Thus, the solution is (2, -1).

7. $\begin{cases} y = -3x+1 & (1) \\ y = \dfrac{1}{2}x - 6 & (2) \end{cases}$

The two equations are in slope-intercept form. Graph each equation and find the point of intersection.

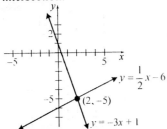

The solution is the ordered pair (2, -5).

8. $\begin{cases} -2x + 3y = -9 & (1) \\ 3x + y = 8 & (2) \end{cases}$

Equation (1) in slope-intercept form is $y = \dfrac{2}{3}x - 3$. Equation (2) in slope-intercept form is $y = -3x + 8$. Graph each equation and find the point of intersection.

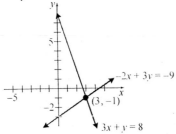

The solution is the ordered pair (3, -1).

9. $\begin{cases} y = -\dfrac{1}{3}x + 2 & (1) \\ x + 3y = -9 & (2) \end{cases}$

Equation (1) is already in slope-intercept form. Equation (2) in slope-intercept form is $y = -\dfrac{1}{3}x - 3$. Graph each equation.

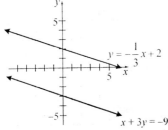

The two lines are parallel. The system is inconsistent. The system has no solution. That is, $\varnothing$ or { }.

10. $\begin{cases} 2x - 3y = 0 & (1) \\ 2x - y = -4 & (2) \end{cases}$

Equation (1) in slope-intercept form is $y = \dfrac{2}{3}x$.

Equation (2) in slope-intercept form is $y = 2x + 4$. Graph each equation and find the point of intersection.

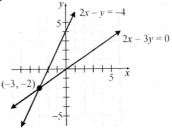

The solution is the ordered pair (-3, -2).

11. $\begin{cases} y = -\dfrac{1}{4}x + 2 & (1) \\ y = 4x - 32 & (2) \end{cases}$

Substituting $-\dfrac{1}{4}x + 2$ for $y$ in equation (2), we obtain

$$-\dfrac{1}{4}x + 2 = 4x - 32$$
$$-4\left(-\dfrac{1}{4}x + 2\right) = -4(4x - 32)$$
$$x - 8 = -16x + 128$$
$$17x = 136$$
$$x = 8$$

Substituting 8 for $x$ in equation (2), we obtain $y = 4(8) - 32 = 32 - 32 = 0$.

The solution is the ordered pair (8, 0).

12. $\begin{cases} y = -\dfrac{3}{4}x + 2 & (1) \\ 3x + 4y = 8 & (2) \end{cases}$

Substituting $-\dfrac{3}{4}x + 2$ for $y$ in equation (2), we obtain

$$3x + 4\left(-\dfrac{3}{4}x + 2\right) = 8$$
$$3x - 3x + 8 = 8$$
$$8 = 8$$

The system is dependent. The solution is
$\left\{(x, y) \mid y = -\dfrac{3}{4}x + 2\right\}$.

13. $\begin{cases} y = 3x - 9 & (1) \\ 4x + 3y = -1 & (2) \end{cases}$

Substituting $3x - 9$ for $y$ in equation (2), we obtain
$4x + 3(3x - 9) = -1$
$4x + 9x - 27 = -1$
$13x - 27 = -1$
$13x = 26$
$x = 2$

Substituting 2 for $x$ in equation (1), we obtain
$y = 3(2) - 9 = 6 - 9 = -3$.

The solution is the ordered pair (2, -3).

14. $\begin{cases} x - 2y = 7 & (1) \\ 3x - y = -4 & (2) \end{cases}$

Equation (1) solved for $x$ is $x = 2y + 7$.

Substituting $2y + 7$ for $x$ in equation (2), we obtain
$3(2y + 7) - y = -4$
$6y + 21 - y = -4$
$5y + 21 = -4$
$5y = -25$
$y = -5$

Substituting -5 for $y$ in equation (1), we obtain
$x - 2(-5) = 7$
$x + 10 = 7$
$x = -3$

The solution is the ordered pair (-3, -5).

15. $\begin{cases} 2x - y = 9 & (1) \\ 3x + y = 11 & (2) \end{cases}$

Add equations (1) and (2).
$2x - y = 9$
$3x + y = 11$
$\overline{\phantom{xxxxxxx}}$
$5x = 20$
$x = 4$

Substituting 4 for $x$ in equation (2), we obtain
$3(4) + y = 11$
$12 + y = 11$
$y = -1$

The solution is the ordered pair (4, -1).

16. $\begin{cases} -x + 3y = 4 & (1) \\ 3x - 4y = -2 & (2) \end{cases}$

Multiply both sides of equation (1) by 3 and add the result to equation (2).
$-3x + 9y = 12$
$3x - 4y = -2$
$\overline{\phantom{xxxxxxx}}$
$5y = 10$
$y = 2$

Substituting 2 for $y$ in equation (1), we obtain
$-x + 3(2) = 4$
$-x + 6 = 4$
$2 = x$

The solution is the ordered pair (2, 2).

17. $\begin{cases} 2x - 4y = 8 & (1) \\ -3x + 6y = -12 & (2) \end{cases}$

Multiply both sides of equation (1) by 3, multiply both sides of equation (2) by 2, and add the results.
$6x - 12y = 24$
$-6x + 12y = -24$
$\overline{\phantom{xxxxxxx}}$
$0 = 0$

The system is dependent. The solution is
$\{(x, y) \mid 2x - 4y = 8\}$.

18. $\begin{cases} 3x - 4y = -11 & (1) \\ 2x - 3y = -7 & (2) \end{cases}$

Multiply both sides of equation (1) by 2, multiply both sides of equation (2) by -3, and add the results.
$6x - 8y = -22$
$-6x + 9y = 21$
$\overline{\phantom{xxxxxxx}}$
$y = -1$

Substituting -1 for $y$ in equation (1), we obtain
$3x - 4(-1) = -11$
$3x + 4 = -11$
$3x = -15$
$x = -5$

The solution is the ordered pair (-5, -1).

19. $\begin{cases} x + y = -4 & (1) \\ 2x - 3y = 12 & (2) \end{cases}$

Because the coefficients of $x$ and $y$ in equation (1) are both 1, we use substitution to solve the system.

Equation (1) solved for $x$ is $x = -y - 4$.

Substituting $-y - 4$ for $x$ in equation (2), we obtain

$$2(-y-4)-3y=12$$
$$-2y-8-3y=12$$
$$-5y-8=12$$
$$-5y=20$$
$$y=-4$$

Substituting -4 for $y$ in equation (1), we obtain
$$x+(-4)=-4$$
$$x-4=-4$$
$$x=0$$

The solution is the ordered pair (0, -4).

20. $\begin{cases} 5x-3y=2 & (1) \\ x+2y=-10 & (2) \end{cases}$

Because the coefficient of $x$ in equation (2) is 1, we use substitution to solve the system.
Equation (2) solved for $x$ is $x=-2y-10$.
Substituting $-2y-10$ for $x$ in equation (1), we obtain
$$5(-2y-10)-3y=2$$
$$-10y-50-3y=2$$
$$-13y-50=2$$
$$-13y=52$$
$$y=-4$$

Substituting -4 for $y$ in equation (2), we obtain
$$x+2(-4)=-10$$
$$x-8=-10$$
$$x=-2$$

The solution is the ordered pair (-2, -4).

21. $\begin{cases} 3x-2y=5 & (1) \\ 4x-5y=9 & (2) \end{cases}$

Because none of variables have a coefficient of 1, we use elimination to solve the system. Multiply both sides of equation (1) by 4, multiply both sides of equation (2) by −3, and add the results.
$$12x-8y=20$$
$$-12x+15y=-27$$
$$\overline{\phantom{xxxxxxxxxx}}$$
$$7y=-7$$
$$y=-1$$

Substituting -1 for $y$ in equation (1), we obtain
$$3x-2(-1)=5$$
$$3x+2=5$$
$$3x=3$$
$$x=1$$

The solution is the ordered pair (1, -1).

22. $\begin{cases} 12x+20y=21 & (1) \\ 3x-2y=0 & (2) \end{cases}$

Because none of variables have a coefficient of 1, we use elimination to solve the system. Multiply both sides of equation (2) by -4 and add the result to equation (1).
$$12x+20y=21$$
$$-12x+8y=0$$
$$\overline{\phantom{xxxxxxxxxx}}$$
$$28y=21$$
$$y=\frac{21}{28}=\frac{3}{4}$$

Substituting $\frac{3}{4}$ for $y$ in equation (1), we obtain
$$12x+20\left(\frac{3}{4}\right)=21$$
$$12x+15=21$$
$$12x=6$$
$$x=\frac{6}{12}=\frac{1}{2}$$

The solution is the ordered pair $\left(\frac{1}{2},\frac{3}{4}\right)$.

23. $\begin{cases} 6x+9y=-3 & (1) \\ 8x+12y=7 & (2) \end{cases}$

Because none of variables have a coefficient of 1, we use elimination to solve the system. Multiply both sides of equation (1) by 4, multiply both sides of equation (2) by −3, and add the results.
$$24x+36y=-12$$
$$-24x-36y=-21$$
$$\overline{\phantom{xxxxxxxxxx}}$$
$$0=-33$$

The system has no solution. The solution set is $\varnothing$ or $\{\ \}$. The system is inconsistent.

24. $\begin{cases} 6x+11y=2 & (1) \\ 5x+8y=-3 & (2) \end{cases}$

Because none of variables have a coefficient of 1, we use elimination to solve the system. Multiply both sides of equation (1) by 5, multiply both sides of equation (2) by -6, and add the results.
$$30x+55y=10$$
$$-30x-48y=18$$
$$\overline{\phantom{xxxxxxxxxx}}$$
$$7y=28$$
$$y=4$$

Substituting 4 for $y$ in equation (1), we obtain

Chapter 4: Systems of Linear Equations and Inequalities

$$6x + 11(4) = 2$$
$$6x + 44 = 2$$
$$6x = -42$$
$$x = -7$$

The solution is the ordered pair (-7, 4).

25. Let $x$ and $y$ represent the two numbers.
$$\begin{cases} x + y = 56 & (1) \\ x - y = 14 & (2) \end{cases}$$
Adding equations (1) and (2) results in
$$2x = 70$$
$$x = 35$$
Substituting 35 for $x$ in equation (1), we obtain
$$35 + y = 56$$
$$y = 21$$
The two numbers are 35 and 21.

26. Let $m$ represent the number of males to be inducted and $f$ represent the number of females to be inducted.
$$\begin{cases} m + f = 73 & (1) \\ f = m + 11 & (2) \end{cases}$$
Substituting $m + 11$ for $f$ in equation (1), we obtain
$$m + (m + 11) = 73$$
$$2m + 11 = 73$$
$$2m = 62$$
$$m = 31$$
Substituting 31 for $m$ in equation (2), we obtain
$$f = 31 + 11 = 42$$
The local chapter of Phi Theta Kappa will induct 31 males and 42 females.

27. Let $p$ represent the number of calories in a slice of pepperoni pizza and $s$ represent the number of calories in a slice of Italian sausage pizza.
$$\begin{cases} 3p + 5s = 2600 & (1) \\ 4p + 2s = 1880 & (2) \end{cases}$$
Multiply both sides of equation (1) by 4, multiply both sides of equation (2) by -3, and add the results.
$$12p + 20s = 10,400$$
$$-12p - 6s = -5640$$
$$\overline{\phantom{12p+20s=10,400}}$$
$$14s = 4760$$
$$s = 340$$
Substituting 340 for $s$ in equation (1), we obtain

$$3p + 5(340) = 2600$$
$$3p + 1700 = 2600$$
$$3p = 900$$
$$p = 300$$
There are 300 calories in a slice of pepperoni pizza and 340 calories in a slice of Italian sausage pizza.

28. Let $l$ represent the length of the rectangle and $w$ represent the width of the rectangle.
$$\begin{cases} l = 2w - 5 & (1) \\ 2l + 2w = 68 & (2) \end{cases}$$
Substituting $2w - 5$ for $l$ in equation (2), we obtain
$$2(2w - 5) + 2w = 68$$
$$4w - 10 + 2w = 68$$
$$6w - 10 = 68$$
$$6w = 78$$
$$w = 13$$
Substituting 13 for $w$ in equation (1), we obtain
$$l = 2(13) - 5 = 26 - 5 = 21$$
The dimensions of the rectangle are 21 inches by 13 inches.

29. Because the sum of the measures of the three angles in a triangle must equal $180°$, we obtain the following system.
$$\begin{cases} x + y + 90 = 180 & (1) \\ y = 2x & (2) \end{cases}$$
Substituting $2x$ for $y$ in equation (1), we obtain
$$x + 2x + 90 = 180$$
$$3x + 90 = 180$$
$$3x = 90$$
$$x = 30$$
Substituting 30 for $x$ in equation (2), we obtain
$$y = 2(30) = 60$$.
Thus, angle $x = 30°$ and angle $y = 60°$.

30. Because angles 1 and 2 are supplemental, and because angles 1 and 3 must be equal, we obtain the following system.
$$\begin{cases} (x + 4y) + (8x + 5y) = 180 & (1) \\ x + 4y = 2x + y & (2) \end{cases}$$
Simplify each equation.
$$\begin{cases} 9x + 9y = 180 & (1) \\ -x + 3y = 0 & (2) \end{cases}$$
Divide both sides of equation (1) by 9 and add the result to equation (2).

ISM − 202

$x + y = 20$
$-x + 3y = 0$
———————
$4y = 20$
$y = 5$

Substituting 5 for $y$ in equation (2), we obtain
$-x + 3(5) = 0$
$-x + 15 = 0$
$15 = x$
Thus, $x = 15$ and $y = 5$.

31. Let $n$ represent the number of nickels and $q$ represent the number of quarters.

| | Number of Coins | Value per Coin | Total Value |
|---|---|---|---|
| Nickels | $n$ | 0.05 | $0.05n$ |
| Quarters | $q$ | 0.25 | $0.25q$ |
| Total | 40 | | 4.40 |

Using the columns for Number of Coins and Total Value, we obtain the following system.
$\begin{cases} n + q = 40 & (1) \\ 0.05n + 0.25q = 4.40 & (2) \end{cases}$
Multiply both sides of equation (1) by -0.05 and add the result to equation (2).
$-0.05n - 0.05q = -2$
$0.05n + 0.25q = 4.40$
———————
$0.20q = 2.40$
$q = 12$

Substituting 12 for $q$ in equation (1), we obtain
$n + 12 = 40$
$n = 28$
Jerome's piggy bank contains 28 nickels and 12 quarters.

32. Let $x$ represent the number of liters of the 25%-hydrochloric-acid (HCl) solution to be used, and let $y$ represent the number of liters of the 40%-HCl solution to be used.

| | Liters of Solution | Percent HCl | Liters of HCl |
|---|---|---|---|
| 25%-HCl Solution | $x$ | 25% = 0.25 | $0.25x$ |
| 40%-HCl Solution | $y$ | 40% = 0.40 | $0.40y$ |
| 30%-HCl Solution | 12 | 30% = 0.30 | 0.30(12) = 3.6 |

Using the columns for Liters of Solution and Liters of HCl, we obtain the following system.
$\begin{cases} x + y = 12 & (1) \\ 0.25x + 0.40y = 3.6 & (2) \end{cases}$
Multiply both sides of equation (1) by -0.25 and add the result to equation (2).
$-0.25x - .25y = -3$
$0.25x + 0.40y = 3.6$
———————
$0.15y = 0.6$
$y = 4$

Substituting 4 for $y$ in equation (1), we obtain
$x + 4 = 12$
$x = 8$
The chemist should mix 8 liters of the 25%-hydrochloric-acid solution with 4 liters of the 40%-hydrochloric-acid solution.

33. Let $s$ represent the amount invested in stocks, and let $b$ represent the amount invested in bonds. We organize the information in the table below, using the interest formula:
Interest = Principal × rate × time or $I = Prt$.

| | $P$ | $r$ | $t$ | $I$ |
|---|---|---|---|---|
| Stocks | $s$ | 0.065 | 1 | $0.065s$ |
| Bonds | $b$ | 0.0425 | 1 | $0.0425b$ |
| Total | 10,000 | | 1 | 582.50 |

Using the columns for Principal and Interest, we obtain the following system.
$\begin{cases} s + b = 10,000 & (1) \\ 0.065s + 0.0425b = 582.50 & (2) \end{cases}$
Solving equation (1) for $b$, we obtain
$b = 10,000 - s$. Substituting $10,000 - s$ for $b$ in equation (2), we obtain
$0.065s + 0.0425(10,000 - s) = 582.50$
$0.065s + 425 - 0.0425s = 582.50$
$0.0225s + 425 = 582.50$
$0.0225s = 157.50$
$s = 7000$
Substituting 7000 for $s$ in equation (1), we obtain
$7000 + b = 10,000$
$b = 3000$
Verna invested $7000 in stocks and $3000 in bonds.

34. Let $p$ represent the speed of the plane and let $w$ represent the speed of the wind.
$\begin{cases} p + w = 160 & (1) \\ p - w = 112 & (2) \end{cases}$

Chapter 4: *Systems of Linear Equations and Inequalities*

Adding equations (1) and (2) results in
$$p + w = 160$$
$$p - w = 112$$
$$\overline{2p \phantom{{}-w} = 272}$$
$$p = 136$$
Substituting 136 for $p$ in equation (1), we obtain
$$136 + w = 160$$
$$w = 24$$
The plane's speed is 136 miles per hour and the wind's speed is 24 miles per hour.

35. Let $d$ represent the distance that both cars have traveled on the turnpike at the point of catch up. Let $t$ represent the Mustang's time on the turnpike at the point of catch up. Then $t + \frac{1}{2}$ represents the Grand Am's time on the turnpike at that point. We obtain the following system by utilizing the formula rate × time = distance :
$$\begin{cases} 80t = d & (1) \\ 50\left(t + \frac{1}{2}\right) = d & (2) \end{cases}$$
Substituting $80t$ for $d$ in equation (2) yields
$$50\left(t + \frac{1}{2}\right) = 80t$$
$$50t + 25 = 80t$$
$$-30t + 25 = 0$$
$$-30t = -25$$
$$t = \frac{-25}{-30} = \frac{5}{6}$$
Substituting $\frac{5}{6}$ for $t$ in equation (1), we obtain
$$80\left(\frac{5}{6}\right) = d$$
$$d = \frac{200}{3} = 66\frac{2}{3}$$
It will take $\frac{5}{6}$ hours (that is, 50 minutes) for the Mustang to catch up to the Grand Am. At that time, the two cars will have traveled $66\frac{2}{3}$ miles on the turnpike.

36. Let $b$ represent the speed of the boat in still water and let $c$ represent the speed of the current. The rate at which the boat will travel downstream will be $b + c$, and the rate upstream will be $b - c$. Using the formula rate × time = distance, we obtain the following system:
$$\begin{cases} (b-c) \cdot 1.5 = 30 & (1) \\ (b+c) \cdot 1 = 30 & (2) \end{cases}$$
Divide both sides of equation (1) by 1.5, simplify equation (2), and add the results.
$$b - c = 20$$
$$b + c = 30$$
$$\overline{2b \phantom{{}+c} = 50}$$
$$b = 25$$
Substituting 25 for $b$ in equation (2), we obtain
$$(25 + c) \cdot 1 = 30$$
$$25 + c = 30$$
$$c = 5$$
The speed of the boat in still water is 25 miles per hour and the speed of the current is 5 miles per hour.

37. **a.** $\begin{cases} M(t) = 0.056t + 3.354 \\ F(t) = 0.103t + 1.842 \end{cases}$

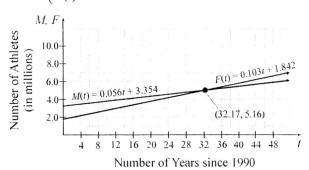

**b.**
$$M(t) = F(t)$$
$$0.056t + 3.354 = 0.103t + 1.842$$
$$-0.047t + 3.354 = 1.842$$
$$-0.047t = -1.512$$
$$t \approx 32.17$$
Male and female participation will be the same approximately 32 years after 1990, which is the year 2022.

38. **a.** $R(x) = 15x$

**b.** $C(x) = 1200 + 2.50x$

**c.**

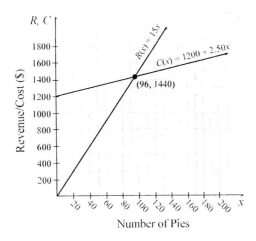

Number of Pies

**d.** $R(x) = C(x)$
$15x = 1200 + 2.50x$
$12.50x = 1200$
$x = 96$
$R(96) = 15(96) = 1440$
$C(96) = 1200 + 2.50(96) = 1440$

The break-even number of pies is 96 pies. The revenue and cost of 96 pies is $1440.

**39.** $\begin{cases} x + y - z = 1 & (1) \\ x - y + z = 7 & (2) \\ x + 2y + z = 1 & (3) \end{cases}$

Multiply both sides of equation (1) by -1 and add the result to equation (2).
$-x - y + z = -1$
$\underline{x - y + z = 7}$
$-2y + 2z = 6 \quad (4)$

Multiply both sides of equation (1) by $-1$ and add the result to equation (3).
$-x - y + z = -1$
$\underline{x + 2y + z = 1}$
$y + 2z = 0 \quad (5)$

Divide both sides of equation (4) by 2 and add the result to equation (5).
$-y + z = 3$
$\underline{y + 2z = 0}$
$3z = 3$
$z = 1$

Substituting 1 for $z$ in equation (4), we obtain

$-2y + 2(1) = 6$
$-2y + 2 = 6$
$-2y = 4$
$y = -2$

Substituting $-2$ for $y$ and 1 for $z$ in equation (1), we obtain
$x + (-2) - 1 = 1$
$x - 3 = 1$
$x = 4$

The solution is the ordered triple (4, -2, 1).

**40.** $\begin{cases} 2x - 2y + z = -10 & (1) \\ 3x + y - 2z = 4 & (2) \\ 5x + 2y - 3z = 7 & (3) \end{cases}$

Multiply both sides of equation (1) by 2 and add the result to equation (2).
$4x - 4y + 2z = -20$
$\underline{3x + y - 2z = 4}$
$7x - 3y = -16 \quad (4)$

Multiply both sides of equation (1) by 3 and add the result to equation (3).
$6x - 6y + 3z = -30$
$\underline{5x + 2y - 3z = 7}$
$11x - 4y = -23 \quad (5)$

Multiply both sides of equation (4) by 4, multiply both sides of equation (5) by -3, and add the results.
$28x - 12y = -64$
$\underline{-33x + 12y = 69}$
$-5x = 5$
$x = -1$

Substituting -1 for $x$ in equation (4), we obtain
$7(-1) - 3y = -16$
$-7 - 3y = -16$
$-3y = -9$
$y = 3$

Substituting $-1$ for $x$ and 3 for $y$ in equation (1), we obtain
$2(-1) - 2(3) + z = -10$
$-2 - 6 + z = -10$
$z = -2$

The solution is the ordered triple (-1, 3, -2).

**41.** $\begin{cases} x + 2y = -1 & (1) \\ 3y + 4z = 7 & (2) \\ 2x - z = 6 & (3) \end{cases}$

Multiply both sides of equation (1) by $-2$ and

## Chapter 4: Systems of Linear Equations and Inequalities

add the result to equation (3).
$$-2x - 4y = 2$$
$$\underline{2x \quad\quad - z = 6}$$
$$-4y - z = 8 \quad (4)$$

Multiply both sides of equation (4) by 4 and add the result to equation (2).
$$-16y - 4z = 32$$
$$\underline{3y + 4z = 7}$$
$$-13y = 39$$
$$y = -3$$

Substituting $-3$ for $y$ in equation (4), we obtain
$$-4(-3) - z = 8$$
$$12 - z = 8$$
$$-z = -4$$
$$z = 4$$

Substituting $-3$ for $y$ in equation (1), we obtain
$$x + 2(-3) = -1$$
$$x - 6 = -1$$
$$x = 5$$

The solution is the ordered triple (5, -3, 4).

42. $\begin{cases} x + 2y - 3z = -4 & (1) \\ x + y + 3z = 5 & (2) \\ 3x + 4y + 3z = 7 & (3) \end{cases}$

Add equations (1) and (2).
$$x + 2y - 3z = -4$$
$$\underline{x + y + 3z = 5}$$
$$2x + 3y = 1 \quad (4)$$

Add equations (1) and (3).
$$x + 2y - 3z = -4$$
$$\underline{3x + 4y + 3z = 7}$$
$$4x + 6y = 3 \quad (5)$$

Multiply both sides of equation (4) by $-2$ and add the result to equation (5).
$$-4x - 6y = -2$$
$$\underline{4x + 6y = 3}$$
$$0 = 1 \quad \text{False}$$

The system has no solution. The solution set is $\varnothing$ or $\{\ \}$. The system is inconsistent.

43. $\begin{cases} 3x + y - 2z = 6 & (1) \\ x + y - z = -2 & (2) \\ -x - 3y + 2z = 14 & (3) \end{cases}$

Multiply both sides of equation (3) by 3 and add the result to equation (1).

$$-3x - 9y + 6z = 42$$
$$\underline{3x + y - 2z = 6}$$
$$-8y + 4z = 48 \quad (4)$$

Add equations (2) and (3).
$$x + y - z = -2$$
$$\underline{-x - 3y + 2z = 14}$$
$$-2y + z = 12 \quad (5)$$

Multiply both sides of equation (5) by $-4$ and add the result to equation (4).
$$8y - 4z = -48$$
$$\underline{-8y + 4z = 48}$$
$$0 = 0 \quad \text{True}$$

Thus, the system is dependent and has an infinite number of solutions.

Solve equation (5) for $y$.
$$-2y + z = 12$$
$$-2y = -z + 12$$
$$y = \frac{-z + 12}{-2}$$
$$y = \frac{1}{2}z - 6$$

Substituting $\frac{1}{2}z - 6$ for $y$ in equation (2), we obtain
$$x + \left(\frac{1}{2}z - 6\right) - z = -2$$
$$x - \frac{1}{2}z - 6 = -2$$
$$x = \frac{1}{2}z + 4$$

The solution to the system is $\left\{(x, y, z) \middle| x = \frac{1}{2}z + 4, y = \frac{1}{2}z - 6, z \text{ is any real number}\right\}$.

44. $\begin{cases} 9x - y + 2z = -5 & (1) \\ -3x - 4y + 4z = 3 & (2) \\ 15x + 3y - 2z = -10 & (3) \end{cases}$

Multiply both sides of equation (2) by 3 and add the result to equation (1).
$$-9x - 12y + 12z = 9$$
$$\underline{9x - y + 2z = -5}$$
$$-13y + 14z = 4 \quad (4)$$

Multiply both sides of equation (2) by 5 and add the result to equation (3).

$$-15x - 20y + 20z = 15$$
$$15x + 3y - 2z = -10$$
$$\overline{\quad -17y + 18z = 5 \quad} \quad (5)$$

Multiply both sides of equation (4) by 17, multiply both sides of equation (5) by -13, and add the results.

$$-221y + 238z = 68$$
$$221y - 234z = -65$$
$$\overline{\quad 4z = 3 \quad}$$
$$z = \frac{3}{4}$$

Substituting $\frac{3}{4}$ for $z$ in equation (4), we obtain

$$-13y + 14\left(\frac{3}{4}\right) = 4$$
$$-13y + \frac{21}{2} = 4$$
$$2\left(-13y + \frac{21}{2}\right) = 2(4)$$
$$-26y + 21 = 8$$
$$-26y = -13$$
$$y = \frac{-13}{-26} = \frac{1}{2}$$

Substituting $\frac{1}{2}$ for $y$ and $\frac{3}{4}$ for $z$ in equation (1), we obtain

$$9x - \frac{1}{2} + 2\left(\frac{3}{4}\right) = -5$$
$$9x - \frac{1}{2} + \frac{3}{2} = -5$$
$$9x + 1 = -5$$
$$9x = -6$$
$$x = \frac{-6}{9} = -\frac{2}{3}$$

The solution is the ordered triple $\left(-\frac{2}{3}, \frac{1}{2}, \frac{3}{4}\right)$.

**45.** $\begin{cases} 4x - 5y + 2z = -8 & (1) \\ 3x + 7y - 3z = 21 & (2) \\ 7x - 4y + 2z = -5 & (3) \end{cases}$

Multiply both sides of equation (1) by 3, multiply both sides of equation (2) by -4, and add the results.

$$12x - 15y + 6z = -24$$
$$-12x - 28y + 12z = -84$$
$$\overline{\quad -43y + 18z = -108 \quad} \quad (4)$$

Multiply both sides of equation (2) by 7, multiply both sides of equation (3) by -3, and add the results.

$$21x + 49y - 21z = 147$$
$$-21x + 12y - 6z = 15$$
$$\overline{\quad 61y - 27z = 162 \quad} \quad (5)$$

Multiply both sides of equation (4) by 3, multiply both sides of equation (5) by 2, and add the results.

$$-129y + 54z = -324$$
$$122y - 54z = 324$$
$$\overline{\quad -7y \quad = 0 \quad}$$
$$y = 0$$

Substituting 0 for $y$ in equation (4), we obtain
$$-43(0) + 18z = -108$$
$$18z = -108$$
$$z = -6$$

Substituting 0 for $y$ and -6 for $z$ in equation (1), we obtain
$$4x - 5(0) + 2(-6) = -8$$
$$4x - 0 - 12 = -8$$
$$4x = 4$$
$$x = 1$$

The solution is the ordered triple (1, 0, -6).

**46.** $\begin{cases} 3x - 2y + 5z = -7 & (1) \\ 4x + y + 3z = -2 & (2) \\ 2x - 3y + 7z = -4 & (3) \end{cases}$

Multiply both sides of equation (1) by 4, multiply both sides of equation (2) by -3, and add the results.

$$12x - 8y + 20z = -28$$
$$-12x - 3y - 9z = 6$$
$$\overline{\quad -11y + 11z = -22 \quad} \quad (4)$$

Multiply both sides of equation (3) by -2 and add the result to equation (2).

$$-4x + 6y - 14z = 8$$
$$4x + y + 3z = -2$$
$$\overline{\quad 7y - 11z = 6 \quad} \quad (5)$$

Add equations (4) and (5).

$$-11y + 11z = -22$$
$$7y - 11z = 6$$
$$\overline{\quad -4y \quad = -16 \quad}$$
$$y = 4$$

Substituting 4 for $y$ in equation (4), we obtain

$-11(4) + 11z = -22$
$-44 + 11z = -22$
$11z = 22$
$z = 2$

Substituting 4 for $y$ and 2 for $z$ in equation (1), we obtain
$3x - 2(4) + 5(2) = -7$
$3x - 8 + 10 = -7$
$3x + 2 = -7$
$3x = -9$
$x = -3$

The solution is the ordered triple (-3, 4, 2).

47. Because the sum of the measures of the three angles in a triangle must equal $180°$, we obtain the following system.
$$\begin{cases} x + y + z = 180 & (1) \\ z = 2y & (2) \\ y = 3x - 10 & (3) \end{cases}$$

Writing the equations in the system in standard form, we obtain
$$\begin{cases} x + y + z = 180 & (1) \\ 2y - z = 0 & (2) \\ 3x - y = 10 & (3) \end{cases}$$

Adding equations (1) and (2), we obtain
$x + y + z = 180$
$2y - z = 0$
$\overline{\phantom{xxxxxxxxxxxxxx}}$
$x + 3y = 180 \quad (4)$

Multiply equation (3) by 3 and add the result to equation (4).
$9x - 3y = 30$
$x + 3y = 180$
$\overline{\phantom{xxxxxxxxxxxxxx}}$
$10x = 210$
$x = 21$

Substituting 21 for $x$ in equation (3), we obtain
$y = 3(21) - 10 = 63 - 10 = 53$

Substituting 53 for $y$ in equation (2), we obtain
$z = 2(53) = 106$

Thus, angle $x = 21°$, angle $y = 53°$, and angle $z = 106°$.

48. Let $b$ represent the number of calories in a cheeseburger, let $f$ represent the number of calories in a medium order of fries, and let $c$ represent the number of calories in a medium Coke.
$$\begin{cases} b + f + c = 1040 & (1) \\ 2b + f + c = 1400 & (2) \\ 2b + f + 2c = 1680 & (3) \end{cases}$$

Multiply equation (1) by -1 and add the result to equation (3).
$-b - f - c = -1040$
$2b + f + 2c = 1680$
$\overline{\phantom{xxxxxxxxxxxxxx}}$
$b \phantom{xx} + c = 640 \quad (4)$

Multiply equation (1) by -1 and add the result to equation (2).
$-b - f - c = -1040$
$2b + f + c = 1400$
$\overline{\phantom{xxxxxxxxxxxxxx}}$
$b \phantom{xxxxxxxxx} = 360$

Substituting 360 for $b$ in equation (4), we obtain
$360 + c = 640$
$c = 280$

Substituting 360 for $b$ and 280 for $c$ in equation (1), we obtain
$360 + f + 280 = 1040$
$f + 640 = 1040$
$f = 400$

At Burger King, a cheeseburger contains 360 calories, a medium order of fries contains 400 calories, and a medium Coke contains 280 calories.

49. $\begin{bmatrix} 3 & 1 & | & 7 \\ 2 & 5 & | & 9 \end{bmatrix}$

50. $\begin{bmatrix} 1 & -5 & | & 14 \\ -1 & 1 & | & -3 \end{bmatrix}$

51. $\begin{bmatrix} 5 & -1 & 4 & | & 6 \\ -3 & 0 & -3 & | & -1 \\ 1 & -2 & 0 & | & 0 \end{bmatrix}$

52. $\begin{bmatrix} 8 & -1 & 3 & | & 14 \\ -3 & 5 & -6 & | & -18 \\ 7 & -4 & 5 & | & 21 \end{bmatrix}$

53. $\begin{cases} x+2y=12 \\ 3y=15 \end{cases}$

54. $\begin{cases} 3x-4y=-5 \\ -x+2y=7 \end{cases}$

55. $\begin{cases} x+3y+4z=20 \\ y-2z=-16 \\ z=7 \end{cases}$

56. $\begin{cases} -3x+7y+9z=1 \\ 4x+10y+7z=5 \\ 2x-5y-6z=-8 \end{cases}$

57. a. $\begin{bmatrix} 1 & -5 & | & 22 \\ -2 & 9 & | & -40 \end{bmatrix}$ $(R_2 = 2r_1 + r_2)$

   $= \begin{bmatrix} 1 & -5 & | & 22 \\ 2(1)+(-2) & 2(-5)+9 & | & 2(22)+(-40) \end{bmatrix}$

   $= \begin{bmatrix} 1 & -5 & | & 22 \\ 0 & -1 & | & 4 \end{bmatrix}$

   b. $\begin{bmatrix} 1 & -5 & | & 22 \\ 0 & -1 & | & 4 \end{bmatrix}$ $(R_2 = -1 \cdot r_2)$

   $= \begin{bmatrix} 1 & -5 & | & 22 \\ -1(0) & -1(-1) & | & -1(4) \end{bmatrix}$

   $= \begin{bmatrix} 1 & -5 & | & 22 \\ 0 & 1 & | & -4 \end{bmatrix}$

58. a. $\begin{bmatrix} 1 & -4 & | & 7 \\ 3 & -7 & | & 6 \end{bmatrix}$ $(R_2 = -3r_1 + r_2)$

   $= \begin{bmatrix} 1 & -4 & | & 7 \\ -3(1)+3 & -3(-4)+(-7) & | & -3(7)+6 \end{bmatrix}$

   $= \begin{bmatrix} 1 & -4 & | & 7 \\ 0 & 5 & | & -15 \end{bmatrix}$

   b. $\begin{bmatrix} 1 & -4 & | & 7 \\ 0 & 5 & | & -15 \end{bmatrix}$ $\left(R_2 = \frac{1}{5}r_2\right)$

   $= \begin{bmatrix} 1 & -4 & | & 7 \\ \frac{1}{5}(0) & \frac{1}{5}(5) & | & \frac{1}{5}(-15) \end{bmatrix}$

   $= \begin{bmatrix} 1 & -4 & | & 7 \\ 0 & 1 & | & -3 \end{bmatrix}$

59. a. $\begin{bmatrix} -1 & 2 & 1 & | & 1 \\ 2 & -1 & 3 & | & -3 \\ -1 & 5 & 6 & | & 2 \end{bmatrix}$ $(R_2 = 2r_1 + r_2)$

   $= \begin{bmatrix} -1 & 2 & 1 & | & 1 \\ 2(-1)+2 & 2(2)+(-1) & 2(1)+3 & | & 2(1)+(-3) \\ -1 & 5 & 6 & | & 2 \end{bmatrix}$

   $= \begin{bmatrix} -1 & 2 & 1 & | & 1 \\ 0 & 3 & 5 & | & -1 \\ -1 & 5 & 6 & | & 2 \end{bmatrix}$

   b. $\begin{bmatrix} -1 & 2 & 1 & | & 1 \\ 0 & 3 & 5 & | & -1 \\ -1 & 5 & 6 & | & 2 \end{bmatrix}$ $(R_3 = -1 \cdot r_1 + r_3)$

   $= \begin{bmatrix} -1 & 2 & 1 & | & 1 \\ 0 & 3 & 5 & | & -1 \\ -1(-1)+(-1) & -1(2)+5 & -1(1)+6 & | & -1(1)+2 \end{bmatrix}$

   $= \begin{bmatrix} -1 & 2 & 1 & | & 1 \\ 0 & 3 & 5 & | & -1 \\ 0 & 3 & 5 & | & 1 \end{bmatrix}$

60. a. $\begin{bmatrix} 1 & 3 & 4 & | & 4 \\ 0 & 5 & 10 & | & -15 \\ 0 & -4 & -7 & | & 7 \end{bmatrix}$ $\left(R_2 = -\frac{1}{5} \cdot r_2\right)$

   $= \begin{bmatrix} 1 & 3 & 4 & | & 4 \\ \frac{1}{5}(0) & \frac{1}{5}(5) & \frac{1}{5}(10) & | & \frac{1}{5}(-15) \\ 0 & -4 & -7 & | & 7 \end{bmatrix}$

   $= \begin{bmatrix} 1 & 3 & 4 & | & 4 \\ 0 & 1 & 2 & | & -3 \\ 0 & -4 & -7 & | & 7 \end{bmatrix}$

   b. $\begin{bmatrix} 1 & 3 & 4 & | & 4 \\ 0 & 1 & 2 & | & -3 \\ 0 & -4 & -7 & | & 7 \end{bmatrix}$ $(R_3 = 4r_2 + r_3)$

   $= \begin{bmatrix} 1 & 3 & 4 & | & 4 \\ 0 & 1 & 2 & | & -3 \\ 4(0)+0 & 4(1)+(-4) & 4(2)+(-7) & | & 4(-3)+7 \end{bmatrix}$

   $= \begin{bmatrix} 1 & 3 & 4 & | & 4 \\ 0 & 1 & 2 & | & -3 \\ 0 & 0 & 1 & | & -5 \end{bmatrix}$

**Chapter 4:** *Systems of Linear Equations and Inequalities*

**61.** Write the augmented matrix of the system and then put it in row echelon form.

$$\begin{bmatrix} 1 & 2 & | & 1 \\ 1 & 3 & | & -2 \end{bmatrix} \quad (R_2 = -1 \cdot r_1 + r_2)$$

$$= \begin{bmatrix} 1 & 2 & | & 1 \\ 0 & 1 & | & -3 \end{bmatrix}$$

Write the system of equations that corresponds to the row-echelon matrix

$$\begin{cases} x + 2y = 1 & (1) \\ y = -3 & (2) \end{cases}$$

This system is consistent and independent.
Substituting -3 for $y$ in equation (1), we obtain
$x + 2(-3) = 1$
$x - 6 = 1$
$x = 7$
The solution is the ordered pair (7, -3).

**62.** Write the augmented matrix of the system and then put it in row echelon form.

$$\begin{bmatrix} 6 & -2 & | & -7 \\ 4 & 3 & | & 17 \end{bmatrix} \quad \left(R_1 = \frac{1}{6}r_1\right)$$

$$= \begin{bmatrix} 1 & -\frac{1}{3} & | & -\frac{7}{6} \\ 4 & 3 & | & 17 \end{bmatrix} \quad (R_2 = -4r_1 + r_2)$$

$$= \begin{bmatrix} 1 & -\frac{1}{3} & | & -\frac{7}{6} \\ 0 & \frac{13}{3} & | & \frac{65}{3} \end{bmatrix} \quad \left(R_2 = \frac{3}{13}r_2\right)$$

$$= \begin{bmatrix} 1 & -\frac{1}{3} & | & -\frac{7}{6} \\ 0 & 1 & | & 5 \end{bmatrix}$$

Write the system of equations that corresponds to the row-echelon matrix

$$\begin{cases} x - \frac{1}{3}y = -\frac{7}{6} & (1) \\ y = 5 & (2) \end{cases}$$

This system is consistent and independent.
Substituting 5 for $y$ in equation (1), we obtain
$x - \frac{1}{3}(5) = -\frac{7}{6}$
$x - \frac{5}{3} = -\frac{7}{6}$
$x = \frac{3}{6} = \frac{1}{2}$
The solution is the ordered pair $\left(\frac{1}{2}, 5\right)$.

**63.** Write the augmented matrix of the system and then put it in row echelon form.

$$\begin{bmatrix} 3 & 2 & | & -10 \\ 2 & -1 & | & -9 \end{bmatrix} \quad \left(R_1 = \frac{1}{3}r_1\right)$$

$$= \begin{bmatrix} 1 & \frac{2}{3} & | & -\frac{10}{3} \\ 2 & -1 & | & -9 \end{bmatrix} \quad (R_2 = -2r_1 + r_2)$$

$$= \begin{bmatrix} 1 & \frac{2}{3} & | & -\frac{10}{3} \\ 0 & -\frac{7}{3} & | & -\frac{7}{3} \end{bmatrix} \quad \left(R_2 = -\frac{3}{7}r_2\right)$$

$$= \begin{bmatrix} 1 & \frac{2}{3} & | & -\frac{10}{3} \\ 0 & 1 & | & 1 \end{bmatrix}$$

Write the system of equations that corresponds to the row-echelon matrix

$$\begin{cases} x + \frac{2}{3}y = -\frac{10}{3} & (1) \\ y = 1 & (2) \end{cases}$$

This system is consistent and independent.
Substituting 1 for $y$ in equation (1), we obtain
$x + \frac{2}{3}(1) = -\frac{10}{3}$
$x + \frac{2}{3} = -\frac{10}{3}$
$x = -\frac{12}{3} = -4$
The solution is the ordered pair (-4, 1).

**64.** Write the augmented matrix of the system and then put it in row echelon form.

$$\begin{bmatrix} -3 & 9 & | & 15 \\ 5 & -15 & | & 11 \end{bmatrix} \quad \left(R_1 = -\frac{1}{3}r_1\right)$$

$$= \begin{bmatrix} 1 & -3 & | & -5 \\ 5 & -15 & | & 11 \end{bmatrix} \quad (R_2 = -5r_1 + r_2)$$

$$= \begin{bmatrix} 1 & -3 & | & -5 \\ 0 & 0 & | & 36 \end{bmatrix}$$

The system is inconsistent. The system has no solution. The solution set is $\varnothing$ or $\{\ \}$.

**65.** Write the augmented matrix of the system and then put it in row echelon form.

$$\begin{bmatrix} 4 & -2 & | & 6 \\ 6 & -3 & | & 9 \end{bmatrix} \quad \left(R_1 = \frac{1}{4}r_1\right)$$

$$= \begin{bmatrix} 1 & -\frac{1}{2} & | & \frac{3}{2} \\ 6 & -3 & | & 9 \end{bmatrix} \quad (R_2 = -6r_1 + r_2)$$

$$= \begin{bmatrix} 1 & -\frac{1}{2} & | & \frac{3}{2} \\ 0 & 0 & | & 0 \end{bmatrix}$$

The system is dependent. The solution to the system is $\{(x, y) \mid 4x - 2y = 6\}$.

66. $\begin{bmatrix} 1 & 4 & | & 4 \\ 4 & -8 & | & 7 \end{bmatrix}$    $(R_2 = -4r_1 + r_2)$

$= \begin{bmatrix} 1 & 4 & | & 4 \\ 0 & -24 & | & -9 \end{bmatrix}$    $\left(R_2 = -\dfrac{1}{24}r_2\right)$

$= \begin{bmatrix} 1 & 4 & | & 4 \\ 0 & 1 & | & \frac{3}{8} \end{bmatrix}$

Write the system of equations that corresponds to the row-echelon matrix

$\begin{cases} x + 4y = 4 & (1) \\ y = \dfrac{3}{8} & (2) \end{cases}$

This system is consistent and independent.

Substituting $\dfrac{3}{8}$ for $y$ in equation (1), we obtain

$x + 4\left(\dfrac{3}{8}\right) = 4$

$x + \dfrac{3}{2} = 4$

$x = \dfrac{5}{2}$

The solution is the ordered pair $\left(\dfrac{5}{2}, \dfrac{3}{8}\right)$.

67. Write the augmented matrix of the system and then put it in row echelon form.

$\begin{bmatrix} 1 & 2 & -2 & | & -11 \\ 2 & 0 & 1 & | & -6 \\ 0 & 5 & -3 & | & -7 \end{bmatrix}$    $(R_2 = -2r_1 + r_2)$

$= \begin{bmatrix} 1 & 2 & -2 & | & -11 \\ 0 & -4 & 5 & | & 16 \\ 0 & 5 & -3 & | & -7 \end{bmatrix}$    $\left(R_2 = -\dfrac{1}{4}r_2\right)$

$= \begin{bmatrix} 1 & 2 & -2 & | & -11 \\ 0 & 1 & -\frac{5}{4} & | & -4 \\ 0 & 5 & -3 & | & -7 \end{bmatrix}$    $(R_3 = -5r_2 + r_3)$

$= \begin{bmatrix} 1 & 2 & -2 & | & -11 \\ 0 & 1 & -\frac{5}{4} & | & -4 \\ 0 & 0 & \frac{13}{4} & | & 13 \end{bmatrix}$    $\left(R_3 = \dfrac{4}{13}r_3\right)$

$= \begin{bmatrix} 1 & 2 & -2 & | & -11 \\ 0 & 1 & -\frac{5}{4} & | & -4 \\ 0 & 0 & 1 & | & 4 \end{bmatrix}$

Write the system of equations that corresponds to the row-echelon matrix

$\begin{cases} x + 2y - 2z = -11 & (1) \\ y - \dfrac{5}{4}z = -4 & (2) \\ z = 4 & (3) \end{cases}$

This system is consistent and independent.
Substituting 4 for $z$ in equation (2), we obtain

$y - \dfrac{5}{4}(4) = -4$

$y - 5 = -4$

$y = 1$

Substituting 4 for $z$ and 1 for $y$ in equation (1), we obtain

$x + 2(1) - 2(4) = -11$

$x + 2 - 8 = -11$

$x - 6 = -11$

$x = -5$

The solution is the ordered triple $(-5, 1, 4)$.

68. Write the augmented matrix of the system and then put it in row echelon form.

$\begin{bmatrix} 2 & -5 & 2 & | & 9 \\ -1 & 1 & -2 & | & -2 \\ -1 & -2 & -4 & | & 8 \end{bmatrix}$    (Interchange $r_1$ and $r_3$)

$= \begin{bmatrix} -1 & -2 & -4 & | & 8 \\ -1 & 1 & -2 & | & -2 \\ 2 & -5 & 2 & | & 9 \end{bmatrix}$    $(R_1 = -1 \cdot r_1)$

$= \begin{bmatrix} 1 & 2 & 4 & | & -8 \\ -1 & 1 & -2 & | & -2 \\ 2 & -5 & 2 & | & 9 \end{bmatrix}$    $\begin{pmatrix} R_2 = r_1 + r_2 \\ R_3 = 2r_2 + r_3 \end{pmatrix}$

$= \begin{bmatrix} 1 & 2 & 4 & | & -8 \\ 0 & 3 & 2 & | & -10 \\ 0 & -3 & -2 & | & 5 \end{bmatrix}$    $(R_3 = r_2 + r_3)$

$= \begin{bmatrix} 1 & 2 & 4 & | & -8 \\ 0 & 3 & 2 & | & -10 \\ 0 & 0 & 0 & | & -5 \end{bmatrix}$

The system is inconsistent. The system has no solution. The solution set is $\emptyset$ or $\{\ \}$.

69. Write the augmented matrix of the system and then put it in row echelon form.

$\begin{bmatrix} 2 & -7 & 11 & | & -5 \\ 4 & -2 & 6 & | & 2 \\ -2 & 19 & -27 & | & 17 \end{bmatrix}$    $\left(R_1 = \dfrac{1}{2} \cdot r_1\right)$

$= \begin{bmatrix} 1 & -\frac{7}{2} & \frac{11}{2} & | & -\frac{5}{2} \\ 4 & -2 & 6 & | & 2 \\ -2 & 19 & -27 & | & 17 \end{bmatrix}$    $\begin{pmatrix} R_2 = -4r_1 + r_2 \\ R_3 = 2r_1 + r_3 \end{pmatrix}$

## Chapter 4: Systems of Linear Equations and Inequalities

$$= \begin{bmatrix} 1 & -\frac{7}{2} & \frac{11}{2} & | & -\frac{5}{2} \\ 0 & 12 & -16 & | & 12 \\ 0 & 12 & -16 & | & 12 \end{bmatrix} \quad \left(R_2 = \frac{1}{12}r_2\right)$$

$$= \begin{bmatrix} 1 & -\frac{7}{2} & \frac{11}{2} & | & -\frac{5}{2} \\ 0 & 1 & -\frac{4}{3} & | & 1 \\ 0 & 12 & -16 & | & 12 \end{bmatrix} \quad (R_3 = -12r_2 + r_3)$$

$$= \begin{bmatrix} 1 & -\frac{7}{2} & \frac{11}{2} & | & -\frac{5}{2} \\ 0 & 1 & -\frac{4}{3} & | & 1 \\ 0 & 0 & 0 & | & 0 \end{bmatrix}$$

The system is dependent and has an infinite number of solutions.

Write the system of equations that corresponds to the row-echelon matrix

$$\begin{cases} x - \frac{7}{2}y + \frac{11}{2}z = -\frac{5}{2} & (1) \\ y - \frac{4}{3}z = 1 & (2) \\ 0 = 0 & (3) \end{cases}$$

Solve equation (2) for $y$.

$$y - \frac{4}{3}z = 1$$

$$y = \frac{4}{3}z + 1$$

Substituting $\frac{4}{3}z + 1$ for $y$ in equation (1), we obtain

$$x - \frac{7}{2}\left(\frac{4}{3}z + 1\right) + \frac{11}{2}z = -\frac{5}{2}$$

$$x - \frac{14}{3}z - \frac{7}{2} + \frac{11}{2}z = -\frac{5}{2}$$

$$6\left(x - \frac{14}{3}z - \frac{7}{2} + \frac{11}{2}z\right) = 6\left(-\frac{5}{2}\right)$$

$$6x - 28z - 21 + 33z = -15$$

$$6x + 5z - 21 = -15$$

$$6x = -5z + 6$$

$$x = \frac{-5z + 6}{6}$$

$$x = -\frac{5}{6}z + 1$$

The solution to the system is

$$\left\{(x, y, z) \mid x = -\frac{5}{6}z + 1, \; y = \frac{4}{3}z + 1, \; z \text{ is any real number}\right\}.$$

70. Write the augmented matrix of the system and then put it in row echelon form.

$$\begin{bmatrix} 5 & 3 & 7 & | & 9 \\ 3 & 5 & 4 & | & 8 \\ 1 & 3 & 3 & | & 9 \end{bmatrix} \quad (\text{Interchange } r_1 \text{ and } r_3)$$

$$= \begin{bmatrix} 1 & 3 & 3 & | & 9 \\ 3 & 5 & 4 & | & 8 \\ 5 & 3 & 7 & | & 9 \end{bmatrix} \quad \begin{pmatrix} R_2 = -3r_1 + r_2 \\ R_3 = -5r_1 + r_3 \end{pmatrix}$$

$$= \begin{bmatrix} 1 & 3 & 3 & | & 9 \\ 0 & -4 & -5 & | & -19 \\ 0 & -12 & -8 & | & -36 \end{bmatrix} \quad \left(R_2 = -\frac{1}{4}r_2\right)$$

$$= \begin{bmatrix} 1 & 3 & 3 & | & 9 \\ 0 & 1 & \frac{5}{4} & | & \frac{19}{4} \\ 0 & -12 & -8 & | & -36 \end{bmatrix} \quad (R_3 = 12r_2 + r_3)$$

$$= \begin{bmatrix} 1 & 3 & 3 & | & 9 \\ 0 & 1 & \frac{5}{4} & | & \frac{19}{4} \\ 0 & 0 & 7 & | & 21 \end{bmatrix} \quad \left(R_3 = \frac{1}{7}r_3\right)$$

$$= \begin{bmatrix} 1 & 3 & 3 & | & 9 \\ 0 & 1 & \frac{5}{4} & | & \frac{19}{4} \\ 0 & 0 & 1 & | & 3 \end{bmatrix}$$

Write the system of equations that corresponds to the row-echelon matrix

$$\begin{cases} x + 3y + 3z = 9 & (1) \\ y + \frac{5}{4}z = \frac{19}{4} & (2) \\ z = 3 & (3) \end{cases}$$

This system is consistent and independent. Substituting 3 for $z$ in equation (2), we obtain

$$y + \frac{5}{4}(3) = \frac{19}{4}$$

$$y + \frac{15}{4} = \frac{19}{4}$$

$$y = 1$$

Substituting 3 for $z$ and 1 for $y$ in equation (1), we obtain

$$x + 3(1) + 3(3) = 9$$

$$x + 3 + 9 = 9$$

$$x + 12 = 9$$

$$x = -3$$

The solution is the ordered triple (-3, 1, 3).

71. $\begin{vmatrix} 3 & 4 \\ -1 & 2 \end{vmatrix} = 3(2) - (-1)(4) = 6 + 4 = 10$

72. $\begin{vmatrix} 2 & -3 \\ -6 & 9 \end{vmatrix} = 2(9)-(-6)(-3) = 18-18 = 0$

73. $\begin{vmatrix} -5 & 7 \\ -4 & 6 \end{vmatrix} = -5(6)-(-4)(7) = -30+28 = -2$

74. $\begin{vmatrix} -7 & 2 \\ 6 & -3 \end{vmatrix} = -7(-3)-6(2) = 21-12 = 9$

75. $\begin{vmatrix} 5 & 0 & 1 \\ 2 & -3 & -1 \\ 3 & 6 & -4 \end{vmatrix} = 5\begin{vmatrix} -3 & -1 \\ 6 & -4 \end{vmatrix} - 0\begin{vmatrix} 2 & -1 \\ 3 & -4 \end{vmatrix} + 1\begin{vmatrix} 2 & -3 \\ 3 & 6 \end{vmatrix}$
$= 5[-3(-4)-6(-1)] - 0[2(-4)-3(-1)]$
$\qquad + 1[2(6)-3(-3)]$
$= 5(18) - 0(-5) + 1(21)$
$= 90 + 0 + 21$
$= 111$

76. $\begin{vmatrix} 1 & -4 & 5 \\ 0 & 1 & -3 \\ 2 & -6 & 4 \end{vmatrix} = 1\begin{vmatrix} 1 & -3 \\ -6 & 4 \end{vmatrix} - (-4)\begin{vmatrix} 0 & -3 \\ 2 & 4 \end{vmatrix} + 5\begin{vmatrix} 0 & 1 \\ 2 & -6 \end{vmatrix}$
$= 1[1(4)-(-6)(-3)] + 4[0(4)-2(-3)]$
$\qquad + 5[0(-6)-2(1)]$
$= 1(-14) + 4(6) + 5(-2)$
$= -14 + 24 - 10$
$= 0$

77. $\begin{vmatrix} 3 & 0 & -1 \\ 2 & 6 & 7 \\ 2 & 5 & 4 \end{vmatrix} = 3\begin{vmatrix} 6 & 7 \\ 5 & 4 \end{vmatrix} - 0\begin{vmatrix} 2 & 7 \\ 2 & 4 \end{vmatrix} + (-1)\begin{vmatrix} 2 & 6 \\ 2 & 5 \end{vmatrix}$
$= 3[6(4)-5(7)] - 0[2(4)-2(7)]$
$\qquad -1[2(5)-2(6)]$
$= 3(-11) - 0(-6) - 1(-2)$
$= -33 + 0 + 2$
$= -31$

78. $\begin{vmatrix} 2 & 3 & 1 \\ 1 & -3 & -7 \\ -5 & 4 & 8 \end{vmatrix}$
$= 2\begin{vmatrix} -3 & -7 \\ 4 & 8 \end{vmatrix} - 3\begin{vmatrix} 1 & -7 \\ -5 & 8 \end{vmatrix} + 1\begin{vmatrix} 1 & -3 \\ -5 & 4 \end{vmatrix}$
$= 2[-3(8)-4(-7)] - 3[1(8)-(-5)(-7)]$
$\qquad + 1[1(4)-(-5)(-3)]$
$= 2(4) - 3(-27) + 1(-11)$
$= 8 + 81 - 11$
$= 78$

79. $D = \begin{vmatrix} 1 & 2 \\ 2 & 3 \end{vmatrix} = 1(3)-2(2) = 3-4 = -1$
$D_x = \begin{vmatrix} -1 & 2 \\ 1 & 3 \end{vmatrix} = -1(3)-1(2) = -3-2 = -5$
$D_y = \begin{vmatrix} 1 & -1 \\ 2 & 1 \end{vmatrix} = 1(1)-2(-1) = 1+2 = 3$
$x = \dfrac{D_x}{D} = \dfrac{-5}{-1} = 5$; $y = \dfrac{D_y}{D} = \dfrac{3}{-1} = -3$
Thus, the solution is the ordered pair (5, -3).

80. $D = \begin{vmatrix} 4 & 9 \\ -5 & 6 \end{vmatrix} = 4(6)-(-5)(9) = 24+45 = 69$
$D_x = \begin{vmatrix} -13 & 9 \\ 22 & 6 \end{vmatrix} = -13(6)-22(9) = -78-198 = -276$
$D_y = \begin{vmatrix} 4 & -13 \\ -5 & 22 \end{vmatrix} = 4(22)-(-5)(-13) = 88-65 = 23$
$x = \dfrac{D_x}{D} = \dfrac{-276}{69} = -4$; $y = \dfrac{D_y}{D} = \dfrac{23}{69} = \dfrac{1}{3}$
Thus, the solution is the ordered pair $\left(-4, \dfrac{1}{3}\right)$.

81. $D = \begin{vmatrix} 4 & -1 \\ 3 & 5 \end{vmatrix} = 4(5)-3(-1) = 20+3 = 23$
$D_x = \begin{vmatrix} -6 & -1 \\ 7 & 5 \end{vmatrix} = -6(5)-7(-1) = -30+7 = -23$
$D_y = \begin{vmatrix} 4 & -6 \\ 3 & 7 \end{vmatrix} = 4(7)-3(-6) = 28+18 = 46$
$x = \dfrac{D_x}{D} = \dfrac{-23}{23} = -1$; $y = \dfrac{D_y}{D} = \dfrac{46}{23} = 2$
Thus, the solution is the ordered pair (-1, 2).

82. $D = \begin{vmatrix} 1 & -1 \\ 2 & 1 \end{vmatrix} = 1(1)-2(-1) = 1+2 = 3$
$D_x = \begin{vmatrix} 2 & -1 \\ 5 & 1 \end{vmatrix} = 2(1)-5(-1) = 2+5 = 7$
$D_y = \begin{vmatrix} 1 & 2 \\ 2 & 5 \end{vmatrix} = 1(5)-2(2) = 5-4 = 1$
$x = \dfrac{D_x}{D} = \dfrac{7}{3}$; $y = \dfrac{D_y}{D} = \dfrac{1}{3}$
Thus, the solution is the ordered pair $\left(\dfrac{7}{3}, \dfrac{1}{3}\right)$.

Chapter 4: Systems of Linear Equations and Inequalities

83. $D = \begin{vmatrix} 6 & 2 \\ 15 & 5 \end{vmatrix} = 6(5) - 15(2) = 30 - 30 = 0$

Because $D = 0$, Cramer's Rule does not apply. Thus, we must use a different method to solve the system. We will use elimination.
$\begin{cases} 6x + 2y = 5 & (1) \\ 15x + 5y = 8 & (2) \end{cases}$

Multiply equation (1) by 5, multiply equation (2) by $-2$, and add the results.
$30x + 10y = 25$
$-30x - 10y = -16$
$\phantom{-30x - 10y =} 0 = 9$

The system has no solution. The solution set is $\varnothing$ or $\{\ \}$. The system is inconsistent.

84. $D = \begin{vmatrix} 12 & 1 \\ -6 & 7 \end{vmatrix} = 12(7) - (-6)(1) = 84 + 6 = 90$

$D_x = \begin{vmatrix} 6 & 1 \\ 15 & 7 \end{vmatrix} = 6(7) - 15(1) = 42 - 15 = 27$

$D_y = \begin{vmatrix} 12 & 6 \\ -6 & 15 \end{vmatrix} = 12(15) - (-6)(6) = 180 + 36 = 216$

$x = \dfrac{D_x}{D} = \dfrac{27}{90} = \dfrac{3}{10}$; $y = \dfrac{D_y}{D} = \dfrac{216}{90} = \dfrac{12}{5}$

Thus, the solution is the ordered pair $\left(\dfrac{3}{10}, \dfrac{12}{5}\right)$.

85. $D = \begin{vmatrix} 1 & -1 & 2 \\ 3 & 2 & -4 \\ 0 & 3 & 5 \end{vmatrix}$

$= 1\begin{vmatrix} 2 & -4 \\ 3 & 5 \end{vmatrix} - (-1)\begin{vmatrix} 3 & -4 \\ 0 & 5 \end{vmatrix} + 2\begin{vmatrix} 3 & 2 \\ 0 & 3 \end{vmatrix}$

$= 1[2(5) - 3(-4)] - (-1)[3(5) - 0(-4)]$
$\phantom{=} + 2[3(3) - 0(2)]$

$= 1(22) - (-1)(15) + 2(9)$
$= 22 + 15 + 18$
$= 55$

$D_x = \begin{vmatrix} 9 & -1 & 2 \\ 7 & 2 & -4 \\ -1 & 3 & 5 \end{vmatrix}$

$= 9\begin{vmatrix} 2 & -4 \\ 3 & 5 \end{vmatrix} - (-1)\begin{vmatrix} 7 & -4 \\ -1 & 5 \end{vmatrix} + 2\begin{vmatrix} 7 & 2 \\ -1 & 3 \end{vmatrix}$

$= 9[2(5) - 3(-4)] - (-1)[7(5) - (-1)(-4)]$
$\phantom{=} + 2[7(3) - (-1)(2)]$

$= 9(22) - (-1)(31) + 2(23)$
$= 198 + 31 + 46$
$= 275$

$D_y = \begin{vmatrix} 1 & 9 & 2 \\ 3 & 7 & -4 \\ 0 & -1 & 5 \end{vmatrix}$

$= 1\begin{vmatrix} 7 & -4 \\ -1 & 5 \end{vmatrix} - 9\begin{vmatrix} 3 & -4 \\ 0 & 5 \end{vmatrix} + 2\begin{vmatrix} 3 & 7 \\ 0 & -1 \end{vmatrix}$

$= 1[7(5) - (-1)(-4)] - 9[3(5) - 0(-4)]$
$\phantom{=} + 2[3(-1) - 0(7)]$

$= 1(31) - 9(15) + 2(-3)$
$= 31 - 135 - 6$
$= -110$

$D_z = \begin{vmatrix} 1 & -1 & 9 \\ 3 & 2 & 7 \\ 0 & 3 & -1 \end{vmatrix}$

$= 1\begin{vmatrix} 2 & 7 \\ 3 & -1 \end{vmatrix} - (-1)\begin{vmatrix} 3 & 7 \\ 0 & -1 \end{vmatrix} + 9\begin{vmatrix} 3 & 2 \\ 0 & 3 \end{vmatrix}$

$= 1[2(-1) - 3(7)] - (-1)[3(-1) - 0(7)]$
$\phantom{=} + 9[3(3) - 0(2)]$

$= 1(-23) - (-1)(-3) + 9(9)$
$= -23 - 3 + 81$
$= 55$

$x = \dfrac{D_x}{D} = \dfrac{275}{55} = 5$; $y = \dfrac{D_y}{D} = \dfrac{-110}{55} = -2$;

$z = \dfrac{D_z}{D} = \dfrac{55}{55} = 1$

Thus, the solution is the ordered triple (5, -2, 1).

86. $D = \begin{vmatrix} 1 & -3 & -3 \\ 7 & 1 & -2 \\ 6 & -5 & -4 \end{vmatrix}$

$= 1\begin{vmatrix} 1 & -2 \\ -5 & -4 \end{vmatrix} - (-3)\begin{vmatrix} 7 & -2 \\ 6 & -4 \end{vmatrix} + (-3)\begin{vmatrix} 7 & 1 \\ 6 & -5 \end{vmatrix}$

$= 1[1(-4) - (-5)(-2)] - (-3)[7(-4) - 6(-2)]$
$\phantom{=} + (-3)[7(-5) - 6(1)]$

$= 1(-14) - (-3)(-16) + (-3)(-41)$
$= -14 - 48 + 123$
$= 61$

$D_x = \begin{vmatrix} -5 & -3 & -3 \\ 24 & 1 & -2 \\ -9 & -5 & -4 \end{vmatrix}$

$= -5\begin{vmatrix} 1 & -2 \\ -5 & -4 \end{vmatrix} - (-3)\begin{vmatrix} 24 & -2 \\ -9 & -4 \end{vmatrix} + (-3)\begin{vmatrix} 24 & 1 \\ -9 & -5 \end{vmatrix}$

$= -5[1(-4) - (-5)(-2)] - (-3)[24(-4) - (-9)(-2)]$
$\phantom{=} + (-3)[24(-5) - (-9)(1)]$

$= -5(-14) - (-3)(-114) + (-3)(-111)$
$= 70 - 342 + 333$
$= 61$

$$D_y = \begin{vmatrix} 1 & -5 & -3 \\ 7 & 24 & -2 \\ 6 & -9 & -4 \end{vmatrix}$$

$$= 1\begin{vmatrix} 24 & -2 \\ -9 & -4 \end{vmatrix} - (-5)\begin{vmatrix} 7 & -2 \\ 6 & -4 \end{vmatrix} + (-3)\begin{vmatrix} 7 & 24 \\ 6 & -9 \end{vmatrix}$$

$$= 1[24(-4)-(-9)(-2)]-(-5)[7(-4)-6(-2)]$$
$$\quad + (-3)[7(-9)-6(24)]$$

$$= 1(-114)-(-5)(-16)+(-3)(-207)$$
$$= -114-80+621$$
$$= 427$$

$$D_z = \begin{vmatrix} 1 & -3 & -5 \\ 7 & 1 & 24 \\ 6 & -5 & -9 \end{vmatrix}$$

$$= 1\begin{vmatrix} 1 & 24 \\ -5 & -9 \end{vmatrix} - (-3)\begin{vmatrix} 7 & 24 \\ 6 & -9 \end{vmatrix} + (-5)\begin{vmatrix} 7 & 1 \\ 6 & -5 \end{vmatrix}$$

$$= 1[1(-9)-(-5)(24)]-(-3)[7(-9)-6(24)]$$
$$\quad + (-5)[7(-5)-6(1)]$$

$$= 1(111)-(-3)(-207)+(-5)(-41)$$
$$= 111-621+205$$
$$= -305$$

$$x = \frac{D_x}{D} = \frac{61}{61} = 1; \ y = \frac{D_y}{D} = \frac{427}{61} = 7;$$
$$z = \frac{D_z}{D} = \frac{-305}{61} = -5$$

Thus, the solution is the ordered triple (1, 7, -5).

87. $D = \begin{vmatrix} 1 & 1 & 1 \\ 3 & 2 & 4 \\ 2 & 2 & 3 \end{vmatrix}$

$$= 1\begin{vmatrix} 2 & 4 \\ 2 & 3 \end{vmatrix} - 1\begin{vmatrix} 3 & 4 \\ 2 & 3 \end{vmatrix} + 1\begin{vmatrix} 3 & 2 \\ 2 & 2 \end{vmatrix}$$

$$= 1[2(3)-2(4)]-1[3(3)-2(4)]$$
$$\quad + 1[3(2)-2(2)]$$

$$= -2-1+2$$
$$= -1$$

$$D_x = \begin{vmatrix} 1 & 1 & 1 \\ -1 & 2 & 4 \\ 0 & 2 & 3 \end{vmatrix}$$

$$= 1\begin{vmatrix} 2 & 4 \\ 2 & 3 \end{vmatrix} - 1\begin{vmatrix} -1 & 4 \\ 0 & 3 \end{vmatrix} + 1\begin{vmatrix} -1 & 2 \\ 0 & 2 \end{vmatrix}$$

$$= 1[2(3)-2(4)]-1[-1(3)-0(4)]$$
$$\quad + 1[-1(2)-0(2)]$$

$$= -2+3-2$$
$$= -1$$

$$D_y = \begin{vmatrix} 1 & 1 & 1 \\ 3 & -1 & 4 \\ 2 & 0 & 3 \end{vmatrix}$$

$$= 1\begin{vmatrix} -1 & 4 \\ 0 & 3 \end{vmatrix} - 1\begin{vmatrix} 3 & 4 \\ 2 & 3 \end{vmatrix} + 1\begin{vmatrix} 3 & -1 \\ 2 & 0 \end{vmatrix}$$

$$= 1[-1(3)-0(4)]-1[3(3)-2(4)]$$
$$\quad + 1[3(0)-2(-1)]$$

$$= -3-1+2$$
$$= -2$$

$$D_z = \begin{vmatrix} 1 & 1 & 1 \\ 3 & 2 & -1 \\ 2 & 2 & 0 \end{vmatrix}$$

$$= 1\begin{vmatrix} 2 & -1 \\ 2 & 0 \end{vmatrix} - 1\begin{vmatrix} 3 & -1 \\ 2 & 0 \end{vmatrix} + 1\begin{vmatrix} 3 & 2 \\ 2 & 2 \end{vmatrix}$$

$$= 1[2(0)-2(-1)]-1[3(0)-2(-1)]$$
$$\quad + 1[3(2)-2(2)]$$

$$= 2-2+2$$
$$= 2$$

$$x = \frac{D_x}{D} = \frac{-1}{-1} = 1; \ y = \frac{D_y}{D} = \frac{-2}{-1} = 2;$$
$$z = \frac{D_z}{D} = \frac{2}{-1} = -2$$

Thus, the solution is the ordered triple (1, 2, -2).

88. $D = \begin{vmatrix} 4 & -3 & 1 \\ 4 & -2 & 3 \\ 8 & -5 & -2 \end{vmatrix}$

$$= 4\begin{vmatrix} -2 & 3 \\ -5 & -2 \end{vmatrix} - (-3)\begin{vmatrix} 4 & 3 \\ 8 & -2 \end{vmatrix} + 1\begin{vmatrix} 4 & -2 \\ 8 & -5 \end{vmatrix}$$

$$= 4[-2(-2)-(-5)(3)]-(-3)[4(-2)-8(3)]$$
$$\quad + 1[4(-5)-8(-2)]$$

$$= 4(19)-(-3)(-32)+1(-4)$$
$$= 76-96-4$$
$$= -24$$

$$D_x = \begin{vmatrix} -6 & -3 & 1 \\ -3 & -2 & 3 \\ -12 & -5 & -2 \end{vmatrix}$$

$$= -6\begin{vmatrix} -2 & 3 \\ -5 & -2 \end{vmatrix} - (-3)\begin{vmatrix} -3 & 3 \\ -12 & -2 \end{vmatrix} + 1\begin{vmatrix} -3 & -2 \\ -12 & -5 \end{vmatrix}$$

$$= -6[-2(-2)-(-5)(3)]-(-3)[-3(-2)-(-12)(3)]$$
$$\quad + 1[-3(-5)-(-12)(-2)]$$

$$= -6(19)-(-3)(42)+1(-9)$$
$$= -114+126-9$$
$$= 3$$

$$D_y = \begin{vmatrix} 4 & -6 & 1 \\ 4 & -3 & 3 \\ 8 & -12 & -2 \end{vmatrix}$$

$$= 4\begin{vmatrix} -3 & 3 \\ -12 & -2 \end{vmatrix} - (-6)\begin{vmatrix} 4 & 3 \\ 8 & -2 \end{vmatrix} + 1\begin{vmatrix} 4 & -3 \\ 8 & -12 \end{vmatrix}$$

$$= 4[-3(-2)-(-12)(3)]-(-6)[4(-2)-8(3)]$$
$$\qquad\qquad +1[4(-12)-8(-3)]$$
$$= 4(42)-(-6)(-32)+1(-24)$$
$$= 168-192-24$$
$$= -48$$

$$D_z = \begin{vmatrix} 4 & -3 & -6 \\ 4 & -2 & -3 \\ 8 & -5 & -12 \end{vmatrix}$$

$$= 4\begin{vmatrix} -2 & -3 \\ -5 & -12 \end{vmatrix} - (-3)\begin{vmatrix} 4 & -3 \\ 8 & -12 \end{vmatrix} + (-6)\begin{vmatrix} 4 & -2 \\ 8 & -5 \end{vmatrix}$$

$$= 4[-2(-12)-(-5)(-3)]-(-3)[4(-12)-8(-3)]$$
$$\qquad\qquad +(-6)[4(-5)-8(-2)]$$
$$= 4(9)-(-3)(-24)+(-6)(-4)$$
$$= 36-72+24$$
$$= -12$$

$$x = \frac{D_x}{D} = \frac{3}{-24} = -\frac{1}{8}; \quad y = \frac{D_y}{D} = \frac{-48}{-24} = 2;$$
$$z = \frac{D_z}{D} = \frac{-12}{-24} = \frac{1}{2}$$

The solution is the ordered triple $\left(-\frac{1}{8}, 2, \frac{1}{2}\right)$.

**89.** $D = \begin{vmatrix} 1 & -1 & -4 \\ 4 & -3 & -3 \\ 3 & -2 & 1 \end{vmatrix}$

$$= 1\begin{vmatrix} -3 & -3 \\ -2 & 1 \end{vmatrix} - (-1)\begin{vmatrix} 4 & -3 \\ 3 & 1 \end{vmatrix} + (-4)\begin{vmatrix} 4 & -3 \\ 3 & -2 \end{vmatrix}$$

$$= 1[-3(1)-(-2)(-3)]-(-1)[4(1)-3(-3)]$$
$$\qquad\qquad +(-4)[4(-2)-3(-3)]$$
$$= 1(-9)-(-1)(13)+(-4)(1)$$
$$= -9+13-4$$
$$= 0$$

Because $D = 0$, Cramer's Rule does not apply. Thus, we must use a different method to solve the system.

$$\begin{cases} x-y-4z=7 & (1) \\ 4x-3y-3z=4 & (2) \\ 3x-2y+z=-3 & (3) \end{cases}$$

Multiply both sides of equation (1) by -4 and add the result to equation (2).
$$-4x+4y+16z=-28$$
$$\underline{4x-3y-3z=4}$$
$$y+13z=-24 \quad (4)$$

Multiply both sides of equation (1) by -3 and add the result to equation (3).
$$-3x+3y+12z=-21$$
$$\underline{3x-2y+z=-3}$$
$$y+13z=-24 \quad (5)$$

Multiply both sides of equation (4) by $-1$ and add the result to equation (5).
$$-y-13z=24$$
$$\underline{y+13z=-24}$$
$$0=0 \quad \text{True}$$

Thus, the system is dependent and has an infinite number of solutions.

Solve equation (5) for $y$.
$$y+13z=-24$$
$$y=-13z-24$$

Substituting $-13z-24$ for $y$ in equation (1), we obtain
$$x-(-13z-24)-4z=7$$
$$x+13z+24-4z=7$$
$$x+9z+24=7$$
$$x=-9z-17$$

The solution to the system is
$$\{(x,y,z) | x=-9z-17, y=-13z-24, z \text{ is any real number}\}$$

**90.** $D = \begin{vmatrix} 1 & 1 & 0 \\ 0 & 2 & -1 \\ 5 & 0 & 1 \end{vmatrix}$

$$= 1\begin{vmatrix} 2 & -1 \\ 0 & 1 \end{vmatrix} - 1\begin{vmatrix} 0 & -1 \\ 5 & 1 \end{vmatrix} + 0\begin{vmatrix} 0 & 2 \\ 5 & 0 \end{vmatrix}$$

$$= 1[2(1)-0(-1)]-1[0(1)-5(-1)]$$
$$\qquad\qquad +0[0(0)-5(2)]$$
$$= 1(2)-1(5)+0(-10)$$
$$= 2-5+0$$
$$= -3$$

$$D_x = \begin{vmatrix} -3 & 1 & 0 \\ -1 & 2 & -1 \\ 1 & 0 & 1 \end{vmatrix}$$

$$= -3\begin{vmatrix} 2 & -1 \\ 0 & 1 \end{vmatrix} - 1\begin{vmatrix} -1 & -1 \\ 1 & 1 \end{vmatrix} + 0\begin{vmatrix} -1 & 2 \\ 1 & 0 \end{vmatrix}$$

$$= -3[2(1)-0(-1)]-1[-1(1)-1(-1)]$$
$$\qquad +0[-1(0)-1(2)]$$
$$= -3(2)-1(0)+0(-2)$$
$$= -6-0+0$$
$$= -6$$

$$D_y = \begin{vmatrix} 1 & -3 & 0 \\ 0 & -1 & -1 \\ 5 & 1 & 1 \end{vmatrix}$$

$$= 1\begin{vmatrix} -1 & -1 \\ 1 & 1 \end{vmatrix} -(-3)\begin{vmatrix} 0 & -1 \\ 5 & 1 \end{vmatrix} +0\begin{vmatrix} 0 & -1 \\ 5 & 1 \end{vmatrix}$$

$$= 1[-1(1)-1(-1)]-(-3)[0(1)-5(-1)]$$
$$\qquad +0[0(1)-5(-1)]$$
$$= 1(0)-(-3)(5)+0(5)$$
$$= 0+15+0$$
$$= 15$$

$$D_z = \begin{vmatrix} 1 & 1 & -3 \\ 0 & 2 & -1 \\ 5 & 0 & 1 \end{vmatrix}$$

$$= 1\begin{vmatrix} 2 & -1 \\ 0 & 1 \end{vmatrix} -1\begin{vmatrix} 0 & -1 \\ 5 & 1 \end{vmatrix} +(-3)\begin{vmatrix} 0 & 2 \\ 5 & 0 \end{vmatrix}$$

$$= 1[2(1)-0(-1)]-1[0(1)-5(-1)]$$
$$\qquad +(-3)[0(0)-5(2)]$$
$$= 1(2)-1(5)+(-3)(-10)$$
$$= 2-5+30$$
$$= 27$$

$$x = \frac{D_x}{D} = \frac{-6}{-3} = 2; \quad y = \frac{D_y}{D} = \frac{15}{-3} = -5;$$
$$z = \frac{D_z}{D} = \frac{27}{-3} = -9$$

Thus, the solution is the ordered triple (2, -5, -9).

**91. a.** Let $x = 5$ and $y = 2$ in each inequality.
$$\begin{cases} x - y = 5 - 2 = 3 > 2 \\ 2x + 3y = 2(5)+3(2) = 16 > 8 \end{cases}$$
Both inequalities are true when when $x = 5$ and $y = 2$, so (5, 2) is a solution.

**b.** Let $x = -3$ and $y = 5$ in each inequality.
$$\begin{cases} x - y = -3 - 5 = -8 \not> 2 \\ 2x + 3y = 2(-3)+3(5) = 9 > 8 \end{cases}$$
The inequality $x - y > 2$ is not true when $x = -3$ and $y = 5$, so (-3, 5) is not a solution.

**92. a.** Let $x = -1$ and $y = 2$ in each inequality.
$$\begin{cases} 7x+3y = 7(-1)+3(2) = -1 \le 21 \\ -2x+y = -2(-1)+2 = 4 \not\ge 5 \end{cases}$$
The inequality $-2x + y \ge 5$ is not true when $x = -1$ and $y = 2$, so (-1, 2) is not a solution.

**b.** Let $x = -3$ and $y = 4$ in each inequality.
$$\begin{cases} 7x+3y = 7(-3)+3(4) = -9 \le 21 \\ -2x+y = -2(-3)+4 = 10 \ge 5 \end{cases}$$
Both inequalities are true when $x = -3$ and $y = 4$, so (-3, 4) is a solution.

**93. a.** Let $x = 2$ and $y = 1$ in each inequality.
$$\begin{cases} x+2y = 2+2(1) = 4 \le 6 \\ 3x-y = 3(2)-1 = 5 \ge 2 \\ x = 2 \ge 0 \\ y = 1 \ge 0 \end{cases}$$
All four inequalities are true when $x = 2$ and $y = 1$, so (2, 1) is a solution.

**b.** Let $x = 1$ and $y = 2$ in each inequality.
$$\begin{cases} x+2y = 1+2(2) = 5 \le 6 \\ 3x-y = 3(1)-2 = 1 \not\ge 2 \\ x = 1 \ge 0 \\ y = 2 \ge 0 \end{cases}$$
The inequality $3x - y \ge 2$ is not true when $x = 1$ and $y = 2$, so (1, 2) is not a solution.

**94. a.** Let $x = -2$ and $y = 3$ in each inequality.
$$\begin{cases} 3x-2y = 3(-2)-2(3) = -12 \le 12 \\ 2x+y = 2(-2)+3 = -1 \le 15 \\ x = -2 \not\ge 0 \\ y = 3 \ge 0 \end{cases}$$
The inequality $x \ge 0$ is not true when $x = -2$ and $y = 3$, so (-2, 3) is not a solution.

**b.** Let $x = 4$ and $y = 1$ in each inequality.
$$\begin{cases} 3x-2y = 3(4)-2(1) = 10 \le 12 \\ 2x+y = 2(4)+1 = 9 \le 15 \\ x = 4 \ge 0 \\ y = 1 \ge 0 \end{cases}$$
All four inequalities are true when $x = 4$ and $y = 1$, so (4, 1) is a solution.

## Chapter 4: Systems of Linear Equations and Inequalities

95. $\begin{cases} x+y \geq 7 \\ 2x-y \geq 5 \end{cases}$

    First, graph the inequality $x+y \geq 7$. To do so, replace the inequality symbol with an equal sign to obtain $x+y=7$. Because the inequality is non-strict, graph $x+y=7$ $(y=-x+7)$ using a solid line.

    Test Point: $(0,0)$: $0+0=0 \not\geq 7$

    Therefore, the half-plane not containing $(0,0)$ is the solution set of $x+y \geq 7$.

    Second, graph the inequality $2x-y \geq 5$. To do so, replace the inequality symbol with an equal sign to obtain $2x-y=5$. Because the inequality is non-strict, graph $2x-y=5$ $(y=2x-5)$ using a solid line.

    Test Point: $(0,0)$: $2(0)-0=0 \not\geq 5$

    Therefore, the half-plane not containing $(0,0)$ is the solution set of $2x-y \geq 5$.

    The overlapping shaded region (that is, the shaded region in the graph below) is the solution to the system of linear inequalities.

    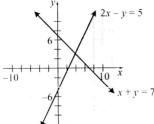

96. $\begin{cases} 2x+3y > 9 \\ x-3y > -18 \end{cases}$

    First, graph the inequality $2x+3y > 9$. To do so, replace the inequality symbol with an equal sign to obtain $2x+3y=9$. Because the inequality is strict, graph $2x+3y=9$ $\left(y=-\dfrac{2}{3}x+3\right)$ using a dashed line.

    Test Point: $(0,0)$: $2(0)+3(0)=0 \not> 9$

    Therefore, the half-plane not containing $(0,0)$ is the solution set of $2x+3y > 9$.

    Second, graph the inequality $x-3y > -18$. To do so, replace the inequality symbol with an equal sign to obtain $x-3y=-18$. Because the inequality is strict, graph $x-3y=-18$ $\left(y=\dfrac{1}{3}x+6\right)$ using a dashed line.

    Test Point: $(0,0)$: $0-3(0)=0 > -18$

    Therefore, the half-plane containing $(0,0)$ is the solution set of $x-3y > -18$.

    The overlapping shaded region (that is, the shaded region in the graph below) is the solution to the system of linear inequalities.

    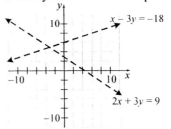

97. $\begin{cases} x-4y > -4 \\ x+2y \leq 8 \end{cases}$

    First, graph the inequality $x-4y > -4$. To do so, replace the inequality symbol with an equal sign to obtain $x-4y=-4$. Because the inequality is strict, graph $x-4y=-4$ $\left(y=\dfrac{1}{4}x+1\right)$ using a dashed line.

    Test Point: $(0,0)$: $0-4(0)=0 > -4$

    Therefore, the half-plane containing $(0,0)$ is the solution set of $x-4y > -4$.

    Second, graph the inequality $x+2y \leq 8$. To do so, replace the inequality symbol with an equal sign to obtain $x+2y=8$. Because the inequality is non-strict, graph $x+2y=8$ $\left(y=-\dfrac{1}{2}x+4\right)$ using a solid line.

    Test Point: $(0,0)$: $0+2(0)=0 \leq 8$

    Therefore, the half-plane containing $(0,0)$ is the solution set of $x+2y \leq 8$.

    The overlapping shaded region (that is, the shaded region in the graph below) is the solution to the system of linear inequalities.

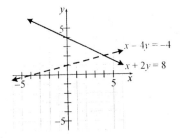

98. $\begin{cases} x - y < -2 \\ 2x - 2y > 6 \end{cases}$

First, graph the inequality $x - y < -2$. To do so, replace the inequality symbol with an equal sign to obtain $x - y = -2$. Because the inequality is strict, graph $x - y = -2$ $(y = x + 2)$ using a dashed line.

Test Point: $(0,0)$: $0 - 0 = 0 \not< -2$

Therefore, the half-plane not containing $(0,0)$ is the solution set of $x - y < -2$.

Second, graph the inequality $2x - 2y > 6$. To do so, replace the inequality symbol with an equal sign to obtain $2x - 2y = 6$. Because the inequality is strict, graph $2x - 2y = 6$ $(y = x - 3)$ using a dashed line.

Test Point: $(0,0)$: $2(0) - 2(0) = 0 \not> 6$

Therefore, the half-plane not containing $(0,0)$ is the solution set of $2x - 2y > 6$.

Notice in the graph below that the two shaded regions do not overlap. Thus, there are no points in the Cartesian plane that satisfy both inequalities. The system has no solution, so the solution set is $\varnothing$ or $\{\ \}$.

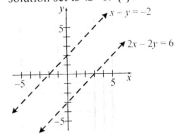

99. $\begin{cases} 2x - y \geq -2 \\ 2x - 3y \leq 6 \end{cases}$

First, graph the inequality $2x - y \geq -2$. To do so, replace the inequality symbol with an equal sign to obtain $2x - y = -2$. Because the inequality is non-strict, graph $2x - y = -2$ $(y = 2x + 2)$ using a solid line.

Test Point: $(0,0)$: $2(0) - 0 = 0 \geq -2$

Therefore, the half-plane containing $(0,0)$ is the solution set of $2x - y \geq -2$.

Second, graph the inequality $2x - 3y \leq 6$. To do so, replace the inequality symbol with an equal sign to obtain $2x - 3y = 6$. Because the inequality is non-strict, graph $2x - 3y = 6$ $\left(y = \dfrac{2}{3}x - 2\right)$ using a solid line.

Test Point: $(0,0)$: $2(0) - 3(0) = 0 \leq 6$

Therefore, the half-plane containing $(0,0)$ is the solution set of $2x - 3y \leq 6$.

The overlapping shaded region (that is, the shaded region in the graph below) is the solution to the system of linear inequalities.

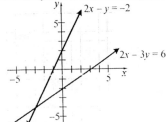

100. $\begin{cases} x \geq 2 \\ y < -3 \end{cases}$

First, graph the inequality $x \geq 2$. To do so, replace the inequality symbol with an equal sign to obtain $x = 2$. Because the inequality is non-strict, graph $x = 2$ using a solid line.

Test Point: $(0,0)$: $0 \not\geq 2$

Therefore, the half-plane not containing $(0,0)$ is the solution set of $x \geq 2$.

Second, graph the inequality $y < -3$. To do so, replace the inequality symbol with an equal sign to obtain $y = -3$. Because the inequality is strict, graph $y = -3$ using a dashed line.

Test Point: $(0,0)$: $0 \not< -3$

Therefore, the half-plane not containing $(0,0)$ is the solution set of $y < -3$.

The overlapping shaded region (that is, the shaded region in the graph below) is the solution to the system of linear inequalities.

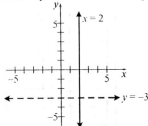

**101.** $\begin{cases} 3x+5y \le 30 \\ 4x-5y \ge 5 \\ x \le 8 \\ y \ge 0 \end{cases}$

First, graph the inequality $3x+5y \le 30$. To do so, replace the inequality symbol with an equal sign to obtain $3x+5y = 30$. Because the inequality is non-strict, graph $3x+5y = 30$ $\left(y = -\dfrac{3}{5}x+6\right)$ using a solid line.

Test Point: $(0,0)$: $3(0)+5(0) = 0 \le 30$

Therefore, the half-plane containing $(0,0)$ is the solution set of $3x+5y \le 30$.

Next, graph the inequality $4x-5y \ge 5$. To do so, replace the inequality symbol with an equal sign to obtain $4x-5y = 5$. Because the inequality is non-strict, graph $4x-5y = 5$ $\left(y = \dfrac{4}{5}x-1\right)$ using a solid line.

Test Point: $(0,0)$: $4(0)-5(0) = 0 \not\ge 5$

Therefore, the half-plane not containing $(0,0)$ is the solution set of $4x-5y \ge 5$.

The inequality $x \le 8$ requires that the graph be to the left of the solid vertical line $x = 8$.

The inequality $y \ge 0$ requires that the graph be above the solid $x$-axis.

The overlapping shaded region (that is, the shaded region in the graph below) is the solution to the system of linear inequalities.

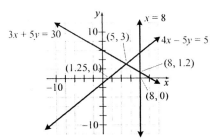

The graph is bounded. The four corner points of the bounded region are $(1.25, 0)$, $(8, 0)$, $(8, 1.2)$, and $(5, 3)$.

**102.** $\begin{cases} 3x+2y \ge 10 \\ x+2y \ge 6 \\ x \ge 0 \\ y \ge 0 \end{cases}$

First, graph the inequality $3x+2y \ge 10$. To do so, replace the inequality symbol with an equal sign to obtain $3x+2y = 10$. Because the inequality is non-strict, graph $3x+2y = 10$ $\left(y = -\dfrac{3}{2}x+5\right)$ using a solid line.

Test Point: $(0,0)$: $3(0)+2(0) = 0 \not\ge 10$

Therefore, the half-plane not containing $(0,0)$ is the solution set of $3x+2y \ge 10$.

Second, graph the inequality $x+2y \ge 6$. To do so, replace the inequality symbol with an equal sign to obtain $x+2y = 6$. Because the inequality is non-strict, graph $x+2y = 6$ $\left(y = -\dfrac{1}{2}x+3\right)$ using a solid line.

Test Point: $(0,0)$: $0+2(0) = 0 \not\ge 6$

Therefore, the half-plane not containing $(0,0)$ is the solution set of $x+2y \ge 6$.

The two inequalities $x \ge 0$ and $y \ge 0$ require that the graph be in quadrant I.

The overlapping shaded region (that is, the shaded region in the graph below) is the solution to the system of linear inequalities.

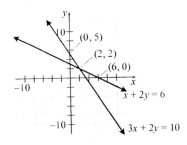

The graph is unbounded. The three corner points of the unbounded region are $(0, 5)$, $(2, 2)$, and $(6, 0)$.

**103. a.** Let $x$ represent the amount to be invested at 6% interest and let $y$ represent the amount to be invested at 8% interest. Now, Anna has at most $4000 to invest, so $x + y \leq 4000$. She wants to earn at least $275 per year in interest, so $0.06x + 0.08y \geq 275$. The amount to be invested at 6% will be at least $500, so $x \geq 500$. The amount to be invested at 8% is at least $2500, so $y \geq 2500$. Thus, we have the following system:
$$\begin{cases} x + y \leq 4000 \\ 0.06x + 0.08y \geq 275 \\ x \geq 500 \\ y \geq 2500 \end{cases}$$

**b.** First, graph the inequality $x + y \leq 4000$. To do so, replace the inequality symbol with an equal sign to obtain $x + y = 4000$. Because the inequality is non-strict, graph $x + y = 4000$ ($y = -x + 4000$) using a solid line.
Test Point: $(0,0)$: $0 + 0 = 0 \leq 4000$
Thus, the solution set of $x + y \leq 4000$ is the half-plane containing $(0,0)$.

Second, graph the inequality $0.06x + 0.08y \geq 275$. To do so, replace the inequality symbol with an equal sign to obtain $0.06x + 0.08y = 275$. Because the inequality is non-strict, graph $0.06x + 0.08y = 275$ $\left(y = -\dfrac{3}{4}x + 3437.5\right)$ using a solid line.

Test Point: $(0,0)$: $0.06(0) + 0.08(0) = 0 \not\geq 275$

Thus, the solution set of $0.06x + 0.08y \geq 275$ is the half-plane not containing $(0,0)$.

The inequality $x \geq 500$ requires that the graph be to the right of the solid vertical line $x = 500$.

The inequality $y \geq 2500$ requires that the graph be above the solid horizontal line $y = 2500$.

The overlapping shaded region (that is, the shaded region in the graph below) is the solution to the system of linear inequalities.

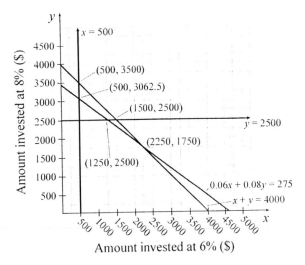

The four corner points are $(500, 3500)$, $(500, 3062.5)$, $(1250, 2500)$, and $(1500, 2500)$.

**104. a.** Let $x$ represent the number of dozens of red tulip bulbs and let $y$ represent the number of dozens of yellow tulip bulbs. Now, Jordan has a maximum of $144 to spend, so $6x + 4y \leq 144$. She wants at least 4 more dozens of red tulip bulbs than yellow, so $x \geq y + 4$. Finally, Jordan cannot buy a negative number of dozens of bulbs, so we have the non-negativity constraints $x \geq 0$ and $y \geq 0$. Thus, we have the following system:
$$\begin{cases} 6x + 4y \leq 144 \\ x \geq y + 4 \\ x \geq 0 \\ y \geq 0 \end{cases}$$

## Chapter 4: Systems of Linear Equations and Inequalities

**b.** First, graph the inequality $6x+4y \leq 144$. To do so, replace the inequality symbol with an equal sign to obtain $6x+4y=144$. Because the inequality is non-strict, graph $6x+4y=144$ $\left(y=-\frac{3}{2}x+36\right)$ using a solid line.

Test Point: $(0,0)$: $6(0)+4(0)=0 \leq 144$

Thus, the solution set of $6x+4y \leq 144$ is the half-plane containing $(0,0)$.

Second, graph the inequality $x \geq y+4$. To do so, replace the inequality symbol with an equal sign to obtain $x=y+4$. Because the inequality is non-strict, graph $x=y+4$ $(y=x-4)$ using a solid line.

Test Point: $(0,0)$: $0 \not\geq 0+4$

Thus, the solution set of $x \geq y+4$ is the half-plane not containing $(0,0)$.

The two inequalities $x \geq 0$ and $y \geq 0$ require that the graph be in quadrant I.

The overlapping shaded region (that is, the shaded region in the graph below) is the solution to the system of linear inequalities.

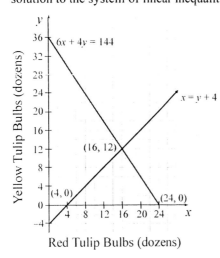

The three corner points are $(4, 0)$, $(24, 0)$, and $(16, 12)$.

## Chapter 4 Test

**1.** $\begin{cases} 2x-y=0 \\ 4x-5y=12 \end{cases}$

The graphs of the linear equations intersect at the point $(-2, -4)$. Thus, the solution is $(-2, -4)$.

**2.** $\begin{cases} 5x+2y=-3 & (1) \\ y=2x-6 & (2) \end{cases}$

Because equation (2) is solved for $y$, we use substitution to solve the system.
Substituting $2x-6$ for $y$ in equation (1), we obtain
$$5x+2(2x-6)=-3$$
$$5x+4x-12=-3$$
$$9x=9$$
$$x=1$$
Substituting 1 for $x$ in equation (2), we obtain
$y=2(1)-6=2-6=-4$.
The solution is the ordered pair $(1, -4)$.

**3.** $\begin{cases} 9x+3y=1 & (1) \\ x-2y=4 & (2) \end{cases}$

Because the coefficient of $x$ in equation (2) is 1, we use substitution to solve the system.
Equation (2) solved for $x$ is $x=2y+4$.
Substituting $2y+4$ for $x$ in equation (1), we obtain
$$9(2y+4)+3y=1$$
$$18y+36+3y=1$$
$$21y=-35$$
$$y=\frac{-35}{21}=-\frac{5}{3}$$
Substituting $-\frac{5}{3}$ for $y$ in equation (1), we obtain
$$9x+3\left(-\frac{5}{3}\right)=1$$
$$9x-5=1$$
$$9x=6$$
$$x=\frac{6}{9}=\frac{2}{3}$$
The solution is the ordered pair $\left(\frac{2}{3}, -\frac{5}{3}\right)$.

**4.** $\begin{cases} 6x - 9y = 5 & (1) \\ 8x - 12y = 7 & (2) \end{cases}$

Because none of variables have a coefficient of 1, we use elimination to solve the system. Multiply both sides of equation (1) by 4, multiply both sides of equation (2) by -3, and add the results.
$$24x - 36y = 20$$
$$-24x + 36y = -21$$
$$\overline{\phantom{aaaaaaaaaa}}$$
$$0 = -1 \quad \text{False}$$

The system has no solution. The solution set is $\emptyset$ or $\{\ \}$. The system is inconsistent.

**5.** $\begin{cases} 2x + y = -4 & (1) \\ \dfrac{1}{3}x + \dfrac{1}{2}y = 2 & (2) \end{cases}$

Because equation (2) contains fractions, we choose to use the elimination method to solve the system. Multiply both sides of equation (2) by -6 and add the result to equation (1).
$$-2x - 3y = -12$$
$$2x + y = -4$$
$$\overline{\phantom{aaaaaaaaaa}}$$
$$-2y = -16$$
$$y = 8$$

Substituting 8 for $y$ in equation (1), we obtain
$$2x + 8 = -4$$
$$2x = -12$$
$$x = -6$$

The solution is the ordered pair (-6, 8).

**6.** $\begin{cases} x - 2y + 3z = 1 & (1) \\ x + y - 3z = 7 & (2) \\ 3x - 4y + 5z = 7 & (3) \end{cases}$

Multiply both sides of equation (1) by -1 and add the result to equation (2).
$$-x + 2y - 3z = -1$$
$$x + y - 3z = 7$$
$$\overline{\phantom{aaaaaaaaaa}}$$
$$3y - 6z = 6 \quad (4)$$

Multiply both sides of equation (1) by -3 and add the result to equation (3).
$$-3x + 6y - 9z = -3$$
$$3x - 4y + 5z = 7$$
$$\overline{\phantom{aaaaaaaaaa}}$$
$$2y - 4z = 4 \quad (5)$$

Multiply both sides of equation (4) by 2, multiply both sides of equation (5) by -3, and add the results.

$$6y - 12z = 12$$
$$-6y + 12z = -12$$
$$\overline{\phantom{aaaaaaaaaa}}$$
$$0 = 0 \quad \text{True}$$

Thus, the system is dependent and has an infinite number of solutions.

Solve equation (5) for $y$.
$$2y - 4z = 4$$
$$2y = 4z + 4$$
$$y = \frac{4z + 4}{2}$$
$$y = 2z + 2$$

Substituting $2z + 2$ for $y$ in equation (1), we obtain
$$x - 2(2z + 2) + 3z = 1$$
$$x - 4z - 4 + 3z = 1$$
$$x - z - 4 = 1$$
$$x = z + 5$$

The solution to the system is
$\{(x, y, z) | x = z + 5, y = 2z + 2, z \text{ is any real number}\}$.

**7.** $\begin{cases} 2x + 4y + 3z = 5 & (1) \\ 3x - y + 2z = 8 & (2) \\ x + y + 2z = 0 & (3) \end{cases}$

Multiply both sides of equation (3) by -2 and add the result to equation (1).
$$-2x - 2y - 4z = 0$$
$$2x + 4y + 3z = 5$$
$$\overline{\phantom{aaaaaaaaaa}}$$
$$2y - z = 5 \quad (4)$$

Multiply both sides of equation (3) by -3 and add the result to equation (2).
$$-3x - 3y - 6z = 0$$
$$3x - y + 2z = 8$$
$$\overline{\phantom{aaaaaaaaaa}}$$
$$-4y - 4z = 8 \quad (5)$$

Multiply both sides of equation (4) by 2 and add the result to equation (5).
$$4y - 2z = 10$$
$$-4y - 4z = 8$$
$$\overline{\phantom{aaaaaaaaaa}}$$
$$-6z = 18$$
$$z = -3$$

Substituting -3 for $z$ in equation (4), we obtain
$$2y - (-3) = 5$$
$$2y + 3 = 5$$
$$2y = 2$$
$$y = 1$$

Substituting -3 for $z$ and 1 for $y$ in equation (1), we obtain

## Chapter 4: Systems of Linear Equations and Inequalities

$2x + 4(1) + 3(-3) = 5$
$2x + 4 - 9 = 5$
$2x - 5 = 5$
$2x = 10$
$x = 5$

The solution is the ordered triple (5, 1, -3).

8. a. $\begin{bmatrix} 1 & -3 & | & -2 \\ 2 & -4 & | & 8 \end{bmatrix}$ $(R_2 = -2r_1 + r_2)$

$= \begin{bmatrix} 1 & -3 & | & -2 \\ -2(1)+2 & -2(-3)+(-4) & | & -2(-2)+8 \end{bmatrix}$

$= \begin{bmatrix} 1 & -3 & | & -2 \\ 0 & 2 & | & 12 \end{bmatrix}$

b. $\begin{bmatrix} 1 & -3 & | & -2 \\ 0 & 2 & | & 12 \end{bmatrix}$ $\left(R_2 = \frac{1}{2}r_2\right)$

$= \begin{bmatrix} 1 & -3 & | & -2 \\ \frac{1}{2}(0) & \frac{1}{2}(2) & | & \frac{1}{2}(12) \end{bmatrix}$

$= \begin{bmatrix} 1 & -3 & | & -2 \\ 0 & 1 & | & 6 \end{bmatrix}$

9. a. $\begin{bmatrix} 1 & -2 & 1 & | & -2 \\ 3 & -5 & 2 & | & 1 \\ 0 & -4 & 5 & | & -32 \end{bmatrix}$ $(R_2 = -3r_1 + r_2)$

$= \begin{bmatrix} 1 & -2 & 1 & | & -2 \\ -3(1)+3 & -3(-2)+(-5) & -3(1)+2 & | & -3(-2)+1 \\ 0 & -4 & 5 & | & -32 \end{bmatrix}$

$= \begin{bmatrix} 1 & -2 & 1 & | & -2 \\ 0 & 1 & -1 & | & 7 \\ 0 & -4 & 5 & | & -32 \end{bmatrix}$

b. $\begin{bmatrix} 1 & -2 & 1 & | & -2 \\ 0 & 1 & -1 & | & 7 \\ 0 & -4 & 5 & | & -32 \end{bmatrix}$ $(R_3 = 4r_2 + r_3)$

$= \begin{bmatrix} 1 & -2 & 1 & | & -2 \\ 0 & 1 & -1 & | & 7 \\ 4(0)+0 & 4(1)+(-4) & 4(-1)+5 & | & 4(7)+(-32) \end{bmatrix}$

$= \begin{bmatrix} 1 & -2 & 1 & | & -2 \\ 0 & 1 & -1 & | & 7 \\ 0 & 0 & 1 & | & -4 \end{bmatrix}$

10. $\begin{cases} x - 5y = 2 & (1) \\ 2x + y = 4 & (2) \end{cases}$

Write the augmented matrix of the system and then put it in row echelon form.

$\begin{bmatrix} 1 & -5 & | & 2 \\ 2 & 1 & | & 4 \end{bmatrix}$ $(R_2 = -2r_1 + r_2)$

$= \begin{bmatrix} 1 & -5 & | & 2 \\ 0 & 11 & | & 0 \end{bmatrix}$ $\left(R_2 = \frac{1}{11}r_2\right)$

$= \begin{bmatrix} 1 & -5 & | & 2 \\ 0 & 1 & | & 0 \end{bmatrix}$

Write the system of equations that corresponds to the row-echelon matrix

$\begin{cases} x - 5y = 2 & (1) \\ y = 0 & (2) \end{cases}$

This system is consistent and independent. Substituting 0 for $y$ in equation (1), we obtain
$x - 5(0) = 2$
$x = 2$

The solution is the ordered pair (2, 0).

11. $\begin{cases} x + 2y + z = 3 & (1) \\ 4y + 3z = 5 & (2) \\ 2x + 3y = 1 & (3) \end{cases}$

Write the augmented matrix of the system and then put it in row echelon form.

$\begin{bmatrix} 1 & 2 & 1 & | & 3 \\ 0 & 4 & 3 & | & 5 \\ 2 & 3 & 0 & | & 1 \end{bmatrix}$ $(R_3 = -2r_1 + r_3)$

$= \begin{bmatrix} 1 & 2 & 1 & | & 3 \\ 0 & 4 & 3 & | & 5 \\ 0 & -1 & -2 & | & -5 \end{bmatrix}$ $\left(R_2 = \frac{1}{4}r_2\right)$

$= \begin{bmatrix} 1 & 2 & 1 & | & 3 \\ 0 & 1 & \frac{3}{4} & | & \frac{5}{4} \\ 0 & -1 & -2 & | & -5 \end{bmatrix}$ $(R_3 = r_3 + r_2)$

$= \begin{bmatrix} 1 & 2 & 1 & | & 3 \\ 0 & 1 & \frac{3}{4} & | & \frac{5}{4} \\ 0 & 0 & -\frac{5}{4} & | & -\frac{15}{4} \end{bmatrix}$ $\left(R_3 = -\frac{4}{5}r_3\right)$

$= \begin{bmatrix} 1 & 2 & 1 & | & 3 \\ 0 & 1 & \frac{3}{4} & | & \frac{5}{4} \\ 0 & 0 & 1 & | & 3 \end{bmatrix}$

Write the system of equations that corresponds to the row-echelon matrix

$\begin{cases} x+2y+z=3 & (1) \\ y+\dfrac{3}{4}z=\dfrac{5}{4} & (2) \\ z=3 & (3) \end{cases}$

This system is consistent and independent.
Substituting 3 for $z$ in equation (2), we obtain

$y+\dfrac{3}{4}(3)=\dfrac{5}{4}$

$y+\dfrac{9}{4}=\dfrac{5}{4}$

$y=\dfrac{-4}{4}=-1$

Substituting 3 for $z$ and -1 for $y$ in equation (1), we obtain

$x+2(-1)+3=3$

$x-2+3=3$

$x=2$

The solution is the ordered triple (2, -1, 3).

12. $\begin{vmatrix} 3 & -5 \\ 4 & -8 \end{vmatrix} = 3(-8)-4(-5)$

$=-24+20$

$=-4$

13. $\begin{vmatrix} 0 & 1 & 2 \\ 3 & 3 & -1 \\ -2 & 1 & 2 \end{vmatrix}$

$= 0\begin{vmatrix} 3 & -1 \\ 1 & 2 \end{vmatrix} - 1\begin{vmatrix} 3 & -1 \\ -2 & 2 \end{vmatrix} + 2\begin{vmatrix} 3 & 3 \\ -2 & 1 \end{vmatrix}$

$= 0[3(2)-1(-1)]-1[3(2)-(-2)(-1)]$
$\qquad +2[3(1)-(-2)(3)]$

$= 0(7)-1(4)+2(9)$

$= 0-4+18$

$= 14$

14. $D = \begin{vmatrix} 1 & -1 \\ 5 & 3 \end{vmatrix} = 1(3)-5(-1)=3+5=8$

$D_x = \begin{vmatrix} -2 & -1 \\ -8 & 3 \end{vmatrix} = -2(3)-(-8)(-1)=-6-8=-14$

$D_y = \begin{vmatrix} 1 & -2 \\ 5 & -8 \end{vmatrix} = 1(-8)-5(-2)=-8+10=2$

$x=\dfrac{D_x}{D}=\dfrac{-14}{8}=-\dfrac{7}{4}$ ; $y=\dfrac{D_y}{D}=\dfrac{2}{8}=\dfrac{1}{4}$

Thus, the solution is the ordered pair $\left(-\dfrac{7}{4},\dfrac{1}{4}\right)$.

15. $D = \begin{vmatrix} 1 & 1 & 1 \\ 1 & 1 & -2 \\ 4 & 2 & 3 \end{vmatrix}$

$= 1\begin{vmatrix} 1 & -2 \\ 2 & 3 \end{vmatrix} - 1\begin{vmatrix} 1 & -2 \\ 4 & 3 \end{vmatrix} + 1\begin{vmatrix} 1 & 1 \\ 4 & 2 \end{vmatrix}$

$= 1[1(3)-2(-2)]-1[1(3)-4(-2)]$
$\qquad +1[1(2)-4(1)]$

$= 1(7)-1(11)+1(-2)$

$= -6$

$D_x = \begin{vmatrix} -2 & 1 & 1 \\ 1 & 1 & -2 \\ -15 & 2 & 3 \end{vmatrix}$

$= -2\begin{vmatrix} 1 & -2 \\ 2 & 3 \end{vmatrix} - 1\begin{vmatrix} 1 & -2 \\ -15 & 3 \end{vmatrix} + 1\begin{vmatrix} 1 & 1 \\ -15 & 2 \end{vmatrix}$

$= -2[1(3)-2(-2)]-1[1(3)-(-15)(-2)]$
$\qquad +1[1(2)-(-15)(1)]$

$= -2(7)-1(-27)+1(17)$

$= -14+27+17$

$= 30$

$D_y = \begin{vmatrix} 1 & -2 & 1 \\ 1 & 1 & -2 \\ 4 & -15 & 3 \end{vmatrix}$

$= 1\begin{vmatrix} 1 & -2 \\ -15 & 3 \end{vmatrix} - (-2)\begin{vmatrix} 1 & -2 \\ 4 & 3 \end{vmatrix} + 1\begin{vmatrix} 1 & 1 \\ 4 & -15 \end{vmatrix}$

$= 1[1(3)-(-15)(-2)]-(-2)[1(3)-4(-2)]$
$\qquad +1[1(-15)-4(1)]$

$= 1(-27)-(-2)(11)+1(-19)$

$= -27+22-19$

$= -24$

$D_z = \begin{vmatrix} 1 & 1 & -2 \\ 1 & 1 & 1 \\ 4 & 2 & -15 \end{vmatrix}$

$= 1\begin{vmatrix} 1 & 1 \\ 2 & -15 \end{vmatrix} - 1\begin{vmatrix} 1 & 1 \\ 4 & -15 \end{vmatrix} + (-2)\begin{vmatrix} 1 & 1 \\ 4 & 2 \end{vmatrix}$

$= 1[1(-15)-2(1)]-1[1(-15)-4(1)]$
$\qquad +(-2)[1(2)-4(1)]$

$= 1(-17)-1(-19)+(-2)(-2)$

$= -17+19+4$

$= 6$

$x=\dfrac{D_x}{D}=\dfrac{30}{-6}=-5$ ; $y=\dfrac{D_y}{D}=\dfrac{-24}{-6}=4$ ;

$z=\dfrac{D_z}{D}=\dfrac{6}{-6}=-1$

Thus, the solution is the ordered triple (-5, 4, -1).

16. $\begin{cases} 2x - y > -2 \\ x - 3y < 9 \end{cases}$

First, graph the inequality $2x - y > -2$. To do so, replace the inequality symbol with an equal sign to obtain $2x - y = -2$. Because the inequality is strict, graph $2x - y = -2$ $(y = 2x + 2)$ using a dashed line.

Test Point: $(0,0)$: $2(0) - 0 = 0 > -2$

Therefore, the half-plane containing $(0,0)$ is the solution set of $2x - y > -2$.

Second, graph the inequality $x - 3y < 9$. To do so, replace the inequality symbol with an equal sign to obtain $x - 3y = 9$. Because the inequality is strict, graph $x - 3y = 9$ $\left(y = \dfrac{1}{3}x - 3\right)$ using a dashed line.

Test Point: $(0,0)$: $0 - 3(0) = 0 < 9$

Therefore, the half-plane containing $(0,0)$ is the solution set of $x - 3y < 9$.

The overlapping shaded region (that is, the shaded region in the graph below) is the solution to the system of linear inequalities.

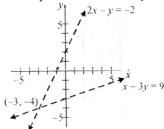

The corner point is $(-3, -4)$.

17. $\begin{cases} 3x + 2y \le 12 \\ x - 2y \ge -4 \\ x \ge 0 \\ y \ge 0 \end{cases}$

First, graph the inequality $3x + 2y \le 12$. To do so, replace the inequality symbol with an equal sign to obtain $3x + 2y = 12$. Because the inequality is non-strict, graph $3x + 2y = 12$ $\left(y = -\dfrac{3}{2}x + 6\right)$ using a solid line.

Test Point: $(0,0)$: $3(0) + 2(0) = 0 \le 12$

Therefore, the half-plane containing $(0,0)$ is the solution set of $3x + 2y \le 12$.

Next, graph the inequality $x - 2y \ge -4$. To do so, replace the inequality symbol with an equal sign to obtain $x - 2y = -4$. Because the inequality is non-strict, graph $x - 2y = -4$ $\left(y = \dfrac{1}{2}x + 2\right)$ using a solid line.

Test Point: $(0,0)$: $0 - 2(0) = 0 \ge -4$

Therefore, the half-plane containing $(0,0)$ is the solution set of $x - 2y \ge -4$.

The two inequalities $x \ge 0$ and $y \ge 0$ require that the graph be in quadrant I.

The overlapping shaded region (that is, the shaded region in the graph below) is the solution to the system of linear inequalities.

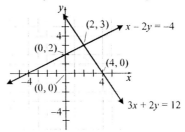

The graph is bounded. The four corner points of the bounded region are $(0, 0)$, $(4, 0)$, $(2, 3)$, and $(0, 2)$.

18. Let $x$ represent the number of twenty-five-ton bins and $y$ represent the number of twenty-ton bins.

$\begin{cases} x + y = 50 & (1) \\ 25x + 20y = 1160 & (2) \end{cases}$

Equation (1) solved for $y$ is $y = -x + 50$.

Substituting $-x + 50$ for $y$ in equation (2), we obtain
$$25x + 20(-x + 50) = 1160$$
$$25x - 20x + 1000 = 1160$$
$$5x + 1000 = 1160$$
$$5x = 160$$
$$x = 32$$

Substituting 32 for $x$ in equation (1), we obtain
$$32 + y = 50$$
$$y = 18$$

The warehouse contains 32 twenty-five-ton bins and 18 twenty-ton bins.

19. Because the sum of the measures of the three angles in a triangle must equal $180°$, we obtain the following system.
$$\begin{cases} x+y+z=180 & (1) \\ z=y+10 & (2) \\ 5x=y+z & (3) \end{cases}$$
Writing the equations in the system in standard form, we obtain
$$\begin{cases} x+y+z=180 & (1) \\ y-z=-10 & (2) \\ 5x-y-z=0 & (3) \end{cases}$$
Adding equations (1) and (2), we obtain
$$\begin{array}{r} x+y+z=180 \\ y-z=-10 \\ \hline x+2y\phantom{+z}=170 \quad (4) \end{array}$$
Adding equations (1) and (3), we obtain.
$$\begin{array}{r} x+y+z=180 \\ 5x-y-z=0 \\ \hline 6x\phantom{+y+z}=180 \\ x=30 \end{array}$$
Substituting 30 for $x$ in equation (4), we obtain
$$30+2y=170$$
$$2y=140$$
$$y=70$$
Substituting 30 for $x$ and 70 for $y$ and in equation (1), we obtain
$$30+70+z=180$$
$$100+z=180$$
$$z=80$$
Thus, angle $x=30°$, angle $y=70°$, and angle $z=80°$.

20. a. Margaret has a maximum of $180 to spend, so $12x+18y \le 180$. She will buy no more than 13 items, so $x+y \le 13$. Finally, Margaret cannot buy a negative number of blouses or sweaters, so we have the non-negativity constraints $x \ge 0$ and $y \ge 0$.
Thus, we have the following system:
$$\begin{cases} 12x+18y \le 180 \\ x+y \le 13 \\ x \ge 0 \\ y \ge 0 \end{cases}$$

b. First, graph the inequality $12x+18y \le 180$. To do so, replace the inequality symbol with an equal sign to obtain $12x+18y=180$. Because the inequality is non-strict, graph $12x+18y=180$ $\left(y=-\dfrac{2}{3}x+10\right)$ using a solid line.
Test Point: $(0,0)$: $12(0)+18(0)=0 \le 180$
Thus, the solution set of $12x+18y \le 180$ is the half-plane containing $(0, 0)$.
Second, graph the inequality $x+y \le 13$. To do so, replace the inequality symbol with an equal sign to obtain $x+y=13$. Because the inequality is non-strict, graph $x+y=13$ $(y=-x+13)$ using a solid line.
Test Point: $(0,0)$: $0+0=0 \le 13$
Thus, the solution set of $x+y \le 13$ is the half-plane containing $(0, 0)$.
The two inequalities $x \ge 0$ and $y \ge 0$ require that the graph be in quadrant I.
The overlapping shaded region (that is, the shaded region in the graph below) is the solution to the system of linear inequalities.

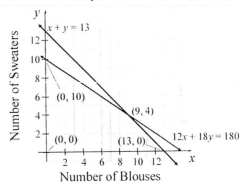

The four corner points are $(0, 0)$, $(13, 0)$, $(9, 4)$, and $(0, 10)$.

# Getting Ready for Chapter 5

2. $a^{m+n}$

4. 3; left

6. True; subtract the exponent in the denominator from the exponent in the numerator.

8. False; to write 0.0001 in scientific notation, we move the decimal point four places to the right. This gives us $1 \times 10^{-4}$.

10. The expression $\left(a^m\right)^n$ can be written as follows:
$$\left(a^m\right)^n = a^{m \cdot n} = a^{n \cdot m} = \left(a^n\right)^m$$
If $n$ is negative, then the base, $a$, cannot be 0 or $a^n$ would yield division by 0. For example, $0^{-3} = \frac{1}{0^3} = \frac{1}{0} = \text{undefined}$. If $n = 0$, then $a$ cannot be 0 or $a^n$ would yield the indeterminate form $0^0$.

12. Written response. Answers may vary.
$$a^n = a^n$$
$$\frac{a^n}{a^n} = 1 \quad \text{(as long as } a \neq 0\text{)}$$
$$a^{n-n} = 1$$
$$a^0 = 1$$

14. Written response. Answers may vary. The idea can be illustrated as follows:
$$\frac{x^7}{x^3} = \frac{x \cdot x \cdot x \cdot x \cdot x \cdot x \cdot x}{x \cdot x \cdot x}$$
$$= \frac{x \cdot x \cdot x \cdot x \cdot \cancel{x \cdot x \cdot x}}{\cancel{x \cdot x \cdot x}}$$
$$= x \cdot x \cdot x \cdot x$$
$$= x^4$$
Notice that the bases are the same. When we divide the two expressions on the left, we will need to subtract the exponents to get the result on the right.

15. Written response. Answers may vary. The idea can be illustrated as follows:
Multiply $x$ together 8 times to get $x^8$.
$$x \cdot x \cdot x \cdot x \cdot x \cdot x \cdot x \cdot x = x^8$$
The Associative Law for Multiplication says that we can group these multiplications however we wish. Group the $x$'s in twos to get
$$(x \cdot x) \cdot (x \cdot x) \cdot (x \cdot x) \cdot (x \cdot x)$$
Remember that exponents indicate multiplication. Since we multiply $x$ twice inside each set of parentheses, we can rewrite this as
$$\left(x^2\right) \cdot \left(x^2\right) \cdot \left(x^2\right) \cdot \left(x^2\right)$$
We again have a series of multiplications, but we are multiplying $x^2$ and not just $x$. Thus, we can write $\left(x^2\right)^4$ since $x^2$ is being multiplied together 4 times. Thus, the end result is that $\left(x^2\right)^4 = x^8$.

16. Written response. Answers may vary. The idea can be illustrated as follows:
Multiply $x \cdot y$ together 4 times to get $(x \cdot y)^4$.
$$(x \cdot y) \cdot (x \cdot y) \cdot (x \cdot y) \cdot (x \cdot y) = (x \cdot y)^4$$
The Commutative and Associative Laws for Multiplication says that we can group these multiplications in any order we wish. Group the $x$'s together and group the $y$'s together to get
$$(x \cdot x \cdot x \cdot x) \cdot (y \cdot y \cdot y \cdot y) = (x \cdot y)^4$$
Using the Product Rule for Exponents, rewrite the expressions inside the parentheses to get
$$x^4 \cdot y^4 = (x \cdot y)^4$$

18. Written response. Answers may vary. See Objective 8.

20. $(-5)^2 = (-5) \cdot (-5) = 25$

22. $5^{-2} = \frac{1}{5^2} = \frac{1}{25}$

24. $-5^0 = -1 \cdot 1 = -1$

26. $\frac{8^7}{8^5} = 8^{7-5} = 8^2 = 64$

28. $\left(\frac{3}{4}\right)^{-3} = \left(\frac{4}{3}\right)^3 = \frac{4^3}{3^3} = \frac{64}{27}$

30. $(-3)^2 \cdot 2^0 = 9 \cdot 1 = 9$

**Getting Ready for Chapter 5**

32. $(-4)^{-5} \cdot (-4)^3 = (-4)^{-5+3} = (-4)^{-2} = \dfrac{1}{(-4)^2} = \dfrac{1}{16}$

34. $\dfrac{(-3)^3}{(-3)^{-2}} = (-3)^{3-(-2)} = (-3)^5 = -243$

36. $\dfrac{3^{-2} \cdot 5^3}{3^2 \cdot 5} = 3^{-2-2} \cdot 5^{3-1}$
    $= 3^{-4} \cdot 5^2$
    $= \dfrac{5^2}{3^4}$
    $= \dfrac{25}{81}$

38. $y^{-3} \cdot y^7 = y^{-3+7} = y^4$

40. $\dfrac{z^5}{z^{-2}} = z^{5-(-2)} = z^{5+2} = z^7$

42. $(5a^2)^0 = 1$ (assuming $a \neq 0$)

44. $10^{-1} = \dfrac{1}{10}$

46. $(6ab) \cdot (3a^3 b^{-4}) = 6 \cdot 3 \cdot a^{1+3} b^{1+(-4)}$
    $= 18a^4 b^{-3}$
    $= \dfrac{18a^4}{b^3}$

48. $(3xy^3) \cdot \left(\dfrac{1}{9}x^2 y\right) = 3 \cdot \dfrac{1}{9} \cdot x^{1+2} y^{3+1}$
    $= \dfrac{1}{3} x^3 y^4$

50. $\dfrac{12 b^6}{18 b^2} = \dfrac{12}{18} b^{6-2}$
    $= \dfrac{2}{3} b^4$

52. $\dfrac{25 a^2 b^3}{5 a b^6} = \dfrac{25}{5} a^{2-1} b^{3-6}$
    $= 5 a^1 b^{-3}$
    $= \dfrac{5a}{b^3}$

54. $\dfrac{25 x^{-2} y}{10 x y^3} = \dfrac{25}{10} x^{-2-1} y^{1-3}$
    $= \dfrac{5}{2} x^{-3} y^{-2}$
    $= \dfrac{5}{2 x^3 y^2}$

56. $\dfrac{25 x y^{-4}}{10 x^{-2} y^2} = \dfrac{25}{10} x^{1-(-2)} y^{-4-2}$
    $= \dfrac{5}{2} x^3 y^{-6}$
    $= \dfrac{5 x^3}{2 y^6}$

58. $(z^2)^{-6} = z^{2 \cdot (-6)} = z^{-12} = \dfrac{1}{z^{12}}$

60. $(5 a^2 b^{-1})^2 = 5^2 (a^2)^2 (b^{-1})^2$
    $= 25 a^{2 \cdot 2} b^{-1 \cdot 2}$
    $= 25 a^4 b^{-2}$
    $= \dfrac{25 a^4}{b^2}$

62. $\left(\dfrac{x}{y}\right)^{-8} = \left(\dfrac{y}{x}\right)^8 = \dfrac{y^8}{x^8}$

64. $(2 y^{-2})^{-4} = 2^{-4} (y^{-2})^{-4}$
    $= \dfrac{1}{2^4} y^{-2 \cdot (-4)}$
    $= \dfrac{y^8}{16}$

66. $(-4 a^{-2} b^2)^{-2} = (-4)^{-2} \cdot (a^{-2})^{-2} \cdot (b^2)^{-2}$
    $= \dfrac{1}{(-4)^2} \cdot a^{-2 \cdot (-2)} \cdot b^{2 \cdot (-2)}$
    $= \dfrac{1}{16} a^4 b^{-4}$
    $= \dfrac{a^4}{16 b^4}$

## Getting Ready for Chapter 5

**68.** $\dfrac{5x^{-3}y^5z}{x^{-2}y^6z^{-2}} = 5x^{-3-(-2)}y^{5-6}z^{1-(-2)}$

$= 5x^{-3+2}y^{-1}z^{1+2}$

$= 5x^{-1}y^{-1}z^3$

$= \dfrac{5z^3}{xy}$

**70.** $\dfrac{3^2 \cdot x^{-3}(y^2)^3}{15x^2y^8} = \dfrac{9x^{-3}y^{2\cdot 3}}{15x^2y^8}$

$= \dfrac{9x^{-3}y^6}{15x^2y^8}$

$= \dfrac{9}{15}x^{-3-2}y^{6-8}$

$= \dfrac{3}{5}x^{-5}y^{-2}$

$= \dfrac{3}{5x^5y^2}$

**72.** $\left(\dfrac{15x^4y^7}{18x^{-3}y}\right)^{-1} = \dfrac{18x^{-3}y}{15x^4y^7}$

$= \dfrac{18}{15}x^{-3-4}y^{1-7}$

$= \dfrac{6}{5}x^{-7}y^{-6}$

$= \dfrac{6}{5x^7y^6}$

**74.** $(9a^2b^{-4})^{-1} \cdot (3ab^{-2})^2$

$= 9^{-1}(a^2)^{-1}(b^{-4})^{-1} \cdot 3^2 a^2 (b^{-2})^2$

$= \dfrac{1}{9}a^{-2}b^4 \cdot 9a^2b^{-4}$

$= \dfrac{9}{9}a^{-2+2}b^{4+(-4)}$

$= 1 \cdot a^0 \cdot b^0$

$= 1 \cdot 1 \cdot 1$

$= 1$

**76.** $(-5a^2b)^2 \cdot (10a^3b^2)^{-1}$

$= (-5)^2(a^2)^2 b^2 \cdot 10^{-1}(a^3)^{-1}(b^2)^{-1}$

$= 25a^4b^2 \cdot \dfrac{1}{10}a^{-3}b^{-2}$

$= \dfrac{25}{10}a^{4+(-3)}b^{2+(-2)}$

$= \dfrac{5}{2}a^1b^0$

$= \dfrac{5}{2}a$

**78.** $\dfrac{(-3)^3 a^3 (ab)^{-2}}{9ab^4} = \dfrac{-27a^3 \cdot a^{-2}b^{-2}}{9ab^4}$

$= \dfrac{-27}{9}a^{3+(-2)-1}b^{-2-4}$

$= -3a^0b^{-6}$

$= \dfrac{-3}{b^6}$

**80.** $\dfrac{(2ab^2c)^{-1}}{(a^{-1}b^3c^2)^{-2}} = \dfrac{2^{-1}a^{-1}(b^2)^{-1}c^{-1}}{(a^{-1})^{-2}(b^3)^{-2}(c^2)^{-2}}$

$= \dfrac{2^{-1}a^{-1}b^{-2}c^{-1}}{a^2b^{-6}c^{-4}}$

$= \dfrac{1}{2}a^{-1-2}b^{-2-(-6)}c^{-1-(-4)}$

$= \dfrac{1}{2}a^{-3}b^4c^3$

$= \dfrac{b^4c^3}{2a^3}$

# Getting Ready for Chapter 5

**82.** $\left(\dfrac{a^{-3}b^{-1}}{2a^4b^{-2}}\right)^2 \cdot \dfrac{(4a^2b)^2}{(2a^{-2}b)^3}$

$= \left(\dfrac{1}{2}a^{-3-4}b^{-1-(-2)}\right)^2 \cdot \dfrac{4^2(a^2)^2 b^2}{2^3(a^{-2})^3 b^3}$

$= \left(\dfrac{1}{2}a^{-7}b^1\right)^2 \cdot \dfrac{16a^4b^2}{8a^{-6}b^3}$

$= \dfrac{1}{4}a^{-14}b^2 \cdot 2a^{4-(-6)}b^{2-3}$

$= \dfrac{1}{4}a^{-14}b^2 \cdot 2a^{10}b^{-1}$

$= \dfrac{2}{4}a^{-14+10}b^{2+(-1)}$

$= \dfrac{1}{2}a^{-4}b^1$

$= \dfrac{b}{2a^4}$

**84.** To write 94,000,000 in scientific notation, we need to move the decimal point to the left 7 places. Thus, we can write 94,000,000 in scientific notation as $9.4 \times 10^7$.

**86.** To write −567,000 in scientific notation, we need to move the decimal point to the left 5 places. Thus, we can write −567,000 in scientific notation as $-5.67 \times 10^5$.

**88.** To write 0.000123 in scientific notation, we need to move the decimal point to the right 4 places. Thus, we can write 0.000123 in scientific notation as $1.23 \times 10^{-4}$.

**90.** To write −0.000004 in scientific notation, we need to move the decimal point to the right 6 places. Thus, we can write −0.000004 in scientific notation as $-4 \times 10^{-6}$.

**92.** $(1.8 \times 10^{-7}) \cdot (3 \times 10^3) = (1.8 \cdot 3) \times (10^{-7} \cdot 10^3)$
$= 5.4 \times 10^{-4}$

**94.** $(6.2 \times 10^3) \cdot (-3.8 \times 10^5)$
$= (6.2)(-3.8) \times (10^3 \cdot 10^5)$
$= -23.56 \times 10^8$
$= (-2.356 \times 10^1) \times 10^8$
$= -2.356 \times 10^9$

**96.** $\dfrac{8.2 \times 10^{-5}}{4.1 \times 10^4} = \dfrac{8.2}{4.1} \times \dfrac{10^{-5}}{10^4}$
$= 2 \times 10^{-5-4}$
$= 2 \times 10^{-9}$

**98.** $(5 \times 10^8)^2 = 5^2 \times (10^8)^2$
$= 25 \times 10^{16}$
$= (2.5 \times 10^1) \times 10^{16}$
$= 2.5 \times 10^{17}$

**100.** $\dfrac{1 \times 10^7}{5 \times 10^{-4}} = \dfrac{1}{5} \times \dfrac{10^7}{10^{-4}}$
$= 0.2 \times 10^{7-(-4)}$
$= (2 \times 10^{-1}) \times 10^{11}$
$= 2 \times 10^{10}$

**102.** $\dfrac{(3 \times 10^9)(8 \times 10^{-6})}{4 \times 10^4} = \dfrac{(3 \cdot 8) \times (10^9 \cdot 10^{-6})}{4 \times 10^4}$
$= \dfrac{24 \times 10^3}{4 \times 10^4}$
$= \dfrac{24}{4} \times \dfrac{10^3}{10^4}$
$= 6 \times 10^{-1}$

## Getting Ready for Chapter 5

**104.**
$$\frac{(-1.5 \times 10^4)(3.2 \times 10^2)}{5 \times 10^5} = \frac{(-1.5)(3.2) \times (10^4 \cdot 10^2)}{5 \times 10^5}$$
$$= \frac{-4.8 \times 10^6}{5 \times 10^5}$$
$$= \frac{-4.8}{5} \times \frac{10^6}{10^5}$$
$$= -0.96 \times 10^1$$
$$= (-9.6 \times 10^{-1}) \times 10^1$$
$$= -9.6$$

**106.**
$$\frac{1.5 \times 10^{13}}{(3 \times 10^4)(5 \times 10^8)} = \frac{1.5 \times 10^{13}}{(3 \cdot 5) \times (10^4 \cdot 10^8)}$$
$$= \frac{1.5 \times 10^{13}}{15 \times 10^{12}}$$
$$= \frac{1.5 \times 10^{13}}{(1.5 \times 10^1) \times 10^{12}}$$
$$= \frac{1.5 \times 10^{13}}{1.5 \times 10^{13}}$$
$$= 1$$

**108.** $(15{,}000) \cdot (3{,}000{,}000) = (1.5 \times 10^4)(3 \times 10^6)$
$$= (1.5 \cdot 3) \times (10^4 \cdot 10^6)$$
$$= 4.5 \times 10^{10}$$
$$= 45{,}000{,}000{,}000$$

**110.** $(12{,}000{,}000)(80{,}000) = (1.2 \times 10^7)(8 \times 10^4)$
$$= (1.2 \cdot 8) \times (10^7 \cdot 10^4)$$
$$= 9.6 \times 10^{11}$$
$$= 960{,}000{,}000{,}000$$

**112.**
$$\frac{0.00012}{0.0000002} = \frac{1.2 \times 10^{-4}}{2 \times 10^{-7}}$$
$$= \frac{1.2}{2} \times 10^{-4-(-7)}$$
$$= 0.6 \times 10^3$$
$$= (6 \times 10^{-1}) \times 10^3$$
$$= 6 \times 10^2$$
$$= 600$$

**114.**
$$\frac{0.00000048}{0.00016} = \frac{4.8 \times 10^{-7}}{1.6 \times 10^{-4}}$$
$$= \frac{4.8}{1.6} \times 10^{-7-(-4)}$$
$$= 3 \times 10^{-3}$$
$$= 0.003$$

**116.**
$$\frac{-5{,}000{,}000 \cdot 0.0015}{25{,}000} = \frac{-5 \times 10^6 \cdot 1.5 \times 10^{-3}}{2.5 \times 10^4}$$
$$= \frac{(-5)(1.5) \times 10^{6+(-3)}}{2.5 \times 10^4}$$
$$= \frac{-7.5 \times 10^3}{2.5 \times 10^4}$$
$$= \frac{-7.5}{2.5} \times \frac{10^3}{10^4}$$
$$= -3 \times 10^{3-4}$$
$$= -3 \times 10^{-1}$$
$$= -0.3$$

**118.**
$$\frac{(0.0001)(3{,}500{,}000)}{(0.0005)(1{,}400{,}000)} = \frac{(1 \times 10^{-4})(3.5 \times 10^6)}{(5 \times 10^{-4})(1.4 \times 10^6)}$$
$$= \frac{(1 \cdot 3.5) \times (10^{-4} \cdot 10^6)}{(5 \cdot 1.4) \times (10^{-4} \cdot 10^6)}$$
$$= \frac{3.5 \times 10^2}{7 \times 10^2}$$
$$= \frac{3.5}{7} \times \frac{10^2}{10^2}$$
$$= 0.5 \times 1$$
$$= 0.5$$

**120.** $A = \pi r^2$
$$= \pi \left(\frac{d}{2}\right)^2$$
$$= \pi \frac{d^2}{2^2}$$
$$= \frac{\pi d^2}{4}$$

The area of a circle is $\dfrac{\pi d^2}{4}$ square units.

**122.** $0.0000000001 = 1 \times 10^{-10}$

The diameter of an atom is about $1 \times 10^{-10}$ meters.

**124.** $7,772,000,000,000 = 7.772 \times 10^{12}$

The national debt as of May 25, 2005 was $\$7.772 \times 10^{12}$.

**126.** $1 \times 10^{16} = 10,000,000,000,000,000$

One lightyear is 10,000,000,000,000,000 meters.

**128.** $1 \times 10^{-8} = 0.00000001$

The human hair grows at a rate of 0.00000001 miles per hour.

$0.00000001 \dfrac{\text{miles}}{\text{hr}} \cdot \dfrac{24 \text{ hr}}{1 \text{ d}} \cdot \dfrac{7 \text{ d}}{1 \text{ wk}} \cdot \dfrac{52 \text{ wk}}{1 \text{ yr}}$

$= (0.00000001)(24 \cdot 7 \cdot 52) \dfrac{\text{miles}}{\text{year}}$

$= 0.00008736 \dfrac{\text{miles}}{\text{year}}$

The human hair grows at a rate of 0.00008736 miles per year.

**130.** $\dfrac{9.9 \times 10^{11}}{2.8 \times 10^{8}} = \dfrac{9.9}{2.8} \times \dfrac{10^{11}}{10^{8}}$

$\approx 3.536 \times 10^{11-8}$

$= 3.536 \times 10^{3}$

$= 3,536$

The per capita tax bill in the United States in 2000 was approximately \$3,536.

**132.** $\dfrac{43.2 \times 10^{9}}{1.74 \times 10^{6}} = \dfrac{43.2}{1.74} \times \dfrac{10^{9}}{10^{6}}$

$\approx 24.828 \times 10^{9-6}$

$= 24.828 \times 10^{3}$

$= 24,828$

The salary per employee in 2000 for the arts and entertainment industry was \$24,828.

**134. a.** $\dfrac{3.52 \times 10^{9}}{2.91 \times 10^{8}} = \dfrac{3.52}{2.91} \times \dfrac{10^{9}}{10^{8}}$

$\approx 1.210 \times 10^{9-8}$

$= 1.210 \times 10^{1} = 12.10$

The per capita number of barrels of oil imported into the United States in 2003 was about 12.10 barrels.

**b.** $(3.52 \times 10^{9})(42) = 147.84 \times 10^{9}$

$= (1.4784 \times 10^{2}) \times 10^{9}$

$= 1.4784 \times 10^{11}$

$1.4784 \times 10^{11}$ gallons of oil were imported into the United States in 2003.

**c.** $\dfrac{1.4784 \times 10^{11}}{2.91 \times 10^{8}} = \dfrac{1.4784}{2.91} \times \dfrac{10^{11}}{10^{8}}$

$\approx 0.508 \times 10^{11-8}$

$= (5.08 \times 10^{-1}) \times 10^{3}$

$= 5.08 \times 10^{2}$

$= 508$

The per capita number of gallons of oil imported into the United States in 2003 was about 508 gallons.

**136.** $\dfrac{3^{2x}}{27} = \dfrac{3^{2x}}{3^{3}} = 3^{2x-3}$

**138.** $9^{-x} \cdot 3^{x+1} = (3^{2})^{-x} \cdot 3^{x+1}$

$= 3^{-2x} \cdot 3^{x+1}$

$= 3^{-2x+x+1}$

$= 3^{-x+1}$

**140.** $4^{x} = 6$

$(4^{x})^{5} = (6)^{5}$

$4^{5x} = 7776$

**142.** $5^{x} = 3$

$(5^{x})^{-3} = (3)^{-3}$

$5^{-3x} = \dfrac{1}{3^{3}}$

$5^{-3x} = \dfrac{1}{27}$

**144.** $x^{0} = 1$ provided $x \neq 0$.

Consider $\dfrac{x^{n}}{x^{n}}$. As long as $x \neq 0$, then $x^{n} \neq 0$ and we have $\dfrac{x^{n}}{x^{n}} = 1$. Using the quotient rule for exponents, we can write

$\dfrac{x^{n}}{x^{n}} = 1$

$x^{n-n} = 1$

$x^{0} = 1$

# Chapter 5

## 5.1 Preparing for Adding and Subtracting Polynomials

1. The coefficient is $-4$, the number multiplied in front of the variable expression.

2. $5x^2 - 3x + 1 - 2x^2 - 6x + 3$
   $= 5x^2 - 2x^2 - 3x - 6x + 1 + 3$
   $= (5-2)x^2 + (-3-6)x + (1+3)$
   $= 3x^2 - 9x + 4$

3. $-4(x-3) = -4 \cdot x - 4 \cdot (-3)$
   $= -4x + 12$

4. $f(x) = -4x + 3$
   $f(3) = -4(3) + 3$
   $= -12 + 3$
   $= -9$

## 5.1 Exercises

2. 0

4. False; a monomial in one variable is the product of a constant and a variable raised to a *nonnegative* integer power.

6. True

8. A linear polynomial is a polynomial of degree 1 or 0.

10. Answers will vary. Examples:
    Monomials: $3$, $4x$, $-7x^2$
    Binomials: $x+2$, $-4x^3 + 3x$
    Trinomials: $x^2 - 7x + 2$, $-2x^5 + 4x^2 - x$
    Linear Polynomial: $x$, $3x+5$, $-7x$.

12. Answers will vary. When adding polynomials we add like terms by adding the coefficients of variables with the same exponent. If the polynomials have the same degree, the sum will have that degree as well (assuming the leading terms do not cancel each other out). If one polynomial has a higher degree, there will be no like term to add to that leading term, so the leading term of the higher-degree polynomial will be the leading term of the sum.

14. Coefficient: 5
    Degree: 4

16. Coefficient: $-12$
    Degree: $1+1=2$

18. Coefficient: $-\dfrac{5}{3}$
    Degree: 5

20. Coefficient: $-7$
    Degree: 0

22. $6p^{-3} - p^{-2} + 3p^{-1}$ is not a polynomial because the exponents are negative.

24. $\dfrac{x^2 + 2}{x}$ is not a polynomial because there is a variable in the denominator.

26. $-3y^2 + 8y + 1$
    Yes; trinomial; degree 2; already in standard form.

28. $\dfrac{1}{x}$ is not a polynomial because there is a variable in the denominator.

30. $8m - 4m^{1/2}$ is not a polynomial because there is a fractional exponent.

32. $-12$
    Yes; monomial; degree 0; already in standard form.

34. $7 - 5p + 2p^2 - p^3$
    Yes; polynomial; degree 3; standard form is $-p^3 + 2p^2 - 5p + 7$

36. $4y^{-2} + 6y - 1$ is not a polynomial because of the negative exponent.

38. $4mn^3 - 2m^2n^3 + mn^8$
    Yes; trinomial; degree 9; standard form is $mn^8 - 2m^2n^3 + 4mn^3$

Section 5.1 Adding and Subtracting Polynomials

40. $-2xyz^2 + 7x^3z - 8y^{1/2}z$ is not a polynomial because there is a fractional exponent.

42. $10y^4 - 6y^4 = (10-6)y^4 = 4y^4$

44. $(x^2 - 4x + 1) + (5x^2 + 2x + 7)$
$= x^2 - 4x + 1 + 5x^2 + 2x + 7$
$= x^2 + 5x^2 - 4x + 2x + 1 + 7$
$= (1+5)x^2 + (-4+2)x + (1+7)$
$= 6x^2 - 2x + 8$

46. $(2w^3 - w^2 + 6w - 5) + (-3w^3 + 5w^2 + 9)$
$= 2w^3 - w^2 + 6w - 5 - 3w^3 + 5w^2 + 9$
$= 2w^3 - 3w^3 - w^2 + 5w^2 + 6w - 5 + 9$
$= (2-3)w^3 + (-1+5)w^2 + 6w + (-5+9)$
$= -w^3 + 4w^2 + 6w + 4$

48. $(7y^2 + 9y + 12) - (4y^2 + 8y - 3)$
$= 7y^2 + 9y + 12 - 4y^2 - 8y + 3$
$= 7y^2 - 4y^2 + 9y - 8y + 12 + 3$
$= (7-4)y^2 + (9-8)y + (12+3)$
$= 3y^2 + y + 15$

50. $(-2x^3y^3 + 7xy - 3) - (x^3y^3 + 5y^2 + xy - 3)$
$= -2x^3y^3 + 7xy - 3 - x^3y^3 - 5y^2 - xy + 3$
$= -2x^3y^3 - x^3y^3 + 7xy - xy - 5y^2 - 3 + 3$
$= (-2-1)x^3y^3 + (7-1)xy - 5y^2 + (-3+3)$
$= -3x^3y^3 + 6xy - 5y^2$

52. $(-3 - 5z + 3z^2) + (1 + 2z + z^2)$
$= -3 - 5z + 3z^2 + 1 + 2z + z^2$
$= 3z^2 + z^2 - 5z + 2z - 3 + 1$
$= (3+1)z^2 + (-5+2)z + (-3+1)$
$= 4z^2 - 3z - 2$

54. $(8 - t^3) - (1 + 3t + 3t^2 + t^3)$
$= 8 - t^3 - 1 - 3t - 3t^2 - t^3$
$= -t^3 - t^3 - 3t^2 - 3t + 8 - 1$
$= (-1-1)t^3 - 3t^2 - 3t + (8-1)$
$= -2t^3 - 3t^2 - 3t + 7$

56. $\left(\dfrac{3}{4}y^3 - \dfrac{1}{8}y + \dfrac{2}{3}\right) + \left(\dfrac{1}{2}y^3 + \dfrac{5}{12}y - \dfrac{5}{6}\right)$
$= \dfrac{3}{4}y^3 - \dfrac{1}{8}y + \dfrac{2}{3} + \dfrac{1}{2}y^3 + \dfrac{5}{12}y - \dfrac{5}{6}$
$= \dfrac{3}{4}y^3 + \dfrac{1}{2}y^3 - \dfrac{1}{8}y + \dfrac{5}{12}y + \dfrac{2}{3} - \dfrac{5}{6}$
$= \left(\dfrac{3}{4} + \dfrac{1}{2}\right)y^3 + \left(-\dfrac{1}{8} + \dfrac{5}{12}\right)y + \left(\dfrac{2}{3} - \dfrac{5}{6}\right)$
$= \dfrac{5}{4}y^3 + \dfrac{7}{24}y - \dfrac{1}{6}$

58. $(7a^3b + 9ab^2 - 4a^2b) + (-4a^3b + 3a^2b - 8ab^2)$
$= 7a^3b + 9ab^2 - 4a^2b - 4a^3b + 3a^2b - 8ab^2$
$= 7a^3b - 4a^3b - 4a^2b + 3a^2b + 9ab^2 - 8ab^2$
$= (7-4)a^3b + (-4+3)a^2b + (9-8)ab^2$
$= 3a^3b - a^2b + ab^2$

60. $(-5xy^2 + 3xy - 9y^2) - (5xy^2 + 7xy - 8y^2)$
$= -5xy^2 + 3xy - 9y^2 - 5xy^2 - 7xy + 8y^2$
$= -5xy^2 - 5xy^2 + 3xy - 7xy - 9y^2 + 8y^2$
$= (-5-5)xy^2 + (3-7)xy + (-9+8)y^2$
$= -10xy^2 - 4xy - y^2$

62. a. $f(x) = x^2 + 5x - 3$
$f(0) = (0)^2 + 5(0) - 3$
$= 0 + 0 - 3$
$= -3$

b. $f(x) = x^2 + 5x - 3$
$f(2) = (2)^2 + 5(2) - 3$
$= 4 + 10 - 3$
$= 11$

## Chapter 5: Polynomials and Polynomial Functions

c. $f(x) = x^2 + 5x - 3$
$f(-3) = (-3)^2 + 5(-3) - 3$
$= 9 - 15 - 3$
$= -9$

64. a. $f(x) = -2x^3 + 3x - 1$
$f(0) = -2(0)^3 + 3(0) - 1$
$= 0 + 0 - 1$
$= -1$

b. $f(x) = -2x^3 + 3x - 1$
$f(2) = -2(2)^3 + 3(2) - 1$
$= -2(8) + 3(2) - 1$
$= -16 + 6 - 1$
$= -11$

c. $f(x) = -2x^3 + 3x - 1$
$f(-3) = -2(-3)^3 + 3(-3) - 1$
$= -2(-27) + 3(-3) - 1$
$= 54 - 9 - 1$
$= 44$

66. a. $f(x) = -2x^3 + x^2 + 5x - 3$
$f(0) = -2(0)^3 + (0)^2 + 5(0) - 3$
$= 0 + 0 + 0 - 3$
$= -3$

b. $f(x) = -2x^3 + x^2 + 5x - 3$
$f(2) = -2(2)^3 + (2)^2 + 5(2) - 3$
$= -2(8) + 4 + 5(2) - 3$
$= -16 + 4 + 10 - 3$
$= -5$

c. $f(x) = -2x^3 + x^2 + 5x - 3$
$f(-3) = -2(-3)^3 + (-3)^2 + 5(-3) - 3$
$= -2(-27) + 9 + 5(-3) - 3$
$= 54 + 9 - 15 - 3$
$= 45$

68. a. $(f+g)(x) = (4x+3) + (2x-3)$
$= 4x + 3 + 2x - 3$
$= 4x + 2x + 3 - 3$
$= 6x$

b. $(f-g)(x) = (4x+3) - (2x-3)$
$= 4x + 3 - 2x + 3$
$= 4x - 2x + 3 + 3$
$= 2x + 6$

c. Since $(f+g)(x) = 6x$, we have
$(f+g)(2) = 6(2) = 12$

d. Since $(f-g)(x) = 2x+6$, we have
$(f-g)(1) = 2(1) + 6$
$= 2 + 6$
$= 8$

70. a. $(f+g)(x) = (3x^2 + x + 2) + (x^2 - 3x - 1)$
$= 3x^2 + x + 2 + x^2 - 3x - 1$
$= 3x^2 + x^2 + x - 3x + 2 - 1$
$= 4x^2 - 2x + 1$

b. $(f-g)(x) = (3x^2 + x + 2) - (x^2 - 3x - 1)$
$= 3x^2 + x + 2 - x^2 + 3x + 1$
$= 3x^2 - x^2 + x + 3x + 2 + 1$
$= 2x^2 + 4x + 3$

c. Since $(f+g)(x) = 4x^2 - 2x + 1$, we have
$(f+g)(2) = 4(2)^2 - 2(2) + 1$
$= 4(4) - 2(2) + 1$
$= 16 - 4 + 1$
$= 13$

d. Since $(f-g)(x) = 2x^2 + 4x + 3$, we have
$(f-g)(1) = 2(1)^2 + 4(1) + 3$
$= 2(1) + 4(1) + 3$
$= 2 + 4 + 3$
$= 9$

## Section 5.1 Adding and Subtracting Polynomials

72. a. $(f+g)(x) = (8x^3+1)+(x^3+3x^2+3x+1)$
$= 8x^3+1+x^3+3x^2+3x+1$
$= 8x^3+x^3+3x^2+3x+1+1$
$= 9x^3+3x^2+3x+2$

b. $(f-g)(x) = (8x^3+1)-(x^3+3x^2+3x+1)$
$= 8x^3+1-x^3-3x^2-3x-1$
$= 8x^3-x^3-3x^2-3x+1-1$
$= 7x^3-3x^2-3x$

c. Since $(f+g)(x) = 9x^3+3x^2+3x+2$, we have $(f+g)(2) = 9(2)^3+3(2)^2+3(2)+2$
$= 9(8)+3(4)+3(2)+2$
$= 72+12+6+2$
$= 92$

d. Since $(f-g)(x) = 7x^3-3x^2-3x$, we have
$(f-g)(1) = 7(1)^3-3(1)^2-3(1)$
$= 7-3-3$
$= 1$

74. $(2x^3-3x^2-5x+7)+(x^3+3x^2-6x-4)$
$= 2x^3-3x^2-5x+7+x^3+3x^2-6x-4$
$= 2x^3+x^3-3x^2+3x^2-5x-6x+7-4$
$= 3x^3-11x+3$

76. $(-5q^3+q^2+2q-1)-(2q^3-3q^2+7q-2)$
$= -5q^3+q^2+2q-1-2q^3+3q^2-7q+2$
$= -5q^3-2q^3+q^2+3q^2+2q-7q-1+2$
$= -7q^3+4q^2-5q+1$

78. a. If the vertex is $(3,4)$, then $x=3$.
$A(x) = -2x^2+10x$
$A(3) = -2(3)^2+10(3)$
$= -2(9)+30$
$= -18+30$
$= 12$
The area of the rectangle is 12 square units.

b. If the vertex is $(4,2)$, then $x=4$.
$A(x) = -2x^2+10x$
$A(4) = -2(4)^2+10(4)$
$= -2(16)+40$
$= -32+40$
$= 8$
The area of the rectangle is 8 square units.

80. a. $I(20) = -42.32(20)^2+4005.76(20)-53062.54$
$= -42.32(400)+80115.20-53062.54$
$= -16,928+27,052.66$
$= 10,124.66$
According to the model, the average income of a 20 year old in 2002 was $10,125.

b. $I(55) = -42.32(55)^2+4005.76(55)-53062.54$
$= -42.32(3025)+220,316.8-53062.54$
$= -128,018+167,254.26$
$= 39,236.26$
According to the model, the average income of a 55 year old in 2002 was $39,236.26.

82. a. Profit is the difference between revenue and costs. Thus,
$P(x) = R(x)-C(x)$
$= (-0.3x^2+30x)-(0.1x^2+7x+400)$
$= -0.3x^2+30x-0.1x^2-7x-400$
$= -0.3x^2-0.1x^2+30x-7x-400$
$= -0.4x^2+23x-400$

b. $P(15) = -0.4(15)^2+23(15)-400$
$= -0.4(225)+23(15)-400$
$= -90+345-400$
$= -145$
A negative profit indicates a loss. Therefore, if 15 clocks were made and sold, there would be a $145 loss.

c. $P(40) = -0.4(40)^2+23(40)-400$
$= -0.4(1600)+920-400$
$= -640+520$
$= -120$
A negative profit indicates a loss. Therefore,

if 40 clocks were made and sold, there would be a $120 loss.

84. $f(x) = -5x+1$; $g(x) = ax-3$

$$(f+g)(x) = (-5x+1)-(ax-3)$$
$$= -5x+1-ax+3$$
$$= (-5-a)x+(1+3)$$
$$= (-5-a)x+4$$

Since $(f-g)(x) = (-5-a)x+4$, we have
$$(-5-a)(2)+4 = 10$$
$$(-5-a)(2) = 6$$
$$-5-a = 3$$
$$-a = 8$$
$$a = -8$$

86. The total number of Americans with work-related disabilities is found by adding the number of males with disabilities to the number of females with disabilities.
$T(a) = M(a) + F(a)$

88. a.-c. $f(x) = -2x^2 + 5x + 3$

$f(4) = -9$
$f(-2) = -15$
$f(6) = -39$

$f(4) = 37$
$f(-2) = 31$
$f(6) = 103$

90. a.-c. $f(x) = -3x^3 + 5x^2 - 8x + 1$

$f(4) = -143$
$f(-2) = 61$
$f(6) = -515$

**5.2 Preparing for Multiplying Polynomials**

1. $4x^2 \cdot 3x^3 = 4 \cdot 3 \cdot x^2 \cdot x^3$
$$= 12x^{2+3}$$
$$= 12x^5$$

2. $(-3x)^2 = (-3)^2 \cdot x^2 = 9x^2$

3. $4(x-5) = 4 \cdot x - 4 \cdot 5$
$$= 4x - 20$$

**5.2 Exercises**

2. $A^2 - 2AB + B^2$

4. False; $(x+a)^2 = x^2 + 2ax + a^2$

6. False; consider $(x-2)(x+2) = x^2 - 4$

8. Answers will vary.

10. $(9a^3b^2)(-3a^2b^5) = (9 \cdot -3)(a^3 \cdot a^2)(b^2 \cdot b^5)$
$$= -27a^{3+2}b^{2+5}$$
$$= -27a^5b^7$$

12. $\left(\frac{12}{5}x^2y\right)\left(\frac{15}{4}x^4y^3\right) = \left(\frac{12}{5} \cdot \frac{15}{4}\right)(x^2 \cdot x^4)(y \cdot y^3)$
$$= 9x^{2+4}y^{1+3}$$
$$= 9x^6y^4$$

14. $6y(y^2 - 4y + 3) = 6y \cdot y^2 - 6y \cdot 4y + 6y \cdot 3$
$$= 6y^3 - 24y^2 + 18y$$

**Section 5.2** *Multiplying Polynomials*

16. $-3mn^3\left(4m^2 - mn + 5n^2\right)$
    $= -3mn^3 \cdot 4m^2 + 3mn^3 \cdot mn - 3mn^3 \cdot 5n^2$
    $= -12m^3n^3 + 3m^2n^4 - 15mn^5$

18. $\dfrac{5}{2}xy\left(\dfrac{4}{15}x^2y - \dfrac{6}{5}xy + \dfrac{3}{10}xy^2\right)$
    $= \dfrac{5}{2}xy \cdot \dfrac{4}{15}x^2y - \dfrac{5}{2}xy \cdot \dfrac{6}{5}xy + \dfrac{5}{2}xy \cdot \dfrac{3}{10}xy^2$
    $= \dfrac{2}{3}x^3y^2 - 3x^2y^2 + \dfrac{3}{4}x^2y^3$

20. $0.8y\left(0.4y^2 + 1.1y - 2.5\right)$
    $= 0.8y \cdot 0.4y^2 + 0.8y \cdot 1.1y - 0.8y \cdot 2.5$
    $= 0.32y^3 + 0.88y^2 - 2y$

22. $(y-2)(y-6) = y \cdot y - y \cdot 6 - 2 \cdot y + 2 \cdot 6$
    $= y^2 - 6y - 2y + 12$
    $= y^2 - 8y + 12$

24. $(z-8)(z+3) = z(z+3) - 8(z+3)$
    $= z \cdot z + z \cdot 3 - 8 \cdot z - 8 \cdot 3$
    $= z^2 + 3z - 8z - 24$
    $= z^2 - 5z - 24$

26. $(5x-3)(x+4) = 5x(x+4) - 3(x+4)$
    $= 5x \cdot x + 5x \cdot 4 - 3 \cdot x - 3 \cdot 4$
    $= 5x^2 + 20x - 3x - 12$
    $= 5x^2 + 17x - 12$

28. $(-x+4)(6x+1) = -x \cdot 6x - x \cdot 1 + 4 \cdot 6x + 4 \cdot 1$
    $= -6x^2 - x + 24x + 4$
    $= -6x^2 + 23x + 4$

30. $(2-7y)(5+2y) = 2 \cdot 5 + 2 \cdot 2y - 7y \cdot 5 - 7y \cdot 2y$
    $= 10 + 4y - 35y - 14y^2$
    $= -14y^2 - 31y + 10$

32. $\left(\dfrac{3}{2}y + 4\right)\left(\dfrac{4}{3}y - 1\right) = \dfrac{3}{2}y\left(\dfrac{4}{3}y - 1\right) + 4\left(\dfrac{4}{3}y - 1\right)$
    $= 2y^2 - \dfrac{3}{2}y + \dfrac{16}{3}y - 4$
    $= 2y^2 + \dfrac{23}{6}y - 4$

34. $(3m - 5n)(m + 2n)$
    $= 3m \cdot m + 3m \cdot 2n - 5n \cdot m - 5n \cdot 2n$
    $= 3m^2 + 6mn - 5mn - 10n^2$
    $= 3m^2 + mn - 10n^2$

36. $(y-2)(y^2 + 5y - 3)$
    $= y(y^2 + 5y - 3) - 2(y^2 + 5y - 3)$
    $= y^3 + 5y^2 - 3y - 2y^2 - 10y + 6$
    $= y^3 + 3y^2 - 13y + 6$

38. $(2b+3)(3b^2 - 2b + 1)$
    $= 2b(3b^2 - 2b + 1) + 3(3b^2 - 2b + 1)$
    $= 6b^3 - 4b^2 + 2b + 9b^2 - 6b + 3$
    $= 6b^3 + 5b^2 - 4b + 3$

40. $(3p^2 - 5p + 3)(7p - 2)$
    $= (7p - 2)(3p^2 - 5p + 3)$
    $= 7p(3p^2 - 5p + 3) - 2(3p^2 - 5p + 3)$
    $= 21p^3 - 35p^2 + 21p - 6p^2 + 10p - 6$
    $= 21p^3 - 41p^2 + 31p - 6$

42. $(w+2)(w^3 - 6w^2 - 3w + 2)$
    $= w(w^3 - 6w^2 - 3w + 2) + 2(w^3 - 6w^2 - 3w + 2)$
    $= w^4 - 6w^3 - 3w^2 + 2w + 2w^3 - 12w^2 - 6w + 4$
    $= w^4 - 4w^3 - 15w^2 - 4w + 4$

**44.** $(3+2z)(z^2+5-3z)$
$= 3(z^2+5-3z)+2z(z^2+5-3z)$
$= 3z^2+15-9z+2z^3+10z-6z^2$
$= 2z^3-3z^2+z+15$

**46.** $(a^2+4a+4)(3a^2-a-2)$
$= a^2(3a^2-a-2)+4a(3a^2-a-2)$
$\quad +4(3a^2-a-2)$
$= 3a^4-a^3-2a^2+12a^3-4a^2-8a$
$\quad +12a^2-4a-8$
$= 3a^4+11a^3+6a^2-12a-8$

**48.** $(a+2b)(2a^2-5ab+3b^2)$
$= a(2a^2-5ab+3b^2)+2b(2a^2-5ab+3b^2)$
$= 2a^3-5a^2b+3ab^2+4a^2b-10ab^2+6b^3$
$= 2a^3-a^2b-7ab^2+6b^3$

**50.** $(xy-2)(x^2+2xy+4y^2)$
$= xy(x^2+2xy+4y^2)-2(x^2+2xy+4y^2)$
$= x^3y+2x^2y^2+4xy^3-2x^2-4xy-8y^2$
$= x^3y+2x^2y^2-2x^2+4xy^3-4xy-8y^2$

**52.** $(y+9)(y-9) = y^2-9^2$
$= y^2-81$

**54.** $(b+3)^2 = b^2+2(b)(3)+3^2$
$= b^2+6b+9$

**56.** $(4z-5)^2 = (4z)^2-2(4z)(5)+5^2$
$= 16z^2-40z+25$

**58.** $(8y+3z)(8y-3z) = (8y)^2-(3z)^2$
$= 64y^2-9z^2$

**60.** $(4a+7b)^2 = (4a)^2+2(4a)(7b)+(7b)^2$
$= 16a^2+56ab+49b^2$

**62.** $(7p-3q)^2 = (7p)^2-2(7p)(3q)+(3q)^2$
$= 49p^2-42pq+9q^2$

**64.** $(m^2-2n^3)(m^2+2n^3) = (m^2)^2-(2n^3)^2$
$= m^4-4n^6$

**66.** $[5-(a+b)][5+(a+b)]$
$= 5^2-(a+b)^2$
$= 25-(a^2+2ab+b^2)$
$= 25-a^2-2ab-b^2$

**68.** $[(m+4)-n]^2$
$= (m+4)^2-2(m+4)(n)+n^2$
$= m^2+2(m)(4)+4^2-2mn-8n+n^2$
$= m^2+8m-2mn-8n+n^2+16$

**70. a.** $(f \cdot g)(x) = f(x) \cdot g(x)$
$= (x+5)(2x+1)$
$= x(2x+1)+5(2x+1)$
$= 2x^2+x+10x+5$
$= 2x^2+11x+5$

**b.** Since $(f \cdot g)(x) = 2x^2+11x+5$, we have
$(f \cdot g)(3) = 2(3)^2+11(3)+5$
$= 2(9)+11(3)+5$
$= 18+33+5$
$= 56$

**72. a.** $(f \cdot g)(x) = f(x) \cdot g(x)$
$= (5x-1)(4x+5)$
$= 5x(4x+5)-1(4x+5)$
$= 20x^2+25x-4x-5$
$= 20x^2+21x-5$

**b.** Since $(f \cdot g)(x) = 20x^2+21x-5$, we have
$(f \cdot g)(3) = 20(3)^2+21(3)-5$
$= 20(9)+21(3)-5$
$= 180+63-5$
$= 238$

## Section 5.2 Multiplying Polynomials

**74. a.** $(f \cdot g)(x) = f(x) \cdot g(x)$
$= (x+5)(x^2 - 2x + 3)$
$= x(x^2 - 2x + 3) + 5(x^2 - 2x + 3)$
$= x^3 - 2x^2 + 3x + 5x^2 - 10x + 15$
$= x^3 + 3x^2 - 7x + 15$

**b.** Since $(f \cdot g)(x) = x^3 + 3x^2 - 7x + 15$, we have
$(f \cdot g)(3) = (3)^3 + 3(3)^2 - 7(3) + 15$
$= 27 + 3(9) - 7(3) + 15$
$= 27 + 27 - 21 + 15$
$= 48$

**76. a.** $f(x) = x^2 - 4$
$f(x+2) = (x+2)^2 - 4$
$= x^2 + 4x + 4 - 4$
$= x^2 + 4x$

**b.** $f(x+h) - f(x)$
$= \left[(x+h)^2 - 4\right] - \left[x^2 - 4\right]$
$= \left[x^2 + 2xh + h^2 - 4\right] - \left[x^2 - 4\right]$
$= x^2 + 2xh + h^2 - 4 - x^2 + 4$
$= 2xh + h^2$

**78. a.** $f(x) = x^2 - 2x + 3$
$f(x+2) = (x+2)^2 - 2(x+2) + 3$
$= x^2 + 4x + 4 - 2x - 4 + 3$
$= x^2 + 2x + 3$

**b.** $f(x+h) - f(x)$
$= \left[(x+h)^2 - 2(x+h) + 3\right] - \left[x^2 - 2x + 3\right]$
$= \left[x^2 + 2xh + h^2 - 2x - 2h + 3\right] - \left[x^2 - 2x + 3\right]$
$= x^2 + 2xh + h^2 - 2x - 2h + 3 - x^2 + 2x - 3$
$= 2xh - 2h + h^2$

**80. a.** $f(x) = -2x^2 + x - 5$
$f(x+2) = -2(x+2)^2 + (x+2) - 5$
$= -2(x^2 + 4x + 4) + (x+2) - 5$
$= -2x^2 - 8x - 8 + x + 2 - 5$
$= -2x^2 - 7x - 11$

**b.** $f(x+h) - f(x)$
$= \left[-2(x+h)^2 + (x+h) - 5\right] - \left[-2x^2 + x - 5\right]$
$= \left[-2(x^2 + 2xh + h^2) + x + h - 5\right] - \left[-2x^2 + x - 5\right]$
$= \left[-2x^2 - 4xh - 2h^2 + x + h - 5\right] - \left[-2x^2 + x - 5\right]$
$= -2x^2 - 4xh - 2h^2 + x + h - 5 + 2x^2 - x + 5$
$= -4xh + h - 2h^2$

**82.** $-3x(x-3)^2 = -3x(x^2 - 2 \cdot x \cdot 3 + 3^2)$
$= -3x(x^2 - 6x + 9)$
$= -3x^3 + 18x^2 - 27x$

**84.** $(9b+2)^2 = (9b)^2 + 2(9b)(2) + 2^2$
$= 81b^2 + 36b + 4$

**86.** $(4x+3)(3x-7) = 4x(3x-7) + 3(3x-7)$
$= 12x^2 - 28x + 9x - 21$
$= 12x^2 - 19x - 21$

**88.** $(6p+q)(5p-2q) - (p+3q)^2$
$= 6p(5p-2q) + q(5p-2q) - \left(p^2 + 2(p)(3q) + (3q)^2\right)$
$= 30p^2 - 12pq + 5pq - 2q^2 - \left(p^2 + 6pq + 9q^2\right)$
$= 30p^2 - 7pq - 2q^2 - p^2 - 6pq - 9q^2$
$= 29p^2 - 13pq - 11q^2$

**90.** $(2y+3)(4y^2 - 6y + 9)$
$= 2y(4y^2 - 6y + 9) + 3(4y^2 - 6y + 9)$
$= 8y^3 - 12y^2 + 18y + 12y^2 - 18y + 27$
$= 8y^3 + 27$

**92.** $\left(3x + \dfrac{1}{3}\right)^2 = (3x)^2 + 2(3x)\left(\dfrac{1}{3}\right) + \left(\dfrac{1}{3}\right)^2$
$= 9x^2 + 2x + \dfrac{1}{9}$

## Chapter 5: Polynomials and Polynomial Functions

94. $(z-3)^3 = (z-3)(z-3)^2$
$= (z-3)\left(z^2 - 2(z)(3) + 3^2\right)$
$= (z-3)\left(z^2 - 6z + 9\right)$
$= z \cdot z^2 - z \cdot 6z + z \cdot 9 - 3 \cdot z^2 + 3 \cdot 6z - 3 \cdot 9$
$= z^3 - 6z^2 + 9z - 3z^2 + 18z - 27$
$= z^3 - 9z^2 + 27z - 27$

96. $(2a+b-5)(4a-2b+1)$
$= 2a(4a-2b+1) + b(4a-2b+1)$
$\quad -5(4a-2b+1)$
$= 8a^2 - 4ab + 2a + 4ab - 2b^2 + b - 20a + 10b - 5$
$= 8a^2 - 18a + 11b - 2b^2 - 5$

98. $(3z+2)(z-2) + (z+2)(z-2)$
$= 3z(z-2) + 2(z-2) + z^2 - 2^2$
$= 3z^2 - 6z + 2z - 4 + z^2 - 4$
$= 4z^2 - 4z - 8$

100. $(a+2)(a-2)(a^2-4) - (a+3)(a^2-3)$
$= (a^2 - 2^2)(a^2 - 4) - \left[a(a^2-3) + 3(a^2-3)\right]$
$= (a^2-4)(a^2-4) - \left[a^3 - 3a + 3a^2 - 9\right]$
$= (a^2)^2 - 2(a^2)(4) + 4^2 - a^3 + 3a - 3a^2 + 9$
$= a^4 - 8a^2 + 16 - a^3 + 3a - 3a^2 + 9$
$= a^4 - a^3 - 11a^2 + 3a + 25$

102.

|  | $2x$ | $5$ |
|---|---|---|
| $x$ | $A_1$ | $A_2$ |
| $2$ | $A_3$ | $A_4$ |

$A = A_1 + A_2 + A_3 + A_4$
$= x(2x) + x(5) + 2(2x) + 2(5)$
$= 2x^2 + 5x + 4x + 10$
$= 2x^2 + 9x + 10$

An alternate method is to note that the length is $(2x+5)$ and the width is $(x+2)$. Since the area of a rectangle is the product of the length and width, we would get:
$A = (2x+5)(x+2)$
$= 2x^2 + 4x + 5x + 10$
$= 2x^2 + 9x + 10$
Notice how the FOIL method gives each of the individual areas.

104.

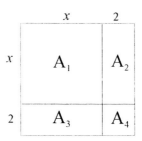

$A = A_1 + A_2 + A_3 + A_4$
$= x(x) + 2(x) + 2(x) + 2(2)$
$= x^2 + 2x + 2x + 4$
$= x^2 + 4x + 4$

An alternate method is to note that the length is $(x+2)$ and the width is $(x+2)$. Since the area of a rectangle is the product of the length and width, we would get:
$A = (x+2)(x+2)$
$= x^2 + 2x + 2x + 2^2$
$= x^2 + 4x + 4$
Notice how the FOIL method gives each of the individual areas.

106. $A = (2x+1)(3x+4) - (x-1)(x+2)$
$= (6x^2 + 8x + 3x + 4) - (x^2 + 2x - x - 2)$
$= (6x^2 + 11x + 4) - (x^2 + x - 2)$
$= 6x^2 + 11x + 4 - x^2 - x + 2$
$= 5x^2 + 10x + 6$
The area of the shaded region is $5x^2 + 10 + 6$ square units.

108. To find the area of the shaded region, consider a larger rectangle with its corners removed.

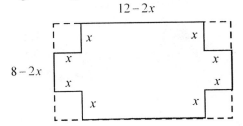

The area of the shaded region is the area of the entire rectangle, minus the area of the four corners. Notice that the length of the rectangle is $12 - 2x + 2x = 12$ units and that the width is $8 - 2x + 2x = 8$ units. Therefore,
$$A = 8(12) - 4(x \cdot x)$$
$$= 96 - 4x^2 \quad \text{or} \quad -4x^2 + 96$$
The area of the shaded region is $-4x^2 + 96$ square units.

110. $(3^x - 1)(3^x - 9) = 3^x \cdot 3^x - 3^x \cdot 9 - 1 \cdot 3^x + 1 \cdot 9$
$$= 3^{x+x} - 9 \cdot 3^x - 3^x + 9$$
$$= 3^{2x} - 10 \cdot 3^x + 9$$

**5.3 Preparing for Dividing Polynomials and Synthetic Division**

1. $\dfrac{15x^5}{12x^3} = \dfrac{15}{12} \cdot \dfrac{x^5}{x^3} = \dfrac{5}{4} x^{5-3} = \dfrac{5}{4} x^2$

2. $\dfrac{2}{7} + \dfrac{5}{7} = \dfrac{2+5}{7} = \dfrac{7}{7} = 1$

**5.3 Exercises**

2. dividend; divisor; remainder

4. False; to use synthetic division, the divisor must be linear and in the form $x - c$.

6. True; this is stated in the Remainder Theorem.

8. Answers will vary.

10. Answers will vary.

12. $\dfrac{6z^3 + 9z^2}{3z^2} = \dfrac{6z^3}{3z^2} + \dfrac{9z^2}{3z^2}$
$$= \dfrac{6}{3} z^{3-2} + \dfrac{9}{3} z^{2-2}$$
$$= 2z + 3$$

14. $\dfrac{4b^3 + 12b^2 + 24b}{6b} = \dfrac{4b^3}{6b} + \dfrac{12b^2}{6b} + \dfrac{24b}{6b}$
$$= \dfrac{4}{6} b^{3-1} + \dfrac{12}{6} b^{2-1} + \dfrac{24}{6} b^{1-1}$$
$$= \dfrac{2}{3} b^2 + 2b + 4$$

16. $\dfrac{3z^4 + 12z^2}{6z^3} = \dfrac{3z^4}{6z^3} + \dfrac{12z^2}{6z^3}$
$$= \dfrac{3}{6} z^{4-3} + \dfrac{12}{6} z^{2-3}$$
$$= \dfrac{1}{2} z + 2z^{-1}$$
$$= \dfrac{z}{2} + \dfrac{2}{z}$$

18. $\dfrac{2x^2y^3 - 9xy^3 + 16x^2y}{2x^2y^2}$
$$= \dfrac{2x^2y^3}{2x^2y^2} - \dfrac{9xy^3}{2x^2y^2} + \dfrac{16x^2y}{2x^2y^2}$$
$$= \dfrac{2}{2} x^{2-2} y^{3-2} - \dfrac{9}{2} x^{1-2} y^{3-2} + \dfrac{16}{2} x^{2-2} y^{1-2}$$
$$= y - \dfrac{9}{2} x^{-1} y + 8 y^{-1}$$
$$= y - \dfrac{9y}{2x} + \dfrac{8}{y}$$

20. 
$$\begin{array}{r} x - 7 \\ x+3 \overline{\smash{\big)}\, x^2 - 4x - 21} \\ \underline{-(x^2 + 3x)\phantom{-21}} \\ -7x - 21 \\ \underline{-(-7x - 21)} \\ 0 \end{array}$$
$$\dfrac{x^2 - 4x - 21}{x + 3} = x - 7$$

## Chapter 5: Polynomials and Polynomial Functions

22. 
$$\begin{array}{r} 3z+5 \\ z-4 \overline{\smash{)}3z^2-7z-28} \\ -\underline{(3z^2-12z)} \\ 5z-28 \\ -\underline{(5z-20)} \\ -8 \end{array}$$

$$\frac{3z^2-7z-28}{z-4} = 3z+5 - \frac{8}{z-4}$$

24. 
$$\begin{array}{r} x-6 \\ 4x+7 \overline{\smash{)}4x^2-17x-33} \\ -\underline{(4x^2+7x)} \\ -24x-33 \\ -\underline{(-24x-42)} \\ 9 \end{array}$$

$$\frac{4x^2-17x-33}{4x+7} = x-6 + \frac{9}{4x+7}$$

26. 
$$\begin{array}{r} x^2-x-20 \\ x+2 \overline{\smash{)}x^3+x^2-22x-40} \\ -\underline{(x^3+2x^2)} \\ -x^2-22x \\ -\underline{(-x^2-2x)} \\ -20x-40 \\ -\underline{(-20x-40)} \\ 0 \end{array}$$

$$\frac{x^3+x^2-22x-40}{x+2} = x^2-x-20$$

28. 
$$\begin{array}{r} a^2-8a+15 \\ a+8 \overline{\smash{)}a^3+0a^2-49a+120} \\ -\underline{(a^3+8a^2)} \\ -8a^2-49a \\ -\underline{(-8a^2-64a)} \\ 15a+120 \\ -\underline{(15a+120)} \\ 0 \end{array}$$

$$\frac{a^3-49a+120}{a+8} = a^2-8a+15$$

30. 
$$\begin{array}{r} 2p^2+5p-3 \\ 2p+5 \overline{\smash{)}4p^3+20p^2+19p-15} \\ -\underline{(4p^3+10p^2)} \\ 10p^2+19p \\ -\underline{(10p^2+25p)} \\ -6p-15 \\ -\underline{(-6p-15)} \\ 0 \end{array}$$

$$\frac{4p^3+20p^2+19p-15}{2p+5} = 2p^2+5p-3$$

32. 
$$\begin{array}{r} x-5 \\ x^2-2 \overline{\smash{)}x^3-5x^2-2x+10} \\ -\underline{(x^3\phantom{-5x^2}-2x)} \\ -5x^2\phantom{-2x}+10 \\ -\underline{(-5x^2\phantom{-2x}+10)} \\ 0 \end{array}$$

$$\frac{x^3-5x^2-2x+10}{x^2-2} = x-5$$

## Section 5.3 Dividing Polynomials: Synthetic Division

**34.**
$$2k^2 - 3 \overline{\smash{\big)}\, 2k^3 + 10k^2 - 6k - 8} \quad \text{quotient: } k+5$$

$$\begin{array}{r}
k+5 \\
2k^2-3 \overline{)2k^3+10k^2-6k-8} \\
-(2k^3 \qquad -3k) \\
\hline
10k^2 - 3k - 8 \\
-(10k^2 \qquad -15) \\
\hline
-3k + 7
\end{array}$$

$$\frac{2k^3+10k^2-6k-8}{2k^2-3} = k+5 + \frac{-3k+7}{2k^2-3}$$

**36.**
$$\begin{array}{r}
2x^2 - 9x - 35 \\
x^2+x-3 \overline{)2x^4-7x^3-50x^2-10x+96} \\
-(2x^4+2x^3-6x^2) \\
\hline
-9x^3 - 44x^2 - 10x \\
-(-9x^3-9x^2+27x) \\
\hline
-35x^2 - 37x + 96 \\
-(-35x^2-35x+105) \\
\hline
-2x - 9
\end{array}$$

$$\frac{2x^4-7x^3-50x^2-10x+96}{x^2+x-3} = 2x^2-9x-35 + \frac{-2x-9}{x^2+x-3}$$

**38.** The divisor is $x-2$ so $c=2$.

$$\begin{array}{r|rrr}
2 & 1 & 4 & -12 \\
  &   & 2 & 12 \\
\hline
  & 1 & 6 & 0
\end{array}$$

$$\frac{x^2+4x-12}{x-2} = x+6$$

**40.** The divisor is $x+8$ so $c=-8$.

$$\begin{array}{r|rrr}
-8 & 3 & 19 & -40 \\
   &   & -24 & 40 \\
\hline
   & 3 & -5 & 0
\end{array}$$

$$\frac{3x^2+19x-40}{x+8} = 3x-5$$

**42.** The divisor is $x-4$ so $c=4$

$$\begin{array}{r|rrr}
4 & 1 & 2 & -17 \\
  &   & 4 & 24 \\
\hline
  & 1 & 6 & 7
\end{array}$$

$$\frac{x^2+2x-17}{x-4} = x+6 + \frac{7}{x-4}$$

**44.** The divisor is $x+3$ so $c=-3$.

$$\begin{array}{r|rrrr}
-3 & 1 & 0 & -13 & -17 \\
   &   & -3 & 9 & 12 \\
\hline
   & 1 & -3 & -4 & -5
\end{array}$$

$$\frac{x^3-13x-17}{x+3} = x^2-3x-4 - \frac{5}{x+3}$$

**46.** The divisor is $x+4$ so $c=-4$.

$$\begin{array}{r|rrrrr}
-4 & 2 & -1 & -38 & 16 & 103 \\
   &   & -8 & 36 & 8 & -96 \\
\hline
   & 2 & -9 & -2 & 24 & 7
\end{array}$$

$$\frac{2x^4-x^3-38x^2+16x+103}{x+4} = 2x^3-9x^2-2x+24 + \frac{7}{x+4}$$

**48.** The divisor is $a-8$ so $c=8$.

$$\begin{array}{r|rrrrr}
8 & 1 & 0 & -65 & 0 & 55 \\
  &   & 8 & 64 & -8 & -64 \\
\hline
  & 1 & 8 & -1 & -8 & -9
\end{array}$$

$$\frac{a^4-65a^2+55}{a-8} = a^3+8a^2-a-8 - \frac{9}{a-8}$$

**50.** The divisor is $x-\frac{2}{3}$ so $c=\frac{2}{3}$.

$$\begin{array}{r|rrrr}
\frac{2}{3} & 3 & 13 & 8 & -12 \\
            &   & 2 & 10 & 12 \\
\hline
            & 3 & 15 & 18 & 0
\end{array}$$

$$\frac{3x^3+13x^2+8x-12}{x-\frac{2}{3}} = 3x^2+15x+18$$

**Chapter 5:** *Polynomials and Polynomial Functions*

**52. a.** $\left(\dfrac{f}{g}\right)(x) = \dfrac{3x^3 - 9x^2 + 12x}{3x}$

$= \dfrac{3x^3}{3x} - \dfrac{9x^2}{3x} + \dfrac{12x}{3x}$

$= \dfrac{3}{3}x^{3-1} - \dfrac{9}{3}x^{2-1} + \dfrac{12}{3}x^{1-1}$

$= x^2 - 3x + 4$

**b.** Using the result from part **a.**, we have:

$\left(\dfrac{f}{g}\right)(2) = (2)^2 - 3(2) + 4$

$= 4 - 6 + 4$

$= 2$

**54. a.** $\left(\dfrac{f}{g}\right)(x) = \dfrac{x^2 + 3x - 4}{x + 4}$

The divisor is $x + 4$ so $c = -4$.

$\begin{array}{r|rrr} -4) & 1 & 3 & -4 \\ & & -4 & 4 \\ \hline & 1 & -1 & 0 \end{array}$

$\left(\dfrac{f}{g}\right)(x) = x - 1$

**b.** $\left(\dfrac{f}{g}\right)(2) = (2) - 1 = 1$

**56. a.** $\left(\dfrac{f}{g}\right)(x) = \dfrac{3x^2 - 6x + 5}{2x + 1}$

$\phantom{2x+1)}\dfrac{3}{2}x - \dfrac{15}{4}$
$2x+1\overline{)3x^2 - 6x + 5}$
$\phantom{2x+1)}-\left(3x^2 + \dfrac{3}{2}x\right)$
$\phantom{2x+1)3x^2}-\dfrac{15}{2}x + 5$
$\phantom{2x+1)3x^2}-\left(-\dfrac{15}{2}x - \dfrac{15}{4}\right)$
$\phantom{2x+1)3x^2-\dfrac{15}{2}x+5}\dfrac{35}{4}$

$\left(\dfrac{f}{g}\right)(x) = \dfrac{3}{2}x - \dfrac{15}{4} + \dfrac{35/4}{2x+1}$

**b.** $\left(\dfrac{f}{g}\right)(2) = \dfrac{3}{2}(2) - \dfrac{15}{4} + \dfrac{35/4}{2(2)+1}$

$= 3 - \dfrac{15}{4} + \dfrac{35/4}{5} = -\dfrac{3}{4} + \dfrac{7}{4}$

$= 1$

**58.** $\left(\dfrac{f}{g}\right)(x) = \dfrac{3x^3 - 2x^2 - 19x - 6}{3x + 1}$

$\phantom{3x+1)}x^2 - x - 6$
$3x+1\overline{)3x^3 - 2x^2 - 19x - 6}$
$\phantom{3x+1)}-(3x^3 + x^2)$
$\phantom{3x+1)3x^3}-3x^2 - 19x$
$\phantom{3x+1)3x^3}-(-3x^2 - x)$
$\phantom{3x+1)3x^3-3x^2}-18x - 6$
$\phantom{3x+1)3x^3-3x^2}-(-18x - 6)$
$\phantom{3x+1)3x^3-3x^2-18x-6}0$

$\left(\dfrac{f}{g}\right)(x) = x^2 - x - 6$

**b.** $\left(\dfrac{f}{g}\right)(2) = (2)^2 - (2) - 6$

$= 4 - 2 - 6$

$= -4$

**60. a.** $\left(\dfrac{f}{g}\right)(x) = \dfrac{x^3 - 19x + 30}{x^2 - x - 6}$

$\phantom{x^2-x-6)}x + 1$
$x^2 - x - 6\overline{)x^3 + 0x^2 - 19x + 30}$
$\phantom{x^2-x-6)}-(x^3 - x^2 - 6x)$
$\phantom{x^2-x-6)x^3}x^2 - 13x + 30$
$\phantom{x^2-x-6)x^3}-(x^2 - x - 6)$
$\phantom{x^2-x-6)x^3-x^2+x}-12x + 36$

$\left(\dfrac{f}{g}\right)(x) = x + 1 + \dfrac{-12x + 36}{x^2 - x - 6}$

Section 5.3 Dividing Polynomials: Synthetic Division

b. $\left(\dfrac{f}{g}\right)(2) = (2) + 1 + \dfrac{-12(2)+36}{(2)^2-(2)-6}$

$= 2 + 1 + \dfrac{-24+36}{4-2-6}$

$= 3 + \dfrac{12}{-4}$

$= 3 - 3$

$= 0$

62. The divisor is $x+2$ so $c = -2$.
$f(-2) = (-2)^2 + 4(-2) - 5$
$= 4 - 8 - 5$
$= -9$
The remainder is $-9$.

64. The divisor is $x-3$ so $c = 3$.
$f(3) = (3)^3 + 3(3)^2 - (3) + 1$
$= 27 + 3(9) - 3 + 1$
$= 27 + 27 - 3 + 1$
$= 52$
The remainder is 52.

66. The divisor is $x+3$ so $c = -3$.
$f(-3) = 3(-3)^3 + 2(-3)^2 - 5$
$= 3(-27) + 2(9) - 5$
$= -81 + 18 - 5$
$= -68$
The remainder is $-68$.

68. The divisor is $x-1$ so $c = 1$.
$f(1) = (1)^4 - 1 = 1 - 1 = 0$
The remainder is 0.

70. $c = 3$
$f(3) = (3)^2 + 5(3) + 6$
$= 9 + 15 + 6$
$= 30$
Since the remainder is not 0, $x-3$ is not a factor of $f(x)$.

72. $c = 2$
$f(2) = 3(2)^2 + (2) - 2$
$= 3(4) + 2 - 2$
$= 12$

Since the remainder is not 0, $x-2$ is not a factor of $f(x)$.

74. $c = 1$
$f(1) = 2(1)^3 - 9(1)^2 - 2(1) + 24$
$= 2(1) - 9(1) - 2(1) + 24$
$= 2 - 9 - 2 + 24$
$= 15$
Since the remainder is not 0, $x-1$ is not a factor of $f(x)$.

76. $c = -2$
$f(-2) = 5(-2)^3 + 8(-2)^2 - 7(-2) - 6$
$= 5(-8) + 8(4) - 7(-2) - 6$
$= -40 + 32 + 14 - 6$
$= 0$
Since the remainder is 0, $x+2$ is a factor.

$\begin{array}{r|rrrr} -2 & 5 & 8 & -7 & -6 \\ & & -10 & 4 & 6 \\ \hline & 5 & -2 & -3 & 0 \end{array}$

$f(x) = (x+2)(5x^2 - 2x - 3)$

78. $\dfrac{5s^4t^3 - 15s^3t^2 + 50s^2t}{5s^2t}$

$= \dfrac{5s^4t^3}{5s^2t} - \dfrac{15s^3t^2}{5s^2t} + \dfrac{50s^2t}{5s^2t}$

$= \dfrac{5}{5}s^{4-2}t^{3-1} - \dfrac{15}{5}s^{3-2}t^{2-1} + \dfrac{50}{5}s^{2-2}t^{1-1}$

$= s^2t^2 - 3st + 10$

80. 
$$\begin{array}{r} a+5 \phantom{xxxx} \\ 4a+3 \overline{) 4a^2 + 23a + 15} \\ -(4a^2 + 3a) \phantom{xxx} \\ \hline 20a + 15 \\ -(20a + 15) \\ \hline 0 \end{array}$$

$\dfrac{4a^2 + 23a + 15}{4a+3} = a + 5$

Chapter 5: Polynomials and Polynomial Functions

82.
$$\begin{array}{r} x-3 \\ x^2+3\overline{\smash{)}x^3-3x^2+5x-12} \\ \underline{-(x^3\phantom{-3x^2}+3x)} \\ -3x^2+2x-12 \\ \underline{-(-3x^2\phantom{+2x}-9)} \\ 2x-3 \end{array}$$

$$\frac{x^3-3x^2+5x-12}{x^2+3} = x-3+\frac{2x-3}{x^2+3}$$

84.
$$\frac{3x^4+6x^2}{9x^3} = \frac{3x^4}{9x^3}+\frac{6x^2}{9x^3}$$
$$= \frac{3}{9}x^{4-3}+\frac{6}{9}x^{2-3}$$
$$= \frac{1}{3}x+\frac{2}{3}x^{-1}$$
$$= \frac{x}{3}+\frac{2}{3x}$$

86. The divisor is $x-5$ so $c=5$.

$$\begin{array}{r|rrrr} 5) & 1 & 5 & -29 & -97 \\ & & 5 & 50 & 105 \\ \hline & 1 & 10 & 21 & 8 \end{array}$$

$$\frac{x^3+5x^2-29x-97}{x-5} = x^2+10x+21+\frac{8}{x-5}$$

88.
$$\begin{array}{r} b^2+2b-\frac{17}{2} \\ 4b+3\overline{\smash{)}4b^3+11b^2-28b-17} \\ \underline{-(4b^3+3b^2)} \\ 8b^2-28b \\ \underline{-(8b^2+6b)} \\ -34b-17 \\ \underline{-\left(-34b-\frac{51}{2}\right)} \\ \frac{17}{2} \end{array}$$

$$\frac{4b^3+11b^2-28b-17}{4b+3} = b^2+2b-\frac{17}{2}+\frac{17/2}{4b+3}$$

90.
$$\begin{array}{r} 4a^2-10a+25 \\ 2a+5\overline{\smash{)}8a^3+0a^2+0a+125} \\ \underline{-(8a^3+20a^2)} \\ -20a^2+0a \\ \underline{-(-20a^2-50a)} \\ 50a+125 \\ \underline{-(50a+125)} \\ 0 \end{array}$$

$$\frac{8a^3+125}{2a+5} = 4a^2-10a+25$$

92. Since the area of a rectangle is the product of the length and width, we can find the length by dividing the area by the width.

$$\begin{array}{r} 5x-3 \\ 2x+3\overline{\smash{)}10x^2+9x-9} \\ \underline{-(10x^2+15x)} \\ -6x-9 \\ \underline{-(-6x-9)} \\ 0 \end{array}$$

The area of the rectangle is given by
$A = 10x^2+9x-9 = (2x+3)(5x-3)$.
The length of the rectangle is $5x-3$ feet.

94. The volume of a rectangular box is the product of the length, width, and height. The product of the width and length is given by
$$(4x+3)(x+1) = 4x^2+4x+3x+3$$
$$= 4x^2+7x+3$$
To find the height, we need to divide the volume by this quantity.

$$\begin{array}{r} x+7 \\ 4x^2+7x+3\overline{\smash{)}4x^3+35x^2+52x+21} \\ \underline{-(4x^3+7x^2+3x)} \\ 28x^2+49x+21 \\ \underline{-(28x^2+49x+21)} \\ 0 \end{array}$$

The volume of the box is given by

## Section 5.3 Dividing Polynomials: Synthetic Division

$V = 4x^3 + 35x^2 + 52x + 21$
$= (4x+3)(x+1)(x+7)$
The height of the box is $x+7$ feet.

96. a. $\bar{C}(x) = \dfrac{0.01x^3 - 0.45x^2 + 16.5x + 600}{x}$

$= \dfrac{0.01x^3}{x} - \dfrac{0.45x^2}{x} + \dfrac{16.5x}{x} + \dfrac{600}{x}$

$= 0.01x^2 - 0.45x + 16.5 + \dfrac{600}{x}$

b. $\bar{C}(50) = 0.01(50)^2 - 0.45(50) + 16.5 + \dfrac{600}{50}$

$= 25 - 22.5 + 16.5 + 12$
$= 31$

The average cost of manufacturing $x = 50$ wagons in a day is $31.

98. $\dfrac{f(x)}{x+3} = 2x + 7$

$f(x) = (x+3)(2x+7)$
$= 2x^2 + 7x + 6x + 21$
$= 2x^2 + 13x + 21$

100. $\dfrac{f(x)}{x-3} = x^2 + 2 + \dfrac{7}{x-3}$

$f(x) = (x-3)\left(x^2 + 2 + \dfrac{7}{x-3}\right)$

$= x^2(x-3) + 2(x-3) + \dfrac{7}{x-3}(x-3)$

$= x^3 - 3x^2 + 2x - 6 + 7$
$= x^3 - 3x^2 + 2x + 1$

**Putting the Concepts Together (Sections 5.1-5.3)**

1. $5m^4 - 2m^3 + 3m + 8$
   Degree: 4

2. $(7a^2 - 4a^3 + 7a - 1) + (2a^2 - 6a - 7)$
   $= 7a^2 - 4a^3 + 7a - 1 + 2a^2 - 6a - 7$
   $= -4a^3 + 7a^2 + 2a^2 + 7a - 6a - 1 - 7$
   $= -4a^3 + (7+2)a^2 + (7-6)a + (-1-7)$
   $= -4a^3 + 9a^2 + a - 8$

3. $\left(\tfrac{1}{5}y^2 + 2y - 6\right) - \left(4y^2 - y + 2\right)$
   $= \tfrac{1}{5}y^2 + 2y - 6 - 4y^2 + y - 2$
   $= \tfrac{1}{5}y^2 - 4y^2 + 2y + y - 6 - 2$
   $= -\tfrac{19}{5}y^2 + 3y - 8$

4. $f(x) = 2x^3 - x^2 + 4x + 9$
   $f(2) = 2(2)^3 - (2)^2 + 4(2) + 9$
   $= 2(8) - 4 + 8 + 9$
   $= 16 - 4 + 8 + 9$
   $= 29$

5. $(f+g)(x) = (6x+5) + (-x^2 + 2x + 3)$
   $= 6x + 5 - x^2 + 2x + 3$
   $= -x^2 + 8x + 8$
   $(f+g)(-3) = -(-3)^2 + 8(-3) + 8$
   $= -9 - 24 + 8$
   $= -25$

6. $(f-g)(x) = (2x^2 + 7) - (x^2 - 4x - 3)$
   $= 2x^2 + 7 - x^2 + 4x + 3$
   $= 2x^2 - x^2 + 4x + 7 + 3$
   $= x^2 + 4x + 10$

7. $2mn^3(m^2n - 4mn + 6)$
   $= 2mn^3 \cdot m^2n - 2mn^3 \cdot 4mn + 2mn^3 \cdot 6$
   $= 2m^{1+2}n^{3+1} - 8m^{1+1}n^{3+1} + 12mn^3$
   $= 2m^3n^4 - 8m^2n^4 + 12mn^3$

8. $(3a - 5b)^2 = (3a)^2 - 2(3a)(5b) + (5b)^2$
   $= 9a^2 - 30ab + 25b^2$

9. $(7n^2 + 3)(7n^2 - 3) = (7n^2)^2 - (3)^2$
   $= 49n^4 - 9$

10. $(3a + 2b)(6a^2 - 2ab + b^2)$
    $= 3a \cdot 6a^2 - 3a \cdot 2ab + 3a \cdot b^2 + 2b \cdot 6a^2$
    $\qquad - 2b \cdot 2ab + 2b \cdot b^2$
    $= 18a^3 - 6a^2b + 3ab^2 + 12a^2b - 4ab^2 + 2b^3$
    $= 18a^3 + 6a^2b - ab^2 + 2b^3$

## Chapter 5: Polynomials and Polynomial Functions

11. $(f \cdot g)(x) = (x+2)(x^2 - 4x + 11)$
    $= x \cdot x^2 - x \cdot 4x + x \cdot 11 + 2 \cdot x^2$
    $\quad - 2 \cdot 4x + 2 \cdot 11$
    $= x^3 - 4x^2 + 11x + 2x^2 - 8x + 22$
    $= x^3 - 2x^2 + 3x + 22$

12. $2z + 7 \overline{\smash{\big)}\,10z^3 + 41z^2 + 7z - 49}$ gives $5z^2 + 3z - 7$

    $\quad\;\;\underline{-(10z^3 + 35z^2)}$
    $\qquad\qquad 6z^2 + 7z$
    $\qquad\;\;\underline{-(6z^2 + 21z)}$
    $\qquad\qquad\qquad -14z - 49$
    $\qquad\qquad\;\;\underline{-(-14z - 49)}$
    $\qquad\qquad\qquad\qquad\;\; 0$

    $\dfrac{10z^3 + 41z^2 + 7z - 49}{2z + 7} = 5z^2 + 3z - 7$

13. The divisor is $x + 9$ so $c = -9$.

    $-9\,\overline{\smash{\big)}\,2 \quad 25 \quad 62 \quad -6}$
    $\qquad\quad\;\; -18 \;\; -63 \;\;\;\; 9$
    $\qquad\;\;\overline{\;\,2 \quad\;\; 7 \quad -1 \quad\;\; 3\;}$

    $\dfrac{2x^3 + 25x^2 + 62x - 6}{x + 9} = 2x^2 + 7x - 1 + \dfrac{3}{x + 9}$

14. $\left(\dfrac{f}{g}\right)(x) = \dfrac{x^3 + 2x^2 - 4x + 5}{x - 1}$

    The divisor is $x - 1$ so $c = 1$.

    $1\,\overline{\smash{\big)}\,1 \quad 2 \quad -4 \quad 5}$
    $\qquad\quad\;\; 1 \quad\;\; 3 \quad -1$
    $\qquad\overline{\;\,1 \quad 3 \quad -1 \quad\;\; 4\;}$

    $\left(\dfrac{f}{g}\right)(x) = x^2 + 3x - 1 + \dfrac{4}{x - 1}$

15. The divisor is $x + 5$ so $c = -5$.
    $f(x) = 3x^3 + 8x^2 - 23x + 60$
    $f(-5) = 3(-5)^3 + 8(-5)^2 - 23(-5) + 60$
    $\quad\;\;\; = 3(-125) + 8(25) + 115 + 60$
    $\quad\;\;\; = -375 + 200 + 115 + 60$
    $\quad\;\;\; = 0$
    Since $f(-5) = 0$, the remainder is 0 and $x + 5$ is a factor of $f(x)$.

    $-5\,\overline{\smash{\big)}\,3 \quad\;\; 8 \quad -23 \quad\;\; 60}$
    $\qquad\qquad -15 \quad\;\; 35 \quad -60$
    $\qquad\;\overline{\;\,3 \quad -7 \quad\;\; 12 \quad\;\;\; 0\;}$

    $f(x) = (x + 5)(3x^2 - 7x + 12)$

### 5.4 Preparing for Greatest Common Factor; Factoring by Grouping

1. $24 = 8 \cdot 3$
   $\quad\;\; = 4 \cdot 2 \cdot 3$
   $\quad\;\; = 2 \cdot 2 \cdot 2 \cdot 3$

2. $4(3x - 5) = 4 \cdot 3x - 4 \cdot 5$
   $\qquad\qquad = 4 \cdot 3 \cdot x - 4 \cdot 5$
   $\qquad\qquad = 12x - 20$

### 5.4 Exercises

2. prime

4. True

6. True

8. Answers will vary.

**Section 5.4** Greatest Common Factor: Factoring by Grouping

10. $3a^2b^3 = 3 \cdot a \cdot a \cdot b \cdot b \cdot b$
    $6ab^4 = 2 \cdot 3 \cdot a \cdot b \cdot b \cdot b \cdot b$
    $12a^3b^2 = 2 \cdot 2 \cdot 3 \cdot a \cdot a \cdot a \cdot b \cdot b$
    GCF: $3 \cdot a \cdot b \cdot b = 3ab^2$

12. $8z + 48$
    GCF: 8
    $8z + 48 = 8 \cdot z + 8 \cdot 6$
    $= 8(z + 6)$

14. $-4b + 32$
    GCF: $-4$
    $-4b + 32 = -4 \cdot b + (-4)(-8)$
    $= -4(b - 8)$

16. $12a^2 + 45a$
    GCF: $3a$
    $12a^2 + 45a = 3a \cdot 4a + 3a \cdot 15$
    $= 3a(4a + 15)$

18. $2w^3 + 10w^2 - 14$
    GCF: 2
    $2w^3 + 10w^2 - 14 = 2 \cdot w^3 + 2 \cdot 5w^2 - 2 \cdot 7$
    $= 2(w^3 + 5w^2 - 7)$

20. $-6q^3 + 36q^2 - 48q$
    GCF: $-6q$
    $-6q^3 + 36q^2 - 48q$
    $= -6q \cdot q^2 + (-6q)(-6q) + (-6q)(8)$
    $= -6q(q^2 - 6q + 8)$

22. $64x^4y^2 - 40x^3y^4 + 96xy^5$
    GCF: $8xy^2$
    $64x^4y^2 - 40x^3y^4 + 96xy^5$
    $= 8xy^2(8x^3) + (8xy^2)(-5x^2y^2) + (8xy^2)(12y^3)$
    $= 8xy^2(8x^3 - 5x^2y^2 + 12y^3)$

24. $-18b^3 + 10b^2 + 6b$
    GCF: $-2b$
    $-18b^3 + 10b^2 + 6b$
    $= -2b(9b^2) + (-2b)(-5b) + (-2b)(-3)$
    $= -2b(9b^2 - 5b - 3)$

26. $6z(5z + 3) + 5(5z + 3)$
    GCF: $5z + 3$
    $6z(5z + 3) + 5(5z + 3)$
    $= (5z + 3) \cdot 6z + (5z + 3) \cdot 5$
    $= (5z + 3)(6z + 5)$

28. $(5b + 3)(b + 4) + (3b + 1)(b + 4)$
    GCF: $b + 4$
    $(5b + 3)(b + 4) + (3b + 1)(b + 4)$
    $= (b + 4) \cdot (5b + 3) + (b + 4)(3b + 1)$
    $= (b + 4)(5b + 3 + 3b + 1)$
    $= (b + 4)(8b + 4)$
    The binomial $8b + 4$ has a greatest common factor of 4. we can write the result as
    $(b + 4)(4 \cdot 2b + 4 \cdot 1) = (b + 4) \cdot 4(2b + 1)$
    $= 4(b + 4)(2b + 1)$

30. $8x - 8y + bx - by = (8x - 8y) + (bx - by)$
    $= 8(x - y) + b(x - y)$
    $= (x - y)(8 + b)$

32. $3y^3 + 9y^2 - 5y - 15 = (3y^3 + 9y^2) + (-5y - 15)$
    $= 3y^2(y + 3) - 5(y + 3)$
    $= (y + 3)(3y^2 - 5)$

34. $p^2 - 3p + 8p - 24 = (p^2 - 3p) + (8p - 24)$
    $= p(p - 3) + 8(p - 3)$
    $= (p - 3)(p + 8)$

## Chapter 5: Polynomials and Polynomial Functions

36. Pull out the GCF first, then factor by grouping.
$$3a^2 - 15a - 9a + 45 = 3(a^2 - 5a - 3a + 15)$$
$$= 3((a^2 - 5a) + (-3a + 15))$$
$$= 3(a(a-5) - 3(a-5))$$
$$= 3(a-5)(a-3)$$

38. Pull out the GCF first, then factor by grouping.
$$2y^3 + 14y^2 - 4y^2 - 28y$$
$$= 2y(y^2 + 7y - 2y - 14)$$
$$= 2y((y^2 + 7y) + (-2y - 14))$$
$$= 2y(y(y+7) - 2(y+7))$$
$$= 2y(y+7)(y-2)$$

40. $15x^2 - 5xy + 18xy - 6y^2$
$$= (15x^2 - 5xy) + (18xy - 6y^2)$$
$$= 5x(3x - y) + 6y(3x - y)$$
$$= (3x - y)(5x + 6y)$$

42. $(x+5)(x-3) - (x-1)(x-3)$
GCF: $x-3$
$(x+5)(x-3) - (x-1)(x-3)$
$$= (x-3) \cdot (x+5) - (x-3) \cdot (x-1)$$
$$= (x-3)(x+5 - x+1)$$
$$= (x-3)(6)$$
$$= 6(x-3)$$

44. $3q^2 + 5q - 12q - 20 = (3q^2 + 5q) + (-12q - 20)$
$$= q(3q+5) - 4(3q+5)$$
$$= (3q+5)(q-4)$$

46. $8a^4b^2 + 12a^3b^3 - 36ab^4$
GCF: $4ab^2$
$8a^4b^2 + 12a^3b^3 - 36ab^4$
$$= 4ab^2 \cdot 2a^3 + 4ab^2 \cdot 3a^2b + 4ab^2 \cdot (-9b^2)$$
$$= 4ab^2(2a^3 + 3a^2b - 9b^2)$$

48. $c^3 - c^2 + 5c - 5 = (c^3 - c^2) + (5c - 5)$
$$= c^2(c-1) + 5(c-1)$$
$$= (c-1)(c^2 + 5)$$

50. $2y(y+4) + 3(y+4)^2$
GCF: $(y+4)$
$2y(y+4) + 3(y+4)^2$
$$= (y+4) \cdot 2y + (y+4) \cdot 3(y+4)$$
$$= (y+4)(2y + 3(y+4))$$
$$= (y+4)(2y + 3y + 12)$$
$$= (y+4)(5y + 12)$$

52. The shaded area is the difference between the area of the rectangle and the area of the circle. The radius of the circle is $x$, and this is also half the length of the width of the rectangle. Thus, we have
$$A = A_{\text{rectangle}} - A_{\text{circle}}$$
$$= (4x)(2x) - \pi x^2$$
$$= 8x^2 - \pi x^2$$
$$= x^2(8 - \pi)$$
The area of the shaded region is $x^2(8-\pi)$ square units.

54. $V = \pi h R^2 - \dfrac{\pi h^3}{4}$
$$= \pi h \cdot R^2 - \pi h \cdot \dfrac{h^2}{4}$$
$$= \pi h \left(R^2 - \dfrac{h^2}{4}\right)$$
The volume of the cylinder is $\pi h\left(R^2 - \dfrac{h^2}{4}\right)$ cubic units.

c. $1.4x - 0.4(1.4x) = 1.4x(1 - 0.4)$
$$= 0.6(1.4x)$$
$$= 0.84x$$

d. No, the sale price is less than the amount that the store paid for the shirt. The 40% increase is on a smaller amount than the 40% decrease. Therefore, the decrease will be larger in magnitude than the increase.

*Section 5.4* **Greatest Common Factor: Factoring by Grouping**

56. a. The selling price of the television is the original price, less the discount.
    $x - 0.2x = (1 - 0.2)x = 0.8x$

    b. The discount is 15% of the originally discounted price. The new sale price is obtained by subtracting this discount from the previous sale price. That is,
    $0.8x - 0.15(0.8x)$

    c. $0.8x - 0.15(0.8x) = 0.8x(1 - 0.15)$
    $= 0.8x(0.85)$
    $= 0.68x$

    d. If the original selling price was $x = 650$, then the sale price after the second discount is $0.68(650) = \$442$.

58. $\frac{1}{2}x + \frac{3}{2} = \frac{1}{2} \cdot x + \frac{1}{2} \cdot 3$
    $= \frac{1}{2}(x + 3)$

60. $\frac{2}{3}x^2 + \frac{4}{9}x = \frac{2}{9}x \cdot 3x + \frac{2}{9}x \cdot 2$
    $= \frac{2}{9}x(3x + 2)$

62. Let $x$ = the original price.

    a) 30% discount:
    Sale price = $x - 0.3x = 0.7x$

    b) 15% discount followed by another 15% discount:
    Price after first discount
    $x - 0.15x = 0.85x$

    Price after second discount
    $0.85x - 0.15(0.85x) = (1 - 0.15)(0.85x)$
    $= (0.85)(0.85x)$
    $= 0.7225x$

    So, with option (a), the sale price will be 70% of the original price. With option (b), the sale price will be 72.25% of the original price. Option (a) is better – take the 30% discount.

**5.5 Preparing for Factoring Trinomials**

1. | factors | 1, 18 | 2, 9 | 3, 6 |
   |---|---|---|---|
   | sum | 19 | 11 | 9 |

   9 and 2 are the factors of 18 that sum to 11.

2. | factors | 1, –24 | 2, –12 | 3, –8 | 4, –6 |
   |---|---|---|---|---|
   | sum | –23 | –10 | –5 | –2 |

   –6 and 4 are the factors of –24 that sum to –2. (note: since the sum is negative, we restrict our choice of factors so that the larger factor (in absolute value) is negative.

3. $4x^2 - 9x + 2$
   The coefficients are the numeric factors of the terms. The coefficients are 4, –9, and 2.

**5.5 Exercises**

2. $2x - 3$

4. True; $4(2x+1)^2 - 3(2x+1) - 1$
   $= (4(2x+1) + 1)((2x+1) - 1)$
   $= (8x + 4 + 1)(2x + 1 - 1)$
   $= (8x + 5)(2x)$
   $= 2x(8x + 5)$

6. When factoring a polynomial, the first step is to pull out the greatest common factor.

8. Answers will vary.

10. $x^2 + 8x + 12$
    We are looking for two factors of $c = 12$ whose sum is $b = 8$. Since both $c$ and $b$ are positive, the two factors must be positive.

    | Factors | 1, 12 | 2, 6 | 3, 4 |
    |---|---|---|---|
    | Sum | 13 | 8 | 7 |

    $x^2 + 8x + 12 = (x + 2)(x + 6)$

## Chapter 5: Polynomials and Polynomial Functions

12. $z^2 + 3z - 28$
    We are looking for two factors of $c = -28$ whose sum is $b = 3$. Since $c$ is negative the two factors have opposite signs, and since $b$ is positive the factor with the larger absolute value must be positive.

    | Factors | −1, 28 | −2, 14 | −4, 7 |
    |---|---|---|---|
    | Sum | 27 | 12 | 3 |

    $z^2 + 3z - 28 = (z - 4)(z + 7)$

14. $y^2 - 12y + 36$
    We are looking for two factors of $c = 36$ whose sum is $b = -12$. Since $c$ is positive the two factors have the same sign, and since $b$ is negative the factors are both negative.

    | Factors | −1, −36 | −2, −18 | −4, −9 | −6, −6 |
    |---|---|---|---|---|
    | Sum | −37 | −20 | −13 | −12 |

    $y^2 - 12y + 36 = (y - 6)(y - 6) = (y - 6)^2$

16. $q^2 + 2q - 80$
    We are looking for two factors of $c = -80$ whose sum is $b = 2$. Since $c$ is negative the two factors have opposite signs, and since $b$ is positive the factor with the larger absolute value must be positive.

    | Factors | −1, 80 | −2, 40 | −4, 20 | −5, 16 | −8, 10 |
    |---|---|---|---|---|---|
    | Sum | 79 | 38 | 16 | 11 | 2 |

    $q^2 + 2q - 80 = (q - 8)(q + 10)$

18. $z^2 - 16z + 48$
    We are looking for two factors of $c = 48$ whose sum is $b = -16$. Since $c$ is positive the two factors have the same sign, and since $b$ is negative the factors are both negative.

    | Factors | −1, −48 | −2, −24 | −4, −12 | −6, −8 |
    |---|---|---|---|---|
    | Sum | −49 | −26 | −16 | −14 |

    $z^2 - 16z + 48 = (z - 4)(z - 12)$

20. $-p^2 + 3p + 54$
    Start by factoring out a $-1$.
    $-p^2 + 3p + 54 = -1(p^2 - 3p - 54)$
    We are looking for two factors of $c = -54$ whose sum is $b = -3$. Since $c$ is negative the two factors have opposite signs, and since $b$ is negative the factor with the larger absolute value must be negative.

    | Factors | 1, −54 | 2, −27 | 3, −18 | 6, −9 |
    |---|---|---|---|---|
    | Sum | −53 | −25 | −15 | −3 |

    $-p^2 + 3p + 54 = -(p - 9)(p + 6)$

22. $m^2 + 7mn + 10n^2$
    We are looking for two factors of $c = 10$ whose sum is $b = 7$. Since both $c$ and $b$ are positive, the two factors will be positive.

    | Factors | 1, 10 | 2, 5 |
    |---|---|---|
    | Sum | 11 | 7 |

    $m^2 + 7mn + 10n^2 = (m + 2n)(m + 5n)$

24. $x^2 - 4xy - 21y^2$
    We are looking for two factors of $c = -21$ whose sum is $b = -4$. Since $c$ is negative, the two factors will have opposite signs. Since $b$ is also negative, the factor with the larger absolute value will be negative.

    | Factors | 1, −21 | 3, −7 |
    |---|---|---|
    | Sum | −20 | −4 |

    $x^2 - 4xy - 21y^2 = (x - 7y)(x + 3y)$

26. $3y^2 - 6y - 189$
    Start by factoring out the GCF.
    $3(y^2 - 2y - 63)$
    We are looking for two factors of $c = -63$ whose sum is $b = -2$. Since $c$ is negative, the two factors will have opposite signs. Since $b$ is also negative, the factor with the larger absolute value will be negative.

    | Factors | 1, −63 | 3, −21 | 7, −9 |
    |---|---|---|---|
    | Sum | −62 | −18 | −2 |

    $3y^2 - 6y - 189 = 3(y - 9)(y + 7)$

28. $-4s^2 - 32s - 48$
    Start by factoring out the GCF.
    $-4(s^2 + 8s + 12)$
    We are looking for two factors of $c = 12$ whose sum is $b = 8$. Since both $c$ and $b$ are positive, the two factors will be positive.

## Section 5.5 Factoring Trinomials

| Factors | 1, 12 | 2, 6 | 3, 4 |
|---------|-------|------|------|
| Sum     | 13    | 8    | 7    |

$-4s^2 - 32s - 48 = -4(s+2)(s+6)$

**30.** <u>ac Method:</u>
$a \cdot c = 3 \cdot -10 = -30$
We are looking for two factors of $-30$ whose sum is $-13$. Since the product is negative, the factors will have opposite signs. The sum is negative so the factor with the largest absolute value will be negative.

| factor 1 | factor 2 | sum |
|----------|----------|-----|
| 1        | −30      | −29 |
| 2        | −15      | −13 ← okay |
| 3        | −10      | −7  |

$3z^2 - 13z - 10 = 3z^2 + 2z - 15z - 10$
$\phantom{3z^2 - 13z - 10} = z(3z+2) - 5(3z+2)$
$\phantom{3z^2 - 13z - 10} = (3z+2)(z-5)$

<u>Trial and Error Method:</u>
First note that there are no common factors and that $a = 3$, $b = -13$, and $c = -10$. Since $c$ is negative, the signs in our factors will be opposites. We will consider factorizations with this form:
$(\underline{\phantom{x}}z + \underline{\phantom{x}})(\underline{\phantom{x}}z - \underline{\phantom{x}})$
If our choice results in a middle term with the wrong sign, we simply switch the signs of the factors.
Since $a = 3$ can be factored as $3 \cdot 1$, we have the following form:
$(3z + \underline{\phantom{x}})(z - \underline{\phantom{x}})$
$|c| = |-10| = 10$ can be factored as $1 \cdot 10$ and $2 \cdot 5$.
$(3z+1)(z-10) \rightarrow 3z^2 - 29z - 10$
$(3z+10)(z-1) \rightarrow 3z^2 + 7z - 10$
$(3z+2)(z-5) \rightarrow 3z^2 - 13z - 10$
$(3z+5)(z-2) \rightarrow 3z^2 - z - 10$
The correct factorization is
$3z^2 - 13z - 10 = (3z+2)(z-5)$

**32.** <u>ac Method:</u>
$a \cdot c = 6 \cdot 6 = 36$
We are looking for two factors of 36 whose sum is $-37$. Since the product is positive, the factors will have the same sign. The sum is negative so the factors will be negative.

| factor 1 | factor 2 | sum |
|----------|----------|-----|
| −1       | −36      | −37 ← okay |
| −2       | −18      | −20 |
| −3       | −12      | −15 |
| −4       | −9       | −13 |
| −6       | −6       | −12 |

$6x^2 - 37x + 6 = 6x^2 - 36x - x + 6$
$\phantom{6x^2 - 37x + 6} = 6x(x-6) - 1(x-6)$
$\phantom{6x^2 - 37x + 6} = (x-6)(6x-1)$

<u>Trial and Error Method:</u>
First note that there are no common factors and that $a = 6$, $b = -37$, and $c = 6$. Since $c$ is positive the signs of our factors will be the same. Since $b$ is negative, the signs in our factors will be negative. We will consider factorizations with this form:
$(\underline{\phantom{x}}x - \underline{\phantom{x}})(\underline{\phantom{x}}x - \underline{\phantom{x}})$
Since $a = 6$ can be factored as $1 \cdot 6$ or $2 \cdot 3$, we have the following forms:
$(x - \underline{\phantom{x}})(6x - \underline{\phantom{x}})$
$(2x - \underline{\phantom{x}})(3x - \underline{\phantom{x}})$
$|c| = |6| = 6$ can be factored as $1 \cdot 6$ and $2 \cdot 3$. Since the original expression had no common factors, the binomials we select cannot have a common factor.
$(x-6)(6x-1) \rightarrow 6x^2 - 37x + 6$
$(2x-3)(3x-2) \rightarrow 6x^2 - 13x + 6$
The correct factorization is
$6x^2 - 37x + 6 = (x-6)(6x-1)$

**34.** <u>ac Method:</u>
$a \cdot c = 12 \cdot -15 = -180$
We are looking for two factors of $-180$ whose sum is 11. Since the product is negative, the factors will have opposite signs. The sum is positive so the factor with the largest absolute value will be positive.

| factor 1 | factor 2 | sum |
|----------|----------|-----|
| −3       | 60       | 57 too large |
| −10      | 18       | 8 too small  |
| −9       | 20       | 11 ← okay |

$12r^2 + 11r - 15 = 12r^2 - 9r + 20r - 15$
$\phantom{12r^2 + 11r - 15} = 3r(4r-3) + 5(4r-3)$
$\phantom{12r^2 + 11r - 15} = (4r-3)(3r+5)$

## Chapter 5: Polynomials and Polynomial Functions

Trial and Error Method:
First note that there are no common factors and that $a = 12$, $b = 11$, and $c = -15$. Since $c$ is negative, the signs in our factors will be opposites. We will consider factorizations with this form:
$$(\underline{\phantom{xx}}r + \underline{\phantom{xx}})(\underline{\phantom{xx}}r - \underline{\phantom{xx}})$$
If our choice results in a middle term with the wrong sign, we simply switch the signs of the factors.
Since $a = 12$ can be factored as $1 \cdot 12$, $2 \cdot 6$, and $3 \cdot 4$, we have the following forms:
$(r + \underline{\phantom{xx}})(12r - \underline{\phantom{xx}})$
$(2r + \underline{\phantom{xx}})(6r - \underline{\phantom{xx}})$
$(3r + \underline{\phantom{xx}})(4r - \underline{\phantom{xx}})$
$|c| = |-15| = 15$ can be factored as $1 \cdot 15$ and $3 \cdot 5$. Since the original expression had no common factors, the binomials we select cannot have a common factor.
$(r+15)(12r-1) \to 12r^2 + 179r - 15$
$(r+3)(12r-5) \to 12r^2 + 31r - 15$
$(2r+15)(6r-1) \to 12r^2 + 88r - 15$
$(2r+3)(6r-5) \to 12r^2 + 8r - 15$
$(3r+1)(4r-15) \to 12r^2 - 41r - 15$
$(3r+5)(4r-3) \to 12r^2 + 11r - 15$
The correct factorization is
$12r^2 + 11r - 15 = (3r+5)(4r-3)$

36. ac Method:
$a \cdot c = 18 \cdot -5 = -90$
We are looking for two factors of $-90$ whose sum is 43. Since the product is negative, the factors will have opposite signs. The sum is positive so the factor with the largest absolute value will be positive.

| factor 1 | factor 2 | sum |
|---|---|---|
| –9 | 10 | 1 too small |
| –3 | 30 | 27 too small |
| –2 | 45 | 43 ← okay |

$18y^2 + 43y - 5 = 18y^2 + 45y - 2y - 5$
$\phantom{18y^2 + 43y - 5} = 9y(2y+5) - 1(2y+5)$
$\phantom{18y^2 + 43y - 5} = (2y+5)(9y-1)$

Trial and Error Method:
First note that there are no common factors and that $a = 18$, $b = 43$, and $c = -5$. Since $c$ is negative, the signs in our factors will be opposites. We will consider factorizations with this form:
$$(\underline{\phantom{xx}}y + \underline{\phantom{xx}})(\underline{\phantom{xx}}y - \underline{\phantom{xx}})$$
If our choice results in a middle term with the wrong sign, we simply switch the signs of the factors.
Since $a = 18$ can be factored as $1 \cdot 18$, $2 \cdot 9$, and $3 \cdot 6$, we have the following forms:
$(y + \underline{\phantom{xx}})(18y - \underline{\phantom{xx}})$
$(2y + \underline{\phantom{xx}})(9y - \underline{\phantom{xx}})$
$(3y + \underline{\phantom{xx}})(6y - \underline{\phantom{xx}})$
$|c| = |-5| = 5$ can be factored as $1 \cdot 5$.
$(y+1)(18y-5) \to 18y^2 + 13y - 5$
$(y+5)(18y-1) \to 18y^2 + 89y - 5$
$(2y+1)(9y-5) \to 18y^2 - y - 5$
$(2y+5)(9y-1) \to 18y^2 + 43y - 5$
$(3y+1)(6y-5) \to 18y^2 - 9y - 5$
$(3y+5)(6y-1) \to 18y^2 + 27y - 5$
The correct factorization is
$18y^2 + 43y - 5 = (2y+5)(9y-1)$

38. ac Method:
$a \cdot c = 20 \cdot 6 = 120$
We are looking for two factors of 120 whose sum is 23. Since the product is positive, the factors will have the same sign. The sum is positive so the factors will both be positive. Since we are adding two positive numbers to get 23, neither factor can exceed 23.

| factor 1 | factor 2 | sum |
|---|---|---|
| 6 | 20 | 26 |
| 8 | 15 | 23 ← okay |
| 10 | 12 | 22 |

$20r^2 + 23r + 6 = 20r^2 + 15r + 8r + 6$
$\phantom{20r^2 + 23r + 6} = 5r(4r+3) + 2(4r+3)$
$\phantom{20r^2 + 23r + 6} = (4r+3)(5r+2)$

Trial and Error Method:
First note that there are no common factors and that $a = 20$, $b = 23$, and $c = 6$. Since $c$ is positive, the signs in our factors will be the same. Since $b$ is also positive, the signs of our factors will be positive. We will consider

### Section 5.5 Factoring Trinomials

factorizations with this form:
$(\_\_r+\_\_)(\_\_r+\_\_)$
Since $a = 20$ can be factored as $1 \cdot 20$, $2 \cdot 10$, and $4 \cdot 5$, we have the following forms:
$(r+\_\_)(20r+\_\_)$
$(2r+\_\_)(10r+\_\_)$
$(4r+\_\_)(5r+\_\_)$
$|c| = |6| = 6$ can be factored as $1 \cdot 6$ and $2 \cdot 3$.
Since the original expression had no common factors, the binomials we select cannot have a common factor.
$(r+6)(20r+1) \to 20r^2 + 121r + 6$
$(r+2)(20r+3) \to 20r^2 + 43r + 6$
$(4r+1)(5r+6) \to 20r^2 + 29r + 6$
$(4r+3)(5r+2) \to 20r^2 + 23r + 6$
The correct factorization is
$20r^2 + 23r + 6 = (4r+3)(5r+2)$

40. <u>ac Method:</u>
Start by factoring out the GCF.
$-24y^2 - 39y + 18 = -3(8y^2 + 13y - 6)$
Now focus on the reduced trinomial.
$a \cdot c = 8 \cdot -6 = -48$
We are looking for two factors of $-48$ whose sum is 13. Since the product is negative, the factors will have opposite signs. The sum is positive so the factor with the larger absolute value will be positive.

| factor 1 | factor 2 | sum |     |
|----------|----------|-----|-----|
| $-2$     | 24       | 22  | too large |
| $-4$     | 12       | 8   | too small |
| $-3$     | 16       | 13  | ← okay |

$-24y^2 - 39y + 18 = -3(8y^2 + 13y - 6)$
$= -3(8y^2 + 16y - 3y - 6)$
$= -3(8y(y+2) - 3(y+2))$
$= -3(y+2)(8y-3)$

<u>Trial and Error Method:</u>
Start by factoring out the GCF.
$-24y^2 - 39y + 18 = -3(8y^2 + 13y - 6)$
Now focus on the reduced trinomial.
$a = 8$, $b = 13$, and $c = -6$. Since $c$ is negative, the signs in our factors will be opposites. We

will consider factorizations with this form:
$(\_\_y+\_\_)(\_\_y-\_\_)$
If our choice results in a middle term with the wrong sign, we simply switch the signs of the factors.
Since $a = 8$ can be factored as $1 \cdot 8$ and $2 \cdot 4$, we have the following forms:
$(y+\_\_)(8y-\_\_)$
$(2y+\_\_)(4y-\_\_)$
$|c| = |-6| = 6$ can be factored as $1 \cdot 6$ and $2 \cdot 3$.
Since the original expression had no common factors, the binomials we select cannot have a common factor.
$(y+1)(8y-6) \to 8y^2 + 2y - 6$
$(y+6)(8y-1) \to 8y^2 + 47y - 6$
$(y+2)(8y-3) \to 8y^2 + 13y - 6$
The third choice is our desired result. Thus, the correct factorization is
$-24y^2 - 39y + 18 = -3(8y^2 + 13y - 6)$
$= -3(y+2)(8y-3)$

42. <u>ac Method:</u>
Start by factoring out the GCF.
$48w^2 + 20w - 42 = 2(24w^2 + 10w - 21)$
Now focus on the reduced trinomial.
$a \cdot c = 24 \cdot -21 = -504$
We are looking for two factors of $-504$ whose sum is 10. Since the product is negative, the factors will have opposite signs. The sum is positive so the factor with the larger absolute value will be positive.

| factor 1 | factor 2 | sum |     |
|----------|----------|-----|-----|
| $-21$    | 24       | 3   | too small |
| $-18$    | 28       | 10  | ← okay |

$48w^2 + 20w - 42 = 2(24w^2 + 10w - 21)$
$= 2(24w^2 + 28w - 18w - 21)$
$= 2(4w(6w+7) - 3(6w+7))$
$= 2(6w+7)(4w-3)$

<u>Trial and Error Method:</u>
Start by factoring out the GCF.
$48w^2 + 20w - 42 = 2(24w^2 + 10w - 21)$
Now focus on the reduced trinomial.

$a = 24$, $b = 10$, and $c = -21$. Since $c$ is negative, the signs in our factors will be opposite. We will consider factorizations with this form:
$(\underline{\phantom{x}}w+\underline{\phantom{x}})(\underline{\phantom{x}}w-\underline{\phantom{x}})$
If our choice results in a middle term with the wrong sign, we simply switch the signs of the factors.
Since $a = 24$ can be factored as $1 \cdot 24$, $2 \cdot 12$, $3 \cdot 8$, and $4 \cdot 6$, we have the following forms:
$(w+\underline{\phantom{x}})(24w-\underline{\phantom{x}})$
$(2w+\underline{\phantom{x}})(12w-\underline{\phantom{x}})$
$(3w+\underline{\phantom{x}})(8w-\underline{\phantom{x}})$
$(4w+\underline{\phantom{x}})(6w-\underline{\phantom{x}})$
$|c| = |-21| = 21$ can be factored as $1 \cdot 21$ and $3 \cdot 7$.
Since the original expression had no common factors, the binomials we select cannot have a common factor.
$(w+21)(24w-1) \to 24w^2 + 503w - 21$
$(w+3)(24w-7) \to 24w^2 + 65w - 21$
$(2w+21)(12w-1) \to 24w^2 + 250w - 21$
$(2w+3)(12w-7) \to 24w^2 + 22w - 21$
$(3w+1)(8w-21) \to 24w^2 - 55w - 21$
$(3w+7)(8w-3) \to 24w^2 + 47w - 21$
$(4w+21)(6w-1) \to 24w^2 + 122w - 21$
$(4w+3)(6w-7) \to 24w^2 - 10w - 21$
Notice that the last choice is correct except for the sign of the middle term. This means we need to switch the signs of our factors.
The correct factorization is
$48w^2 + 20w - 42 = 2(24w^2 + 10w - 21)$
$= 2(4w-3)(6w+7)$

44. ac Method:
$a \cdot c = 3 \cdot -6 = -18$
We are looking for two factors of $-18$ whose sum is 7. Since the product is negative, the factors will have opposite signs. The sum is positive so the factor with the larger absolute value will be positive.

| factor 1 | factor 2 | sum | |
|---|---|---|---|
| 18 | $-1$ | 17 | too large |
| 6 | $-3$ | 3 | too small |
| 9 | $-2$ | 7 | ← okay |

$3m^2 + 7mn - 6n^2 = 3m^2 + 9mn - 2mn - 6n^2$
$= 3m(m+3n) - 2n(m+3n)$
$= (m+3n)(3m-2n)$

Trial and Error Method:
$a = 3$, $b = 7$, and $c = -6$. Since $c$ is negative, the signs in our factors will be opposite. We will consider factorizations with this form:
$(\underline{\phantom{x}}m+\underline{\phantom{x}}n)(\underline{\phantom{x}}m-\underline{\phantom{x}}n)$
If our choice results in a middle term with the wrong sign, we simply switch the signs of the factors.
Since $a = 3$ can be factored as $1 \cdot 3$, we have the following form:
$(m+\underline{\phantom{x}}n)(3m-\underline{\phantom{x}}n)$
$|c| = |-6| = 6$ can be factored as $1 \cdot 6$ and $2 \cdot 3$.
Since the original expression had no common factors, the binomials we select cannot have a common factor.
$(m+6n)(3m-n) \to 3m^2 + 17mn - 6n^2$
$(m+3n)(3m-2n) \to 3m^2 + 7mn - 6n^2$
The correct factorization is
$3m^2 + 7mn - 6n^2 = (m+3n)(3m-2n)$

46. ac Method:
$a \cdot c = 6 \cdot 4 = 24$
We are looking for two factors of 24 whose sum is $-25$. Since the product is positive, the factors will have the same sign. The sum is negative so the factors will be negative. Since we are adding two negatives to get $-25$, neither factor can be less than $-25$.

| factor 1 | factor 2 | sum | |
|---|---|---|---|
| $-3$ | $-8$ | $-11$ | too large |
| $-2$ | $-12$ | $-14$ | too large |
| $-1$ | $-24$ | $-25$ | ← okay |

$6r^2 - 25rs + 4s^2 = 6r^2 - 24rs - rs + 4s^2$
$= 6r(r-4s) - s(r-4s)$
$= (r-4s)(6r-s)$

Trial and Error Method:
$a = 6$, $b = -25$, and $c = 4$. Since $c$ is positive, the signs in our factors will be the same. Since $b$ is negative, the two factors will be negative. We will consider factorizations with this form:
$(\underline{\phantom{x}}r-\underline{\phantom{x}}s)(\underline{\phantom{x}}r-\underline{\phantom{x}}s)$

### Section 5.5 Factoring Trinomials

Since $a = 6$ can be factored as $1 \cdot 6$ and $2 \cdot 3$, we have the following forms:
$(r - \underline{\phantom{x}}s)(6r - \underline{\phantom{x}}s)$
$(2r - \underline{\phantom{x}}s)(3r - \underline{\phantom{x}}s)$
$|c| = |4| = 4$ can be factored as $1 \cdot 4$ and $2 \cdot 2$.
Since the original expression had no common factors, the binomials we select cannot have a common factor.
$(r - 4s)(6r - s) \to 6r^2 - 25rs + 4s^2$
$(2r - s)(3r - 4s) \to 6r^2 - 11rs + 4s^2$
The correct factorization is
$6r^2 - 25rs + 4s^2 = (r - 4s)(6r - s)$

48. <u>ac Method:</u>
First note that there are no common factors and that $a = 18$, $b = 37$, and $c = -20$.
$a \cdot c = 18 \cdot -20 = -360$. We want to determine two factors whose product is $-360$ and whose sum is 37. Since the product is negative we know the two factors have opposite signs. Since the sum is positive we know that the factor with the larger absolute value must be positive.

| factor 1 | factor 2 | sum |
|---|---|---|
| $-6$ | $60$ | $54$ too large |
| $-10$ | $36$ | $26$ too small |
| $-8$ | $45$ | $37$ ← okay |

$18x^2 + 37xy - 20y^2 = 18x^2 + 45xy - 8xy - 20y^2$
$= 9x(2x + 5y) - 4y(2x + 5y)$
$= (2x + 5y)(9x - 4y)$

<u>Trial and Error Method:</u>
First note that there are no common factors and that $a = 18$, $b = 37$, and $c = -20$.
Since $c$ is negative the signs in our factors will be opposite. We will consider factorizations with this form:
$(\underline{\phantom{x}}x + \underline{\phantom{x}}y)(\underline{\phantom{x}}x - \underline{\phantom{x}}y)$.
If our choice results in a middle term with the wrong sign, we simply switch the signs of the factors.
Since $a = 18$ can be factored as $1 \cdot 18$, $2 \cdot 9$, and $3 \cdot 6$, we have the following forms:
$(x + \underline{\phantom{x}}y)(18x - \underline{\phantom{x}}y)$
$(2x + \underline{\phantom{x}}y)(9x - \underline{\phantom{x}}y)$
$(3x + \underline{\phantom{x}}y)(6x - \underline{\phantom{x}}y)$
$|c| = |-20| = 20$ can be factored as $1 \cdot 20$, $2 \cdot 10$,

and $4 \cdot 5$. Since the original expression had no common factors, the binomials we select cannot have a common factor.
$(x + 20y)(18x - y) \to 18x^2 + 359xy - 20y^2$
$(x + 4y)(18x - 5y) \to 18x^2 + 67xy - 20y^2$
$(2x + y)(9x - 20y) \to 18x^2 - 31xy - 20y^2$
$(2x + 5y)(9x - 4y) \to 18x^2 + 37xy - 20y^2$
$(3x + 20y)(6x - y) \to 18x^2 + 117xy - 20y^2$
$(3x + 4y)(6x - 5y) \to 18x^2 - xy - 20y^2$
The correct factorization is
$18x^2 + 37xy - 20y^2 = (2x + 5y)(9x - 4y)$

50. Start by factoring out the GCF.
$4x^3 - 52x^2 + 144x = 4x(x^2 - 13x + 36)$
Now we focus on the reduced trinomial.
We need two factors of 36 whose sum is $-13$. Since the product is positive, the factors will have the same sign. Since the sum is negative, the factors will be negative.

| Factors | $-1, -36$ | $-2, -18$ | $-3, -12$ | $-4, -9$ | $-6, -6$ |
|---|---|---|---|---|---|
| Sum | $-37$ | $-20$ | $-15$ | $-13$ | $-12$ |

$4x^3 - 52x^2 + 144x = 4x(x^2 - 13x + 36)$
$= 4x(x - 4)(x - 9)$

52. Start by factoring out the GCF.
$-24m^3n - 18m^2n + 27mn$
$= -3mn(8m^2 + 6m - 9)$
Now we focus on the reduced trinomial.
$a \cdot c = 8 \cdot -9 = -72$
We need two factors of $-72$ whose sum is 6. Since the product is negative, the two factors will have opposite signs. Since the sum is positive, the factor with the larger absolute value will be positive.

| factor 1 | factor 2 | sum |
|---|---|---|
| $-2$ | $36$ | $34$ too large |
| $-8$ | $9$ | $1$ too small |
| $-6$ | $12$ | $6$ ← okay |

54. Start by factoring out the GCF.
$54x^3y + 33x^2y - 72xy$
$= 3xy(18x^2 + 11x - 24)$

## Chapter 5: Polynomials and Polynomial Functions

Now we focus on the reduced trinomial.
$a \cdot c = 18 \cdot -24 = -432$
We need two factors of $-432$ whose sum is 11. Since the product is negative, the two factors will have opposite signs. Since the sum is positive, the factor with the larger absolute value will be positive.

| factor 1 | factor 2 | sum | |
|---|---|---|---|
| $-18$ | 24 | 6 | too small |
| $-8$ | 54 | 46 | too large |
| $-16$ | 27 | 11 | ← okay |

$54x^3y + 33x^2y - 72xy$
$= 3xy(18x^2 + 11x - 24)$
$= 3xy(18x^2 + 27x - 16x - 24)$
$= 3xy(9x(2x+3) - 8(2x+3))$
$= 3xy(2x+3)(9x-8)$

**56.** $y^4 + 5y^2 + 6$

Let $u = y^2$. This gives us $u^2 + 5u + 6$.

We need two factors of 6 whose sum is 5. Since the product is positive, the two factors will have the same sign. Since the sum is positive, the two factors will both be positive.
$u^2 + 5u + 6 = (u+2)(u+3)$
Now get back in terms of $y$:
$y^4 + 5y^2 + 6 = (y^2+2)(y^2+3)$

**58.** $r^2s^2 + 8rs - 48$

Let $u = rs$. This gives us $u^2 + 8u - 48$.
We need two factors of $-48$ whose sum is 8. Since the product is negative, the factors will have opposite signs. Since the sum is positive, the factor with the larger absolute value will be positive.
$u^2 + 8u - 48 = (u+12)(u-4)$
Now get back in terms of $r$ and $s$.
$r^2s^2 + 8rs - 48 = (rs+12)(rs-4)$

**60.** $(y-3)^2 + 3(y-3) + 2$

Let $u = y-3$. This gives us $u^2 + 3u + 2$.
We need two factors of 2 whose sum is 3. Since the product is positive, the two factors will have the same sign. Since the sum is also positive, the two factors are positive. The only possibility is

$u^2 + 3u + 2 = (u+1)(u+2)$
Now get back in terms of $y$.
$(y-3)^2 + 3(y-3) + 2 = (y-3+1)(y-3+2)$
$= (y-2)(y-1)$

**62.** $(5z-3)^2 - 12(5z-3) + 32$

Let $u = 5z-3$. This gives us $u^2 - 12u + 32$.
We need two factors of 32 whose sum is $-12$. Since the product is positive, the two factors will have the same sign. Since the sum is negative, the two factors will be negative.
$u^2 - 12u + 32 = (u-8)(u-4)$
Now get back in terms of $z$.
$(5z-3)^2 - 12(5z-3) + 32$
$= (5z-3-8)(5z-3-4)$
$= (5z-11)(5z-7)$

**64.** $3(z+3)^2 + 14(z+3) + 8$

Let $u = z+3$. This gives us $3u^2 + 14u + 8$.
$a \cdot c = 3 \cdot 8 = 24$
We need two factors of 24 whose sum is 14. Since the product is positive and the sum is positive, both factors will be positive.

| factor 1 | factor 2 | sum | |
|---|---|---|---|
| 1 | 24 | 25 | |
| 2 | 12 | 14 | ← okay |
| 3 | 8 | 11 | |
| 4 | 6 | 10 | |

$3u^2 + 14u + 8 = 3u^2 + 12u + 2u + 8$
$= 3u(u+4) + 2(u+4)$
$= (u+4)(3u+2)$
Now get back in terms of $z$.
$3(z+3)^2 + 14(z+3) + 8$
$= (z+3+4)(3(z+3)+2)$
$= (z+7)(3z+9+2)$
$= (z+7)(3z+11)$

**66.** *ac* Method:
$a \cdot c = 8 \cdot 11 = 88$
We are looking for two factors of 88 whose sum is 26. Since the product is positive, the factors will have the same sign. The sum is positive so the factors will both be positive.

| factor 1 | factor 2 | sum |
|---|---|---|
| 1 | 88 | 89 |
| 2 | 44 | 46 |
| 4 | 22 | 26 ← okay |
| 8 | 11 | 19 |

$8q^2 + 26q + 11 = 8q^2 + 4q + 22q + 11$
$\phantom{8q^2 + 26q + 11} = 4q(2q+1) + 11(2q+1)$
$\phantom{8q^2 + 26q + 11} = (2q+1)(4q+11)$

Trial and Error Method:
First note that there are no common factors and that $a = 8$, $b = 26$, and $c = 11$. Since $c$ is positive, the signs in our factors will be the same. Since $b$ is also positive, the signs of our factors will be positive. We will consider factorizations with this form:

$(\_\_q + \_\_)(\_\_q + \_\_)$

Since $a = 8$ can be factored as $1 \cdot 8$ and $2 \cdot 4$, we have the following forms:

$(q + \_\_)(8q + \_\_)$
$(2q + \_\_)(4q + \_\_)$

$|c| = |11| = 11$ can be factored as $1 \cdot 11$. This gives the following possibilities:

$(q+1)(8q+11) \to 8q^2 + 19q + 11$
$(q+11)(8q+1) \to 8q^2 + 89q + 11$
$(2q+1)(4w+11) \to 8q^2 + 26q + 11$
$(2q+11)(4w+1) \to 8q^2 + 46q + 11$

The correct factorization is
$8q^2 + 26q + 11 = (2q+1)(4q+11)$

**68.** $2(3z-1)^2 + 3(3z-1) + 1$

Let $u = 3z - 1$. This gives us $2u^2 + 3u + 1$.
$a \cdot c = 2 \cdot 1 = 2$
We need two factors of 2 whose sum is 3. Since the product is positive, the factors will have the same sign. Since the sum is also positive, the factors will be positive.

$2u^2 + u + 2u + 1 = u(2u+1) + 1(2u+1)$
$\phantom{2u^2 + u + 2u + 1} = (2u+1)(u+1)$

Now get back in terms of $z$.
$2(3z-1)^2 + 3(3z-1) + 1$
$= (2(3z-1)+1)(3z-1+1)$
$= (6z-2+1)(3z)$
$= 3z(6z-1)$

**70.** ac Method:
First note that there are no common factors and that $a = 24$, $b = 58$, and $c = 9$.
$a \cdot c = 24 \cdot 9 = 216$. We want to determine two factors whose product is 216 and whose sum is 58. Since the product is positive we know the two integers have the same sign. Since the sum is positive we know that they must both be positive and neither factor can be larger than 58.

| factor 1 | factor 2 | sum |
|---|---|---|
| 9 | 24 | 33 too small |
| 6 | 36 | 42 too small |
| 4 | 54 | 58 ← okay |

$24m^2 + 54mn + 4mn + 9n^2$
$= 6m(4m+9n) + n(4m+9n)$
$= (4m+9n)(6m+n)$

Trial and Error Method:
First note that there are no common factors and that $a = 24$, $b = 58$, and $c = 9$.
Since $c$ is positive the signs in our factors will be the same, and since $b$ is positive we know that the signs will both be positive. We will consider factorizations with this form:

$(\_\_m + \_\_n)(\_\_m + \_\_n)$.

Since $a = 24$ can be factored as $1 \cdot 24$, $2 \cdot 12$, $3 \cdot 8$ and $4 \cdot 6$, we have the following forms:

$(m + \_\_n)(24m + \_\_n)$
$(2m + \_\_n)(12m + \_\_n)$
$(3m + \_\_n)(8m + \_\_n)$
$(4m + \_\_n)(6m + \_\_n)$

$|c| = |9| = 9$ can be factored as $1 \cdot 9$ or $3 \cdot 3$.
Since the original expression had no common factors, the binomials we select cannot have a common factor.

## Chapter 5: Polynomials and Polynomial Functions

$(m+9n)(24m+n) \to 24m^2 + 217mn + 9n^2$

$(2m+9n)(12m+n) \to 24m^2 + 110mn + 9n^2$

$(3m+9n)(8m+n) \to 24m^2 + 75mn + 9n^2$

$(4m+9n)(6m+n) \to 24m^2 + 58mn + 9n^2$

The correct factorization is
$24m^2 + 58mn + 9n^2 = (4m+9n)(6m+n)$

72. $t^2 - 5t + 8$

We are looking for two factors of $c = 8$ whose sum is $b = -5$. Since $c$ is positive, the two factors will have the same sign. Since $b$ is negative, the two factors will be negative.

| Factors | $-1, -8$ | $-2, -4$ |
|---|---|---|
| Sum | $-9$ | $-6$ |

Since none of the possibilities work, $t^2 - 5t + 8$ is prime.

74. $a^6 + 7a^3 + 12$

Let $u = a^3$. This gives us $u^2 + 7u + 12$.
We need two factors of $c = 12$ whose sum is $b = 7$. Since the product is positive, the two factors will have the same sign. Since the sum is also positive, the factors will both be positive.
$u^2 + 7u + 12 = (u+3)(u+4)$
Now get back in terms of $a$.
$a^6 + 7a^3 + 12 = (a^3 + 3)(a^3 + 4)$

76. $r^6 - 6r^3 + 8$

Let $u = r^3$. This gives us $u^2 - 6u + 8$.
We need two factors of 8 whose sum is $-6$. Since the product is positive, the two factors will have the same sign. Since the sum is negative, the two factors will be negative.
$u^2 - 6u + 8 = (u-2)(u-4)$
Now get back in terms of $r$.
$r^6 - 6r^3 + 8 = (r^3 - 2)(r^3 - 4)$

78. $p^2 - 14pq + 45q^2$

We are looking for two factors of $c = 45$ whose sum is $b = -14$. Since $c$ is positive, the two factors must have the same sign. Since $b$ is negative, the two factors are negative.

| Factors | $-1, -45$ | $-3, -15$ | $-5, -9$ |
|---|---|---|---|
| Sum | $-46$ | $-18$ | $-14$ |

$p^2 - 14pq + 45q^2 = (p - 5q)(p - 9q)$

80. $9(a+2)^2 - 10(a+2) + 1$

Let $u = a + 2$. This gives us $9u^2 - 10u + 1$.
$a \cdot c = 9 \cdot 1 = 9$
We need two factors of 9 whose sum is $-10$. Since the product is positive, the two factors will have the same sign. Since the sum is negative, the two factors will be negative.

| factor 1 | factor 2 | sum |
|---|---|---|
| $-1$ | $-9$ | $-10 \leftarrow$ okay |
| $-3$ | $-3$ | $-6$ |

$9u^2 - 10u + 1 = 9u^2 - 9u - u + 1$
$= 9u(u-1) - 1(u-1)$
$= (u-1)(9u-1)$

Now get back in terms of $w$.
$9(a+2)^2 - 10(a+2) + 1$
$= (a+2-1)(9(a+2)-1)$
$= (a+1)(9a+18-1)$
$= (a+1)(9a+17)$

82. $5w^2 - 10w + 12$

We need two factors of $c = 12$ whose sum is $b = -10$. Since $c$ is positive, the two factors will have the same sign. Since $b$ is negative, the factors will both be negative.

| Factors | $-1, -12$ | $-2, -6$ | $-3, -4$ |
|---|---|---|---|
| Sum | $-13$ | $-8$ | $-7$ |

Since none of the possibilities work,
$5w^2 - 10w + 12$ is prime.

84. a. $s(3) = -16(3)^2 + 32(3) + 240$
$= -16(9) + 32(3) + 240$
$= -144 + 96 + 240$
$= 192$

Three seconds after the rock is thrown, it is 192 feet above the ocean.

b. $s(t) = -16t^2 + 32t + 240$
$= -16(t^2 - 2t - 15)$
$= -16(t-5)(t+3)$

c. $s(3) = -16(3-5)(3+3)$
$= -16(-2)(6)$
$= 32(6)$
$= 192$

d. Answers may vary. Often it will be easier in factored form.

86. If we know one factor, we can find the other by dividing the polynomial by the known factor.

$$\begin{array}{r} 4x-3 \\ 2x+7 \overline{\smash{\big)}\, 8x^2 + 22x - 21} \\ \underline{-(8x^2 + 28x)} \\ -6x - 21 \\ \underline{-(-6x - 21)} \\ 0 \end{array}$$

The other factor is $4x - 3$.
$8x^2 + 22x - 21 = (2x+7)(4x-3)$.

88. $\frac{1}{4}x^2 + 2x + 3 = \frac{1}{4}x^2 + \frac{8}{4}x + \frac{12}{4}$
$= \frac{1}{4}(x^2 + 8x + 12)$
$= \frac{1}{4}(x+2)(x+6)$

90. $\frac{1}{4}z^2 + \frac{1}{2}z - \frac{15}{4} = \frac{1}{4}z^2 + \frac{2}{4}z - \frac{15}{4}$
$= \frac{1}{4}(z^2 + 2z - 15)$
$= \frac{1}{4}(z-3)(z+5)$

91. $\frac{4}{3}a^2 - \frac{8}{3}a - 32 = \frac{4}{3}a^2 - \frac{8}{3}a - \frac{96}{3}$
$= \frac{4}{3}a^2 - \frac{4}{3} \cdot 2a - \frac{4}{3} \cdot 24$
$= \frac{4}{3}(a^2 - 2a - 24)$
$= \frac{4}{3}(a-6)(a+4)$

92. $\frac{3}{8}b^2 - \frac{15}{8}b - 9 = \frac{3}{8}b^2 - \frac{15}{8}b - \frac{72}{8}$
$= \frac{3}{8}b^2 - \frac{3}{8} \cdot 5b - \frac{3}{8} \cdot 24$
$= \frac{3}{8}(b^2 - 5b - 24)$
$= \frac{3}{8}(b-8)(b+3)$

## 5.6 Preparing for Factoring Special Products

1. $1^2 = 1 \cdot 1 = 1;$ $2^2 = 2 \cdot 2 = 4;$
$3^2 = 3 \cdot 3 = 9;$ $4^2 = 4 \cdot 4 = 16;$
$5^2 = 5 \cdot 5 = 25$

2. $\left(-\frac{3}{2}\right)^2 = \left(-\frac{3}{2}\right)\left(-\frac{3}{2}\right) = \frac{9}{4}$

## 5.6 Exercises

2. $(A+B)(A^2 - AB + B^2)$

4. False; $x^2 + 25$ is prime over the real numbers.

6. Answers will vary. One example would be $x^2 + 8x + 16 = (x+4)^2$

8. $y^2 + 6y + 9 = y^2 + 2 \cdot 3 \cdot y + 3^2$
$= (y+3)^2$

10. $49 - 14d + d^2 = 7^2 - 2 \cdot 7 \cdot d + d^2$
$= (7-d)^2$ or $(d-7)^2$

12. $9z^2 - 6z + 1 = (3z)^2 - 2 \cdot 3z \cdot 1 + 1^2$
$= (3z-1)^2$

14. $16y^2 - 24y + 9 = (4y)^2 - 2 \cdot 4y \cdot 3 + 3^2$
$= (4y-3)^2$

16. $36b^2 + 84b + 49 = (6b)^2 + 2 \cdot 6b \cdot 7 + 7^2$
$= (6b+7)^2$

## Chapter 5: Polynomials and Polynomial Functions

18. $4a^2 + 20ab + 25b^2 = (2a)^2 + 2 \cdot 2a \cdot 5b + (5b)^2$
$= (2a + 5b)^2$

20. $4c^2 - 24c + 36 = 4(c^2 - 6c + 9)$
$= 4(c^2 - 2 \cdot 3 \cdot c + 3^2)$
$= 4(c - 3)^2$

22. $-2a^2 - 32a - 128 = -2(a^2 + 16a + 64)$
$= -2(a^2 + 2 \cdot 8 \cdot a + 8^2)$
$= -2(a + 8)^2$

24. $12x^2 - 84xy + 147y^2$
$= 3(4x^2 - 24xy + 49y^2)$
$= 3((2x)^2 - 2 \cdot 2x \cdot 7y + (7y)^2)$
$= 3(2x - 7y)^2$

26. $b^4 + 8b^2 + 16 = (b^2)^2 + 2 \cdot 4 \cdot b^2 + 4^2$
$= (b^2 + 4)^2$

28. $z^2 - 64 = z^2 - 8^2$
$= (z - 8)(z + 8)$

30. $81 - a^2 = 9^2 - a^2$
$= (9 - a)(9 + a)$

32. $16y^2 - 81 = (4y)^2 - 9^2$
$= (4y - 9)(4y + 9)$

35. $m^4 - 36n^2 = (m^2)^2 - (6n)^2$
$= (m^2 - 6n)(m^2 + 6n)$

37. $8p^2 - 18q^2 = 2(4p^2 - 9q^2)$
$= 2((2p)^2 - (3q)^2)$
$= 2(2p - 3q)(2p + 3q)$

39. $80p^2r - 245b^2r = 5r(16p^2 - 49b^2)$
$= 5r((4p)^2 - (7b)^2)$
$= 5r(4p - 7b)(4p + 7b)$

41. $(x + y)^2 - 9 = (x + y)^2 - 3^2$
$= (x + y - 3)(x + y + 3)$

43. $x^3 - 8 = x^3 - 2^3$
$= (x - 2)(x^2 + x(2) + 2^2)$
$= (x - 2)(x^2 + 2x + 4)$

45. $125 + m^3 = 5^3 + m^3$
$= (5 + m)(5^2 - 5m + m^2)$
$= (5 + m)(25 - 5m + m^2)$
$= (m + 5)(m^2 - 5m + 25)$

46. $216 - n^3 = 6^3 - n^3$
$= (6 - n)(6^2 + 6 \cdot n + n^2)$
$= (6 - n)(36 + 6n + n^2)$
$= (6 - n)(n^2 + 6n + 36)$

48. $m^6 - 27n^3 = (m^2)^3 - (3n)^3$
$= (m^2 - 3n)((m^2)^2 + m^2 \cdot 3n + (3n)^2)$
$= (m^2 - 3n)(m^4 + 3m^2n + 9n^2)$

50. $16m^3 + 54n^3$
$= 2(8m^3 + 27n^3)$
$= 2((2m)^3 + (3n)^3)$
$= 2(2m + 3n)((2m)^2 - 2m \cdot 3n + (3n)^2)$
$= 2(2m + 3n)(4m^2 - 6mn + 9n^2)$

**Section 5.6** Factoring Special Products

52. $(y-2)^3 - 8$
$= (y-2)^3 - 2^3$
$= ((y-2)-2)((y-2)^2 + (y-2)\cdot 2 + 2^2)$
$= (y-2-2)(y^2 - 4y + 4 + 2y - 4 + 4)$
$= (y-4)(y^2 - 2y + 4)$

54. $(2z+3)^3 + 27z^3$
$= (2z+3)^3 + (3z)^3$
$= (2z+3+3z) + ((2z+3)^2 - (2z+3)(3z) + (3z)^2)$
$= (5z+3)(4z^2 + 12z + 9 - 6z^2 - 9z + 9z^2)$
$= (5z+3)(7z^2 + 3z + 9)$

56. $(y+5)^3 + (y-5)^3$
$= (y+5+y-5)((y+5)^2 - (y+5)(y-5) + (y-5)^2)$
$= (2y)(y^2 + 10y + 25 - (y^2 - 25) + y^2 - 10y + 25)$
$= 2y(2y^2 + 50 - y^2 + 25)$
$= 2y(y^2 + 75)$

57. $y^6 + z^9 = (y^2)^3 + (z^3)^3$
$= (y^2 + z^3)((y^2)^2 - y^2 z^3 + (z^3)^2)$
$= (y^2 + z^3)(y^4 - y^2 z^3 + z^6)$

58. $m^9 + n^{12} = (m^3)^3 + (n^4)^3$
$= (m^3 + n^4)((m^3)^2 - m^3 n^4 + (n^4)^2)$
$= (m^3 + n^4)(m^6 - m^3 n^4 + n^8)$

59. $y^9 - 1 = (y^3)^3 - 1^3$
$= (y^3 - 1)((y^3)^2 + y^3 \cdot 1 + 1^2)$
$= (y^3 - 1)(y^6 + y^3 + 1)$
$= (y^3 - 1^3)(y^6 + y^3 + 1)$
$= (y-1)(y^2 + y + 1)(y^6 + y^3 + 1)$

60.
$x^{12} - 1 = (x^6)^2 - (1)^2$
$= (x^6 + 1)(x^6 - 1)$
$= ((x^2)^3 + 1^3)((x^3)^2 - 1^2)$
$= (x^2 + 1)((x^2)^2 - x^2 + 1)(x^3 - 1)(x^3 + 1)$
$= (x^2 + 1)(x^4 - x^2 + 1)(x-1)(x^2 + x + 1)(x+1)(x^2 - x + 1)$
$= (x+1)(x-1)(x^2 + 1)(x^2 - x + 1)(x^2 + x + 1)(x^4 - x^2 + 1)$

62. $9a^2 - b^2 = (3a)^2 - b^2$
$= (3a - b)(3a + b)$

64. $64x^3 - 125 = (4x)^3 - 5^3$
$= (4x - 5)((4x)^2 + 4x \cdot 5 + 5^2)$
$= (4x - 5)(16x^2 + 20x + 25)$

66. $p^2 - 20p + 100 = p^2 - 2 \cdot 10 \cdot p + 10^2$
$= (p - 10)^2$

68. $3m^4 - 81mn^3 = 3m(m^3 - 27n^3)$
$= 3m(m^3 - (3n)^3)$
$= 3m(m - 3n)(m^2 + m \cdot 3n + (3n)^2)$
$= 3m(m - 3n)(m^2 + 3mn + 9n^2)$

70. $81p^2 - 72pq + 16q^2 = (9p)^2 - 2 \cdot 9p \cdot 4q + (4q)^2$
$= (9p - 4q)^2$

72. $p^4 - 18p^2 + 81 = (p^2)^2 - 2 \cdot 9 \cdot p^2 + 9^2$
$= (p^2 - 9)^2$
$= (p^2 - 3^2)^2$
$= [(p-3)(p+3)]^2$
$= (p-3)^2 (p+3)^2$

74. $9m^2n^2 - 30mn + 25 = (3mn)^2 - 2 \cdot 5 \cdot 3mn + 5^2$
$= (3mn - 5)^2$

76. $p^2 + 8p + 16 - q^2 = (p^2 + 8p + 16) - q^2$
$= (p^2 + 2 \cdot 4 \cdot p + 4^2) - q^2$
$= (p+4)^2 - q^2$
$= (p+4-q)(p+4+q)$
$= (p-q+4)(p+q+4)$

78. $2a^2 - 2b^2 - 24b - 72$
$= 2(a^2 - b^2 - 12b - 36)$
$= 2[a^2 - (b^2 + 12b + 36)]$
$= 2[a^2 - (b^2 + 2 \cdot 6 \cdot b + 6^2)]$
$= 2[a^2 - (b+6)^2]$
$= 2(a - (b+6))(a + (b+6))$
$= 2(a - b - 6)(a + b + 6)$

80. $-5a^3 - 40 = -5(a^3 + 8)$
$= -5(a^3 + 2^3)$
$= -5(a+2)(a^2 - 2a + 4)$

82. $36m^2 + 12mn + n^2 - 81$
$= (36m^2 + 12mn + n^2) - 81$
$= ((6m)^2 + 2 \cdot 6m \cdot n + n^2) - 81$
$= (6m + n)^2 - 81$
$= (6m + n)^2 - 9^2$
$= (6m + n - 9)(6m + n + 9)$

84. The area of the shaded region is the area of the larger square minus the area of the smaller square.
$x^2 - 4^2 = (x-4)(x+4)$
The area of the shaded region is $(x-4)(x+4)$ square units.

86. The area of the shaded region is the area of the large square minus the total area of the four square corners.
$x^2 - 4(3^2) = x^2 - 36$
$= x^2 - 6^2$
$= (x-6)(x+6)$
The area of the shaded region is $(x-6)(x+6)$ square units..

88. The volume of the shaded region is the volume of the larger cylinder minus the volume of the smaller cylinder in the middle.
$V = \pi R^2 h - \pi r^2 h$
$= \pi h(R^2 - r^2)$
$= \pi h(R - r)(R + r)$
The volume of the shaded region is $\pi h(R-r)(R+r)$ cubic units.

90. The volume of the shaded region is the volume of the larger sphere minus the volume of the smaller sphere in the middle.
$V = \frac{4}{3}\pi R^3 - \frac{4}{3}\pi r^3$
$= \frac{4}{3}\pi(R^3 - r^3)$
$= \frac{4}{3}\pi(R - r)(R^2 + R \cdot r + r^2)$
The volume of the shaded region is
$\frac{4}{3}\pi(R-r)(R^2 + R \cdot r + r^2)$ cubic units.

92. $A = (a-b) \cdot b + (a-b) \cdot a$
$= ab - b^2 + a^2 - ab$
$= a^2 - b^2$
$A = (a-b) \cdot b + (a-b) \cdot a$
$= (a-b)(b+a)$
$= (a-b)(a+b)$

Therefore, $a^2 - b^2 = (a-b)(a+b)$.

94. $16y^2 + 24y + c$

To be a perfect square trinomial of this form, the middle term must equal twice the product of the quantities that are squared to get the first and last terms.

$16y^2 = (4y)^2$ and $c = (\sqrt{c})^2$

Thus, the middle term must equal twice the product of $4y$ and $\sqrt{c}$.

$24y = 2 \cdot 4y \cdot \sqrt{c}$

$3 = \sqrt{c}$

$3^2 = c$

$9 = c$

Therefore, we need $c = 9$ to make the expression a perfect square trinomial.

96. $4x^2 + 36x$

Note that $4x^2 = (2x)^2$.

To be a perfect square trinomial, we need to have the form $a^2 + 2ab + b^2$. Here we know that $a = 2x$ and we want $2ab = 36x$.

$2ab = 36x$

$2(2x)b = 36x$

$4bx = 36x$

$4b = 36$

$b = 9$

Since $b = 9$, we need to add $9^2 = 81$ to make a perfect square trinomial.

$4x^2 + 36x + 81 = (2x+9)^2$

98. $x^2 + 0.6x + 0.09 = x^2 + 2 \cdot 0.3 \cdot x + (0.3)^2$
$= (x+0.3)^2$

100. $100x^2 - \dfrac{1}{81} = (10x)^2 - \left(\dfrac{1}{9}\right)^2$
$= \left(10x - \dfrac{1}{9}\right)\left(10x + \dfrac{1}{9}\right)$

102. $\dfrac{a^2}{36} - \dfrac{b^2}{49} = \left(\dfrac{a}{6}\right)^2 - \left(\dfrac{b}{7}\right)^2$
$= \left(\dfrac{a}{6} - \dfrac{b}{7}\right)\left(\dfrac{a}{6} + \dfrac{b}{7}\right)$

104. $\dfrac{a^3}{27} + \dfrac{b^3}{64} = \left(\dfrac{a}{3}\right)^3 + \left(\dfrac{b}{4}\right)^3$
$= \left(\dfrac{a}{3} + \dfrac{b}{4}\right)\left(\left(\dfrac{a}{3}\right)^2 - \dfrac{a}{3} \cdot \dfrac{b}{4} + \left(\dfrac{b}{4}\right)^2\right)$
$= \left(\dfrac{a}{3} + \dfrac{b}{4}\right)\left(\dfrac{a^2}{9} - \dfrac{ab}{12} + \dfrac{b^2}{16}\right)$

## 5.7 Exercises

2. Answers will vary. "Factored completely" means that there are no common factors remaining.

4. $3x^2 + 6x - 105 = 3(x^2 + 2x - 35)$
$= 3(x-5)(x+7)$

6. $-5a^2 + 80 = -5(a^2 - 16)$
$= -5(a-4)(a+4)$

8. $a \cdot c = 8 \cdot 49 = 392$

We need two factors of 392 that add to get $-42$. Since the product is positive, they must have the same sign, and since the sum is negative, the factors will both be negative with neither being less than $-42$.

| factor 1 | factor 2 | sum |
|---|---|---|
| $-7$ | $-56$ | $-63$ too small |
| $-8$ | $-49$ | $-57$ too small |
| $-14$ | $-28$ | $-42$ okay |

$8m^2 - 42m + 49 = 8m^2 - 28m - 14m + 49$
$= 4m(2m-7) - 7(2m-7)$
$= (2m-7)(4m-7)$

10. $54p^6 - 2q^3 = 2(27p^6 - q^3)$
$= 2\left((3p^2)^3 - q^3\right)$
$= 2(3p^2 - q)\left((3p^2)^2 + 3p^2q + q^2\right)$
$= 2(3p^2 - q)(9p^4 + 3p^2q + q^2)$

12. $-4c^3 + 16c^2 - 28c = -4c(c^2 - 4c + 7)$

14. $a \cdot c = 18 \cdot -20 = -360$
    We need two factors of $-360$ that add to get $-9$. Since the product is negative, they must have different signs, and since the sum is negative, the factor with the larger absolute value must be negative. The sum is also not too far from 0 so we want the factors to be near each other in magnitude.

    | factor 1 | factor 2 | sum |
    |---|---|---|
    | 12 | −30 | −18 too small |
    | 18 | −20 | −2 too large |
    | 15 | −24 | −9 okay |

    $18t^2 - 9t - 20 = 18t^2 + 15t - 24t - 20$
    $= 3t(6t + 5) - 4(6t + 5)$
    $= (6t + 5)(3t - 4)$

16. $p^3 + 7p^2 - 3p - 21 = (p^3 + 7p^2) + (-3p - 21)$
    $= p^2(p + 7) - 3(p + 7)$
    $= (p + 7)(p^2 - 3)$

18. $12p^2 + 50q^2 = 2(6p^2 + 25q^2)$

20. $16w^4 - 1 = (4w^2)^2 - 1^2$
    $= (4w^2 - 1)(4w^2 + 1)$
    $= ((2w)^2 - 1^2)(4w^2 + 1)$
    $= (2w - 1)(2w + 1)(4w^2 + 1)$

22. $a \cdot c = 4 \cdot -6 = -24$
    We need two factors of $-24$ that add to get $-3$. Since the product is negative, they must have different signs, and since the sum is negative, the factor with the larger absolute value must be negative.

    | factor 1 | factor 2 | sum |
    |---|---|---|
    | 1 | −24 | −23 |
    | 2 | −12 | −10 |
    | 3 | −8 | −5 |
    | 4 | −6 | −2 |

    None of the possibilities work. The expression is prime.

24. $20k^3 - 60k^2 + 45k = 5k(4k^2 - 12k + 9)$
    $= 5k((2k)^2 - 2 \cdot 2k \cdot 3 + 3^2)$
    $= 5k(2k - 3)^2$

26. $20p^3q - 2p^2q - 4pq = 2pq(10p^2 - p - 2)$
    Focusing on the reduced trinomial we get
    $a \cdot c = 10 \cdot -2 = -20$
    We need two factors of $-20$ that sum to $-1$. Since the product is negative, the factors will have opposite signs. Since the sum is negative, the factor with the larger absolute value is negative.

    | factor 1 | factor 2 | sum |
    |---|---|---|
    | 1 | −20 | −19 too small |
    | 2 | −10 | −8 too small |
    | 4 | −5 | −1 okay |

    $20p^3q - 2p^2q - 4pq$
    $= 2pq(10p^2 - p - 2)$
    $= 2pq(10p^2 - 5p + 4p - 2)$
    $= 2pq(5p(2p - 1) + 2(2p - 1))$
    $= 2pq(2p - 1)(5p + 2)$

28. $54p^5 + 16p^2q^3$
    $= 2p^2(27p^3 + 8q^3)$
    $= 2p^2((3p)^3 + (2q)^3)$
    $= 2p^2(3p + 2q)((3p)^2 - (3p)(2q) + (2q)^2)$
    $= 2p^2(3p + 2q)(9p^2 - 6pq + 4q^2)$

Section 5.7 Factoring: A General Strategy

30. $3x^3 - 6x^2 - 48x + 96$
$= 3(x^3 - 2x^2 - 16x + 32)$
$= 3((x^3 - 2x^2) + (-16x + 32))$
$= 3(x^2(x-2) - 16(x-2))$
$= 3(x-2)(x^2 - 16)$
$= 3(x-2)(x-4)(x+4)$

32. $4z^4 - 25 = (2z^2)^2 - 5^2$
$= (2z^2 - 5)(2z^2 + 5)$

34. $4b^4 + 4b^2 - 15$
Let $u = b^2$. This gives us $4u^2 + 4u - 15$.
$a \cdot c = 4 \cdot -15 = -60$
We need two factors of $-60$ whose sum is 4. Since the product is negative, the two factors will have opposite signs. Since the sum is positive, the factor with the larger absolute value will be positive.

| factor 1 | factor 2 | sum |     |
|----------|----------|-----|-----|
| $-2$     | 30       | 28  | too large |
| $-6$     | 10       | 4   | okay |

$4u^2 + 4u - 15 = 4u^2 + 10u - 6u - 15$
$= 2u(2u+5) - 3(2u+5)$
$= (2u+5)(2u-3)$

Now get back in terms of $b$.
$4b^4 + 4b^2 - 15 = (2b^2 + 5)(2b^2 - 3)$

36. $(3x+5)^2 + 4(3x+5) - 21$
Let $u = 3x+5$. This gives us
$u^2 + 4u - 21 = (u-3)(u+7)$
Now get back in terms of $x$.
$(3x+5)^2 + 4(3x+5) - 21$
$= (3x+5-3)(3x+5+7)$
$= (3x+2)(3x+12)$
Notice that the second factor here has a common factor of 3 than can be pulled out. Therefore,
$(3x+5)^2 + 4(3x+5) - 21 = 3(3x+2)(x+4)$

38. $a^2 + 12a + 36 - 4b^2 = (a^2 + 12a + 36) - 4b^2$
$= (a^2 + 2 \cdot 6 \cdot a + 6^2) - 4b^2$
$= (a+6)^2 - (2b)^2$
$= (a+6-2b)(a+6+2b)$
$= (a-2b+6)(a+2b+6)$

40. $w^6 + 4w^3 - 5$
Let $u = w^3$. This gives us
$u^2 + 4u - 5 = (u-1)(u+5)$
Now get back in terms of $w$.
$w^6 + 4w^3 - 5 = (w^3 - 1)(w^3 + 5)$
$= (w^3 - 1^3)(w^3 + 5)$
$= (w-1)(w^2 + w + 1)(w^3 + 5)$

42. $q^6 + 1 = (q^2)^3 + 1^3$
$= (q^2 + 1)((q^2)^2 - (q^2)(1) + 1^2)$
$= (q^2 + 1)(q^4 - q^2 + 1)$

44. $-2y^3 - 4y^2 + 32y + 64$
$= -2(y^3 + 2y^2 - 16y - 32)$
$= -2(y^2(y+2) - 16(y+2))$
$= -2(y+2)(y^2 - 16)$
$= -2(y+2)(y-4)(y+4)$

46. $-5z - 20z^3 = -5z(1 + 4z^2) = -5z(4z^2 + 1)$

48. $18h^5 + 154h^3 - 72h = 2h(9h^4 + 77h^2 - 36)$
Focusing on the reduced trinomial we get:
$a \cdot c = 9 \cdot -36 = -324$
We need two factors of $-324$ that add to get 77. Since the product is negative, they must have different signs, and since the sum is positive, the factor with the larger absolute value must be positive.

*Chapter 5: Polynomials and Polynomial Functions*

| factor 1 | factor 2 | sum | |
|---|---|---|---|
| −9 | 36 | 27 | too small |
| −2 | 162 | 160 | too large |
| −4 | 81 | 77 | okay |

$$18h^5 + 154h^3 - 72h = 2h(9h^4 + 77h^2 - 36)$$
$$= 2h(9h^4 + 81h^2 - 4h^2 - 36)$$
$$= 2h(9h^2(h^2 + 9) - 4(h^2 + 9))$$
$$= 2h(h^2 + 9)(9h^2 - 4)$$
$$= 2h(h^2 + 9)((3h)^2 - 2^2)$$
$$= 2h(h^2 + 9)(3h - 2)(3h + 2)$$

50. $2p^4q + 14p^3q - 32p^2q - 224pq$
$$= 2pq(p^3 + 7p^2 - 16p - 112)$$
$$= 2pq(p^2(p + 7) - 16(p + 7))$$
$$= 2pq(p + 7)(p^2 - 16)$$
$$= 2pq(p + 7)(p - 4)(p + 4)$$

52. The area of the shaded region is the area of the large rectangle minus the area of the small rectangle.
$$(2x + 1)(x + 2) - (x + 2)(x - 1)$$
$$= (x + 2)[(2x + 1) - (x - 1)]$$
$$= (x + 2)(2x + 1 - x + 1)$$
$$= (x + 2)(x + 2)$$
$$= (x + 2)^2$$
The area of the shaded region is $(x + 2)^2$ square units.

54. We can get the area of the entire shaded region by finding the areas of the smaller square and rectangular regions and then adding the results together.
$$x \cdot y + x^2 + y^2 + x \cdot y = x^2 + 2xy + y^2$$
$$= (x + y)^2$$
The total area is $(x + y)^2$ square units.

56. $(5x)^3 - (2y)^3$
$$= (5x - 2y)((5x)^2 + (5x)(2y) + (2y)^2)$$
$$= (5x - 2y)(25x^2 + 10xy + 4y^2)$$
The volumes of the boxes differ by
$(5x - 2y)(25x^2 + 10xy + 4y^2)$ cubic units.

58. Answers will vary. There is no pair of integer factors of $-8$ whose sum is $-4$.

60. $x^{3/2} - 4x^{-1/2} = x^{-1/2}(x^2 - 4)$
$$= x^{-1/2}(x^2 - 2^2)$$
$$= x^{-1/2}(x - 2)(x + 2)$$

62. $3 - 2x^{-1} - x^{-2} = x^{-2}(3x^2 - 2x - 1)$
$$= x^{-2}(3x + 1)(x - 1)$$

### 5.8 Preparing for Polynomial Equations

1. $x + 4 = 0$
$$x + 4 - 4 = 0 - 4$$
$$x = -4$$
Solution set: $\{-4\}$

2. $3(x - 2) - 12 = 0$
$$3x - 6 - 12 = 0$$
$$3x - 18 = 0$$
$$3x - 18 + 18 = 0 + 18$$
$$3x = 18$$
$$\frac{3x}{3} = \frac{18}{3}$$
$$x = 6$$
Solution set: $\{6\}$

3. $2x^2 + 3x + 1$

   a. When $x = 2$, we get
   $$2x^2 + 3x + 1 = 2(2)^2 + 3(2) + 1$$
   $$= 2(4) + 3(2) + 1$$
   $$= 8 + 6 + 1$$
   $$= 15$$

b. When $x = -1$, we get
$$\begin{aligned} 2x^2 + 3x + 1 &= 2(-1)^2 + 3(-1) + 1 \\ &= 2(1) + 3(-1) + 1 \\ &= 2 - 3 + 1 \\ &= 0 \end{aligned}$$

4. $f(x) = 4x + 3$
$$11 = 4x + 3$$
$$11 - 3 = 4x + 3 - 3$$
$$8 = 4x$$
$$\frac{8}{4} = \frac{4x}{4}$$
$$2 = x$$
Solution set: $\{2\}$
The point $(2,11)$ is on the graph of $f$.

5. $f(x) = -2x + 8$
$$f(4) = -2(4) + 8$$
$$= -8 + 8$$
$$= 0$$
The point $(4,0)$ is on the graph of $f$.

## 5.8 Exercises

2. quadratic equation

4. True

6. True

8. Answers will vary. The degree of the equation indicates the maximum number of solutions that are possible.

10. $(x+3)(x-8) = 0$
$x + 3 = 0$ or $x - 8 = 0$
$x = -3$ $\quad\quad x = 8$
The solution set is $\{-3, 8\}$.

12. $4x(2x-3) = 0$
$4x = 0$ or $2x - 3 = 0$
$x = 0$ $\quad\quad 2x = 3$
$\quad\quad\quad\quad x = \frac{3}{2}$
The solution set is $\left\{0, \frac{3}{2}\right\}$.

14. $3a(a-9)(a+11) = 0$
$3a = 0$ or $a - 9 = 0$ or $a + 11 = 0$
$a = 0$ $\quad\quad a = 9$ $\quad\quad a = -11$
The solution set is $\{-11, 9, 0\}$.

16. $5c^2 + 15c = 0$
$5c(c+3) = 0$
$5c = 0$ or $c + 3 = 0$
$c = 0$ $\quad\quad c = -3$
The solution set is $\{-3, 0\}$

18. $\quad\quad 4t^2 = -20t$
$4t^2 + 20t = 0$
$4t(t+5) = 0$
$4t = 0$ or $t + 5 = 0$
$t = 0$ $\quad\quad t = -5$
The solution set is $\{-5, 0\}$.

20. $x^2 + 3x - 40 = 0$
$(x+8)(x-5) = 0$
$x + 8 = 0$ or $x - 5 = 0$
$x = -8$ $\quad\quad x = 5$
The solution set is $\{-8, 5\}$.

22. $\quad\quad y^2 + 13y = -42$
$y^2 + 13y + 42 = 0$
$(y+6)(y+7) = 0$
$y + 6 = 0$ or $y + 7 = 0$
$y = -6$ $\quad\quad y = -7$
The solution set is $\{-7, -6\}$.

## Chapter 5: Polynomials and Polynomial Functions

**24.** $a^2 + 12a + 36 = 0$
$(a+6)^2 = 0$
$a + 6 = 0$
$a = -6$
The solution set is $\{-6\}$.

**26.** $4c^2 + 6 = 25c$
$4c^2 - 25c + 6 = 0$
$(4c-1)(c-6) = 0$
$4c - 1 = 0$ or $c - 6 = 0$
$4c = 1 \quad\quad c = 6$
$c = \dfrac{1}{4}$
The solution set is $\left\{\dfrac{1}{4}, 6\right\}$.

**28.** $6z^2 + 17z = -5$
$6z^2 + 17z + 5 = 0$
$(3z+1)(2z+5) = 0$
$3z + 1 = 0$ or $2z + 5 = 0$
$3z = -1 \quad\quad 2z = -5$
$z = -\dfrac{1}{3} \quad\quad z = -\dfrac{5}{2}$
The solution set is $\left\{-\dfrac{5}{2}, -\dfrac{1}{3}\right\}$.

**30.** $4y^2 - 20y - 56 = 0$
$4(y^2 - 5y - 14) = 0$
$4(y-7)(y+2) = 0$
$y - 7 = 0$ or $y + 2 = 0$
$y = 7 \quad\quad y = -2$
The solution set is $\{-2, 7\}$.

**32.** $-6n^2 - 9n + 60 = 0$
$-3(2n^2 + 3n - 20) = 0$
$-3(2n-5)(n+4) = 0$
$2n - 5 = 0$ or $n + 4 = 0$
$2n = 5 \quad\quad n = -4$
$n = \dfrac{5}{2}$
The solution set is $\left\{-4, \dfrac{5}{2}\right\}$.

**34.** $\dfrac{1}{2}t^2 - 3t - 8 = 0$
$2\left(\dfrac{1}{2}t^2 - 3t - 8\right) = 2(0)$
$t^2 - 6t - 16 = 0$
$(t-8)(t+2) = 0$
$t - 8 = 0$ or $t + 2 = 0$
$t = 8 \quad\quad t = -2$
The solution set is $\{-2, 8\}$.

**36.** $\dfrac{2}{3}x^2 + \dfrac{7}{3}x = 5$
$\dfrac{2}{3}x^2 + \dfrac{7}{3}x - 5 = 0$
$3\left(\dfrac{2}{3}x^2 + \dfrac{7}{3}x - 5\right) = 3(0)$
$2x^2 + 7x - 15 = 0$
$(2x-3)(x+5) = 0$
$2x - 3 = 0$ or $x + 5 = 0$
$2x = 3 \quad\quad x = -5$
$x = \dfrac{3}{2}$
The solution set is $\left\{-5, \dfrac{3}{2}\right\}$.

The solution set is $\{-11, 3\}$.

## Section 5.8 Polynomial Equations

**38.**
$y(y+4) = 45$
$y^2 + 4y = 45$
$y^2 + 4y - 45 = 0$
$(y+9)(y-5) = 0$
$y+9 = 0$ or $y-5 = 0$
$y = -9 \qquad y = 5$
The solution set is $\{-9, 5\}$.

**40.** $(y-4)(y-1)(3y+2) = 0$
$y-4 = 0$ or $y-1 = 0$ or $3y+2 = 0$
$y = 4 \qquad y = 1 \qquad 3y = -2$
$\qquad\qquad\qquad\qquad\qquad y = -\dfrac{2}{3}$

The solution set is $\left\{-\dfrac{2}{3}, 1, 4\right\}$.

**42.**
$7q^3 + 31q^2 = -12q$
$7q^3 + 31q^2 + 12q = 0$
$q(7q^2 + 31q + 12) = 0$
$q(7q+3)(q+4) = 0$
$q = 0$ or $7q+3 = 0$ or $q+4 = 0$
$\qquad\qquad 7q = -3 \qquad\quad q = -4$
$\qquad\qquad q = -\dfrac{3}{7}$

The solution set is $\left\{-4, -\dfrac{3}{7}, 0\right\}$.

**44.**
$w^3 + 5w^2 - 16w - 80 = 0$
$w^2(w+5) - 16(w+5) = 0$
$(w+5)(w^2 - 16) = 0$
$(w+5)(w-4)(w+4) = 0$
$w+5 = 0$ or $w-4 = 0$ or $w+4 = 0$
$w = -5 \qquad w = 4 \qquad w = -4$
The solution set is $\{-5, -4, 4\}$.

**46.**
$-24b^3 + 27b = 18b^2$
$-24b^3 - 18b^2 + 27b = 0$
$-3b(8b^2 + 6b - 9) = 0$
$-3b(4b-3)(2b+3) = 0$
$-3b = 0$ or $4b-3 = 0$ or $2b+3 = 0$
$b = 0 \qquad 4b = 3 \qquad\quad 2b = -3$
$\qquad\qquad b = \dfrac{3}{4} \qquad\quad b = -\dfrac{3}{2}$

The solution set is $\left\{-\dfrac{3}{2}, 0, \dfrac{3}{4}\right\}$.

**48.**
$(x+2)^3 = x^3 - 2x$
$(x+2)(x+2)(x+2) = x^3 - 2x$
$(x^2 + 4x + 4)(x+2) = x^3 - 2x$
$x^3 + 4x^2 + 4x + 2x^2 + 8x + 8 = x^3 - 2x$
$6x^2 + 14x + 8 = 0$
$2(3x^2 + 7x + 4) = 0$
$2(3x+4)(x+1) = 0$
$3x+4 = 0$ or $x+1 = 0$
$3x = -4 \qquad\quad x = -1$
$x = -\dfrac{4}{3}$

The solution set is $\left\{-\dfrac{4}{3}, -1\right\}$.

**50. a.**
$f(x) = 3$
$x^2 + 5x + 3 = 3$
$x^2 + 5x = 0$
$x(x+5) = 0$
$x = 0$ or $x+5 = 0$
$\qquad\qquad x = -5$
The solution set is $\{-5, 0\}$.

## Chapter 5: Polynomials and Polynomial Functions

**b.** $f(x) = 17$
$x^2 + 5x + 3 = 17$
$x^2 + 5x - 14 = 0$
$(x+7)(x-2) = 0$
$x + 7 = 0$ or $x - 2 = 0$
$x = -7 \qquad x = 2$
The solution set is $\{-7, 2\}$.

The points $(-5, 3)$, $(0, 3)$, $(-7, 17)$, and $(2, 17)$ are on the graph of $f$.

**52. a.** $h(x) = -8$
$3x^2 - 9x - 8 = -8$
$3x^2 - 9x = 0$
$3x(x - 3) = 0$
$3x = 0$ or $x - 3 = 0$
$x = 0 \qquad x = 3$
The solution set is $\{0, 3\}$.

**b.** $h(x) = 22$
$3x^2 - 9x - 8 = 22$
$3x^2 - 9x - 30 = 0$
$3(x^2 - 3x - 10) = 0$
$3(x - 5)(x + 2) = 0$
$x - 5 = 0$ or $x + 2 = 0$
$x = 5 \qquad x = -2$
The solution set is $\{-2, 5\}$.

The points $(0, -8)$, $(3, -8)$, $(-2, 22)$, and $(5, 22)$ are on the graph of $h$.

**54. a.** $G(x) = 1$
$-x^2 + 4x + 6 = 1$
$-x^2 + 4x + 5 = 0$
$-(x^2 - 4x - 5) = 0$
$-(x - 5)(x + 1) = 0$
$x - 5 = 0$ or $x + 1 = 0$
$x = 5 \qquad x = -1$
The solution set is $\{-1, 5\}$.

**b.** $G(x) = 9$
$-x^2 + 4x + 6 = 9$
$-x^2 + 4x - 3 = 0$
$-(x^2 - 4x + 3) = 0$
$-(x - 1)(x - 3) = 0$
$x - 1 = 0$ or $x - 3 = 0$
$x = 1 \qquad x = 3$
The solution set is $\{1, 3\}$.

The points $(-1, 1)$, $(5, 1)$, $(1, 9)$, and $(3, 9)$ are on the graph of $G$.

**56.** $f(x) = x^2 - 13x + 42$
$0 = x^2 - 13x + 42$
$0 = (x - 7)(x - 6)$
$x - 7 = 0$ or $x - 6 = 0$
$x = 7 \qquad x = 6$
The zeros of $f(x)$ are 6 and 7.
The $x$-intercepts are 6 and 7.

**58.** $h(p) = 8p^2 - 18p - 35$
$0 = 8p^2 - 18p - 35$
$0 = (4p + 5)(2p - 7)$
$4p + 5 = 0$ or $2p - 7 = 0$
$4p = -5 \qquad 2p = 7$
$p = -\frac{5}{4} \qquad p = \frac{7}{2}$
The zeros of $h(p)$ are $-\frac{5}{4}$ and $\frac{7}{2}$.
The $x$-intercepts are $-\frac{5}{4}$ and $\frac{7}{2}$.

**60.** $f(a) = 3a^3 - 15a^2 - 42a$
$0 = 3a^3 - 15a^2 - 42a$
$0 = 3a(a^2 - 5a - 14)$
$0 = 3a(a - 7)(a + 2)$
$3a = 0$ or $a - 7 = 0$ or $a + 2 = 0$
$a = 0 \qquad a = 7 \qquad a = -2$
The zeros of $f(a)$ are $-2$, 0, and 7.
The $x$-intercepts are $-2$, 0, and 7.

**Section 5.8** Polynomial Equations

62. $(x+7)(x-3) = 11$
$x^2 - 3x + 7x - 21 = 11$
$x^2 + 4x - 21 = 11$
$x^2 + 4x - 32 = 0$
$(x+8)(x-4) = 0$
$x+8 = 0$ or $x-4 = 0$
$x = -8 \qquad x = 4$
The solution set is $\{-8, 4\}$.

64. $3t^2 + 7t - 20 = 0$
$(3t-5)(t+4) = 0$
$3t - 5 = 0$ or $t + 4 = 0$
$3t = 5 \qquad t = -4$
$t = \frac{5}{3}$
The solution set is $\left\{-4, \frac{5}{3}\right\}$.

66. $-7z^2 + 42z = 0$
$-7z(z-6) = 0$
$-7z = 0$ or $z - 6 = 0$
$z = 0 \qquad z = 6$
The solution set is $\{0, 6\}$.

68. $(x+7)(x-6) = 0$
$x+7 = 0$ or $x-6 = 0$
$x = -7 \qquad x = 6$
The solution set is $\{-7, 6\}$.

70. $2c^3 + 3c^2 - 8c - 12 = 0$
$c^2(2c+3) - 4(2c+3) = 0$
$(2c+3)(c^2-4) = 0$
$(2c+3)(c-2)(c+2) = 0$
$2c+3 = 0$ or $c-2 = 0$ or $c+2 = 0$
$2c = -3 \qquad c = 2 \qquad c = -2$
$c = -\frac{3}{2}$
The solution set is $\left\{-2, -\frac{3}{2}, 2\right\}$.

72. Let $x$ = width, then $2(x+3)$ = length.
$A$ = width · length
$A = x \cdot 2(x+3) = 2x(x+3) = 2x^2 + 6x$
$2x^2 + 6x = 216$
$2x^2 + 6x - 216 = 0$
$2(x^2 + 3x - 108) = 0$
$2(x-9)(x+12) = 0$
$x - 9 = 0$ or $x + 12 = 0$
$x = 9 \qquad \cancel{x = -12}$
The width of the rectangle is 9 inches and the length is 24 inches.

74. Let $x$ = height, then $x - 4$ = base.
$A = \frac{1}{2} \cdot \text{base} \cdot \text{height} = \frac{1}{2}(x-4) \cdot x = \frac{1}{2}x^2 - 2x$
$\frac{1}{2}x^2 - 2x = 48$
$2\left(\frac{1}{2}x^2 - 2x\right) = 2(48)$
$x^2 - 4x = 96$
$x^2 - 4x - 96 = 0$
$(x-12)(x+8) = 0$
$x - 12 = 0$ or $x + 8 = 0$
$x = 12 \qquad \cancel{x = -8}$
The height is 12 meters and the base is 8 meters.

76. $S = 36$
$\frac{n(n+1)}{2} = 36$
$2 \cdot \frac{n(n+1)}{2} = 2 \cdot 36$
$n(n+1) = 72$
$n^2 + n = 72$
$n^2 + n - 72 = 0$
$(n+9)(n-8) = 0$
$n+9 = 0$ or $n-8 = 0$
$\cancel{n = -9} \qquad n = 8$
The first 8 consecutive integers must be added for a sum of 36.

78. Let $x$ = width of the corral. Then the length must be $300 - 3x$.

ISM - 275

## Chapter 5: Polynomials and Polynomial Functions

$$x(300-3x) = 4800$$
$$-3x^2 + 300x = 4800$$
$$-3x^2 + 300x - 4800 = 0$$
$$-3(x^2 - 100x + 1600) = 0$$
$$-3(x-80)(x-20) = 0$$
$$x - 80 = 0 \quad \text{or} \quad x - 20 = 0$$
$$x = 80 \qquad\qquad x = 20$$
$$300 - 3(80) \qquad 300 - 3(20)$$
$$= 300 - 240 \qquad = 300 - 60$$
$$= 60 \qquad\qquad = 240$$

The dimensions of the corral are 60 feet (length) by 80 feet (width) or 240 feet (length) by 20 feet (width).

**80.** Let $x$ = width of the frame. The dimensions of the picture must be $40 - 2x$ by $32 - 2x$.
$$(40-2x)(32-2x) = 1008$$
$$1280 - 80x - 64x + 4x^2 = 1008$$
$$4x^2 - 144x + 272 = 0$$
$$4(x^2 - 36x + 68) = 0$$
$$4(x-2)(x-34) = 0$$
$$x - 2 = 0 \quad \text{or} \quad x - 34 = 0$$
$$x = 2 \qquad\qquad \cancel{x = 34}$$

The picture frame is 2 inches wide.

**82.** Let $x$ = the width of the cardboard. Then $x + 8$ = length. The height is 3 inches (when the sides are folded up).
To obtain the length and width of the box, we subtract 6 from both the length and width of the cardboard because a corner is cut from each end.
$$V = l \cdot w \cdot h$$
$$315 = (x+8-6)(x-6)(3)$$
$$315 = 3(x-6)(x+2)$$
$$105 = (x-6)(x+2)$$
$$105 = x^2 - 4x - 12$$
$$0 = x^2 - 4x - 117$$
$$0 = (x-13)(x+9)$$
$$x - 13 = 0 \quad \text{or} \quad x + 9 = 0$$
$$x = 13 \qquad\qquad \cancel{x = -9}$$

The cardboard measures 13 inches by 21 inches.

**84. a.** $C(x) = \dfrac{1}{2}x^2 - 30x + 475$
$$C(30) = \dfrac{1}{2}(30)^2 - 30(30) + 475$$
$$= \dfrac{1}{2}(900) - 900 + 475$$
$$= 450 - 900 + 475$$
$$= 25$$

The marginal cost for producing the 30th cell phone is $25.

**b.** $C(x) = 75$
$$\dfrac{1}{2}x^2 - 30x + 475 = 75$$
$$2\left(\dfrac{1}{2}x^2 - 30x + 475\right) = 2(75)$$
$$x^2 - 60x + 950 = 150$$
$$x^2 - 60x + 800 = 0$$
$$(x-20)(x-40) = 0$$
$$x - 20 = 0 \quad \text{or} \quad x - 40 = 0$$
$$x = 20 \qquad\qquad x = 40$$

The marginal cost will be $75 for the 20th and 40th cell phones manufactured.

**c.** $C(x) = 97$
$$\dfrac{1}{2}x^2 - 30x + 475 = 97$$
$$2\left(\dfrac{1}{2}x^2 - 30x + 475\right) = 2(97)$$
$$x^2 - 60x + 950 = 194$$
$$x^2 - 60x + 756 = 0$$
$$(x-18)(x-42) = 0$$
$$x - 18 = 0 \quad \text{or} \quad x - 42 = 0$$
$$x = 18 \qquad\qquad x = 42$$

Since production would continue until marginal revenue equals marginal cost, the company should manufacture 18 cell phones.

## Section 5.8 Polynomial Equations

**86. a.**
$$s(t) = 320$$
$$-16t^2 + 64t + 260 = 320$$
$$-16t^2 + 64t - 60 = 0$$
$$-4(4t^2 - 16t + 15) = 0$$
$$4t^2 - 16t + 15 = 0$$
$$(2t-3)(2t-5) = 0$$
$$2t - 3 = 0 \quad \text{or} \quad 2t - 5 = 0$$
$$2t = 3 \qquad \qquad 2t = 5$$
$$t = \frac{3}{2} \qquad \qquad t = \frac{5}{2}$$

The cannonball will be at a height of 320 feet after 1.5 seconds and again after 2.5 seconds.

**b.** When the cannonball hits the ground, its height will be 0. Therefore, we need to solve $s(t) = 0$.
$$s(t) = 0$$
$$-16t^2 + 64t + 260 = 0$$
$$-4(4t^2 - 16t - 65) = 0$$
$$4t^2 - 16t - 65 = 0$$
$$(2t+5)(2t-13) = 0$$
$$2t + 5 = 0 \quad \text{or} \quad 2t - 13 = 0$$
$$2t = -5 \qquad \qquad 2t = 13$$
$$t = -\frac{5}{2} \qquad \qquad t = \frac{13}{2}$$

The cannonball will strike the ground after 6.5 seconds.

**88.**
$$9z^4 - 13z^2 + 4 = 0$$
$$(9z^2 - 4)(z^2 - 1) = 0$$
$$((3z)^2 - 2^2)(z^2 - 1^2) = 0$$
$$(3z - 2)(3z + 2)(z - 1)(z + 1) = 0$$
$$3z - 2 = 0 \text{ or } 3z + 2 = 0 \text{ or } z - 1 = 0 \text{ or } z + 1 = 0$$
$$3z = 2 \qquad 3z = -2 \qquad z = 1 \qquad z = -1$$
$$z = \frac{2}{3} \qquad z = -\frac{2}{3}$$

The solution set is $\left\{-1, -\frac{2}{3}, \frac{2}{3}, 1\right\}$.

**90.**
$$(2b+1)^2 + 7(2b+1) = -12$$
$$(2b+1)^2 + 7(2b+1) + 12 = 0$$
$$((2b+1)+4)((2b+1)+3) = 0$$
$$(2b+5)(2b+4) = 0$$
$$2(2b+5)(b+2) = 0$$
$$2b + 5 = 0 \quad \text{or} \quad b + 2 = 0$$
$$2b = -5 \qquad \qquad b = -2$$
$$b = -\frac{5}{2}$$

The solution set is $\left\{-\frac{5}{2}, -2\right\}$.

**92.** $f(x) = \dfrac{-9}{x^2 + 6x + 5}$

The domain is all real numbers except where the denominator equals 0.
$$x^2 + 6x + 5 = 0$$
$$(x+5)(x+1) = 0$$
$$x + 5 = 0 \quad \text{or} \quad x + 1 = 0$$
$$x = -5 \qquad \qquad x = -1$$

The domain is all real numbers except $-5$ and $-1$. That is, $\{x \mid x \neq -5, x \neq -1\}$.

**94.** $h(x) = \dfrac{x+4}{3x^2 - 7x - 6}$

The domain is all real numbers except where the denominator equals 0.
$$3x^2 - 7x - 6 = 0$$
$$(3x+2)(x-3) = 0$$
$$3x + 2 = 0 \quad \text{or} \quad x - 3 = 0$$
$$3x = -2 \qquad \qquad x = 3$$
$$x = -\frac{2}{3}$$

The domain is all real numbers except $-\frac{2}{3}$ and 3. That is, $\left\{x \mid x \neq -\frac{2}{3}, x \neq 3\right\}$.

**96.** $2x^2 - x - 10 = 0$

## Chapter 5: Polynomials and Polynomial Functions

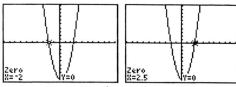

The solution set is $\{-2, 2.5\}$.

98. $0.4x^2 - 2.7x + 1 = 0$

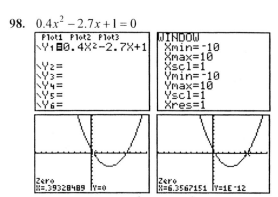

The solution set is $\{0.39, 6.36\}$.

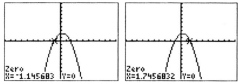

The solution set is $\{-1.15, 1.75\}$.

100. $-3.1x^2 - 0.4x = -3$ or $-3.1x^2 - 0.4x + 3 = 0$

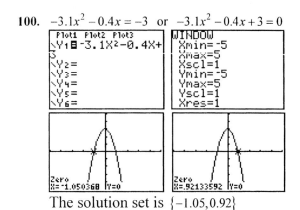

The solution set is $\{-1.05, 0.92\}$

## Chapter 5 Review

1. Coefficient: $-7$
   Degree: 4

2. Coefficient: $\dfrac{1}{9}$
   Degree: 3

3. $7x^3 - 2x^2 + x - 8$
   Degree: 3

4. $y^4 - 3y^2 + 2y + 3$
   Degree: 4

5. $(x^2 + 2x - 7) + (3x^2 - x - 4)$
   $= x^2 + 2x - 7 + 3x^2 - x - 4$
   $= x^2 + 3x^2 + 2x - x - 7 - 4$
   $= (1+3)x^2 + (2-1)x + (-7-4)$
   $= 4x^2 + x - 11$

6. $(4x^3 - 3x^2 + x - 5) - (x^4 + 2x^2 - 7x + 1)$
   $= 4x^3 - 3x^2 + x - 5 - x^4 - 2x^2 + 7x - 1$
   $= -x^4 + 4x^3 - 3x^2 - 2x^2 + x + 7x - 5 - 1$
   $= -x^4 + 4x^3 + (-3-2)x^2 + (1+7)x + (-5-1)$
   $= -x^4 + 4x^3 - 5x^2 + 8x - 6$

7. $\left(\dfrac{1}{4}x^2 - \dfrac{1}{2}x\right) - \left(4x - \dfrac{1}{6}\right)$
   $= \dfrac{1}{4}x^2 - \dfrac{1}{2}x - 4x + \dfrac{1}{6}$
   $= \dfrac{1}{4}x^2 + \left(-\dfrac{1}{2} - 4\right)x + \dfrac{1}{6}$
   $= \dfrac{1}{4}x^2 - \dfrac{9}{2}x + \dfrac{1}{6}$

8. $\left(\dfrac{1}{2}x^2 - x + \dfrac{1}{4}\right) + \left(\dfrac{1}{3}x^2 + \dfrac{2}{5}\right)$
   $= \dfrac{1}{2}x^2 - x + \dfrac{1}{4} + \dfrac{1}{3}x^2 + \dfrac{2}{5}$
   $= \dfrac{1}{2}x^2 + \dfrac{1}{3}x^2 - x + \dfrac{1}{4} + \dfrac{2}{5}$
   $= \left(\dfrac{1}{2} + \dfrac{1}{3}\right)x^2 - x + \left(\dfrac{1}{4} + \dfrac{2}{5}\right)$
   $= \dfrac{5}{6}x^2 - x + \dfrac{13}{20}$

9. $(x^3y^2 + 6x^2y^2 - xy) + (-x^3y^2 + 4x^2y^2 + xy)$
   $= x^3y^2 + 6x^2y^2 - xy - x^3y^2 + 4x^2y^2 + xy$
   $= x^3y^2 - x^3y^2 + 6x^2y^2 + 4x^2y^2 - xy + xy$
   $= (1-1)x^3y^2 + (6+4)x^2y^2 + (-1+1)xy$
   $= 10x^2y^2$

10. $\left(a^2b - 4ab^2 + 3\right) - \left(2a^2b + 2ab^2 + 7\right)$
$= a^2b - 4ab^2 + 3 - 2a^2b - 2ab^2 - 7$
$= a^2b - 2a^2b - 4ab^2 - 2ab^2 + 3 - 7$
$= (1-2)a^2b + (-4-2)ab^2 + (3-7)$
$= -a^2b - 6ab^2 - 4$

11. a. $f(x) = -3x^2 + 2x - 8$
$f(-2) = -3(-2)^2 + 2(-2) - 8$
$= -3(4) - 4 - 8$
$= -12 - 4 - 8$
$= -24$

b. $f(x) = -3x^2 + 2x - 8$
$f(0) = -3(0)^2 + 2(0) - 8$
$= 0 + 0 - 8$
$= -8$

c. $f(x) = -3x^2 + 2x - 8$
$f(3) = -3(3)^2 + 2(3) - 8$
$= -3(9) + 6 - 8$
$= -27 + 6 - 8$
$= -29$

12. a. $f(x) = x^3 - 5x^2 + 3x - 1$
$f(-3) = (-3)^3 - 5(-3)^2 + 3(-3) - 1$
$= -27 - 5(9) - 9 - 1$
$= -27 - 45 - 9 - 1$
$= -82$

b. $f(x) = x^3 - 5x^2 + 3x - 1$
$f(0) = (0)^3 - 5(0)^2 + 3(0) - 1$
$= 0 - 0 + 0 - 1$
$= -1$

c. $f(x) = x^3 - 5x^2 + 3x - 1$
$f(2) = (2)^3 - 5(2)^2 + 3(2) - 1$
$= 8 - 5(4) + 6 - 1$
$= 8 - 20 + 6 - 1$
$= -7$

13. a. $(f+g)(x) = (4x - 3) + (x^2 + 3x + 2)$
$= 4x - 3 + x^2 + 3x + 2$
$= x^2 + 4x + 3x - 3 + 2$
$= x^2 + 7x - 1$

b. Since $(f+g)(x) = x^2 + 7x - 1$, we have
$(f+g)(3) = (3)^2 + 7(3) - 1$
$= 9 + 21 - 1$
$= 29$

14. a. $(f-g)(x) = (2x^3 + x^2 - 7) - (3x^2 - x + 5)$
$= 2x^3 + x^2 - 7 - 3x^2 + x - 5$
$= 2x^3 + x^2 - 3x^2 + x - 7 - 5$
$= 2x^3 - 2x^2 + x - 12$

b. Since $(f-g)(x) = 2x^3 - 2x^2 + x - 12$, we have
$(f-g)(2) = 2(2)^3 - 2(2)^2 + (2) - 12$
$= 2(8) - 2(4) + 2 - 12$
$= 16 - 8 + 2 - 12$
$= -2$

15. a. Profit is the difference between revenue and costs. Thus,
$P(x) = R(x) - C(x)$
$= (-1.5x^2 + 180x) - (x^2 - 100x + 3290)$
$= -1.5x^2 + 180x - x^2 + 100x - 3290$
$= -1.5x^2 - x^2 + 180x + 100x - 3290$
$= -2.5x^2 + 280x - 3290$
The profit function is:
$P(x) = -2.5x^2 + 280x - 3290$

b. $P(25) = -2.5(25)^2 + 280(25) - 3290$
$= -2.5(625) + 7000 - 3290$
$= -1562.5 + 7000 - 3290$
$= 2147.5$
If 25 graphing calculators are sold, there would be a profit of $2147.50.

*Chapter 5: Polynomials and Polynomial Functions*

16. a. If the point is $(2,1)$, then $x=2$.
$$A(x) = -x^2 + 5x$$
$$A(2) = -(2)^2 + 5(2)$$
$$= -4 + 10$$
$$= 6$$
The area of the region would be 6 square units.

   b. If the point is $(1,3)$, then $x=1$.
$$A(x) = -x^2 + 5x$$
$$A(1) = -(1)^2 + 5(1)$$
$$= -1 + 5$$
$$= 4$$
The area of the region would be 4 square units.

17. $(-3x^3y)(4xy^2) = (-3 \cdot 4)(x^3 \cdot x)(y \cdot y^2)$
$$= -12x^{3+1}y^{1+2}$$
$$= -12x^4y^3$$

18. $\left(\frac{1}{3}mn^4\right)(18m^3n^3) = \left(\frac{1}{3} \cdot 18\right)(m \cdot m^3)(n^4 \cdot n^3)$
$$= \frac{18}{3}m^{1+3}n^{4+3}$$
$$= 6m^4n^7$$

19. $5ab(-2a^2b + ab^2 - 3ab)$
$$= -5ab \cdot 2a^2b + 5ab \cdot ab^2 - 5ab \cdot 3ab$$
$$= -10a^3b^2 + 5a^2b^3 - 15a^2b^2$$

20. $0.5c(1.7c^2 + 4.3c + 8.9)$
$$= 0.5c \cdot 1.7c^2 + 0.5c \cdot 4.3c + 0.5c \cdot 8.9$$
$$= 0.85c^3 + 2.15c^2 + 4.45c$$

21. $(x+2)(x-9) = x \cdot x - x \cdot 9 + 2 \cdot x - 2 \cdot 9$
$$= x^2 - 9x + 2x - 18$$
$$= x^2 - 7x - 18$$

22. $(-3x+1)(2x-8)$
$$= -3x \cdot 2x - 3x \cdot (-8) + 1 \cdot 2x - 1 \cdot 8$$
$$= -6x^2 + 24x + 2x - 8$$
$$= -6x^2 + 26x - 8$$

23. $(m-4n)(2m+n)$
$$= m \cdot 2m + m \cdot n - 4n \cdot 2m - 4n \cdot n$$
$$= 2m^2 + mn - 8mn - 4n^2$$
$$= 2m^2 - 7mn - 4n^2$$

24. $(2a+15)(-a+3)$
$$= 2a \cdot (-a) + 2a \cdot 3 + 15 \cdot (-a) + 15 \cdot 3$$
$$= -2a^2 + 6a - 15a + 45$$
$$= -2a^2 - 9a + 45$$

25. $(x+2)(3x^2 - 5x + 1)$
$$= x \cdot 3x^2 - x \cdot 5x + x \cdot 1 + 2 \cdot 3x^2 - 2 \cdot 5x + 2 \cdot 1$$
$$= 3x^3 - 5x^2 + x + 6x^2 - 10x + 2$$
$$= 3x^3 + x^2 - 9x + 2$$

26. $(w-4)(w^2 + w - 8)$
$$= w \cdot w^2 + w \cdot w - w \cdot 8 - 4 \cdot w^2 - 4 \cdot w - 4 \cdot (-8)$$
$$= w^3 + w^2 - 8w - 4w^2 - 4w + 32$$
$$= w^3 - 3w^2 - 12w + 32$$

27. $(m^2 - 2m + 3)(2m^2 + 5m - 7)$
$$= m^2 \cdot 2m^2 + m^2 \cdot 5m - m^2 \cdot 7 - 2m \cdot 2m^2 - 2m \cdot 5m$$
$$\quad - 2m \cdot (-7) + 3 \cdot 2m^2 + 3 \cdot 5m - 3 \cdot 7$$
$$= 2m^4 + 5m^3 - 7m^2 - 4m^3 - 10m^2 + 14m$$
$$\quad + 6m^2 + 15m - 21$$
$$= 2m^4 + m^3 - 11m^2 + 29m - 21$$

28. $(2p-3q)(p^2 + 7pq - 4q^2)$
$$= 2p \cdot p^2 + 2p \cdot 7pq - 2p \cdot 4q^2 - 3q \cdot p^2$$
$$\quad - 3q \cdot 7pq - 3q \cdot (-4q^2)$$
$$= 2p^3 + 14p^2q - 8pq^2 - 3p^2q - 21pq^2 + 12q^3$$
$$= 2p^3 + 11p^2q - 29pq^2 + 12q^3$$

29. $(3w+1)(3w-1) = (3w)^2 - (1)^2$
$$= 9w^2 - 1$$

30. $(2x-5y)(2x+5y) = (2x)^2 - (5y)^2$
$$= 4x^2 - 25y^2$$

**31.** $(6k-5)^2 = (6k)^2 - 2(6k)(5) + (5)^2$
$= 36k^2 - 60k + 25$

**32.** $(3a+2b)^2 = (3a)^2 + 2(3a)(2b) + (2b)^2$
$= 9a^2 + 12ab + 4b^2$

**33.** $(x+2)(x^2 - 2x + 4)$
$= x \cdot x^2 - x \cdot 2x + x \cdot 4 + 2 \cdot x^2 - 2 \cdot 2x + 2 \cdot 4$
$= x^3 - 2x^2 + 4x + 2x^2 - 4x + 8$
$= x^3 + 8$

**34.** $(2x-3)(4x^2 + 6x + 9)$
$= 2x \cdot 4x^2 + 2x \cdot 6x + 2x \cdot 9 - 3 \cdot 4x^2 - 3 \cdot 6x - 3 \cdot 9$
$= 8x^3 + 12x^2 + 18x - 12x^2 - 18x - 27$
$= 8x^3 - 27$

**35. a.** $(f \cdot g)(x) = f(x) \cdot g(x)$
$= (3x-7)(6x+5)$
$= 3x \cdot 6x + 3x \cdot 5 - 7 \cdot 6x - 7 \cdot 5$
$= 18x^2 + 15x - 42x - 35$
$= 18x^2 - 27x - 35$

**b.** Since $(f \cdot g)(x) = 18x^2 - 27x - 35$, we have
$(f \cdot g)(-2) = 18(-2)^2 - 27(-2) - 35$
$= 18(4) + 54 - 35$
$= 72 + 54 - 35$
$= 91$

**36. a.** $(f \cdot g)(x) = f(x) \cdot g(x)$
$= (x+2)(3x^2 - x + 1)$
$= x \cdot 3x^2 - x \cdot x + x \cdot 1 + 2 \cdot 3x^2$
$\quad - 2 \cdot x + 2 \cdot 1$
$= 3x^3 - x^2 + x + 6x^2 - 2x + 2$
$= 3x^3 + 5x^2 - x + 2$

**b.** Since $(f \cdot g)(x) = 3x^3 + 5x^2 - x + 2$, we have
$(f \cdot g)(4) = 3(4)^3 + 5(4)^2 - (4) + 2$
$= 3(64) + 5(16) - 4 + 2$
$= 192 + 80 - 4 + 2$
$= 270$

**37.** $f(x) = 5x^2 + 8$
$f(x-3) = 5(x-3)^2 + 8$
$= 5(x^2 - 2 \cdot x \cdot 3 + 3^2) + 8$
$= 5(x^2 - 6x + 9) + 8$
$= 5x^2 - 30x + 45 + 8$
$= 5x^2 - 30x + 53$

**38.** $f(x+h) - f(x) = \left[-(x+h)^2 + 3(x+h) - 5\right]$
$\quad - \left[-x^2 + 3x - 5\right]$
$= \left[-x^2 - 2xh - h^2 + 3x + 3h - 5\right]$
$\quad - \left[-x^2 + 3x - 5\right]$
$= -x^2 - 2xh - h^2 + 3x + 3h - 5$
$\quad + x^2 - 3x + 5$
$= -2xh + 3h - h^2$

**39.** $\dfrac{12x^3 - 6x^2}{3x} = \dfrac{12x^3}{3x} - \dfrac{6x^2}{3x}$
$= \dfrac{12}{3}x^{3-1} - \dfrac{6}{3}x^{2-1}$
$= 4x^2 - 2x$

**40.** $\dfrac{15w^5 - 5w^3 + 25w^2 + 10w}{5w}$
$= \dfrac{15w^5}{5w} - \dfrac{5w^3}{5w} + \dfrac{25w^2}{5w} + \dfrac{10w}{5w}$
$= \dfrac{15}{5}w^{5-1} - \dfrac{5}{5}w^{3-1} + \dfrac{25}{5}w^{2-1} + \dfrac{10}{5}w^{1-1}$
$= 3w^4 - w^2 + 5w + 2$

**41.** $\dfrac{7y^3 + 12y^2 - 6y}{2y} = \dfrac{7y^3}{2y} + \dfrac{12y^2}{2y} - \dfrac{6y}{2y}$
$= \dfrac{7}{2}y^{3-1} + \dfrac{12}{2}y^{2-1} - \dfrac{6}{2}y^{1-1}$
$= \dfrac{7}{2}y^2 + 6y - 3$

## Chapter 5: Polynomials and Polynomial Functions

42. $\dfrac{2m^3n^2 + 8m^2n^2 - 14mn^3}{4m^2n^3}$

$= \dfrac{2m^3n^2}{4m^2n^3} + \dfrac{8m^2n^2}{4m^2n^3} - \dfrac{14mn^3}{4m^2n^3}$

$= \dfrac{2}{4}m^{3-2}n^{2-3} + \dfrac{8}{4}m^{2-2}n^{2-3} - \dfrac{14}{4}m^{1-2}n^{3-3}$

$= \dfrac{1}{2}mn^{-1} + 2n^{-1} - \dfrac{7}{2}m^{-1}$

$= \dfrac{m}{2n} + \dfrac{2}{n} - \dfrac{7}{2m}$

43. 
$$\begin{array}{r} 3x+4 \\ x-2\overline{\smash{)}3x^2-2x-8} \\ \underline{-(3x^2-6x)} \\ 4x-8 \\ \underline{-(4x-8)} \\ 0 \end{array}$$

$\dfrac{3x^2 - 2x - 8}{x - 2} = 3x + 4$

44. 
$$\begin{array}{r} -2x+7 \\ x+5\overline{\smash{)}-2x^2-3x+40} \\ \underline{-(-2x^2-10x)} \\ 7x+40 \\ \underline{-(7x+35)} \\ 5 \end{array}$$

$\dfrac{-2x^2 - 3x + 40}{x + 5} = -2x + 7 + \dfrac{5}{x+5}$

45. 
$$\begin{array}{r} 3z^2+2 \\ 2z+3\overline{\smash{)}6z^3+9z^2+4z-6} \\ \underline{-(6z^3+9z^2)} \\ 4z-6 \\ \underline{-(4z+6)} \\ -12 \end{array}$$

$\dfrac{6z^3 + 9z^2 + 4z - 6}{2z + 3} = 3z^2 + 2 - \dfrac{12}{2z+3}$

46.
$$\begin{array}{r} 4k^2+k-2 \\ 3k-8\overline{\smash{)}12k^3-29k^2-14k+16} \\ \underline{-(12k^3-32k^2)} \\ 3k^2-14k \\ \underline{-(3k^2-8k)} \\ -6k+16 \\ \underline{-(-6k+16)} \\ 0 \end{array}$$

$\dfrac{12k^3 - 29k^2 - 14k + 16}{3k - 8} = 4k^2 + k - 2$

47.
$$\begin{array}{r} 8x^3+12x^2+18x+27 \\ 2x-3\overline{\smash{)}16x^4+0x^3+0x^2+0x-81} \\ \underline{-(16x^4-24x^3)} \\ 24x^3+0x^2 \\ \underline{-(24x^3-36x^2)} \\ 36x^2+0x \\ \underline{-(36x^2-54x)} \\ 54x-81 \\ \underline{-(54x-81)} \\ 0 \end{array}$$

$\dfrac{16x^4 - 81}{2x - 3} = 8x^3 + 12x^2 + 18x + 27$

48.
$$\begin{array}{r} 2x^2-5x+12 \\ x^2-3x+4\overline{\smash{)}2x^4-11x^3+35x^2-54x+55} \\ \underline{-(2x^4-6x^3+8x^2)} \\ -5x^3+27x^2-54x \\ \underline{-(-5x^3+15x^2-20x)} \\ 12x^2-34x+55 \\ \underline{-(12x^2-36x+48)} \\ 2x+7 \end{array}$$

$\dfrac{2x^4 - 11x^3 + 35x^2 - 54x + 55}{x^2 - 3x + 4}$

$= 2x^2 - 5x + 12 + \dfrac{2x+7}{x^2-3x+4}$

49. The divisor is $x+2$ so $c=-2$.

$$\begin{array}{r|rrr} -2 & 5 & 11 & 8 \\ & & -10 & -2 \\ \hline & 5 & 1 & 6 \end{array}$$

$$\frac{5x^2+11x+8}{x+2}=5x+1+\frac{6}{x+2}$$

50. The divisor is $a-2$ so $c=2$.

$$\begin{array}{r|rrr} 2 & 9 & -14 & -8 \\ & & 18 & 8 \\ \hline & 9 & 4 & 0 \end{array}$$

$$\frac{9a^2-14a-8}{a-2}=9a+4$$

51. The divisor is $m+3$ so $c=-3$.

$$\begin{array}{r|rrrr} -3 & 3 & 11 & -5 & -33 \\ & & -9 & -6 & 33 \\ \hline & 3 & 2 & -11 & 0 \end{array}$$

$$\frac{3m^3+11m^2-5m-33}{m+3}=3m^2+2m-11$$

52. The divisor is $n-4$ so $c=4$.

$$\begin{array}{r|rrrr} 4 & 1 & 2 & -39 & 67 \\ & & 4 & 24 & -60 \\ \hline & 1 & 6 & -15 & 7 \end{array}$$

$$\frac{n^3+2n^2-39n+67}{n-4}=n^2+6n-15+\frac{7}{n-4}$$

53. The divisor is $x+1$ so $c=-1$.

$$\begin{array}{r|rrrrr} -1 & 1 & 0 & 6 & 0 & -7 \\ & & -1 & 1 & -7 & 7 \\ \hline & 1 & -1 & 7 & -7 & 0 \end{array}$$

$$\frac{x^4+6x^2-7}{x+1}=x^3-x^2+7x-7$$

54. The divisor is $x+2$ so $c=-2$.

$$\begin{array}{r|rrrr} -2 & 2 & 0 & 5 & -8 \\ & & -4 & 8 & -26 \\ \hline & 2 & -4 & 13 & -34 \end{array}$$

$$\frac{2x^3+5x-8}{x+2}=2x^2-4x+13-\frac{34}{x+2}$$

55. a. $\left(\dfrac{f}{g}\right)(x)=\dfrac{5x^3+25x^2-15x}{5x}$

$$=\frac{5x^3}{5x}+\frac{25x^2}{5x}-\frac{15x}{5x}$$

$$=\frac{5}{5}x^{3-1}+\frac{25}{5}x^{2-1}-\frac{15}{5}x^{1-1}$$

$$=x^2+5x-3$$

$\left(\dfrac{f}{g}\right)(x)=x^2+5x-3$

b. Since $\left(\dfrac{f}{g}\right)(x)=x^2+5x-3$, we have

$$\left(\frac{f}{g}\right)(2)=(2)^2+5(2)-3$$

$$=4+10-3$$

$$=11$$

56. a. $\left(\dfrac{f}{g}\right)(x)=\dfrac{9x^2+54x-31}{3x-2}$

$$\begin{array}{r} 3x+20\phantom{)} \\ 3x-2{\overline{\smash{\big)}\,9x^2+54x-31\phantom{)}}} \\ \underline{-(9x^2-6x)\phantom{-31)}} \\ 60x-31\phantom{)} \\ \underline{-(60x-40)} \\ 9\phantom{)} \end{array}$$

$\left(\dfrac{f}{g}\right)(x)=3x+20+\dfrac{9}{3x-2}$

b. Since $\left(\dfrac{f}{g}\right)(x)=3x+20+\dfrac{9}{3x-2}$, we have

$$\left(\frac{f}{g}\right)(-3)=3(-3)+20+\frac{9}{3(-3)-2}$$

$$=-9+20+\frac{9}{-11}$$

$$=11-\frac{9}{11}$$

$$=\frac{112}{11}$$

57. a. $\left(\dfrac{f}{g}\right)(x)=\dfrac{2x^3+12x^2+9x-28}{x+4}$

The divisor is $x+4$ so $c=-4$.

*Chapter 5:* *Polynomials and Polynomial Functions*

$$\begin{array}{r|rrrr} -4) & 2 & 12 & 9 & -28 \\ & & -8 & -16 & 28 \\ \hline & 2 & 4 & -7 & 0 \end{array}$$

$\left(\dfrac{f}{g}\right)(x) = 2x^2 + 4x - 7$

b. Since $\left(\dfrac{f}{g}\right)(x) = 2x^2 + 4x - 7$, we have

$\left(\dfrac{f}{g}\right)(-2) = 2(-2)^2 + 4(-2) - 7$

$= 2(4) - 8 - 7$
$= 8 - 8 - 7$
$= -7$

58. a. $\left(\dfrac{f}{g}\right)(x) = \dfrac{3x^4 - 14x^3 + 31x^2 - 58x + 22}{x^2 - x + 5}$

$$\begin{array}{r} 3x^2 - 11x + 5 \\ x^2 - x + 5 \overline{\smash{)}3x^4 - 14x^3 + 31x^2 - 58x + 22} \\ -\left(3x^4 - 3x^3 + 15x^2\right) \\ \hline -11x^3 + 16x^2 - 58x \\ -\left(-11x^3 + 11x^2 - 55x\right) \\ \hline 5x^2 - 3x + 22 \\ -\left(5x^2 - 5x + 25\right) \\ \hline 2x - 3 \end{array}$$

$\left(\dfrac{f}{g}\right)(x) = 3x^2 - 11x + 5 + \dfrac{2x-3}{x^2 - x + 5}$

b. Since $\left(\dfrac{f}{g}\right)(x) = 3x^2 - 11x + 5 + \dfrac{2x-3}{x^2-x+5}$, we have

$\left(\dfrac{f}{g}\right)(4) = 3(4)^2 - 11(4) + 5 + \dfrac{2(4)-3}{(4)^2 - (4) + 5}$

$= 3(16) - 44 + 5 + \dfrac{8-3}{16-4+5}$

$= 48 - 44 + 5 + \dfrac{5}{17}$

$= 9 + \dfrac{5}{17}$

$= \dfrac{158}{17}$

59. The divisor is $x - 4$ so $c = 4$
$f(4) = 4(4)^2 - 7(4) + 23$
$= 4(16) - 28 + 23$
$= 64 - 28 + 23$
$= 59$
The remainder is 59.

60. The divisor is $x + 2$ so $c = -2$
$f(-2) = (-2)^3 - 2(-2)^2 + 12(-2) - 5$
$= -8 - 2(4) - 24 - 5$
$= -8 - 8 - 24 - 5 = -45$
The remainder is $-45$.

61. $c = 2$
$f(2) = 3(2)^2 + (2) - 14$
$= 3(4) + 2 - 14$
$= 12 + 2 - 14$
$= 0$
Since the remainder is 0, $x - 2$ is a factor.

$$\begin{array}{r|rrr} 2) & 3 & 1 & -14 \\ & & 6 & 14 \\ \hline & 3 & 7 & 0 \end{array}$$

$f(x) = (x - 2)(3x + 7)$

62. $c = -4$
$f(-4) = 2(-4)^2 + 13(-4) + 22$
$= 2(16) - 52 + 22$
$= 32 - 52 + 22 = 2$
Since the remainder is not 0, $x + 4$ is not a factor of $f(x)$.

63. Since $A = L \cdot W$ we get $L = \dfrac{A}{W}$

$$\begin{array}{r} 5x + 1 \\ 4x - 3 \overline{\smash{)}20x^2 - 11x - 3} \\ -\left(20x^2 - 15x\right) \\ \hline 4x - 3 \\ -(4x - 3) \\ \hline 0 \end{array}$$

The length of the rectangle is $5x + 1$ meters.

64. We have $V = L \cdot W \cdot H$ and the area of the top of the box is $L \cdot W$. We can rewrite to get

$L \cdot W = \dfrac{V}{H}$

$$\begin{array}{r|rrr} 2) & 2 & -7 & -6 \\ & & 4 & 10 & 6 \\ \hline & 2 & 5 & 3 & 0 \end{array}$$

The area of the top of the box is $2x^2 + 5x + 3$ square centimeters.

65. $4z + 24$
    GCF: 4
    $4z + 24 = 4 \cdot z + 4 \cdot 6$
    $\phantom{4z + 24} = 4(z + 6)$

66. $-7y^2 + 91y$
    GCF: $-7y$
    $-7y^2 + 91y = -7y \cdot y - 7y \cdot (-13)$
    $\phantom{-7y^2 + 91y} = -7y(y - 13)$

67. $14x^3y^2 + 2xy^2 - 8x^2y$
    GCF: $2xy$
    $14x^3y^2 + 2xy^2 - 8x^2y$
    $= 2xy \cdot 7x^2y + 2xy \cdot y + 2xy \cdot (-4x)$
    $= 2xy(7x^2y + y - 4x)$
    $= 2xy(7x^2y - 4x + y)$

68. $30a^4b^3 + 15a^3b - 25a^2b^2$
    GCF: $5a^2b$
    $30a^4b^3 + 15a^3b - 25a^2b^2$
    $= 5a^2b \cdot 6a^2b^2 + 5a^2b \cdot 3a + 5a^2b \cdot (-5b)$
    $= 5a^2b(6a^2b^2 + 3a - 5b)$

69. $3x(x+5) - 4(x+5)$
    GCF: $x + 5$
    $3x(x+5) - 4(x+5) = 3x \cdot (x+5) - 4 \cdot (x+5)$
    $\phantom{3x(x+5) - 4(x+5)} = (x+5)(3x - 4)$

70. $-4c(2c+9) + 3(2c+9)$
    GCF: $2c + 9$
    $-4c(2c+9) + 3(2c+9)$
    $= -4c \cdot (2c+9) + 3 \cdot (2c+9)$
    $= (2c+9)(-4c+3)$
    $= (2c+9)(3-4c)$

71. $(5x+3)(x-5y) + (x+2)(x-5y)$
    GCF: $x - 5y$
    $(5x+3)(x-5y) + (x+2)(x-5y)$
    $= (x-5y)((5x+3) + (x+2))$
    $= (x-5y)(5x+3+x+2)$
    $= (x-5y)(6x+5)$

72. $(3a-b)(a+7) - (a+1)(a+7)$
    GCF: $a + 7$
    $(3a-b)(a+7) - (a+1)(a+7)$
    $= (a+7)((3a-b) - (a+1))$
    $= (a+7)(3a-b-a-1)$
    $= (a+7)(2a-b-1)$

73. $x^2 + 6x - 3x - 18 = (x^2 + 6x) + (-3x - 18)$
    $= x(x+6) - 3(x+6)$
    $= (x+6)(x-3)$

74. $c^2 + 2c - 5c - 10 = (c^2 + 2c) + (-5c - 10)$
    $= c(c+2) - 5(c+2)$
    $= (c+2)(c-5)$

75. $14z^2 + 16z - 21z - 24$
    $= (14z^2 + 16z) + (-21z - 24)$
    $= 2z(7z+8) - 3(7z+8)$
    $= (7z+8)(2z-3)$

76. $21w^2 - 28w + 6w - 8 = (21w^2 - 28w) + (6w - 8)$
    $= 7w(3w-4) + 2(3w-4)$
    $= (3w-4)(7w+2)$

## Chapter 5: Polynomials and Polynomial Functions

77. Pull out the GCF first and then factor by grouping.
$$2x^3 + 2x^2 - 18x^2 - 18x$$
$$= 2x(x^2 + x - 9x - 9)$$
$$= 2x((x^2 + x) + (-9x - 9))$$
$$= 2x(x(x+1) - 9(x+1))$$
$$= 2x(x+1)(x-9)$$

78. Pull out the GCF first and then factor by grouping.
$$10a^4 + 15a^3 + 70a^3 + 105a^2$$
$$= 5a^2(2a^2 + 3a + 14a + 21)$$
$$= 5a^2((2a^2 + 3a) + (14a + 21))$$
$$= 5a^2(a(2a+3) + 7(2a+3))$$
$$= 5a^2(2a+3)(a+7)$$

79. a. $\frac{1}{2}n^2 + \frac{1}{2}n = \frac{1}{2}n \cdot n + \frac{1}{2}n \cdot 1$
$$= \frac{1}{2}n(n+1)$$

   b. $n = 32$
   $$\frac{1}{2}(32)(32+1) = 16(33) = 528$$

80. $R(x) = 5200x - 2x^3$
$$= 2x \cdot 2600 - 2x \cdot x^2$$
$$= 2x(2600 - x^2)$$
The revenue function is $R(x) = 2x(2600 - x^2)$.

81. $w^2 - 11w - 26$
We are looking for two factors of $c = -26$ whose sum is $b = -11$. Since $c$ is negative, the factors will have opposite signs, and since $b$ is negative the factor with the larger absolute value must be negative.

| Factors | 1, −26 | 2, −13 |
|---|---|---|
| Sum | −25 | −11 |

$w^2 - 11w - 26 = (w+2)(w-13)$

82. $x^2 - 9x + 15$
We are looking for two factors of $c = 15$ whose sum is $b = -9$. Since $c$ is positive the two factors have the same sign, and since $b$ is negative the factors are both negative.

| Factors | −1, −15 | −3, −5 |
|---|---|---|
| Sum | −16 | −8 |

Since none of the possibilities work, $x^2 - 9x + 15$ is prime.

83. $-t^2 + 6t + 72$
Start by factoring out $-1$.
$$-t^2 + 6t + 72 = -1(t^2 - 6t - 72)$$
We are looking for two factors of $c = -72$ whose sum is $b = -6$. Since $c$ is negative, the two factors will have opposite signs. Since $b$ is negative the factor with the larger absolute value will be negative.

| Factors | 1, −72 | 2, −36 | 3, −24 | 4, −18 | 6, −12 | 8, −9 |
|---|---|---|---|---|---|---|
| Sum | −71 | −34 | −21 | −14 | −6 | −1 |

$$-t^2 + 6t + 72 = -1(t^2 - 6t - 72)$$
$$= -1(t-12)(t+6)$$

84. $m^2 + 10m + 21$
We are looking for two factors of $c = 21$ whose sum is $b = 10$. Since both $c$ and $b$ are positive, the two factors must be positive.

| Factors | 1, 21 | 3, 7 |
|---|---|---|
| Sum | 22 | 10 |

$m^2 + 10m + 21 = (m+3)(m+7)$

85. $x^2 + 4xy - 320y^2$
We are looking for two factors of $c = -320$ whose sum is $b = 4$. Since $c$ is negative, the two factors have opposite signs, and since $b$ is positive the factor with the larger absolute value must be positive.

| Factors | −1, 320 | −2, 160 | −4, 80 | −5, 64 |
|---|---|---|---|---|
| Sum | 319 | 158 | 76 | 59 |

| Factors | −8, 40 | −10, 32 | −16, 20 |
|---|---|---|---|
| Sum | 32 | 22 | 4 |

$x^2 + 4xy - 320y^2 = (x+20y)(x-16y)$

86. $r^2 - 5rs + 6s^2$

We are looking for two factors of $c = 6$ whose sum is $b = -5$. Since $c$ is positive, the factors will have the same sign, and since $b$ is negative the factors will both be negative.

| Factors | −1,−6 | −2,−3 |
|---------|-------|-------|
| Sum     | −7    | −5    |

$r^2 - 5rs + 6s^2 = (r - 2s)(r - 3s)$

87. <u>ac Method:</u>
$a \cdot c = 5 \cdot -6 = -30$
We are looking for two factors of $-30$ whose sum is 13. Since the product is negative, the factors will have opposite signs. The sum is positive so the factor with the larger absolute value will be positive.

| factor 1 | factor 2 | sum |
|----------|----------|-----|
| −1       | 30       | 29  |
| −2       | 15       | 13 ← okay |
| −3       | 10       | 7   |
| −5       | 6        | 1   |

$5x^2 + 13x - 6 = 5x^2 - 2x + 15x - 6$
$= x(5x - 2) + 3(5x - 2)$
$= (x + 3)(5x - 2)$

<u>Trial and Error Method:</u>
First note that there are no common factors and that $a = 5$, $b = 13$, and $c = -6$. Since $c$ is negative, the signs in our factors will be opposites. We will consider factorizations with this form:
$(\_x + \_)(\_x - \_)$
If our choice results in a middle term with the wrong sign, we simply switch the signs of the factors.
Since $a = 5$ can be factored as $1 \cdot 5$, we have the following form:
$(x + \_)(5x - \_)$
$|c| = |-6| = 6$ can be factored as $1 \cdot 6$ or $2 \cdot 3$.
$(x + 1)(5x - 6) \to 5x^2 - x - 6$
$(x + 6)(5x - 1) \to 5x^2 + 29x - 6$
$(x + 2)(5x - 3) \to 5x^2 + 7x - 6$
$(x + 3)(5x - 2) \to 5x^2 + 13x - 6$

The correct factorization is
$5x^2 + 13x - 6 = (x + 3)(5x - 2)$

88. <u>ac Method:</u>
$a \cdot c = 6 \cdot 44 = 264$
We are looking for two factors of 264 whose sum is 41. Since the product is positive, the factors will have the same sign. The sum is positive so the factors will both be positive, and both must be less than 41.

| factor 1 | factor 2 | sum |
|----------|----------|-----|
| 8        | 33       | 41 ← okay |
| 11       | 24       | 35  |
| 12       | 22       | 34  |

$6m^2 + 41m + 44 = 6m^2 + 8m + 33m + 44$
$= 2m(3m + 4) + 11(3m + 4)$
$= (2m + 11)(3m + 4)$

<u>Trial and Error Method:</u>
First note that there are no common factors and that $a = 6$, $b = 41$, and $c = 44$. Since $c$ is positive, the signs in our factors will be the same. Since $b$ is positive, both factors will be positive. We will consider factorizations with this form:
$(\_m + \_)(\_m + \_)$
Since $a = 6$ can be factored as $1 \cdot 6$ or $2 \cdot 3$, we have the following forms:
$(m + \_)(6m + \_)$
$(2m + \_)(3m + \_)$
$c = 44$ can be factored as $1 \cdot 44$, $2 \cdot 22$, or $4 \cdot 11$. Since the original expression had no common factors, the binomials we select cannot have a common factor.
$(m + 44)(6m + 1) \to 6m^2 + 265m + 44$
$(2m + 11)(3m + 4) \to 6m^2 + 41m + 44$
$(2m + 1)(3m + 44) \to 6m^2 + 91m + 44$
$(m + 4)(6m + 11) \to 6m^2 + 35m + 44$

The correct factorization is
$6m^2 + 41m + 44 = (2m + 11)(3m + 4)$

89. <u>ac Method:</u>
$a \cdot c = 4 \cdot 7 = 28$
We are looking for two factors of 28 whose sum is $-5$. Since the product is positive, the factors will have the same sign. The sum is negative so

## Chapter 5: Polynomials and Polynomial Functions

the factors will both be negative.

| factor 1 | factor 2 | sum |
|---|---|---|
| −1 | −28 | −29 |
| −2 | −14 | −16 |
| −4 | −7 | −11 |

Since none of the possibilities work, the expression is prime.

Trial and Error Method:
First note that there are no common factors and that $a = 4$, $b = -5$, and $c = 7$. Since $c$ is positive, the signs in our factors will be the same. Since $b$ is negative, the signs will both be negative. We will consider factorizations with this form:
$(\_y-\_)(\_y-\_)$
Since $a = 4$ can be factored as $1 \cdot 4$ or $2 \cdot 2$, we have the following forms:
$(y-\_)(4y-\_)$
$(2y-\_)(2y-\_)$
$c = 7$ can be factored as $1 \cdot 7$.
$(y-1)(4y-7) \to 4y^2 - 11y + 7$
$(y-7)(4y-1) \to 4y^2 - 29y + 7$
$(2y-1)(2y-7) \to 4y^2 - 16y + 7$
None of the possibilities yield the given expression. Therefore, the expression is prime.

90. ac Method:
Start by factoring out the GCF.
$8t^2 + 22t - 6 = 2(4t^2 + 11t - 3)$
Now focus on the reduced trinomial.
$a \cdot c = 4 \cdot -3 = -12$
We are looking for two factors of −12 whose sum is 11. Since the product is negative, the two factors will have opposite signs. The sum is positive so the factor with the larger absolute value will be positive.

| factor 1 | factor 2 | sum |  |
|---|---|---|---|
| −1 | 12 | 11 | ← okay |
| −2 | 6 | 4 |  |
| −3 | 4 | 1 |  |

$8t^2 + 22t - 6 = 2(4t^2 + 11t - 3)$
$= 2(4t^2 + 12t - t - 3)$
$= 2(4t(t+3) - 1(t+3))$
$= 2(t+3)(4t-1)$

Trial and Error Method:
Start by factoring out the GCF.
$8t^2 + 22t - 6 = 2(4t^2 + 11t - 3)$
Now focus on the reduced trinomial with $a = 4$, $b = 11$, and $c = -3$. Since $c$ is negative, our factors will have opposite signs. We will consider factorizations of this form:
$(\_t+\_)(\_t-\_)$
If our choice results in a middle term with the wrong sign, we simply switch the signs in our factors to obtain the correct result.
Since $a = 4$ can be factored as $1 \cdot 4$ or $2 \cdot 2$, we have the following forms:
$(t+\_)(4t-\_)$
$(2t+\_)(2t-\_)$
$|c| = |-3| = 3$ can be factored as $1 \cdot 3$.
$(t+1)(4t-3) \to 4t^2 + t - 3$
$(t+3)(4t-1) \to 4t^2 + 11t - 3$
$(2t+1)(2t-3) \to 4t^2 - 4t - 3$
The correct factorization is
$8t^2 + 22t - 6 = 2(t+3)(4t-1)$

91. ac Method:
$a \cdot c = 6 \cdot 5 = 30$
We are looking for two factors of 30 whose sum is −13. Since the product is positive, the factors will have the same sign. The sum is negative so the factors will both be negative.

| factor 1 | factor 2 | sum |  |
|---|---|---|---|
| −1 | −30 | −31 |  |
| −2 | −15 | −17 |  |
| −3 | −10 | −13 | ← okay |
| −5 | −6 | −11 |  |

$6x^2 - 13x + 5 = 6x^2 - 3x - 10x + 5$
$= 3x(2x-1) - 5(2x-1)$
$= (2x-1)(3x-5)$

Trial and Error Method:

First note that there are no common factors and that $a = 6$, $b = -13$, and $c = 5$. Since $c$ is positive, the signs in our factors will be the same. Since $b$ is negative, the signs will both be negative. We will consider factorizations with this form:
$$(\_x - \_)(\_x - \_)$$
Since $a = 6$ can be factored as $1 \cdot 6$ or $2 \cdot 3$, we have the following forms:
$$(x - \_)(6x - \_)$$
$$(2x - \_)(3x - \_)$$
$c = 5$ can be factored as $1 \cdot 5$.
$$(x-1)(6x-5) \to 6x^2 - 11x + 5$$
$$(x-5)(6x-1) \to 6x^2 - 31x + 5$$
$$(2x-1)(3x-5) \to 6x^2 - 13x + 5$$
$$(2x-5)(3x-1) \to 6x^2 - 17x + 5$$
The correct factorization is
$$6x^2 - 13x + 5 = (2x-1)(3x-5)$$

**92.** ac Method:
$a \cdot c = 21 \cdot -2 = -42$
We are looking for two factors of $-42$ whose sum is $-1$. Since the product is negative, the two factors will have opposite signs. Since the sum is negative, the factor with the larger absolute value will be negative.

| factor 1 | factor 2 | sum | |
|---|---|---|---|
| 1 | −42 | −41 | |
| 2 | −21 | −19 | |
| 3 | −14 | −11 | |
| 6 | −7 | −1 | ← okay |

$$21r^2 - rs - 2s^2 = 21r^2 + 6rs - 7rs - 2s^2$$
$$= 3r(7r + 2s) - s(7r + 2s)$$
$$= (7r + 2s)(3r - s)$$

Trial and Error Method:
First note that there are no common factors and that $a = 21$, $b = -1$, and $c = -2$. Since $c$ is negative, the factors will have opposite signs. We will consider factorizations with this form:
$$(\_r + \_s)(\_r - \_s)$$
If our choice results in a middle term with the wrong sign, we simply switch the signs in our factors to obtain the correct result.
Since $a = 21$ can be factored as $1 \cdot 21$ or $3 \cdot 7$,

we have the following forms:
$$(r + \_s)(21r - \_s)$$
$$(3r + \_s)(7r - \_s)$$
$|c| = |-2| = 2$ can be factored as $1 \cdot 2$.
$$(r+s)(21r-2s) \to 21r^2 + 19rs - 2s^2$$
$$(r+2s)(21r-s) \to 21r^2 + 41rs - 2s^2$$
$$(3r+s)(7r-2s) \to 21r^2 + rs - 2s^2$$
$$(3r+2s)(7r-s) \to 21r^2 + 11rs - 2s^2$$
The third result is almost what we need, except the sign on the middle term is wrong. We simply change the signs of the factors to get the correct result.
$$(3r-s)(7r+2s) \to 21r^2 - rs - 2s^2$$
The correct factorization is
$$21r^2 - rs - 2s^2 = (3r-s)(7r+2s)$$

**93.** ac Method:
$a \cdot c = 20 \cdot 27 = 540$
We are looking for two factors of 540 whose sum is $-57$. Since the product is positive, the two factors will have the same sign. Since the sum is negative, the factors will both be negative. We are adding two negative numbers to get $-57$ so neither can be less than $-57$.

| factor 1 | factor 2 | sum | |
|---|---|---|---|
| −10 | −54 | −64 | too small |
| −15 | −36 | −51 | too large |
| −12 | −45 | −57 | ← okay |

$$20x^2 - 57xy + 27y^2 = 20x^2 - 12xy - 45xy + 27y^2$$
$$= 4x(5x - 3y) - 9y(5x - 3y)$$
$$= (5x - 3y)(4x - 9y)$$

Trial and Error Method
First note that there are no common factors and that $a = 20$, $b = -57$, and $c = 27$. Since $c$ is positive, the signs in our factors will be the same. Since $b$ is negative, the signs will both be negative. We will consider factorizations with this form:
$$(\_x - \_y)(\_x - \_y)$$
Since $a = 20$ can be factored as $1 \cdot 20$, $2 \cdot 10$, or $4 \cdot 5$, we have the following forms:

*Chapter 5: Polynomials and Polynomial Functions*

$(x - \underline{\phantom{y}}y)(20x - \underline{\phantom{y}}y)$
$(2x - \underline{\phantom{y}}y)(10x - \underline{\phantom{y}}y)$
$(4x - \underline{\phantom{y}}y)(5x - \underline{\phantom{y}}y)$
$c = 27$ can be factored as $1 \cdot 27$ or $3 \cdot 9$.
$(x - y)(20x - 27y) \to 20x^2 - 47xy + 27y^2$
$(x - 27y)(20x - y) \to 20x^2 - 541xy + 27y^2$
$(2x - y)(10x - 27y) \to 20x^2 - 64xy + 27y^2$
$(2x - 27y)(10x - y) \to 20x^2 - 272xy + 27y^2$
$(4x - y)(5x - 27y) \to 20x^2 - 113xy + 27y^2$
$(4x - 27y)(5x - y) \to 20x^2 - 139xy + 27y^2$
$(x - 3y)(20x - 9y) \to 20x^2 - 69xy + 27y^2$
$(x - 9y)(20x - 3y) \to 20x^2 - 183xy + 27y^2$
$(2x - 3y)(10x - 9y) \to 20x^2 - 48xy + 27y^2$
$(2x - 9y)(10x - 3y) \to 20x^2 - 96xy + 27y^2$
$(4x - 3y)(5x - 9y) \to 20x^2 - 51xy + 27y^2$
$(4x - 9y)(5x - 3y) \to 20x^2 - 57xy + 27y^2$
The correct factorization is
$20x^2 - 57xy + 27y^2 = (4x - 9y)(5x - 3y)$

94. Begin by factoring out the GCF.
$-2s^2 + 12s + 14 = -2(s^2 - 6s - 7)$
Now focus on the reduced trinomial.
We are looking for two factors of $-7$ whose sum is $-6$. Since the product is negative, the factors have opposite signs. Since the sum is negative, the factor with the larger absolute value must be negative.
$-2s^2 + 12s + 14 = -2(s^2 - 6s - 7)$
$= -2(s - 7)(s + 1)$

95. $x^4 - 10x^2 - 11$
Let $u = x^2$. This gives $u^2 - 10u - 11$.
We need two factors of $-11$ whose sum is $-10$. Since the product is negative, the two factors will have opposite signs. Since the sum is negative, the factor with the larger absolute value will be negative.
$u^2 - 10u - 11 = (u + 1)(u - 11)$
Now get back in terms of $x$.
$x^4 - 10x^2 - 11 = (x^2 + 1)(x^2 - 11)$

96. $10x^2y^2 + 41xy + 4$
Let $u = xy$. This gives $10u^2 + 41u + 4$.
$a \cdot c = 10 \cdot 4 = 40$
We need two factors of 40 whose sum is 41. Since the product is positive, the two factors will have the same sign. Since the sum is also positive, the two factors are both positive.

| factor 1 | factor 2 | sum |
|---|---|---|
| 1 | 40 | 41 ← okay |
| 2 | 20 | 22 |
| 4 | 10 | 14 |
| 5 | 8 | 13 |

$10u^2 + 41u + 4 = 10u^2 + 40u + u + 4$
$= 10u(u + 4) + 1(u + 4)$
$= (u + 4)(10u + 1)$
Now get back in terms of $x$ and $y$.
$10x^2y^2 + 41xy + 4 = (xy + 4)(10xy + 1)$

97. $(a + 4)^2 - 9(a + 4) - 36$
Let $u = a + 4$. This gives $u^2 - 9u - 36$.
We need two factors of $-36$ whose sum is $-9$. Since the product is negative, the two factors will have opposite signs. Since the sum is negative, the factor with the larger absolute value will be negative.

| factor 1 | factor 2 | sum |
|---|---|---|
| 1 | −36 | −35 |
| 2 | −18 | −16 |
| 3 | −12 | −9 ← okay |
| 4 | −9 | −5 |
| 6 | −6 | 0 |

$u^2 - 9u - 36 = (u + 3)(u - 12)$
Now get back in terms of $a$.
$(a + 4)^2 - 9(a + 4) - 36 = (a + 4 + 3)(a + 4 - 12)$
$= (a + 7)(a - 8)$

98. $2(w-1)^2 + 11(w-1) + 9$

Let $u = w-1$. This gives $2u^2 + 11u + 9$.
$a \cdot c = 2 \cdot 9 = 18$
We are looking for two factors of 18 whose sum is 11. Since the product is positive, the factors will have the same sign. Since the sum is also positive, the two factors will both be positive.

| factor 1 | factor 2 | sum |
|---|---|---|
| 1 | 18 | 19 |
| 2 | 9 | 11 ← okay |
| 3 | 6 | 9 |

$2u^2 + 11u + 9 = 2u^2 + 2u + 9u + 9$
$= 2u(u+1) + 9(u+1)$
$= (u+1)(2u+9)$

Now go back in terms of $w$.
$2(w-1)^2 + 11(w-1) + 9$
$= (w-1+1)(2(w-1)+9)$
$= w(2w-2+9)$
$= w(2w+7)$

99. $x^2 + 22x + 121 = x^2 + 2 \cdot 11 \cdot x + 11^2$
$= (x+11)^2$

100. $w^2 - 34w + 289 = w^2 - 2 \cdot 17 \cdot w + 17^2$
$= (w-17)^2$

101. $144 - 24c + c^2 = 12^2 - 2 \cdot 12 \cdot c + c^2$
$= (12-c)^2$

102. $x^2 - 8x + 16 = x^2 - 2 \cdot (4) \cdot x + (4)^2$
$= (x-4)^2$

103. $64y^2 + 80y + 25 = (8y)^2 + 2 \cdot (8y) \cdot 5 + 5^2$
$= (8y+5)^2$

104. $12z^2 + 48z + 48 = 12(z^2 + 4z + 4)$
$= 12(z^2 + 2 \cdot 2 \cdot z + 2^2)$
$= 12(z+2)^2$

105. $x^2 - 196 = x^2 - 14^2$
$= (x-14)(x+14)$

106. $49 - y^2 = 7^2 - y^2$
$= (7-y)(7+y)$

107. $t^2 - 225 = t^2 - (15)^2$
$= (t-15)(t+15)$

108. $4w^2 - 81 = (2w)^2 - 9^2$
$= (2w-9)(2w+9)$

109. $36x^4 - 25y^2 = (6x^2)^2 - (5y)^2$
$= (6x^2 - 5y)(6x^2 + 5y)$

110. $80mn^2 - 20m = 20m(4n^2 - 1)$
$= 20m((2n)^2 - 1^2)$
$= 20m(2n-1)(2n+1)$

111. $x^3 - 343 = x^3 - 7^3$
$= (x-7)(x^2 + (7)x + 7^2)$
$= (x-7)(x^2 + 7x + 49)$

112. $729 - y^3 = 9^3 - y^3$
$= (9-y)(9^2 + 9(y) + y^2)$
$= (9-y)(81 + 9y + y^2)$
$= (9-y)(y^2 + 9y + 81)$

113. $27x^3 - 125y^3$
$= (3x)^3 - (5y)^3$
$= (3x - 5y)((3x)^2 + (3x)(5y) + (5y)^2)$
$= (3x - 5y)(9x^2 + 15xy + 25y^2)$

## Chapter 5: Polynomials and Polynomial Functions

114. $8m^6 + 27n^3$
$= (2m^2)^3 + (3n)^3$
$= (2m^2 + 3n)\left((2m^2)^2 - (2m^2)(3n) + (3n)^2\right)$
$= (2m^2 + 3n)(4m^4 - 6m^2n + 9n^2)$

115. $2a^6 - 2b^6$
$= 2(a^6 - b^6)$
$= 2\left((a^3)^2 - (b^3)^2\right)$
$= 2(a^3 - b^3)(a^3 + b^3)$
$= 2(a-b)(a^2+ab+b^2)(a+b)(a^2-ab+b^2)$
$= 2(a-b)(a+b)(a^2+ab+b^2)(a^2-ab+b^2)$

116. $(y-1)^3 + 64$
$= (y-1)^3 + 4^3$
$= (y-1+4)\left((y-1)^2 - (y-1)(4) + 4^2\right)$
$= (y+3)(y^2 - 2y + 1 - 4y + 4 + 16)$
$= (y+3)(y^2 - 6y + 21)$

117. $x^2 + 7x + 6 = (x+1)(x+6)$

118. $c^2 - 24c + 144 = c^2 - 2(12)c + (12)^2$
$= (c-12)^2$

119. $z^2 - 9z - 112$
We need two factors of $-112$ whose sum is $-9$. Since the product is negative, the factors will have opposite signs. Since the sum is negative, the factor with the larger absolute value will be negative.

| factor 1 | factor 2 | sum | |
|---|---|---|---|
| 2 | −56 | −54 | too small |
| 8 | −14 | −6 | too big |
| 7 | −16 | −9 | ← okay |

$z^2 - 9z - 112 = (z+7)(z-16)$

120. $-8x^2y^3 + 12xy^3 = -4xy^3(2x - 3)$

121. $7x^3 - 28x^2 + 63x = 7x(x^2 - 4x + 9)$

122. $3x^2 - 3x - 18 = 3(x^2 - x - 6)$
$= 3(x-3)(x+2)$

123. $4z^2 - 60z + 225 = (2z)^2 - 2(2z)(15) + 15^2$
$= (2z - 15)^2$

124. $12x^2 + 7x - 49$
$a \cdot c = 12 \cdot -49 = -588$
We are looking for two factors of $-588$ whose sum is $7$. Since the product is negative, the factors will have opposite signs. Since the sum is positive, the factor with the larger absolute value will be positive. The difference is relatively small, so we should select factors whose absolute values are close to each other.

| factor 1 | factor 2 | sum | |
|---|---|---|---|
| −14 | 42 | 28 | |
| −21 | 28 | 7 | ← okay |

$12x^2 + 7x - 49 = 12x^2 - 21x + 28x - 49$
$= 3x(4x - 7) + 7(4x - 7)$
$= (3x + 7)(4x - 7)$

125. $45 + 6x - 3x^2 = -3(x^2 - 2x - 15)$
$= -3(x-5)(x+3)$

126. $10n^2 - 33n - 7$
$a \cdot c = 10 \cdot -7 = -70$
We are looking for two factors of $-70$ whose sum is $-33$. Since the product is negative, the factors will have opposite signs. Since the sum is also negative, the factor with the larger absolute value will be negative.

| factor 1 | factor 2 | sum | |
|---|---|---|---|
| 1 | −70 | −69 | |
| 2 | −35 | −33 | ← okay |

$10n^2 - 33n - 7 = 10n^2 + 2n - 35n - 7$
$= 2n(5n + 1) - 7(5n + 1)$
$= (2n - 7)(5n + 1)$

127. $8 - 2y - y^2 = -(y^2 + 2y - 8)$
$= -(y-2)(y+4)$

**128.** $2x^3 - 10x^2 + 6x - 30 = 2(x^3 - 5x^2 + 3x - 15)$
$= 2(x^2(x-5) + 3(x-5))$
$= 2(x-5)(x^2 + 3)$

**129.** $(w-3z)^2 - (z+2)^2$
$= ((w-3z)-(z+2))((w-3z)+(z+2))$
$= (w-3z-z-2)(w-3z+z+2)$
$= (w-4z-2)(w-2z+2)$

**130.** $(3h+2)^3 + 64$
$= (3h+2)^3 + 4^3$
$= (3h+2+4)((3h+2)^2 - (3h+2)(4) + 4^2)$
$= (3h+6)((3h)^2 + 2(3h)(2) + 2^2 - 12h - 8 + 16)$
$= 3(h+2)(9h^2 + 12h + 4 - 12h + 8)$
$= 3(h+2)(9h^2 + 12)$
$= 3(h+2) \cdot 3(3h^2 + 4)$
$= 9(h+2)(3h^2 + 4)$

**131.** $5p^3q^2 - 80p = 5p(p^2q^2 - 16)$
$= 5p((pq)^2 - 4^2)$
$= 5p(pq-4)(pq+4)$

**132.** $36a^2 - 20a - 27a + 15 = 4a(9a-5) - 3(9a-5)$
$= (4a-3)(9a-5)$

**133.** $m^4 - 5m^2 + 4 = (m^2)^2 - 5(m^2) + 4$
Let $u = m^2$. This gives $u^2 - 5u + 4$.
$u^2 - 5u + 4 = (u-1)(u-4)$
Now go back in terms of $m$.
$m^4 - 5m^2 + 4 = (m^2 - 1)(m^2 - 4)$
$= (m-1)(m+1)(m-2)(m+2)$

**134.** $686 - 16m^6$
$= -2(8m^6 - 343)$
$= -2((2m^2)^3 - 7^3)$
$= -2(2m^2 - 7)((2m^2)^2 + (2m^2)(7) + 7^2)$
$= -2(2m^2 - 7)(4m^4 + 14m^2 + 49)$

**135.** $h^3 + 2h^2 - h - 2 = h^2(h+2) - 1(h+2)$
$= (h^2 - 1)(h+2)$
$= (h-1)(h+1)(h+2)$

**136.** $108x^3 + 4y^3 = 4(27x^3 + y^3)$
$= 4((3x)^3 + y^3)$
$= 4(3x+y)((3x)^2 - (3x)(y) + y^2)$
$= 4(3x+y)(9x^2 - 3xy + y^2)$

**137.** The area of the larger square is $x \cdot x = x^2$ and the area of the smaller square is $5 \cdot 5 = 25$. The shaded area is the difference of the areas of the two squares.
$x^2 - 25 = x^2 - 5^2$
$= (x-5)(x+5)$
The area of the shaded region is $(x-5)(x+5)$ square units.

**138.** The area of the shaded region is the difference between the area of the square and the area of the circle. The length of each side of the square is twice the radius of the circle, or $2x$. Thus, the area of the square is $(2x)^2 = 4x^2$ and the area of the circle is $\pi x^2$. The difference is
$4x^2 - \pi x^2 = (4-\pi)x^2$.
The area of the shaded region is $(4-\pi)x^2$ square units.

**139.** $(w+5)(w-13) = 0$
$w+5 = 0$ or $w-13 = 0$
$w = -5$ or $w = 13$
The solution set is $\{-5, 13\}$.

## Chapter 5: Polynomials and Polynomial Functions

**140.** $x(2x+1)(3x-5) = 0$

$x = 0$ or $2x+1 = 0$ or $3x-5 = 0$

$x = 0$ or $2x = -1$ or $3x = 5$

$x = 0$ or $x = -\frac{1}{2}$ or $x = \frac{5}{3}$

The solution set is $\left\{-\frac{1}{2}, 0, \frac{5}{3}\right\}$.

**141.** $5a^2 = -20a$

$5a^2 + 20a = 0$

$5a(a+4) = 0$

$5a = 0$ or $a+4 = 0$

$a = 0$ or $a = -4$

The solution set is $\{-4, 0\}$.

**142.** $y^2 + 2y = 15$

$y^2 + 2y - 15 = 0$

$(y+5)(y-3) = 0$

$y+5 = 0$ or $y-3 = 0$

$y = -5$ or $y = 3$

The solution set is $\{-5, 3\}$.

**143.** $x^2 + 21x + 54 = 0$

$(x+18)(x+3) = 0$

$x+18 = 0$ or $x+3 = 0$

$x = -18$ or $x = -3$

The solution set is $\{-18, -3\}$.

**144.** $15x^2 + 29x - 14 = 0$

$(5x-2)(3x+7) = 0$

$5x-2 = 0$ or $3x+7 = 0$

$5x = 2$ or $3x = -7$

$x = \frac{2}{5}$ or $x = -\frac{7}{3}$

The solution set is $\left\{-\frac{7}{3}, \frac{2}{5}\right\}$.

**145.** $x(x+1) = 110$

$x^2 + x = 110$

$x^2 + x - 110 = 0$

$(x+11)(x-10) = 0$

$x+11 = 0$ or $x-10 = 0$

$x = -11$ or $x = 10$

The solution set is $\{-11, 10\}$.

**146.** $\frac{1}{2}x^2 + 5x + 12 = 0$

$2\left(\frac{1}{2}x^2 + 5x + 12\right) = 2 \cdot 0$

$x^2 + 10x + 24 = 0$

$(x+6)(x+4) = 0$

$x+6 = 0$ or $x+4 = 0$

$x = -6$ or $x = -4$

The solution set is $\{-6, -4\}$.

**147.** $(b+1)(b-3) = 5$

$b^2 + b - 3b - 3 = 5$

$b^2 - 2b - 3 = 5$

$b^2 - 2b - 8 = 0$

$(b-4)(b+2) = 0$

$b-4 = 0$ or $b+2 = 0$

$b = 4$ or $b = -2$

The solution set is $\{-2, 4\}$.

**148.** $(x+7)^3 = x^3 + 133$

$(x+7)(x+7)^2 = x^3 + 133$

$(x+7)(x^2 + 14x + 49) = x^3 + 133$

$x^3 + 14x^2 + 49x + 7x^2 + 98x + 343 = x^3 + 133$

$x^3 + 21x^2 + 147x + 343 = x^3 + 133$

$21x^2 + 147x + 210 = 0$

$21(x^2 + 7x + 10) = 0$

$x^2 + 7x + 10 = 0$

$(x+5)(x+2) = 0$

$x+5 = 0$ or $x+2 = 0$

$x = -5$ or $x = -2$

The solution set is $\{-5, -2\}$.

149. a. $f(x) = x^2 + 5x - 18$
$6 = x^2 + 5x - 18$
$0 = x^2 + 5x - 24$
$0 = (x+8)(x-3)$
$x + 8 = 0$ or $x - 3 = 0$
$x = -8$ or $x = 3$
The solution set is $\{-8, 3\}$.

b. $f(x) = x^2 + 5x - 18$
$-4 = x^2 + 5x - 18$
$0 = x^2 + 5x - 14$
$0 = (x+7)(x-2)$
$x + 7 = 0$ or $x - 2 = 0$
$x = -7$ or $x = 2$
The solution set is $\{-7, 2\}$.
The points $(-8, 6)$, $(3, 6)$, $(-7, -4)$, and $(2, -4)$ are on the graph.

150. a. $f(x) = 5x^2 - 4x + 3$
$3 = 5x^2 - 4x + 3$
$0 = 5x^2 - 4x$
$0 = x(5x - 4)$
$x = 0$ or $5x - 4 = 0$
$x = 0$ or $x = \frac{4}{5}$
The solution set is $\{0, \frac{4}{5}\}$.

b. $f(x) = 5x^2 - 4x + 3$
$4 = 5x^2 - 4x + 3$
$0 = 5x^2 - 4x - 1$
$0 = (5x+1)(x-1)$
$5x + 1 = 0$ or $x - 1 = 0$
$x = -\frac{1}{5}$ or $x = 1$
The solution set is $\{-\frac{1}{5}, 1\}$.
The points $(0, 3)$, $(\frac{4}{5}, 3)$, $(-\frac{1}{5}, 4)$, and $(1, 4)$ are on the graph of $f$.

151. $f(x) = 3x^3 + 18x^2 + 24x$
$0 = 3x(x^2 + 6x + 8)$
$0 = 3x(x+4)(x+2)$
$3x = 0$ or $x + 4 = 0$ or $x + 2 = 0$
$x = 0$ or $x = -4$ or $x = -2$
The zeros of $f(x)$ are $-4$, $-2$, and 0. The x-intercepts are also $-4$, $-2$, and 0.

152. $f(x) = -4x^2 + 22x + 42$
$0 = -4x^2 + 22x + 42$
$0 = 2x^2 - 11x - 21$
$0 = (2x+3)(x-7)$
$2x + 3 = 0$ or $x - 7 = 0$
$x = -\frac{3}{2}$ or $x = 7$
The zeros of $f(x)$ are $-\frac{3}{2}$ and 7. The x-intercepts are also $-\frac{3}{2}$ and 7.

153. $s(t) = -16t^2 + 2064$
$1280 = -16t^2 + 2064$
$0 = -16t^2 + 784$
$0 = t^2 - 49$
$0 = (t-7)(t+7)$
$t - 7 = 0$ or $t + 7 = 0$
$t = 7$ or $\cancel{t = -7}$
The object will be 1280 feet above the ground after 7 seconds.

*Chapter 5:* Polynomials and Polynomial Functions

**154. a.** $R = 1-(1-r)^2$
$0.96 = 1-(1-r)^2$
$0 = (1-r)^2 - 0.04$
$0 = ((1-r)-0.2)((1-r)+0.2)$
$0 = (0.8-r)(1.2-r)$
$0.8-r = 0$ or $1.2-r = 0$
$r = 0.8$ or $\cancel{r=1.2}$
The component reliability is 0.80.
(Reliability cannot exceed 1)

**b.** $R = 1-(1-r)^2$
$0.99 = 1-(1-r)^2$
$0 = (1-r)^2 - 0.01$
$0 = ((1-r)-0.1)((1-r)+0.1)$
$0 = (0.9-r)(1.1-r)$
$0.9-r = 0$ or $1.1-r = 0$
$r = 0.9$ or $\cancel{r=1.1}$
The component reliability is 0.90.
(Reliability cannot exceed 1)

**Chapter 5 Test**

**1.** $-5x^7 + x^4 + 7x^2 - x + 1$
Degree: 7

**2.** $\left(-2a^3b^2 + 5a^2b + ab + 1\right) + \left(\frac{1}{3}a^3b^2 + 4a^2b - 6ab - 5\right)$
$= -2a^3b^2 + 5a^2b + ab + 1 + \frac{1}{3}a^3b^2 + 4a^2b - 6ab - 5$
$= -2a^3b^2 + \frac{1}{3}a^3b^2 + 5a^2b + 4a^2b + ab - 6ab + 1 - 5$
$= -\frac{5}{3}a^3b^2 + 9a^2b - 5ab - 4$

**3.** $f(x) = x^3 + 3x^2 - x + 1$
$f(-2) = (-2)^3 + 3(-2)^2 - (-2) + 1$
$= -8 + 3(4) + 2 + 1$
$= -8 + 12 + 2 + 1$
$= 7$

**4.** $(f-g)(x) = (7x^3 - 1) - (4x^2 + 3x - 2)$
$= 7x^3 - 1 - 4x^2 - 3x + 2$
$= 7x^3 - 4x^2 - 3x + 1$

**5.** $\frac{1}{2}a^2b(4ab^2 - 6ab + 8)$
$= \frac{1}{2}a^2b \cdot 4ab^2 - \frac{1}{2}a^2b \cdot 6ab + \frac{1}{2}a^2b \cdot 8$
$= 2a^{2+1}b^{1+2} - 3a^{2+1}b^{1+1} + 4a^2b$
$= 2a^3b^3 - 3a^3b^2 + 4a^2b$

**6.** $(3x-1)(4x+17) = 3x \cdot 4x + 3x \cdot 17 - 1 \cdot 4x - 1 \cdot 17$
$= 12x^2 + 51x - 4x - 17$
$= 12x^2 + 47x - 17$

**7.** $(2m-n)^2 = (2m)^2 - 2(2m)(n) + n^2$
$= 4m^2 - 4mn + n^2$

**8.** 
$$\require{enclose} \begin{array}{r} 3z-7 \\ 2z^2+1 \enclose{longdiv}{6z^3 - 14z^2 + z + 4} \end{array}$$
$\phantom{xx}-(6z^3 \phantom{xxx} + 3z)$
$\phantom{xxxxxxx}-14z^2 - 2z + 4$
$\phantom{xxxxxx}-(-14z^2 \phantom{xxx} -7)$
$\phantom{xxxxxxxxxxxx} -2z + 11$

$\dfrac{6z^3 - 14z^2 + z + 4}{2z^2 + 1} = 3z - 7 + \dfrac{-2z+11}{2z^2+1}$

**9.** 
$$6\overline{)5 \quad -27 \quad -18}$$
$\phantom{xxx} \underline{\phantom{xx}30 \quad \phantom{x}18}$
$\phantom{xxx} 5 \quad \phantom{xx}3 \quad \phantom{xx}0$

$\dfrac{5x^2 - 27x - 18}{x-6} = 5x + 3$

10. 
$$\phantom{2x+3)}\;\;3x-4$$
$$2x+3\overline{)6x^2+x-12}$$
$$\phantom{2x+3)}\underline{-(6x^2+9x)}$$
$$\phantom{2x+3)00}-8x-12$$
$$\phantom{2x+3)00}\underline{-(-8x-12)}$$
$$\phantom{2x+3)0000000}0$$

$\left(\frac{f}{g}\right)(x) = 3x - 4$

$\left(\frac{f}{g}\right)(2) = 3(2) - 4 = 6 - 4 = 2$

11. The divisor is $x - 3$ so $c = 3$.
$$f(3) = 2(3)^3 - 3(3)^2 - 4(3) + 7$$
$$= 2(27) - 3(9) - 12 + 7$$
$$= 54 - 27 - 12 + 7$$
$$= 22$$

12. GCF: $4ab^2$
$$12a^3b^2 + 8a^2b^2 - 16ab^3$$
$$= 4ab^2 \cdot 3a^2 + 4ab^2 \cdot 2a - 4ab^2 \cdot 4b$$
$$= 4ab^2(3a^2 + 2a - 4b)$$

13. $6c^2 + 21c - 4c - 14 = 3c(2c + 7) - 2(2c + 7)$
$$= (2c + 7)(3c - 2)$$

14. $x^2 - 13x - 48$
We need two factors of $-48$ whose sum is $-13$. Since the product is negative, the factors will have opposite signs. Since the sum is also negative, the factor with the larger absolute value will be negative.

| Factors | 1, −48 | 2, −24 | 3, −16 | 4, −12 | 6, −8 |
|---|---|---|---|---|---|
| Sum | −47 | −22 | −13 | −8 | −2 |

$x^2 - 13x - 48 = (x - 16)(x + 3)$

15. $-14p^2 - 17p + 6$
$a \cdot c = -14 \cdot 6 = -84$
We need two factors of $-84$ whose sum is $-17$. The product is negative, so the two factors will have opposite signs. Since the sum is also negative, the factor with the larger absolute value will be negative.

| factor 1 | factor 2 | sum | |
|---|---|---|---|
| 3 | −28 | −25 | ← too small |
| 6 | −14 | −8 | ← too large |
| 4 | −21 | −17 | ← okay |

$-14p^2 - 17p + 6 = -14p^2 - 21p + 4p + 6$
$= -7p(2p + 3) + 2(2p + 3)$
$= (2p + 3)(-7p + 2)$
or
$= (2p + 3)(2 - 7p)$

16. $5(z - 1)^2 + 17(z - 1) - 12$
Let $u = z - 1$. This gives $5u^2 + 17u - 12$.
$a \cdot c = 5 \cdot -12 = -60$
We need two factors of $-60$ whose sum is $17$. The product is negative, so the two factors will have opposite signs. Since the sum is positive, the factor with the larger absolute value will be positive.

| factor 1 | factor 2 | sum | |
|---|---|---|---|
| 15 | −4 | 11 | ← too small |
| 30 | −2 | 28 | ← too large |
| 20 | −3 | 17 | ← okay |

$5u^2 + 17u - 12 = 5u^2 + 20u - 3u - 12$
$= 5u(u + 4) - 3(u + 4)$
$= (u + 4)(5u - 3)$

Now go back in terms of $z$.
$5(z - 1)^2 + 17(z - 1) - 12$
$= (z - 1 + 4)(5(z - 1) - 3)$
$= (z + 3)(5z - 5 - 3)$
$= (z + 3)(5z - 8)$

17. $-98x^2 + 112x - 32 = -2(49x^2 - 56x + 16)$
$= -2((7x)^2 - 2(7x)(4) + 4^2)$
$= -2(7x - 4)^2$

## Chapter 5: Polynomials and Polynomial Functions

18. $16x^2 - 196 = 4(4x^2 - 49)$
$= 4((2x)^2 - 7^2)$
$= 4(2x - 7)(2x + 7)$

19. $3m^2 - 5m = 5m - 7$
$3m^2 - 5m - 5m + 7 = 0$
$3m^2 - 10m + 7 = 0$
$(3m - 7)(m - 1) = 0$
$3m - 7 = 0$ or $m - 1 = 0$
$m = \dfrac{7}{3}$ or $m = 1$
The solution set is $\left\{1, \dfrac{7}{3}\right\}$.

20. Let $x$ = the width (shorter side). Then $x + 3$ is the length (longer side).
From the formula for the area of a rectangle, we have:
$A = (\text{length})(\text{width})$
$= (x + 3)(x)$
$= x^2 + 3x$
We know the area is 108 square meters, so we have:
$108 = x^2 + 3x$
$0 = x^2 + 3x - 108$
$0 = (x - 9)(x + 12)$
$x - 9 = 0$ or $x + 12 = 0$
$x = 9$ or $\cancel{x = -12}$
The patio is 9 meters wide and $9 + 3 = 12$ meters long.

## Cumulative Review Chapters R-5

1. $(-3)^2 + 4 - 16 \div 2 = 9 + 4 - 8$
$= 13 - 8$
$= 5$

2. $|2 - 3^2| + 7 = |2 - 9| + 7$
$= |-7| + 7$
$= 7 + 7$
$= 14$

3. $2(5x + 1) - 4(x - 3) = 10x + 2 - 4x + 12$
$= 10x - 4x + 2 + 12$
$= 6x + 14$

4. $3(x + 2) - 10 = 4x$
$3x + 6 - 10 = 4x$
$3x - 4 = 4x$
$-4 = x$
The solution set is $\{-4\}$.

5. $2|x - 5| + 3 = 5$
$2|x - 5| = 2$
$|x - 5| = 1$
$x - 5 = 1$ or $x - 5 = -1$
$x = 6$ or $x = 4$
The solution set is $\{4, 6\}$.

6. $4x - 5y = 30$
$-5y = -4x + 30$
$\dfrac{-5y}{-5} = \dfrac{-4x + 30}{-5}$
$y = \dfrac{4}{5}x - 6$

7. $4x - 3(x + 1) \geq 7 - 4x$
$4x - 3x - 3 \geq 7 - 4x$
$x - 3 \geq 7 - 4x$
$5x - 3 \geq 7$
$5x \geq 10$
$\dfrac{5x}{5} \geq \dfrac{10}{5}$
$x \geq 2$
Solution set: $\{x \mid x \geq 2\}$
Interval: $[2, \infty)$

8. Let $x =$ liters of 70% sulfuric acid solution. There will be $x+12$ liters in the final solution.

total acid = acid in 30% sol. + acid added

(%)(liters) = (%)(liters) + (%)(liters)

$0.40(x+12) = 0.30(12) + 0.70(x)$

$0.4x + 4.8 = 3.6 + 0.7x$

$4.8 = 3.6 + 0.3x$

$1.2 = 0.3x$

$4 = x$

4 liters of the 70% solution should be added to obtain a 40% sulfuric acid solution.

9. To find x-intercepts, we locate the points where the graph crosses the x-axis. From the graph, we see that the x-intercepts are $x = -1$ and $x = 2$. To find the y-intercept, we locate the points where the graph crosses the y-axis. From the graph, we see that the y-intercept is $y = -2$.

The intercepts are $(-1,0), (2,0)$, and $(0,-2)$.

10. $y = 2x^2 + 4x - 1$

| $x$ | $y = 2x^2 + 4x - 1$ | Point |
|---|---|---|
| $-3$ | $y = 2(-3)^2 + 4(-3) - 1 = 5$ | $(-3,5)$ |
| $-2$ | $y = 2(-2)^2 + 4(-2) - 1 = -1$ | $(-2,-1)$ |
| $-1$ | $y = 2(-1)^2 + 4(-1) - 1 = -3$ | $(-1,-3)$ |
| $0$ | $y = 2(0)^2 + 4(0) - 1 = -1$ | $(0,-1)$ |
| $1$ | $y = 2(1)^2 + 4(1) - 1 = 5$ | $(1,5)$ |

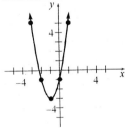

Domain: $\{x \mid x \text{ is any real number}\}$ or $(-\infty, \infty)$

Range: $\{y \mid y \geq -3\}$ or $[-3, \infty)$

11. a. Function. For each value of the independent variable (input) there is exactly one corresponding value for the dependent variable (output).

Domain: $\{-2, 3, 5, 7, 10\}$

Range: $\{-3, 1, 4, 13\}$

b. Not a function. The graph fails the vertical line test. For a given input, it is possible to have more than one corresponding output.

Domain: $\{x \mid x \geq 2\}$ or $[2, \infty)$

Range: $\{y \mid y \text{ is any real number}\}$ or $(-\infty, \infty)$

12. Let $x = -4, -2, 0$, and $2$.

$x = -4$: $3(-4) - 2y = 6$

$-12 - 2y = 6$

$-2y = 18$

$y = -9$

$x = -2$: $3(-2) - 2y = 6$

$-6 - 2y = 6$

$-2y = 12$

$y = -6$

$x = 0$: $3(0) - 2y = 6$

$0 - 2y = 6$

$-2y = 6$

$y = -3$

$x = 2$: $3(2) - 2y = 6$

$6 - 2y = 6$

$-2y = 0$

$y = 0$

So the points $(-4,-9)$, $(-2,-6)$, $(0,-3)$, and $(2,0)$ are on the graph.

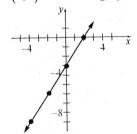

13. $m = \frac{2}{5}$, $(x_1, y_1) = (10, -4)$

    $y - y_1 = m(x - x_1)$

    $y - (-4) = \frac{2}{5}(x - 10)$

    $y + 4 = \frac{2}{5}x - 4$

    $y = \frac{2}{5}x - 8$ or $2x - 5y = 40$

14. a. For roughly the first four months shown, the average price per gallon steadily decreased. Beginning in mid-December, the price began to steadily increase for the last two months.

    b. During this time period it appears, according to the graph, that gas was cheapest around December 14th.

    c. According to the graph, on January 1, 2004 the average price per gallon was roughly $1.49.

15. Replace the inequality symbol with an equal sign to obtain $4x - 2y = 5$. Because the inequality is strict, graph $4x - 2y = 5$ $\left(y = 2x - \frac{5}{2}\right)$ using a dashed line.

    Test Point $(0, 0)$: $4(0) - 2(0) \overset{?}{<} 5$

    $0 < 5$  True

    Therefore, $(0, 0)$ is a solution to $4x - 2y < 5$. Shade the half-plane that contains $(0, 0)$.

    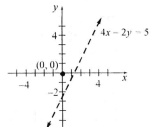

16. $\begin{cases} y = x - 3 & (1) \\ 3x - 2y = 4 & (2) \end{cases}$

    Since equation (1) is already solved for $y$, we will use substitution to solve the system. Substituting $x - 3$ for $y$ in equation (2), we obtain

    $3x - 2(x - 3) = 4$

    $3x - 2x + 6 = 4$

    $x + 6 = 4$

    $x = -2$

    Substituting $-2$ for $x$ in equation (1), we obtain $y = (-2) - 3 = -5$

    The solution is the ordered pair $(-2, -5)$.

17. $\begin{cases} 6x + 2y = 2 \\ 3x - y = 11 \end{cases}$

    Each equation is in standard form so we can write the augmented matrix:

    $\begin{bmatrix} 6 & 2 & | & 2 \\ 3 & -1 & | & 11 \end{bmatrix}$

    Now we need to put this matrix into row echelon form.

    $\begin{bmatrix} 6 & 2 & | & 2 \\ 3 & -1 & | & 11 \end{bmatrix} R_1 = \frac{1}{6}r_1$

    $\begin{bmatrix} 1 & \frac{1}{3} & | & \frac{1}{3} \\ 3 & -1 & | & 11 \end{bmatrix} R_2 = -3r_1 + r_2$

    $\begin{bmatrix} 1 & \frac{1}{3} & | & \frac{1}{3} \\ 0 & -2 & | & 10 \end{bmatrix} R_2 = -\frac{1}{2}r_2$

    $\begin{bmatrix} 1 & \frac{1}{3} & | & \frac{1}{3} \\ 0 & 1 & | & -5 \end{bmatrix}$

    Write the system that corresponds to the row echelon matrix.

    $x + \frac{1}{3}y = \frac{1}{3}$

    $y = -5$

    The system is consistent and independent. Substitute $-5$ for $y$ in the first equation to get

    $x + \frac{1}{3}(-5) = \frac{1}{3}$

    $x - \frac{5}{3} = \frac{1}{3}$

    $x = \frac{6}{3}$

    $x = 2$

    The solution is the ordered pair $(2, -5)$.

18. From the problem statement, we can write two ordered pairs, $(x, y)$, where $x$ is the year and $y$ is the average cost of a 30 second ad (in millions) during the Super Bowl. The ordered pairs are: $(1994, 0.9)$ and $(2004, 2.3)$.

The average rate of change is the slope of the line connecting these ordered pairs.
$$m = \frac{2.3 - 0.9}{2004 - 1994}$$
$$= \frac{1.4}{10}$$
$$= 0.14$$
The average rate of change in Super Bowl ad price is $140,000 per year.

19. Define the following:
    $x$ = price of a cheese pizza
    $y$ = price of a sausage pizza
    $z$ = price of a pepperoni pizza

From the problem statement we need to write three equations, one for each seller.
Brandon: $4x + 5y + 2z = 77.50$
Matt: $8x + 3y + 5z = 110.50$
Ethan: $1x + 4y + 7z = 92.00$

We can solve the system using matrices. Begin with the augmented matrix.
$$\begin{bmatrix} 4 & 5 & 2 & | & 77.50 \\ 8 & 3 & 5 & | & 110.50 \\ 1 & 4 & 7 & | & 92.00 \end{bmatrix}$$
Now get the matrix in row echelon form.
$$\begin{bmatrix} 4 & 5 & 2 & | & 77.50 \\ 8 & 3 & 5 & | & 110.50 \\ 1 & 4 & 7 & | & 92.00 \end{bmatrix} \text{Switch rows 1 and 3}$$
$$\begin{bmatrix} 1 & 4 & 7 & | & 92.00 \\ 8 & 3 & 5 & | & 110.50 \\ 4 & 5 & 2 & | & 77.50 \end{bmatrix} \begin{matrix} R_2 = -8r_1 + r_2 \\ R_3 = -4r_1 + r_3 \end{matrix}$$
$$\begin{bmatrix} 1 & 4 & 7 & | & 92.00 \\ 0 & -29 & -51 & | & -625.50 \\ 0 & -11 & -26 & | & -290.50 \end{bmatrix} R_2 = r_2 \div -29$$

$$\begin{bmatrix} 1 & 4 & 7 & | & 92.00 \\ 0 & 1 & \frac{51}{29} & | & \frac{625.50}{29} \\ 0 & -11 & -26 & | & -290.50 \end{bmatrix} R_3 = 11r_2 + r_3$$
$$\begin{bmatrix} 1 & 4 & 7 & | & 92.00 \\ 0 & 1 & \frac{51}{29} & | & \frac{625.50}{29} \\ 0 & 0 & -\frac{193}{29} & | & -\frac{1544}{29} \end{bmatrix} R_3 = -\frac{29}{193} r_3$$
$$\begin{bmatrix} 1 & 4 & 7 & | & 92.00 \\ 0 & 1 & \frac{51}{29} & | & \frac{625.50}{29} \\ 0 & 0 & 1 & | & 8 \end{bmatrix}$$

Write the system that corresponds to the row echelon matrix.
$x + 4y + 7z = 92.00$
$y + \frac{51}{29} z = \frac{625.50}{29}$
$z = 8$

The system is consistent and independent.
Substitute 8 for $z$ in the second equation to get
$$y + \frac{51}{29}(8) = \frac{625.50}{29}$$
$$y + \frac{408}{29} = \frac{625.50}{29}$$
$$y = 7.50$$

Substitute 8 for $z$ and 7.5 for $y$ in the first equation to get
$$x + 4(7.5) + 7(8) = 92.00$$
$$x + 30 + 56 = 92$$
$$x + 86 = 92$$
$$x = 6$$
Therefore, the cheese pizzas cost $6, the sausage pizzas cost $7.50, and the pepperoni pizzas cost $8.

20. a. Cost: $C(x) = 0.22x + 50$
    Revenue: $R(x) = 0.75x$

    b. To make a profit, revenue must exceed costs. That is, we need
    $$R(x) > C(x)$$
    $$0.75x > 0.22x + 50$$
    $$0.75x - 0.22x > 0.22x + 50 - 0.22x$$
    $$0.53x > 50$$
    $$\frac{0.53x}{0.53} > \frac{50}{0.53}$$
    $$x > 94.34$$
    The band must sell at least 95 candy bars to make a profit.

## Chapter 5: Polynomials and Polynomial Functions

c. To purchase the saxophone, the band needs to profit at least $1396. The profit function is
$$P(x) = R(x) - C(x)$$
$$= 0.75x - (0.22x + 50)$$
$$= 0.75x - 0.22x - 50$$
$$= 0.53x - 50$$
Therefore, the band needs
$$P(x) \geq 1396$$
$$0.53x - 50 \geq 1396$$
$$0.53x - 50 + 50 \geq 1396 + 50$$
$$0.53x \geq 1446$$
$$\frac{0.53x}{0.53} \geq \frac{1446}{0.53}$$
$$x \geq 2728.30$$
To purchase the saxophone, the band needs to sell at least 2729 candy bars.

21. $(12x^3 + 5x^2 - 3x + 1) - (2x^3 - 4x + 8)$
$$= 12x^3 + 5x^2 - 3x + 1 - 2x^3 + 4x - 8$$
$$= 10x^3 + 5x^2 + x - 7$$

22. $(2x+1)(x^2 - 3x + 5)$
$$= 2x(x^2 - 3x + 5) + 1(x^2 - 3x + 5)$$
$$= 2x^3 - 6x^2 + 10x + x^2 - 3x + 5$$
$$= 2x^3 - 5x^2 + 7x + 5$$

23. The divisor is $x + 5$ so $c = -5$.

$$\begin{array}{r|rrrr} -5 & 3 & 10 & -23 & 1 \\ & & -15 & 25 & -10 \\ \hline & 3 & -5 & 2 & -9 \end{array}$$

$$\frac{3x^3 + 10x^2 - 23x + 1}{x+5} = 3x^2 - 5x + 2 - \frac{9}{x+5}$$

24. a. $\frac{1}{3}x^2 + 2x + 3 = \frac{1}{3}(x^2 + 6x + 9)$
$$= \frac{1}{3}(x^2 + 2(3)x + 3^2)$$
$$= \frac{1}{3}(x+3)^2$$

b. $w^2 - 7w - 60 = (w-12)(w+5)$

c. $32a^4 - 128a^2 = 32a^2(a^2 - 4)$
$$= 32a^2(a-2)(a+2)$$

25. $f(x) = 4x^2 - 2x - 30$
$$0 = 4x^2 - 2x - 30$$
$$0 = 2x^2 - x - 15$$
$$0 = (2x+5)(x-3)$$
$$2x + 5 = 0 \text{ or } x - 3 = 0$$
$$x = -\frac{5}{2} \text{ or } x = 3$$
The zeros of $f(x)$ are $-\frac{5}{2}$ and $3$.

# Getting Ready for Chapter 6

2. least common denominator

4. Answers may vary.
   Factor each denominator completely into their prime factors. List each factor that is common to the denominators, then list the factors that are not common. The LCD is the product of the listed factors.

6. Answers may vary.
   To divide two rational numbers, take the reciprocal of the divisor and multiply by the dividend.

8. $\dfrac{6}{18} = \dfrac{\not{2} \cdot \not{3}}{\not{2} \cdot \not{3} \cdot 3} = \dfrac{1}{3}$

10. $\dfrac{-12}{28} = \dfrac{-1 \cdot \not{2} \cdot \not{2} \cdot 3}{\not{2} \cdot \not{2} \cdot 7} = \dfrac{-3}{7} = -\dfrac{3}{7}$

12. $\dfrac{81}{-27} = \dfrac{3 \cdot \not{3} \cdot \not{3} \cdot \not{3}}{-1 \cdot \not{3} \cdot \not{3} \cdot \not{3}} = \dfrac{3}{-1} = -3$

14. $\dfrac{2}{3} \cdot \dfrac{15}{6} = \dfrac{\not{2} \cdot \not{3} \cdot 5}{\not{3} \cdot \not{2} \cdot 3} = \dfrac{5}{3}$

16. $-\dfrac{9}{8} \cdot \dfrac{16}{3} = -\dfrac{3 \cdot \not{3}}{\not{2} \cdot \not{2} \cdot \not{2}} \cdot \dfrac{\not{2} \cdot \not{2} \cdot \not{2} \cdot 2}{\not{3}} = -\dfrac{6}{1} = -6$

18. $\dfrac{3}{14} \cdot \dfrac{7}{12} = \dfrac{\not{3}}{2 \cdot \not{7}} \cdot \dfrac{\not{7}}{2 \cdot 2 \cdot \not{3}} = \dfrac{1}{8}$

20. $\dfrac{2}{3} \div \dfrac{8}{9} = \dfrac{2}{3} \cdot \dfrac{9}{8} = \dfrac{\not{2}}{\not{3}} \cdot \dfrac{3 \cdot \not{3}}{2 \cdot 2 \cdot \not{2}} = \dfrac{3}{4}$

22. $\dfrac{\tfrac{12}{7}}{-\tfrac{18}{21}} = \dfrac{12}{7} \cdot \left(-\dfrac{21}{18}\right) = \dfrac{2 \cdot 2 \cdot \not{3}}{\not{7}} \cdot \dfrac{-1 \cdot \not{3} \cdot \not{7}}{\not{2} \cdot \not{3} \cdot \not{3}} = -2$

24. $\dfrac{\tfrac{10}{-7}}{\tfrac{-5}{14}} = \dfrac{10}{-7} \cdot \dfrac{14}{-5} = \dfrac{2 \cdot \not{5}}{-1 \cdot \not{7}} \cdot \dfrac{2 \cdot \not{7}}{-1 \cdot \not{5}} = \dfrac{4}{1} = 4$

26. $\dfrac{3}{4} + \dfrac{9}{4} = \dfrac{3+9}{4} = \dfrac{12}{4} = \dfrac{3 \cdot \not{4}}{\not{4}} = 3$

28. $\dfrac{19}{6} - \dfrac{5}{6} = \dfrac{19-5}{6} = \dfrac{14}{6} = \dfrac{\not{2} \cdot 7}{\not{2} \cdot 3} = \dfrac{7}{3}$

30. 7 is prime
    $9 = 3 \cdot 3$
    The LCD $= 3 \cdot 3 \cdot 7 = 63$
    $\dfrac{-1}{7} + \dfrac{4}{9} = \dfrac{-1}{7} \cdot \dfrac{9}{9} + \dfrac{4}{9} \cdot \dfrac{7}{7} = \dfrac{-9}{63} + \dfrac{28}{63} = \dfrac{-9+28}{63} = \dfrac{19}{63}$

32. $8 = 2 \cdot 2 \cdot 2$
    $12 = 2 \cdot 2 \cdot 3$
    The LCD $= 2 \cdot 2 \cdot 2 \cdot 3 = 24$
    $\dfrac{5}{8} + \dfrac{5}{12} = \dfrac{5}{8} \cdot \dfrac{3}{3} + \dfrac{5}{12} \cdot \dfrac{2}{2} = \dfrac{15}{24} + \dfrac{10}{24} = \dfrac{15+10}{24} = \dfrac{25}{24}$

34. $16 = 2 \cdot 2 \cdot 2 \cdot 2$
    $20 = 2 \cdot 2 \cdot 5$
    The LCD $= 2 \cdot 2 \cdot 2 \cdot 2 \cdot 5 = 80$
    $\dfrac{7}{16} - \dfrac{9}{20} = \dfrac{7}{16} \cdot \dfrac{5}{5} - \dfrac{9}{20} \cdot \dfrac{4}{4}$
    $= \dfrac{35}{80} - \dfrac{36}{80}$
    $= \dfrac{35-36}{80}$
    $= -\dfrac{1}{80}$

36. $24 = 2 \cdot 2 \cdot 2 \cdot 3$
    $32 = 2 \cdot 2 \cdot 2 \cdot 2 \cdot 2$
    The LCD $= 2 \cdot 2 \cdot 2 \cdot 2 \cdot 2 \cdot 3 = 96$
    $\dfrac{5}{24} + \dfrac{7}{32} = \dfrac{5}{24} \cdot \dfrac{4}{4} + \dfrac{7}{32} \cdot \dfrac{3}{3}$
    $= \dfrac{20}{96} + \dfrac{21}{96}$
    $= \dfrac{20+21}{96}$
    $= \dfrac{41}{96}$

38. $12 = 2 \cdot 2 \cdot 3$
    $15 = 3 \cdot 5$
    The LCD $= 2 \cdot 2 \cdot 3 \cdot 5 = 60$

## Getting Ready for Chapter 6

$$-\frac{7}{12}+\frac{2}{15} = -\frac{7}{12}\cdot\frac{5}{5}+\frac{2}{15}\cdot\frac{4}{4}$$
$$= -\frac{35}{60}+\frac{8}{60}$$
$$= \frac{-35+8}{60}$$
$$= -\frac{27}{60} = -\frac{\cancel{3}\cdot 9}{\cancel{3}\cdot 20}$$
$$= -\frac{9}{20}$$

**40.** $28 = \phantom{3\cdot}2\cdot 2\cdot 7$
$12 = 3\cdot 2\cdot 2$
The LCD = $3\cdot 2\cdot 2\cdot 7 = 84$

$$\frac{3}{28}+\frac{5}{12} = \frac{3}{28}\cdot\frac{3}{3}+\frac{5}{12}\cdot\frac{7}{7}$$
$$= \frac{9}{84}+\frac{35}{84}$$
$$= \frac{9+35}{84}$$
$$= \frac{44}{84} = \frac{\cancel{4}\cdot 11}{\cancel{4}\cdot 21}$$
$$= \frac{11}{21}$$

**42.** $18 = 2\cdot 3\cdot 3$
$45 = \phantom{2\cdot}3\cdot 3\cdot 5$
The LCD = $2\cdot 3\cdot 3\cdot 5 = 90$

$$-\frac{5}{18}-\frac{1}{45} = -\frac{5}{18}\cdot\frac{5}{5}-\frac{1}{45}\cdot\frac{2}{2}$$
$$= -\frac{25}{90}-\frac{2}{90}$$
$$= \frac{-25-2}{90}$$
$$= -\frac{27}{90} = -\frac{3\cdot\cancel{9}}{10\cdot\cancel{9}}$$
$$= -\frac{3}{10}$$

# Chapter 6

## 6.1 Preparing for Multiplying and Dividing Rational Expressions

1. $a \cdot c = 2(-21) = -42$

   The factors of $-42$ that add to $-11$ (the linear coefficient) are $-14$ and $3$. Thus,
   $$2x^2 - 11x - 21 = 2x^2 - 14x + 3x - 21$$
   $$= 2x(x-7) + 3(x-7)$$
   $$= (2x+3)(x-7)$$

2. $q^2 - 16 = 0$
   $(q+4)(q-4) = 0$
   $q+4 = 0$ or $q-4 = 0$
   $q = -4$ or $q = 4$

   The solution set is $\{-4, 4\}$.

3. The reciprocal of $\dfrac{5}{2}$ is $\dfrac{2}{5}$ because $\dfrac{5}{2} \cdot \dfrac{2}{5} = 1$.

4. The domain is the set of all inputs for which the relation (or function) is defined.

## 6.1 Exercises

2. numerator; denominator

4. FALSE. The domain of rational functions is all real numbers except those that cause the denominator to equal 0.

6. Answers will vary. One possibility follows: A rational expression is in lowest terms if the numerator and denominator contain no common factors.

8. The expression $\dfrac{\sqrt{x}}{x+1}$ is not a rational expression because the numerator is not a polynomial.

10. We need to find all values of $x$ that cause the denominator $x - 7$ to equal 0.
    $x - 7 = 0$
    $x = 7$

    Thus, the domain of $\dfrac{4}{x-7}$ is $\{x \mid x \neq 7\}$.

12. We need to find all values of $x$ that cause the denominator $x^2 + 4x - 45$ to equal 0.
    $x^2 + 4x - 45 = 0$
    $(x+9)(x-5) = 0$
    $x+9 = 0$ or $x-5 = 0$
    $x = -9$ or $x = 5$

    Thus, the domain of $\dfrac{2x+1}{x^2+4x-45}$ is $\{x \mid x \neq -9, x \neq 5\}$.

14. We need to find all values of $m$ that cause the denominator $m^2 + 5m + 6$ to equal 0.
    $3m^2 + 4m - 4 = 0$
    $(3m-2)(m+2) = 0$
    $3m-2 = 0$ or $m+2 = 0$
    $m = \dfrac{2}{3}$ or $m = -2$

    Thus, the domain of $\dfrac{m^2+5m+6}{3m^2+4m-4}$ is $\left\{m \mid m \neq -2, m \neq \dfrac{2}{3}\right\}$.

16. We need to find all values of $x$ that cause the denominator $x^2 + 4$ to equal 0. However, if $x$ is a real number, then $x^2 + 4$ can never equal 0.

    Thus, the domain of $\dfrac{x-2}{x^2+4}$ is $\{x \mid x \text{ is any real number}\}$.

18. We need to find all values of $x$ that cause the denominator $x^2 + 8x + 16$ to equal 0.
    $x^2 + 8x + 16 = 0$
    $(x+4)^2 = 0$
    $x+4 = 0$
    $x = -4$

    Thus, the domain of $\dfrac{x+5}{x^2+8x+16}$ is $\{x \mid x \neq -4\}$.

20. $\dfrac{x^2 - 3x}{x^2 - 9} = \dfrac{x(x-3)}{(x+3)(x-3)} = \dfrac{x}{x+3}$

## Chapter 6: Rational Expressions and Rational Functions

22. $\dfrac{a^2 - 2a - 24}{a+4} = \dfrac{(a-6)(a+4)}{a+4} = \dfrac{a-6}{1} = a-6$

24. $\dfrac{6x-42}{x^3-7x^2} = \dfrac{6(x-7)}{x^2(x-7)} = \dfrac{6}{x^2}$

26. $\dfrac{w^2+5w-14}{w^2+6w-16} = \dfrac{(w+7)(w-2)}{(w+8)(w-2)} = \dfrac{w+7}{w+8}$

28. $\dfrac{3n^2+n-2}{3n^2-20n+12} = \dfrac{(3n-2)(n+1)}{(3n-2)(n-6)} = \dfrac{n+1}{n-6}$

30. $\dfrac{25-k^2}{k^2+2k-35} = \dfrac{-1(k-5)(k+5)}{(k-5)(k+7)} = -\dfrac{k+5}{k+7}$

32. $\dfrac{2z^2-10z-28}{4z^3-32z^2+28z} = \dfrac{2(z-7)(z+2)}{4z(z-7)(z-1)} = \dfrac{z+2}{2z(z-1)}$

34. $\dfrac{a^2+5ab+4b^2}{a^2+8ab+16b^2} = \dfrac{(a+4b)(a+b)}{(a+4b)(a+4b)} = \dfrac{a+b}{a+4b}$

36. $\dfrac{v^3+3v^2-5v-15}{v^2+6v+9} = \dfrac{(v+3)(v^2-5)}{(v+3)(v+3)} = \dfrac{v^2-5}{v+3}$

38. $\dfrac{27q^3+1}{6q^2-7q-3} = \dfrac{(3q+1)(9q^2-3q+1)}{(3q+1)(2q-3)} = \dfrac{9q^2-3q+1}{2q-3}$

40. $\dfrac{5x^2}{x+3} \cdot \dfrac{x^2+7x+12}{20x} = \dfrac{x \cdot 5x}{x+3} \cdot \dfrac{(x+4)(x+3)}{4 \cdot 5x}$
$= \dfrac{x \cdot 5x}{x+3} \cdot \dfrac{(x+4)(x+3)}{4 \cdot 5x}$
$= \dfrac{x(x+4)}{4}$

42. $\dfrac{3x^2+14x-5}{x^2+x-30} \cdot \dfrac{x^2-2x-15}{3x^2+8x-3}$
$= \dfrac{(3x-1)(x+5)}{(x+6)(x-5)} \cdot \dfrac{(x-5)(x+3)}{(3x-1)(x+3)}$
$= \dfrac{(3x-1)(x+5)}{(x+6)(x-5)} \cdot \dfrac{(x-5)(x+3)}{(3x-1)(x+3)} = \dfrac{x+5}{x+6}$

44. $\dfrac{p^2-16}{p^2-25} \cdot \dfrac{p^2+2p-24}{p^2+3p-4}$
$= \dfrac{(p-4)(p+4)}{(p-5)(p+5)} \cdot \dfrac{(p-4)(p+6)}{(p-1)(p+4)}$
$= \dfrac{(p-4)(p+4)}{(p-5)(p+5)} \cdot \dfrac{(p-4)(p+6)}{(p-1)(p+4)}$
$= \dfrac{(p-4)^2(p+6)}{(p-5)(p+5)(p-1)}$

46. $\dfrac{2y^2-5y-12}{2y^2-y-6} \cdot \dfrac{4y^2-5y-6}{4-y}$
$= \dfrac{(2y+3)(y-4)}{(2y+3)(y-2)} \cdot \dfrac{(4y+3)(y-2)}{-1(y-4)}$
$= \dfrac{(2y+3)(y-4)}{(2y+3)(y-2)} \cdot \dfrac{(4y+3)(y-2)}{-1(y-4)}$
$= \dfrac{4y+3}{-1} = -(4y+3)$ or $-4y-3$

48. $\dfrac{p^2-4p-5}{p^2-5p-6} \cdot (p-6) = \dfrac{(p-5)(p+1)}{(p-6)(p+1)} \cdot (p-6)$
$= \dfrac{(p-5)(p+1)}{(p-6)(p+1)} \cdot (p-6)$
$= p-5$

50. $\dfrac{a^2+2ab+b^2}{3a+3b} \cdot \dfrac{b-a}{a^2-b^2}$
$= \dfrac{(a+b)(a+b)}{3(a+b)} \cdot \dfrac{-1(a-b)}{(a-b)(a+b)}$
$= \dfrac{(a+b)(a+b)}{3(a+b)} \cdot \dfrac{-1(a-b)}{(a-b)(a+b)}$
$= \dfrac{-1}{3} = -\dfrac{1}{3}$

Section 6.1 Multiplying and Dividing Rational Expressions

52. $\dfrac{\dfrac{x-2}{3x}}{\dfrac{5x-10}{x}} = \dfrac{x-2}{3x} \cdot \dfrac{x}{5x-10}$

$= \dfrac{x-2}{3 \cdot x} \cdot \dfrac{x}{5(x-2)}$

$= \dfrac{\cancel{x-2}}{3 \cdot \cancel{x}} \cdot \dfrac{\cancel{x}}{5\cancel{(x-2)}} = \dfrac{1}{15}$

54. $\dfrac{\dfrac{9m^3}{2n^2}}{\dfrac{3m}{8n^4}} = \dfrac{9m^3}{2n^2} \cdot \dfrac{8n^4}{3m} = \dfrac{\cancel{9m^3}^{3m^2}}{\cancel{2n^2}} \cdot \dfrac{\cancel{8n^4}^{4n^2}}{\cancel{3m}} = 12m^2n^2$

56. $\dfrac{\dfrac{y^2-9}{2y^2-y-15}}{\dfrac{3y^2+10y+3}{2y^2+y-10}} = \dfrac{y^2-9}{2y^2-y-15} \cdot \dfrac{2y^2+y-10}{3y^2+10y+3}$

$= \dfrac{(y-3)(y+3)}{(2y+5)(y-3)} \cdot \dfrac{(2y+5)(y-2)}{(3y+1)(y+3)}$

$= \dfrac{\cancel{(y-3)}\cancel{(y+3)}}{\cancel{(2y+5)}\cancel{(y-3)}} \cdot \dfrac{\cancel{(2y+5)}(y-2)}{(3y+1)\cancel{(y+3)}}$

$= \dfrac{y-2}{3y+1}$

58. $\dfrac{\dfrac{8x^3+1}{2x}}{\dfrac{x^3+2x^2-15x}{2x^2-5x-3}}$

$= \dfrac{8x^3+1}{2x} \cdot \dfrac{2x^2-5x-3}{x^3+2x^2-15x}$

$= \dfrac{(2x+1)(4x^2-2x+1)}{2x} \cdot \dfrac{(2x+1)(x-3)}{x(x+5)(x-3)}$

$= \dfrac{(2x+1)(4x^2-2x+1)}{2x} \cdot \dfrac{(2x+1)\cancel{(x-3)}}{x(x+5)\cancel{(x-3)}}$

$= \dfrac{(2x+1)^2(4x^2-2x+1)}{2x^2(x+5)}$

60. We need to find all values of $x$ that cause the denominator $x+3$ to equal 0.

$x+3 = 0$
$x = -3$

Thus, the domain of $R(x) = \dfrac{5}{x+3}$ is $\{x \mid x \neq -3\}$.

62. We need to find all values of $x$ that cause the denominator $(4x-1)(x+5)$ to equal 0.

$(4x-1)(x+5) = 0$
$4x-1 = 0$ or $x+5 = 0$
$x = \dfrac{1}{4}$ or $x = -5$

Thus, the domain of $R(x) = \dfrac{3x+2}{(4x-1)(x+5)}$ is $\left\{x \mid x \neq -5, x \neq \dfrac{1}{4}\right\}$.

64. We need to find all values of $x$ that cause the denominator $x^2-6x-16$ to equal 0.

$x^2-6x-16 = 0$
$(x+2)(x-8) = 0$
$x+2 = 0$ or $x-8 = 0$
$x = -2$ or $x = 8$

Thus, the domain of $R(x) = \dfrac{5x-2}{x^2-6x-16}$ is $\{x \mid x \neq -2, x \neq 8\}$.

66. We need to find all values of $x$ that cause the denominator $3x^2+7x-6$ to equal 0.

$3x^2+7x-6 = 0$
$(3x-2)(x+3) = 0$
$3x-2 = 0$ or $x+3 = 0$
$x = \dfrac{2}{3}$ or $x = -3$

Thus, the domain of $R(x) = \dfrac{x+3}{3x^2+7x-6}$ is $\left\{x \mid x \neq -3, x \neq \dfrac{2}{3}\right\}$.

68. We need to find all values of $x$ that cause the denominator $4x^2+1$ to equal 0. However, $4x^2+1$ cannot equal 0 if $x$ is a real number, so the domain of $R(x) = \dfrac{4x}{4x^2+1}$ is $\{x \mid x$ is any real number$\}$.

ISM - 307

**70. a.** $R(x) = f(x) \cdot g(x)$
$$= \frac{x^2 - 7x - 8}{2x - 5} \cdot \frac{2x^2 + 3x - 20}{x^2 - 10x + 16}$$
$$= \frac{(x+1)(x-8)}{2x-5} \cdot \frac{(2x-5)(x+4)}{(x-2)(x-8)}$$
$$= \frac{(x+1)\cancel{(x-8)}}{\cancel{2x-5}} \cdot \frac{\cancel{(2x-5)}(x+4)}{(x-2)\cancel{(x-8)}}$$
$$= \frac{(x+1)(x+4)}{x-2}$$

The domain of $f(x)$ is $\left\{x \mid x \neq \frac{5}{2}\right\}$. The domain of $g(x)$ is $\{x \mid x \neq 2, x \neq 8\}$. Therefore, the domain of $R(x)$ is $\left\{x \mid x \neq 2, x \neq \frac{5}{2}, x \neq 8\right\}$.

**b.** $R(x) = \dfrac{f(x)}{h(x)} = \dfrac{\dfrac{x^2 - 7x - 8}{2x - 5}}{\dfrac{x^2 - 3x - 40}{x + 9}}$
$$= \frac{x^2 - 7x - 8}{2x - 5} \cdot \frac{x + 9}{x^2 - 3x - 40}$$
$$= \frac{(x+1)(x-8)}{2x-5} \cdot \frac{x+9}{(x+5)(x-8)}$$
$$= \frac{(x+1)\cancel{(x-8)}}{2x-5} \cdot \frac{x+9}{(x+5)\cancel{(x-8)}}$$
$$= \frac{(x+1)(x+9)}{(2x-5)(x+5)}$$

The domain of $f(x)$ is $\left\{x \mid x \neq \frac{5}{2}\right\}$. The domain of $h(x)$ is $\{x \mid x \neq -9\}$. Because the denominator of $R(x)$ cannot equal 0, we must exclude those values of $x$ such that the numerator of $h(x)$ is 0. That is, we must exclude the values of $x$ such that $x^2 - 3x - 40 = 0$. These values are $x = -5$ and $x = 8$. Therefore, the domain of $R(x)$ is $\left\{x \mid x \neq -9, x \neq -5, x \neq \frac{5}{2}, x \neq 8\right\}$.

**72. a.** $R(x) = (f \cdot g)(x) = \dfrac{4x^2 - 9x - 9}{x^3 - 8} \cdot \dfrac{x^2 + 7x - 18}{5x^2 - 14x - 3}$
$$= \frac{(4x+3)(x-3)}{(x-2)(x^2+2x+4)} \cdot \frac{(x+9)(x-2)}{(5x+1)(x-3)}$$
$$= \frac{(4x+3)\cancel{(x-3)}}{\cancel{(x-2)}(x^2+2x+4)} \cdot \frac{(x+9)\cancel{(x-2)}}{(5x+1)\cancel{(x-3)}}$$
$$= \frac{(4x+3)(x+9)}{(x^2+2x+4)(5x+1)}$$

The domain of $f(x)$ is $\{x \mid x \neq 2\}$. (Note that $x^2 + 2x + 4 = 0$ has no real solution.) The domain of $g(x)$ is $\left\{x \mid x \neq -\dfrac{1}{5}, x \neq 3\right\}$.
Therefore, the domain of $R(x)$ is $\left\{x \mid x \neq -\dfrac{1}{5}, x \neq 2, x \neq 3\right\}$.

**b.** $R(x) = \left(\dfrac{f}{h}\right)(x) = \dfrac{\dfrac{4x^2 - 9x - 9}{x^3 - 8}}{\dfrac{x^2 - 6x + 9}{x^2 - 4}}$
$$= \frac{4x^2 - 9x - 9}{x^3 - 8} \cdot \frac{x^2 - 4}{x^2 - 6x + 9}$$
$$= \frac{(4x+3)(x-3)}{(x-2)(x^2+2x+4)} \cdot \frac{(x-2)(x+2)}{(x-3)(x-3)}$$
$$= \frac{(4x+3)\cancel{(x-3)}}{\cancel{(x-2)}(x^2+2x+4)} \cdot \frac{\cancel{(x-2)}(x+2)}{\cancel{(x-3)}(x-3)}$$
$$= \frac{(4x+3)(x+2)}{(x^2+2x+4)(x-3)}$$

The domain of $f(x)$ is $\{x \mid x \neq 2\}$. (Note that $x^2 + 2x + 4 = 0$ has no real solution.) The domain of $h(x)$ is $\{x \mid x \neq -2, x \neq 2\}$. Because the denominator of $R(x)$ cannot equal 0, we must exclude those values of $x$ such that the numerator of $h(x)$ is 0. That is, we must exclude the values of $x$ such that $x^2 - 6x + 9 = 0$. The only such value is $x = 3$. Therefore, the domain of $R(x)$ is $\{x \mid x \neq -2, x \neq 2, x \neq 3\}$.

**Section 6.1** *Multiplying and Dividing Rational Expressions*

74. $\dfrac{x^3-27}{2x^2+5x-25} \cdot \dfrac{x^2+2x-15}{x^3+3x^2+9x}$

$= \dfrac{(x-3)(x^2+3x+9)}{(2x-5)(x+5)} \cdot \dfrac{(x+5)(x-3)}{x(x^2+3x+9)}$

$= \dfrac{(x-3)\cancel{(x^2+3x+9)}}{(2x-5)\cancel{(x+5)}} \cdot \dfrac{\cancel{(x+5)}(x-3)}{x\cancel{(x^2+3x+9)}}$

$= \dfrac{(x-3)^2}{x(2x-5)}$

76. $\dfrac{\dfrac{x^2+2xy+y^2}{x^2+3xy+2y^2}}{\dfrac{x^2-y^2}{x+2y}}$

$= \dfrac{(m-2n)(m+2n)}{(m-n)(m^2+mn+n^2)} \cdot \dfrac{m^2+mn+n^2}{2(m+2n)}$

$= \dfrac{(m-2n)\cancel{(m+2n)}}{(m-n)\cancel{(m^2+mn+n^2)}} \cdot \dfrac{\cancel{m^2+mn+n^2}}{2\cancel{(m+2n)}}$

$= \dfrac{m-2n}{2(m-n)}$

78. $\dfrac{5m-5}{m^2+6m} \cdot \dfrac{m^2+2m-24}{m^2+3m-4}$

$= \dfrac{5(m-1)}{m(m+6)} \cdot \dfrac{(m+6)(m-4)}{(m+4)(m-1)}$

$= \dfrac{5\cancel{(m-1)}}{m\cancel{(m+6)}} \cdot \dfrac{\cancel{(m+6)}(m-4)}{(m+4)\cancel{(m-1)}}$

$= \dfrac{5(m-4)}{m(m+4)}$

80. $\dfrac{\dfrac{3x+15}{2x+4}}{\dfrac{x+5}{x^2-4}} = \dfrac{3x+15}{2x+4} \cdot \dfrac{x^2-4}{x+5}$

$= \dfrac{3(x+5)}{2(x+2)} \cdot \dfrac{(x+2)(x-2)}{x+5}$

$= \dfrac{3\cancel{(x+5)}}{2\cancel{(x+2)}} \cdot \dfrac{\cancel{(x+2)}(x-2)}{\cancel{x+5}}$

$= \dfrac{3(x-2)}{2}$

82. Answers will vary. Any rational expression with $x+2$ as a factor of the denominator, but not the numerator, will be undefined at $x=-2$. One possible expression is $\dfrac{1}{x+2}$.

84. Answers will vary. Any rational expression with both $x+6$ and $x$ as factors of the denominator, but not the numerator, will be undefined at $x=-6$ and $x=0$. One possible rational expression $\dfrac{1}{x(x+6)} = \dfrac{1}{x^2+6x}$.

86. The rational function will have both $x+4$ and $x-3$ as factors of the denominator. Consider the function $R(x) = \dfrac{k}{(x+4)(x-3)} = \dfrac{k}{x^2+x-12}$, where $k$ is a real number. To determine the value of $k$, we solve the following for $k$:

$R(4) = 1$

$\dfrac{k}{4^2+4-12} = 1$

$\dfrac{k}{8} = 1$

$k = 8$

Thus, a rational function $R$ that is undefined at $x=-4$ and $x=3$ such that $R(4)=1$ will be

$R(x) = \dfrac{8}{x^2+x-12}$.

88. **a.** We evaluate $G(s) = \dfrac{s}{1-b}$ for $b=0.9$ and $s=100$ million: $G(100) = \dfrac{100}{1-0.9} = 1000$.

The change in GDP will be $1000 million, or $1 billion.

**b.** We evaluate $G(s) = \dfrac{s}{1-b}$ for $b=0.95$ and $s=100$ million: $G(100) = \dfrac{100}{1-0.95} = 2000$.

The change in GDP will be $2000 million, or $2 billion.

90. a. $R(x) = \dfrac{2x+1}{x-2}$

| $x$ | $R(x)$ |
|---|---|
| 5 | $\dfrac{11}{3} \approx 3.67$ |
| 10 | $\dfrac{21}{8} \approx 2.63$ |
| 50 | $\dfrac{101}{48} \approx 2.10$ |
| 100 | $\dfrac{201}{98} \approx 2.05$ |
| 1000 | $\dfrac{2001}{998} \approx 2.01$ |

As $x$ gets larger in the positive direction, $R$ approaches 2.

b.

| $x$ | $R(x)$ |
|---|---|
| $-5$ | $\dfrac{9}{7} \approx 1.29$ |
| $-10$ | $\dfrac{19}{12} \approx 1.58$ |
| $-50$ | $\dfrac{99}{52} \approx 1.90$ |
| $-100$ | $\dfrac{199}{102} \approx 1.95$ |
| $-1000$ | $\dfrac{1999}{1002} \approx 2.00$ |

As $x$ gets larger in the negative direction, $R$ approaches 2.

c. The term of highest degree in the numerator is $2x$. The term of highest degree in the denominator is $x$. The ratio of the coefficients of these terms is $\dfrac{2}{1} = 2$. The ratio agrees with the results in parts (a) and (b).

d. As $x$ approaches $\infty$, the graph approaches 2 from above. As $x$ approaches $-\infty$, the graph approaches 2 from below. These results agree with those in parts (a) and (b).

92. a. $R(x) = \dfrac{2x^2 - 3x - 20}{x^2 + 4x - 32}$
$= \dfrac{(2x+5)(x-4)}{(x+8)(x-4)}$
$= \dfrac{2x+5}{x+8}$

b. $R(x) = 0$
$\dfrac{2x^2 - 3x - 20}{x^2 + 4x - 32} = 0$
$\dfrac{2x+5}{x+8} = 0$
$2x + 5 = 0$
$2x = -5$
$x = -\dfrac{5}{2}$

The $x$-intercept of the graph of $R$ is $-\dfrac{5}{2}$.

94. Answers may vary. One possibility follows:
$f(x) = \dfrac{x^2 - 3x + 2}{x - 1} = \dfrac{(x-1)(x-2)}{x-1} = x - 2$.
$f(x)$ and $g(x)$ are equal for all values of $x$ expect 1. When $x = 1$, $g(x)$ is $-1$, but $f(x)$ is indeterminate. That is, $g(1) = 1 - 2 = -1$, but
$f(1) = \dfrac{1^2 - 3(1) + 2}{1 - 1} = \dfrac{0}{0} = $ indeterminate.

96. For $f(x) = 3x - 6$, the slope is 3 and the $y$-intercept is $-6$. Begin at $(0, -6)$ and move to the right 1 unit and up 3 units to find the point $(1, -3)$.

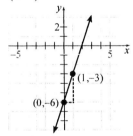

98. For $G(x) = -5x + 10$, the slope is $-5$ and the $y$-intercept is 10. Begin at $(0, 10)$ and move to the right 1 unit and down 5 units to find the point $(1, 5)$.

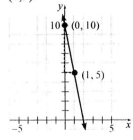

**100.** $H(x) = -x^2 + 4$

Let $x = -3, -2, -1, 0, 1, 2,$ and $3$.

$H(-3) = -(-3)^2 + 4 = -9 + 4 = -5$
$H(-2) = -(-2)^2 + 4 = -4 + 4 = 0$
$H(-1) = -(-1)^2 + 4 = -1 + 4 = 3$
$H(0) = -(0)^2 + 4 = 0 + 4 = 4$
$H(1) = -(1)^2 + 4 = -1 + 4 = 3$
$H(2) = -(2)^2 + 4 = -4 + 4 = 0$
$H(3) = -(3)^2 + 4 = -9 + 4 = -5$

Thus, the points $(-3,-5)$, $(-2,0)$, $(-1,3)$, $(0,4)$, $(1,3)$, $(2,0)$, and $(3,-5)$ are on the graph. We plot these points and connect them with a smooth curve.

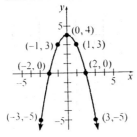

## 6.2 Preparing for Adding and Subtracting Rational Expressions

**1.** $\dfrac{1}{6} - \dfrac{5}{8} = \dfrac{4}{24} - \dfrac{15}{24} = \dfrac{4-15}{24} = -\dfrac{11}{24}$

**2. a.** The additive inverse of 5 is $-5$ because $5 + (-5) = 0$.

**b.** The additive inverse of $x - 2$ is $-(x-2) = -x + 2 = 2 - x$ because $(x-2) + (2-x) = 0$.

## 6.2 Exercises

**2.** $\dfrac{a-b}{c}$

**4.** TRUE

**6.** Answers may vary. To add or subtract rational expressions with unlike denominators, first find the common denominator and rewrite each rational expression with that common denominator, multiplying out the numerators but leaving the denominators in factored form. Second, add or subtract the common rational expressions and write the result in lowest terms, if necessary.

**8.** $\dfrac{5x}{x-3} + \dfrac{2}{x-3} = \dfrac{5x+2}{x-3}$

**10.** $\dfrac{9x}{6x-5} - \dfrac{2}{6x-5} = \dfrac{9x-2}{6x-5}$

**12.** $\dfrac{x}{3x^2+8x-3} + \dfrac{3}{3x^2+8x-3}$
$= \dfrac{x+3}{3x^2+8x-3} = \dfrac{x+3}{(3x-1)(x+3)} = \dfrac{1}{3x-1}$

**14.** $\dfrac{3x^2+8x-1}{x^2-3x-28} - \dfrac{2x^2+2x-9}{x^2-3x-28}$
$= \dfrac{3x^2+8x-1-(2x^2+2x-9)}{x^2-3x-28}$
$= \dfrac{3x^2+8x-1-2x^2-2x+9}{x^2-3x-28}$
$= \dfrac{x^2+6x+8}{x^2-3x-28} = \dfrac{(x+4)(x+2)}{(x+4)(x-7)} = \dfrac{x+2}{x-7}$

**16.** $\dfrac{3x}{x-6} + \dfrac{2}{6-x} = \dfrac{3x}{x-6} + \dfrac{2}{-1(x-6)}$
$= \dfrac{3x}{x-6} + \dfrac{-2}{x-6} = \dfrac{3x-2}{x-6}$

**18.** $\dfrac{x^2+2x-5}{x-4} - \dfrac{x^2-5x-15}{4-x}$
$= \dfrac{x^2+2x-5}{x-4} - \dfrac{x^2-5x-15}{-1(x-4)}$
$= \dfrac{x^2+2x-5}{x-4} + \dfrac{x^2-5x-15}{x-4}$
$= \dfrac{x^2+2x-5+x^2-5x-15}{x-4}$
$= \dfrac{2x^2-3x-20}{x-4} = \dfrac{(2x+5)(x-4)}{x-4} = 2x+5$

## Chapter 6: Rational Expressions and Rational Functions

20. $3a^3 = 3 \cdot a^3$
$9a^2 = 3^2 \cdot a^2$
$\text{LCD} = 3^2 \cdot a^3 = 9a^3$

22. $8a^3b = 2^3 \cdot a^3 \cdot b$
$12ab^2 = 2^2 \cdot 3 \cdot a \cdot b^2$
$\text{LCD} = 2^3 \cdot a^3 \cdot b^2 \cdot 3 = 24a^3b^2$

24. $x+2$
$x-5$
$\text{LCD} = (x+2)(x-5)$

26. $m^2 + 3m - 18 = (m+6)(m-3)$
$m^2 - 7m + 12 = (m-3)(m-4)$
$\text{LCD} = (m-3)(m+6)(m-4)$

28. $x^2 - 9 = (x+3)(x-3)$
$x^3 - 3x^2 = x^2(x-3)$
$\text{LCD} = x^2(x-3)(x+3)$

30. $9x = 3^2 \cdot x$
$3x^2 = 3 \cdot x^2$
$\text{LCD} = 3^2 \cdot x^2 = 9x^2$
$\dfrac{2}{9x} + \dfrac{5}{3x^2} = \dfrac{2}{9x} \cdot \dfrac{x}{x} + \dfrac{5}{3x^2} \cdot \dfrac{3}{3} = \dfrac{2x}{9x^2} + \dfrac{15}{9x^2} = \dfrac{2x+15}{9x^2}$

32. $14mn^3 = 2 \cdot 7 \cdot m \cdot n^3$
$21m^2n = 3 \cdot 7 \cdot m^2 \cdot n$
$\text{LCD} = 7 \cdot m^2 \cdot n^3 \cdot 2 \cdot 3 = 42m^2n^3$
$\dfrac{3}{14mn^3} - \dfrac{2}{21m^2n} = \dfrac{3}{14mn^3} \cdot \dfrac{3m}{3m} - \dfrac{2}{21m^2n} \cdot \dfrac{2n^2}{2n^2}$
$= \dfrac{9m}{42m^2n^3} - \dfrac{4n^2}{42m^2n^3}$
$= \dfrac{9m - 4n^2}{42m^2n^3}$

34. $x-3$
$x+1$
$\text{LCD} = (x-3)(x+1)$
$\dfrac{x+2}{x-3} - \dfrac{x+2}{x+1} = \dfrac{x+2}{x-3} \cdot \dfrac{x+1}{x+1} - \dfrac{x+2}{x+1} \cdot \dfrac{x-3}{x-3}$
$= \dfrac{x^2+3x+2}{(x-3)(x+1)} - \dfrac{x^2-x-6}{(x-3)(x+1)}$
$= \dfrac{x^2+3x+2-(x^2-x-6)}{(x-3)(x+1)}$
$= \dfrac{x^2+3x+2-x^2+x+6}{(x-3)(x+1)}$
$= \dfrac{4x+8}{(x-3)(x+1)}$ or $\dfrac{4(x+2)}{(x-3)(x+1)}$

36. $z+3$
$z^2 - z - 12 = (z+3)(z-4)$
$\text{LCD} = (z+3)(z-4)$
$\dfrac{z+1}{z+3} - \dfrac{z+17}{z^2-z-12}$
$= \dfrac{z+1}{z+3} \cdot \dfrac{z-4}{z-4} - \dfrac{z+17}{(z+3)(z-4)}$
$= \dfrac{z^2-3z-4}{(z+3)(z-4)} - \dfrac{z+17}{(z+3)(z-4)}$
$= \dfrac{z^2-3z-4-(z+17)}{(z+3)(z-4)} = \dfrac{z^2-3z-4-z-17}{(z+3)(z-4)}$
$= \dfrac{z^2-4z-21}{(z+3)(z-4)} = \dfrac{(z+3)(z-7)}{(z+3)(z-4)} = \dfrac{z-7}{z-4}$

38. $(x-1)(x+3)$
$(x+1)(x-1)$
$\text{LCD} = (x-1)(x+1)(x+3)$
$\dfrac{1}{(x-1)(x+3)} - \dfrac{5}{(x+1)(x-1)}$
$= \dfrac{1}{(x-1)(x+3)} \cdot \dfrac{x+1}{x+1} - \dfrac{5}{(x+1)(x-1)} \cdot \dfrac{x+3}{x+3}$
$= \dfrac{x+1}{(x-1)(x+1)(x+3)} - \dfrac{5x+15}{(x-1)(x+1)(x+3)}$
$= \dfrac{x+1-(5x+15)}{(x-1)(x+1)(x+3)}$
$= \dfrac{x+1-5x-15}{(x-1)(x+1)(x+3)}$
$= \dfrac{-4x-14}{(x-1)(x+1)(x+3)}$ or $\dfrac{-2(2x+7)}{(x-1)(x+1)(x+3)}$

## Section 6.2 Adding and Subtracting Rational Expressions

**40.** $x^2 + 4x + 3 = (x+1)(x+3)$
$x^2 - 1 = (x+1)(x-1)$
$\text{LCD} = (x+1)(x-1)(x+3)$

$\dfrac{x-5}{x^2+4x+3} + \dfrac{x-2}{x^2-1}$

$= \dfrac{x-5}{(x+1)(x+3)} \cdot \dfrac{x-1}{x-1} + \dfrac{x-2}{(x+1)(x-1)} \cdot \dfrac{x+3}{x+3}$

$= \dfrac{x^2-6x+5}{(x+1)(x-1)(x+3)} + \dfrac{x^2+x-6}{(x+1)(x-1)(x+3)}$

$= \dfrac{x^2-6x+5+x^2+x-6}{(x+1)(x-1)(x+3)}$

$= \dfrac{2x^2-5x-1}{(x+1)(x-1)(x+3)}$

**42.** $3y^2 - y - 2 = (3y+2)(y-1)$
$3y^2 + 14y + 8 = (3y+2)(y+4)$
$\text{LCD} = (3y+2)(y-1)(y+4)$

$\dfrac{y+4}{3y^2-y-2} - \dfrac{1}{3y^2+14y+8}$

$= \dfrac{y+4}{(3y+2)(y-1)} \cdot \dfrac{y+4}{y+4} - \dfrac{1}{(3y+2)(y+4)} \cdot \dfrac{y-1}{y-1}$

$= \dfrac{y^2+8y+16}{(3y+2)(y-1)(y+4)} - \dfrac{y-1}{(3y+2)(y-1)(y+4)}$

$= \dfrac{y^2+8y+16-(y-1)}{(3y+2)(y-1)(y+4)} = \dfrac{y^2+8y+16-y+1}{(3y+2)(y-1)(y+4)}$

$= \dfrac{y^2+7y+17}{(3y+2)(y-1)(y+4)}$

**44.** $m^2 + 4mn + 4n^2 = (m+2n)^2$
$m^2 - mn - 6n^2 = (m+2n)(m-3n)$
$\text{LCD} = (m+2n)^2(m-3n)$

$\dfrac{m-2n}{m^2+4mn+4n^2} + \dfrac{m-n}{m^2-mn-6n^2}$

$= \dfrac{m-2n}{(m+2n)^2} \cdot \dfrac{m-3n}{m-3n} + \dfrac{m-n}{(m+2n)(m-3n)} \cdot \dfrac{m+2n}{m+2n}$

$= \dfrac{m^2-5mn+6n^2}{(m+2n)^2(m-3n)} + \dfrac{m^2+mn-2n^2}{(m+2n)^2(m-3n)}$

$= \dfrac{m^2-5mn+6n^2+m^2+mn-2n^2}{(m+2n)^2(m-3n)}$

$= \dfrac{2m^2-4mn+4n^2}{(m+2n)^2(m-3n)}$ or $\dfrac{2(m^2-2mn+2n^2)}{(m+2n)^2(m-3n)}$

**46.** $x^2 + 7x + 10 = (x+5)(x+2)$
$x^2 + 6x + 5 = (x+5)(x+1)$
$\text{LCD} = (x+5)(x+2)(x+1)$

$\dfrac{3}{x^2+7x+10} - \dfrac{4}{x^2+6x+5}$

$= \dfrac{3}{(x+5)(x+2)} \cdot \dfrac{x+1}{x+1} - \dfrac{4}{(x+5)(x+1)} \cdot \dfrac{x+2}{x+2}$

$= \dfrac{3x+3}{(x+5)(x+2)(x+1)} - \dfrac{4x+8}{(x+5)(x+2)(x+1)}$

$= \dfrac{3x+3-(4x+8)}{(x+5)(x+2)(x+1)} = \dfrac{3x+3-4x-8}{(x+5)(x+2)(x+1)}$

$= \dfrac{-x-5}{(x+5)(x+2)(x+1)} = \dfrac{-(x+5)}{(x+5)(x+2)(x+1)}$

$= \dfrac{-1}{(x+2)(x+1)}$

**48.** $y^2 - 9 = (y-3)(y+3)$
$9 - y^2 = -1(y-3)(y+3)$
$\text{LCD} = (y-3)(y+3)$

$\dfrac{y^2+4y+4}{y^2-9} + \dfrac{y^2+4y+4}{9-y^2}$

$= \dfrac{y^2+4y+4}{(y-3)(y+3)} + \dfrac{y^2+4y+4}{-1(y-3)(y+3)} \cdot \dfrac{-1}{-1}$

$= \dfrac{y^2+4y+4}{(y-3)(y+3)} + \dfrac{-y^2-4y-4}{(y-3)(y+3)}$

$= \dfrac{y^2+4y+4-y^2-4y-4}{(y-3)(y+3)} = \dfrac{0}{(y-3)(y+3)} = 0$

*Chapter 6: Rational Expressions and Rational Functions*

**50.** $m-3$
$m$
$m^2 - 9 = (m-3)(m+3)$
LCD $= m(m-3)(m+3)$

$$\frac{7}{m-3} - \frac{5}{m} - \frac{2m+6}{m^2-9}$$

$$= \frac{7}{m-3} \cdot \frac{m(m+3)}{m(m+3)} - \frac{5}{m} \cdot \frac{(m+3)(m-3)}{(m+3)(m-3)} - \frac{2m+6}{m^2-9} \cdot \frac{m}{m}$$

$$= \frac{7m^2+21m}{m(m-3)(m+3)} - \frac{5m^2-45}{m(m-3)(m+3)} - \frac{2m^2+6m}{m(m-3)(m+3)}$$

$$= \frac{7m^2+21m-(5m^2-45)-(2m^2+6m)}{m(m-3)(m+3)}$$

$$= \frac{7m^2+21m-5m^2+45-2m^2-6m}{m(m-3)(m+3)}$$

$$= \frac{15m+45}{m(m-3)(m+3)}$$

$$= \frac{15(m+3)}{m(m-3)(m+3)} = \frac{15}{m(m-3)}$$

**52.** $x^2 - 16 = (x+4)(x-4)$
$x+4$
$3x^2 - 7x - 20 = (3x+5)(x-4)$
LCD $= (x+4)(x-4)(3x+5)$

$$\frac{x-1}{x^2-16} + \frac{1}{x+4} \cdot \frac{4x+1}{3x^2-7x-20}$$

$$= \frac{x-1}{(x+4)(x-4)} \cdot \frac{3x+5}{3x+5} + \frac{1}{x+4} \cdot \frac{(x-4)(3x+5)}{(x-4)(3x+5)} - \frac{4x+1}{(x-4)(3x+5)} \cdot \frac{x+4}{x+4}$$

$$= \frac{3x^2+2x-5}{(x+4)(x-4)(3x+5)} + \frac{3x^2-7x-20}{(x+4)(x-4)(3x+5)} - \frac{4x^2+17x+4}{(x+4)(x-4)(3x+5)}$$

$$= \frac{3x^2+2x-5+3x^2-7x-20-(4x^2+17x+4)}{(x+4)(x-4)(3x+5)}$$

$$= \frac{3x^2+2x-5+3x^2-7x-20-4x^2-17x-4}{(x+4)(x-4)(3x+5)}$$

$$= \frac{2x^2-22x-29}{(x+4)(x-4)(3x+5)}$$

**54.** $x$
$x+2$
$(x+2)^2$
LCD $= x(x+2)^2$

$$\frac{2}{x} - \frac{2}{x+2} + \frac{2}{(x+2)^2}$$

$$= \frac{2}{x} \cdot \frac{(x+2)^2}{(x+2)^2} - \frac{2}{x+2} \cdot \frac{x(x+2)}{x(x+2)} + \frac{2}{(x+2)^2} \cdot \frac{x}{x}$$

$$= \frac{2x^2+8x+8}{x(x+2)^2} - \frac{2x^2+4x}{x(x+2)^2} + \frac{2x}{x(x+2)^2}$$

$$= \frac{2x^2+8x+8-(2x^2+4x)+2x}{x(x+2)^2}$$

$$= \frac{2x^2+8x+8-2x^2-4x+2x}{x(x+2)^2}$$

$$= \frac{6x+8}{x(x+2)^2} \text{ or } \frac{2(3x+4)}{x(x+2)^2}$$

**56. a.** $x+2$
$x-1$
LCD $= (x+2)(x-1)$

$$R(x) = f(x) + g(x) = \frac{5}{x+2} + \frac{3}{x-1}$$

$$= \frac{5}{x+2} \cdot \frac{x-1}{x-1} + \frac{3}{x-1} \cdot \frac{x+2}{x+2}$$

$$= \frac{5x-5}{(x+2)(x-1)} + \frac{3x+6}{(x+2)(x-1)}$$

$$= \frac{5x-5+3x+6}{(x+2)(x-1)}$$

$$= \frac{8x+1}{(x+2)(x-1)}$$

**b.** Since $R$ is the sum of $f$ and $g$, the restrictions for the domain of $R$ will consist of all the restrictions for the domains of both $f$ and $g$. Since $-2$ is restricted from the domain of $f$ and since 1 is restricted from the domain of $g$, both $-2$ and 1 are restricted from the domain of $R$. That is, the domain of $R(x)$ is
$\{x \mid x \neq -2, x \neq 1\}$.

**58.** $x^2 - 5x - 6 = (x-6)(x+1)$
$x^2 - 4x - 12 = (x-6)(x+2)$
LCD $= (x-6)(x+1)(x+2)$

## Section 6.2 Adding and Subtracting Rational Expressions

a.  $R(x) = f(x) + g(x)$

$= \dfrac{x+5}{x^2 - 5x - 6} + \dfrac{x+1}{x^2 - 4x - 12}$

$= \dfrac{x+5}{(x-6)(x+1)} \cdot \dfrac{x+2}{x+2} + \dfrac{x+1}{(x-6)(x+2)} \cdot \dfrac{x+1}{x+1}$

$= \dfrac{x^2 + 7x + 10}{(x-6)(x+1)(x+2)} + \dfrac{x^2 + 2x + 1}{(x-6)(x+1)(x+2)}$

$= \dfrac{x^2 + 7x + 10 + x^2 + 2x + 1}{(x-6)(x+1)(x+2)}$

$= \dfrac{2x^2 + 9x + 11}{(x-6)(x+1)(x+2)}$

b.  Since $R$ is the sum of $f$ and $g$, the restrictions for the domain of $R$ will consist of all the restrictions for the domains of both $f$ and $g$. Since 6 and $-1$ are restricted from the domain of $f$ and since 6 and $-2$ are restricted from the domain of $g$, then 6, $-1$, and $-2$ are restricted from the domain of $R$. That is, the domain of $R(x)$ is $\{x \mid x \neq -2, x \neq -1, x \neq 6\}$.

c.  $H(x) = f(x) - g(x)$

$= \dfrac{x+5}{x^2 - 5x - 6} - \dfrac{x+1}{x^2 - 4x - 12}$

$= \dfrac{x+5}{(x-6)(x+1)} \cdot \dfrac{x+2}{x+2} - \dfrac{x+1}{(x-6)(x+2)} \cdot \dfrac{x+1}{x+1}$

$= \dfrac{x^2 + 7x + 10}{(x-6)(x+1)(x+2)} - \dfrac{x^2 + 2x + 1}{(x-6)(x+1)(x+2)}$

$= \dfrac{x^2 + 7x + 10 - (x^2 + 2x + 1)}{(x-6)(x+1)(x+2)}$

$= \dfrac{x^2 + 7x + 10 - x^2 - 2x - 1}{(x-6)(x+1)(x+2)}$

$= \dfrac{5x + 9}{(x-6)(x+1)(x+2)}$

d.  Since $H$ is the difference of $f$ and $g$, the restrictions for the domain of $H$ will consist of all the restrictions for the domains of both $f$ and $g$. Since 6 and $-1$ are restricted from the domain of $f$ and since 6 and $-2$ are restricted from the domain of $g$, then 6, $-1$, and $-2$ are restricted from the domain of $H$. That is, the domain of $H(x)$ is $\{x \mid x \neq -2, x \neq -1, x \neq 6\}$.

60. $x + 2$

$x^3 + 8 = (x+2)(x^2 - 2x + 4)$

$\text{LCD} = (x+2)(x^2 - 2x + 4)$

$\dfrac{1}{x+2} + \dfrac{x-10}{x^3 + 8}$

$= \dfrac{1}{x+2} \cdot \dfrac{x^2 - 2x + 4}{x^2 - 2x + 4} + \dfrac{x-10}{(x+2)(x^2 - 2x + 4)}$

$= \dfrac{x^2 - 2x + 4}{(x+2)(x^2 - 2x + 4)} + \dfrac{x-10}{(x+2)(x^2 - 2x + 4)}$

$= \dfrac{x^2 - 2x + 4 + x - 10}{(x+2)(x^2 - 2x + 4)} = \dfrac{x^2 - x - 6}{(x+2)(x^2 - 2x + 4)}$

$= \dfrac{(x+2)(x-3)}{(x+2)(x^2 - 2x + 4)} = \dfrac{x-3}{x^2 - 2x + 4}$

62. $\dfrac{2x^2}{x-1} - \dfrac{x^2 - 2x}{x-1} + \dfrac{x-4}{x-1}$

$= \dfrac{2x^2 - (x^2 - 2x) + x - 4}{x-1}$

$= \dfrac{2x^2 - x^2 + 2x + x - 4}{x-1}$

$= \dfrac{x^2 + 3x - 4}{x-1} = \dfrac{(x-1)(x+4)}{x-1} = x+4$

64. $\dfrac{1}{x-4}$

$\text{LCD} = x - 4$

$3 + \dfrac{x+4}{x-4} = \dfrac{3}{1} \cdot \dfrac{x-4}{x-4} + \dfrac{x+4}{x-4} = \dfrac{3x-12}{x-4} + \dfrac{x+4}{x-4}$

$= \dfrac{3x - 12 + x + 4}{x-4}$

$= \dfrac{4x - 8}{x-4}$ or $\dfrac{4(x-2)}{x-4}$

66. $a^2 - 8a + 15 = (a-3)(a-5)$

$a^2 - 9 = (a-3)(a+3)$

$\text{LCD} = (a-3)(a-5)(a+3)$

ISM – 315

## Chapter 6: Rational Expressions and Rational Functions

$$\frac{a+3}{a^2-8a+15}+\frac{a+3}{a^2-9}$$
$$=\frac{a+3}{(a-3)(a-5)}\cdot\frac{a+3}{a+3}+\frac{a+3}{(a-3)(a+3)}\cdot\frac{a-5}{a-5}$$
$$=\frac{a^2+6a+9}{(a-3)(a-5)(a+3)}+\frac{a^2-2a-15}{(a-3)(a-5)(a+3)}$$
$$=\frac{a^2+6a+9+a^2-2a-15}{(a-3)(a-5)(a+3)}$$
$$=\frac{2a^2+4a-6}{(a-3)(a-5)(a+3)}=\frac{2(a+3)(a-1)}{(a-3)(a-5)(a+3)}$$
$$=\frac{2(a-1)}{(a-3)(a-5)}$$

**68.** $z-6$
$z-2$
$z^2-8z+12=(z-6)(z-2)$
LCD $=(z-6)(z-2)$
$$\frac{z+3}{z-6}+\frac{z-1}{z-2}-\frac{6z}{z^2-8z+12}$$
$$=\frac{z+3}{z-6}\cdot\frac{z-2}{z-2}+\frac{z-1}{z-2}\cdot\frac{z-6}{z-6}-\frac{6z}{(z-6)(z-2)}$$
$$=\frac{z^2+z-6}{(z-6)(z-2)}+\frac{z^2-7z+6}{(z-6)(z-2)}-\frac{6z}{(z-6)(z-2)}$$
$$=\frac{z^2+z-6+z^2-7z+6-6z}{(z-6)(z-2)}$$
$$=\frac{2z^2-12z}{(z-6)(z-2)}=\frac{2z(z-6)}{(z-6)(z-2)}=\frac{2z}{z-2}$$

**70.** $x^2+4x-5=(x+5)(x-1)$
$x^2+3x-10=(x+5)(x-2)$
$x^2-3x+2=(x-2)(x-1)$
LCD $=(x+5)(x-1)(x-2)$
$$\frac{x-1}{x^2+4x-5}+\frac{3x-1}{x^2+3x-10}-\frac{4x+1}{x^2-3x+2}$$
$$=\frac{x-1}{(x+5)(x-1)}\cdot\frac{x-2}{x-2}+\frac{3x-1}{(x+5)(x-2)}\cdot\frac{x-1}{x-1}$$
$$-\frac{4x+1}{(x-1)(x-2)}\cdot\frac{x+5}{x+5}$$
$$=\frac{x^2-3x+2}{(x+5)(x-1)(x-2)}+\frac{3x^2-4x+1}{(x+5)(x-1)(x-2)}$$
$$-\frac{4x^2+21x+5}{(x+5)(x-1)(x-2)}$$
$$=\frac{x^2-3x+2+(3x^2-4x+1)-(4x^2+21x+5)}{(x+5)(x-1)(x-2)}$$
$$=\frac{x^2-3x+2+3x^2-4x+1-4x^2-21x-5}{(x+5)(x-1)(x-2)}$$
$$=\frac{-28x-2}{(x+5)(x-1)(x-2)} \text{ or } \frac{-2(14x+1)}{(x+5)(x-1)(x-2)}$$

**72. a.** $S(r)=2\pi r^2+\dfrac{400}{r}$    (Note: LCD $=r$)
$$=\frac{2\pi r^2}{1}\cdot\frac{r}{r}+\frac{400}{r}$$
$$=\frac{2\pi r^3}{r}+\frac{400}{r}$$
$$=\frac{2\pi r^3+400}{r}$$

**b.** $S(4)=\dfrac{2\pi(4)^3+400}{4}$
$$=\frac{128\pi+400}{4}$$
$$=32\pi+100\approx 200.53$$
For a cylindrical can with volume 200 cubic centimeters, if the radius is 4 centimeters, then the surface area of the can will be $32\pi+100\approx 200.53$ square centimeters.

Section 6.2 Adding and Subtracting Rational Expressions

74. a. $s$
$s+5$
LCD $= s(s+5)$
$$T(s) = \frac{600}{s} + \frac{600}{s+5}$$
$$= \frac{600}{s} \cdot \frac{s+5}{s+5} + \frac{600}{s+5} \cdot \frac{s}{s}$$
$$= \frac{600s + 3000}{s(s+5)} + \frac{600s}{s(s+5)}$$
$$= \frac{1200s + 3000}{s(s+5)} \text{ or } \frac{600(2s+5)}{s(s+5)}$$

b. $T(50) = \dfrac{1200(50)+3000}{50(50+5)} = \dfrac{63,000}{2750} \approx 22.91$

If the Sullivan family's average speed between Chicago and Atlanta is 50 miles per hour, then it will take a total of about 22.91 hours for the trip.

c. Using the result from part (b), the average speed for the entire trip will be
$\dfrac{1200 \text{ miles}}{22.91 \text{ hours}} \approx 52.38$ miles per hour.
Answers may vary, but this result should not be very surprising. If the average speed of the first half of the trip (Chicago to Atlanta) is 50 miles per hour, then the average speed of the second half of the trip (Atlanta to Naples) is 55 miles per hour. It seems logical that the average speed for the entire trip should be the average of the two halves.

76. $\left(\dfrac{a}{b}\right)^{-1} + \left(\dfrac{b}{a}\right)^{-1} = \dfrac{b}{a} + \dfrac{a}{b}$ (Note: LCD $= ab$)
$$= \dfrac{b}{a} \cdot \dfrac{b}{b} + \dfrac{a}{b} \cdot \dfrac{a}{a}$$
$$= \dfrac{b^2}{ab} + \dfrac{a^2}{ab}$$
$$= \dfrac{b^2 + a^2}{ab} \text{ or } \dfrac{a^2 + b^2}{ab}$$

78. $-5z(z+4) = -5z \cdot z + (-5z) \cdot 4 = -5z^2 - 20z$

80. $(3q+1)(3q-1) = 9q^2 - 3q + 3q - 1 = 9q^2 - 1$

82. $(2v+1)(4v^2 - 2v + 1)$
$= 8v^3 - 4v^2 + 2v + 4v^2 - 2v + 1$
$= 8v^3 + 1$

6.3 Preparing for Complex Rational Expressions

1. $a \cdot c = 6(-6) = -36$
The factors of $-36$ that add to $-5$ (the linear coefficient) are $-9$ and $4$. Thus,
$6y^2 - 5y - 6 = 6y^2 - 9y + 4y - 6$
$= 3y(2y-3) + 2(2y-3)$
$= (3y+2)(2y-3)$

2. $\left(\dfrac{3ab^2}{2a^{-1}b^5}\right)^{-2} = \left(\dfrac{2a^{-1}b^5}{3ab^2}\right)^2 = \left(\dfrac{2b^{5-2}}{3a^{1+1}}\right)^2$
$= \left(\dfrac{2b^3}{3a^2}\right)^2 = \dfrac{2^2 b^{3 \cdot 2}}{3^2 a^{2 \cdot 2}} = \dfrac{4b^6}{9a^4}$

6.3 Exercises

2. TRUE.

4. Answers will vary.

6. Method 1: Write the numerator of the complex rational expression as a single rational expression. Note that the LCD of the rational expressions in the numerator is $x^2$.
$1 + \dfrac{1}{x^2} = \dfrac{1}{1} \cdot \dfrac{x^2}{x^2} + \dfrac{1}{x^2} = \dfrac{x^2}{x^2} + \dfrac{1}{x^2} = \dfrac{x^2+1}{x^2}$
Write the denominator of the complex rational expression as a single rational expression. Note that the LCD of the rational expressions in the denominator is $x^2$.
$1 - \dfrac{1}{x^2} = \dfrac{1}{1} \cdot \dfrac{x^2}{x^2} - \dfrac{1}{x^2} = \dfrac{x^2}{x^2} - \dfrac{1}{x^2} = \dfrac{x^2-1}{x^2}$
Rewrite the complex rational expression using the single rational expressions and simplify.
$$\dfrac{1 + \dfrac{1}{x^2}}{1 - \dfrac{1}{x^2}} = \dfrac{\dfrac{x^2+1}{x^2}}{\dfrac{x^2-1}{x^2}} = \dfrac{x^2+1}{x^2} \cdot \dfrac{x^2}{x^2-1}$$
$$= \dfrac{x^2+1}{x^2} \cdot \dfrac{x^2}{x^2-1} \text{ or } \dfrac{x^2+1}{(x+1)(x-1)}$$

Method 2: The LCD of all the denominators in the complex rational expression is $x^2$.

$$\frac{1+\dfrac{1}{x^2}}{1-\dfrac{1}{x^2}} = \frac{1+\dfrac{1}{x^2}}{1-\dfrac{1}{x^2}} \cdot \frac{x^2}{x^2} = \frac{1 \cdot x^2 + \dfrac{1}{x^2} \cdot x^2}{1 \cdot x^2 - \dfrac{1}{x^2} \cdot x^2}$$

$$= \frac{x^2+1}{x^2-1} \text{ or } \frac{x^2+1}{(x+1)(x-1)}$$

8. Method 1: Write the numerator of the complex rational expression as a single rational expression. Note that the LCD of the rational expressions in the numerator is $wx$.

$$\frac{7}{w} + \frac{9}{x} = \frac{7}{w} \cdot \frac{x}{x} + \frac{9}{x} \cdot \frac{w}{w} = \frac{7x}{wx} + \frac{9w}{wx} = \frac{7x+9w}{wx}$$

Write the denominator of the complex rational expression as a single rational expression. Note that the LCD of the rational expressions in the denominator is $wx$.

$$\frac{9}{w} - \frac{7}{x} = \frac{9}{w} \cdot \frac{x}{x} - \frac{7}{x} \cdot \frac{w}{w} = \frac{9x}{wx} - \frac{7w}{wx} = \frac{9x-7w}{wx}$$

Rewrite the complex rational expression using the single rational expressions and simplify.

$$\frac{\dfrac{7}{w}+\dfrac{9}{x}}{\dfrac{9}{w}-\dfrac{7}{x}} = \frac{\dfrac{7x+9w}{wx}}{\dfrac{9x-7w}{wx}} = \frac{7x+9w}{wx} \cdot \frac{wx}{9x-7w}$$

$$= \frac{7x+9w}{\cancel{wx}} \cdot \frac{\cancel{wx}}{9x-7w}$$

$$= \frac{7x+9w}{9x-7w}$$

Method 2: The LCD of all the denominators in the complex rational expression is $wx$.

$$\frac{\dfrac{7}{w}+\dfrac{9}{x}}{\dfrac{9}{w}-\dfrac{7}{x}} = \frac{\dfrac{7}{w}+\dfrac{9}{x}}{\dfrac{9}{w}-\dfrac{7}{x}} \cdot \frac{wx}{wx}$$

$$= \frac{\dfrac{7}{w} \cdot wx + \dfrac{9}{x} \cdot wx}{\dfrac{9}{w} \cdot wx - \dfrac{7}{x} \cdot wx}$$

$$= \frac{7x+9w}{9x-7w}$$

10. Method 1: Write the numerator of the complex rational expression as a single rational expression. Note that the LCD of the rational

expressions in the numerator is $a+1$.

$$\frac{a}{a+1} - 1 = \frac{a}{a+1} - \frac{1}{1} \cdot \frac{a+1}{a+1} = \frac{a}{a+1} - \frac{a+1}{a+1}$$

$$= \frac{a-(a+1)}{a+1} = \frac{a-a-1}{a+1} = \frac{-1}{a+1}$$

Write the denominator of the complex rational expression as a single rational expression. Note that the LCD of the rational expressions in the denominator is $a$.

$$\frac{a+3}{a} - 2 = \frac{a+3}{a} - \frac{2}{1} \cdot \frac{a}{a} = \frac{a+3}{a} - \frac{2a}{a}$$

$$= \frac{a+3-2a}{a} = \frac{-a+3}{a} = \frac{-1(a-3)}{a}$$

Rewrite the complex rational expression using the single rational expressions and simplify.

$$\frac{\dfrac{a}{a+1}-1}{\dfrac{a+3}{a}-2} = \frac{\dfrac{-1}{a+1}}{\dfrac{-1(a-3)}{a}} = \frac{-1}{a+1} \cdot \frac{a}{-1(a-3)}$$

$$= \frac{\cancel{-1}}{a+1} \cdot \frac{a}{\cancel{-1}(a-3)} = \frac{a}{(a+1)(a-3)}$$

Method 2: The LCD of all the denominators in the complex rational expression is $a(a+1)$.

$$\frac{\dfrac{a}{a+1}-1}{\dfrac{a+3}{a}-2} = \frac{\dfrac{a}{a+1}-1}{\dfrac{a+3}{a}-2} \cdot \frac{a(a+1)}{a(a+1)}$$

$$= \frac{\dfrac{a}{a+1} \cdot a(a+1) - 1 \cdot a(a+1)}{\dfrac{a+3}{a} \cdot a(a+1) - 2 \cdot a(a+1)}$$

$$= \frac{\dfrac{a}{\cancel{a+1}} \cdot a\cancel{(a+1)} - 1 \cdot a(a+1)}{\dfrac{a+3}{\cancel{a}} \cdot \cancel{a}(a+1) - 2 \cdot a(a+1)}$$

$$= \frac{a^2 - a^2 - a}{a^2+4a+3-2a^2-2a} = \frac{-a}{-a^2+2a+3}$$

$$= \frac{-a}{-(a+1)(a-3)} = \frac{a}{(a+1)(a-3)}$$

12. Method 1: Write the numerator of the complex rational expression as a single rational expression. Note that the LCD of the rational expressions in the numerator is $(x-2)(x-1)$.

$$\frac{\dfrac{x+5}{x-2}-\dfrac{x+3}{x-1}}{} = \frac{x+5}{x-2}\cdot\frac{x-1}{x-1}-\frac{x+3}{x-1}\cdot\frac{x-2}{x-2}$$

$$=\frac{x^2+4x-5}{(x-2)(x-1)}-\frac{x^2+x-6}{(x-2)(x-1)}$$

$$=\frac{x^2+4x-5-(x^2+x-6)}{(x-2)(x-1)}$$

$$=\frac{x^2+4x-5-x^2-x+6}{(x-2)(x-1)}=\frac{3x+1}{(x-2)(x-1)}$$

The denominator is already a single rational expression: $\dfrac{3x+1}{1}$.

Rewrite the complex rational expression using the single rational expressions and simplify.

$$\frac{\dfrac{x+5}{x-2}-\dfrac{x+3}{x-1}}{3x+1}=\frac{\dfrac{3x+1}{(x-2)(x-1)}}{\dfrac{3x+1}{1}}$$

$$=\frac{3x+1}{(x-2)(x-1)}\cdot\frac{1}{3x+1}$$

$$=\frac{\cancel{3x+1}}{(x-2)(x-1)}\cdot\frac{1}{\cancel{3x+1}}=\frac{1}{(x-2)(x-1)}$$

Method 2: The LCD of all the denominators in the complex rational expression is $(x-2)(x-1)$.

$$\frac{\dfrac{x+5}{x-2}-\dfrac{x+3}{x-1}}{3x+1}=\frac{\dfrac{x+5}{x-2}-\dfrac{x+3}{x-1}}{3x+1}\cdot\frac{(x-2)(x-1)}{(x-2)(x-1)}$$

$$=\frac{\dfrac{x+5}{x-2}\cdot(x-2)(x-1)-\dfrac{x+3}{x-1}\cdot(x-2)(x-1)}{(3x+1)\cdot(x-2)(x-1)}$$

$$=\frac{\dfrac{x+5}{\cancel{x-2}}\cdot\cancel{(x-2)}(x-1)-\dfrac{x+3}{\cancel{x-1}}\cdot(x-2)\cancel{(x-1)}}{(3x+1)\cdot(x-2)(x-1)}$$

$$=\frac{(x+5)(x-1)-(x+3)(x-2)}{(3x+1)(x-2)(x-1)}$$

$$=\frac{x^2+4x-5-(x^2+x-6)}{(3x+1)(x-2)(x-1)}$$

$$=\frac{x^2+4x-5-x^2-x+6}{(3x+1)(x-2)(x-1)}$$

$$=\frac{3x+1}{(3x+1)(x-2)(x-1)}=\frac{1}{(x-2)(x-1)}$$

14. Method 1: Write the numerator of the complex rational expression as a single rational expression. Note that the LCD of the rational expressions in the numerator is $(x-1)(x-3)$.

$$\frac{x-4}{x-1}-\frac{x}{x-3}=\frac{x-4}{x-1}\cdot\frac{x-3}{x-3}-\frac{x}{x-3}\cdot\frac{x-1}{x-1}$$

$$=\frac{x^2-7x+12}{(x-1)(x-3)}-\frac{x^2-x}{(x-1)(x-3)}$$

$$=\frac{x^2-7x+12-(x^2-x)}{(x-1)(x-3)}$$

$$=\frac{x^2-7x+12-x^2+x}{(x-1)(x-3)}$$

$$=\frac{-6x+12}{(x-1)(x-3)}$$

$$=\frac{-6(x-2)}{(x-1)(x-3)}$$

Write the denominator of the complex rational expression as a single rational expression. Note that the LCD of the rational expressions in the denominator is $x-3$.

$$3+\frac{12}{x-3}=\frac{3}{1}\cdot\frac{x-3}{x-3}+\frac{12}{x-3}$$

$$=\frac{3x-9}{x-3}+\frac{12}{x-3}$$

$$=\frac{3x-9+12}{x-3}$$

$$=\frac{3x+3}{x-3} \text{ or } \frac{3(x+1)}{x-3}$$

Rewrite the complex rational expression using the single rational expressions and simplify.

$$\frac{\dfrac{x-4}{x-1}-\dfrac{x}{x-3}}{3+\dfrac{12}{x-3}}=\frac{\dfrac{-6(x-2)}{(x-1)(x-3)}}{\dfrac{3(x+1)}{x-3}}$$

$$=\frac{-6(x-2)}{(x-1)(x-3)}\cdot\frac{x-3}{3(x+1)}$$

$$=\frac{\overset{-2}{\cancel{-6}}(x-2)}{(x-1)\cancel{(x-3)}}\cdot\frac{\cancel{x-3}}{\cancel{3}(x+1)}$$

$$=\frac{-2(x-2)}{(x-1)(x+1)}$$

ISM – 319

## Chapter 6: Rational Expressions and Rational Functions

Method 2: The LCD of all the denominators in the complex rational expression is $(x-1)(x-3)$.

$$\frac{\dfrac{x-4}{x-1} - \dfrac{x}{x-3}}{3 + \dfrac{12}{x-3}} = \frac{\dfrac{x-4}{x-1} - \dfrac{x}{x-3}}{3 + \dfrac{12}{x-3}} \cdot \frac{(x-1)(x-3)}{(x-1)(x-3)}$$

$$= \frac{\dfrac{x-4}{x-1} \cdot (x-1)(x-3) - \dfrac{x}{x-3} \cdot (x-1)(x-3)}{3 \cdot (x-1)(x-3) + \dfrac{12}{x-3} \cdot (x-1)(x-3)}$$

$$= \frac{(x-4)(x-3) - x(x-1)}{3(x-1)(x-3) + 12(x-1)}$$

$$= \frac{x^2 - 7x + 12 - x^2 + x}{3(x-1)\left[(x-3) + 4\right]}$$

$$= \frac{-6x + 12}{3(x-1)(x+1)}$$

$$= \frac{-6(x-2)}{3(x-1)(x+1)}$$

$$= \frac{-2(x-2)}{(x-1)(x+1)}$$

**16.** Using Method 2, the LCD of all the denominators in the complex rational expression is $z$.

$$\frac{1 - \dfrac{4}{z}}{z - \dfrac{16}{z}} = \frac{1 - \dfrac{4}{z}}{z - \dfrac{16}{z}} \cdot \frac{z}{z} = \frac{1 \cdot z - \dfrac{4}{z} \cdot z}{z \cdot z - \dfrac{16}{z} \cdot z} = \frac{z - 4}{z^2 - 16}$$

$$= \frac{z - 4}{(z-4)(z+4)} = \frac{1}{z+4}$$

**18.** Using Method 2, the LCD of all the denominators in the complex rational expression is $mn$.

$$\frac{\dfrac{n^2}{m} - \dfrac{m^2}{n}}{\dfrac{1}{m} - \dfrac{1}{n}} = \frac{\dfrac{n^2}{m} - \dfrac{m^2}{n}}{\dfrac{1}{m} - \dfrac{1}{n}} \cdot \frac{mn}{mn} = \frac{\dfrac{n^2}{m} \cdot mn - \dfrac{m^2}{n} \cdot mn}{\dfrac{1}{m} \cdot mn - \dfrac{1}{n} \cdot mn}$$

$$= \frac{n^3 - m^3}{n - m}$$

$$= \frac{(n-m)(n^2 + nm + m^2)}{2(n+m)}$$

$$= \frac{(n-m)(n^2 + nm + m^2)}{2(n+m)}$$

$$= n^2 + nm + m^2 \quad \text{or} \quad m^2 + mn + n^2$$

**20.** Using Method 1, write the numerator of the complex rational expression as a single rational expression. Note that the LCD of the rational expressions in the numerator is $x$.

$$1 + \frac{5}{x} = \frac{1}{1} \cdot \frac{x}{x} + \frac{5}{x} = \frac{x}{x} + \frac{5}{x} = \frac{x+5}{x}$$

Write the denominator of the complex rational expression as a single rational expression. Note that the LCD of the rational expressions in the denominator is $x+4$.

$$1 + \frac{1}{x+4} = \frac{1}{1} \cdot \frac{x+4}{x+4} + \frac{1}{x+4}$$

$$= \frac{x+4}{x+4} + \frac{1}{x+4} = \frac{x+4+1}{x+4} = \frac{x+5}{x+4}$$

Rewrite the complex rational expression using the single rational expressions and simplify.

$$\frac{1 + \dfrac{5}{x}}{1 + \dfrac{1}{x+4}} = \frac{\dfrac{x+5}{x}}{\dfrac{x+5}{x+4}}$$

$$= \frac{x+5}{x} \cdot \frac{x+4}{x+5} = \frac{x+5}{x} \cdot \frac{x+4}{x+5} = \frac{x+4}{x}$$

**Section 6.3** Complex Rational Expressions

22. Using Method 2, the LCD of all the denominators in the complex rational expression is $5x^2$.

$$\frac{\frac{5}{x}-\frac{x}{5}}{\frac{1}{5}-\frac{5}{x^2}} = \frac{\frac{5}{x}-\frac{x}{5}}{\frac{1}{5}-\frac{5}{x^2}} \cdot \frac{5x^2}{5x^2}$$

$$= \frac{\frac{5}{x}\cdot 5x^2 - \frac{x}{5}\cdot 5x^2}{\frac{1}{5}\cdot 5x^2 - \frac{5}{x^2}\cdot 5x^2}$$

$$= \frac{\frac{5}{\cancel{x}}\cdot 5x^{\cancel{2}} - \frac{x}{\cancel{5}}\cdot \cancel{5}x^2}{\frac{1}{\cancel{5}}\cdot \cancel{5}x^2 - \frac{5}{\cancel{x^2}}\cdot 5\cancel{x^2}}$$

$$= \frac{25x - x^3}{x^2 - 25}$$

$$= \frac{-x(x-5)(x+5)}{(x-5)(x+5)}$$

$$= \frac{-x\cancel{(x-5)}\cancel{(x+5)}}{\cancel{(x-5)}\cancel{(x+5)}} = -x$$

24. Using Method 1, write the numerator of the complex rational expression as a single rational expression. Note that the LCD of the rational expressions in the numerator is $(x-3)(x+4)$.

$$\frac{x+5}{x-3} - \frac{x}{x+4} = \frac{x+5}{x-3}\cdot\frac{x+4}{x+4} - \frac{x}{x+4}\cdot\frac{x-3}{x-3}$$

$$= \frac{x^2+9x+20}{(x-3)(x+4)} - \frac{x^2-3x}{(x-3)(x+4)}$$

$$= \frac{x^2+9x+20-(x^2-3x)}{(x-3)(x+4)}$$

$$= \frac{x^2+9x+20-x^2+3x}{(x-3)(x+4)}$$

$$= \frac{12x+20}{(x-3)(x+4)}$$

$$= \frac{4(3x+5)}{(x-3)(x+4)}$$

The denominator is already a single rational expression: $3x^2 - 4x - 15 = (3x+5)(x-3)$.

Rewrite the complex rational expression using the single rational expressions and simplify.

$$\frac{\frac{x+5}{x-3} - \frac{x}{x+4}}{3x^2-4x-15} = \frac{\frac{4(3x+5)}{(x-3)(x+4)}}{\frac{(3x+5)(x-3)}{1}}$$

$$= \frac{4(3x+5)}{(x-3)(x+4)} \cdot \frac{1}{(3x+5)(x-3)}$$

$$= \frac{4\cancel{(3x+5)}}{(x-3)(x+4)} \cdot \frac{1}{\cancel{(3x+5)}(x-3)}$$

$$= \frac{4}{(x-3)^2(x+4)}$$

26. Using Method 2, the LCD of all the denominators in the complex rational expression is $(x+3)(x-4)$.

$$\frac{\frac{x-3}{x+3} + \frac{x-3}{x-4}}{1 + \frac{x+3}{x-4}} = \frac{\frac{x-3}{x+3} + \frac{x-3}{x-4}}{1 + \frac{x+3}{x-4}} \cdot \frac{(x+3)(x-4)}{(x+3)(x-4)}$$

$$= \frac{\frac{x-3}{x+3}\cdot(x+3)(x-4) + \frac{x-3}{x-4}\cdot(x+3)(x-4)}{1\cdot(x+3)(x-4) + \frac{x+3}{x-4}\cdot(x+3)(x-4)}$$

$$= \frac{\frac{x-3}{\cancel{x+3}}\cdot\cancel{(x+3)}(x-4) + \frac{x-3}{\cancel{x-4}}\cdot(x+3)\cancel{(x-4)}}{1\cdot(x+3)(x-4) + \frac{x+3}{\cancel{x-4}}\cdot(x+3)\cancel{(x-4)}}$$

$$= \frac{(x-3)(x-4)+(x-3)(x+3)}{(x+3)(x-4)+(x+3)^2}$$

$$= \frac{(x-3)[(x-4)+(x+3)]}{(x+3)[(x-4)+(x+3)]}$$

$$= \frac{(x-3)(2x-1)}{(x+3)(2x-1)}$$

$$= \frac{(x-3)\cancel{(2x-1)}}{(x+3)\cancel{(2x-1)}} = \frac{x-3}{x+3}$$

28. Using Method 1, write the numerator of the complex rational expression as a single rational expression. Note that the LCD of the rational

## Chapter 6: Rational Expressions and Rational Functions

expressions in the numerator is $(m-3)(m+3)$.

$$\frac{m+3}{m-3} - \frac{m-3}{m+3} = \frac{m+3}{m-3} \cdot \frac{m+3}{m+3} - \frac{m-3}{m+3} \cdot \frac{m-3}{m-3}$$
$$= \frac{m^2+6m+9}{(m-3)(m+3)} - \frac{m^2-6m+9}{(m-3)(m+3)}$$
$$= \frac{m^2+6m+9-(m^2-6m+9)}{(m-3)(m+3)}$$
$$= \frac{m^2+6m+9-m^2+6m-9}{(m-3)(m+3)}$$
$$= \frac{12m}{(m-3)(m+3)}$$

Write the denominator of the complex rational expression as a single rational expression. Note that the LCD of the rational expressions in the denominator is $(m-3)(m+3)$.

$$\frac{m+3}{m-3} + \frac{m-3}{m+3} = \frac{m+3}{m-3} \cdot \frac{m+3}{m+3} + \frac{m-3}{m+3} \cdot \frac{m-3}{m-3}$$
$$= \frac{m^2+6m+9}{(m-3)(m+3)} + \frac{m^2-6m+9}{(m-3)(m+3)}$$
$$= \frac{m^2+6m+9+m^2-6m+9}{(m-3)(m+3)}$$
$$= \frac{2m^2+18}{(m-3)(m+3)}$$
$$= \frac{2(m^2+9)}{(m-3)(m+3)}$$

Rewrite the complex rational expression using the single rational expressions and simplify.

$$\frac{\frac{m+3}{m-3} - \frac{m-3}{m+3}}{\frac{m+3}{m-3} + \frac{m-3}{m+3}} = \frac{\frac{12m}{(m-3)(m+3)}}{\frac{2(m^2+9)}{(m-3)(m+3)}}$$
$$= \frac{12m}{(m-3)(m+3)} \cdot \frac{(m-3)(m+3)}{2(m^2+9)}$$
$$= \frac{\cancel{12}^6 m}{\cancel{(m-3)(m+3)}} \cdot \frac{\cancel{(m-3)(m+3)}}{\cancel{2}(m^2+9)}$$
$$= \frac{6m}{m^2+9}$$

**30.** Using Method 1, the numerator of the complex rational expression is already a single rational expression: $\dfrac{-6}{x^2+5x+6} = \dfrac{-6}{(x+3)(x+2)}$.

Write the denominator of the complex rational expression as a single rational expression. Note that the LCD of the rational expressions in the denominator is $(x+3)(x+2)$.

$$\frac{2}{x+3} - \frac{3}{x+2} = \frac{2}{x+3} \cdot \frac{x+2}{x+2} - \frac{3}{x+2} \cdot \frac{x+3}{x+3}$$
$$= \frac{2x+4}{(x+3)(x+2)} - \frac{3x+9}{(x+3)(x+2)}$$
$$= \frac{2x+4-(3x+9)}{(x+3)(x+2)}$$
$$= \frac{2x+4-3x-9}{(x+3)(x+2)}$$
$$= \frac{-x-5}{(x+3)(x+2)}$$
$$= \frac{-(x+5)}{(x+3)(x+2)}$$

Rewrite the complex rational expression using the single rational expressions and simplify.

$$\frac{\frac{-6}{x^2+5x+6}}{\frac{2}{x+3} - \frac{3}{x+2}} = \frac{\frac{-6}{(x+3)(x+2)}}{\frac{-(x+5)}{(x+3)(x+2)}}$$
$$= \frac{-6}{(x+3)(x+2)} \cdot \frac{(x+3)(x+2)}{-(x+5)}$$
$$= \frac{\cancel{-6}^6}{\cancel{(x+3)(x+2)}} \cdot \frac{\cancel{(x+3)(x+2)}}{\cancel{-}(x+5)}$$
$$= \frac{6}{x+5}$$

**32.** Rewrite the expression so that it does not contain any negative exponents.

$$\frac{2x^{-1}+2y^{-1}}{xy^{-1}-x^{-1}y} = \frac{\frac{2}{x}+\frac{2}{y}}{\frac{x}{y}-\frac{y}{x}}$$

Using Method 2, the LCD of all the denominators in the complex rational expression is $xy$.

$$\frac{2x^{-1}+2y^{-1}}{xy^{-1}-x^{-1}y} = \frac{\dfrac{2}{x}+\dfrac{2}{y}}{\dfrac{x}{y}-\dfrac{y}{x}} = \frac{\dfrac{2}{x}+\dfrac{2}{y}}{\dfrac{x}{y}-\dfrac{y}{x}} \cdot \frac{xy}{xy}$$

$$= \frac{\dfrac{2}{x} \cdot xy + \dfrac{2}{y} \cdot xy}{\dfrac{x}{y} \cdot xy - \dfrac{y}{x} \cdot xy}$$

$$= \frac{2y+2x}{x^2-y^2} = \frac{2(y+x)}{(x-y)(x+y)} = \frac{2}{x-y}$$

**34.** Rewrite the expression so that it does not contain any negative exponents.

$$\frac{(x-y)^{-1}}{x^{-1}-y^{-1}} = \frac{\dfrac{1}{x-y}}{\dfrac{1}{x}-\dfrac{1}{y}}$$

Using Method 1, note the numerator of the complex rational expression is already a single rational expression.

Write the denominator of the complex rational expression as a single rational expression. Note that the LCD of the rational expressions in the denominator is $xy$.

$$\frac{1}{x}-\frac{1}{y} = \frac{1}{x} \cdot \frac{y}{y} - \frac{1}{y} \cdot \frac{x}{x} = \frac{y}{xy}-\frac{x}{xy} = \frac{y-x}{xy} = \frac{-(x-y)}{xy}$$

Rewrite the complex rational expression using the single rational expressions and simplify.

**36.** Rewrite the expression so that it does not contain any negative exponents.

$$\frac{a^{-3}+8b^{-3}}{a^{-2}-4b^{-2}} = \frac{\dfrac{1}{a^3}+\dfrac{8}{b^3}}{\dfrac{1}{a^2}-\dfrac{4}{b^2}}$$

Using Method 2, the LCD of all the denominators in the complex rational expression is $a^3b^3$.

$$\frac{a^{-3}+8b^{-3}}{a^{-2}-4b^{-2}} = \frac{\dfrac{1}{a^3}+\dfrac{8}{b^3}}{\dfrac{1}{a^2}-\dfrac{4}{b^2}}$$

$$= \frac{\dfrac{1}{a^3}+\dfrac{8}{b^3}}{\dfrac{1}{a^2}-\dfrac{4}{b^2}} \cdot \frac{a^3b^3}{a^3b^3}$$

$$= \frac{\dfrac{1}{a^3} \cdot a^3b^3 + \dfrac{8}{b^3} \cdot a^3b^3}{\dfrac{1}{a^2} \cdot a^3b^3 - \dfrac{4}{b^2} \cdot a^3b^3}$$

$$= \frac{b^3+8a^3}{ab^3-4a^3b}$$

$$= \frac{(b+2a)(b^2-2ba+4a^2)}{ab(b-2a)(b+2a)}$$

$$= \frac{b^2-2ba+4a^2}{ab(b-2a)} \text{ or } \frac{4a^2-2ab+b^2}{ab(b-2a)}$$

**38. a.** Using Method 2, the LCD of all the denominators in the complex rational expression is $R_1R_2R_3$.

$$R = \frac{1}{\dfrac{1}{R_1}+\dfrac{1}{R_2}+\dfrac{1}{R_3}} = \frac{1}{\dfrac{1}{R_1}+\dfrac{1}{R_2}+\dfrac{1}{R_3}} \cdot \frac{R_1R_2R_3}{R_1R_2R_3}$$

$$= \frac{1 \cdot R_1R_2R_3}{\dfrac{1}{R_1} \cdot R_1R_2R_3 + \dfrac{1}{R_2} \cdot R_1R_2R_3 + \dfrac{1}{R_3} \cdot R_1R_2R_3}$$

$$= \frac{1 \cdot R_1R_2R_3}{\dfrac{1}{R_1} \cdot R_1R_2R_3 + \dfrac{1}{R_2} \cdot R_1R_2R_3 + \dfrac{1}{R_3} \cdot R_1R_2R_3}$$

$$= \frac{R_1R_2R_3}{R_2R_3+R_1R_3+R_1R_2}$$

**b.** For $R_1 = 4$ ohms, $R_2 = 6$ ohms, and $R_2 = 10$ ohms, then

$$R = \frac{R_1R_2R_3}{R_2R_3+R_1R_3+R_1R_2}$$

$$= \frac{4 \cdot 6 \cdot 10}{6 \cdot 10 + 4 \cdot 10 + 4 \cdot 6}$$

$$= \frac{240}{124} = \frac{60}{31} \approx 1.935 \text{ ohms}$$

**39. a.** Using Method 2, the LCD of all the

*Chapter 6: Rational Expressions and Rational Functions*

**40. a.** Using Method 2, the LCD of all the denominators in the complex rational expression is $R_1 R_2$.

$$f = \dfrac{1}{(n-1)\left[\dfrac{1}{R_1}+\dfrac{1}{R_2}\right]}$$

$$= \dfrac{1}{(n-1)\left[\dfrac{1}{R_1}+\dfrac{1}{R_2}\right]} \cdot \dfrac{R_1 R_2}{R_1 R_2}$$

$$= \dfrac{1 \cdot R_1 R_2}{(n-1)\left[\dfrac{1}{R_1}\cdot R_1 R_2 + \dfrac{1}{R_2}\cdot R_1 R_2\right]}$$

$$= \dfrac{R_1 R_2}{(n-1)[R_2+R_1]} \text{ or } \dfrac{R_1 R_2}{nR_2+nR_1-R_2-R_1}$$

**b.** If $n=1.5$, $R_1=0.5$ meter, and $R_2=0.3$ meter, then

$$f = \dfrac{R_1 R_2}{(n-1)[R_2+R_1]}$$

$$= \dfrac{(0.5)(0.3)}{(1.5-1)(0.3+0.5)} = 0.375 \text{ meter}$$

**42. a.**

$$1+\dfrac{1}{1+\dfrac{1}{x}} = 1+\dfrac{1}{\dfrac{x}{x}+\dfrac{1}{x}}$$

$$= 1+\dfrac{1}{\dfrac{x+1}{x}}$$

$$= 1+1\cdot\dfrac{x}{x+1}$$

$$= 1+\dfrac{x}{x+1}$$

$$= \dfrac{x+1}{x+1}+\dfrac{x}{x+1}$$

$$= \dfrac{x+1+x}{x+1}$$

$$= \dfrac{2x+1}{x+1}$$

**b.** $1+\dfrac{1}{1+\dfrac{1}{1+\dfrac{1}{x}}} = 1+\dfrac{1}{\dfrac{2x+1}{x+1}}$ (from part (a))

$$= 1+1\cdot\dfrac{x+1}{2x+1}$$

$$= 1+\dfrac{x+1}{2x+1}$$

$$= \dfrac{2x+1}{2x+1}+\dfrac{x+1}{2x+1}$$

$$= \dfrac{2x+1+x+1}{2x+1}$$

$$= \dfrac{3x+2}{2x+1}$$

**c.** $1+\dfrac{1}{1+\dfrac{1}{1+\dfrac{1}{1+\dfrac{1}{x}}}} = 1+\dfrac{1}{\dfrac{3x+2}{2x+1}}$ (from part (b))

$$= 1+1\cdot\dfrac{2x+1}{3x+2}$$

$$= 1+\dfrac{2x+1}{3x+2}$$

$$= \dfrac{3x+2}{3x+2}+\dfrac{2x+1}{3x+2}$$

$$= \dfrac{3x+2+2x+1}{3x+2}$$

$$= \dfrac{5x+3}{3x+2}$$

**d.** $1+\dfrac{1}{1+\dfrac{1}{1+\dfrac{1}{1+\dfrac{1}{1+\dfrac{1}{x}}}}} = 1+\dfrac{1}{\dfrac{5x+3}{3x+2}}$ (from part (c))

$$= 1+1\cdot\dfrac{3x+2}{5x+3}$$

$$= 1+\dfrac{3x+2}{5x+3}$$

$$= \dfrac{5x+3}{5x+3}+\dfrac{3x+2}{5x+3}$$

$$= \dfrac{5x+3+3x+2}{5x+3}$$

$$= \dfrac{8x+5}{5x+3}$$

e. From part (a), 2, 1, 1;
   From part (b), 3, 2, 1;
   From part (c), 5, 3, 2;
   From part (d), 8, 5, 3.
   Pattern: Each subsequent term is the sum of the previous two.
   The first 6 terms in the Fibonacci sequence are: 1, 1, 2, 3, 5, 8.

44. $-5a + 2 = 22$
    $-5a = 20$
    $a = -4$
    The solution set is $\{-4\}$.

46. $\frac{2}{3}x - \frac{5}{7}(x+21) = \frac{11}{21}x + \frac{4}{7}$
    $21\left[\frac{2}{3}x - \frac{5}{7}(x+21)\right] = 21\left[\frac{11}{21}x + \frac{4}{7}\right]$
    $21 \cdot \frac{2}{3}x - 21 \cdot \frac{5}{7}(x+21) = 21 \cdot \frac{11}{21}x + 21 \cdot \frac{4}{7}$
    $14x - 15(x+21) = 11x + 12$
    $14x - 15x - 315 = 11x + 12$
    $-x - 315 = 11x + 12$
    $-12x - 315 = 12$
    $-12x = 327$
    $\frac{-12x}{-12} = \frac{327}{-12}$
    $x = -\frac{109}{4}$
    The solution set is $\left\{-\frac{109}{4}\right\}$.

48. $3p^2 + 19p = 14$
    $3p^2 + 19p - 14 = 0$
    $(3p-2)(p+7) = 0$
    $3p - 2 = 0$ or $p + 7 = 0$
    $p = \frac{2}{3}$ or $p = -7$
    The solution set is $\left\{-7, \frac{2}{3}\right\}$.

**Putting the Concepts Together (Sections 6.1 – 6.3)**

1. We need to find all values of $x$ that cause the denominator $3x^2 - 17x - 6$ to equal 0.

   $3x^2 - 17x - 6 = 0$
   $(3x+1)(x-6) = 0$
   $3x + 1 = 0$ or $x - 6 = 0$
   $3x = -1$ or $x = 6$
   $x = -\frac{1}{3}$

   Thus, the domain of $g(x) = \frac{3x+1}{3x^2 - 17x - 6}$ is $\left\{x \,\middle|\, x \neq -\frac{1}{3},\ x \neq 6\right\}$.

2. $\frac{24n - 4n^2}{2n^2 - 9n - 18} = \frac{-4n(n-6)}{(2n+3)(n-6)}$
   $= \frac{-4n\cancel{(n-6)}}{(2n+3)\cancel{(n-6)}} = \frac{-4n}{2n+3}$

3. $\frac{2p^2 - pq - 10q^2}{3p^2 + 2pq - 8q^2} = \frac{(2p-5q)(p+2q)}{(3p-4q)(p+2q)}$
   $= \frac{(2p-5q)\cancel{(p+2q)}}{(3p-4q)\cancel{(p+2q)}}$
   $= \frac{2p-5q}{3p-4q}$

4. $\frac{a^2 - 16}{12a^2 + 48a} \cdot \frac{6a^3 - 30a^2}{a^2 + 2a - 24}$
   $= \frac{(a-4)(a+4)}{12a(a+4)} \cdot \frac{6a^2(a-5)}{(a-4)(a+6)}$
   $= \frac{\cancel{(a-4)}\cancel{(a+4)}}{2 \cdot \cancel{6a}\cancel{(a+4)}} \cdot \frac{\cancel{6a} \cdot a(a-5)}{\cancel{(a-4)}(a+6)} = \frac{a(a-5)}{2(a+6)}$

5. $\dfrac{\dfrac{x^2+x-2}{3x^2-5x-2}}{\dfrac{3x^2-2x-1}{x^2-9x+14}} = \dfrac{x^2+x-2}{3x^2-5x-2} \cdot \dfrac{x^2-9x+14}{3x^2-2x-1}$
   $= \frac{(x-1)(x+2)}{(3x+1)(x-2)} \cdot \frac{(x-2)(x-7)}{(3x+1)(x-1)}$
   $= \frac{\cancel{(x-1)}(x+2)}{(3x+1)\cancel{(x-2)}} \cdot \frac{\cancel{(x-2)}(x-7)}{(3x+1)\cancel{(x-1)}}$
   $= \frac{(x+2)(x-7)}{(3x+1)^2}$

*Chapter 6: Rational Expressions and Rational Functions*

6. $\dfrac{x^2-10}{x^2-4} - \dfrac{3x}{x^2-4} = \dfrac{x^2-10-3x}{x^2-4} = \dfrac{x^2-3x-10}{x^2-4}$
$= \dfrac{(x-5)(x+2)}{(x-2)(x+2)} = \dfrac{(x-5)\cancel{(x+2)}}{(x-2)\cancel{(x+2)}} = \dfrac{x-5}{x-2}$

7. $n^2-7n+10 = (n-2)(n-5)$
$n^2-8n+15 = (n-3)(n-5)$
LCD $= (n-2)(n-3)(n-5)$

$\dfrac{3n}{n^2-7n+10} - \dfrac{2n}{n^2-8n+15}$
$= \dfrac{3n}{(n-2)(n-5)} - \dfrac{2n}{(n-3)(n-5)}$
$= \dfrac{3n}{(n-2)(n-5)} \cdot \dfrac{n-3}{n-3} - \dfrac{2n}{(n-2)(n-3)(n-5)} \cdot \dfrac{n-2}{n-2}$
$= \dfrac{3n^2-9n}{(n-2)(n-3)(n-5)} - \dfrac{2n^2-4n}{(n-2)(n-3)(n-5)}$
$= \dfrac{3n^2-9n-(2n^2-4n)}{(n-2)(n-3)(n-5)} = \dfrac{3n^2-9n-2n^2+4n}{(n-2)(n-3)(n-5)}$
$= \dfrac{n^2-5n}{(n-2)(n-3)(n-5)} = \dfrac{n(n-5)}{(n-2)(n-3)(n-5)}$
$= \dfrac{n}{(n-2)(n-3)}$

8. $y^2+5y-24 = (y-3)(y+8)$
$y^2+4y-32 = (y-4)(y+8)$
LCD $= (y-3)(y-4)(y+8)$

$\dfrac{3y+2}{y^2+5y-24} + \dfrac{7}{y^2+4y-32}$
$= \dfrac{3y+2}{(y-3)(y+8)} \cdot \dfrac{y-4}{y-4} + \dfrac{7}{(y-4)(y+8)} \cdot \dfrac{y-3}{y-3}$
$= \dfrac{3y^2-10y-8}{(y-3)(y-4)(y+8)} + \dfrac{7y-21}{(y-3)(y-4)(y+8)}$
$= \dfrac{3y^2-10y-8+7y-21}{(y-3)(y-4)(y+8)}$
$= \dfrac{3y^2-3y-29}{(y-3)(y-4)(y+8)}$

9. $P(x) = f(x) \cdot g(x)$
$= \dfrac{2x+1}{x^2-11x+28} \cdot \dfrac{3x-12}{4x^2+4x+1}$
$= \dfrac{2x+1}{(x-4)(x-7)} \cdot \dfrac{3(x-4)}{(2x+1)(2x+1)}$
$= \dfrac{\cancel{2x+1}}{\cancel{(x-4)}(x-7)} \cdot \dfrac{3\cancel{(x-4)}}{\cancel{(2x+1)}(2x+1)}$
$= \dfrac{3}{(x-7)(2x+1)}$

10. $x-7 = x-7$
$x^2-11x+28 = (x-4)(x-7)$
LCD $= (x-4)(x-7)$
$D(x) = h(x) - f(x)$
$= \dfrac{3x}{x-7} - \dfrac{2x+1}{x^2-11x+28}$
$= \dfrac{3x}{x-7} - \dfrac{2x+1}{(x-4)(x-7)}$
$= \dfrac{3x}{x-7} \cdot \dfrac{x-4}{x-4} - \dfrac{2x+1}{(x-4)(x-7)}$
$= \dfrac{3x^2-12x}{(x-4)(x-7)} - \dfrac{2x+1}{(x-4)(x-7)}$
$= \dfrac{3x^2-12x-(2x+1)}{(x-4)(x-7)}$
$= \dfrac{3x^2-12x-2x-1}{(x-4)(x-7)} = \dfrac{3x^2-14x-1}{(x-4)(x-7)}$

11. Using Method 2, the LCD of all the denominators in the complex rational expression is $m^2n^2$.

$\dfrac{\dfrac{1}{m^2}-\dfrac{1}{n^2}}{\dfrac{1}{m}-\dfrac{1}{n}} = \dfrac{\dfrac{1}{m^2}-\dfrac{1}{n^2}}{\dfrac{1}{m}-\dfrac{1}{n}} \cdot \dfrac{m^2n^2}{m^2n^2}$
$= \dfrac{\dfrac{1}{m^2}\cdot m^2n^2 - \dfrac{1}{n^2}\cdot m^2n^2}{\dfrac{1}{m}\cdot m^2n^2 - \dfrac{1}{n}\cdot m^2n^2}$
$= \dfrac{n^2-m^2}{mn^2-m^2n}$
$= \dfrac{(n-m)(n+m)}{mn(n-m)}$
$= \dfrac{n+m}{mn}$

**Putting the Concepts Together (Sections 6.1 – 6.3)**

12. Using Method 2, the LCD of all the denominators in the complex rational expression is $(z-2)(z+2)$.

$$\frac{\frac{z^2-2}{z^2-4}+\frac{7}{z-2}}{\frac{z^2+z-24}{z^2-4}-\frac{2}{z+2}} = \frac{\frac{z^2-2}{(z-2)(z+2)}+\frac{7}{z-2}}{\frac{z^2+z-24}{(z-2)(z+2)}-\frac{2}{z+2}}$$

$$= \frac{\frac{z^2-2}{(z-2)(z+2)}+\frac{7}{z-2}}{\frac{z^2+z-24}{(z-2)(z+2)}-\frac{2}{z+2}} \cdot \frac{(z-2)(z+2)}{(z-2)(z+2)}$$

$$= \frac{\frac{z^2-2}{(z-2)(z+2)}\cdot(z-2)(z+2)+\frac{7}{z-2}\cdot(z-2)(z+2)}{\frac{z^2+z-24}{(z-2)(z+2)}\cdot(z-2)(z+2)-\frac{2}{z+2}\cdot(z-2)(z+2)}$$

$$= \frac{z^2-2+7(z+2)}{z^2+z-24-2(z-2)}$$

$$= \frac{z^2-2+7z+14}{z^2+z-24-2z+4}$$

$$= \frac{z^2+7z+12}{z^2-z-20} = \frac{(z+3)(z+4)}{(z+4)(z-5)} = \frac{z+3}{z-5}$$

## 6.4 Preparing for Rational Equations

1. $\frac{2}{3}x + \frac{1}{2} = \frac{3}{4}$

   $12\left(\frac{2}{3}x + \frac{1}{2}\right) = 12\left(\frac{3}{4}\right)$

   $8x + 6 = 9$

   $8x = 3$

   $x = \frac{3}{8}$

   The solution set is $\left\{\frac{3}{8}\right\}$.

2. $a \cdot c = 3(-4) = -12$

   The factors of $-12$ that add to 11 (the linear coefficient) are $-1$ and 12. Thus,

   $3z^2 + 11z - 4 = 3z^2 + 12z - z - 4$

   $= 3z(z+4) - 1(z+4)$

   $= (3z-1)(z+4)$

3. $6y^2 - y - 12 = 0$

   $(3y+4)(2y-3) = 0$

   $3y+4 = 0$ or $2y-3 = 0$

   $3y = -4$ or $2y = 3$

   $y = -\frac{4}{3}$ or $y = \frac{3}{2}$

   The solution set is $\left\{-\frac{4}{3}, \frac{3}{2}\right\}$.

4. a. If $x = -4$, then

   $\frac{x+4}{x^2-5x-24} = \frac{-4+4}{(-4)^2-5(-4)-24}$

   $= \frac{-4+4}{16+20-24} = \frac{0}{12} = 0$

   Since $\frac{x+4}{x^2-5x-24}$ is defined for $x = -4$,

   $-4$ is in the domain.

   b. If $x = 8$, then

   $\frac{x+4}{x^2-5x-24} = \frac{8+4}{8^2-5(8)-24}$

   $= \frac{8+4}{64-40-24} = \frac{12}{0} =$ undefined

   Since $\frac{x+4}{x^2-5x-24}$ is not defined for $x = 8$,

   8 is not in the domain.

5. $f(x) = 3$

   $x^2 - 3x - 15 = 3$

   $x^2 - 3x - 18 = 0$

   $(x+3)(x-6) = 0$

   $x+3 = 0$ or $x-6 = 0$

   $x = -3$ or $x = 6$

   The solution set is $\{-3, 6\}$.

6. If $g(4) = 3$, means that $y = 3$ when $x = 4$.
   Thus, the point $(4, 3)$ is on the graph of $g$.

## 6.4 Exercises

2. least common denominator

4. TRUE.

Chapter 6: Rational Expressions and Rational Functions

6. No. The solution set to the equation $\dfrac{x-6}{x-6}=1$ is all real numbers except 6. That is, the solution set is $\{x \mid x \neq 6\}$ or, using interval notation, $(-\infty, 6) \cup (6, \infty)$.

8.
$$\dfrac{8}{p}+\dfrac{1}{4p}=\dfrac{11}{8}$$
$$8p\left(\dfrac{8}{p}+\dfrac{1}{4p}\right)=8p\left(\dfrac{11}{8}\right)$$
$$64+2=11p$$
$$66=11p$$
$$6=p$$

Check:
$$\dfrac{8}{6}+\dfrac{1}{4(6)} \stackrel{?}{=} \dfrac{11}{8}$$
$$\dfrac{32}{24}+\dfrac{1}{24} \stackrel{?}{=} \dfrac{11}{8}$$
$$\dfrac{33}{24} \stackrel{?}{=} \dfrac{11}{8}$$
$$\dfrac{11}{8}=\dfrac{11}{8} \leftarrow \text{True}$$

The solution checks, so the solution set is $\{6\}$.

10.
$$\dfrac{3}{x-5}=\dfrac{-2}{x+2}$$
$$(x-5)(x+2)\left(\dfrac{3}{x-5}\right)=(x-5)(x+2)\left(\dfrac{-2}{x+2}\right)$$
$$3(x+2)=-2(x-5)$$
$$3x+6=-2x+10$$
$$5x=4$$
$$x=\dfrac{4}{5}$$

Check:
$$\dfrac{3}{\frac{4}{5}-5} \stackrel{?}{=} \dfrac{-2}{\frac{4}{5}+2}$$
$$\dfrac{3}{\frac{4}{5}-5}\cdot\dfrac{5}{5} \stackrel{?}{=} \dfrac{-2}{\frac{4}{5}+2}\cdot\dfrac{5}{5}$$
$$\dfrac{15}{4-25} \stackrel{?}{=} \dfrac{-10}{4+10}$$
$$\dfrac{15}{-21} \stackrel{?}{=} \dfrac{-10}{14}$$
$$-\dfrac{5}{7}=-\dfrac{5}{7} \leftarrow \text{True}$$

The solution checks, so the solution set is $\left\{\dfrac{4}{5}\right\}$.

12.
$$\dfrac{w-4}{w+1}=\dfrac{w-3}{w+3}$$
$$(w+1)(w+3)\left(\dfrac{w-4}{w+1}\right)=(w+1)(w+3)\left(\dfrac{w-3}{w+3}\right)$$
$$(w+3)(w-4)=(w+1)(w-3)$$
$$w^2-w-12=w^2-2w-3$$
$$w=9$$

Check:
$$\dfrac{9-4}{9+1} \stackrel{?}{=} \dfrac{9-3}{9+3}$$
$$\dfrac{5}{10} \stackrel{?}{=} \dfrac{6}{12}$$
$$\dfrac{1}{2}=\dfrac{1}{2} \leftarrow \text{True}$$

The solution checks, so the solution set is $\{9\}$.

14.
$$\dfrac{2x+1}{x+3}=\dfrac{4(x-1)}{2x+3}$$
$$(x+3)(2x+3)\left(\dfrac{2x+1}{x+3}\right)=(x+3)(2x+3)\left(\dfrac{4(x-1)}{2x+3}\right)$$
$$(2x+3)(2x+1)=(x+3)(4x-4)$$
$$4x^2+8x+3=4x^2+8x-12$$
$$3=-12$$

The final statement is a contradiction, so the equation has no solution. The solution set is $\{\ \}$ or $\varnothing$.

16.
$$m+\dfrac{8}{m}=6$$
$$m\left(m+\dfrac{8}{m}\right)=m\cdot 6$$
$$m^2+8=6m$$
$$m^2-6m+8=0$$
$$(m-2)(m-4)=0$$
$$m-2=0 \text{ or } m-4=0$$
$$m=2 \text{ or } \quad m=4$$

Check:
$$2+\dfrac{8}{2} \stackrel{?}{=} 6$$
$$2+4 \stackrel{?}{=} 6$$
$$6=6 \leftarrow \text{True}$$

Check:
$$4+\dfrac{8}{4} \stackrel{?}{=} 6$$
$$4+2 \stackrel{?}{=} 6$$
$$6=6 \leftarrow \text{True}$$

Both solutions check, so the solution set is $\{2,4\}$.

**Section 6.4** *Rational Equations*

18. $$8b - \frac{3}{b} = 2$$
$$b\left(8b - \frac{3}{b}\right) = b \cdot 2$$
$$8b^2 - 3 = 2b$$
$$8b^2 - 2b - 3 = 0$$
$$(4b-3)(2b+1) = 0$$
$$4b - 3 = 0 \quad \text{or} \quad 2b + 1 = 0$$
$$4b = 3 \quad \text{or} \quad 2b = -1$$
$$b = \frac{3}{4} \quad \text{or} \quad b = -\frac{1}{2}$$

Check:
$$8\left(-\frac{1}{2}\right) - \frac{3}{\left(-\frac{1}{2}\right)} \stackrel{?}{=} 2$$
$$-4 + 6 \stackrel{?}{=} 2$$
$$2 = 2$$
↑ True

Check:
$$8\left(-\frac{1}{2}\right) - \frac{3}{\left(-\frac{1}{2}\right)} \stackrel{?}{=} 2$$
$$-4 + 6 \stackrel{?}{=} 2$$
$$2 = 2$$
↑ True

Both solutions check, so the solution set is $\left\{-\frac{1}{2}, \frac{3}{4}\right\}$.

20. $$2 - \frac{3}{p+2} = \frac{6}{p}$$
$$p(p+2)\left(2 - \frac{3}{p+2}\right) = p(p+2)\left(\frac{6}{p}\right)$$
$$2p(p+2) - 3p = 6(p+2)$$
$$2p^2 + 4p - 3p = 6p + 12$$
$$2p^2 - 5p - 12 = 0$$
$$(2p+3)(p-4) = 0$$
$$2p + 3 = 0 \quad \text{or} \quad p - 4 = 0$$
$$p = -\frac{3}{2} \quad \text{or} \quad p = 4$$

Check:
$$2 - \frac{3}{-\frac{3}{2}+2} \stackrel{?}{=} \frac{6}{-\frac{3}{2}}$$
$$2 - \frac{3}{\frac{1}{2}} \stackrel{?}{=} -4$$
$$2 - 6 \stackrel{?}{=} -4$$
$$-4 = -3$$
↑ True

Check:
$$2 - \frac{3}{4+2} \stackrel{?}{=} \frac{6}{4}$$
$$2 - \frac{3}{6} \stackrel{?}{=} \frac{3}{2}$$
$$\frac{4}{2} - \frac{1}{2} \stackrel{?}{=} \frac{3}{2}$$
$$\frac{3}{2} = \frac{3}{2}$$
↑ True

Both solutions check, so the solution set is $\left\{-\frac{3}{2}, 4\right\}$.

22. $$\frac{3-y}{y-3} + 2 = \frac{2}{y}$$
$$y(y-3)\left(\frac{3-y}{y-3} + 2\right) = y(y-3)\left(\frac{2}{y}\right)$$
$$y(3-y) + 2y(y-3) = 2(y-3)$$
$$3y - y^2 + 2y^2 - 6y = 2y - 6$$
$$y^2 - 5y + 6 = 0$$
$$(y-2)(y-3) = 0$$
$$y - 2 = 0 \quad \text{or} \quad y - 3 = 0$$
$$y = 2 \quad \text{or} \quad y = 3$$

Since $y = 3$ is not in the domain of the variable, it is an extraneous solution.

Check $y = 2$:
$$\frac{3-2}{2-3} + 2 \stackrel{?}{=} \frac{2}{2}$$
$$\frac{1}{-1} + 2 \stackrel{?}{=} 1$$
$$-1 + 2 \stackrel{?}{=} 1$$
$$1 = 1 \leftarrow \text{True}$$

The solution set is $\{2\}$.

24. $$\frac{4}{3} + \frac{7}{x-4} = \frac{x-1}{3x-12}$$
$$\frac{4}{3} + \frac{7}{x-4} = \frac{x-1}{3(x-4)}$$
$$3(x-4)\left(\frac{4}{3} + \frac{7}{x-4}\right) = 3(x-4)\left(\frac{x-1}{3(x-4)}\right)$$
$$4(x-4) + 3(7) = x - 1$$
$$4x - 16 + 21 = x - 1$$
$$3x = -6$$
$$x = -2$$

Check: $$\frac{4}{3} + \frac{7}{-2-4} \stackrel{?}{=} \frac{-2-1}{3(-2)-12}$$
$$\frac{4}{3} + \frac{7}{-6} \stackrel{?}{=} \frac{-3}{-6-12}$$
$$\frac{8}{6} - \frac{7}{6} \stackrel{?}{=} \frac{-3}{-18}$$
$$\frac{1}{6} = \frac{1}{6} \leftarrow \text{True}$$

The solution set is $\{-2\}$.

Chapter 6: *Rational Expressions and Rational Functions*

**26.**
$$\frac{5}{x+2} = 1 - \frac{3}{x-2}$$
$$(x+2)(x-2)\left(\frac{5}{x+2}\right) = (x+2)(x-2)\left(1 - \frac{3}{x-2}\right)$$
$$5(x-2) = (x+2)(x-2) - 3(x+2)$$
$$5x - 10 = x^2 + 2x - 2x - 4 - 3x - 6$$
$$0 = x^2 - 8x$$
$$0 = x(x-8)$$
$$x = 0 \text{ or } x - 8 = 0$$
$$x = 8$$

Check:
$$\frac{5}{0+2} \stackrel{?}{=} 1 - \frac{3}{0-2}$$
$$\frac{5}{2} \stackrel{?}{=} 1 - \frac{3}{-2}$$
$$\frac{5}{2} \stackrel{?}{=} \frac{2}{2} + \frac{3}{2}$$
$$\frac{5}{2} = \frac{5}{2} \leftarrow \text{True}$$

Check:
$$\frac{5}{8+2} \stackrel{?}{=} 1 - \frac{3}{8-2}$$
$$\frac{5}{10} \stackrel{?}{=} 1 - \frac{3}{6}$$
$$\frac{1}{2} \stackrel{?}{=} 1 - \frac{1}{2}$$
$$\frac{1}{2} = \frac{1}{2} \leftarrow \text{True}$$

Both solutions check, so the solution set is $\{0, 8\}$.

**28.**
$$\frac{4}{x+3} + \frac{5}{x-6} = \frac{4x+1}{x^2 - 3x - 18}$$
$$\frac{4}{x+3} + \frac{5}{x-6} = \frac{4x+1}{(x-6)(x+3)}$$
$$(x+3)(x-6)\left(\frac{4}{x+3} + \frac{5}{x-6}\right) =$$
$$(x+3)(x-6)\left(\frac{4x+1}{(x-6)(x+3)}\right)$$
$$4(x-6) + 5(x+3) = 4x+1$$
$$4x - 24 + 5x + 15 = 4x + 1$$
$$9x - 9 = 4x + 1$$
$$5x = 10$$
$$x = 2$$

Check:
$$\frac{4}{2+3} + \frac{5}{2-6} \stackrel{?}{=} \frac{4(2)+1}{2^2 - 3(2) - 18}$$
$$\frac{4}{5} + \frac{5}{-4} \stackrel{?}{=} \frac{8+1}{4 - 6 - 18}$$
$$\frac{16}{20} - \frac{25}{20} \stackrel{?}{=} \frac{9}{-20}$$
$$-\frac{9}{20} = -\frac{9}{20} \leftarrow \text{True}$$

The solution checks, so the solution set is $\{2\}$.

**30.**
$$\frac{3}{x-4} = \frac{5x+4}{x^2 - 16} - \frac{4}{x+4}$$
$$\frac{3}{x-4} = \frac{5x+4}{(x+4)(x-4)} - \frac{4}{x+4}$$
$$(x+4)(x-4)\left(\frac{3}{x-4}\right) =$$
$$(x+4)(x-4)\left(\frac{5x+4}{(x+4)(x-4)} - \frac{4}{x+4}\right)$$
$$3(x+4) = 5x + 4 - 4(x-4)$$
$$3x + 12 = 5x + 4 - 4x + 16$$
$$3x + 12 = x + 20$$
$$2x = 8$$
$$x = 4$$

Since $x = 4$ is not in the domain of the variable, it is an extraneous solution. Thus, the equation has no solution. The solution set is $\{\ \}$ or $\varnothing$.

**32.**
$$\frac{5}{z-4} + \frac{3}{z-2} = \frac{z^2 - z - 2}{z^2 - 6z + 8}$$
$$\frac{5}{z-4} + \frac{3}{z-2} = \frac{z^2 - z - 2}{(z-4)(z-2)}$$
$$(z-4)(z-2)\left(\frac{5}{z-4} + \frac{3}{z-2}\right) =$$
$$(z-4)(z-2)\left(\frac{z^2 - z - 2}{(z-4)(z-2)}\right)$$
$$5(z-2) + 3(z-4) = z^2 - z - 2$$
$$5z - 10 + 3z - 12 = z^2 - z - 2$$
$$8z - 22 = z^2 - z - 2$$
$$0 = z^2 - 9z + 20$$
$$0 = (z-4)(z-5)$$
$$z - 4 = 0 \text{ or } z - 5 = 0$$
$$z = 4 \text{ or } z = 5$$

Since $z = 4$ is not in the domain of the variable, it is an extraneous solution.
Check $z = 5$:
$$\frac{5}{5-4} + \frac{3}{5-2} \stackrel{?}{=} \frac{5^2 - 5 - 2}{5^2 - 6(5) + 8}$$
$$\frac{5}{1} + \frac{3}{3} \stackrel{?}{=} \frac{25 - 5 - 2}{25 - 30 + 8}$$
$$5 + 1 \stackrel{?}{=} \frac{18}{3}$$
$$6 = 6 \leftarrow \text{True}$$
The solution set is $\{5\}$.

**Section 6.4** Rational Equations

34.
$$\frac{3}{x^2-5x-6}+\frac{3}{x^2-7x+6}=\frac{6}{x^2-1}$$
$$\frac{3}{(x+1)(x-6)}+\frac{3}{(x-1)(x-6)}=\frac{6}{(x+1)(x-1)}$$
$$(x+1)(x-1)(x-6)\left(\frac{3}{(x+1)(x-6)}+\frac{3}{(x-1)(x-6)}\right)=$$
$$(x+1)(x-1)(x-6)\left(\frac{6}{(x+1)(x-1)}\right)$$
$$3(x-1)+3(x+1)=6(x-6)$$
$$3x-3+3x+3=6x-36$$
$$6x=6x-36$$
$$0=-36$$

The final statement is a contradiction. Thus, the equation has no solution. The solution set is { } or $\varnothing$.

36.
$$\frac{x+5}{x+1}+1=\frac{x-5}{x-2}$$
$$(x+1)(x-2)\left(\frac{x+5}{x+1}+1\right)=(x+1)(x-2)\left(\frac{x-5}{x-2}\right)$$
$$(x-2)(x+5)+(x+1)(x-2)=(x+1)(x-5)$$
$$x^2+3x-10+x^2-x-2=x^2-4x-5$$
$$2x^2+2x-12=x^2-4x-5$$
$$x^2+6x-7=0$$
$$(x+7)(x-1)=0$$
$$x+7=0 \text{ or } x-1=0$$
$$x=-7 \text{ or } x=1$$

Check $x=-7$:
$$\frac{(-7)+5}{(-7)+1}+1\stackrel{?}{=}\frac{(-7)-5}{(-7)-2}$$
$$\frac{-2}{-6}+1\stackrel{?}{=}\frac{-12}{-9}$$
$$\frac{1}{3}+1\stackrel{?}{=}\frac{4}{3}$$
$$\frac{4}{3}=\frac{4}{3} \leftarrow \text{True}$$

Check $x=1$:
$$\frac{1+5}{1+1}+1\stackrel{?}{=}\frac{1-5}{1-2}$$
$$\frac{6}{2}+1\stackrel{?}{=}\frac{-4}{-1}$$
$$3+1\stackrel{?}{=}4$$
$$4=4 \leftarrow \text{True}$$

Both solutions check, so the solution set is $\{-7,1\}$.

38.
$$f(x)=8$$
$$x+\frac{7}{x}=8$$
$$x\left(x+\frac{7}{x}\right)=x\cdot 8$$
$$x^2+7=8x$$
$$x^2-8x+7=0$$
$$(x-1)(x-7)=0$$
$$x-1=0 \text{ or } x-7=0$$
$$x=1 \text{ or } x=7$$

Thus, $f(1)=8$ and $f(7)=8$, so the points $(1,8)$ and $(7,8)$ are on the graph of $f$.

40.
$$f(x)=-10$$
$$2x+\frac{8}{x}=-10$$
$$x\left(2x+\frac{8}{x}\right)=x\cdot(-10)$$
$$2x^2+8=-10x$$
$$2x^2+10x+8=0$$
$$2(x^2+5x+4)=0$$
$$2(x+1)(x+4)=0$$
$$x+1=0 \text{ or } x+4=0$$
$$x=-1 \text{ or } x=-4$$

Thus, $f(-1)=-10$ and $f(-4)=-10$, so the points $(-1,-10)$ and $(-4,-10)$ are on the graph of $f$.

42.
$$f(x)=\frac{1}{5}$$
$$\frac{x+5}{x-3}=\frac{1}{5}$$
$$5(x-3)\left(\frac{x+5}{x-3}\right)=5(x-3)\cdot\left(\frac{1}{5}\right)$$
$$5(x+5)=x-3$$
$$5x+25=x-3$$
$$4x=-28$$
$$x=-7$$

Thus, $f(-7)=\frac{1}{5}$, so the point $\left(-7,\frac{1}{5}\right)$ is on the graph of $f$.

## Chapter 6: Rational Expressions and Rational Functions

**44.**
$$f(x) = g(x)$$
$$\frac{4x+1}{8x+5} = \frac{x-4}{2x-7}$$
$$(8x+5)(2x-7)\left(\frac{4x+1}{8x+5}\right) = (8x+5)(2x-7)\left(\frac{x-4}{2x-7}\right)$$
$$(2x-7)(4x+1) = (8x+5)(x-4)$$
$$8x^2 - 26x - 7 = 8x^2 - 27x - 20$$
$$-26x - 7 = -27x - 20$$
$$x = -13$$

Now, $f(-13) = \frac{4(-13)+1}{8(-13)+5} = \frac{-52+1}{-104+5} = \frac{-51}{-99} = \frac{17}{33}$

and $g(-13) = \frac{(-13)-4}{2(-13)-7} = \frac{-13-4}{-26-7} = \frac{-17}{-33} = \frac{17}{33}$, so

the intersection point of $f$ and $g$ is $\left(-13, \frac{17}{33}\right)$.

The solution checks, so the solution set is $\left\{\frac{1}{4}\right\}$.

**46.**
$$\frac{2b-1}{b+5} - \frac{2}{3} = \frac{1}{3b+15}$$
$$\frac{2b-1}{b+5} - \frac{2}{3} = \frac{1}{3(b+5)}$$
$$3(b+5)\left(\frac{2b-1}{b+5} - \frac{2}{3}\right) = 3(b+5)\left(\frac{1}{3(b+5)}\right)$$
$$3(2b-1) - 2(b+5) = 1$$
$$6b - 3 - 2b - 10 = 1$$
$$4b - 13 = 1$$
$$4b = 14$$
$$z = \frac{14}{4} = \frac{7}{2}$$

Check:
$$\frac{2\left(\frac{7}{2}\right)-1}{\frac{7}{2}+5} - \frac{2}{3} \stackrel{?}{=} \frac{1}{3\left(\frac{7}{2}\right)+15}$$
$$\frac{7-1}{\frac{7}{2}+5} - \frac{2}{3} \stackrel{?}{=} \frac{1}{\frac{21}{2}+15}$$
$$\frac{7-1}{\frac{7}{2}+5} \cdot \frac{2}{2} - \frac{2}{3} \stackrel{?}{=} \frac{1}{\frac{21}{2}+15} \cdot \frac{2}{2}$$
$$\frac{14-2}{7+10} - \frac{2}{3} \stackrel{?}{=} \frac{2}{21+30}$$
$$\frac{12}{17} - \frac{2}{3} \stackrel{?}{=} \frac{2}{51}$$
$$\frac{36}{51} - \frac{34}{51} \stackrel{?}{=} \frac{2}{51}$$
$$\frac{2}{51} = \frac{2}{51} \leftarrow \text{True}$$

The solution checks, so the solution set is $\left\{\frac{7}{2}\right\}$.

**48.**
$$p + \frac{25}{p} = 10$$
$$p\left(p + \frac{25}{p}\right) = p \cdot 10$$
$$p^2 + 25 = 10p$$
$$p^2 - 10p + 25 = 0$$
$$(p-5)(p-5) = 0$$
$$p - 5 = 0$$
$$p = 5$$

Check:
$$5 + \frac{25}{5} \stackrel{?}{=} 10$$
$$5 + 5 \stackrel{?}{=} 10$$
$$5 = 5 \leftarrow \text{True}$$

The solution checks, so the solution set is $\{5\}$.

**50.**
$$\frac{3}{a^2+3a-10} + \frac{2}{a^2+7a+10} = \frac{4}{a^2-4}$$
$$\frac{3}{(a+5)(a-2)} + \frac{2}{(a+5)(a+2)} = \frac{4}{(a+2)(a-2)}$$
$$(a+5)(a+2)(a-2)\left(\frac{3}{(a+5)(a-2)} + \frac{2}{(a+5)(a+2)}\right) =$$
$$(a+5)(a+2)(a-2)\left(\frac{4}{(a+2)(a-2)}\right)$$
$$3(a+2) + 2(a-2) = 4(a+5)$$
$$3a + 6 + 2a - 4 = 4a + 20$$
$$5a + 2 = 4a + 20$$
$$a = 18$$

Check:
$$\frac{3}{18^2+3(18)-10}+\frac{2}{18^2+7(18)+10}\stackrel{?}{=}\frac{4}{18^2-4}$$
$$\frac{3}{324+54-10}+\frac{2}{324+126+10}\stackrel{?}{=}\frac{4}{324-4}$$
$$\frac{3}{368}+\frac{2}{460}\stackrel{?}{=}\frac{4}{320}$$
$$\frac{15}{1840}+\frac{8}{1840}\stackrel{?}{=}\frac{1}{80}$$
$$\frac{23}{1840}\stackrel{?}{=}\frac{1}{80}$$
$$\frac{1}{80}=\frac{1}{80}\leftarrow\text{True}$$

The solution set is $\{18\}$.

**52.** $\dfrac{9}{b}+\dfrac{4}{5b}=\dfrac{7}{10}$

$10b\left(\dfrac{9}{b}+\dfrac{4}{5b}\right)=10b\cdot\dfrac{7}{10}$

$90+8=7b$
$98=7b$
$14=b$

Check:
$$\frac{9}{14}+\frac{4}{5(14)}\stackrel{?}{=}\frac{7}{10}$$
$$\frac{45}{70}+\frac{4}{70}\stackrel{?}{=}\frac{7}{10}$$
$$\frac{49}{70}\stackrel{?}{=}\frac{7}{10}$$
$$\frac{7}{10}=\frac{7}{10}\leftarrow\text{True}$$

The solution checks, so the solution set is $\{14\}$.

**54.** $\dfrac{x+3}{x-2}+4=\dfrac{x+2}{x+1}$

$(x-2)(x+1)\left(\dfrac{x+3}{x-2}+4\right)=(x-2)(x+1)\left(\dfrac{x+2}{x+1}\right)$

$(x+1)(x+3)+4(x-2)(x+1)=(x-2)(x+2)$
$x^2+4x+3+4x^2-4x-8=x^2-4$
$5x^2-5=x^2-4$
$4x^2=1$
$x^2=\dfrac{1}{4}$
$x=\pm\sqrt{\dfrac{1}{4}}=\pm\dfrac{1}{2}$

Check: $x=-\dfrac{1}{2}$     Check: $x=-\dfrac{1}{2}$

$\dfrac{-\frac{1}{2}+3}{-\frac{1}{2}-2}+4\stackrel{?}{=}\dfrac{-\frac{1}{2}+2}{-\frac{1}{2}+1}$     $\dfrac{\frac{1}{2}+3}{\frac{1}{2}-2}+4\stackrel{?}{=}\dfrac{\frac{1}{2}+2}{\frac{1}{2}+1}$

$\dfrac{\frac{5}{2}}{-\frac{5}{2}}+4\stackrel{?}{=}\dfrac{\frac{3}{2}}{\frac{1}{2}}$     $\dfrac{\frac{7}{2}}{-\frac{3}{2}}+4\stackrel{?}{=}\dfrac{\frac{5}{2}}{\frac{3}{2}}$

$-1+4\stackrel{?}{=}3$     $-\dfrac{7}{3}+\dfrac{12}{3}\stackrel{?}{=}\dfrac{5}{3}$

$3=3$     $\dfrac{5}{3}=\dfrac{5}{3}$

↑     ↑
True     True

Both check, so the solution set is $\left\{-\dfrac{1}{2},\dfrac{1}{2}\right\}$.

**56.** $P(x)=1350$

$\dfrac{200(1+0.4t)}{2(1+0.01t)}=1350$

$2(1+0.01t)\left(\dfrac{200(1+0.4t)}{2(1+0.01t)}\right)=2(1+0.01t)\cdot 1350$

$200(1+0.4t)=2700(1+0.01t)$
$200+80t=2700+27t$
$53t=2500$
$t=\dfrac{2500}{53}\approx 47.17$

The population will be 1350 insects after about 47.17 months (almost 4 years).

*Chapter 6: Rational Expressions and Rational Functions*

**58. a.**
$$P(x) = 0.7$$
$$\frac{0.8x - 0.8}{0.8x + 0.1} = 0.7$$
$$(0.8x + 0.1)\left(\frac{0.8x - 0.8}{0.8x + 0.1}\right) = (0.8x + 0.1) \cdot 0.7$$
$$0.8x - 0.8 = 0.56x + 0.07$$
$$0.24x = 0.87$$
$$x = 3.625$$

Thus, the student would need to study 3.625 hours (or 3 hours 37.5 minutes) to learn 70% of the vocabulary words.

**b.**
$$P(x) = \frac{400}{500}$$
$$\frac{0.8x - 0.8}{0.8x + 0.1} = 0.8$$
$$(0.8x + 0.1)\left(\frac{0.8x - 0.8}{0.8x + 0.1}\right) = (0.8x + 0.1) \cdot 0.8$$
$$0.8x - 0.8 = 0.64x + 0.08$$
$$0.16x = 0.88$$
$$x = 5.5$$

Thus, the student would need to study 5.5 hours (or 5 hours 30 minutes) to learn 400 of the 500 vocabulary words.

**60.** Substitute $I = 135°$ into the given formula and solve for $n$.
$$I = \frac{180°(n-2)}{n}$$
$$135° = \frac{180°(n-2)}{n}$$
$$135° = \frac{180°n - 360°}{n}$$
$$n(135°) = n\left(\frac{180°n - 360°}{n}\right)$$
$$135°n = 180°n - 360°$$
$$-45°n = -360°$$
$$n = \frac{-360°}{-45°} = 8$$

The regular polygon has 8 sides. In other words, it must be an octagon.

**62.** Answers will vary. One possibility follows:

The equation $\dfrac{x+1}{x^2 - 2x - 8} = \dfrac{1}{x-4}$ has no real solution.

**64.**
$$2 + 11a^{-1} = -12a^{-2}$$
$$2 + \frac{11}{a} = -\frac{12}{a^2}$$
$$a^2\left(2 + \frac{11}{a}\right) = a^2\left(-\frac{12}{a^2}\right)$$
$$2a^2 + 11a = -12$$
$$2a^2 + 11a + 12 = 0$$
$$(2a+3)(a+4) = 0$$
$$2a + 3 = 0 \text{ or } a + 4 = 0$$
$$a = -\frac{3}{2} \text{ or } a = -4$$

Both solutions check. The solution set is $\left\{-4, -\dfrac{3}{2}\right\}$.

**66.**
$$\left(\frac{z^{-2}}{2z^{-3}}\right)^{-1} + 3(z-1)^{-1} = \left(\frac{z^{-2-(-3)}}{2}\right)^{-1} + 3(z-1)^{-1}$$
$$= \left(\frac{z}{2}\right)^{-1} + 3(z-1)^{-1}$$
$$= \frac{2}{z} + \frac{3}{z-1}$$
$$= \frac{2}{z} \cdot \frac{z-1}{z-1} + \frac{3}{z-1} \cdot \frac{z}{z}$$
$$= \frac{2z - 2 + 3z}{z(z-1)}$$
$$= \frac{5z - 2}{z(z-1)}$$

**68.**
$$\frac{5}{x-6} + \frac{2}{x+2} = \frac{1}{x^2 - 4x - 12}$$
$$\frac{5}{x-6} + \frac{2}{x+2} = \frac{1}{(x-6)(x+2)}$$
$$(x-6)(x+2)\left(\frac{5}{x-6} + \frac{2}{x+2}\right) = (x-6)(x+2)\left(\frac{1}{(x-6)(x+2)}\right)$$
$$5(x+2) + 2(x-6) = 1$$
$$5x + 10 + 2x - 12 = 1$$
$$7x - 2 = 1$$
$$7x = 3$$
$$x = \frac{3}{7}$$

This solution checks, so the solution set is $\left\{\dfrac{3}{7}\right\}$.

70. Answers will vary. One possibility follows: The directions "simplify" indicates that an algebraic expression should written in a more concise (i.e., simpler) manner. Here, we cannot determine a specific value for the variable(s) because there is no statement of equality. On the other hand, the directions "solve" indicates that we have an equation. The goal, then, is to determine the value(s) for the variable that make the equation true. In summary, we simplify an expression and we solve an equation.

72. $\dfrac{x-6}{x+1} = \dfrac{2}{3}$

74. $\dfrac{4}{3} + \dfrac{7}{x-4} = \dfrac{-7}{3x-12}$

 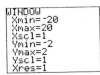

The solution set is $\{-3\}$.

76. $\dfrac{3x^2 + 10x + 3}{x+3} = -8$

 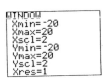

The apparent solution is $x = -3$. However, notice that $-3$ is not in the domain of the left side of the equation. Thus, the equation has no solution. The solution set is $\varnothing$ or $\{\ \}$.

## 6.5 Preparing for Rational Inequalities

1. The inequality $-1 < x \leq 8$ in interval notation is $(-1, 8]$.

The solution set is $\{20\}$.

2. $2x + 3 > 4x - 9$
$-4x + 2x + 3 > -4x + 4x - 9$
$-2x + 3 > -9$
$-2x + 3 - 3 > -9 - 3$
$-2x > -12$
$\dfrac{-2x}{-2} < \dfrac{-12}{-2}$
$x < 6$

The solution set is $\{x \mid x < 6\}$ or, using interval notation, $(-\infty, 6)$.

## 6.5 Exercises

2. TRUE. Since $x = 2$ will make the inequality true, it is a solution. Please recognize, however, that it is not the only solution to the inequality.

4. Answers will vary. One possibility follows: A rational function is continuous (i.e., it has no breaks) everywhere except where it is undefined. Therefore, the only possible places on the real numbers that a rational function can change signs (from positive to negative, or from negative to positive) is either where the function equals zero or where the function is undefined. So, if we use these numbers to form the intervals when solving the inequality, we are guaranteed that a sign change cannot occur anywhere else.

6. $\dfrac{x+5}{x-2} > 0$

The rational expression will equal 0 when $x = -5$. It is undefined when $x = 2$. Determine where the numerator and denominator are positive and negative and where the quotient is positive and negative.

## Chapter 6: Rational Expressions and Rational Functions

| Interval | $(-\infty, -5)$ | $-5$ | $(-5, 2)$ | $2$ | $(2, \infty)$ |
|---|---|---|---|---|---|
| $x-5$ | $----$ | $0$ | $++++$ | $+$ | $++++$ |
| $x-2$ | $----$ | $-$ | $----$ | $0$ | $++++$ |
| $\dfrac{x+5}{x-2}$ | $++++$ | $0$ | $----$ | $\varnothing$ | $++++$ |

The rational function is undefined at $x = 2$, so 2 is not part of the solution. The inequality is strict, so $-5$ is not part of the solution. Now, $\dfrac{x+5}{x-2}$ is greater than zero where the quotient is positive. The solution is $\{x \mid x < -5 \text{ or } x > 2\}$ in set-builder notation; the solution is $(-\infty, -5) \cup (2, \infty)$ in interval notation.

8. $\dfrac{x+8}{x+2} > 0$

   The rational expression will equal 0 when $x = -8$. It is undefined when $x = -2$. Determine where the numerator and denominator are positive and negative and where the quotient is positive and negative.

| Interval | $(-\infty, -8)$ | $-8$ | $(-8, -2)$ | $-2$ | $(-2, \infty)$ |
|---|---|---|---|---|---|
| $x+8$ | $----$ | $0$ | $++++$ | $+$ | $++++$ |
| $x+2$ | $----$ | $-$ | $----$ | $0$ | $++++$ |
| $\dfrac{x+8}{x+2}$ | $++++$ | $0$ | $----$ | $\varnothing$ | $++++$ |

   The rational function is undefined at $x = -2$, so $-2$ is not part of the solution. The inequality is strict, so 8 is not part of the solution. Now, $\dfrac{x+8}{x+2}$ is greater than zero where the quotient is positive. The solution is $\{x \mid x < -8 \text{ or } x > -2\}$ in set-builder notation; the solution is $(-\infty, -8) \cup (-2, \infty)$ in interval notation.

10. $\dfrac{x+12}{x-2} \geq 0$

    The rational expression will equal 0 when $x = -12$. It is undefined when $x = 2$. Determine where the numerator and denominator are positive and negative and where the quotient is positive and negative.

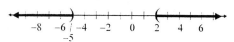

| Interval | $(-\infty, -12)$ | $-12$ | $(-12, 2)$ | $2$ | $(2, \infty)$ |
|---|---|---|---|---|---|
| $x+12$ | $----$ | $0$ | $++++$ | $+$ | $++++$ |
| $x-2$ | $----$ | $-$ | $----$ | $0$ | $++++$ |
| $\dfrac{x+12}{x-2}$ | $++++$ | $0$ | $----$ | $\varnothing$ | $++++$ |

The rational function is undefined at $x = 2$, so 2 is not part of the solution. The inequality is non-strict, so $-12$ is part of the solution. Now, $\dfrac{x+12}{x-2}$ is greater than zero where the quotient is positive. The solution is $\{x \mid x \leq -12 \text{ or } x > 2\}$ in set-builder notation; the solution is $(-\infty, -12] \cup (2, \infty)$ in interval notation.

12. $\dfrac{x-10}{x+5} \leq 0$

    The rational expression will equal 0 when $x = 10$. It is undefined when $x = -5$. Determine where the numerator and denominator are positive and negative and where the quotient is positive and negative.

| Interval | $(-\infty, -5)$ | $-5$ | $(-5, 10)$ | $10$ | $(10, \infty)$ |
|---|---|---|---|---|---|
| $x-10$ | $----$ | $-$ | $----$ | $0$ | $++++$ |
| $x+5$ | $----$ | $0$ | $++++$ | $+$ | $++++$ |
| $\dfrac{x-10}{x+5}$ | $++++$ | $\varnothing$ | $----$ | $0$ | $++++$ |

    The rational function is undefined at $x = -5$, so $-5$ is not part of the solution. The inequality is non-strict, so 10 is part of the solution. Now, $\dfrac{x-10}{x+5}$ is less than zero where the quotient is negative. The solution is $\{x \mid -5 < x \leq 10\}$ in set-builder notation; the solution is $(-5, 10]$ in interval notation.

14. $\dfrac{(5x-2)(x+4)}{x-5} < 0$

    The rational expression will equal 0 when $x = \dfrac{2}{5}$ and when $x = -4$. It is undefined when $x = 5$.

ISM - 336

Determine where the factors of the numerator and denominator are positive and negative and where the quotient is positive and negative.

| Interval | $(-\infty, -4)$ | $-4$ | $\left(-4, \frac{2}{5}\right)$ | $\frac{2}{5}$ | $\left(\frac{2}{5}, 5\right)$ | $5$ | $(5, \infty)$ |
|---|---|---|---|---|---|---|---|
| $5x-2$ | $---$ | $-$ | $---$ | $0$ | $+++$ | $+$ | $+++$ |
| $x+4$ | $---$ | $0$ | $+++$ | $+$ | $+++$ | $+$ | $+++$ |
| $x-5$ | $---$ | $-$ | $---$ | $-$ | $---$ | $0$ | $+++$ |
| $\frac{(5x-2)(x+4)}{x-5}$ | $---$ | $0$ | $+++$ | $0$ | $---$ | $\emptyset$ | $+++$ |

The rational function is undefined at $x = 5$, so 5 is not part of the solution. The inequality is strict, so $-4$ and $\frac{2}{5}$ are not part of the solution.

Now, $\frac{(5x-2)(x+4)}{x-5}$ is less than zero where the quotient is negative. The solution is $\left\{x \mid x < -4 \text{ or } \frac{2}{5} < x < 5\right\}$ in set-builder notation; the solution is $(-\infty, -4) \cup \left(\frac{2}{5}, 5\right)$ in interval notation.

16. $\frac{(3x-2)(x-6)}{x+1} \geq 0$

The rational expression will equal 0 when $x = \frac{2}{3}$ and when $x = 6$. It is undefined when $x = -1$. Determine where the factors of the numerator and denominator are positive and negative and where the quotient is positive and negative.

| Interval | $(-\infty, -1)$ | $-1$ | $\left(-1, \frac{2}{3}\right)$ | $\frac{2}{3}$ | $\left(\frac{2}{3}, 6\right)$ | $6$ | $(6, \infty)$ |
|---|---|---|---|---|---|---|---|
| $3x-2$ | $---$ | $-$ | $---$ | $0$ | $+++$ | $+$ | $+++$ |
| $x-6$ | $---$ | $-$ | $---$ | $-$ | $---$ | $0$ | $---$ |
| $x+1$ | $---$ | $0$ | $+++$ | $+$ | $+++$ | $+$ | $+++$ |
| $\frac{(3x-2)(x-6)}{x+1}$ | $---$ | $0$ | $+++$ | $0$ | $---$ | $\emptyset$ | $+++$ |

The rational function is undefined at $x = -1$, so $-1$ is not part of the solution. The inequality is non-strict, so $\frac{2}{3}$ and 8 are part of the solution.

Now, $\frac{(3x-2)(x-6)}{x+1}$ is greater than zero where the quotient is positive. The solution is $\left\{x \mid -1 < x \leq \frac{2}{3} \text{ or } x \geq 6\right\}$ in set-builder notation; the solution is $\left(-1, \frac{2}{3}\right] \cup [6, \infty)$ in interval notation.

18. $\frac{x+3}{x-4} > 1$

$\frac{x+3}{x-4} - 1 > 0$

$\frac{x+3}{x-4} - \frac{x-4}{x-4} > 0$

$\frac{x+3-x+4}{x-4} > 0$

$\frac{7}{x-4} > 0$

The numerator is always positive, so the rational expression will never equal 0. However, it is undefined when $x = 4$.
Determine where the denominator is positive and negative and where the quotient is positive and negative.

| Interval | $(-\infty, -4)$ | $4$ | $(4, \infty)$ |
|---|---|---|---|
| $7$ | $++++$ | $+$ | $++++$ |
| $x-4$ | $----$ | $0$ | $++++$ |
| $\frac{7}{x-4}$ | $----$ | $\emptyset$ | $++++$ |

The rational function is undefined at $x = 4$, so 4 is not part of the solution. Now, $\frac{7}{x-4}$ is greater than zero where the quotient is positive. The solution is $\{x \mid x > 4\}$ in set-builder notation; the solution is $(4, \infty)$ in interval notation.

ISM – 337

## Chapter 6: Rational Expressions and Rational Functions

**20.**
$$\frac{3x-7}{x+2} \leq 2$$
$$\frac{3x-7}{x+2} - 2 \leq 0$$
$$\frac{3x-7}{x+2} - \frac{2(x+2)}{x+2} \leq 0$$
$$\frac{3x-7-2x-4}{x+2} \leq 0$$
$$\frac{x-11}{x+2} \leq 0$$

The rational expression will equal 0 when $x = 11$. It is undefined when $x = -2$. Determine where the numerator and denominator are positive and negative and where the quotient is positive and negative.

| Interval | $(-\infty, -2)$ | $-2$ | $(-2, 11)$ | $11$ | $(11, \infty)$ |
|---|---|---|---|---|---|
| $x - 11$ | $----$ | $-$ | $----$ | $0$ | $++++$ |
| $x + 2$ | $----$ | $0$ | $++++$ | $+$ | $++++$ |
| $\frac{x-11}{x+2}$ | $++++$ | $\varnothing$ | $----$ | $0$ | $++++$ |

The rational function is undefined at $x = -2$, so $-2$ is not part of the solution. The inequality is non-strict, so 11 is part of the solution. Now, $\frac{x-11}{x+2}$ is less than zero where the quotient is negative. The solution is $\{x \mid -2 < x \leq 11\}$ in set-builder notation; the solution is $(-2, 11]$ in interval notation.

**22.**
$$\frac{3x+20}{x+6} < 5$$
$$\frac{3x+20}{x+6} - 5 < 0$$
$$\frac{3x+20}{x+6} - \frac{5(x+6)}{x+6} < 0$$
$$\frac{3x+20-5x-30}{x+6} < 0$$
$$\frac{-2x-10}{x+6} < 0$$

The rational expression will equal 0 when $x = -5$. It is undefined when $x = -6$. Determine where the numerator and denominator are positive and negative and where the quotient is positive and negative.

| Interval | $(-\infty, -6)$ | $-6$ | $(-6, -5)$ | $-5$ | $(-5, \infty)$ |
|---|---|---|---|---|---|
| $-2x - 10$ | $++++$ | $+$ | $++++$ | $0$ | $----$ |
| $x + 6$ | $----$ | $0$ | $++++$ | $+$ | $++++$ |
| $\frac{-2x-10}{x+6}$ | $----$ | $\varnothing$ | $++++$ | $0$ | $----$ |

The rational function is undefined at $x = -6$, so $-6$ is not part of the solution. The inequality is strict, so $-5$ is not part of the solution. Now, $\frac{-2x-10}{x+6}$ is less than zero where the quotient is negative. The solution is $\{x \mid x < -6 \text{ or } x > -5\}$ in set-builder notation; the solution is $(-\infty, -6) \cup (-5, \infty)$ in interval notation.

**24.**
$$\frac{2}{x+3} + \frac{2}{x} \leq 0$$
$$\frac{2x}{x(x+3)} + \frac{2(x+3)}{x(x+3)} \leq 0$$
$$\frac{2x+2x+6}{x(x+3)} \leq 0$$
$$\frac{4x+6}{x(x+3)} \leq 0$$

The rational expression will equal 0 when $x = -\frac{3}{2}$. It is undefined when $x = 0$ and when $x = -3$. Determine where the numerator and the factors of the denominator are positive and negative and where the quotient is positive and negative.

| Interval | $(-\infty, -3)$ | $-3$ | $\left(-3, -\frac{3}{2}\right)$ | $-\frac{3}{2}$ | $\left(-\frac{3}{2}, 0\right)$ | $0$ | $(0, \infty)$ |
|---|---|---|---|---|---|---|---|
| $4x + 6$ | $---$ | $-$ | $---$ | $0$ | $+++$ | $+$ | $+++$ |
| $x$ | $---$ | $-$ | $---$ | $-$ | $---$ | $0$ | $+++$ |
| $x + 3$ | $---$ | $0$ | $+++$ | $+$ | $+++$ | $+$ | $+++$ |
| $\frac{4x+6}{x(x+3)}$ | $---$ | $\varnothing$ | $+++$ | $0$ | $---$ | $\varnothing$ | $+++$ |

The rational function is undefined at $x = 0$ and $x = -3$, so 0 and $-3$ are not part of the solution. The inequality is non-strict, so $-\frac{3}{2}$ is part of the solution. Now, $\frac{4x+6}{x(x+3)}$ is less than zero where the quotient is negative. The solution is

## Section 6.5: Rational Inequalities

$\{x \mid x < -3 \text{ or } -\frac{3}{2} \leq x < 0\}$ in set-builder notation; the solution is $(-\infty, -3) \cup \left[-\frac{3}{2}, 0\right)$ in interval notation.

26.
$$\frac{1}{x-4} \geq \frac{3}{2x+1}$$
$$\frac{1}{x-4} - \frac{3}{2x+1} \geq 0$$
$$\frac{2x+1}{(x-4)(2x+1)} - \frac{3(x-4)}{(x-4)(2x+1)} \geq 0$$
$$\frac{2x+1-3(x-4)}{(x-4)(2x+1)} \geq 0$$
$$\frac{2x+1-3x+12}{(x-4)(2x+1)} \geq 0$$
$$\frac{13-x}{(x-4)(2x+1)} \geq 0$$

The rational expression will equal 0 when $x = 13$. It is undefined when $x = 4$ and when $x = -\frac{1}{2}$.

Determine where the numerator and the factors of the denominator are positive and negative and where the quotient is positive and negative.

| Interval | $\left(-\infty, -\frac{1}{2}\right)$ | $-\frac{1}{2}$ | $\left(-\frac{1}{2}, 0\right)$ | 4 | (4, 13) | 13 | (13, ∞) |
|---|---|---|---|---|---|---|---|
| $13-x$ | +++ | + | +++ | + | +++ | 0 | --- |
| $x-4$ | --- | - | --- | 0 | +++ | + | +++ |
| $2x+1$ | --- | 0 | +++ | + | +++ | + | +++ |
| $\frac{13-x}{(x-4)(2x+1)}$ | +++ | ∅ | --- | ∅ | +++ | 0 | --- |

The rational function is undefined at $x = -\frac{1}{2}$ and $x = 4$, so $-\frac{1}{2}$ and 4 are not part of the solution. The inequality is non-strict, so 13 is part of the solution. Now, $\frac{13-x}{(x-4)(2x+1)}$ is greater than zero where the quotient is positive. The solution is $\left\{x \mid x < -\frac{1}{2} \text{ or } 4 < x \leq 13\right\}$ in set-builder notation; the solution is $\left(-\infty, -\frac{1}{2}\right) \cup (4, 13]$ in interval notation.

28. $R(x) \geq 0$
$$\frac{x+3}{x-8} \geq 0$$

The rational expression will equal 0 when $x = -3$. It is undefined when $x = 8$.
Determine where the numerator and denominator are positive and negative and where the quotient is positive and negative.

| Interval | (-∞, -3) | -3 | (-3, 8) | 8 | (8, ∞) |
|---|---|---|---|---|---|
| $x+3$ | ---- | 0 | ++++ | + | ++++ |
| $x-8$ | ---- | - | ---- | 0 | ++++ |
| $\frac{x+3}{x-8}$ | ++++ | 0 | ---- | ∅ | ++++ |

The rational function is undefined at $x = 8$, so 8 is not part of the solution. The inequality is non-strict, so $-3$ is part of the solution. Now, $\frac{x+3}{x-8}$ is greater than zero where the quotient is positive. The solution is $\{x \mid x \leq -3 \text{ or } x > 8\}$ in set-builder notation; the solution is $(-\infty, -3] \cup (8, \infty)$ in interval notation.

30. $R(x) < 0$
$$\frac{3x+2}{x-4} < 0$$

The rational expression will equal 0 when $x = -\frac{2}{3}$. It is undefined when $x = 4$.
Determine where the numerator and denominator are positive and negative and where the quotient is positive and negative.

| Interval | $\left(-\infty, -\frac{2}{3}\right)$ | $-\frac{2}{3}$ | $\left(-\frac{2}{3}, 4\right)$ | 4 | (4, ∞) |
|---|---|---|---|---|---|
| $3x+2$ | ---- | 0 | ++++ | + | ++++ |
| $x-4$ | ---- | - | ---- | 0 | ++++ |
| $\frac{3x+2}{x-4}$ | ++++ | 0 | ---- | ∅ | ++++ |

The rational function is undefined at $x = 4$, so 4 is not part of the solution. The inequality is strict, so $-\frac{2}{3}$ is not part of the solution. Now, $\frac{3x+2}{x-4}$ is less than zero where the quotient is negative. The solution is $\left\{x \mid -\frac{2}{3} < x < 4\right\}$ in set-builder notation; the solution is $\left(-\frac{2}{3}, 4\right)$ in interval notation.

32. The average cost will be no more than $130 when $\overline{C}(x) \le 130$.

$$\frac{90x + 5000}{x} \le 130$$

$$\frac{90x + 5000}{x} - 130 \le 0$$

$$\frac{90x + 5000}{x} - \frac{130x}{x} \le 0$$

$$\frac{5000 - 40x}{x} \le 0$$

The rational expression will equal 0 when $x = 125$. It is undefined when $x = 0$.
Determine where the numerator and denominator are positive and negative and where the quotient is positive and negative.

| Interval | $(-\infty, 0)$ | 0 | $(0, 125)$ | 125 | $(125, \infty)$ |
|---|---|---|---|---|---|
| $5000 - 40x$ | $++++$ | $+$ | $++++$ | 0 | $----$ |
| $x$ | $----$ | 0 | $++++$ | $+$ | $++++$ |
| $\frac{5000-40x}{x}$ | $----$ | $\varnothing$ | $++++$ | 0 | $----$ |

The rational function is undefined at $x = 0$, so 0 is not part of the solution. The inequality is non-strict, so 120 is part of the solution. Now, $\frac{5000 - 40x}{x}$ is less than zero where the quotient is negative. The solution is $\{x \mid x < 0 \text{ or } x \ge 125\}$ in set-builder notation; the solution is $(-\infty, 0) \cup [125, \infty)$ in interval notation. However, for this problem, $x < 0$ is not meaningful. Thus, the average cost will be no more than $130 when 125 or more bicycles are produced each day.

34. Answers may vary. One possibility follows: The left endpoint of the interval is $x = -2$ and the right endpoint of the interval is $x = 5$. We notice that the endpoint $-2$ is not included in the solution, but the endpoint 5 is included in the solution. Thus, we can use $x + 2$ as a factor of the denominator and $x - 5$ as a factor of the numerator of the rational function. Now, $x + 2$ will be negative for $x < -2$ and positive for $x > -2$. Likewise, $x - 5$ will be negative for $x < 5$ and positive for $x > 5$. Thus, the quotient $\frac{x-5}{x+2}$ will be positive when $x < -2$ or $x > 5$. It will be negative for $-2 < x < 5$. Thus, the rational inequality $\frac{x-5}{x+2} \le 0$ will have $(-2, 5]$ as the solution set.
NOTE: Because the solution set contains the endpoint 5, the inequality must be non-strict.

36. To find the $x$-intercept(s), we solve $G(x) = 0$:

$$G(x) = 0$$
$$5x + 30 = 0$$
$$5x = -30$$
$$x = -6$$

The $x$-intercept of $G$ is $-6$.

38. To find the $x$-intercept(s), we solve $h(x) = 0$:

$$h(x) = 0$$
$$-3x^2 - 7x + 20 = 0$$
$$-1(3x - 5)(x + 4) = 0$$
$$3x - 5 = 0 \quad \text{or} \quad x + 4 = 0$$
$$x = \frac{5}{3} \quad \text{or} \quad x = -4$$

The $x$-intercepts of $h$ are $-4$ and $\frac{5}{3}$.

40. To find the $x$-intercept(s), we solve $R(x) = 0$:

$$R(x) = 0$$
$$\frac{x^2 + 5x + 6}{x + 2} = 0$$
$$(x+2)\left(\frac{x^2 + 5x + 6}{x + 2}\right) = (x+2)0$$
$$x^2 + 5x + 6 = 0$$
$$(x+3)(x+2) = 0$$

$x+3=0$ or $x+2=0$
$x=-3$       $x=-2$

Notice that the apparent solution $x=-2$ does not belong to the domain $R$. Therefore, $-2$ is extraneous and cannot be an $x$-intercept. Thus, the only $x$-intercept of $R$ is $-3$.

42. $\dfrac{x+2}{x-5} > -2$

Let $Y_1 = \dfrac{x+2}{x-5}$ and $Y_2 = -2$. Graph the functions. Use the INTERSECT feature to find the $x$-coordinates of the point(s) of intersection.

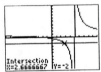

The rational function is undefined at $x=5$, so 5 is not part of the solution. The inequality is strict, so $2.\overline{6} = \dfrac{8}{3}$ is not part of the solution.

From the graph, we can see that $\dfrac{x+2}{x-5} > -2$ where $x < \dfrac{8}{3}$ and where $x > 5$. Thus, the solution set is $\left\{x \middle| x < \dfrac{8}{3} \text{ or } x > 5\right\}$ in set-builder notation; the solution is $\left(-\infty, \dfrac{8}{3}\right) \cup (5, \infty)$ in interval notation.

44. $\dfrac{2x-1}{x+5} \le 4$

Let $Y_1 = \dfrac{2x-1}{x+5}$ and $Y_2 = 4$. Graph the functions. Use the INTERSECT feature to find the $x$-coordinates of the point(s) of intersection.

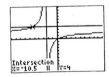

The rational function is undefined at $x=-5$, so $-5$ is not part of the solution. The inequality is non-strict, so $-10.5 = -\dfrac{21}{2}$ is part of the

solution. From the graph, we can see that $\dfrac{2x-1}{x+5} \le 4$ where $x \le -\dfrac{21}{2}$ and where $x > -5$.

Thus, the solution set is $\left\{x \middle| x \le -\dfrac{21}{2} \text{ or } x > -5\right\}$ in set-builder notation; the solution is $\left(-\infty, -\dfrac{21}{2}\right] \cup (-5, \infty)$ in interval notation.

## 6.6 Preparing for Models Involving Rational Expressions

1. $4x - 2y = 10$
   $-2y = -4x + 10$
   $y = \dfrac{-4x + 10}{-2}$
   $y = 2x - 5$

2. Since $y = 4$ when $x = 12$, then
   $y = kx$
   $4 = k \cdot 12$
   $k = \dfrac{4}{12} = \dfrac{1}{3}$
   So, $y = \dfrac{1}{3}x$.

   Thus, when $x = 24$, $y = \dfrac{1}{3}(24) = 8$.

## 6.6 Exercises

2. similar

4. TRUE

6. Answers will vary. The assumption "no gain or loss of efficiency" means that we assume each individual will work at constant rates. That is, the speed at which an individual works does not increase or decrease as more individuals are added to the task.

   This assumption is reasonable for machines, but it is probably not reasonable for humans. As more and more individuals are added to a particular chore (such as mowing a lawn) they would eventually begin to get into each others way, causing the time of completion to increase actually increase.

## Chapter 6: Rational Expressions and Rational Functions

We make this assumption in order to make the mathematics more manageable.

8.  $$\frac{V_1}{V_2} = \frac{P_2}{P_1}$$
    $$V_2 P_1 \left(\frac{V_1}{V_2}\right) = V_2 P_1 \left(\frac{P_2}{P_1}\right)$$
    $$P_1 V_1 = V_2 P_2$$
    $$\frac{V_1 P_1}{P_2} = V_2 \text{ or } V_2 = \frac{V_1 P_1}{P_2}$$

10. $$P = \frac{A}{1+r}$$
    $$(1+r) \cdot P = (1+r)\left(\frac{A}{1+r}\right)$$
    $$P + Pr = A$$
    $$Pr = A - P$$
    $$r = \frac{A-P}{P}$$

12. $$m = \frac{y - y_1}{x - x_1}$$
    $$(x - x_1) \cdot m = (x - x_1)\left(\frac{y - y_1}{x - x_2}\right)$$
    $$mx - mx_1 = y - y_1$$
    $$mx - y + y_1 = mx_1$$
    $$\frac{mx - y + y_1}{m} = x_1 \text{ or } x_1 = \frac{mx - y + y_1}{m}$$

14. $$\omega = \frac{rmv}{I + mr^2}$$
    $$(I + mr^2) \cdot \omega = (I + mr^2)\left(\frac{rmv}{I + mr^2}\right)$$
    $$\omega(I + mr^2) = rmv$$
    $$\omega I + \omega mr^2 = rmv$$
    $$\omega I = rmv - \omega mr^2$$
    $$I = \frac{rmv - \omega mr^2}{\omega}$$

16. $$v_2 = \frac{2m_1}{m_1 + m_2} v_1$$
    $$(m_1 + m_2) \cdot v_2 = (m_1 + m_2)\left(\frac{2m_1}{m_1 + m_2} v_1\right)$$
    $$m_1 v_2 + m_2 v_2 = 2m_1 v_1$$
    $$m_2 v_2 = 2m_1 v_1 - m_1 v_2$$
    $$m_2 v_2 = m_1 (2v_1 - v_2)$$
    $$\frac{m_2 v_2}{2v_1 - v_2} = m_1 \text{ or } m_1 = \frac{m_2 v_2}{2v_1 - v_2}$$

18. a. $$y = \frac{k}{x}$$
    $$15 = \frac{k}{3}$$
    $$3 \cdot 15 = 3\left(\frac{k}{3}\right)$$
    $$45 = k$$

    b. $$y = \frac{45}{x}$$

    c. When $x = 5$, $y = \frac{45}{5} = 9$.

20. a. $$y = \frac{k}{x}$$
    $$4 = \frac{k}{20}$$
    $$20 \cdot 4 = 20\left(\frac{k}{20}\right)$$
    $$80 = k$$

    b. $$y = \frac{80}{x}$$

    c. When $x = 35$, $y = \frac{80}{35} = \frac{16}{7}$.

22. a. $y = kxz$
    $20 = k \cdot 6 \cdot 10$
    $20 = 60k$
    $$k = \frac{20}{60} = \frac{1}{3}$$

    b. $y = \frac{1}{3} xz$

    c. When $x = 8$ and $z = 15$, $y = \frac{1}{3} \cdot 8 \cdot 15 = 40$.

Section 6.6 Models Involving Rational Expressions

24. a. $Q = \dfrac{kx}{y}$

$$\dfrac{14}{5} = \dfrac{k \cdot 4}{3}$$

$$15\left(\dfrac{14}{5}\right) = 15\left(\dfrac{k \cdot 4}{3}\right)$$

$$42 = 20k$$

$$k = \dfrac{42}{20} = \dfrac{21}{10}$$

b. $Q = \dfrac{(21/10)x}{y} = \dfrac{21x}{10y}$

c. When $x = 8$ and $y = 3$,

$$Q = \dfrac{21 \cdot 8}{10 \cdot 3} = \dfrac{168}{30} = \dfrac{28}{5}.$$

26. $$\dfrac{AB}{AC} = \dfrac{DE}{DF}$$

$$\dfrac{2x+3}{10} = \dfrac{6}{x-2}$$

$$10(x-2)\left(\dfrac{2x+3}{10}\right) = 10(x-2)\left(\dfrac{6}{x-2}\right)$$

$$(x-2)(2x+3) = 10 \cdot 6$$

$$2x^2 - x - 6 = 60$$

$$2x^2 - x - 66 = 0$$

$$(2x+11)(x-6) = 0$$

$$2x+11 = 0 \quad \text{or} \quad x-6 = 0$$

$$x = -\dfrac{11}{2} \quad \text{or} \quad x = 6$$

Since the length of DF cannot be negative, we discard $x = -\dfrac{11}{2}$ and keep $x = 6$. Thus,
$AB = 2x+3 = 2(6)+3 = 15$ and
$DF = x-2 = 6-2 = 4$.

28. Let $x$ represent the number of flight hours in 2001.

$$\dfrac{1.22}{100{,}000} = \dfrac{321}{x}$$

$$100{,}000x\left(\dfrac{1.22}{100{,}000}\right) = 100{,}000x\left(\dfrac{321}{x}\right)$$

$$1.22x = 32{,}100{,}000$$

$$x = \dfrac{321{,}100{,}000}{1.22}$$

$$x \approx 26{,}311{,}475$$

In 2001, approximately 26,311,475 flight hours were flown.

30. Let $x$ represent the amount Eduardo borrowed.

$$\dfrac{0.0191}{1} = \dfrac{340}{x}$$

$$x\left(\dfrac{0.0191}{1}\right) = x\left(\dfrac{340}{x}\right)$$

$$0.0191x = 340$$

$$x = \dfrac{340}{0.0191}$$

$$x \approx 17{,}801$$

Eduardo borrowed approximately $17,801.

32. Utilize Pascal's Principal.

$$\dfrac{F_1}{A_1} = \dfrac{F_2}{A_2}$$

$$\dfrac{40}{15} = \dfrac{F_2}{8}$$

$$F_2 = 8\left(\dfrac{40}{15}\right) = \dfrac{64}{3} \approx 21.33$$

Thus, the force exerted by the right pipe is approximately 21.33 pounds.

34. Let $t$ represent the time required if Latoya and Lisa paint the five 10-foot-by-14-foot rooms together.

$$\begin{pmatrix}\text{Part done}\\ \text{by Latoya}\\ \text{in 1 hour}\end{pmatrix} + \begin{pmatrix}\text{Part done}\\ \text{by Lisa}\\ \text{in 1 hour}\end{pmatrix} = \begin{pmatrix}\text{Part done}\\ \text{together}\\ \text{in 1 hour}\end{pmatrix}$$

$$\dfrac{1}{14} + \dfrac{1}{10} = \dfrac{1}{t}$$

$$140t\left(\dfrac{1}{14} + \dfrac{1}{10}\right) = 140t\left(\dfrac{1}{t}\right)$$

$$10t + 14t = 140$$

$$24t = 140$$

$$t = \dfrac{140}{24} \approx 5.83$$

Working together, it will take Latoya and Lisa about 5.83 hours (or 5 hours and 50 minutes) to paint the five 10-foot-by-14-foot rooms.

36. Let $t$ represent the time required for Frank to assemble a King Kong swingset when working

## Chapter 6: Rational Expressions and Rational Functions

alone.

$$\begin{pmatrix} \text{Part done} \\ \text{by Alexandria} \\ \text{in 1 hour} \end{pmatrix} + \begin{pmatrix} \text{Part done} \\ \text{by Frank} \\ \text{in 1 hour} \end{pmatrix} = \begin{pmatrix} \text{Part done} \\ \text{together} \\ \text{in 1 hour} \end{pmatrix}$$

$$\frac{1}{10} + \frac{1}{t} = \frac{1}{6}$$

$$30t\left(\frac{1}{10} + \frac{1}{t}\right) = 30t\left(\frac{1}{6}\right)$$

$$3t + 30 = 5t$$
$$30 = 2t$$
$$t = \frac{30}{2} = 15$$

Working alone, it will take Frank 15 hours to assemble a King Kong swingset.

**38.** Let $t$ represent the time required for 10-horsepower pump to empty the pond when working alone.

Then $t + 4$ will represent the time required for the 4-horsepower pump to empty the pond when working alone.

$$\begin{pmatrix} \text{Part done by} \\ \text{10-horse pump} \\ \text{in 1 hour} \end{pmatrix} + \begin{pmatrix} \text{Part done by} \\ \text{4-horse pump} \\ \text{in 1 hour} \end{pmatrix} = \begin{pmatrix} \text{Part done} \\ \text{together} \\ \text{in 1 hour} \end{pmatrix}$$

$$\frac{1}{t} + \frac{1}{t+4} = \frac{1}{15/4}$$

$$\frac{1}{t} + \frac{1}{t+4} = \frac{4}{15}$$

$$15(t+4) + 15t = 4t(t+4)$$
$$15t + 60 + 15t = 4t^2 + 16t$$
$$30t + 60 = 4t^2 + 16t$$
$$0 = 4t^2 - 14t - 60$$
$$0 = 2(2t+5)(t-6)$$
$$2t + 5 = 0 \quad \text{or} \quad t - 6 = 0$$
$$t = -\frac{5}{2} \quad \text{or} \quad t = 6$$

Since time cannot be negative, we discard $t = -\frac{5}{2}$ and keep $t = 6$.

Working alone, it will take 10-horsepower pump 6 hours to empty the pond.

**40.** Let $x$ represent the average speed going to the waterfall.

Then $x - 4$ will represent the average speed on the return trip.

| | Distance (miles) | Rate (mph) | Time (hours) |
|---|---|---|---|
| Going to waterfall | 20 | $x$ | $\dfrac{20}{x}$ |
| Return trip | 12 | $x - 4$ | $\dfrac{12}{x-4}$ |

$$\frac{20}{x} = \frac{12}{x-4}$$

$$x(x-4)\left(\frac{20}{x}\right) = x(x-4)\left(\frac{12}{x-4}\right)$$

$$20(x-4) = 12x$$
$$20x - 80 = 12x$$
$$8x = 80$$
$$x = 10$$

The average speed going to the waterfall is 10 miles per hour.

**42.** Let $x$ represent Karli's speed up the stairs.

| | Distance (feet) | Rate (ft/s) | Time (sec) |
|---|---|---|---|
| Escalator | 50 | $x + 1.5$ | $\dfrac{50}{x+1.5}$ |
| Stairs | 30 | $x$ | $\dfrac{30}{x}$ |

$$\frac{50}{x+1.5} = \frac{30}{x}$$

$$x(x+1.5)\left(\frac{50}{x+1.5}\right) = x(x+1.5)\left(\frac{30}{x}\right)$$

$$50x = 30(x+1.5)$$
$$50x = 30x + 45$$
$$20x = 45$$
$$x = \frac{45}{20} = 2.25$$

Karli's speed up the stairs is 2.25 feet per second.

### Section 6.6 Models Involving Rational Expressions

**44.** Let $x$ represent the distance Roger has run when Jeff catches up to him.

|  | Distance (miles) | Rate (mi/min) | Time (min) |
|---|---|---|---|
| Jeff | $x$ | $\frac{1}{6}$ | $\frac{x}{1/6} = 6x$ |
| Roger | $x$ | $\frac{1}{8}$ | $\frac{x}{1/8} = 8x$ |

$t_{\text{Jeff}} = t_{\text{Rodger}} - 1$
$6x = 8x - 1$
$-2x = -1$
$x = \frac{-1}{-2} = \frac{1}{2}$

Both men will have run $\frac{1}{2}$ mile when Jeff catches up to Roger. Jeff will have bee running for $6 \cdot \frac{1}{2} = 3$ minutes while Roger will have been running for $8 \cdot \frac{1}{2} = 4$ minutes.

**46.** Let $x$ represent the speed of the plane in still air.

|  | Distance (miles) | Rate (mph) | Time (hours) |
|---|---|---|---|
| With wind | 600 | $x+15$ | $\frac{600}{x+15}$ |
| Against wind | 600 | $x-15$ | $\frac{600}{x-15}$ |

$\frac{600}{x+15} + \frac{600}{x-15} = 9$

$(x+15)(x-15)\left(\frac{600}{x+15} + \frac{600}{x-15}\right)$
$\qquad = (x+15)(x-15)(9)$
$600(x-15) + 600(x+15) = 9(x+15)(x-15)$
$600x - 9000 + 600x + 9000 = 9x^2 - 2025$
$0 = 9x^2 - 1200x - 2025$
$0 = 3(3x+5)(x-135)$
$3x+5 = 0 \quad \text{or} \quad x - 135 = 0$
$x = -\frac{5}{3} \quad \text{or} \quad x = 135$

Since speed cannot be negative, we discard $x = -\frac{5}{3}$ and keep $x = 135$.

The speed of the plane in still air is 135 miles per hour.

**48.** Let $x$ represent Dirk's average speed.
Then $x + 2$ will represent Garrett's average speed.

|  | Distance (miles) | Rate (mph) | Time (hours) |
|---|---|---|---|
| Dirk | 40 | $x+2$ | $\frac{40}{x+2}$ |
| Garrett | 40 | $x$ | $\frac{40}{x}$ |

$\frac{40}{x+2} + \frac{2}{3} = \frac{40}{x}$

$3x(x+2)\left(\frac{40}{x+2} + \frac{2}{3}\right) = 3x(x+2)\left(\frac{40}{x}\right)$
$3x(40) + 2x(x+2) = 3(x+2) \cdot 40$
$120x + 2x^2 + 4x = 120x + 240$
$2x^2 + 4x - 240 = 0$
$x^2 + 2x - 120 = 0$
$(x+12)(x-10) = 0$
$x + 12 = 0 \quad \text{or} \quad x - 10 = 0$
$x = -12 \quad \text{or} \quad x = 10$

Since speed cannot be negative, we discard $x = -12$ and keep $x = 10$. Thus, Garrett's average speed was 10 miles per hour, and Dirks average speed was $x + 2 = 10 + 2 = 12$ miles per hour.

**50. a.** $t = \frac{k}{s}$

$\frac{1}{2} = \frac{k}{35}$

$35 \cdot \frac{1}{2} = 35\left(\frac{k}{35}\right)$

$\frac{35}{2} = k$

Thus, the driving time to school (in minutes) as a function of speed (in miles per hour) is

$t(s) = \frac{35}{2s}$.

## Chapter 6: Rational Expressions and Rational Functions

**b.** $t(30) = \dfrac{35}{2 \cdot 30} = \dfrac{35}{60} = \dfrac{7}{12}$

If the average speed to school is 30 miles per hour, then the driving time will be $\dfrac{7}{12}$ hours, or 35 minutes.

**52.** 
$$i = \dfrac{k}{R}$$
$$30 = \dfrac{k}{8}$$
$$8 \cdot 30 = 8\left(\dfrac{k}{8}\right)$$
$$240 = k$$
$$i(R) = \dfrac{240}{R}$$
$$i(10) = \dfrac{240}{10} = 24$$

When the resistance is 10 ohms, the current is 24 amperes.

**54.** Let $I$ represent the intensity of light that is a distance $d$ from the bulb.
$$I = \dfrac{k}{d^2}$$
$$0.075 = \dfrac{k}{2^2}$$
$$2^2 \cdot 0.075 = 2^2\left(\dfrac{k}{2^2}\right)$$
$$0.3 = k$$
$$I(d) = \dfrac{0.3}{d^2}$$
$$I(3) = \dfrac{0.3}{3^2} \approx 0.033$$

The intensity of the 100-watt light bulb at a distance of 3 meters is approximately 0.033 foot-candles.

**56.** 
$$K = kmv^2$$
$$3520 = k \cdot 110 \cdot 8^2$$
$$3520 = 7040k$$
$$k = \dfrac{3520}{7040} = \dfrac{1}{2}$$
$$K = \dfrac{1}{2}mv^2$$

For $m = 90$ and $v = 10$,
$$K = \dfrac{1}{2}mv^2 = \dfrac{1}{2}(90)(10)^2 = 4500$$

The kinetic energy of a wide receiver weighing 90 kilograms and running at a speed of 10 meters per second is 4500 joules.

**58.** Let $R$ represent the electrical resistance of a wire of length $x$ and diameter $d$.
$$R = \dfrac{kx}{d^2}$$
$$0.255 = \dfrac{k \cdot 50}{3^2}$$
$$0.255 = \dfrac{50}{9}k$$
$$\dfrac{9}{50}(0.255) = \dfrac{9}{50}\left(\dfrac{50}{9}k\right)$$
$$0.0459 = k$$
$$R = \dfrac{0.0459x}{d^2}$$

For $R = 0.147$ and $d = 2.5$,
$$0.147 = \dfrac{0.0459x}{(2.5)^2}$$
$$0.147 = 0.007344x$$
$$20.02 \approx x$$

The length of a wire with diameter 2.5 mm and resistance 0.147 ohms is approximately 20.02 feet.

**60.** 
$$V = \dfrac{kT}{P}$$
$$100 = \dfrac{k \cdot 300}{15}$$
$$100 = 20k$$
$$5 = k$$

The constant of proportionality is 5, and the equation relating the variable is $V = \dfrac{5T}{P}$.

Section 6.6 Models Involving Rational Expressions

For $V = 70$ and $T = 315$,
$$70 = \frac{5(315)}{P}$$
$$70 = \frac{1575}{P}$$
$$P(70) = P\left(\frac{1575}{P}\right)$$
$$70P = 1575$$
$$P = \frac{1575}{70} = 22.5$$

The pressure is 22.5 atmospheres when the volume is 70 liters and the temperature is $315°K$.

62. Carl Lewis would require $9.99 - 9.79 = 0.2$ seconds more to finish the race than Maurice Greene. Let $d$ represent the distance that Carl Lewis can run in 0.2 seconds. Assuming Lewis' speed will remain constant,
$$\frac{100}{9.99} = \frac{x}{0.2}$$
$$\frac{100}{9.99} = \frac{x}{0.2}$$
$$1.998\left(\frac{100}{9.99}\right) = 1.998\left(\frac{x}{0.2}\right)$$
$$20 = 9.99x$$
$$x \approx 2.002$$
Maurice Green would win by approximately 2 meters.

64. $a\left(a^2 b^{-3}\right)^3 = a \cdot a^6 b^{-9} = a^7 b^{-9} = \frac{a^7}{b^9}$

66. $\left(\frac{13a^5 b^2}{ab^{-5}}\right)^0 = 1$

68. $\left(\frac{12pq^{-3}}{3p^4 q^{-4}}\right)^2 = \left(4p^{1-4} q^{-3-(-4)}\right)^2$
$= \left(4p^{-3} q^1\right)^2$
$= 4^2 p^{-3 \cdot 2} q^{1 \cdot 2}$
$= 16 p^{-6} q^2$
$= \frac{16q^2}{p^6}$

**Chapter 6 Review**

1. We need to find all values of $x$ that cause the denominator $3x - 2$ to equal 0.
$$3x - 2 = 0$$
$$3x = 2$$
$$x = \frac{2}{3}$$
Thus, the domain of $\frac{x-5}{3x-2}$ is $\left\{x \mid x \neq \frac{2}{3}\right\}$

2. We need to find all values of $x$ that cause the denominator $a^2 - 3a - 28$ to equal 0.
$$a^2 - 3a - 28 = 0$$
$$(a+4)(a-7) = 0$$
$$a + 4 = 0 \quad \text{or} \quad a - 7 = 0$$
$$a = -4 \quad \text{or} \quad a = 7$$
Thus, the domain of $\frac{a^2 - 16}{a^2 - 3a - 28}$ is $\{a \mid a \neq -4, a \neq 7\}$

3. We need to find all values of $x$ that cause the denominator $m^2 + 9$ to equal 0. However, if $x$ is a real number, then $m^2 + 9$ can never equal 0. The domain of $\frac{m-3}{m^2+9}$ is $\{m \mid m \text{ is any real number}\}$.

4. We need to find all values of $p$ that cause the denominator $n^2 - 2n - 8$ to equal 0.
$$n^2 - 2n - 8 = 0$$
$$(n+2)(n-4) = 0$$
$$n + 2 = 0 \quad \text{or} \quad n - 4 = 0$$
$$n = -2 \quad \text{or} \quad n = 4$$
Thus, the domain of $\frac{n^2 + 7n + 10}{n^2 - 2n - 8}$ is $\{n \mid n \neq -2, n \neq 4\}$.

5. $\frac{6x + 30}{x^2 - 25} = \frac{6(x+5)}{(x+5)(x-5)} = \frac{6\cancel{(x+5)}}{\cancel{(x+5)}(x-5)} = \frac{6}{x-5}$

**Chapter 6:** *Rational Expressions and Rational Functions*

6. $\dfrac{4y^2 - 28y}{2y^5 - 14y^4} = \dfrac{4y(y-7)}{2y^4(y-7)}$
$= \dfrac{2 \cdot 2y(y-7)}{2y \cdot y^3(y-7)}$
$= \dfrac{2 \cdot \cancel{2y}\,\cancel{(y-7)}}{\cancel{2y} \cdot y^3 \cancel{(y-7)}}$
$= \dfrac{2}{y^3}$

7. $\dfrac{w^2 - 4w - 21}{w^2 + 7w + 12} = \dfrac{(w+3)(w-7)}{(w+3)(w+4)}$
$= \dfrac{\cancel{(w+3)}(w-7)}{\cancel{(w+3)}(w+4)}$
$= \dfrac{w-7}{w+4}$

8. $\dfrac{6a^2 - 7ab - 3b^2}{10a^2 - 11ab - 6b^2} = \dfrac{(3a+b)(2a-3b)}{(5a+2b)(2a-3b)}$
$= \dfrac{(3a+b)\cancel{(2a-3b)}}{(5a+2b)\cancel{(2a-3b)}}$
$= \dfrac{3a+b}{5a+2b}$

9. $\dfrac{7-m}{3m^2 - 20m - 7} = \dfrac{-1(m-7)}{(3m+1)(m-7)}$
$= \dfrac{-1\cancel{(m-7)}}{(3m+1)\cancel{(m-7)}} = \dfrac{-1}{3m+1}$

10. $\dfrac{n^3 - 4n^2 + 3n - 12}{n^2 - 8n + 16} = \dfrac{(n-4)(n^2+3)}{(n-4)(n-4)}$
$= \dfrac{\cancel{(n-4)}(n^2+3)}{\cancel{(n-4)}(n-4)} = \dfrac{n^2+3}{n-4}$

11. $\dfrac{4p^2}{p^2 - 3p - 18} \cdot \dfrac{p+3}{8p} = \dfrac{4p \cdot p}{(p+3)(p-6)} \cdot \dfrac{p+3}{2 \cdot 4p}$
$= \dfrac{\cancel{4p} \cdot p}{\cancel{(p+3)}(p-6)} \cdot \dfrac{\cancel{p+3}}{2 \cdot \cancel{4p}}$
$= \dfrac{p}{2(p-6)}$

12. $\dfrac{q^2 + 6q}{6q + 12} \cdot \dfrac{4q+8}{q^2 + q - 30} = \dfrac{q(q+6)}{6(q+2)} \cdot \dfrac{4(q+2)}{(q-5)(q+6)}$
$= \dfrac{q\cancel{(q+6)}}{\underset{3}{\cancel{6}}\cancel{(q+2)}} \cdot \dfrac{\overset{2}{\cancel{4}}\cancel{(q+2)}}{(q-5)\cancel{(q+6)}}$
$= \dfrac{2q}{3(q-5)}$

13. $\dfrac{x^3 - 4x^2}{x^2 - 4} \cdot \dfrac{x^2 + 4x - 12}{x^3 + 2x^2}$
$= \dfrac{x^2(x-4)}{(x+2)(x-2)} \cdot \dfrac{(x-2)(x+6)}{x^2(x+2)}$
$= \dfrac{\cancel{x^2}(x-4)}{(x+2)\cancel{(x-2)}} \cdot \dfrac{\cancel{(x-2)}(x+6)}{\cancel{x^2}(x+2)} = \dfrac{(x-4)(x+6)}{(x+2)^2}$

14. $\dfrac{y^2 - 3y - 28}{y^3 + 4y^2} \cdot \dfrac{2y^2 + 10y}{y^2 - 12y + 35}$
$= \dfrac{(y+4)(y-7)}{y^2(y+4)} \cdot \dfrac{2y(y+5)}{(y-5)(y-7)}$
$= \dfrac{(y+4)(y-7)}{y \cdot y \cdot (y+4)} \cdot \dfrac{2 \cdot y \cdot (y+5)}{(y-5)(y-7)}$
$= \dfrac{\cancel{(y+4)}\cancel{(y-7)}}{\cancel{y} \cdot y \cdot \cancel{(y+4)}} \cdot \dfrac{2 \cdot \cancel{y} \cdot (y+5)}{(y-5)\cancel{(y-7)}} = \dfrac{2(y+5)}{y(y-5)}$

15. $\dfrac{6a^2 + ab - b^2}{3a^2 + 2ab - b^2} \cdot \dfrac{3a^2 + 4ab + b^2}{4a^2 - b^2}$
$= \dfrac{(3a-b)(2a+b)}{(3a-b)(a+b)} \cdot \dfrac{(3a+b)(a+b)}{(2a+b)(2a-b)}$
$= \dfrac{\cancel{(3a-b)}\cancel{(2a+b)}}{\cancel{(3a-b)}\cancel{(a+b)}} \cdot \dfrac{(3a+b)\cancel{(a+b)}}{\cancel{(2a+b)}(2a-b)}$
$= \dfrac{3a+b}{2a-b}$

16. $\dfrac{m^2 + m - 20}{m^3 - 64} \cdot \dfrac{3m^2 + 12m + 48}{m^2 + 3m - 10}$
$= \dfrac{(m-4)(m+5)}{(m-4)(m^2+4m+16)} \cdot \dfrac{3(m^2+4m+16)}{(m-2)(m+5)}$
$= \dfrac{\cancel{(m-4)}\cancel{(m+5)}}{\cancel{(m-4)}\cancel{(m^2+4m+16)}} \cdot \dfrac{3\cancel{(m^2+4m+16)}}{(m-2)\cancel{(m+5)}}$
$= \dfrac{3}{m-2}$

17. $\dfrac{\dfrac{4c^2}{3d^4}}{\dfrac{8c}{27d}} = \dfrac{4c^2}{3d^4} \cdot \dfrac{27d}{8c} = \dfrac{\cancel{4c}\cdot c}{3d\cdot d^3}\cdot\dfrac{9\cdot \cancel{3d}}{2\cdot \cancel{4c}} = \dfrac{9c}{2d^3}$

18. $\dfrac{\dfrac{6z-24}{7z+21}}{\dfrac{z-4}{z^2-9}} = \dfrac{6z-24}{7z+21}\cdot\dfrac{z^2-9}{z-4}$

    $= \dfrac{6(z-4)}{7(z+3)}\cdot\dfrac{(z-3)(z+3)}{z-4}$

    $= \dfrac{6\cancel{(z-4)}}{7\cancel{(z+3)}}\cdot\dfrac{(z-3)\cancel{(z+3)}}{\cancel{z-4}} = \dfrac{6(z-3)}{7}$

19. $\dfrac{\dfrac{x^2-11x+30}{x^2-8x+15}}{\dfrac{x^2-5x-6}{x^2+8x+7}} = \dfrac{x^2-11x+30}{x^2-8x+15}\cdot\dfrac{x^2+8x+7}{x^2-5x-6}$

    $= \dfrac{(x-5)(x-6)}{(x-3)(x-5)}\cdot\dfrac{(x+1)(x+7)}{(x+1)(x-6)}$

    $= \dfrac{\cancel{(x-5)}\cancel{(x-6)}}{(x-3)\cancel{(x-5)}}\cdot\dfrac{\cancel{(x+1)}(x+7)}{\cancel{(x+1)}\cancel{(x-6)}}$

    $= \dfrac{x+7}{x-3}$

20. $\dfrac{\dfrac{m^2+mn-12n^2}{m^3-27n^3}}{\dfrac{m+5n}{m^2+3mn+9n^2}}$

    $= \dfrac{m^2+mn-12n^2}{m^3-27n^3}\cdot\dfrac{m^2+3mn+9n^2}{m+5n}$

    $= \dfrac{(m-3n)(m+4n)}{(m-3n)(m^2+3mn+9n^2)}\cdot\dfrac{m^2+3mn+9n^2}{m+5n}$

    $= \dfrac{\cancel{(m-3n)}(m+4n)}{\cancel{(m-3n)}\cancel{(m^2+3mn+9n^2)}}\cdot\dfrac{\cancel{m^2+3mn+9n^2}}{m+5n}$

    $= \dfrac{m+4n}{m+5n}$

21. $\dfrac{\dfrac{4p^3-4pq^2}{p^2-5pq-24q^2}}{\dfrac{2p^3+4p^2q+2pq^2}{p^2-7pq-8q^2}}$

    $= \dfrac{4p^3-4pq^2}{p^2-5pq-24q^2}\cdot\dfrac{p^2-7pq-8q^2}{2p^3+4p^2q+2pq^2}$

    $= \dfrac{4p(p-q)(p+q)}{(p-8q)(p+3q)}\cdot\dfrac{(p+q)(p-8q)}{2p(p+q)(p+q)}$

    $= \dfrac{2\cdot\cancel{2p}(p-q)\cancel{(p+q)}}{\cancel{(p-8q)}(p+3q)}\cdot\dfrac{\cancel{(p+q)}\cancel{(p-8q)}}{\cancel{2p}\cancel{(p+q)}\cancel{(p+q)}}$

    $= \dfrac{2(p-q)}{p+3q}$

22. $\dfrac{\dfrac{15a^2+11a-14}{25a^2-49}}{\dfrac{27a^3-8}{10a^2+11a-35}}$

    $= \dfrac{15a^2+11a-14}{25a^2-49}\cdot\dfrac{10a^2+11a-35}{27a^3-8}$

    $= \dfrac{(3a-2)(5a+7)}{(5a-7)(5a+7)}\cdot\dfrac{(2a+5)(5a-7)}{(3a-2)(9a^2+6a+4)}$

    $= \dfrac{\cancel{(3a-2)}\cancel{(5a+7)}}{\cancel{(5a-7)}\cancel{(5a+7)}}\cdot\dfrac{(2a+5)\cancel{(5a-7)}}{\cancel{(3a-2)}(9a^2+6a+4)}$

    $= \dfrac{2a+5}{9a^2+6a+4}$

23. $P(x) = f(x)\cdot g(x)$

    $= \dfrac{2x^2+3x-2}{x-5}\cdot\dfrac{x^2-3x-10}{2x-1}$

    $= \dfrac{(2x-1)(x+2)}{x-5}\cdot\dfrac{(x+2)(x-5)}{2x-1}$

    $= \dfrac{\cancel{(2x-1)}(x+2)}{\cancel{x-5}}\cdot\dfrac{(x+2)\cancel{(x-5)}}{\cancel{2x-1}}$

    $= (x+2)^2$

    The domain of $f(x)$ is $\{x\,|\,x\ne 5\}$. The domain of $g(x)$ is $\left\{x\,\middle|\,x\ne\dfrac{1}{2}\right\}$. Therefore, the domain of $P(x)$ is $\left\{x\,\middle|\,x\ne\dfrac{1}{2}, x\ne 5\right\}$.

## Chapter 6: Rational Expressions and Rational Functions

24. $R(x) = g(x) \cdot h(x)$
$= \dfrac{x^2 - 3x - 10}{2x - 1} \cdot \dfrac{2x - 1}{x^2 + 9x + 14}$
$= \dfrac{(x+2)(x-5)}{2x-1} \cdot \dfrac{2x-1}{(x+2)(x+7)}$
$= \dfrac{\cancel{(x+2)}(x-5)}{\cancel{2x-1}} \cdot \dfrac{\cancel{2x-1}}{\cancel{(x+2)}(x+7)}$
$= \dfrac{x-5}{x+7}$

The domain of $g(x)$ is $\left\{x \mid x \neq \dfrac{1}{2}\right\}$. The domain of $h(x)$ is $\{x \mid x \neq -7, x \neq -2\}$. Therefore, the domain of $R(x)$ is $\left\{x \mid x \neq -7, x \neq -2, x \neq \dfrac{1}{2}\right\}$.

25. $Q(x) = \dfrac{g(x)}{f(x)} = \dfrac{\dfrac{x^2 - 3x - 10}{2x - 1}}{\dfrac{2x^2 + 3x - 2}{x - 5}}$
$= \dfrac{x^2 - 3x - 10}{2x - 1} \cdot \dfrac{x - 5}{2x^2 + 3x - 2}$
$= \dfrac{(x+2)(x-5)}{2x-1} \cdot \dfrac{x-5}{(2x-1)(x+2)}$
$= \dfrac{\cancel{(x+2)}(x-5)}{2x-1} \cdot \dfrac{x-5}{(2x-1)\cancel{(x+2)}}$
$= \dfrac{(x-5)^2}{(2x-1)^2}$

The domain of $g(x)$ is $\left\{x \mid x \neq \dfrac{1}{2}\right\}$. The domain of $f(x)$ is $\{x \mid x \neq 5\}$. Because the denominator of $Q(x)$ cannot equal 0, we must exclude those values of $x$ such that the numerator of $f(x)$ is 0. That is, we must exclude the values of $x$ such that $2x^2 + 3x - 2 = 0$. These values are $x = -2$ and $x = \dfrac{1}{2}$. Therefore, the domain of $Q(x)$ is $\left\{x \mid x \neq -2, x \neq \dfrac{1}{2}, x \neq 5\right\}$.

26. $T(x) = \dfrac{f(x)}{h(x)} = \dfrac{\dfrac{2x^2 + 3x - 2}{x - 5}}{\dfrac{2x - 1}{x^2 + 9x + 14}}$
$= \dfrac{2x^2 + 3x - 2}{x - 5} \cdot \dfrac{x^2 + 9x + 14}{2x - 1}$
$= \dfrac{(2x-1)(x+2)}{x-5} \cdot \dfrac{(x+2)(x+7)}{2x-1}$
$= \dfrac{\cancel{(2x-1)}(x+2)}{x-5} \cdot \dfrac{(x+2)(x+7)}{\cancel{2x-1}}$
$= \dfrac{(x+2)^2 (x+7)}{x-5}$

The domain of $f(x)$ is $\{x \mid x \neq 5\}$. The domain of $h(x)$ is $\{x \mid x \neq -7, x \neq -2\}$. Because the denominator of $T(x)$ cannot equal 0, we must exclude those values of $x$ such that the numerator of $h(x)$ is 0. That is, we must exclude the value of $x$ such that $2x - 1 = 0$. This value is $x = \dfrac{1}{2}$. Therefore, the domain of $T(x)$ is $\left\{x \mid x \neq -7, x \neq -2, x \neq \dfrac{1}{2}, x \neq 5\right\}$.

27. $\dfrac{4x}{x-5} + \dfrac{3}{x-5} = \dfrac{4x+3}{x-5}$

28. $\dfrac{4y}{y-3} - \dfrac{12}{y-3} = \dfrac{4y-12}{y-3} = \dfrac{4(y-3)}{y-3} = 4$

29. $\dfrac{a^2 - 2a - 4}{a^2 - 6a + 8} + \dfrac{4a - 20}{a^2 - 6a + 8} = \dfrac{a^2 - 2a - 4 + 4a - 20}{a^2 - 6a + 8}$
$= \dfrac{a^2 + 2a - 24}{a^2 - 6a + 8}$
$= \dfrac{(a-4)(a+6)}{(a-4)(a-2)}$
$= \dfrac{a+6}{a-2}$

30. $\dfrac{3b^2+8b-5}{2b^2-5b-12}-\dfrac{2b^2+7b+15}{2b^2-5b-12}$
$=\dfrac{3b^2+8b-5-(2b^2+7b+15)}{2b^2-5b-12}$
$=\dfrac{3b^2+8b-5-2b^2-7b-15}{2b^2-5b-12}$
$=\dfrac{b^2+b-20}{2b^2-5b-12}=\dfrac{(b+5)(b-4)}{(2b+3)(b-4)}=\dfrac{b+5}{2b+3}$

31. $\dfrac{5c^2-8c}{c-8}+\dfrac{2c^2+16c}{8-c}=\dfrac{5c^2-8c}{c-8}+\dfrac{2c^2+16c}{-1(c-8)}$
$=\dfrac{5c^2-8c}{c-8}-\dfrac{2c^2+16c}{c-8}$
$=\dfrac{5c^2-8c-(2c^2+16c)}{c-8}$
$=\dfrac{5c^2-8c-2c^2-16c}{c-8}$
$=\dfrac{3c^2-24c}{c-8}$
$=\dfrac{3c(c-8)}{c-8}$
$=3c$

32. $\dfrac{2d^2+d}{d^2-1}-\dfrac{d^2+1}{d^2-1}+\dfrac{d-2}{d^2-1}=\dfrac{2d^2+d-(d^2+1)+d-2}{d^2-1}$
$=\dfrac{2d^2+d-d^2-1+d-2}{d^2-1}$
$=\dfrac{d^2+2d-3}{d^2-1}$
$=\dfrac{(d-1)(d+3)}{(d+1)(d-1)}$
$=\dfrac{d+3}{d+1}$

33. $9x^4=3^2\cdot x^4$
$12x^2=2^2\cdot 3\cdot x$
LCD $=2^2\cdot 3^2\cdot x^4=36x^4$

34. $y-9$
$y+2$
LCD $=(y+2)(y-9)$

35. $2p^2-3p-20=(2p+5)(p-4)$
$2p^3+5p^2=p^2(2p+5)$
LCD $=p^2(2p+5)(p-4)$

36. $q^2+4q-5=(q-1)(q+5)$
$q^2+2q-15=(q-3)(q+5)$
LCD $=(q+5)(q-1)(q-3)$

37. $mn^4$
$m^3n^2$
LCD $=m^3n^4$
$\dfrac{1}{mn^4}+\dfrac{4}{m^3n^2}=\dfrac{1}{mn^4}\cdot\dfrac{m^2}{m^2}+\dfrac{4}{m^3n^2}\cdot\dfrac{n^2}{n^2}$
$=\dfrac{m^2}{m^3n^4}+\dfrac{4n^2}{m^3n^4}$
$=\dfrac{m^2+4n^2}{m^3n^4}$

38. $2xy^3=2\cdot x\cdot y^3$
$6x^2y=2\cdot 3\cdot x^2\cdot y$
LCD $=2\cdot 3\cdot x^2\cdot y^3=6x^2y^3$
$\dfrac{3}{2xy^3}-\dfrac{7}{6x^2y}=\dfrac{3}{2xy^3}\cdot\dfrac{3x}{3x}-\dfrac{7}{6x^2y}\cdot\dfrac{y^2}{y^2}$
$=\dfrac{9x}{6x^2y^3}-\dfrac{7y^2}{6x^2y^3}$
$=\dfrac{9x-7y^2}{6x^2y^3}$

39. $p-q$
$p+q$
LCD $=(p-q)(p+q)$
$\dfrac{p}{p-q}-\dfrac{q}{p+q}=\dfrac{p}{p-q}\cdot\dfrac{p+q}{p+q}-\dfrac{q}{p+q}\cdot\dfrac{p-q}{p-q}$
$=\dfrac{p^2+pq}{(p-q)(p+q)}-\dfrac{pq-q^2}{(p-q)(p+q)}$
$=\dfrac{p^2+pq-(pq-q^2)}{(p-q)(p+q)}$
$=\dfrac{p^2+pq-pq+q^2}{(p-q)(p+q)}=\dfrac{p^2+q^2}{(p-q)(p+q)}$

## Chapter 6: Rational Expressions and Rational Functions

**40.** $x^2 - 10x + 21 = (x-3)(x-7)$
$x^2 - 3x - 28 = (x+4)(x-7)$
LCD $= (x-7)(x-3)(x+4)$

$\dfrac{x+8}{x^2-10x+21} - \dfrac{x-5}{x^2-3x-28}$

$= \dfrac{x+8}{(x-3)(x-7)} \cdot \dfrac{x+4}{x+4} - \dfrac{x-5}{(x+4)(x-7)} \cdot \dfrac{x-3}{x-3}$

$= \dfrac{x^2+12x+32}{(x-7)(x-3)(x+4)} - \dfrac{x^2-8x+15}{(x-7)(x-3)(x+4)}$

$= \dfrac{x^2+12x+32-(x^2-8x+15)}{(x-7)(x-3)(x+4)}$

$= \dfrac{x^2+12x+32-x^2+8x-15}{(x-7)(x-3)(x+4)}$

$= \dfrac{20x+17}{(x-7)(x-3)(x+4)}$

**41.** $y^2 - 2y + 1 = (y-1)^2$
$y^2 + y - 2 = (y-1)(y+2)$
LCD $= (y-1)^2(y+2)$

$\dfrac{3}{y^2-2y+1} - \dfrac{2}{y^2+y-2}$

$= \dfrac{3}{(y-1)^2} \cdot \dfrac{y+2}{y+2} - \dfrac{2}{(y-1)(y+2)} \cdot \dfrac{y-1}{y-1}$

$= \dfrac{3y+6}{(y-1)^2(y+2)} - \dfrac{2y-2}{(y-1)^2(y+2)}$

$= \dfrac{3y+6-(2y-2)}{(y-1)^2(y+2)}$

$= \dfrac{3y+6-2y+2}{(y-1)^2(y+2)} = \dfrac{y+8}{(y-1)^2(y+2)}$

**42.** $4a^2 - 9b^2 = (2a-3b)(2a+3b)$
$2a - 3b$
LCD $= (2a-3b)(2a+3b)$

$\dfrac{3a-5b}{4a^2-9b^2} + \dfrac{4}{2a-3b}$

$= \dfrac{3a-5b}{(2a-3b)(2a+3b)} + \dfrac{4}{2a-3b} \cdot \dfrac{2a+3b}{2a+3b}$

$= \dfrac{3a-5b}{(2a-3b)(2a+3b)} + \dfrac{8a+12b}{(2a-3b)(2a+3b)}$

$= \dfrac{3a-5b+8a+12b}{(2a-3b)(2a+3b)} = \dfrac{11a+7b}{(2a-3b)(2a+3b)}$

**43.** $x^2 - 9 = (x-3)(x+3)$
$9 - x^2 = -1(x-3)(x+3)$
LCD $= (x-3)(x+3)$

$\dfrac{4x^2-10x}{x^2-9} + \dfrac{8x-2x^2}{9-x^2}$

$= \dfrac{4x^2-10x}{(x-3)(x+3)} + \dfrac{8x-2x^2}{-1(x-3)(x+3)} \cdot \dfrac{-1}{-1}$

$= \dfrac{4x^2-10x}{(x-3)(x+3)} + \dfrac{-8x+2x^2}{(x-3)(x+3)}$

$= \dfrac{4x^2-10x-8x+2x^2}{(x-3)(x+3)}$

$= \dfrac{6x^2-18x}{(x-3)(x+3)} = \dfrac{6x(x-3)}{(x-3)(x+3)} = \dfrac{6x}{x+3}$

**44.** $n+5$
$n^3 + 125 = (n+5)(n^2-5n+25)$
LCD $= (n+5)(n^2-5n+25)$

$\dfrac{1}{n+5} - \dfrac{n^2-10n}{n^3+125}$

$= \dfrac{1}{n+5} \cdot \dfrac{n^2-5n+25}{n^2-5n+25} - \dfrac{n^2-10n}{(n+5)(n^2-5n+25)}$

$= \dfrac{n^2-5n+25}{(n+5)(n^2-5n+25)} - \dfrac{n^2-10n}{(n+5)(n^2-5n+25)}$

$= \dfrac{n^2-5n+25-(n^2-10n)}{(n+5)(n^2-5n+25)}$

$= \dfrac{n^2-5n+25-n^2+10n}{(n+5)(n^2-5n+25)} = \dfrac{5n+25}{(n+5)(n^2-5n+25)}$

$= \dfrac{5(n+5)}{(n+5)(n^2-5n+25)} = \dfrac{5}{n^2-5n+25}$

**45.** $m+3n$
$m-7n$
$m^2 - 4mn - 21n^2 = (m+3n)(m-7n)$
LCD $= (m+3n)(m-7n)$

$$\frac{m+n}{m+3n} - \frac{m-4n}{m-7n} + \frac{7mn+n^2}{m^2-4mn-21n^2}$$
$$= \frac{m+n}{m+3n} \cdot \frac{m-7n}{m-7n} - \frac{m-4n}{m-7n} \cdot \frac{m+3n}{m+3n} + \frac{7mn+n^2}{(m+3n)(m-7n)}$$
$$= \frac{m^2-6mn-7n^2}{(m+3n)(m-7n)} - \frac{m^2-mn-12n^2}{(m+3n)(m-7n)} + \frac{7mn+n^2}{(m+3n)(m-7n)}$$
$$= \frac{m^2-6mn-7n^2-(m^2-mn-12n^2)+7mn+n^2}{(m+3n)(m-7n)}$$
$$= \frac{m^2-6mn-7n^2-m^2+mn+12n^2+7mn+n^2}{(m+3n)(m-7n)}$$
$$= \frac{2mn+6n^2}{(m+3n)(m-7n)}$$
$$= \frac{2n(m+3n)}{(m+3n)(m-7n)} = \frac{2n}{m-7n}$$

46. $z^2 - 9 = (z-3)(z+3)$
    $z-3$
    $z+3$
    LCD $= (z-3)(z+3)$
$$\frac{z^2+10z+3}{z^2-9} - \frac{2z}{z-3} + \frac{z}{z+3}$$
$$= \frac{z^2+10z+3}{(z-3)(z+3)} - \frac{2z}{z-3} \cdot \frac{z+3}{z+3} + \frac{z}{z+3} \cdot \frac{z-3}{z-3}$$
$$= \frac{z^2+10z+3}{(z-3)(z+3)} - \frac{2z^2+6z}{(z-3)(z+3)} + \frac{z^2-3z}{(z-3)(z+3)}$$
$$= \frac{z^2+10z+3-(2z^2+6z)+z^2-3z}{(z-3)(z+3)}$$
$$= \frac{z^2+10z+3-2z^2-6z+z^2-3z}{(z-3)(z+3)}$$
$$= \frac{z+3}{(z-3)(z+3)}$$
$$= \frac{1}{z-3}$$

47. $y-2$
    $y+2$
    $y^2-4 = (y-2)(y+2)$
    LCD $= (y-2)(y+2)$

$$\frac{y-1}{y-2} - \frac{y+1}{y+2} + \frac{y-6}{y^2-4}$$
$$= \frac{y-1}{y-2} \cdot \frac{y+2}{y+2} - \frac{y+1}{y+2} \cdot \frac{y-2}{y-2} + \frac{y-6}{(y-2)(y+2)}$$
$$= \frac{y^2+y-2}{(y-2)(y+2)} - \frac{y^2-y-2}{(y-2)(y+2)} + \frac{y-6}{(y-2)(y+2)}$$
$$= \frac{y^2+y-2-(y^2-y-2)+y-6}{(y-2)(y+2)}$$
$$= \frac{y^2+y-2-y^2+y+2+y-6}{(y-2)(y+2)}$$
$$= \frac{3y-6}{(y-2)(y+2)}$$
$$= \frac{3(y-2)}{(y-2)(y+2)} = \frac{3}{y+2}$$

48. $a^2 - 16 = (a-4)(a+4)$
    $a-4$
    $a^2+2a-8 = (a-2)(a+4)$
    LCD $= (a-4)(a+4)(a-2)$
$$\frac{2a}{a^2-16} - \frac{1}{a-4} - \frac{1}{a^2+2a-8}$$
$$= \frac{2a}{(a-4)(a+4)} \cdot \frac{a-2}{a-2} - \frac{1}{a-4} \cdot \frac{(a+4)(a-2)}{(a+4)(a-2)} - \frac{1}{(a+4)(a-2)} \cdot \frac{a-4}{a-4}$$
$$= \frac{2a^2-4a}{(a-4)(a+4)(a-2)} - \frac{a^2+2a-8}{(a-4)(a+4)(a-2)} - \frac{a-4}{(a-4)(a+4)(a-2)}$$
$$= \frac{2a^2-4a-(a^2+2a-8)-(a-4)}{(a-4)(a+4)(a-2)}$$
$$= \frac{2a^2-4a-a^2-2a+8-a+4}{(a-4)(a+4)(a-2)}$$
$$= \frac{a^2-7a+12}{(a-4)(a+4)(a-2)}$$
$$= \frac{(a-3)(a-4)}{(a-4)(a+4)(a-2)} = \frac{a-3}{(a+4)(a-2)}$$

Chapter 6: Rational Expressions and Rational Functions

49. **a.**  $x-4$
    $x+2$
    $\text{LCD} = (x-4)(x+2)$

    $S(x) = f(x) + g(x) = \dfrac{5}{x-4} + \dfrac{x}{x+2}$

    $= \dfrac{5}{x-4} \cdot \dfrac{x+2}{x+2} + \dfrac{x}{x+2} \cdot \dfrac{x-4}{x-4}$

    $= \dfrac{5x+10}{(x-4)(x+2)} + \dfrac{x^2-4x}{(x-4)(x+2)}$

    $= \dfrac{5x+10+x^2-4x}{(x-4)(x+2)}$

    $= \dfrac{x^2+x+10}{(x-4)(x+2)}$

    **b.** Since $S$ is the sum of $f$ and $g$, the restrictions for the domain of $S$ will consist of all the restrictions for the domains of both $f$ and $g$. Since 4 is restricted from the domain of $f$ and since $-2$ is restricted from the domain of $g$, both 4 and $-2$ are restricted from the domain of $S$. That is, the domain of $S(x)$ is

    $\{x \mid x \neq -2, x \neq 4\}$.

50. **a.**  $2x^2 + x - 15 = (2x-5)(x+3)$
    $4x^2 - 8x - 5 = (2x-5)(2x+1)$
    $\text{LCD} = (2x-5)(2x+1)(x+3)$

    $D(x) = f(x) - g(x)$

    $= \dfrac{x+3}{2x^2+x-15} - \dfrac{x-7}{4x^2-8x-5}$

    $= \dfrac{x+3}{(2x-5)(x+3)} \cdot \dfrac{2x+1}{2x+1} - \dfrac{x-7}{(2x-5)(2x+1)} \cdot \dfrac{x+3}{x+3}$

    $= \dfrac{2x^2+7x+3}{(2x-5)(2x+1)(x+3)} - \dfrac{x^2-4x-21}{(2x-5)(2x+1)(x+3)}$

    $= \dfrac{2x^2+7x+3-(x^2-4x-21)}{(2x-5)(2x+1)(x+3)}$

    $= \dfrac{2x^2+7x+3-x^2+4x+21}{(2x-5)(2x+1)(x+3)}$

    $= \dfrac{x^2+11x+24}{(2x-5)(2x+1)(x+3)}$

    $= \dfrac{(x+3)(x+8)}{(2x-5)(2x+1)(x+3)}$

    $= \dfrac{x+8}{(2x-5)(2x+1)}$

    **b.** Since $D$ is the difference of $f$ and $g$, the restrictions for the domain of $D$ will consist of all the restrictions for the domains of both $f$ and $g$. Since $-3$ and $\dfrac{5}{2}$ are restricted from the domain of $f$ and since $-\dfrac{1}{2}$ and $\dfrac{5}{2}$ are restricted from the domain of $g$, then $-3$, $-\dfrac{1}{2}$, and $\dfrac{5}{2}$ are restricted from the domain of $D$. That is, the domain of $D(x)$ is

    $\left\{ x \mid x \neq -3, x \neq -\dfrac{1}{2}, x \neq \dfrac{5}{2} \right\}$.

51. Write the numerator of the complex rational expression as a single rational expression. Note that the LCD of the rational expressions in the numerator is $x$.

    $x - \dfrac{1}{x} = \dfrac{x}{1} \cdot \dfrac{x}{x} - \dfrac{1}{x} = \dfrac{x^2}{x} - \dfrac{1}{x} = \dfrac{x^2-1}{x}$

    Write the denominator of the complex rational expression as a single rational expression. Note that the LCD of the rational expressions in the denominator is $x$.

    $1 - \dfrac{1}{x} = \dfrac{1}{1} \cdot \dfrac{x}{x} - \dfrac{1}{x} = \dfrac{x}{x} - \dfrac{1}{x} = \dfrac{x-1}{x}$

    Rewrite the complex rational expression using the single rational expressions and simplify.

    $\dfrac{x - \dfrac{1}{x}}{1 - \dfrac{1}{x}} = \dfrac{\dfrac{x^2-1}{x}}{\dfrac{x-1}{x}} = \dfrac{x^2-1}{x} \cdot \dfrac{x}{x-1}$

    $= \dfrac{(x-1)(x+1)}{x} \cdot \dfrac{x}{x-1}$

    $= \dfrac{\cancel{(x-1)}(x+1)}{\cancel{x}} \cdot \dfrac{\cancel{x}}{\cancel{x-1}}$

    $= x+1$

52. Write the numerator of the complex rational expression as a single rational expression. Note that the LCD of the rational expressions in the numerator is $xy$.

    $\dfrac{1}{x} - \dfrac{1}{y} = \dfrac{1}{x} \cdot \dfrac{y}{y} - \dfrac{1}{y} \cdot \dfrac{x}{x} = \dfrac{y}{xy} - \dfrac{x}{xy} = \dfrac{y-x}{xy}$

    Write the denominator of the complex rational expression as a single rational expression. Note that the LCD of the rational expressions in the denominator is $x^2y^2$.

$$\frac{\dfrac{1}{x^2}-\dfrac{1}{y^2}}=\dfrac{1}{x^2}\cdot\dfrac{y^2}{y^2}-\dfrac{1}{y^2}\cdot\dfrac{x^2}{x^2}$$

$$=\dfrac{y^2}{x^2y^2}-\dfrac{x^2}{x^2y^2}=\dfrac{y^2-x^2}{x^2y^2}$$

Rewrite the complex rational expression using the single rational expressions and simplify.

$$\dfrac{\dfrac{1}{x}-\dfrac{1}{y}}{\dfrac{1}{x^2}-\dfrac{1}{y^2}}=\dfrac{\dfrac{y-x}{xy}}{\dfrac{y^2-x^2}{x^2y^2}}=\dfrac{y-x}{xy}\cdot\dfrac{x^2y^2}{y^2-x^2}$$

$$=\dfrac{\cancel{y-x}}{\cancel{x\cdot y}}\cdot\dfrac{\cancel{x}\cdot x\cdot\cancel{y}\cdot y}{\cancel{(y-x)}(y+x)}$$

$$=\dfrac{xy}{y+x}\quad\text{or}\quad\dfrac{xy}{x+y}$$

**53.** Write the numerator of the complex rational expression as a single rational expression. Note that the LCD of the rational expressions in the numerator is $b(a+b)$.

$$\dfrac{a}{b}-\dfrac{a-b}{a+b}=\dfrac{a}{b}\cdot\dfrac{a+b}{a+b}-\dfrac{a-b}{a+b}\cdot\dfrac{b}{b}$$

$$=\dfrac{a^2+ab}{b(a+b)}-\dfrac{ab-b^2}{b(a+b)}$$

$$=\dfrac{a^2+ab-(ab-b^2)}{b(a+b)}$$

$$=\dfrac{a^2+ab-ab+b^2}{b(a+b)}$$

$$=\dfrac{a^2+b^2}{b(a+b)}$$

Write the denominator of the complex rational expression as a single rational expression. Note that the LCD of the rational expressions in the denominator is $b(a-b)$.

$$\dfrac{a}{b}+\dfrac{a+b}{a-b}=\dfrac{a}{b}\cdot\dfrac{a-b}{a-b}+\dfrac{a+b}{a-b}\cdot\dfrac{b}{b}$$

$$=\dfrac{a^2-ab}{b(a-b)}+\dfrac{ab+b^2}{b(a-b)}$$

$$=\dfrac{a^2+b^2}{b(a-b)}$$

Rewrite the complex rational expression using the single rational expressions and simplify.

$$\dfrac{\dfrac{a}{b}-\dfrac{a-b}{a+b}}{\dfrac{a}{b}+\dfrac{a+b}{a-b}}=\dfrac{\dfrac{a^2+b^2}{b(a+b)}}{\dfrac{a^2+b^2}{b(a-b)}}=\dfrac{a^2+b^2}{b(a+b)}\cdot\dfrac{b(a-b)}{a^2+b^2}$$

$$=\dfrac{\cancel{a^2+b^2}}{\cancel{b}(a+b)}\cdot\dfrac{\cancel{b}(a-b)}{\cancel{a^2+b^2}}$$

$$=\dfrac{a-b}{a+b}$$

**54.** Write the numerator of the complex rational expression as a single rational expression. Note that the LCD of the rational expressions in the numerator is $a+2$.

$$\dfrac{2}{a+2}-1=\dfrac{2}{a+2}-\dfrac{1}{1}\cdot\dfrac{a+2}{a+2}=\dfrac{2}{a+2}-\dfrac{a+2}{a+2}$$

$$=\dfrac{2-(a+2)}{a+2}=\dfrac{2-a-2}{a+2}=\dfrac{-a}{a+2}$$

Write the denominator of the complex rational expression as a single rational expression. Note that the LCD of the rational expressions in the denominator is $a+2$.

$$\dfrac{1}{a+2}+1=\dfrac{1}{a+2}+\dfrac{1}{1}\cdot\dfrac{a+2}{a+2}$$

$$=\dfrac{1}{a+2}+\dfrac{a+2}{a+2}=\dfrac{1+a+2}{a+2}=\dfrac{a+3}{a+2}$$

Rewrite the complex rational expression using the single rational expressions and simplify.

$$\dfrac{\dfrac{2}{a+2}-1}{\dfrac{1}{a+2}+1}=\dfrac{\dfrac{-a}{a+2}}{\dfrac{a+3}{a+2}}=\dfrac{-a}{a+2}\cdot\dfrac{a+2}{a+3}$$

$$=\dfrac{-a}{\cancel{a+2}}\cdot\dfrac{\cancel{a+2}}{a+3}=\dfrac{-a}{a+3}$$

**55.** The LCD of all the denominators in the complex rational expression is $t^2$.

$$\dfrac{\dfrac{3}{t}+\dfrac{4}{t^2}}{5+\dfrac{1}{t^2}}=\dfrac{\dfrac{3}{t}+\dfrac{4}{t^2}}{5+\dfrac{1}{t^2}}\cdot\dfrac{t^2}{t^2}=\dfrac{\dfrac{3}{t}\cdot t^2+\dfrac{4}{t^2}\cdot t^2}{5\cdot t^2+\dfrac{1}{t^2}\cdot t^2}$$

$$=\dfrac{\dfrac{3}{\cancel{t}}\cdot t^{\cancel{2}}+\dfrac{4}{\cancel{t^2}}\cdot\cancel{t^2}}{5\cdot t^2+\dfrac{1}{\cancel{t^2}}\cdot\cancel{t^2}}=\dfrac{3t+4}{5t^2+1}$$

## Chapter 6: Rational Expressions and Rational Functions

**56.** The LCD of all the denominators in the complex rational expression is $ab$.

$$\frac{\frac{1}{a}-\frac{1}{b}}{\frac{b}{a}-\frac{a}{b}} = \frac{\frac{1}{a}-\frac{1}{b}}{\frac{b}{a}-\frac{a}{b}} \cdot \frac{ab}{ab} = \frac{\frac{1}{a}\cdot ab - \frac{1}{b}\cdot ab}{\frac{b}{a}\cdot ab - \frac{a}{b}\cdot ab}$$

$$= \frac{\frac{1}{\cancel{a}}\cdot \cancel{a}b - \frac{1}{\cancel{b}}\cdot a\cancel{b}}{\frac{b}{\cancel{a}}\cdot \cancel{a}b - \frac{a}{\cancel{b}}\cdot a\cancel{b}} = \frac{b-a}{b^2-a^2}$$

$$= \frac{b-a}{(b-a)(b+a)} = \frac{\cancel{b-a}}{\cancel{(b-a)}(b+a)}$$

$$= \frac{1}{b+a} \text{ or } \frac{1}{a+b}$$

**57.** The LCD of all the denominators in the complex rational expression is $z(z-1)(z+1)$.

$$\frac{\frac{1}{z-1}-\frac{1}{z}}{\frac{1}{z}-\frac{1}{z+1}} = \frac{\frac{1}{z-1}-\frac{1}{z}}{\frac{1}{z}-\frac{1}{z+1}} \cdot \frac{z(z-1)(z+1)}{z(z-1)(z+1)}$$

$$= \frac{\frac{1}{z-1}\cdot z(z-1)(z+1) - \frac{1}{z}\cdot z(z-1)(z+1)}{\frac{1}{z}\cdot z(z-1)(z+1) - \frac{1}{z+1}\cdot z(z-1)(z+1)}$$

$$= \frac{\frac{1}{\cancel{z-1}}\cdot z\cancel{(z-1)}(z+1) - \frac{1}{\cancel{z}}\cdot \cancel{z}(z-1)(z+1)}{\frac{1}{\cancel{z}}\cdot \cancel{z}(z-1)(z+1) - \frac{1}{\cancel{z+1}}\cdot z(z-1)\cancel{(z+1)}}$$

$$= \frac{z(z+1)-(z-1)(z+1)}{(z-1)(z+1)-z(z-1)}$$

$$= \frac{z^2+z-(z^2-1)}{z^2-1-z^2+z}$$

$$= \frac{z^2+z-z^2+1}{z-1} = \frac{z+1}{z-1}$$

**58.** The LCD of all the denominators in the complex rational expression is $(x-1)(x+1)$.

$$\frac{1+\frac{x}{x+1}}{\frac{2x+1}{x-1}} = \frac{1+\frac{x}{x+1}}{\frac{2x+1}{x-1}} \cdot \frac{(x-1)(x+1)}{(x-1)(x+1)}$$

$$= \frac{1\cdot(x-1)(x+1)+\frac{x}{x+1}\cdot(x-1)(x+1)}{\frac{2x+1}{x-1}\cdot(x-1)(x+1)}$$

$$= \frac{1\cdot(x-1)(x+1)+\frac{x}{\cancel{x+1}}\cdot(x-1)\cancel{(x+1)}}{\frac{2x+1}{\cancel{x-1}}\cdot \cancel{(x-1)}(x+1)}$$

$$= \frac{(x-1)(x+1)+x(x-1)}{(2x+1)(x+1)}$$

$$= \frac{x^2-1+x^2-x}{(2x+1)(x+1)}$$

$$= \frac{2x^2-x-1}{(2x+1)(x+1)}$$

$$= \frac{(2x+1)(x-1)}{(2x+1)(x+1)}$$

$$= \frac{\cancel{(2x+1)}(x-1)}{\cancel{(2x+1)}(x+1)} = \frac{x-1}{x+1}$$

**59.** Using Method 2, the LCD of all the denominators in the complex rational expression is $y$.

$$\frac{\frac{x}{y}+1}{\frac{x}{y}-1} = \frac{\frac{x}{y}+1}{\frac{x}{y}-1} \cdot \frac{y}{y} = \frac{\frac{x}{y}\cdot y + 1\cdot y}{\frac{x}{y}\cdot y - 1\cdot y}$$

$$= \frac{\frac{x}{\cancel{y}}\cdot \cancel{y}+1\cdot y}{\frac{x}{\cancel{y}}\cdot \cancel{y}-1\cdot y} = \frac{x+y}{x-y}$$

**60.** Using Method 2, the LCD of all the denominators in the complex rational expression is $(a-b)(a+b)$.

$$\frac{\frac{a}{a-b}-\frac{b}{a+b}}{\frac{b}{a-b}+\frac{a}{a+b}} = \frac{\frac{a}{a-b}-\frac{b}{a+b}}{\frac{b}{a-b}+\frac{a}{a+b}} \cdot \frac{(a-b)(a+b)}{(a-b)(a+b)}$$

$$= \frac{\frac{a}{a-b}\cdot(a-b)(a+b)-\frac{b}{a+b}\cdot(a-b)(a+b)}{\frac{b}{a-b}\cdot(a-b)(a+b)+\frac{a}{a+b}\cdot(a-b)(a+b)}$$

$$= \frac{\frac{a}{\cancel{a-b}}\cdot \cancel{(a-b)}(a+b)-\frac{b}{\cancel{a+b}}\cdot(a-b)\cancel{(a+b)}}{\frac{b}{\cancel{a-b}}\cdot \cancel{(a-b)}(a+b)+\frac{a}{\cancel{a+b}}\cdot(a-b)\cancel{(a+b)}}$$

$$= \frac{a(a+b)-b(a-b)}{b(a+b)+a(a-b)}$$

$$= \frac{a^2+ab-ab+b^2}{ab+b^2+a^2-ab} = \frac{a^2+b^2}{a^2+b^2} = 1$$

ISM – 356

**61.** Using Method 1, write the numerator of the complex rational expression as a single rational expression. Note that the LCD of the rational expressions in the numerator is $(x-2)(x+2)$.

$$\frac{1}{x-2} - \frac{x}{x^2-4} = \frac{1}{x-2} - \frac{x}{(x-2)(x+2)}$$
$$= \frac{1}{x-2} \cdot \frac{x+2}{x+2} - \frac{x}{(x-2)(x+2)}$$
$$= \frac{x+2}{(x-2)(x+2)} - \frac{x}{(x-2)(x+2)}$$
$$= \frac{x+2-x}{(x-2)(x+2)} = \frac{2}{(x-2)(x+2)}$$

Write the denominator of the complex rational expression as a single rational expression. Note that the LCD of the rational expressions in the denominator is $x+2$.

$$1 - \frac{2}{x+2} = \frac{1}{1} \cdot \frac{x+2}{x+2} - \frac{2}{x+2}$$
$$= \frac{x+2}{x+2} - \frac{2}{x+2} = \frac{x+2-2}{x+2} = \frac{x}{x+2}$$

Rewrite the complex rational expression using the single rational expressions and simplify.

$$\frac{\frac{1}{x-2} - \frac{x}{x^2-4}}{1 - \frac{2}{x+2}} = \frac{\frac{2}{(x-2)(x+2)}}{\frac{x}{x+2}}$$
$$= \frac{2}{(x-2)(x+2)} \cdot \frac{x+2}{x}$$
$$= \frac{2}{(x-2)(x\!\!\!/+\!\!\!/2)} \cdot \frac{x\!\!\!/+\!\!\!/2}{x}$$
$$= \frac{2}{x(x-2)}$$

**62.** Using Method 1, write the numerator of the complex rational expression as a single rational expression. Note that the LCD of the rational expressions in the numerator is $z+5$.

$$z - \frac{5z}{z+5} = \frac{z}{1} \cdot \frac{z+5}{z+5} - \frac{5z}{z+5}$$
$$= \frac{z^2+5z}{z+5} - \frac{5z}{z+5}$$
$$= \frac{z^2+5z-5z}{z+5} = \frac{z^2}{z+5}$$

Write the denominator of the complex rational expression as a single rational expression. Note that the LCD of the rational expressions in the denominator is $z-5$.

$$z + \frac{5z}{z-5} = \frac{z}{1} \cdot \frac{z-5}{z-5} + \frac{5z}{z-5}$$
$$= \frac{z^2-5z}{z-5} + \frac{5z}{z-5}$$
$$= \frac{z^2-5z+5z}{z-5} = \frac{z^2}{z-5}$$

Rewrite the complex rational expression using the single rational expressions and simplify.

$$\frac{z - \frac{5z}{z+5}}{z + \frac{5z}{z-5}} = \frac{\frac{z^2}{z+5}}{\frac{z^2}{z-5}} = \frac{z^2}{z+5} \cdot \frac{z-5}{z^2} = \frac{z\!\!\!/^2}{z+5} \cdot \frac{z-5}{z\!\!\!/^2} = \frac{z-5}{z+5}$$

**63.** Using Method 2, the LCD of all the denominators in the complex rational expression is $mn(m+n)$.

$$\frac{\frac{m-n}{m+n} + \frac{n}{m}}{\frac{m}{n} - \frac{m-n}{m+n}} = \frac{\frac{m-n}{m+n} + \frac{n}{m}}{\frac{m}{n} - \frac{m-n}{m+n}} \cdot \frac{mn(m+n)}{mn(m+n)}$$
$$= \frac{\frac{m-n}{m+n} \cdot mn(m+n) + \frac{n}{m} \cdot mn(m+n)}{\frac{m}{n} \cdot mn(m+n) - \frac{m-n}{m+n} \cdot mn(m+n)}$$
$$= \frac{\frac{m-n}{m\!\!\!/+\!\!\!/n} \cdot mn(m\!\!\!/+\!\!\!/n) + \frac{n}{m\!\!\!/} \cdot m\!\!\!/n(m+n)}{\frac{m}{n\!\!\!/} \cdot m n\!\!\!/(m+n) - \frac{m-n}{m\!\!\!/+\!\!\!/n} \cdot mn(m\!\!\!/+\!\!\!/n)}$$
$$= \frac{mn(m-n) + n^2(m+n)}{m^2(m+n) - mn(m-n)}$$
$$= \frac{m^2n - mn^2 + mn^2 + n^3}{m^3 + m^2n - m^2n + mn^2}$$
$$= \frac{m^2n + n^3}{m^3 + mn^2} = \frac{n(m^2+n^2)}{m(m^2+n^2)}$$
$$= \frac{n(m^2\!\!\!/+\!\!\!/n^2)}{m(m^2\!\!\!/+\!\!\!/n^2)} = \frac{n}{m}$$

**64.** Using Method 1, write the numerator of the complex rational expression as a single rational expression. Note that the LCD of the rational expressions in the numerator is $(x+1)(x-2)$.

$$\frac{x+4}{x-2} - \frac{x-3}{x+1} = \frac{x+4}{x-2} \cdot \frac{x+1}{x+1} - \frac{x-3}{x+1} \cdot \frac{x-2}{x-2}$$
$$= \frac{x^2+5x+4}{(x+1)(x-2)} - \frac{x^2-5x+6}{(x+1)(x-2)}$$

## Chapter 6: Rational Expressions and Rational Functions

$$= \frac{x^2+5x+4-\left(x^2-5x+6\right)}{(x+1)(x-2)}$$
$$= \frac{x^2+5x+4-x^2+5x-6}{(x+1)(x-2)}$$
$$= \frac{10x-2}{(x+1)(x-2)}$$
$$= \frac{2(5x-1)}{(x+1)(x-2)}$$

The denominator is already a single rational expression: $5x^2+4x-1=(5x-1)(x+1)$.

Rewrite the complex rational expression using the single rational expressions and simplify.

$$\frac{\dfrac{x+4}{x-2}-\dfrac{x-3}{x+1}}{5x^2+4x-1} = \frac{\dfrac{2(5x-1)}{(x+1)(x-2)}}{(5x-1)(x+1)}$$
$$= \frac{2(5x-1)}{(x+1)(x-2)} \cdot \frac{1}{(5x-1)(x+1)}$$
$$= \frac{2\cancel{(5x-1)}}{(x+1)(x-2)} \cdot \frac{1}{\cancel{(5x-1)}(x+1)}$$
$$= \frac{2}{(x+1)^2(x-2)}$$

**65.** Rewrite the expression so that it does not contain any negative exponents.

$$\frac{3x^{-1}-3y^{-1}}{(x+y)^{-1}} = \frac{\dfrac{3}{x}-\dfrac{3}{y}}{\dfrac{1}{x+y}}$$

Using Method 2, the LCD of all the denominators in the complex rational expression is $xy(x+y)$.

$$\frac{3x^{-1}-3y^{-1}}{(x+y)^{-1}} = \frac{\dfrac{3}{x}-\dfrac{3}{y}}{\dfrac{1}{x+y}} = \frac{\dfrac{3}{x}-\dfrac{3}{y}}{\dfrac{1}{x+y}} \cdot \frac{xy(x+y)}{xy(x+y)}$$
$$= \frac{\dfrac{3}{x}\cdot xy(x+y)-\dfrac{3}{y}\cdot xy(x+y)}{\dfrac{1}{x+y}\cdot xy(x+y)}$$
$$= \frac{\dfrac{3}{\cancel{x}}\cdot \cancel{x}y(x+y)-\dfrac{3}{\cancel{y}}\cdot x\cancel{y}(x+y)}{\dfrac{1}{\cancel{x+y}}\cdot xy\cancel{(x+y)}}$$

$$= \frac{3y(x+y)-3x(x+y)}{xy}$$
$$= \frac{3xy+3y^2-3x^2-3xy}{xy}$$
$$= \frac{3y^2-3x^2}{xy} \text{ or } \frac{3(y-x)(y+x)}{xy}$$

**66.** Rewrite the expression so that it does not contain any negative exponents.

$$\frac{2c^{-1}-(3d)^{-1}}{(6d)^{-1}} = \frac{\dfrac{2}{c}-\dfrac{1}{3d}}{\dfrac{1}{6d}}$$

Using Method 2, the LCD of all the denominators in the complex rational expression is $6cd$.

$$\frac{2c^{-1}-(3d)^{-1}}{(6d)^{-1}} = \frac{\dfrac{2}{c}-\dfrac{1}{3d}}{\dfrac{1}{6d}} = \frac{\dfrac{2}{c}-\dfrac{1}{3d}}{\dfrac{1}{6d}} \cdot \frac{6cd}{6cd}$$
$$= \frac{\dfrac{2}{c}\cdot 6cd-\dfrac{1}{3d}\cdot 6cd}{\dfrac{1}{6d}\cdot 6cd} = \frac{\dfrac{2}{\cancel{c}}\cdot 6\cancel{c}d-\dfrac{1}{\cancel{3d}}\cdot \overset{2}{\cancel{6}}c\cancel{d}}{\dfrac{1}{\cancel{6d}}\cdot \cancel{6}c\cancel{d}}$$
$$= \frac{12d-2c}{c} \text{ or } \frac{2(6d-c)}{c}$$

**67.**
$$\frac{2}{z}-\frac{1}{3z}=\frac{1}{6}$$
$$6z\left(\frac{2}{z}-\frac{1}{3z}\right)=6z\left(\frac{1}{6}\right)$$
$$12-2=z$$
$$10=z$$

Check:
$$\frac{2}{10}-\frac{1}{3\cdot 10}\overset{?}{=}\frac{1}{6}$$
$$\frac{6}{30}-\frac{1}{30}\overset{?}{=}\frac{1}{6}$$
$$\frac{5}{30}\overset{?}{=}\frac{1}{6}$$
$$\frac{1}{6}=\frac{1}{6} \leftarrow \text{True}$$

The solution checks, so the solution set is $\{10\}$.

**68.**
$$\frac{4}{m-4}=\frac{-5}{m+2}$$
$$(m-4)(m+2)\left(\frac{4}{m-4}\right)=(m-4)(m+2)\left(\frac{-5}{m+2}\right)$$
$$4(m+2)=-5(m-4)$$
$$4m+8=-5m+20$$
$$9m=12$$
$$m=\frac{12}{9}=\frac{4}{3}$$

ISM − 358

Check:
$$\frac{4}{\frac{4}{3}-4} \stackrel{?}{=} \frac{-5}{\frac{4}{3}+2}$$
$$\frac{4}{\frac{4}{3}-\frac{12}{3}} \stackrel{?}{=} \frac{-5}{\frac{4}{3}+\frac{6}{3}}$$
$$\frac{4}{-\frac{8}{3}} \stackrel{?}{=} \frac{-5}{\frac{10}{3}}$$
$$4 \cdot \left(-\frac{3}{8}\right) \stackrel{?}{=} -5 \cdot \frac{3}{10}$$
$$-\frac{3}{2} = -\frac{3}{2} \leftarrow \text{True}$$

The solution checks, so the solution set is $\left\{\frac{4}{3}\right\}$.

**69.**
$$m - \frac{14}{m} = 5$$
$$m\left(m - \frac{14}{m}\right) = m(5)$$
$$m^2 - 14 = 5m$$
$$m^2 - 5m - 14 = 0$$
$$(m+2)(m-7) = 0$$
$$m+2 = 0 \quad \text{or} \quad m-7 = 0$$
$$m = -2 \quad \text{or} \quad m = 7$$

Check:
$$-2 - \frac{14}{-2} \stackrel{?}{=} 5$$
$$-2 + 7 \stackrel{?}{=} 5$$
$$5 = 5 \leftarrow \text{True}$$

Check:
$$7 - \frac{14}{7} \stackrel{?}{=} 5$$
$$7 - 2 \stackrel{?}{=} 5$$
$$5 = 5 \leftarrow \text{True}$$

Both solutions check, so the solution set is $\{-2, 7\}$.

**70.**
$$\frac{2}{n+3} = \frac{1}{n-3}$$
$$(n-3)(n+3)\left(\frac{2}{n+3}\right) = (n-3)(n+3)\left(\frac{1}{n-3}\right)$$
$$2(n-3) = n+3$$
$$2n - 6 = n + 3$$
$$n = 9$$

Check:
$$\frac{2}{9+3} \stackrel{?}{=} \frac{1}{9-3}$$
$$\frac{2}{12} \stackrel{?}{=} \frac{1}{6}$$
$$\frac{1}{6} = \frac{1}{6} \leftarrow \text{True}$$

The solution checks, so the solution set is $\{9\}$.

**71.**
$$\frac{s}{s-1} = 1 + \frac{2}{s}$$
$$s(s-1)\left(\frac{s}{s-1}\right) = s(s-1)\left(1 + \frac{2}{s}\right)$$
$$s^2 = s(s-1) + 2(s-1)$$
$$s^2 = s^2 - s + 2s - 2$$
$$s^2 = s^2 + s - 2$$
$$0 = s - 2$$
$$2 = s$$

Check:
$$\frac{2}{2-1} \stackrel{?}{=} 1 + \frac{2}{2}$$
$$\frac{2}{1} \stackrel{?}{=} 1 + 1$$
$$2 = 2 \leftarrow \text{True}$$

The solution checks, so the solution set is $\{2\}$.

**72.**
$$\frac{3}{x^2 - 7x + 10} + 2 = \frac{x-4}{x-5}$$
$$\frac{3}{(x-2)(x-5)} + 2 = \frac{x-4}{x-5}$$
$$(x-2)(x-5)\left(\frac{3}{(x-2)(x-5)} + 2\right) = (x-2)(x-5)\left(\frac{x-4}{x-5}\right)$$
$$3 + 2(x-2)(x-5) = (x-4)(x-2)$$
$$3 + 2x^2 - 14x + 20 = x^2 - 6x + 8$$
$$2x^2 - 14x + 23 = x^2 - 6x + 8$$
$$x^2 - 8x + 15 = 0$$
$$(x-3)(x-5) = 0$$
$$x - 3 = 0 \quad \text{or} \quad x - 5 = 0$$
$$x = 3 \quad \text{or} \quad x = 5$$

Since $x = 5$ is not in the domain of the variable, it is an extraneous solution.

## Chapter 6: Rational Expressions and Rational Functions

Check $x = 3$:  $\dfrac{3}{3^2 - 7 \cdot 3 + 10} + 2 \stackrel{?}{=} \dfrac{3-4}{3-5}$

$\dfrac{3}{9 - 21 + 10} + 2 \stackrel{?}{=} \dfrac{-1}{-2}$

$\dfrac{3}{-2} + 2 \stackrel{?}{=} \dfrac{1}{2}$

$\dfrac{1}{2} = \dfrac{1}{2}$ ← True

The solution set is $\{3\}$.

**73.**
$\dfrac{1}{k-1} + \dfrac{1}{k+2} = \dfrac{3}{k^2 + k - 2}$

$\dfrac{1}{k-1} + \dfrac{1}{k+2} = \dfrac{3}{(k-1)(k+2)}$

$(k-1)(k+2)\left(\dfrac{1}{k-1} + \dfrac{1}{k+2}\right) =$

$(k-1)(k+2)\left(\dfrac{3}{(k-1)(k+2)}\right)$

$(k+2) + (k-1) = 3$

$2k + 1 = 3$

$2k = 2$

$k = 1$

Since $k = 1$ is not in the domain of the variable, it is an extraneous solution. The solution set is $\varnothing$ or $\{\ \}$.

**74.**
$x + \dfrac{3x}{x-3} = \dfrac{9}{x-3}$

$(x-3)\left(x + \dfrac{3x}{x-3}\right) = (x-3)\left(\dfrac{9}{x-3}\right)$

$x(x-3) + 3x = 9$

$x^2 - 3x + 3x = 9$

$x^2 = 9$

$x^2 - 9 = 0$

$(x-3)(x+3) = 0$

$x - 3 = 0$  or  $x + 3 = 0$

$x = 3$  or  $x = -3$

Since $x = 3$ is not in the domain of the variable, it is an extraneous solution.

Check $x = -3$:  $-3 + \dfrac{3(-3)}{-3-3} \stackrel{?}{=} \dfrac{9}{-3-3}$

$-3 + \dfrac{-9}{-6} \stackrel{?}{=} \dfrac{9}{-6}$

$-3 + \dfrac{3}{2} \stackrel{?}{=} -\dfrac{3}{2}$

$-\dfrac{3}{2} = -\dfrac{3}{2}$ ← True

The solution set is $\{-3\}$.

**75.**
$\dfrac{2}{a+3} - \dfrac{4}{a^2 - 4} = \dfrac{a+1}{a^2 + 5a + 6}$

$\dfrac{2}{a+3} - \dfrac{4}{(a-2)(a+2)} = \dfrac{a+1}{(a+2)(a+3)}$

$(a-2)(a+2)(a+3)\left(\dfrac{2}{a+3} - \dfrac{4}{(a-2)(a+2)}\right) =$

$(a-2)(a+2)(a+3)\left(\dfrac{a+1}{(a+2)(a+3)}\right)$

$2(a-2)(a+2) - 4(a+3) = (a+1)(a-2)$

$2a^2 - 8 - 4a - 12 = a^2 - a - 2$

$2a^2 - 4a - 20 = a^2 - a - 2$

$a^2 - 3a - 18 = 0$

$(a+3)(a-6) = 0$

$a + 3 = 0$  or  $a - 6 = 0$

$a = -3$  or  $a = 6$

Since $a = -3$ is not in the domain of the variable, it is an extraneous solution.

Check $a = 6$:

$\dfrac{2}{6+3} - \dfrac{4}{6^2 - 4} \stackrel{?}{=} \dfrac{6+1}{6^2 + 5 \cdot 6 + 6}$

$\dfrac{2}{9} - \dfrac{4}{36-4} \stackrel{?}{=} \dfrac{7}{36 + 30 + 6}$

$\dfrac{2}{9} - \dfrac{4}{32} \stackrel{?}{=} \dfrac{7}{72}$

$\dfrac{64}{288} - \dfrac{36}{288} \stackrel{?}{=} \dfrac{7}{72}$

$\dfrac{28}{288} \stackrel{?}{=} \dfrac{7}{72}$

$\dfrac{7}{72} = \dfrac{7}{72}$ ← True

The solution set is $\{6\}$.

**76.**
$\dfrac{2}{z^2 + 2z - 8} = \dfrac{1}{z^2 + 9z + 20} + \dfrac{4}{z^2 + 3z - 10}$

$\dfrac{2}{(z-2)(z+4)} = \dfrac{1}{(z+4)(z+5)} + \dfrac{4}{(z-2)(z+5)}$

$(z-2)(z+4)(z+5)\left(\dfrac{2}{(z-2)(z+4)}\right) =$

$(z-2)(z+4)(z+5)\left(\dfrac{1}{(z+4)(z+5)} + \dfrac{4}{(z-2)(z+5)}\right)$

$2(z+5) = (z-2) + 4(z+4)$

$2z + 10 = z - 2 + 4z + 16$

$2z + 10 = 5z + 14$

$-3z = 4$

$z = -\dfrac{4}{3}$

Check:

$$\frac{2}{\left(-\frac{4}{3}\right)^2 + 2\left(-\frac{4}{3}\right) - 8} \stackrel{?}{=}$$

$$\frac{1}{\left(-\frac{4}{3}\right)^2 + 9\left(-\frac{4}{3}\right) + 20} + \frac{4}{\left(-\frac{4}{3}\right)^2 + 3\left(-\frac{4}{3}\right) - 10}$$

$$\frac{2}{\frac{16}{9} - \frac{8}{3} - 8} \stackrel{?}{=} \frac{1}{\frac{16}{9} - 12 + 20} + \frac{4}{\frac{16}{9} - 4 - 10}$$

$$\frac{2}{-\frac{80}{9}} \stackrel{?}{=} \frac{1}{\frac{88}{9}} + \frac{4}{-\frac{110}{9}}$$

$$-\frac{9}{40} \stackrel{?}{=} \frac{9}{88} - \frac{18}{55}$$

$$-\frac{9}{40} \stackrel{?}{=} \frac{45}{440} - \frac{144}{440}$$

$$-\frac{9}{40} \stackrel{?}{=} -\frac{99}{440}$$

$$-\frac{9}{40} = -\frac{9}{40} \quad \leftarrow \text{True}$$

The solution set is $\left\{-\frac{4}{3}\right\}$.

**77.**
$$\frac{x-3}{x+4} = \frac{14}{x^2 + 6x + 8}$$

$$\frac{x-3}{x+4} = \frac{14}{(x+2)(x+4)}$$

$$(x+2)(x+4)\left(\frac{x-3}{x+4}\right) = (x+2)(x+4)\left(\frac{14}{(x+2)(x+4)}\right)$$

$$(x+2)(x-3) = 14$$
$$x^2 - x - 6 = 14$$
$$x^2 - x - 20 = 0$$
$$(x+4)(x-5) = 0$$
$$x+4 = 0 \quad \text{or} \quad x-5 = 0$$
$$x = -4 \quad \text{or} \quad x = 5$$

Since $x = -4$ is not in the domain of the variable, it is an extraneous solution.

Check $x = 5$:
$$\frac{5-3}{5+4} \stackrel{?}{=} \frac{14}{5^2 + 6\cdot 5 + 8}$$
$$\frac{2}{9} \stackrel{?}{=} \frac{14}{25 + 30 + 8}$$
$$\frac{2}{9} \stackrel{?}{=} \frac{14}{63}$$
$$\frac{2}{9} = \frac{2}{9} \quad \leftarrow \text{True}$$

The solution set is $\{5\}$.

**78.**
$$\frac{5}{y-5} + 4 = \frac{3y-10}{y-5}$$

$$(y-5)\left(\frac{5}{y-5} + 4\right) = (y-5)\left(\frac{3y-10}{y-5}\right)$$

$$5 + 4(y-5) = 3y - 10$$
$$5 + 4y - 20 = 3y - 10$$
$$4y - 15 = 3y - 10$$
$$y = 5$$

Since $y = 5$ is not in the domain of the variable, it is an extraneous solution. The solution set is $\varnothing$ or $\{\ \}$.

**79.**
$$f(x) = 2$$
$$\frac{6}{x-2} = 2$$
$$(x-2)\left(\frac{6}{x-2}\right) = (x-2)\cdot 2$$
$$6 = 2x - 4$$
$$10 = 2x$$
$$5 = x$$

Thus, $f(5) = 2$, so the point $(5, 2)$ is on the graph of $f$.

**80.**
$$g(x) = 4$$
$$x - \frac{21}{x} = 4$$
$$x\left(x - \frac{21}{x}\right) = x \cdot 4$$
$$x^2 - 21 = 4x$$
$$x^2 - 4x - 21 = 0$$
$$(x+3)(x-7) = 0$$
$$x+3 = 0 \quad \text{or} \quad x-7 = 0$$
$$x = -3 \quad \text{or} \quad x = 7$$

Thus, $g(-3) = 4$ and $g(7) = 4$, so the points $(-3, 4)$ and $(7, 4)$ are on the graph of $f$.

**81.** $\dfrac{x-4}{x+2} \geq 0$

The rational expression will equal 0 when $x = 4$. It is undefined when $x = -2$.
Determine where the numerator and denominator are positive and negative and where the quotient is positive and negative.

| Interval | $(-\infty, -2)$ | $-2$ | $(-2, 4)$ | $4$ | $(4, \infty)$ |
|---|---|---|---|---|---|
| $x - 4$ | $----$ | $-$ | $----$ | $0$ | $++++$ |
| $x + 2$ | $----$ | $0$ | $++++$ | $+$ | $++++$ |
| $\dfrac{x-4}{x+2}$ | $++++$ | $\varnothing$ | $----$ | $0$ | $++++$ |

ISM − 361

## Chapter 6: Rational Expressions and Rational Functions

The rational function is undefined at $x = -2$, so -2 is not part of the solution. The inequality is non-strict, so 4 is part of the solution. Now, $\dfrac{x-4}{x+2}$ is greater than zero where the quotient is positive. The solution is $\{x \mid x < -2 \text{ or } x \geq 4\}$ in set-builder notation; the solution is $(-\infty, -2) \cup [4, \infty)$ in interval notation.

82. $\dfrac{y-5}{y+4} < 0$

The rational expression will equal 0 when $y = 5$. It is undefined when $y = -4$. Determine where the numerator and denominator are positive and negative and where the quotient is positive and negative.

| Interval | $(-\infty, -4)$ | $-4$ | $(-4, 5)$ | $5$ | $(5, \infty)$ |
|---|---|---|---|---|---|
| $y - 5$ | $----$ | $-$ | $----$ | $0$ | $++++$ |
| $y + 4$ | $----$ | $0$ | $++++$ | $+$ | $++++$ |
| $\dfrac{y-5}{y+4}$ | $++++$ | $\varnothing$ | $----$ | $0$ | $++++$ |

The rational function is undefined at $y = -4$, so -4 is not part of the solution. The inequality is strict, so 5 is not part of the solution. Now, $\dfrac{y-5}{y+4}$ is less than zero where the quotient is negative. The solution is $\{y \mid -4 < y < 5\}$ in set-builder notation; the solution is $(-4, 5)$ in interval notation.

83. $\dfrac{4}{z^2 - 9} \leq 0$

$\dfrac{4}{(z-3)(z+3)} \leq 0$

Because the numerator is a constant, this rational expression cannot equal 0. However, it is undefined when $z = 3$ and when $z = -3$. Determine where the factors of the denominator are positive and negative and where the quotient is positive and negative.

| Interval | $(-\infty, -3)$ | $-3$ | $(-3, 3)$ | $3$ | $(3, \infty)$ |
|---|---|---|---|---|---|
| $4$ | $++++$ | $+$ | $++++$ | $+$ | $++++$ |
| $z - 3$ | $----$ | $-$ | $----$ | $0$ | $++++$ |
| $z + 3$ | $----$ | $0$ | $++++$ | $+$ | $++++$ |
| $\dfrac{4}{(z-3)(z+3)}$ | $++++$ | $\varnothing$ | $----$ | $\varnothing$ | $++++$ |

The rational function is undefined at $z = -3$ and $z = 3$, so -3 and 3 are not part of the solution. Now, $\dfrac{4}{(z-3)(z+3)}$ is less than zero where the quotient is negative. The solution is $\{z \mid -3 < z < 3\}$ in set-builder notation; the solution is $(-3, 3)$ in interval notation.

84. $\dfrac{w^2 + 5w - 14}{w - 4} < 0$

$\dfrac{(w-2)(w+7)}{w-4} < 0$

The rational expression will equal 0 when $w = 2$ and when $w = -7$. It is undefined when $w = 4$. Determine where the factors of the numerator and the denominator are positive and negative and where the quotient is positive and negative.

| Interval | $(-\infty, -7)$ | $-7$ | $(-7, 2)$ | $2$ | $(2, 4)$ | $4$ | $(4, \infty)$ |
|---|---|---|---|---|---|---|---|
| $w - 2$ | $---$ | $-$ | $---$ | $0$ | $+++$ | $+$ | $+++$ |
| $w + 7$ | $---$ | $0$ | $+++$ | $+$ | $+++$ | $+$ | $+++$ |
| $w - 4$ | $---$ | $-$ | $---$ | $-$ | $---$ | $0$ | $+++$ |
| $\dfrac{(w-2)(w+7)}{w-4}$ | $---$ | $0$ | $+++$ | $0$ | $---$ | $\varnothing$ | $+++$ |

The rational function is undefined at $w = 4$, so 4 is not part of the solution. The inequality is strict, so -7 and 2 are not part of the solution. Now, $\dfrac{(w-2)(w+7)}{w-4}$ is less than zero where the quotient is negative. The solution is $\{w \mid w < -7 \text{ or } 2 < w < 4\}$ in set-builder notation; the solution is $(-\infty, -7) \cup (2, 4)$ in interval notation.

85. $$\frac{m-5}{m^2+3m-10} \geq 0$$
$$\frac{m-5}{(m-2)(m+5)} \geq 0$$

The rational expression will equal 0 when $m=5$. It is undefined when $m=2$ and when $m=-5$.

Determine where the numerator and the factors of the denominator are positive and negative and where the quotient is positive and negative.

| Interval | $(-\infty, -5)$ | $-5$ | $(-5, 2)$ | $2$ | $(2, 5)$ | $5$ | $(5, \infty)$ |
|---|---|---|---|---|---|---|---|
| $m-5$ | $---$ | $-$ | $---$ | $-$ | $---$ | $0$ | $+++$ |
| $m-2$ | $---$ | $-$ | $---$ | $0$ | $+++$ | $+$ | $+++$ |
| $m+5$ | $---$ | $0$ | $+++$ | $+$ | $+++$ | $+$ | $+++$ |
| $\frac{m-5}{(m-2)(m+5)}$ | $---$ | $\varnothing$ | $+++$ | $\varnothing$ | $---$ | $0$ | $+++$ |

The rational function is undefined at $m=-5$ and when $m=2$, so -5 and 2 are not part of the solution. The inequality is non-strict, so 5 is part of the solution. Now, $\frac{m-5}{(m-2)(m+5)}$ is greater than zero where the quotient is positive. The solution is $\{m \mid -5 < m < 2 \text{ or } m \geq 5\}$ in set-builder notation; the solution is $(-5, 2) \cup [5, \infty)$ in interval notation.

86. $$\frac{4}{n-2} \leq -2$$
$$\frac{4}{n-2} + 2 \leq 0$$
$$\frac{4}{n-2} + \frac{2(n-2)}{n-2} \leq 0$$
$$\frac{4+2n-4}{n-2} \leq 0$$
$$\frac{2n}{n-2} \leq 0$$

The rational expression will equal 0 when $n=0$. It is undefined when $n=2$.

Determine where the numerator and denominator are positive and negative and where the quotient is positive and negative.

| Interval | $(-\infty, 0)$ | $0$ | $(0, 2)$ | $2$ | $(2, \infty)$ |
|---|---|---|---|---|---|
| $2n$ | $----$ | $0$ | $++++$ | $+$ | $++++$ |
| $n-2$ | $----$ | $-$ | $----$ | $0$ | $++++$ |
| $\frac{2n}{n-2}$ | $++++$ | $0$ | $----$ | $\varnothing$ | $++++$ |

The rational function is undefined at $n=2$, so 2 is not part of the solution. The inequality is non-strict, so 0 is part of the solution. Now, $\frac{2n}{n-2}$ is less than zero where the quotient is negative. The solution is $\{n \mid 0 \leq n < 2\}$ in set-builder notation; the solution is $[0, 2)$ in interval notation.

87. $$\frac{a+1}{a-2} > 3$$
$$\frac{a+1}{a-2} - 3 > 0$$
$$\frac{a+1}{a-2} - \frac{3(a-2)}{a-2} > 0$$
$$\frac{a+1-3a+6}{a-2} > 0$$
$$\frac{-2a+7}{a-2} > 0$$

The rational expression will equal 0 when $a = \frac{7}{2}$. It is undefined when $a=2$.

Determine where the numerator and denominator are positive and negative and where the quotient is positive and negative.

| Interval | $(-\infty, 2)$ | $2$ | $\left(2, \frac{7}{2}\right)$ | $\frac{7}{2}$ | $\left(\frac{7}{2}, \infty\right)$ |
|---|---|---|---|---|---|
| $-2a+7$ | $++++$ | $+$ | $++++$ | $0$ | $----$ |
| $a-2$ | $----$ | $0$ | $++++$ | $+$ | $++++$ |
| $\frac{-2a+7}{a-2}$ | $----$ | $\varnothing$ | $++++$ | $0$ | $----$ |

The rational function is undefined at $a=2$, so 2 is not part of the solution. The inequality is strict, so $\frac{7}{2}$ is not part of the solution. Now, $\frac{-2a+7}{a-2}$ is greater than zero where the quotient is positive. The solution is $\left\{a \,\middle|\, 2 < a < \frac{7}{2}\right\}$ in set-builder notation; the solution is $\left(2, \frac{7}{2}\right)$ in interval notation.

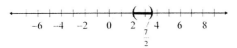

**88.**
$$\frac{4}{c-2}-\frac{3}{c}<0$$
$$\frac{4c}{c(c-2)}-\frac{3(c-2)}{c(c-2)}<0$$
$$\frac{4c-3c+6}{c(c-2)}<0$$
$$\frac{c+6}{c(c-2)}<0$$

The rational expression will equal 0 when $c=-6$. It is undefined when $c=0$ and when $c=2$.

Determine where the numerator and the factors of the denominator are positive and negative and where the quotient is positive and negative.

| Interval | $(-\infty,-6)$ | $-6$ | $(-6,0)$ | $0$ | $(0,2)$ | $2$ | $(2,\infty)$ |
|---|---|---|---|---|---|---|---|
| $c+6$ | $---$ | $0$ | $+++$ | $+$ | $+++$ | $+$ | $+++$ |
| $c$ | $---$ | $-$ | $---$ | $0$ | $+++$ | $+$ | $+++$ |
| $c-2$ | $---$ | $-$ | $---$ | $-$ | $---$ | $0$ | $+++$ |
| $\frac{c+6}{c(c-2)}$ | $---$ | $0$ | $+++$ | $\varnothing$ | $---$ | $\varnothing$ | $+++$ |

The rational function is undefined at $c=0$ and $c=2$, so 0 and 2 are not part of the solution. The inequality is strict, so -6 is not part of the solution. Now, $\frac{c+6}{c(c-2)}$ is less than zero where the quotient is negative. The solution is $\{c \mid c<-6 \text{ or } 0<c<2\}$ in set-builder notation; the solution is $(-\infty,-6)\cup(0,2)$ in interval notation.

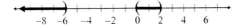

**89.** $Q(x)<0$
$$\frac{2x+3}{x-4}<0$$

The rational expression will equal 0 when $x=-\frac{3}{2}$. It is undefined when $x=4$.

Determine where the numerator and denominator are positive and negative and where the quotient is positive and negative.

| Interval | $\left(-\infty,-\frac{3}{2}\right)$ | $-\frac{3}{2}$ | $\left(-\frac{3}{2},4\right)$ | $4$ | $(4,\infty)$ |
|---|---|---|---|---|---|
| $2x+3$ | $---$ | $0$ | $++++$ | $+$ | $++++$ |
| $x-4$ | $---$ | $-$ | $----$ | $0$ | $++++$ |
| $\frac{2x+3}{x-4}$ | $++++$ | $0$ | $----$ | $\varnothing$ | $++++$ |

The rational function is undefined at $x=4$, so 4 is not part of the solution. The inequality is strict, so $-\frac{3}{2}$ is not part of the solution. Now, $\frac{2x+3}{x-4}$ is less than zero where the quotient is negative. The solution is $\left\{x \mid -\frac{3}{2}<x<4\right\}$ in set-builder notation; the solution is $\left(-\frac{3}{2},4\right)$ in interval notation.

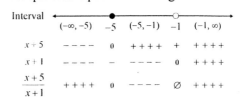

**90.** $R(x)\geq 0$
$$\frac{x+5}{x+1}\geq 0$$

The rational expression will equal 0 when $x=-5$. It is undefined when $x=-1$. Determine where the numerator and the denominator are positive and negative and where the quotient is positive and negative.

| Interval | $(-\infty,-5)$ | $-5$ | $(-5,-1)$ | $-1$ | $(-1,\infty)$ |
|---|---|---|---|---|---|
| $x+5$ | $----$ | $0$ | $++++$ | $+$ | $++++$ |
| $x+1$ | $----$ | $-$ | $----$ | $0$ | $++++$ |
| $\frac{x+5}{x+1}$ | $++++$ | $0$ | $----$ | $\varnothing$ | $++++$ |

The rational function is undefined at $x=-1$, so -1 is not part of the solution. The inequality is non-strict, so -5 is part of the solution. Now,

$\dfrac{x+5}{x+1}$ is greater than zero where the quotient is positive. The solution is $\{x \mid x \le -5 \text{ or } x > -1\}$ in set-builder notation; the solution is $(-\infty,\ -5] \cup (-1,\ \infty)$ in interval notation.

91.
$$\dfrac{1}{C_1}+\dfrac{1}{C_2}=\dfrac{1}{C}$$
$$CC_1C_2\left(\dfrac{1}{C_1}+\dfrac{1}{C_2}\right)=CC_1C_2\left(\dfrac{1}{C}\right)$$
$$CC_2+CC_1=C_1C_2$$
$$C(C_2+C_1)=C_1C_2$$
$$C=\dfrac{C_1C_2}{C_2+C_1}$$

92.
$$\dfrac{P_1V_1}{T_1}=\dfrac{P_2V_2}{T_2}$$
$$T_1T_2\left(\dfrac{P_1V_1}{T_1}\right)=T_1T_2\left(\dfrac{P_2V_2}{T_2}\right)$$
$$T_2P_1V_1=T_1P_2V_2$$
$$T_2=\dfrac{T_1P_2V_2}{P_1V_1}$$

93.
$$T=\dfrac{4\pi^2a^2}{MG}$$
$$MG\cdot T=MG\left(\dfrac{4\pi^2a^2}{MG}\right)$$
$$MGT=4\pi^2a^2$$
$$G=\dfrac{4\pi^2a^2}{MT}$$

94.
$$z=\dfrac{x-\mu}{\sigma}$$
$$z\cdot\sigma=\left(\dfrac{x-\mu}{\sigma}\right)\sigma$$
$$z\cdot\sigma=x-\mu$$
$$z\cdot\sigma+\mu=x$$

95.
$$\dfrac{AB}{AC}=\dfrac{DE}{DF}$$
$$\dfrac{3x-6}{16}=\dfrac{15}{x}$$
$$16x\left(\dfrac{3x-6}{16}\right)=16x\left(\dfrac{15}{x}\right)$$
$$x(3x-6)=16\cdot 15$$
$$3x^2-6x=240$$
$$3x^2-6x-240=0$$
$$3(x^2-2x-80)=0$$
$$3(x+8)(x-10)=0$$
$$x+8=0 \quad \text{or} \quad x-10=0$$
$$x=-8 \quad \text{or} \quad x=10$$

Since the length of DF cannot be negative, we discard $x=-8$ and keep $x=10$. The length of $AB$ is $3x-6=3\cdot 10-6=30-6=24$.

In summary, $AB=24$ and $DF=10$.

96. Let $x$ represent the height of the tree.
$$\dfrac{\text{tree's height}}{\text{tree's shadow length}}=\dfrac{\text{post's height}}{\text{post's shadow length}}$$
$$\dfrac{x}{30}=\dfrac{5}{8}$$
$$120\left(\dfrac{x}{30}\right)=120\left(\dfrac{5}{8}\right)$$
$$4x=75$$
$$x=\dfrac{75}{4}=18.75$$

The height of the pine tree is 18.75 feet.

97. Let $x$ represent the number of grams of total carbohydrates in the 3-cup bowl.
$$\dfrac{\tfrac{3}{4}}{26}=\dfrac{3}{x}$$
$$26x\left(\dfrac{\tfrac{3}{4}}{26}\right)=26x\left(\dfrac{3}{x}\right)$$
$$\dfrac{3}{4}x=78$$
$$\dfrac{4}{3}\left(\dfrac{3}{4}x\right)=\dfrac{4}{3}(78)$$
$$x=104$$

Three bowls of Honey Nut Chex® will contain 104 grams of total carbohydrates.

## Chapter 6: Rational Expressions and Rational Functions

**98.** Let $x$ represent the amount Jeri will earn working an 8-hour day.

$$\frac{48.75}{5} = \frac{x}{8}$$

$$40\left(\frac{48.75}{5}\right) = 40\left(\frac{x}{8}\right)$$

$$8(48.75) = 5x$$

$$390 = 5x$$

$$78 = x$$

Jeri will earn $78.00 for working an 8-hour day.

**99.** Let $t$ represent the time required to fill the tank if both pipes are used.
NOTE: 1 hour and 12 minutes = 72 minutes

$$\begin{pmatrix}\text{Part filled by} \\ \text{first pipe} \\ \text{in 1 minute}\end{pmatrix} + \begin{pmatrix}\text{Part filled by} \\ \text{second pipe} \\ \text{in 1 minute}\end{pmatrix} = \begin{pmatrix}\text{Part filled} \\ \text{together} \\ \text{in 1 minute}\end{pmatrix}$$

$$\frac{1}{48} + \frac{1}{72} = \frac{1}{t}$$

$$144t\left(\frac{1}{48} + \frac{1}{72}\right) = 144t\left(\frac{1}{t}\right)$$

$$3t + 2t = 144$$

$$5t = 144$$

$$t = \frac{144}{5} = 28.8$$

If both pipes are used, it will take 28.8 minutes (or 28 minutes and 48 seconds) to fill the tank.

**100.** Let $t$ represent the time required for Craig to mow the lawn when working alone.

NOTE: 1 hour and 10 minutes = $\frac{7}{6}$ hours

$$\begin{pmatrix}\text{Part done} \\ \text{by Diane} \\ \text{in 1 hour}\end{pmatrix} + \begin{pmatrix}\text{Part done} \\ \text{by Craig} \\ \text{in 1 hour}\end{pmatrix} = \begin{pmatrix}\text{Part done} \\ \text{together} \\ \text{in 1 hour}\end{pmatrix}$$

$$\frac{1}{2} + \frac{1}{t} = \frac{1}{\frac{7}{6}}$$

$$\frac{1}{2} + \frac{1}{t} = \frac{6}{7}$$

$$14t\left(\frac{1}{2} + \frac{1}{t}\right) = 14t\left(\frac{6}{7}\right)$$

$$7t + 14 = 12t$$

$$14 = 5t$$

$$t = \frac{14}{5} = 2.8$$

Working alone, it will take Craig 2.8 hours (or 2 hours and 48 minutes) to mow the lawn.

**101.** Let $t$ represent the time required for John to carpet the room when working alone.
Then $t - 7$ will represent the time required for Rick to carpet the room when working alone.

$$\begin{pmatrix}\text{Part done} \\ \text{by John} \\ \text{in 1 hour}\end{pmatrix} + \begin{pmatrix}\text{Part done} \\ \text{by Rick} \\ \text{in 1 hour}\end{pmatrix} = \begin{pmatrix}\text{Part done} \\ \text{together} \\ \text{in 1 hour}\end{pmatrix}$$

$$\frac{1}{t} + \frac{1}{t-7} = \frac{1}{12}$$

$$12t(t-7)\left(\frac{1}{t} + \frac{1}{t-7}\right) = 12t(t-7)\left(\frac{1}{12}\right)$$

$$12(t-7) + 12t = t(t-7)$$

$$12t - 84 + 12t = t^2 - 7t$$

$$24t - 84 = t^2 - 7t$$

$$0 = t^2 - 31t + 84$$

$$0 = (t-3)(t-28)$$

$$t - 3 = 0 \quad \text{or} \quad t - 28 = 0$$

$$t = 3 \quad \text{or} \quad t = 28$$

If John's time is $t = 3$, then Rick's time will be $t - 7 = 3 - 7 = -4$. Since time cannot be negative, we discard $t = 3$ and keep $t = 28$.

Working alone, it would take John 28 hours to carpet the room. It would take Rick $28 - 7 = 21$ hours to carpet the room alone.

**102.** Let $t$ represent the time required for the drain to empty a full sink.
NOTE: 1 minute 30 seconds = 1.5 minutes.

$$\begin{pmatrix}\text{Part filled in} \\ \text{1 minute if} \\ \text{the drain is} \\ \text{plugged}\end{pmatrix} - \begin{pmatrix}\text{Part emptied} \\ \text{in 1 minute} \\ \text{if the drain} \\ \text{is unplugged}\end{pmatrix} = \begin{pmatrix}\text{Part filled in} \\ \text{1 minute if} \\ \text{the drain is} \\ \text{unplugged}\end{pmatrix}$$

$$\frac{1}{1} - \frac{1}{t} = \frac{1}{1.5}$$

$$3t\left(1 - \frac{1}{t}\right) = 3t\left(\frac{1}{1.5}\right)$$

$$3t - 3 = 2t$$

$$t = 3$$

It will take 3 minutes for the drain to empty a full sink.

103. Let $x$ represent the speed of the wind.

    NOTE: 1 hour and 15 minutes = $\frac{5}{4}$ hours.

    |  | Distance (miles) | Rate (mph) | Time (hours) |
    |---|---|---|---|
    | With the wind | 100 | $180 + x$ | $\frac{100}{180+x}$ |
    | Against the wind | 100 | $180 - x$ | $\frac{100}{180-x}$ |
    | Total | 200 |  | $\frac{5}{4}$ |

    $$\frac{100}{180+x} + \frac{100}{180-x} = \frac{5}{4}$$
    $$4(180+x)(180-x)\left(\frac{100}{180+x} + \frac{100}{180-x}\right) =$$
    $$4(180+x)(180-x)\left(\frac{5}{4}\right)$$
    $$400(180-x) + 400(180+x) = 5(180+x)(180-x)$$
    $$72{,}000 - 400x + 72{,}000 + 400x = 162{,}000 - 5x^2$$
    $$144{,}000 = 162{,}000 - 5x^2$$
    $$\frac{144{,}000}{5} = \frac{162{,}000 - 5x^2}{5}$$
    $$28{,}800 = 32{,}400 - x^2$$
    $$x^2 - 3600 = 0$$
    $$(x-60)(x+60) = 0$$
    $$x - 60 = 0 \quad \text{or} \quad x + 60 = 0$$
    $$x = 60 \quad \text{or} \quad x = -60$$

    Because the rate cannot be negative, we discard $x = -60$ and keep $x = 60$. The speed of the wind was 60 miles per hour.

104. Let $x$ represent the speed of Jesse's boat in still water. Then $x + 5$ will represent the speed of his boat downstream and $x - 5$ will be the speed of his boat up stream.

    |  | Distance (miles) | Rate (mph) | Time (hours) |
    |---|---|---|---|
    | Down | 20 | $x + 5$ | $\frac{20}{x+5}$ |
    | Up | 10 | $x - 5$ | $\frac{10}{x-5}$ |

    $$\frac{20}{x+5} = \frac{10}{x-5}$$
    $$(x+5)(x-5)\left(\frac{20}{x+5}\right) = (x+5)(x-5)\left(\frac{10}{x-5}\right)$$
    $$20(x-5) = 10(x+5)$$
    $$20x - 100 = 10x + 50$$
    $$10x = 150$$
    $$x = 15$$

    Jesse's boat can travel 15 miles per hour in still water.

105. Let $x$ represent the average speed at which Todd walks. Then $4x$ will represent the average speed at which he runs.

    NOTE: 35 minutes = $\frac{7}{12}$ hours.

    |  | Distance (miles) | Rate (mph) | Time (hours) |
    |---|---|---|---|
    | Walking | 1 | $x$ | $\frac{1}{x}$ |
    | Running | 3 | $4x$ | $\frac{3}{4x}$ |

    $$\frac{1}{x} + \frac{3}{4x} = \frac{7}{12}$$
    $$12x\left(\frac{1}{x} + \frac{3}{4x}\right) = 12x\left(\frac{7}{12}\right)$$
    $$12 + 9 = 7x$$
    $$21 = 7x$$
    $$3 = x$$

    Todd's average walking speed is 3 miles per hour. His average running speed is $4(3) = 12$ miles per hour.

106. Let $x$ represent the average speed for the last 80 miles of the trip.

    |  | Distance (miles) | Rate (mph) | Time (hours) |
    |---|---|---|---|
    | First 20 miles | 20 | 30 | $\frac{20}{30} = \frac{2}{3}$ |
    | Last 80 miles | 80 | $x$ | $\frac{80}{x}$ |
    | Whole trip | 100 | 50 | $\frac{100}{50} = 2$ |

## Chapter 6: Rational Expressions and Rational Functions

$$\frac{2}{3} + \frac{80}{x} = 2$$

$$3x\left(\frac{2}{3} + \frac{80}{x}\right) = 3x \cdot 2$$

$$2x + 240 = 6x$$

$$240 = 4x$$

$$60 = x$$

Danielle's average speed for the last 80 miles of her trip was 60 miles per hour.

**107.**
a. $y = \dfrac{k}{x}$

$15 = \dfrac{k}{4}$

$60 = k$

b. $y = \dfrac{60}{x}$

c. If $x = 5$, then $y = \dfrac{60}{5} = 12$.

**108.**
a. $y = kxz$

$45 = k \cdot 6 \cdot 10$

$45 = 60k$

$k = \dfrac{45}{60} = \dfrac{3}{4}$

b. $y = \dfrac{3}{4}xz$

c. If $x = 8$ and $z = 7$, then $y = \dfrac{3}{4} \cdot 8 \cdot 7 = 42$.

**109.**
a. $s = \dfrac{k}{t^2}$

$18 = \dfrac{k}{2^2}$

$18 = \dfrac{k}{4}$

$72 = k$

b. $s = \dfrac{72}{t^2}$

c. If $t = 3$, then $s = \dfrac{72}{3^2} = \dfrac{72}{9} = 8$.

**110.**
a. $w = k \cdot \dfrac{x}{z}$

$\dfrac{4}{3} = k \cdot \dfrac{10}{12}$

$k = \dfrac{4}{3} \cdot \dfrac{12}{10} = \dfrac{8}{5}$

b. $w = \dfrac{8}{5} \cdot \dfrac{x}{z}$ or $w = \dfrac{8x}{5z}$

c. If $x = 9$ and $z = 16$, then $w = \dfrac{8 \cdot 9}{5 \cdot 16} = \dfrac{9}{10}$.

**111.** Let $f$ represent the frequency of a radio signal with wavelength $l$.

$$f = \dfrac{k}{l}$$

$$800 = \dfrac{k}{375}$$

$$375 \cdot 800 = 375\left(\dfrac{k}{375}\right)$$

$$300{,}000 = k$$

$$f(l) = \dfrac{300{,}000}{l}$$

$$f(250) = \dfrac{300{,}000}{250} = 1200$$

When the wavelength is 250 meters, the frequency is 1200 kilohertz.

**112.** Let $C$ represent the electrical current when the resistance is $R$.

$$C = \dfrac{k}{R}$$

$$8 = \dfrac{k}{15}$$

$$15 \cdot 8 = 15\left(\dfrac{k}{15}\right)$$

$$120 = k$$

$$C(R) = \dfrac{120}{R}$$

$$C(10) = \dfrac{120}{10} = 12$$

If the electrical current is 10 amperes, then the resistance is 12 ohms.

**113.** $V = khd^2$

$231 = k \cdot 6 \cdot 7^2$

$231 = 294k$

$k = \dfrac{231}{294} = \dfrac{11}{14}$

$V = \dfrac{11}{14}hd^2$

For $h = 14$ and $d = 8$, $V = \dfrac{11}{14}(14)(8)^2 = 704$.

If the diameter is 8 centimeters and the height is 14 centimeters, then the volume of the cylinder is 704 cubic centimeters.

**114.**
$$V = kBh$$
$$270 = k(81)(10)$$
$$270 = 810k$$
$$k = \frac{270}{810} = \frac{1}{3}$$
$$V = \frac{1}{3}Bh$$

For $B = 125$ and $h = 9$, $V = \frac{1}{3}(125)(9) = 375$.

When the area of the base is 125 square inches and the height is 9 inches, the volume of the pyramid is 375 cubic inches.

## Chapter 6 Test

1. We need to find all values of $x$ that cause the denominator $2x^2 - 13x - 7$ to equal 0.
$$2x^2 - 13x - 7 = 0$$
$$(2x+1)(x-7) = 0$$
$$2x + 1 = 0 \quad \text{or} \quad x - 7 = 0$$
$$2x = -1 \quad \text{or} \quad x = 7$$
$$x = -\frac{1}{2}$$

Thus, the domain of $f(x) = \dfrac{2x+1}{2x^2 - 13x - 7}$ is $\left\{ x \,\middle|\, x \ne -\dfrac{1}{2},\ x \ne 7 \right\}$.

2. $\dfrac{2m^2 + 5m - 12}{3m^2 + 11m - 4} = \dfrac{(2m-3)(m+4)}{(3m-1)(m+4)}$
$$= \dfrac{(2m-3)\cancel{(m+4)}}{(3m-1)\cancel{(m+4)}}$$
$$= \dfrac{2m-3}{3m-1}$$

3. $\dfrac{2b - 3a}{3a^2 + 10ab - 8b^2} = \dfrac{-1(3a-2b)}{(3a-2b)(a+4b)}$
$$= \dfrac{-1\cancel{(3a-2b)}}{\cancel{(3a-2b)}(a+4b)}$$
$$= \dfrac{-1}{a+4b}$$

4. $\dfrac{4x^2 - 12x}{x^2 - 9} \cdot \dfrac{2x^2 + 11x + 15}{8x^3 - 32x^2}$
$$= \dfrac{4x(x-3)}{(x+3)(x-3)} \cdot \dfrac{(2x+5)(x+3)}{8x^2(x-4)}$$
$$= \dfrac{\cancel{4x}\cancel{(x-3)}}{\cancel{(x+3)}\cancel{(x-3)}} \cdot \dfrac{(2x+5)\cancel{(x+3)}}{2x \cdot \cancel{4x}(x-4)}$$
$$= \dfrac{2x+5}{2x(x-4)}$$

5. $\dfrac{\dfrac{y^2 + 2y - 8}{4y^2 - 5y - 6}}{\dfrac{3y^2 - 14y - 5}{4y^2 - 17y - 15}}$
$$= \dfrac{y^2 + 2y - 8}{4y^2 - 5y - 6} \cdot \dfrac{4y^2 - 17y - 15}{3y^2 - 14y - 5}$$
$$= \dfrac{(y-2)(y+4)}{(4y+3)(y-2)} \cdot \dfrac{(4y+3)(y-5)}{(3y+1)(y-5)}$$
$$= \dfrac{\cancel{(y-2)}(y+4)}{\cancel{(4y+3)}\cancel{(y-2)}} \cdot \dfrac{\cancel{(4y+3)}\cancel{(y-5)}}{(3y+1)\cancel{(y-5)}}$$
$$= \dfrac{y+4}{3y+1}$$

6. $p^2 - q^2 = (p-q)(p+q)$
$p - q$
$\text{LCD} = (p-q)(p+q)$
$$\dfrac{3p^2 + 3pq}{p^2 - q^2} - \dfrac{3p - 2q}{p - q}$$
$$= \dfrac{3p^2 + 3pq}{(p-q)(p+q)} - \dfrac{3p - 2q}{p - q} \cdot \dfrac{p+q}{p+q}$$
$$= \dfrac{3p^2 + 3pq}{(p-q)(p+q)} - \dfrac{3p^2 + pq - 2q^2}{(p-q)(p+q)}$$
$$= \dfrac{3p^2 + 3pq - (3p^2 + pq - 2q^2)}{(p-q)(p+q)}$$
$$= \dfrac{3p^2 + 3pq - 3p^2 - pq + 2q^2}{(p-q)(p+q)}$$
$$= \dfrac{2pq + 2q^2}{(p-q)(p+q)}$$
$$= \dfrac{2q(p+q)}{(p-q)(p+q)}$$
$$= \dfrac{2q}{p-q}$$

## Chapter 6: Rational Expressions and Rational Functions

7. $3c^2 - 2c - 8 = (3c+4)(c-2)$
$3c^2 + c - 4 = (3c+4)(c-1)$
LCD $= (3c+4)(c-2)(c-1)$

$$\frac{9c+2}{3c^2-2c-8} + \frac{7}{3c^2+c-4}$$
$$= \frac{9c+2}{(3c+4)(c-2)} \cdot \frac{c-1}{c-1} + \frac{7}{(3c+4)(c-1)} \cdot \frac{c-2}{c-2}$$
$$= \frac{9c^2-7c-2}{(3c+4)(c-2)(c-1)} + \frac{7c-14}{(3c+4)(c-2)(c-1)}$$
$$= \frac{9c^2-7c-2+7c-14}{(3c+4)(c-2)(c-1)}$$
$$= \frac{9c^2-16}{(3c+4)(c-2)(c-1)}$$
$$= \frac{(3c+4)(3c-4)}{(3c+4)(c-2)(c-1)}$$
$$= \frac{3c-4}{(c-2)(c-1)}$$

8. $Q(x) = \frac{f(x)}{h(x)} = \dfrac{\dfrac{3x}{x^2-4}}{\dfrac{9x^2-45x}{x^2-2x-8}}$

$$= \frac{3x}{x^2-4} \cdot \frac{x^2-2x-8}{9x^2-45x}$$
$$= \frac{3x}{(x+2)(x-2)} \cdot \frac{(x-4)(x+2)}{9x(x-5)}$$
$$= \frac{\cancel{3x}}{\cancel{(x+2)}(x-2)} \cdot \frac{(x-4)\cancel{(x+2)}}{3 \cdot \cancel{3x}(x-5)}$$
$$= \frac{x-4}{3(x-2)(x-5)}$$

The domain of $f(x)$ is $\{x \mid x \neq -2, x \neq 2\}$. The domain of $h(x)$ is $\{x \mid x \neq -2, x \neq 4\}$. Because the denominator of $Q(x)$ cannot equal 0, we must exclude those values of $x$ such that the numerator of $h(x)$ is 0. That is, we must exclude the values of $x$ such that $9x^2 - 45x$. These values are $x = 0$ and $x = 5$. Therefore, the domain of $Q(x)$ is $\{x \mid x \neq -2, x \neq 0, x \neq 2, x \neq 4, x \neq 5\}$.

9. $x^2 - 4 = (x+2)(x-2)$
$x^2 + 2x = x(x+2)$
LCD $= x(x+2)(x-2)$

$S(x) = f(x) + g(x)$
$$= \frac{3x}{x^2-4} + \frac{6}{x^2+2x}$$
$$= \frac{3x}{(x+2)(x-2)} \cdot \frac{x}{x} + \frac{6}{x(x+2)} \cdot \frac{x-2}{x-2}$$
$$= \frac{3x^2}{x(x+2)(x-2)} + \frac{6x-12}{x(x+2)(x-2)}$$
$$= \frac{3x^2+6x-12}{x(x+2)(x-2)} \text{ or } \frac{3(x^2+2x-4)}{x(x+2)(x-2)}$$

The domain of $f(x)$ is $\{x \mid x \neq -2, x \neq 2\}$. The domain of $g(x)$ is $\{x \mid x \neq -2, x \neq 0\}$. Therefore, the domain of $S(x)$ is $\{x \mid x \neq -2, x \neq 0, x \neq 2\}$.

10. Using Method 2, the LCD of all the denominators in the complex rational expression is $a^2$.

$$\frac{1-\dfrac{1}{a}}{1-\dfrac{1}{a^2}} = \frac{1-\dfrac{1}{a}}{1-\dfrac{1}{a^2}} \cdot \frac{a^2}{a^2} = \frac{1 \cdot a^2 - \dfrac{1}{a} \cdot a^2}{1 \cdot a^2 - \dfrac{1}{a^2} \cdot a^2}$$
$$= \frac{a^2-a}{a^2-1} = \frac{a(a-1)}{(a+1)(a-1)} = \frac{a}{a+1}$$

11. Using Method 2, the LCD of all the denominators in the complex rational expression is $(d+2)(d-2)$.

$$\frac{\dfrac{5}{d+2}-\dfrac{1}{d-2}}{\dfrac{3}{d+2}-\dfrac{6}{d-2}} = \frac{\dfrac{5}{d+2}-\dfrac{1}{d-2}}{\dfrac{3}{d+2}-\dfrac{6}{d-2}} \cdot \frac{(d+2)(d-2)}{(d+2)(d-2)}$$
$$= \frac{\dfrac{5}{d+2}\cdot(d+2)(d-2) - \dfrac{1}{d-2}\cdot(d+2)(d-2)}{\dfrac{3}{d+2}\cdot(d+2)(d-2) - \dfrac{6}{d-2}\cdot(d+2)(d-2)}$$
$$= \frac{5(d-2)-(d+2)}{3(d-2)-6(d+2)}$$
$$= \frac{5d-10-d-2}{3d-6-6d-12} = \frac{4d-12}{-3d-18} \text{ or } \frac{4(d-3)}{-3(d+6)}$$

12. 
$$\frac{1}{6x} - \frac{1}{3} = \frac{5}{4x} + \frac{3}{4}$$
$$12x\left(\frac{1}{6x} - \frac{1}{3}\right) = 12x\left(\frac{5}{4x} + \frac{3}{4}\right)$$
$$12x\left(\frac{1}{6x}\right) - 12x\left(\frac{1}{3}\right) = 12x\left(\frac{5}{4x}\right) + 12x\left(\frac{3}{4}\right)$$
$$2 - 4x = 15 + 9x$$
$$-13x = 13$$
$$x = -1$$

Check:
$$\frac{1}{6(-1)} - \frac{1}{3} \stackrel{?}{=} \frac{5}{4(-1)} + \frac{3}{4}$$
$$\frac{1}{-6} - \frac{1}{3} \stackrel{?}{=} \frac{5}{-4} + \frac{3}{4}$$
$$-\frac{1}{6} - \frac{2}{6} \stackrel{?}{=} -\frac{5}{4} + \frac{3}{4}$$
$$\frac{-3}{6} \stackrel{?}{=} \frac{-2}{4}$$
$$-\frac{1}{2} = -\frac{1}{2} \quad \leftarrow \text{True}$$

The solution checks, so the solution set is $\{-1\}$.

13.
$$\frac{7n}{n+3} + \frac{21}{n-3} = \frac{126}{n^2 - 9}$$
$$\frac{7n}{n+3} + \frac{21}{n-3} = \frac{126}{(n+3)(n-3)}$$
$$(n+3)(n-3)\left(\frac{7n}{n+3} + \frac{21}{n-3}\right) =$$
$$(n+3)(n-3)\left(\frac{126}{(n+3)(n-3)}\right)$$
$$7n(n-3) + 21(n+3) = 126$$
$$7n^2 - 21n + 21n + 63 = 126$$
$$7n^2 + 63 = 126$$
$$7n^2 - 63 = 0$$
$$\frac{7n^2 - 63}{7} = \frac{0}{7}$$
$$n^2 - 9 = 0$$
$$(n+3)(n-3) = 0$$
$$n+3 = 0 \quad \text{or} \quad n-3 = 0$$
$$n = -3 \quad \text{or} \quad n = 3$$

Since $n = -3$ and $n = 3$ are neither in the domain of the variable, they are both extraneous solutions. The solution set is $\varnothing$ or $\{\ \}$.

14.
$$\frac{x+5}{x-2} \geq 3$$
$$\frac{x+5}{x-2} - 3 \geq 0$$
$$\frac{x+5}{x-2} - \frac{3(x-2)}{x-2} \geq 0$$
$$\frac{x+5-3x+6}{x-2} \geq 0$$
$$\frac{-2x+11}{x-2} \geq 0$$

The rational expression will equal 0 when $x = \frac{11}{2}$.
It is undefined when $x = 2$.
Determine where the numerator and denominator are positive and negative and where the quotient is positive and negative.

| Interval | $(-\infty, 2)$ | $2$ | $\left(2, \frac{11}{2}\right)$ | $\frac{11}{2}$ | $\left(\frac{11}{2}, \infty\right)$ |
|---|---|---|---|---|---|
| $-2x+11$ | $++++$ | $+$ | $++++$ | $0$ | $----$ |
| $x-2$ | $----$ | $0$ | $++++$ | $+$ | $++++$ |
| $\frac{-2x+11}{x-2}$ | $----$ | $\varnothing$ | $++++$ | $0$ | $----$ |

The rational function is undefined at $x = 2$, so 2 is not part of the solution. The inequality is non-strict, so $\frac{11}{2}$ is part of the solution. Now, $\frac{-2x+11}{x-2}$ is greater than zero where the quotient is positive. The solution is $\left\{x \mid 2 < x \leq \frac{11}{2}\right\}$ in set-builder notation; the solution is $\left(2, \frac{11}{2}\right]$ in interval notation.

15.
$$\frac{1}{F} = \frac{D^2}{kq_1q_2}$$
$$Fkq_1q_2\left(\frac{1}{F}\right) = Fkq_1q_2\left(\frac{D^2}{kq_1q_2}\right)$$
$$kq_1q_2 = FD^2$$
$$k = \frac{FD^2}{q_1q_2}$$

16. Let $t$ represent the time to print out the 48-page document.

$$\frac{\text{Number of pages}}{\text{time to print document}} = \frac{\text{Number of pages}}{\text{time to print document}}$$

$$\frac{10}{25} = \frac{48}{t}$$

$$25t\left(\frac{10}{25}\right) = 25t\left(\frac{48}{t}\right)$$

$$10t = 1200$$

$$t = 120$$

The printer will take 120 seconds (or 2 minutes) to print out the 48-page document.

17. Let $t$ represent the time required if Linnette and Darrell work together to clean their house.

$$\begin{pmatrix}\text{Part done}\\ \text{by Linnette}\\ \text{in 1 hour}\end{pmatrix} + \begin{pmatrix}\text{Part done}\\ \text{by Darrell}\\ \text{in 1 hour}\end{pmatrix} = \begin{pmatrix}\text{Part done}\\ \text{together}\\ \text{in 1 hour}\end{pmatrix}$$

$$\frac{1}{4} + \frac{1}{6} = \frac{1}{t}$$

$$24t\left(\frac{1}{4}+\frac{1}{6}\right) = 24t\left(\frac{1}{t}\right)$$

$$6t + 4t = 24$$

$$10t = 24$$

$$t = \frac{24}{10} = 2.4$$

Working together, it will take Linnette and Darrell 2.4 hours (or 2 hours and 24 minutes) to clean their house.

18. Let $x$ represent the speed of the current.

|  | Distance (miles) | Rate (mph) | Time (hours) |
|---|---|---|---|
| Up | 4 | $7-x$ | $\frac{4}{7-x}$ |
| Down | 10 | $7+x$ | $\frac{10}{7+x}$ |

$$\frac{4}{7-x} = \frac{10}{7+x}$$

$$(7-x)(7+x)\left(\frac{4}{7-x}\right) = (7-x)(7+x)\left(\frac{10}{7+x}\right)$$

$$4(7+x) = 10(7-x)$$

$$28 + 4x = 70 - 10x$$

$$14x = 42$$

$$x = 3$$

The rate of the current is 3 miles per hour.

19. $F = \dfrac{k}{l}$

$50 = \dfrac{k}{4}$

$200 = k$

$F = \dfrac{200}{l}$

If $l = 10$, then $F = \dfrac{200}{10} = 20$.

If the length of the length of the force arm of the lever is 10 feet, the force required to lift the boulder will be 20 pounds.

20. $L = krh$

$528 = k(7)(12)$

$528 = 84k$

$k = \dfrac{528}{84} = \dfrac{44}{7}$

$L = \dfrac{44}{7}rh$

For $r = 9$ and $h = 14$, $L = \dfrac{44}{7}(9)(14) = 792$.

When the radius is 9 centimeters and the height is 14 centimeters, the lateral surface area of the right circular cylinder is 792 square centimeters.

# Getting Ready for Chapter 7

2. principal square root

4. True

6. False; the square root of a negative number is not a real number.

8. Answers may vary. When taking a square root, we are looking for a number that we can multiply by itself to obtain the radicand. Since $a \cdot a = a^2$, it should follow that $\sqrt{a^2} = a$. However, since the product of two negative numbers is positive, it is impossible to determine from $a^2$ whether $a$ was positive or negative. We define $\sqrt{a^2} = |a|$ to ensure that the principle square root is positive even if $a < 0$.
For example:
$2^2 = 4$ and $\sqrt{4} = 2$
$(-2)^2 = 4$ and $\sqrt{4} = 2$ not $-2$ since
$\sqrt{(-2)^2} = |-2| = 2$.

10. $\sqrt{9} = \sqrt{3^2} = 3$

12. $-\sqrt{144} = -\sqrt{12^2} = -12$

14. $\sqrt{\frac{4}{81}} = \sqrt{\left(\frac{2}{9}\right)^2} = \frac{2}{9}$

16. $\sqrt{0.16} = \sqrt{(0.4)^2} = 0.4$

18. $\left(\sqrt{3.7}\right)^2 = 3.7$

20. $\sqrt{-50}$ is not a real number because the radicand of the square root is negative.

22. $\sqrt{121}$ is rational because 121 is a perfect square.
$\sqrt{121} = \sqrt{(11)^2} = 11$

24. $\sqrt{\frac{49}{100}}$ is rational because $\frac{49}{100}$ is a perfect square.
$\sqrt{\frac{49}{100}} = \sqrt{\left(\frac{7}{10}\right)^2} = \frac{7}{10}$

26. $\sqrt{24}$ is irrational because 24 is not a perfect square.
$\sqrt{24} \approx 4.90$

28. $\sqrt{12}$ is irrational because 12 is not a perfect square.
$\sqrt{12} \approx 3.46$

30. $\sqrt{-64}$ is not a real number because the radicand of the square root is negative.

32. $\sqrt{5^2} = 5$

34. $\sqrt{(-13)^2} = |-13| = 13$

36. $\sqrt{w^2} = |w|$

38. $\sqrt{(x-8)^2} = |x-8|$

40. $\sqrt{(5x+2)^2} = |5x+2|$

42. $\sqrt{9z^2 - 24z + 16} = \sqrt{(3z-4)^2} = |3z-4|$

44. $\sqrt{9+16} = \sqrt{25} = \sqrt{5^2} = 5$

46. $\sqrt{9} + \sqrt{16} = \sqrt{3^2} + \sqrt{4^2} = 3 + 4 = 7$

48. $\sqrt{-36}$ is not a real number.

50. $-10\sqrt{16} = -10\sqrt{4^2} = -10 \cdot 4 = -40$

52. $2\sqrt{\frac{9}{4}} - \sqrt{4} = 2\sqrt{\left(\frac{3}{2}\right)^2} - \sqrt{2^2}$
$= 2 \cdot \frac{3}{2} - 2 = 3 - 2 = 1$

# Getting Ready for Chapter 7

**54.** $\sqrt{9^2 - 4 \cdot 1 \cdot 20} = \sqrt{81 - 80} = \sqrt{1} = 1$

**56.** $\sqrt{(-3)^2 - 4 \cdot 3 \cdot 2} = \sqrt{9 - 24} = \sqrt{-15}$ is not a real number.

**58.** 
$$\frac{-7 + \sqrt{7^2 - 4 \cdot 2 \cdot 6}}{2 \cdot 2} = \frac{-7 + \sqrt{49 - 48}}{4}$$
$$= \frac{-7 + \sqrt{1}}{4}$$
$$= \frac{-7 + 1}{4}$$
$$= \frac{-6}{4}$$
$$= -\frac{3}{2}$$

**60.** 
$$\sqrt{(2-(-1))^2 + (6-2)^2} = \sqrt{(3)^2 + (4)^2}$$
$$= \sqrt{9 + 16}$$
$$= \sqrt{25}$$
$$= \sqrt{5^2}$$
$$= 5$$

**62.** The square roots of 64 are 8 and −8 because $(8)^2 = 64$ and $(-8)^2 = 64$.
$\sqrt{64} = 8$ (the principal square root).

**64.** 
$$Z = \frac{X - \mu}{\frac{\sigma}{\sqrt{n}}} = \frac{40 - 50}{\frac{10}{\sqrt{5}}}$$
$$= \frac{-10}{\frac{10}{\sqrt{5}}} = (-10) \cdot \frac{\sqrt{5}}{10}$$
$$= -\sqrt{5} \approx -2.24$$

# Chapter 7

## 7.1 Preparing for *n*th Root and Rational Exponents

1. $\left(\dfrac{x^2 y}{xy^{-2}}\right)^{-3} = \left(x^{2-1} y^{1-(-2)}\right)^{-3}$
$= \left(xy^3\right)^{-3}$
$= \left(\dfrac{1}{xy^3}\right)^3$
$= \dfrac{1}{x^3 y^9}$

2. $\left(\sqrt{7}\right)^2 = 7$

3. $\sqrt{64} = 8$ since $8^2 = 64$.

## 7.1 Exercises

2. $a^{-\frac{m}{n}} = \dfrac{1}{a^{\frac{m}{n}}} = \dfrac{1}{\sqrt[n]{a^m}}$ or $\dfrac{1}{\left(\sqrt[n]{a}\right)^m}$
provided $\sqrt[n]{a}$ exists.

4. True, provided $\sqrt[n]{a}$ exists.

6. Answers will vary.
$(-9)^{1/2}$ and $-9^{1/2}$ are different because of order of operations. This is easier seen in a different form:
$(-9)^{1/2} = \sqrt{-9}$ (radicand is negative, so the result is not a real number)
$-9^{1/2} = -\left(9^{1/2}\right) = -\sqrt{9} = -3$ (radicand is positive, so the result is a real number)

8. If $\dfrac{m}{n}$, in lowest terms, is positive, then $a^{\frac{m}{n}}$ is a real number provided $\sqrt[n]{a}$ is a real number.
If $\dfrac{m}{n}$, in lowest terms, is negative, then $a^{\frac{m}{n}}$ is a real number provided $a \neq 0$ and $\sqrt[n]{a}$ is a real number.

10. $\sqrt[3]{216} = \sqrt[3]{6^3} = 6$

12. $\sqrt[3]{-64} = \sqrt[3]{(-4)^3} = -4$

14. $-\sqrt[4]{256} = -\sqrt[4]{(4)^4} = -4$

16. $\sqrt[3]{\dfrac{8}{125}} = \sqrt[3]{\left(\dfrac{2}{5}\right)^3} = \dfrac{2}{5}$

18. $-\sqrt[5]{-1024} = -\sqrt[5]{(-4)^5} = -(-4) = 4$

20. $\sqrt[4]{6^4} = 6$

22. $\sqrt[5]{n^5} = n$

24. $\sqrt[6]{(2x-3)^6} = |2x-3|$

26. $-\sqrt[3]{(6z-5)^3} = -(6z-5) = -6z+5$

28. $16^{1/2} = \sqrt{16} = \sqrt{4^2} = 4$

30. $-25^{1/2} = -\sqrt{25} = -\sqrt{5^2} = -5$

32. $27^{1/3} = \sqrt[3]{27} = \sqrt[3]{3^3} = 3$

34. $-81^{1/4} = -\sqrt[4]{81} = -\sqrt[4]{3^4} = -3$

36. $\left(\dfrac{8}{27}\right)^{1/3} = \sqrt[3]{\dfrac{8}{27}} = \sqrt[3]{\left(\dfrac{2}{3}\right)^3} = \dfrac{2}{3}$

38. $(-216)^{1/3} = \sqrt[3]{-216} = \sqrt[3]{(-6)^3} = -6$

40. $(-81)^{1/2} = \sqrt{-81}$ is not a real number.

42. $25^{3/2} = \left(\sqrt{25}\right)^3 = (5)^3 = 125$

44. $-100^{5/2} = -\left(\sqrt{100}\right)^5 = -10^5 = -100,000$

46. $27^{4/3} = \left(\sqrt[3]{27}\right)^4 = 3^4 = 81$

## Chapter 7: Radicals and Rational Exponents

**48.** $(-125)^{2/3} = \left(\sqrt[3]{-125}\right)^2 = (-5)^2 = 25$

**50.** $-(-216)^{2/3} = -\left(\sqrt[3]{-216}\right)^2 = -(-6)^2 = -36$

**52.** $121^{-1/2} = \dfrac{1}{121^{1/2}} = \dfrac{1}{\sqrt{121}} = \dfrac{1}{11}$

**54.** $\dfrac{1}{49^{-3/2}} = 49^{3/2} = \left(\sqrt{49}\right)^3 = 7^3 = 343$

**56.** $27^{-4/3} = \dfrac{1}{27^{4/3}} = \dfrac{1}{\left(\sqrt[3]{27}\right)^4} = \dfrac{1}{3^4} = \dfrac{1}{81}$

**58.** $\sqrt[5]{2y} = (2y)^{1/5}$

**60.** $\sqrt{\dfrac{w}{2}} = \left(\dfrac{w}{2}\right)^{1/2}$

**62.** $\sqrt[3]{p^5} = p^{5/3}$

**64.** $\left(\sqrt[4]{6z}\right)^3 = (6z)^{3/4}$

**66.** $\sqrt[6]{\left(\dfrac{2a}{b}\right)^5} = \left(\dfrac{2a}{b}\right)^{5/6}$

**68.** $\sqrt[4]{(3pq)^7} = (3pq)^{7/4}$

**70.** $\sqrt[3]{85} = 85\wedge(1/3) \approx 4.40$

**72.** $\sqrt[4]{2} = 2\wedge(1/4) \approx 1.19$

**76.** $100^{3/4} = 100\wedge(3/4) \approx 31.62$

**78.** $100^{0.25} = 100\wedge(0.25) \approx 3.16$

**80.** $\sqrt[3]{-125} = \sqrt[3]{(-5)^3} = -5$

**82.** $100^{3/2} = \left(\sqrt{100}\right)^3 = 10^3 = 1000$

**84.** $\sqrt[4]{-1}$ is not a real number because the index is even and the radicand is negative.

**86.** $125^{-1/3} = \dfrac{1}{125^{1/3}} = \dfrac{1}{\sqrt[3]{125}} = \dfrac{1}{5}$

**88.** $\sqrt[4]{0.0081} = \sqrt[4]{(0.3)^4} = 0.3$

**90.** $100^{1/2} - 4^{3/2} = \sqrt{100} - \left(\sqrt{4}\right)^3 = 10 - 2^3$
$= 10 - 8 = 2$

**92.** $(-125)^{-1/3} = \dfrac{1}{(-125)^{1/3}} = \dfrac{1}{\sqrt[3]{-125}} = \dfrac{1}{-5} = -\dfrac{1}{5}$

**94.** 9 is the only cube root of 729.
$\sqrt[3]{729} = \sqrt[3]{9^3} = 9$

**96. a.** $P = 100, A = 144, t = 2$
$r = \left(\dfrac{144}{100}\right)^{1/2} - 1 = \sqrt{\dfrac{144}{100}} - 1 = \dfrac{12}{10} - 1$
$= \dfrac{12}{10} - \dfrac{10}{10} = \dfrac{2}{10} = \dfrac{1}{5} = 0.2$
The annual interest rate is 20%.

**b.** $P = 100, A = 337.5, t = 3$
$r = \left(\dfrac{337.5}{100}\right)^{1/3} - 1 = (3.375)^{1/3} - 1$
$= \sqrt[3]{3.375} - 1 = \sqrt[3]{(1.5)^3} - 1 = 1.5 - 1$
$= 0.50$
The annual interest rate is 50%.

**c.** $P = 1000, A = 2000, t = 8$
$r = \left(\dfrac{2000}{1000}\right)^{1/8} - 1 = 2^{1/8} - 1 \approx 0.0905$
The annual interest rate is about 0.0905 or about 9.05%.
$\dfrac{72}{100(0.0905)} = \dfrac{72}{9.05} \approx 7.96$ years
This is very close to the actual time of 8 years.

**98. a.** $T = 0.241, r = 5.79 \times 10^{10}$
$0.241 = k \cdot \left(5.79 \times 10^{10}\right)^{3/2}$
$k = \dfrac{0.241}{\left(5.79 \times 10^{10}\right)^{3/2}} \approx 1.73 \times 10^{-17}$
$T = \left(1.73 \times 10^{-17}\right) r^{3/2}$

b. $T = \left(1.73 \times 10^{-17}\right)\left(2.28 \times 10^{11}\right)^{3/2} \approx 1.88$

It will take Mars about 1.88 years to complete one orbit around the sun.

100. $g(x) = x^{-3/2}$

$g(16) = 16^{-3/2} = \dfrac{1}{16^{3/2}} = \dfrac{1}{\left(\sqrt{16}\right)^3} = \dfrac{1}{4^3} = \dfrac{1}{64}$

102. $G(a) = a^{5/3}$

$G(-8) = (-8)^{5/3} = \left(\sqrt[3]{-8}\right)^5 = (-2)^5 = -32$

104. $\dfrac{(x+2)^2 (x-1)^4}{(x+2)(x-1)} = (x+2)^{2-1}(x-1)^{4-1}$

$= (x+2)(x-1)^3$

106. $\dfrac{(3a^2 + 5a - 3) - (a^2 - 2a - 9)}{4a^2 + 12a + 9}$

$= \dfrac{3a^2 + 5a - 3 - a^2 + 2a + 9}{4a^2 + 12a + 9} = \dfrac{2a^2 + 7a + 6}{4a^2 + 12a + 9}$

$= \dfrac{(2a+3)(a+2)}{(2a+3)^2} = \dfrac{(a+2)}{(2a+3)^{2-1}}$

$= \dfrac{a+2}{2a+3}$

## 7.2 Preparing for Simplifying Expressions Using the Laws of Exponents

1. $z^{-3} = \dfrac{1}{z^3}$

2. $x^{-2} \cdot x^5 = x^{-2+5} = x^3$

3. $\left(\dfrac{2a^2}{b^{-1}}\right)^3 = \left(2a^2 b\right)^3 = 2^3 \left(a^2\right)^3 b^3 = 8a^6 b^3$

4. $\sqrt{64} = 8$ because $8^2 = 64$.

## 7.2 Exercises

2. $a^{r+s}$

4. $3^{1/3} \cdot 3^{5/3} = 3^{\frac{1}{3} + \frac{5}{3}} = 3^{\frac{6}{3}} = 3^2 = 9$

6. $\dfrac{10^{7/5}}{10^{2/5}} = 10^{\frac{7}{5} - \frac{2}{5}} = 10^{\frac{5}{5}} = 10^1 = 10$

8. $9^{-5/4} \cdot 9^{1/3} = 9^{-\frac{5}{4} + \frac{1}{3}} = 9^{-\frac{15}{12} + \frac{4}{12}} = 9^{-\frac{11}{12}}$

$= \dfrac{1}{9^{11/12}} = \dfrac{1}{\left(3^2\right)^{11/12}} = \dfrac{1}{3^{2 \cdot \frac{11}{12}}} = \dfrac{1}{3^{11/6}}$

10. $\dfrac{y^{1/5}}{y^{9/10}} = y^{\frac{1}{5} - \frac{9}{10}} = y^{\frac{2}{10} - \frac{9}{10}} = y^{-\frac{7}{10}} = \dfrac{1}{y^{7/10}}$

12. $\left(9^{3/5}\right)^{5/6} = 9^{\frac{3}{5} \cdot \frac{5}{6}} = 9^{1/2} = \sqrt{9} = 3$

14. $\left(36^{-1/4} \cdot 9^{3/4}\right)^{-2} = \left(36^{-1/4}\right)^{-2} \cdot \left(9^{3/4}\right)^{-2}$

$= 36^{-\frac{1}{4} \cdot (-2)} \cdot 9^{\frac{3}{4} \cdot (-2)}$

$= 36^{1/2} \cdot 9^{-3/2}$

$= \dfrac{\sqrt{36}}{\left(\sqrt{9}\right)^3} = \dfrac{6}{3^3} = \dfrac{6}{27} = \dfrac{2}{9}$

16. $\left(a^{5/4} \cdot b^{3/2}\right)^{2/5} = \left(a^{5/4}\right)^{2/5} \cdot \left(b^{3/2}\right)^{2/5}$

$= a^{\frac{5}{4} \cdot \frac{2}{5}} \cdot b^{\frac{3}{2} \cdot \frac{2}{5}}$

$= a^{1/2} b^{3/5}$

18. $\left(a^{4/3} \cdot b^{-1/2}\right)\left(a^{-2} \cdot b^{5/2}\right) = a^{\frac{4}{3} + (-2)} b^{\left(-\frac{1}{2} + \frac{5}{2}\right)}$

$= a^{\frac{4}{3} - \frac{6}{3}} b^{\frac{4}{2}} = a^{-2/3} b^2$

$= \dfrac{b^2}{a^{2/3}}$

20. $\left(25 p^{2/5} q^{-1}\right)^{1/2} = 25^{1/2} \cdot \left(p^{2/5}\right)^{1/2} \cdot \left(q^{-1}\right)^{1/2}$

$= \sqrt{25} \cdot p^{\frac{2}{5} \cdot \frac{1}{2}} \cdot q^{-1 \cdot \frac{1}{2}}$

$= 5 p^{1/5} q^{-1/2}$

$= \dfrac{5 p^{1/5}}{q^{1/2}}$

## Chapter 7: Radicals and Rational Exponents

**22.** $\left(\dfrac{64m^{1/2}n}{m^{-2}n^{4/3}}\right)^{1/2} = \left(64m^{\frac{1}{2}-(-2)}n^{1-\frac{4}{3}}\right)^{1/2}$

$= \left(64m^{5/2}n^{-1/3}\right)^{1/2}$

$= 64^{1/2}\left(m^{5/2}\right)^{1/2}\left(n^{-1/3}\right)^{1/2}$

$= 8m^{\frac{5}{2}\cdot\frac{1}{2}}n^{-\frac{1}{3}\cdot\frac{1}{2}} = 8m^{5/4}n^{-1/6}$

$= \dfrac{8m^{5/4}}{n^{1/6}}$

**24.** $\left(\dfrac{27x^{1/2}y^{-1}}{y^{-2/3}x^{-1/2}}\right)^{1/3} - \left(\dfrac{4x^{1/3}y^{4/9}}{x^{-1/3}y^{2/3}}\right)^{1/2}$

$= \left(27x^{\frac{1}{2}-\left(-\frac{1}{2}\right)}y^{-1-\left(-\frac{2}{3}\right)}\right)^{1/3} - \left(4x^{\frac{1}{3}-\left(-\frac{1}{3}\right)}y^{\frac{4}{9}-\frac{2}{3}}\right)^{1/2}$

$= \left(27x^1 y^{-1/3}\right)^{1/3} - \left(4x^{2/3}y^{-2/9}\right)^{1/2}$

$= 27^{1/3}x^{1/3}\left(y^{-1/3}\right)^{1/3} - 4^{1/2}\left(x^{2/3}\right)^{1/2}\left(y^{-2/9}\right)^{1/2}$

$= 3x^{1/3}y^{-1/9} - 2x^{1/3}y^{-1/9}$

$= x^{1/3}y^{-1/9}$

$= \dfrac{x^{1/3}}{y^{1/9}}$

**26.** $x^{1/3}\left(x^{5/3}+4\right) = x^{1/3}\cdot x^{5/3} + x^{1/3}\cdot 4$

$= x^{6/3} + 4x^{1/3}$

$= x^2 + 4x^{1/3}$

**28.** $3a^{-1/2}(2-a) = 3a^{-1/2}\cdot 2 - 3a^{-1/2}\cdot a$

$= 6a^{-1/2} - 3a^{1/2}$

$= \dfrac{6}{a^{1/2}} - 3a^{1/2}$

**30.** $8p^{2/3}\left(p^{4/3} - 4p^{-2/3}\right)$

$= 8p^{2/3}\cdot p^{4/3} - 8p^{2/3}\cdot 4p^{-2/3}$

$= 8p^{6/3} - 32p^0$

$= 8p^2 - 32$

$= 8\left(p^2 - 4\right)$

$= 8(p-2)(p+2)$

**32.** $\sqrt[3]{x^6} = x^{6/3} = x^2$

**34.** $\sqrt[9]{125^6} = 125^{6/9} = 125^{2/3} = \left(\sqrt[3]{125}\right)^2 = 5^2 = 25$

**36.** $\sqrt{25x^4y^6} = \left(25x^4y^6\right)^{1/2}$

$= 25^{1/2}\left(x^4\right)^{1/2}\left(y^6\right)^{1/2}$

$= 5x^2y^3$

**38.** $\dfrac{\sqrt[3]{y^2}}{\sqrt{y}} = \dfrac{y^{2/3}}{y^{1/2}} = y^{\frac{2}{3}-\frac{1}{2}} = y^{1/6} = \sqrt[6]{y}$

**40.** $\sqrt[4]{p^3}\cdot\sqrt[3]{p} = p^{3/4}\cdot p^{1/3} = p^{\frac{3}{4}+\frac{1}{3}}$

$= p^{13/12} = \sqrt[12]{p^{13}}$

**42.** $\sqrt[3]{\sqrt{x^3}} = \left(x^{3/2}\right)^{1/3} = x^{1/2} = \sqrt{x}$

**44.** $\sqrt{5}\cdot\sqrt[3]{25} = \sqrt{5}\cdot\sqrt[3]{5^2} = 5^{1/2}\cdot 5^{2/3}$

$= 5^{\frac{1}{2}+\frac{2}{3}} = 5^{7/6}$

$= \sqrt[6]{5^7}$

**46.** $\dfrac{\sqrt[4]{49}}{\sqrt{7}} = \dfrac{\sqrt[4]{7^2}}{\sqrt{7}} = \dfrac{7^{2/4}}{7^{1/2}} = 7^{\frac{2}{4}-\frac{1}{2}} = 7^0 = 1$

**48.** $6x^{4/3} + 4x^{1/3}(2x-3)$

$= x^{1/3}\cdot 6x^{3/3} + x^{1/3}\cdot 4(2x-3)$

$= x^{1/3}\left(6x + 4(2x-3)\right)$

$= x^{1/3}(6x + 8x - 12)$

$= x^{1/3}(14x - 12)$

$= 2x^{1/3}(7x - 6)$

**50.** $3(x-5)^{1/2}(3x+1) + 6(x-5)^{3/2}$

$= (x-5)^{1/2}\cdot 3(3x+1) + (x-5)^{1/2}\cdot 6(x-5)^{2/2}$

$= (x-5)^{1/2}\left(3(3x+1) + 6(x-5)\right)$

$= (x-5)^{1/2}(9x + 3 + 6x - 30)$

$= (x-5)^{1/2}(15x - 27)$

$= 3(x-5)^{1/2}(5x - 9)$

**Section 7.2** Simplify Expressions Using the Laws of Exponents

52. $x^{-2/3}(3x+2)+9x^{1/3}$
$= x^{-2/3}(3x+2)+x^{-2/3}\cdot 9x^{3/3}$
$= x^{-2/3}(3x+2+9x)$
$= x^{-2/3}(12x+2)$
$= 2x^{-2/3}(6x+1)$
$= \dfrac{2(6x+1)}{x^{2/3}}$

54. $4(x+3)^{1/2}+(x+3)^{-1/2}(2x+1)$
$= (x+3)^{-1/2}\cdot 4(x+3)+(x+3)^{-1/2}(2x+1)$
$= (x+3)^{-1/2}(4(x+3)+(2x+1))$
$= (x+3)^{-1/2}(4x+12+2x+1)$
$= (x+3)^{-1/2}(6x+13)$
$= \dfrac{6x+13}{(x+3)^{1/2}}$

56. $24x(x^2-1)^{1/3}+9(x^2-1)^{4/3}$
$= 3(x^2-1)^{1/3}\cdot 8x+3(x^2-1)^{1/3}\cdot 3(x^2-1)^{3/3}$
$= 3(x^2-1)^{1/3}(8x+3(x^2-1))$
$= 3(x^2-1)^{1/3}(3x^2+8x-3)$
$= 3(x^2-1)^{1/3}(3x-1)(x+3)$

58. $\sqrt[6]{27^2}=27^{2/6}=27^{1/3}=\sqrt[3]{27}=3$

60. $25^{3/4}\cdot 25^{3/4}=25^{\frac{3}{4}+\frac{3}{4}}=25^{6/4}$
$=25^{3/2}=(\sqrt{25})^3=125$

62. $(8^4)^{5/12}=8^{4\cdot\frac{5}{12}}=8^{5/3}=(\sqrt[3]{8})^5=2^5=32$

64. $(\sqrt[6]{27})^2=27^{2/6}=27^{1/3}=\sqrt[3]{27}=3$

66. $\sqrt[9]{a^6}-\dfrac{\sqrt[6]{a^5}}{\sqrt[6]{a}}=a^{6/9}-\dfrac{a^{5/6}}{a^{1/6}}$
$= a^{2/3}-a^{\frac{5}{6}-\frac{1}{6}}$
$= a^{2/3}-a^{2/3}$
$= 0$

68. $(4^{-1}\cdot 81^{1/2})^{1/2}=\left(\dfrac{1}{4}\cdot\sqrt{81}\right)^{1/2}$
$=\left(\dfrac{9}{4}\right)^{1/2}=\sqrt{\dfrac{9}{4}}=\dfrac{3}{2}$

70. $5^x=64$
$5^{x/3}=5^{x\cdot\frac{1}{3}}=(5^x)^{1/3}=(64)^{1/3}=\sqrt[3]{64}=4$

72. $5^x=27$
$\sqrt[3]{5^x}=5^{x/3}=5^{x\cdot\frac{1}{3}}=(5^x)^{1/3}=27^{1/3}=\sqrt[3]{27}=3$

74. $\sqrt[5]{\sqrt[3]{\sqrt{x^2}}}=\left(\left((x^2)^{1/2}\right)^{1/3}\right)^{1/5}=\left(\left(x^{2\cdot\frac{1}{2}}\right)^{1/3}\right)^{1/5}$
$=(x^{1/3})^{1/5}=x^{\frac{1}{3}\cdot\frac{1}{5}}$
$= x^{1/15}=\sqrt[15]{x}$

76. $g(x)=(x-3)^{1/2}(x-1)^{-1/2}=\dfrac{\sqrt{x-3}}{\sqrt{x-1}}$

We need $x-3\ge 0$ and $x-1>0$.
$x-3\ge 0$ and $x-1>0$
$x\ge 3$ $\qquad$ $x>1$
The domain is the set of values that satisfy both inequalities. Thus, the domain is $x\ge 3$.
Domain: $\{x\,|\,x\ge 3\}$ or $[3,\infty)$
We can check this graphically:

*Chapter 7: Radicals and Rational Exponents*

78. $(2x-1)(x+4)-(x+1)(x-1)$
    $= (2x^2 + 8x - x - 4) - (x^2 - 1)$
    $= 2x^2 + 7x - 4 - x^2 + 1$
    $= x^2 + 7x - 3$

80. $\dfrac{\sqrt{x^2 + 4x + 4}}{x+2} = \dfrac{\sqrt{(x+2)^2}}{x+2} = \dfrac{x+2}{x+2} = 1$
    (since $x + 2 > 0$)

## 7.3 Preparing for Simplifying Radical Expressions

1. $1^2 = 1$, $2^2 = 4$, $3^2 = 9$, $4^2 = 16$, $5^2 = 25$,
   $6^2 = 36$, $7^2 = 49$, $8^2 = 64$, $9^2 = 81$,
   $10^2 = 100$, $11^2 = 121$, $12^2 = 144$, $13^2 = 169$,
   and $14^2 = 196$.

2. $1^3 = 1$, $2^3 = 8$, $3^3 = 27$, $4^3 = 64$, and $5^3 = 125$.

3. a. $\sqrt{16} = \sqrt{4^2} = 4$
   b. $\sqrt{p^2} = |p|$

## 7.3 Exercises

2. perfect square; perfect cube

4. True

6. $1 = 1^3$, $8 = 2^3$, $27 = 3^3$, $64 = 4^3$, $125 = 5^3$, $216 = 6^3$

8. The indexes must be the same.

10. $\sqrt{3} \cdot \sqrt{10} = \sqrt{3 \cdot 10} = \sqrt{30}$

12. $\sqrt[3]{-5} \cdot \sqrt[3]{7} = \sqrt[3]{-5 \cdot 7} = \sqrt[3]{-35}$

14. $\sqrt[4]{6a^2} \cdot \sqrt[4]{7b^2} = \sqrt[4]{6a^2 \cdot 7b^2} = \sqrt[4]{42a^2 b^2}$

16. $\sqrt{p-5} \cdot \sqrt{p+5} = \sqrt{(p-5)(p+5)} = \sqrt{p^2 - 25}$
    if $|p| \geq 5$, otherwise the result is not a real number.

18. $\sqrt[3]{\dfrac{-9x^2}{4}} \cdot \sqrt[3]{\dfrac{4}{3x}} = \sqrt[3]{\dfrac{-9x^2}{4} \cdot \dfrac{4}{3x}} = \sqrt[3]{-3x} = -\sqrt[3]{3x}$

20. $\sqrt{32} = \sqrt{16 \cdot 2} = \sqrt{16} \cdot \sqrt{2} = 4\sqrt{2}$

22. $\sqrt[4]{162} = \sqrt[4]{81 \cdot 2} = \sqrt[4]{81} \cdot \sqrt[4]{2} = 3\sqrt[4]{2}$

24. $\sqrt{20a^2} = \sqrt{4a^2 \cdot 5} = \sqrt{4a^2} \cdot \sqrt{5} = 2|a|\sqrt{5}$

26. $\sqrt[3]{-64p^3} = -4p$

28. $\sqrt[4]{48z^4} = \sqrt[4]{16z^4 \cdot 3} = \sqrt[4]{16z^4} \cdot \sqrt[4]{3} = 2|z|\sqrt[4]{3}$

30. $\sqrt{45m^2 n} = \sqrt{9m^2 \cdot 5n} = \sqrt{9m^2} \cdot \sqrt{5n} = 3|m|\sqrt{5n}$

32. $4\sqrt{27b} = 4\sqrt{9 \cdot 3b} = 4\sqrt{9} \cdot \sqrt{3b}$
    $= 4 \cdot 3\sqrt{3b} = 12\sqrt{3b}$

34. $\sqrt{98w^8} = \sqrt{49w^8 \cdot 2} = \sqrt{49w^8} \cdot \sqrt{2} = 7w^4\sqrt{2}$

36. $\sqrt{s^9} = \sqrt{s^8 \cdot s} = \sqrt{s^8} \cdot \sqrt{s} = s^4\sqrt{s}$

38. $\sqrt[5]{x^{12}} = \sqrt[5]{x^{10} \cdot x^2} = \sqrt[5]{x^{10}} \cdot \sqrt[5]{x^2} = x^2 \sqrt[5]{x^2}$

40. $\sqrt{x^7 y^2} = \sqrt{x^6 y^2 \cdot x} = \sqrt{x^6 y^2} \cdot \sqrt{x} = |x^3 y|\sqrt{x}$

42. $\sqrt{243ab^5} = \sqrt{81b^4 \cdot 3ab} = \sqrt{81b^4} \cdot \sqrt{3ab}$
    $= 9b^2\sqrt{3ab}$

44. $\sqrt[3]{-54q^{12}} = \sqrt[3]{-27q^{12} \cdot 2} = \sqrt[3]{-27q^{12}} \cdot \sqrt[3]{2}$
    $= -3q^4 \sqrt[3]{2}$

46. $\sqrt[4]{x^6 y^9 z^4} = \sqrt[4]{x^4 y^8 z^4 \cdot x^2 y}$
    $= \sqrt[4]{x^4 y^8 z^4} \cdot \sqrt[4]{x^2 y}$
    $= |xz| y^2 \sqrt[4]{x^2 y}$

48. $\sqrt{75x^6 y} = \sqrt{25x^6 \cdot 3y}$
    $= \sqrt{25x^6} \cdot \sqrt{3y}$
    $= 5|x^3|\sqrt{3y}$

### Section 7.3 Simplifying Radical Expressions

50. $\sqrt[3]{(a+b)^5} = \sqrt[3]{(a+b)^3 \cdot (a+b)^2}$
$= \sqrt[3]{(a+b)^3} \cdot \sqrt[3]{(a+b)^2}$
$= (a+b)\sqrt[3]{(a+b)^2}$

52. $\sqrt[3]{8a^3 + 8b^3} = \sqrt[3]{8(a^3 + b^3)}$
$= \sqrt[3]{8} \cdot \sqrt[3]{a^3 + b^3}$
$= 2\sqrt[3]{a^3 + b^3}$

54. $\dfrac{5-\sqrt{100}}{5} = \dfrac{5-10}{5} = \dfrac{-5}{5} = -1$

56. $\dfrac{10-\sqrt{75}}{5} = \dfrac{10-\sqrt{25 \cdot 3}}{5} = \dfrac{10-\sqrt{25}\cdot\sqrt{3}}{5}$
$= \dfrac{10-5\sqrt{3}}{5} = \dfrac{5(2-\sqrt{3})}{5}$
$= 2 - \sqrt{3}$

58. $\dfrac{-6+\sqrt{48}}{8} = \dfrac{-6+\sqrt{16 \cdot 3}}{8} = \dfrac{-6+\sqrt{16}\cdot\sqrt{3}}{8}$
$= \dfrac{-6+4\sqrt{3}}{8} = \dfrac{2(-3+2\sqrt{3})}{2 \cdot 4}$
$= \dfrac{-3+2\sqrt{3}}{4}$

60. $\dfrac{-6+\sqrt{108}}{6} = \dfrac{-6+\sqrt{36\cdot 3}}{6} = \dfrac{-6+\sqrt{36}\cdot\sqrt{3}}{6}$
$= \dfrac{-6+6\sqrt{3}}{6} = \dfrac{6(-1+\sqrt{3})}{6}$
$= -1+\sqrt{3}$

62. $\sqrt{6}\cdot\sqrt{6} = \sqrt{6\cdot 6} = \sqrt{36} = 6$

64. $\sqrt{3}\cdot\sqrt{12} = \sqrt{3\cdot 12} = \sqrt{36} = 6$

66. $\sqrt[3]{9}\cdot\sqrt[3]{3} = \sqrt[3]{9\cdot 3} = \sqrt[3]{27} = 3$

67. $\sqrt{5x}\cdot\sqrt{15x} = \sqrt{5x\cdot 15x}$
$= \sqrt{75x^2}$
$= \sqrt{25x^2 \cdot 3}$
$= \sqrt{25x^2}\cdot\sqrt{3}$
$= 5x\sqrt{3}$

69. $\sqrt[3]{4b^2}\cdot\sqrt[3]{6b^2} = \sqrt[3]{4b^2\cdot 6b^2}$
$= \sqrt[3]{24b^4}$
$= \sqrt[3]{8b^3\cdot 3b}$
$= \sqrt[3]{8b^3}\cdot\sqrt[3]{3b}$
$= 2b\sqrt[3]{3b}$

71. $2\sqrt{6ab}\cdot 3\sqrt{15ab^3} = 6\sqrt{6ab\cdot 15ab^3}$
$= 6\sqrt{90a^2b^4}$
$= 6\sqrt{9a^2b^4\cdot 10}$
$= 6\sqrt{9a^2b^4}\cdot\sqrt{10}$
$= 18ab^2\sqrt{10}$

73. $\sqrt[4]{27p^3q^2}\cdot\sqrt[4]{12p^2q^2} = \sqrt[4]{27p^3q^2\cdot 12p^2q^2}$
$= \sqrt[4]{324p^5q^4}$
$= \sqrt[4]{81p^4q^4\cdot 4p}$
$= \sqrt[4]{81p^4q^4}\cdot\sqrt[4]{4p}$
$= 3pq\sqrt[4]{4p}$

75. $\sqrt[5]{-8a^3b^4}\cdot\sqrt[5]{12a^3b} = \sqrt[5]{-8a^3b^4\cdot 12a^3b}$
$= \sqrt[5]{-96a^6b^5}$
$= \sqrt[5]{-32a^5b^5\cdot 3a}$
$= \sqrt[5]{-32a^5b^5}\cdot\sqrt[5]{3a}$
$= -2ab\sqrt[5]{3a}$

## Chapter 7: Radicals and Rational Exponents

78. $\sqrt[3]{9(a+b)^2} \cdot \sqrt[3]{6(a+b)^5}$
    $= \sqrt[3]{9(a+b)^2 \cdot 6(a+b)^5}$
    $= \sqrt[3]{54(a+b)^7}$
    $= \sqrt[3]{27(a+b)^6 \cdot 2(a+b)}$
    $= \sqrt[3]{27(a+b)^6} \cdot \sqrt[3]{2(a+b)}$
    $= 3(a+b)^2 \sqrt[3]{2(a+b)}$

80. $\sqrt{\dfrac{5}{36}} = \dfrac{\sqrt{5}}{\sqrt{36}} = \dfrac{\sqrt{5}}{6}$

82. $\sqrt{\dfrac{81}{64}} = \dfrac{\sqrt{81}}{\sqrt{64}} = \dfrac{9}{8}$

84. $\sqrt[4]{\dfrac{2a^8}{81}} = \dfrac{\sqrt[4]{2a^8}}{\sqrt[4]{81}} = \dfrac{a^2 \sqrt[4]{2}}{3}$

86. $\sqrt{\dfrac{4a^4}{81b^2}} = \dfrac{\sqrt{4a^4}}{\sqrt{81b^2}} = \dfrac{2a^2}{9b}$

88. $\sqrt[5]{\dfrac{-32a^{15}}{243b^{10}}} = \dfrac{\sqrt[5]{-32a^{15}}}{\sqrt[5]{243b^{10}}} = \dfrac{-2a^3}{3b^2} = -\dfrac{2a^3}{3b^2}$

90. $\dfrac{\sqrt{27}}{\sqrt{3}} = \sqrt{\dfrac{27}{3}} = \sqrt{9} = 3$

92. $\dfrac{\sqrt[4]{64}}{\sqrt[4]{4}} = \sqrt[4]{\dfrac{64}{4}} = \sqrt[4]{16} = 2$

94. $\dfrac{\sqrt{54y^5}}{\sqrt{3y}} = \sqrt{\dfrac{54y^5}{3y}} = \sqrt{18y^4} = 3y^2\sqrt{2}$

96. $\dfrac{\sqrt{360m^7n^3}}{\sqrt{5mn^5}} = \sqrt{\dfrac{360m^7n^3}{5mn^5}} = \sqrt{\dfrac{72m^6}{n^2}}$
    $= \dfrac{\sqrt{72m^6}}{\sqrt{n^2}} = \dfrac{6m^3\sqrt{2}}{n}$

98. $\dfrac{\sqrt{375x^2y^7}}{10\sqrt{3y}} = \dfrac{1}{10}\sqrt{\dfrac{375x^2y^7}{3y}}$
    $= \dfrac{1}{10}\sqrt{125x^2y^6}$
    $= \dfrac{1}{10} \cdot 5xy^3\sqrt{5}$
    $= \dfrac{xy^3\sqrt{5}}{2}$

100. $\dfrac{\sqrt[3]{-128x^8}}{\sqrt[3]{2x^{-1}}} = \sqrt[3]{\dfrac{-128x^8}{2x^{-1}}} = \sqrt[3]{-64x^9} = -4x^3$

102. $\dfrac{\sqrt{96a^5b^{-3}}}{\sqrt{3a^{-5}b}} = \sqrt{\dfrac{96a^5b^{-3}}{3a^{-5}b}} = \sqrt{\dfrac{32a^{10}}{b^4}}$
     $= \dfrac{\sqrt{32a^{10}}}{\sqrt{b^4}} = \dfrac{4a^5\sqrt{2}}{b^2}$

104. $\sqrt{2} \cdot \sqrt[3]{7} = 2^{1/2} \cdot 7^{1/3}$
     $= 2^{3/6} \cdot 7^{2/6}$
     $= (2^3)^{1/6} \cdot (7^2)^{1/6}$
     $= (2^3 \cdot 7^2)^{1/6}$
     $= (392)^{1/6}$
     $= \sqrt[6]{392}$

106. $\sqrt[4]{3} \cdot \sqrt[8]{5} = 3^{1/4} \cdot 5^{1/8}$
     $= 3^{2/8} \cdot 5^{1/8}$
     $= (3^2)^{1/8} \cdot (5^1)^{1/8}$
     $= (3^2 \cdot 5)^{1/8}$
     $= (45)^{1/8}$
     $= \sqrt[8]{45}$

## Section 7.3 Simplifying Radical Expressions

108. $\sqrt{6} \cdot \sqrt[3]{9} = 6^{1/2} \cdot 9^{1/3}$
$= 6^{1/2} \cdot (3^2)^{1/3}$
$= 6^{1/2} \cdot 3^{2/3}$
$= 6^{3/6} \cdot 3^{4/6}$
$= (6^3)^{1/6} \cdot (3^4)^{1/6}$
$= (6^3 \cdot 3^4)^{1/6}$
$= ((2 \cdot 3)^3 \cdot 3^4)^{1/6}$
$= (2^3 \cdot 3^3 \cdot 3^4)^{1/6}$
$= (2^3 \cdot 3^7)^{1/6}$
$= 3(2^3 \cdot 3)^{1/6}$
$= 3(24)^{1/6}$
$= 3\sqrt[6]{24}$

110. $\sqrt[5]{8} \cdot \sqrt[10]{16} = 8^{1/5} \cdot 16^{1/10}$
$= 8^{2/10} \cdot 16^{1/10}$
$= (8^2)^{1/10} \cdot (16^1)^{1/10}$
$= (8^2 \cdot 16)^{1/10}$
$= ((2^3)^2 \cdot 2^4)^{1/10}$
$= (2^6 \cdot 2^4)^{1/10}$
$= (2^{10})^{1/10}$
$= 2$

112. $\sqrt[3]{\dfrac{7a^2}{64}} = \dfrac{\sqrt[3]{7a^2}}{\sqrt[3]{64}} = \dfrac{\sqrt[3]{7a^2}}{4}$

114. $\sqrt[5]{8b^2} \cdot \sqrt[5]{3b} = \sqrt[5]{8b^2 \cdot 3b} = \sqrt[5]{24b^3}$

116. $\sqrt{24b^6} = \sqrt{4b^6 \cdot 6} = \sqrt{4b^6} \cdot \sqrt{6} = 2b^3\sqrt{6}$

118. $\sqrt[4]{8x^3y^2} \cdot \sqrt[4]{4x^2y^3} = \sqrt[4]{8x^3y^2 \cdot 4x^2y^3}$
$= \sqrt[4]{32x^5y^5}$
$= \sqrt[4]{16x^4y^4 \cdot 2xy}$
$= \sqrt[4]{16x^4y^4} \cdot \sqrt[4]{2xy}$
$= 2xy\sqrt[4]{2xy}$

120. $\dfrac{\sqrt[3]{-250p^2}}{\sqrt[3]{2p^5}} = \sqrt[3]{\dfrac{-250p^2}{2p^5}} = \sqrt[3]{\dfrac{-125}{p^3}}$
$= \dfrac{\sqrt[3]{-125}}{\sqrt[3]{p^3}} = \dfrac{-5}{p}$
$= -\dfrac{5}{p}$

122. $-7\sqrt[3]{250p^3} = -7\sqrt[3]{125p^3 \cdot 2}$
$= -7\sqrt[3]{125p^3} \cdot \sqrt[3]{2}$
$= -7 \cdot 5p \cdot \sqrt[3]{2}$
$= -35p\sqrt[3]{2}$

124. $\sqrt[5]{32p^7q^{11}} = \sqrt[5]{32p^5q^{10} \cdot p^2q}$
$= \sqrt[5]{32p^5q^{10}} \cdot \sqrt[5]{p^2q}$
$= 2pq^2\sqrt[5]{p^2q}$

126. $\sqrt[4]{8} \cdot \sqrt[4]{18} = \sqrt[4]{8 \cdot 18} = \sqrt[4]{144}$
$= \sqrt[4]{16 \cdot 9} = \sqrt[4]{16} \cdot \sqrt[4]{9}$
$= 2\sqrt[4]{9}$ or $2\sqrt{3}$

128. a.

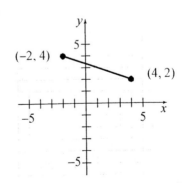

## Chapter 7: Radicals and Rational Exponents

b. $\sqrt{(4-2)^2 + (-2-4)^2} = \sqrt{2^2 + (-6)^2}$
$= \sqrt{4+36}$
$= \sqrt{40}$
$= 2\sqrt{10}$

The line segment has a length of $2\sqrt{10}$ units.

130. a. $r = \sqrt[3]{\dfrac{3(9)}{4\pi}} = \dfrac{\sqrt[3]{27}}{\sqrt[3]{4\pi}} = \dfrac{3}{\sqrt[3]{4\pi}} \approx 1.29$ cm

b. $r = \sqrt[3]{\dfrac{3(32\pi)}{4\pi}} = \sqrt[3]{24} = 2\sqrt[3]{3} \approx 2.88$ m

132. $a=1, b=4, c=1$

$x = \dfrac{-4 \pm \sqrt{4^2 - 4(1)(1)}}{2(1)}$

$= \dfrac{-4 \pm \sqrt{16-4}}{2} = \dfrac{-4 \pm \sqrt{12}}{2}$

$= \dfrac{-4 \pm 2\sqrt{3}}{2} = \dfrac{2(-2 \pm \sqrt{3})}{2}$

$= -2 \pm \sqrt{3}$

$x = -2 - \sqrt{3}$ or $x = -2 + \sqrt{3}$.

134. $a=2, b=1, c=-1$

$x = \dfrac{-1 \pm \sqrt{1^2 - 4(2)(-1)}}{2(2)}$

$= \dfrac{-1 \pm \sqrt{1+8}}{4} = \dfrac{-1 \pm \sqrt{9}}{4}$

$= \dfrac{-1 \pm 3}{4}$

$x = \dfrac{-1-3}{4} = \dfrac{-4}{4} = -1$ or $x = \dfrac{-1+3}{4} = \dfrac{2}{4} = \dfrac{1}{2}$.

136. $4x+3 = 13$
$4x = 10$
$x = \dfrac{10}{4}$
$x = \dfrac{5}{2}$

The solution set is $\left\{\dfrac{5}{2}\right\}$.

138. $\dfrac{3}{5}x + 2 \le 28$

$\dfrac{3}{5}x \le 26$

$x \le \dfrac{130}{3}$

The solution set is $\left\{x \mid x \le \dfrac{130}{3}\right\}$, or in interval notation we would write $\left(-\infty, \dfrac{130}{3}\right]$.

140. Answers will vary.

### 7.4 Preparing for Adding, Subtracting, and Multiplying Radical Expressions

1. $4y^3 - 2y^2 + 8y - 1 + (-2y^3 + 7y^2 - 3y + 9)$
$= 4y^3 - 2y^3 - 2y^2 + 7y^2 + 8y - 3y - 1 + 9$
$= 2y^3 + 5y^2 + 5y + 8$

2. $5z^2 + 6 - (3z^2 - 8z - 3)$
$= 5z^2 + 6 - 3z^2 + 8z + 3$
$= 5z^2 - 3z^2 + 8z + 6 + 3$
$= 2z^2 + 8z + 9$

3. $(4x+3)(x-5) = 4x \cdot x - 4x \cdot 5 + 3 \cdot x - 3 \cdot 5$
$= 4x^2 - 20x + 3x - 15$
$= 4x^2 - 17x - 15$

4. $(2y-3)(2y+3) = (2y)^2 - 3^2$
$= 4y^2 - 9$

### 7.4 Exercises

2. conjugates

4. False; the conjugate of $-5 + \sqrt{2}$ is $-5 - \sqrt{2}$.

6. $(\sqrt{a} - \sqrt{b})(\sqrt{a} + \sqrt{b}) = (\sqrt{a})^2 - (\sqrt{b})^2$
$= a - b$

Answers will vary. The product of conjugates involving square roots will not have square roots in the resulting expression.

**Section 7.4** *Adding, Subtracting, and Multiplying Radical Expressions*

8. $6\sqrt{3} + 8\sqrt{3} = (6+8)\sqrt{3} = 14\sqrt{3}$

10. $12\sqrt[4]{z} - 5\sqrt[4]{z} = (12-5)\sqrt[4]{z} = 7\sqrt[4]{z}$

12. $4\sqrt[3]{3y} + 8\sqrt[3]{3y} - 10\sqrt[3]{3y} = (4+8-10)\sqrt[3]{3y}$
    $= 2\sqrt[3]{3y}$

14. $12\sqrt{7} + 5\sqrt[4]{7} - 5\sqrt{7} + 6\sqrt[4]{7}$
    $= (12-5)\sqrt{7} + (5+6)\sqrt[4]{7}$
    $= 7\sqrt{7} + 11\sqrt[4]{7}$

16. $6\sqrt{3} + \sqrt{12} = 6\sqrt{3} + \sqrt{4}\cdot\sqrt{3}$
    $= 6\sqrt{3} + 2\sqrt{3}$
    $= (6+2)\sqrt{3}$
    $= 8\sqrt{3}$

18. $\sqrt[3]{32} - 5\sqrt[3]{4} = \sqrt[3]{8}\cdot\sqrt[3]{4} - 5\sqrt[3]{4}$
    $= 2\sqrt[3]{4} - 5\sqrt[3]{4}$
    $= (2-5)\sqrt[3]{4}$
    $= -3\sqrt[3]{4}$

20. $7\sqrt[4]{48} - 4\sqrt[4]{243} = 7\sqrt[4]{16}\cdot\sqrt[4]{3} - 4\sqrt[4]{81}\cdot\sqrt[4]{3}$
    $= 7\cdot 2\sqrt[4]{3} - 4\cdot 3\sqrt[4]{3}$
    $= 14\sqrt[4]{3} - 12\sqrt[4]{3}$
    $= (14-12)\sqrt[4]{3}$
    $= 2\sqrt[4]{3}$

22. $2\sqrt{48z} - \sqrt{75z} = 2\sqrt{16}\cdot\sqrt{3z} - \sqrt{25}\cdot\sqrt{3z}$
    $= 2\cdot 4\sqrt{3z} - 5\sqrt{3z}$
    $= 8\sqrt{3z} - 5\sqrt{3z}$
    $= (8-5)\sqrt{3z}$
    $= 3\sqrt{3z}$

24. $4\sqrt{12} + 2\sqrt{20} = 4\sqrt{4}\cdot\sqrt{3} + 2\sqrt{4}\cdot\sqrt{5}$
    $= 4\cdot 2\sqrt{3} + 2\cdot 2\sqrt{5}$
    $= 8\sqrt{3} + 4\sqrt{5}$

26. $3\sqrt{63z^3} + 2z\sqrt{28z} = 3\sqrt{9z^2}\cdot\sqrt{7z} + 2z\sqrt{4}\cdot\sqrt{7z}$
    $= 3\cdot 3z\sqrt{7z} + 2z\cdot 2\sqrt{7z}$
    $= 9z\sqrt{7z} + 4z\sqrt{7z}$
    $= (9+4)z\sqrt{7z}$
    $= 13z\sqrt{7z}$

28. $\sqrt{48y^2} - 4y\sqrt{12} + \sqrt{108y^2}$
    $= \sqrt{16y^2}\cdot\sqrt{3} - 4y\sqrt{4}\cdot\sqrt{3} + \sqrt{36y^2}\cdot\sqrt{3}$
    $= 4y\sqrt{3} - 4y\cdot 2\sqrt{3} + 6y\sqrt{3}$
    $= 4y\sqrt{3} - 8y\sqrt{3} + 6y\sqrt{3}$
    $= (4-8+6)y\sqrt{3}$
    $= 2y\sqrt{3}$

30. $2\sqrt[3]{-5x^3} + 4x\sqrt[3]{40} - \sqrt[3]{135}$
    $= 2\sqrt[3]{-x^3}\cdot\sqrt[3]{5} + 4x\sqrt[3]{8}\cdot\sqrt[3]{5} - \sqrt[3]{27}\cdot\sqrt[3]{5}$
    $= -2x\sqrt[3]{5} + 4x\cdot 2\sqrt[3]{5} - 3\sqrt[3]{5}$
    $= -2x\sqrt[3]{5} + 8x\sqrt[3]{5} - 3\sqrt[3]{5}$
    $= (-2x + 8x - 3)\sqrt[3]{5}$
    $= (6x - 3)\sqrt[3]{5}$ or $3(2x-1)\sqrt[3]{5}$

32. $\sqrt{4x+12} - \sqrt{9x+27}$
    $= \sqrt{4(x+3)} - \sqrt{9(x+3)}$
    $= \sqrt{4}\cdot\sqrt{x+3} - \sqrt{9}\cdot\sqrt{x+3}$
    $= 2\sqrt{x+3} - 3\sqrt{x+3}$
    $= (2-3)\sqrt{x+3}$
    $= -\sqrt{x+3}$

34. $\sqrt{25x} - \sqrt[4]{x^2} = \sqrt{25x} - x^{2/4}$
    $= \sqrt{25}\cdot\sqrt{x} - x^{1/2}$
    $= 5\sqrt{x} - \sqrt{x}$
    $= (5-1)\sqrt{x}$
    $= 4\sqrt{x}$

## Chapter 7: Radicals and Rational Exponents

36. $\sqrt[4]{16y} + \sqrt[8]{y^2} = \sqrt[4]{16y} + y^{2/8}$
$= \sqrt[4]{16} \cdot \sqrt[4]{y} + y^{1/4}$
$= 2\sqrt[4]{y} + \sqrt[4]{y}$
$= (2+1)\sqrt[4]{y}$
$= 3\sqrt[4]{y}$

38. $\sqrt{5}(5 + 3\sqrt{3}) = \sqrt{5} \cdot 5 + \sqrt{5} \cdot 3\sqrt{3}$
$= 5\sqrt{5} + 3\sqrt{5 \cdot 3}$
$= 5\sqrt{5} + 3\sqrt{15}$

40. $\sqrt{2}(\sqrt{5} - 2\sqrt{10}) = \sqrt{2} \cdot \sqrt{5} - \sqrt{2} \cdot 2\sqrt{10}$
$= \sqrt{2 \cdot 5} - 2\sqrt{2 \cdot 10}$
$= \sqrt{10} - 2\sqrt{20}$
$= \sqrt{10} - 2\sqrt{4} \cdot \sqrt{5}$
$= \sqrt{10} - 2 \cdot 2\sqrt{5}$
$= \sqrt{10} - 4\sqrt{5}$

42. $\sqrt[3]{6}(\sqrt[3]{2} + \sqrt[3]{12}) = \sqrt[3]{6} \cdot \sqrt[3]{2} + \sqrt[3]{6} \cdot \sqrt[3]{12}$
$= \sqrt[3]{6 \cdot 2} + \sqrt[3]{6 \cdot 12}$
$= \sqrt[3]{12} + \sqrt[3]{72}$
$= \sqrt[3]{12} + \sqrt[3]{8} \cdot \sqrt[3]{9}$
$= \sqrt[3]{12} + 2\sqrt[3]{9}$

44. $\sqrt{5x}(6 + \sqrt{15x}) = \sqrt{5x} \cdot 6 + \sqrt{5x} \cdot \sqrt{15x}$
$= 6\sqrt{5x} + \sqrt{5x \cdot 15x}$
$= 6\sqrt{5x} + \sqrt{75x^2}$
$= 6\sqrt{5x} + \sqrt{25x^2} \cdot \sqrt{3}$
$= 6\sqrt{5x} + 5x\sqrt{3}$

46. $(5 + \sqrt{5})(3 + \sqrt{6})$
$= 5 \cdot 3 + +5 \cdot \sqrt{6} + \sqrt{5} \cdot 3 + \sqrt{5} \cdot \sqrt{6}$
$= 15 + 5\sqrt{6} + 3\sqrt{5} + \sqrt{30}$

48. $(7 - \sqrt{3})(6 + \sqrt{5})$
$= 7 \cdot 6 + 7 \cdot \sqrt{5} - \sqrt{3} \cdot 6 - \sqrt{3} \cdot \sqrt{5}$
$= 42 + 7\sqrt{5} - 6\sqrt{3} - \sqrt{15}$

50. $(9 + 5\sqrt{10})(1 - 3\sqrt{10})$
$= 9 \cdot 1 - 9 \cdot 3\sqrt{10} + 5\sqrt{10} \cdot 1 - 5\sqrt{10} \cdot 3\sqrt{10}$
$= 9 - 27\sqrt{10} + 5\sqrt{10} - 15\sqrt{100}$
$= 9 + (-27 + 5)\sqrt{10} - 15 \cdot 10$
$= 9 - 22\sqrt{10} - 150$
$= -141 - 22\sqrt{10}$

52. $(2\sqrt{3} + \sqrt{10})(\sqrt{5} - 2\sqrt{2})$
$= 2\sqrt{3} \cdot \sqrt{5} - 2\sqrt{3} \cdot 2\sqrt{2} + \sqrt{10} \cdot \sqrt{5} - \sqrt{10} \cdot 2\sqrt{2}$
$= 2\sqrt{15} - 4\sqrt{6} + \sqrt{50} - 2\sqrt{20}$
$= 2\sqrt{15} - 4\sqrt{6} + 5\sqrt{2} - 2 \cdot 2\sqrt{5}$
$= 2\sqrt{15} - 4\sqrt{6} + 5\sqrt{2} - 4\sqrt{5}$

54. $(\sqrt{6} - 2\sqrt{2})(2\sqrt{6} + 3\sqrt{2})$
$= \sqrt{6} \cdot 2\sqrt{6} + \sqrt{6} \cdot 3\sqrt{2} - 2\sqrt{2} \cdot 2\sqrt{6} - 2\sqrt{2} \cdot 3\sqrt{2}$
$= 2\sqrt{36} + 3\sqrt{12} - 4\sqrt{12} - 6\sqrt{4}$
$= 2 \cdot 6 + 3 \cdot 2\sqrt{3} - 4 \cdot 2\sqrt{3} - 6 \cdot 2$
$= 12 + 6\sqrt{3} - 8\sqrt{3} - 12$
$= 12 - 12 + (6 - 8)\sqrt{3}$
$= -2\sqrt{3}$

56. $(2 - \sqrt{3})^2 = 2^2 - 2(2)(\sqrt{3}) + (\sqrt{3})^2$
$= 4 - 4\sqrt{3} + \sqrt{9}$
$= 4 - 4\sqrt{3} + 3$
$= 7 - 4\sqrt{3}$

58. $(\sqrt{7} - \sqrt{3})^2 = (\sqrt{7})^2 - 2(\sqrt{7})(\sqrt{3}) + (\sqrt{3})^2$
$= \sqrt{49} - 2\sqrt{21} + \sqrt{9}$
$= 7 - 2\sqrt{21} + 3$
$= 10 - 2\sqrt{21}$

60. $(\sqrt{z} + \sqrt{5})^2 = (\sqrt{z})^2 + 2(\sqrt{z})(\sqrt{5}) + (\sqrt{5})^2$
$= \sqrt{z^2} + 2\sqrt{5z} + \sqrt{25}$
$= z + 2\sqrt{5z} + 5$

**Section 7.4** Adding, Subtracting, and Multiplying Radical Expressions

62. $(\sqrt{3}-1)(\sqrt{3}+1) = (\sqrt{3})^2 - 1^2$
$\phantom{(\sqrt{3}-1)(\sqrt{3}+1)} = \sqrt{9}-1$
$\phantom{(\sqrt{3}-1)(\sqrt{3}+1)} = 3-1$
$\phantom{(\sqrt{3}-1)(\sqrt{3}+1)} = 2$

64. $(6+3\sqrt{2})(6-3\sqrt{2}) = 6^2 - (3\sqrt{2})^2$
$\phantom{(6+3\sqrt{2})(6-3\sqrt{2})} = 36 - 9\sqrt{4}$
$\phantom{(6+3\sqrt{2})(6-3\sqrt{2})} = 36 - 9 \cdot 2$
$\phantom{(6+3\sqrt{2})(6-3\sqrt{2})} = 36 - 18$
$\phantom{(6+3\sqrt{2})(6-3\sqrt{2})} = 18$

66. $(\sqrt{5a}+\sqrt{7b})(\sqrt{5a}-\sqrt{7b}) = (\sqrt{5a})^2 - (\sqrt{7b})^2$
$\phantom{(\sqrt{5a}+\sqrt{7b})(\sqrt{5a}-\sqrt{7b})} = \sqrt{25a^2} - \sqrt{49b^2}$
$\phantom{(\sqrt{5a}+\sqrt{7b})(\sqrt{5a}-\sqrt{7b})} = 5a - 7b$

68. $(\sqrt[3]{y}-6)(\sqrt[3]{y}+3)$
$= \sqrt[3]{y} \cdot \sqrt[3]{y} + \sqrt[3]{y} \cdot 3 - 6 \cdot \sqrt[3]{y} - 6 \cdot 3$
$= \sqrt[3]{y^2} + 3\sqrt[3]{y} - 6\sqrt[3]{y} - 18$
$= \sqrt[3]{y^2} + (3-6)\sqrt[3]{y} - 18$
$= \sqrt[3]{y^2} - 3\sqrt[3]{y} - 18$

70. $(\sqrt[3]{4p}-1)(\sqrt[3]{4p}+3)$
$= \sqrt[3]{4p} \cdot \sqrt[3]{4p} + \sqrt[3]{4p} \cdot 3 - 1 \cdot \sqrt[3]{4p} - 1 \cdot 3$
$= \sqrt[3]{16p^2} + 3\sqrt[3]{4p} - \sqrt[3]{4p} - 3$
$= 2\sqrt[3]{2p^2} + (3-1)\sqrt[3]{4p} - 3$
$= 2\sqrt[3]{2p^2} + 2\sqrt[3]{4p} - 3$

72. $\sqrt{7}(\sqrt{14}+\sqrt{3}) = \sqrt{7} \cdot \sqrt{14} + \sqrt{7} \cdot \sqrt{3}$
$\phantom{\sqrt{7}(\sqrt{14}+\sqrt{3})} = \sqrt{98} + \sqrt{21}$
$\phantom{\sqrt{7}(\sqrt{14}+\sqrt{3})} = \sqrt{49} \cdot \sqrt{2} + \sqrt{21}$
$\phantom{\sqrt{7}(\sqrt{14}+\sqrt{3})} = 7\sqrt{2} + \sqrt{21}$

74. $\sqrt{180a^5} + a^2\sqrt{20} - a\sqrt{80a^3}$
$= \sqrt{36a^4} \cdot \sqrt{5a} + a^2\sqrt{4} \cdot \sqrt{5} - a\sqrt{16a^2} \cdot \sqrt{5a}$
$= 6a^2\sqrt{5a} + a^2 \cdot 2\sqrt{5} - a \cdot 4a\sqrt{5a}$
$= 6a^2\sqrt{5a} + 2a^2\sqrt{5} - 4a^2\sqrt{5a}$
$= (6-4)a^2\sqrt{5a} + 2a^2\sqrt{5}$
$= 2a^2\sqrt{5a} + 2a^2\sqrt{5}$ or $2a^2\sqrt{5}(\sqrt{a}+1)$

76. $(4\sqrt{2}-2)(4\sqrt{2}+2) = (4\sqrt{2})^2 - 2^2$
$\phantom{(4\sqrt{2}-2)(4\sqrt{2}+2)} = 16\sqrt{4} - 4$
$\phantom{(4\sqrt{2}-2)(4\sqrt{2}+2)} = 16 \cdot 2 - 4$
$\phantom{(4\sqrt{2}-2)(4\sqrt{2}+2)} = 32 - 4$
$\phantom{(4\sqrt{2}-2)(4\sqrt{2}+2)} = 28$

78. $\sqrt[3]{9}(5+2\sqrt[3]{2}) = \sqrt[3]{9} \cdot 5 + \sqrt[3]{9} \cdot 2\sqrt[3]{2}$
$\phantom{\sqrt[3]{9}(5+2\sqrt[3]{2})} = 5\sqrt[3]{9} + 2\sqrt[3]{9 \cdot 2}$
$\phantom{\sqrt[3]{9}(5+2\sqrt[3]{2})} = 5\sqrt[3]{9} + 2\sqrt[3]{18}$

80. $(5\sqrt{5}-3)(3\sqrt{5}-4)$
$= 5\sqrt{5} \cdot 3\sqrt{5} - 5\sqrt{5} \cdot 4 - 3 \cdot 3\sqrt{5} + 3 \cdot 4$
$= 15\sqrt{25} - 20\sqrt{5} - 9\sqrt{5} + 12$
$= 15 \cdot 5 + (-20-9)\sqrt{5} + 12$
$= 75 - 29\sqrt{5} + 12$
$= 87 - 29\sqrt{5}$

82. $5\sqrt{20} + 2\sqrt{80} = 5\sqrt{4} \cdot \sqrt{5} + 2\sqrt{16} \cdot \sqrt{5}$
$\phantom{5\sqrt{20} + 2\sqrt{80}} = 5 \cdot 2\sqrt{5} + 2 \cdot 4\sqrt{5}$
$\phantom{5\sqrt{20} + 2\sqrt{80}} = 10\sqrt{5} + 8\sqrt{5}$
$\phantom{5\sqrt{20} + 2\sqrt{80}} = (10+8)\sqrt{5}$
$\phantom{5\sqrt{20} + 2\sqrt{80}} = 18\sqrt{5}$

84. $(\sqrt{2}-\sqrt{7})^2 = (\sqrt{2})^2 - 2(\sqrt{2})(\sqrt{7}) + (\sqrt{7})^2$
$\phantom{(\sqrt{2}-\sqrt{7})^2} = \sqrt{4} - 2\sqrt{14} + \sqrt{49}$
$\phantom{(\sqrt{2}-\sqrt{7})^2} = 2 - 2\sqrt{14} + 7$
$\phantom{(\sqrt{2}-\sqrt{7})^2} = 9 - 2\sqrt{14}$

## Chapter 7: Radicals and Rational Exponents

86. $5\sqrt[3]{3m^3n} + \sqrt[3]{81n} = 5\sqrt[3]{m^3} \cdot \sqrt[3]{3n} + \sqrt[3]{27} \cdot \sqrt[3]{3n}$
$= 5m\sqrt[3]{3n} + 3\sqrt[3]{3n}$
$= (5m+3)\sqrt[3]{3n}$

88. $(\sqrt{3a} - \sqrt{4b})(\sqrt{3a} + \sqrt{4b}) = (\sqrt{3a})^2 - (\sqrt{4b})^2$
$= \sqrt{9a^2} - \sqrt{16b^2}$
$= 3a - 4b$

90. $\dfrac{4}{5} \cdot \left(-\dfrac{\sqrt{5}}{5}\right) + \left(-\dfrac{3}{5}\right) \cdot \left(-\dfrac{2\sqrt{5}}{5}\right)$
$= -\dfrac{4\sqrt{5}}{25} + \dfrac{6\sqrt{5}}{25}$
$= \dfrac{-4\sqrt{5} + 6\sqrt{5}}{25}$
$= \dfrac{(-4+6)\sqrt{5}}{25}$
$= \dfrac{2\sqrt{5}}{25}$

92. $A = l \cdot w$
$= \sqrt{148} \cdot \sqrt{48}$
$= \sqrt{4} \cdot \sqrt{37} \cdot \sqrt{16} \cdot \sqrt{3}$
$= 2\sqrt{37} \cdot 4\sqrt{3}$
$= 8\sqrt{111}$
The area is $8\sqrt{111} \approx 84.3$ square units.

$P = 2l + 2w$
$= 2(\sqrt{148}) + 2(\sqrt{48})$
$= 2\sqrt{4} \cdot \sqrt{37} + 2\sqrt{16} \cdot \sqrt{3}$
$= 2 \cdot 2\sqrt{37} + 2 \cdot 4\sqrt{3}$
$= 4\sqrt{37} + 8\sqrt{3}$
The perimeter is $4\sqrt{37} + 8\sqrt{3} \approx 38.2$ units.

94. Area of larger triangle:
$s = \dfrac{1}{2}(20 + 16 + 12) = \dfrac{1}{2}(48) = 24$

$A = \sqrt{24(24-20)(24-16)(24-12)}$
$= \sqrt{24(4)(8)(12)}$
$= \sqrt{96 \cdot 96}$
$= \sqrt{(96)^2}$
$= 96$

Area of smaller triangle:
$s = \dfrac{1}{2}(10 + 8 + 6) = \dfrac{1}{2}(24) = 12$
$A = \sqrt{12(12-10)(12-8)(12-6)}$
$= \sqrt{12(2)(4)(6)}$
$= \sqrt{24 \cdot 24}$
$= \sqrt{(24)^2}$
$= 24$

Area of shaded region:
area of large − area of smaller
$= 96 - 24$
$= 72$
The area of the shaded region is 72 square units.

96. a. $(f+g)(x) = \sqrt{4x-4} + \sqrt{25x-25}$
$= \sqrt{4(x-1)} + \sqrt{25(x-1)}$
$= \sqrt{4} \cdot \sqrt{x-1} + \sqrt{25} \cdot \sqrt{x-1}$
$= 2\sqrt{x-1} + 5\sqrt{x-1}$
$= (2+5)\sqrt{x-1}$
$= 7\sqrt{x-1}$

b. $(f+g)(10) = 7\sqrt{10-1} = 7\sqrt{9} = 7 \cdot 3 = 21$

c. $(f \cdot g)(x) = \sqrt{4x-4} \cdot \sqrt{25x-25}$
$= \sqrt{(4x-4)(25x-25)}$
$= \sqrt{4(x-1) \cdot 25(x-1)}$
$= \sqrt{100(x-1)^2}$
$= \sqrt{10^2(x-1)^2}$
$= 10(x-1), \ x \geq 1$

98. $(3a^3b)(4a^2b^4) = 3 \cdot 4 a^{3+2} b^{1+4} = 12a^5b^5$

## Section 7.4 Adding, Subtracting, and Multiplying Radical Expressions

100. $(3y+2)(2y-1) = 3y(2y-1) + 2(2y-1)$
$= 6y^2 - 3y + 4y - 2$
$= 6y^2 + y - 2$

102. $(5w+2)(5w-2) = (5w)^2 - 2^2 = 25w^2 - 4$

### 7.5 Preparing for Rationalizing Radical Expressions

1. Start by finding the prime factors of 12.
$12 = 3 \cdot 4 = 3 \cdot 2 \cdot 2$
To make a perfect square, we need to have each unique prime factor occur an even number of times. To make the smallest perfect square, we want to use the smallest even number possible for each factor. In this case, the factor 2 occurs twice so no additional factors of 2 are needed. However, the factor 3 only occurs once. Therefore, we need to multiply by one more factor of 3.
The smallest perfect square that is a multiple of 12 is $12 \cdot 3 = 36$.

2. $\sqrt{25x^2} = \sqrt{(5x)^2} = 5x$
(since $x > 0$, we do not need $|x|$)

### 7.5 Exercises

2. $\sqrt{7}$

4. Answers will vary.
To maintain an equivalent expression, we need to multiply the numerator and denominator by the same nonzero quantity. To eliminate the radical in the denominator, we multiply by the conjugate of the denominator.

6. $\dfrac{2}{\sqrt{3}} = \dfrac{2}{\sqrt{3}} \cdot \dfrac{\sqrt{3}}{\sqrt{3}} = \dfrac{2\sqrt{3}}{\sqrt{9}} = \dfrac{2\sqrt{3}}{3}$

8. $-\dfrac{3}{2\sqrt{3}} = -\dfrac{3}{2\sqrt{3}} \cdot \dfrac{\sqrt{3}}{\sqrt{3}}$
$= -\dfrac{3\sqrt{3}}{2\sqrt{9}}$
$= -\dfrac{3\sqrt{3}}{2 \cdot 3}$
$= -\dfrac{\sqrt{3}}{2}$

10. $\dfrac{5}{\sqrt{20}} = \dfrac{5}{2\sqrt{5}} = \dfrac{5}{2\sqrt{5}} \cdot \dfrac{\sqrt{5}}{\sqrt{5}}$
$= \dfrac{5\sqrt{5}}{2\sqrt{25}}$
$= \dfrac{5\sqrt{5}}{2 \cdot 5}$
$= \dfrac{\sqrt{5}}{2}$

12. $\dfrac{\sqrt{3}}{\sqrt{11}} = \dfrac{\sqrt{3}}{\sqrt{11}} \cdot \dfrac{\sqrt{11}}{\sqrt{11}} = \dfrac{\sqrt{33}}{\sqrt{121}} = \dfrac{\sqrt{33}}{11}$

14. $\sqrt{\dfrac{5}{z}} = \dfrac{\sqrt{5}}{\sqrt{z}} = \dfrac{\sqrt{5}}{\sqrt{z}} \cdot \dfrac{\sqrt{z}}{\sqrt{z}} = \dfrac{\sqrt{5z}}{\sqrt{z^2}} = \dfrac{\sqrt{5z}}{z}$

16. $\dfrac{\sqrt{32}}{\sqrt{a^5}} = \dfrac{4\sqrt{2}}{a^2\sqrt{a}} = \dfrac{4\sqrt{2}}{a^2\sqrt{a}} \cdot \dfrac{\sqrt{a}}{\sqrt{a}}$
$= \dfrac{4\sqrt{2a}}{a^2\sqrt{a^2}} = \dfrac{4\sqrt{2a}}{a^2 \cdot a}$
$= \dfrac{4\sqrt{2a}}{a^3}$

18. $\dfrac{5}{\sqrt[3]{3}} = \dfrac{5}{\sqrt[3]{3}} \cdot \dfrac{\sqrt[3]{9}}{\sqrt[3]{9}} = \dfrac{5\sqrt[3]{9}}{\sqrt[3]{27}} = \dfrac{5\sqrt[3]{9}}{3}$

20. $\sqrt[3]{\dfrac{-4}{p}} = \dfrac{\sqrt[3]{-4}}{\sqrt[3]{p}} = \dfrac{\sqrt[3]{-4}}{\sqrt[3]{p}} \cdot \dfrac{\sqrt[3]{p^2}}{\sqrt[3]{p^2}}$
$= \dfrac{\sqrt[3]{-4p^2}}{\sqrt[3]{p^3}}$
$= \dfrac{-\sqrt[3]{4p^2}}{p}$

## Chapter 7: Radicals and Rational Exponents

**22.** $\sqrt[3]{\dfrac{-5}{72}} = \dfrac{\sqrt[3]{-5}}{\sqrt[3]{72}} = \dfrac{-\sqrt[3]{5}}{2\sqrt[3]{9}}$

$= \dfrac{-\sqrt[3]{5}}{2\sqrt[3]{9}} \cdot \dfrac{\sqrt[3]{3}}{\sqrt[3]{3}} = \dfrac{-\sqrt[3]{15}}{2\sqrt[3]{27}}$

$= \dfrac{-\sqrt[3]{15}}{2 \cdot 3}$

$= -\dfrac{\sqrt[3]{15}}{6}$

**24.** $\dfrac{8}{\sqrt[3]{36z^2}} = \dfrac{8}{\sqrt[3]{36z^2}} \cdot \dfrac{\sqrt[3]{6z}}{\sqrt[3]{6z}}$

$= \dfrac{8\sqrt[3]{6z}}{\sqrt[3]{216z^3}} = \dfrac{8\sqrt[3]{6z}}{6z}$

$= \dfrac{4\sqrt[3]{6z}}{3z}$

**26.** $\dfrac{6}{\sqrt[4]{9b^2}} = \dfrac{6}{\sqrt[4]{9b^2}} \cdot \dfrac{\sqrt[4]{9b^2}}{\sqrt[4]{9b^2}}$

$= \dfrac{6\sqrt[4]{9b^2}}{\sqrt[4]{81b^4}} = \dfrac{6\sqrt[4]{9b^2}}{3b}$

$= \dfrac{2\sqrt[4]{9b^2}}{b} = \dfrac{2\sqrt[4]{(3b)^2}}{b} = \dfrac{2\sqrt{3b}}{b}$

**28.** $\dfrac{-3}{\sqrt[5]{ab^3}} = \dfrac{-3}{\sqrt[5]{ab^3}} \cdot \dfrac{\sqrt[5]{a^4b^2}}{\sqrt[5]{a^4b^2}}$

$= \dfrac{-3\sqrt[5]{a^4b^2}}{\sqrt[5]{a^5b^5}}$

$= -\dfrac{3\sqrt[5]{a^4b^2}}{ab}$

**30.** $\dfrac{6}{\sqrt{7}-2} = \dfrac{6}{\sqrt{7}-2} \cdot \dfrac{\sqrt{7}+2}{\sqrt{7}+2} = \dfrac{6(\sqrt{7}+2)}{(\sqrt{7})^2 - 2^2}$

$= \dfrac{6(\sqrt{7}+2)}{7-4} = \dfrac{6(\sqrt{7}+2)}{3}$

$= 2(\sqrt{7}+2)$

**32.** $\dfrac{10}{\sqrt{10}+3} = \dfrac{10}{\sqrt{10}+3} \cdot \dfrac{\sqrt{10}-3}{\sqrt{10}-3} = \dfrac{10(\sqrt{10}-3)}{(\sqrt{10})^2 - 3^2}$

$= \dfrac{10(\sqrt{10}-3)}{10-9}$

$= 10(\sqrt{10}-3)$

**34.** $\dfrac{12}{\sqrt{11}-\sqrt{7}} = \dfrac{12}{\sqrt{11}-\sqrt{7}} \cdot \dfrac{\sqrt{11}+\sqrt{7}}{\sqrt{11}+\sqrt{7}}$

$= \dfrac{12(\sqrt{11}+\sqrt{7})}{(\sqrt{11})^2 - (\sqrt{7})^2}$

$= \dfrac{12(\sqrt{11}+\sqrt{7})}{11-7}$

$= \dfrac{12(\sqrt{11}+\sqrt{7})}{4}$

$= 3(\sqrt{11}+\sqrt{7})$

**36.** $\dfrac{\sqrt{3}}{\sqrt{15}-\sqrt{6}} = \dfrac{\sqrt{3}}{\sqrt{15}-\sqrt{6}} \cdot \dfrac{\sqrt{15}+\sqrt{6}}{\sqrt{15}+\sqrt{6}}$

$= \dfrac{\sqrt{3}(\sqrt{15}+\sqrt{6})}{(\sqrt{15})^2 - (\sqrt{6})^2} = \dfrac{\sqrt{45}+\sqrt{18}}{15-6}$

$= \dfrac{3\sqrt{5}+3\sqrt{2}}{9} = \dfrac{3(\sqrt{5}+\sqrt{2})}{9}$

$= \dfrac{\sqrt{5}+\sqrt{2}}{3}$

**38.** $\dfrac{\sqrt{a}}{\sqrt{a}+\sqrt{b}} = \dfrac{\sqrt{a}}{\sqrt{a}+\sqrt{b}} \cdot \dfrac{\sqrt{a}-\sqrt{b}}{\sqrt{a}-\sqrt{b}}$

$= \dfrac{\sqrt{a}(\sqrt{a}-\sqrt{b})}{(\sqrt{a})^2 - (\sqrt{b})^2}$

$= \dfrac{\sqrt{a^2}-\sqrt{ab}}{a-b}$

$= \dfrac{a-\sqrt{ab}}{a-b}$

**Section 7.5** *Rationalizing Radical Expressions*

40. $\dfrac{15}{3\sqrt{5}+4\sqrt{3}} = \dfrac{15}{3\sqrt{5}+4\sqrt{3}} \cdot \dfrac{3\sqrt{5}-4\sqrt{3}}{3\sqrt{5}-4\sqrt{3}}$
$= \dfrac{15(3\sqrt{5}-4\sqrt{3})}{(3\sqrt{5})^2-(4\sqrt{3})^2}$
$= \dfrac{15(3\sqrt{5}-4\sqrt{3})}{45-48}$
$= \dfrac{15(3\sqrt{5}-4\sqrt{3})}{-3}$
$= -5(3\sqrt{5}-4\sqrt{3})$
$= 5(4\sqrt{3}-3\sqrt{5})$

42. $\dfrac{\sqrt{5}+3}{\sqrt{5}-3} = \dfrac{\sqrt{5}+3}{\sqrt{5}-3} \cdot \dfrac{\sqrt{5}+3}{\sqrt{5}+3}$
$= \dfrac{(\sqrt{5})^2+2(\sqrt{5})(3)+3^2}{(\sqrt{5})^2-3^2}$
$= \dfrac{5+6\sqrt{5}+9}{5-9}$
$= \dfrac{14+6\sqrt{5}}{-4}$
$= -\dfrac{7+3\sqrt{5}}{2}$

44. $\dfrac{3\sqrt{6}+5\sqrt{7}}{2\sqrt{6}-3\sqrt{7}} = \dfrac{3\sqrt{6}+5\sqrt{7}}{2\sqrt{6}-3\sqrt{7}} \cdot \dfrac{2\sqrt{6}+3\sqrt{7}}{2\sqrt{6}+3\sqrt{7}}$
$= \dfrac{(3\sqrt{6}+5\sqrt{7})(2\sqrt{6}+3\sqrt{7})}{(2\sqrt{6})^2-(3\sqrt{7})^2}$
$= \dfrac{6\sqrt{36}+9\sqrt{42}+10\sqrt{42}+15\sqrt{49}}{24-63}$
$= \dfrac{36+19\sqrt{42}+105}{-39}$
$= \dfrac{141+19\sqrt{42}}{-39}$
$= -\dfrac{141+19\sqrt{42}}{39}$

46. $\dfrac{\sqrt{x}-4}{\sqrt{x}+4} = \dfrac{\sqrt{x}-4}{\sqrt{x}+4} \cdot \dfrac{\sqrt{x}-4}{\sqrt{x}-4}$
$= \dfrac{(\sqrt{x})^2-2(\sqrt{x})(4)+4^2}{(\sqrt{x})^2-4^2}$
$= \dfrac{x-8\sqrt{x}+16}{x-16}$

48. $\dfrac{2\sqrt{3}+3}{\sqrt{12}-\sqrt{3}}$
$= \dfrac{2\sqrt{3}+3}{2\sqrt{3}-\sqrt{3}}$
$= \dfrac{2\sqrt{3}+3}{\sqrt{3}}$
$= \dfrac{2\sqrt{3}+3}{\sqrt{3}} \cdot \dfrac{\sqrt{3}}{\sqrt{3}}$
$= \dfrac{6+3\sqrt{3}}{3}$
$= 2+\sqrt{3}$

50. $\sqrt{5}-\dfrac{1}{\sqrt{5}} = \sqrt{5} - \dfrac{1}{\sqrt{5}} \cdot \dfrac{\sqrt{5}}{\sqrt{5}}$
$= \sqrt{5} - \dfrac{\sqrt{5}}{5}$
$= \dfrac{5\sqrt{5}}{5} - \dfrac{\sqrt{5}}{5}$
$= \dfrac{4\sqrt{5}}{5}$

52. $\dfrac{\sqrt{5}}{2}+\dfrac{3}{\sqrt{5}} = \dfrac{\sqrt{5}}{2} + \dfrac{3}{\sqrt{5}} \cdot \dfrac{\sqrt{5}}{\sqrt{5}}$
$= \dfrac{\sqrt{5}}{2} + \dfrac{3\sqrt{5}}{5}$
$= \dfrac{5\sqrt{5}}{10} + \dfrac{6\sqrt{5}}{10}$
$= \dfrac{11\sqrt{5}}{10}$

*Chapter 7:* *Radicals and Rational Exponents*

54. $\sqrt{\dfrac{2}{5}} + \sqrt{20} - \sqrt{45} = \dfrac{\sqrt{2}}{\sqrt{5}} + 2\sqrt{5} - 3\sqrt{5}$

$= \dfrac{\sqrt{2}}{\sqrt{5}} - \sqrt{5}$

$= \dfrac{\sqrt{2}}{\sqrt{5}} \cdot \dfrac{\sqrt{5}}{\sqrt{5}} - \sqrt{5}$

$= \dfrac{\sqrt{10}}{5} - \sqrt{5}$

$= \dfrac{\sqrt{10}}{5} - \dfrac{5\sqrt{5}}{5}$

$= \dfrac{\sqrt{10} - 5\sqrt{5}}{5}$

56. $\sqrt{\dfrac{4}{3}} + \dfrac{4}{\sqrt{48}} = \dfrac{\sqrt{4}}{\sqrt{3}} + \dfrac{4}{4\sqrt{3}} = \dfrac{2}{\sqrt{3}} + \dfrac{1}{\sqrt{3}}$

$= \dfrac{3}{\sqrt{3}} = \dfrac{3}{\sqrt{3}} \cdot \dfrac{\sqrt{3}}{\sqrt{3}} = \dfrac{3\sqrt{3}}{3}$

$= \sqrt{3}$

58. $\dfrac{\sqrt{2}}{\sqrt{18}} = \dfrac{\sqrt{2}}{3\sqrt{2}} = \dfrac{1}{3}$

60. $\dfrac{7}{\sqrt{98}} = \dfrac{7}{7\sqrt{2}} = \dfrac{1}{\sqrt{2}} = \dfrac{1}{\sqrt{2}} \cdot \dfrac{\sqrt{2}}{\sqrt{2}} = \dfrac{\sqrt{2}}{2}$

62. $\sqrt{\dfrac{9}{5}} = \dfrac{\sqrt{9}}{\sqrt{5}} = \dfrac{3}{\sqrt{5}} = \dfrac{3}{\sqrt{5}} \cdot \dfrac{\sqrt{5}}{\sqrt{5}} = \dfrac{3\sqrt{5}}{5}$

64. $\dfrac{\sqrt{2} - 5}{\sqrt{2} + 5} = \dfrac{\sqrt{2} - 5}{\sqrt{2} + 5} \cdot \dfrac{\sqrt{2} - 5}{\sqrt{2} - 5}$

$= \dfrac{(\sqrt{2})^2 - 2(\sqrt{2})(5) + 5^2}{(\sqrt{2})^2 - 5^2}$

$= \dfrac{2 - 10\sqrt{2} + 25}{2 - 25}$

$= \dfrac{27 - 10\sqrt{2}}{-23}$

$= \dfrac{10\sqrt{2} - 27}{23}$

66. $\dfrac{5}{\sqrt{6}+4} = \dfrac{5}{\sqrt{6}+4} \cdot \dfrac{\sqrt{6}-4}{\sqrt{6}-4}$

$= \dfrac{5(\sqrt{6}-4)}{(\sqrt{6})^2 - 4^2} = \dfrac{5(\sqrt{6}-4)}{6-16}$

$= \dfrac{5(\sqrt{6}-4)}{-10} = \dfrac{\sqrt{6}-4}{-2}$

$= \dfrac{4-\sqrt{6}}{2}$

68. $\dfrac{\sqrt{75}}{\sqrt{3}} = \dfrac{5\sqrt{3}}{\sqrt{3}} = 5$

70. $\dfrac{1}{\sqrt{7}} = \dfrac{1}{\sqrt{7}} \cdot \dfrac{\sqrt{7}}{\sqrt{7}} = \dfrac{\sqrt{7}}{7}$

72. $\dfrac{1}{\sqrt[3]{18}} = \dfrac{1}{\sqrt[3]{18}} \cdot \dfrac{\sqrt[3]{12}}{\sqrt[3]{12}}$

$= \dfrac{\sqrt[3]{12}}{\sqrt[3]{216}}$

$= \dfrac{\sqrt[3]{12}}{6}$

(Since $18 = 3^2 \cdot 2$, we need to multiply by $2^2 \cdot 3$ to get powers of 3 on the factors.)

74. $\dfrac{1}{7-\sqrt{2}} = \dfrac{1}{7-\sqrt{2}} \cdot \dfrac{7+\sqrt{2}}{7+\sqrt{2}}$

$= \dfrac{7+\sqrt{2}}{7^2 - (\sqrt{2})^2} = \dfrac{7+\sqrt{2}}{49-2}$

$= \dfrac{7+\sqrt{2}}{47}$

**76.** $-\sqrt{\dfrac{2}{3}} \cdot \left(-\dfrac{2}{\sqrt{5}}\right) + \dfrac{1}{\sqrt{3}} \cdot \dfrac{1}{\sqrt{5}}$

$= -\dfrac{\sqrt{2}}{\sqrt{3}} \cdot \left(-\dfrac{2}{\sqrt{5}}\right) + \dfrac{1}{\sqrt{3}} \cdot \dfrac{1}{\sqrt{5}}$

$= \dfrac{2\sqrt{2}}{\sqrt{15}} + \dfrac{1}{\sqrt{15}}$

$= \dfrac{2\sqrt{2}+1}{\sqrt{15}}$

$= \dfrac{2\sqrt{2}+1}{\sqrt{15}} \cdot \dfrac{\sqrt{15}}{\sqrt{15}}$

$= \dfrac{2\sqrt{30}+\sqrt{15}}{15}$

**78.** $\dfrac{\sqrt{3}+2}{2} = \dfrac{\sqrt{3}+2}{2} \cdot \dfrac{\sqrt{3}-2}{\sqrt{3}-2} = \dfrac{\left(\sqrt{3}\right)^2 - 2^2}{2\left(\sqrt{3}-2\right)}$

$= \dfrac{3-4}{2\left(\sqrt{3}-2\right)} = \dfrac{-1}{2\left(\sqrt{3}-2\right)}$

$= \dfrac{1}{2\left(2-\sqrt{3}\right)}$

**80.** $\dfrac{\sqrt{a}-\sqrt{b}}{\sqrt{2}} = \dfrac{\sqrt{a}-\sqrt{b}}{\sqrt{2}} \cdot \dfrac{\sqrt{a}+\sqrt{b}}{\sqrt{a}+\sqrt{b}}$

$= \dfrac{\left(\sqrt{a}\right)^2 - \left(\sqrt{b}\right)^2}{\sqrt{2}\left(\sqrt{a}+\sqrt{b}\right)}$

$= \dfrac{a-b}{\sqrt{2}\left(\sqrt{a}+\sqrt{b}\right)}$

**82.** $\dfrac{2}{\sqrt{2}+\sqrt{3}-\sqrt{9}} = \dfrac{2}{\left(\sqrt{2}+\sqrt{3}\right)-3}$

$= \dfrac{2}{\left(\sqrt{2}+\sqrt{3}\right)-3} \cdot \dfrac{\left(\sqrt{2}+\sqrt{3}\right)+3}{\left(\sqrt{2}+\sqrt{3}\right)+3}$

$= \dfrac{2\left[\left(\sqrt{2}+\sqrt{3}\right)+3\right]}{\left(\sqrt{2}+\sqrt{3}\right)^2 - 3^2}$

$= \dfrac{2\left[\left(\sqrt{2}+\sqrt{3}\right)+3\right]}{\left(\sqrt{2}\right)^2 + 2\left(\sqrt{2}\right)\left(\sqrt{3}\right) + \left(\sqrt{3}\right)^2 - 9}$

$= \dfrac{2\left[\left(\sqrt{2}+\sqrt{3}\right)+3\right]}{2 + 2\sqrt{6} + 3 - 9}$

$= \dfrac{2\left[\left(\sqrt{2}+\sqrt{3}\right)+3\right]}{2\sqrt{6}-4}$

$= \dfrac{\left(\sqrt{2}+\sqrt{3}\right)+3}{\sqrt{6}-2}$

$= \dfrac{\left(\sqrt{2}+\sqrt{3}\right)+3}{\sqrt{6}-2} \cdot \dfrac{\sqrt{6}+2}{\sqrt{6}+2}$

$= \dfrac{\left[\left(\sqrt{2}+\sqrt{3}\right)+3\right]\left[\sqrt{6}+2\right]}{\left(\sqrt{6}\right)^2 - 2^2}$

$= \dfrac{\left(\sqrt{2}+\sqrt{3}\right)\left(\sqrt{6}+2\right) + 3\left(\sqrt{6}+2\right)}{6-4}$

$= \dfrac{\sqrt{12}+2\sqrt{2}+\sqrt{18}+2\sqrt{3}+3\sqrt{6}+6}{2}$

$= \dfrac{2\sqrt{3}+2\sqrt{2}+3\sqrt{2}+2\sqrt{3}+3\sqrt{6}+6}{2}$

$= \dfrac{4\sqrt{3}+5\sqrt{2}+3\sqrt{6}+6}{2}$

## Chapter 7: Radicals and Rational Exponents

**84.** $g(x) = -3x + 9$

| $x$ | $y = -3x + 9$ | $(x, y)$ |
|---|---|---|
| $-1$ | $y = -3(-1) + 9 = 12$ | $(-1, 12)$ |
| $0$ | $y = -3(0) + 9 = 9$ | $(0, 9)$ |
| $1$ | $y = -3(1) + 9 = 6$ | $(1, 6)$ |
| $2$ | $y = -3(2) + 9 = 3$ | $(2, 3)$ |
| $3$ | $y = -3(3) + 9 = 0$ | $(3, 0)$ |

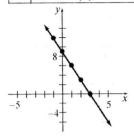

**86.** $F(x) = x^3$

| $x$ | $y = x^3$ | $(x, y)$ |
|---|---|---|
| $-2$ | $y = (-2)^3 = -8$ | $(-2, -8)$ |
| $-1$ | $y = (-1)^3 = -1$ | $(-1, -1)$ |
| $0$ | $y = 0^3 = 0$ | $(0, 0)$ |
| $1$ | $y = 1^3 = 1$ | $(1, 1)$ |
| $2$ | $y = 2^3 = 8$ | $(2, 8)$ |

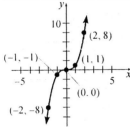

**Putting the Concepts Together (Sections 7.1-7.5)**

**1.** $-25^{1/2} = -\sqrt{25} = -5$

**2.** $(-64)^{-2/3} = \dfrac{1}{(-64)^{2/3}} = \dfrac{1}{\left(\sqrt[3]{-64}\right)^2} = \dfrac{1}{(-4)^2} = \dfrac{1}{16}$

**3.** $\sqrt[4]{3x^3} = \left(3x^3\right)^{1/4}$

**4.** $7z^{4/5} = 7\left(z^4\right)^{1/5} = 7\sqrt[5]{z^4}$

**5.** $\sqrt[3]{\sqrt{64x^3}} = \left(\left(64x^3\right)^{1/2}\right)^{1/3}$
$= \left(64x^3\right)^{1/6}$
$= \left(2^6 x^3\right)^{1/6}$
$= 2^{6/6} x^{3/6}$
$= 2x^{1/2}$ or $2\sqrt{x}$

**6.** $c^{1/2}\left(c^{3/2} + c^{5/2}\right) = c^{1/2} \cdot c^{3/2} + c^{1/2} \cdot c^{5/2}$
$= c^{\frac{1}{2}+\frac{3}{2}} + c^{\frac{1}{2}+\frac{5}{2}}$
$= c^{\frac{4}{2}} + c^{\frac{6}{2}}$
$= c^2 + c^3$

**7.** $\left(a^{2/3} b^{-1/3}\right)\left(a^{4/3} b^{-5/3}\right) = a^{\frac{2}{3}+\frac{4}{3}} b^{-\frac{1}{3}+\left(-\frac{5}{3}\right)}$
$= a^{6/3} b^{-6/3}$
$= a^2 b^{-2}$
$= \dfrac{a^2}{b^2}$

**8.** $\dfrac{x^{3/4}}{x^{1/8}} = x^{\frac{3}{4}-\frac{1}{8}} = x^{\frac{6}{8}-\frac{1}{8}} = x^{5/8}$ or $\sqrt[8]{x^5}$

**9.** $\left(x^{3/4} y^{-1/8}\right)^8 = \left(x^{3/4}\right)^8 \left(y^{-1/8}\right)^8$
$= x^{\frac{3}{4} \cdot 8} y^{-\frac{1}{8} \cdot 8}$
$= x^6 y^{-1}$
$= \dfrac{x^6}{y}$

**10.** $\sqrt{15a} \cdot \sqrt{2b} = \sqrt{15a \cdot 2b} = \sqrt{30ab}$

**11.** $\sqrt{10m^3 n^2} \cdot \sqrt{20mn} = \sqrt{10m^3 n^2 \cdot 20mn}$
$= \sqrt{200m^4 n^3}$
$= \sqrt{100m^4 n^2 \cdot 2n}$
$= \sqrt{100m^4 n^2} \cdot \sqrt{2n}$
$= 10m^2 n \sqrt{2n}$

**12.** $\sqrt[3]{\dfrac{-32xy^4}{4x^{-2}y}} = \sqrt[3]{-8x^3 y^3} = \sqrt[3]{(-2)^3 x^3 y^3} = -2xy$

ISM − 394

Putting the Concepts Together (Sections 7.1 – 7.5)

13. $2\sqrt{108} - 3\sqrt{75} + \sqrt{48}$
$= 2 \cdot \sqrt{36 \cdot 3} - 3\sqrt{25 \cdot 3} + \sqrt{16 \cdot 3}$
$= 2\sqrt{36} \cdot \sqrt{3} - 3\sqrt{25} \cdot \sqrt{3} + \sqrt{16} \cdot \sqrt{3}$
$= 2 \cdot 6\sqrt{3} - 3 \cdot 5\sqrt{3} + 4\sqrt{3}$
$= 12\sqrt{3} - 15\sqrt{3} + 4\sqrt{3}$
$= (12 - 15 + 4)\sqrt{3}$
$= \sqrt{3}$

14. $-5b\sqrt{8b} + 7\sqrt{18b^3} = -5b\sqrt{4 \cdot 2b} + 7\sqrt{9b^2 \cdot 2b}$
$= -5b\sqrt{4} \cdot \sqrt{2b} + 7\sqrt{9b^2} \cdot \sqrt{2b}$
$= -5b \cdot 2\sqrt{2b} + 7 \cdot 3b\sqrt{2b}$
$= -10b\sqrt{2b} + 21b\sqrt{2b}$
$= (-10 + 21)b\sqrt{2b}$
$= 11b\sqrt{2b}$

15. $\sqrt[3]{16y^4} - y\sqrt[3]{2y} = \sqrt[3]{8y^3 \cdot 2y} - y\sqrt[3]{2y}$
$= \sqrt[3]{8y^3} \cdot \sqrt[3]{2y} - y\sqrt[3]{2y}$
$= 2y\sqrt[3]{2y} - y\sqrt[3]{2y}$
$= (2-1)y\sqrt[3]{2y}$
$= y\sqrt[3]{2y}$

16. $(3\sqrt{x})(4\sqrt{x}) = 3 \cdot 4 \cdot \sqrt{x \cdot x}$
$= 12\sqrt{x^2}$
$= 12x$

17. $3\sqrt{x} + 4\sqrt{x} = (3+4)\sqrt{x} = 7\sqrt{x}$

18. $(2 - 3\sqrt{2})(10 + \sqrt{2})$
$= 2 \cdot 10 + 2 \cdot \sqrt{2} - 3\sqrt{2} \cdot 10 - 3\sqrt{2} \cdot \sqrt{2}$
$= 20 + 2\sqrt{2} - 30\sqrt{2} - 3 \cdot 2$
$= 20 - 28\sqrt{2} - 6$
$= 14 - 28\sqrt{2} = 14(1 - 2\sqrt{2})$

19. $(4\sqrt{2} - 3)^2 = (4\sqrt{2})^2 - 2(4\sqrt{2})(3) + (3)^2$
$= 16 \cdot 2 - 24\sqrt{2} + 9$
$= 32 - 24\sqrt{2} + 9$
$= 41 - 24\sqrt{2}$

20. $\dfrac{3}{2\sqrt{32}} = \dfrac{3}{2 \cdot 4\sqrt{2}} = \dfrac{3}{8\sqrt{2}}$
$= \dfrac{3}{8\sqrt{2}} \cdot \dfrac{\sqrt{2}}{\sqrt{2}} = \dfrac{3\sqrt{2}}{8 \cdot 2}$
$= \dfrac{3\sqrt{2}}{16}$

21. $\dfrac{4}{\sqrt{3} - 8} = \dfrac{4}{\sqrt{3} - 8} \cdot \dfrac{\sqrt{3} + 8}{\sqrt{3} + 8}$
$= \dfrac{4(\sqrt{3} + 8)}{(\sqrt{3} - 8)(\sqrt{3} + 8)}$
$= \dfrac{4(\sqrt{3} + 8)}{(\sqrt{3})^2 - 8^2}$
$= \dfrac{4\sqrt{3} + 32}{3 - 64}$
$= \dfrac{4\sqrt{3} + 32}{-61}$
$= -\dfrac{4\sqrt{3} + 32}{61} = -\dfrac{4(\sqrt{3} + 8)}{61}$

**7.6 Preparing for Functions Involving Radicals**

1. $\sqrt{121} = \sqrt{11^2} = 11$

2. $\sqrt{p^2} = |p|$

3. $f(x) = x^2 - 4$
$f(3) = (3)^2 - 4$
$= 9 - 4$
$= 5$

## Chapter 7: Radicals and Rational Exponents

4. $-2x + 3 \geq 0$
$-2x + 3 - 3 \geq 0 - 3$
$-2x \geq -3$
$\dfrac{-2x}{-2} \leq \dfrac{-3}{-2}$
$x \leq \dfrac{3}{2}$

The solution set is $\left\{ x \mid x \leq \dfrac{3}{2} \right\}$, or $\left( -\infty, \dfrac{3}{2} \right]$ in interval notation.

5. $f(x) = x^2 + 1$

| $x$ | $y = x^2 + 1$ | $(x, y)$ |
|---|---|---|
| $-2$ | $y = (-2)^2 + 1 = 5$ | $(-2, 5)$ |
| $-1$ | $y = (-1)^2 + 1 = 2$ | $(-1, 2)$ |
| $0$ | $y = (0)^2 + 1 = 1$ | $(0, 1)$ |
| $1$ | $y = (1)^2 + 1 = 2$ | $(1, 2)$ |
| $2$ | $y = (2)^2 + 1 = 5$ | $(2, 5)$ |

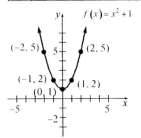

### 7.6 Exercises

2. odd

4. True

6. $f(x) = \sqrt{x + 10}$

   a. $f(6) = \sqrt{6 + 10} = \sqrt{16} = 4$

   b. $f(2) = \sqrt{2 + 10} = \sqrt{12} = 2\sqrt{3}$

   c. $f(-6) = \sqrt{-6 + 10} = \sqrt{4} = 2$

8. $g(x) = -\sqrt{4x + 5}$

   a. $g(1) = -\sqrt{4(1) + 5} = -\sqrt{9} = -3$

   b. $g(10) = -\sqrt{4(10) + 5} = -\sqrt{45} = -3\sqrt{5}$

   c. $g\left( \dfrac{1}{8} \right) = -\sqrt{4\left( \dfrac{1}{8} \right) + 5}$
   $= -\sqrt{\dfrac{1}{2} + 5} = -\sqrt{\dfrac{11}{2}}$
   $= -\dfrac{\sqrt{11}}{\sqrt{2}} = -\dfrac{\sqrt{11}}{\sqrt{2}} \cdot \dfrac{\sqrt{2}}{\sqrt{2}}$
   $= -\dfrac{\sqrt{22}}{2}$

10. $G(p) = 3\sqrt{4p + 1}$

    a. $G(2) = 3\sqrt{4(2) + 1} = 3\sqrt{9} = 3 \cdot 3 = 9$

    b. $G(11) = 3\sqrt{4(11) + 1} = 3\sqrt{45}$
    $= 3 \cdot 3\sqrt{5} = 9\sqrt{5}$

    c. $G\left( \dfrac{1}{8} \right) = 3\sqrt{4\left( \dfrac{1}{8} \right) + 1} = 3\sqrt{\dfrac{3}{2}}$
    $= 3\dfrac{\sqrt{3}}{\sqrt{2}} \cdot \dfrac{\sqrt{2}}{\sqrt{2}} = \dfrac{3\sqrt{6}}{2}$

12. $G(t) = \sqrt[3]{t - 6}$

    a. $G(7) = \sqrt[3]{7 - 6} = \sqrt[3]{1} = 1$

    b. $G(-21) = \sqrt[3]{-21 - 6} = \sqrt[3]{-27} = -3$

    c. $G(22) = \sqrt[3]{22 - 6} = \sqrt[3]{16} = 2\sqrt[3]{2}$

14. $f(x) = \sqrt{\dfrac{x - 4}{x + 4}}$

    a. $f(5) = \sqrt{\dfrac{5 - 4}{5 + 4}} = \sqrt{\dfrac{1}{9}} = \dfrac{1}{3}$

    b. $f(8) = \sqrt{\dfrac{8 - 4}{8 + 4}} = \sqrt{\dfrac{4}{12}} = \sqrt{\dfrac{1}{3}}$
    $= \dfrac{\sqrt{1}}{\sqrt{3}} = \dfrac{1}{\sqrt{3}} \cdot \dfrac{\sqrt{3}}{\sqrt{3}} = \dfrac{\sqrt{3}}{3}$

Section 7.6 Functions Involving Radicals

c. $f(12) = \sqrt{\dfrac{12-4}{12+4}} = \sqrt{\dfrac{8}{16}} = \sqrt{\dfrac{1}{2}}$

$= \dfrac{\sqrt{1}}{\sqrt{2}} = \dfrac{1}{\sqrt{2}} \cdot \dfrac{\sqrt{2}}{\sqrt{2}} = \dfrac{\sqrt{2}}{2}$

16. $H(z) = \sqrt[3]{\dfrac{3z}{z+5}}$

   a. $H(3) = \sqrt[3]{\dfrac{3(3)}{3+5}} = \sqrt[3]{\dfrac{9}{8}} = \dfrac{\sqrt[3]{9}}{\sqrt[3]{8}} = \dfrac{\sqrt[3]{9}}{2}$

   b. $H(4) = \sqrt[3]{\dfrac{3(4)}{4+5}} = \sqrt[3]{\dfrac{12}{9}} = \sqrt[3]{\dfrac{4}{3}}$

   $= \dfrac{\sqrt[3]{4}}{\sqrt[3]{3}} \cdot \dfrac{\sqrt[3]{9}}{\sqrt[3]{9}} = \dfrac{\sqrt[3]{36}}{\sqrt[3]{27}} = \dfrac{\sqrt[3]{36}}{3}$

   c. $H(-1) = \sqrt[3]{\dfrac{3(-1)}{-1+5}} = \sqrt[3]{\dfrac{-3}{4}} = \dfrac{\sqrt[3]{-3}}{\sqrt[3]{4}}$

   $= \dfrac{-\sqrt[3]{3}}{\sqrt[4]{4}} \cdot \dfrac{\sqrt[3]{2}}{\sqrt[3]{2}} = \dfrac{-\sqrt[3]{6}}{2}$

18. $f(x) = \sqrt{x+4}$

   $x + 4 \geq 0$

   $x \geq -4$

   The domain of the function is $\{x \mid x \geq -4\}$ or the interval $[-4, \infty)$.

20. $g(x) = \sqrt{3x+7}$

   $3x + 7 \geq 0$

   $3x \geq -7$

   $x \geq -\dfrac{7}{3}$

   The domain of the function is $\left\{x \mid x \geq -\dfrac{7}{3}\right\}$ or the interval $\left[-\dfrac{7}{3}, \infty\right)$.

22. $G(x) = \sqrt{5-2x}$

   $5 - 2x \geq 0$

   $-2x \geq -5$

   $x \leq \dfrac{5}{2}$

The domain of the function is $\left\{x \mid x \leq \dfrac{5}{2}\right\}$ or the interval $\left(-\infty, \dfrac{5}{2}\right]$.

24. $G(z) = \sqrt[3]{5z-3}$

   Since the index is odd, the domain of the function is all real numbers.

   $\{z \mid z \text{ is any real number}\}$ or $(-\infty, \infty)$.

26. $C(y) = \sqrt[4]{3y-2}$

   $3y - 2 \geq 0$

   $3y \geq 2$

   $y \geq \dfrac{2}{3}$

   The domain of the function is $\left\{y \mid y \geq \dfrac{2}{3}\right\}$ or the interval $\left[\dfrac{2}{3}, \infty\right)$.

28. $g(x) = \sqrt[5]{x+9}$

   Since the index is odd, the domain of the function is all real numbers.

   $\{x \mid x \text{ is any real number}\}$ or $(-\infty, \infty)$.

30. $f(x) = \sqrt{\dfrac{3}{x-3}}$

   The index is even so we need the radicand to be nonnegative. The numerator is positive which means we need the denominator to be positive as well.

   $x - 3 > 0$

   $x > 3$

   The domain of the function is $\{x \mid x > 3\}$ or the interval $(3, \infty)$.

32. $H(x) = \sqrt{\dfrac{x-5}{x}}$

   The index is even so we need the radicand to be non-negative. That is, we need to solve

ISM – 397

## Chapter 7: Radicals and Rational Exponents

$\dfrac{x-5}{x} \geq 0$

The radicand equals 0 when $x = 5$ and it is undefined when $x = 0$. We can use these two values to split up the real number line into subintervals.

| Interval | $(-\infty, 0)$ | $(0, 5)$ | $(5, \infty)$ |
|---|---|---|---|
| Num. chosen | $-1$ | $1$ | $6$ |
| Value of radicand | $6$ | $-4$ | $\frac{1}{6}$ |
| Conclusion | positive | negative | positive |

Since we need the radicand to be positive or 0, the domain is $\{x \mid x < 0 \text{ or } x \geq 5\}$ or the interval $(-\infty, 0) \cup [5, \infty)$.

**34.** $f(x) = \sqrt{x-1}$

a. $x - 1 \geq 0$
   $x \geq 1$
   The domain is $\{x \mid x \geq 1\}$ or the interval $[1, \infty)$.

b. 
| $x$ | $f(x) = \sqrt{x-1}$ | $(x, y)$ |
|---|---|---|
| 1 | $f(1) = \sqrt{1-1} = 0$ | $(1, 0)$ |
| 2 | $f(2) = 1$ | $(2, 1)$ |
| 5 | $f(5) = 2$ | $(5, 2)$ |
| 10 | $f(10) = 3$ | $(10, 3)$ |
| 17 | $f(17) = 4$ | $(17, 4)$ |

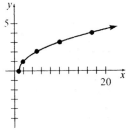

c. Based on the graph, the range is $[0, \infty)$.

**36.** $g(x) = \sqrt{x+5}$

a. $x + 5 \geq 0$
   $x \geq -5$
   The domain is $\{x \mid x \geq -5\}$ or the interval $[-5, \infty)$

b.
| $x$ | $g(x) = \sqrt{x+5}$ | $(x, y)$ |
|---|---|---|
| $-5$ | $g(-5) = \sqrt{-5+5} = 0$ | $(-5, 0)$ |
| $-4$ | $g(-4) = 1$ | $(-4, 1)$ |
| $-1$ | $g(-1) = 2$ | $(-1, 2)$ |
| 4 | $g(4) = 3$ | $(4, 3)$ |
| 11 | $g(11) = 4$ | $(11, 4)$ |

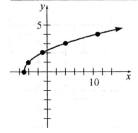

c. Based on the graph, the range is $[0, \infty)$

**38.** $F(x) = \sqrt{4-x}$

a. $4 - x \geq 0$
   $-x \geq -4$
   $x \leq 4$
   The domain is $\{x \mid x \leq 4\}$ or the interval $(-\infty, 4]$.

b.
| $x$ | $F(x) = \sqrt{4-x}$ | $(x, y)$ |
|---|---|---|
| $-12$ | $F(-12) = \sqrt{4-(-12)} = 4$ | $(-12, 4)$ |
| $-5$ | $F(-5) = 3$ | $(-5, 3)$ |
| 0 | $F(0) = 2$ | $(0, 2)$ |
| 3 | $F(3) = 1$ | $(3, 1)$ |
| 4 | $F(4) = 0$ | $(4, 0)$ |

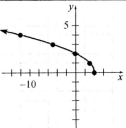

c. Based on the graph, the range is $[0, \infty)$.

**40.** $f(x) = \sqrt{x} + 1$

a. The domain is $\{x \mid x \geq 0\}$ or the interval $[0, \infty)$.

ISM − 398

b.

| $x$ | $f(x)=\sqrt{x}+1$ | $(x,y)$ |
|---|---|---|
| 0 | $f(0)=\sqrt{0}+1=1$ | $(0,1)$ |
| 1 | $f(1)=2$ | $(1,2)$ |
| 4 | $f(4)=3$ | $(4,3)$ |
| 9 | $f(9)=4$ | $(9,4)$ |

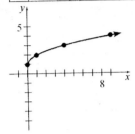

c. Based on the graph, the range is $[1,\infty)$.

42. $g(x)=\sqrt{x}-2$

   a. The domain is $\{x \mid x \geq 0\}$ or the interval $[0,\infty)$.

   b.

| $x$ | $g(x)=\sqrt{x}-2$ | $(x,y)$ |
|---|---|---|
| 0 | $g(0)=\sqrt{0}-2=-2$ | $(0,-2)$ |
| 1 | $g(1)=-1$ | $(1,-1)$ |
| 4 | $g(4)=0$ | $(4,0)$ |
| 9 | $g(9)=1$ | $(9,1)$ |

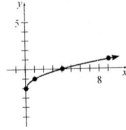

   c. Based on the graph, the range is $[-2,\infty)$.

44. $h(x)=3\sqrt{x}$

   a. The domain is $\{x \mid x \geq 0\}$ or the interval $[0,\infty)$.

b.

| $x$ | $h(x)=3\sqrt{x}$ | $(x,y)$ |
|---|---|---|
| 0 | $h(0)=3\sqrt{0}=0$ | $(0,0)$ |
| 1 | $h(1)=3$ | $(1,3)$ |
| 4 | $h(4)=6$ | $(4,6)$ |
| 9 | $h(9)=9$ | $(9,9)$ |

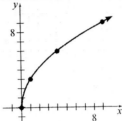

   c. Based on the graph, the range is $[0,\infty)$.

46. $g(x)=\dfrac{1}{4}\sqrt{x}$

   a. The domain is $\{x \mid x \geq 0\}$ or the interval $[0,\infty)$.

   b.

| $x$ | $g(x)=\tfrac{1}{4}\sqrt{x}$ | $(x,y)$ |
|---|---|---|
| 0 | $g(0)=\tfrac{1}{4}\sqrt{0}=0$ | $(0,0)$ |
| 1 | $g(1)=\tfrac{1}{4}$ | $\left(1,\tfrac{1}{4}\right)$ |
| 4 | $g(4)=\tfrac{1}{2}$ | $\left(4,\tfrac{1}{2}\right)$ |
| 9 | $g(9)=\tfrac{3}{4}$ | $\left(9,\tfrac{3}{4}\right)$ |
| 16 | $g(16)=1$ | $(16,1)$ |

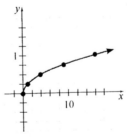

   c. Based on the graph, the range is $[0,\infty)$.

48. $F(x)=\sqrt{-x}$

   a. The domain is $\{x \mid x \leq 0\}$ or the interval $(-\infty,0]$.

**b.**

| $x$ | $F(x)=\sqrt{-x}$ | $(x,y)$ |
|---|---|---|
| 0 | $F(0)=\sqrt{-0}=0$ | $(0,0)$ |
| $-1$ | $F(-1)=1$ | $(-1,1)$ |
| $-4$ | $F(-4)=2$ | $(-4,2)$ |
| $-9$ | $F(-9)=3$ | $(-9,3)$ |
| $-16$ | $F(-16)=4$ | $(-16,4)$ |

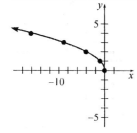

**c.** Based on the graph, the range is $[0,\infty)$.

**50.** $g(x)=\sqrt[3]{x-4}$

**a.** The index is odd so the domain is all real numbers, $\{x \mid x \text{ is any real number}\}$ or $(-\infty,\infty)$.

**b.**

| $x$ | $g(x)=\sqrt[3]{x-4}$ | $(x,y)$ |
|---|---|---|
| $-4$ | $g(-4)=\sqrt[3]{-4-4}=-2$ | $(-4,-2)$ |
| 3 | $g(3)=-1$ | $(3,-1)$ |
| 4 | $g(4)=0$ | $(4,0)$ |
| 5 | $g(5)=1$ | $(5,1)$ |
| 12 | $g(12)=2$ | $(12,2)$ |

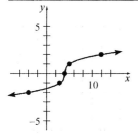

**c.** Based on the graph, the range is $(-\infty,\infty)$.

**52.** $H(x)=\sqrt[3]{x}+3$

**a.** The index is odd so the domain is all real numbers, $\{x \mid x \text{ is any real number}\}$ or $(-\infty,\infty)$.

**b.**

| $x$ | $H(x)=\sqrt[3]{x}+3$ | $(x,y)$ |
|---|---|---|
| $-8$ | $H(-8)=\sqrt[3]{-8}+3=1$ | $(-8,1)$ |
| $-1$ | $H(-1)=2$ | $(-1,2)$ |
| 0 | $H(0)=3$ | $(0,3)$ |
| 1 | $H(1)=4$ | $(1,4)$ |
| 8 | $H(8)=5$ | $(8,5)$ |

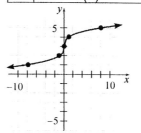

**c.** Based on the graph, the range is $(-\infty,\infty)$.

**54.** $F(x)=3\sqrt[3]{x}$

**a.** The index is odd so the domain is all real numbers, $\{x \mid x \text{ is any real number}\}$ or $(-\infty,\infty)$.

**b.**

| $x$ | $F(x)=3\sqrt[3]{x}$ | $(x,y)$ |
|---|---|---|
| $-8$ | $F(-8)=3\sqrt[3]{-8}=-6$ | $(-8,-6)$ |
| $-1$ | $F(-1)=-3$ | $(-1,-3)$ |
| 0 | $F(0)=0$ | $(0,0)$ |
| 1 | $F(1)=3$ | $(1,3)$ |
| 8 | $F(8)=6$ | $(8,6)$ |

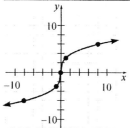

**c.** Based on the graph, the range is $(-\infty,\infty)$.

**56. a.** $d(0)=\sqrt{0^4-7(0)^2+16}=\sqrt{16}=4$

The distance is 4 units.

**b.** $d(1)=\sqrt{1^4-7(1)^2+16}=\sqrt{10}$

The distance is $\sqrt{10}\approx 3.162$ units.

Section 7.6 Functions Involving Radicals

c. $d(4) = \sqrt{4^4 - 7(4)^2 + 16} = \sqrt{160} = 4\sqrt{10}$

The distance is $4\sqrt{10} \approx 12.649$ units.

58. a. $x = 1$

$A = 2(1)\sqrt{16 - 1^2} = 2\sqrt{15}$

The area is $2\sqrt{15} \approx 7.746$ square units.

b. $x = 2$

$A = 2(2)\sqrt{16 - 2^2} = 4\sqrt{12} = 4 \cdot 2\sqrt{3} = 8\sqrt{3}$

The area is $8\sqrt{3} \approx 13.856$ square units.

c. $x = 2\sqrt{2}$

$A = 2(2\sqrt{2})\sqrt{16 - (2\sqrt{2})^2}$

$= 4\sqrt{2} \cdot \sqrt{16 - 8} = 4\sqrt{2} \cdot \sqrt{8}$

$= 4\sqrt{16} = 4 \cdot 4 = 16$

The area is 16 square units.

60. The graph of $g(x) = \sqrt{x} + c$ can be obtained from the graph of $f(x) = \sqrt{x}$ by shifting the graph of $f(x)$ $c$ units up (if $c > 0$) or down (if $c < 0$).

62. $\dfrac{1}{5} + \dfrac{3}{4} = \dfrac{4}{20} + \dfrac{15}{20} = \dfrac{4+15}{20} = \dfrac{19}{20}$

64. $\dfrac{5}{x-3} + \dfrac{2}{x+1} = \dfrac{5(x+1)}{(x-3)(x+1)} + \dfrac{2(x-3)}{(x-3)(x+1)}$

$= \dfrac{5(x+1) + 2(x-3)}{(x-3)(x+1)}$

$= \dfrac{5x + 5 + 2x - 6}{(x-3)(x+1)}$

$= \dfrac{7x - 1}{(x-3)(x+1)}$

66. Answers will vary.
With no common factors, the LCD is the product of the denominators.

68. $f(x) = \sqrt{x-1}$

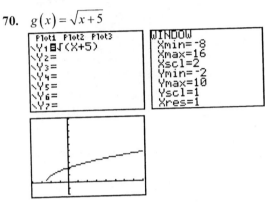

70. $g(x) = \sqrt{x+5}$

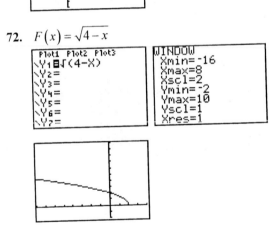

72. $F(x) = \sqrt{4-x}$

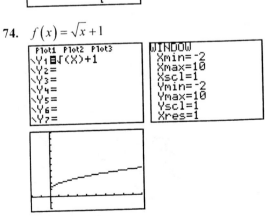

74. $f(x) = \sqrt{x} + 1$

*Chapter 7: Radicals and Rational Exponents*

76. $g(x) = \sqrt{x} - 2$

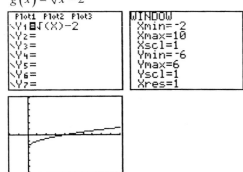

78. $h(x) = 3\sqrt{x}$

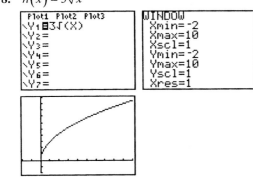

80. $g(x) = \dfrac{1}{4}\sqrt{x}$

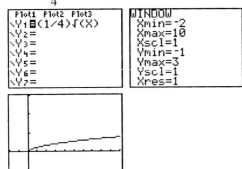

82. $F(x) = \sqrt{-x}$

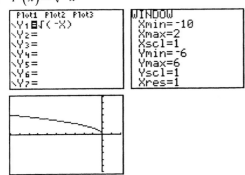

84. $g(x) = \sqrt[3]{x-4}$

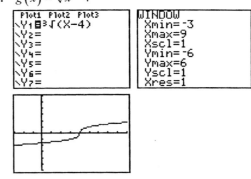

86. $H(x) = \sqrt[3]{x} + 3$

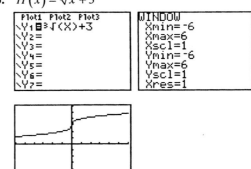

88. $F(x) = 3\sqrt[3]{x}$

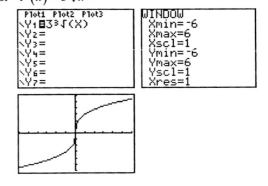

**7.7 Preparing for Radical Equations and Their Applications**

1. $3x - 5 = 0$
   $3x = 5$
   $x = \dfrac{5}{3}$

   The solution set is $\left\{\dfrac{5}{3}\right\}$.

**Section 7.7** *Radical Equations and Their Applications*

2. $2p^2 + 4p - 6 = 0$
$p^2 + 2p - 3 = 0$
$(p+3)(p-1) = 0$
$p + 3 = 0$ or $p - 1 = 0$
$p = -3 \qquad p = 1$
The solution set is $\{-3, 1\}$.

3. $\left(\sqrt[3]{x-5}\right)^3 = x - 5$

**7.7 Exercises**

2. extraneous

4. True

6. Answers will vary.
Since the principal square root is never negative, the left side of the equation will always be at least 5. Since it will never be 0 (the right side), the equation has no real solution.

8. $\sqrt{p} = 6$
$\left(\sqrt{p}\right)^2 = 6^2$
$p = 36$

Check: $\sqrt{36} = 6$ ?
$6 = 6$ T
The solution set is $\{36\}$.

10. $\sqrt{y-5} = 3$
$\left(\sqrt{y-5}\right)^2 = 3^2$
$y - 5 = 9$
$y = 14$

Check: $\sqrt{14-5} = 3$ ?
$\sqrt{9} = 3$ ?
$3 = 3$ T
The solution set is $\{14\}$.

12. $\sqrt{3w-2} = 4$
$\left(\sqrt{3w-2}\right)^2 = 4^2$
$3w - 2 = 16$
$3w = 18$
$w = 6$

Check: $\sqrt{3(6)-2} = 4$ ?
$\sqrt{18-2} = 4$ ?
$\sqrt{16} = 4$ ?
$4 = 4$ T
The solution set is $\{6\}$.

14. $\sqrt{6p-5} = -5$
Since the principal square root is never negative, this equation has no real solution. Solving this equation by squaring both sides will yield an extraneous solution.
$\sqrt{6p-5} = -5$ n
$\left(\sqrt{6p-5}\right)^2 = (-5)^2$
$6p - 5 = 25$
$6p = 30$
$p = 5$

Check: $\sqrt{6(5)-5} = -5$ ?
$\sqrt{25} = -5$ ?
$5 = -5$ False
The solution does not check so the problem has no real solution.

16. $\sqrt[3]{9w} = 3$
$\left(\sqrt[3]{9w}\right)^3 = 3^3$
$9w = 27$
$w = 3$

Check: $\sqrt[3]{9(3)} = 3$ ?
$\sqrt[3]{27} = 3$ ?
$3 = 3$ T
The solution set is $\{3\}$.

## Chapter 7: Radicals and Rational Exponents

18. $\sqrt[3]{7m+20} = 5$
$\left(\sqrt[3]{7m+20}\right)^3 = 5^3$
$7m + 20 = 125$
$7m = 105$
$m = 15$

Check: $\sqrt[3]{7(15)+20} = 5$ ?
$\sqrt[3]{125} = 5$ ?
$5 = 5$ T

The solution set is $\{15\}$.

20. $\sqrt{q} - 5 = 2$
$\sqrt{q} = 7$
$\left(\sqrt{q}\right)^2 = 7^2$
$q = 49$

Check: $\sqrt{49} - 5 = 2$ ?
$7 - 5 = 2$ ?
$2 = 2$ T

The solution set is $\{49\}$.

22. $\sqrt{x-4} + 4 = 7$
$\sqrt{x-4} = 3$
$\left(\sqrt{x-4}\right)^2 = 3^2$
$x - 4 = 9$
$x = 13$

Check: $\sqrt{13-4} + 4 = 7$ ?
$\sqrt{9} + 4 = 7$ ?
$3 + 4 = 7$ ?
$7 = 7$ T

The solution set is $\{13\}$.

24. $\sqrt{4x+1} - 2 = 3$
$\sqrt{4x+1} = 5$
$\left(\sqrt{4x+1}\right)^2 = 5^2$
$4x + 1 = 25$
$4x = 24$
$x = 6$

Check: $\sqrt{4(6)+1} - 2 = 3$ ?
$\sqrt{25} - 2 = 3$ ?
$5 - 2 = 3$ ?
$3 = 3$ T

The solution set is $\{6\}$.

26. $4\sqrt{t} - 2 = 10$
$4\sqrt{t} = 12$
$\sqrt{t} = 3$
$\left(\sqrt{t}\right)^2 = 3^2$
$t = 9$

Check: $4\sqrt{9} - 2 = 10$ ?
$4(3) - 2 = 10$ ?
$12 - 2 = 10$ ?
$10 = 10$ T

The solution set is $\{9\}$.

28. $\sqrt{6-w} - 3 = 1$
$\sqrt{6-w} = 4$
$\left(\sqrt{6-w}\right)^2 = 4^2$
$6 - w = 16$
$-w = 10$
$w = -10$

Check: $\sqrt{6-(-10)} - 3 = 1$ ?
$\sqrt{16} - 3 = 1$ ?
$4 - 3 = 1$ ?
$1 = 1$ T

The solution set is $\{-10\}$.

**Section 7.7** Radical Equations and Their Applications

30. $\sqrt{q} = 3q$
$(\sqrt{q})^2 = (3q)^2$
$q = 9q^2$
$0 = 9q^2 - q$
$0 = q(9q - 1)$
$q = 0$ or $9q - 1 = 0$
$\phantom{q = 0 \text{ or }} 9q = 1$
$\phantom{q = 0 \text{ or }} q = \dfrac{1}{9}$

Check:
$\sqrt{0} = 3(0)$ ? $\qquad \sqrt{\dfrac{1}{9}} = 3\left(\dfrac{1}{9}\right)$ ?
$0 = 0$ T $\qquad\qquad \dfrac{1}{3} = \dfrac{1}{3}$ T

The solution set is $\left\{0, \dfrac{1}{9}\right\}$.

32. $\sqrt{2p + 8} = p$
$(\sqrt{2p + 8})^2 = p^2$
$2p + 8 = p^2$
$0 = p^2 - 2p - 8$
$0 = (p - 4)(p + 2)$
$p - 4 = 0$ or $p + 2 = 0$
$p = 4 \qquad\qquad p = -2$

Check:
$\sqrt{2(4) + 8} = 4$ ? $\qquad \sqrt{2(-2) + 8} = 4$ ?
$\sqrt{16} = 4$ ? $\qquad\qquad \sqrt{4} = 4$ ?
$4 = 4$ T $\qquad\qquad\quad 2 = 4$ False

The second solution does not check. Therefore, the solution set is $\{4\}$.

34. $\sqrt{m} = m - 12$
$(\sqrt{m})^2 = (m - 12)^2$
$m = m^2 - 24m + 144$
$0 = m^2 - 25m + 144$
$0 = (m - 16)(m - 9)$
$m - 16 = 0$ or $m - 9 = 0$
$m = 16 \qquad\qquad m = 9$

Check:
$\sqrt{16} = 16 - 12$ ? $\qquad \sqrt{9} = 9 - 12$ ?
$4 = 4$ T $\qquad\qquad\quad 3 = -3$ False

The second solution does not check. Therefore, the solution set is $\{16\}$.

36. $\sqrt{1 - 4x} - 5 = x$
$\sqrt{1 - 4x} = x + 5$
$(\sqrt{1 - 4x})^2 = (x + 5)^2$
$1 - 4x = x^2 + 10x + 25$
$0 = x^2 + 14x + 24$
$0 = (x + 12)(x + 2)$
$x + 12 = 0$ or $x + 2 = 0$
$x = -12 \qquad\qquad x = -2$

Check:
$\sqrt{1 - 4(-12)} - 5 = -12$ ?
$\sqrt{49} - 5 = -12$ ?
$7 - 5 = -12$ ?
$2 = -12$ False

$\sqrt{1 - 4(-2)} - 5 = -2$ ?
$\sqrt{9} - 5 = -2$ ?
$3 - 5 = -2$ ?
$-2 = -2$ T

The first solution does not check. Therefore, the solution set is $\{-2\}$.

38. $\sqrt{z^2 - z - 7} + 3 = z + 2$
$\sqrt{z^2 - z - 7} = z - 1$
$(\sqrt{z^2 - z - 7})^2 = (z - 1)^2$
$z^2 - z - 7 = z^2 - 2z + 1$
$z - 8 = 0$
$z = 8$

## Chapter 7: Radicals and Rational Exponents

Check: $\sqrt{8^2 - 8 - 7} + 3 = 8 + 2$ ?
$\sqrt{49} + 3 = 10$ ?
$7 + 3 = 10$ ?
$10 = 10$ T

The solution set is $\{8\}$.

**40.** $\sqrt{3x+1} = \sqrt{2x+7}$
$\left(\sqrt{3x+1}\right)^2 = \left(\sqrt{2x+7}\right)^2$
$3x + 1 = 2x + 7$
$x + 1 = 7$
$x = 6$

Check: $\sqrt{3(6)+1} = \sqrt{2(6)+7}$ ?
$\sqrt{19} = \sqrt{19}$ T

The solution set is $\{6\}$.

**42.** $\sqrt[3]{3y-2} = \sqrt[3]{5y+8}$
$\left(\sqrt[3]{3y-2}\right)^3 = \left(\sqrt[3]{5y+8}\right)^3$
$3y - 2 = 5y + 8$
$-2y - 2 = 8$
$-2y = 10$
$y = -5$

Check: $\sqrt[3]{3(-5)-2} = \sqrt[3]{5(-5)+8}$ ?
$\sqrt[3]{-17} = \sqrt[3]{-17}$ T

The solution set is $\{-5\}$.

**44.** $\sqrt{2x^2 + 7x - 10} = \sqrt{x^2 + 4x + 8}$
$\left(\sqrt{2x^2 + 7x - 10}\right)^2 = \left(\sqrt{x^2 + 4x + 8}\right)^2$
$2x^2 + 7x - 10 = x^2 + 4x + 8$
$x^2 + 3x - 18 = 0$
$(x+6)(x-3) = 0$
$x + 6 = 0$ or $x - 3 = 0$
$x = -6$ $\qquad x = 3$

Check:
$\sqrt{2(-6)^2 + 7(-6) - 10} = \sqrt{(-6)^2 + 4(-6) + 8}$ ?
$\sqrt{2(36) - 42 - 10} = \sqrt{36 - 24 + 8}$ ?
$\sqrt{20} = \sqrt{20}$ T

$\sqrt{2(3)^2 + 7(3) - 10} = \sqrt{(3)^2 + 4(3) + 8}$ ?
$\sqrt{2(9) + 21 - 10} = \sqrt{9 + 12 + 8}$ ?
$\sqrt{29} = \sqrt{29}$ T

The solution set is $\{-6, 3\}$.

**46.** $\sqrt{3y-2} = 2 + \sqrt{y}$
$\left(\sqrt{3y-2}\right)^2 = \left(2 + \sqrt{y}\right)^2$
$3y - 2 = 2^2 + 2(2)\left(\sqrt{y}\right) + \left(\sqrt{y}\right)^2$
$3y - 2 = 4 + 4\sqrt{y} + y$
$2y - 6 = 4\sqrt{y}$
$(2y-6)^2 = \left(4\sqrt{y}\right)^2$
$4y^2 - 24y + 36 = 16y$
$4y^2 - 40y + 36 = 0$
$y^2 - 10y + 9 = 0$
$(y-9)(y-1) = 0$
$y - 9 = 0$ or $y - 1 = 0$
$y = 9$ $\qquad y = 1$

**Section 7.7** *Radical Equations and Their Applications*

Check:
$\sqrt{3(9)-2} = 2+\sqrt{9}$ ?
$\sqrt{25} = 2+3$ ?
$5 = 5$ T

$\sqrt{3(1)-2} = 2+\sqrt{1}$ ?
$\sqrt{1} = 2+1$ ?
$1 = 3$ False

The second solution does not check. The solution set is $\{9\}$.

48. $\sqrt{2x-1} - \sqrt{x-1} = 1$
$\sqrt{2x-1} = 1+\sqrt{x-1}$
$\left(\sqrt{2x-1}\right)^2 = \left(1+\sqrt{x-1}\right)^2$
$2x-1 = 1+2\sqrt{x-1}+x-1$
$x-1 = 2\sqrt{x-1}$
$(x-1)^2 = \left(2\sqrt{x-1}\right)^2$
$x^2 - 2x+1 = 4(x-1)$
$x^2 - 2x+1 = 4x-4$
$x^2 - 6x+5 = 0$
$(x-5)(x-1) = 0$
$x-5 = 0$  or  $x-1 = 0$
$x = 5$    $x = 1$

Check:
$\sqrt{2(5)-1} - \sqrt{5-1} = 1$ ?
$\sqrt{9} - \sqrt{4} = 1$ ?
$3-2 = 1$ ?
$1 = 1$ T

$\sqrt{2(1)-1} - \sqrt{1-1} = 1$ ?
$\sqrt{1} - \sqrt{0} = 1$ ?
$1 = 1$ T

The solution set is $\{1, 5\}$.

50. $\sqrt{2x+6} - \sqrt{x-6} = 3$
$\sqrt{2x+6} = 3+\sqrt{x-6}$
$\left(\sqrt{2x+6}\right)^2 = \left(3+\sqrt{x-6}\right)^2$
$2x+6 = 9+6\sqrt{x-6}+x-6$
$x+3 = 6\sqrt{x-6}$
$(x+3)^2 = \left(6\sqrt{x-6}\right)^2$
$x^2+6x+9 = 36(x-6)$
$x^2+6x+9 = 36x-216$
$x^2-30x+225 = 0$
$(x-15)^2 = 0$
$x = 15$

Check: $\sqrt{2(15)+6} - \sqrt{15-6} = 3$ ?
$\sqrt{36} - \sqrt{9} = 3$ ?
$6-3 = 3$ ?
$3 = 3$ T

The solution set is $\{15\}$.

52. $\sqrt{4x+1} - \sqrt{2x+1} = 2$
$\sqrt{4x+1} = 2+\sqrt{2x+1}$
$\left(\sqrt{4x+1}\right)^2 = \left(2+\sqrt{2x+1}\right)^2$
$4x+1 = 4+4\sqrt{2x+1}+2x+1$
$2x-4 = 4\sqrt{2x+1}$
$x-2 = 2\sqrt{2x+1}$
$(x-2)^2 = \left(2\sqrt{2x+1}\right)^2$
$x^2-4x+4 = 4(2x+1)$
$x^2-4x+4 = 8x+4$
$x^2-12x = 0$
$x(x-12) = 0$
$x = 0$  or  $x = 12$

Check:
$\sqrt{4(12)+1} - \sqrt{2(12)+1} = 2$ ?
$\sqrt{49} - \sqrt{25} = 2$ ?
$7-5 = 2$ ?
$2 = 2$ T

## Chapter 7: Radicals and Rational Exponents

$\sqrt{4(0)+1} - \sqrt{2(0)+1} = 2$ ?

$\sqrt{1} - \sqrt{1} = 2$ ?

$1 - 1 = 2$ ?

$0 = 2$  False

The second solution does not check. The solution set is $\{12\}$.

**54.** $(4x+1)^{1/2} = 5$

$\left[(4x+1)^{1/2}\right]^2 = 5^2$

$4x+1 = 25$

$4x = 24$

$x = 6$

Check: $(4(6)+1)^{1/2} = 5$ ?

$(25)^{1/2} = 5$ ?

$5 = 5$  T

The solution set is $\{6\}$.

**56.** $(6p+3)^{1/5} = (4p-9)^{1/5}$

$\left[(6p+3)^{1/5}\right]^5 = \left[(4p-9)^{1/5}\right]^5$

$6p+3 = 4p-9$

$2p+3 = -9$

$2p = -12$

$p = -6$

Check: $(6(-6)+3)^{1/5} = (4(-6)-9)^{1/5}$ ?

$(-36+3)^{1/5} = (-24-9)^{1/5}$ ?

$(-33)^{1/5} = (-33)^{1/5}$  T

The solution set is $\{-6\}$.

**58.** $(3x+1)^{1/2} - (x-1)^{1/2} = 2$

$(3x+1)^{1/2} = (x-1)^{1/2} + 2$

$\left[(3x+1)^{1/2}\right]^2 = \left[(x-1)^{1/2} + 2\right]^2$

$3x+1 = x-1 + 4(x-1)^{1/2} + 4$

$2x-2 = 4(x-1)^{1/2}$

$x-1 = 2(x-1)^{1/2}$

$(x-1)^2 = \left[2(x-1)^{1/2}\right]^2$

$x^2 - 2x + 1 = 4(x-1)$

$x^2 - 2x + 1 = 4x - 4$

$x^2 - 6x + 5 = 0$

$(x-5)(x-1) = 0$

$x-5 = 0$  or  $x-1 = 0$

$x = 5$           $x = 1$

Check:

$(3(5)+1)^{1/2} - (5-1)^{1/2} = 2$ ?

$(16)^{1/2} - (4)^{1/2} = 2$ ?

$4 - 2 = 2$ ?

$2 = 2$  T

$(3(1)+1)^{1/2} - (1-1)^{1/2} = 2$ ?

$(4)^{1/2} - (0)^{1/2} = 2$ ?

$2 - 0 = 2$ ?

$2 = 2$  T

The solution set is $\{1, 5\}$.

**60.** $v = \sqrt{ar}$

$v^2 = \left(\sqrt{ar}\right)^2$

$v^2 = ar$

$\dfrac{v^2}{r} = a$  or  $a = \dfrac{v^2}{r}$

### Section 7.7 Radical Equations and Their Applications

**62.**
$$r = \sqrt{\frac{S}{4\pi}}$$
$$r^2 = \left(\sqrt{\frac{S}{4\pi}}\right)^2$$
$$r^2 = \frac{S}{4\pi}$$
$$4\pi r^2 = S \quad \text{or} \quad S = 4\pi r^2$$

**64.**
$$V = \sqrt{\frac{2U}{C}}$$
$$V^2 = \left(\sqrt{\frac{2U}{C}}\right)^2$$
$$V^2 = \frac{2U}{C}$$
$$CV^2 = 2U$$
$$\frac{CV^2}{2} = U \quad \text{or} \quad U = \frac{CV^2}{2}$$

**66.**
$$\sqrt{3b-2} + 8 = 5$$
$$\sqrt{3b-2} = -3$$

At this point we see that there is no real solution. Since the principal square root is never negative, $\sqrt{3b-2} = -3$ has no real solution. Solving the usual way yields the same result.

$$\left(\sqrt{3b-2}\right)^2 = (-3)^2$$
$$3b - 2 = 9$$
$$3b = 11$$
$$b = \frac{11}{3}$$

Check: $\sqrt{3\left(\frac{11}{3}\right) - 2} + 8 = 5$ ?
$\sqrt{11-2} + 8 = 5$ ?
$\sqrt{9} + 8 = 5$ ?
$3 + 8 = 5$ ?
$11 = 5$ False

The solution does not check, so the equation has no real solution.

**68.**
$$\sqrt{x+20} = x$$
$$\left(\sqrt{x+20}\right)^2 = x^2$$
$$x + 20 = x^2$$
$$0 = x^2 - x - 20$$
$$0 = (x-5)(x+4)$$
$$x - 5 = 0 \quad \text{or} \quad x + 4 = 0$$
$$x = 5 \qquad\qquad x = -4$$

Check:
$\sqrt{5+20} = 5$ ?   $\sqrt{-4+20} = -4$ ?
$\sqrt{25} = 5$ ?     $\sqrt{16} = -4$ ?
$5 = 5$ T             $4 = -4$ False

The second solution does not check. The solution set is $\{5\}$.

**70.**
$$\sqrt{3a-5} = 2$$
$$\left(\sqrt{3a-5}\right)^2 = 2^2$$
$$3a - 5 = 4$$
$$3a = 9$$
$$a = 3$$

Check: $\sqrt{3(3)-5} = 2$ ?
$\sqrt{4} = 2$ ?
$2 = 2$ T

The solution set is $\{3\}$.

**72.**
$$\sqrt[5]{x+23} = 2$$
$$\left(\sqrt[5]{x+23}\right)^5 = 2^5$$
$$x + 23 = 32$$
$$x = 9$$

Check: $\sqrt[5]{9+23} = 2$ ?
$\sqrt[5]{32} = 2$ ?
$2 = 2$ T

The solution set is $\{9\}$.

## Chapter 7: Radicals and Rational Exponents

**74.** $(5x-2)^{1/3} + 3 = 0$

$(5x-2)^{1/3} = -3$

$\left[(5x-2)^{1/3}\right]^3 = (-3)^3$

$5x - 2 = -27$

$5x = -25$

$x = -5$

Check: $(5(-5)-2)^{1/3} + 3 = 0$ ?

$(-27)^{1/3} + 3 = 0$ ?

$0 = 0$ T

The solution set is $\{-5\}$.

**76.** $\sqrt{a} - 5 = -2$

$\sqrt{a} = 3$

$\left(\sqrt{a}\right)^2 = 3^2$

$a = 9$

Check: $\sqrt{9} - 5 = -2$ ?

$3 - 5 = -2$ ?

$-2 = -2$ T

The solution set is $\{9\}$.

**78.** $\sqrt{4c-5} = \sqrt{3c+1}$

$\left(\sqrt{4c-5}\right)^2 = \left(\sqrt{3c+1}\right)^2$

$4c - 5 = 3c + 1$

$c - 5 = 1$

$c = 6$

Check: $\sqrt{4(6)-5} = \sqrt{3(6)+1}$ ?

$\sqrt{19} = \sqrt{19}$ T

The solution set is $\{6\}$.

**80.** $\sqrt{x-3} + \sqrt{x+4} = 7$

$\sqrt{x-3} = 7 - \sqrt{x+4}$

$\left(\sqrt{x-3}\right)^2 = \left(7 - \sqrt{x+4}\right)^2$

$x - 3 = 49 - 14\sqrt{x+4} + x + 4$

$14\sqrt{x+4} = 56$

$\sqrt{x+4} = 4$

$\left(\sqrt{x+4}\right)^2 = 4^2$

$x + 4 = 16$

$x = 12$

Check: $\sqrt{12-3} + \sqrt{12+4} = 7$ ?

$\sqrt{9} + \sqrt{16} = 7$ ?

$3 + 4 = 7$ ?

$7 = 7$ T

The solution set is $\{12\}$.

**82.** $g(x) = \sqrt{x+3}$

**a.** $g(x) = 0$

$\sqrt{x+3} = 0$

$\left(\sqrt{x+3}\right)^2 = 0^2$

$x + 3 = 0$

$x = -3$

The point $(-3, 0)$ is on the graph of $g$.

**b.** $g(x) = 1$

$\sqrt{x+3} = 1$

$\left(\sqrt{x+3}\right)^2 = 1^2$

$x + 3 = 1$

$x = -2$

The point $(-2, 1)$ is on the graph of $g$.

**c.** $g(x) = 2$

$\sqrt{x+3} = 2$

$\left(\sqrt{x+3}\right)^2 = 2^2$

$x + 3 = 4$

$x = 1$

The point $(1, 2)$ is on the graph of $g$.

d. The points $(-3,0)$, $(-2,1)$, and $(1,2)$ are on the graph of $g$.

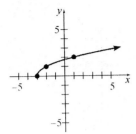

e. The equation $g(x) = -1$ has no solution because the graph of the function does not go below the x-axis. Therefore, the value of the function will never be negative.

84. a.
$$\sqrt{x^2 + 4^2} = 5$$
$$\left(\sqrt{x^2 + 4^2}\right)^2 = 5^2$$
$$x^2 + 4^2 = 5^2$$
$$x^2 + 16 = 25$$
$$x^2 = 9$$
$$x = \pm 3$$

b. The points are $(0,3)$, $(-3,-1)$, and $(3,-1)$.

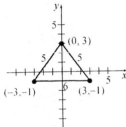

The figure is an isosceles triangle whose common side has a length of 5 units. The base has a length of 6 units.

86. $r = \sqrt[3]{\dfrac{3V}{4\pi}}$

$r^3 = \left(\sqrt[3]{\dfrac{3V}{4\pi}}\right)^3$

$r^3 = \dfrac{3V}{4\pi}$

$4\pi r^3 = 3V$

$\dfrac{4}{3}\pi r^3 = V$ or $V = \dfrac{4}{3}\pi r^3$

a. When $r = 3$ meters, we get
$$V = \dfrac{4}{3}\pi(3)^3 = 36\pi \approx 113.10 \text{ m}^3.$$

b. Use the result from part (a). When $r = 2$ meters, we get
$$V = \dfrac{4}{3}\pi(2)^3 = \dfrac{32}{3}\pi \approx 33.51 \text{ m}^3.$$

88. a. $P = 1000, r = 0.05, t = 2$

$$0.05 = \sqrt{\dfrac{A}{1000}} - 1$$

$$1.05 = \sqrt{\dfrac{A}{1000}}$$

$$(1.05)^2 = \left(\sqrt{\dfrac{A}{1000}}\right)^2$$

$$1.1025 = \dfrac{A}{1000}$$

$$1{,}102.50 = A$$

After 2 years, you would have $1,102.50.

b. $P = 1000, r = 0.05, t = 3$

$$0.05 = \sqrt[3]{\dfrac{A}{1000}} - 1$$

$$1.05 = \sqrt[3]{\dfrac{A}{1000}}$$

$$(1.05)^3 = \left(\sqrt[3]{\dfrac{A}{1000}}\right)^3$$

$$1.157625 = \dfrac{A}{1000}$$

$$1{,}157.63 = A$$

After 3 years, you would have $1,157.63.

90. $\sqrt[3]{2\sqrt{x-2}} = \sqrt[3]{x-1}$

$\left(\sqrt[3]{2\sqrt{x-2}}\right)^3 = \left(\sqrt[3]{x-1}\right)^3$

$2\sqrt{x-2} = x-1$

$\left(2\sqrt{x-2}\right)^2 = (x-1)^2$

$4(x-2) = x^2 - 2x + 1$

$4x - 8 = x^2 - 2x + 1$

$0 = x^2 - 6x + 9$

$0 = (x-3)^2$

$0 = x - 3$

$3 = x$

Check:
$\sqrt[3]{2\sqrt{3-2}} = \sqrt[3]{3-1}$ ?

$\sqrt[3]{2\sqrt{1}} = \sqrt[3]{2}$ ?

$\sqrt[3]{2} = \sqrt[3]{2}$ T

The solution set is $\{3\}$.

92. $0$, $-4$, and $12$ are the integers.

94. $\sqrt{2^3}$, $\pi$, and $\sqrt[3]{-4}$ are irrational numbers.

96. Answers will vary.
A rational number is a real number that can be expressed as the ratio of two integers. An irrational number is a real number that cannot be written as the ratio of two integers. In decimal form, a rational number will terminate or repeat (e.g. 3.7 or $1.\overline{3}$) while an irrational number will be non-terminating and non-repeating (e.g. $\sqrt{2} = 1.414213562...$).

98.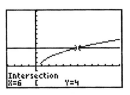

100.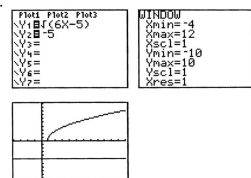

## 7.8 Preparing for The Complex Number System

1. a. 8 is the only natural number in the set.

   b. 8 and 0 are the whole numbers in the set.

   c. $-23$, $-\dfrac{12}{3}$, 0, and 8 are the integers in the set.

   d. $-23$, $-\dfrac{12}{3}$, $-\dfrac{1}{3}$, 0, $1.\overline{26}$, and 8 are the rational numbers in the set.

   e. $\sqrt{2}$ is the only irrational number in the set.

   f. $-23$, $-\dfrac{12}{3}$, $-\dfrac{1}{3}$, 0, $1.\overline{26}$, $\sqrt{2}$, and 8 are the real numbers in the set.

2. $3x(4x - 3) = 3x \cdot 4x - 3x \cdot 3 = 12x^2 - 9x$

3. $(z+4)(3z-2) = z \cdot 3z + 4 \cdot 3z - 2 \cdot z - 4 \cdot 2$
   $= 3z^2 + 12z - 2z - 8$
   $= 3z^2 + 10z - 8$

4. $(2y+5)(2y-5) = (2y)^2 - 5^2 = 4y^2 - 25$

**Section 7.8** *The Complex Number System*

**7.8 Exercises**

2. pure imaginary number

4. True; $\sqrt{-N} = \sqrt{N \cdot -1} = \sqrt{N} \cdot \sqrt{-1} = \sqrt{N}\, i$

6. True; the set of real numbers is a subset of the set of complex numbers. Thus, every real number is a complex number.

8. Answers will vary.

10. Answers will vary.
    In both cases, we multiply the numerator and denominator by the same quantity to change the form of the expression. To rationalize a denominator, we multiply the numerator and denominator by an appropriate radical expression. To write the quotient of two complex numbers in standard form, we multiply the numerator and denominator by the conjugate of the denominator.

12. $\sqrt{-25} = \sqrt{-1} \cdot \sqrt{25} = i \cdot 5 = 5i$

14. $-\sqrt{-100} = -\sqrt{-1} \cdot \sqrt{100} = -i \cdot 10 = -10i$

16. $\sqrt{-48} = \sqrt{-1} \cdot \sqrt{16} \cdot \sqrt{3} = i \cdot 4\sqrt{3} = 4\sqrt{3}\, i$

18. $\sqrt{-162} = \sqrt{-1} \cdot \sqrt{81} \cdot \sqrt{2} = i \cdot 9\sqrt{2} = 9\sqrt{2}\, i$

20. $\sqrt{-13} = \sqrt{-1} \cdot \sqrt{13} = \sqrt{13}\, i$

22. $4 - \sqrt{-36} = 4 - \sqrt{-1} \cdot \sqrt{36}$
    $= 4 - 6i$

24. $10 + \sqrt{-32} = 10 + \sqrt{-1} \cdot \sqrt{16} \cdot \sqrt{2}$
    $= 10 + i \cdot 4\sqrt{2}$
    $= 10 + 4\sqrt{2}\, i$

26. $\dfrac{10 - \sqrt{-25}}{5} = \dfrac{10 - \sqrt{-1} \cdot \sqrt{25}}{5}$
    $= \dfrac{10 - 5i}{5}$
    $= 2 - i$

28. $\dfrac{15 - \sqrt{-50}}{5} = \dfrac{15 - \sqrt{-1} \cdot \sqrt{25} \cdot \sqrt{2}}{5}$
    $= \dfrac{15 - i \cdot 5\sqrt{2}}{5}$
    $= 3 - \sqrt{2}\, i$

30. $(-6 + 2i) + (3 + 12i) = -6 + 2i + 3 + 12i$
    $= (-6 + 3) + (2 + 12)i$
    $= -3 + 14i$

32. $(-7 + 3i) - (-3 + 2i) = -7 + 3i + 3 - 2i$
    $= (-7 + 3) + (3 - 2)i$
    $= -4 + i$

34. $(-4 + \sqrt{-25}) + (1 - \sqrt{-16}) = (-4 + 5i) + (1 - 4i)$
    $= -4 + 5i + 1 - 4i$
    $= (-4 + 1) + (5 - 4)i$
    $= -3 + i$

36. $(-10 + \sqrt{-20}) - (-6 + \sqrt{-45})$
    $= (-10 + 2\sqrt{5}\, i) - (-6 + 3\sqrt{5}\, i)$
    $= -10 + 2\sqrt{5}\, i + 6 - 3\sqrt{5}\, i$
    $= (-10 + 6) + (2\sqrt{5} - 3\sqrt{5})i$
    $= -4 - \sqrt{5}\, i$

38. $3i(-2 - 6i) = 3i \cdot (-2) - 3i \cdot 6i$
    $= -6i - 18i^2$
    $= -6i - 18(-1)$
    $= 18 - 6i$

40. $\dfrac{1}{3}i(12 + 15i) = \dfrac{1}{3}i \cdot 12 + \dfrac{1}{3}i \cdot 15i$
    $= 4i + 5i^2$
    $= 4i + 5(-1)$
    $= -5 + 4i$

42. $(3 - i)(1 + 2i) = 3 \cdot 1 + 3 \cdot 2i - i \cdot 1 - i \cdot 2i$
    $= 3 + 6i - i - 2i^2$
    $= 3 + 5i + 2$
    $= 5 + 5i$

## Chapter 7: Radicals and Rational Exponents

**44.** $(5-2i)(-1+2i) = 5\cdot(-1)+5\cdot 2i+2i\cdot 1-2i\cdot 2i$
$= -5+10i+2i-4i^2$
$= -5+12i+4$
$= -1+12i$

**46.** $(2+8i)(-3-i) = 2\cdot(-3)-2\cdot i-8i\cdot 3-8i\cdot i$
$= -6-2i-24i-8i^2$
$= -6-26i+8$
$= 2-26i$

**48.** $(1+\sqrt{3}i)(-4-\sqrt{3}i)$
$= 1\cdot(-4)-1\cdot\sqrt{3}i+\sqrt{3}i\cdot(-4)-\sqrt{3}i\cdot\sqrt{3}i$
$= -4-\sqrt{3}i-4\sqrt{3}i-3i^2$
$= -4-5\sqrt{3}i+3$
$= -1-5\sqrt{3}i$

**50.** $\left(-\dfrac{2}{3}+\dfrac{4}{3}i\right)\left(\dfrac{1}{2}-\dfrac{3}{2}i\right)$
$= -\dfrac{2}{3}\cdot\dfrac{1}{2}+\dfrac{2}{3}\cdot\dfrac{3}{2}i+\dfrac{4}{3}\cdot\dfrac{1}{2}i-\dfrac{4}{3}\cdot\dfrac{3}{2}i\cdot i$
$= -\dfrac{1}{3}+i+\dfrac{2}{3}i-2i^2$
$= -\dfrac{1}{3}+\dfrac{5}{3}i+2$
$= \dfrac{5}{3}+\dfrac{5}{3}i$

**52.** $(2+5i)^2 = 2^2+2(2)(5i)+(5i)^2$
$= 4+20i+25i^2$
$= 4+20i-25$
$= -21+20i$

**54.** $(2-7i)^2 = 2^2-2(2)(7i)+(7i)^2$
$= 4-28i+49i^2$
$= 4-28i-49$
$= -45-28i$

**56.** $\sqrt{-36}\cdot\sqrt{-4} = 6i\cdot 2i = 12i^2 = -12$

**58.** $\sqrt{-12}\cdot\sqrt{-15} = 2\sqrt{3}i\cdot\sqrt{15}i$
$= 2\sqrt{45}i^2$
$= 2\cdot 3\sqrt{5}\cdot(-1)$
$= -6\sqrt{5}$

**60.** $(1-\sqrt{-64})(-2+\sqrt{-49})$
$= (1-8i)(-2+7i)$
$= 1\cdot(-2)+1\cdot 7i+8i\cdot 2-8i\cdot 7i$
$= -2+7i+16i-56i^2$
$= -2+23i+56$
$= 54+23i$

**62. a.** The conjugate of $5+2i$ is $5-2i$.

**b.** $(5+2i)(5-2i) = 5^2-(2i)^2$
$= 25-4i^2$
$= 25-4(-1)$
$= 25+4$
$= 29$

**64. a.** The conjugate of $9-i$ is $9+i$.

**b.** $(9-i)(9+i) = 9^2-i^2$
$= 81-(-1)$
$= 81+1$
$= 82$

**66. a.** The conjugate of $-1-4i$ is $-1+4i$.

**b.** $(-1-4i)(-1+4i) = (-1)^2-(4i)^2$
$= 1-16i^2$
$= 1-16(-1)$
$= 1+16$
$= 17$

## Section 7.8 The Complex Number System

**68.** 
$$\frac{2-i}{2i} = \frac{2-i}{2i} \cdot \frac{i}{i}$$
$$= \frac{2i - i^2}{2i^2}$$
$$= \frac{2i - (-1)}{2(-1)}$$
$$= \frac{2i + 1}{-2}$$
$$= -\frac{1}{2} - i$$

**70.**
$$\frac{-4+5i}{6i} = \frac{-4+5i}{6i} \cdot \frac{i}{i}$$
$$= \frac{-4i + 5i^2}{6i^2}$$
$$= \frac{-4i + 5(-1)}{6(-1)}$$
$$= \frac{-4i - 5}{-6}$$
$$= \frac{5}{6} + \frac{2}{3}i$$

**72.**
$$\frac{2}{4+i} = \frac{2}{4+i} \cdot \frac{4-i}{4-i}$$
$$= \frac{8-2i}{4^2 - i^2}$$
$$= \frac{8-2i}{16-(-1)}$$
$$= \frac{8-2i}{17}$$
$$= \frac{8}{17} - \frac{2}{17}i$$

**74.**
$$\frac{-4}{-5-3i} = \frac{-4}{-5-3i} \cdot \frac{-5+3i}{-5+3i}$$
$$= \frac{20-12i}{(-5)^2 - (3i)^2}$$
$$= \frac{20-12i}{25-(-9)}$$
$$= \frac{20-12i}{34}$$
$$= \frac{10}{17} - \frac{6}{17}i$$

**76.**
$$\frac{2+5i}{5-2i} = \frac{2+5i}{5-2i} \cdot \frac{5+2i}{5+2i}$$
$$= \frac{2 \cdot 5 + 2 \cdot 2i + 5i \cdot 5 + 5i \cdot 2i}{5^2 - (2i)^2}$$
$$= \frac{10 + 4i + 25i + 10i^2}{25 - 4i^2}$$
$$= \frac{10 + 29i - 10}{25 + 4}$$
$$= \frac{29i}{29} = i$$

**78.**
$$\frac{-6+2i}{1+i} = \frac{-6+2i}{1+i} \cdot \frac{1-i}{1-i}$$
$$= \frac{-6 \cdot 1 + 6 \cdot i + 2i \cdot 1 - 2i \cdot i}{1^2 - i^2}$$
$$= \frac{-6 + 6i + 2i - 2i^2}{1 - i^2}$$
$$= \frac{-6 + 8i + 2}{1+1}$$
$$= \frac{-4 + 8i}{2}$$
$$= -2 + 4i$$

**80.**
$$\frac{5-3i}{2+4i} = \frac{5-3i}{2+4i} \cdot \frac{2-4i}{2-4i}$$
$$= \frac{5 \cdot 2 - 5 \cdot 4i - 3i \cdot 2 + 3i \cdot 4i}{2^2 - (4i)^2}$$
$$= \frac{10 - 20i - 6i + 12i^2}{4 - 16i^2}$$
$$= \frac{10 - 26i - 12}{4 + 16}$$
$$= \frac{-2 - 26i}{20}$$
$$= -\frac{1}{10} - \frac{13}{10}i$$

**82.** $i^{72} = \left(i^4\right)^{18} = 1^{18} = 1$

**84.** $i^{110} = i^{108} \cdot i^2 = \left(i^4\right)^{27} \cdot i^2 = 1^{27} \cdot (-1) = -1$

**86.** $i^{131} = i^{128} \cdot i^3 = \left(i^4\right)^{32} \cdot i^3 = 1^{32} \cdot -i = -i$

## Chapter 7: Radicals and Rational Exponents

**88.** $i^{-26} = i^{-28} \cdot i^2 = (i^4)^{-7} \cdot i^2 = 1^{-7} \cdot (-1) = -1$

**90.** $(-5+2i)(5-2i) = -5 \cdot 5 + 5 \cdot 2i + 2i \cdot 5 - 2i \cdot 2i$
$= -25 + 10i + 10i - 4i^2$
$= -25 + 20i + 4$
$= -21 + 20i$

**92.** $(-3+2i)^2 = (-3)^2 + 2(-3)(2i) + (2i)^2$
$= 9 - 12i + 4i^2$
$= 9 - 12i - 4$
$= 5 - 12i$

**94.** $\dfrac{5-3i}{4i} = \dfrac{5-3i}{4i} \cdot \dfrac{i}{i}$
$= \dfrac{5i - 3i^2}{4i^2}$
$= \dfrac{5i + 3}{-4}$
$= -\dfrac{3}{4} - \dfrac{5}{4}i$

**96.** $\dfrac{-4+6i}{-5-i} = \dfrac{-4+6i}{-5-i} \cdot \dfrac{-5+i}{-5+i}$
$= \dfrac{-4(-5) - 4 \cdot i + 6i(-5) + 6i \cdot i}{(-5)^2 - (i)^2}$
$= \dfrac{20 - 4i - 30i + 6i^2}{25 - i^2}$
$= \dfrac{20 - 34i - 6}{25 + 1}$
$= \dfrac{14 - 34i}{26}$
$= \dfrac{7}{13} - \dfrac{17}{13}i$

**98.** $(-4+5i) + (4-2i) = -4 + 5i + 4 - 2i$
$= (-4+4) + (5-2)i$
$= 3i$

**100.** $2i(3-4i) = 2i \cdot 3 - 2i \cdot 4i$
$= 6i - 8i^2$
$= 6i + 8$
$= 8 + 6i$

**102.** $\sqrt{-8} \cdot \sqrt{-12} = \sqrt{-1} \cdot \sqrt{4} \cdot \sqrt{2} \cdot \sqrt{-1} \cdot \sqrt{4} \cdot \sqrt{3}$
$= i \cdot 2\sqrt{2} \cdot i \cdot 2\sqrt{3}$
$= i^2 \cdot 4\sqrt{6}$
$= -1 \cdot 4\sqrt{6}$
$= -4\sqrt{6}$

**104.** $\dfrac{1}{7i} = \dfrac{1}{7i} \cdot \dfrac{i}{i} = \dfrac{i}{7i^2} = \dfrac{i}{-7} = -\dfrac{1}{7}i$

**106.** $\dfrac{1}{3-5i} = \dfrac{1}{3-5i} \cdot \dfrac{3+5i}{3+5i}$
$= \dfrac{3+5i}{3^2 - (5i)^2} = \dfrac{3+5i}{9 - 25i^2}$
$= \dfrac{3+5i}{9+25} = \dfrac{3+5i}{34}$
$= \dfrac{3}{34} + \dfrac{5}{34}i$

**108.** $\dfrac{1}{-6+2i} = \dfrac{1}{-6+2i} \cdot \dfrac{-6-2i}{-6-2i}$
$= \dfrac{-6-2i}{(-6)^2 - (2i)^2} = \dfrac{-6-2i}{36 - 4i^2}$
$= \dfrac{-6-2i}{36+4} = \dfrac{-6-2i}{40}$
$= -\dfrac{3}{20} - \dfrac{1}{20}i$

**110.** $f(x) = x^2 + x$

  **a.** $f(i) = i^2 + i = -1 + i$

  **b.** $f(1+i) = (1+i)^2 + (1+i)$
  $= 1 + 2i + i^2 + 1 + i$
  $= 2 + 3i + i^2$
  $= 2 + 3i - 1$
  $= 1 + 3i$

**112.** $f(x) = x^2 + x - 1$

  **a.** $f(2i) = (2i)^2 + (2i) - 1$
  $= 4i^2 + 2i - 1$
  $= -4 + 2i - 1$
  $= -5 + 2i$

**b.** $f(2+i) = (2+i)^2 + (2+i) - 1$
$\quad = 4 + 4i + i^2 + 2 + i - 1$
$\quad = 5 + 5i + i^2$
$\quad = 5 + 5i - 1$
$\quad = 4 + 5i$

**114.** $\dfrac{1}{Z} = \dfrac{1}{Z_1} + \dfrac{1}{Z_2}$

$= \dfrac{1}{5} + \dfrac{1}{1-2i}$

$= \dfrac{1}{5} \cdot \dfrac{1-2i}{1-2i} + \dfrac{1}{1-2i} \cdot \dfrac{5}{5}$

$= \dfrac{1-2i+5}{5(1-2i)}$

$= \dfrac{6-2i}{5(1-2i)}$

Therefore,

$Z = \dfrac{5(1-2i)}{6-2i} = \dfrac{5(1-2i)}{6-2i} \cdot \dfrac{6+2i}{6+2i}$

$= \dfrac{5(6-10i-4i^2)}{36-4i^2}$

$= \dfrac{5(6-10i+4)}{36+4}$

$= \dfrac{5(10-10i)}{40}$

$= \dfrac{50(1-i)}{40}$

$= \dfrac{5(1-i)}{4}$

$= \dfrac{5}{4} - \dfrac{5}{4}i$

The total impedance for the circuit is $\dfrac{5}{4} - \dfrac{5}{4}i$ ohms.

**116.** $f(x) = x^2 - 2x + 2$

**a.** $f(1+i) = (1+i)^2 - 2(1+i) + 2$
$\quad = (1)^2 + 2(1)i + i^2 - 2 - 2i + 2$
$\quad = 1 + 2i - 1 - 2 - 2i + 2$
$\quad = 0$

**b.** $f(1-i) = (1-i)^2 - 2(1-i) + 2$
$\quad = (1)^2 - 2(1)i + i^2 - 2 + 2i + 2$
$\quad = 1 - 2i - 1 - 2 + 2i + 2$
$\quad = 0$

**118.** $f(x) = x^3 - 1$

**a.** $f(1) = (1)^3 - 1 = 1 - 1 = 0$

**b.** $f\left(-\dfrac{1}{2} + \dfrac{\sqrt{3}}{2}i\right) = \left(-\dfrac{1}{2} + \dfrac{\sqrt{3}}{2}i\right)^3 - 1$

$= \left(-\dfrac{1}{2} + \dfrac{\sqrt{3}}{2}i\right)\left(-\dfrac{1}{2} + \dfrac{\sqrt{3}}{2}i\right)^2 - 1$

$= \left(-\dfrac{1}{2} + \dfrac{\sqrt{3}}{2}i\right)\left(\dfrac{1}{4} - \dfrac{\sqrt{3}}{2}i + \dfrac{3}{4}i^2\right) - 1$

$= \left(-\dfrac{1}{2} + \dfrac{\sqrt{3}}{2}i\right)\left(\dfrac{1}{4} - \dfrac{\sqrt{3}}{2}i - \dfrac{3}{4}\right) - 1$

$= \left(-\dfrac{1}{2} + \dfrac{\sqrt{3}}{2}i\right)\left(-\dfrac{1}{2} - \dfrac{\sqrt{3}}{2}i\right) - 1$

$= \left(-\dfrac{1}{2}\right)^2 - \left(\dfrac{\sqrt{3}}{2}i\right)^2 - 1$

$= \dfrac{1}{4} - \dfrac{3}{4}i^2 - 1$

$= -\dfrac{3}{4} + \dfrac{3}{4}$

$= 0$

Chapter 7: Radicals and Rational Exponents

c. $f\left(-\dfrac{1}{2}-\dfrac{\sqrt{3}}{2}i\right) = \left(-\dfrac{1}{2}-\dfrac{\sqrt{3}}{2}i\right)^3 - 1$

$= \left(-\dfrac{1}{2}-\dfrac{\sqrt{3}}{2}i\right)\left(-\dfrac{1}{2}-\dfrac{\sqrt{3}}{2}i\right)^2 - 1$

$= \left(-\dfrac{1}{2}-\dfrac{\sqrt{3}}{2}i\right)\left(\dfrac{1}{4}+\dfrac{\sqrt{3}}{2}i+\dfrac{3}{4}i^2\right) - 1$

$= \left(-\dfrac{1}{2}-\dfrac{\sqrt{3}}{2}i\right)\left(\dfrac{1}{4}+\dfrac{\sqrt{3}}{2}i-\dfrac{3}{4}\right) - 1$

$= \left(-\dfrac{1}{2}-\dfrac{\sqrt{3}}{2}i\right)\left(-\dfrac{1}{2}+\dfrac{\sqrt{3}}{2}i\right) - 1$

$= \left(-\dfrac{1}{2}\right)^2 - \left(\dfrac{\sqrt{3}}{2}i\right)^2 - 1$

$= \dfrac{1}{4} - \dfrac{3}{4}i^2 - 1$

$= -\dfrac{3}{4} + \dfrac{3}{4}$

$= 0$

120. $(x+2)^3 = x^3 + 3x^2 \cdot 2 + 3x \cdot 2^2 + 2^3$
$= x^3 + 6x^2 + 12x + 8$

122. $(3+i)^3 = 3^3 + 3 \cdot 3^2 \cdot i + 3 \cdot 3 \cdot i^2 + i^3$
$= 27 + 27i + 9i^2 + i^3$
$= 27 + 27i - 9 - i$
$= 18 + 26i$

124. Answers will vary.

126.
```
(-3.4+1.9i)-(6.5
-5.3i)
         -9.9+7.2i
```

128.
```
(-4.3+0.2i)(7.2-
0.5i)
         -30.86+3.59i
```

130.
```
(1-7i)/(-4+i)▶Fr
ac
         -11/17+27/17i
```

132.
```
-6i^14+5i^6-(2+i
)(3+11i)
              6-25i
```

**Chapter 7 Review**

1. $\sqrt[3]{343} = \sqrt[3]{7^3} = 7$

2. $\sqrt[3]{-125} = \sqrt[3]{(-5)^3} = -5$

3. $\sqrt[3]{\dfrac{8}{27}} = \sqrt[3]{\left(\dfrac{2}{3}\right)^3} = \dfrac{2}{3}$

4. $\sqrt[4]{81} = \sqrt[4]{3^4} = 3$

5. $-\sqrt[5]{-243} = -\sqrt[5]{(-3)^5} = -(-3) = 3$

6. $\sqrt[3]{10^3} = 10$

7. $\sqrt[5]{z^5} = z$

8. $\sqrt[4]{(5p-3)^4} = |5p-3|$

9. $81^{1/2} = \sqrt{81} = \sqrt{9^2} = 9$

10. $(-256)^{1/4} = \sqrt[4]{-256}$ is not a real number.

11. $-4^{1/2} = -\sqrt{4} = -\sqrt{2^2} = -2$

12. $729^{1/3} = \sqrt[3]{729} = \sqrt[3]{9^3} = 9$

13. $16^{7/4} = \left(\sqrt[4]{16}\right)^7 = \left(\sqrt[4]{2^4}\right)^7 = 2^7 = 128$

## Chapter 7 Review

14. $-(-27)^{2/3} = -\left(\sqrt[3]{-27}\right)^2$
$= -\left(\sqrt[3]{(-3)^3}\right)^2$
$= -(-3)^2$
$= -9$

15. $-121^{3/2} = -\left(\sqrt{121}\right)^3$
$= -\left(\sqrt{11^2}\right)^3$
$= -(11)^3$
$= -1331$

16. $\dfrac{1}{36^{-1/2}} = 36^{1/2} = \sqrt{36} = \sqrt{6^2} = 6$

17. $(-65)^{1/3} \approx -4.02$

18. $4^{3/5} \approx 2.30$

19. $\sqrt[3]{100} \approx 4.64$

20. $\sqrt[4]{10} \approx 1.78$

21. $\sqrt[3]{5a} = (5a)^{1/3}$

22. $\sqrt[5]{p^7} = p^{7/5}$

23. $\left(\sqrt[4]{10z}\right)^3 = (10z)^{3/4}$

24. $\sqrt[6]{(2ab)^5} = (2ab)^{5/6}$

25. $4^{2/3} \cdot 4^{7/3} = 4^{\frac{2}{3}+\frac{7}{3}} = 4^{\frac{9}{3}} = 4^3 = 64$

26. $\dfrac{k^{1/2}}{k^{3/4}} = k^{\frac{1}{2}-\frac{3}{4}} = k^{\frac{2}{4}-\frac{3}{4}} = k^{-1/4} = \dfrac{1}{k^{1/4}}$

27. $\left(p^{4/3} \cdot q^4\right)^{3/2} = \left(p^{4/3}\right)^{3/2} \cdot \left(q^4\right)^{3/2}$
$= p^{\frac{4}{3}\cdot\frac{3}{2}} \cdot q^{4\cdot\frac{3}{2}}$
$= p^2 \cdot q^6 \quad \text{or} \quad \left(p \cdot q^3\right)^2$

28. $\left(32a^{-3/2} \cdot b^{1/4}\right)^{1/5} = (32)^{1/5} \cdot \left(a^{-3/2}\right)^{1/5} \cdot \left(b^{1/4}\right)^{1/5}$
$= 2 \cdot a^{-3/10} \cdot b^{1/20}$
$= \dfrac{2b^{1/20}}{a^{3/10}} \quad \text{or} \quad 2\left(\dfrac{b}{a^6}\right)^{1/20}$

29. $5m^{-2/3}\left(2m + m^{-1/3}\right)$
$= 5m^{-2/3} \cdot 2m + 5m^{-2/3} \cdot m^{-1/3}$
$= 10m^{-2/3} \cdot m^{3/3} + 5m^{-3/3}$
$= 10m^{1/3} + 5m^{-1}$
$= 10m^{1/3} + \dfrac{5}{m}$

30. $\left(\dfrac{16x^{1/3}}{x^{-1/3}}\right)^{-1/2} + \left(\dfrac{x^{-3/2}}{64x^{-1/2}}\right)^{1/3}$
$= \left(16x^{\frac{1}{3}-\left(-\frac{1}{3}\right)}\right)^{-1/2} + \left(64^{-1}x^{-\frac{3}{2}-\left(-\frac{1}{2}\right)}\right)^{1/3}$
$= \left(16x^{2/3}\right)^{-1/2} + \left(\dfrac{1}{64}x^{-1}\right)^{1/3}$
$= \left(\dfrac{1}{16}x^{-2/3}\right)^{1/2} + \left(\dfrac{1}{64}x^{-1}\right)^{1/3}$
$= \left(\dfrac{1}{16}\right)^{1/2}\left(x^{-2/3}\right)^{1/2} + \left(\dfrac{1}{64}\right)^{1/3}\left(x^{-1}\right)^{1/3}$
$= \dfrac{1}{4}x^{-1/3} + \dfrac{1}{4}x^{-1/3}$
$= \dfrac{1}{2}x^{-1/3}$
$= \dfrac{1}{2x^{1/3}}$

31. $\sqrt[8]{x^6} = x^{6/8} = x^{3/4} = \sqrt[4]{x^3}$

32. $\sqrt{121x^4y^{10}} = \left(121x^4y^{10}\right)^{1/2}$
$= 121^{1/2}\left(x^4\right)^{1/2}\left(y^{10}\right)^{1/2}$
$= 11x^{4/2}y^{10/2}$
$= 11x^2y^5$

## Chapter 7: Radicals and Rational Exponents

33. $\sqrt[3]{m^2} \cdot \sqrt{m^3} = m^{2/3} \cdot m^{3/2}$
$= m^{\frac{2}{3}+\frac{3}{2}}$
$= m^{\frac{4}{6}+\frac{9}{6}}$
$= m^{13/6}$
$= \sqrt[6]{m^{13}}$
$= m^2 \sqrt[6]{m}$

34. $\dfrac{\sqrt[3]{c}}{\sqrt[6]{c^4}} = \dfrac{c^{1/3}}{c^{4/6}} = \dfrac{c^{1/3}}{c^{2/3}} = c^{\frac{1}{3}-\frac{2}{3}} = c^{-1/3} = \dfrac{1}{c^{1/3}} = \dfrac{1}{\sqrt[3]{c}}$

35. $2(3m-1)^{1/4} + (m-7)(3m-1)^{5/4}$
$= 2 \cdot (3m-1)^{1/4} + (m-7)(3m-1)^{1/4} \cdot (3m-1)$
$= (3m-1)^{1/4} \left(2 + (m-7)(3m-1)\right)$
$= (3m-1)^{1/4} \left(2 + 3m^2 - 21m - m + 7\right)$
$= (3m-1)^{1/4} \left(3m^2 - 22m + 9\right)$

36. $3(x^2-5)^{1/3} - 4x(x^2-5)^{-2/3}$
$= 3(x^2-5)^{-2/3} \cdot (x^2-5) - 4x(x^2-5)^{-2/3}$
$= (x^2-5)^{-2/3} \left(3(x^2-5) - 4x\right)$
$= (x^2-5)^{-2/3} \left(3x^2 - 15 - 4x\right)$
$= (x^2-5)^{-2/3} \left(3x^2 - 4x - 15\right)$
$= \dfrac{3x^2 - 4x - 15}{(x^2-5)^{2/3}} = \dfrac{(3x+5)(x-3)}{(x^2-5)^{2/3}}$

37. $\sqrt{15} \cdot \sqrt{7} = \sqrt{15 \cdot 7} = \sqrt{105}$

38. $\sqrt[4]{2ab^2} \cdot \sqrt[4]{6a^2 b} = \sqrt[4]{2ab^2 \cdot 6a^2 b} = \sqrt[4]{12a^3 b^3}$

39. $\sqrt{80} = \sqrt{16 \cdot 5} = \sqrt{16} \cdot \sqrt{5} = 4\sqrt{5}$

40. $\sqrt[3]{-500} = \sqrt[3]{-125 \cdot 4} = \sqrt[3]{-125} \cdot \sqrt[3]{4} = -5\sqrt[3]{4}$

41. $\sqrt[3]{162 m^6 n^4} = \sqrt[3]{27 m^6 n^3 \cdot 6n}$
$= \sqrt[3]{27 m^6 n^3} \cdot \sqrt[3]{6n}$
$= 3m^2 n \sqrt[3]{6n}$

42. $\sqrt[4]{50 p^8 q^4} = \sqrt[4]{p^8 q^4 \cdot 50}$
$= \sqrt[4]{p^8 q^4} \cdot \sqrt[4]{50}$
$= p^2 |q| \sqrt[4]{50}$

43. $2\sqrt{16 x^6 y} = 2\sqrt{16 x^6 \cdot y}$
$= 2\sqrt{16 x^6} \cdot \sqrt{y}$
$= 2 \cdot 4|x^3| \sqrt{y}$
$= 8|x^3| \sqrt{y}$

as long as $y \geq 0$. Otherwise the result is not a real number.

44. $\sqrt{(2x+1)^3} = \sqrt{(2x+1)^2 \cdot (2x+1)}$
$= \sqrt{(2x+1)^2} \cdot \sqrt{2x+1}$
$= (2x+1)\sqrt{2x+1}$

as long as $2x+1 \geq 0$. Otherwise the result is not a real number.

45. $\sqrt{w^3 z^2} = \sqrt{w^2 z^2 \cdot w} = \sqrt{w^2 z^2} \cdot \sqrt{w} = wz\sqrt{w}$

46. $\sqrt{45 x^4 y z^3} = \sqrt{9x^4 z^2 \cdot 5yz}$
$= \sqrt{9x^4 z^2} \cdot \sqrt{5yz}$
$= 3x^2 z \sqrt{5yz}$

47. $\sqrt[3]{16 a^{12} b^5} = \sqrt[3]{8 a^{12} b^3 \cdot 2b^2}$
$= \sqrt[3]{8 a^{12} b^3} \cdot \sqrt[3]{2b^2}$
$= 2a^4 b \sqrt[3]{2b^2}$

48. $\sqrt{4x^2 + 8x + 4} = \sqrt{4(x^2 + 2x + 1)}$
$= \sqrt{4(x+1)^2}$
$= 2(x+1)$

49. $\sqrt{15} \cdot \sqrt{18} = \sqrt{15 \cdot 18}$
$= \sqrt{270}$
$= \sqrt{9 \cdot 30}$
$= \sqrt{9} \cdot \sqrt{30}$
$= 3\sqrt{30}$

50. $\sqrt[3]{20} \cdot \sqrt[3]{30} = \sqrt[3]{20 \cdot 30}$
$= \sqrt[3]{600}$
$= \sqrt[3]{8 \cdot 75}$
$= \sqrt[3]{8} \cdot \sqrt[3]{75}$
$= 2\sqrt[3]{75}$

51. $\sqrt[3]{-3x^4y^7} \cdot \sqrt[3]{24x^3y^2} = \sqrt[3]{-3x^4y^7 \cdot 24x^3y^2}$
$= \sqrt[3]{-72x^7y^9}$
$= \sqrt[3]{-8x^6y^9 \cdot 9x}$
$= \sqrt[3]{-8x^6y^9} \cdot \sqrt[3]{9x}$
$= -2x^2y^3 \sqrt[3]{9x}$

52. $3\sqrt{4xy^2} \cdot 5\sqrt{3x^2y} = 15\sqrt{4xy^2 \cdot 3x^2y}$
$= 15\sqrt{12x^3y^3}$
$= 15\sqrt{4x^2y^2 \cdot 3xy}$
$= 15\sqrt{4x^2y^2} \cdot \sqrt{3xy}$
$= 15 \cdot 2xy\sqrt{3xy}$
$= 30xy\sqrt{3xy}$

53. $\sqrt{\dfrac{121}{25}} = \dfrac{\sqrt{121}}{\sqrt{25}} = \dfrac{11}{5}$

54. $\sqrt{\dfrac{5a^4}{64b^2}} = \dfrac{\sqrt{5a^4}}{\sqrt{64b^2}} = \dfrac{a^2\sqrt{5}}{8b}$

55. $\sqrt[3]{\dfrac{54k^2}{9k^5}} = \sqrt[3]{\dfrac{6}{k^3}} = \dfrac{\sqrt[3]{6}}{\sqrt[3]{k^3}} = \dfrac{\sqrt[3]{6}}{k}$

56. $\sqrt[3]{\dfrac{-160w^{11}}{343w^{-4}}} = \sqrt[3]{\dfrac{-160w^{15}}{343}}$
$= \dfrac{\sqrt[3]{-160w^{15}}}{\sqrt[3]{343}}$
$= \dfrac{\sqrt[3]{-8w^{15} \cdot 20}}{\sqrt[3]{7^3}}$
$= \dfrac{-2w^5 \sqrt[3]{20}}{7}$

57. $\dfrac{\sqrt{12h^3}}{\sqrt{3h}} = \sqrt{\dfrac{12h^3}{3h}} = \sqrt{4h^2} = 2h$

58. $\dfrac{\sqrt{50a^3b^3}}{\sqrt{8a^5b^{-3}}} = \sqrt{\dfrac{50a^3b^3}{8a^5b^{-3}}} = \sqrt{\dfrac{25b^6}{4a^2}} = \dfrac{\sqrt{25b^6}}{\sqrt{4a^2}} = \dfrac{5b^3}{2a}$

59. $\dfrac{\sqrt[3]{-8x^7y}}{\sqrt[3]{27xy^4}} = \sqrt[3]{\dfrac{-8x^7y}{27xy^4}} = \sqrt[3]{\dfrac{-8x^6}{27y^3}} = \dfrac{\sqrt[3]{-8x^6}}{\sqrt[3]{27y^3}} = \dfrac{-2x^2}{3y}$

60. $\dfrac{\sqrt[4]{48m^2n^7}}{\sqrt[4]{3m^6n}} = \sqrt[4]{\dfrac{48m^2n^7}{3m^6n}}$
$= \sqrt[4]{\dfrac{16n^6}{m^4}}$
$= \dfrac{\sqrt[4]{16n^6}}{\sqrt[4]{m^4}}$
$= \dfrac{2n\sqrt[4]{n^2}}{m}$
$= \dfrac{2n\sqrt{n}}{m}$

61. $\sqrt{5} \cdot \sqrt[3]{2} = 5^{1/2} \cdot 2^{1/3}$
$= 5^{3/6} \cdot 2^{2/6}$
$= \left(5^3\right)^{1/6} \cdot \left(2^2\right)^{1/6}$
$= \left(5^3 \cdot 2^2\right)^{1/6}$
$= (125 \cdot 4)^{1/6}$
$= 500^{1/6}$
$= \sqrt[6]{500}$

62. $\sqrt[4]{8} \cdot \sqrt[6]{4} = 8^{1/4} \cdot 4^{1/6}$
$= \left(2^3\right)^{1/4} \cdot \left(2^2\right)^{1/6}$
$= 2^{3/4} \cdot 2^{1/3} = 2^{\frac{3}{4}+\frac{1}{3}}$
$= 2^{13/12} = 2^{\frac{12}{12}} \cdot 2^{\frac{1}{12}}$
$= 2\sqrt[12]{2}$

63. $2\sqrt[4]{x} + 6\sqrt[4]{x} = (2+6)\sqrt[4]{x} = 8\sqrt[4]{x}$

64. $7\sqrt[3]{4y} + 2\sqrt[3]{4y} - 3\sqrt[3]{4y} = (7+2-3)\sqrt[3]{4y}$
$= 6\sqrt[3]{4y}$

65. $5\sqrt{2} - 2\sqrt{12} = 5\sqrt{2} - 2\cdot\sqrt{4}\cdot\sqrt{3}$
$= 5\sqrt{2} - 2\cdot 2\cdot\sqrt{3}$
$= 5\sqrt{2} - 4\sqrt{3}$
Cannot be simplified further.

66. $\sqrt{18} + 2\sqrt{50} = \sqrt{9}\cdot\sqrt{2} + 2\sqrt{25}\cdot\sqrt{2}$
$= 3\cdot\sqrt{2} + 2\cdot 5\cdot\sqrt{2}$
$= 3\sqrt{2} + 10\sqrt{2}$
$= (3+10)\sqrt{2}$
$= 13\sqrt{2}$

67. $\sqrt[3]{-16z} + \sqrt[3]{54z} = \sqrt[3]{-8}\cdot\sqrt[3]{2z} + \sqrt[3]{27}\cdot\sqrt[3]{2z}$
$= -2\sqrt[3]{2z} + 3\sqrt[3]{2z}$
$= (-2+3)\sqrt[3]{2z}$
$= \sqrt[3]{2z}$

68. $7\sqrt[3]{8x^2} - \sqrt[3]{-27x^2} = 7\cdot\sqrt[3]{8}\cdot\sqrt[3]{x^2} - \sqrt[3]{-27}\cdot\sqrt[3]{x^2}$
$= 7\cdot 2\cdot\sqrt[3]{x^2} - (-3)\cdot\sqrt[3]{x^2}$
$= 14\sqrt[3]{x^2} + 3\sqrt[3]{x^2}$
$= (14+3)\sqrt[3]{x^2}$
$= 17\sqrt[3]{x^2}$

69. $\sqrt{16a} + \sqrt[6]{729a^3} = \sqrt{16}\cdot\sqrt{a} + \sqrt[6]{729}\cdot\sqrt[6]{a^3}$
$= \sqrt{16}\cdot\sqrt{a} + \sqrt[6]{729}\cdot a^{3/6}$
$= 4\sqrt{a} + 3\cdot a^{1/2}$
$= 4\sqrt{a} + 3\sqrt{a}$
$= (4+3)\sqrt{a}$
$= 7\sqrt{a}$

70. $\sqrt{27x^2} - x\sqrt{48} + 2\sqrt{75x^2}$
$= \sqrt{9x^2}\cdot\sqrt{3} - x\cdot\sqrt{16}\cdot\sqrt{3} + 2\cdot\sqrt{25x^2}\cdot\sqrt{3}$
$= 3x\sqrt{3} - 4x\sqrt{3} + 10x\sqrt{3}$
$= (3+10-4)x\sqrt{3}$
$= 9x\sqrt{3}$

71. $5\sqrt[3]{4m^5y^2} - \sqrt[6]{16m^{10}y^4}$
$= 5\cdot\sqrt[3]{m^3}\cdot\sqrt[3]{4m^2y^2} - \sqrt[6]{m^6}\cdot\sqrt[6]{16m^4y^4}$
$= 5m\sqrt[3]{4m^2y^2} - m\cdot\sqrt[6]{(4m^2y^2)^2}$
$= 5m\sqrt[3]{4m^2y^2} - m\cdot(4m^2y^2)^{2/6}$
$= 5m\sqrt[3]{4m^2y^2} - m\cdot(4m^2y^2)^{1/3}$
$= 5m\sqrt[3]{4m^2y^2} - m\sqrt[3]{4m^2y^2}$
$= (5-1)m\sqrt[3]{4m^2y^2}$
$= 4m\sqrt[3]{4m^2y^2}$

72. $\sqrt{y^3 - 4y^2} - 2\sqrt{y-4} + \sqrt[4]{y^2 - 8y + 16}$
$= \sqrt{y^2}\cdot\sqrt{(y-4)} - 2\sqrt{y-4} + \sqrt[4]{(y-4)^2}$
$= y\sqrt{y-4} - 2\sqrt{y-4} + (y-4)^{2/4}$
$= y\sqrt{y-4} - 2\sqrt{y-4} + (y-4)^{1/2}$
$= y\sqrt{y-4} - 2\sqrt{y-4} + \sqrt{y-4}$
$= (y-2+1)\sqrt{y-4}$
$= (y-1)\sqrt{y-4}$

73. $\sqrt{3}(\sqrt{5} - \sqrt{15}) = \sqrt{3}\cdot\sqrt{5} - \sqrt{3}\cdot\sqrt{15}$
$= \sqrt{3}\cdot\sqrt{5} - \sqrt{3\cdot 15}$
$= \sqrt{3}\cdot\sqrt{5} - \sqrt{45}$
$= \sqrt{3}\cdot\sqrt{5} - \sqrt{9}\cdot\sqrt{5}$
$= \sqrt{3}\cdot\sqrt{5} - 3\cdot\sqrt{5}$
$= \sqrt{15} - 3\sqrt{5}$

74. $\sqrt[3]{5}(3 + \sqrt[3]{4}) = \sqrt[3]{5}\cdot 3 + \sqrt[3]{5}\cdot\sqrt[3]{4} = 3\sqrt[3]{5} + \sqrt[3]{20}$

75. $(3+\sqrt{5})(4-\sqrt{5}) = 3\cdot 4 - 3\cdot\sqrt{5} + 4\cdot\sqrt{5} - \sqrt{5}\cdot\sqrt{5}$
$= 12 - 3\sqrt{5} + 4\sqrt{5} - \sqrt{25}$
$= 12 + \sqrt{5} - 5$
$= 7 + \sqrt{5}$

76. $(7+\sqrt{3})(6+\sqrt{2})$
$= 7\cdot 6 + 7\cdot\sqrt{2} + \sqrt{3}\cdot 6 + \sqrt{3}\cdot\sqrt{2}$
$= 42 + 7\sqrt{2} + 6\sqrt{3} + \sqrt{6}$

Chapter 7 Review

77. $(1-3\sqrt{5})(1+3\sqrt{5}) = (1)^2 - (3\sqrt{5})^2$
$= 1 - 9\sqrt{25}$
$= 1 - 9 \cdot 5$
$= 1 - 45$
$= -44$

78. $(\sqrt[3]{x}+1)(9\sqrt[3]{x}-4)$
$= \sqrt[3]{x} \cdot 9\sqrt[3]{x} + \sqrt[3]{x} \cdot (-4) + 1 \cdot 9\sqrt[3]{x} + 1 \cdot (-4)$
$= 9\sqrt[3]{x^2} - 4\sqrt[3]{x} + 9\sqrt[3]{x} - 4$
$= 9\sqrt[3]{x^2} + 5\sqrt[3]{x} - 4$

79. $(\sqrt{x}-\sqrt{5})^2 = (\sqrt{x})^2 - 2(\sqrt{x})(\sqrt{5}) + (\sqrt{5})^2$
$= \sqrt{x^2} - 2\sqrt{5x} + \sqrt{25}$
$= x - 2\sqrt{5x} + 5$

80. $(11\sqrt{2}+\sqrt{5})^2$
$= (11\sqrt{2})^2 + 2(11\sqrt{2})(\sqrt{5}) + (\sqrt{5})^2$
$= 121\sqrt{4} + 22\sqrt{10} + \sqrt{25}$
$= 121 \cdot 2 + 22\sqrt{10} + 5$
$= 242 + 22\sqrt{10} + 5$
$= 247 + 22\sqrt{10}$

81. $(\sqrt{2a}-b)(\sqrt{2a}+b) = (\sqrt{2a})^2 - (b)^2$
$= \sqrt{4a^2} - b^2$
$= 2a - b^2$

82. $(\sqrt[3]{6s}+2)(\sqrt[3]{6s}-7)$
$= \sqrt[3]{6s} \cdot \sqrt[3]{6s} + \sqrt[3]{6s} \cdot (-7) + 2 \cdot \sqrt[3]{6s} + 2 \cdot (-7)$
$= \sqrt[3]{36s^2} - 7\sqrt[3]{6s} + 2\sqrt[3]{6s} - 14$
$= \sqrt[3]{36s^2} - 5\sqrt[3]{6s} - 14$

83. $\dfrac{2}{\sqrt{6}} = \dfrac{2}{\sqrt{6}} \cdot \dfrac{\sqrt{6}}{\sqrt{6}} = \dfrac{2\sqrt{6}}{\sqrt{36}} = \dfrac{2\sqrt{6}}{6} = \dfrac{\sqrt{6}}{3}$

84. $\dfrac{6}{\sqrt{3}} = \dfrac{6}{\sqrt{3}} \cdot \dfrac{\sqrt{3}}{\sqrt{3}} = \dfrac{6\sqrt{3}}{\sqrt{9}} = \dfrac{6\sqrt{3}}{3} = 2\sqrt{3}$

85. $\dfrac{\sqrt{48}}{\sqrt{p^3}} = \dfrac{4\sqrt{3}}{p\sqrt{p}} = \dfrac{4\sqrt{3}}{p\sqrt{p}} \cdot \dfrac{\sqrt{p}}{\sqrt{p}}$
$= \dfrac{4\sqrt{3p}}{p\sqrt{p^2}} = \dfrac{4\sqrt{3p}}{p \cdot p}$
$= \dfrac{4\sqrt{3p}}{p^2}$

86. $\dfrac{5}{\sqrt{2a}} = \dfrac{5}{\sqrt{2a}} \cdot \dfrac{\sqrt{2a}}{\sqrt{2a}} = \dfrac{5\sqrt{2a}}{\sqrt{4a^2}} = \dfrac{5\sqrt{2a}}{2a}$

87. $\dfrac{-2}{\sqrt{6y^3}} = \dfrac{-2}{y\sqrt{6y}} = \dfrac{-2}{y\sqrt{6y}} \cdot \dfrac{\sqrt{6y}}{\sqrt{6y}}$
$= \dfrac{-2\sqrt{6y}}{y\sqrt{36y^2}} = \dfrac{-2\sqrt{6y}}{y \cdot 6y}$
$= \dfrac{-2\sqrt{6y}}{6y^2}$
$= -\dfrac{\sqrt{6y}}{3y^2}$

88. $\dfrac{3}{\sqrt[3]{5}} = \dfrac{3}{\sqrt[3]{5}} \cdot \dfrac{\sqrt[3]{25}}{\sqrt[3]{25}} = \dfrac{3\sqrt[3]{25}}{\sqrt[3]{125}} = \dfrac{3\sqrt[3]{25}}{5}$

89. $\sqrt[3]{\dfrac{-4}{45}} = \dfrac{\sqrt[3]{-4}}{\sqrt[3]{45}} = \dfrac{\sqrt[3]{-4}}{\sqrt[3]{45}} \cdot \dfrac{\sqrt[3]{75}}{\sqrt[3]{75}} = \dfrac{\sqrt[3]{-300}}{\sqrt[3]{3375}} = -\dfrac{\sqrt[3]{300}}{15}$

90. $\dfrac{27}{\sqrt[5]{8p^3q^4}} = \dfrac{27}{\sqrt[5]{8p^3q^4}} \cdot \dfrac{\sqrt[5]{4p^2q}}{\sqrt[5]{4p^2q}}$
$= \dfrac{27\sqrt[5]{4p^2q}}{\sqrt[5]{32p^5q^5}}$
$= \dfrac{27\sqrt[5]{4p^2q}}{2pq}$

*Chapter 7:* Radicals and Rational Exponents

91. $\dfrac{6}{7-\sqrt{6}} = \dfrac{6}{7-\sqrt{6}} \cdot \dfrac{7+\sqrt{6}}{7+\sqrt{6}}$
$= \dfrac{6(7+\sqrt{6})}{7^2 - (\sqrt{6})^2}$
$= \dfrac{42 + 6\sqrt{6}}{49 - 6}$
$= \dfrac{42 + 6\sqrt{6}}{43}$

92. $\dfrac{3}{\sqrt{3}-9} = \dfrac{3}{\sqrt{3}-9} \cdot \dfrac{\sqrt{3}+9}{\sqrt{3}+9}$
$= \dfrac{3(\sqrt{3}+9)}{(\sqrt{3})^2 - 9^2}$
$= \dfrac{3\sqrt{3}+27}{3 - 81}$
$= \dfrac{3\sqrt{3}+27}{-78}$
$= -\dfrac{3\sqrt{3}+27}{78}$
$= -\dfrac{\sqrt{3}+9}{26}$

93. $\dfrac{\sqrt{3}}{3+\sqrt{2}} = \dfrac{\sqrt{3}}{3+\sqrt{2}} \cdot \dfrac{3-\sqrt{2}}{3-\sqrt{2}}$
$= \dfrac{\sqrt{3}(3-\sqrt{2})}{3^2 - (\sqrt{2})^2}$
$= \dfrac{\sqrt{3}(3-\sqrt{2})}{9 - 2}$
$= \dfrac{\sqrt{3}(3-\sqrt{2})}{7}$ or $\dfrac{3\sqrt{3}-\sqrt{6}}{7}$

94. $\dfrac{\sqrt{k}}{\sqrt{k}-\sqrt{m}} = \dfrac{\sqrt{k}}{\sqrt{k}-\sqrt{m}} \cdot \dfrac{\sqrt{k}+\sqrt{m}}{\sqrt{k}+\sqrt{m}}$
$= \dfrac{\sqrt{k}(\sqrt{k}+\sqrt{m})}{(\sqrt{k})^2 - (\sqrt{m})^2}$
$= \dfrac{k + \sqrt{km}}{k - m}$

95. $\dfrac{\sqrt{10}+2}{\sqrt{10}-2} = \dfrac{\sqrt{10}+2}{\sqrt{10}-2} \cdot \dfrac{\sqrt{10}+2}{\sqrt{10}+2}$
$= \dfrac{(\sqrt{10})^2 + 2 \cdot \sqrt{10} \cdot 2 + 2^2}{(\sqrt{10})^2 - 2^2}$
$= \dfrac{10 + 4\sqrt{10} + 4}{10 - 4}$
$= \dfrac{14 + 4\sqrt{10}}{6}$
$= \dfrac{2(7 + 2\sqrt{10})}{2(3)}$
$= \dfrac{7 + 2\sqrt{10}}{3}$

96. $\dfrac{3-\sqrt{y}}{3+\sqrt{y}} = \dfrac{3-\sqrt{y}}{3+\sqrt{y}} \cdot \dfrac{3-\sqrt{y}}{3-\sqrt{y}}$
$= \dfrac{3^2 - 2 \cdot 3 \cdot \sqrt{y} + (\sqrt{y})^2}{3^2 - (\sqrt{y})^2}$
$= \dfrac{9 - 6\sqrt{y} + y}{9 - y}$ or $\dfrac{y - 6\sqrt{y} + 9}{9 - y}$

97. $\dfrac{4}{2\sqrt{3}+5\sqrt{2}} = \dfrac{4}{2\sqrt{3}+5\sqrt{2}} \cdot \dfrac{2\sqrt{3}-5\sqrt{2}}{2\sqrt{3}-5\sqrt{2}}$
$= \dfrac{4(2\sqrt{3}-5\sqrt{2})}{(2\sqrt{3})^2 - (5\sqrt{2})^2}$
$= \dfrac{8\sqrt{3} - 20\sqrt{2}}{4 \cdot 3 - 25 \cdot 2}$
$= \dfrac{8\sqrt{3} - 20\sqrt{2}}{12 - 50}$
$= \dfrac{8\sqrt{3} - 20\sqrt{2}}{-38}$
$= \dfrac{-2(-4\sqrt{3} + 10\sqrt{2})}{-2(19)}$
$= \dfrac{10\sqrt{2} - 4\sqrt{3}}{19}$

## Chapter 7 Review

**98.** $\dfrac{\sqrt{5}-\sqrt{6}}{\sqrt{10}+\sqrt{3}} = \dfrac{\sqrt{5}-\sqrt{6}}{\sqrt{10}+\sqrt{3}} \cdot \dfrac{\sqrt{10}-\sqrt{3}}{\sqrt{10}-\sqrt{3}}$

$= \dfrac{(\sqrt{5}-\sqrt{6})(\sqrt{10}-\sqrt{3})}{(\sqrt{10}+\sqrt{3})(\sqrt{10}-\sqrt{3})}$

$= \dfrac{\sqrt{50}-\sqrt{15}-\sqrt{60}+\sqrt{18}}{(\sqrt{10})^2-(\sqrt{3})^2}$

$= \dfrac{5\sqrt{2}-\sqrt{15}-2\sqrt{15}+3\sqrt{2}}{10-3}$

$= \dfrac{8\sqrt{2}-3\sqrt{15}}{7}$

**99.** $\dfrac{\sqrt{7}}{3} + \dfrac{6}{\sqrt{7}} = \dfrac{\sqrt{7}}{3} \cdot \dfrac{\sqrt{7}}{\sqrt{7}} + \dfrac{6}{\sqrt{7}} \cdot \dfrac{3}{3}$

$= \dfrac{7}{3\sqrt{7}} + \dfrac{18}{3\sqrt{7}}$

$= \dfrac{25}{3\sqrt{7}}$

$= \dfrac{25}{3\sqrt{7}} \cdot \dfrac{\sqrt{7}}{\sqrt{7}}$

$= \dfrac{25\sqrt{7}}{3 \cdot 7}$

$= \dfrac{25\sqrt{7}}{21}$

**100.** $(4-\sqrt{7})^{-1} = \dfrac{1}{4-\sqrt{7}}$

$= \dfrac{1}{4-\sqrt{7}} \cdot \dfrac{4+\sqrt{7}}{4+\sqrt{7}}$

$= \dfrac{4+\sqrt{7}}{4^2-(\sqrt{7})^2}$

$= \dfrac{4+\sqrt{7}}{16-7}$

$= \dfrac{4+\sqrt{7}}{9}$

**101.** $f(x) = \sqrt{x+4}$

a. $f(-3) = \sqrt{-3+4} = \sqrt{1} = 1$

b. $f(0) = \sqrt{0+4} = \sqrt{4} = 2$

c. $f(5) = \sqrt{5+4} = \sqrt{9} = 3$

**102.** $g(x) = \sqrt{3x-2}$

a. $g\left(\dfrac{2}{3}\right) = \sqrt{3\left(\dfrac{2}{3}\right)-2} = \sqrt{2-2} = \sqrt{0} = 0$

b. $g(2) = \sqrt{3(2)-2} = \sqrt{6-2} = \sqrt{4} = 2$

c. $g(6) = \sqrt{3(6)-2} = \sqrt{18-2} = \sqrt{16} = 4$

**103.** $H(t) = \sqrt[3]{t+3}$

a. $H(-2) = \sqrt[3]{-2+3} = \sqrt[3]{1} = 1$

b. $H(-4) = \sqrt[3]{-4+3} = \sqrt[3]{-1} = -1$

c. $H(5) = \sqrt[3]{5+3} = \sqrt[3]{8} = 2$

**104.** $G(z) = \sqrt{\dfrac{z-1}{z+2}}$

a. $G(1) = \sqrt{\dfrac{1-1}{1+2}} = \sqrt{\dfrac{0}{3}} = \sqrt{0} = 0$

b. $G(-3) = \sqrt{\dfrac{-3-1}{-3+2}} = \sqrt{\dfrac{-4}{-1}} = \sqrt{4} = 2$

c. $G(2) = \sqrt{\dfrac{2-1}{2+2}} = \sqrt{\dfrac{1}{4}} = \dfrac{1}{2}$

**105.** $f(x) = \sqrt{3x-5}$

$3x - 5 \geq 0$

$3x \geq 5$

$x \geq \dfrac{5}{3}$

The domain of the function is $\left\{x \mid x \geq \dfrac{5}{3}\right\}$ or the interval $\left[\dfrac{5}{3}, \infty\right)$.

**106.** $g(x) = \sqrt[3]{2x-7}$

Since the index is odd, the domain of the function is all real numbers, $\{x \mid x \text{ is any real number}\}$ or the interval $(-\infty, \infty)$.

## Chapter 7: Radicals and Rational Exponents

**107.** $h(x) = \sqrt[4]{6x+1}$

$6x + 1 \geq 0$

$6x \geq -1$

$x \geq -\dfrac{1}{6}$

The domain of the function is $\left\{x \mid x \geq -\dfrac{1}{6}\right\}$ or the interval $\left[-\dfrac{1}{6}, \infty\right)$.

**108.** $F(x) = \sqrt[5]{2x-9}$

Since the index is odd, the domain of the function is all real numbers, $\{x \mid x \text{ is any real number}\}$, or the interval $(-\infty, \infty)$.

**109.** $G(x) = \sqrt{\dfrac{4}{x-2}}$

The index is even so we need the radicand to be nonnegative. The numerator is positive, so we need the denominator to be positive as well.

$x - 2 > 0$

$x > 2$

The domain of the function is $\{x \mid x > 2\}$ or the interval $(2, \infty)$.

**110.** $H(x) = \sqrt{\dfrac{x-3}{x}}$

The index is even so we need the radicand to be nonnegative. The radicand will equal zero when the numerator equals zero. The radicand will be positive when the numerator and denominator have the same sign.

$x - 3 = 0$

$x = 3$

$x - 3 > 0 \text{ and } x > 0$

$\quad x > 3 \text{ and } x > 0 \;\to\; x > 3$

$x - 3 < 0 \text{ and } x < 0$

$\quad x < 3 \text{ and } x < 0 \;\to\; x < 0$

The domain of the function is $\{x \mid x < 0 \text{ or } x \geq 3\}$ or the interval $(-\infty, 0) \cup [3, \infty)$.

**111.** $f(x) = \dfrac{1}{2}\sqrt{1-x}$

a. $1 - x \geq 0$

$x \leq 1$

The domain of the function is $\{x \mid x \leq 1\}$ or the interval $(-\infty, 1]$.

b. 
| $x$ | $f(x) = \dfrac{1}{2}\sqrt{1-x}$ | $(x, y)$ |
|---|---|---|
| $-15$ | $f(-15) = \dfrac{1}{2}\sqrt{1-(-15)} = 2$ | $(-15, 2)$ |
| $-8$ | $f(-8) = \dfrac{3}{2}$ | $\left(-8, \dfrac{3}{2}\right)$ |
| $-3$ | $f(-3) = 1$ | $(-3, 1)$ |
| $0$ | $f(0) = \dfrac{1}{2}$ | $\left(0, \dfrac{1}{2}\right)$ |
| $1$ | $f(1) = 0$ | $(1, 0)$ |

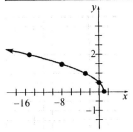

c. Based on the graph, the range is $[0, \infty)$.

**112.** $g(x) = \sqrt{x+1} - 2$

a. $x + 1 \geq 0 \text{ or } x \geq -1$

The domain of the function is $\{x \mid x \geq -1\}$ or the interval $[-1, \infty)$.

b.
| $x$ | $g(x) = \sqrt{x+1} - 2$ | $(x, y)$ |
|---|---|---|
| $-1$ | $g(-1) = \sqrt{-1+1} - 2 = -2$ | $(-1, -2)$ |
| $0$ | $g(0) = -1$ | $(0, -1)$ |
| $3$ | $g(3) = 0$ | $(3, 0)$ |
| $8$ | $g(8) = 1$ | $(8, 1)$ |
| $15$ | $g(15) = 2$ | $(15, 2)$ |

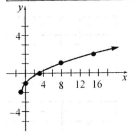

c. Based on the graph, the range is $[-2, \infty)$.

113. $h(x) = -\sqrt{x+3}$

   a. $x + 3 \geq 0$
   
   $x \geq -3$
   
   The domain of the function is $\{x \mid x \geq -3\}$ or the interval $[-3, \infty)$.

   b.
   | $x$ | $h(x) = -\sqrt{x+3}$ | $(x, y)$ |
   |---|---|---|
   | $-3$ | $h(-3) = -\sqrt{-3+3} = 0$ | $(-3, 0)$ |
   | $-2$ | $h(-2) = -1$ | $(-2, -1)$ |
   | $1$ | $h(1) = -2$ | $(1, -2)$ |
   | $6$ | $h(6) = -3$ | $(6, -3)$ |
   | $13$ | $h(13) = -4$ | $(13, -4)$ |

   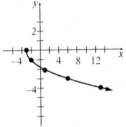

   c. Based on the graph, the range is $(-\infty, 0]$.

114. $F(x) = \sqrt[3]{x+1}$

   a. Since the index is odd, the domain of the function is all real numbers, $\{x \mid x \text{ is any real number}\}$, or the interval $(-\infty, \infty)$.

   b.
   | $x$ | $F(x) = \sqrt[3]{x+1}$ | $(x, y)$ |
   |---|---|---|
   | $-9$ | $F(-9) = \sqrt[3]{-9+1} = -2$ | $(-9, -2)$ |
   | $-2$ | $F(-2) = -1$ | $(-2, -1)$ |
   | $-1$ | $F(-1) = 0$ | $(-1, 0)$ |
   | $0$ | $F(0) = 1$ | $(0, 1)$ |
   | $7$ | $F(7) = 2$ | $(7, 2)$ |

   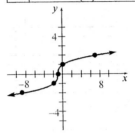

   c. Based on the graph, the range is $(-\infty, \infty)$.

115. $\sqrt{m} = 13$

   $\left(\sqrt{m}\right)^2 = 13^2$
   
   $m = 169$

   Check: $\sqrt{169} = 13$ ?
   
   $13 = 13$  T
   
   The solution set is $\{169\}$.

116. $\sqrt[3]{3t+1} = -2$

   $\left(\sqrt[3]{3t+1}\right)^3 = (-2)^3$
   
   $3t + 1 = -8$
   
   $3t = -9$
   
   $t = -3$

   Check: $\sqrt[3]{3(-3)+1} = -2$ ?
   
   $\sqrt[3]{-9+1} = -2$ ?
   
   $\sqrt[3]{-8} = -2$ ?
   
   $-2 = -2$  T
   
   The solution set is $\{-3\}$.

*Chapter 7:* Radicals and Rational Exponents

**117.** $\sqrt[4]{3x-8} = 3$

$\left(\sqrt[4]{3x-8}\right)^4 = 3^4$

$3x - 8 = 81$

$3x = 89$

$x = \dfrac{89}{3}$

Check: $\sqrt[4]{3\left(\dfrac{89}{3}\right) - 8} = 3$ ?

$\sqrt[4]{89 - 8} = 3$ ?

$\sqrt[4]{81} = 3$ ?

$3 = 3$ ✓

The solution set is $\left\{\dfrac{89}{3}\right\}$.

**118.** $\sqrt{2x+5} + 4 = 2$

$\sqrt{2x+5} = -2$

$\left(\sqrt{2x+5}\right)^2 = (-2)^2$

$2x + 5 = 4$

$2x = -1$

$x = -\dfrac{1}{2}$

Check: $\sqrt{2\left(-\dfrac{1}{2}\right) + 5} + 4 = 2$ ?

$\sqrt{-1 + 5} + 4 = 2$ ?

$\sqrt{4} + 4 = 2$ ?

$2 + 4 = 2$ ?

$6 = 2$ False

The solution does not check so the problem has no real solution.

**119.** $\sqrt{k+4} - 3 = -1$

$\sqrt{k+4} = 2$

$\left(\sqrt{k+4}\right)^2 = 2^2$

$k + 4 = 4$

$k = 0$

Check: $\sqrt{0+4} - 3 = -1$ ?

$\sqrt{4} - 3 = -1$ ?

$2 - 3 = -1$ ?

$-1 = -1$ ✓

The solution set is $\{0\}$.

**120.** $3\sqrt{t} - 4 = 11$

$3\sqrt{t} = 15$

$\sqrt{t} = 5$

$\left(\sqrt{t}\right)^2 = 5^2$

$t = 25$

Check: $3\sqrt{25} - 4 = 11$ ?

$3 \cdot 5 - 4 = 11$ ?

$15 - 4 = 11$ ?

$11 = 11$ ✓

The solution set is $\{25\}$.

**121.** $2\sqrt[3]{m} + 5 = -11$

$2\sqrt[3]{m} = -16$

$\sqrt[3]{m} = -8$

$\left(\sqrt[3]{m}\right)^3 = (-8)^3$

$m = -512$

Check: $2\sqrt[3]{-512} + 5 = -11$ ?

$2(-8) + 5 = -11$ ?

$-16 + 5 = -11$ ?

$-11 = -11$ ✓

The solution set is $\{-512\}$.

**122.** $\sqrt{q+2} = q$

$\left(\sqrt{q+2}\right)^2 = q^2$

$q + 2 = q^2$

$q^2 - q - 2 = 0$

$(q-2)(q+1) = 0$

$q - 2 = 0$ or $q + 1 = 0$

$q = 2$ or $q = -1$

Check: $\sqrt{2+2} = 2$ ?

$\sqrt{4} = 2$ ?

$2 = 2$ ✓

$\sqrt{-1+2} = -1$ ?

$\sqrt{1} = -1$ ?

$1 = -1$ False

The second solution is extraneous, so the solution set is $\{2\}$.

123. $\sqrt{w+11}+3 = w+2$
$\sqrt{w+11} = w-1$
$\left(\sqrt{w+11}\right)^2 = (w-1)^2$
$w+11 = w^2 - 2w + 1$
$w^2 - 3w - 10 = 0$
$(w-5)(w+2) = 0$
$w-5 = 0$ or $w+2 = 0$
$w = 5$ or $w = -2$

Check: $\sqrt{5+11}+3 = 5+2$ ?
$\sqrt{16}+3 = 7$ ?
$4+3 = 7$ ?
$7 = 7$ T

$\sqrt{-2+11}+3 = -2+2$ ?
$\sqrt{9}+3 = 0$ ?
$3+3 = 0$ ?
$6 = 0$ False

The second solution is extraneous, so the solution set is $\{5\}$.

124. $\sqrt{p^2 - 2p + 9} = p+1$
$\left(\sqrt{p^2 - 2p + 9}\right)^2 = (p+1)^2$
$p^2 - 2p + 9 = p^2 + 2p + 1$
$-4p + 8 = 0$
$-4p = -8$
$p = 2$

Check: $\sqrt{2^2 - 2(2) + 9} = 2+1$ ?
$\sqrt{4 - 4 + 9} = 3$ ?
$\sqrt{9} = 3$ ?
$3 = 3$ T

The solution set is $\{2\}$.

125. $\sqrt{a+10} = \sqrt{2a-1}$
$\left(\sqrt{a+10}\right)^2 = \left(\sqrt{2a-1}\right)^2$
$a+10 = 2a - 1$
$11 = a$

Check: $\sqrt{11+10} = \sqrt{2(11)-1}$ ?
$\sqrt{21} = \sqrt{22-1}$ ?
$\sqrt{21} = \sqrt{21}$ T

The solution set is $\{11\}$.

126. $\sqrt{5x+9} = \sqrt{7x-3}$
$\left(\sqrt{5x+9}\right)^2 = \left(\sqrt{7x-3}\right)^2$
$5x + 9 = 7x - 3$
$12 = 2x$
$6 = x$

Check: $\sqrt{5(6)+9} = \sqrt{7(6)-3}$ ?
$\sqrt{30+9} = \sqrt{42-3}$ ?
$\sqrt{39} = \sqrt{39}$ T

The solution set is $\{6\}$.

127. $\sqrt{c-8} + \sqrt{c} = 4$
$\sqrt{c-8} = 4 - \sqrt{c}$
$\left(\sqrt{c-8}\right)^2 = \left(4 - \sqrt{c}\right)^2$
$c - 8 = 4^2 - 2(4)\sqrt{c} + \left(\sqrt{c}\right)^2$
$c - 8 = 16 - 8\sqrt{c} + c$
$8\sqrt{c} = 24$
$\sqrt{c} = 3$
$\left(\sqrt{c}\right)^2 = 3^2$
$c = 9$

Check: $\sqrt{9-8} + \sqrt{9} = 4$ ?
$\sqrt{1} + \sqrt{9} = 4$ ?
$1 + 3 = 4$ ?
$4 = 4$ T

The solution set is $\{9\}$.

## Chapter 7: Radicals and Rational Exponents

**128.** $\sqrt{x+2} - \sqrt{x+9} = 7$

$\sqrt{x+2} = \sqrt{x+9} + 7$

$\left(\sqrt{x+2}\right)^2 = \left(\sqrt{x+9} + 7\right)^2$

$x+2 = \left(\sqrt{x+9}\right)^2 + 2(7)\sqrt{x+9} + 7^2$

$x+2 = x+9 + 14\sqrt{x+9} + 49$

$x+2 = x + 14\sqrt{x+9} + 58$

$-56 = 14\sqrt{x+9}$

$-4 = \sqrt{x+9}$

$(-4)^2 = \left(\sqrt{x+9}\right)^2$

$16 = x+9$

$7 = x$

Check: $\sqrt{7+2} - \sqrt{7+9} = 7$ ?

$\sqrt{9} - \sqrt{16} = 7$ ?

$3 - 4 = 7$ ?

$-1 = 7$ False

The solution does not check so the equation has no real solution.

**129.** $(4x-3)^{1/3} - 3 = 0$

$(4x-3)^{1/3} = 3$

$\left((4x-3)^{1/3}\right)^3 = 3^3$

$4x - 3 = 27$

$4x = 30$

$x = \dfrac{30}{4} = \dfrac{15}{2}$

Check: $\left(4\left(\dfrac{15}{2}\right) - 3\right)^{1/3} - 3 = 0$ ?

$(30 - 3)^{1/3} - 3 = 0$ ?

$27^{1/3} - 3 = 0$ ?

$3 - 3 = 0$ ?

$0 = 0$ T

The solution set is $\left\{\dfrac{15}{2}\right\}$.

**130.** $\left(x^2 - 9\right)^{1/4} = 2$

$\left(\left(x^2 - 9\right)^{1/4}\right)^4 = 2^4$

$x^2 - 9 = 16$

$x^2 = 25$

$x = \pm 5$

Check:

$\left(5^2 - 9\right)^{1/4} = 2$ ?

$(25 - 9)^{1/4} = 2$ ?

$16^{1/4} = 2$ ?

$2 = 2$ T

$\left((-5)^2 - 9\right)^{1/4} = 2$ ?

$(25 - 9)^{1/4} = 2$ ?

$16^{1/4} = 2$ ?

$2 = 2$ T

The solution set is $\{-5, 5\}$.

**131.** $r = \sqrt{\dfrac{3V}{\pi h}}$

$(r)^2 = \left(\sqrt{\dfrac{3V}{\pi h}}\right)^2$

$r^2 = \dfrac{3V}{\pi h}$

$h \cdot r^2 = \dfrac{3V}{\pi}$

$h = \dfrac{3V}{\pi r^2}$

**132.** $f_s = \sqrt[3]{\dfrac{30}{v}}$

$(f_s)^3 = \left(\sqrt[3]{\dfrac{30}{v}}\right)^3$

$f_s^3 = \dfrac{30}{v}$

$v \cdot f_s^3 = 30$

$v = \dfrac{30}{f_s^3}$

**133.** $\sqrt{-29} = \sqrt{-1} \cdot \sqrt{29} = i \cdot \sqrt{29} = \sqrt{29}\, i$

**134.** $\sqrt{-54} = \sqrt{-1} \cdot \sqrt{9} \cdot \sqrt{6} = i \cdot 3 \cdot \sqrt{6} = 3\sqrt{6}\,i$

**135.** $14 - \sqrt{-162} = 14 - \sqrt{-1} \cdot \sqrt{81} \cdot \sqrt{2}$
$= 14 - i \cdot 9 \cdot \sqrt{2}$
$= 14 - 9\sqrt{2}\,i$

**136.** $\dfrac{6 + \sqrt{-45}}{3} = \dfrac{6 + \sqrt{-1} \cdot \sqrt{9} \cdot \sqrt{5}}{3}$
$= \dfrac{6 + i \cdot 3 \cdot \sqrt{5}}{3}$
$= \dfrac{6 + 3\sqrt{5}\,i}{3}$
$= \dfrac{3(2 + \sqrt{5}\,i)}{3}$
$= 2 + \sqrt{5}\,i$

**137.** $(3 - 7i) + (-2 + 5i) = 3 - 2 - 7i + 5i$
$= 1 - 2i$

**138.** $(4 + 2i) - (9 - 8i) = 4 + 2i - 9 + 8i$
$= 4 - 9 + 2i + 8i$
$= -5 + 10i$

**139.** $\left(8 - \sqrt{-45}\right) - \left(3 + \sqrt{-80}\right)$
$= \left(8 - 3\sqrt{5}\,i\right) - \left(3 + 4\sqrt{5}\,i\right)$
$= 8 - 3\sqrt{5}\,i - 3 - 4\sqrt{5}\,i$
$= 8 - 3 - 3\sqrt{5}\,i - 4\sqrt{5}\,i$
$= 5 - 7\sqrt{5}\,i$

**140.** $\left(1 + \sqrt{-9}\right) + \left(-6 + \sqrt{-16}\right) = (1 + 3i) + (-6 + 4i)$
$= 1 + 3i - 6 + 4i$
$= 1 - 6 + 3i + 4i$
$= -5 + 7i$

**141.** $(4 - 5i)(3 + 7i) = 4 \cdot 3 + 4 \cdot 7i - 5i \cdot 3 - 5i \cdot 7i$
$= 12 + 28i - 15i - 35i^2$
$= 12 + 13i - 35(-1)$
$= 12 + 13i + 35$
$= 47 + 13i$

**142.** $\left(\dfrac{1}{2} + \dfrac{2}{3}i\right)(4 - 9i)$
$= \dfrac{1}{2} \cdot 4 + \dfrac{1}{2}(-9i) + \dfrac{2}{3}i \cdot 4 + \dfrac{2}{3}i(-9i)$
$= 2 - \dfrac{9}{2}i + \dfrac{8}{3}i - 6i^2$
$= 2 - \dfrac{11}{6}i - 6(-1)$
$= 2 - \dfrac{11}{6}i + 6$
$= 8 - \dfrac{11}{6}i$

**143.** $\sqrt{-3} \cdot \sqrt{-27} = \sqrt{3}\,i \cdot 3\sqrt{3}\,i$
$= 3\left(\sqrt{3}\right)^2 \cdot i^2$
$= 3 \cdot 3 \cdot (-1)$
$= -9$
Note: It is necessary to write each radical in terms of $i$ prior to performing the operation. Otherwise the sign of the answer will be wrong.
$\sqrt{(-3)(-27)} = \sqrt{81} = 9 \neq -9 = \sqrt{-3} \cdot \sqrt{-27}$

**144.** $\left(1 + \sqrt{-36}\right)\left(-5 - \sqrt{-144}\right)$
$= (1 + 6i)(-5 - 12i)$
$= 1 \cdot (-5) + 1 \cdot (-12i) + 6i \cdot (-5) + 6i \cdot (-12i)$
$= -5 - 12i - 30i - 72i^2$
$= -5 - 42i - 72(-1)$
$= -5 - 42i + 72$
$= 67 - 42i$

**145.** $(1 + 12i)(1 - 12i) = 1^2 - (12i)^2$
$= 1 - 144i^2$
$= 1 - 144(-1)$
$= 1 + 144$
$= 145$

**146.** $(7 + 2i)(5 + 4i) = 7 \cdot 5 + 7 \cdot 4i + 2i \cdot 5 + 2i \cdot 4i$
$= 35 + 28i + 10i + 8i^2$
$= 35 + 38i + 8(-1)$
$= 35 + 38i - 8$
$= 27 + 38i$

## Chapter 7: Radicals and Rational Exponents

**147.** $\dfrac{4}{3+5i} = \dfrac{4}{3+5i} \cdot \dfrac{3-5i}{3-5i}$

$= \dfrac{4(3-5i)}{3^2 - (5i)^2}$

$= \dfrac{4(3-5i)}{9 - 25i^2}$

$= \dfrac{4(3-5i)}{9+25}$

$= \dfrac{4(3-5i)}{34}$

$= \dfrac{2 \cdot 2(3-5i)}{2 \cdot 17}$

$= \dfrac{2(3-5i)}{17}$

$= \dfrac{6-10i}{17}$

$= \dfrac{6}{17} - \dfrac{10}{17}i$

**148.** $\dfrac{-3}{7-2i} = \dfrac{-3}{7-2i} \cdot \dfrac{7+2i}{7+2i}$

$= \dfrac{-3(7+2i)}{7^2 - (2i)^2}$

$= \dfrac{-21-6i}{49 - 4i^2}$

$= \dfrac{-21-6i}{49 - 4(-1)}$

$= \dfrac{-21-6i}{49+4}$

$= \dfrac{-21-6i}{53}$

$= -\dfrac{21}{53} - \dfrac{6}{53}i$

**149.** $\dfrac{2-3i}{5+2i} = \dfrac{2-3i}{5+2i} \cdot \dfrac{5-2i}{5-2i}$

$= \dfrac{(2-3i)(5-2i)}{(5+2i)(5-2i)}$

$= \dfrac{10 - 4i - 15i + 6i^2}{25 - 4i^2}$

$= \dfrac{10 - 19i + 6(-1)}{25 - 4(-1)}$

$= \dfrac{10 - 19i - 6}{25 + 4}$

$= \dfrac{4 - 19i}{29}$

$= \dfrac{4}{29} - \dfrac{19}{29}i$

**150.** $\dfrac{4+3i}{1-i} = \dfrac{4+3i}{1-i} \cdot \dfrac{1+i}{1+i}$

$= \dfrac{(4+3i)(1+i)}{(1-i)(1+i)}$

$= \dfrac{4 + 4i + 3i + 3i^2}{1 - i^2}$

$= \dfrac{4 + 7i + 3(-1)}{1 - (-1)}$

$= \dfrac{4 + 7i - 3}{1+1}$

$= \dfrac{1 + 7i}{2}$

$= \dfrac{1}{2} + \dfrac{7}{2}i$

**151.** $i^{59} = i^{56} \cdot i^3 = (i^4)^{14} \cdot i^3 = 1 \cdot i^3 = i^2 \cdot i = -1 \cdot i = -i$

**152.** $i^{173} = i^{172} \cdot i = (i^4)^{43} \cdot i = 1 \cdot i = i$

## Chapter 7 Test

1. $49^{-1/2} = \dfrac{1}{49^{1/2}} = \dfrac{1}{\sqrt{49}} = \dfrac{1}{7}$

2. $\sqrt[3]{8x^{1/2}y^3} \cdot \sqrt{9xy^{1/2}}$
   $= \left(8x^{1/2}y^3\right)^{1/3} \cdot \left(9xy^{1/2}\right)^{1/2}$
   $= 8^{1/3}\left(x^{1/2}\right)^{1/3}\left(y^3\right)^{1/3} \cdot 9^{1/2} x^{1/2} \left(y^{1/2}\right)^{1/2}$
   $= 2x^{1/6}y \cdot 3x^{1/2}y^{1/4}$
   $= 6x^{\frac{1}{6}+\frac{1}{2}} y^{1+\frac{1}{4}}$
   $= 6x^{2/3} y^{5/4}$
   $= 6x^{8/12} y^{15/12}$
   $= 6y\sqrt[12]{x^8 y^3}$

3. $\sqrt[5]{\left(2a^4 b^3\right)^7} = \left(2a^4 b^3\right)^{7/5} = 2^{7/5} a^{28/5} b^{21/5}$
   $= 2a^5 b^4 \cdot 2^{2/5} a^{3/5} b^{1/5}$
   $= 2a^5 b^4 \sqrt[5]{4a^3 b}$

4. $\sqrt{3m} \cdot \sqrt{13n} = \sqrt{3m \cdot 13n} = \sqrt{39mn}$

5. $\sqrt{32x^7 y^4} = \sqrt{16x^6 y^4 \cdot 2x}$
   $= \sqrt{16x^6 y^4} \cdot \sqrt{2x}$
   $= 4x^3 y^2 \sqrt{2x}$

6. $\dfrac{\sqrt{9a^3 b^{-3}}}{\sqrt{4ab}} = \sqrt{\dfrac{9a^3 b^{-3}}{4ab}}$
   $= \sqrt{\dfrac{9a^2}{4b^4}}$
   $= \dfrac{\sqrt{9a^2}}{\sqrt{4b^4}}$
   $= \dfrac{3a}{2b^2}$

7. $\sqrt{5x^3} + 2\sqrt{45x} = \sqrt{x^2 \cdot 5x} + 2\sqrt{9 \cdot 5x}$
   $= \sqrt{x^2} \cdot \sqrt{5x} + 2\sqrt{9} \cdot \sqrt{5x}$
   $= x\sqrt{5x} + 2 \cdot 3 \cdot \sqrt{5x}$
   $= x\sqrt{5x} + 6\sqrt{5x}$
   $= (x+6)\sqrt{5x}$

8. $\sqrt{9a^2 b} - \sqrt[4]{16a^4 b^2} = \sqrt{9a^2} \cdot \sqrt{b} - \sqrt[4]{16a^4} \cdot \sqrt[4]{b^2}$
   $= 3a\sqrt{b} - 2a \cdot b^{2/4}$
   $= 3a\sqrt{b} - 2a \cdot b^{1/2}$
   $= 3a\sqrt{b} - 2a\sqrt{b}$
   $= (3-2)a\sqrt{b}$
   $= a\sqrt{b}$

9. $\left(11 + 2\sqrt{x}\right)\left(3 - \sqrt{x}\right)$
   $= 11 \cdot 3 + 11\left(-\sqrt{x}\right) + 2\sqrt{x} \cdot 3 + 2\sqrt{x}\left(-\sqrt{x}\right)$
   $= 33 - 11\sqrt{x} + 6\sqrt{x} - 2\left(\sqrt{x}\right)^2$
   $= 33 + (-11+6)\sqrt{x} - 2x$
   $= 33 - 5\sqrt{x} - 2x$

10. $\dfrac{-2}{3\sqrt{72}} = \dfrac{-2}{3 \cdot 6\sqrt{2}} = \dfrac{-2}{18\sqrt{2}}$
    $= \dfrac{-1}{9\sqrt{2}} = \dfrac{-1}{9\sqrt{2}} \cdot \dfrac{\sqrt{2}}{\sqrt{2}}$
    $= \dfrac{-\sqrt{2}}{9 \cdot 2}$
    $= \dfrac{-\sqrt{2}}{18}$

11. $\dfrac{\sqrt{5}}{\sqrt{5}+2} = \dfrac{\sqrt{5}}{\sqrt{5}+2} \cdot \dfrac{\sqrt{5}-2}{\sqrt{5}-2}$
    $= \dfrac{\sqrt{5}\left(\sqrt{5}-2\right)}{\left(\sqrt{5}\right)^2 - 2^2}$
    $= \dfrac{5 - 2\sqrt{5}}{5 - 4}$
    $= 5 - 2\sqrt{5}$

*Chapter 7*: Radicals and Rational Exponents

12. $f(x) = \sqrt{-2x+3}$

    a. $f(1) = \sqrt{-2(1)+3} = \sqrt{-2+3} = \sqrt{1} = 1$

    b. $f(-3) = \sqrt{-2(-3)+3} = \sqrt{6+3} = \sqrt{9} = 3$

13. $g(x) = \sqrt{-3x+5}$

    $-3x + 5 \geq 0$
    $-3x \geq -5$
    $x \leq \dfrac{5}{3}$

    The domain of the function is $\left\{x \mid x \leq \dfrac{5}{3}\right\}$ or the interval $\left(-\infty, \dfrac{5}{3}\right]$.

14. $f(x) = \sqrt{x} - 3$

    a. $x \geq 0$

    The domain of the function is $\{x \mid x \geq 0\}$ or the interval $[0, \infty)$.

    b. 
    | $x$ | $f(x) = \sqrt{x} - 3$ | $(x,y)$ |
    |---|---|---|
    | 0 | $f(0) = \sqrt{0} - 3 = -3$ | $(0,-3)$ |
    | 1 | $f(1) = -2$ | $(1,-2)$ |
    | 4 | $f(4) = -1$ | $(4,-1)$ |
    | 9 | $f(9) = 0$ | $(9,0)$ |

    c. Based on the graph, the range is $[-3, \infty)$.

15. $\sqrt{x+3} = 4$

    $\left(\sqrt{x+3}\right)^2 = 4^2$
    $x + 3 = 16$
    $x = 13$

    Check: $\sqrt{13+3} = 4$ ?
    $\sqrt{16} = 4$ ?
    $4 = 4$ T
    The solution set is $\{13\}$.

16. $\sqrt{x+13} - 4 = x - 3$

    $\sqrt{x+13} = x + 1$
    $\left(\sqrt{x+13}\right)^2 = (x+1)^2$
    $x + 13 = x^2 + 2x + 1$
    $x^2 + x - 12 = 0$
    $(x+4)(x-3) = 0$
    $x + 4 = 0$ or $x - 3 = 0$
    $x = -4$ or $x = 3$

    Check: $\sqrt{-4+13} - 4 = -4 - 3$ ?
    $\sqrt{9} - 4 = -7$ ?
    $3 - 4 = -7$ ?
    $-1 = -7$ False

    $\sqrt{3+13} - 4 = 3 - 3$ ?
    $\sqrt{16} - 4 = 0$ ?
    $4 - 4 = 0$ ?
    $0 = 0$ T

    The first solution is extraneous, so the solution set is $\{3\}$.

17. $\sqrt{x-1} + \sqrt{x+2} = 3$

    $\sqrt{x-1} = 3 - \sqrt{x+2}$
    $\left(\sqrt{x-1}\right)^2 = \left(3 - \sqrt{x+2}\right)^2$
    $x - 1 = 3^2 - 2(3)\sqrt{x+2} + \left(\sqrt{x+2}\right)^2$
    $x - 1 = 9 - 6\sqrt{x+2} + x + 2$
    $-12 = -6\sqrt{x+2}$
    $2 = \sqrt{x+2}$
    $2^2 = \left(\sqrt{x+2}\right)^2$
    $4 = x + 2$
    $2 = x$

    Check: $\sqrt{2-1} + \sqrt{2+2} = 3$ ?
    $\sqrt{1} + \sqrt{4} = 3$ ?
    $1 + 2 = 3$ ?

$3 = 3$ T

The solution set is $\{2\}$.

18. $(13+2i)+(4-15i) = 13+2i+4-15i$
$= 13+4+2i-15i$
$= 17-13i$

19. $(4-7i)(2+3i) = 4 \cdot 2 + 4 \cdot 3i - 7i \cdot 2 - 7i \cdot 3i$
$= 8+12i-14i-21i^2$
$= 8-2i-21(-1)$
$= 8-2i+21$
$= 29-2i$

20. $\dfrac{7-i}{12+11i} = \dfrac{7-i}{12+11i} \cdot \dfrac{12-11i}{12-11i}$
$= \dfrac{(7-i)(12-11i)}{(12+11i)(12-11i)}$
$= \dfrac{84-77i-12i+11i^2}{144-121i^2}$
$= \dfrac{84-89i+11(-1)}{144-121(-1)}$
$= \dfrac{84-89i-11}{144+121}$
$= \dfrac{73-89i}{265}$
$= \dfrac{73}{265} - \dfrac{89}{265}i$

**Cumulative Review R-7**

1. $6-3^2 \div (9-3) = 6-3^2 \div 6$
$= 6-9 \div 6$
$= 6 - \dfrac{3}{2}$
$= \dfrac{9}{2}$

2. $(3x+2y)-(2x-5y+3)+9$
$= 3x+2y-2x+5y-3+9$
$= 3x-2x+2y+5y-3+9$
$= x+7y+6$

3. $(3x+5)-2 = 7x-13$
$3x+5-2 = 7x-13$
$3x+3 = 7x-13$
$3x+16 = 7x$
$16 = 4x$
$\dfrac{16}{4} = \dfrac{4x}{4}$
$4 = x$

The solution set is $\{4\}$.

4. $6x + \dfrac{1}{2}(4x-2) \le 3x+9$
$6x+2x-1 \le 3x+9$
$8x-1 \le 3x+9$
$5x-1 \le 9$
$5x \le 10$
$x \le 2$
Interval: $(-\infty, 2]$

5. $f(x) = 3x^2 - x + 5$

   a. $f(-2) = 3(-2)^2 - (-2) + 5$
   $= 3(4)+2+5 = 12+7 = 19$

   b. $f(3) = 3(3)^2 - (3) + 5$
   $= 3(9)-3+5 = 27+2 = 29$

6. The function $g$ tells us to divide $x^2 - 9$ by $x^2 - 2x - 8$. Since division by 0 is not defined, the denominator can never be 0.
$x^2 - 2x - 8 = 0$
$(x-4)(x+2) = 0$
$x-4 = 0$ or $x+2 = 0$
$x = 4$ or $x = -2$
Thus, the domain of the function is $\{x \mid x \ne 4, -2\}$. In interval notation we would write $(-\infty, -2) \cup (-2, 4) \cup (4, \infty)$.

Chapter 7: Radicals and Rational Exponents

7.  a. $n(50) = -50(50) + 6,000$
    $= -2,500 + 6,000$
    $= 3,500$
    If the price of the game were \$50, you would sell 3500 games per year.

    b. We want to solve $n(p) = 0$.
    $-50p + 6,000 = 0$
    $6,000 = 50p$
    $\dfrac{6,000}{50} = \dfrac{50p}{50}$
    $120 = p$
    When the price of the game reaches \$120, no games will be sold.

8.  First we use the two points to find the slope of the line.
    $m = \dfrac{-2-6}{3-(-1)} = \dfrac{-8}{4} = -2$
    Next, we use the slope and one of the points to determine the y-intercept.
    $y = mx + b$
    $6 = -2(-1) + b$
    $6 = 2 + b$
    $4 = b$
    The slope is $m = -2$ and the y-intercept is $b = 4$. Therefore, the equation of the line is $y = -2x + 4$.

9.  Start by solving the inequality for $y$.
    $6x + 3y > 24$
    $3y > -6x + 24$
    $\dfrac{3y}{3} > \dfrac{-6x+24}{3}$
    $y > -2x + 8$
    Since the inequality is strict, graph the line $y = -2x + 8$ with a dashed line. We will use the point $(0,0)$ as our test point.
    $6(0) + 3(0) > 24$ ?
    $0 > 24$ false
    Since we obtained a contradiction, we shade the region that does not contain the point $(0,0)$.

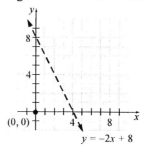

10. $4x - y = 17$
    $5x + 6y = 14$
    Solve the first equation for $y$.
    $4x - y = 17$
    $-y = -4x + 17$
    $y = 4x - 17$
    Substitute this result for $y$ in the second equation and solve for $x$.
    $5x + 6(4x - 17) = 14$
    $5x + 24x - 102 = 14$
    $29x - 102 = 14$
    $29x = 116$
    $x = 4$
    Use this result to solve for $y$.
    $y = 4x - 17$
    $y = 4(4) - 17$
    $= 16 - 17$
    $= -1$
    The ordered pair solution is $(4, -1)$.

    Check: $4(4) - (-1) = 17$ ?
    $16 + 1 = 17$ ?
    $17 = 17$ T

    $5(4) + 6(-1) = 14$ ?
    $20 - 6 = 14$ ?
    $14 = 14$ T

11. Let $x$ = pounds of dried fruit and $y$ = pounds of nuts. Since the total pounds will be 10, we have the equation
    $x + y = 10$
    We want the total revenue to be the same. Thus, our second equation will be
    $3.45x + 2.10y = 2.64(10)$
    $3.45x + 2.10y = 26.40$

ISM – 436

Putting the two equations together gives us the following system:
$$x + y = 10$$
$$3.45x + 2.10y = 26.40$$
We can solve this system by using substitution. Start by solving the first equation for $y$.
$$x + y = 10$$
$$y = -x + 10$$
Substitute this result for $y$ in the second equation and solve for $x$.
$$3.45x + 2.10(-x + 10) = 26.40$$
$$3.45x - 2.10x + 21.00 = 26.40$$
$$1.35x + 21.00 = 26.40$$
$$1.35x = 5.40$$
$$x = 4$$
Use this result to solve for $y$.
$$y = -x + 10$$
$$= -4 + 10$$
$$= 6$$
The trail mix should contain 4 pounds of dried fruit and 6 pounds of nuts.

12. $\begin{vmatrix} 5 & -2 & 3 \\ 3 & -3 & 4 \\ -2 & 4 & 1 \end{vmatrix} = 5\begin{vmatrix} -3 & 4 \\ 4 & 1 \end{vmatrix} - (-2)\begin{vmatrix} 3 & 4 \\ -2 & 1 \end{vmatrix} + 3\begin{vmatrix} 3 & -3 \\ -2 & 4 \end{vmatrix}$
$$= 5(-3 \cdot 1 - 4 \cdot 4) + 2(3 \cdot 1 - (-2)4)$$
$$+ 3(3 \cdot 4 - (-2)(-3))$$
$$= 5(-3 - 16) + 2(3 + 8) + 3(12 - 6)$$
$$= 5(-19) + 2(11) + 3(6)$$
$$= -95 + 22 + 18$$
$$= -55$$

13. $(8x^3 - 4x^2 + 5x + 3) + (2x^2 - 8x + 7)$
$$= 8x^3 - 4x^2 + 5x + 3 + 2x^2 - 8x + 7$$
$$= 8x^3 - 4x^2 + 2x^2 + 5x - 8x + 3 + 7$$
$$= 8x^3 - 2x^2 - 3x + 10$$

14. $(2x - 1)(4x^2 + 2x - 9)$
$$= 8x^3 + 4x^2 - 18x - 4x^2 - 2x + 9$$
$$= 8x^3 + 4x^2 - 4x^2 - 18x - 2x + 9$$
$$= 8x^3 - 20x + 9$$

15. 
$$\begin{array}{r} 3x^2 + 2x - 6 \\ 2x^2 + 3x - 5 \overline{\smash{\big)} 6x^4 + 13x^3 - 21x^2 - 28x + 37} \\ \underline{-(6x^4 + 9x^3 - 15x^2)} \\ 4x^3 - 6x^2 - 28x \\ \underline{-(4x^3 + 6x^2 - 10x)} \\ -12x^2 - 18x + 37 \\ \underline{-(-12x^2 - 18x + 30)} \\ 7 \end{array}$$

$$\frac{6x^4 + 13x^3 - 21x^2 - 28x + 37}{2x^2 + 3x - 5}$$
$$= 3x^2 + 2x - 6 + \frac{7}{2x^2 + 3x - 5}$$

16. $8x^2 - 44x - 84$
Begin by factoring out the greatest common factor.
$$4(2x^2 - 11x - 21)$$
For the reduced polynomial, we can use the AC method.
$$AC = 2(-21) = -42$$
We need two factors of $-42$ whose sum is $-11$. Since the product is negative, the factors will have opposite signs. Since the sum is also negative, the factor with the larger absolute value will be negative.

| factor 1 | factor 2 | sum |     |
|----------|----------|-----|-----|
| 2        | -21      | -19 | too small |
| 6        | -7       | -1  | too large |
| 3        | -14      | -11 | okay |

$$8x^2 - 44x - 84 = 4(2x^2 - 11x - 21)$$
$$= 4(2x^2 + 3x - 14x - 21)$$
$$= 4(x(2x + 3) + (-7)(2x + 3))$$
$$= 4(2x + 3)(x - 7)$$

## Chapter 7: Radicals and Rational Exponents

17. $\dfrac{2x}{x-3} - \dfrac{x+1}{x+2}$

    To subtract, we need a common denominator. Since both denominators are prime, the LCD will be the product of the denominators.
    $LCD = (x-3)(x+2)$

    Next we write equivalent fractions using the LCD.
    $$\dfrac{2x(x+2)}{(x-3)(x+2)} - \dfrac{(x+1)(x-3)}{(x-3)(x+2)}$$

    Now we can combine the numerators and simplify.
    $$\dfrac{2x(x+2) - (x+1)(x-3)}{(x-3)(x+2)}$$
    $$= \dfrac{2x^2 + 4x - (x^2 - 2x - 3)}{(x-3)(x+2)}$$
    $$= \dfrac{2x^2 + 4x - x^2 + 2x + 3}{(x-3)(x+2)}$$
    $$= \dfrac{x^2 + 6x + 3}{(x-3)(x+2)}$$

18. $\dfrac{9}{k-2} = \dfrac{6}{k} + 3$

    $LCD = k(k-2)$

    The solution set cannot contain the restricted values 0 or 2 since these values will make one of the terms have a 0 in its denominator. Multiply each term by the LCD.
    $$k(k-2)\left(\dfrac{9}{k-2}\right) = k(k-2)\dfrac{6}{k} + k(k-2)(3)$$
    $$k\cancel{(k-2)}\left(\dfrac{9}{\cancel{k-2}}\right) = \cancel{k}(k-2)\dfrac{6}{\cancel{k}} + k(k-2)(3)$$
    $$9k = 6(k-2) + 3k(k-2)$$
    $$9k = 6k - 12 + 3k^2 - 6k$$
    $$9k = 3k^2 - 12$$
    $$0 = 3k^2 - 9k - 12$$
    $$0 = k^2 - 3k - 4$$
    $$0 = (k-4)(k+1)$$
    $k - 4 = 0$ or $k + 1 = 0$
    $k = 4$ or $k = -1$

    Neither of these are restricted values, so the solution set is $\{-1, 4\}$.

19. $\dfrac{x+8}{x-4} \leq 3$

    $\dfrac{x+8}{x-4} - 3 \leq 0$

    $\dfrac{x+8}{x-4} - \dfrac{3(x-4)}{x-4} \leq 0$

    $\dfrac{x+8 - 3x + 12}{x-4} \leq 0$

    $\dfrac{-2x + 20}{x-4} \leq 0$

    $\dfrac{-2(x-10)}{x-4} \leq 0$

    $\dfrac{x-10}{x-4} \geq 0$

    The rational expression will equal 0 when $x = 10$. It is undefined when $x = 4$. Determine where the numerator and denominator are positive and negative, and where the quotient is positive and negative.

    | Interval | $(-\infty, 4)$ | 4 | $(4, 10)$ | 10 | $(10, \infty)$ |
    |---|---|---|---|---|---|
    | $x - 10$ | $----$ | $-$ | $----$ | 0 | $++++$ |
    | $x - 4$ | $----$ | 0 | $++++$ | $+$ | $++++$ |
    | $\dfrac{x-10}{x-4}$ | $++++$ | $\emptyset$ | $----$ | 0 | $++++$ |

    The rational function is undefined at $x = 4$, so 4 is not part of the solution. The inequality is non-strict, so $x = 10$ is part of the solution. Now, $\dfrac{x-10}{x-4}$ is greater than zero where the quotient is positive. The solution is $\{x \mid x < 4 \text{ or } x \geq 10\}$.
    Interval: $(-\infty, 4) \cup [10, \infty)$
    Graph:

20. Let $x$ = hours needed to paint the room together.
    $$\dfrac{1 \text{ room}}{4 \text{ hours}} + \dfrac{1 \text{ room}}{6 \text{ hours}} = \dfrac{1 \text{ room}}{x \text{ hours}}$$
    $$12x\left(\dfrac{1}{4}\right) + 12x\left(\dfrac{1}{6}\right) = 12x\left(\dfrac{1}{x}\right)$$
    $$3x + 2x = 12$$
    $$5x = 12$$
    $$x = \dfrac{12}{5} = 2.4$$

    It will take Shawn and Payton 2.4 hours to paint the room together.

21. $\dfrac{\sqrt{50a^3b}}{\sqrt{2a^{-1}b^3}} = \sqrt{\dfrac{50a^3b}{2a^{-1}b^3}} = \sqrt{\dfrac{25a^4}{b^2}} = \dfrac{\sqrt{25a^4}}{\sqrt{b^2}} = \dfrac{5a^2}{b}$

22. $f(x) = \sqrt[4]{8-3x}$

Since the index is even, we need the radicand to be greater than or equal to zero.

$8 - 3x \geq 0$

$-3x \geq -8$

$\dfrac{-3x}{-3} \leq \dfrac{-8}{-3}$

$x \leq \dfrac{8}{3}$

The domain of the function is $\left\{x \mid x \leq \dfrac{8}{3}\right\}$ or the interval $\left(-\infty, \dfrac{8}{3}\right]$.

23. $\sqrt{x+7} - 8 = x - 7$

$\sqrt{x+7} = x + 1$

$\left(\sqrt{x+7}\right)^2 = (x+1)^2$

$x + 7 = x^2 + 2x + 1$

$0 = x^2 + x - 6$

$0 = (x+3)(x-2)$

$x + 3 = 0$ or $x - 2 = 0$

$x = -3$ or $x = 2$

Check: $\sqrt{-3+7} - 8 = -3 - 7$ ?

$\sqrt{4} - 8 = -10$ ?

$2 - 8 = -10$ ?

$-6 = -10$ False

$\sqrt{2+7} - 8 = 2 - 7$ ?

$\sqrt{9} - 8 = -5$ ?

$3 - 8 = -5$ ?

$-5 = -5$ T

The first solution is extraneous, so the solution set is $\{2\}$.

24. $\dfrac{3i}{1-7i} = \dfrac{3i}{1-7i} \cdot \dfrac{1+7i}{1+7i}$

$= \dfrac{3i + 21i^2}{1 - 49i^2}$

$= \dfrac{3i + 21(-1)}{1 - 49(-1)}$

$= \dfrac{-21 + 3i}{1 + 49}$

$= \dfrac{-21 + 3i}{50}$

$= -\dfrac{21}{50} + \dfrac{3}{50}i$

25. a. $\dfrac{2}{4-\sqrt{11}} = \dfrac{2}{4-\sqrt{11}} \cdot \dfrac{4+\sqrt{11}}{4+\sqrt{11}}$

$= \dfrac{2(4+\sqrt{11})}{4^2 - (\sqrt{11})^2}$

$= \dfrac{8 + 2\sqrt{11}}{16 - 11}$

$= \dfrac{8 + 2\sqrt{11}}{5}$

# Chapter 8

**8.1 Preparing for Solving Quadratic Equations by Completing the Square**

1. $(2p+3)^2 = (2p)^2 + 2(2p)(3) + 3^2$
   $= 4p^2 + 12p + 9$

2. $y^2 - 8y + 16 = y^2 - 2 \cdot y \cdot 4 + 4^2 = (y-4)^2$

3. $x^2 + 5x - 14 = 0$
   $(x+7)(x-2) = 0$
   $x + 7 = 0$ or $x - 2 = 0$
   $x = -7$ or $x = 2$
   The solution set is $\{-7, 2\}$.

4. $x^2 - 16 = 0$
   $(x+4)(x-4) = 0$
   $x + 4 = 0$ or $x - 4 = 0$
   $x = -4$ or $x = 4$
   The solution set is $\{-4, 4\}$.

5. a. $\sqrt{36} = 6$ because $6^2 = 36$.
   b. $\sqrt{45} = \sqrt{9 \cdot 5} = 3\sqrt{5}$
   c. $\sqrt{-12} = \sqrt{-1 \cdot 4 \cdot 3} = 2\sqrt{3}\,i$

6. The complex conjugate is $-3 - 2i$.

**8.1 Exercises**

2. perfect square trinomial

4. TRUE.

6. FALSE. The Pythagorean Theorem states that, in a *right* triangle, the *square* of the length of the hypotenuse is equal to the sum of the squares of the length of the legs.

8. Answers may vary. One possibility follows: The first step in solving the quadratic equation $3x^2 - 6x + 12 = 0$ should be to divide both sides of the equation by 3. This action will make the coefficient of the square term equal to 1, which is necessary to complete the square.

10. $x^2 = 81$
    $x = \pm\sqrt{81}$
    $x = \pm 9$
    The solution set is $\{-9, 9\}$.

12. $z^2 = 48$
    $z = \pm\sqrt{48}$
    $z = \pm 4\sqrt{3}$
    The solution set is $\{-4\sqrt{3}, 4\sqrt{3}\}$.

14. $n^2 = -49$
    $n = \pm\sqrt{-49}$
    $n = \pm 7i$
    The solution set is $\{-7i, 7i\}$.

16. $z^2 = \dfrac{8}{9}$
    $z = \pm\sqrt{\dfrac{8}{9}}$
    $z = \pm\dfrac{\sqrt{8}}{3}$
    $z = \pm\dfrac{2\sqrt{2}}{3}$
    The solution set is $\left\{-\dfrac{2\sqrt{2}}{3}, \dfrac{2\sqrt{2}}{3}\right\}$.

18. $w^2 - 6 = 14$
    $w^2 = 20$
    $w = \pm\sqrt{20}$
    $w = \pm 2\sqrt{5}$
    The solution set is $\{-2\sqrt{5}, 2\sqrt{5}\}$.

20. $4y^2 = 100$
    $y^2 = 25$
    $y = \pm\sqrt{25}$
    $y = \pm 5$
    The solution set is $\{-5, 5\}$.

**Section 8.1** *Solving Quadratic Equations by Completing the Square*

22. $5x^2 = 32$

$x^2 = \dfrac{32}{5}$

$x = \pm\sqrt{\dfrac{32}{5}}$

$x = \pm\dfrac{\sqrt{32}}{\sqrt{5}}$

$x = \pm\dfrac{4\sqrt{2}}{\sqrt{5}} \cdot \dfrac{\sqrt{5}}{\sqrt{5}}$

$x = \pm\dfrac{4\sqrt{10}}{5}$

The solution set is $\left\{-\dfrac{4\sqrt{10}}{5}, \dfrac{4\sqrt{10}}{5}\right\}$.

24. $-3x^2 - 5 = 22$

$-3x^2 = 27$

$x^2 = -9$

$x = \pm\sqrt{-9}$

$x = \pm 3i$

The solution set is $\{-3i, 3i\}$.

26. $(y-2)^2 = 9$

$y - 2 = \pm\sqrt{9}$

$y - 2 = \pm 3$

$y = 2 \pm 3$

$y = 2 - 3$ or $y = 2 + 3$

$z = -1$ or $y = 5$

The solution set is $\{-1, 5\}$.

28. $(z+4)^2 = -24$

$z + 4 = \pm\sqrt{-24}$

$z + 4 = \pm 2\sqrt{6}\,i$

$z = -4 \pm 2\sqrt{6}\,i$

The solution set is $\{-4 - 2\sqrt{6}\,i,\ -4 + 2\sqrt{6}\,i\}$.

30. $5(x-3)^2 + 2 = 27$

$5(x-3)^2 = 25$

$(x-3)^2 = 5$

$x - 3 = \pm\sqrt{5}$

$x = 3 \pm \sqrt{5}$

The solution set is $\{3 - \sqrt{5},\ 3 + \sqrt{5}\}$.

32. $(2p+3)^2 = 16$

$2p + 3 = \pm\sqrt{16}$

$2p + 3 = \pm 4$

$2p = -3 \pm 4$

$2p = -3 - 4$ or $2p = -3 + 4$

$2p = -7$ or $2p = 1$

$p = -\dfrac{7}{2}$ or $p = \dfrac{1}{2}$

The solution set is $\left\{-\dfrac{7}{2}, \dfrac{1}{2}\right\}$.

34. $\left(y + \dfrac{3}{2}\right)^2 = \dfrac{3}{4}$

$y + \dfrac{3}{2} = \pm\sqrt{\dfrac{3}{4}}$

$y + \dfrac{3}{2} = \pm\dfrac{\sqrt{3}}{2}$

$y = -\dfrac{3}{2} \pm \dfrac{\sqrt{3}}{2}$

The solution set is $\left\{-\dfrac{3}{2} - \dfrac{\sqrt{3}}{2},\ -\dfrac{3}{2} + \dfrac{\sqrt{3}}{2}\right\}$.

36. $q^2 - 6q + 9 = 16$

$(q-3)^2 = 16$

$q - 3 = \pm\sqrt{16}$

$q - 3 = \pm 4$

$q = 3 \pm 4$

$q = 3 - 4$ or $q = 3 + 4$

$q = -1$ or $q = 7$

The solution set is $\{-1, 7\}$.

38. Start: $y^2 + 16y$

Add: $\left[\dfrac{1}{2} \cdot 16\right]^2 = 64$

Result: $y^2 + 16y + 64$

Factored Form: $(y+8)^2$

40. Start: $p^2 - 4p$

Add: $\left[\dfrac{1}{2} \cdot (-4)\right]^2 = 4$

Result: $p^2 - 4p + 4$

Factored Form: $(p-2)^2$

## Chapter 8: Quadratic Equations and Functions

42. Start: $x^2 + x$

    Add: $\left(\dfrac{1}{2} \cdot 1\right)^2 = \dfrac{1}{4}$

    Result: $x^2 + x + \dfrac{1}{4}$

    Factored Form: $\left(x + \dfrac{1}{2}\right)^2$

44. Start: $z^2 - \dfrac{1}{3}z$

    Add: $\left[\dfrac{1}{2}\left(-\dfrac{1}{3}\right)\right]^2 = \dfrac{1}{36}$

    Result: $z^2 - \dfrac{1}{3}z + \dfrac{1}{36}$

    Factored Form: $\left(z - \dfrac{1}{6}\right)^2$

46. Start: $m^2 + \dfrac{5}{2}m$

    Add: $\left(\dfrac{1}{2} \cdot \dfrac{5}{2}\right)^2 = \dfrac{25}{16}$

    Result: $m^2 + \dfrac{5}{2}m + \dfrac{25}{16}$

    Factored Form: $\left(m + \dfrac{5}{4}\right)^2$

48. $$z^2 - 6z = 7$$
    $$z^2 - 6z + \left(\dfrac{1}{2} \cdot (-6)\right)^2 = 7 + \left(\dfrac{1}{2} \cdot (-6)\right)^2$$
    $$z^2 - 6z + 9 = 7 + 9$$
    $$(z-3)^2 = 16$$
    $$z - 3 = \pm\sqrt{16}$$
    $$z - 3 = \pm 4$$
    $$z = 3 \pm 4$$
    $$z = -1 \text{ or } z = 7$$
    The solution set is $\{-1,\ 7\}$.

50. $$y^2 + 3y - 18 = 0$$
    $$y^2 + 3y = 18$$
    $$y^2 + 3y + \left(\dfrac{1}{2} \cdot 3\right)^2 = 18 + \left(\dfrac{1}{2} \cdot 3\right)^2$$
    $$y^2 + 3y + \dfrac{9}{4} = 18 + \dfrac{9}{4}$$
    $$\left(y + \dfrac{3}{2}\right)^2 = \dfrac{81}{4}$$
    $$y + \dfrac{3}{2} = \pm\sqrt{\dfrac{81}{4}}$$
    $$y + \dfrac{3}{2} = \pm\dfrac{9}{2}$$
    $$y = -\dfrac{3}{2} \pm \dfrac{9}{2}$$
    $$y = -6 \text{ or } y = 3$$
    The solution set is $\{-6,\ 3\}$.

52. $$p^2 - 6p + 4 = 0$$
    $$p^2 - 6p = -4$$
    $$p^2 - 6p + \left(\dfrac{1}{2} \cdot (-6)\right)^2 = -4 + \left(\dfrac{1}{2} \cdot (-6)\right)^2$$
    $$p^2 - 6p + 9 = -4 + 9$$
    $$(p-3)^2 = 5$$
    $$p - 3 = \pm\sqrt{5}$$
    $$p = 3 \pm \sqrt{5}$$
    The solution set is $\{3 - \sqrt{5},\ 3 + \sqrt{5}\}$.

54. $$b^2 + 10b + 19 = 0$$
    $$b^2 + 10b = -19$$
    $$b^2 + 10b + \left(\dfrac{1}{2} \cdot 10\right)^2 = -19 + \left(\dfrac{1}{2} \cdot 10\right)^2$$
    $$b^2 + 10b + 25 = -19 + 25$$
    $$(b+5)^2 = 6$$
    $$b + 5 = \pm\sqrt{6}$$
    $$b = -5 \pm \sqrt{6}$$
    The solution set is $\{-5 - \sqrt{6},\ -5 + \sqrt{6}\}$.

## Section 8.1 Solving Quadratic Equations by Completing the Square

**56.**
$$m^2 - 2m + 5 = 0$$
$$m^2 - 2m = -5$$
$$m^2 - 2m + \left(\frac{1}{2} \cdot (-2)\right)^2 = -5 + \left(\frac{1}{2} \cdot (-2)\right)^2$$
$$m^2 - 2m + 1 = -5 + 1$$
$$(m-1)^2 = -4$$
$$m - 1 = \pm\sqrt{-4}$$
$$m - 1 = \pm 2i$$
$$m = 1 \pm 2i$$
The solution set is $\{1 - 2i,\ 1 + 2i\}$.

**58.**
$$q^2 + 7q + 7 = 0$$
$$q^2 + 7q = -7$$
$$q^2 + 7q + \left(\frac{1}{2} \cdot 7\right)^2 = -7 + \left(\frac{1}{2} \cdot 7\right)^2$$
$$q^2 + 7q + \frac{49}{4} = -7 + \frac{49}{4}$$
$$\left(q + \frac{7}{2}\right)^2 = \frac{21}{4}$$
$$q + \frac{7}{2} = \pm\sqrt{\frac{21}{4}}$$
$$q + \frac{7}{2} = \pm\frac{\sqrt{21}}{2}$$
$$q = -\frac{7}{2} \pm \frac{\sqrt{21}}{2}$$
The solution set is $\left\{-\frac{7}{2} - \frac{\sqrt{21}}{2},\ -\frac{7}{2} + \frac{\sqrt{21}}{2}\right\}$.

**60.**
$$x^2 - 5x - 3 = 0$$
$$x^2 - 5x = 3$$
$$x^2 - 5x + \left(\frac{1}{2} \cdot (-5)\right)^2 = 3 + \left(\frac{1}{2} \cdot (-5)\right)^2$$
$$x^2 - 5x + \frac{25}{4} = 3 + \frac{25}{4}$$
$$\left(x - \frac{5}{2}\right)^2 = \frac{37}{4}$$
$$x - \frac{5}{2} = \pm\sqrt{\frac{37}{4}}$$
$$x - \frac{5}{2} = \pm\frac{\sqrt{37}}{2}$$
$$x = \frac{5}{2} \pm \frac{\sqrt{37}}{2}$$
The solution set is $\left\{\frac{5}{2} - \frac{\sqrt{37}}{2},\ \frac{5}{2} + \frac{\sqrt{37}}{2}\right\}$.

**62.**
$$n^2 = 10n + 5$$
$$n^2 - 10n = 5$$
$$n^2 - 10n + \left(\frac{1}{2} \cdot (-10)\right)^2 = 5 + \left(\frac{1}{2} \cdot (-10)\right)^2$$
$$n^2 - 10n + 25 = 5 + 25$$
$$(n - 5)^2 = 30$$
$$n - 5 = \pm\sqrt{30}$$
$$n = 5 \pm \sqrt{30}$$
The solution set is $\{5 - \sqrt{30},\ 5 + \sqrt{30}\}$.

**64.**
$$z^2 - 3z + 5 = 0$$
$$z^2 - 3z = -5$$
$$z^2 - 3z + \left(\frac{1}{2} \cdot (-3)\right)^2 = -5 + \left(\frac{1}{2} \cdot (-3)\right)^2$$
$$z^2 - 3z + \frac{9}{4} = -5 + \frac{9}{4}$$
$$\left(z - \frac{3}{2}\right)^2 = -\frac{11}{4}$$
$$z - \frac{3}{2} = \pm\sqrt{-\frac{11}{4}}$$
$$z - \frac{3}{2} = \pm\frac{\sqrt{11}}{2}i$$
$$z = \frac{3}{2} \pm \frac{\sqrt{11}}{2}i$$
The solution set is $\left\{\frac{3}{2} - \frac{\sqrt{11}}{2}i,\ \frac{3}{2} + \frac{\sqrt{11}}{2}i\right\}$.

The solution set is $\left\{-\frac{3}{2},\ 4\right\}$.

## Chapter 8: Quadratic Equations and Functions

**66.**
$$3a^2 - 4a - 4 = 0$$
$$\frac{3a^2 - 4a - 4}{3} = \frac{0}{3}$$
$$a^2 - \frac{4}{3}a - \frac{4}{3} = 0$$
$$a^2 - \frac{4}{3}a = \frac{4}{3}$$
$$a^2 - \frac{4}{3}a + \left[\frac{1}{2} \cdot \left(-\frac{4}{3}\right)\right]^2 = \frac{4}{3} + \left[\frac{1}{2} \cdot \left(-\frac{4}{3}\right)\right]^2$$
$$a^2 - \frac{4}{3}a + \frac{4}{9} = \frac{4}{3} + \frac{4}{9}$$
$$\left(a - \frac{2}{3}\right)^2 = \frac{16}{9}$$
$$a - \frac{2}{3} = \pm\sqrt{\frac{16}{9}}$$
$$a - \frac{2}{3} = \pm\frac{4}{3}$$
$$a = \frac{2}{3} \pm \frac{4}{3}$$
$$y = -\frac{2}{3} \text{ or } y = 2$$

The solution set is $\left\{-\frac{2}{3}, 2\right\}$.

**68.**
$$2y^2 - 2y - 1 = 0$$
$$\frac{2y^2 - 2y - 1}{2} = \frac{0}{2}$$
$$y^2 - y - \frac{1}{2} = 0$$
$$y^2 - y = \frac{1}{2}$$
$$y^2 - y + \left(\frac{1}{2} \cdot (-1)\right)^2 = \frac{1}{2} + \left(\frac{1}{2} \cdot (-1)\right)^2$$
$$y^2 - y + \frac{1}{4} = \frac{1}{2} + \frac{1}{4}$$
$$\left(y - \frac{1}{2}\right)^2 = \frac{3}{4}$$
$$y - \frac{1}{2} = \pm\sqrt{\frac{3}{4}}$$
$$y - \frac{1}{2} = \pm\frac{\sqrt{3}}{2}$$
$$y = \frac{1}{2} \pm \frac{\sqrt{3}}{2}$$

The solution set is $\left\{\frac{1}{2} - \frac{\sqrt{3}}{2}, \frac{1}{2} + \frac{\sqrt{3}}{2}\right\}$.

**70.**
$$2x^2 - 7x + 2 = 0$$
$$\frac{2x^2 - 7x + 2}{2} = \frac{0}{2}$$
$$x^2 - \frac{7}{2}x + 1 = 0$$
$$x^2 - \frac{7}{2}x = -1$$
$$x^2 - \frac{7}{2}x + \left[\frac{1}{2} \cdot \left(-\frac{7}{2}\right)\right]^2 = -1 + \left[\frac{1}{2} \cdot \left(-\frac{7}{2}\right)\right]^2$$
$$x^2 - \frac{7}{2}x + \frac{49}{16} = -1 + \frac{49}{16}$$
$$\left(x - \frac{7}{4}\right)^2 = \frac{33}{16}$$
$$x - \frac{7}{4} = \pm\sqrt{\frac{33}{16}}$$
$$x - \frac{7}{4} = \pm\frac{\sqrt{33}}{4}$$
$$x = \frac{7}{4} \pm \frac{\sqrt{33}}{4}$$

The solution set is $\left\{\frac{7}{4} - \frac{\sqrt{33}}{4}, \frac{7}{4} + \frac{\sqrt{33}}{4}\right\}$.

**72.**
$$2z^2 + 6z + 5 = 0$$
$$\frac{2z^2 + 6z + 5}{2} = \frac{0}{2}$$
$$z^2 + 3z + \frac{5}{2} = 0$$
$$z^2 + 3z = -\frac{5}{2}$$
$$z^2 + 3z + \left(\frac{1}{2} \cdot 3\right)^2 = -\frac{5}{2} + \left(\frac{1}{2} \cdot 3\right)^2$$
$$z^2 + 3z + \frac{9}{4} = -\frac{5}{2} + \frac{9}{4}$$
$$\left(z + \frac{3}{2}\right)^2 = -\frac{1}{4}$$
$$z + \frac{3}{2} = \pm\sqrt{-\frac{1}{4}}$$
$$z + \frac{3}{2} = \pm\frac{1}{2}i$$
$$z = -\frac{3}{2} \pm \frac{1}{2}i$$

The solution set is $\left\{-\frac{3}{2} - \frac{1}{2}i, -\frac{3}{2} + \frac{1}{2}i\right\}$.

**Section 8.1** Solving Quadratic Equations by Completing the Square

74. $3m^2 + 2m - 7 = 0$

$$\frac{3m^2 + 2m - 7}{3} = \frac{0}{3}$$

$$m^2 + \frac{2}{3}m - \frac{7}{3} = 0$$

$$m^2 + \frac{2}{3}m = \frac{7}{3}$$

$$m^2 + \frac{2}{3}m + \left[\frac{1}{2}\cdot\left(\frac{2}{3}\right)\right]^2 = \frac{7}{3} + \left[\frac{1}{2}\cdot\left(\frac{2}{3}\right)\right]^2$$

$$m^2 + \frac{2}{3}m + \frac{1}{9} = \frac{7}{3} + \frac{1}{9}$$

$$\left(m + \frac{1}{3}\right)^2 = \frac{22}{9}$$

$$m + \frac{1}{3} = \pm\sqrt{\frac{22}{9}}$$

$$m + \frac{1}{3} = \pm\frac{\sqrt{22}}{3}$$

$$m = -\frac{1}{3} \pm \frac{\sqrt{22}}{3}$$

The solution set is $\left\{-\frac{1}{3} - \frac{\sqrt{22}}{3}, -\frac{1}{3} + \frac{\sqrt{22}}{3}\right\}$.

76. $c^2 = 7^2 + 24^2$
$= 49 + 576$
$= 625$
$c = \sqrt{625} = 25$

78. $c^2 = 15^2 + 8^2$
$= 225 + 64$
$= 289$
$c = \sqrt{289} = 17$

80. $c^2 = 3^2 + 3^2$
$= 9 + 9$
$= 18$
$c = \sqrt{18} = 3\sqrt{2} \approx 4.24$

82. $c^2 = 2^2 + \left(\sqrt{5}\right)^2$
$= 4 + 5$
$= 9$
$c = \sqrt{9} = 3$

84. $c^2 = 8^2 + 10^2$
$= 64 + 100$
$= 164$
$c = \sqrt{164} = 2\sqrt{41} \approx 12.81$

86. $c^2 = a^2 + b^2$
$10^2 = 4^2 + b^2$
$100 = 16 + b^2$
$84 = b^2$
$b = \sqrt{84}$
$b = 2\sqrt{21} \approx 9.17$

88. $c^2 = a^2 + b^2$
$10^2 = a^2 + 2^2$
$100 = a^2 + 4$
$96 = a^2$
$a = \sqrt{96}$
$a = 4\sqrt{6} \approx 9.80$

90. $f(x) = 49$
$(x-5)^2 = 49$
$x - 5 = \pm\sqrt{49}$
$x - 5 = \pm 7$
$x = 5 \pm 7$
$x = 5 - 7$ or $x = 5 + 7$
$x = -2$ or $x = 12$
The solution set is $\{-2, 12\}$.

92. $h(x) = 32$
$(x+1)^2 = 32$
$x + 1 = \pm\sqrt{32}$
$x + 1 = \pm 4\sqrt{2}$
$x = -1 \pm 4\sqrt{2}$
The solution set is $\left\{-1 - 4\sqrt{2}, -1 + 4\sqrt{2}\right\}$.

94. Let $x$ represent the diagonal.
$x^2 = 6^2 + 9^2$
$x^2 = 36 + 81$
$x^2 = 117$
$x = \sqrt{117} = 3\sqrt{13} \approx 10.817$

96. Let $x$ represent the distance from home plate.
$x^2 = 310^2 + 40^2$
$x^2 = 96,100 + 1600$
$x^2 = 97,700$
$x = \sqrt{97,700} \approx 312.570$

Jermaine Dye approximately 312.570 feet from home plate.

**98.** Let $x$ represent the length of the guy wire.
$$x^2 = 8^2 + 40^2$$
$$x^2 = 64 + 1600$$
$$x^2 = 1664$$
$$x = \sqrt{1664} \approx 40.792$$
The guy wire is approximately 40.792 feet long.

**100.** Let $x$ represent the height the ladder can reach above the truck (see figure).

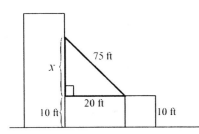

$$x^2 + 20^2 = 75^2$$
$$x^2 + 400 = 5625$$
$$x^2 = 5225$$
$$x = \sqrt{5225} \approx 72.284$$
So, the ladder can reach approximately 72.284 feet above the truck. This means that the ladder can reach a total of 72.284 + 10 = 82.284 feet up the building.

**102. a.**
$$A = \frac{\sqrt{3}}{4}x^2$$
$$\frac{8\sqrt{3}}{9} = \frac{\sqrt{3}}{4}x^2$$
$$\frac{4}{\sqrt{3}}\left(\frac{8\sqrt{3}}{9}\right) = \frac{4}{\sqrt{3}}\left(\frac{\sqrt{3}}{4}x^2\right)$$
$$\frac{32}{9} = x^2$$
$$x = \sqrt{\frac{32}{9}} = \frac{\sqrt{32}}{\sqrt{9}} = \frac{4\sqrt{2}}{3} \approx 1.886$$
The length of the side is approximately 1.886 feet.

**b.**
$$A = \frac{\sqrt{3}}{4}x^2$$
$$\frac{25\sqrt{3}}{4} = \frac{\sqrt{3}}{4}x^2$$
$$\frac{4}{\sqrt{3}}\left(\frac{25\sqrt{3}}{4}\right) = \frac{4}{\sqrt{3}}\left(\frac{\sqrt{3}}{4}x^2\right)$$
$$25 = x^2$$
$$x = \sqrt{25} = 5$$
The length of the side is 5 feet.

**104.**
$$6200 = 5000(1+r)^2$$
$$1.24 = (1+r)^2$$
$$1+r = \sqrt{1.24}$$
$$r = -1 + \sqrt{1.24}$$
$$r \approx .1136$$
The required rate of interest is approximately 11.36%.

**106.** $8^2 \stackrel{?}{=} 4^2 + 6^2$
$64 \stackrel{?}{=} 16 + 36$
$64 \stackrel{?}{=} 52$ ← False
Because $c^2 \neq a^2 + b^2$, the triangle is not a right triangle.

**108.** $52^2 \stackrel{?}{=} 20^2 + 48^2$
$2704 \stackrel{?}{=} 400 + 2304$
$2704 \stackrel{?}{=} 2704$ ← True
Because $c^2 = a^2 + b^2$, the triangle is a right triangle. The hypotenuse is 52.

**110.** Assuming $a \neq 0$,
$$ax^2 + bx + c = 0$$
$$\frac{ax^2 + bx + c}{a} = \frac{0}{a}$$
$$x^2 + \frac{b}{a}x + \frac{c}{a} = 0$$
$$x^2 + \frac{b}{a}x = -\frac{c}{a}$$
$$x^2 + \frac{b}{a}x + \left(\frac{1}{2} \cdot \frac{b}{a}\right)^2 = -\frac{c}{a} + \left(\frac{1}{2} \cdot \frac{b}{a}\right)^2$$
$$x^2 + \frac{b}{a}x + \frac{b^2}{4a^2} = -\frac{c}{a} + \frac{b^2}{4a^2}$$
$$\left(x + \frac{b}{2a}\right)^2 = \frac{b^2 - 4ac}{4a^2}$$

*Section 8.1* Solving Quadratic Equations by Completing the Square

$$x + \frac{b}{2a} = \pm\sqrt{\frac{b^2 - 4ac}{4a^2}}$$

$$x + \frac{b}{2a} = \pm\frac{\sqrt{b^2 - 4ac}}{2a}$$

$$x = -\frac{b}{2a} \pm \frac{\sqrt{b^2 - 4ac}}{2a}$$

$$x = \frac{-b \pm \sqrt{b^2 - 4ac}}{2a}$$

112.  $p^2 + 4p = 32$
$p^2 + 4p - 32 = 0$
$(p+8)(p-4) = 0$
$p + 8 = 0$ or $p - 4 = 0$
$p = -8$ or $p = 4$
The solution set is $\{-8, 4\}$.

114.  $\left|\frac{3}{4}w - \frac{2}{3}\right| = \frac{5}{2}$

$\frac{3}{4}w - \frac{2}{3} = \frac{5}{2}$ or $\frac{3}{4}w - \frac{2}{3} = -\frac{5}{2}$

$12\left(\frac{3}{4}w - \frac{2}{3}\right) = 12\left(\frac{5}{2}\right)$ or $12\left(\frac{3}{4}w - \frac{2}{3}\right) = 12\left(-\frac{5}{2}\right)$

$9w - 8 = 30$ or $9w - 8 = -30$
$9w = 38$ or $9w = -22$
$w = \frac{38}{9}$ or $w = -\frac{22}{9}$

The solution set is $\left\{-\frac{22}{9}, \frac{38}{9}\right\}$.

115.  In both cases, the simpler equations are linear.

## 8.2 Preparing for Solving Quadratic Equations by the Quadratic Formula

1.  a.  $\sqrt{54} = \sqrt{9 \cdot 6} = 3\sqrt{6}$

    b.  $\sqrt{121} = 11$ because $11^2 = 121$.

2.  a.  $\sqrt{-9} = \sqrt{-1 \cdot 9} = 3i$

    b.  $\sqrt{-72} = \sqrt{-1 \cdot 36 \cdot 2} = 6\sqrt{2}\,i$

3.  $\frac{3 + \sqrt{18}}{6} = \frac{3 + 3\sqrt{2}}{6}$
$= \frac{3(1 + \sqrt{2})}{6}$
$= \frac{1 + \sqrt{2}}{2}$ or $\frac{1}{2} + \frac{\sqrt{2}}{2}$

## 8.2 Exercises

2.  discriminant

4.  TRUE.

6.  TRUE.

8.  Answers may vary. One possibility follows: If the equation is either of the form $x^2 = p$ or $ax^2 + c = 0$ (or if it can be easily put in one of these forms), then the Square Root Property would be a good choice for solving the equation.

10.  Putting the equation $3x^2 - x = 5$ in standard form, we obtain $3x^2 - x - 5 = 0$. Thus, $a = 3$, $b = -1$, and $c = -5$.

12.  $p^2 - 4p - 32 = 0$
For this equation, $a = 1$, $b = -4$, and $c = -32$.

$p = \frac{-(-4) \pm \sqrt{(-4)^2 - 4(1)(-32)}}{2(1)}$

$= \frac{4 \pm \sqrt{16 + 128}}{2}$

$= \frac{4 \pm \sqrt{144}}{2}$

$= \frac{4 \pm 12}{2}$

$p = \frac{4 - 12}{2}$ or $p = \frac{4 + 12}{2}$

$= \frac{-8}{2}$ or $= \frac{16}{2}$

$= -4$ or $= 8$

The solution set is $\{-4, 8\}$.

## Chapter 8: Quadratic Equations and Functions

**14.** $10x^2 + x - 2 = 0$

For this equation, $a = 10$, $b = 1$, and $c = -2$.

$$x = \frac{-1 \pm \sqrt{1^2 - 4(10)(-2)}}{2(10)}$$

$$= \frac{-1 \pm \sqrt{1 + 80}}{20}$$

$$= \frac{-1 \pm \sqrt{81}}{20}$$

$$= \frac{-1 \pm 9}{20}$$

$x = \frac{-1-9}{20}$ or $x = \frac{-1+9}{20}$

$= \frac{-10}{20}$ or $= \frac{8}{20}$

$= -\frac{1}{2}$ or $= \frac{2}{5}$

The solution set is $\left\{-\frac{1}{2}, \frac{2}{5}\right\}$.

**16.** $2q^2 - 4q + 1 = 0$

For this equation, $a = 2$, $b = -4$, and $c = 1$.

$$q = \frac{-(-4) \pm \sqrt{(-4)^2 - 4(2)(1)}}{2(2)}$$

$$= \frac{4 \pm \sqrt{16 - 8}}{4}$$

$$= \frac{4 \pm \sqrt{8}}{4}$$

$$= \frac{4 \pm 2\sqrt{2}}{4}$$

$$= \frac{4}{4} \pm \frac{2\sqrt{2}}{4} = 1 \pm \frac{\sqrt{2}}{2}$$

The solution set is $\left\{1 - \frac{\sqrt{2}}{2}, 1 + \frac{\sqrt{2}}{2}\right\}$.

**18.** $x + \frac{1}{x} = 3$

$x\left(x + \frac{1}{x}\right) = x(3)$

$x^2 + 1 = 3x$

$x^2 - 3x + 1 = 0$

For this equation, $a = 1$, $b = -3$, and $c = 1$.

$$x = \frac{-(-3) \pm \sqrt{(-3)^2 - 4(1)(1)}}{2(1)}$$

$$= \frac{3 \pm \sqrt{9 - 4}}{2}$$

$$= \frac{3 \pm \sqrt{5}}{2} = \frac{3}{2} \pm \frac{\sqrt{5}}{2}$$

The solution set is $\left\{\frac{3}{2} - \frac{\sqrt{5}}{2}, \frac{3}{2} + \frac{\sqrt{5}}{2}\right\}$.

**20.** $5w^2 = -3w + 1$

$5w^2 + 3w - 1 = 0$

For this equation, $a = 5$, $b = 3$, and $c = -1$.

$$w = \frac{-3 \pm \sqrt{3^2 - 4(5)(-1)}}{2(5)}$$

$$= \frac{-3 \pm \sqrt{9 + 20}}{10}$$

$$= \frac{-3 \pm \sqrt{29}}{10} = -\frac{3}{10} \pm \frac{\sqrt{29}}{10}$$

The solution set is $\left\{-\frac{3}{10} - \frac{\sqrt{29}}{10}, -\frac{3}{10} + \frac{\sqrt{29}}{10}\right\}$.

**22.** $y^2 - 4y + 5 = 0$

For this equation, $a = 1$, $b = -4$, and $c = 5$.

$$y = \frac{-(-4) \pm \sqrt{(-4)^2 - 4(1)(5)}}{2(1)}$$

$$= \frac{4 \pm \sqrt{16 - 20}}{2}$$

$$= \frac{4 \pm \sqrt{-4}}{2}$$

$$= \frac{4 \pm 2i}{2}$$

$$= \frac{4}{2} \pm \frac{2i}{2} = 2 \pm i$$

The solution set is $\{2 - i, 2 + i\}$.

**24.** $2z^2 + 7 = 4z$

$2z^2 - 4z + 7 = 0$

For this equation, $a = 2$, $b = -4$, and $c = 7$.

## Section 8.2 Solving Quadratic Equations by the Quadratic Formula

$$z = \frac{-(-4) \pm \sqrt{(-4)^2 - 4(2)(7)}}{2(2)}$$

$$= \frac{4 \pm \sqrt{16 - 56}}{4}$$

$$= \frac{4 \pm \sqrt{-40}}{4}$$

$$= \frac{4 \pm 2\sqrt{10}\, i}{4} = \frac{4}{4} \pm \frac{2\sqrt{10}\, i}{4} = 1 \pm \frac{\sqrt{10}}{2} i$$

The solution set is $\left\{ 1 - \frac{\sqrt{10}}{2} i,\, 1 + \frac{\sqrt{10}}{2} i \right\}$.

26. $6p^2 = 4p + 1$
$6p^2 - 4p - 1 = 0$
For this equation, $a = 6$, $b = -4$, and $c = -1$.

$$p = \frac{-(-4) \pm \sqrt{(-4)^2 - 4(6)(-1)}}{2(6)}$$

$$= \frac{4 \pm \sqrt{16 + 24}}{12}$$

$$= \frac{4 \pm \sqrt{40}}{12}$$

$$= \frac{4 \pm 2\sqrt{10}}{12}$$

$$= \frac{4}{12} \pm \frac{2\sqrt{10}}{12} = \frac{1}{3} \pm \frac{\sqrt{10}}{6}$$

The solution set is $\left\{ \frac{1}{3} - \frac{\sqrt{10}}{6},\, \frac{1}{3} + \frac{\sqrt{10}}{6} \right\}$.

28. $1 = 5w^2 + 6w$
$0 = 5w^2 + 6w - 1$
For this equation, $a = 5$, $b = 6$, and $c = -1$.

$$w = \frac{-6 \pm \sqrt{6^2 - 4(5)(-1)}}{2(5)}$$

$$= \frac{-6 \pm \sqrt{36 + 20}}{10}$$

$$= \frac{-6 \pm \sqrt{56}}{10}$$

$$= \frac{-6 \pm 2\sqrt{14}}{10} = -\frac{6}{10} \pm \frac{2\sqrt{14}}{10} = -\frac{3}{5} \pm \frac{\sqrt{14}}{5}$$

The solution set is $\left\{ -\frac{3}{5} - \frac{\sqrt{14}}{5},\, -\frac{3}{5} + \frac{\sqrt{14}}{5} \right\}$.

30. $p^2 + 4p - 2 = 0$
For this equation, $a = 1$, $b = 4$, and $c = -2$.

$b^2 - 4ac = 4^2 - 4(1)(-2) = 16 + 8 = 24$

Because $b^2 - 4ac = 24$ is positive, but not a perfect square, the quadratic equation will have two irrational solutions.

32. $2y^2 - 3y + 5 = 0$
For this equation, $a = 2$, $b = -3$, and $c = 5$.

$b^2 - 4ac = (-3)^2 - 4(2)(5) = 9 - 40 = -31$

Because $b^2 - 4ac = -31$ is negative, the quadratic equation will have two complex solutions that are not real. The solutions will be complex conjugates of each other.

34. $16x^2 + 24x + 9 = 0$
For this equation, $a = 16$, $b = 24$, and $c = 9$.

$b^2 - 4ac = (24)^2 - 4(16)(9) = 576 - 576 = 0$

Because $b^2 - 4ac = 0$, the quadratic equation will have one repeated real solution.

36. $6x^2 - x = -4$
$6x^2 - x + 4 = 0$
For this equation, $a = 6$, $b = -1$, and $c = 4$.

$b^2 - 4ac = (-1)^2 - 4(6)(4) = 1 - 96 = -95$

Because $b^2 - 4ac = -95$ is negative, the quadratic equation will have two complex solutions that are not real. The solutions will be complex conjugates of each other.

38. $10w^2 = 3$
$10w^2 - 3 = 0$
For this equation, $a = 10$, $b = 0$, and $c = -3$.

$b^2 - 4ac = 0^2 - 4(10)(-3) = 0 + 120 = 120$

Because $b^2 - 4ac = 120$ is positive, but not a perfect square, the quadratic equation will have two irrational solutions.

40. $q^2 - 7q + 7 = 0$
Because this equation does not easily factor, solve by using the quadratic formula. For this equation, $a = 1$, $b = -7$, and $c = 7$.

ISM – 449

## Chapter 8: Quadratic Equations and Functions

$$q = \frac{-(-7) \pm \sqrt{(-7)^2 - 4(1)(7)}}{2(1)}$$

$$= \frac{7 \pm \sqrt{49 - 28}}{2} = \frac{7 \pm \sqrt{21}}{2} = \frac{7}{2} \pm \frac{\sqrt{21}}{2}$$

The solution set is $\left\{\dfrac{7}{2} - \dfrac{\sqrt{21}}{2}, \dfrac{7}{2} + \dfrac{\sqrt{21}}{2}\right\}$.

42. $4p^2 + 5p = 9$
$4p^2 + 5p - 9 = 0$
Because this equation factors easily, solve by factoring.
$(4p + 9)(p - 1) = 0$
$4p + 9 = 0 \quad$ or $\quad p - 1 = 0$
$\quad 4p = -9 \quad$ or $\quad p = 1$
$\quad p = -\dfrac{9}{4}$

The solution set is $\left\{-\dfrac{9}{4}, 1\right\}$.

44. $3x^2 + 5x = 2$
$3x^2 + 5x - 2 = 0$
Because this equation factors easily, solve by factoring.
$(3x - 1)(x + 2) = 0$
$3x - 1 = 0 \quad$ or $\quad x + 2 = 0$
$\quad 3x = 1 \quad$ or $\quad x = -2$
$\quad x = \dfrac{1}{3}$

The solution set is $\left\{-2, \dfrac{1}{3}\right\}$.

46. $w^2 + 4w + 9 = 0$
Because this equation does not easily factor, solve by using the quadratic formula. For this equation, $a = 1$, $b = 4$, and $c = 9$.

$$w = \frac{-4 \pm \sqrt{4^2 - 4(1)(9)}}{2(1)}$$

$$= \frac{-4 \pm \sqrt{16 - 36}}{2}$$

$$= \frac{-4 \pm \sqrt{-20}}{2}$$

$$= \frac{-4 \pm 2\sqrt{5}\,i}{2}$$

$$= \frac{-4}{2} \pm \frac{2\sqrt{5}\,i}{2} = -2 \pm \sqrt{5}\,i$$

The solution set is $\left\{-2 - \sqrt{5}\,i,\ -2 + \sqrt{5}\,i\right\}$.

48. $6x^2 = 2x + 4$
$6x^2 - 2x - 4 = 0$
$\dfrac{6x^2 - 2x - 4}{2} = \dfrac{0}{2}$
$3x^2 - x - 2 = 0$
Because this equation factors easily, solve by factoring.
$(3x + 2)(x - 1) = 0$
$3x + 2 = 0 \quad$ or $\quad x - 1 = 0$
$\quad 3x = -2 \quad$ or $\quad x = 1$
$\quad x = -\dfrac{2}{3}$

The solution set is $\left\{-\dfrac{2}{3}, 1\right\}$.

50. $5m - 4 = \dfrac{5}{m}$
$(5m - 4)m = \left(\dfrac{5}{m}\right)m$
$5m^2 - 4m = 5$
$5m^2 - 4m - 5 = 0$
Because this equation does not easily factor, solve by using the quadratic formula. For this equation, $a = 5$, $b = -4$, and $c = -5$.

## Section 8.2 Solving Quadratic Equations by the Quadratic Formula

$$m = \frac{-(-4) \pm \sqrt{(-4)^2 - 4(5)(-5)}}{2(5)}$$

$$= \frac{4 \pm \sqrt{16 + 100}}{10}$$

$$= \frac{4 \pm \sqrt{116}}{10}$$

$$= \frac{4 \pm 2\sqrt{29}}{10}$$

$$= \frac{4}{10} \pm \frac{2\sqrt{29}}{10} = \frac{2}{5} \pm \frac{\sqrt{29}}{5}$$

The solution set is $\left\{ \dfrac{2}{5} - \dfrac{\sqrt{29}}{5}, \dfrac{2}{5} + \dfrac{\sqrt{29}}{5} \right\}$.

**52.** $4p^2 - 100 = 0$

Because this equation has no linear term, solve by using the square root method.

$$4p^2 = 100$$
$$p^2 = 25$$
$$p = \pm\sqrt{25} = \pm 5$$

The solution set is $\{-5, 5\}$.

**54.** $4q^2 + 1 = 2q$

$4q^2 - 2q + 1 = 0$

Because this equation does not easily factor, solve by using the quadratic formula. For this equation, $a = 4$, $b = -2$, and $c = 1$.

$$q = \frac{-(-2) \pm \sqrt{(-2)^2 - 4(4)(1)}}{2(4)}$$

$$= \frac{2 \pm \sqrt{4 - 16}}{8}$$

$$= \frac{2 \pm \sqrt{-12}}{8}$$

$$= \frac{2 \pm 2\sqrt{3}\,i}{8}$$

$$= \frac{2}{8} \pm \frac{2\sqrt{3}\,i}{8} = \frac{1}{4} \pm \frac{\sqrt{3}}{4}i$$

The solution set is $\left\{ \dfrac{1}{4} - \dfrac{\sqrt{3}}{4}i, \dfrac{1}{4} + \dfrac{\sqrt{3}}{4}i \right\}$.

**56.** $8p^2 - 40p + 50 = 0$

$$\frac{8p^2 - 40p + 50}{2} = \frac{0}{2}$$

$4p^2 - 20p + 25 = 0$

Because this equation factors easily, solve by factoring.

$(2p - 5)(2p - 5) = 0$

$2p - 5 = 0$ or $2p - 5 = 0$
$2p = 5$ or $2p = 5$
$p = \dfrac{5}{2}$ or $p = \dfrac{5}{2}$

The solution set is $\left\{ \dfrac{5}{2} \right\}$.

**58.** $\dfrac{1}{2}x^2 + \dfrac{3}{4}x - 1 = 0$

$$4\left( \dfrac{1}{2}x^2 + \dfrac{3}{4}x - 1 \right) = 4(0)$$

$2x^2 + 3x - 4 = 0$

Because this equation does not easily factor, solve by using the quadratic formula. For this equation, $a = 2$, $b = 3$, and $c = -4$.

$$x = \frac{-3 \pm \sqrt{3^2 - 4(2)(-4)}}{2(2)}$$

$$= \frac{-3 \pm \sqrt{9 + 32}}{4}$$

$$= \frac{-3 \pm \sqrt{41}}{4} = -\frac{3}{4} \pm \frac{\sqrt{41}}{4}$$

The solution set is $\left\{ -\dfrac{3}{4} - \dfrac{\sqrt{41}}{4}, -\dfrac{3}{4} + \dfrac{\sqrt{41}}{4} \right\}$.

**60.** $(a - 3)(a + 1) = 2$

$a^2 - 2a - 3 = 2$
$a^2 - 2a - 5 = 0$

Because this equation does not easily factor, solve by using the quadratic formula. For this equation, $a = 1$, $b = -2$, and $c = -5$.

## Chapter 8: Quadratic Equations and Functions

$$a = \frac{-(-2) \pm \sqrt{(-2)^2 - 4(1)(-5)}}{2(1)}$$

$$= \frac{2 \pm \sqrt{4+20}}{2}$$

$$= \frac{2 \pm \sqrt{24}}{2}$$

$$= \frac{2 \pm 2\sqrt{6}}{2}$$

$$= \frac{2}{2} \pm \frac{2\sqrt{6}}{2} = 1 \pm \sqrt{6}$$

The solution set is $\{1-\sqrt{6}, 1+\sqrt{6}\}$.

**62.** Note: $x \neq -3$.

$$\frac{x-5}{x+3} = x-3$$

$$(x+3)\left(\frac{x-5}{x+3}\right) = (x+3)(x-3)$$

$$x-5 = x^2 - 9$$

$$0 = x^2 - x - 4$$

Because this equation does not easily factor, solve by using the quadratic formula. For this equation, $a=1$, $b=-1$, and $c=-4$.

$$x = \frac{-(-1) \pm \sqrt{(-1)^2 - 4(1)(-4)}}{2(1)}$$

$$= \frac{1 \pm \sqrt{1+16}}{2}$$

$$= \frac{1 \pm \sqrt{17}}{2} = \frac{1}{2} \pm \frac{\sqrt{17}}{2}$$

The solution set is $\left\{\frac{1}{2} - \frac{\sqrt{17}}{2}, \frac{1}{2} + \frac{\sqrt{17}}{2}\right\}$.

**64.**

$$\frac{x-1}{x^2+4} = 1$$

$$(x^2+4)\left(\frac{x-1}{x^2+4}\right) = (x^2+4)(1)$$

$$x-1 = x^2 + 4$$

$$0 = x^2 - x + 5$$

Because this equation does not easily factor, solve by using the quadratic formula. For this equation, $a=1$, $b=-1$, and $c=5$.

$$x = \frac{-(-1) \pm \sqrt{(-1)^2 - 4(1)(5)}}{2(1)}$$

$$= \frac{1 \pm \sqrt{1-20}}{2}$$

$$= \frac{1 \pm \sqrt{-19}}{2}$$

$$= \frac{1 \pm \sqrt{19}\, i}{2} = \frac{1}{2} \pm \frac{\sqrt{19}}{2} i$$

The solution set is $\left\{\frac{1}{2} - \frac{\sqrt{19}}{2}i, \frac{1}{2} + \frac{\sqrt{19}}{2}i\right\}$.

**66. a.**

$$f(x) = 0$$

$$x^2 + 2x - 8 = 0$$

$$(x+4)(x-2) = 0$$

$$x+4 = 0 \quad \text{or} \quad x-2 = 0$$

$$x = -4 \quad \text{or} \quad x = 2$$

The solution set is $\{-4, 2\}$.

**b.**

$$f(x) = -8$$

$$x^2 + 2x - 8 = -8$$

$$x^2 + 2x = 0$$

$$x(x+2) = 0$$

$$x = 0 \quad \text{or} \quad x+2 = 0$$

$$x = -2$$

The solution set is $\{-2, 0\}$.

**68. a.**

$$g(x) = 0$$

$$3x^2 + x - 1 = 0$$

For this equation, $a=3$, $b=1$, and $c=-1$.

$$x = \frac{-1 \pm \sqrt{1^2 - 4(3)(-1)}}{2(3)}$$

$$= \frac{-1 \pm \sqrt{1+12}}{6}$$

$$= \frac{-1 \pm \sqrt{13}}{6} = -\frac{1}{6} \pm \frac{\sqrt{13}}{6}$$

The solution set is $\left\{-\frac{1}{6} - \frac{\sqrt{13}}{6}, -\frac{1}{6} + \frac{\sqrt{13}}{6}\right\}$.

**b.**

$$g(x) = 4$$

$$3x^2 + x - 1 = 4$$

$$3x^2 + x - 5 = 0$$

For this equation, $a=3$, $b=1$, and $c=-5$.

### Section 8.2 Solving Quadratic Equations by the Quadratic Formula

$$x = \frac{-1 \pm \sqrt{1^2 - 4(3)(-5)}}{2(3)}$$

$$= \frac{-1 \pm \sqrt{1 + 60}}{6}$$

$$= \frac{-1 \pm \sqrt{61}}{6} = -\frac{1}{6} \pm \frac{\sqrt{61}}{6}$$

The solution set is $\left\{-\frac{1}{6} - \frac{\sqrt{61}}{6}, -\frac{1}{6} + \frac{\sqrt{61}}{6}\right\}$.

**70. a.** 
$$F(x) = 0$$
$$-x^2 + 3x - 3 = 0$$
$$x^2 - 3x + 3 = 0$$

For this equation, $a = 1$, $b = -3$, and $c = 3$.

$$x = \frac{-(-3) \pm \sqrt{(-3)^2 - 4(1)(3)}}{2(1)}$$

$$= \frac{3 \pm \sqrt{9 - 12}}{2}$$

$$= \frac{3 \pm \sqrt{-3}}{2}$$

$$= \frac{3 \pm \sqrt{3}\,i}{2} = \frac{3}{2} \pm \frac{\sqrt{3}}{2}i$$

The solution set is $\left\{\frac{3}{2} - \frac{\sqrt{3}}{2}i, \frac{3}{2} + \frac{\sqrt{3}}{2}i\right\}$.

**b.** 
$$F(x) = -2$$
$$-x^2 + 3x - 3 = -2$$
$$-x^2 + 3x - 1 = 0$$
$$x^2 - 3x + 1 = 0$$

For this equation, $a = 1$, $b = -3$, and $c = 1$.

$$x = \frac{-(-3) \pm \sqrt{(-3)^2 - 4(1)(1)}}{2(1)}$$

$$= \frac{3 \pm \sqrt{9 - 4}}{2} = \frac{3 \pm \sqrt{5}}{2} = \frac{3}{2} \pm \frac{\sqrt{5}}{2}$$

The solution set is $\left\{\frac{3}{2} - \frac{\sqrt{5}}{2}, \frac{3}{2} + \frac{\sqrt{5}}{2}\right\}$.

**72.**
$$x^2 + (x+7)^2 = (2x+3)^2$$
$$x^2 + x^2 + 14x + 49 = 4x^2 + 12x + 9$$
$$2x^2 + 14x + 49 = 4x^2 + 12x + 9$$
$$0 = 2x^2 - 2x - 40$$
$$0 = x^2 - x - 20$$
$$0 = (x+4)(x-5)$$
$$x + 4 = 0 \quad \text{or} \quad x - 5 = 0$$
$$x = -4 \quad \text{or} \quad x = 5$$

Disregard $x = -4$ because $x$ represents the length of one leg of the triangle. Thus, $x = 5$ is the only viable answer. Now, $x + 7 = 5 + 7 = 12$ and $2x + 3 = 2(5) + 3 = 13$. The three measurements are 5, 12, and 13.

**74.**
$$(4x)^2 + (x-1)^2 = (4x+1)^2$$
$$16x^2 + x^2 - 2x + 1 = 16x^2 + 8x + 1$$
$$17x^2 - 2x + 1 = 16x^2 + 8x + 1$$
$$x^2 - 10x = 0$$
$$x(x - 10) = 0$$
$$x = 0 \quad \text{or} \quad x - 10 = 0$$
$$x = 10$$

Disregard $x = 0$ because $x - 1 = 0 - 1 = -1$ cannot represents the length of one leg of the triangle. Thus, $x = 10$ is the only viable answer. Now, $4x = 4(10) = 40$, $x - 1 = 10 - 1 = 9$, and $4x + 1 = 4(10) + 1 = 41$. The three measurements are 9, 40, and 41.

**76.** Let $x$ represent the length of the rectangle. Then $x + 6$ will represent the width.

$$x(x + 6) = 60$$
$$x^2 + 6x = 60$$
$$x^2 + 6x - 60 = 0$$

For this equation, $a = 1$, $b = 6$, and $c = -60$.

$$x = \frac{-6 \pm \sqrt{6^2 - 4(1)(-60)}}{2(1)}$$

$$= \frac{-6 \pm \sqrt{36 + 240}}{2}$$

$$= \frac{-6 \pm \sqrt{276}}{2}$$

$$= \frac{-6 \pm 2\sqrt{69}}{2} = -\frac{6}{2} \pm \frac{2\sqrt{69}}{2} = -3 \pm \sqrt{69}$$

Disregard $x = -3 - \sqrt{69} \approx -11.307$ because $x$ represents the length of the rectangle, which must be positive. Thus, $x = -3 + \sqrt{69} \approx 5.307$ is

the only viable answer. Now,
$x+6 = -3+\sqrt{69}+6 = 3+\sqrt{69} \approx 11.307$. Thus, the dimensions of the rectangle are $-3+\sqrt{69}$ inches by $3+\sqrt{69}$ inches, which is approximately 5.307 inches by 11.307 inches.

78. Let $x$ represent the base of the triangle. Then $x-2$ will represent the height.
$$\frac{1}{2}x(x-2) = 35$$
$$\frac{1}{2}x^2 - x - 35 = 0$$
$$2\left(\frac{1}{2}x^2 - x - 35\right) = 2(0)$$
$$x^2 - 2x - 70 = 0$$
For this equation, $a=1$, $b=-2$, and $c=-70$.
$$x = \frac{-(-2) \pm \sqrt{(-2)^2 - 4(1)(-70)}}{2(1)}$$
$$= \frac{2 \pm \sqrt{4+280}}{2}$$
$$= \frac{2 \pm \sqrt{284}}{2}$$
$$= \frac{2 \pm 2\sqrt{71}}{2} = \frac{2}{2} \pm \frac{2\sqrt{71}}{2} = 1 \pm \sqrt{71}$$
Disregard $x = 1-\sqrt{71} \approx -7.426$ because $x$ represents the base of the triangle, which must be positive. Thus, $x = 1+\sqrt{71} \approx 9.426$ is the only viable answer. Now,
$x-2 = 1+\sqrt{71}-2 = -1+\sqrt{71} \approx 7.426$. Thus, the base of the triangle is $1+\sqrt{71}$ inches, which is approximately 9.426 inches. The height of the triangle is $-1+\sqrt{71}$ inches, which is approximately 7.426 inches.

80. a. $R(300) = -0.02(300)^2 + 24(300) = 5400$
If 300 all-day passes are sold per day, then the company's revenue will be $5400.

$R(800) = -0.02(800)^2 + 24(800) = 6400$
If 800 all-day passes are sold per day, then the company's revenue will be $6400.

b. $R(x) = 4000$
$-0.02x^2 + 24x = 4000$
$0 = 0.02x^2 - 24x + 4000$
For this equation, $a = 0.02$, $b = -24$, and $c = 4000$.
$$x = \frac{-(-24) \pm \sqrt{(-24)^2 - 4(0.02)(4000)}}{2(0.02)}$$
$$= \frac{24 \pm \sqrt{576-320}}{0.04}$$
$$= \frac{24 \pm \sqrt{256}}{0.04}$$
$$= \frac{24 \pm 16}{0.04}$$
$$x = \frac{24-16}{0.04} \text{ or } x = \frac{24+16}{0.04}$$
$$= \frac{8}{0.04} \text{ or } = \frac{40}{0.04}$$
$$= 200 \text{ or } = 1000$$
The revenue will be $4000 per day if either 200 or 1000 all-day passes are sold.

c. $R(x) = 7200$
$-0.02x^2 + 24x = 7200$
$0 = 0.02x^2 - 24x + 7200$
For this equation, $a = 0.02$, $b = -24$, and $c = 7200$.
$$x = \frac{-(-24) \pm \sqrt{(-24)^2 - 4(0.02)(7200)}}{2(0.02)}$$
$$= \frac{24 \pm \sqrt{576-576}}{0.04}$$
$$= \frac{24 \pm \sqrt{0}}{0.04}$$
$$= \frac{24 \pm 0}{0.04}$$
$$= \frac{24}{0.04} = 600$$
The revenue will be $7200 per day if 600 all-day passes are sold.

Section 8.2 Solving Quadratic Equations by the Quadratic Formula

82. a.
$$s(t) = 200$$
$$-16t^2 + 150t + 2 = 200$$
$$-16t^2 + 150t - 198 = 0$$
$$\frac{-16t^2 + 150t - 198}{-2} = \frac{0}{-2}$$
$$8t^2 - 75t + 99 = 0$$
For this equation, $a = 8$, $b = -75$, and $c = 99$.
$$x = \frac{-(-75) \pm \sqrt{(-75)^2 - 4(8)(99)}}{2(8)}$$
$$= \frac{75 \pm \sqrt{5625 - 3168}}{16}$$
$$= \frac{75 \pm \sqrt{2457}}{16}$$
$$= \frac{75 \pm 3\sqrt{273}}{16}$$
$$x = \frac{75 - 3\sqrt{273}}{16} \text{ or } x = \frac{75 + 3\sqrt{273}}{16}$$
$$\approx 1.589 \quad \text{or} \quad \approx 7.786$$
Rounding to the nearest tenth, the height of the rocket will be 200 feet after approximately 1.6 seconds and after approximately 7.8 seconds.

b.
$$s(t) = 300$$
$$-16t^2 + 150t + 2 = 300$$
$$-16t^2 + 150t - 298 = 0$$
$$\frac{-16t^2 + 150t - 298}{-2} = \frac{0}{-2}$$
$$8t^2 - 75t + 149 = 0$$
For this equation, $a = 8$, $b = -75$, and $c = 149$.
$$x = \frac{-(-75) \pm \sqrt{(-75)^2 - 4(8)(149)}}{2(8)}$$
$$= \frac{75 \pm \sqrt{5625 - 4768}}{16}$$
$$= \frac{75 \pm \sqrt{857}}{16}$$
$$x = \frac{75 - \sqrt{857}}{16} \text{ or } x = \frac{75 + \sqrt{857}}{16}$$
$$\approx 2.858 \quad \text{or} \quad \approx 6.517$$
Rounding to the nearest tenth, the height of the rocket will be 300 feet after approximately 2.9 seconds and after approximately 6.5 seconds.

c.
$$s(t) = 500$$
$$-16t^2 + 150t + 2 = 500$$
$$-16t^2 + 150t - 498 = 0$$
$$\frac{-16t^2 + 150t - 498}{-2} = \frac{0}{-2}$$
$$8t^2 - 75t + 249 = 0$$
For this equation, $a = 8$, $b = -75$, and $c = 249$.
$$x = \frac{-(-75) \pm \sqrt{(-75)^2 - 4(8)(249)}}{2(8)}$$
$$= \frac{75 \pm \sqrt{5625 - 7968}}{16}$$
$$= \frac{75 \pm \sqrt{-2343}}{16}$$
$$= \frac{75 \pm \sqrt{2343}\, i}{16}$$
$$x = \frac{75}{16} - \frac{\sqrt{2343}}{16} i \text{ or } x = \frac{75}{16} + \frac{\sqrt{2343}}{16} i$$
The rocket will never reach a height of 500 feet. This is clear because the solutions to the equation above are complex solutions that are not real.

84. Let $x$ represent the number.
$$3x^2 = 2x + 5$$
$$3x^2 - 2x - 5 = 0$$
$$(3x - 5)(x + 1) = 0$$
$$3x - 5 = 0 \quad \text{or} \quad x + 1 = 0$$
$$3x = 5 \qquad\qquad x = -1$$
$$x = \frac{5}{3}$$
The number can be either $-1$ or $\frac{5}{3}$.

86. a.
$$P(a) = 200$$
$$0.015a^2 - 4.962a + 290.580 = 200$$
$$0.015a^2 - 4.962a + 90.580 = 0$$
For this equation, $a = 0.015$, $b = -4.962$, and $c = 90.580$.

## Chapter 8: Quadratic Equations and Functions

$$x = \frac{-(-4.962) \pm \sqrt{(-4.962)^2 - 4(0.015)(90.580)}}{2(0.015)}$$

$$= \frac{4.962 \pm \sqrt{24.621444 - 5.4348}}{0.03}$$

$$= \frac{4.962 \pm \sqrt{19.186644}}{0.03}$$

$$x = \frac{4.962 - \sqrt{19.186644}}{0.03} \approx 19.391$$

or

$$x = \frac{4.962 + \sqrt{19.186644}}{0.03} \approx 311.409$$

We disregard 311.409 since this is not a valid choice for human age. Thus, the only viable answer is 19.391. Rounding to the nearest year, there were 200 million Americans in 2001 who were age 19 years or older.

**b.** $\quad P(a) = 50$
$0.015a^2 - 4.962a + 290.580 = 50$
$0.015a^2 - 4.962a + 240.580 = 0$
For this equation, $a = 0.015$, $b = -4.962$, and $c = 240.580$.

$$x = \frac{-(-4.962) \pm \sqrt{(-4.962)^2 - 4(0.015)(240.580)}}{2(0.015)}$$

$$= \frac{4.962 \pm \sqrt{24.621444 - 14.4348}}{0.03}$$

$$= \frac{4.962 \pm \sqrt{10.186644}}{0.03}$$

$$x = \frac{4.962 - \sqrt{10.186644}}{0.03} \approx 59.012$$

or

$$x = \frac{4.962 + \sqrt{10.186644}}{0.03} \approx 271.788$$

We disregard 271.788 since this is not a valid choice for human age. Thus, the only viable answer is 59.012. Rounding to the nearest year, there were 50 million Americans in 2001 who were age 59 years or older.

**88.** Let $x$ represent the effect (i.e., speed) of the jet stream.

|  | Distance | Rate | Time |
|---|---|---|---|
| With jet stream | 200 | $120 - x$ | $\dfrac{200}{120 - x}$ |
| Against jet stream | 200 | $120 + x$ | $\dfrac{200}{120 + x}$ |

$$\frac{200}{120 - x} + \frac{200}{120 + x} = 4$$

$$(120 - x)(120 + x)\left(\frac{200}{120 - x} + \frac{200}{120 + x}\right)$$
$$= (120 - x)(120 + x)(4)$$

$200(120 + x) + 200(120 - x) = (14{,}400 - x^2)(4)$
$24{,}000 + 200x + 24{,}000 - 200x = 57{,}600 - 4x^2$
$48{,}000 = 57{,}600 - 4x^2$
$4x^2 = 9600$
$x^2 = 2400$
$x = \pm\sqrt{2400} = \pm 20\sqrt{6}$
$x \approx \pm 48.990$

Disregard $-48.990$. The effect of the jet stream is approximately 49.0 miles per hour.

**90.** Let $t$ represent the time required for house hose to fill the pool alone. Then $t - 8$ will represent the time required for tanker hose to fill the pool alone.

$$\begin{pmatrix}\text{Part done by}\\ \text{house hose}\\ \text{in 1 hour}\end{pmatrix} + \begin{pmatrix}\text{Part done by}\\ \text{tanker hose}\\ \text{in 1 hour}\end{pmatrix} = \begin{pmatrix}\text{Part done}\\ \text{together}\\ \text{in 1 hour}\end{pmatrix}$$

$$\frac{1}{t} + \frac{1}{t - 8} = \frac{1}{5}$$

$$5t(t-8)\left(\frac{1}{t} + \frac{1}{t-8}\right) = 5t(t-8)\left(\frac{1}{5}\right)$$

$5(t-8) + 5t = t(t-8)$
$5t - 40 + 5t = t^2 - 8t$
$10t - 40 = t^2 - 8t$
$0 = t^2 - 18t + 40$

For this equation, $a = 1$, $b = -18$, and $c = 40$.

$$t = \frac{-(-18) \pm \sqrt{(-18)^2 - 4(1)(40)}}{2(1)}$$

$$= \frac{18 \pm \sqrt{324 - 160}}{2}$$

$$= \frac{18 \pm \sqrt{164}}{2}$$

$$= \frac{18 \pm 2\sqrt{41}}{2} = 9 \pm \sqrt{41}$$

$t = 9 - \sqrt{41}$ or $t = 9 + \sqrt{41}$
$\approx 2.597$ or $\approx 15.403$

Disregard $t \approx 2.597$ because this value makes the tanker hose's time negative:
$t - 1 = 2.597 - 8 = -5.403$. The only viable answer is $t \approx 15.403$ hours. Working alone, it will take the house hose approximately 15.4 hours to fill the tank.

### Section 8.2 Solving Quadratic Equations by the Quadratic Formula

92. By the quadratic formula, the solutions of the equation $ax^2 + bx + c = 0$ are
$$x = \frac{-b - \sqrt{b^2 - 4ac}}{2a} \text{ and } x = \frac{-b + \sqrt{b^2 - 4ac}}{2a}.$$
The product of these two solutions is:
$$\left(\frac{-b - \sqrt{b^2 - 4ac}}{2a}\right)\left(\frac{-b + \sqrt{b^2 - 4ac}}{2a}\right)$$
$$= \frac{b^2 - \sqrt{b^2 - 4ac} + \sqrt{b^2 - 4ac} - (b^2 - 4ac)}{4a^2}$$
$$= \frac{b^2 - b^2 + 4ac}{4a^2} = \frac{4ac}{4a^2} = \frac{c}{a}$$

94. Assume $b^2 - 4ac \geq 0$. The solutions of $ax^2 + bx + c = 0$ are $x = \frac{-b \pm \sqrt{b^2 - 4ac}}{2a}$.

The solutions of $cx^2 + bx + a = 0$ are
$$x = \frac{-b \pm \sqrt{b^2 - 4ca}}{2c} = \frac{-b \pm \sqrt{b^2 - 4ac}}{2c}.$$
Recall that the product of reciprocals is 1. Now,
$$\left(\frac{-b + \sqrt{b^2 - 4ac}}{2a}\right)\left(\frac{-b - \sqrt{b^2 - 4ac}}{2c}\right)$$
$$= \frac{b^2 - (b^2 - 4ac)}{4ac} = \frac{b^2 - b^2 + 4ac}{4ac} = \frac{4ac}{4ac} = 1.$$
Thus, $\frac{-b + \sqrt{b^2 - 4ac}}{2a}$ and $\frac{-b - \sqrt{b^2 - 4ac}}{2c}$ are reciprocals. Likewise,
$$\left(\frac{-b - \sqrt{b^2 - 4ac}}{2a}\right)\left(\frac{-b + \sqrt{b^2 - 4ac}}{2c}\right)$$
$$= \frac{b^2 - (b^2 - 4ac)}{4ac} = \frac{b^2 - b^2 + 4ac}{4ac} = \frac{4ac}{4ac} = 1.$$
Thus, $\frac{-b - \sqrt{b^2 - 4ac}}{2a}$ and $\frac{-b + \sqrt{b^2 - 4ac}}{2c}$ are reciprocals. Therefore, the real solutions of $ax^2 + bx + c = 0$ are the reciprocals of the real solution of $cx^2 + bx + a = 0$.

96. a. $f(x) = x^2 - x - 6$
Let $x = -2, -1, 0, 0.5, 1, 2,$ and $3$.
$f(-2) = (-2)^2 - (-2) - 6 = 4 + 2 - 6 = 0$
$f(-1) = (-1)^2 - (-1) - 6 = 1 + 1 - 6 = -4$
$f(0) = 0^2 - 0 - 6 = -6$
$f(0.5) = (0.5)^2 - (0.5) - 6$
$= 0.25 - 0.5 - 6 = -6.25$
$f(1) = 1^2 - 1 - 6 = 1 - 1 - 6 = -6$
$f(2) = 2^2 - 2 - 6 = 4 - 2 - 6 = -4$
$f(3) = 3^2 - 3 - 6 = 9 - 3 - 6 = 0$
Thus, the points $(-2, 0)$, $(-1, -4)$, $(0, -6)$, $(0.5, -6.25)$, $(1, -6)$, $(2, -4)$, and $(3, 0)$ are on the graph of $f$. We plot the points and connect them with a smooth curve.

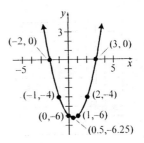

b. $x^2 - x - 6 = 0$
$(x + 2)(x - 3) = 0$
$x + 2 = 0$ or $x - 3 = 0$
$x = -2$ or $x = 3$
The solution set is $\{-2, 3\}$.

c. From the graph in part (a), the x-intercepts of the function $f(x) = x^2 - x - 6$ are $-2$ and 3, which are the same as the solutions of the equation $x^2 - x - 6 = 0$. To find the x-intercepts of a function, set the function equal to zero and solve for x.

98. a. $g(x) = x^2 + 4x + 4$
Let $x = -4, -3, -2, -1,$ and $0$.
$g(-4) = (-4)^2 + 4(-4) + 4 = 16 - 16 + 4 = 4$
$g(-3) = (-3)^2 + 4(-3) + 4 = 9 - 12 + 4 = 1$
$g(-2) = (-2)^2 + 4(-2) + 4 = 4 - 8 + 4 = 0$
$g(-1) = (-1)^2 + 4(-1) + 4 = 1 - 4 + 4 = 1$
$g(0) = 0^2 + 4(0) + 4 = 0 + 0 + 4 = 4$
Thus, the points $(-4, 4)$, $(-3, 1)$, $(-2, 0)$, $(-1, 1)$, and $(0, 4)$ are on the graph of $g$. We plot the points and connect them with a smooth curve.

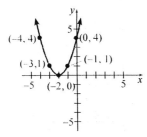

**b.** $x^2 + 4x + 4 = 0$
$(x+2)^2 = 0$
$x + 2 = 0$
$x = -2$
The solution set is $\{-2\}$.

**c.** From the graph in part (a), the x-intercept of the function $g(x) = x^2 + 4x + 4$ is $-2$, which is the same as the solution of the equation $x^2 + 4x + 4 = 0$. To find the x-intercepts of a function, set the function equal to zero and solve for x.

**100.** $f(x) = 0$
$-x^2 - 5x + 1 = 0$
For this equation, $a = -1$, $b = -5$, and $c = 1$.
$b^2 - 4ac = (-5)^2 - 4(-1)(1) = 25 + 4 = 29$

Because $b^2 - 4ac = 29$ is positive, but not a perfect square, the equation has two irrational solutions. This conclusion based on the discriminant is apparent in the graph because the graph crosses the x-axis in two places. That is, the graph has two x-intercepts.

**102.** $f(x) = 0$
$x^2 - 6x + 9 = 0$
For this equation, $a = 1$, $b = -6$, and $c = 9$.
$b^2 - 4ac = (-6)^2 - 4(1)(9) = 36 - 36 = 0$

Because $b^2 - 4ac = 0$, the quadratic equation will have one repeated real solution. This conclusion based on the discriminant is apparent in the graph because the graph touches the x-axis at only one place. That is, the graph has only one x-intercept.

**104. a.** $x^2 - 4x - 45 = 0$
$(x+5)(x-9) = 0$
$x + 5 = 0$ or $x - 9 = 0$
$x = -5$ or $x = 9$
The solution set is $\{-5, 9\}$.

**b.** The x-intercepts are $-5$ and $9$ (see graph).

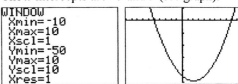

The x-intercepts of $y = x^2 - 4x - 45$ are the same as the solutions of $x^2 - 4x - 45 = 0$.

**106. a.** $x^2 + 10x + 25 = 0$
$(x+5)(x+5) = 0$
$x + 5 = 0$ or $x + 5 = 0$
$x = -5$ or $x = -5$
The solution set is $\{-5\}$.

**b.** The x-intercept is $-5$ (see graph).

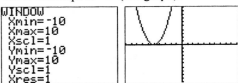

The x-intercept of $y = x^2 + 10x + 25$ are the same as the solutions of $x^2 + 10x + 25 = 0$.

**108. a.** $x^2 + 2x + 5 = 0$
For this equation, $a = 1$, $b = 2$, and $c = 5$.
$x = \dfrac{-2 \pm \sqrt{2^2 - 4(1)(5)}}{2(1)}$
$= \dfrac{-2 \pm \sqrt{4 - 21}}{2}$
$= \dfrac{-2 \pm \sqrt{-16}}{2}$
$= \dfrac{-2 \pm 4i}{2} = -1 \pm 2i$

The solution set is $\{-1 - 2i,\ -1 + 2i\}$.

**b.** The graph has no x-intercepts (see graph).

## Section 8.2 Solving Quadratic Equations by the Quadratic Formula

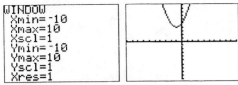

$y = x^2 + 2x + 5$ has no $x$-intercepts, and the solutions of $x^2 + 2x + 5 = 0$ are not real.

### 8.3 Preparing for Solving Equations Quadratic in Form

1. $x^4 - 5x^2 - 6 = (x^2 - 6)(x^2 + 1)$

2. $2(p+3)^2 + 3(p+3) - 5$
   $= [2(p+3)+5][(p+3)-1]$
   $= (2p+6+5)(p+3-1)$
   $= (2p+11)(p+2)$

3. a. $(x^2)^2 = x^{2 \cdot 2} = x^4$

   b. $(p^{-1})^2 = p^{-1 \cdot 2} = p^{-2} = \dfrac{1}{p^2}$

### 8.3 Exercises

2. $3x + 1$

4. TRUE.

6. Answers will vary. One possibility follows: Extraneous solutions may be introduced when we raise both side of the equation to an even power. They may also occur when the equation involves rational expressions.

8. The appropriate choice for $u$ is $\dfrac{1}{x}$. That is, let $u = \dfrac{1}{x}$.

10. $x^4 - 10x^2 + 9 = 0$
    $(x^2)^2 - 10(x^2) + 9 = 0$
    Let $u = x^2$.

    $u^2 - 10u + 9 = 0$
    $(u-1)(u-9) = 0$
    $u - 1 = 0$ or $u - 9 = 0$
    $u = 1$ or $u = 9$
    $x^2 = 1$ or $x^2 = 9$
    $x = \pm\sqrt{1} \pm 1$ or $x = \pm\sqrt{9} = \pm 3$

    Check:
    $x = -1$: $(-1)^4 - 10(-1)^2 + 9 \stackrel{?}{=} 0$
    $1 - 10 + 9 \stackrel{?}{=} 0$
    $0 = 0$ ✓

    $x = 1$: $1^4 - 10(1)^2 + 9 \stackrel{?}{=} 0$
    $1 - 10 + 9 \stackrel{?}{=} 0$
    $0 = 0$ ✓

    $x = -3$: $(-3)^4 - 10(-3)^2 + 9 \stackrel{?}{=} 0$
    $81 - 10 \cdot 9 + 9 \stackrel{?}{=} 0$
    $81 - 90 + 9 \stackrel{?}{=} 0$
    $0 = 0$ ✓

    $x = 3$: $3^4 - 10(3)^2 + 9 \stackrel{?}{=} 0$
    $81 - 10 \cdot 9 + 9 \stackrel{?}{=} 0$
    $81 - 90 + 9 \stackrel{?}{=} 0$
    $0 = 0$ ✓

    All check; the solution set is $\{-3, -1, 1, 3\}$.

12. $z^4 + 10z^2 + 9 = 0$
    $(z^2)^2 + 10(z^2) + 9 = 0$
    Let $u = z^2$.
    $u^2 + 10u + 9 = 0$
    $(u+1)(u+9) = 0$

    $z^2 = -1$ or $z^2 = -9$
    $z = \pm\sqrt{-1}$ or $z = \pm\sqrt{-9}$
    $z = \pm i$ or $z = \pm 3i$

    Check:
    $z = -i$: $(-i)^4 + 10(-i)^2 + 9 \stackrel{?}{=} 0$
    $i^4 + 10 \cdot i^2 + 9 \stackrel{?}{=} 0$
    $1 - 10 + 9 \stackrel{?}{=} 0$
    $0 = 0$ ✓

## Chapter 8: Quadratic Equations and Functions

$z = i$: $(i)^4 + 10(i)^2 + 9 \overset{?}{=} 0$
$$1 - 10 + 9 \overset{?}{=} 0$$
$$0 = 0 \checkmark$$

$z = -3i$: $(-3i)^4 + 10(-3i)^2 + 9 \overset{?}{=} 0$
$$81i^4 + 10 \cdot 9i^2 + 9 \overset{?}{=} 0$$
$$81(1) + 10 \cdot 9 \cdot (-1) + 9 \overset{?}{=} 0$$
$$81 - 90 + 9 \overset{?}{=} 0$$
$$0 = 0 \checkmark$$

$z = 3i$: $(3i)^4 + 10(3i)^2 + 9 \overset{?}{=} 0$
$$81i^4 + 10 \cdot 9i^2 + 9 \overset{?}{=} 0$$
$$81(1) + 10 \cdot 9 \cdot (-1) + 9 \overset{?}{=} 0$$
$$81 - 90 + 9 \overset{?}{=} 0$$
$$0 = 0 \checkmark$$

All check; the solution set is $\{-i, i, -3i, 3i\}$.

**14.** $4b^4 - 5b^2 + 1 = 0$
$$4(b^2)^2 - 5(b^2) + 1 = 0$$
Let $u = b^2$.
$$4u^2 - 5u + 1 = 0$$
$$(4u - 1)(u - 1) = 0$$
$4u - 1 = 0$ or $u - 1 = 0$
$4u = 1$ or $u = 1$
$u = \dfrac{1}{4}$

$b^2 = \dfrac{1}{4}$ or $b^2 = 4$

$b = \pm\sqrt{\dfrac{1}{4}}$ or $b = \pm\sqrt{4}$

$b = \pm\dfrac{1}{2}$ or $b = \pm 2$

Check:

$b = -\dfrac{1}{2}$: $4\left(-\dfrac{1}{2}\right)^4 - 5\left(-\dfrac{1}{2}\right)^2 + 1 \overset{?}{=} 0$
$$4\left(\dfrac{1}{16}\right) - 5\left(\dfrac{1}{4}\right) + 1 \overset{?}{=} 0$$
$$\dfrac{1}{4} - \dfrac{5}{4} + 1 \overset{?}{=} 0$$
$$0 = 0 \checkmark$$

$b = \dfrac{1}{2}$: $4\left(\dfrac{1}{2}\right)^4 - 5\left(\dfrac{1}{2}\right)^2 + 1 \overset{?}{=} 0$
$$4\left(\dfrac{1}{16}\right) - 5\left(\dfrac{1}{4}\right) + 1 \overset{?}{=} 0$$
$$\dfrac{1}{4} - \dfrac{5}{4} + 1 \overset{?}{=} 0$$
$$0 = 0 \checkmark$$

$b = -1$: $4(-1)^4 - 5(-1)^2 + 1 \overset{?}{=} 0$
$$4 \cdot 1 - 5 \cdot 1 + 1 \overset{?}{=} 0$$
$$4 - 5 + 1 \overset{?}{=} 0$$
$$0 = 0 \checkmark$$

$b = 1$: $4(1)^4 - 5(1)^2 + 1 \overset{?}{=} 0$
$$4 \cdot 1 - 5 \cdot 1 + 1 \overset{?}{=} 0$$
$$4 - 5 + 1 \overset{?}{=} 0$$
$$0 = 0 \checkmark$$

All check; the solution set is $\left\{-1, -\dfrac{1}{2}, \dfrac{1}{2}, 1\right\}$.

**16.** $q^4 + 15 = 8q^2$
$$q^4 - 8q^2 + 15 = 0$$
$$(q^2)^2 - 8(q^2) + 15 = 0$$
Let $u = q^2$.
$$u^2 - 8u + 15 = 0$$
$$(u - 3)(u - 5) = 0$$
$u - 3 = 0$ or $u - 5 = 0$
$u = 3$ or $u = 5$
$q^2 = 3$ or $q^2 = 5$
$q = \pm\sqrt{3}$ or $q = \pm\sqrt{5}$

Check:

$q = -\sqrt{3}$: $(-\sqrt{3})^4 + 15 \overset{?}{=} 8(-\sqrt{3})^2$
$$9 + 15 \overset{?}{=} 8 \cdot 3$$
$$24 = 24 \checkmark$$

$q = \sqrt{3}$: $(\sqrt{3})^4 + 15 \overset{?}{=} 8(\sqrt{3})^2$
$$9 + 15 \overset{?}{=} 8 \cdot 3$$
$$24 = 24 \checkmark$$

$q = -\sqrt{5}$: $(-\sqrt{5})^4 + 15 \overset{?}{=} 8(-\sqrt{5})^2$
$$25 + 15 \overset{?}{=} 8 \cdot 5$$
$$40 = 40 \checkmark$$

## Section 8.3 Solving Equations Quadratic in Form

$q = \sqrt{5}$: $(\sqrt{5})^4 + 15 \stackrel{?}{=} 8(\sqrt{5})^2$
$\qquad\qquad 25 + 15 \stackrel{?}{=} 8 \cdot 5$
$\qquad\qquad\qquad 40 = 40$ ✓

All check; the solution set is $\{-\sqrt{5}, -\sqrt{3}, \sqrt{3}, \sqrt{5}\}$.

**18.** $(x+2)^2 - 3(x+2) - 10 = 0$

Let $u = x + 2$.
$u^2 - 3u - 10 = 0$
$(u-5)(u+2) = 0$
$u - 5 = 0$ or $u + 2 = 0$
$\quad u = 5$ or $\quad u = -2$
$x + 2 = 5$ or $x + 2 = -2$
$\quad x = 3$ or $\quad x = -4$

Check:
$x = 3$: $(3+2)^2 - 3(3+2) - 10 \stackrel{?}{=} 0$
$\qquad\quad 5^2 - 3(5) - 10 \stackrel{?}{=} 0$
$\qquad\quad 25 - 15 - 10 \stackrel{?}{=} 0$
$\qquad\qquad\qquad 0 = 0$ ✓

$x = -4$: $(-4+2)^2 - 3(-4+2) - 10 \stackrel{?}{=} 0$
$\qquad\qquad (-2)^2 - 3(-2) - 10 \stackrel{?}{=} 0$
$\qquad\qquad\quad 4 + 6 - 10 \stackrel{?}{=} 0$
$\qquad\qquad\qquad\qquad 0 = 0$ ✓

Both check; the solution set is $\{-4, 3\}$.

**20.** $(p^2 - 2)^2 - 8(p^2 - 2) + 12 = 0$

Let $u = p^2 - 2$.
$u^2 - 8u + 12 = 0$
$(u-2)(u-6) = 0$
$u - 2 = 0$ or $u - 6 = 0$
$\quad u = 2$ or $\quad u = 6$
$p^2 - 2 = 2$ or $p^2 - 2 = 6$
$\quad p^2 = 4$ or $\quad p^2 = 8$
$p = \pm\sqrt{4}$ or $p = \pm\sqrt{8}$
$p = \pm 2$ or $p = \pm 2\sqrt{2}$

Check:
$p = -2$: $((-2)^2 - 2)^2 - 8((-2)^2 - 2) + 12 \stackrel{?}{=} 0$
$\qquad\qquad (4-2)^2 - 8(4-2) + 12 \stackrel{?}{=} 0$
$\qquad\qquad\quad 2^2 - 8 \cdot 2 + 12 \stackrel{?}{=} 0$
$\qquad\qquad\quad 4 - 16 + 12 \stackrel{?}{=} 0$
$\qquad\qquad\qquad\qquad 0 = 0$ ✓

$p = 2$: $(2^2 - 2)^2 - 8(2^2 - 2) + 12 \stackrel{?}{=} 0$
$\qquad\quad (4-2)^2 - 8(4-2) + 12 \stackrel{?}{=} 0$
$\qquad\quad\quad 2^2 - 8 \cdot 2 + 12 \stackrel{?}{=} 0$
$\qquad\quad\quad 4 - 16 + 12 \stackrel{?}{=} 0$
$\qquad\qquad\qquad\quad 0 = 0$ ✓

$p = -2\sqrt{2}$:
$((-2\sqrt{2})^2 - 2)^2 - 8((-2\sqrt{2})^2 - 2) + 12 \stackrel{?}{=} 0$
$\qquad (8-2)^2 - 8(8-2) + 12 \stackrel{?}{=} 0$
$\qquad\quad 6^2 - 8 \cdot 6 + 12 \stackrel{?}{=} 0$
$\qquad\quad 36 - 48 + 12 \stackrel{?}{=} 0$
$\qquad\qquad\qquad 0 = 0$ ✓

$p = 2\sqrt{2}$:
$((2\sqrt{2})^2 - 2)^2 - 8((2\sqrt{2})^2 - 2) + 12 \stackrel{?}{=} 0$
$\qquad (8-2)^2 - 8(8-2) + 12 \stackrel{?}{=} 0$
$\qquad\quad 6^2 - 8 \cdot 6 + 12 \stackrel{?}{=} 0$
$\qquad\quad 36 - 48 + 12 \stackrel{?}{=} 0$
$\qquad\qquad\qquad 0 = 0$ ✓

All check; the solution set is $\{-2\sqrt{2}, -2, 2, 2\sqrt{2}\}$.

**22.** $(q^2 + 4)^2 + 3(q^2 + 4) - 4 = 0$

Let $u = q^2 + 4$.
$u^2 + 3u - 4 = 0$
$(u+4)(u-1) = 0$
$u + 4 = 0$ or $u - 1 = 0$
$\quad u = -4$ or $\quad u = 1$
$q^2 + 4 = -4$ or $q^2 + 4 = 1$
$\quad q^2 = -8$ or $\quad q^2 = -3$
$q = \pm\sqrt{-8}$ or $q = \pm\sqrt{-3}$
$q = \pm 2\sqrt{2}\,i$ or $q = \pm\sqrt{3}\,i$

## Chapter 8: Quadratic Equations and Functions

Check:
$q = -2\sqrt{2}\ i$:

$$\left(\left(-2\sqrt{2}\ i\right)^2 + 4\right)^2 + 3\left(\left(-2\sqrt{2}\ i\right)^2 + 4\right) - 4 \stackrel{?}{=} 0$$

$$\left(4 \cdot 2 \cdot i^2 + 4\right)^2 + 3\left(4 \cdot 2 \cdot i^2 + 4\right) - 4 \stackrel{?}{=} 0$$

$$\left(8i^2 + 4\right)^2 + 3\left(8i^2 + 4\right) - 4 \stackrel{?}{=} 0$$

$$(-8+4)^2 + 3(-8+4) - 4 \stackrel{?}{=} 0$$

$$(-4)^2 + 3(-4) - 4 \stackrel{?}{=} 0$$

$$16 - 12 - 4 \stackrel{?}{=} 0$$

$$0 = 0 \checkmark$$

$q = 2\sqrt{2}\ i$:

$$\left(\left(2\sqrt{2}\ i\right)^2 + 4\right)^2 + 3\left(\left(2\sqrt{2}\ i\right)^2 + 4\right) - 4 \stackrel{?}{=} 0$$

$$\left(8i^2 + 4\right)^2 + 3\left(8i^2 + 4\right) - 4 \stackrel{?}{=} 0$$

$$(-8+4)^2 + 3(-8+4) - 4 \stackrel{?}{=} 0$$

$$(-4)^2 + 3(-4) - 4 \stackrel{?}{=} 0$$

$$16 - 12 - 4 \stackrel{?}{=} 0$$

$$0 = 0 \checkmark$$

$q = -\sqrt{3}\ i$:

$$\left(\left(-\sqrt{3}\ i\right)^2 + 4\right)^2 + 3\left(\left(-\sqrt{3}\ i\right)^2 + 4\right) - 4 \stackrel{?}{=} 0$$

$$\left(3i^2 + 4\right)^2 + 3\left(3i^2 + 4\right) - 4 \stackrel{?}{=} 0$$

$$(-3+4)^2 + 3(-3+4) - 4 \stackrel{?}{=} 0$$

$$(1)^2 + 3(1) - 4 \stackrel{?}{=} 0$$

$$1 + 3 - 4 \stackrel{?}{=} 0$$

$$0 = 0 \checkmark$$

$q = \sqrt{3}\ i$:

$$\left(\left(\sqrt{3}\ i\right)^2 + 4\right)^2 + 3\left(\left(\sqrt{3}\ i\right)^2 + 4\right) - 4 \stackrel{?}{=} 0$$

$$\left(3i^2 + 4\right)^2 + 3\left(3i^2 + 4\right) - 4 \stackrel{?}{=} 0$$

$$(-3+4)^2 + 3(-3+4) - 4 \stackrel{?}{=} 0$$

$$(1)^2 + 3(1) - 4 \stackrel{?}{=} 0$$

$$1 + 3 - 4 \stackrel{?}{=} 0$$

$$0 = 0 \checkmark$$

All check; the solution set is
$\left\{-2\sqrt{2}\ i,\ 2\sqrt{2}\ i, -\sqrt{3}\ i,\ \sqrt{3}\ i\right\}$.

24. $\qquad x - 5\sqrt{x} - 6 = 0$

$$\left(\sqrt{x}\right)^2 - 5\left(\sqrt{x}\right) - 6 = 0$$

Let $u = \sqrt{x}$.

$$u^2 - 5u - 6 = 0$$
$$(u-6)(u+1) = 0$$
$$u - 6 = 0 \quad \text{or} \quad u + 1 = 0$$
$$u = 6 \quad \text{or} \quad u = -1$$
$$\sqrt{x} = 6 \quad \text{or} \quad \sqrt{x} = -1$$
$$x = 6^2 \quad \text{or} \quad x = (-1)^2$$
$$x = 36 \quad \text{or} \quad x = 1$$

Check:

$x = 36$: $\quad 36 - 5\sqrt{36} - 6 \stackrel{?}{=} 0$

$$36 - 5 \cdot 6 - 6 \stackrel{?}{=} 0$$

$$36 - 30 - 6 \stackrel{?}{=} 0$$

$$0 = 0 \checkmark$$

$x = 1$: $\quad 1 + 5\sqrt{1} - 6 \stackrel{?}{=} 0$

$$1 + 5 \cdot 1 - 6 \stackrel{?}{=} 0$$

$$1 + 5 - 6 \stackrel{?}{=} 0$$

$$-6 \ne 0 \ \text{✗}$$

$x = 1$ does not check; the solution set is $\{36\}$.

26. $\qquad z + 7\sqrt{z} + 6 = 0$

$$\left(\sqrt{z}\right)^2 + 7\left(\sqrt{z}\right) + 6 = 0$$

Let $u = \sqrt{z}$.

$$u^2 + 7u + 6 = 0$$
$$(u+6)(u+1) = 0$$
$$u + 6 = 0 \quad \text{or} \quad u + 1 = 0$$
$$u = -6 \quad \text{or} \quad u = -1$$
$$\sqrt{z} = -6 \quad \text{or} \quad \sqrt{z} = -1$$
$$z = (-6)^2 \quad \text{or} \quad z = (-1)^2$$
$$z = 36 \quad \text{or} \quad z = 1$$

Check:

$z = 36$: $36 + 7\sqrt{36} + 6 \stackrel{?}{=} 0$
$36 + 7 \cdot 6 + 6 \stackrel{?}{=} 0$
$36 + 42 + 6 \stackrel{?}{=} 0$
$84 \neq 0$ ✗

$z = 1$: $1 + 7\sqrt{1} + 6 \stackrel{?}{=} 0$
$1 + 7 \cdot 1 + 6 \stackrel{?}{=} 0$
$1 + 7 + 6 \stackrel{?}{=} 0$
$14 \neq 0$ ✗

Neither possibility checks; the equation has no solution. The solution set is $\{\ \}$ or $\emptyset$.

28. $\qquad 3x = 11\sqrt{x} + 4$
$3x - 11\sqrt{x} - 4 = 0$
$3\left(\sqrt{x}\right)^2 - 11\left(\sqrt{x}\right) - 4 = 0$

Let $u = \sqrt{x}$.
$3u^2 - 11u - 4 = 0$
$(3u + 1)(u - 4) = 0$
$3u + 1 = 0 \quad$ or $\quad u - 4 = 0$
$u = -\dfrac{1}{3} \quad$ or $\quad u = 4$
$\sqrt{x} = -\dfrac{1}{3} \quad$ or $\quad \sqrt{x} = 4$
$x = \left(-\dfrac{1}{3}\right)^2 \quad$ or $\quad x = 4^2$
$x = \dfrac{1}{9} \quad$ or $\quad x = 16$

Check:

$x = \dfrac{1}{9}$: $3 \cdot \dfrac{1}{9} \stackrel{?}{=} 11\sqrt{\dfrac{1}{9}} + 4$
$\dfrac{1}{3} \stackrel{?}{=} 11 \cdot \dfrac{1}{3} + 4$
$\dfrac{1}{3} \stackrel{?}{=} \dfrac{11}{3} + 4$
$\dfrac{1}{3} \neq \dfrac{23}{3}$ ✗

$x = 16$: $3 \cdot 16 \stackrel{?}{=} 11\sqrt{16} + 4$
$48 \stackrel{?}{=} 11 \cdot 4 + 4$
$48 \stackrel{?}{=} 44 + 4$
$48 = 48$ ✓

$x = \dfrac{1}{9}$ does not check; the solution set is $\{16\}$.

30. $\qquad q^{-2} + 2q^{-1} = 15$
$q^{-2} + 2q^{-1} - 15 = 0$
$\left(q^{-1}\right)^2 + 2\left(q^{-1}\right) - 15 = 0$

Let $u = q^{-1}$.
$u^2 + 2u - 15 = 0$
$(u + 5)(u - 3) = 0$
$u = -5 \quad$ or $\quad u = 3$
$q^{-1} = -5 \quad$ or $\quad q^{-1} = 3$
$\dfrac{1}{q} = -5 \quad$ or $\quad \dfrac{1}{q} = 3$
$q = -\dfrac{1}{5} \quad$ or $\quad q = \dfrac{1}{3}$

Check:

$q = -\dfrac{1}{5}$: $\left(-\dfrac{1}{5}\right)^{-2} + 2\left(-\dfrac{1}{5}\right)^{-1} \stackrel{?}{=} 15$
$(-5)^2 + 2(-5) \stackrel{?}{=} 15$
$25 - 10 \stackrel{?}{=} 15$
$15 = 15$ ✓

$q = \dfrac{1}{3}$: $\left(\dfrac{1}{3}\right)^{-2} + 2\left(\dfrac{1}{3}\right)^{-1} \stackrel{?}{=} 15$
$(3)^2 + 2(3) \stackrel{?}{=} 15$
$9 + 6 \stackrel{?}{=} 15$
$15 = 15$ ✓

Both check; the solution set is $\left\{-\dfrac{1}{5}, \dfrac{1}{3}\right\}$

32. $\qquad 10a^{-2} + 23a^{-1} = 5$
$10a^{-2} + 23a^{-1} - 5 = 0$
$10\left(a^{-1}\right)^2 + 23\left(a^{-1}\right) - 5 = 0$

Let $u = a^{-1}$.
$10u^2 + 23u - 5 = 0$
$(5u - 1)(2u + 5) = 0$
$5u - 1 = 0 \quad$ or $\quad 2u + 5 = 0$
$5u = 1 \quad$ or $\quad 2u = -5$
$u = \dfrac{1}{5} \quad$ or $\quad u = -\dfrac{5}{2}$

## Chapter 8: Quadratic Equations and Functions

$a^{-1} = \dfrac{1}{5}$ or $a^{-1} = -\dfrac{5}{2}$

$\dfrac{1}{a} = \dfrac{1}{5}$ or $\dfrac{1}{a} = -\dfrac{5}{2}$

$a = 5$ or $a = -\dfrac{2}{5}$

Check:

$a = 5: \ 10(5)^{-2} + 23(5)^{-1} \stackrel{?}{=} 5$

$10\left(\dfrac{1}{5}\right)^2 + 23\left(\dfrac{1}{5}\right) \stackrel{?}{=} 5$

$10\left(\dfrac{1}{25}\right) + 23\left(\dfrac{1}{5}\right) \stackrel{?}{=} 5$

$\dfrac{2}{5} + \dfrac{23}{5} \stackrel{?}{=} 5$

$\dfrac{25}{5} \stackrel{?}{=} 5$

$5 = 5 \ \checkmark$

$a = -\dfrac{2}{5}: \ 10\left(-\dfrac{2}{5}\right)^{-2} + 23\left(-\dfrac{2}{5}\right)^{-1} \stackrel{?}{=} 5$

$10\left(-\dfrac{5}{2}\right)^2 + 23\left(-\dfrac{5}{2}\right) \stackrel{?}{=} 5$

$10\left(\dfrac{25}{4}\right) + 23\left(-\dfrac{5}{2}\right) \stackrel{?}{=} 5$

$\dfrac{125}{2} - \dfrac{115}{2} \stackrel{?}{=} 5$

$\dfrac{10}{2} \stackrel{?}{=} 5$

$5 = 5 \ \checkmark$

Both check; the solution set is $\left\{-\dfrac{2}{5},\ 5\right\}$.

**34.** $y^{2/3} - 2y^{1/3} - 3 = 0$

$\left(y^{1/3}\right)^2 - 2\left(y^{1/3}\right) - 3 = 0$

Let $u = y^{1/3}$.

$u^2 - 2u - 3 = 0$

$(u - 3)(u + 1) = 0$

$u - 3 = 0$ or $u + 1 = 0$
$u = 3$ or $u = -1$
$y^{1/3} = 3$ or $y^{1/3} = -1$
$\left(y^{1/3}\right)^3 = 3^3$ or $\left(y^{1/3}\right)^3 = (-1)^3$
$y = 27$ or $y = -1$

Check:

$y = 27: \ (27)^{2/3} - 2 \cdot (27)^{1/3} - 3 \stackrel{?}{=} 0$

$\left(\sqrt[3]{27}\right)^2 - 2\left(\sqrt[3]{27}\right) - 3 \stackrel{?}{=} 0$

$(3)^2 - 2(3) - 3 \stackrel{?}{=} 0$

$9 - 6 - 3 \stackrel{?}{=} 0$

$0 = 0 \ \checkmark$

$y = -1: \ (-1)^{2/3} - 2 \cdot (-1)^{1/3} - 3 \stackrel{?}{=} 0$

$\left(\sqrt[3]{-1}\right)^2 - 2\left(\sqrt[3]{-1}\right) - 3 \stackrel{?}{=} 0$

$(-1)^2 - 2(-1) - 3 \stackrel{?}{=} 0$

$1 + 2 - 3 \stackrel{?}{=} 0$

$0 = 0 \ \checkmark$

Both check; the solution set is $\{-1,\ 27\}$.

**36.** $w^{2/3} + 2w^{1/3} = 3$

$\left(w^{1/3}\right)^2 + 2\left(w^{1/3}\right) - 3 = 0$

Let $u = w^{1/3}$.

$u^2 + 2u - 3 = 0$

$(u - 1)(u + 3) = 0$

$u - 1 = 0$ or $u + 3 = 0$
$u = 1$ or $u = -3$
$w^{1/3} = 1$ or $w^{1/3} = -3$
$\left(w^{1/3}\right)^3 = 1^3$ or $\left(w^{1/3}\right)^3 = (-3)^3$
$w = 1$ or $w = -27$

Check:

$w = 1: \ (1)^{2/3} + 2(1)^{1/3} \stackrel{?}{=} 3$

$\left(\sqrt[3]{1}\right)^2 + 2\left(\sqrt[3]{1}\right) \stackrel{?}{=} 3$

$(1)^2 + 2 \cdot 1 \stackrel{?}{=} 3$

$1 + 2 \stackrel{?}{=} 3$

$3 = 3 \ \checkmark$

$w = -27: \ (-27)^{2/3} + 2(-27)^{1/3} \stackrel{?}{=} 3$

$\left(\sqrt[3]{-27}\right)^2 + 2\left(\sqrt[3]{-27}\right) \stackrel{?}{=} 3$

$(-3)^2 + 2(-3) \stackrel{?}{=} 3$

$9 - 6 \stackrel{?}{=} 3$

$3 = 3 \ \checkmark$

Both check; the solution set is $\{-27, 1\}$.

**38.** $\quad b + 3b^{1/2} = 28$

$\left(b^{1/2}\right)^2 + 3\left(b^{1/2}\right) - 28 = 0$

Let $u = b^{1/2}$.

$u^2 + 3u - 28 = 0$

$(u-4)(u+7) = 0$

$\begin{aligned} u - 4 &= 0 &\text{or}& & u + 7 &= 0 \\ u &= 4 &\text{or}& & u &= -7 \\ b^{1/2} &= 4 &\text{or}& & b^{1/2} &= -7 \\ \left(b^{1/2}\right)^2 &= 4^2 &\text{or}& & \left(b^{1/2}\right)^2 &= (-7)^2 \\ b &= 16 &\text{or}& & b &= 49 \end{aligned}$

Check:

$b = 16: \quad 16 + 3(16)^{1/2} \stackrel{?}{=} 28$
$\qquad\qquad 16 + 3\sqrt{16} \stackrel{?}{=} 28$
$\qquad\qquad 16 + 3(4) \stackrel{?}{=} 28$
$\qquad\qquad 16 + 12 \stackrel{?}{=} 28$
$\qquad\qquad 28 = 28 \checkmark$

$b = 49: \quad 49 + 3(49)^{1/2} \stackrel{?}{=} 28$
$\qquad\qquad 49 + 3\sqrt{49} \stackrel{?}{=} 28$
$\qquad\qquad 49 + 3(7) \stackrel{?}{=} 28$
$\qquad\qquad 49 + 21 \stackrel{?}{=} 28$
$\qquad\qquad 70 \neq 28 \text{ ✗}$

$b = 49$ does not check; the solution set is $\{16\}$.

**40.** $\quad \dfrac{1}{x^2} - \dfrac{7}{x} + 12 = 0$

$\left(\dfrac{1}{x}\right)^2 - 7\left(\dfrac{1}{x}\right) + 12 = 0$

Let $u = \dfrac{1}{x}$.

$u^2 - 7u + 12 = 0$

$(u-4)(u-3) = 0$

$\begin{aligned} u - 4 &= 0 &\text{or}& & u - 3 &= 0 \\ u &= 4 &\text{or}& & u &= 3 \\ \dfrac{1}{x} &= 4 &\text{or}& & \dfrac{1}{x} &= 3 \\ x &= \dfrac{1}{4} &\text{or}& & x &= \dfrac{1}{3} \end{aligned}$

Check:

$x = \dfrac{1}{4}: \quad \dfrac{1}{\left(\frac{1}{4}\right)^2} - \dfrac{7}{\frac{1}{4}} + 12 \stackrel{?}{=} 0$
$\qquad\qquad \dfrac{1}{\frac{1}{16}} - \dfrac{7}{\frac{1}{4}} + 12 \stackrel{?}{=} 0$
$\qquad\qquad 16 - 28 + 12 \stackrel{?}{=} 0$
$\qquad\qquad 0 = 0 \checkmark$

$x = \dfrac{1}{3}: \quad \dfrac{1}{\left(\frac{1}{3}\right)^2} - \dfrac{7}{\frac{1}{3}} + 12 \stackrel{?}{=} 0$
$\qquad\qquad \dfrac{1}{\frac{1}{9}} - \dfrac{7}{\frac{1}{3}} + 12 \stackrel{?}{=} 0$
$\qquad\qquad 9 - 21 + 12 \stackrel{?}{=} 0$
$\qquad\qquad 0 = 0 \checkmark$

Both check; the solution set is $\left\{\dfrac{1}{4}, \dfrac{1}{3}\right\}$.

**42.** $\quad \left(\dfrac{1}{x+2}\right)^2 + \dfrac{6}{x+2} = 7$

$\left(\dfrac{1}{x+2}\right)^2 + 6\left(\dfrac{1}{x+2}\right) - 7 = 0$

Let $u = \dfrac{1}{x+2}$.

$u^2 + 6u - 7 = 0$

$(u-1)(u+7) = 0$

$\begin{aligned} u - 1 &= 0 &\text{or}& & u + 7 &= 0 \\ u &= 1 &\text{or}& & u &= -7 \\ \dfrac{1}{x+2} &= 1 &\text{or}& & \dfrac{1}{x+2} &= -7 \\ x + 2 &= 1 &\text{or}& & x + 2 &= -\dfrac{1}{7} \\ x &= -1 &\text{or}& & x &= -\dfrac{15}{7} \end{aligned}$

Check:

$x = -1: \quad \left(\dfrac{1}{-1+2}\right)^2 + \dfrac{6}{-1+2} \stackrel{?}{=} 7$
$\qquad\qquad \left(\dfrac{1}{1}\right)^2 + \dfrac{6}{1} \stackrel{?}{=} 7$
$\qquad\qquad 1 + 6 \stackrel{?}{=} 7$
$\qquad\qquad 7 = 7 \checkmark$

## Chapter 8: Quadratic Equations and Functions

$x = -\dfrac{15}{7}: \left(\dfrac{1}{-\frac{15}{7}+2}\right)^2 + \dfrac{6}{-\frac{15}{7}+2} \stackrel{?}{=} 7$

$\left(\dfrac{1}{-\frac{1}{7}}\right)^2 + \dfrac{6}{-\frac{1}{7}} \stackrel{?}{=} 7$

$(-7)^2 + 6(-7) \stackrel{?}{=} 7$

$49 - 42 \stackrel{?}{=} 7$

$7 = 7$ ✓

Both check; the solution set is $\left\{-\dfrac{15}{7}, -1\right\}$.

**44.** $y^6 - 7y^3 - 8 = 0$

$(y^3)^2 - 7(y^3) - 8 = 0$

Let $u = y^3$.

$u^2 - 7u - 8 = 0$

$(u-8)(u+1) = 0$

$u - 8 = 0 \quad \text{or} \quad u + 1 = 0$

$y^3 - 8 = 0 \quad \text{or} \quad y^3 + 1 = 0$

First consider $y^3 - 8 = 0$. Note that the expression on the left is a difference of cubes.

$y^3 - 8 = 0$

$(y-2)(y^2 + 2y + 4) = 0$

$y - 2 = 0 \quad \text{or} \quad y^2 + 2y + 4 = 0$

$y = 2 \quad \text{or} \quad y = \dfrac{-2 \pm \sqrt{2^2 - 4(1)(4)}}{2(1)}$

$= \dfrac{-2 \pm \sqrt{-12}}{2}$

$= \dfrac{-2 \pm 2\sqrt{3}\,i}{2}$

$= -1 \pm \sqrt{3}\,i$

Now consider $y^3 + 1 = 0$. Note that the expression on the left is a difference of cubes.

$y^3 + 1 = 0$

$(y+1)(y^2 - y + 1) = 0$

$y + 1 = 0 \quad \text{or} \quad y^2 - y + 1 = 0$

$y = -1 \quad \text{or} \quad y = \dfrac{-(-1) \pm \sqrt{(-1)^2 - 4(1)(1)}}{2(1)}$

$= \dfrac{1 \pm \sqrt{-3}}{2}$

$= \dfrac{1 \pm \sqrt{3}\,i}{2}$

$= \dfrac{1}{2} \pm \dfrac{\sqrt{3}}{2}i$

All will check; the solution set is

$\left\{-1,\, 2,\, -1 - \sqrt{3}i,\, -1 + \sqrt{3}i,\, \dfrac{1}{2} - \dfrac{\sqrt{3}}{2}i,\, \dfrac{1}{2} + \dfrac{\sqrt{3}}{2}i\right\}$.

**46.** $6b^{-2} - b^{-1} = 1$

$6b^{-2} - b^{-1} - 1 = 0$

$6(b^{-1})^2 - (b^{-1}) - 1 = 0$

Let $u = b^{-1}$.

$6u^2 - u - 1 = 0$

$(3u+1)(2u-1) = 0$

$3u + 1 = 0 \quad \text{or} \quad 2u - 1 = 0$

$3u = -1 \quad \text{or} \quad 2u = 1$

$u = -\dfrac{1}{3} \quad \text{or} \quad u = \dfrac{1}{2}$

$b^{-1} = -\dfrac{1}{3} \quad \text{or} \quad b^{-1} = \dfrac{1}{2}$

$\dfrac{1}{b} = -\dfrac{1}{3} \quad \text{or} \quad \dfrac{1}{b} = \dfrac{1}{2}$

$b = -3 \quad \text{or} \quad b = 2$

Check:

$b = -3:\ 6(-3)^{-2} - (-3)^{-1} \stackrel{?}{=} 1$

$6\left(-\dfrac{1}{3}\right)^2 - \left(-\dfrac{1}{3}\right) \stackrel{?}{=} 1$

$6\left(\dfrac{1}{9}\right) + \dfrac{1}{3} \stackrel{?}{=} 1$

$\dfrac{2}{3} + \dfrac{1}{3} \stackrel{?}{=} 1$

$1 = 1$ ✓

### Section 8.3 Solving Equations Quadratic in Form

$b = 2:$ $6(2)^{-2} - (2)^{-1} \stackrel{?}{=} 1$

$\qquad 6\left(\dfrac{1}{2}\right)^2 - \dfrac{1}{2} \stackrel{?}{=} 1$

$\qquad 6\left(\dfrac{1}{4}\right) - \dfrac{1}{2} \stackrel{?}{=} 1$

$\qquad \dfrac{3}{2} - \dfrac{1}{2} \stackrel{?}{=} 1$

$\qquad 1 = 1$ ✓

Both check; the solution set is $\{-3, 2\}$.

**48.** $\quad x^4 + 3x^2 = 4$

$\quad x^4 + 3x^2 - 4 = 0$

$\quad (x^2)^2 + 3(x^2) - 4 = 0$

Let $u = x^2$.

$u^2 + 3u - 4 = 0$

$(u-1)(u+4) = 0$

$u - 1 = 0 \quad$ or $\quad u + 4 = 0$

$u = 1 \quad$ or $\quad u = -4$

$x^2 = 1 \quad$ or $\quad x^2 = -4$

$x = \pm\sqrt{1} \quad$ or $\quad x = \pm\sqrt{-4}$

$x = \pm 1 \quad$ or $\quad x = \pm 2i$

Check:

$x = -1:$ $(-1)^4 + 3(-1)^2 \stackrel{?}{=} 4$

$\qquad 1 + 3(1) \stackrel{?}{=} 4$

$\qquad 1 + 3 \stackrel{?}{=} 4$

$\qquad 4 = 4$ ✓

$x = 1:$ $(1)^4 + 3(1)^2 \stackrel{?}{=} 4$

$\qquad 1 + 3(1) \stackrel{?}{=} 4$

$\qquad 1 + 3 \stackrel{?}{=} 4$

$\qquad 4 = 4$ ✓

$x = -2i:$ $(-2i)^4 + 3(-2i)^2 \stackrel{?}{=} 4$

$\qquad 16i^4 + 3 \cdot 4i^2 \stackrel{?}{=} 4$

$\qquad 16(1) + 3(-4) \stackrel{?}{=} 4$

$\qquad 16 - 12 \stackrel{?}{=} 4$

$\qquad 4 = 4$ ✓

$x = 2i:$ $(2i)^4 + 3(2i)^2 \stackrel{?}{=} 4$

$\qquad 16i^4 + 3 \cdot 4i^2 \stackrel{?}{=} 4$

$\qquad 16(1) + 3(-4) \stackrel{?}{=} 4$

$\qquad 16 - 12 \stackrel{?}{=} 4$

$\qquad 4 = 4$ ✓

All check; the solution set is $\{-1, 1, -2i, 2i\}$.

**50.** $\quad c^{1/2} + c^{1/4} - 12 = 0$

$\quad (c^{1/4})^2 + (c^{1/4}) - 12 = 0$

Let $u = c^{1/4}$.

$u^2 + u - 12 = 0$

$(u-3)(u+4) = 0$

$u - 3 = 0 \quad$ or $\quad u + 4 = 0$

$u = 3 \quad$ or $\quad u = -4$

$c^{1/4} = 3 \quad$ or $\quad c^{1/4} = -4$

$(c^{1/4})^4 = 3^4 \quad$ or $\quad (c^{1/4})^4 = (-4)^4$

$c = 81 \quad$ or $\quad c = 256$

Check:

$c = 81:$ $(81)^{1/2} + (81)^{1/4} - 12 \stackrel{?}{=} 0$

$\qquad \sqrt{81} + \sqrt[4]{81} - 12 \stackrel{?}{=} 0$

$\qquad 9 + 3 - 12 \stackrel{?}{=} 0$

$\qquad 0 = 0$ ✓

$c = 256:$ $(256)^{1/2} + (256)^{1/4} - 12 \stackrel{?}{=} 0$

$\qquad \sqrt{256} + \sqrt[4]{256} - 12 \stackrel{?}{=} 0$

$\qquad 16 + 4 - 12 \stackrel{?}{=} 0$

$\qquad 8 \neq 0$ ✗

$c = 256$ does not check; the solution set is $\{81\}$.

**52.** $\quad p^4 - 15p^2 - 16 = 0$

$\quad (p^2)^2 - 15(p^2) - 16 = 0$

Let $u = p^2$.

$u^2 - 15u - 16 = 0$

$(u-16)(u+1) = 0$

$u - 16 = 0 \quad$ or $\quad u + 1 = 0$

$u = 16 \quad$ or $\quad u = -1$

$p^2 = 16 \quad$ or $\quad p^2 = -1$

$p = \pm\sqrt{16} \quad$ or $\quad p = \pm\sqrt{-1}$

$p = \pm 4 \quad$ or $\quad p = \pm i$

Check:
$p = -4:\ (-4)^4 - 15(-4)^2 - 16 \stackrel{?}{=} 0$
$256 - 15 \cdot 16 - 16 \stackrel{?}{=} 0$
$256 - 240 - 16 \stackrel{?}{=} 0$
$0 = 0\ \checkmark$

$p = 4:\ (4)^4 - 15(4)^2 - 16 \stackrel{?}{=} 0$
$256 - 15 \cdot 16 - 16 \stackrel{?}{=} 0$
$256 - 240 - 16 \stackrel{?}{=} 0$
$0 = 0\ \checkmark$

$p = -i:\ (-i)^4 - 15(-i)^2 - 16 \stackrel{?}{=} 0$
$i^4 - 15i^2 - 16 \stackrel{?}{=} 0$
$1 - 15(-1) - 16 \stackrel{?}{=} 0$
$1 + 15 - 16 \stackrel{?}{=} 0$
$0 = 0\ \checkmark$

$p = i:\ (i)^4 - 15(i)^2 - 16 \stackrel{?}{=} 0$
$1 - 15(-1) - 16 \stackrel{?}{=} 0$
$1 + 15 - 16 \stackrel{?}{=} 0$
$0 = 0\ \checkmark$

All check; the solution set is $\{-4, 4, -i, i\}$.

**54.** $\left(\dfrac{1}{x-1}\right)^2 + \dfrac{7}{x-1} = 8$

$\left(\dfrac{1}{x-1}\right)^2 + 7\left(\dfrac{1}{x-1}\right) - 8 = 0$

Let $u = \dfrac{1}{x-1}$.

$u^2 + 7u - 8 = 0$
$(u - 1)(u + 8) = 0$
$u - 1 = 0$ or $u + 8 = 0$
$u = 1$ or $u = -8$
$\dfrac{1}{x-1} = 1$ or $\dfrac{1}{x-1} = -8$
$x - 1 = 1$ or $x - 1 = -\dfrac{1}{8}$
$x = 2$ or $x = \dfrac{7}{8}$

Check:
$x = 2:\ \left(\dfrac{1}{2-1}\right)^2 + \dfrac{7}{2-1} \stackrel{?}{=} 8$
$\left(\dfrac{1}{1}\right)^2 + \dfrac{7}{1} \stackrel{?}{=} 8$
$1 + 7 \stackrel{?}{=} 8$
$8 = 8\ \checkmark$

$x = \dfrac{7}{8}:\ \left(\dfrac{1}{\frac{7}{8} - 1}\right)^2 + \dfrac{7}{\frac{7}{8} - 1} \stackrel{?}{=} 8$
$\left(\dfrac{1}{-\frac{1}{8}}\right)^2 + \dfrac{7}{-\frac{1}{8}} \stackrel{?}{=} 8$
$(-8)^2 + 7(-8) \stackrel{?}{=} 8$
$64 - 56 \stackrel{?}{=} 8$
$8 = 8\ \checkmark$

Both check; the solution set is $\left\{\dfrac{7}{8},\ 2\right\}$.

**56.** $x - 8\sqrt{x} + 12 = 0$
$(\sqrt{x})^2 - 8(\sqrt{x}) + 12 = 0$

Let $u = \sqrt{x}$.
$u^2 - 8u + 12 = 0$
$(u - 2)(u - 6) = 0$
$u - 2 = 0$ or $u - 6 = 0$
$u = 2$ or $u = 6$
$\sqrt{x} = 2$ or $\sqrt{x} = 6$
$x = 2^2$ or $x = 6^2$
$x = 4$ or $x = 36$

Check:
$x = 4:\ 4 - 8\sqrt{4} + 12 \stackrel{?}{=} 0$
$4 - 8 \cdot 2 + 12 \stackrel{?}{=} 0$
$4 - 16 + 12 \stackrel{?}{=} 0$
$0 = 0\ \checkmark$

$x = 36:\ 36 - 8\sqrt{36} + 12 \stackrel{?}{=} 0$
$36 - 8 \cdot 6 + 12 \stackrel{?}{=} 0$
$36 - 48 + 12 \stackrel{?}{=} 0$
$0 = 0\ \checkmark$

Both check; the solution set is $\{4,\ 36\}$.

**58.** $3(y-2)^2 - 4(y-2) = 4$
$3(y-2)^2 - 4(y-2) - 4 = 0$
Let $u = y-2$.
$$3u^2 - 4u - 4 = 0$$
$$(3u+2)(u-2) = 0$$
$3u + 2 = 0 \quad \text{or} \quad u - 2 = 0$
$u = -\dfrac{2}{3} \quad \text{or} \quad u = 2$
$y - 2 = -\dfrac{2}{3} \quad \text{or} \quad y - 2 = 2$
$y = \dfrac{4}{3} \quad \text{or} \quad y = 4$

Check:
$y = \dfrac{4}{3}: \quad 3\left(\dfrac{4}{3} - 2\right)^2 - 4\left(\dfrac{4}{3} - 2\right) \overset{?}{=} 4$
$3\left(-\dfrac{2}{3}\right)^2 - 4\left(-\dfrac{2}{3}\right) \overset{?}{=} 4$
$\dfrac{4}{3} + \dfrac{8}{3} \overset{?}{=} 4$
$4 = 4 \checkmark$

$y = 4: \quad 3(4-2)^2 - 4(4-2) \overset{?}{=} 4$
$3(2)^2 - 4(2) \overset{?}{=} 4$
$12 - 8 \overset{?}{=} 4$
$4 = 4 \checkmark$

Both check; the solution set is $\left\{\dfrac{4}{3}, 4\right\}$.

**60. a.** $f(x) = 3$
$x^4 + 5x^2 + 3 = 3$
$(x^2)^2 + 5(x^2) = 0$
Let $u = x^2$.
$u^2 + 5u = 0$
$u(u+5) = 0$
$u + 5 = 0 \quad \text{or} \quad u = 0$
$u = -5$
$x^2 = -5 \quad \text{or} \quad x^2 = 0$
$x = \pm\sqrt{-5} \quad \text{or} \quad x = \pm\sqrt{0}$
$x = \pm\sqrt{5}\, i \quad \text{or} \quad x = 0$

Check:
$f(-\sqrt{5}\, i) = (-\sqrt{5}\, i)^4 + 5(-\sqrt{5}\, i)^2 + 3$
$= 25i^4 + 5 \cdot 5i^2 + 3$
$= 25(1) + 5 \cdot 5(-1) + 3$
$= 25 - 25 + 3$
$= 3 \checkmark$

$f(\sqrt{5}\, i) = (\sqrt{5}\, i)^4 + 5(\sqrt{5}\, i)^2 + 3$
$= 25i^4 + 5 \cdot 5i^2 + 3$
$= 25(1) + 5 \cdot 5(-1) + 3$
$= 25 - 25 + 3$
$= 3 \checkmark$

$f(0) = 0^4 + 5 \cdot 0^2 + 3$
$= 3 \checkmark$

All check; the values that make $f(x) = 3$ are $\{-\sqrt{5}\, i,\ \sqrt{5}\, i,\ 0\}$.

**b.** $f(x) = 17$
$x^4 + 5x^2 + 3 = 17$
$(x^2)^2 + 5(x^2) - 14 = 0$
Let $u = x^2$.
$u^2 + 5u - 14 = 0$
$(u+7)(u-2) = 0$
$u + 7 = 0 \quad \text{or} \quad u - 2 = 0$
$u = -7 \quad \text{or} \quad u = 2$
$x^2 = -7 \quad \text{or} \quad x^2 = 2$
$x = \pm\sqrt{-7} \quad \text{or} \quad x = \pm\sqrt{2}$
$x = \pm\sqrt{7}\, i$

Check:
$f(-\sqrt{7}\, i) = (-\sqrt{7}\, i)^4 + 5(-\sqrt{7}\, i)^2 + 3$
$= 49i^4 + 5 \cdot 7i^2 + 3$
$= 49(1) + 5 \cdot 7(-1) + 3$
$= 49 - 35 + 3$
$= 17 \checkmark$

$f(\sqrt{7}\, i) = (\sqrt{7}\, i)^4 + 5(\sqrt{7}\, i)^2 + 3$
$= 49i^4 + 5 \cdot 7i^2 + 3$
$= 49(1) + 5 \cdot 7(-1) + 3$
$= 49 - 35 + 3$
$= 17 \checkmark$

## Chapter 8: Quadratic Equations and Functions

$f(-\sqrt{2}) = (-\sqrt{2})^4 + 5(-\sqrt{2})^2 + 3$
$\phantom{f(-\sqrt{2})} = 4 + 5 \cdot 2 + 3$
$\phantom{f(-\sqrt{2})} = 4 + 10 + 3$
$\phantom{f(-\sqrt{2})} = 17 \checkmark$

$f(\sqrt{2}) = (\sqrt{2})^4 + 5(\sqrt{2})^2 + 3$
$\phantom{f(\sqrt{2})} = 4 + 5 \cdot 2 + 3$
$\phantom{f(\sqrt{2})} = 4 + 10 + 3$
$\phantom{f(\sqrt{2})} = 17 \checkmark$

All check; the values that make $f(x) = 17$ are $\{-\sqrt{7}\,i,\ \sqrt{7}\,i, -\sqrt{2},\ \sqrt{2}\}$.

**62. a.**
$h(x) = -8$
$3x^4 - 9x^2 - 8 = -8$
$3x^4 - 9x^2 = 0$
$3(x^2)^2 - 9(x^2) = 0$

Let $u = x^2$.
$3u^2 - 9u = 0$
$3u(u - 3) = 0$
$3u = 0$ or $u - 3 = 0$
$u = 0$ or $u = 3$
$x^2 = 0$ or $x^2 = 3$
$x = \pm\sqrt{0}$ or $x = \pm\sqrt{3}$
$x = 0$

Check:
$h(0) = 3 \cdot 0^4 - 9 \cdot 0^2 - 8$
$\phantom{h(0)} = -8 \checkmark$

$h(-\sqrt{3}) = 3(-\sqrt{3})^4 - 9(-\sqrt{3})^2 - 8$
$\phantom{h(-\sqrt{3})} = 3 \cdot 9 - 9 \cdot 3 - 8$
$\phantom{h(-\sqrt{3})} = 27 - 27 - 8$
$\phantom{h(-\sqrt{3})} = -8 \checkmark$

$h(\sqrt{3}) = 3(\sqrt{3})^4 - 9(\sqrt{3})^2 - 8$
$\phantom{h(\sqrt{3})} = 3 \cdot 9 - 9 \cdot 3 - 8$
$\phantom{h(\sqrt{3})} = 27 - 27 - 8$
$\phantom{h(\sqrt{3})} = -8 \checkmark$

All check; the values that make $h(x) = -8$ are $\{-\sqrt{3},\ 0,\ \sqrt{3}\}$.

**b.**
$h(x) = 22$
$3x^4 - 9x^2 - 8 = 22$
$3x^4 - 9x^2 - 30 = 0$
$\dfrac{3x^4 - 9x^2 - 30}{3} = \dfrac{0}{3}$
$x^4 - 3x^2 - 10 = 0$
$(x^2)^2 - 3(x^2) - 10 = 0$

Let $u = x^2$.
$u^2 - 3u - 10 = 0$
$(u + 2)(u - 5) = 0$
$u + 2 = 0$ or $u - 5 = 0$
$u = -2$ or $u = 5$
$x^2 = -2$ or $x^2 = 5$
$x = \pm\sqrt{-2}$ or $x = \pm\sqrt{5}$
$x = \pm\sqrt{2}\,i$

Check:
$h(-\sqrt{2}\,i) = 3(-\sqrt{2}\,i)^4 - 9(-\sqrt{2}\,i)^2 - 8$
$\phantom{h(-\sqrt{2}\,i)} = 3 \cdot 4i^4 - 9 \cdot 2i^2 - 8$
$\phantom{h(-\sqrt{2}\,i)} = 3 \cdot 4(1) - 9 \cdot 2(-1) - 8$
$\phantom{h(-\sqrt{2}\,i)} = 12 + 18 - 8$
$\phantom{h(-\sqrt{2}\,i)} = 22 \checkmark$

$h(\sqrt{2}\,i) = 3(\sqrt{2}\,i)^4 - 9(\sqrt{2}\,i)^2 - 8$
$\phantom{h(\sqrt{2}\,i)} = 3 \cdot 4i^4 - 9 \cdot 2i^2 - 8$
$\phantom{h(\sqrt{2}\,i)} = 3 \cdot 4(1) - 9 \cdot 2(-1) - 8$
$\phantom{h(\sqrt{2}\,i)} = 12 + 18 - 8$
$\phantom{h(\sqrt{2}\,i)} = 22 \checkmark$

$h(-\sqrt{5}) = 3(-\sqrt{5})^4 - 9(-\sqrt{5})^2 - 8$
$\phantom{h(-\sqrt{5})} = 3 \cdot 25 - 9 \cdot 5 - 8$
$\phantom{h(-\sqrt{5})} = 75 - 45 - 8$
$\phantom{h(-\sqrt{5})} = 22 \checkmark$

$h(\sqrt{5}) = 3(\sqrt{5})^4 - 9(\sqrt{5})^2 - 8$
$\phantom{h(\sqrt{5})} = 3 \cdot 25 - 9 \cdot 5 - 8$
$\phantom{h(\sqrt{5})} = 75 - 45 - 8$
$\phantom{h(\sqrt{5})} = 22 \checkmark$

All check; the values that make $h(x) = 22$ are $\{-\sqrt{2}\,i,\ \sqrt{2}\,i,\ -\sqrt{5},\ \sqrt{5}\}$.

Section 8.3 Solving Equations Quadratic in Form

**64. a.**
$$f(x) = 4$$
$$x^{-2} - 3x^{-1} = 4$$
$$(x^{-1})^2 - 3(x^{-1}) - 4 = 0$$
Let $u = x^{-1}$.
$$u^2 - 3u - 4 = 0$$
$$(u-4)(u+1) = 0$$
$u - 4 = 0$ or $u + 1 = 0$
$u = 4$ or $u = -1$
$x^{-1} = 4$ or $x^{-1} = -1$
$\dfrac{1}{x} = 4$ or $\dfrac{1}{x} = -1$
$x = \dfrac{1}{4}$ or $x = -1$

Check:
$$f\left(\dfrac{1}{4}\right) = \left(\dfrac{1}{4}\right)^{-2} - 3\left(\dfrac{1}{4}\right)^{-1}$$
$$= 4^2 - 3 \cdot 4^1$$
$$= 16 - 12$$
$$= 4 \checkmark$$
$$f(-1) = (-1)^{-2} - 3(-1)^{-1}$$
$$= (-1)^2 - 3(-1)^1$$
$$= 1 + 3$$
$$= 4 \checkmark$$
Both check; the values that make $f(x) = 4$ are $\left\{-1, \dfrac{1}{4}\right\}$.

**b.**
$$f(x) = 18$$
$$x^{-2} - 3x^{-1} = 18$$
$$(x^{-1})^2 - 3(x^{-1}) - 18 = 0$$
Let $u = x^{-1}$.
$$u^2 - 3u - 18 = 0$$
$$(u-6)(u+3) = 0$$
$u - 6 = 0$ or $u + 3 = 0$
$u = 6$ or $u = -3$
$x^{-1} = 6$ or $x^{-1} = -3$
$\dfrac{1}{x} = 6$ or $\dfrac{1}{x} = -3$
$x = \dfrac{1}{6}$ or $x = -\dfrac{1}{3}$

Check:
$$f\left(\dfrac{1}{6}\right) = \left(\dfrac{1}{6}\right)^{-2} - 3\left(\dfrac{1}{6}\right)^{-1}$$
$$= 6^2 - 3 \cdot 6^1$$
$$= 36 - 18$$
$$= 18 \checkmark$$
$$f\left(-\dfrac{1}{3}\right) = \left(-\dfrac{1}{3}\right)^{-2} - 3\left(-\dfrac{1}{3}\right)^{-1}$$
$$= (-3)^2 - 3 \cdot (-3)^1$$
$$= 9 + 9$$
$$= 18 \checkmark$$
Both check; the values that make $f(x) = 18$ are $\left\{-\dfrac{1}{3}, \dfrac{1}{6}\right\}$.

**66.**
$$x^4 - 13x^2 + 42 = 0$$
$$(x^2)^2 - 13(x^2) + 42 = 0$$
Let $u = x^2$.
$$u^2 - 13u + 42 = 0$$
$$(u-6)(u-7) = 0$$
$u - 6 = 0$ or $u - 7 = 0$
$u = 6$ or $u = 7$
$x^2 = 6$ or $x^2 = 7$
$x = \pm\sqrt{6}$ or $x = \pm\sqrt{7}$

Check:
$$f(-\sqrt{6}) = (-\sqrt{6})^4 - 13(-\sqrt{6})^2 + 42$$
$$= 36 - 13 \cdot 6 + 42$$
$$= 36 - 78 + 42$$
$$= 0 \checkmark$$
$$f(\sqrt{6}) = (\sqrt{6})^4 - 13(\sqrt{6})^2 + 42$$
$$= 36 - 13 \cdot 6 + 42$$
$$= 36 - 78 + 42$$
$$= 0 \checkmark$$
$$f(-\sqrt{7}) = (-\sqrt{7})^4 - 13(-\sqrt{7})^2 + 42$$
$$= 49 - 13 \cdot 7 + 42$$
$$= 49 - 91 + 42$$
$$= 0 \checkmark$$
$$f(\sqrt{7}) = (\sqrt{7})^4 - 13(\sqrt{7})^2 + 42$$
$$= 49 - 13 \cdot 7 + 42$$
$$= 49 - 91 + 42$$
$$= 0 \checkmark$$

All check; the zeros of $f$ are $\{-\sqrt{7}, -\sqrt{6}, \sqrt{6}, \sqrt{7}\}$.

**68.** $8p - 18\sqrt{p} - 35 = 0$
$8(\sqrt{p})^2 - 18(\sqrt{p}) - 35 = 0$
Let $u = \sqrt{p}$.
$8u^2 - 18u - 35 = 0$
$(2u - 7)(4u + 5) = 0$
$2u - 7 = 0$ or $4u + 5 = 0$
$2u = 7$ or $4u = -5$
$u = \dfrac{7}{2}$ or $u = -\dfrac{5}{4}$
$\sqrt{p} = \dfrac{7}{2}$ or $\sqrt{p} = -\dfrac{5}{4}$
$p = \left(\dfrac{7}{2}\right)^2$ or $p = \left(-\dfrac{5}{4}\right)^2$
$p = \dfrac{49}{4}$ or $p = \dfrac{25}{16}$

Check:
$h\left(\dfrac{49}{4}\right) = 8 \cdot \dfrac{49}{4} - 18\sqrt{\dfrac{49}{4}} - 35$
$= 8 \cdot \dfrac{49}{4} - 18 \cdot \dfrac{7}{2} - 35$
$= 98 - 63 - 35$
$= 0$ ✓

$h\left(\dfrac{25}{16}\right) = 8 \cdot \dfrac{25}{16} - 18\sqrt{\dfrac{25}{16}} - 35$
$= \dfrac{25}{2} - 18 \cdot \dfrac{5}{4} - 35$
$= \dfrac{25}{2} - \dfrac{45}{2} - 35$
$= -45 \neq 0$ ✗

$p = \dfrac{25}{16}$ does not check; the zero of $h$ is $\left\{\dfrac{49}{4}\right\}$.

**70.** $\dfrac{1}{(a-2)^2} + \dfrac{3}{a-2} - 4 = 0$
$\left(\dfrac{1}{a-2}\right)^2 + 3\left(\dfrac{1}{a-2}\right) - 4 = 0$
Let $u = \dfrac{1}{a-2}$.

$u^2 + 3u - 4 = 0$
$(u - 1)(u + 4) = 0$
$u - 1 = 0$ or $u + 4 = 0$
$u = 1$ or $u = -4$
$\dfrac{1}{a-2} = 1$ or $\dfrac{1}{a-2} = -4$
$a - 2 = 1$ or $a - 2 = -\dfrac{1}{4}$
$a = 3$ or $a = \dfrac{7}{4}$

Check:
$f(3) = \dfrac{1}{(3-2)^2} + \dfrac{3}{3-2} - 4$
$= \dfrac{1}{1^2} + \dfrac{3}{1} - 4$
$= 1 + 3 - 4$
$= 0$ ✓

$f\left(\dfrac{7}{4}\right) = \dfrac{1}{\left(\frac{7}{4} - 2\right)^2} + \dfrac{3}{\frac{7}{4} - 2} - 4$
$= \dfrac{1}{\left(-\frac{1}{4}\right)^2} + \dfrac{3}{-\frac{1}{4}} - 4$
$= \dfrac{1}{\frac{1}{16}} + \dfrac{3}{-\frac{1}{4}} - 4$
$= 16 - 12 - 4$
$= 0$ ✓

Both check; the zeros of $s$ are $\left\{\dfrac{7}{4}, 3\right\}$.

**72. a.** $x^2 + 3x - 18 = 0$
$(x + 6)(x - 3) = 0$
$x + 6 = 0$ or $x - 3 = 0$
$x = -6$ or $x = 3$
Both check; the solution set is $\{-6, 3\}$.

**b.** $(x-1)^2 + 3(x-1) - 18 = 0$
Let $u = x - 1$.
$u^2 + 3u - 18 = 0$
$(u + 6)(u - 3) = 0$
$u + 6 = 0$ or $u - 3 = 0$
$u = -6$ or $u = 3$
$x - 1 = -6$ or $x - 1 = 3$
$x = -5$ or $x = 4$
Both check; the solution set is $\{-5, 4\}$.
Comparing these solutions to those in part (a), we note that $-5 = -6 + 1$ and $4 = 3 + 1$.

c. $(x+5)^2 + 3(x+5) - 18 = 0$
Let $u = x+5$.
$u^2 + 3u - 18 = 0$
$(u+6)(u-3) = 0$
$u+6 = 0$ or $u-3 = 0$
$u = -6$ or $u = 3$
$x+5 = -6$ or $x+5 = 3$
$x = -11$ or $x = -2$
Both check; the solution set is $\{-11, -2\}$.
Comparing these solutions to those in part (a), we note that $-11 = -6 - 5$ and $-2 = 3 - 5$.

d. $(x-3)^2 + 3(x-3) - 18 = 0$
Let $u = x-3$.
$u^2 + 3u - 18 = 0$
$(u+6)(u-3) = 0$
$u+6 = 0$ or $u-3 = 0$
$u = -6$ or $u = 3$
$x-3 = -6$ or $x-3 = 3$
$x = -3$ or $x = 6$
Both check; the solution set is $\{-3, 6\}$.
Comparing these solutions to those in part (a), we note that $-3 = -6 + 3$ and $6 = 3 + 3$.

e. Conjecture: The solution set of the equation $(x-a)^2 + 3(x-a) - 18 = 0$ is $\{-6+a, 3+a\}$.
NOTE: This conjecture can be shown to be true by using the techniques from parts (b) through (d).

74. a. $f(x) = 3x^2 - 5x - 2$
$3x^2 - 5x - 2 = 0$
$(3x+1)(x-2) = 0$
$3x+1 = 0$ or $x-2 = 0$
$3x = -1$ or $x = 2$
$x = -\dfrac{1}{3}$
Both check; the zeros of $f(x)$ are $\left\{-\dfrac{1}{3}, 2\right\}$.

b. $f(x-1) = 3(x-1)^2 - 5(x-1) - 2$
Let $u = x-1$.
$3u^2 - 5u - 2 = 0$
$(3u+1)(u-2) = 0$
$3u+1 = 0$ or $u-2 = 0$
$3u = -1$ or $u = 2$
$u = -\dfrac{1}{3}$
$x-1 = -\dfrac{1}{3}$ or $x-1 = 2$
$x = \dfrac{2}{3}$ or $x = 3$
Both check; the zeros of $f(x-1)$ are $\left\{\dfrac{2}{3}, 3\right\}$.
Comparing these solutions to those in part (a), we note that $\dfrac{2}{3} = -\dfrac{1}{3} + 1$ and $3 = 2 + 1$.

c. $f(x-4) = 3(x-4)^2 - 5(x-4) - 2$
Let $u = x-4$.
$3u^2 - 5u - 2 = 0$
$(3u+1)(u-2) = 0$
$3u+1 = 0$ or $u-2 = 0$
$3u = -1$ or $u = 2$
$u = -\dfrac{1}{3}$
$x-4 = -\dfrac{1}{3}$ or $x-4 = 2$
$x = \dfrac{11}{3}$ or $x = 6$
Both check; the zeros of $f(x-1)$ are $\left\{\dfrac{11}{3}, 6\right\}$. Comparing these solutions to those in part (a), we note that $\dfrac{11}{3} = -\dfrac{1}{3} + 4$ and $6 = 2 + 4$.

d. Conjecture: For $f(x) = 3x^2 - 5x - 2$, the zeros of the $f(x-a)$ are $\left\{-\dfrac{1}{3}+a, 2+a\right\}$.
NOTE: This conjecture can be shown to be true by using the techniques from parts (b) and (c).

## Chapter 8: Quadratic Equations and Functions

**76.**

**a.** $R(2000)$
$$= \frac{(2000-2000)^2}{3} + \frac{5(2000-20000)}{3} + 2000$$
$$= \frac{0^2}{3} + \frac{5(0)}{3} + 2000$$
$$= 2000$$

Interpretation: The revenue in 2000 was $2,000 thousand (or $2,000,000).

**b.** $\frac{(x-2000)^2}{3} + \frac{5(x-2000)}{3} + 2000 = 2250$

$\frac{1}{3}(x-2000)^2 + \frac{5}{3}(x-2000) - 250 = 0$

Let $u = x - 2000$

$\frac{1}{3}u^2 + \frac{5}{3}u - 250 = 0$

$3\left(\frac{1}{3}u^2 + \frac{5}{3}u - 250\right) = 3(0)$

$u^2 + 5u - 750 = 0$

$(u-25)(u+30) = 0$

$u - 25 = 0$ or $u + 30 = 0$
$u = 25$ or $u = -30$
$x - 2000 = 25$ or $x - 2000 = -30$
$x = 2025$ or $x = 1970$

Since 1970 is before 2000, disregard it. The solution set is $\{2025\}$.

Interpretation: In the year 2025, revenue will be $2,250 thousand (or $2,250,000).

**c.** $\frac{(x-2000)^2}{3} + \frac{5(x-2000)}{3} + 2000 = 2350$

$\frac{1}{3}(x-2000)^2 + \frac{5}{3}(x-2000) - 350 = 0$

Let $u = x - 2000$

$\frac{1}{3}u^2 + \frac{5}{3}u - 350 = 0$

$3\left(\frac{1}{3}u^2 + \frac{5}{3}u - 350\right) = 3(0)$

$u^2 + 5u - 1050 = 0$

$(u-30)(u+35) = 0$

$u - 30 = 0$ or $u + 35 = 0$
$u = 30$ or $u = -35$
$x - 2000 = 30$ or $x - 2000 = -35$
$x = 2030$ or $x = 1965$

Since 1965 is before 2000, disregard it. The solution set is $\{2030\}$. The model predicts that revenue will be $2,350 thousand (or $2,350,000) in the year 2030.

**78.** $x^4 + 7x^2 + 4 = 0$

$(x^2)^2 + 7(x^2) + 4 = 0$

Let $u = x^2$.

$u^2 + 7u + 4 = 0$

For this equation, $a = 1$, $b = 7$, and $c = 4$.

$u = \frac{-7 \pm \sqrt{7^2 - 4(1)(4)}}{2(1)}$

$= \frac{-7 \pm \sqrt{49-16}}{2}$

$= \frac{-7 \pm \sqrt{33}}{2}$

$u = \frac{-7-\sqrt{33}}{2}$ or $u = \frac{-7+\sqrt{33}}{2}$

$x^2 = \frac{-7-\sqrt{33}}{2}$ or $x^2 = \frac{-7+\sqrt{33}}{2}$

$x = \pm\sqrt{\frac{-7-\sqrt{33}}{2}}$ or $x = \pm\sqrt{\frac{-7+\sqrt{33}}{2}}$

$x = \pm\sqrt{\frac{-1(7+\sqrt{33})}{2}}$   $x = \pm\sqrt{\frac{-1(7-\sqrt{33})}{2}}$

$x = \pm i\sqrt{\frac{7+\sqrt{33}}{2}}$   $x = \pm i\sqrt{\frac{7-\sqrt{33}}{2}}$

$x = \pm i\sqrt{\frac{7+\sqrt{33}}{2} \cdot \frac{2}{2}}$   $x = \pm i\sqrt{\frac{7-\sqrt{33}}{2} \cdot \frac{2}{2}}$

$x = \pm\frac{\sqrt{14+2\sqrt{33}}}{2}i$   $x = \pm\frac{\sqrt{14-2\sqrt{33}}}{2}i$

All check, the solution set is $\left\{-\frac{\sqrt{14+2\sqrt{33}}}{2}i,\right.$

$\left.\frac{\sqrt{14+2\sqrt{33}}}{2}i, -\frac{\sqrt{14-2\sqrt{33}}}{2}i, \frac{\sqrt{14-2\sqrt{33}}}{2}i\right\}$.

**80.** $3(x+1)^2 + 6(x+1) - 1 = 0$

Let $u = x + 1$.

$3u^2 + 6u - 1 = 0$

For this equation, $a = 3$, $b = 6$, and $c = -1$.

$u = \frac{-6 \pm \sqrt{6^2 - 4(3)(-1)}}{2(3)}$

$= \frac{-6 \pm \sqrt{36+12}}{6}$

$= \frac{-6 \pm \sqrt{48}}{6}$

$= \frac{-6 \pm 4\sqrt{3}}{6} = -1 \pm \frac{2\sqrt{3}}{3}$

Section 8.3 Solving Equations Quadratic in Form

$$u = -1 - \frac{2\sqrt{3}}{3} \quad \text{or} \quad u = -1 + \frac{2\sqrt{3}}{3}$$

$$x + 1 = -1 - \frac{2\sqrt{3}}{3} \quad \text{or} \quad x + 1 = -1 + \frac{2\sqrt{3}}{3}$$

$$x = -2 - \frac{2\sqrt{3}}{3} \quad \text{or} \quad x = -2 + \frac{2\sqrt{3}}{2}$$

Both check, the solution set is
$$\left\{ -2 - \frac{2\sqrt{3}}{3},\ -2 + \frac{2\sqrt{3}}{3} \right\}.$$

82. $(5y^3 - 2y^2 + y + 4) - (2y^3 + 6y^2 - 3)$
$= 5y^3 - 2y^2 + y + 4 - 2y^3 - 6y^2 + 3$
$= 3y^3 - 8y^2 + y + 7$

84. $3\sqrt{2x} - \sqrt{8x} + \sqrt{50x}$
$= 3\sqrt{2x} - \sqrt{4 \cdot 2x} + \sqrt{25 \cdot 2x}$
$= 3\sqrt{2x} - 2\sqrt{2x} + 5\sqrt{2x}$
$= 6\sqrt{2x}$

86. Answers will vary. One possibility follows: To add algebraic expressions, we combine like terms or expressions. To subtract algebraic expression, we must first distribute the subtraction and then combine the like terms or expressions.

88. Let $Y_1 = x^4 - 4x^2 - 12$.

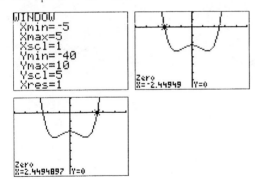

The solution set is approximately $\{-2.45, 2.45\}$.

90. Let $Y_1 = 3(x+3)^2$ and $Y_2 = 2(x+3) + 6$.

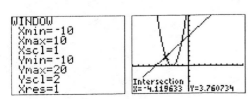

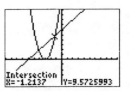

The solution set is approximately $\{-4.12, -1.21\}$.

92. Let $Y_1 = x + 4\sqrt{x}$ and $Y_2 = 5$.

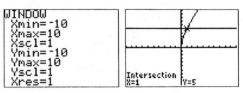

The solution set is $\{1\}$.

93. For the graphs shown in parts (a) through (c) the WINDOW setting is the following:

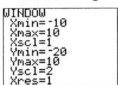

a. $Y_1 = x^2 - 5x - 6$

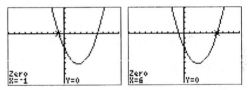

The x-intercepts are $-1$ and $6$.

b. $Y_1 = (x+2)^2 - 5(x+2) - 6$

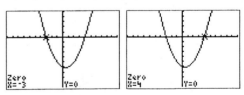

The x-intercepts are $-3$ and $4$.

ISM − 475

Chapter 8: Quadratic Equations and Quadratic Functions

c. $Y_1 = (x+5)^2 - 5(x+5) - 6$

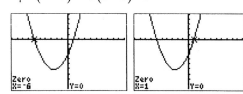

The x-intercepts are −6 and 1.

d. The x-intercepts of the graph of
$y = f(x) = x^2 - 5x - 6$ are −1 and 6.
The x-intercepts of the graph of
$y = f(x+a) = (x+a)^2 - 5(x+a) - 6$ are
$-1-a$ and $6-a$.

94. For the graphs shown in parts (a) through (c) the WINDOW setting is the following:

```
WINDOW
 Xmin=-10
 Xmax=10
 Xscl=1
 Ymin=-10
 Ymax=10
 Yscl=1
 Xres=1
```

a. $Y_1 = x^2 + 4x + 3$

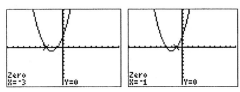

The x-intercepts are −3 and −1.

b. $Y_1 = (x-3)^2 + 4(x-3) + 3$

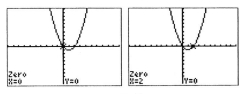

The x-intercepts are 0 and 2.

c. $Y_1 = (x-6)^2 + 4(x-6) + 3$

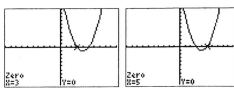

The x-intercepts are 3 and 5.

d. The x-intercepts of the graph of
$y = f(x) = x^2 + 4x + 3$ are −3 and −1.
The x-intercepts of the graph of
$y = f(x-a) = (x-a)^2 + 4(x-a) + 3$ are
$-3+a$ and $-1+a$.

**Putting the Concepts Together (8.1 – 8.3)**

1. Start: $z^2 + 10z$

   Add: $\left[\dfrac{1}{2} \cdot 10\right]^2 = 25$

   Result: $z^2 + 10z + 25$

   Factored Form: $(z+5)^2$

2. Start: $x^2 + 7x$

   Add: $\left[\dfrac{1}{2} \cdot 7\right]^2 = \dfrac{49}{4}$

   Result: $x^2 + 7x + \dfrac{49}{4}$

   Factored Form: $\left(x + \dfrac{7}{2}\right)^2$

3. Start: $n^2 - \dfrac{1}{4}n$

   Add: $\left[\dfrac{1}{2} \cdot \left(-\dfrac{1}{4}\right)\right]^2 = \dfrac{1}{64}$

   Result: $n^2 - \dfrac{1}{4}n + \dfrac{1}{64}$

   Factored Form: $\left(n - \dfrac{1}{8}\right)^2$

4. $(2x-3)^2 - 5 = -1$
   $(2x-3)^2 = 4$
   $2x - 3 = \pm\sqrt{4}$
   $2x - 3 = \pm 2$
   $2x = 3 \pm 2$
   $2x = 1$ or $2x = 5$
   $x = \dfrac{1}{2}$ or $x = \dfrac{5}{2}$

   The solution set is $\left\{\dfrac{1}{2}, \dfrac{5}{2}\right\}$.

**Putting the Concepts Together (Sections 8.1 – 8.3)**

5.  $x^2 + 8x + 4 = 0$
    $x^2 + 8x = -4$
    $x^2 + 8x + \left(\frac{1}{2} \cdot 8\right)^2 = -4 + \left(\frac{1}{2} \cdot 8\right)^2$
    $x^2 + 8x + 16 = -4 + 16$
    $(x+4)^2 = 12$
    $x + 4 = \pm\sqrt{12}$
    $x + 4 = \pm 2\sqrt{3}$
    $x = -4 \pm 2\sqrt{3}$
    The solution set is $\{-4 - 2\sqrt{3},\ -4 + 2\sqrt{3}\}$.

6.  $x(x-6) = -7$
    $x^2 - 6x + 7 = 0$
    For this equation, $a = 1$, $b = -6$, and $c = 7$.
    $x = \dfrac{-(-6) \pm \sqrt{(-6)^2 - 4(1)(7)}}{2(1)}$
    $= \dfrac{6 \pm \sqrt{36 - 28}}{2}$
    $= \dfrac{6 \pm \sqrt{8}}{2} = \dfrac{6 \pm 2\sqrt{2}}{2} = \dfrac{6}{2} \pm \dfrac{2\sqrt{2}}{2} = 3 \pm \sqrt{2}$
    The solution set is $\{3 - \sqrt{2},\ 3 + \sqrt{2}\}$.

7.  $49x^2 - 80 = 0$
    $49x^2 = 80$
    $x^2 = \dfrac{80}{49}$
    $x = \pm\sqrt{\dfrac{80}{49}} = \pm\dfrac{\sqrt{80}}{\sqrt{49}} = \pm\dfrac{4\sqrt{5}}{7}$
    The solution set is $\left\{-\dfrac{4\sqrt{5}}{7},\ \dfrac{4\sqrt{5}}{7}\right\}$.

8.  $p^2 - 8p + 6 = 0$
    Because this equation does not easily factor, solve by using the quadratic formula. For this equation, $a = 1$, $b = -8$, and $c = 6$.

    $p = \dfrac{-(-8) \pm \sqrt{(-8)^2 - 4(1)(6)}}{2(1)}$
    $= \dfrac{8 \pm \sqrt{64 - 24}}{2}$
    $= \dfrac{8 \pm \sqrt{40}}{2}$
    $= \dfrac{8 \pm 2\sqrt{10}}{2}$
    $= \dfrac{8}{2} \pm \dfrac{2\sqrt{10}}{2} = 4 \pm \sqrt{10}$
    The solution set is $\{4 - \sqrt{10},\ 4 + \sqrt{10}\}$.

9.  $3y^2 + 6y + 4 = 0$
    Because this equation does not easily factor, solve by using the quadratic formula. For this equation, $a = 3$, $b = 6$, and $c = 4$.
    $y = \dfrac{-6 \pm \sqrt{6^2 - 4(3)(4)}}{2(3)}$
    $= \dfrac{-6 \pm \sqrt{36 - 48}}{6}$
    $= \dfrac{-6 \pm \sqrt{-12}}{6}$
    $= \dfrac{-6 \pm 2\sqrt{3}\,i}{6}$
    $= \dfrac{-6}{6} \pm \dfrac{2\sqrt{3}}{6}i = -1 \pm \dfrac{\sqrt{3}}{3}i$
    The solution set is $\left\{-1 - \dfrac{\sqrt{3}}{3}i,\ -1 + \dfrac{\sqrt{3}}{3}i\right\}$.

**Chapter 8:** Quadratic Equations and Quadratic Functions

10. $\frac{1}{4}n^2 + n = \frac{1}{6}$

$12\left(\frac{1}{4}n^2 + n\right) = 12\left(\frac{1}{6}\right)$

$3n^2 + 12n = 2$

$3n^2 + 12n - 2 = 0$

Because this equation does not easily factor, solve by using the quadratic formula. For this equation, $a = 3$, $b = 12$, and $c = -2$.

$n = \frac{-12 \pm \sqrt{12^2 - 4(3)(-2)}}{2(3)}$

$= \frac{-12 \pm \sqrt{144 + 24}}{6}$

$= \frac{-12 \pm \sqrt{168}}{6}$

$= \frac{-12 \pm 2\sqrt{42}}{6}$

$= -\frac{12}{6} \pm \frac{2\sqrt{42}}{6} = -2 \pm \frac{\sqrt{42}}{3}$

The solution set is $\left\{-2 - \frac{\sqrt{42}}{3},\ -2 + \frac{\sqrt{42}}{3}\right\}$.

11. $9x^2 + 12x + 4 = 0$

For this equation, $a = 9$, $b = 12$, and $c = 4$.

$b^2 - 4ac = 12^2 - 4(9)(4) = 144 - 144 = 0$

Because $b^2 - 4ac = 0$, the quadratic equation will have one repeated real solution.

12. $3x^2 + 6x - 2 = 0$

For this equation, $a = 3$, $b = 6$, and $c = -2$.

$b^2 - 4ac = 6^2 - 4(3)(-2) = 36 + 24 = 60$

Because $b^2 - 4ac = 60$ is positive, but not a perfect square, the quadratic equation will have two irrational solutions.

13. $2x^2 + 6x + 5 = 0$

For this equation, $a = 2$, $b = 6$, and $c = 5$.

$b^2 - 4ac = 6^2 - 4(2)(5) = 36 - 40 = -4$

Because $b^2 - 4ac = -4$ is negative, the quadratic equation will have two complex solutions that are not real. The solutions will be complex conjugates of each other.

14. $c^2 = 4^2 + 10^2 = 16 + 100 = 116$

$c = \sqrt{116} = 2\sqrt{29}$

15. $2m + 7\sqrt{m} - 15 = 0$

$2(\sqrt{m})^2 + 7(\sqrt{m}) - 15 = 0$

Let $u = \sqrt{m}$.

$2u^2 + 7u - 15 = 0$

$(2u - 3)(u + 5) = 0$

$2u - 3 = 0$ or $u + 5 = 0$

$2u = 3$ or $u = -5$

$u = \frac{3}{2}$

$\sqrt{m} = \frac{3}{2}$ or $\sqrt{m} = -5$

$m = \left(\frac{3}{2}\right)^2$ or $m = (-5)^2$

$m = \frac{9}{4}$ or $m = 25$

Check:

$m = \frac{9}{4}$: $2 \cdot \frac{9}{4} + 7\sqrt{\frac{9}{4}} - 15 \stackrel{?}{=} 0$

$2 \cdot \frac{9}{4} + 7 \cdot \frac{3}{2} - 15 \stackrel{?}{=} 0$

$\frac{9}{2} + \frac{21}{2} - 15 \stackrel{?}{=} 0$

$0 = 0$ ✓

$m = 25$: $2 \cdot 25 + 7\sqrt{25} - 15 \stackrel{?}{=} 0$

$2 \cdot 25 + 7 \cdot 5 - 15 \stackrel{?}{=} 0$

$50 + 35 - 15 \stackrel{?}{=} 0$

$70 \neq 0$ ✗

$m = 25$ does not check; the solution set is $\left\{\frac{9}{4}\right\}$.

16. $p^{-2} - 3p^{-1} - 18 = 0$

$(p^{-1})^2 - 3(p^{-1}) - 18 = 0$

Let $u = p^{-1}$.

$u^2 - 3u - 18 = 0$

$(u + 3)(u - 6) = 0$

$u + 3 = 0$ or $u - 6 = 0$

$u = -3$ or $u = 6$

$p^{-1} = -3$ or $p^{-1} = 6$

$\frac{1}{p} = -3$ or $\frac{1}{p} = 6$

$p = -\frac{1}{3}$ or $p = \frac{1}{6}$

Section 8.4 Graphing Quadratic Functions Using Transformations

Check:

$p = -\dfrac{1}{3}$: $\left(-\dfrac{1}{3}\right)^{-2} - 3\left(-\dfrac{1}{3}\right)^{-1} - 18 \stackrel{?}{=} 0$

$(-3)^2 - 3(-3) - 18 \stackrel{?}{=} 0$

$9 + 9 - 18 \stackrel{?}{=} 0$

$0 = 0$ ✓

$p = \dfrac{1}{6}$: $\left(\dfrac{1}{6}\right)^{-2} - 3\left(\dfrac{1}{6}\right)^{-1} - 18 \stackrel{?}{=} 0$

$(6)^2 - 3(6) - 18 \stackrel{?}{=} 0$

$36 - 18 - 18 \stackrel{?}{=} 0$

$0 = 0$ ✓

Both check; the solution set is $\left\{\dfrac{1}{6}, -\dfrac{1}{3}\right\}$.

17.  $12,000 = -0.4x^2 + 140x$
$0.4x^2 - 140x + 12,000 = 0$

For this equation, $a = 0.4$, $b = -140$, and $c = 12,000$.

$x = \dfrac{-(-140) \pm \sqrt{(-140)^2 - 4(0.4)(12,000)}}{2(0.4)}$

$= \dfrac{140 \pm \sqrt{400}}{0.8}$

$= \dfrac{140 \pm 20}{0.8}$

$x = \dfrac{140 + 20}{0.8}$ or $x = \dfrac{140 - 20}{0.8}$

$= 200$ or $= 150$

Revenue will be $12,000 when either 150 microwaves or 200 microwaves are sold.

18. Let $x$ represent the speed of the wind.

|  | Distance | Rate | Time |
| --- | --- | --- | --- |
| Against Wind | 300 | $140 - x$ | $\dfrac{300}{140 - x}$ |
| With Wind | 300 | $140 + x$ | $\dfrac{300}{140 + x}$ |

$\dfrac{300}{140 - x} + \dfrac{300}{140 + x} = 5$

$(140 - x)(140 + x)\left(\dfrac{300}{140 - x} + \dfrac{300}{140 + x}\right)$
$= (140 - x)(140 + x)(5)$

$300(140 + x) + 300(140 - x) = (19,600 - x^2)(5)$

$42,000 + 300x + 42,000 - 300x = 98,000 - 5x^2$

$84,000 = 98,000 - 5x^2$

$-14,000 = -5x^2$

$2800 = x^2$

$x = \pm\sqrt{2800}$

$= \pm 20\sqrt{7}$

$\approx \pm 52.915$

Because the speed should be positive, we disregard $x \approx -52.915$. Thus, the only viable answer is $x \approx 52.915$. Rounding to the nearest tenth, the speed of the wind was approximately 52.9 miles per hour.

## 8.4 Preparing for Graphs of Quadratic Functions Using Transfromations

1. Locate some points on the graph of $f(x) = x^2$.

| $x$ | $f(x) = x^2$ | $(x, f(x))$ |
| --- | --- | --- |
| $-3$ | $f(-3) = (-3)^2 = 9$ | $(-3, 9)$ |
| $-2$ | $f(-2) = (-2)^2 = 4$ | $(-2, 4)$ |
| $-1$ | $f(-1) = (-1)^2 = 1$ | $(-1, 1)$ |
| $0$ | $f(0) = 0^2 = 0$ | $(0, 0)$ |
| $1$ | $f(1) = 1^2 = 1$ | $(1, 1)$ |
| $2$ | $f(2) = 2^2 = 4$ | $(2, 4)$ |
| $3$ | $f(3) = 3^2 = 9$ | $(3, 9)$ |

Plot the points and connect them with a smooth curve.

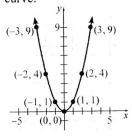

2. Locate some points on the graph of $f(x) = x^2 - 3$.

| $x$ | $f(x) = x^2 - 3$ | $(x, f(x))$ |
|---|---|---|
| $-3$ | $f(-3) = (-3)^2 - 3 = 6$ | $(-3, 6)$ |
| $-2$ | $f(-2) = (-2)^2 - 3 = 1$ | $(-2, 1)$ |
| $-1$ | $f(-1) = (-1)^2 - 3 = -2$ | $(-1, -2)$ |
| 0 | $f(0) = 0^2 - 3 = -3$ | $(0, -3)$ |
| 1 | $f(1) = 1^2 - 3 = -2$ | $(1, -2)$ |
| 2 | $f(2) = 2^2 - 3 = 1$ | $(2, 1)$ |
| 3 | $f(3) = 3^2 - 3 = 6$ | $(3, 6)$ |

Plot the points and connect them with a smooth curve.

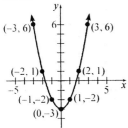

3. For $f(x) = 2x^2 + 5x + 1$, the independent variable $x$ can take on any value on the real number line. That is, $f$ is defined for all real $x$ values. Thus, the domain is the set of all real number, or $(-\infty, \infty)$.

## 8.4 Exercises

2. up; down

4. TRUE.

6. TRUE.

8. Answers will vary. One possibility follows: They $y$-coordinates of the graph of $y = ax^2$ are $a$ times the $y$-coordinates to the graph of $y = x^2$. Since the graph of $y = x^2$ opens upward, the graph of $y = ax^2$ will continue to open upward when $a > 0$ because all of the $y$-coordinates will be multiplied by a positive value. On the other hand, if $a < 0$, then $y = ax^2$ will open downward because all of the $y$-coordinates will be multiplied by a negative value. When we generalize to a quadratic function of the form $y = a(x - h)^2 + k$ or $y = ax^2 + bx + c$, the coefficient $a$ continues to have the same effect on the graph.

10. (I) The graph of $f(x) = (x - 2)^2 - 4$ is the graph of $y = x^2$ shifted 2 units to the right and 4 units down. Thus, the graph of the function is graph (B).

(II) The graph of $f(x) = -(x - 2)^2 + 4$ is the graph of $y = x^2$ shifted 2 units to the right, reflected over the $x$-axis (so that it opens downward), and shifted 4 units up. Thus, the graph of the function is graph (D).

(III) The graph of $f(x) = -(x + 2)^2 - 4$ is the graph of $y = x^2$ shifted 2 units to the left, reflected over the $x$-axis (so that it opens downward), and shifted 4 units up. Thus, the graph of the function is graph (C).

(IV) The graph of $f(x) = 2(x - 2)^2 - 4$ is the graph of $y = x^2$ shifted 2 units to the right, stretched vertically by a factor of 2, and shifted 4 units down. Thus, the graph of the function is graph (A).

12. To obtain the graph of $G(x) = (x - 9)^2$, begin with the graph of $y = x^2$ and shift it 9 units to the right.

14. To obtain the graph of $g(x) = x^2 - 8$, begin with the graph of $y = x^2$ and shift it 8 units down.

16. To obtain the graph of $h(x) = 4(x + 7)^2$, begin with the graph of $y = x^2$, shift it 7 units to the left, and vertically stretch it by a factor of 4 (multiply the $y$-coordinates by 4).

18. To obtain the graph of $F(x) = -\frac{1}{2}(x - 3)^2 - 5$, begin with the graph of $y = x^2$, shift it 3 units to the right, multiply the $y$-coordinates by $-\frac{1}{2}$ (which means it opens down and is compressed vertically by a factor of $\frac{1}{2}$), and shift the graph down 5 units.

Section 8.4 Graphing Quadratic Functions Using Transformations

20. Begin with the graph of $y = x^2$, then shift the graph down 1 unit to obtain the graph of $f(x) = x^2 - 1$.

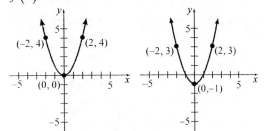

22. Begin with the graph of $y = x^2$, then shift the graph down 7 units to obtain the graph of $g(x) = x^2 - 7$.

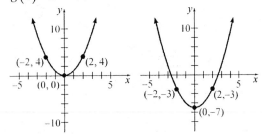

24. Begin with the graph of $y = x^2$, then shift the graph to the right 2 units to obtain the graph of $F(x) = (x - 2)^2$.

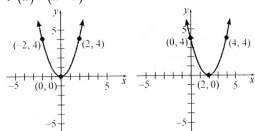

26. Begin with the graph of $y = x^2$, then shift the graph to the left 4 units to obtain the graph of $f(x) = (x + 4)^2$.

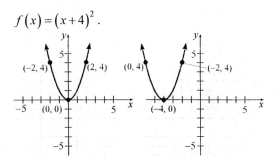

28. Begin with the graph of $y = x^2$, then vertically stretch the graph by a factor of 5 (multiply each $y$-coordinate by 5) to obtain the graph of $G(x) = 5x^2$.

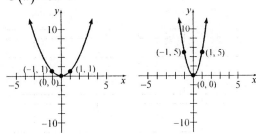

30. Begin with the graph of $y = x^2$, then vertically stretch the graph by a factor of $\frac{3}{2}$ (multiply each $y$-coordinate by $\frac{3}{2}$) to obtain the graph of $h(x) = \frac{3}{2}x^2$.

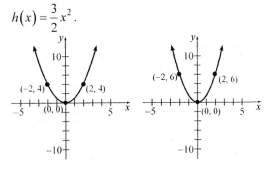

32. Begin with the graph of $y = x^2$, then multiply each $y$-coordinate by $-3$ to obtain the graph of

$P(x) = -3x^2$.

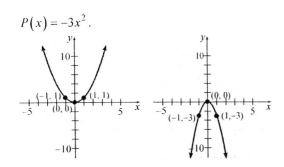

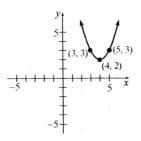

**34.** Begin with the graph of $y = x^2$, then shift the graph 2 units to the left to obtain the graph of $y = (x+2)^2$. Shift this graph down 1 unit to obtain the graph of $g(x) = (x+2)^2 - 1$.

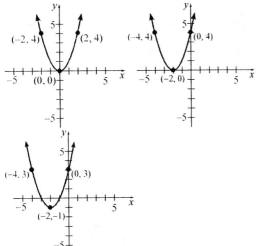

**36.** Begin with the graph of $y = x^2$, then shift the graph 4 units to the right to obtain the graph of $y = (x-4)^2$. Shift this graph up 2 units to obtain the graph of $G(x) = (x-4)^2 + 2$.

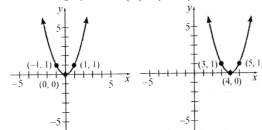

**38.** Begin with the graph of $y = x^2$, then shift the graph 3 units to the right to obtain the graph of $y = (x-3)^2$. Multiply the y-coordinates by $-1$ to obtain the graph of $y = -(x-3)^2$. Lastly, shift the graph up 5 units to obtain the graph of $H(x) = -(x-3)^2 + 5$.

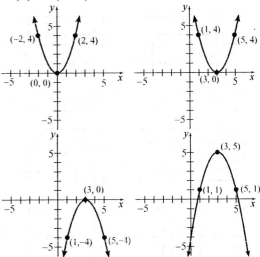

**40.** Begin with the graph of $y = x^2$, then shift the graph 2 unit to the right to obtain the graph of $y = (x-2)^2$. Vertically stretch this graph by a factor of 3 (multiply the y-coordinates by 3) to obtain the graph of $y = 3(x-2)^2$. Lastly, shift the graph down 1 unit to obtain the graph of $F(x) = 3(x-2)^2 - 1$.

**Section 8.4** *Graphing Quadratic Functions Using Transformations*

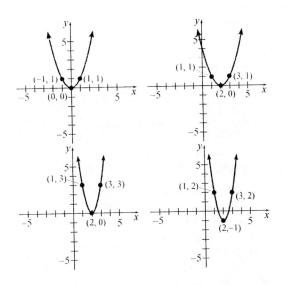

42. Begin with the graph of $y = x^2$, then shift the graph 6 units to the left to obtain the graph of $y = (x+6)^2$. Multiply the y-coordinates by $-\frac{1}{2}$ to obtain the graph of $y = -\frac{1}{2}(x+6)^2$. Lastly, shift the graph up 2 units to obtain the graph of $f(x) = -\frac{1}{2}(x+6)^2 + 2$.

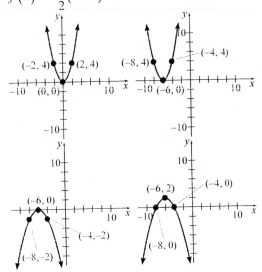

44. Use completing the square to write the function in the form $y = a(x-h)^2 + k$.

$$f(x) = x^2 + 4x - 1$$
$$= (x^2 + 4x) - 1$$
$$= (x^2 + 4x + 4) - 1 - 4$$
$$= (x+2)^2 - 5$$

Begin with the graph of $y = x^2$, the shift the graph 2 units to the left to obtain the graph of $y = (x+2)^2$. Shift this graph down 5 units to obtain the graph of $f(x) = (x+2)^2 - 5$.

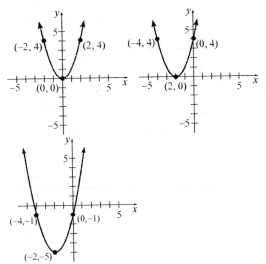

The vertex is $(h, k) = (-2, -5)$ and the axis of symmetry is $x = -2$. The domain is the set of all real numbers or, using interval notation, $(-\infty, \infty)$. The range is $\{y \mid y \geq -5\}$ or, using interval notation, $[-5, \infty)$.

46. Use completing the square to write the function in the form $y = a(x-h)^2 + k$.

$$G(x) = x^2 - 2x + 7$$
$$= (x^2 - 2x) + 7$$
$$= (x^2 - 2x + 1) + 7 - 1$$
$$= (x-1)^2 + 6$$

Begin with the graph of $y = x^2$, then shift the graph right 1 unit to obtain the graph of $y = (x-1)^2$. Shift this graph up 6 units to obtain

ISM − 483

## Chapter 8: Quadratic Equations and Quadratic Functions

the graph of $G(x) = (x-1)^2 + 6$.

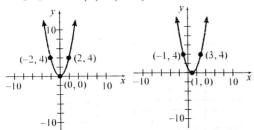

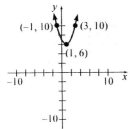

The vertex is $(h,k) = (1,6)$ and the axis of symmetry is $x = 1$. The domain is the set of all real numbers or, using interval notation, $(-\infty, \infty)$. The range is $\{y \mid y \geq 6\}$ or, using interval notation, $[6, \infty)$.

48. Use completing the square to write the function in the form $y = a(x-h)^2 + k$.

$$f(x) = x^2 + 4x + 5$$
$$= (x^2 + 4x) + 5$$
$$= (x^2 + 4x + 4) + 5 - 4$$
$$= (x+2)^2 + 1$$

Begin with the graph of $y = x^2$, then shift the graph 2 units left to obtain the graph of $y = (x+2)^2$. Shift this result up 1 unit to obtain the graph of $f(x) = (x+2)^2 + 1$.

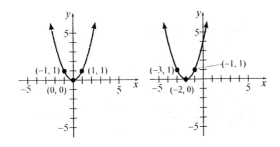

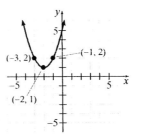

The vertex is $(h,k) = (-2, 1)$ and the axis of symmetry is $x = -2$. The domain is the set of all real numbers or, using interval notation, $(-\infty, \infty)$. The range is $\{y \mid y \geq 1\}$ or, using interval notation, $[1, \infty)$.

50. Use completing the square to write the function in the form $y = a(x-h)^2 + k$.

$$h(x) = x^2 - 7x + 10$$
$$= (x^2 - 7x) + 10$$
$$= \left(x^2 - 7x + \frac{49}{4}\right) + 10 - \frac{49}{4}$$
$$= \left(x - \frac{7}{2}\right)^2 - \frac{9}{4}$$

Begin with the graph of $y = x^2$, then shift the graph right $\frac{7}{2}$ unit to obtain the graph of $y = \left(x - \frac{7}{2}\right)^2$. Shift this graph down $\frac{9}{4}$ units to obtain the graph of $h(x) = \left(x - \frac{7}{2}\right)^2 - \frac{9}{4}$.

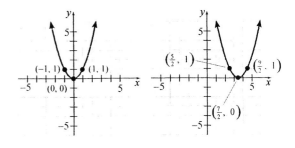

Section 8.4 Graphing Quadratic Functions Using Transformations

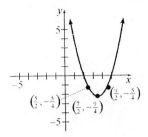

The vertex is $(h,k) = \left(\dfrac{7}{2}, -\dfrac{9}{4}\right)$ and the axis of symmetry is $x = \dfrac{7}{2}$. The domain is the set of all real numbers or, using interval notation, $(-\infty, \infty)$. The range is $\left\{y \mid y \geq -\dfrac{9}{4}\right\}$ or, using interval notation, $\left[-\dfrac{9}{4}, \infty\right)$.

52. Use completing the square to write the function in the form $y = a(x-h)^2 + k$.

$g(x) = 2x^2 + 4x - 3$
$= (2x^2 + 4x) - 3$
$= 2(x^2 + 2x) - 3$
$= 2(x^2 + 2x + 1) - 3 - 2$
$= 2(x+1)^2 - 5$

Begin with the graph of $y = x^2$, then shift the graph left 1 unit to obtain the graph of $y = (x+1)^2$. Vertically stretch this graph by a factor of 2 (multiply the y-coordinates by 2) to obtain the graph of $y = 2(x+1)^2$. Lastly, shift the graph down 5 units to obtain the graph of $g(x) = 2(x+1)^2 - 5$.

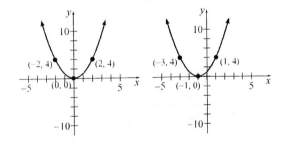

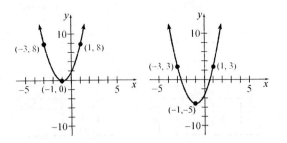

The vertex is $(h,k) = (-1,-5)$ and the axis of symmetry is $x = -1$. The domain is the set of all real numbers or, using interval notation, $(-\infty, \infty)$. The range is $\{y \mid y \geq -5\}$ or, using interval notation, $[-5, \infty)$.

54. Use completing the square to write the function in the form $y = a(x-h)^2 + k$.

$f(x) = 3x^2 + 18x + 25$
$= (3x^2 + 18x) + 25$
$= 3(x^2 + 6x) + 25$
$= 3(x^2 + 6x + 9) + 25 - 27$
$= 3(x+3)^2 - 2$

Begin with the graph of $y = x^2$, then shift the graph left 3 units to obtain the graph of $y = (x+3)^2$. Vertically stretch this graph by a factor of 3 (multiply the y-coordinates by 3) to obtain the graph of $y = 3(x+3)^2$. Lastly, shift the graph down 2 units to obtain the graph of $f(x) = 3(x+3)^2 - 2$.

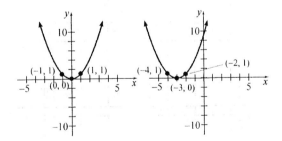

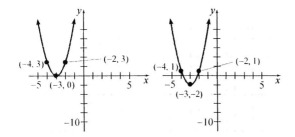

The vertex is $(h,k)=(-3,-2)$ and the axis of symmetry is $x=-3$. The domain is the set of all real numbers or, using interval notation, $(-\infty,\infty)$. The range is $\{y \mid y \geq -2\}$ or, using interval notation, $[-2,\infty)$.

56. Use completing the square to write the function in the form $y = a(x-h)^2 + k$.

$$g(x) = -x^2 - 8x - 14$$
$$= (-x^2 - 8x) - 14$$
$$= -(x^2 + 8x) - 14$$
$$= -(x^2 + 8x + 16) - 14 + 16$$
$$= -(x+4)^2 + 2$$

Begin with the graph of $y = x^2$, then shift the graph left 4 units to obtain the graph of $y = (x+4)^2$. Multiply the y-coordinates by $-1$ to obtain the graph of $y = -(x+4)^2$. Lastly, shift the graph up 2 units to obtain the graph of $g(x) = -(x+4)^2 + 2$.

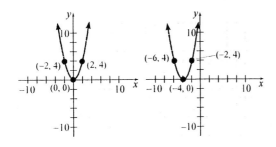

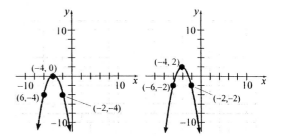

The vertex is $(h,k)=(-4,2)$ and the axis of symmetry is $x=-4$. The domain is the set of all real numbers or, using interval notation, $(-\infty,\infty)$. The range is $\{y \mid y \leq 2\}$ or, using interval notation, $(-\infty,2]$.

58. Use completing the square to write the function in the form $y = a(x-h)^2 + k$.

$$f(x) = -x^2 + 10x - 17$$
$$= (-x^2 + 10x) - 17$$
$$= -(x^2 - 10x) - 17$$
$$= -(x^2 - 10x + 25) - 17 + 25$$
$$= -(x-5)^2 + 8$$

Begin with the graph of $y = x^2$, then shift the graph right 5 units to obtain the graph of $y = (x-5)^2$. Multiply the y-coordinates by $-1$ to obtain the graph of $y = -(x-5)^2$. Lastly, shift the graph up 8 units to obtain the graph of $f(x) = -(x-5)^2 + 8$.

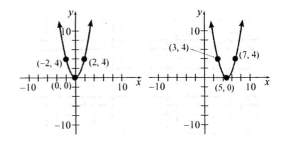

## Section 8.4 Graphing Quadratic Functions Using Transformations

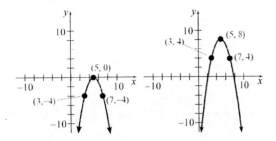

The vertex is $(h,k) = (5,8)$ and the axis of symmetry is $x = 5$. The domain is the set of all real numbers or, using interval notation, $(-\infty, \infty)$. The range is $\{y \mid y \leq 8\}$ or, using interval notation, $(-\infty, 8]$.

**60.** Use completing the square to write the function in the form $y = a(x-h)^2 + k$.

$$h(x) = -2x^2 + 12x - 17$$
$$= (-2x^2 + 12x) - 17$$
$$= -2(x^2 - 6x) - 17$$
$$= -2(x^2 - 6x + 9) - 17 + 18$$
$$= -2(x-3)^2 + 1$$

Begin with the graph of $y = x^2$, then shift the graph right 3 units to obtain the graph of $y = (x-3)^2$. Multiply the y-coordinates by $-2$ to obtain the graph of $y = -2(x-3)^2$. Lastly, shift the graph up 1 unit to obtain the graph of $h(x) = -2(x-3)^2 + 1$.

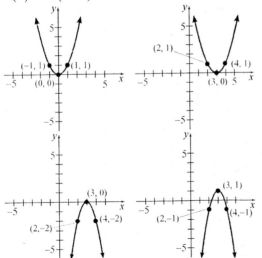

The vertex is $(h,k) = (3,1)$ and the axis of symmetry is $x = 3$. The domain is the set of all real numbers or, using interval notation, $(-\infty, \infty)$. The range is $\{y \mid y \leq 1\}$ or, using interval notation, $(-\infty, 1]$.

**62.** Use completing the square to write the function in the form $y = a(x-h)^2 + k$.

$$f(x) = \frac{1}{2}x^2 + 2x - 1$$
$$= \left(\frac{1}{2}x^2 + 2x\right) - 1$$
$$= \frac{1}{2}(x^2 + 4x) - 1$$
$$= \frac{1}{2}(x^2 + 4x + 4) - 1 - 2$$
$$= \frac{1}{2}(x+2)^2 - 3$$

Begin with the graph of $y = x^2$, then shift the graph left 2 units to obtain the graph of $y = (x+2)^2$. Vertically compress the graph by a factor of $\frac{1}{2}$ (multiply the y-coordinates by $\frac{1}{2}$) to obtain the graph of $y = \frac{1}{2}(x+2)^2$. Lastly, shift the graph down 3 units to obtain the graph of $f(x) = \frac{1}{2}(x+2)^2 - 3$.

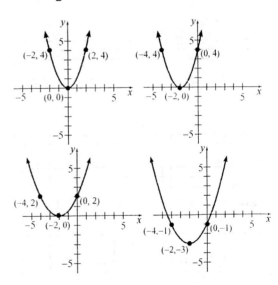

ISM – 487

## Chapter 8: Quadratic Equations and Quadratic Functions

The vertex is $(h,k)=(-2,-3)$ and the axis of symmetry is $x=-2$. The domain is the set of all real numbers or, using interval notation, $(-\infty,\infty)$. The range is $\{y \mid y \geq -3\}$ or, using interval notation, $[-3,\infty)$.

64. Use completing the square to write the function in the form $y = a(x-h)^2 + k$.

$$h(x) = -4x^2 + 4x$$
$$= -4(x^2 - x)$$
$$= -4\left(x^2 - x + \frac{1}{4}\right) + 1$$
$$= -4\left(x - \frac{1}{2}\right)^2 + 1$$

Begin with the graph of $y = x^2$, then shift right $\frac{1}{2}$ unit to obtain the graph of $y = \left(x - \frac{1}{2}\right)^2$. Multiply the y-coordinates by $-4$ to obtain the graph of $y = -4\left(x - \frac{1}{2}\right)^2$. Lastly, shift the graph up 1 unit to obtain the graph of $h(x) = -4\left(x - \frac{1}{2}\right)^2 + 1$.

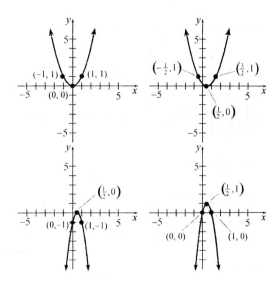

The vertex is $(h,k)=\left(\frac{1}{2}, 1\right)$ and the axis of symmetry is $x = \frac{1}{2}$. The domain is the set of all real numbers or, using interval notation, $(-\infty,\infty)$. The range is $\{y \mid y \leq 1\}$ or, using interval notation, $(-\infty,1]$.

66. Answers may vary. Since the graph opens up, one possibility is to let $a = 1$. The vertex is $(0,2)$ so we have $h = 0$ and $k = 2$. Substituting these values into the form $y = a(x-h)^2 + k$ gives $y = f(x) = x^2 + 2$.

68. Answers may vary. Since the graph opens up, one possibility is to let $a = 1$. The vertex is $(4,-2)$ so we have $h = 4$ and $k = -2$. Substituting these values into the form $y = a(x-h)^2 + k$ give $y = f(x) = (x-4)^2 - 2$.

70. Answers may vary. Since the graph opens down, one possibility is to let $a = -1$. The vertex is $(-4,-7)$ so we have $h = -4$ and $k = -7$. Substituting these values into the form $y = a(x-h)^2 + k$ gives $y = f(x) = -(x+4)^2 - 7$.

72. Consider the form $y = a(x-h)^2 + k$. Since the graph opens up, we know that $a > 0$. The graph is vertically compressed by a factor of $\frac{1}{2}$ so we know that $a = \frac{1}{2}$. The vertex is $(-5,0)$ so we have $h = -5$ and $k = 0$. Substituting these values gives $y = f(x) = \frac{1}{2}(x+5)^2$.

74. Consider the form $y = a(x-h)^2 + k$. Since the graph opens down, we know that $a < 0$. The graph is vertically stretched by a factor of 5 so we know that $a = -5$. The vertex is $(5,8)$ so we have $h = 5$ and $k = 8$. Substituting these values gives $y = f(x) = -5(x-5)^2 + 8$.

**Section 8.4** *Graphing Quadratic Functions Using Transformations*

76. Consider the form $y = a(x-h)^2 + k$. From the graph we know that the vertex is $(2,1)$ so we have $h = 2$ and $k = 1$. The graph also passes through the point $(x, y) = (0, 5)$. Substituting these values for $x$, $y$, $h$, and $k$, we can solve for $a$:
$5 = a(0-2)^2 + 1$
$5 = a(-2)^2 + 1$
$5 = 4a + 1$
$4 = 4a$
$1 = a$
The quadratic function is $f(x) = (x-2)^2 + 1$.

78. Consider the form $y = a(x-h)^2 + k$. From the graph we know that the vertex is $(-2, 5)$ so we have $h = -2$ and $k = 5$. The graph also passes through the point $(x, y) = (0, -7)$. Substituting these values for $x$, $y$, $h$, and $k$, we can solve for $a$:
$-7 = a(0-(-2))^2 + 5$
$-7 = a(2)^2 + 5$
$-7 = 4a + 5$
$-12 = 4a$
$-3 = a$
The quadratic function is $f(x) = -3(x+2)^2 + 5$.

80. Consider the form $y = a(x-h)^2 + k$. From the graph we know that the vertex is $(0, -5)$ so we have $h = 0$ and $k = -5$. The graph also passes through the point $(x, y) = (1, -4)$. Substituting these values for $x$, $y$, $h$, and $k$, we can solve for $a$:

$-4 = a(1-0)^2 + (-5)$
$-4 = a(1)^2 - 5$
$-4 = a - 5$
$1 = a$
The quadratic function is $f(x) = x^2 - 5$.

82. No. A quadratic function cannot have more than one $y$-intercept. In fact, no function can have more than one $y$-intercept. If a graph were to intersect the $y$-axis in more than one place, then it would fail the vertical line test and would therefore not be a function.

84. 
$$\begin{array}{r} 4x-1 \\ x+5{\overline{\smash{\big)}\,4x^2+19x-1}} \\ \underline{4x^2+20x} \\ -x-1 \\ \underline{-x-5} \\ 4 \end{array}$$

So, $\dfrac{4x^2 + 19x - 1}{x+5} = 4x - 1 + \dfrac{4}{x+5}$.

86. Answers will vary. One possibility follows. Both in division of real number and in division of polynomials, we can use the "long division" process to determine the quotient.

88. $f(x) = x^2 - 3.5$

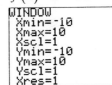 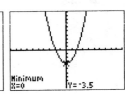

Vertex: $(0, -3.5)$
Axis of symmetry: $x = 0$
Range: $\{y \mid y \geq -3.5\} = [-3.5, \infty)$

89. $g(x) = (x - 2.5)^2$

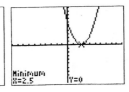 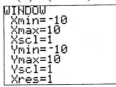

Vertex: $(2.5, 0)$
Axis of symmetry: $x = 2.5$
Range: $\{y \mid y \geq 0\} = [0, \infty)$

90. $G(x) = (x + 4.5)^2$

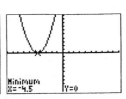

Vertex: $(-4.5, 0)$
Axis of symmetry: $x = -4.5$
Range: $\{y \mid y \geq 0\} = [0, \infty)$

ISM - 489

92. $H(x) = 1.2(x+0.4)^2 - 1.3$

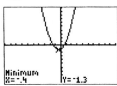

Vertex: $(-0.4, -1.3)$
Axis of symmetry: $x = -0.4$
Range: $\{y \mid y \geq -1.3\} = [-1.3, \infty)$

94. $f(x) = 0.3(x+3.8)^2 - 8.9$

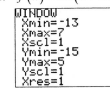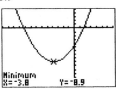

Vertex: $(-3.8, -8.9)$
Axis of symmetry: $x = -3.8$
Range: $\{y \mid y \geq -8.9\} = [-8.9, \infty)$

## 8.5 Preparing for Graphing Quadratic Functions Using Properties

1. To find the $y$-intercept, let $x = 0$ and solve for $y$:
$2(0) + 5y = 20$
$5y = 20$
$y = 4$
To find the $x$-intercept, let $y = 0$ and solve for $x$:
$2x + 5(0) = 20$
$2x = 20$
$x = 10$
The intercepts are $(0, 4)$ and $(10, 0)$.

2. $2x^2 - 3x - 20 = 0$
$(2x+5)(x-4) = 0$
$2x + 5 = 0$ or $x - 4 = 0$
$x = -\dfrac{5}{2}$ or $x = 4$
The solution set is $\left\{-\dfrac{5}{2}, 4\right\}$.

3. $f(x) = 0$
$x^2 - 3x - 4 = 0$
$(x+1)(x-4) = 0$
$x + 1 = 0$ or $x - 4 = 0$
$x = -1$ or $x = 4$
The zeros are $-1$ and $4$.

## Section 8.5

2. $>$

4. TRUE. For $f(x) = -2x^2 - 3x + 6$, we have $a = -2$, $b = -3$, and $c = 6$. Thus,
$b^2 - 4ac = (-3)^2 - 4(-2)(6) = 9 + 72 = 81$.
Because the discriminant is positive, the parabola will have two distinct $x$-intercepts.

6. TRUE.

8. Answers may vary. One possibility follows:
The vertex of $f(x) = ax^2 + bx + c$ can be found by using either of the following formulas:
$\left(-\dfrac{b}{2a}, \dfrac{4ac-b^2}{4a}\right)$ or $\left(-\dfrac{b}{2a}, f\left(-\dfrac{b}{2a}\right)\right)$.

10. $g(x) = 2x^2 - 7x - 4$;
$a = 2, b = -7, c = -4$;
$b^2 - 4ac = (-7)^2 - 4(2)(-4) = 49 + 32 = 81$.
Because the discriminant is positive, the parabola will have two distinct $x$-intercepts.
The $x$-intercepts are:
$g(x) = 0$
$2x^2 - 7x - 4 = 0$
$(2x+1)(x-4) = 0$
$2x + 1 = 0$ or $x - 4 = 0$
$x = -\dfrac{1}{2}$   $x = 4$

12. $H(x) = x^2 - 3x + 5$;
$a = 1, b = -3, c = 5$;
$b^2 - 4ac = (-3)^2 - 4(1)(5) = 9 - 20 = -11$.
Because the discriminant is negative, the parabola will have no $x$-intercepts.

14. $f(x) = x^2 - 6x + 9$;
$a = 1, b = -6, c = 9$;
$b^2 - 4ac = (-6)^2 - 4(1)(9) = 36 - 36 = 0$.
Because the discriminant is zero, the parabola will have one $x$-intercept.

## Section 8.5 Graphing Quadratic Functions Using Properties

The $x$-intercept is:
$$f(x) = 0$$
$$x^2 - 6x + 9 = 0$$
$$(x-3)^2 = 0$$
$$x - 3 = 0$$
$$x = 3$$

16. $P(x) = -2x^2 + 3x + 1$;
$a = -2, b = 3, c = 1$;
$b^2 - 4ac = 3^2 - 4(-2)(1) = 9 + 8 = 17$.

Because the discriminant is positive, the parabola will have two distinct $x$-intercepts.

The $x$-intercepts are:
$$P(x) = 0$$
$$-2x^2 + 3x + 1 = 0$$
$$x = \frac{-3 \pm \sqrt{3^2 - 4(-2)(1)}}{2(-2)}$$
$$= \frac{-3 \pm \sqrt{17}}{-4}$$
$$x \approx -0.28 \quad \text{or} \quad x \approx 1.78$$

18. $f(x) = x^2 - 2x - 8$
$a = 1, b = -2, c = -8$
The graph opens up because $a > 0$.

vertex:
$$x = -\frac{b}{2a} = -\frac{(-2)}{2(1)} = 1$$
$$f(1) = (1)^2 - 2(1) - 8 = -9$$
The vertex is $(1, -9)$ and the axis of symmetry is $x = 1$.

$y$-intercept:
$$f(0) = (0)^2 - 2(0) - 8 = -8$$

$x$-intercepts:
$$b^2 - 4ac = (-2)^2 - 4(1)(-8) = 36 > 0$$
There are two distinct $x$-intercepts. We find these by solving
$$f(x) = 0$$
$$x^2 - 2x - 8 = 0$$
$$(x-4)(x+2) = 0$$

$x - 4 = 0$ or $x + 2 = 0$
$x = 4$ or $x = -2$

Graph:
The $y$-intercept point, $(0, -8)$, is one unit to the left of the axis of symmetry. Therefore, if we move one unit to the right of the axis of symmetry, we obtain the point $(2, -8)$ which must also be on the graph.

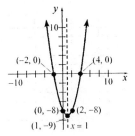

The domain is the set of all real numbers or, using interval notation, $(-\infty, \infty)$. The range is $\{y \mid y \geq -9\}$ or, using interval notation, $[-9, \infty)$.

20. $g(x) = x^2 - 12x + 27$
$a = 1, b = -12, c = 27$
The graph opens up because $a > 0$.

vertex:
$$x = -\frac{b}{2a} = -\frac{(-12)}{2(1)} = 6$$
$$g(6) = (6)^2 - 12(6) + 27 = -9$$
The vertex is $(6, -9)$ and the axis of symmetry is $x = 6$.

$y$-intercept:
$$g(0) = (0)^2 - 12(0) + 27 = 27$$

$x$-intercepts:
$$b^2 - 4ac = (-12)^2 - 4(1)(27) = 36 > 0$$
There are two distinct $x$-intercepts. We find these by solving
$$g(x) = 0$$
$$x^2 - 12x + 27 = 0$$
$$(x-3)(x-9) = 0$$
$x - 3 = 0$ or $x - 9 = 0$
$x = 3$ or $x = 9$

Graph:
The $y$-intercept point, $(0, 27)$, is six units to the left of the axis of symmetry. Therefore, if we move six units to the right of the axis of symmetry, we obtain the point $(12, 27)$ which must also be on the graph.

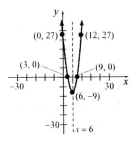

The domain is the set of all real numbers or, using interval notation, $(-\infty, \infty)$. The range is $\{y \mid y \geq -9\}$ or, using interval notation, $[-9, \infty)$.

22. $g(x) = -x^2 + 2x + 15$
$a = -1, b = 2, c = 15$
The graph opens down because $a < 0$.

vertex:
$$x = -\frac{b}{2a} = -\frac{(2)}{2(-1)} = 1$$
$$g(1) = -(1)^2 + 2(1) + 15 = 16$$
The vertex is $(1, 16)$ and the axis of symmetry is $x = 1$.

$y$-intercept:
$$g(0) = -(0)^2 + 2(0) + 15 = 15$$

$x$-intercepts:
$b^2 - 4ac = (2)^2 - 4(-1)(15) = 64 > 0$
There are two distinct $x$-intercepts. We find these by solving
$$g(x) = 0$$
$$-x^2 + 2x + 15 = 0$$
$$x^2 - 2x - 15 = 0$$
$$(x+3)(x-5) = 0$$
$x + 3 = 0$ or $x - 5 = 0$
$x = -3$ or $x = 5$

Graph:
The $y$-intercept point, $(0, 15)$, is one unit to the left of the axis of symmetry. Therefore, if we move one unit to the right of the axis of symmetry, we obtain the point $(2, 15)$ which must also be on the graph.

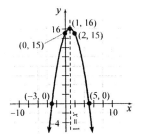

The domain is the set of all real numbers or, using interval notation, $(-\infty, \infty)$. The range is $\{y \mid y \leq 16\}$ or, using interval notation, $(-\infty, 16]$.

24. $h(x) = x^2 + 6x + 9$
$a = 1, b = 6, c = 9$
The graph opens up because $a > 0$.

vertex:
$$x = -\frac{b}{2a} = -\frac{6}{2(1)} = -3$$
$$h(-3) = (-3)^2 + 6(-3) + 9 = 0$$
The vertex is $(-3, 0)$ and the axis of symmetry is $x = -3$.

$y$-intercept:
$$h(0) = (0)^2 + 6(0) + 9 = 9$$

$x$-intercepts:
Since the discriminant is 0, the $x$-coordinate of the vertex is the only $x$-intercept, $x = -3$.

Graph:
The $y$-intercept point, $(0, 9)$, is three units to the right of the axis of symmetry. Therefore, if we move three units to the left of the axis of symmetry, we obtain the point $(-6, 9)$ which must also be on the graph.

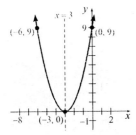

The domain is the set of all real numbers or, using interval notation, $(-\infty,\infty)$. The range is $\{y \mid y \geq 0\}$ or, using interval notation, $[0,\infty)$.

26. $f(x) = x^2 - 4x + 7$
$a = 1, b = -4, c = 7$
The graph opens up because $a > 0$.

vertex:
$x = -\dfrac{b}{2a} = -\dfrac{(-4)}{2(1)} = 2$

$f(2) = (2)^2 - 4(2) + 7 = 3$

The vertex is $(2,3)$ and the axis of symmetry is $x = 2$.

y-intercept:
$f(0) = (0)^2 - 4(0) + 7 = 7$

x-intercepts:
$b^2 - 4ac = (-4)^2 - 4(1)(7) = -12 < 0$
There are no x-intercepts since the discriminant is negative.

Graph:
The y-intercept point, $(0,7)$, is two units to the left of the axis of symmetry. Therefore, if we move two units to the right of the axis of symmetry, we obtain the point $(4,7)$ which must also be on the graph.

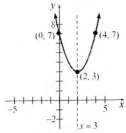

The domain is the set of all real numbers or, using interval notation, $(-\infty,\infty)$. The range is $\{y \mid y \geq 3\}$ or, using interval notation, $[3,\infty)$.

28. $P(x) = -x^2 - 12x - 36$
$a = -1, b = -12, c = -36$
The graph opens down because $a < 0$.

vertex:
$x = -\dfrac{b}{2a} = -\dfrac{(-12)}{2(-1)} = -6$

$P(-6) = -(-6)^2 - 12(-6) - 36 = 0$

The vertex is $(-6,0)$ and the axis of symmetry is $x = -6$.

y-intercept:
$P(0) = -(0)^2 - 12(0) - 36 = -36$

x-intercepts:
$b^2 - 4ac = (-12)^2 - 4(-1)(-36) = 144 - 144 = 0$
Since the discriminant is 0, the x-coordinate of the vertex is the only x-intercept. $x = -6$.

Graph:
The y-intercept point, $(0,-36)$, is six units to the right of the axis of symmetry. Therefore, if we move six units to the left of the axis of symmetry, we obtain the point $(-12,-36)$ which must also be on the graph.

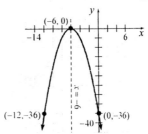

The domain is the set of all real numbers or, using interval notation, $(-\infty,\infty)$. The range is $\{y \mid y \leq 0\}$ or, using interval notation, $(-\infty,0]$.

30. $f(x) = -x^2 + 4x - 6$
$a = -1, b = 4, c = -6$
The graph opens down because $a < 0$.

vertex:
$x = -\dfrac{b}{2a} = -\dfrac{4}{2(-1)} = 2$

$f(2) = -(2)^2 + 4(2) - 6 = -2$

## Chapter 8: Quadratic Equations and Quadratic Functions

The vertex is $(2,-2)$ and the axis of symmetry is $x = 2$.

y-intercept:
$$f(0) = -(0)^2 + 4(0) - 6 = -6$$

x-intercepts:
$$b^2 - 4ac = (4)^2 - 4(-1)(-6) = -8 < 0$$
There are no x-intercepts since the discriminant is negative.

Graph:
The y-intercept point, $(0,-6)$, is two units to the left of the axis of symmetry. Therefore, if we move two units to the right of the axis of symmetry, we obtain the point $(4,-6)$ which must also be on the graph.

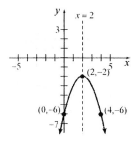

The domain is the set of all real numbers or, using interval notation, $(-\infty, \infty)$. The range is $\{y \mid y \leq -2\}$ or, using interval notation, $(-\infty, -2]$.

**32.** $f(x) = 4x^2 - 8x - 21$
$a = 4, b = -8, c = -21$
The graph opens up because $a > 0$.

vertex:
$$x = -\frac{b}{2a} = -\frac{(-8)}{2(4)} = 1$$
$$f(1) = 4(1)^2 - 8(1) - 21 = -25$$
The vertex is $(1,-25)$ and the axis of symmetry is $x = 1$.

y-intercept:
$$f(0) = 4(0)^2 - 8(0) - 21 = -21$$

x-intercepts:
$$b^2 - 4ac = (-8)^2 - 4(4)(-21) = 400 > 0$$
There are two distinct x-intercepts. We find these

by solving
$$f(x) = 0$$
$$4x^2 - 8x - 21 = 0$$
$$(2x+3)(2x-7) = 0$$
$$2x+3 = 0 \quad \text{or} \quad 2x-7 = 0$$
$$2x = -3 \quad \text{or} \quad 2x = 7$$
$$x = -\frac{3}{2} \quad \text{or} \quad x = \frac{7}{2}$$

Graph:
The y-intercept point, $(0,-21)$, is 1 unit to the left of the axis of symmetry. Therefore, if we move 1 unit to the right of the axis of symmetry, we obtain the point $(2,-21)$ which must also be on the graph.

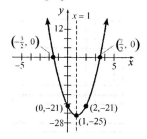

The domain is the set of all real numbers or, using interval notation, $(-\infty, \infty)$. The range is $\{y \mid y \geq -25\}$ or, using interval notation, $[-25, \infty)$.

**34.** $g(x) = -9x^2 - 36x - 20$
$a = -9, b = -36, c = -20$
The graph opens down because $a < 0$.

vertex:
$$x = -\frac{b}{2a} = -\frac{(-36)}{2(-9)} = -2$$
$$g(1) = -9(-2)^2 - 36(-2) - 20 = 16$$
The vertex is $(-2, 16)$ and the axis of symmetry is $x = -2$.

y-intercept:
$$g(0) = -9(0)^2 - 36(0) - 20 = -20$$

x-intercepts:
$$b^2 - 4ac = (-36)^2 - 4(-9)(-20) = 576 > 0$$
There are two distinct x-intercepts. We find these by solving

**Section 8.5** Graphing Quadratic Functions Using Properties

$$g(x) = 0$$
$$-9x^2 - 36x - 20 = 0$$
$$9x^2 + 36x + 20 = 0$$
$$(3x+10)(3x+2) = 0$$
$$3x+10 = 0 \quad \text{or} \quad 3x+2 = 0$$
$$3x = -10 \quad \text{or} \quad 3x = -2$$
$$x = -\frac{10}{3} \quad \text{or} \quad x = -\frac{2}{3}$$

Graph:
The $y$-intercept point, $(0, -20)$, is two units to the right of the axis of symmetry. Therefore, if we move two units to the left of the axis of symmetry, we obtain the point $(-4, -20)$ which must also be on the graph.

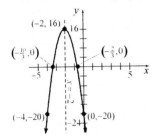

The domain is the set of all real numbers or, using interval notation, $(-\infty, \infty)$. The range is $\{y \mid y \leq 16\}$ or, using interval notation, $(-\infty, 16]$.

36. $h(x) = 9x^2 + 12x + 4$
$a = 9, b = 12, c = 4$
The graph opens up because $a > 0$.

vertex:
$$x = -\frac{b}{2a} = -\frac{12}{2(9)} = -\frac{2}{3}$$
$$h\left(-\frac{2}{3}\right) = -9\left(-\frac{2}{3}\right)^2 + 12\left(-\frac{2}{3}\right) + 4 = 0$$
The vertex is $\left(-\frac{2}{3}, 0\right)$ and the axis of symmetry is $x = -\frac{2}{3}$.

$y$-intercept:
$$h(0) = 9(0)^2 + 12(0) + 4 = 4$$

$x$-intercepts:
$$b^2 - 4ac = (12)^2 - 4(9)(4) = 0$$
Since the discriminant is 0, the $x$-coordinate of the vertex is the only $x$-intercept, $x = -\frac{2}{3}$.

Graph:
The $y$-intercept point, $(0, 4)$, is $\frac{2}{3}$ unit to the right of the axis of symmetry. If we move $\frac{2}{3}$ unit to the left of the axis of symmetry, we obtain the point $\left(-\frac{4}{3}, 4\right)$ which must also be on the graph.

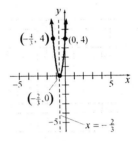

The domain is the set of all real numbers or, using interval notation, $(-\infty, \infty)$. The range is $\{y \mid y \geq 0\}$ or, using interval notation, $[0, \infty)$.

38. $F(x) = -4x^2 - 20x - 25$
$a = -4, b = -20, c = -25$
The graph opens down because $a < 0$.

vertex:
$$x = -\frac{b}{2a} = -\frac{(-20)}{2(-4)} = -\frac{5}{2}$$
$$F\left(-\frac{5}{2}\right) = -4\left(-\frac{5}{2}\right)^2 - 20\left(-\frac{5}{2}\right) - 25 = 0$$
The vertex is $\left(-\frac{5}{2}, 0\right)$ and the axis of symmetry is $x = -\frac{5}{2}$.

$y$-intercept:
$$F(0) = -4(0)^2 - 20(0) - 25 = -25$$

$x$-intercepts:
$$b^2 - 4ac = (-20)^2 - 4(-4)(-25) = 0$$
Since the discriminant is 0, the only $x$-intercept is the $x$-coordinate of the vertex, $x = -\frac{5}{2}$.

Graph:
The $y$-intercept point, $(0, -25)$, is $\frac{5}{2}$ units to the right of the axis of symmetry. Therefore, if we move $\frac{5}{2}$ units to the left of the axis of symmetry,

we obtain the point $(-5, -25)$ which must also be on the graph.

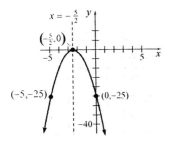

The domain is the set of all real numbers or, using interval notation, $(-\infty, \infty)$. The range is $\{y \mid y \leq 0\}$ or, using interval notation, $(-\infty, 0]$.

40. $F(x) = 3x^2 + 6x + 7$
$a = 3, b = 6, c = 7$
The graph opens up because $a > 0$.

vertex:
$$x = -\frac{b}{2a} = -\frac{(6)}{2(3)} = -1$$
$$F(-1) = 3(-1)^2 + 6(-1) + 7 = 4$$
The vertex is $(-1, 4)$ and the axis of symmetry is $x = -1$.

y-intercept:
$$F(0) = 3(0)^2 + 6(0) + 7 = 7$$

x-intercepts:
$$b^2 - 4ac = (6)^2 - 4(3)(7) = -48 < 0$$
There are no x-intercepts since the discriminant is negative.

Graph:
The y-intercept point, $(0, 7)$, is one unit to the right of the axis of symmetry. Therefore, if we move one unit to the left of the axis of symmetry, we obtain the point $(-2, 7)$ which must also be on the graph.

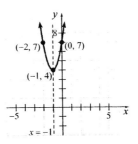

The domain is the set of all real numbers or, using interval notation, $(-\infty, \infty)$. The range is $\{y \mid y \geq 4\}$ or, using interval notation, $[4, \infty)$.

42. $p(x) = -2x^2 + 6x + 5$
$a = -2, b = 6, c = 5$
The graph opens down because $a < 0$.

vertex:
$$x = -\frac{b}{2a} = -\frac{(6)}{2(-2)} = \frac{3}{2}$$
$$p\left(\tfrac{3}{2}\right) = -2\left(\tfrac{3}{2}\right)^2 + 6\left(\tfrac{3}{2}\right) + 5 = \tfrac{19}{2}$$
The vertex is $\left(\tfrac{3}{2}, \tfrac{19}{2}\right)$ and the axis of symmetry is $x = \tfrac{3}{2}$.

y-intercept:
$$p(0) = -2(0)^2 + 6(0) + 5 = 5$$

x-intercepts:
$$b^2 - 4ac = (6)^2 - 4(-2)(5) = 76 > 0$$
There are two distinct x-intercepts. We find these by solving
$$p(x) = 0$$
$$-2x^2 + 6x + 5 = 0$$
$$x = \frac{-6 \pm \sqrt{76}}{-4} = \frac{-6 \pm 2\sqrt{19}}{-4} = \frac{3 \pm \sqrt{19}}{2}$$
$$x \approx -0.68 \text{ or } x \approx 3.68$$

Graph:
The y-intercept point, $(0, 5)$, is $\tfrac{3}{2}$ units to the left of the axis of symmetry. Therefore, if we move $\tfrac{3}{2}$ unit to the right of the axis of symmetry, we obtain the point $(3, 5)$ which must also be on the graph.

### Section 8.5 Graphing Quadratic Functions Using Properties

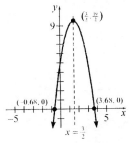

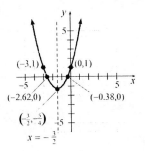

The domain is the set of all real numbers or, using interval notation, $(-\infty, \infty)$. The range is $\left\{y \mid y \leq \frac{19}{2}\right\}$ or, using interval notation, $\left(-\infty, \frac{19}{2}\right]$.

The domain is the set of all real numbers or, using interval notation, $(-\infty, \infty)$. The range is $\left\{y \mid y \geq -\frac{5}{4}\right\}$ or, using interval notation, $\left[-\frac{5}{4}, \infty\right)$.

**44.** $H(x) = x^2 + 3x + 1$
$a = 1, b = 3, c = 1$
The graph opens up because $a > 0$.

vertex:
$$x = -\frac{b}{2a} = -\frac{(3)}{2(1)} = -\frac{3}{2}$$
$$H\left(-\frac{3}{2}\right) = \left(-\frac{3}{2}\right)^2 + 3\left(-\frac{3}{2}\right) + 1 = -\frac{5}{4}$$
The vertex is $\left(-\frac{3}{2}, -\frac{5}{4}\right)$ and the axis of symmetry is $x = -\frac{3}{2}$.

y-intercept:
$$H(0) = (0)^2 + 3(0) + 1 = 1$$

x-intercepts:
$$b^2 - 4ac = (3)^2 - 4(1)(1) = 5 > 0$$
There are two distinct x-intercepts. We find these by solving
$$H(x) = 0$$
$$x^2 + 3x + 1 = 0$$
$$x = \frac{-3 \pm \sqrt{5}}{2}$$
$$x \approx -2.62 \text{ or } x \approx -0.38$$

Graph:
The y-intercept point, $(0,1)$, is $\frac{3}{2}$ units to the right of the axis of symmetry. Therefore, if we move $\frac{3}{2}$ units to the left of the axis of symmetry, we obtain the point $(-3,1)$ which must also be on the graph.

**46.** $F(x) = -2x^2 + 6x + 1$
$a = -2, b = 6, c = 1$
The graph opens down because $a < 0$.

vertex:
$$x = -\frac{b}{2a} = -\frac{(6)}{2(-2)} = \frac{3}{2}$$
$$F\left(\frac{3}{2}\right) = -2\left(\frac{3}{2}\right)^2 + 6\left(\frac{3}{2}\right) + 1 = \frac{11}{2}$$
The vertex is $\left(\frac{3}{2}, \frac{11}{2}\right)$ and the axis of symmetry is $x = \frac{3}{2}$.

y-intercept:
$$F(0) = -2(0)^2 + 6(0) + 1 = 1$$

x-intercepts:
$$b^2 - 4ac = (6)^2 - 4(-2)(1) = 44 > 0$$
There are two distinct x-intercepts. We find these by solving
$$F(x) = 0$$
$$-2x^2 + 6x + 1 = 0$$
$$x = \frac{-6 \pm \sqrt{44}}{2(-2)} = \frac{-6 \pm 2\sqrt{11}}{-4} = \frac{3 \pm \sqrt{11}}{2}$$
$$x \approx -0.16 \text{ or } x \approx 3.16$$

Graph:
The y-intercept point, $(0,1)$, is $\frac{3}{2}$ units to the left of the axis of symmetry. Therefore, if we move $\frac{3}{2}$ units to the right of the axis of symmetry, we obtain the point $(3,1)$ which must also be on the graph.

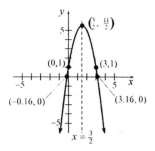

The domain is the set of all real numbers or, using interval notation, $(-\infty, \infty)$. The range is $\left\{y \mid y \leq \dfrac{11}{2}\right\}$ or, using interval notation, $\left(-\infty, \dfrac{11}{2}\right]$.

48. $F(x) = 4x^2 + 4x - 1$
$a = 4, b = 4, c = -1$
The graph opens up because $a > 0$.

vertex:
$$x = -\dfrac{b}{2a} = -\dfrac{(4)}{2(4)} = -\dfrac{1}{2}$$
$$F\left(\tfrac{1}{2}\right) = 4\left(-\tfrac{1}{2}\right)^2 + 4\left(-\tfrac{1}{2}\right) - 1 = -2$$
The vertex is $\left(-\tfrac{1}{2}, -2\right)$ and the axis of symmetry is $x = -\tfrac{1}{2}$.

y-intercept:
$F(0) = 4(0)^2 + 4(0) - 1 = -1$

x-intercepts:
$b^2 - 4ac = (4)^2 - 4(4)(-1) = 32 > 0$
There are two distinct x-intercepts. We find these by solving
$$F(x) = 0$$
$$4x^2 + 4x - 1 = 0$$
$$x = \dfrac{-4 \pm \sqrt{32}}{2(4)} = \dfrac{-4 \pm 4\sqrt{2}}{8} = \dfrac{-1 \pm \sqrt{2}}{2}$$
$x \approx -1.21$ or $x \approx 0.21$

Graph:
The y-intercept point, $(0, -1)$, is $\tfrac{1}{2}$ unit to the right of the axis of symmetry. Therefore, if we move one unit to the left of the axis of

symmetry, we obtain the point $(-1, -1)$ which must also be on the graph.

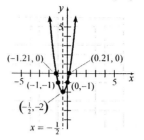

The domain is the set of all real numbers or, using interval notation, $(-\infty, \infty)$. The range is $\{y \mid y \geq -2\}$ or, using interval notation, $[-2, \infty)$.

50. $h(x) = -4x^2 + 8x$
$a = -4, b = 8, c = 0$
The graph opens down because $a < 0$.

vertex:
$$x = -\dfrac{b}{2a} = -\dfrac{(8)}{2(-4)} = 1$$
$$h(1) = -4(1)^2 + 8(1) = 4$$
The vertex is $(1, 4)$ and the axis of symmetry is $x = 1$.

y-intercept:
$h(0) = -4(0)^2 + 8(0) = 0$

x-intercepts:
$b^2 - 4ac = (8)^2 - 4(-4)(0) = 64 > 0$
There are two distinct x-intercepts. We find these by solving
$$h(x) = 0$$
$$-4x^2 + 8x = 0$$
$$-4x(x - 2) = 0$$
$-4x = 0$ or $x - 2 = 0$
$x = 0$ or $x = 2$

Graph:
The y-intercept point, $(0, 0)$, is one unit to the left of the axis of symmetry. Therefore, if we move one unit to the right of the axis of symmetry, we obtain the point $(2, 0)$ which must also be on the graph (these points are actually the x-intercept points).

### Section 8.5 Graphing Quadratic Functions Using Properties

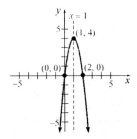

The domain is the set of all real numbers or, using interval notation, $(-\infty, \infty)$. The range is $\{y \mid y \leq 4\}$ or, using interval notation, $(-\infty, 4]$.

52. $g(x) = x^2 + \dfrac{5}{2}x - 6$

$a = 1, b = \dfrac{5}{2}, c = -6$

The graph opens up because $a > 0$.

vertex:
$$x = -\dfrac{b}{2a} = -\dfrac{\left(\dfrac{5}{2}\right)}{2(1)} = -\dfrac{5}{4}$$

$$g\left(-\dfrac{5}{4}\right) = \left(-\dfrac{5}{4}\right)^2 + \dfrac{5}{2}\left(-\dfrac{5}{4}\right) - 6 = -\dfrac{121}{16}$$

The vertex is $\left(-\dfrac{5}{4}, -\dfrac{121}{16}\right)$ and the axis of symmetry is $x = -\dfrac{5}{4}$.

y-intercept:
$$g(0) = (0)^2 + \dfrac{5}{2}(0) - 6 = -6$$

x-intercepts:
$$b^2 - 4ac = \left(\dfrac{5}{2}\right)^2 - 4(1)(-6) = \dfrac{121}{4} > 0$$

There are two distinct x-intercepts. We find these by solving
$$g(x) = 0$$
$$x^2 + \dfrac{5}{2}x - 6 = 0$$
$$2x^2 + 5x - 12 = 0$$
$$(2x - 3)(x + 4) = 0$$
$$2x - 3 = 0 \quad \text{or} \quad x + 4 = 0$$
$$x = \dfrac{3}{2} \quad \text{or} \quad x = -4$$

Graph:
The y-intercept point, $(0, -6)$, is $\dfrac{5}{4}$ units to the right of the axis of symmetry. Therefore, if we move $\dfrac{5}{4}$ units to the left of the axis of symmetry,

we obtain the point $\left(-\dfrac{5}{2}, -6\right)$ which must also be on the graph.

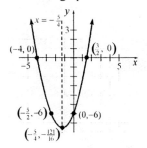

The domain is the set of all real numbers or, using interval notation, $(-\infty, \infty)$. The range is $\left\{y \mid y \geq -\dfrac{121}{16}\right\}$ or, using interval notation, $\left[-\dfrac{121}{16}, \infty\right)$.

54. $H(x) = \dfrac{1}{4}x^2 + x - 8$

$a = \dfrac{1}{4}, b = 1, c = -8$

The graph opens up because $a > 0$.

vertex:
$$x = -\dfrac{b}{2a} = -\dfrac{(1)}{2\left(\dfrac{1}{4}\right)} = -2$$

$$H(-2) = \dfrac{1}{4}(-2)^2 + (-2) - 8 = -9$$

The vertex is $(-2, -9)$ and the axis of symmetry is $x = -2$.

y-intercept:
$$H(0) = \dfrac{1}{4}(0)^2 + (0) - 8 = -8$$

x-intercepts:
$$b^2 - 4ac = (1)^2 - 4\left(\dfrac{1}{4}\right)(-8) = 9 > 0$$

There are two distinct x-intercepts. We find these by solving
$$H(x) = 0$$
$$\dfrac{1}{4}x^2 + x - 8 = 0$$
$$x^2 + 4x - 32 = 0$$
$$(x + 8)(x - 4) = 0$$
$$x + 8 = 0 \quad \text{or} \quad x - 4 = 0$$
$$x = -8 \quad \text{or} \quad x = 4$$

ISM - 499

## Chapter 8: Quadratic Equations and Quadratic Functions

Graph:
The $y$-intercept point, $(0,-8)$, is two units to the right of the axis of symmetry. Therefore, if we move two units to the left of the axis of symmetry, we obtain the point $(-4,-8)$ which must also be on the graph.

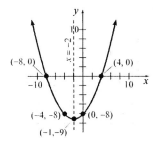

The domain is the set of all real numbers or, using interval notation, $(-\infty, \infty)$. The range is $\{y \mid y \geq -9\}$ or, using interval notation, $[-9, \infty)$.

**56.** $G(x) = -\dfrac{1}{2}x^2 - 8x - 24$

$a = -\dfrac{1}{2}, b = -8, c = -24$

The graph opens down because $a < 0$.

vertex:
$$x = -\dfrac{b}{2a} = -\dfrac{(-8)}{2\left(-\frac{1}{2}\right)} = -8$$

$$G(-8) = -\dfrac{1}{2}(-8)^2 - 8(-8) - 24 = 8$$

The vertex is $(-8, 8)$ and the axis of symmetry is $x = -8$.

$y$-intercept:
$$G(0) = -\dfrac{1}{2}(0)^2 - 8(0) - 24 = -24$$

$x$-intercepts:
$$b^2 - 4ac = (-8)^2 - 4\left(-\dfrac{1}{2}\right)(-24) = 16 > 0$$

There are two distinct $x$-intercepts. We find these by solving

$G(x) = 0$
$-\dfrac{1}{2}x^2 - 8x - 24 = 0$
$x^2 + 16x + 48 = 0$
$(x+12)(x+4) = 0$
$x + 12 = 0$ or $x + 4 = 0$
$x = -12$ or $x = -4$

Graph:
The $y$-intercept point, $(0, -24)$, is eight units to the right of the axis of symmetry. Therefore, if we move eight units to the left of the axis of symmetry, we obtain the point $(-16, -24)$ which must also be on the graph.

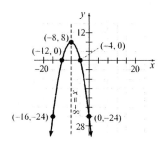

The domain is the set of all real numbers or, using interval notation, $(-\infty, \infty)$. The range is $\{y \mid y \leq 8\}$ or, using interval notation, $(-\infty, 8]$.

**58.** If we compare $f(x) = x^2 - 6x + 3$ to $f(x) = ax^2 + bx + c$, we find that $a = 1$, $b = -6$, and $c = 3$. Because $a > 0$, we know the graph will open up, so the function will have a minimum value.
The minimum value occurs at
$$x = -\dfrac{b}{2a} = -\dfrac{(-6)}{2(1)} = 3.$$
The minimum value is
$f(3) = 3^2 - 6(3) + 3 = -6$.
So, the minimum value is $-6$ and it occurs when $x = 3$.

**60.** If we compare $g(x) = -x^2 + 4x + 12$ to $g(x) = ax^2 + bx + c$, we find that $a = -1$, $b = 4$, and $c = 12$. Because $a < 0$, we know the graph will open down, so the function will have a maximum value.
The maximum value occurs at

ISM − 500

Section 8.5 Graphing Quadratic Functions Using Properties

$x = -\dfrac{b}{2a} = -\dfrac{4}{2(-1)} = 2$.

The maximum value is

$g(2) = -(2)^2 + 4(2) + 12 = 16$.

So, the maximum value is 16 and it occurs when $x = 2$.

62. If we compare $H(x) = -3x^2 + 12x - 1$ to $H(x) = ax^2 + bx + c$, we find that $a = -3$, $b = 12$, and $c = -1$. Because $a < 0$, we know the graph will open down, so the function will have a maximum value.
The maximum value occurs at

$x = -\dfrac{b}{2a} = -\dfrac{12}{2(-3)} = 2$.

The maximum value is

$H(2) = -3(2)^2 + 12(2) - 1 = 11$.

So, the maximum value is 11 and it occurs when $x = 2$.

64. If we compare $G(x) = 5x^2 + 10x - 1$ to $G(x) = ax^2 + bx + c$, we find that $a = 5$, $b = 10$, and $c = -1$. Because $a > 0$, we know the graph will open up, so the function will have a minimum value.
The minimum value occurs at

$x = -\dfrac{b}{2a} = -\dfrac{10}{2(5)} = -1$.

The minimum value is

$G(-1) = 5(-1)^2 + 10(-1) - 1 = -6$.

So, the minimum value is $-6$ and it occurs when $x = -1$.

66. If we compare $F(x) = 3x^2 + 4x - 3$ to $F(x) = ax^2 + bx + c$, we find that $a = 3$, $b = 4$, and $c = -3$. Because $a > 0$, we know the graph will open up, so the function will have a minimum value.
The minimum value occurs at

$x = -\dfrac{b}{2a} = -\dfrac{4}{2(3)} = -\dfrac{2}{3}$.

The minimum value is

$F\left(-\dfrac{2}{3}\right) = 3\left(-\dfrac{2}{3}\right)^2 + 4\left(-\dfrac{2}{3}\right) - 3 = -\dfrac{13}{3}$.

So, the minimum value is $-\dfrac{13}{3}$ and it occurs when $x = -\dfrac{2}{3}$.

68. If we compare $h(x) = -4x^2 - 6x + 1$ to $h(x) = ax^2 + bx + c$, we find that $a = -4$, $b = -6$, and $c = 1$. Because $a < 0$, we know the graph will open down, so the function will have a maximum value.
The maximum value occurs at

$x = -\dfrac{b}{2a} = -\dfrac{(-6)}{2(-4)} = -\dfrac{3}{4}$.

The maximum value is

$h\left(-\dfrac{3}{4}\right) = -4\left(-\dfrac{3}{4}\right)^2 - 6\left(-\dfrac{3}{4}\right) + 1 = \dfrac{13}{4}$.

So, the maximum value is $\dfrac{13}{4}$ and it occurs when $x = -\dfrac{3}{4}$.

70. a. We first recognize that the quadratic function has the leading coefficient $a = -5 < 0$. This means that the graph will open down and the function indeed has a maximum value. The maximum value occurs when $p = -\dfrac{b}{2a} = -\dfrac{600}{2(-5)} = 60$.

The revenue will be maximized when the cellular phones are sold at a price of $60.

b. The maximum revenue is obtained by evaluating the revenue function at the price found in part (a).
$R(60) = -5(60)^2 + 600(60) = 18{,}000$
The maximum revenue is $18,000.

72. First we recognize that the quadratic function has the leading coefficient $a = 0.05 > 0$. This means that the graph will open up and the function will indeed have a minimum value. The minimum value occurs when $x = -\dfrac{b}{2a} = -\dfrac{(-9)}{2(0.05)} = 90$

The marginal cost will be minimized when 90 portable CD players are produced.
To find the minimum marginal cost, we evaluate the marginal cost function for $x = 90$.
$C(90) = 0.05(90)^2 - 9(90) + 435 = 30$
The minimum marginal cost is $30.

74. a. First we recognize that the quadratic function has the leading coefficient $a = -16 < 0$. This means the graph will

ISM − 501

*Chapter 8: Quadratic Equations and Quadratic Functions*

open down and the function will indeed have a maximum.
The maximum height will occur when
$$t = -\frac{b}{2a} = -\frac{155}{2(-16)} \approx 4.84.$$
The pumpkin will reach a maximum height after about 4.84 seconds.

b. The maximum height can be found by evaluating $s(t)$ for the value of $t$ found in part (a).
$$s(4.84) = -16(4.84)^2 + 155(4.84) + 8 \approx 383.39$$
The pumpkin will reach a maximum height of 383.39 feet.

c. When the pumpkin is on the ground, it will have a height of 0 feet. Thus, we need to solve $s(t) = 0$.
$$-16t^2 + 155t + 8 = 0$$
$$16t^2 - 155t - 8 = 0$$
$$a = 16, b = -155, c = -8$$
$$t = \frac{-(-155) \pm \sqrt{(-155)^2 - 4(16)(-8)}}{2(16)}$$
$$= \frac{155 \pm \sqrt{24537}}{32}$$
$t \approx -0.051$ or $t \approx 9.739$
Since the time of flight cannot be negative, we discard the negative solution. The pumpkin will hit the ground in about 9.739 seconds.

76. a. We first recognize that the quadratic function has the leading coefficient $a = \frac{-32}{220^2} < 0$. This means the graph will open down and the function will indeed have a maximum. The maximum height will occur when
$$x = -\frac{b}{2a} = -\frac{1}{2(-32/220^2)} = 756.25.$$
The pumpkin will reach a maximum height when it is 756.25 feet from the cannon.

b. The maximum height is obtained by evaluating $h(x)$ for the value of $x$ found in part (a).
$$h(756.25) = \frac{-32}{220^2}(756.25)^2 + (756.25) + 8$$
$$= 386.125$$
The pumpkin will reach a maximum height of 386.125 feet.

c. When the pumpkin is on the ground, it will have a height of 0 feet. Thus, we need to solve $h(x) = 0$.
$$\frac{-32}{220^2}x^2 + x + 8 = 0$$
$$a = \frac{-32}{220^2}, b = 1, c = 8$$
$$x = \frac{-1 \pm \sqrt{1^2 - 4\left(\frac{-32}{220^2}\right)(8)}}{2\left(\frac{-32}{220^2}\right)}$$
$x \approx -7.96$ or $x \approx 1520.46$
Since the horizontal distance traveled cannot be negative, we discard the negative result. The pumpkin will hit the ground at a distance of about 1520.46 feet from the cannon.

d. The two answers are close. The difference is because the initial velocity component used in the formula for problem 74 was an approximation while the value used in the formula for this problem was exact. The exact formula for problem 74 would be
$$s(t) = -16t^2 + \frac{220}{\sqrt{2}}t + 8.$$
If we had used this formula, the two results would be the same.

78. a. We first recognize that the quadratic function has the leading coefficient $a = -0.008 < 0$. This means the graph will open down and the function will indeed have a maximum.
The maximum value occurs when
$$x = -\frac{0.868}{2(-0.008)} = 54.25.$$
Advanced degrees will be maximized at an age of about 54.25 years.

b. $P(54.25)$
$$= -0.008(54.25)^2 + 0.868(54.25) - 11.884$$
$$= 11.6605$$
According to the model, about 11.66% of Americans aged 54.25 will have earned an advanced degree.

80. Let $x$ represent the first number. Then the second number must be $50 - x$. We can express the product of the two numbers as the function
$$p(x) = x(50 - x) = -x^2 + 50x$$

### Section 8.5 Graphing Quadratic Functions Using Properties

This is a quadratic function with $a = -1$, $b = 50$, and $c = 0$. The function is maximized when
$$x = -\frac{b}{2a} = -\frac{50}{2(-1)} = 25.$$
The maximum product can be obtained by evaluating $p(x)$ when $x = 25$.
$$p(x) = 25(50-25) = 25(25) = 625$$
Two numbers that sum to 50 have a maximum product of 625 when both numbers are 25.

**82.** Let $x$ represent the smaller number. Then the larger number must be $x + 10$. We can express the product of the two numbers as the function
$$p(x) = x(x+10) = x^2 + 10x.$$
This is a quadratic function with $a = 1$, $b = 10$, and $c = 0$. The product will be a minimum when
$$x = -\frac{b}{2a} = -\frac{10}{2(1)} = -5.$$
The minimum product can be found by evaluating $p(x)$ when $x = -5$.
$$p(-5) = -5(-5+10) = -5(5) = -25$$
Two numbers whose difference is 10 have a minimum product of $-25$ when the smaller number is $-5$ and the larger number is 5.

**83.** Let $l$ = length and $w$ = width.

**84.** Let $l$ = length and $w$ = width.
The area of a rectangle is given by $A = l \cdot w$. Before we can work on maximizing area, we need to get the equation in terms of one independent variable.
The 800 yards of fencing will form the perimeter of the rectangle. That is, we have
$2l + 2w = 800$.
We can solve this equation for $l$ and substitute the result in the area equation.
$2l + 2w = 800$
$l + w = 400$
$l = 400 - w$
Thus, the area equation becomes
$A = l \cdot w = (400-w) \cdot w = -w^2 + 400w$
Since $a = -1 < 0$, we know the graph opens down, so there will be a maximum area. This occurs when $w = -\frac{b}{2a} = -\frac{400}{2(-1)} = 200$.
The maximum area can be found by substituting this value for $w$ in the area equation.
$A = 200(400 - 200) = 200(200) = 40{,}000$
The rectangular field will have a maximum area of 40,000 square yards when the field measures 200 yards × 200 yards.

**86.** Let $x$ represent the length of the side of the pen with the dividing fence parallel to it. Then $\frac{8000 - 3x}{2} = 4000 - 1.5x$ will be the length of the other side. (see figure).

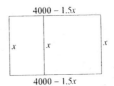

The area of a rectangular region is the product of the length and width.
$A = x(4000 - 1.5x) = -1.5x^2 + 4000x$
Since $a = -1.5 < 0$, we know the graph opens down and there will be a maximum area. This value occurs when
$$x = -\frac{b}{2a} = -\frac{4000}{2(-1.5)} = \frac{4000}{3} = 1333\tfrac{1}{3}.$$
The maximum area can be found by substituting this value for $x$ into the area equation.
$$A = \frac{4000}{3}\left(4000 - 1.5\left(\frac{4000}{3}\right)\right) = \frac{4000}{3}(2000)$$
$$= \frac{8{,}000{,}000}{3} = 2{,}666{,}666\tfrac{2}{3}$$
The rectangular field will have a maximum area of $\frac{8{,}000{,}000}{3} = 2{,}666{,}666\tfrac{2}{3}$ square meters when the field measures $\frac{4000}{3}$ m × 2000 m or $1333\tfrac{1}{3}$ m × 2000 m.

**88.** Let $l$ = length and $w$ = width of the rectangular base. The volume of a rectangular box is given by $V = l \cdot w \cdot h$. We are given that $h = 15$ inches. Before we can work on maximizing volume, we need to get the equation in terms of one independent variable. Since the perimeter of the base is 40 inches, we have $2l + 2w = 40$. We can solve this equation for $l$ and substitute the result in the volume equation.

$2l + 2w = 40$
$l + w = 20$
$l = 20 - w$
$V = (20 - w) \cdot w \cdot 15 = -15w^2 + 300w$

Since $a = -15 < 0$, the graph will open down and the function will have a maximum value. This value occurs when
$$w = -\frac{b}{2a} = -\frac{300}{2(-15)} = 10$$

The maximum volume can be found by substituting this value for $x$ in the area equation.
$V = (20 - 10) \cdot 10 \cdot 15 = 10 \cdot 10 \cdot 15 = 1500$

The box will have a maximum volume of 1500 cubic inches. The dimensions of the box will be 10 inches $\times$ 10 inches $\times$ 15 inches.

90. a. Since $R = x \cdot p$, we have
$R = (-800p + 8000) \cdot p = -800p^2 + 8000p$

b. The revenue function is quadratic with $a = -800 < 0$. This means the graph will open down and the function will have a maximum value. This value occurs when
$$p = -\frac{b}{2a} = -\frac{8000}{2(-800)} = 5.$$

The maximum revenue can be found by substituting this value for $p$ in the revenue equation.
$R = (-800(5) + 8000) \cdot 5 = 4000 \cdot 5 = 20,000$

There will be a maximum revenue of \$20,000 if the price is set at \$5 for each hotdog.

c. To determine how many hotdogs will be sold, we substitute the maximizing price into the demand equation.
$x = -800p + 8000$
$\phantom{x} = -800(5) + 8000$
$\phantom{x} = 4000$

If the price is \$5, then 4000 hotdogs will be sold.

92. a. $a = 1$:
$f(x) = 1(x + 1)(x - 5) = x^2 - 4x - 5$

$a = 2$:
$f(x) = 2(x + 1)(x - 5) = 2x^2 - 8x - 10$

$a = -2$:
$f(x) = -2(x + 1)(x - 5) = -2x^2 + 8x + 10$

b. The value of $a$ has no effect on the $x$-intercepts. These depend only on the factors and are given in the problem.
The value of $a$ does have an effect on the $y$-intercept which can be expressed as $c = a \cdot r_1 \cdot r_2 = -5a$.

c. The value of $a$ has no effect on the axis of symmetry. The axis of symmetry lies halfway between the two $x$-intercepts which are fixed. Note that in this case we have
$f(x) = ax^2 - 4ax - 5a$
The axis of symmetry would be
$$x = -\frac{(-4a)}{2(a)} = \frac{4a}{2a} = 2$$
which does not depend on $a$.

d. Consider the general function in this case written in the form $f(x) = a(x - h)^2 + k$.
$f(x) = ax^2 - 4ax - 5a$
$\phantom{f(x)} = a(x^2 - 4x) - 5a$
$\phantom{f(x)} = a(x^2 - 4x + 4) - 5a - 4a$
$\phantom{f(x)} = a(x - 2)^2 - 9a$

The $x$-coordinate of the vertex is 2, which does not depend on $a$. However, the $y$-coordinate is $-9a$ which does depend on $a$.

94. $f(x) = -2x + 12$
Let $x = -2$, 0, and 2.
$f(-2) = -2(-2) + 12 = 4 + 12 = 16$
$f(0) = -2(0) + 12 = 0 + 12 = 12$
$f(2) = -2(2) + 12 = -4 + 12 = 8$
Thus, the points $(-4, -3)$, $(0, -2)$, and $(4, -1)$ are on the graph.

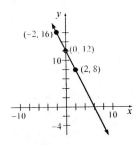

96. $f(x) = x^2 - 5$

Let $x = -3, -2, -1, 0, 1, 2,$ and $3$.

$f(-3) = (-3)^2 - 5 = 9 - 5 = 4$

$f(-2) = (-2)^2 - 5 = 4 - 5 = -1$

$f(-1) = (-1)^2 - 5 = 1 - 5 = -4$

$f(0) = 0^2 - 5 = 0 - 5 = -5$

$f(1) = 1^2 - 5 = 1 - 5 = -4$

$f(2) = 2^2 - 5 = 4 - 5 = -1$

$f(3) = 3^2 - 5 = 9 - 5 = 4$

Thus, the points $(-3, 4)$, $(-2, -1)$, $(-1, -4)$, $(0, -5)$, $(1, -4)$, $(2, -1)$, and $(3, 4)$ are on the graph.

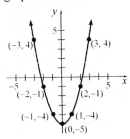

98. Answers may vary. Possibilities follow:
For problem 94, we could have used the slope-intercept form to graph the line $f(x) = -2x + 12$. Doing so, we would begin at the $y$-intercept point $(0, 12)$. Using the slope $-2$, we can move to the right 1 unit and down 2 units to find the point $(1, 10)$. We could also have moved to the left 1 unit and up 2 units to find the point $(-2, 14)$. Connect the points with a line to obtain the graph.
For problem 95, we could have used the slope-intercept form to graph the line $y = \dfrac{1}{4}x - 2$.

Doing so, we would begin at the $y$-intercept point $(0, -2)$. Using the slope $\dfrac{1}{4}$, we can move to the right 4 units and up 1 unit to find the point $(4, -1)$. We could also have moved to the left 4 units and down 1 unit to find the point $(-4, -3)$. Connect the points with a line to obtain the graph.

For problem 96, begin with the graph $y = x^2$. Shift the graph down 5 units to obtain the graph of $f(x) = x^2 - 5$.

For problem 97, begin with the graph $y = x^2$. Shift the graph to the left 2 units to obtain the graph of $y = (x+2)^2$. Then shift up 4 units to obtain the graph of $f(x) = (x+2)^2 + 4$.

100. Vertex: $(-1.5, 5.75)$

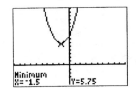

102. Vertex: $(-0.125, 11.0625)$

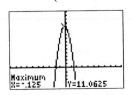

104. Vertex: $(-0.33, -21.33)$

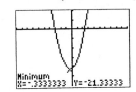

*Chapter 8: Quadratic Equations and Quadratic Functions*

**106.** Vertex: $(-0.89, -1.59)$

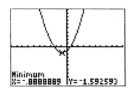

**108.** $b$ has an effect on the location of the vertex of the parabola. In terms of transformations, we can write: $f(x) = x^2 + bx + 1$

$$= (x^2 + bx) + 1$$
$$= \left(x^2 + bx + \left(\frac{b}{2a}\right)^2\right) + 1 - \left(\frac{b}{2a}\right)^2$$
$$= \left(x + \frac{b}{2a}\right)^2 + 1 - \frac{b^2}{4a^2}$$
$$= \left(x + \frac{b}{2a}\right)^2 + \frac{4a^2 - b^2}{4a^2}$$

The vertex for this family of parabolas will be $\left(-\frac{b}{2a}, \frac{4a^2 - b^2}{4a^2}\right)$.

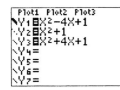 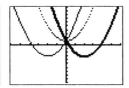

### 8.6 Preparing for Quadratic Inequalities

**1.** The inequality $-4 \leq x < 5$ in interval notation is $[-4, 5)$.

**2.**
$$3x + 5 > 5x - 3$$
$$3x + 5 - 5x > 5x - 3 - 5x$$
$$-2x + 5 > -3$$
$$-2x + 5 - 5 > -3 - 5$$
$$-2x > -8$$
$$\frac{-2x}{-2} < \frac{-8}{-2}$$
$$x < 4$$

The solution set is $\{x \mid x < 4\}$ or, using interval notation, $(-\infty, 4)$.

### 8.6 Exercises

**2.** negative

**4.** TRUE.

**6.** Answers may vary. One possibility follows: $x^2 \geq 0$ for all real values of $x$, so $x^2 - 1 \geq -1$ for all real values of $x$. That is, $x^2 - 1$ is always $-1$ or larger.

**8.** No. If $x = 0$, we obtain the following:
$$x^2 + 1 > 1$$
$$0^2 + 1 > 1$$
$$1 > 1 \leftarrow \text{False}$$

Thus, 0 is not a solution to the inequality. The inequality is true for all other real numbers. Thus, the solution set to the inequality $x^2 + 1 > 1$ is all real numbers except 0. That is, the solution set is $\{x \mid x \neq 0\}$ or, using interval notation, $(-\infty, 0) \cup (0, \infty)$.

**10. a.** The graph is greater than 0 for $x < -7$ or $x > 4$. The solution is $\{x \mid x < -7 \text{ or } x > 4\}$ using set-builder notation; the solution is $(-\infty, -7) \cup (4, \infty)$ using interval notation.

**b.** The graph is 0 or less for $-7 \leq x \leq 4$. The solution is $\{x \mid -7 \leq x \leq 4\}$ using set-builder notation; the solution is $[-7, 4]$ using interval notation.

**12. a.** The graph is greater than 0 for $-4 < x < \frac{7}{2}$. The solution is $\left\{x \mid -4 < x < \frac{7}{2}\right\}$ using set-builder notation; the solution is $\left(-4, \frac{7}{2}\right)$ using interval notation.

**b.** The graph is 0 or less for $x \leq -4$ or $x \geq \frac{7}{2}$. The solution is $\left\{x \mid x \leq -4 \text{ or } x \geq \frac{7}{2}\right\}$ using set-builder notation; the solution is

$(-\infty, -4] \cup \left[\dfrac{7}{2}, \infty\right)$ using interval notation.

14. $(x-8)(x+1) \geq 0$

Solve: $(x-8)(x+1) = 0$

$x - 8 = 0$ or $x + 1 = 0$
$x = 8$ or $x = -1$

Determine where each factor is positive and negative and where the product of these factors is positive and negative.

| Interval | $(-\infty, -1)$ | $-1$ | $(-1, 8)$ | $8$ | $(8, \infty)$ |
|---|---|---|---|---|---|
| $x - 8$ | $----$ | $-$ | $----$ | $0$ | $++++$ |
| $x + 1$ | $----$ | $0$ | $++++$ | $+$ | $++++$ |
| $(x-8)(x+1)$ | $++++$ | $0$ | $----$ | $0$ | $++++$ |

The inequality is non-strict, so $-1$ and $8$ are part of the solution. Now, $(x-8)(x+1)$ is less than zero where the product is negative. The solution is $\{x \mid -1 \leq x \leq 8\}$ in set-builder notation; the solution is $[-1, 8]$ in interval notation.

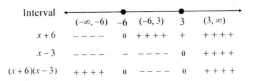

16. $(x-4)(x-10) > 0$

Solve: $(x-4)(x-10) = 0$

$x - 4 = 0$ or $x - 10 = 0$
$x = 4$ or $x = 10$

Determine where each factor is positive and negative and where the product of these factors is positive and negative.

| Interval | $(-\infty, 4)$ | $4$ | $(4, 10)$ | $10$ | $(10, \infty)$ |
|---|---|---|---|---|---|
| $x - 4$ | $----$ | $0$ | $++++$ | $+$ | $++++$ |
| $x - 10$ | $----$ | $-$ | $----$ | $0$ | $++++$ |
| $(x-4)(x-10)$ | $++++$ | $0$ | $----$ | $0$ | $++++$ |

The inequality is strict, so $4$ and $10$ are not part of the solution. Now, $(x-4)(x-10)$ is greater than zero where the product is positive. The solution is $\{x \mid x < 4 \text{ or } x > 10\}$ in set-builder notation; the solution is $(-\infty, 4) \cup (10, \infty)$ in interval notation.

18. $x^2 + 3x - 18 \geq 0$

Solve: $x^2 + 3x - 18 = 0$
$(x+6)(x-3) = 0$
$x + 6 = 0$ or $x - 3 = 0$
$x = -6$ or $x = 3$

Determine where each factor is positive and negative and where the product of these factors is positive and negative.

| Interval | $(-\infty, -6)$ | $-6$ | $(-6, 3)$ | $3$ | $(3, \infty)$ |
|---|---|---|---|---|---|
| $x + 6$ | $----$ | $0$ | $++++$ | $+$ | $++++$ |
| $x - 3$ | $----$ | $-$ | $----$ | $0$ | $++++$ |
| $(x+6)(x-3)$ | $++++$ | $0$ | $----$ | $0$ | $++++$ |

The inequality is non-strict, so $-6$ and $3$ are part of the solution. Now, $(x+6)(x-3)$ is greater than zero where the product is positive. The solution is $\{x \mid x \leq -6 \text{ or } x \geq 3\}$ in set-builder notation; the solution is $(-\infty, -6] \cup [3, \infty)$ in interval notation.

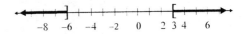

20. $p^2 + 5p + 4 < 0$

Solve: $p^2 + 5p + 4 = 0$
$(p+4)(p+1) = 0$
$p + 4 = 0$ or $p + 1 = 0$
$p = -4$ or $p = -1$

Determine where each factor is positive and negative and where the product of these factors is positive and negative.

| Interval | $(-\infty, -4)$ | $-4$ | $(-4, -1)$ | $-1$ | $(-1, \infty)$ |
|---|---|---|---|---|---|
| $p + 4$ | $----$ | $0$ | $++++$ | $+$ | $++++$ |
| $p + 1$ | $----$ | $-$ | $----$ | $0$ | $++++$ |
| $(p+4)(p+1)$ | $++++$ | $0$ | $----$ | $0$ | $++++$ |

The inequality is strict, so $-4$ and $-1$ are not part of the solution. Now, $(p+4)(p+1)$ is less than zero where the product is negative. The solution is $\{p \mid -4 < p < -1\}$ in set-builder

## Chapter 8: Quadratic Equations and Quadratic Functions

notation; the solution is $(-4, -1)$ in interval notation.

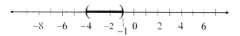

22. $z^2 > 7z + 8$
$z^2 - 7z - 8 > 0$
Solve: $z^2 - 7z - 8 = 0$
$(z + 1)(z - 8) = 0$
$z + 1 = 0$ or $z - 8 = 0$
$z = -1$ or $z = 8$

Determine where each factor is positive and negative and where the product of these factors is positive and negative.

| Interval | $(-\infty, -1)$ | $-1$ | $(-1, 8)$ | $8$ | $(8, \infty)$ |
|---|---|---|---|---|---|
| $z + 1$ | $----$ | $0$ | $++++$ | $+$ | $++++$ |
| $z - 8$ | $----$ | $-$ | $----$ | $0$ | $++++$ |
| $(z+1)(z-8)$ | $++++$ | $0$ | $----$ | $0$ | $++++$ |

The inequality is strict, so $-1$ and $8$ are not part of the solution. Now, $(z+1)(z-8)$ is greater than zero where the product is positive. The solution is $\{z \mid z < -1 \text{ or } z > 8\}$ in set-builder notation; the solution is $(-\infty, -1) \cup (8, \infty)$ in interval notation.

24. $2b^2 + 5b < 7$
$2b^2 + 5b - 7 < 0$
Solve: $2b^2 + 5b - 7 = 0$
$(2b + 7)(b - 1) = 0$
$2b + 7 = 0$ or $b - 1 = 0$
$b = -\dfrac{7}{2}$ or $b = 1$

Determine where each factor is positive and negative and where the product of these factors is positive and negative.

| Interval | $\left(-\infty, -\dfrac{7}{2}\right)$ | $-\dfrac{7}{2}$ | $\left(-\dfrac{7}{2}, 1\right)$ | $1$ | $(1, \infty)$ |
|---|---|---|---|---|---|
| $2p + 7$ | $----$ | $0$ | $++++$ | $+$ | $++++$ |
| $p - 1$ | $----$ | $-$ | $----$ | $0$ | $++++$ |
| $(2p+7)(p-1)$ | $++++$ | $0$ | $----$ | $0$ | $++++$ |

The inequality is non-strict, so $-\dfrac{7}{2}$ and $1$ are part of the solution. Now, $(2b+7)(b-1)$ is greater than zero where the product is positive. The solution is $\left\{b \mid -\dfrac{7}{2} < b < 1\right\}$ in set-builder notation; the solution is $\left(-\dfrac{7}{2}, 1\right)$ in interval notation.

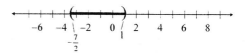

26. $x + 6 < x^2$
$0 < x^2 - x - 6$
$x^2 - x - 6 > 0$
Solve: $x^2 - x - 6 = 0$
$(x + 2)(x - 3) = 0$
$x + 2 = 0$ or $x - 3 = 0$
$x = -2$ or $x = 3$

Determine where each factor is positive and negative and where the product of these factors is positive and negative.

| Interval | $(-\infty, -2)$ | $-2$ | $(-2, 3)$ | $3$ | $(3, \infty)$ |
|---|---|---|---|---|---|
| $x + 2$ | $----$ | $0$ | $++++$ | $+$ | $++++$ |
| $x - 3$ | $----$ | $-$ | $----$ | $0$ | $++++$ |
| $(x+2)(x-3)$ | $++++$ | $0$ | $----$ | $0$ | $++++$ |

The inequality is strict, so $-2$ and $3$ are not part of the solution. Now, $(x+2)(x-3)$ is greater than zero where the product is positive. The solution is $\{x \mid x < -2 \text{ or } x > 3\}$ in set-builder notation; the solution is $(-\infty, -2) \cup (3, \infty)$ in interval notation.

28. $-x^2 > 4x - 21$
$0 > x^2 + 4x - 21$
$x^2 + 4x - 21 < 0$
Solve: $x^2 + 4x - 21 = 0$
$(x + 7)(x - 3) = 0$
$x + 7 = 0$ or $x - 3 = 0$
$x = -7$ or $x = 3$

Determine where each factor is positive and

negative and where the product of these factors is positive and negative.

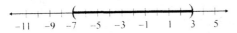

The inequality is strict, so $-7$ and $3$ are not part of the solution. Now, $(x+7)(x-3)$ is less than zero where the product is negative. The solution is $\{x\,|\,-7<x<3\}$ in set-builder notation; the solution is $(-7,\,3)$ in interval notation.

30. $-3m^2 \geq 16m+5$
$0 \geq 3m^2 + 16m + 5$
$3m^2 + 16m + 5 \leq 0$
Solve: $3m^2 + 16m + 5 = 0$
$(3m+1)(m+5) = 0$
$3m+1 = 0$ or $m+5 = 0$
$m = -\dfrac{1}{3}$ or $m = -5$

Determine where each factor is positive and negative and where the product of these factors is positive and negative.

The inequality is non-strict, so $-\dfrac{1}{3}$ and $-5$ are part of the solution. Now, $(3m+1)(m+5)$ is less than zero where the product is negative. The solution is $\left\{m\,|\,-5 \leq m \leq -\dfrac{1}{3}\right\}$ in set-builder notation; the solution is $\left[-5,\,-\dfrac{1}{3}\right]$ in interval notation.

32. $x^2 - 3x - 5 \geq 0$
Solve: $x^2 - 3x - 5 = 0$
$x = \dfrac{-(-3) \pm \sqrt{(-3)^2 - 4(1)(-5)}}{2(1)}$
$= \dfrac{3 \pm \sqrt{29}}{2}$
$x = \dfrac{3 - \sqrt{29}}{2}$ or $x = \dfrac{3 + \sqrt{29}}{2}$
$\approx -1.19$ or $\approx 4.19$

Determine where each factor is positive and negative and where the product of these factors is positive and negative.

The inequality is non-strict, so $\dfrac{3-\sqrt{29}}{2}$ and $\dfrac{3+\sqrt{29}}{2}$ are part of the solution. Now, $\left[x-\left(\dfrac{3-\sqrt{29}}{2}\right)\right]\left[x-\left(\dfrac{3+\sqrt{29}}{2}\right)\right]$ is greater than zero where the product is positive. The solution is $\left\{x\,\Big|\,x \leq \dfrac{3-\sqrt{29}}{2} \text{ or } x \geq \dfrac{3+\sqrt{29}}{2}\right\}$ in set-builder notation; the solution is $\left(-\infty,\,\dfrac{3-\sqrt{29}}{2}\right] \cup \left[\dfrac{3+\sqrt{29}}{2},\,\infty\right)$ in interval notation.

34. $-3p^2 < 3p - 5$
$0 < 3p^2 + 3p - 5$
$3p^2 + 3p - 5 > 0$
Solve: $3p^2 + 3p - 5 = 0$

$$p = \frac{-3 \pm \sqrt{3^2 - 4(3)(-5)}}{2(3)}$$
$$= \frac{-3 \pm \sqrt{69}}{6}$$
$$p = \frac{-3 - \sqrt{69}}{6} \text{ or } p = \frac{-3 + \sqrt{69}}{6}$$
$$\approx -1.88 \text{ or } \approx 0.88$$

Determine where each factor is positive and negative and where the product of these factors is positive and negative.

The inequality is strict, so $\frac{-3 - \sqrt{69}}{6}$ and $\frac{-3 + \sqrt{69}}{6}$ are not part of the solution. Now,
$$\left[p - \left(\frac{-3 - \sqrt{69}}{6}\right)\right]\left[p - \left(\frac{-3 + \sqrt{69}}{6}\right)\right] \text{ is greater than}$$
zero where the product is positive. The solution is
$\left\{p \mid p < \frac{-3 - \sqrt{69}}{6} \text{ or } p > \frac{-3 + \sqrt{69}}{6}\right\}$ in set-builder notation; the solution is $\left(-\infty, \frac{-3 - \sqrt{69}}{6}\right) \cup \left(\frac{-3 + \sqrt{69}}{6}, \infty\right)$ in interval notation.

36. $y^2 + 3y + 5 \geq 0$
Solve: $y^2 + 3y + 5 = 0$

$$y = \frac{-3 \pm \sqrt{3^2 - 4(1)(5)}}{2(1)}$$
$$= \frac{-3 \pm \sqrt{-11}}{2}$$
$$= \frac{-3 \pm \sqrt{11}\, i}{2}$$
$$= -\frac{3}{2} \pm \frac{\sqrt{11}}{2} i$$

The solutions to the equation are non-real. This means that $y^2 + 3y + 5$ will not divide the number line into positive and negative intervals. Instead, $y^2 + 3y + 5$ will either be positive on the entire number line or be negative on the entire number line. The graph below shows that $f(y) = y^2 + 3y + 5$ is always positive.

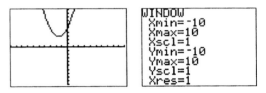

This means that $y^2 + 3y + 5$ is always greater than zero. The solution is $\{y \mid y \text{ is any real number}\}$; the solution is $(-\infty, \infty)$ in interval notation.

38. $3w^2 + w < -2$
$3w^2 + w + 2 < 0$
Solve: $3w^2 + w + 2 = 0$
$$w = \frac{-1 \pm \sqrt{1^2 - 4(3)(2)}}{2(3)}$$
$$= \frac{-1 \pm \sqrt{-23}}{6} = \frac{-1 \pm \sqrt{23}\, i}{6} = -\frac{1}{6} \pm \frac{\sqrt{23}}{6} i$$

The solutions to the equation are non-real. This means that $3w^2 + w + 2$ will not divide the number line into positive and negative intervals. Instead, $3w^2 + w + 2$ will either be positive on the entire number line or be negative on the entire number line. The graph below shows that $f(w) = 3w^2 + w + 2$ is always positive.

### Section 8.6 Quadratic Inequalities

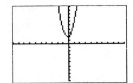

This means that $3w^2 + w + 2$ is never less than or equal to zero. The quadratic inequality has no solution: { } or $\emptyset$.

**40.** $p^2 - 8p + 16 \leq 0$

Solve: $p^2 - 8p - 16 = 0$
$(p-4)(p-4) = 0$
$p - 4 = 0$ or $p - 4 = 0$
$p = 4$ or $p = 4$

Determine where each factor is positive and negative and where the product of these factors is positive and negative.

| Interval | $(-\infty, 4)$ | 4 | $(4, \infty)$ |
|---|---|---|---|
| $p - 4$ | $----$ | 0 | $++++$ |
| $p - 4$ | $----$ | 0 | $++++$ |
| $(p-4)(p-4)$ | $++++$ | 0 | $++++$ |

The inequality is non-strict, so 4 is part of the solution. Now, $(p-4)(p-4)$ is less than zero where the product is negative. Thus, $p^2 - 8p + 16$ is never less than zero. However, $p^2 - 8p + 16$ does equal zero when $p = 4$. Thus, the solution is $\{4\}$.

**42.** $f(x) > 0$
$x^2 + 4x > 0$

Solve: $x^2 + 4x = 0$
$x(x+4) = 0$
$x = 0$ or $x + 4 = 0$
or $x = -4$

Determine where each factor is positive and negative and where the product of these factors is positive and negative.

| Interval | $(-\infty, -4)$ | $-4$ | $(-4, 0)$ | 0 | $(0, \infty)$ |
|---|---|---|---|---|---|
| $x$ | $----$ | $----$ | $----$ | 0 | $++++$ |
| $x + 4$ | $----$ | 0 | $++++$ | $+$ | $++++$ |
| $x(x+4)$ | $++++$ | 0 | $----$ | 0 | $++++$ |

The inequality is strict, so 0 and $-4$ are not part of the solution. Now, $x(x+4)$ is greater than zero where the product is positive. The solution is $\{x \mid x < -4 \text{ or } x > 0\}$ in set-builder notation; the solution is $(-\infty, -4) \cup (0, \infty)$ in interval notation.

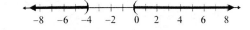

**44.** $f(x) \leq 0$
$x^2 + 2x - 48 \leq 0$

Solve: $x^2 + 2x - 48 = 0$
$(x+8)(x-6) = 0$
$x + 8 = 0$ or $x - 6 = 0$
$x = -8$ or $x = 6$

Determine where each factor is positive and negative and where the product of these factors is positive and negative.

| Interval | $(-\infty,-8)$ | $-8$ | $(-8, 6)$ | 6 | $(6, \infty)$ |
|---|---|---|---|---|---|
| $x + 8$ | $----$ | 0 | $++++$ | $+$ | $++++$ |
| $x - 6$ | $----$ | $----$ | $----$ | 0 | $++++$ |
| $(x+8)(x+4)$ | $++++$ | 0 | $----$ | 0 | $++++$ |

The inequality is non-strict, so $-8$ and 6 are part of the solution. Now, $(x+8)(x-6)$ is less than zero where the product is negative. The solution is $\{x \mid -8 \leq x \leq 6\}$ in set-builder notation; the solution is $[\,8, 6]$ in interval notation.

**46.** $F(x) < 0$
$2x^2 + 7x - 15 < 0$

Solve: $2x^2 + 7x - 15 = 0$
$(2x-3)(x+5) = 0$
$2x - 3 = 0$ or $x + 5 = 0$
$2x = 3$ or $x = -5$
$x = \dfrac{3}{2}$

Determine where each factor is positive and negative and where the product of these factors is positive and negative.

ISM – 511

## Chapter 8: Quadratic Equations and Quadratic Functions

| Interval | $(-\infty, -5)$ | $-5$ | $\left(-5, \frac{3}{2}\right)$ | $\frac{3}{2}$ | $\left(\frac{3}{2}, \infty\right)$ |
|---|---|---|---|---|---|
| $2x-3$ | $----$ | $-$ | $----$ | $0$ | $++++$ |
| $x+5$ | $----$ | $0$ | $++++$ | $+$ | $++++$ |
| $(2x-3)(x+5)$ | $++++$ | $0$ | $----$ | $0$ | $++++$ |

The inequality is strict, so $-5$ and $\frac{3}{2}$ are not part of the solution. Now, $(2x-3)(x+5)$ is less than zero where the product is negative. The solution is $\left\{x \mid -5 < x < \frac{3}{2}\right\}$ in set-builder notation; the solution is $\left(-5, \frac{3}{2}\right)$ in interval notation.

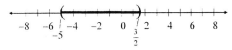

48. The domain of $f(x) = \sqrt{x^2 - 5x}$ will be the solution set of $x^2 - 5x \geq 0$.
Solve: $x^2 - 5x = 0$
$x(x-5) = 0$
$x = 0$ or $x - 5 = 0$
or $x = 5$

Determine where each factor is positive and negative and where the product of these factors is positive and negative.

| Interval | $(-\infty, 0)$ | $0$ | $(0, 5)$ | $5$ | $(5, \infty)$ |
|---|---|---|---|---|---|
| $x$ | $----$ | $0$ | $++++$ | $+$ | $++++$ |
| $x-5$ | $----$ | $-$ | $----$ | $0$ | $++++$ |
| $x(x-5)$ | $++++$ | $0$ | $----$ | $0$ | $++++$ |

The inequality is non-strict, so $0$ and $5$ are part of the solution. Now, $x(x-5)$ is greater than zero where the product is positive. Thus, the domain of $f$ is $\{x \mid x \leq 0 \text{ or } x \geq 5\}$ in set-builder notation; the domain is $(-\infty, 0] \cup [5, \infty)$ in interval notation.

50. The domain of $G(x) = \sqrt{x^2 + 2x - 63}$ will be the solution set of $x^2 + 2x - 63 \geq 0$.
Solve: $x^2 + 2x - 63 = 0$
$(x+9)(x-7) = 0$
$x + 9 = 0$ or $x - 7 = 0$
$x = -9$ or $x = 7$

Determine where each factor is positive and negative and where the product of these factors is positive and negative.

| Interval | $(-\infty, -9)$ | $-9$ | $(-9, 7)$ | $7$ | $(7, \infty)$ |
|---|---|---|---|---|---|
| $x+9$ | $----$ | $0$ | $++++$ | $+$ | $++++$ |
| $x-7$ | $----$ | $-$ | $----$ | $0$ | $++++$ |
| $(x+9)(x-7)$ | $++++$ | $0$ | $----$ | $0$ | $++++$ |

The inequality is non-strict, so $-9$ and $7$ are part of the solution. Now, $(x+9)(x-7)$ is greater than zero where the product is positive. Thus, the domain of $G$ is $\{x \mid x \leq -9 \text{ or } x \geq 7\}$ in set-builder notation; the domain is $(-\infty, -9] \cup [7, \infty)$ in interval notation.

52. The water balloon will be more than $248$ feet above sea level when $s(t) > 248$.

$-16t^2 + 64t + 200 > 248$
$0 > 16t^2 - 64t + 48$
$16t^2 - 64t + 48 < 0$
$\dfrac{16t^2 - 64t + 48}{16} < \dfrac{0}{16}$
$t^2 - 4t + 3 < 0$

Solve: $t^2 - 4t + 3 = 0$
$(t-1)(t-3) = 0$
$t - 1 = 0$ or $t - 3 = 0$
$t = 1$ or $t = 3$

Determine where each factor is positive and negative and where the product of these factors is positive and negative.

| Interval | $(-\infty, 1)$ | $1$ | $(1, 3)$ | $3$ | $(3, \infty)$ |
|---|---|---|---|---|---|
| $t-1$ | $----$ | $0$ | $++++$ | $+$ | $++++$ |
| $t-3$ | $----$ | $-$ | $----$ | $0$ | $++++$ |
| $(t-1)(t-3)$ | $++++$ | $0$ | $----$ | $0$ | $++++$ |

The inequality is strict, so $1$ and $3$ are not part of the solution. Now, $(t-1)(t-3)$ is less than zero where the product is negative. The solution is

$\{t \mid 1 < t < 3\}$ in set-builder notation; the solution is $(1, 3)$ in interval notation. Thus, the water balloon will be more than 248 feet above sea level when the time is between 1 and 3 seconds after it is thrown.

54. The revenue will exceed $17,500 when $R(p) > 17,500$.

$$-5p^2 + 600p > 17,500$$
$$0 > 5p^2 - 600p + 17,500$$
$$5p^2 - 600p + 17,500 < 0$$
$$\frac{5p^2 - 600p + 17,500}{5} < \frac{0}{5}$$
$$p^2 - 120t + 3500 < 0$$

Solve: $p^2 - 120t + 3500 = 0$
$$(p - 50)(p - 70) = 0$$
$$p - 50 = 0 \quad \text{or} \quad p - 70 = 0$$
$$p = 50 \quad \text{or} \quad p = 70$$

Determine where each factor is positive and negative and where the product of these factors is positive and negative.

| Interval | $(-\infty, 50)$ | 50 | $(50, 70)$ | 70 | $(70, \infty)$ |
|---|---|---|---|---|---|
| $p - 50$ | $----$ | 0 | $++++$ | $+$ | $++++$ |
| $p - 70$ | $----$ | $-$ | $----$ | 0 | $++++$ |
| $(p-50)(p-70)$ | $++++$ | 0 | $----$ | 0 | $++++$ |

The inequality is strict, so 50 and 70 are not part of the solution. Now, $(p-50)(p-70)$ is less than zero where the product is negative. The solution is $\{p \mid 50 < p < 70\}$ in set-builder notation; the solution is $(50, 70)$ in interval notation. Thus, the revenue will exceed $17,500 when the cellular telephone is sold for a price between $50 and $70.

56. By inspection, the only solution is all real numbers except 4. That is, the solution set is $\{x \mid x \neq 4\}$.

Explanation: The expression on the left side of the inequality is a perfect square. A perfect square is always zero or greater. Therefore, the solution will be all real numbers except for those that cause the perfect square expression to equal zero, which is 4.

58. By inspection, the inequality has no solution. That is, the solution set is $\varnothing$ or $\{\ \}$.

Explanation: The expression on the left side of the inequality is a perfect square. A perfect square must always be zero or greater. Therefore, it can never be less than –2. Thus, no values of $x$ will make the inequality true. The inequality has no solution.

60. Answers may vary. One possibility follows: We want $x > 0$ and $x < 5$. Now, $x < 5$ means $x - 5 < 0$ (negative). If we multiply a positive ($x > 0$) by a negative, we get a negative result. Thus, $x(x-5) \leq 0$. Multiplying out the expression on the left, we get $x^2 - 5x < 0$. The solution set of $x^2 - 5x < 0$ is $(0, 5)$.

NOTE: Because the solution does not contain the endpoints 0 and 5, the inequality must be strict.

62. $(x+1)(x-2)(x-5) > 0$

Solve: $(x+1)(x-2)(x-5) = 0$
$$x + 1 = 0 \quad \text{or} \quad x - 2 = 0 \quad \text{or} \quad x - 5 = 0$$
$$x = -1 \quad \text{or} \quad x = 2 \quad \text{or} \quad x = 5$$

Determine where each factor is positive and negative and where the product of these factors is positive and negative.

| Interval | $(-\infty, -1)$ | $-1$ | $(-1, 2)$ | 2 | $(2, 5)$ | 5 | $(5, \infty)$ |
|---|---|---|---|---|---|---|---|
| $x + 1$ | $---$ | 0 | $+++$ | $+$ | $+++$ | $+$ | $+++$ |
| $x - 2$ | $---$ | $-$ | $---$ | 0 | $+++$ | $+$ | $+++$ |
| $x - 5$ | $---$ | $-$ | $---$ | $-$ | $---$ | 0 | $+++$ |
| $(x+1)(x-2)(x-5)$ | $---$ | 0 | $+++$ | 0 | $---$ | 0 | $+++$ |

The inequality is strict, so $-1$, 2, and 5 are not part of the solution. Now, $(x+1)(x-2)(x-5)$ is greater than zero where the product is positive. The solution is $\{x \mid -1 < x < 2 \text{ or } x > 5\}$ in set-builder notation; the solution is $(-1, 2) \cup (5, \infty)$ in interval notation.

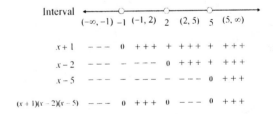

## Chapter 8: Quadratic Equations and Quadratic Functions

**64.** $(2x+1)(x-4)(x-9) \leq 0$

Solve: $(2x+1)(x-4)(x-9) = 0$

$2x+1 = 0$ or $x-4 = 0$ or $x-9 = 0$

$x = -\dfrac{1}{2}$ or $x = 4$ or $x = 9$

Determine where each factor is positive and negative and where the product of these factors is positive and negative.

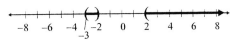

The inequality is non-strict, so $-\dfrac{1}{2}$, 4, and 9 are part of the solution. Now, $(2x+1)(x-4)(x-9)$ is less than zero where the product is negative. The solution is $\left\{x \mid x \leq -\dfrac{1}{2} \text{ or } 4 \leq x \leq 9\right\}$ in set-builder notation; the solution is $\left(-\infty, -\dfrac{1}{2}\right] \cup [4, 9]$ in interval notation.

**66.** $\dfrac{x^2 + 5x + 6}{x - 2} > 0$

$\dfrac{(x+3)(x+2)}{x-2} > 0$

The rational expression will equal 0 when $x = -3$ and when $x = -2$. It is undefined when $x = 2$. Determine where the factors of the numerator and the denominator are positive and negative and where the quotient is positive and negative.

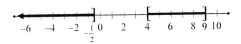

The inequality is strict, so $-3$, $-2$, and 2 are not part of the solution. Now, $\dfrac{(x+3)(x+2)}{x-2}$ is greater than zero where the quotient is positive. The solution is $\{x \mid -3 < x < -2 \text{ or } x > 2\}$ in set-builder notation; the solution is $(-3, -2) \cup (2, \infty)$ in interval notation.

**68.** $\dfrac{3a^4 b}{12a^{-3}b^5} = \dfrac{1 a^{4-(-3)}}{4 b^{5-1}} = \dfrac{a^7}{4b^4}$

**70.** $\left(\dfrac{3x^4 y}{6x^{-2} y^5}\right)^{\frac{1}{2}} = \left(\dfrac{1 x^{4-(-2)}}{2 y^{5-1}}\right)^{\frac{1}{2}}$

$= \left(\dfrac{x^6}{2y^4}\right)^{\frac{1}{2}} = \dfrac{(x^6)^{\frac{1}{2}}}{(2y^4)^{\frac{1}{2}}}$

$= \dfrac{x^3}{2^{\frac{1}{2}} y^2} = \dfrac{x^3}{y^2 \sqrt{2}} \cdot \dfrac{\sqrt{2}}{\sqrt{2}} = \dfrac{x^3 \sqrt{2}}{2y^2}$

**72.** Yes, the Laws of Exponents presented in the "Getting Ready: Integer Exponents" section also apply to the Laws of Exponents for ration exponents presented in Section 7.1.

**74.** $2x^2 + 3x - 27 < 0$

Let $Y_1 = 2x^2 + 7x - 49$. Graph the quadratic function. Use the ZERO feature to find the x-intercepts.

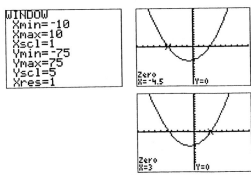

The inequality is strict, so $-4.5$ and 3 are not part of the solution. From the graph, we can see that $2x^2 + 3x - 27 < 0$ for $-4.5 < x < 3$. Thus, the solution set is $\{x \mid -4.5 < x < 3\}$ in set-builder notation; the solution is $(-4.5, 3)$ in interval notation.

76. $8x^2 + 18x \geq 81$
$8x^2 + 18x - 81 \geq 0$

Let $Y_1 = 8x^2 + 18x - 81$. Graph the quadratic function. Use the ZERO feature to find the $x$-intercepts.

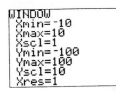
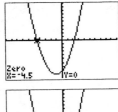

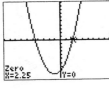

The inequality is non-strict, so $-4.5$ and $2.25$ are part of the solution. From the graph, we can see that $8x^2 + 18x - 81 \geq 0$ for $x \leq -4.5$ or $x \geq 2.25$. Thus, the solution set is $\{x \mid x \leq -4.5 \text{ or } x \geq 2.25\}$ in set-builder notation; the solution is $(-\infty, -4.5] \cup [2.25, \infty)$ in interval notation.

## Chapter 8 Review

1. $m^2 = 169$
$m = \pm\sqrt{169}$
$m = \pm 13$
The solution set is $\{-13, 13\}$.

2. $n^2 = 75$
$n = \pm\sqrt{75}$
$n = \pm 5\sqrt{3}$
The solution set is $\{-5\sqrt{3}, 5\sqrt{3}\}$.

3. $a^2 = -16$
$a = \pm\sqrt{-16}$
$a = \pm 4i$
The solution set is $\{-4i, 4i\}$.

4. $b^2 = \dfrac{8}{9}$
$b = \pm\sqrt{\dfrac{8}{9}}$
$b = \pm\dfrac{\sqrt{8}}{\sqrt{9}}$
$b = \pm\dfrac{2\sqrt{2}}{3}$

The solution set is $\left\{-\dfrac{2\sqrt{2}}{3}, \dfrac{2\sqrt{2}}{3}\right\}$.

5. $(x-8)^2 = 81$
$x - 8 = \pm\sqrt{81}$
$x - 8 = \pm 9$
$x = 8 \pm 9$
$x = 8 - 9$ or $x = 8 + 9$
$x = -1$ or $x = 17$
The solution set is $\{-1, 17\}$.

6. $(y-2)^2 - 62 = 88$
$(y-2)^2 = 150$
$y - 2 = \pm\sqrt{150}$
$y - 2 = \pm 5\sqrt{6}$
$y = 2 \pm 5\sqrt{6}$
The solution set is $\{2 - 5\sqrt{6},\ 2 + 5\sqrt{6}\}$.

7. $(3z+5)^2 = 100$
$3z + 5 = \pm\sqrt{100}$
$3z + 5 = \pm 10$
$3z = -5 \pm 10$
$3z = -5 - 10$ or $3z = -5 + 10$
$3z = -15$ or $3z = 5$
$z = -5$ or $z = \dfrac{5}{3}$
The solution set is $\left\{-5, \dfrac{5}{3}\right\}$.

## Chapter 8: Quadratic Equations and Quadratic Functions

8. $7p^2 = 18$

   $p^2 = \dfrac{18}{7}$

   $p = \pm\sqrt{\dfrac{18}{7}}$

   $p = \pm\dfrac{\sqrt{18}}{\sqrt{7}} \cdot \dfrac{\sqrt{7}}{\sqrt{7}}$

   $p = \pm\dfrac{\sqrt{126}}{\sqrt{49}}$

   $p = \pm\dfrac{3\sqrt{14}}{7}$

   The solution set is $\left\{-\dfrac{3\sqrt{14}}{7}, \dfrac{3\sqrt{14}}{7}\right\}$.

9. $3q^2 + 251 = 11$

   $3q^2 = -240$

   $q^2 = -80$

   $q = \pm\sqrt{-80}$

   $q = \pm 4\sqrt{5}\, i$

   The solution set is $\{-4\sqrt{5}\, i,\, 4\sqrt{5}\, i\}$.

10. $\left(x + \dfrac{3}{4}\right)^2 = \dfrac{13}{16}$

    $x + \dfrac{3}{4} = \pm\sqrt{\dfrac{13}{16}}$

    $x + \dfrac{3}{4} = \pm\dfrac{\sqrt{13}}{4}$

    $x = -\dfrac{3}{4} \pm \dfrac{\sqrt{13}}{4}$

    The solution set is $\left\{-\dfrac{3}{4} - \dfrac{\sqrt{13}}{4},\, -\dfrac{3}{4} + \dfrac{\sqrt{13}}{4}\right\}$.

11. Start: $a^2 + 30a$

    Add: $\left(\dfrac{1}{2} \cdot 30\right)^2 = 225$

    Result: $a^2 + 30a + 225$

    Factored Form: $(a + 15)^2$

12. Start: $b^2 - 14b$

    Add: $\left[\dfrac{1}{2} \cdot (-14)\right]^2 = 49$

    Result: $b^2 - 14b + 49$

    Factored Form: $(b - 7)^2$

13. Start: $c^2 - 11c$

    Add: $\left[\dfrac{1}{2} \cdot (-11)\right]^2 = \dfrac{121}{4}$

    Result: $c^2 - 11c + \dfrac{121}{4}$

    Factored Form: $\left(c - \dfrac{11}{2}\right)^2$

14. Start: $d^2 + 9d$

    Add: $\left(\dfrac{1}{2} \cdot 9\right)^2 = \dfrac{81}{4}$

    Result: $d^2 + 9d + \dfrac{81}{4}$

    Factored Form: $\left(d + \dfrac{9}{2}\right)^2$

15. Start: $m^2 - \dfrac{1}{4}m$

    Add: $\left[\dfrac{1}{2} \cdot \left(-\dfrac{1}{4}\right)\right]^2 = \dfrac{1}{64}$

    Result: $m^2 - \dfrac{1}{4}m + \dfrac{1}{64}$

    Factored Form: $\left(m - \dfrac{1}{8}\right)^2$

16. Start: $n^2 + \dfrac{6}{7}n$

    Add: $\left(\dfrac{1}{2} \cdot \dfrac{6}{7}\right)^2 = \dfrac{9}{49}$

    Result: $n^2 + \dfrac{6}{7}n + \dfrac{9}{49}$

    Factored Form: $\left(n + \dfrac{3}{7}\right)^2$

17. 
$$x^2 - 10x + 16 = 0$$
$$x^2 - 10x = -16$$
$$x^2 - 10x + \left(\frac{1}{2} \cdot (-10)\right)^2 = -16 + \left(\frac{1}{2} \cdot (-10)\right)^2$$
$$x^2 - 10x + 25 = -16 + 25$$
$$(x-5)^2 = 9$$
$$x - 5 = \pm\sqrt{9}$$
$$x - 5 = \pm 3$$
$$x = 5 \pm 3$$
$$x = 2 \text{ or } x = 8$$
The solution set is $\{2, 8\}$.

18. 
$$y^2 - 3y - 28 = 0$$
$$y^2 - 3y = 28$$
$$y^2 - 3y + \left(\frac{1}{2} \cdot (-3)\right)^2 = 28 + \left(\frac{1}{2} \cdot (-3)\right)^2$$
$$y^2 - 3y + \frac{9}{4} = 28 + \frac{9}{4}$$
$$\left(y - \frac{3}{2}\right)^2 = \frac{121}{4}$$
$$y - \frac{3}{2} = \pm\sqrt{\frac{121}{4}}$$
$$y - \frac{3}{2} = \pm\frac{11}{2}$$
$$y = \frac{3}{2} \pm \frac{11}{2}$$
$$y = -4 \text{ or } y = 7$$
The solution set is $\{-4, 7\}$.

19. 
$$z^2 - 6z - 3 = 0$$
$$z^2 - 6z = 3$$
$$z^2 - 6z + \left(\frac{1}{2} \cdot (-6)\right)^2 = 3 + \left(\frac{1}{2} \cdot (-6)\right)^2$$
$$z^2 - 6z + 9 = 3 + 9$$
$$(z-3)^2 = 12$$
$$z - 3 = \pm\sqrt{12}$$
$$z - 3 = \pm 2\sqrt{3}$$
$$z = 3 \pm 2\sqrt{3}$$
The solution set is $\{3 - 2\sqrt{3},\ 3 + 2\sqrt{3}\}$.

20. 
$$a^2 - 5a - 7 = 0$$
$$a^2 - 5a = 7$$
$$a^2 - 5a + \left(\frac{1}{2} \cdot (-5)\right)^2 = 7 + \left(\frac{1}{2} \cdot (-5)\right)^2$$
$$a^2 - 5a + \frac{25}{4} = 7 + \frac{25}{4}$$
$$\left(a - \frac{5}{2}\right)^2 = \frac{53}{4}$$
$$a - \frac{5}{2} = \pm\sqrt{\frac{53}{4}}$$
$$a - \frac{5}{2} = \pm\frac{\sqrt{53}}{2}$$
$$a = \frac{5}{2} \pm \frac{\sqrt{53}}{2}$$
The solution set is $\left\{\frac{5}{2} - \frac{\sqrt{53}}{2},\ \frac{5}{2} + \frac{\sqrt{53}}{2}\right\}$.

21. 
$$b^2 + b + 7 = 0$$
$$b^2 + b = -7$$
$$b^2 + b + \left(\frac{1}{2} \cdot 1\right)^2 = -7 + \left(\frac{1}{2} \cdot 1\right)^2$$
$$b^2 + b + \frac{1}{4} = -7 + \frac{1}{4}$$
$$\left(b + \frac{1}{2}\right)^2 = -\frac{27}{4}$$
$$b + \frac{1}{2} = \pm\sqrt{-\frac{27}{4}}$$
$$b + \frac{1}{2} = \pm\frac{3\sqrt{3}}{2}i$$
$$b = -\frac{1}{2} \pm \frac{3\sqrt{3}}{2}i$$
The solution set is $\left\{-\frac{1}{2} - \frac{3\sqrt{3}}{2}i,\ -\frac{1}{2} + \frac{3\sqrt{3}}{2}i\right\}$.

## Chapter 8: Quadratic Equations and Quadratic Functions

**22.**
$$c^2 - 6c + 17 = 0$$
$$c^2 - 6c = -17$$
$$c^2 - 6c + \left(\frac{1}{2} \cdot (-6)\right)^2 = -17 + \left(\frac{1}{2} \cdot (-6)\right)^2$$
$$c^2 - 6c + 9 = -17 + 9$$
$$(c-3)^2 = -8$$
$$c - 3 = \pm\sqrt{-8}$$
$$c - 3 = \pm 2\sqrt{2}\, i$$
$$c = 3 \pm 2\sqrt{2}\, i$$
The solution set is $\{3 - 2\sqrt{2}\, i,\ 3 + 2\sqrt{2}\, i\}$.

**23.**
$$2d^2 - 7d + 3 = 0$$
$$\frac{2d^2 - 7d + 3}{2} = \frac{0}{2}$$
$$d^2 - \frac{7}{2}d + \frac{3}{2} = 0$$
$$d^2 - \frac{7}{2}d = -\frac{3}{2}$$
$$d^2 - \frac{7}{2}d + \left[\frac{1}{2} \cdot \left(-\frac{7}{2}\right)\right]^2 = -\frac{3}{2} + \left[\frac{1}{2} \cdot \left(-\frac{7}{2}\right)\right]^2$$
$$d^2 - \frac{7}{2}d + \frac{49}{16} = -\frac{3}{2} + \frac{49}{16}$$
$$\left(d - \frac{7}{4}\right)^2 = \frac{25}{16}$$
$$d - \frac{7}{4} = \pm\sqrt{\frac{25}{16}}$$
$$d - \frac{7}{4} = \pm\frac{5}{4}$$
$$d = \frac{7}{4} \pm \frac{5}{4}$$
$$d = \frac{1}{2}\ \text{or}\ d = 3$$
The solution set is $\left\{\frac{1}{2},\ 3\right\}$.

**24.**
$$2w^2 + 2w + 5 = 0$$
$$\frac{2w^2 + 2w + 5}{2} = \frac{0}{2}$$
$$w^2 + w + \frac{5}{2} = 0$$
$$w^2 + w = -\frac{5}{2}$$
$$w^2 + w + \left(\frac{1}{2} \cdot 1\right)^2 = -\frac{5}{2} + \left(\frac{1}{2} \cdot 1\right)^2$$
$$w^2 + w + \frac{1}{4} = -\frac{5}{2} + \frac{1}{4}$$
$$\left(w + \frac{1}{2}\right)^2 = -\frac{9}{4}$$
$$w + \frac{1}{2} = \pm\sqrt{-\frac{9}{4}}$$
$$w + \frac{1}{2} = \pm\frac{3}{2}i$$
$$w = -\frac{1}{2} \pm \frac{3}{2}i$$
The solution set is $\left\{-\frac{1}{2} - \frac{3}{2}i,\ -\frac{1}{2} + \frac{3}{2}i\right\}$.

25. $3x^2 - 9x + 8 = 0$

$\dfrac{3x^2 - 9x + 8}{3} = \dfrac{0}{3}$

$x^2 - 3x + \dfrac{8}{3} = 0$

$x^2 - 3x = -\dfrac{8}{3}$

$x^2 - 3x + \left(\dfrac{1}{2}\cdot(-3)\right)^2 = -\dfrac{8}{3} + \left(\dfrac{1}{2}\cdot(-3)\right)^2$

$x^2 - 3x + \dfrac{9}{4} = -\dfrac{8}{3} + \dfrac{9}{4}$

$\left(x - \dfrac{3}{2}\right)^2 = -\dfrac{5}{12}$

$x - \dfrac{3}{2} = \pm\sqrt{-\dfrac{5}{12}}$

$x - \dfrac{3}{2} = \pm\dfrac{\sqrt{5}}{\sqrt{12}}i \cdot \dfrac{\sqrt{3}}{\sqrt{3}}$

$x - \dfrac{3}{2} = \pm\dfrac{\sqrt{15}}{\sqrt{36}}i$

$x - \dfrac{3}{2} = \pm\dfrac{\sqrt{15}}{6}i$

$x = \dfrac{3}{2} \pm \dfrac{\sqrt{15}}{6}i$

The solution set is $\left\{\dfrac{3}{2} - \dfrac{\sqrt{15}}{6}i,\ \dfrac{3}{2} + \dfrac{\sqrt{15}}{6}i\right\}$.

26. $3x^2 + 4x - 2 = 0$

$x^2 + \dfrac{4}{3}x - \dfrac{2}{3} = 0$

$x^2 + \dfrac{4}{3}x = \dfrac{2}{3}$

$x^2 + \dfrac{4}{3}x + \left(\dfrac{1}{2}\cdot\dfrac{4}{3}\right)^2 = \dfrac{2}{3} + \left(\dfrac{1}{2}\cdot\dfrac{4}{3}\right)^2$

$x^2 + \dfrac{4}{3}x + \dfrac{4}{9} = \dfrac{2}{3} + \dfrac{4}{9}$

$\left(x + \dfrac{2}{3}\right)^2 = \dfrac{10}{9}$

$x + \dfrac{2}{3} = \pm\sqrt{\dfrac{10}{9}}$

$x + \dfrac{2}{3} = \pm\dfrac{\sqrt{10}}{3}$

$x = -\dfrac{2}{3} \pm \dfrac{\sqrt{10}}{3}$

The solution set is $\left\{-\dfrac{2}{3} - \dfrac{\sqrt{10}}{3},\ -\dfrac{2}{3} + \dfrac{\sqrt{10}}{3}\right\}$.

27. $c^2 = 9^2 + 12^2$
$= 81 + 144$
$= 225$
$c = \sqrt{225}$
$= 15$

28. $c^2 = 8^2 + 8^2$
$= 64 + 64$
$= 128$
$c = \sqrt{128}$
$= 8\sqrt{2}$

29. $c^2 = 3^2 + 6^2$
$= 9 + 36$
$= 45$
$c = \sqrt{45}$
$= 3\sqrt{5}$

30. $c^2 = 10^2 + 24^2$
$= 100 + 576$
$= 676$
$c = \sqrt{676}$
$= 26$

## Chapter 8: Quadratic Equations and Quadratic Functions

**31.** $c^2 = 5^2 + \left(\sqrt{11}\right)^2$
$= 25 + 11$
$= 36$
$c = \sqrt{36}$
$= 6$

**32.** $c^2 = 6^2 + \left(\sqrt{13}\right)^2$
$= 36 + 13$
$= 49$
$c = \sqrt{49}$
$= 7$

**33.** $c^2 = a^2 + b^2$
$12^2 = 9^2 + b^2$
$144 = 81 + b^2$
$63 = b^2$
$b = \sqrt{63}$
$b = 3\sqrt{7}$

**34.** $c^2 = a^2 + b^2$
$10^2 = a^2 + 5^2$
$100 = a^2 + 25$
$75 = a^2$
$a = \sqrt{75}$
$a = 5\sqrt{3}$

**35.** $c^2 = a^2 + b^2$
$17^2 = a^2 + 6^2$
$289 = a^2 + 36$
$253 = a^2$
$a = \sqrt{253}$

**36.** For this problem, we are actually looking for the hypotenuse of a right triangle that has legs 90 feet and 90 feet. We use the Pythagorean Theorem to find the desired distance:
$c^2 = 90^2 + 90^2$
$= 8100 + 8100$
$= 16,200$
$c = \sqrt{16,200}$
$= 90\sqrt{2}$
$\approx 127.3$
The distance from home plate to 2$^{nd}$ base is exactly $90\sqrt{2}$ feet or approximately 127.3 feet.

**37.** $x^2 - x - 20 = 0$
For this equation, $a = 1$, $b = -1$, and $c = -20$.

$x = \dfrac{-(-1) \pm \sqrt{(-1)^2 - 4(1)(-20)}}{2(1)}$
$= \dfrac{1 \pm \sqrt{1 + 80}}{2}$
$= \dfrac{1 \pm \sqrt{81}}{2}$
$= \dfrac{1 \pm 9}{2}$

$x = \dfrac{1 - 9}{2}$ or $x = \dfrac{1 + 9}{2}$
$= \dfrac{-8}{2}$ or $= \dfrac{10}{2}$
$= -4$ or $= 5$
The solution set is $\{-4, 5\}$.

**38.** $4y^2 = 8y + 21$
$4y^2 - 8y - 21 = 0$
For this equation, $a = 4$, $b = -8$, and $c = -21$.

$y = \dfrac{-(-8) \pm \sqrt{(-8)^2 - 4(4)(-21)}}{2(4)}$
$= \dfrac{8 \pm \sqrt{64 + 336}}{8}$
$= \dfrac{8 \pm \sqrt{400}}{8}$
$= \dfrac{8 \pm 20}{8}$

$y = \dfrac{8 - 20}{8}$ or $y = \dfrac{8 + 20}{8}$
$= \dfrac{-12}{8}$ or $= \dfrac{28}{8}$
$= -\dfrac{3}{2}$ or $= \dfrac{7}{2}$
The solution set is $\left\{-\dfrac{3}{2}, \dfrac{7}{2}\right\}$.

**39.** $3p^2 + 8p = -3$
$3p^2 + 8p + 3 = 0$
For this equation, $a = 3$, $b = 8$, and $c = 3$.

$$p = \frac{-8 \pm \sqrt{8^2 - 4(3)(3)}}{2(3)}$$
$$= \frac{-8 \pm \sqrt{64 - 36}}{6}$$
$$= \frac{-8 \pm \sqrt{28}}{6}$$
$$= \frac{-8 \pm 2\sqrt{7}}{6}$$
$$= \frac{-8}{6} \pm \frac{2\sqrt{7}}{6}$$
$$= -\frac{4}{3} \pm \frac{\sqrt{7}}{3}$$

The solution set is $\left\{-\frac{4}{3} - \frac{\sqrt{7}}{3}, -\frac{4}{3} + \frac{\sqrt{7}}{3}\right\}$.

40. $2q^2 - 3 = 4q$
$2q^2 - 4q - 3 = 0$
For this equation, $a = 2$, $b = -4$, and $c = -3$.
$$q = \frac{-(-4) \pm \sqrt{(-4)^2 - 4(2)(-3)}}{2(2)}$$
$$= \frac{4 \pm \sqrt{16 + 24}}{4}$$
$$= \frac{4 \pm \sqrt{40}}{4}$$
$$= \frac{4 \pm 2\sqrt{10}}{4}$$
$$= \frac{4}{4} \pm \frac{2\sqrt{10}}{4}$$
$$= 1 \pm \frac{\sqrt{10}}{2}$$

The solution set is $\left\{1 - \frac{\sqrt{10}}{2}, 1 + \frac{\sqrt{10}}{2}\right\}$.

41. $3w^2 + w = -3$
$3w^2 + w + 3 = 0$
For this equation, $a = 3$, $b = 1$, and $c = 3$.

$$w = \frac{-1 \pm \sqrt{1^2 - 4(3)(3)}}{2(3)}$$
$$= \frac{-1 \pm \sqrt{1 - 36}}{6}$$
$$= \frac{-1 \pm \sqrt{-35}}{6}$$
$$= \frac{-1 \pm \sqrt{35}\, i}{6}$$
$$= -\frac{1}{6} \pm \frac{\sqrt{35}}{6} i$$

The solution set is $\left\{-\frac{1}{6} - \frac{\sqrt{35}}{6}i, -\frac{1}{6} + \frac{\sqrt{35}}{6}i\right\}$.

42. $9z^2 + 16 = 24z$
$9z^2 - 24z + 16 = 0$
For this equation, $a = 9$, $b = -24$, and $c = 16$.
$$z = \frac{-(-24) \pm \sqrt{(-24)^2 - 4(9)(16)}}{2(9)}$$
$$= \frac{24 \pm \sqrt{576 - 576}}{18}$$
$$= \frac{24 \pm \sqrt{0}}{18}$$
$$= \frac{24}{18}$$
$$= \frac{4}{3}$$

The solution set is $\left\{\frac{4}{3}\right\}$. It is a double root.

43. $m^2 - 4m + 2 = 0$
For this equation, $a = 1$, $b = -4$, and $c = 2$.
$$m = \frac{-(-4) \pm \sqrt{(-4)^2 - 4(1)(2)}}{2(1)}$$
$$= \frac{4 \pm \sqrt{16 - 8}}{2}$$
$$= \frac{4 \pm \sqrt{8}}{2}$$
$$= \frac{4 \pm 2\sqrt{2}}{2}$$
$$= \frac{4}{2} \pm \frac{2\sqrt{2}}{2}$$
$$= 2 \pm \sqrt{2}$$

The solution set is $\left\{2 - \sqrt{2},\ 2 + \sqrt{2}\right\}$.

**Chapter 8**: Quadratic Equations and Quadratic Functions

44. $5n^2 + 4n + 1 = 0$
    For this equation, $a = 5$, $b = 4$, and $c = 1$.
    $$n = \frac{-4 \pm \sqrt{4^2 - 4(5)(1)}}{2(5)}$$
    $$= \frac{-4 \pm \sqrt{16 - 20}}{10}$$
    $$= \frac{-4 \pm \sqrt{-4}}{10}$$
    $$= \frac{-4 \pm 2i}{10}$$
    $$= -\frac{4}{10} \pm \frac{2i}{10} = -\frac{2}{5} \pm \frac{1}{5}i$$
    The solution set is $\left\{ -\frac{2}{5} - \frac{1}{5}i,\ -\frac{2}{5} + \frac{1}{5}i \right\}$.

45. $5x + 13 = -x^2$
    $x^2 + 5x + 13 = 0$
    For this equation, $a = 1$, $b = 5$, and $c = 13$.
    $$x = \frac{-5 \pm \sqrt{5^2 - 4(1)(13)}}{2(1)}$$
    $$= \frac{-5 \pm \sqrt{25 - 52}}{2}$$
    $$= \frac{-5 \pm \sqrt{-27}}{2}$$
    $$= \frac{-5 \pm 3\sqrt{3}\,i}{2} = -\frac{5}{2} \pm \frac{3\sqrt{3}}{2}i$$
    The solution set is $\left\{ -\frac{5}{2} - \frac{3\sqrt{3}}{2}i,\ -\frac{5}{2} + \frac{3\sqrt{3}}{2}i \right\}$.

46. $-2y^2 = 6y + 7$
    $0 = 2y^2 + 6y + 7$
    For this equation, $a = 2$, $b = 6$, and $c = 7$.
    $$y = \frac{-6 \pm \sqrt{6^2 - 4(2)(7)}}{2(2)}$$
    $$= \frac{-6 \pm \sqrt{36 - 56}}{4}$$
    $$= \frac{-6 \pm \sqrt{-20}}{4}$$
    $$= \frac{-6 \pm 2\sqrt{5}\,i}{4}$$
    $$= -\frac{6}{4} \pm \frac{2\sqrt{5}}{4}i = -\frac{3}{2} \pm \frac{\sqrt{5}}{2}i$$
    The solution set is $\left\{ -\frac{3}{2} - \frac{\sqrt{5}}{2}i,\ -\frac{3}{2} + \frac{\sqrt{5}}{2}i \right\}$.

47. $p^2 - 5p - 8 = 0$
    For this equation, $a = 1$, $b = -5$, and $c = -8$.
    $b^2 - 4ac = (-5)^2 - 4(1)(-8) = 25 + 32 = 57$
    Because $b^2 - 4ac = 57$ is positive, but not a perfect square, the quadratic equation will have two irrational solutions.

48. $m^2 + 8m + 16 = 0$
    For this equation, $a = 1$, $b = 8$, and $c = 16$.
    $b^2 - 4ac = 8^2 - 4(1)(16) = 64 - 64 = 0$
    Because $b^2 - 4ac = 0$, the quadratic equation will have one repeated real solution.

49. $3n^2 + n = -4$
    $3n^2 + n + 4 = 0$
    For this equation, $a = 3$, $b = 1$, and $c = 4$.
    $b^2 - 4ac = 1^2 - 4(3)(4) = 1 - 48 = -47$
    Because $b^2 - 4ac = -47$ is negative, the quadratic equation will have two complex solutions that are not real. The solutions will be complex conjugates of each other.

50. $7w^2 + 3 = 8w$
    $7w^2 - 8w + 3 = 0$
    For this equation, $a = 7$, $b = -8$, and $c = 3$.
    $b^2 - 4ac = (-8)^2 - 4(7)(3) = 64 - 84 = -20$
    Because $b^2 - 4ac = -20$ is negative, the quadratic equation will have two complex solutions that are not real. The solutions will be complex conjugates of each other.

51. $4x^2 + 49 = 28x$
    $4x^2 - 28x + 49 = 0$
    For this equation, $a = 4$, $b = -28$, and $c = 49$.
    $b^2 - 4ac = (-28)^2 - 4(4)(49) = 784 - 784 = 0$
    Because $b^2 - 4ac = 0$, the quadratic equation will have one repeated real solution.

52. $11z - 12 = 2z^2$
    $0 = 2z^2 - 11z + 12$
    For this equation, $a = 2$, $b = -11$, and $c = 12$.
    $b^2 - 4ac = (-11)^2 - 4(2)(12) = 121 - 96 = 25$
    Because $b^2 - 4ac = 25$ is positive and a perfect square, the quadratic equation will have two rational solutions.

**53.** $x^2 + 8x - 9 = 0$
Because this equation factors easily, solve by factoring.
$(x+9)(x-1) = 0$
$x + 9 = 0$ or $x - 1 = 0$
$x = -9$ or $x = 1$
The solution set is $\{-9, 1\}$.

**54.** $6p^2 + 13p = 5$
$6p^2 + 13p - 5 = 0$
Because this equation factors easily, solve by factoring.
$(3p-1)(2p+5) = 0$
$3p - 1 = 0$ or $2p + 5 = 0$
$3p = 1$ or $2p = -5$
$p = \dfrac{1}{3}$ or $p = -\dfrac{5}{2}$
The solution set is $\left\{-\dfrac{5}{2}, \dfrac{1}{3}\right\}$.

**55.** $n^2 + 13 = -4n$
$n^2 + 4n + 13 = 0$
Because this equation does not easily factor, solve by using the quadratic formula. For this equation, $a = 1$, $b = 4$, and $c = 13$.
$n = \dfrac{-4 \pm \sqrt{4^2 - 4(1)(13)}}{2(1)}$
$= \dfrac{-4 \pm \sqrt{16 - 52}}{2}$
$= \dfrac{-4 \pm \sqrt{-36}}{2}$
$= \dfrac{-4 \pm 6i}{2}$
$= -\dfrac{4}{2} \pm \dfrac{6i}{2}$
$= -2 \pm 3i$
The solution set is $\{-2 - 3i, -2 + 3i\}$.

**56.** $5y^2 - 60 = 0$
Because this equation has no linear term, solve by using the square root method.

$5y^2 = 60$
$y^2 = 12$
$y = \pm\sqrt{12}$
$y = \pm 2\sqrt{3}$
The solution set is $\{-2\sqrt{3}, 2\sqrt{3}\}$.

**57.** $\dfrac{1}{4}q^2 - \dfrac{1}{2}q - \dfrac{3}{8} = 0$
$8\left(\dfrac{1}{4}q^2 - \dfrac{1}{2}q - \dfrac{3}{8}\right) = 8(0)$
$2q^2 - 4q - 3 = 0$
Because this equation does not easily factor, solve by using the quadratic formula. For this equation, $a = 2$, $b = -4$, and $c = -3$.
$q = \dfrac{-(-4) \pm \sqrt{(-4)^2 - 4(2)(-3)}}{2(2)}$
$= \dfrac{4 \pm \sqrt{16 + 24}}{4}$
$= \dfrac{4 \pm \sqrt{40}}{4}$
$= \dfrac{4 \pm 2\sqrt{10}}{4}$
$= \dfrac{4}{4} \pm \dfrac{2\sqrt{10}}{4}$
$= 1 \pm \dfrac{\sqrt{10}}{2}$
The solution set is $\left\{1 - \dfrac{\sqrt{10}}{2}, 1 + \dfrac{\sqrt{10}}{2}\right\}$.

**58.** $\dfrac{1}{8}m^2 + m + \dfrac{5}{2} = 0$
$8\left(\dfrac{1}{8}m^2 + m + \dfrac{5}{2}\right) = 8(0)$
$m^2 + 8m + 20 = 0$
Because this equation does not easily factor, solve by using the quadratic formula. For this equation, $a = 1$, $b = 8$, and $c = 20$.
$m = \dfrac{-8 \pm \sqrt{8^2 - 4(1)(20)}}{2(1)}$
$= \dfrac{-8 \pm \sqrt{64 - 80}}{2}$
$= \dfrac{-8 \pm \sqrt{-16}}{2}$

## Chapter 8: Quadratic Equations and Quadratic Functions

$$= \frac{-8 \pm 4i}{2}$$
$$= -\frac{8}{2} \pm \frac{4i}{2}$$
$$= -4 \pm 2i$$

The solution set is $\{-4-2i,\ -4+2i\}$.

**59.** $(w-8)(w+6) = -33$
$w^2 - 2w - 48 = -33$
$w^2 - 2w - 15 = 0$
Because this equation factors easily, solve by factoring.
$(w-5)(w+3) = 0$
$w - 5 = 0$ or $w + 3 = 0$
$w = 5$ or $w = -3$
The solution set is $\{-3,\ 5\}$.

**60.** $(x-3)(x+1) = -2$
$x^2 - 2x - 3 = -2$
$x^2 - 2x - 1 = 0$
Because this equation does not easily factor, solve by using the quadratic formula. For this equation, $a = 1$, $b = -2$, and $c = -1$.

$$x = \frac{-(-2) \pm \sqrt{(-2)^2 - 4(1)(-1)}}{2(1)}$$
$$= \frac{2 \pm \sqrt{4+4}}{2}$$
$$= \frac{2 \pm \sqrt{8}}{2}$$
$$= \frac{2 \pm 2\sqrt{2}}{2}$$
$$= \frac{2}{2} \pm \frac{2\sqrt{2}}{2}$$
$$= 1 \pm \sqrt{2}$$

The solution set is $\{1-\sqrt{2},\ 1+\sqrt{2}\}$.

**61.** $9z^2 = 16$
Because this equation has no linear term, solve by using the square root method.

$$z^2 = \frac{16}{9}$$
$$z = \pm\sqrt{\frac{16}{9}}$$
$$z = \pm\frac{4}{3}$$

The solution set is $\left\{-\frac{4}{3},\ \frac{4}{3}\right\}$.

**62.** $\dfrac{1-2x}{x^2+5} = 1$

$(x^2+5)\left(\dfrac{1-2x}{x^2+5}\right) = (x^2+5)(1)$
$1 - 2x = x^2 + 5$
$0 = x^2 + 2x + 4$

Because this equation does not easily factor, solve by using the quadratic formula. For this equation, $a = 1$, $b = 2$, and $c = 4$.

$$x = \frac{-2 \pm \sqrt{2^2 - 4(1)(4)}}{2(1)}$$
$$= \frac{-2 \pm \sqrt{4-16}}{2}$$
$$= \frac{-2 \pm \sqrt{-12}}{2}$$
$$= \frac{-2 \pm 2\sqrt{3}\ i}{2}$$
$$= -1 \pm \sqrt{3}\ i$$

The solution set is $\{-1-\sqrt{3}\ i,\ -1+\sqrt{3}\ i\}$.

**63.** $(x+2)^2 + (x-5)^2 = (x+3)^2$
$x^2 + 4x + 4 + x^2 - 10x + 25 = x^2 + 6x + 9$
$2x^2 - 6x + 29 = x^2 + 6x + 9$
$x^2 - 12x + 20 = 0$
$(x-2)(x-10) = 0$
$x - 2 = 0$ or $x - 10 = 0$
$x = 2$ or $x = 10$
Disregard $x = 2$ because this value will cause the length of one of the legs to be negative: $x - 5 = 2 - 5 = -3$. Thus, $x = 10$ is the only viable answer. Now, $x - 5 = 10 - 5 = 5$, $x + 2 = 10 + 2 = 12$, and $x + 3 = 10 + 3 = 13$. The three measurements are 5, 12, and 13.

**64.** Let $x$ represent the length of the rectangle. Then $x - 3$ will represent the width.

$$x(x-3) = 108$$
$$x^2 - 3x = 108$$
$$x^2 - 3x - 108 = 0$$
$$(x-12)(x+9) = 0$$
$$x - 12 = 0 \quad \text{or} \quad x + 9 = 0$$
$$x = 12 \quad \text{or} \quad x = -9$$

Disregard $x = -9$ because $x$ represents the length of the rectangle, which must be positive. Thus, $x = 12$ is the only viable answer. Now, $x - 3 = 12 - 3 = 9$. Thus, the dimensions of the rectangle are 12 centimeters by 9 centimeters.

**65. a.**
$$-0.2x^2 + 180x = 36{,}000$$
$$0 = 0.2x^2 - 180x + 36{,}000$$

For this equation, $a = 0.2$, $b = -180$, and $c = 36{,}000$.

$$x = \frac{-(-180) \pm \sqrt{(-180)^2 - 4(0.2)(36{,}000)}}{2(0.2)}$$
$$= \frac{180 \pm \sqrt{32{,}400 - 28{,}800}}{0.4}$$
$$= \frac{180 \pm \sqrt{3600}}{0.4}$$
$$= \frac{180 \pm 60}{0.4}$$
$$x = \frac{180 - 60}{0.4} \quad \text{or} \quad x = \frac{180 + 60}{0.4}$$
$$= 300 \quad \text{or} \quad = 600$$

The revenue will be \$36,000 per week if either 300 or 600 cellular phones are sold per week.

**b.**
$$-0.2x^2 + 180x = 40{,}500$$
$$0 = 0.2x^2 - 180x + 40{,}500$$

For this equation, $a = 0.2$, $b = -180$, and $c = 40{,}500$.

$$x = \frac{-(-180) \pm \sqrt{(-180)^2 - 4(0.2)(40{,}500)}}{2(0.2)}$$
$$= \frac{180 \pm \sqrt{32{,}400 - 32{,}400}}{0.4}$$
$$= \frac{180 \pm \sqrt{0}}{0.4}$$
$$= \frac{180}{0.4}$$
$$= 450$$

The revenue will be \$40,500 per week if 450 cellular phones are sold per week.

**66. a.**
$$200 = -16t^2 + 50t + 180$$
$$16t^2 - 50t + 20 = 0$$

For this equation, $a = 16$, $b = -50$, and $c = 20$.

$$t = \frac{-(-50) \pm \sqrt{(-50)^2 - 4(16)(20)}}{2(16)}$$
$$= \frac{50 \pm \sqrt{2500 - 1280}}{32}$$
$$= \frac{50 \pm \sqrt{1220}}{32}$$
$$= \frac{50 \pm 2\sqrt{305}}{32}$$
$$= \frac{50}{32} \pm \frac{2\sqrt{305}}{32}$$
$$= \frac{25}{16} \pm \frac{\sqrt{305}}{16}$$
$$t = \frac{25}{16} - \frac{\sqrt{305}}{16} \quad \text{or} \quad t = \frac{25}{16} + \frac{\sqrt{305}}{16}$$
$$\approx 0.471 \quad \text{or} \quad \approx 2.654$$

Rounding to the nearest tenth, the height of the ball will be 200 feet after approximately 0.5 seconds and after approximately 2.7 seconds.

**b.**
$$100 = -16t^2 + 50t + 180$$
$$16t^2 - 50t - 80 = 0$$

For this equation, $a = 16$, $b = -50$, and $c = -80$.

$$t = \frac{-(-50) \pm \sqrt{(-50)^2 - 4(16)(-80)}}{2(16)}$$
$$= \frac{50 \pm \sqrt{2500 + 5120}}{32}$$
$$= \frac{50 \pm \sqrt{7620}}{32}$$
$$= \frac{50 \pm 2\sqrt{1905}}{32}$$
$$= \frac{50}{32} \pm \frac{2\sqrt{1905}}{32}$$
$$= \frac{25}{16} \pm \frac{\sqrt{1905}}{16}$$
$$t = \frac{25}{16} - \frac{\sqrt{1905}}{16} \quad \text{or} \quad t = \frac{25}{16} + \frac{\sqrt{1905}}{16}$$
$$\approx -1.165 \quad \text{or} \quad \approx 4.290$$

Because time cannot be negative, we disregard $-1.165$. Thus, $t \approx 4.290$ is the only viable answer. Rounding to the nearest tenth, the height of the ball will be 100 feet after approximately 4.3 seconds.

c. $$300 = -16t^2 + 50t + 180$$
$$16t^2 - 50t + 120 = 0$$
For this equation, $a = 16$, $b = -50$, and $c = 120$.
$$t = \frac{-(-50) \pm \sqrt{(-50)^2 - 4(16)(120)}}{2(16)}$$
$$= \frac{50 \pm \sqrt{2500 - 7680}}{32}$$
$$= \frac{50 \pm \sqrt{-5180}}{32}$$
$$= \frac{50 \pm 2\sqrt{1295}\, i}{32}$$
$$= \frac{50}{32} \pm \frac{2\sqrt{1295}}{32} i$$
$$= \frac{25}{16} \pm \frac{\sqrt{1295}}{16} i$$
$$t = \frac{25}{16} - \frac{\sqrt{1295}}{16} i \text{ or } t = \frac{25}{16} + \frac{\sqrt{1295}}{16} i$$

The ball will never reach a height of 300 feet. This is clear because the solutions to the equation above are complex solutions that are not real.

67. Let $x$ represent the speed the boat would travel in still water.

|  | Distance | Rate | Time |
|---|---|---|---|
| Up Stream | 10 | $x - 3$ | $\frac{10}{x-3}$ |
| Down Stream | 10 | $x + 3$ | $\frac{10}{x+3}$ |

$$\frac{10}{x-3} + \frac{10}{x+3} = 2$$
$$(x-3)(x+3)\left(\frac{10}{x-3} + \frac{10}{x+3}\right) = (x-3)(x+3)(2)$$
$$10(x+3) + 10(x-3) = (x^2 - 9)(2)$$
$$10x + 30 + 10x - 30 = 2x^2 - 18$$
$$20x = 2x^2 - 18$$
$$0 = 2x^2 - 20x - 18$$

For this equation, $a = 2$, $b = -20$, and $c = -18$.

$$x = \frac{-(-20) \pm \sqrt{(-20)^2 - 4(2)(-18)}}{2(2)}$$
$$= \frac{20 \pm \sqrt{400 + 144}}{4}$$
$$= \frac{20 \pm \sqrt{544}}{4}$$
$$= \frac{20 \pm 4\sqrt{34}}{4}$$
$$= 5 \pm \sqrt{34}$$
$$x = 5 - \sqrt{34} \text{ or } x = 5 + \sqrt{34}$$
$$\approx -0.831 \text{ or } \approx 10.831$$

Because the speed should be positive, we disregard $x \approx -0.831$. Thus, the only viable answer is $x \approx 10.831$. Rounding to the nearest tenth, the boat would travel approximately 10.8 miles per hour in still water.

68. Let $t$ represent the time required for Beth to wash the car alone. Then $t - 14$ will represent the time required for Tom to wash the car alone.

$$\begin{pmatrix} \text{Part done} \\ \text{by Beth in} \\ \text{1 minute} \end{pmatrix} + \begin{pmatrix} \text{Part done} \\ \text{by Tom in} \\ \text{1 minute} \end{pmatrix} = \begin{pmatrix} \text{Part done} \\ \text{together in} \\ \text{1 minute} \end{pmatrix}$$

$$\frac{1}{t} + \frac{1}{t-14} = \frac{1}{30}$$
$$30t(t-14)\left(\frac{1}{t} + \frac{1}{t-14}\right) = 30t(t-14)\left(\frac{1}{30}\right)$$
$$30(t-14) + 30t = t(t-14)$$
$$30t - 420 + 30t = t^2 - 14t$$
$$60t - 420 = t^2 - 14t$$
$$0 = t^2 - 74t + 420$$

For this equation, $a = 1$, $b = -74$, and $c = 420$.

$$t = \frac{-(-74) \pm \sqrt{(-74)^2 - 4(1)(420)}}{2(1)}$$
$$= \frac{74 \pm \sqrt{5476 - 1680}}{2}$$
$$= \frac{74 \pm \sqrt{3796}}{2}$$
$$= \frac{74 \pm 2\sqrt{949}}{2}$$
$$= 37 \pm \sqrt{949}$$
$$t = 37 - \sqrt{949} \text{ or } t = 37 + \sqrt{949}$$
$$\approx 6.194 \text{ or } \approx 67.806$$

Disregard $t \approx 6.194$ because this value makes Tom's time negative: $t - 14 = 6.194 - 14 = -7.806$. The only viable answer is $t \approx 67.806$ minutes. Working alone, it will take Beth approximately 67.8 minutes to wash the car.

69. $x^4 + 7x^2 - 144 = 0$

$(x^2)^2 + 7(x^2) - 144 = 0$

Let $u = x^2$.

$u^2 + 7u - 144 = 0$

$(u - 9)(u + 16) = 0$

$u - 9 = 0$ or $u + 16 = 0$

$u = 9$ or $u = -16$

$x^2 = 9$ or $x^2 = -16$

$x = \pm\sqrt{9}$ or $x = \pm\sqrt{-16}$

$x = \pm 3$ or $x = \pm 4i$

Check:

$x = -3$: $(-3)^4 + 7(-3)^2 - 144 \stackrel{?}{=} 0$

$81 + 7 \cdot 9 - 144 \stackrel{?}{=} 0$

$81 + 63 - 144 \stackrel{?}{=} 0$

$0 = 0$ ✓

$x = 3$: $3^4 + 7(3)^2 - 144 \stackrel{?}{=} 0$

$81 + 7 \cdot 9 - 144 \stackrel{?}{=} 0$

$81 + 63 - 144 \stackrel{?}{=} 0$

$0 = 0$ ✓

$x = -4i$: $(-4i)^4 + 7(-4i)^2 - 144 \stackrel{?}{=} 0$

$256i^4 + 7 \cdot 16i^2 - 144 \stackrel{?}{=} 0$

$256(1) + 7 \cdot 16(-1) - 144 \stackrel{?}{=} 0$

$256 - 112 - 144 \stackrel{?}{=} 0$

$0 = 0$ ✓

$x = 4i$: $(4i)^4 + 7(4i)^2 - 144 \stackrel{?}{=} 0$

$256i^4 + 7 \cdot 16i^2 - 144 \stackrel{?}{=} 0$

$256(1) + 7 \cdot 16(-1) - 144 \stackrel{?}{=} 0$

$256 - 112 - 144 \stackrel{?}{=} 0$

$0 = 0$ ✓

All check; the solution set is $\{-3, 3, -4i, 4i\}$.

70. $4w^4 + 5w^2 - 6 = 0$

$4(w^2)^2 + 5(w^2) - 6 = 0$

Let $u = w^2$.

$4u^2 + 5u - 6 = 0$

$(4u - 3)(u + 2) = 0$

$4u - 3 = 0$ or $u + 2 = 0$

$4u = 3$ or $u = -2$

$u = \dfrac{3}{4}$

$w^2 = \dfrac{3}{4}$ or $w^2 = -2$

$w = \pm\sqrt{\dfrac{3}{4}}$ or $w = \pm\sqrt{-2}$

$w = \pm\dfrac{\sqrt{3}}{2}$ or $w = \pm\sqrt{2}\,i$

Check:

$w = -\dfrac{\sqrt{3}}{2}$: $4\left(-\dfrac{\sqrt{3}}{2}\right)^4 + 5\left(-\dfrac{\sqrt{3}}{2}\right)^2 - 6 \stackrel{?}{=} 0$

$4\left(\dfrac{9}{16}\right) + 5\left(\dfrac{3}{4}\right) - 6 \stackrel{?}{=} 0$

$\dfrac{9}{4} + \dfrac{15}{4} - 6 \stackrel{?}{=} 0$

$0 = 0$ ✓

$w = \dfrac{\sqrt{3}}{2}$: $4\left(\dfrac{\sqrt{3}}{2}\right)^4 + 5\left(\dfrac{\sqrt{3}}{2}\right)^2 - 6 \stackrel{?}{=} 0$

$4\left(\dfrac{9}{16}\right) + 5\left(\dfrac{3}{4}\right) - 6 \stackrel{?}{=} 0$

$\dfrac{9}{4} + \dfrac{15}{4} - 6 \stackrel{?}{=} 0$

$0 = 0$ ✓

$w = -\sqrt{2}\,i$: $4\left(-\sqrt{2}\,i\right)^4 + 5\left(-\sqrt{2}\,i\right)^2 - 6 \stackrel{?}{=} 0$

$4 \cdot 4i^4 + 5 \cdot 2i^2 - 6 \stackrel{?}{=} 0$

$4 \cdot 4(1) + 5 \cdot 2(-1) - 6 \stackrel{?}{=} 0$

$16 - 10 - 6 \stackrel{?}{=} 0$

$0 = 0$ ✓

$w = \sqrt{2}\,i$: $4\left(\sqrt{2}\,i\right)^4 + 5\left(\sqrt{2}\,i\right)^2 - 6 \stackrel{?}{=} 0$

$4 \cdot 4i^4 + 5 \cdot 2i^2 - 6 \stackrel{?}{=} 0$

$4 \cdot 4(1) + 5 \cdot 2(-1) - 6 \stackrel{?}{=} 0$

$16 - 10 - 6 \stackrel{?}{=} 0$

$0 = 0$ ✓

All check; the solution set is

$\left\{-\dfrac{\sqrt{3}}{2}, \dfrac{\sqrt{3}}{2}, -\sqrt{2}\,i, \sqrt{2}\,i\right\}$.

## Chapter 8: Quadratic Equations and Quadratic Functions

**71.** $3(a+4)^2 - 11(a+4) + 6 = 0$

Let $u = a+4$.

$3u^2 - 11u + 6 = 0$
$(3u-2)(u-3) = 0$
$3u - 2 = 0 \quad \text{or} \quad u - 3 = 0$
$3u = 2 \quad \text{or} \quad u = 3$
$u = \dfrac{2}{3}$
$a + 4 = \dfrac{2}{3} \quad \text{or} \quad a + 4 = 3$
$a = -\dfrac{10}{3} \quad \text{or} \quad a = -1$

Check:

$a = -\dfrac{10}{3}: \ 3\left(-\dfrac{10}{3}+4\right)^2 - 11\left(-\dfrac{10}{3}+4\right) + 6 \stackrel{?}{=} 0$

$3\left(\dfrac{2}{3}\right)^2 - 11\left(\dfrac{2}{3}\right) + 6 \stackrel{?}{=} 0$

$3\left(\dfrac{4}{9}\right) - \dfrac{22}{3} + 6 \stackrel{?}{=} 0$

$\dfrac{4}{3} - \dfrac{22}{3} + 6 \stackrel{?}{=} 0$

$0 = 0 \ \checkmark$

$a = -1: \ 3(-1+4)^2 - 11(-1+4) + 6 \stackrel{?}{=} 0$

$3(3)^2 - 11(3) + 6 \stackrel{?}{=} 0$

$3(9) - 33 + 6 \stackrel{?}{=} 0$

$27 - 33 + 6 \stackrel{?}{=} 0$

$0 = 0 \ \checkmark$

Both check; the solution set is $\left\{-\dfrac{10}{3},\ -1\right\}$.

**72.** $(q^2 - 11)^2 - 2(q^2 - 11) - 15 = 0$

Let $u = q^2 - 11$.

$u^2 - 2u - 15 = 0$
$(u+3)(u-5) = 0$
$u + 3 = 0 \quad \text{or} \quad u - 5 = 0$
$u = -3 \quad \text{or} \quad u = 5$
$q^2 - 11 = -3 \quad \text{or} \quad q^2 - 11 = 5$
$q^2 = 8 \quad \text{or} \quad q^2 = 16$
$q = \pm\sqrt{8} \quad \text{or} \quad q = \pm\sqrt{16}$
$q = \pm 2\sqrt{2} \quad \text{or} \quad q = \pm 4$

Check:

$q = -2\sqrt{2}$:

$\left((-2\sqrt{2})^2 - 11\right)^2 - 2\left((-2\sqrt{2})^2 - 11\right) - 15 \stackrel{?}{=} 0$

$(8-11)^2 - 2(8-11) - 15 \stackrel{?}{=} 0$

$(-3)^2 - 2(-3) - 15 \stackrel{?}{=} 0$

$9 + 6 - 15 \stackrel{?}{=} 0$

$0 = 0 \ \checkmark$

$q = 2\sqrt{2}$:

$\left((2\sqrt{2})^2 - 11\right)^2 - 2\left((2\sqrt{2})^2 - 11\right) - 15 \stackrel{?}{=} 0$

$(8-11)^2 - 2(8-11) - 15 \stackrel{?}{=} 0$

$(-3)^2 - 2(-3) - 15 \stackrel{?}{=} 0$

$9 + 6 - 15 \stackrel{?}{=} 0$

$0 = 0 \ \checkmark$

$q = -4: \ \left((-4)^2 - 11\right)^2 - 2\left((-4)^2 - 11\right) - 15 \stackrel{?}{=} 0$

$(16-11)^2 - 2(16-11) - 15 \stackrel{?}{=} 0$

$(5)^2 - 2(5) - 15 \stackrel{?}{=} 0$

$25 - 10 - 15 \stackrel{?}{=} 0$

$0 = 0 \ \checkmark$

$q = 4: \ (4^2 - 11)^2 - 2(4^2 - 11) - 15 \stackrel{?}{=} 0$

$(16-11)^2 - 2(16-11) - 15 \stackrel{?}{=} 0$

$(5)^2 - 2(5) - 15 \stackrel{?}{=} 0$

$25 - 10 - 15 \stackrel{?}{=} 0$

$0 = 0 \ \checkmark$

All check; the solution set is $\left\{-4,\ -2\sqrt{2},\ 2\sqrt{2},\ 4\right\}$.

**73.** $y - 13\sqrt{y} + 36 = 0$

$(\sqrt{y})^2 - 13(\sqrt{y}) + 36 = 0$

Let $u = \sqrt{y}$.

$u^2 - 13u + 36 = 0$
$(u-4)(u-9) = 0$
$u - 4 = 0 \quad \text{or} \quad u - 9 = 0$
$u = 4 \quad \text{or} \quad u = 9$

$\sqrt{y} = 4$ or $\sqrt{y} = 9$
$y = 4^2$ or $y = 9^2$
$y = 16$ or $y = 81$

Check:
$y = 16$: $16 - 13\sqrt{16} + 36 \stackrel{?}{=} 0$
$16 - 13 \cdot 4 + 36 \stackrel{?}{=} 0$
$16 - 52 + 36 \stackrel{?}{=} 0$
$0 = 0$ ✓

$y = 16$: $81 - 13\sqrt{81} + 36 \stackrel{?}{=} 0$
$81 - 13 \cdot 9 + 36 \stackrel{?}{=} 0$
$81 - 117 + 36 \stackrel{?}{=} 0$
$0 = 0$ ✓

Both check, the solution set is $\{16, 81\}$.

**74.** $5z + 2\sqrt{z} - 3 = 0$
$5(\sqrt{z})^2 + 2(\sqrt{z}) - 3 = 0$
Let $u = \sqrt{z}$.
$5u^2 + 2u - 3 = 0$
$(5u - 3)(u + 1) = 0$
$5u - 3 = 0$ or $u + 1 = 0$
$5u = 3$ or $u = -1$
$u = \dfrac{3}{5}$

$\sqrt{z} = \dfrac{3}{5}$ or $\sqrt{z} = -1$
$z = \left(\dfrac{3}{5}\right)^2$ or $z = (-1)^2$
$z = \dfrac{9}{25}$ or $z = 1$

Check:
$z = \dfrac{9}{25}$: $5 \cdot \dfrac{9}{25} + 2\sqrt{\dfrac{9}{25}} - 3 \stackrel{?}{=} 0$
$\dfrac{9}{5} + 2 \cdot \dfrac{3}{5} - 3 \stackrel{?}{=} 0$
$\dfrac{9}{5} + \dfrac{6}{5} - 3 \stackrel{?}{=} 0$
$0 = 0$ ✓

$z = 1$: $5 \cdot 1 + 2\sqrt{1} - 3 \stackrel{?}{=} 0$
$5 + 2 \cdot 1 - 3 \stackrel{?}{=} 0$
$5 + 2 - 3 \stackrel{?}{=} 0$
$4 \neq 0$ ✗

$z = 1$ does not check; the solution set is $\left\{\dfrac{9}{25}\right\}$.

**75.** $p^{-2} - 4p^{-1} - 21 = 0$
$(p^{-1})^2 - 4(p^{-1}) - 21 = 0$
Let $u = p^{-1}$.
$u^2 - 4u - 21 = 0$
$(u + 3)(u - 7) = 0$
$u + 3 = 0$ or $u - 7 = 0$
$u = -3$ or $u = 7$
$p^{-1} = -3$ or $p^{-1} = 7$
$\dfrac{1}{p} = -3$ or $\dfrac{1}{p} = 7$
$p = -\dfrac{1}{3}$ or $p = \dfrac{1}{7}$

Check:
$p = -\dfrac{1}{3}$: $\left(-\dfrac{1}{3}\right)^{-2} - 4\left(-\dfrac{1}{3}\right)^{-1} - 21 \stackrel{?}{=} 0$
$(-3)^2 - 4(-3) - 21 \stackrel{?}{=} 0$
$9 + 12 - 21 \stackrel{?}{=} 0$
$0 = 0$ ✓

$p = \dfrac{1}{7}$: $\left(\dfrac{1}{7}\right)^{-2} - 4\left(\dfrac{1}{7}\right)^{-1} - 21 \stackrel{?}{=} 0$
$(7)^2 - 4(7) - 21 \stackrel{?}{=} 0$
$49 - 28 - 21 \stackrel{?}{=} 0$
$0 = 0$ ✓

Both check; the solution set is $\left\{-\dfrac{1}{3}, \dfrac{1}{7}\right\}$.

**76.** $2b^{2/3} + 13b^{1/3} - 7 = 0$
$2(b^{1/3})^2 + 13(b^{1/3}) - 7 = 0$
Let $u = b^{1/3}$.

*Chapter 8:* Quadratic Equations and Quadratic Functions

$$2u^2 + 13u - 7 = 0$$
$$(2u-1)(u+7) = 0$$
$$2u-1 = 0 \quad \text{or} \quad u+7 = 0$$
$$2u = 1 \quad \text{or} \quad u = -7$$
$$u = \frac{1}{2}$$
$$b^{1/3} = \frac{1}{2} \quad \text{or} \quad b^{1/3} = -7$$
$$\left(b^{1/3}\right)^3 = \left(\frac{1}{2}\right)^3 \quad \text{or} \quad \left(b^{1/3}\right)^3 = (-7)^3$$
$$b = \frac{1}{8} \quad \text{or} \quad b = -343$$

Check:
$$b = \frac{1}{8}: \quad 2\left(\frac{1}{8}\right)^{2/3} + 13\left(\frac{1}{8}\right)^{1/3} - 7 \stackrel{?}{=} 0$$
$$2\left(\sqrt[3]{\frac{1}{8}}\right)^2 + 13\left(\sqrt[3]{\frac{1}{8}}\right) - 7 \stackrel{?}{=} 0$$
$$2\left(\frac{1}{2}\right)^2 + 13\left(\frac{1}{2}\right) - 7 \stackrel{?}{=} 0$$
$$2\left(\frac{1}{4}\right) + 13\left(\frac{1}{2}\right) - 7 \stackrel{?}{=} 0$$
$$\frac{1}{2} + \frac{13}{2} - 7 \stackrel{?}{=} 0$$
$$0 = 0 \checkmark$$

$$b = -343: \quad 2(-343)^{2/3} + 13(-343)^{1/3} - 7 \stackrel{?}{=} 0$$
$$2\left(\sqrt[3]{-343}\right)^2 + 13\left(\sqrt[3]{-343}\right) - 7 \stackrel{?}{=} 0$$
$$2(-7)^2 + 13(-7) - 7 \stackrel{?}{=} 0$$
$$2(49) - 91 - 7 \stackrel{?}{=} 0$$
$$98 - 91 - 7 \stackrel{?}{=} 0$$
$$0 = 0 \checkmark$$

Both check; the solution set is $\left\{-343, \frac{1}{8}\right\}$.

**77.**
$$m^{1/2} + 2m^{1/4} - 8 = 0$$
$$\left(m^{1/4}\right)^2 + 2\left(m^{1/4}\right) - 8 = 0$$
Let $u = m^{1/4}$.

$$u^2 + 2u - 8 = 0$$
$$(u-2)(u+4) = 0$$
$$u-2 = 0 \quad \text{or} \quad u+4 = 0$$
$$u = 2 \quad \text{or} \quad u = -4$$
$$m^{1/4} = 2 \quad \text{or} \quad m^{1/4} = -4$$
$$\left(m^{1/4}\right)^4 = 2^4 \quad \text{or} \quad \left(m^{1/4}\right)^4 = (-4)^4$$
$$m = 16 \quad \text{or} \quad m = 256$$

Check:
$$m = 16: \quad (16)^{1/2} + 2(16)^{1/4} - 8 \stackrel{?}{=} 0$$
$$\sqrt{16} + 2\sqrt[4]{16} - 8 \stackrel{?}{=} 0$$
$$4 + 2 \cdot 2 - 8 \stackrel{?}{=} 0$$
$$4 + 4 - 8 \stackrel{?}{=} 0$$
$$0 = 0 \checkmark$$

$$m = 256: \quad (256)^{1/2} + 2(256)^{1/4} - 8 \stackrel{?}{=} 0$$
$$\sqrt{256} + 2\sqrt[4]{256} - 8 \stackrel{?}{=} 0$$
$$16 + 2 \cdot 4 - 8 \stackrel{?}{=} 0$$
$$16 + 8 - 8 \stackrel{?}{=} 0$$
$$16 \neq 0 \; \times$$

$m = 256$ does not check; the solution set is $\{16\}$.

**78.**
$$\left(\frac{1}{x+5}\right)^2 + \frac{3}{x+5} = 28$$
$$\left(\frac{1}{x+5}\right)^2 + 3\left(\frac{1}{x+5}\right) - 28 = 0$$
Let $u = \frac{1}{x+5}$.

$$u^2 + 3u - 28 = 0$$
$$(u-4)(u+7) = 0$$
$$u-4 = 0 \quad \text{or} \quad u+7 = 0$$
$$u = 4 \quad \text{or} \quad u = -7$$
$$\frac{1}{x+5} = 4 \quad \text{or} \quad \frac{1}{x+5} = -7$$
$$x+5 = \frac{1}{4} \quad \text{or} \quad x+5 = -\frac{1}{7}$$
$$x = -\frac{19}{4} \quad \text{or} \quad x = -\frac{36}{7}$$

Check:

$x = -\dfrac{19}{4}: \left(\dfrac{1}{-\frac{19}{4}+5}\right)^2 + \dfrac{3}{-\frac{19}{4}+5} \stackrel{?}{=} 28$

$\left(\dfrac{1}{\frac{1}{4}}\right)^2 + \dfrac{3}{\frac{1}{4}} \stackrel{?}{=} 28$

$(4)^2 + 12 \stackrel{?}{=} 28$

$16 + 12 \stackrel{?}{=} 28$

$28 = 28$ ✓

$x = -\dfrac{36}{7}: \left(\dfrac{1}{-\frac{36}{7}+5}\right)^2 + \dfrac{3}{-\frac{36}{7}+5} \stackrel{?}{=} 28$

$\left(\dfrac{1}{-\frac{1}{7}}\right)^2 + \dfrac{3}{-\frac{1}{7}} \stackrel{?}{=} 28$

$(-7)^2 - 21 \stackrel{?}{=} 28$

$49 - 21 \stackrel{?}{=} 28$

$28 = 28$ ✓

Both check; the solution set is $\left\{-\dfrac{36}{7}, -\dfrac{19}{4}\right\}$.

79. $4x - 20\sqrt{x} + 21 = 0$

$4(\sqrt{x})^2 - 20(\sqrt{x}) + 21 = 0$

Let $u = \sqrt{x}$.

$4u^2 - 20u + 21 = 0$

$(2u - 3)(2u - 7) = 0$

$2u - 3 = 0 \quad \text{or} \quad 2u - 7 = 0$

$2u = 3 \quad \text{or} \quad 2u = 7$

$u = \dfrac{3}{2} \quad \text{or} \quad u = \dfrac{7}{2}$

$\sqrt{x} = \dfrac{3}{2} \quad \text{or} \quad \sqrt{x} = \dfrac{7}{2}$

$x = \left(\dfrac{3}{2}\right)^2 \quad \text{or} \quad x = \left(\dfrac{7}{2}\right)^2$

$x = \dfrac{9}{4} \quad \text{or} \quad x = \dfrac{49}{4}$

Check:

$f\left(\dfrac{9}{4}\right) = 4 \cdot \dfrac{9}{4} - 20\sqrt{\dfrac{9}{4}} + 21$

$= 4 \cdot \dfrac{9}{4} - 20 \cdot \dfrac{3}{2} + 21$

$= 9 - 30 + 21$

$= 0$ ✓

$f\left(\dfrac{49}{4}\right) = 4 \cdot \dfrac{49}{4} - 20\sqrt{\dfrac{49}{4}} + 21$

$= 4 \cdot \dfrac{49}{4} - 20 \cdot \dfrac{7}{2} + 21$

$= 49 - 70 + 21$

$= 0$ ✓

Both check; the zeros of $f$ are $\left\{\dfrac{9}{4}, \dfrac{49}{4}\right\}$.

80. $x^4 - 17x^2 + 60 = 0$

$(x^2)^2 - 17(x^2) + 60 = 0$

Let $u = x^2$.

$u^2 - 17u + 60 = 0$

$(u - 12)(u - 5) = 0$

$u - 12 = 0 \quad \text{or} \quad u - 5 = 0$

$u = 12 \quad \text{or} \quad u = 5$

$x^2 = 12 \quad \text{or} \quad x^2 = 5$

$x = \pm\sqrt{12} \quad \text{or} \quad x = \pm\sqrt{5}$

$x = \pm 2\sqrt{3}$

Check:

$g(-2\sqrt{3}) = (-2\sqrt{3})^4 - 17(-2\sqrt{3})^2 + 60$

$= 144 - 17 \cdot 12 + 60$

$= 144 - 204 + 60$

$= 0$ ✓

$g(2\sqrt{3}) = (2\sqrt{3})^4 - 17(2\sqrt{3})^2 + 60$

$= 144 - 17 \cdot 12 + 60$

$= 144 - 204 + 60$

$= 0$ ✓

$g(-\sqrt{5}) = (-\sqrt{5})^4 - 17(-\sqrt{5})^2 + 60$

$= 25 - 17 \cdot 5 + 60$

$= 25 - 85 + 60$

$= 0$ ✓

$g(\sqrt{5}) = (\sqrt{5})^4 - 17(\sqrt{5})^2 + 60$

$= 25 - 17 \cdot 5 + 60$

$= 25 - 85 + 60$

$= 0$ ✓

All check; the zeros of $g$ are $\left\{-2\sqrt{3}, -\sqrt{5}, \sqrt{5}, 2\sqrt{3}\right\}$.

81. Begin with the graph of $y = x^2$, then shift the graph up 4 units to obtain the graph of

$f(x) = x^2 + 4$.

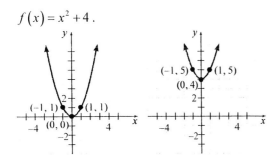

82. Begin with the graph of $y = x^2$, then shift the graph down 5 units to obtain the graph of $g(x) = x^2 - 5$.

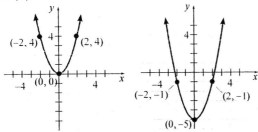

83. Begin with the graph of $y = x^2$, then shift the graph to the left 1 unit to obtain the graph of $h(x) = (x+1)^2$.

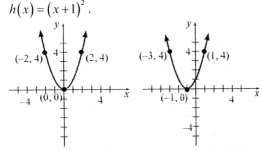

84. Begin with the graph of $y = x^2$, then shift the graph to the right 4 units to obtain the graph of $F(x) = (x-4)^2$.

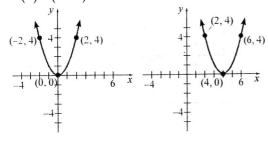

85. Begin with the graph of $y = x^2$, then multiply each y-coordinate by $-4$ to obtain the graph of

$G(x) = -4x^2$.

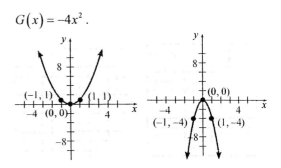

86. Begin with the graph of $y = x^2$, then vertically compress the graph by a factor of $\frac{1}{5}$ (multiply each y-coordinate by $\frac{1}{5}$) to obtain the graph of $H(x) = \frac{1}{5}x^2$.

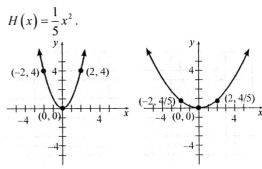

87. Begin with the graph of $y = x^2$, then shift the graph 4 units to the right to obtain the graph of $y = (x-4)^2$. Shift this graph down 3 units to obtain the graph of $p(x) = (x-4)^2 - 3$.

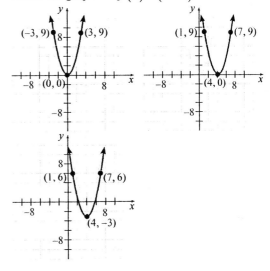

88. Begin with the graph of $y = x^2$, then shift the graph 4 units to the left to obtain the graph of

$y = (x+4)^2$. Shift this graph up 2 units to obtain the graph of $P(x) = (x+4)^2 + 2$.

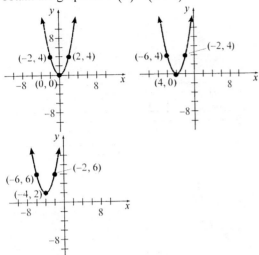

89. Begin with the graph of $y = x^2$, then shift the graph 1 unit to the right to obtain the graph of $y = (x-1)^2$. Multiply the y-coordinates by $-1$ to obtain the graph of $y = -(x-1)^2$. Shift this graph up 4 units to obtain the graph of $f(x) = -(x-1)^2 + 4$.

90. Begin with the graph of $y = x^2$, then shift the graph 2 unit to the left to obtain the graph of $y = (x+2)^2$. Multiply the y-coordinates by $\dfrac{1}{2}$ to obtain the graph of $y = \dfrac{1}{2}(x+2)^2$. Shift this graph down 1 unit to obtain the graph of $F(x) = \dfrac{1}{2}(x+2)^2 - 1$.

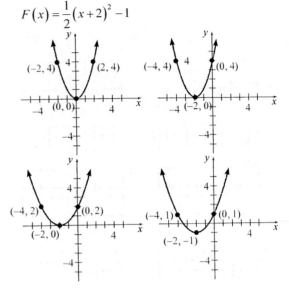

91. Use completing the square to write the function in the form $y = a(x-h)^2 + k$.

$g(x) = x^2 - 6x + 10$
$= (x^2 - 6x) + 10$
$= (x^2 - 6x + 9) + 10 - 9$
$= (x-3)^2 + 1$

Begin with the graph of $y = x^2$, then shift the graph right 3 units to obtain the graph of $y = (x-3)^2$. Shift this result up 1 unit to obtain the graph of $g(x) = (x-3)^2 + 1$.

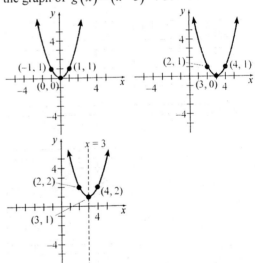

## Chapter 8: Quadratic Equations and Quadratic Functions

The vertex is $(3, 1)$; The axis of symmetry is $x = 3$.

92. Use completing the square to write the function in the form $y = a(x-h)^2 + k$.

$$G(x) = x^2 + 8x + 11$$
$$= (x^2 + 8x) + 11$$
$$= (x^2 + 8x + 16) + 11 - 16$$
$$= (x+4)^2 - 5$$

Begin with the graph of $y = x^2$, then shift the graph left 4 units to obtain the graph of $y = (x+4)^2$. Shift this result down 5 units to obtain the graph of $G(x) = (x+4)^2 - 5$.

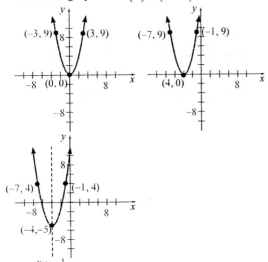

The vertex is $(-4, -5)$. The axis of symmetry is $x = -4$.

93. Use completing the square to write the function in the form $y = a(x-h)^2 + k$.

$$h(x) = 2x^2 - 4x - 3$$
$$= (2x^2 - 4x) - 3$$
$$= 2(x^2 - 2x) - 3$$
$$= 2(x^2 - 2x + 1) - 3 - 2$$
$$= 2(x-1)^2 - 5$$

Begin with the graph of $y = x^2$, then shift the graph right 1 unit to obtain the graph of $y = (x-1)^2$. Vertically stretch this graph by a factor of 2 (multiply the y-coordinates by 2) to obtain the graph of $y = 2(x-1)^2$. Lastly, shift the graph down 5 units to obtain the graph of $H(x) = 2(x-1)^2 - 5$.

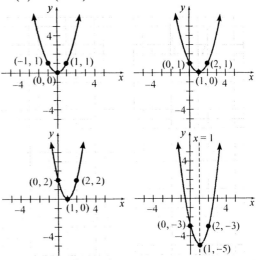

The vertex is $(1, -5)$. The axis of symmetry is $x = 1$.

94. Use completing the square to write the function in the form $y = a(x-h)^2 + k$.

$$H(x) = -x^2 - 6x - 10$$
$$= (-x^2 - 6x) - 10$$
$$= -(x^2 + 6x) - 10$$
$$= -(x^2 + 6x + 9) - 10 + 9$$
$$= -(x+3)^2 - 1$$

Begin with the graph of $y = x^2$, then shift the graph left 3 units to obtain the graph of $y = (x+3)^2$. Multiply the y-coordinates by $-1$ to obtain the graph of $y = -(x+3)^2$. Lastly, shift the graph down 1 unit to obtain the graph of $H(x) = -(x+3)^2 - 1$.

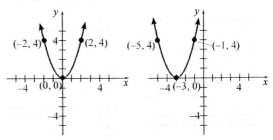

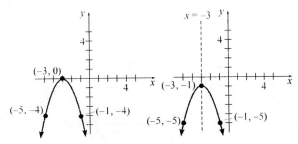

The vertex is $(-3, -1)$. The axis of symmetry is $x = -3$.

95. Use completing the square to write the function in the form $y = a(x-h)^2 + k$.

$$p(x) = -3x^2 + 12x - 8$$
$$= (-3x^2 + 12x) - 8$$
$$= -3(x^2 - 4x) - 8$$
$$= -3(x^2 - 4x + 4) - 8 + 12$$
$$= -3(x-2)^2 + 4$$

Begin with the graph of $y = x^2$, then shift the graph right 2 units to obtain the graph of $y = (x-2)^2$. Multiply the y-coordinates by $-3$ to obtain the graph of $y = -3(x-2)^2$. Lastly, shift the graph up 4 units to obtain the graph of $p(x) = -3(x-2)^2 + 4$.

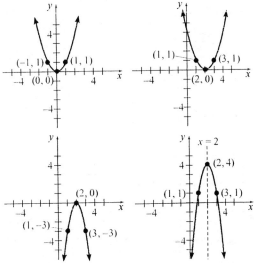

The vertex is $(2, 4)$. The axis of symmetry is $x = 2$.

96. Use completing the square to write the function in the form $y = a(x-h)^2 + k$.

$$P(x) = \frac{1}{2}x^2 - 2x + 5$$
$$= \left(\frac{1}{2}x^2 - 2x\right) + 5$$
$$= \frac{1}{2}(x^2 - 4x) + 5$$
$$= \frac{1}{2}(x^2 - 4x + 4) + 5 - 2$$
$$= \frac{1}{2}(x-2)^2 + 3$$

Begin with the graph of $y = x^2$, then shift the graph right 2 units to obtain the graph of $y = (x-2)^2$. Vertically compress this graph by a factor of $\frac{1}{2}$ (multiply the y-coordinates by $\frac{1}{2}$) to obtain the graph of $y = \frac{1}{2}(x-2)^2$. Lastly, shift the graph up 3 units to obtain the graph of $P(x) = \frac{1}{2}(x-2)^2 + 3$.

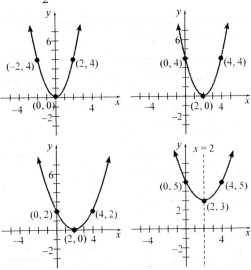

The vertex is $(2, 3)$. The axis of symmetry is $x = 2$.

97. Consider the form $y = a(x-h)^2 + k$. From the graph we know that the vertex is $(2, -4)$, so we have $h = 2$ and $k = -4$. The graph also passes through the point $(x, y) = (0, 4)$. Substituting values for $x$, $y$, $h$, and $k$, we can solve for $a$:

$4 = a(0-2)^2 - 4$
$4 = a(-2)^2 - 4$
$8 = 4a$
$2 = a$
The quadratic function is $f(x) = 2(x-2)^2 - 4$ or $f(x) = 2x^2 - 8x + 4$.

98. Consider the form $y = a(x-h)^2 + k$. From the graph we know that the vertex is $(4,3)$, so we have $h = 4$ and $k = 3$. The graph also passes through the point $(x, y) = (3, 2)$. Substituting values for x, y, h, and k, we can solve for a:
$2 = a(3-4)^2 + 3$
$2 = a(-1)^2 + 3$
$2 = a + 3$
$-1 = a$
The quadratic function is $f(x) = -(x-4)^2 + 3$ or $f(x) = -x^2 + 8x - 13$.

99. Consider the form $y = a(x-h)^2 + k$. From the graph we know that the vertex is $(-2, -1)$, so we have $h = -2$ and $k = -1$. The graph also passes through the point $(x, y) = (0, -3)$. Substituting values for x, y, h, and k, we can solve for a:
$-3 = a(0-(-2))^2 - 1$
$-3 = a(0+2)^2 - 1$
$-3 = a(2)^2 - 1$
$-2 = 4a$
$-\frac{1}{2} = a$
The quadratic function is $f(x) = -\frac{1}{2}(x+2)^2 - 1$ or $f(x) = -\frac{1}{2}x^2 - 2x - 3$.

100. Consider the form $y = a(x-h)^2 + k$. From the graph we know that the vertex is $(-2, 0)$, so we have $h = -2$ and $k = 0$. The graph also passes through the point $(x, y) = (-3, 3)$. Substituting values for x, y, h, and k, we can solve for a:

$3 = a(-3-(-2))^2 - 0$
$3 = a(-3+2)^2$
$3 = a(-1)^2$
$3 = a$
The quadratic function is $f(x) = 3(x+2)^2$ or $f(x) = 3x^2 + 12x + 12$.

101. $f(x) = x^2 + 2x - 8$
$a = 1, b = 2, c = -8$
The graph opens up because the coefficient on $x^2$ is positive.

vertex:
$x = -\frac{b}{2a} = -\frac{2}{2(1)} = -1$
$f(-1) = (-1)^2 + 2(-1) - 8 = -9$
The vertex is $(-1, -9)$ and the axis of symmetry is $x = -1$.

y-intercept:
$f(0) = (0)^2 + 2(0) - 8 = -8$

x-intercepts:
$b^2 - 4ac = 2^2 - 4(1)(-8) = 36 > 0$
There are two distinct x-intercepts. We find these by solving
$f(x) = 0$
$x^2 + 2x - 8 = 0$
$(x-2)(x+4) = 0$
$x - 2 = 0 \text{ or } x + 4 = 0$
$x = 2 \text{ or } x = -4$

Graph:
The y-intercept point, $(0, -8)$, is one unit to the right of the axis of symmetry. Therefore, if we move one unit to the left of the axis of symmetry, we obtain the point $(-2, -8)$ which must also be on the graph.

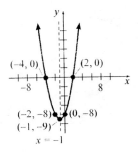

**102.** $F(x) = 2x^2 - 5x + 3$

$a = 2, b = -5, c = 3$

The graph opens up because the coefficient on $x^2$ is positive.

vertex:
$$x = -\frac{b}{2a} = -\frac{(-5)}{2(2)} = \frac{5}{4}$$

$$F\left(\frac{5}{4}\right) = 2\left(\frac{5}{4}\right)^2 - 5\left(\frac{5}{4}\right) + 3 = -\frac{1}{8}$$

The vertex is $\left(\frac{5}{4}, -\frac{1}{8}\right)$ and the axis of symmetry is $x = \frac{5}{4}$.

y-intercept:
$$F(0) = 2(0)^2 - 5(0) + 3 = 3$$

x-intercepts:
$$b^2 - 4ac = (-5)^2 - 4(2)(3) = 1 > 0$$

There are two distinct x-intercepts. We find these by solving
$$F(x) = 0$$
$$2x^2 - 5x + 3 = 0$$
$$(2x - 3)(x - 1) = 0$$
$$2x - 3 = 0 \text{ or } x - 1 = 0$$
$$2x = 3 \text{ or } x = 1$$
$$x = \frac{3}{2}$$

Graph:
The y-intercept point, $(0, 3)$, is five-fourths units to the left of the axis of symmetry. Therefore, if we move five-fourths units to the right of the axis of symmetry, we obtain the point $\left(\frac{5}{2}, 3\right)$ which must also be on the graph.

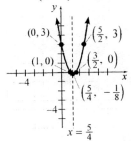

**103.** $g(x) = -x^2 + 6x - 7$

$a = -1, b = 6, c = -7$

The graph opens down because the coefficient on $x^2$ is negative.

vertex:
$$x = -\frac{b}{2a} = -\frac{6}{2(-1)} = 3$$

$$g(3) = -(3)^2 + 6(3) - 7 = 2$$

The vertex is $(3, 2)$ and the axis of symmetry is $x = 3$.

y-intercept:
$$g(0) = -(0)^2 + 6(0) - 7 = -7$$

x-intercepts:
$$b^2 - 4ac = 6^2 - 4(-1)(-7) = 8 > 0$$

There are two distinct x-intercepts. We find these by solving
$$g(x) = 0$$
$$-x^2 + 6x - 7 = 0$$
$$x = \frac{-6 \pm \sqrt{8}}{2(-1)}$$
$$= \frac{-6 \pm 2\sqrt{2}}{-2}$$
$$= 3 \pm \sqrt{2}$$
$$x \approx 1.59 \text{ or } x \approx 4.41$$

Graph:
The y-intercept point, $(0, -7)$, is three units to the left of the axis of symmetry. Therefore, if we move three units to the right of the axis of

symmetry, we obtain the point $(6,-7)$ which must also be on the graph.

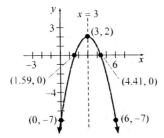

**104.** $G(x) = -2x^2 + 4x + 3$

$a = -2, b = 4, c = 3$

The graph opens down because the coefficient on $x^2$ is negative.

vertex:
$$x = -\frac{b}{2a} = -\frac{4}{2(-2)} = 1$$

$$G(1) = -2(1)^2 + 4(1) + 3 = 5$$

The vertex is $(1,5)$ and the axis of symmetry is $x = 1$.

y-intercept:
$$G(0) = -2(0)^2 + 4(0) + 3 = 3$$

x-intercepts:
$$b^2 - 4ac = 4^2 - 4(-2)(3) = 40 > 0$$

There are two distinct x-intercepts. We find these by solving
$$G(x) = 0$$
$$-2x^2 + 4x + 3 = 0$$
$$x = \frac{-4 \pm \sqrt{40}}{2(-2)}$$
$$= \frac{-4 \pm 2\sqrt{10}}{-4}$$
$$= 1 \pm \frac{\sqrt{10}}{2}$$

$x \approx -0.58$ or $x \approx 2.58$

Graph:
The y-intercept point, $(0,3)$, is one unit to the left of the axis of symmetry. Therefore, if we move one unit to the right of the axis of

symmetry, we obtain the point $(2,3)$ which must also be on the graph.

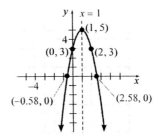

**105.** $h(x) = 4x^2 - 12x + 9$

$a = 4, b = -12, c = 9$

The graph opens up because the coefficient on $x^2$ is positive.

vertex:
$$x = -\frac{b}{2a} = -\frac{(-12)}{2(4)} = \frac{3}{2}$$

$$h\left(\frac{3}{2}\right) = 4\left(\frac{3}{2}\right)^2 - 12\left(\frac{3}{2}\right) + 9 = 0$$

The vertex is $\left(\frac{3}{2}, 0\right)$ and the axis of symmetry is $x = \frac{3}{2}$.

y-intercept:
$$h(0) = 4(0)^2 - 12(0) + 9 = 9$$

x-intercepts:
$$b^2 - 4ac = (-12)^2 - 4(4)(9) = 0$$

Since the discriminant is 0, the x-coordinate of the vertex is the only x-intercept, $x = \frac{3}{2}$.

Graph:
The y-intercept point, $(0,9)$, is three-halves units to the left of the axis of symmetry. Therefore, if we move three-halves units to the right of the axis of symmetry, we obtain the point $(3,9)$ which must also be on the graph.

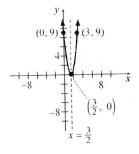

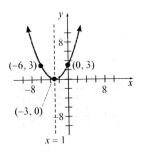

**106.** $H(x) = \frac{1}{3}x^2 + 2x + 3$

$a = \frac{1}{3}, b = 2, c = 3$

The graph opens up because the coefficient on $x^2$ is positive.

vertex:

$x = -\frac{b}{2a} = -\frac{2}{2\left(\frac{1}{3}\right)} = -3$

$H(-3) = \frac{1}{3}(-3)^2 + 2(-3) + 3 = 0$

The vertex is $(-3, 0)$ and the axis of symmetry is $x = -3$.

y-intercept:

$H(0) = \frac{1}{3}(0)^2 + 2(0) + 3 = 3$

x-intercepts:

$b^2 - 4ac = 2^2 - 4\left(\frac{1}{3}\right)(3) = 0$

Since the discriminant is 0, the x-coordinate of the vertex is the only x-intercept, $x = -3$.

Graph:
The y-intercept point, $(0, 3)$, is three units to the right of the axis of symmetry. Therefore, if we move three units to the left of the axis of symmetry, we obtain the point $(-6, 3)$ which must also be on the graph.

**107.** $p(x) = \frac{1}{4}x^2 + 3x + 10$

$a = \frac{1}{4}, b = 3, c = 10$

The graph opens up because the coefficient on $x^2$ is positive.

vertex:

$x = -\frac{b}{2a} = -\frac{3}{2\left(\frac{1}{4}\right)} = -6$

$p(-6) = \frac{1}{4}(-6)^2 + 3(-6) + 10 = 1$

The vertex is $(-6, 1)$ and the axis of symmetry is $x = -6$.

y-intercept:

$p(0) = \frac{1}{4}(0)^2 + 3(0) + 10 = 10$

x-intercepts:

$b^2 - 4ac = 3^2 - 4\left(\frac{1}{4}\right)(10) = -1 < 0$

There are no x-intercepts since the discriminant is negative.

Graph:
The y-intercept point, $(0, 10)$, is six units to the right of the axis of symmetry. Therefore, if we move six unit to the left of the axis of symmetry, we obtain the point $(-12, 10)$ which must also be on the graph.

## Chapter 8: Quadratic Equations and Quadratic Functions

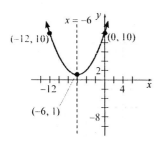

**108.** $P(x) = -x^2 + 4x - 9$

$a = -1, b = 4, c = -9$

The graph opens down because the coefficient on $x^2$ is negative.

vertex:
$$x = -\frac{b}{2a} = -\frac{4}{2(-1)} = 2$$
$$P(2) = -(2)^2 + 4(2) - 9 = -5$$

The vertex is $(2, -5)$ and the axis of symmetry is $x = 2$.

y-intercept:
$$P(0) = -(0)^2 + 4(0) - 9 = -9$$

x-intercepts:
$$b^2 - 4ac = 4^2 - 4(-1)(-9) = -20 < 0$$

There are no x-intercepts since the discriminant is negative.

Graph:
The y-intercept point, $(0, -9)$, is 2 units to the left of the axis of symmetry. Therefore, if we move 2 units to the right of the axis of symmetry, we obtain the point $(4, -9)$ which must also be on the graph.

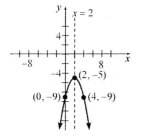

**109.** If we compare $f(x) = -2x^2 + 16x - 10$ to $f(x) = ax^2 + bx + c$, we find that $a = -2$, $b = 16$, and $c = -10$. Because $a < 0$, we know the graph will open down, so the function will have a maximum value.
The maximum value occurs at
$$x = -\frac{b}{2a} = -\frac{16}{2(-2)} = 4.$$
The maximum value is
$$f(4) = -2(4)^2 + 16(4) - 10 = 22.$$
So, the maximum value is 22, and it occurs when $x = 4$.

**110.** If we compare $g(x) = 6x^2 - 3x - 1$ to $g(x) = ax^2 + bx + c$, we find that $a = 6$, $b = -3$, and $c = -1$. Because $a > 0$, we know the graph will open up, so the function will have a minimum value.
The minimum value occurs at
$$x = -\frac{b}{2a} = -\frac{(-3)}{2(6)} = \frac{1}{4}.$$
The minimum value is
$$g\left(\frac{1}{4}\right) = 6\left(\frac{1}{4}\right)^2 - 3\left(\frac{1}{4}\right) - 1 = -\frac{11}{8}.$$
So, the minimum value is $-\frac{11}{8}$, and it occurs when $x = \frac{1}{4}$.

**111.** If we compare $h(x) = -4x^2 + 8x + 3$ to $h(x) = ax^2 + bx + c$, we find that $a = -4$, $b = 8$, and $c = 3$. Because $a < 0$, we know the graph will open down, so the function will have a maximum value.
The maximum value occurs at
$$x = -\frac{b}{2a} = -\frac{8}{2(-4)} = 1.$$
The maximum value is
$$h(4) = -4(1)^2 + 8(1) + 3 = 7.$$
So, the maximum value is 7, and it occurs when $x = 1$.

112. If we compare $F(x) = -\frac{1}{3}x^2 + 4x - 7$ to $F(x) = ax^2 + bx + c$, we find that $a = -\frac{1}{3}$, $b = 4$, and $c = -7$. Because $a < 0$, we know the graph will open down, so the function will have a maximum value.
The maximum value occurs at
$$x = -\frac{b}{2a} = -\frac{4}{2\left(-\frac{1}{3}\right)} = 6.$$
The maximum value is
$$F(6) = -\frac{1}{3}(6)^2 + 4(6) - 7 = 5.$$
So, the maximum value is 5, and it occurs when $x = 6$.

113. a. We first recognize that the quadratic function has a negative leading coefficient. This means that the graph will open down and the function indeed has a maximum value. The maximum value occurs when
$$p = -\frac{b}{2a} = -\frac{150}{2\left(-\frac{1}{3}\right)} = 225.$$
The revenue will be maximized when the televisions are sold at a price of $225.

   b. The maximum revenue is obtained by evaluating the revenue function at the price found in part (a).
$$R(225) = -\frac{1}{3}(225)^2 + 150(225) = 16,875$$
The maximum revenue is $16,875.

114. a. We first recognize that the quadratic function has a negative leading coefficient. This means that the graph will open down and the function indeed has a maximum value. The maximum value occurs when
$$I = -\frac{b}{2a} = -\frac{120}{2(-16)} = 3.75.$$
The power will be maximized when the current is 3.75 amperes.

   b. The maximum power is obtained by evaluating the power function for the current found in part (a).
$$P(3.75) = -16(3.75)^2 + 120(3.75) = 225$$
The maximum power is 225 watts.

115. Let $x$ represent the first number. Then the second number must be $24 - x$. We can express the product of the two numbers as the function
$$p(x) = x(24 - x) = -x^2 + 24x$$
This is a quadratic function with $a = -1$, $b = 24$, and $c = 0$. The function is maximized when
$$x = -\frac{b}{2a} = -\frac{24}{2(-1)} = 12.$$
The maximum product can be obtained by evaluating $p(x)$ when $x = 12$.
$$p(x) = 12(24 - 12) = 12(12) = 144$$
Two numbers that sum to 24 have a maximum product of 144 when both numbers are 12.

116. a. Let $x$ represent the width of the rectangular kennel (the side that is not parallel to the garage). Then $15 - 2x$ is the length of the kennel (the side that is parallel to the garage). The area is the product of the length and width:
$$A = (15 - 2x) \cdot x$$
$$= -2x^2 + 15x$$
The leading coefficient is negative so we know the graph opens down and there will be a maximum area. This value occurs when the width is
$$x = -\frac{b}{2a} = -\frac{15}{2(-2)} = 3.75.$$
Then the length is $15 - 2(3.75) = 7.5$.
The dimensions that maximize the area of the kennel are 3.75 yards by 7.5 yards.

   b. The maximum area can be found by substituting 3.75 for $x$ into the area function.
$$A = 3.75(15 - 2(3.75)) = 28.125$$
The maximum area of the kennel is 28.125 square yards.

117. a. First we recognize that the quadratic function has a negative leading coefficient. This means the graph will open down and the function will indeed have a maximum. The maximum height will occur when
$$x = -\frac{b}{2a} = -\frac{1}{2(-0.005)} = 100.$$
The ball will reach a maximum height when it is 100 feet from Ted.

## Chapter 8: Quadratic Equations and Quadratic Functions

b. The maximum height can be found by evaluating $h(x)$ for the value of $x$ found in part (a).
$h(100) = -0.005(100)^2 + 100 = 50$
The ball will reach a maximum height of 50 feet.

c. When the ball is on the ground, it will have a height of 0 feet. Thus, we need to solve $h(x) = 0$.
$$-0.005x^2 + x = 0$$
$$200(-0.005x^2 + x) = 200(0)$$
$$-x^2 + 200x = 0$$
$$-x(x - 200) = 0$$
$$-x = 0 \quad \text{or} \quad x - 200 = 0$$
$$x = 0 \quad \text{or} \quad x = 200$$

Zero (0) represents the distance from Ted before he kicks the ball. The ball will strike the ground again when it is 200 feet away from Ted.

118. a. Since $R = x \cdot p$, we have
$R = x \cdot p$
$= (-0.002p + 60) \cdot p$
$= -0.002p^2 + 60p$

b. The revenue function is quadratic with a negative leading coefficient. This means the graph will open down and the function will have a maximum value. This value occurs when
$$p = -\frac{b}{2a} = -\frac{60}{2(-0.002)} = 15,000.$$
The maximum revenue can be found by substituting this value for $p$ in the revenue equation.
$R = -0.002(15,000)^2 + 60(15,000) = 450,000$

The maximum revenue of $450,000 will occur if the price of each automobile is set at $15,000.

c. To determine how many automobiles will be sold, we substitute the maximizing price into the demand equation.

$x = -0.002p + 60$
$= -0.002(15,000) + 60$
$= 30$

When the price per automobile is $15,000, the dealership will sell 30 automobiles per month.

119. a. The graph is greater than 0 for $x < -2$ or $x > 3$. The solution is $\{x | x < -2 \text{ or } x > 3\}$ in set-builder notation. The solution is $(-\infty, -2) \cup (3, \infty)$ using interval notation.

b. The graph is less than 0 for $-2 < x < 3$. The solution is $\{x | -2 < x < 3\}$ using set-builder notation. The solution is $(-2, 3)$ using interval notation.

120. a. The graph is 0 or greater for $-\frac{7}{2} \leq x \leq 1$. The solution is $\left\{x \left| -\frac{7}{2} \leq x \leq 1 \right.\right\}$ using set-builder notation. The solution is $\left[-\frac{7}{2}, 1\right]$ using interval notation.

b. The graph is 0 or less for $x \leq -\frac{7}{2}$ or $x \geq 1$. The solution is $\left\{x \left| x \leq -\frac{7}{2} \text{ or } x \geq 1 \right.\right\}$ using set-builder notation. The solution is $\left(-\infty, -\frac{7}{2}\right] \cup [1, \infty)$ using interval notation.

121. $x^2 - 2x - 24 \leq 0$
Solve: $x^2 - 2x - 24 = 0$
$(x - 6)(x + 4) = 0$
$x - 6 = 0 \quad \text{or} \quad x + 4 = 0$
$x = 6 \quad \text{or} \quad x = -4$

Determine where each factor is positive and negative and where the product of these factors is positive and negative.

| Interval | $(-\infty, -4)$ | $-4$ | $(-4, 6)$ | $6$ | $(6, \infty)$ |
|---|---|---|---|---|---|
| $x - 6$ | $----$ | $-$ | $----$ | $0$ | $++++$ |
| $x + 4$ | $----$ | $0$ | $++++$ | $+$ | $++++$ |
| $(x-6)(x+4)$ | $++++$ | $0$ | $----$ | $0$ | $++++$ |

The inequality is non-strict, so $-4$ and $6$ are part of the solution. Now, $(x-6)(x+4)$ is less than zero where the product is negative. The solution is $\{x \mid -4 \leq x \leq 6\}$ or, using interval notation, $[-4, 6]$.

122. $y^2 + 7y - 8 \geq 0$

Solve: $y^2 + 7y - 8 = 0$
$(y-1)(y+8) = 0$
$y-1 = 0$ or $y+8 = 0$
$y = 1$ or $y = -8$

Determine where each factor is positive and negative and where the product of these factors is positive and negative.

The inequality is non-strict, so $-8$ and $1$ are part of the solution. Now, $(y-1)(y+8)$ is greater than zero where the product is positive. The solution is $\{y \mid y \leq -8 \text{ or } y \geq 1\}$ or, using interval notation; $(-\infty, -8] \cup [1, \infty)$.

123. $3z^2 - 19z + 20 > 0$

Solve: $3z^2 - 19z + 20 = 0$
$(3z-4)(z-5) = 0$
$3z - 4 = 0$ or $z - 5 = 0$
$3z = 4$ or $z = 5$
$z = \dfrac{4}{3}$

Determine where each factor is positive and negative and where the product of these factors is positive and negative.

The inequality is strict, so $\dfrac{4}{3}$ and $5$ are not part of the solution. Now, $(3z-4)(z-5)$ is greater than zero where the product is positive. The solution is $\left\{z \mid z < \dfrac{4}{3} \text{ or } z > 5\right\}$ or, using interval notation, $\left(-\infty, \dfrac{4}{3}\right) \cup (5, \infty)$.

124. $p^2 + 4p - 2 < 0$

Solve: $p^2 + 4p - 2 = 0$
$p = \dfrac{-4 \pm \sqrt{4^2 - 4(1)(-2)}}{2(1)}$
$= \dfrac{-4 \pm \sqrt{24}}{2}$
$= \dfrac{-4 \pm 2\sqrt{6}}{2}$
$= -2 \pm \sqrt{6}$
$x = -2 - \sqrt{6}$ or $x = -2 + \sqrt{6}$
$\approx -4.45$ or $\approx 0.45$

Determine where each factor is positive and negative and where the product of these factors is positive and negative.

The inequality is strict, so $-2 - \sqrt{6}$ and $-2 + \sqrt{6}$ are not part of the solution. Now, $\left[p - \left(-2 - \sqrt{6}\right)\right]\left[p - \left(-2 + \sqrt{6}\right)\right]$ is less than

## Chapter 8: Quadratic Equations and Quadratic Functions

zero where the product is negative. The solution is $\{p \mid -2-\sqrt{6} < p < -2+\sqrt{6}\}$ or, using interval notation, $(-2-\sqrt{6},\ -2+\sqrt{6})$.

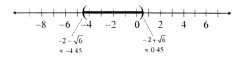

**125.** $4m^2 - 20m + 25 \geq 0$

Solve: $4m^2 - 20m + 25 = 0$
$(2m-5)(2m-5) = 0$

The inequality is non-strict, so $\dfrac{5}{2}$ is part of the solution. Now, $(2m-5)(2m-5)$ is greater than zero where the product is positive. Thus, $4m^2 - 20m + 25$ is always greater than or equal to zero. The solution is $\{m \mid m$ is any real number$\}$ or, using interval notation $(-\infty,\ \infty)$.

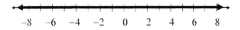

**126.** $6w^2 - 19w - 7 \leq 0$

Solve: $6w^2 - 19w - 7 = 0$
$(3w+1)(2w-7) = 0$

$3w+1 = 0$ or $2w-7 = 0$
$3w = -1$ or $2w = 7$
$w = -\dfrac{1}{3}$ or $w = \dfrac{7}{2}$

Determine where each factor is positive and negative and where the product of these factors is positive and negative.

| Interval | $\left(-\infty, -\dfrac{1}{3}\right)$ | $-\dfrac{1}{3}$ | $\left(-\dfrac{1}{3}, \dfrac{7}{2}\right)$ | $\dfrac{7}{2}$ | $\left(\dfrac{7}{2}, \infty\right)$ |
|---|---|---|---|---|---|
| $3w+1$ | $----$ | 0 | $++++$ | $+$ | $++++$ |
| $2w-7$ | $----$ | $-$ | $----$ | 0 | $++++$ |
| $(3w+1)(2w-7)$ | $++++$ | 0 | $----$ | 0 | $++++$ |

The inequality is non-strict, so $-\dfrac{1}{3}$ and $\dfrac{7}{2}$ are part of the solution. Now, $(3w+1)(2w-7)$ is less than zero where the product is negative. The solution is $\left\{w \mid -\dfrac{1}{3} \leq w \leq \dfrac{7}{2}\right\}$ or, using interval

$2m-5 = 0$ or $2m-5 = 0$
$2m = 5$ or $2m = 5$
$m = \dfrac{5}{2}$ or $m = \dfrac{5}{2}$

Determine where each factor is positive and negative and where the product of these factors is positive and negative.

| Interval | $\left(-\infty, \dfrac{5}{2}\right)$ | $\dfrac{5}{2}$ | $\left(\dfrac{5}{2}, \infty\right)$ |
|---|---|---|---|
| $2m-5$ | $----$ | 0 | $++++$ |
| $2m-5$ | $----$ | 0 | $++++$ |
| $(2m-5)(2m-5)$ | $++++$ | 0 | $++++$ |

notation, $\left[-\dfrac{1}{3},\ \dfrac{7}{2}\right]$.

## Chapter 8 Test

**1.** Start: $x^2 - 3x$

Add: $\left[\dfrac{1}{2}\cdot(-3)\right]^2 = \dfrac{9}{4}$

Result: $x^2 - 3x + \dfrac{9}{4}$

Factored Form: $\left(x - \dfrac{3}{2}\right)^2$

**2.** Start: $m^2 + \dfrac{2}{5}m$

Add: $\left[\dfrac{1}{2}\cdot\left(\dfrac{2}{5}\right)\right]^2 = \dfrac{1}{25}$

Result: $m^2 + \dfrac{2}{5}m + \dfrac{1}{25}$

Factored Form: $\left(m + \dfrac{1}{5}\right)^2$

3. $9\left(x+\dfrac{4}{3}\right)^2 = 1$

$\left(x+\dfrac{4}{3}\right)^2 = \dfrac{1}{9}$

$x + \dfrac{4}{3} = \pm\sqrt{\dfrac{1}{9}}$

$x + \dfrac{4}{3} = \pm\dfrac{1}{3}$

$x = -\dfrac{4}{3} \pm \dfrac{1}{3}$

$x = -\dfrac{4}{3} - \dfrac{1}{3}$ or $x = -\dfrac{4}{3} + \dfrac{1}{3}$

$x = -\dfrac{5}{3}$ or $x = -1$

The solution set is $\left\{-\dfrac{5}{3},\ -1\right\}$.

4. $m^2 - 6m + 4 = 0$

Because this equation does not easily factor, solve by using the quadratic formula. For this equation, $a = 1$, $b = -6$, and $c = 4$.

$m = \dfrac{-(-6) \pm \sqrt{(-6)^2 - 4(1)(4)}}{2(1)}$

$= \dfrac{6 \pm \sqrt{36 - 16}}{2}$

$= \dfrac{6 \pm \sqrt{20}}{2}$

$= \dfrac{6 \pm 2\sqrt{5}}{2}$

$= \dfrac{6}{2} \pm \dfrac{2\sqrt{5}}{2}$

$= 3 \pm \sqrt{5}$

The solution set is $\{3 - \sqrt{5},\ 3 + \sqrt{5}\}$.

5. $2w^2 - 4w + 3 = 0$

Because this equation does not easily factor, solve by using the quadratic formula. For this equation, $a = 2$, $b = -4$, and $c = 3$.

$w = \dfrac{-(-4) \pm \sqrt{(-4)^2 - 4(2)(3)}}{2(2)}$

$= \dfrac{4 \pm \sqrt{16 - 24}}{4}$

$= \dfrac{4 \pm \sqrt{-8}}{4}$

$= \dfrac{4 \pm 2\sqrt{2}\,i}{4}$

$= \dfrac{4}{4} \pm \dfrac{2\sqrt{2}}{4}i$

$= 1 \pm \dfrac{\sqrt{2}}{2}i$

The solution set is $\left\{1 - \dfrac{\sqrt{2}}{2}i,\ 1 + \dfrac{\sqrt{2}}{2}i\right\}$.

6. $\dfrac{1}{2}z^2 - \dfrac{3}{2}z = -\dfrac{7}{6}$

$6\left(\dfrac{1}{2}z^2 - \dfrac{3}{2}z\right) = 6\left(-\dfrac{7}{6}\right)$

$3z^2 - 9z = -7$

$3z^2 - 9z + 7 = 0$

Because this equation does not easily factor, solve by using the quadratic formula. For this equation, $a = 3$, $b = -9$, and $c = 7$.

$z = \dfrac{-(-9) \pm \sqrt{(-9)^2 - 4(3)(7)}}{2(3)}$

$= \dfrac{9 \pm \sqrt{81 - 84}}{6}$

$= \dfrac{9 \pm \sqrt{-3}}{6}$

$= \dfrac{9 \pm \sqrt{3}\,i}{6}$

$= \dfrac{9}{6} \pm \dfrac{\sqrt{3}}{6}i$

$= \dfrac{3}{2} \pm \dfrac{\sqrt{3}}{6}i$

The solution set is $\left\{\dfrac{3}{2} - \dfrac{\sqrt{3}}{6}i,\ \dfrac{3}{2} + \dfrac{\sqrt{3}}{6}i\right\}$.

7. $2x^2 + 5x = 4$

$2x^2 + 5x - 4 = 0$

For this equation, $a = 2$, $b = 5$, and $c = -4$.

$b^2 - 4ac = 5^2 - 4(2)(-4) = 25 + 32 = 57$

Because $b^2 - 4ac = 57$, but not a perfect square, the quadratic equation will have two irrational solutions.

## Chapter 8: Quadratic Equations and Quadratic Functions

8. $c^2 = a^2 + b^2$
   $11^2 = a^2 + 7^2$
   $121 = a^2 + 49$
   $72 = a^2$
   $a = \sqrt{72}$
   $a = 6\sqrt{2}$

9. $x^4 - 5x^2 - 36 = 0$
   $(x^2)^2 - 5(x^2) - 36 = 0$
   Let $u = x^2$.
   $u^2 - 5u - 36 = 0$
   $(u-9)(u+4) = 0$
   $u - 9 = 0$ or $u + 4 = 0$
   $u = 9$ or $u = -4$
   $x^2 = 9$ or $x^2 = -4$
   $x = \pm\sqrt{9}$ or $x = \pm\sqrt{-4}$
   $x = \pm 3$ or $x = \pm 2i$

   Check:
   $x = -3$: $(-3)^4 - 5(-3)^2 - 36 \stackrel{?}{=} 0$
   $81 - 5 \cdot 9 - 36 \stackrel{?}{=} 0$
   $81 - 45 - 36 \stackrel{?}{=} 0$
   $0 = 0$ ✓

   $x = 3$: $3^4 - 5(3)^2 - 36 \stackrel{?}{=} 0$
   $81 - 5 \cdot 9 - 36 \stackrel{?}{=} 0$
   $81 - 45 - 36 \stackrel{?}{=} 0$
   $0 = 0$ ✓

   $x = -2i$: $(-2i)^4 - 5(-2i)^2 - 36 \stackrel{?}{=} 0$
   $16i^4 - 5 \cdot 4i^2 - 36 \stackrel{?}{=} 0$
   $16(1) - 5 \cdot 4(-1) - 36 \stackrel{?}{=} 0$
   $16 + 20 - 36 \stackrel{?}{=} 0$
   $0 = 0$ ✓

   $x = 2i$: $(2i)^4 - 5(2i)^2 - 36 \stackrel{?}{=} 0$
   $16i^4 - 5 \cdot 4i^2 - 36 \stackrel{?}{=} 0$
   $16(1) - 5 \cdot 4(-1) - 36 \stackrel{?}{=} 0$
   $16 + 20 - 36 \stackrel{?}{=} 0$
   $0 = 0$ ✓

   All check; the solution set is $\{-3, 3, -2i, 2i\}$.

10. $6y^{1/2} + 13y^{1/4} - 5 = 0$
    $6\left(y^{1/4}\right)^2 + 13\left(y^{1/4}\right) - 5 = 0$
    Let $u = y^{1/4}$.
    $6u^2 + 13u - 5 = 0$
    $(3u - 1)(2u + 5) = 0$
    $3u - 1 = 0$ or $2u + 5 = 0$
    $3u = 1$ or $2u = -5$
    $u = \dfrac{1}{3}$ or $u = -\dfrac{5}{2}$
    $y^{1/4} = \dfrac{1}{3}$ or $y^{1/4} = -\dfrac{5}{2}$
    $\left(y^{1/4}\right)^4 = \left(\dfrac{1}{3}\right)^4$ or $\left(y^{1/4}\right)^4 = \left(-\dfrac{5}{2}\right)^4$
    $y = \dfrac{1}{81}$ or $y = \dfrac{625}{16}$

    Check:
    $y = \dfrac{1}{81}$: $6\left(\dfrac{1}{81}\right)^{1/2} + 13\left(\dfrac{1}{81}\right)^{1/4} - 5 \stackrel{?}{=} 0$
    $6\sqrt{\dfrac{1}{81}} + 13\sqrt[4]{\dfrac{1}{81}} - 5 \stackrel{?}{=} 0$
    $6\left(\dfrac{1}{9}\right) + 13\left(\dfrac{1}{3}\right) - 5 \stackrel{?}{=} 0$
    $\dfrac{2}{3} + \dfrac{13}{3} - 5 \stackrel{?}{=} 0$
    $0 = 0$ ✓

    $y = \dfrac{625}{16}$: $6\left(\dfrac{625}{16}\right)^{1/2} + 13\left(\dfrac{625}{16}\right)^{1/4} - 5 \stackrel{?}{=} 0$
    $6\sqrt{\dfrac{625}{16}} + 13\sqrt[4]{\dfrac{625}{16}} - 5 \stackrel{?}{=} 0$
    $6\left(\dfrac{25}{4}\right) + 13\left(\dfrac{5}{2}\right) - 5 \stackrel{?}{=} 0$
    $\dfrac{75}{2} + \dfrac{65}{2} - 5 \stackrel{?}{=} 0$
    $65 \neq 0$ ✗

    $y = \dfrac{625}{16}$ does not check; the solution set is $\left\{\dfrac{1}{81}\right\}$.

11. Begin with the graph of $y = x^2$, then shift the graph 2 units to the left to obtain the graph of $y = (x+2)^2$. Shift this graph down 5 units to obtain the graph of $f(x) = (x+2)^2 - 5$.

# Chapter 8 Test

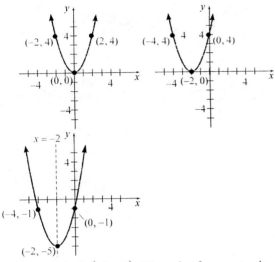

The vertex is $(-2, -5)$. The axis of symmetry is $x = -2$.

12. $g(x) = -2x^2 - 8x - 3$
    $a = -2, b = -8, c = -3$
    The graph opens down because the coefficient on $x^2$ is negative.

    vertex:
    $$x = -\frac{b}{2a} = -\frac{(-8)}{2(-2)} = -2$$
    $$g(-2) = -2(-2)^2 - 8(-2) - 3 = 5$$
    The vertex is $(-2, 5)$. The axis of symmetry is $x = -2$.

    y-intercept:
    $$g(0) = -2(0)^2 - 8(0) - 3 = -3$$

    x-intercepts:
    $$b^2 - 4ac = (-8)^2 - 4(-2)(-3) = 40 > 0$$
    There are two distinct x-intercepts. We find these by solving
    $$g(x) = 0$$
    $$-2x^2 - 8x - 3 = 0$$
    $$x = \frac{-(-8) \pm \sqrt{40}}{2(-2)}$$
    $$= \frac{8 \pm 2\sqrt{10}}{-4}$$
    $$= -2 \pm \frac{\sqrt{10}}{2}$$
    $x \approx -3.58$ or $x \approx -0.42$

Graph:
The y-intercept point, $(0, -3)$, is two units to the right of the axis of symmetry. Therefore, if we move two units to the left of the axis of symmetry, we obtain the point $(-4, -3)$ which must also be on the graph.

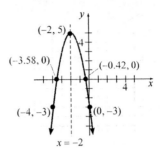

13. Consider the form $y = a(x - h)^2 + k$. From the graph we know that the vertex is $(-3, -5)$, so we have $h = -3$ and $k = -5$. The graph also passes through the point $(x, y) = (0, -2)$. Substituting values for x, y, h, and k, we can solve for a:
    $$-2 = a(0 - (-3))^2 - 5$$
    $$-2 = a(0 + 3)^2 - 5$$
    $$-2 = a(3)^2 - 5$$
    $$-2 = 9a - 5$$
    $$3 = 9a$$
    $$\frac{3}{9} = a$$
    $$\frac{1}{3} = a$$

    The quadratic function is $f(x) = \frac{1}{3}(x + 3)^2 - 5$ or $f(x) = \frac{1}{3}x^2 + 2x - 2$.

14. If we compare $h(x) = -\frac{1}{4}x^2 + x + 5$ to $f(x) = ax^2 + bx + c$, we find that $a = -\frac{1}{4}$, $b = 1$, and $c = 5$. Because $a < 0$, we know the graph will open down, so the function will have a maximum value.
    The maximum value occurs at
    $$x = -\frac{b}{2a} = -\frac{1}{2(-\frac{1}{4})} = -\frac{1}{-\frac{1}{2}} = 2.$$

## Chapter 8: Quadratic Equations and Quadratic Functions

The maximum value is
$h(2) = -\frac{1}{4}(2)^2 + 2 + 5 = 6$.
So, the maximum value is 6, and it occurs when $x = 2$.

**15.** $2m^2 + m - 15 > 0$

Solve: $2m^2 + m - 15 = 0$
$(2m - 5)(m + 3) = 0$
$2m - 5 = 0$ or $m + 3 = 0$
$2m = 5$ or $m = -3$
$m = \frac{5}{2}$

Determine where each factor is positive and negative and where the product of these factors is positive and negative.

The inequality is strict, so $-3$ and $\frac{5}{2}$ are not part of the solution. Now, $(2m-5)(m+3)$ is greater than zero where the product is positive. The solution is $\left\{ m \mid m < -3 \text{ or } m > \frac{5}{2} \right\}$ or, using interval notation, $(-\infty, -3) \cup \left(\frac{5}{2}, \infty\right)$.

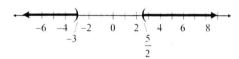

**16.** $z^2 + 6z - 1 \leq 0$

Solve: $z^2 + 6z - 1 = 0$
$z = \frac{-6 \pm \sqrt{6^2 - 4(1)(-1)}}{2(1)}$
$= \frac{-6 \pm \sqrt{40}}{2}$
$= \frac{-6 \pm 2\sqrt{10}}{2}$
$= -3 \pm \sqrt{10}$
$z = -3 - \sqrt{10}$ or $z = -3 + \sqrt{10}$
$\approx -6.16$ or $\approx 0.16$

Determine where each factor is positive and negative and where the product of these factors is positive and negative.

The inequality is non-strict, so $-3 - \sqrt{10}$ and $-3 + \sqrt{10}$ are part of the solution. Now, $\left[z - \left(-3 - \sqrt{10}\right)\right]\left[z - \left(-3 + \sqrt{10}\right)\right]$ is less than zero where the product is negative. The solution is $\left\{z \mid -3 - \sqrt{10} \leq z \leq -3 + \sqrt{10}\right\}$ or, using interval notation, $\left[-3 - \sqrt{10},\; -3 + \sqrt{10}\right]$.

**17.** 
$50 = -16t^2 + 80t + 20$
$16t^2 - 80t + 30 = 0$
For this equation, $a = 16$, $b = -80$, and $c = 30$.
$t = \frac{-(-80) \pm \sqrt{(-80)^2 - 4(16)(30)}}{2(16)}$
$= \frac{80 \pm \sqrt{6400 - 1920}}{32}$
$= \frac{80 \pm \sqrt{4480}}{32}$
$= \frac{80 \pm 8\sqrt{70}}{32}$
$= \frac{80}{32} \pm \frac{8\sqrt{70}}{32}$
$= \frac{5}{2} \pm \frac{\sqrt{70}}{4}$
$t = \frac{5}{2} - \frac{\sqrt{70}}{4}$ or $t = \frac{5}{2} - \frac{\sqrt{70}}{4}$
$\approx 0.408$ or $\approx 4.592$

Rounding to the nearest tenth, the height of the rock will be 50 feet at both 0.4 seconds and 4.6 seconds.

18. Let $t$ represent the time required for Rupert to roof the house alone. Then $t-4$ will represent the time required for Lex to roof the house alone.

$$\begin{pmatrix} \text{Part done} \\ \text{by Rupert} \\ \text{in 1 hour} \end{pmatrix} + \begin{pmatrix} \text{Part done} \\ \text{by Lex} \\ \text{in 1 hour} \end{pmatrix} = \begin{pmatrix} \text{Part done} \\ \text{together} \\ \text{in 1 hour} \end{pmatrix}$$

$$\frac{1}{t} + \frac{1}{t-4} = \frac{1}{16}$$

$$16t(t-4)\left(\frac{1}{t} + \frac{1}{t-4}\right) = 16t(t-4)\left(\frac{1}{16}\right)$$

$$16(t-4) + 16t = t(t-4)$$

$$16t - 64 + 16t = t^2 - 4t$$

$$32t - 64 = t^2 - 4t$$

$$0 = t^2 - 36t + 64$$

For this equation, $a=1$, $b=-36$, and $c=64$.

$$t = \frac{-(-36) \pm \sqrt{(-36)^2 - 4(1)(64)}}{2(1)}$$

$$= \frac{36 \pm \sqrt{1296 - 256}}{2}$$

$$= \frac{36 \pm \sqrt{1040}}{2}$$

$$= \frac{36 \pm 4\sqrt{65}}{2}$$

$$= 18 \pm 2\sqrt{65}$$

$t = 18 - 2\sqrt{65}$ or $t = 18 + 2\sqrt{65}$
$\approx 1.875$ or $\approx 34.125$

Disregard $t \approx 1.875$ because this value makes Lex's time negative: $t - 4 = 1.875 - 4 = -2.125$. The only viable answer is $t \approx 34.125$ hours. Rounding to the nearest tenth, Rupert can roof the house in 34.1 hours when working alone.

19. a. We first recognize that the quadratic function has a negative leading coefficient. This means that the graph will open down and the function indeed has a maximum value. The maximum value occurs when

$$p = -\frac{b}{2a} = -\frac{170}{2(-0.25)} = 340.$$

The revenue will be maximized when the product is sold at a price of $340.

b. The maximum revenue is obtained by evaluating the revenue function at the price found in part (a).

$$R(340) = -0.25(340)^2 + 170(340) = 28,900$$

The maximum weekly revenue is $28,900.

20. a. Let $x$ represent the width of the base of the box. Then $\frac{50-2x}{2} = 25 - x$ is the length of the base. The volume is the product of the length, width, and height:
$$V = (25-x) \cdot x \cdot 12$$
$$= -12x^2 + 300x$$
The leading coefficient is negative so we know the graph opens down and there will be a maximum volume. This value occurs when the width is
$$x = -\frac{b}{2a} = -\frac{300}{2(-12)} = 12.5.$$
Then the length is $25 - 12.5 = 12.5$.
The dimensions that maximize the volume of the box are 12.5 in. by 12.5 in. by 12 in.

b. The maximum volume can be found by substituting 12.5 for $x$ into the volume function.
$$V = (25-12.5) \cdot 12.5 \cdot 12 = 1875$$
The maximum volume of the box is 1875 cubic inches.

# Chapter 9

## 9.1 Preparing for Composite Functions and Inverse Functions

1. We need to find all values of $x$ that cause the denominator $x^2 + 3x - 28$ to equal 0.
$$x^2 + 3x - 28 = 0$$
$$(x+7)(x-4) = 0$$
$$x + 7 = 0 \quad \text{or} \quad x - 4 = 0$$
$$x = -7 \quad \text{or} \quad x = 4$$
Thus, the domain of $R(x) = \dfrac{x^2 - 9}{x^2 + 3x - 28}$ is $\{x \mid x \neq -7, x \neq 4\}$.

2. a. $f(-2) = 2(-2)^2 - (-2) + 1$
$= 8 + 2 + 1$
$= 11$

   b. $f(a+1) = 2(a+1)^2 - (a+1) + 1$
$= 2(a^2 + 2a + 1) - (a+1) + 1$
$= 2a^2 + 4a + 2 - a - 1 + 1$
$= 2a^2 + 3a + 2$

3. The graph shown is not a function because it fails the vertical line test.

## 9.1 Exercises

2. one-to-one

4. TRUE.

6. FALSE. $(f \circ g)(x) = f(g(x))$

8. Answers may vary. One possibility follows: By definition, the inverse of a function with ordered pairs of the form $(a, b)$ is the set of ordered pairs of the form $(b, a)$. This means that the set of $y$-coordinates (i.e., the range) of the original function will become the set of $x$-coordinates (i.e., the domain) of the inverse function. Likewise, the set of $x$-coordinates (i.e., the domain) of the original function will become the set of $y$-coordinates (i.e., the range) of the inverse function.

10. $g(x) = 4x - 3$

12. $f(x) = 4x - 3$; $g(x) = x + 2$

    a. $g(3) = 3 + 2 = 5$
    $f(5) = 4(5) - 3 = 20 - 3 = 17$
    $(f \circ g)(3) = f(g(3)) = f(5) = 17$

    b. $f(-2) = 4(-2) - 3 = -8 - 3 = -11$
    $g(-11) = -11 + 2 = -9$
    $(g \circ f)(-2) = g(f(-2)) = g(-11) = -9$

    c. $f(1) = 4(1) - 3 = 4 - 3 = 1$
    $f(1) = 4(1) - 3 = 4 - 3 = 1$
    $(f \circ f)(1) = f(f(1)) = f(1) = 1$

    d. $g(-4) = -4 + 2 = -2$
    $g(-2) = -2 + 2 = 0$
    $(g \circ g)(-4) = g(g(-4)) = g(-2) = 0$

14. $f(x) = x^2 - 3$; $g(x) = 5x + 1$

    a. $g(3) = 5(3) + 1 = 15 + 1 = 16$
    $f(16) = (16)^2 - 3 = 256 - 3 = 253$
    $(f \circ g)(3) = f(g(3)) = f(16) = 253$

    b. $f(-2) = (-2)^2 - 3 = 4 - 3 = 1$
    $g(1) = 5(1) + 1 = 5 + 1 = 6$
    $(g \circ f)(-2) = g(f(-2)) = g(1) = 6$

    c. $f(1) = (1)^2 - 3 = 1 - 3 = -2$
    $f(-2) = (-2)^2 - 3 = 4 - 3 = 1$
    $(f \circ f)(1) = f(f(1)) = f(-2) = 1$

    d. $g(-4) = 5(-4) + 1 = -20 + 1 = -19$
    $g(-19) = 5(-19) + 1 = -95 + 1 = -94$
    $(g \circ g)(-4) = g(g(-4)) = g(-19) = -94$

16. $f(x) = -2x^3$; $g(x) = x^2 + 1$

    a. $g(3) = (3)^2 + 1 = 10$
    $f(10) = -2(10)^3 = -2000$
    $(f \circ g)(3) = f(g(3)) = f(10) = -2000$

**Section 9.1** Composite Functions and Inverse Functions

b. $f(-2) = -2(-2)^3 = -2(-8) = 16$
$g(16) = (16)^2 + 1 = 256 + 1 = 257$
$(g \circ f)(-2) = g(f(-2)) = g(16) = 257$

c. $f(1) = 2(1)^3 = 2$
$f(2) = 2(2)^3 = 16$
$(f \circ f)(1) = f(f(1)) = f(2) = 16$

d. $g(-4) = (-4)^2 + 1 = 16 + 1 = 17$
$g(17) = (17)^2 + 1 = 289 + 1 = 290$
$(g \circ g)(-4) = g(g(-4)) = g(17) = 290$

18. $f(x) = \sqrt{x+8}$; $g(x) = x^2 - 4$

    a. $g(3) = (3)^2 - 4 = 9 - 4 = 5$
    $f(5) = \sqrt{5+8} = \sqrt{13}$
    $(f \circ g)(3) = f(g(3)) = f(5) = \sqrt{13}$

    b. $f(-2) = \sqrt{-2+8} = \sqrt{6}$
    $g(\sqrt{6}) = (\sqrt{6})^2 - 4 = 6 - 4 = 2$
    $(g \circ f)(-2) = g(f(-2)) = g(\sqrt{6}) = 2$

    c. $f(1) = \sqrt{1+8} = \sqrt{9} = 3$
    $f(3) = \sqrt{3+8} = \sqrt{11}$
    $(f \circ f)(1) = f(f(1)) = f(3) = \sqrt{11}$

    d. $g(-4) = (-4)^2 - 4 = 16 - 4 = 12$
    $g(12) = (12)^2 - 4 = 144 - 4 = 140$

20. $f(x) = x - 3$; $g(x) = 4x$

    a. $(f \circ g)(x) = f(g(x)) = (4x) - 3 = 4x - 3$

    b. $(g \circ f)(x) = g(f(x)) = 4(x-3) = 4x - 12$

    c. $(f \circ f)(x) = f(f(x)) = (x-3) - 3 = x - 6$

    d. $(g \circ g)(x) = g(g(x)) = 4(4x) = 16x$

22. $f(x) = 3x - 1$; $g(x) = -2x + 5$

    a. $(f \circ g)(x) = f(g(x))$
    $= 3(-2x+5) - 1$
    $= -6x + 15 - 1$
    $= -6x + 14$

    b. $(g \circ f)(x) = g(f(x))$
    $= -2(3x-1) + 5$
    $= -6x + 2 + 5$
    $= -6x + 7$

    c. $(f \circ f)(x) = f(f(x))$
    $= 3(3x-1) - 1$
    $= 9x - 3 - 1$
    $= 9x - 4$

    d. $(g \circ g)(x) = g(g(x))$
    $= -2(-2x+5) + 5$
    $= 4x - 10 + 5$
    $= 4x - 5$

24. $f(x) = x^2 + 1$; $g(x) = x + 1$

    a. $(f \circ g)(x) = f(g(x))$
    $= (x+1)^2 + 1$
    $= x^2 + 2x + 1 + 1$
    $= x^2 + 2x + 2$

    b. $(g \circ f)(x) = g(f(x)) = (x^2 + 1) + 1 = x^2 + 2$

    c. $(f \circ f)(x) = f(f(x))$
    $= (x^2 + 1)^2 + 1$
    $= x^4 + 2x^2 + 1 + 1$
    $= x^4 + 2x^2 + 2$

    d. $(g \circ g)(x) = g(g(x)) = (x+1) + 1 = x + 2$

26. $f(x) = \sqrt{x+2}$; $g(x) = x - 2$

    a. $(f \circ g)(x) = f(g(x)) = \sqrt{(x-2)+2} = \sqrt{x}$

    b. $(g \circ f)(x) = g(f(x)) = \sqrt{x+2} - 2$

    c. $(f \circ f)(x) = f(f(x)) = \sqrt{\sqrt{x+2}+2}$

**d.** $(g \circ g)(x) = g(g(x)) = (x-2) - 2 = x - 4$

28. $f(x) = |x-3|$; $g(x) = x^3 + 3$

   **a.** $(f \circ g)(x) = f(g(x)) = |(x^3+3) - 3| = |x^3|$

   **b.** $(g \circ f)(x) = g(f(x))$
   $= (|x-3|)^3 + 3$
   $= |(x-3)^3| + 3$
   $= |x^3 - 9x^2 + 27x - 27| + 3$

   **c.** $(f \circ f)(x) = f(f(x)) = ||x-3|-3|$

   **d.** $(g \circ g)(x) = g(g(x))$
   $= (x^3+3)^3 + 3$
   $= x^9 + 9x^6 + 27x^3 + 27 + 3$
   $= x^9 + 9x^6 + 27x^3 + 30$

30. $f(x) = \dfrac{2}{x-1}$; $g(x) = \dfrac{4}{x}$

   **a.** $(f \circ g)(x) = f(g(x))$
   $= \dfrac{2}{\frac{4}{x} - 1} = \dfrac{2}{\frac{4-x}{x}} = \dfrac{2x}{4-x}$
   where $x \neq 0, 4$.

   **b.** $(g \circ f)(x) = g(f(x))$
   $= \dfrac{4}{\frac{2}{x-1}} = 4 \cdot \dfrac{x-1}{2}$
   $= 2(x-1)$ or $2x - 2$
   where $x \neq 1$.

   **c.** $(f \circ f)(x) = f(f(x))$
   $= \dfrac{2}{\frac{2}{x-1} - 1} = \dfrac{2}{\frac{2-(x-1)}{x-1}}$
   $= \dfrac{2}{\frac{3-x}{x-1}} = 2 \cdot \dfrac{x-1}{3-x}$
   $= \dfrac{2(x-1)}{3-x}$ or $\dfrac{-2(x-1)}{x-3}$
   where $x \neq 1, 3$.

   **d.** $(g \circ g)(x) = g(g(x)) = \dfrac{4}{\frac{4}{x}} = 4 \cdot \dfrac{x}{4} = x$
   where $x \neq 0$.

32. The function is one-to-one. Each element in the range corresponds to exactly one element in the domain.

34. The function is not one-to-one. There is an element in the range (3) that corresponds to more than one element in the domain (Nevada and New Mexico).

36. The function is not one-to-one. There is an element in the range (6) that corresponds to more than one element in the domain (–2 and 2).

38. The function is one-to-one. Each element in the range corresponds to exactly one element in the domain.

40. The function is not one-to-one. There is an element in the range (0) that corresponds to more than one element in the domain (–3 and –1).

42. The graph passes the horizontal line test, so the graph is that of a one-to-one function.

44. The graph fails the horizontal line test. Therefore, the function is not one-to-one.

46. The graph fails the horizontal line test. Therefore, the function is not one-to-one.

48. Inverse:
   | Quantity Demanded | Price ($) |
   |---|---|
   | 152 | 2300 |
   | 159 | 2000 |
   | 164 | 1700 |
   | 171 | 1500 |
   | 176 | 1300 |

50. To obtain the inverse, we switch the $x$- and $y$-coordinates.
   Inverse: $\{(4,-1),(1,0),(-2,1),(-5,2)\}$

52. To obtain the inverse, we switch the $x$- and $y$-coordinates.
   Inverse: $\{(1,-10),(4,-5),(3,0),(2,-5)\}$

54. To plot the inverse, switch the $x$- and $y$-coordinates in each point and connect the corresponding points. The graph of the function

(shaded) and the line $y = x$ (dashed) are included for reference.

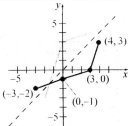

56. To plot the inverse, switch the $x$- and $y$-coordinates in each point and connect the corresponding points. The graph of the function (shaded) and the line $y = x$ (dashed) are included for reference.

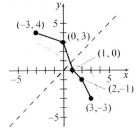

58. To plot the inverse, switch the $x$- and $y$-coordinates in each point and connect the corresponding points. The graph of the function (shaded) and the line $y = x$ (dashed) are included for reference.

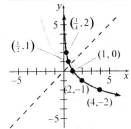

60. $f(g(x)) = 10\left(\dfrac{x}{10}\right) = x$

$g(f(x)) = \dfrac{10x}{10} = x$

Since $f(g(x)) = g(f(x)) = x$, the two functions are inverses of each other.

62. $f(g(x)) = 3\left(\dfrac{x+5}{3}\right) - 5 = x + 5 - 5 = x$

$g(f(x)) = \dfrac{(3x-5)+5}{3} = \dfrac{3x}{3} = x$

Since $f(g(x)) = g(f(x)) = x$, the two functions are inverses of each other.

64. $f(g(x)) = 8\left(\dfrac{x-3}{8}\right) + 3 = x - 3 + 3 = x$

$g(f(x)) = \dfrac{(8x+3)-3}{8} = \dfrac{8x}{8} = x$

Since $f(g(x)) = g(f(x)) = x$, the two functions are inverses of each other.

66. $f(g(x)) = \dfrac{2}{\left(\dfrac{2}{x}-4\right)+4} = \dfrac{2}{\dfrac{2}{x}} = 2 \cdot \dfrac{x}{2} = x$

$g(f(x)) = \dfrac{2}{\dfrac{2}{x+4}} - 4 = 2 \cdot \dfrac{x+4}{2} - 4 = x + 4 - 4 = x$

Since $f(g(x)) = g(f(x)) = x$, the two functions are inverses of each other.

68. $f(g(x)) = \sqrt[3]{2\left(\dfrac{x^3-1}{2}\right)+1} = \sqrt[3]{x^3-1+1} = \sqrt[3]{x^3} = x$

$g(f(x)) = \dfrac{\left(\sqrt[3]{2x+1}\right)^3 - 1}{2} = \dfrac{2x+1-1}{2} = \dfrac{2x}{2} = x$

Since $f(g(x)) = g(f(x)) = x$, the two functions are inverses of each other.

70. $f(x) = 12x$

$y = 12x$

$x = 12y$

$\dfrac{x}{12} = y$

$f^{-1}(x) = \dfrac{x}{12}$

Check:

$f(f^{-1}(x)) = 12\left(\dfrac{x}{12}\right) = x$

$f^{-1}(f(x)) = \dfrac{12x}{12} = x$

## Chapter 9: Exponential and Logarithmic Functions

**72.** $g(x) = x + 6$
$y = x + 6$
$x = y + 6$
$x - 6 = y$
$g^{-1}(x) = x - 6$

Check:
$g(g^{-1}(x)) = (x - 6) + 6 = x$
$g^{-1}(g(x)) = (x + 6) - 6 = x$

**74.** $H(x) = 3x + 8$
$y = 3x + 8$
$x = 3y + 8$
$x - 8 = 3y$
$\dfrac{x - 8}{3} = y$
$H^{-1}(x) = \dfrac{x - 8}{3}$

Check:
$H(H^{-1}(x)) = 3\left(\dfrac{x - 8}{3}\right) + 8 = x - 8 + 8 = x$
$H^{-1}(H(x)) = \dfrac{(3x + 8) - 8}{3} = \dfrac{3x}{3} = x$

**76.** $F(x) = 1 - 6x$
$y = 2 - 5x$
$x = 1 - 6y$
$x - 1 = -6y$
$\dfrac{x - 1}{-6} = y$
$\dfrac{1 - x}{6} = y$
$F^{-1}(x) = \dfrac{1 - x}{6}$

Check:
$F(F^{-1}(x)) = 1 - 6\left(\dfrac{1 - x}{6}\right) = 1 - (1 - x) = x$
$F^{-1}(F(x)) = \dfrac{1 - (1 - 6x)}{6} = \dfrac{1 - 1 + 6x}{6} = \dfrac{6x}{6} = x$

**78.** $f(x) = x^3 - 2$
$y = x^3 - 2$
$x = y^3 - 2$
$x + 2 = y^3$
$\sqrt[3]{x + 2} = y$
$f^{-1}(x) = \sqrt[3]{x + 2}$

Check:
$f(f^{-1}(x)) = \left(\sqrt[3]{x + 2}\right)^3 - 2 = x + 2 - 2 = x$
$f^{-1}(f(x)) = \sqrt[3]{(x^3 - 2) + 2} = \sqrt[3]{x^3} = x$

**80.** $P(x) = \dfrac{1}{x + 1}$
$y = \dfrac{1}{x + 1}$
$x = \dfrac{1}{y + 1}$
$y + 1 = \dfrac{1}{x}$
$y = \dfrac{1}{x} - 1$
$P^{-1}(x) = \dfrac{1}{x} - 1$

Check:
$P(P^{-1}(x)) = \dfrac{1}{\left(\dfrac{1}{x} - 1\right) + 1} = \dfrac{1}{\dfrac{1}{x}} = 1 \cdot \dfrac{x}{1} = x$
$P^{-1}(P(x)) = \dfrac{1}{\dfrac{1}{x + 1}} - 1 = 1 \cdot \dfrac{x + 1}{1} - 1 = x + 1 - 1 = x$

## Section 9.1 Composite Functions and Inverse Functions

**82.** $G(x) = \dfrac{2}{3-x}$

$y = \dfrac{2}{3-x}$

$x = \dfrac{2}{3-y}$

$3 - y = \dfrac{2}{x}$

$-y = -3 + \dfrac{2}{x}$

$y = 3 - \dfrac{2}{x}$

$G^{-1}(x) = 3 - \dfrac{2}{x}$

Check:

$G(G^{-1}(x)) = \dfrac{2}{3 - \left(3 - \dfrac{2}{x}\right)} = \dfrac{2}{\dfrac{2}{x}} = 2 \cdot \dfrac{x}{2} = x$

$G^{-1}(G(x)) = 3 - \dfrac{2}{\dfrac{2}{3-x}} = 3 - (3-x) = x$

**84.** $f(x) = \sqrt[5]{x+5}$

$y = \sqrt[5]{x+5}$

$x = \sqrt[5]{y+5}$

$x^5 = y + 5$

$x^5 - 5 = y$

$f^{-1}(x) = x^5 - 5$

Check:

$f(f^{-1}(x)) = \sqrt[5]{x^5 - 5 + 5} = \sqrt[5]{x^5} = x$

$f^{-1}(f(x)) = \left(\sqrt[5]{x+5}\right)^5 - 5 = x + 5 - 5 = x$

**86.** $R(x) = \dfrac{2x}{x+4}$

$y = \dfrac{2x}{x+4}$

$x = \dfrac{2y}{y+4}$

$x(y+4) = 2y$

$xy + 4x = 2y$

$4x = 2y - xy$

$4x = y(2-x)$

$\dfrac{4x}{2-x} = y$

$R^{-1}(x) = \dfrac{4x}{2-x}$

Check:

$R(R^{-1}(x)) = \dfrac{2\left(\dfrac{4x}{2-x}\right)}{\dfrac{4x}{2-x} + 4} = \dfrac{\dfrac{8x}{2-x}}{\dfrac{4x+8-4x}{2-x}} = \dfrac{\dfrac{8x}{2-x}}{\dfrac{8}{2-x}}$

$= \dfrac{8x}{2-x} \cdot \dfrac{2-x}{8} = x$

$R^{-1}(R(x)) = \dfrac{4\left(\dfrac{2x}{x+4}\right)}{2 - \dfrac{2x}{x+4}} = \dfrac{\dfrac{8x}{x+4}}{\dfrac{2x+8-2x}{x+4}} = \dfrac{\dfrac{8x}{x+4}}{\dfrac{8}{x+4}}$

$= \dfrac{8x}{x+4} \cdot \dfrac{x+4}{8} = x$

**88.** $g(x) = \sqrt[3]{x+2} - 3$

$y = \sqrt[3]{x+2} - 3$

$x = \sqrt[3]{y+2} - 3$

$x + 3 = \sqrt[3]{y+2}$

$(x+3)^3 = y + 2$

$(x+3)^3 - 2 = y$

$g^{-1}(x) = (x+3)^3 - 2$

Check:

$g(g^{-1}(x)) = \sqrt[3]{\left[(x+3)^3 - 2\right] + 2} - 3$

$= \sqrt[3]{(x+3)^3} - 3$

$= x + 3 - 3$

$= x$

ISM – 555

## Chapter 9: Exponential and Logarithmic Functions

$$g^{-1}(g(x)) = \left[\left(\sqrt[3]{x+2}-3\right)+3\right]^3 - 2$$
$$= \left(\sqrt[3]{x+2}\right)^3 - 2$$
$$= x+2-2$$
$$= x$$

**90.** $V(r) = \dfrac{4}{3}\pi r^3$; $r(t) = 3\sqrt[3]{t}$

$V(t) = V(r(t)) = \dfrac{4}{3}\pi\left(3\sqrt[3]{t}\right)^3 = \dfrac{4}{3}\pi(27t) = 36\pi t$

$V(30) = 36\pi(30) = 1080\pi \approx 3392.92 \text{ m}^3$

**92.** $G(h) = 20h$; $T(G) = 0.18G$

**a.** $T(h) = T(G(h)) = 0.18(20h) = 3.6h$

**b.** $T(28) = 3.6(28) = 100.8$
The federal tax withholding will be $100.80.

**94.** $g^{-1}(7) = g^{-1}(g(-2)) = -2$

**96.** To find the domain and range of the inverse, we simply switch the values for the domain and range of the function.
Domain of $f^{-1}$: $[0, \infty)$
Range of $f^{-1}$: $[5, \infty)$

**98.** To find the domain and range of the inverse, we simply switch the values for the domain and range of the function.
Domain of $g^{-1}$: $(0, 8)$
Range of $g^{-1}$: $[0, 15]$

**100.** $H(a) = 22.8a - 117.5$ for $15 \leq a \leq 90$

$H = 22.8a - 117.5$
$H + 117.5 = 22.8a$
$\dfrac{H+117.5}{22.8} = a$
$a(H) = \dfrac{H+117.5}{22.8}$

To determine the restrictions, we find
$H(15) = 22.8(15) - 117.5 = 224.5$ and
$H(90) = 22.8(90) - 117.5 = 1934.5$. Thus,
$a(H) = \dfrac{H+117.5}{22.8}$ for $224.5 \leq H \leq 1934.5$.

**102.** $f(x) = x^2 - 3x + 1$; $g(x) = x - a$

$(f \circ g)(x) = f(g(x))$
$= (x-a)^2 - 3(x-a) + 1$
$= x^2 - 2ax + a^2 - 3x + 3a + 1$

The y-intercept is found by letting $x = 0$ and solving for y. We know the y-intercept of $(f \circ g)(x)$ is $-1$. Therefore, we get
$(f \circ g)(0) = x^2 - 2ax + a^2 - 3x + 3a + 1$
$(0)^2 - 2a(0) + a^2 - 3(0) + 3a + 1$
$-1 = a^2 + 3a + 1$
$0 = a^2 + 3a + 2$
$0 = (a+2)(a+1)$
$a + 2 = 0$ or $a + 1 = 0$
$a = -2$ $\qquad a = -1$

The solution set for $a$ is $\{-2, -1\}$.

**104.** $f(x) = 4x - 3$; $g(x) = x + 2$

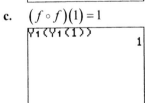

**a.** $(f \circ g)(3) = 17$

**b.** $(g \circ f)(-2) = -9$

**c.** $(f \circ f)(1) = 1$

**Section 9.1** Composite Functions and Inverse Functions

d. $(g \circ g)(-4) = 0$

a. $(f \circ g)(3) = -2000$

106. $f(x) = x^2 + 4$; $g(x) = 2x + 3$

b. $(g \circ f)(-2) = 257$

a. $(f \circ g)(3) = 253$

c. $(f \circ f)(1) = 16$

b. $(g \circ f)(-2) = 6$

d. $(g \circ g)(-4) = 290$

c. $(f \circ f)(1) = 1$

110. $f(x) = \sqrt{x+8}$; $g(x) = x^2 - 4$

a. $(f \circ g)(3) = \sqrt{13}$

d. $(g \circ g)(-4) = -94$

108. $f(x) = 2x^3$; $g(x) = -2x^2 + 5$

b. $(g \circ f)(-2) = 2$

ISM – 557

## Chapter 9: Exponential and Logarithmic Functions

c. $(f \circ f)(1) = \sqrt{11}$

d. $(g \circ g)(-4) = 140$

112. $f(x) = 10x$ ; $g(x) = \dfrac{x}{10}$

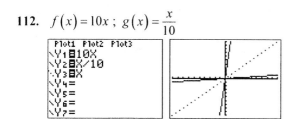

114. $f(x) = 3x - 5$ ; $g(x) = \dfrac{x+5}{3}$

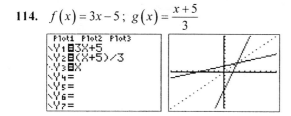

116. $f(x) = 8x + 3$ ; $g(x) = \dfrac{x-3}{8}$

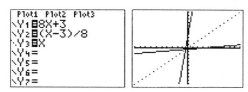

### 9.2 Preparing for Exponential Functions

1.  a. $2^3 = 2 \cdot 2 \cdot 2 = 8$

    b. $2^{-1} = \dfrac{1}{2^1} = \dfrac{1}{2}$

    c. $3^4 = 3 \cdot 3 \cdot 3 \cdot 3 = 81$

2. Locate some points on the graph of $f(x) = x^2$.

| $x$ | $f(x) = x^2$ | $(x, f(x))$ |
|---|---|---|
| $-3$ | $f(-3) = (-3)^2 = 9$ | $(-3, 9)$ |
| $-2$ | $f(-2) = (-2)^2 = 4$ | $(-2, 4)$ |
| $-1$ | $f(-1) = (-1)^2 = 1$ | $(-1, 1)$ |
| $0$ | $f(0) = 0^2 = 0$ | $(0, 0)$ |
| $1$ | $f(1) = 1^2 = 1$ | $(1, 1)$ |
| $2$ | $f(2) = 2^2 = 4$ | $(2, 4)$ |
| $3$ | $f(3) = 3^2 = 9$ | $(3, 9)$ |

Plot the points and connect them with a smooth curve.

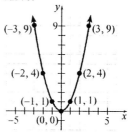

3. A **rational number** is a number that can be expressed as a quotient $\dfrac{p}{q}$ of two integers. The integer $p$ is called the *numerator*, and the integer $q$, which cannot be 0, is called the **denominator**. The set of rational numbers is the numbers
$$\mathbb{Q} = \left\{ x \mid x = \dfrac{p}{q}, \text{ where } p, q \text{ are integers and } q \neq 0 \right\}.$$

4. An irrational number has a decimal representation that neither repeats nor terminates.

5.  a. Rounding to 4 decimal places, we obtain $3.20349193 \approx 3.2035$.

    b. Truncating to 4 decimal places, we obtain $3.20349193 \approx 3.2034$.

6.  a. $m^3 \cdot m^5 = m^{3+5} = m^8$

    b. $\dfrac{a^7}{a^2} = a^{7-2} = a^5$

    c. $\left(z^3\right)^4 = z^{3 \cdot 4} = z^{12}$

## Section 9.2 Exponential Functions

7. $x^2 - 5x = 14$
$x^2 - 5x - 14 = 0$
$(x-7)(x+2) = 0$
$x - 7 = 0$ or $x + 2 = 0$
$x = 7$ or $x = -2$
The solution set is $\{-2, 7\}$.

### 9.2 Exercises

2. $\left(-1, \dfrac{1}{a}\right); (0,1); (1, a)$

4. TRUE.
$a^{\log_a x} = x$. Using this property, it follows that $2^x = 2^{\log_2 12}$, and we <u>can</u> solve the equation by using the fact that if $a^u = a^v$ then $u = v$:
$2^x = 12$
$2^x = 2^{\log_2 12}$
$x = \log_2 12$

12. a. $5^{1.4} \approx 9.518$
b. $5^{1.41} \approx 9.673$
c. $5^{1.414} \approx 9.735$
d. $5^{1.4142} \approx 9.738$
e. $5^{\sqrt{2}} \approx 9.739$

14. a. $10^{2.7} \approx 501.187$
b. $10^{2.72} \approx 524.807$
c. $10^{2.718} \approx 522.396$
d. $10^{2.7183} \approx 522.757$
e. $10^e \approx 522.735$

16. a. $2.7^{3.1} \approx 21.738$
b. $2.72^{3.14} \approx 23.150$
c. $2.718^{3.142} \approx 23.143$
d. $2.7183^{3.1416} \approx 23.141$
e. $e^\pi \approx 23.141$

18. $e^3 \approx 20.086$

20. $e^{-3} \approx 0.050$

22. $e^{1.5} \approx 4.482$

6. TRUE.

8. Answers may vary. One possibility follows:
$g(x) = \left(\dfrac{1}{2}\right)^x = \left(2^{-1}\right)^x = 2^{-1 \cdot x} = 2^{-x} = f(x)$.

10. Answers may vary. One possibility follows: Using the information we have learned in this textbook so far, we cannot solve the equation $2^x = 12$ using the fact that if $a^u = a^v$, then $u = v$, because we do not have the information required to write 12 as a power of 2.
Note: Later in this chapter, when we study logarithms, we will discuss the property

24. c. $f(x) = 2^{x+1}$ because the following points are on the graph:

| $x$ | $f(x) = 2^{x+1}$ | $(x, f(x))$ |
|---|---|---|
| $-2$ | $f(-2) = 2^{-2+1} = 2^{-1} = \dfrac{1}{2}$ | $\left(-2, \dfrac{1}{2}\right)$ |
| $-1$ | $f(-1) = 2^{-1} = 2^0 = 1$ | $(-1, 1)$ |
| 0 | $f(0) = 2^{0+1} = 2^1 = 2$ | $(0, 2)$ |

26. b. $f(x) = 2^{-x}$ because the following points are on the graph:

| $x$ | $f(x) = 2^{-x}$ | $(x, f(x))$ |
|---|---|---|
| $-1$ | $f(-1) = 2^{-(-1)} = 2^1 = 2$ | $(-1, 2)$ |
| 0 | $f(0) = 2^{-0} = 2^0 = 1$ | $(0, 1)$ |
| 1 | $f(1) = 2^{-1} = \dfrac{1}{2}$ | $\left(1, \dfrac{1}{2}\right)$ |

28. a. $f(x) = 2^x$ because the following points are on the graph:

| $x$ | $f(x) = 2^x$ | $(x, f(x))$ |
|---|---|---|
| $-1$ | $f(-1) = 2^{-1} = \dfrac{1}{2}$ | $\left(-1, \dfrac{1}{2}\right)$ |
| 0 | $f(0) = 2^0 = 1$ | $(0, 1)$ |
| 1 | $f(1) = 2^1 = 2$ | $(1, 2)$ |

## Chapter 9: Exponential and Logarithmic Functions

**30. d.** $f(x) = 2^{x-1}$ because the following points are on the graph:

| $x$ | $f(x) = 2^{x-1}$ | $(x, f(x))$ |
|---|---|---|
| 0 | $f(0) = 2^{0-1} = 2^{-1} = \dfrac{1}{2}$ | $\left(0, \dfrac{1}{2}\right)$ |
| 1 | $f(1) = 2^{1-1} = 2^0 = 1$ | $(1, 1)$ |
| 2 | $f(2) = 2^{2-1} = 2^1 = 2$ | $(0, 2)$ |

**32.** Locate some points on the graph of $f(x) = 7^x$.

| $x$ | $f(x) = 7^x$ | $(x, f(x))$ |
|---|---|---|
| $-2$ | $f(-2) = 7^{-2} = \dfrac{1}{7^2} = \dfrac{1}{49}$ | $\left(-2, \dfrac{1}{49}\right)$ |
| $-1$ | $f(-1) = 7^{-1} = \dfrac{1}{7^1} = \dfrac{1}{7}$ | $\left(-1, \dfrac{1}{7}\right)$ |
| 0 | $f(0) = 7^0 = 1$ | $(0, 1)$ |
| 1 | $f(1) = 7^1 = 7$ | $(1, 7)$ |

Plot the points and connect them with a smooth curve.

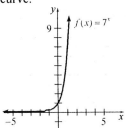

The domain of $f$ is all real numbers or, using interval notation, $(-\infty, \infty)$. The range of $f$ is $\{y \mid y > 0\}$ or, using interval notation, $(0, \infty)$.

**34.** Locate some points on the graph of $G(x) = 8^x$.

| $x$ | $G(x) = 8^x$ | $(x, G(x))$ |
|---|---|---|
| $-2$ | $G(-2) = 8^{-2} = \dfrac{1}{8^2} = \dfrac{1}{64}$ | $\left(-2, \dfrac{1}{64}\right)$ |
| $-1$ | $G(-1) = 8^{-1} = \dfrac{1}{8^1} = \dfrac{1}{8}$ | $\left(-1, \dfrac{1}{8}\right)$ |
| 0 | $G(0) = 8^0 = 1$ | $(0, 1)$ |
| 1 | $G(1) = 8^1 = 7$ | $(1, 8)$ |

Plot the points and connect them with a smooth curve.

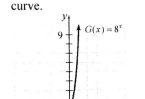

The domain of $G$ is all real numbers or, using interval notation, $(-\infty, \infty)$. The range of $G$ is $\{y \mid y > 0\}$ or, using interval notation, $(0, \infty)$.

**36.** Locate some points on the graph of $F(x) = \left(\dfrac{1}{7}\right)^x$.

| $x$ | $F(x) = \left(\dfrac{1}{7}\right)^x$ | $(x, F(x))$ |
|---|---|---|
| $-1$ | $F(-1) = \left(\dfrac{1}{7}\right)^{-1} = 7^1 = 7$ | $(-1, 7)$ |
| 0 | $F(0) = \left(\dfrac{1}{7}\right)^0 = 1$ | $(0, 1)$ |
| 1 | $F(1) = \left(\dfrac{1}{7}\right)^1 = \dfrac{1}{7}$ | $\left(1, \dfrac{1}{7}\right)$ |
| 2 | $F(2) = \left(\dfrac{1}{7}\right)^2 = \dfrac{1}{49}$ | $\left(2, \dfrac{1}{49}\right)$ |

Plot the points and connect them with a smooth curve.

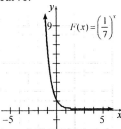

The domain of $F$ is all real numbers or, using interval notation, $(-\infty, \infty)$. The range of $F$ is $\{y \mid y > 0\}$ or, using interval notation, $(0, \infty)$.

**38.** Locate some points on the graph of $g(x) = \left(\dfrac{1}{8}\right)^x$.

| $x$ | $g(x) = \left(\dfrac{1}{8}\right)^x$ | $(x, g(x))$ |
|---|---|---|
| $-1$ | $g(-1) = \left(\dfrac{1}{8}\right)^{-1} = 8^1 = 8$ | $(-1, 8)$ |
| $0$ | $g(0) = \left(\dfrac{1}{8}\right)^0 = 1$ | $(0, 8)$ |
| $1$ | $g(1) = \left(\dfrac{1}{8}\right)^1 = \dfrac{1}{7}$ | $\left(1, \dfrac{1}{8}\right)$ |
| $2$ | $g(2) = \left(\dfrac{1}{8}\right)^2 = \dfrac{1}{64}$ | $\left(2, \dfrac{1}{64}\right)$ |

Plot the points and connect them with a smooth curve.

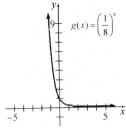

The domain of $g$ is all real numbers or, using interval notation, $(-\infty, \infty)$. The range of $g$ is $\{y \mid y > 0\}$ or, using interval notation, $(0, \infty)$.

**40.** Locate some points on the graph of $H(x) = 2^{x-2}$.

| $x$ | $H(x) = 2^{x-2}$ | $(x, H(x))$ |
|---|---|---|
| $-1$ | $H(-1) = 2^{-1-2} = 2^{-3} = \dfrac{1}{2^3} = \dfrac{1}{8}$ | $\left(-1, \dfrac{1}{8}\right)$ |
| $0$ | $H(0) = 2^{0-2} = 2^{-2} = \dfrac{1}{2^2} = \dfrac{1}{4}$ | $\left(0, \dfrac{1}{4}\right)$ |
| $1$ | $H(1) = 2^{1-2} = 2^{-1} = \dfrac{1}{2^1} = \dfrac{1}{2}$ | $\left(1, \dfrac{1}{2}\right)$ |
| $2$ | $H(2) = 2^{2-2} = 2^0 = 1$ | $(2, 1)$ |
| $3$ | $H(3) = 2^{3-2} = 2^1 = 2$ | $(3, 2)$ |
| $4$ | $H(4) = 2^{4-2} = 2^2 = 4$ | $(4, 4)$ |
| $5$ | $H(5) = 2^{5-2} = 2^3 = 8$ | $(5, 8)$ |

Plot the points and connect them with a smooth curve.

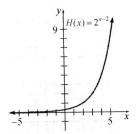

The domain of $H$ is all real numbers or, using interval notation, $(-\infty, \infty)$. The range of $H$ is $\{y \mid y > 0\}$ or, using interval notation, $(0, \infty)$.

**42.** Locate some points on the graph of $F(x) = 2^x - 3$.

| $x$ | $F(x) = 2^x - 3$ | $(x, F(x))$ |
|---|---|---|
| $-2$ | $F(-2) = 2^{-2} - 3 = \dfrac{1}{4} - 3 = -\dfrac{11}{4}$ | $\left(-2, -\dfrac{11}{4}\right)$ |
| $-1$ | $F(-1) = 2^{-1} - 3 = \dfrac{1}{2} - 3 = -\dfrac{5}{2}$ | $\left(-1, -\dfrac{5}{2}\right)$ |
| $0$ | $F(0) = 2^0 - 3 = 1 - 3 = -2$ | $(0, -2)$ |
| $1$ | $F(1) = 2^1 - 3 = 2 - 3 = -1$ | $(1, -1)$ |
| $2$ | $F(2) = 2^2 - 3 = 4 - 3 = 1$ | $(2, 1)$ |

Plot the points and connect them with a smooth curve.

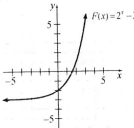

The domain of $F$ is all real numbers or, using interval notation, $(-\infty, \infty)$. The range of $F$ is $\{y \mid y > -3\}$ or, using interval notation, $(-3, \infty)$.

**44.** Locate some points on the graph of

$G(x) = \left(\dfrac{1}{2}\right)^x + 2$.

| $x$ | $G(x) = \left(\dfrac{1}{2}\right)^x + 2$ | $(x, G(x))$ |
|---|---|---|
| $-2$ | $G(-1) = \left(\dfrac{1}{2}\right)^{-2} + 2 = 4 + 2 = 6$ | $(-1, 6)$ |
| $-1$ | $G(-1) = \left(\dfrac{1}{2}\right)^{-1} + 2 = 2 + 2 = 4$ | $(-1, 4)$ |
| $0$ | $G(0) = \left(\dfrac{1}{2}\right)^{0} + 2 = 1 + 2 = 3$ | $(0, 3)$ |
| $1$ | $G(1) = \left(\dfrac{1}{2}\right)^{1} + 2 = \dfrac{1}{2} + 2 = \dfrac{5}{2}$ | $\left(1, \dfrac{5}{2}\right)$ |
| $2$ | $G(2) = \left(\dfrac{1}{2}\right)^{2} + 2 = \dfrac{1}{4} + 2 = \dfrac{9}{4}$ | $\left(2, \dfrac{9}{4}\right)$ |
| $3$ | $G(3) = \left(\dfrac{1}{2}\right)^{3} + 2 = \dfrac{1}{8} + 2 = \dfrac{17}{8}$ | $\left(3, \dfrac{17}{8}\right)$ |

Plot the points and connect them with a smooth curve.

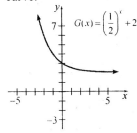

The domain of $G$ is all real numbers or, using interval notation, $(-\infty, \infty)$. The range of $G$ is $\{y \mid y > 2\}$ or, using interval notation, $(2, \infty)$.

**46.** Locate some points on the graph of $p(x) = \left(\dfrac{1}{3}\right)^{x+2}$.

| $x$ | $p(x) = \left(\dfrac{1}{3}\right)^{x+2}$ | $(x, p(x))$ |
|---|---|---|
| $-4$ | $p(-4) = \left(\dfrac{1}{3}\right)^{-4+2} = \left(\dfrac{1}{3}\right)^{-2} = 9$ | $(-4, 9)$ |
| $-3$ | $p(-3) = \left(\dfrac{1}{3}\right)^{-3+2} = \left(\dfrac{1}{3}\right)^{-1} = 3$ | $(-3, 3)$ |
| $-2$ | $p(-2) = \left(\dfrac{1}{3}\right)^{-2+2} = \left(\dfrac{1}{3}\right)^{0} = 1$ | $(-2, 1)$ |
| $-1$ | $p(-1) = \left(\dfrac{1}{3}\right)^{-1+2} = \left(\dfrac{1}{3}\right)^{1} = \dfrac{1}{3}$ | $\left(-1, \dfrac{1}{3}\right)$ |
| $0$ | $p(0) = \left(\dfrac{1}{3}\right)^{0+2} = \left(\dfrac{1}{3}\right)^{2} = \dfrac{1}{9}$ | $\left(0, \dfrac{1}{9}\right)$ |

Plot the points and connect them with a smooth curve.

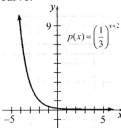

The domain of $p$ is all real numbers or, using interval notation, $(-\infty, \infty)$. The range of $p$ is $\{y \mid y > 0\}$ or, using interval notation, $(0, \infty)$.

**48.** Locate some points on the graph of $f(x) = e^x - 1$.

| $x$ | $f(x) = e^x - 1$ | $(x, f(x))$ |
|---|---|---|
| $-2$ | $f(-2) = e^{-2} - 1 \approx -0.865$ | $\left(-2, e^{-2} - 1\right)$ |
| $-1$ | $f(-1) = e^{-1} - 1 \approx -0.632$ | $\left(-1, e^{-1} - 1\right)$ |
| $0$ | $f(0) = e^0 - 1 = 1 - 1 = 0$ | $(0, 1)$ |
| $1$ | $f(1) = e^1 - 1 = e - 1 \approx 1.718$ | $(1, e - 1)$ |
| $2$ | $f(2) = e^2 - 1 \approx 6.389$ | $\left(3, e^2 - 1\right)$ |

Plot the points and connect them with a smooth curve.

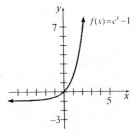

The domain of $f$ is all real numbers or, using interval notation, $(-\infty, \infty)$. The range of $f$ is $\{y \mid y > -1\}$ or, using interval notation, $(-1, \infty)$.

50. $3^x = 3^{-2}$
    $x = -2$
    The solution set is $\{-2\}$.

52. $4^{-x} = 64$
    $4^{-x} = 4^3$
    $-x = 3$
    $x = -3$
    The solution set is $\{-3\}$.

54. $\left(\dfrac{1}{3}\right)^x = \dfrac{1}{243}$
    $\left(\dfrac{1}{3}\right)^x = \left(\dfrac{1}{3}\right)^5$
    $x = 5$
    The solution set is $\{5\}$.

56. $2^{x+3} = 128$
    $2^{x+3} = 2^7$
    $x + 3 = 7$
    $x = 4$
    The solution set is $\{4\}$.

58. $9^x = 27$
    $(3^2)^x = 3^3$
    $3^{2x} = 3^3$
    $2x = 3$
    $x = \dfrac{3}{2}$
    The solution set is $\left\{\dfrac{3}{2}\right\}$.

60. $3^{-x+4} = 27^x$
    $3^{-x+4} = (3^3)^x$
    $3^{-x+4} = 3^{3x}$
    $-x + 4 = 3x$
    $4 = 4x$
    $1 = x$
    The solution set is $\{1\}$.

62. $5^{x^2-10} = 125^x$
    $5^{x^2-10} = (5^3)^x$
    $5^{x^2-10} = 5^{3x}$
    $x^2 - 10 = 3x$
    $x^2 - 3x - 10 = 0$
    $(x-5)(x+2) = 0$
    $x - 5 = 0$ or $x + 2 = 0$
    $x = 5$ or $x = -2$
    The solution set is $\{-2, 5\}$.

64. $9^{2x} \cdot 27^{x^2} = 3^{-1}$
    $(3^2)^{2x} \cdot (3^3)^{x^2} = 3^{-1}$
    $3^{4x} \cdot 3^{3x^2} = 3^{-1}$
    $3^{3x^2 + 4x} = 3^{-1}$
    $3x^2 + 4x = -1$
    $3x^2 + 4x + 1 = 0$
    $(3x+1)(x+1) = 0$
    $3x + 1 = 0$ or $x + 1 = 0$
    $x = -\dfrac{1}{3}$ or $x = -1$
    The solution set is $\left\{-1, -\dfrac{1}{3}\right\}$.

66. $3^x \cdot 9 = 27^x$
    $3^x \cdot 3^2 = (3^3)^x$
    $3^{x+2} = 3^{3x}$
    $x + 2 = 3x$
    $2 = 2x$
    $1 = x$
    The solution set is $\{1\}$.

## Chapter 9: Exponential and Logarithmic Functions

**68.** $\left(\dfrac{1}{6}\right)^x - 36 = 0$

$\left(\dfrac{1}{6}\right)^x = 36$

$\left(6^{-1}\right)^x = 6^2$

$6^{-x} = 6^2$

$-x = 2$

$x = -2$

The solution set is $\{-2\}$.

**70.** $\left(3^x\right)^x = 81$

$3^{x^2} = 3^4$

$x^2 = 4$

$x = \pm\sqrt{4}$

$x = \pm 2$

The solution set is $\{-2, 2\}$.

**72.** $e^{3x} = e^2$

$3x = 2$

$x = \dfrac{2}{3}$

The solution set is $\left\{\dfrac{2}{3}\right\}$.

**74.** $\left(e^3\right)^x = e^2 \cdot e^x$

$e^{3x} = e^{2+x}$

$3x = 2 + x$

$2x = 2$

$x = 1$

The solution set is $\{1\}$.

**76. a.** $f(2) = 3^2 = 9$.

The point $(2, 9)$ is on the graph of $f$.

**b.** $f(x) = \dfrac{1}{81}$

$3^x = \dfrac{1}{81}$

$3^x = \dfrac{1}{3^4} = 3^{-4}$

$x = -4$

The point $\left(-4, \dfrac{1}{81}\right)$ is on the graph of $f$.

**78. a.** $g(-1) = 5^{-1} + 1 = \dfrac{1}{5} + 1 = \dfrac{6}{5}$.

The point $\left(-1, \dfrac{6}{5}\right)$ is on the graph of $g$.

**b.** $g(x) = 126$

$5^x + 1 = 126$

$5^x = 125$

$5^x = 5^3$

$x = 3$

The point $(3, 126)$ is on the graph of $g$.

**80. a.** $F(-1) = -2 \cdot \left(\dfrac{1}{3}\right)^{-1} = -2 \cdot 3^1 = -2 \cdot 3 = -6$.

The point $(-1, -6)$ is on the graph of $F$.

**b.** $F(x) = -18$

$-2 \cdot \left(\dfrac{1}{3}\right)^x = -18$

$\left(\dfrac{1}{3}\right)^x = 9$

$3^{-x} = 3^2$

$-x = 2$

$x = -2$

The point $(-2, -18)$ is on the graph of $F$.

**82. a.** $P(2008) = 6{,}448(1.0126)^{2008-2005} \approx 6{,}694.818$

According to the model, the population of the world in 2008 will be approximately 6,695 million people (or 6.7 billion people).

**b.** $P(2027) = 6{,}448(1.0126)^{2027-2005} \approx 8{,}492.965$

According to the model, the population of the world in 2015 will be approximately 8,493 million people (or 8.5 billion people).

**c.** $8493 - 8000 = 493$. The U.S. Census Bureau's prediction for the population in 2042 is 493 million people fewer than that of the model.

Reasons given for the differences may vary. One possibility is that perhaps the Census Bureau expects the rate of population growth to decline below 1.26% per year over time.

**84.** We use the compound interest formula with $P = \$8000$, $r = 0.04$, and $n = 4$, so that
$$A = 8000\left(1 + \frac{0.04}{4}\right)^{4t} = 8000(1.01)^{4t}$$

  **a.** The value of the account after $t = 1$ years is
  $A = 8000(1.01)^{4(1)} \approx \$8,324.83$.

  **b.** The value of the account after $t = 3$ years is
  $A = 8000(1.01)^{4(3)} \approx \$9,014.60$.

  **c.** The value of the account after $t = 5$ years is
  $A = 8000(1.01)^{4(5)} \approx \$9,761.52$.

**86. a.** We use the compound interest formula with $P = \$1000$, $r = 0.06$, $t = 3$, and $n = 1$, so that $A = 1000\left(1 + \frac{0.06}{1}\right)^{1(3)} \approx \$1,191.02$.

  **b.** We use the compound interest formula with $P = \$1000$, $r = 0.06$, $t = 3$, and $n = 4$, so that $A = 1000\left(1 + \frac{0.06}{4}\right)^{4(3)} \approx \$1,195.62$.

  **c.** We use the compound interest formula with $P = \$1000$, $r = 0.06$, $t = 3$, and $n = 12$, so that $A = 1000\left(1 + \frac{0.06}{12}\right)^{12(3)} \approx \$1,196.68$.

  **d.** We use the compound interest formula with $P = \$1000$, $r = 0.06$, $t = 3$, and $n = 365$, so that $A = 1000\left(1 + \frac{0.06}{365}\right)^{365(3)} \approx \$1,197.20$.

  **e.** Answers may vary. One possibility follows: All other things equal, the number of compounding periods does not have a very significant impact on the future value. In this case, for example, the difference between compounding annually ($n = 1$) and daily ($n = 365$) is only $6.18 over 3 years.

**88. a.** $V(0) = 19,282(0.84)^0 = 19,282$.
  According to the model, the value of a brand-new Dodge Stratus is $19,282.

  **b.** $V(2) = 19,282(0.84)^2 \approx 13,605.38$.
  According to the model, the value of a 2-year-old Dodge Stratus is $13,605.38.

  **c.** $V(5) = 19,282(0.84)^5 \approx 8,063.96$.
  According to the model, the value of a 5-year-old Dodge Stratus is $8,063.96.

**90. a.** $A(1) = 100\left(\frac{1}{2}\right)^{1/19.255} \approx 96.464$.
  After 1 second, approximately 96.464 grams of carbon-10 will be left in the sample.

  **b.** $A(19.255) = 100\left(\frac{1}{2}\right)^{19.255/19.255} = 50$.
  After 19.255 seconds, 50 grams of carbon-10 will be left in the sample.

  **c.** $A(38.51) = 100\left(\frac{1}{2}\right)^{38.51/19.255} = 25$.
  After 38.51 seconds, 25 grams of carbon-10 will be left in the sample.

  **d.** $A(100) = 100\left(\frac{1}{2}\right)^{100/19.255} \approx 2.733$.
  After 100 seconds, approximately 2.733 grams of carbon-10 will be left in the sample.

**92. a.** $u(5) = 70 + 100e^{-0.045(5)} \approx 149.852$.
  According to the model, the temperature of the coffee after 5 minutes will be approximately $149.852°F$.

  **b.** $u(5) = 70 + 100e^{-0.045(10)} \approx 133.763$.
  According to the model, the temperature of the coffee after 10 minutes will be approximately $133.763°F$.

  **c.** $u(20) = 70 + 100e^{-0.045(20)} \approx 110.657$.
  According to the model, the temperature of the coffee after 20 minutes will be approximately $110.657°F$. Since this is below $120°F$, the coffee will be bad after cooling for 20 minutes.

**94. a.** $L(45) = 50\left(1 - e^{-0.0223(45)}\right) \approx 31.67$.
  According to the model, the student will learn approximately 32 words after 45 minutes.

  **b.** $L(60) = 50\left(1 - e^{-0.0223(60)}\right) \approx 36.88$.
  According to the model, the student will learn approximately 37 words after 60 minutes.

**96. a.** We use the equation with $E = 120$, $R = 2500$, $C = 100$, and $t = 0$, so that
$$I = \frac{120}{2500}e^{-0/(2500 \cdot 100)} = 0.048 \text{ amperes.}$$

We use the equation with $E = 120$, $R = 2500$, $C = 100$, and $t = 50$, so that

$$I = \frac{120}{2500} e^{-50/(2500 \cdot 100)} \approx 0.048 \text{ ampere.}$$

b. We use the equation with $E = 240$, $R = 2500$, $C = 100$, and $t = 0$, so that

$$I = \frac{240}{2500} e^{-0/(2500 \cdot 100)} = 0.096 \text{ ampere.}$$

We use the equation with $E = 240$, $R = 2500$, $C = 100$, and $t = 50$, so that

$$I = \frac{240}{2500} e^{-50/(2500 \cdot 100)} \approx 0.096 \text{ ampere.}$$

**98.** The exponential function will be of the form $y = a^x$. Now the function contains the points $\left(-1, \frac{1}{4}\right)$, $(0,1)$, and $(1,4)$, so $a = 4$. Thus, the equation of the function is $y = 4^x$.

**100. a.** $x = -4$: $x^3 - 5x + 2 = (-4)^3 - 5(-4) + 2$
$= -64 + 20 + 2$
$= -42$

**b.** $x = 2$: $x^3 - 5x + 2 = (2)^3 - 5(2) + 2$
$= 8 - 10 + 2$
$= 0$

**102. a.** $x = -4$: $\dfrac{2x}{x+1} = \dfrac{2(-4)}{-4+1} = \dfrac{-8}{-3} = \dfrac{8}{3}$

**b.** $x = 5$: $\dfrac{2x}{x+1} = \dfrac{2(5)}{5+1} = \dfrac{10}{6} = \dfrac{5}{3}$

**104. a.** $x = 3$: $\sqrt[3]{3x-1} = \sqrt[3]{3(3)-1}$
$= \sqrt[3]{9-1} = \sqrt[3]{8} = 2$

**b.** $x = 0$: $\sqrt[3]{3x-1} = \sqrt[3]{3(0)-1}$
$= \sqrt[3]{0-1} = \sqrt[3]{-1} = -1$

**106.** Let $Y_1 = 3.1^x$.

 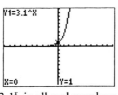

The domain of $G(x) = 3.1^x$ is all real numbers, or using interval notation $(-\infty, \infty)$. The range is $\{y \mid y > 0\}$, or using interval notation $(0, \infty)$.

**108.** Let $Y_1 = 0.3^x$.

 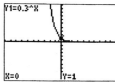

The domain of $F(x) = 0.3^x$ is all real numbers, or using interval notation $(-\infty, \infty)$. The range is $\{y \mid y > 0\}$, or using interval notation $(0, \infty)$.

**110.** Let $Y_1 = 1.7^x - 2$.

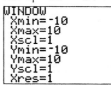 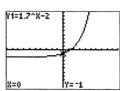

The domain of $f(x) = 1.7^x - 2$ is all real numbers, or using interval notation $(-\infty, \infty)$. The range is $\{y \mid y > -2\}$, or using interval notation $(-2, \infty)$.

**112.** Let $Y_1 = 0.3^{x+2}$.

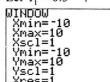 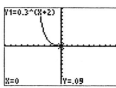

The domain of $g(x) = 0.3^{x+2}$ is all real numbers, or using interval notation $(-\infty, \infty)$. The range is $\{y \mid y > 0\}$, or using interval notation $(0, \infty)$.

## Section 9.3 Logarithmic Functions

### 9.3 Preparing for Logarithmic Functions

1. $3x + 2 > 0$
$3x + 2 - 2 > 0 - 2$
$3x > -2$
$\dfrac{3x}{3} > \dfrac{-2}{3}$
$x > -\dfrac{2}{3}$

The solution set is $\left\{x \mid x > -\dfrac{2}{3}\right\}$ or, using interval notation, $\left(-\dfrac{2}{3}, \infty\right)$.

2. $\sqrt{x+2} = x$
$\left(\sqrt{x+2}\right)^2 = (x)^2$
$x + 2 = x^2$
$0 = x^2 - x - 2$
$0 = (x-2)(x+1)$
$x - 2 = 0$ or $x + 1 = 0$
$x = 2$ or $x = -1$

Check:
$x = 2$: $\sqrt{2+2} \stackrel{?}{=} 2$
$\sqrt{4} \stackrel{?}{=} 2$
$2 = 2$ ✓

$x = -1$: $\sqrt{-1+2} \stackrel{?}{=} -1$
$\sqrt{1} \stackrel{?}{=} -1$
$1 \neq -1$ ✗

The potential solution $x = -1$ does not check; the solution set is $\{2\}$.

3. $x^2 = 6x + 7$
$x^2 - 6x - 7 = 0$
$(x-7)(x+1) = 0$
$x - 7 = 0$ or $x + 1 = 0$
$x = 7$ or $x = -1$
The solution set is $\{-1, 7\}$.

### 9.3 Exercises

2. $(0, \infty)$ or $\{x \mid x > 0\}$

4. FALSE. See Exercise 87.

6. TRUE.

8. The domain of $f(x) = \log_a\left(x^2 + 1\right)$ is all values of $x$ that make that $x^2 + 1$ is positive. Now $x^2 + 1 > 0$ is equivalent to $x^2 > -1$, which must be true for all real numbers (since $x^2$ is always positive and, therefore, greater than $-1$). Thus, since $x^2 + 1$ is positive for all real numbers $x$, the domain of $f(x) = \log_a\left(x^2 + 1\right)$ is all real numbers.

10. If $64 = 8^2$, then $2 = \log_8 64$.

12. If $16 = 2^4$, then $4 = \log_2 16$.

14. If $\dfrac{1}{9} = 3^{-2}$, then $-2 = \log_3 \left(\dfrac{1}{9}\right)$.

16. If $e^{4.2} = M$, then $\ln M = 4.2$.

18. If $b^4 = 23$, then $\log_b 23 = 4$.

20. If $10^{-3} = z$, then $\log z = -3$.

22. If $\log_3 81 = 4$, then $3^4 = 81$.

24. If $\log_2 \dfrac{1}{32} = -5$, then $2^{-5} = \dfrac{1}{32}$.

26. If $\ln(x-1) = 3$, then $e^3 = x - 1$.

28. If $\log_6 x = -4$, then $6^{-4} = x$.

30. If $\log_a 16 = 2$, then $a^2 = 16$.

32. If $\log_{1/2} 18 = z$, then $\left(\dfrac{1}{2}\right)^z = 18$.

34. Let $y = \log_5 5$. Then,
$5^y = 5$
$5^y = 5^1$
$y = 1$
Thus, $\log_5 5 = 1$.

36. Let $y = \log_4 16$. Then,
$4^y = 16$
$4^y = 4^2$
$y = 2$
Thus, $\log_4 16 = 2$.

38. Let $y = \log_5\left(\dfrac{1}{125}\right)$. Then,
$5^y = \dfrac{1}{125}$
$5^y = \dfrac{1}{5^3} = 5^{-3}$
$y = -3$
Thus, $\log_5 \dfrac{1}{125} = -3$.

40. Let $y = \log_{\sqrt{3}} 3$. Then,
$\left(\sqrt{3}\right)^y = 3$
$\left(3^{\frac{1}{2}}\right)^y = 3^1$
$3^{\frac{1}{2}y} = 3^1$
$\dfrac{1}{2}y = 1$
$y = 2$
Thus, $\log_{\sqrt{3}} 3 = 2$.

42. $f(9) = \log_3 9$. To determine the value, let $y = \log_3 9$. Then,
$3^y = 9$
$3^y = 3^2$
$y = 2$
Therefore, $f(9) = 2$.

44. $g\left(\sqrt[3]{5}\right) = \log_5 \sqrt[3]{5}$. To determine the value, let $y = \log_5 \sqrt[3]{5}$. Then,
$5^y = \sqrt[3]{5}$
$5^y = 5^{\frac{1}{3}}$
$y = \dfrac{1}{3}$
Therefore, $g\left(\sqrt[3]{5}\right) = \dfrac{1}{3}$.

46. $H(100,000) = \log(100,000)$. To determine the value, let $y = \log(100,000)$. Then,
$10^y = 100,000$
$10^y = 10^5$
$y = 5$
Therefore, $H(100,000) = 5$.

48. $P(e^{-3}) = \ln e^{-3}$. To determine the value, let $y = \ln e^{-3}$. Then,
$e^y = e^{-3}$
$y = -3$
Therefore, $P(e^{-3}) = -3$.

50. The domain of $f(x) = \log_3(x-2)$ is the set of all real numbers $x$ such that
$x - 2 > 0$
$x > 2$
Thus, the domain of $f(x) = \log_3(x-2)$ is $\{x \mid x > 2\}$, or using interval notation, $(2, \infty)$.

52. The domain of $g(x) = \log_4(x+10)$ is the set of all real numbers $x$ such that
$x + 10 > 0$
$x > -10$
Thus, the domain of $g(x) = \log_4(x+10)$ is $\{x \mid x > -10\}$, or using interval notation, $(-10, \infty)$.

54. The domain of $h(x) = \log_4(5x)$ is the set of all real numbers $x$ such that
$5x > 0$
$x > 0$
Thus, the domain of $h(x) = \log_4(5x)$ is $\{x \mid x > 0\}$, or using interval notation, $(0, \infty)$.

56. The domain of $F(x) = \log_2(4x-3)$ is the set of all real numbers $x$ such that
$4x - 3 > 0$
$4x > 3$
$x > \dfrac{3}{4}$
Thus, the domain of $F(x) = \log_2(4x-3)$ is $\left\{x \mid x > \dfrac{3}{4}\right\}$, or using interval notation, $\left(\dfrac{3}{4}, \infty\right)$.

**58.** The domain of $f(x) = \log_3(5x+3)$ is the set of all real numbers $x$ such that
$$5x+3 > 0$$
$$5x > -3$$
$$x > -\frac{3}{5}$$
Thus, the domain of $f(x) = \log_3(5x+3)$ is $\left\{x \mid x > -\frac{3}{5}\right\}$, or using interval notation, $\left(-\frac{3}{5}, \infty\right)$.

**60.** The domain of $G(x) = \log_4(3-5x)$ is the set of all real numbers $x$ such that
$$3-5x > 0$$
$$-5x > -3$$
$$x < \frac{-3}{-5}$$
$$x < \frac{3}{5}$$
Thus, the domain of $G(x) = \log_4(3-5x)$ is $\left\{x \mid x < \frac{3}{5}\right\}$, or using interval notation, $\left(-\infty, \frac{3}{5}\right)$.

**62.** Rewrite $y = f(x) = \log_7 x$ as $x = 7^y$. Locate some points on the graph of $x = 7^y$.

| $y$ | $x = 7^y$ | $(x, y)$ |
|---|---|---|
| $-1$ | $x = 7^{-1} = \dfrac{1}{7^1} = \dfrac{1}{7}$ | $\left(\dfrac{1}{7}, -1\right)$ |
| $0$ | $x = 7^0 = 1$ | $(1, 0)$ |
| $1$ | $x = 7^1 = 5$ | $(7, 1)$ |

Plot the points and connect them with a smooth curve.

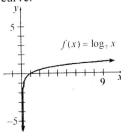

The domain of $f$ is $\{x \mid x > 0\}$ or, using interval notation, $(0, \infty)$. The range of $f$ is all real numbers or, using interval notation, $(-\infty, \infty)$.

**64.** Rewrite $y = G(x) = \log_6 x$ as $x = 8^y$. Locate some points on the graph of $x = 8^y$.

| $y$ | $x = 8^y$ | $(x, y)$ |
|---|---|---|
| $-1$ | $x = 8^{-1} = \dfrac{1}{8^1} = \dfrac{1}{8}$ | $\left(\dfrac{1}{8}, -1\right)$ |
| $0$ | $x = 8^0 = 1$ | $(1, 0)$ |
| $1$ | $x = 8^1 = 8$ | $(8, 1)$ |

Plot the points and connect them with a smooth curve.

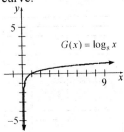

The domain of $G$ is $\{x \mid x > 0\}$ or, using interval notation, $(0, \infty)$. The range of $G$ is all real numbers or, using interval notation, $(-\infty, \infty)$.

**66.** Rewrite $y = F(x) = \log_{1/7} x$ as $x = \left(\dfrac{1}{7}\right)^y$. Locate some points on the graph of $x = \left(\dfrac{1}{7}\right)^y$.

| $y$ | $x = \left(\dfrac{1}{7}\right)^y$ | $(x, y)$ |
|---|---|---|
| $-1$ | $x = \left(\dfrac{1}{7}\right)^{-1} = 7^1 = 7$ | $(7, -1)$ |
| $0$ | $x = \left(\dfrac{1}{7}\right)^0 = 1$ | $(1, 0)$ |
| $1$ | $x = \left(\dfrac{1}{7}\right)^1 = \dfrac{1}{7}$ | $\left(\dfrac{1}{7}, 1\right)$ |

Plot the points and connect them with a smooth curve.

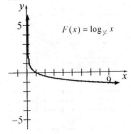

The domain of F is $\{x \mid x > 0\}$ or, using interval notation, $(0, \infty)$. The range of F is all real numbers or, using interval notation, $(-\infty, \infty)$.

68. Rewrite $y = g(x) = \log_{1/8} x$ as $x = \left(\dfrac{1}{8}\right)^y$. Locate some points on the graph of $x = \left(\dfrac{1}{8}\right)^y$.

| $y$ | $x = \left(\dfrac{1}{8}\right)^y$ | $(x, y)$ |
|---|---|---|
| $-1$ | $x = \left(\dfrac{1}{8}\right)^{-1} = 8^1 = 8$ | $(8, -1)$ |
| $0$ | $x = \left(\dfrac{1}{8}\right)^0 = 1$ | $(1, 0)$ |
| $1$ | $x = \left(\dfrac{1}{8}\right)^1 = \dfrac{1}{8}$ | $\left(\dfrac{1}{8}, 1\right)$ |
| $2$ | $x = \left(\dfrac{1}{8}\right)^2 = \dfrac{1}{64}$ | $\left(\dfrac{1}{64}, 1\right)$ |

Plot the points and connect them with a smooth curve.

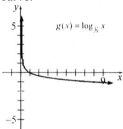

The domain of g is $\{x \mid x > 0\}$ or, using interval notation, $(0, \infty)$. The range of g is all real numbers or, using interval notation, $(-\infty, \infty)$.

70. Rewrite $y = f(x) = \log x$ as $x = 10^y$. Locate some points on the graph of $x = 10^y$.

| $y$ | $x = 10^y$ | $(x, y)$ |
|---|---|---|
| $-2$ | $x = 10^{-2} = \dfrac{1}{10^2} = \dfrac{1}{100}$ | $\left(\dfrac{1}{100}, -2\right)$ |
| $-1$ | $x = 10^{-1} = \dfrac{1}{10^1} = \dfrac{1}{10}$ | $\left(\dfrac{1}{10}, -1\right)$ |
| $0$ | $x = 10^0 = 1$ | $(1, 0)$ |
| $1$ | $x = 10^1 = 10$ | $(10, 1)$ |

Plot the points and connect them with a smooth curve.

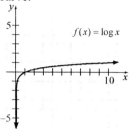

The domain of f is $\{x \mid x > 0\}$ or, using interval notation, $(0, \infty)$. The range of f is all real numbers or, using interval notation, $(-\infty, \infty)$.

72. $\log 106 \approx 2.025$

74. $\ln 10.4 \approx 2.342$

76. $\log 0.78 \approx -0.108$

78. $\ln 0.4 \approx -0.916$

80. $\log \dfrac{10}{7} \approx 0.155$

82. $\ln \dfrac{1}{2} \approx -0.693$

84. $\log_3(5x - 3) = 3$
$5x - 3 = 3^3$
$5x - 3 = 27$
$5x = 30$
$x = 6$
The solution set is $\{6\}$.

86. $\log_4(8x + 10) = 3$
$8x + 10 = 4^3$
$8x + 10 = 64$
$8x = 54$
$x = \dfrac{54}{8} = \dfrac{27}{4}$
The solution set is $\left\{\dfrac{27}{4}\right\}$.

88. $\log_a 81 = 2$
$a^2 = 81$
$a = \pm\sqrt{81}$
$a = \pm 9$
Since the base of a logarithm must always be

Section 9.3 Logarithmic Functions

positive, we know that $a = -9$ is extraneous.
The solution set is $\{9\}$.

90. $\log_a 28 = 2$
$a^2 = 28$
$a = \pm\sqrt{28}$
$a = \pm 2\sqrt{7}$

Since the base of a logarithm must always be positive, we know that $a = -2\sqrt{7}$ is extraneous.
The solution set is $\{2\sqrt{7}\}$.

92. $\log_a 243 = 5$
$a^5 = 243$
$a = \sqrt[5]{243}$
$a = 3$
The solution set is $\{3\}$.

94. $\ln x = 10$
$x = e^{10} \approx 22{,}026.466$
The solution set is $\{e^{10}\}$.

96. $\log(2x+3) = 1$
$2x+3 = 10^1$
$2x+3 = 10$
$2x = 7$
$x = \dfrac{7}{2}$
The solution set is $\left\{\dfrac{7}{2}\right\}$.

98. $\ln e^{2x} = 8$
$e^{2x} = e^8$
$2x = 8$
$x = 4$
The solution set is $\{4\}$.

100. $\log_4(16) = x+1$
$4^{x+1} = 16$
$4^{x+1} = 4^2$
$x+1 = 2$
$x = 1$
The solution set is $\{1\}$.

102. $\log_3(x^2+1) = 2$
$x^2+1 = 3^2$
$x^2+1 = 9$
$x^2 = 8$
$x = \pm\sqrt{8}$
$x = \pm 2\sqrt{2}$
The solution set is $\{-2\sqrt{2},\ 2\sqrt{2}\}$.

104. a. $f(5) = \log_5 5$. To determine the value, let $y = \log_5 5$. Then,
$5^y = 5$
$5^y = 5^1$
$y = 1$
Therefore, $f(5) = 1$, and the point $(5,1)$ is on the graph of $f$.

b. $f(x) = -2$
$\log_5 x = -2$
$x = 5^{-2}$
$x = \dfrac{1}{5^2} = \dfrac{1}{25}$
The point $\left(\dfrac{1}{25}, -2\right)$ is on the graph of $f$.

106. a. $F(8) = \log_2(8) - 3$. To determine the value, let $y = \log_2(8) - 3$. Then,
$y+3 = \log_2(8)$
$2^{y+3} = 8$
$2^{y+3} = 2^3$
$y+3 = 3$
$y = 0$
Therefore, $F(8) = 0$, and the point $(8, 0)$ is on the graph of $F$.

b. $F(x) = -1$
$\log_2 x - 3 = -1$
$\log_2 x = 2$
$x = 2^2$
$x = 4$
The point $(4, -1)$ is on the graph of $F$.

108. For the front row, we evaluate
$L(x) = 10\log\dfrac{x}{10^{-12}}$ at $x = 10^{-1}$.

ISM - 571

## Chapter 9: Exponential and Logarithmic Functions

$$L(10^{-1}) = 10\log\frac{10^{-1}}{10^{-12}}$$
$$= 10\log 10^{-1-(-12)}$$
$$= 10\log 10^{11} = 10(11) = 110$$

The loudness of a concert in the front row is 110 decibels.

For the 15$^{th}$ row, we evaluate

$L(x) = 10\log\dfrac{x}{10^{-12}}$ at $x = 10^{-2}$.

$$L(10^{-2}) = 10\log\frac{10^{-2}}{10^{-12}}$$
$$= 10\log 10^{-2-(-12)}$$
$$= 10\log 10^{10} = 10(10) = 100$$

The loudness of a concert in the 15$^{th}$ row is 100 decibels.

**110.** We evaluate $L(x) = 10\log\dfrac{x}{10^{-12}}$ at $x = 10^4$.

$$L(10^4) = 10\log\frac{10^4}{10^{-12}}$$
$$= 10\log 10^{4-(-12)}$$
$$= 10\log 10^{16} = 10(16) = 160$$

Instant perforation of the ear drum will occur at 160 decibels.

**112.** We evaluate $M(x) = \log\left(\dfrac{x}{10^{-3}}\right)$ at $x = 1{,}584{,}893$.

$$M(1{,}584{,}893) = \log\left(\frac{1{,}584{,}893}{10^{-3}}\right)$$
$$= \log(1{,}584{,}893{,}000) \approx 9.2$$

The magnitude of the 1964 Alaska earthquake was approximately 9.2 on the Richter scale.

**114.** We solve $M(x) = \log\left(\dfrac{x}{10^{-3}}\right)$ for $M(x) = 7.3$.

$$M(x) = 7.3$$
$$\log\left(\frac{x}{10^{-3}}\right) = 7.3$$
$$\frac{x}{10^{-3}} = 10^{7.3}$$
$$x = 10^{7.3} \cdot 10^{-3}$$
$$= 10^{7.3+(-3)}$$
$$= 10^{4.3}$$
$$\approx 19{,}953$$

The seismographic reading of the 1886 South Carolina earthquake 100 kilometers from the epicenter was approximately 19,953 millimeters.

**116. a.** We solve $\log E_S = 11.8 + 1.5M$ for $E_S$ when $M = 5.8$.
$$\log E_S = 11.8 + 1.5(5.8)$$
$$\log E_S = 20.5$$
$$E_S = 10^{20.5} \approx 3.16 \times 10^{20}$$
An earthquake with magnitude 5.8 on the Richter scale will give off approximately $3.16 \times 10^{20}$ ergs of energy.

**b.** We solve $\log E_S = 11.8 + 1.5M$ for $E_S$ when $M = 8.7$.
$$\log E_S = 11.8 + 1.5(8.7)$$
$$\log E_S = 24.85$$
$$E_S = 10^{24.85} \approx 7.08 \times 10^{24}$$
The 1965 Rat Islands earthquake gave off approximately $7.08 \times 10^{24}$ ergs of energy.

**118.** The domain of $f(x) = \log_5(x^2 + 2x - 24)$ is the set of all real numbers $x$ such that $x^2 + 2x - 24 > 0$.

Solve:  $x^2 + 2x - 24 = 0$
$(x-4)(x+6) = 0$
$x - 4 = 0$ or $x + 6 = 0$
$x = 4$ or $x = -6$

Determine where each factor is positive and negative and where the product of these factors is positive and negative.

| Interval | $(-\infty, -6)$ | $-6$ | $(-6, 4)$ | $4$ | $(4, \infty)$ |
|---|---|---|---|---|---|
| $x - 4$ | $----$ | $-$ | $----$ | $0$ | $++++$ |
| $x + 6$ | $----$ | $0$ | $++++$ | $+$ | $++++$ |
| $(x-4)(x+6)$ | $++++$ | $0$ | $----$ | $0$ | $++++$ |

The inequality is strict, so $-6$ and $4$ are not part of the solution. Now, $(x-4)(x+6)$ is greater than zero where the product is positive. Thus, the solution of the inequality and the domain of $f$ is $\{x \mid x < -6 \text{ or } x > 4\}$, or using interval notation, $(-\infty, -6) \cup (4, \infty)$.

**120.** The domain of $f(x) = \log\left(\dfrac{x+4}{x-3}\right)$ is the set of all real numbers $x$ such that $\dfrac{x+4}{x-3} > 0$.

The rational expression will equal 0 when $x = -4$. It is undefined when $x = 3$. Determine where the numerator and the denominator are positive and negative and where the quotient is positive and negative:

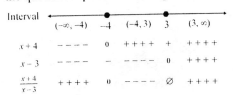

The rational function is undefined at $x = 3$, so 3 is not part of the solution. The inequality is strict, so $-4$ is not part of the solution. Now, $\dfrac{x+4}{x-3}$ is greater than zero where the quotient is positive. Thus, the solution of the inequality and the domain of $f$ is $\{x \mid x < -4 \text{ or } x > 3\}$, or using interval notation, $(-\infty, -4) \cup (3, \infty)$.

122. If the graph of $f(x) = \log_a x$ contains the point $\left(\dfrac{1}{4}, -2\right)$, then

$$\log_a\left(\dfrac{1}{4}\right) = -2$$

$$a^{-2} = \dfrac{1}{4}$$

$$\dfrac{1}{a^2} = \dfrac{1}{4}$$

$$a^2 = 4$$

$$a = \pm\sqrt{4} = \pm 2$$

Since the base of a logarithm must always be positive, we know that $a = -2$ is extraneous. Thus, $a = 2$.

124. $(x^3 - 2x^2 + 10x + 2) + (3x^3 - 4x - 5)$
$= x^3 - 2x^2 + 10x + 2 + 3x^3 - 4x - 5$
$= 4x^3 - 2x^2 + 6x - 3$

126. $x^2 + 5x + 6 = (x+3)(x+2)$
$x^2 + 2x - 3 = (x+3)(x-1)$
LCD $= (x+3)(x+2)(x-1)$

$$\dfrac{x+5}{x^2+5x+6} - \dfrac{2}{x^2+2x-3}$$

$$= \dfrac{x+5}{(x+3)(x+2)} \cdot \dfrac{x-1}{x-1} - \dfrac{2}{(x+3)(x-1)} \cdot \dfrac{x+2}{x+2}$$

$$= \dfrac{x^2 - x + 5x - 5}{(x+3)(x+2)(x-1)} - \dfrac{2x+4}{(x+3)(x+2)(x-1)}$$

$$= \dfrac{x^2 - x + 5x - 5 - (2x+4)}{(x+3)(x+2)(x-1)}$$

$$= \dfrac{x^2 - x + 5x - 5 - 2x - 4}{(x+3)(x+2)(x-1)}$$

$$= \dfrac{x^2 + 2x - 9}{(x+3)(x+2)(x-1)}$$

128. $\sqrt[3]{27a} + 2\sqrt[3]{a} - \sqrt[3]{8a} = 3\sqrt[3]{a} + 2\sqrt[3]{a} - 2\sqrt[3]{a} = 3\sqrt[3]{a}$

130. Let $Y_1 = \log(x-2)$.

The domain of $g(x) = \log(x-2)$ is $\{x \mid x > 2\}$ or, using interval notation, $(2, \infty)$. The range is all real numbers or, using interval notation, $(-\infty, \infty)$.

Note: The graph shown above is a bit misleading. The curve does not terminate at $x = 2$. Instead, the curve has an asymptote at $x = 2$. More specifically, as $x$ approaches 2 (but stays larger than 2), $g$ goes to $-\infty$.

132. Let $Y_1 = \ln(x) - 4$.

The domain of $F(x) = \ln(x) - 4$ is $\{x \mid x > 0\}$ or, using interval notation, $(0, \infty)$. The range is all real numbers or, using interval notation, $(-\infty, \infty)$.

Note: The graph shown above is a bit misleading. The curve does not terminate at $x = 0$. Instead, the curve has an asymptote at

## Chapter 9: Exponential and Logarithmic Functions

$x = 0$. More specifically, as $x$ approaches 0 (but stays larger than 0), $F$ goes to $-\infty$.

134. Let $Y_1 = -\log(x+1) - 3$.

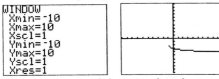

The domain of $G(x) = -\log(x+1) - 3$ is $\{x \mid x > -1\}$ or, using interval notation, $(-1, \infty)$. The range is all real numbers or, using interval notation, $(-\infty, \infty)$.

Note: The graph shown above is a bit misleading. The curve does not terminate at $x = -1$. Instead, the curve has an asymptote at $x = -1$. More specifically, as $x$ approaches $-1$ (but stays larger than $-1$), $G$ goes to $\infty$.

## Putting the Concepts Together (Sections 9.1 – 9.3)

1. $f(x) = 2x + 3$; $g(x) = 2x^2 - 4x$

   a. $(f \circ g)(x) = f(g(x))$
   $= 2(2x^2 - 4x) + 3$
   $= 4x^2 - 8x + 3$

   b. $(g \circ f)(x) = g(f(x))$
   $= 2(2x+3)^2 - 4(2x+3)$
   $= 2(4x^2 + 12x + 9) - 4(2x+3)$
   $= 8x^2 + 24x + 18 - 8x - 12$
   $= 8x^2 + 16x + 6$

   c. Using the result from part (a):
   $(f \circ g)(3) = 4(3)^2 - 8(3) + 3$
   $= 4(9) - 8(3) + 3$
   $= 36 - 24 + 3$
   $= 15$

   d. Using the result from part (b):
   $(g \circ f)(-2) = 8(-2)^2 + 16(-2) + 6$
   $= 8(4) + 16(-2) + 6$
   $= 32 - 32 + 6$
   $= 6$

   e. $f(1) = 2(1) + 3 = 2 + 3 = 5$
   $f(5) = 2(5) + 3 = 10 + 3 = 13$
   $(f \circ f)(1) = f(f(1)) = f(5) = 13$

2. a. $f(x) = 3x + 4$
   $y = 3x + 4$
   $x = 3y + 4$
   $x - 4 = 3y$
   $\dfrac{x-4}{3} = y$
   $f^{-1}(x) = \dfrac{x-4}{3}$

   Check:
   $f(f^{-1}(x)) = 3\left(\dfrac{x-4}{3}\right) + 4 = x - 4 + 4 = x$
   $f^{-1}(f(x)) = \dfrac{(3x+4) - 4}{3} = \dfrac{3x}{3} = x$

   b. $g(x) = x^3 - 4$
   $y = x^3 - 4$
   $x = y^3 - 4$
   $x + 4 = y^3$
   $\sqrt[3]{x+4} = y$
   $g^{-1}(x) = \sqrt[3]{x+4}$

   Check:
   $g(g^{-1}(x)) = \left(\sqrt[3]{x+4}\right)^3 - 4 = x + 4 - 4 = x$
   $g^{-1}(g(x)) = \sqrt[3]{(x^3 - 4) + 4} = \sqrt[3]{x^3} = x$

3. To plot the inverse, switch the $x$- and $y$-coordinates in each point and connect the corresponding points. The graphs of the function (shaded) and the line $y = x$ (dashed) are included for reference.

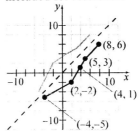

## Putting the Concepts Together (Sections 9.1 – 9.3)

4. a. $2.7^{2.7} \approx 14.611$
   b. $2.72^{2.72} \approx 15.206$
   c. $2.718^{2.718} \approx 15.146$
   d. $2.7183^{2.7183} \approx 15.155$
   e. $e^e \approx 15.154$

5. a. If $a^4 = 6.4$, then $\log_a 6.4 = 4$.
   b. If $10^x = 278$, then $\log 278 = x$.

6. a. If $\log_2 x = 7$, then $2^7 = x$.
   b. If $\ln 16 = M$, then $e^M = 16$.

7. a. Let $y = \log_5 625$. Then,
   $5^y = 625$
   $5^y = 5^4$
   $y = 4$
   Thus, $\log_5 625 = 4$.

   b. Let $y = \log_{\frac{2}{3}}\left(\frac{9}{4}\right)$. Then,
   $\left(\frac{2}{3}\right)^y = \frac{9}{4}$
   $\left(\frac{2}{3}\right)^y = \left(\frac{3}{2}\right)^2$
   $\left(\frac{2}{3}\right)^y = \left(\frac{2}{3}\right)^{-2}$
   $y = -2$
   Thus, $\log_{\frac{2}{3}}\left(\frac{9}{4}\right) = -2$.

8. The domain of $f(x) = \log_{13}(2x+12)$ is the set of all real numbers $x$ such that
   $2x + 12 > 0$
   $2x > -12$
   $x > -6$
   Thus, the domain of $f(x) = \log_{13}(2x+12)$ is $\{x \mid x > -6\}$, or using interval notation, $(-6, \infty)$.

9. Locate some points on the graph of $f(x) = \left(\frac{1}{6}\right)^x$.

| $x$ | $f(x) = \left(\frac{1}{6}\right)^x$ | $(x, f(x))$ |
|---|---|---|
| $-1$ | $f(-1) = \left(\frac{1}{6}\right)^{-1} = \left(\frac{6}{1}\right)^1 = 6$ | $(-1, 6)$ |
| $0$ | $f(0) = \left(\frac{1}{6}\right)^0 = 1$ | $(0, 1)$ |
| $1$ | $f(1) = \left(\frac{1}{6}\right)^1 = \frac{1}{6}$ | $\left(1, \frac{1}{6}\right)$ |
| $2$ | $f(2) = \left(\frac{1}{6}\right)^2 = \frac{1}{36}$ | $\left(2, \frac{1}{36}\right)$ |

Plot the points and connect them with a smooth curve.

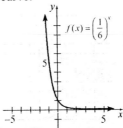

The domain of $f$ is all real numbers or, using interval notation, $(-\infty, \infty)$. The range of $f$ is $\{y \mid y > 0\}$ or, using interval notation, $(0, \infty)$.

10. Rewrite $y = g(x) = \log_{\frac{3}{2}} x$ as $x = \left(\frac{3}{2}\right)^y$. Locate some points on the graph of $x = \left(\frac{3}{2}\right)^y$.

## Chapter 9: Exponential and Logarithmic Functions

| $y$ | $x = \left(\dfrac{3}{2}\right)^y$ | $(x, y)$ |
|---|---|---|
| $-2$ | $x = \left(\dfrac{3}{2}\right)^{-2} = \left(\dfrac{2}{3}\right)^2 = \dfrac{4}{9}$ | $\left(\dfrac{4}{9}, -2\right)$ |
| $-1$ | $x = \left(\dfrac{3}{2}\right)^{-1} = \left(\dfrac{2}{3}\right)^1 = \dfrac{2}{3}$ | $\left(\dfrac{2}{3}, -1\right)$ |
| $0$ | $x = \left(\dfrac{3}{2}\right)^0 = 1$ | $(1, 0)$ |
| $1$ | $x = \left(\dfrac{3}{2}\right)^1 = \dfrac{3}{2}$ | $\left(\dfrac{3}{2}, 1\right)$ |
| $2$ | $x = \left(\dfrac{3}{2}\right)^2 = \dfrac{9}{4}$ | $\left(\dfrac{9}{4}, 2\right)$ |

Plot the points and connect them with a smooth curve.

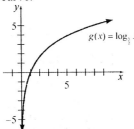

The domain of $g$ is $\{x \mid x > 0\}$ or, using interval notation, $(0, \infty)$. The range of $g$ is all real numbers or, using interval notation, $(-\infty, \infty)$.

11. $3^{-x+2} = 27$
$3^{-x+2} = 3^3$
$-x + 2 = 3$
$-x = 1$
$x = -1$
The solution set is $\{-1\}$.

12. $e^x = e^{2x+5}$
$x = 2x + 5$
$-x = 5$
$x = -5$
The solution set is $\{-5\}$.

13. $\log_2(2x + 5) = 4$
$2x + 5 = 2^4$
$2x + 5 = 16$
$2x = 11$
$x = \dfrac{11}{2}$
The solution set is $\left\{\dfrac{11}{2}\right\}$.

14. $\ln x = 7$
$x = e^7 \approx 1096.633$
The solution set is $\{e^7\}$.

15. $L(90) = 150\left(1 - e^{-0.0052(90)}\right) \approx 56.06$.
According to the model, the student will learn approximately 56 terms after 90 minutes.

### 9.4 Preparing for Properties of Logarithms

1. $3.03468 \approx 3.035$

2. If $a \neq 0$, then $a^0 = 1$.

### 9.4 Exercises

2. $\log_a x + \log_a y$

4. FALSE. $\log(x \cdot 4) = \log x + \log 4$, but there is no such rule for $\log(x + 4)$.

6. TRUE.

8. Answers may vary. One possibility follows: The log of a quotient is equal to the difference of the logs.

10. Answers may vary. One example follows: Let $x = 8$, $r = 3$, and $a = 2$. Then $(\log_a x)^r = (\log_2 8)^3 = 3^3 = 27$, and $r \log_a x = 3 \log_2 8 = 3(3) = 9$. Since $27 \neq 9$, $(\log_a x)^r \neq r \log_a x$.

12. $\log_5 5^{-3} = -3$

14. $\ln e^9 = 9$

**Section 9.4:** *Properties of Logarithms*

16. $5^{\log_5 \sqrt{2}} = \sqrt{2}$

18. $e^{\ln 10} = 10$

20. $\log_6 2 + \log_6 3 = \log_6 (2 \cdot 3) = \log_6 6 = 1$

22. $\log_4 20 - \log_4 5 = \log_4 \left(\dfrac{20}{5}\right) = \log_4 4 = 1$

24. $e^{\ln 24 - \ln 3} = e^{\ln(24/3)} = e^{\ln 8} = 8$

26. $\ln \dfrac{3}{2} = \ln 3 - \ln 2 = b - a$

28. $\ln 4 = \ln 2^2 = 2 \ln 2 = 2a$

30. $\ln 18 = \ln(2 \cdot 3^2)$
$= \ln 2 + \ln 3^2$
$= \ln 2 + 2 \ln 3$
$= a + 2b$

32. $\ln \sqrt[4]{3} = \ln 3^{1/4} = \dfrac{1}{4} \ln 3 = \dfrac{1}{4} b$

34. $\log_4 \left(\dfrac{a}{b}\right) = \log_4 a - \log_4 b$

36. $\log_3 z^{-2} = -2 \log_3 z$

38. $\log_3 (a^3 b) = \log_3 a^3 + \log_3 b = 3 \log_3 a + \log_3 b$

40. $\log_2 (8z) = \log_2 8 + \log_2 z$
$= \log_2 2^3 + \log_2 z$
$= 3 + \log_2 z$

42. $\log_2 \left(\dfrac{16}{p}\right) = \log_2 16 - \log_2 p$
$= \log_2 2^4 - \log_2 p$
$= 4 - \log_2 p$

44. $\ln \left(\dfrac{x}{e^3}\right) = \ln x - \ln e^3 = \ln x - 3$

46. $\log_2 (32 \sqrt[4]{z}) = \log_2 32 + \log_2 \sqrt[4]{z}$
$= \log_2 2^5 + \log_2 z^{1/4}$
$= 5 + \dfrac{1}{4} \log_2 z$

48. $\log_3 \left(x^3 \sqrt{x^2 - 1}\right)$
$= \log_3 x^3 + \log_3 \sqrt{x^2 - 1}$
$= \log_3 x^3 + \log_3 (x^2 - 1)^{1/2}$
$= 3 \log_3 x + \dfrac{1}{2} \log_3 (x^2 - 1)$
$= 3 \log_3 x + \dfrac{1}{2} \log_3 [(x+1)(x-1)]$
$= 3 \log_3 x + \dfrac{1}{2} [\log_3 (x+1) + \log_3 (x-1)]$
$= 3 \log_3 x + \dfrac{1}{2} \log_3 (x+1) + \dfrac{1}{2} \log_3 (x-1)$

50. $\ln \left(\dfrac{\sqrt[5]{x}}{(x+2)^2}\right) = \ln \sqrt[5]{x} - \ln (x+2)^2$
$= \ln x^{1/5} - \ln (x+2)^2$
$= \dfrac{1}{5} \ln x - 2 \ln (x+2)$

52. $\log_6 \sqrt[3]{\dfrac{x-2}{x+1}} = \log_6 \left(\dfrac{x-2}{x+1}\right)^{1/3}$
$= \dfrac{1}{3} \log_6 \left(\dfrac{x-2}{x+1}\right)$
$= \dfrac{1}{3} [\log_6 (x-2) - \log_6 (x+1)]$
$= \dfrac{1}{3} \log_6 (x-2) - \dfrac{1}{3} \log_6 (x+1)$

54. $\log_4 \left[\dfrac{x^3 (x-3)}{\sqrt[3]{x+1}}\right]$
$= \log_4 [x^3 (x-3)] - \log_4 \sqrt[3]{x+1}$
$= \log_4 x^3 + \log_4 (x-3) - \log_4 \sqrt[3]{x+1}$
$= \log_4 x^3 + \log_4 (x-3) - \log_4 (x+1)^{1/3}$
$= 3 \log_4 x + \log_4 (x-3) - \dfrac{1}{3} \log_4 (x+1)$

**56.** $\log_4 32 + \log_4 2 = \log_4(32 \cdot 2)$
$= \log_4 64$
$= \log_4 4^3$
$= 3$

**58.** $\log_2 6 + \log_2 z = \log_2(6z)$

**60.** $\log_2 48 - \log_2 3 = \log_2\left(\dfrac{48}{3}\right)$
$= \log_2 16$
$= \log_2 2^4$
$= 4$

**62.** $8\log_2 z = \log_2 z^8$

**64.** $\log_5(2y-1) - \log_5 y = \log_5\left(\dfrac{2y-1}{y}\right)$

**66.** $4\log_2 a + 2\log_2 b = \log_2 a^4 + \log_2 b^2$
$= \log_2(a^4 b^2)$

**68.** $\dfrac{1}{3}\log_4 z + 2\log_4(2z+1) = \log_4 z^{1/3} + \log_4(2z+1)^2$
$= \log_4 \sqrt[3]{z} + \log_4(2z+1)^2$
$= \log_4\left[\sqrt[3]{z}(2z+1)^2\right]$

**70.** $\log_7 x^4 - 2\log_7 x = \log_7 x^4 - \log_7 x^2$
$= \log_7\left(\dfrac{x^4}{x^2}\right)$
$= \log_7(x^2)$

**72.** $\dfrac{1}{3}\left[\ln(x-1) + \ln(x+1)\right] = \dfrac{1}{3}\ln\left[(x-1)(x+1)\right]$
$= \dfrac{1}{3}\ln(x^2-1)$
$= \ln(x^2-1)^{1/3}$
$= \ln\sqrt[3]{x^2-1}$

**74.** $\log_5(x^2+3x+2) - \log_5(x+2)$
$= \log_5\left(\dfrac{x^2+3x+2}{x+2}\right)$
$= \log_5\left[\dfrac{(x+2)(x+1)}{x+2}\right]$
$= \log_5(x+1)$

**76.** $10\log_4 \sqrt[5]{x} + 4\log_4 \sqrt{x} - \log_4 16$
$= 10\log_4 x^{1/5} + 4\log_4 x^{1/2} - \log_4 16$
$= \log_4\left(x^{1/5}\right)^{10} + \log_4\left(x^{1/2}\right)^4 - \log_4 16$
$= \log_4 x^2 + \log_4 x^2 - \log_4 16$
$= \log_4\left(\dfrac{x^2 \cdot x^2}{16}\right)$
$= \log_4\left(\dfrac{x^4}{16}\right)$

**78.** Using common logarithms:
$\log_3 10 = \dfrac{\log 18}{\log 3} \approx 2.631$

**80.** Using common logarithms:
$\log_7 5 = \dfrac{\log 5}{\log 7} \approx 0.827$

**82.** Using natural logarithms:
$\log_{1/4} 3 = \dfrac{\ln 3}{\ln\left(\dfrac{1}{4}\right)} \approx -0.792$

**84.** Using natural logarithms:
$\log_{\sqrt{3}} \sqrt{6} = \dfrac{\ln\sqrt{6}}{\ln\sqrt{3}} \approx 1.631$

**86.** $\log_2 4 \cdot \log_4 6 \cdot \log_6 8 = \dfrac{\log 4}{\log 2} \cdot \dfrac{\log 6}{\log 4} \cdot \dfrac{\log 8}{\log 6}$
$= \dfrac{\log 4}{\log 2} \cdot \dfrac{\log 6}{\log 4} \cdot \dfrac{\log 8}{\log 6}$
$= \dfrac{\log 8}{\log 2}$
$= \log_2 8$
$= \log_2 2^3$
$= 3$

**Section 9.4: Properties of Logarithms**

88. $\log_3 3 \cdot \log_3 9 \cdot \log_3 27 \cdot ... \cdot \log_3 3^n$
$= \log_3 3 \cdot \log_3 3^2 \cdot \log_3 3^3 \cdot ... \cdot \log_3 3^n$
$= 1 \cdot 2 \cdot 3 \cdot ... \cdot n$
$= n!$

90. $\log_a\left(\sqrt{x}+\sqrt{x-1}\right)+\log_a\left(\sqrt{x}-\sqrt{x-1}\right)$
$= \log_a\left[\left(\sqrt{x}+\sqrt{x-1}\right)\left(\sqrt{x}-\sqrt{x-1}\right)\right]$
$= \log_a\left[x-\sqrt{x}\sqrt{x-1}+\sqrt{x}\sqrt{x-1}-(x-1)\right]$
$= \log_a(x-x+1)$
$= \log_a 1$
$= 0$

92. The domain of $f(x) = \log_a x^2$ is all real numbers. The domain of $g(x) = 2\log_a x$ is $\{x \mid x > 0\}$. Explanations for differences will vary. One possibility follows: From the restrictions placed on the power rule, the statement $\log_a x^2 = 2\log_a x$ is only true when $x$ is a positive real number. Thus, if $x$ is negative, then $\log_a x^2 \neq 2\log_a x$. This accounts for the difference in the domains of the two functions.

100. Since $\log_5 x = \dfrac{\log x}{\log 5}$, let $Y_1 = \dfrac{\log x}{\log 5}$.

 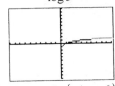

The domain of $f(x) = \log_5 x$ is $\{x \mid x > 0\}$, or using interval notation, $(0, \infty)$; the range is all real numbers or, using interval notation, $(-\infty, \infty)$.

Note: The graph shown above is a bit misleading. The curve does not terminate at $x = 0$. Instead, the curve has an asymptote at $x = 0$. More specifically, as $x$ approaches 0 (but stays larger than 0), $f$ goes to $-\infty$.

102. Since $\log_{1/3} x = \dfrac{\log x}{\log\left(\frac{1}{3}\right)}$, let $Y_1 = \dfrac{\log x}{\log\left(\frac{1}{3}\right)}$.

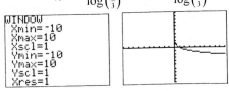

94. $-3x+10 = 4$
$-3x = -6$
$x = 2$
The solution set is $\{2\}$.

96. $3x^2 = 2x+1$
$3x^2 - 2x - 1 = 0$
$(3x+1)(x-1) = 0$
$3x+1 = 0$ or $x-1 = 0$
$x = -\dfrac{1}{3}$ or $x = 1$
The solution set is $\left\{-\dfrac{1}{3}, 1\right\}$.

98. $\sqrt[3]{2x} - 2 = -5$
$\sqrt[3]{2x} = -3$
$\left(\sqrt[3]{2x}\right)^3 = (-3)^3$
$2x = -27$
$x = -\dfrac{27}{2}$
The solution set is $\left\{-\dfrac{27}{2}\right\}$.

The domain of $G(x) = \log_{1/3} x$ is $\{x \mid x > 0\}$, or using interval notation, $(0, \infty)$; the range is all real numbers or, using interval notation, $(-\infty, \infty)$.

Note: The graph shown above is a bit misleading. The curve does not terminate at $x = 0$. Instead, the curve has an asymptote at $x = 0$. More specifically, as $x$ approaches 0 (but stays larger than 0), $G$ goes to $\infty$.

**9.5 Preparing for Exponential and Logarithmic Equations**

1. $2x+5 = 13$
$2x+5-5 = 13-5$
$2x = 8$
$\dfrac{2x}{2} = \dfrac{8}{2}$
$x = 4$
The solution set is $\{4\}$.

## Chapter 9: Exponential and Logarithmic Functions

2.  $x^2 - 4x = -3$
    $x^2 - 4x + 3 = 0$
    $(x-1)(x-3) = 0$
    $x - 1 = 0$ or $x - 3 = 0$
    $x = 1$ or $x = 3$
    The solution set is $\{1, 3\}$.

3.  $3a^2 = a + 5$
    $3a^2 - a - 5 = 0$
    For this equation, $a = 3$, $b = -1$, and $c = -5$.
    $a = \dfrac{-(-1) \pm \sqrt{(-1)^2 - 4(3)(-5)}}{2(3)}$
    $= \dfrac{1 \pm \sqrt{1 + 60}}{6}$
    $= \dfrac{1 \pm \sqrt{61}}{6}$
    The solution set is $\left\{ \dfrac{1 - \sqrt{61}}{6}, \dfrac{1 + \sqrt{61}}{6} \right\}$.

4.  $(x+3)^2 + 2(x+3) - 8 = 0$
    Let $u = x + 3$.
    $u^2 + 2u - 8 = 0$
    $(u+4)(u-2) = 0$
    $u + 4 = 0$ or $u - 2 = 0$
    $u = -4$ or $u = 2$
    $x + 3 = -4$ or $x + 3 = 2$
    $x = -7$ or $x = -1$

Check:
$x = -7$: $(-7+3)^2 + 2(-7+3) - 8 \stackrel{?}{=} 0$
$(-4)^2 + 2(-4) - 8 \stackrel{?}{=} 0$
$16 - 8 - 8 \stackrel{?}{=} 0$
$0 = 0$ ✓

$x = -1$: $(-1+3)^2 + 2(-1+3) - 8 \stackrel{?}{=} 0$
$(2)^2 + 2(2) - 8 \stackrel{?}{=} 0$
$4 + 4 - 8 \stackrel{?}{=} 0$
$0 = 0$ ✓

Both check; the solution set is $\{-7, -1\}$.

### 9.5 Exercises

2. natural logarithm; common logarithm

4. FALSE. For example, see Exercise 9.

Note: Technically, we <u>can</u> use the property $a^{\log_a x} = x$ along with the fact that if $a^u = a^v$, then $u = v$, to solve the equation:
$2^x = 7$
$2^x = 2^{\log_2 7}$
$x = \log_2 7$

6. From the expression $\log_3(x+3)$, the apparent solution $x = -2$ does not appear to be extraneous because the argument $x + 3$ is positive when $x = -2$.

8.  $\log_5 x = \log_5 13$
    $x = 13$
    The solution set is $\{13\}$.

10. $2\log_3 x = \log_3 4$
    $\log_3 x^2 = \log_3 4$
    $x^2 = 4$
    $x = \pm\sqrt{4} = \pm 2$
    The apparent solution $x = -2$ is extraneous because the argument of a logarithm must be positive. The solution set is $\{2\}$.

## Section 9.5 Exponential and Logarithmic Equations

12. $\log(2x-3) = \log 11$
$2x - 3 = 11$
$2x = 14$
$x = 7$
The solution set is $\{7\}$.

14. $\dfrac{1}{2}\log_2 x = 2\log_2 2$
$\log_2 x = 4\log_2 2$
$\log_2 x = \log_2 2^4$
$\log_2 x = \log_2 16$
$x = 16$
The solution set is $\{16\}$.

16. $\log_2(x-7) + \log_2 x = 3$
$\log_2[x(x-7)] = 3$
$\log_2(x^2 - 7x) = 3$
$x^2 - 7x = 2^3$
$x^2 - 7x = 8$
$x^2 - 7x - 8 = 0$
$(x+1)(x-8) = 0$
$x + 1 = 0$ or $x - 8 = 0$
$x = -1$ or $x = 8$
The apparent solution $x = -1$ is extraneous because it causes the argument of a logarithm to be negative. The solution set is $\{8\}$.

18. $\log_2(x-2) + \log_2(x+4) = 4$
$\log_2[(x-2)(x+4)] = 4$
$\log_2(x^2 + 2x - 8) = 4$
$x^2 + 2x - 8 = 2^4$
$x^2 + 2x - 8 = 16$
$x^2 + 2x - 24 = 0$
$(x+6)(x-4) = 0$
$x + 6 = 0$ or $x - 4 = 0$
$x = -6$ or $x = 4$
The apparent solution $x = -6$ is extraneous because it causes the argument of a logarithm to be negative. The solution set is $\{4\}$.

20. $\log_3(x+5) - \log_3 x = 2$
$\log_3\left(\dfrac{x+5}{x}\right) = 2$
$\dfrac{x+5}{x} = 3^2$
$x\left(\dfrac{x+5}{x}\right) = x(9)$
$x + 5 = 9x$
$5 = 8x$
$x = \dfrac{5}{8}$
The solution set is $\left\{\dfrac{5}{8}\right\}$.

22. $\log_3(x+2) - \log_3(x-2) = 4$
$\log_3\left(\dfrac{x+2}{x-2}\right) = 4$
$\dfrac{x+2}{x-2} = 3^4$
$(x-2)\left(\dfrac{x+2}{x-2}\right) = (x-2)81$
$x + 2 = 81x - 162$
$-80x = -164$
$x = \dfrac{-164}{-80} = \dfrac{41}{20}$
The solution set is $\left\{\dfrac{41}{20}\right\}$.

24. $\log_5(x+3) + \log_5(x-4) = \log_5 8$
$\log_5[(x+3)(x-4)] = \log_5 8$
$\log_5(x^2 - x - 12) = \log_5 8$
$x^2 - x - 12 = 8$
$x^2 - x - 20 = 0$
$(x+4)(x-5) = 0$
$x + 4 = 0$ or $x - 5 = 0$
$x = -4$ or $x = 5$
The apparent solution $x = -4$ is extraneous because it causes the argument of a logarithm to be negative. The solution set is $\{5\}$.

## Chapter 9: Exponential and Logarithmic Functions

**26.** $3^x = 8$
$\log 3^x = \log 8$
$x \log 3 = \log 8$
$x = \dfrac{\log 8}{\log 3} \approx 1.893$

The solution set is $\left\{\dfrac{\log 8}{\log 3}\right\} \approx \{1.893\}$. If we had taken the natural logarithm of both sides, the solution set would be $\left\{\dfrac{\ln 8}{\ln 3}\right\} \approx \{1.893\}$.

**28.** $4^x = 20$
$\log 4^x = \log 20$
$x \log 4 = \log 20$
$x = \dfrac{\log 20}{\log 4} \approx 2.161$

The solution set is $\left\{\dfrac{\log 20}{\log 4}\right\} \approx \{2.161\}$. If we had taken the natural logarithm of both sides, the solution set would be $\left\{\dfrac{\ln 20}{\ln 4}\right\} \approx \{2.161\}$.

**30.** $\left(\dfrac{1}{2}\right)^x = 10$
$\log\left(\dfrac{1}{2}\right)^x = \log 10$
$x \log\left(\dfrac{1}{2}\right) = 1$
$x = \dfrac{1}{\log\left(\frac{1}{2}\right)} \approx -3.322$

The solution set is $\left\{\dfrac{1}{\log\left(\frac{1}{2}\right)}\right\} \approx \{-3.322\}$. If we had taken the natural logarithm of both sides, the solution set would be $\left\{\dfrac{\ln 10}{\ln\left(\frac{1}{2}\right)}\right\} \approx \{-3.322\}$.

**32.** $e^x = 3$
$\ln e^x = \ln 3$
$x = \ln 3 \approx 1.099$
The solution set is $\{\ln 3\} \approx \{1.099\}$.

**34.** $10^x = 0.2$
$\log 10^x = \log 0.2$
$x = \log 0.2 \approx -0.699$
The solution set is $\{\log 0.2\} \approx \{-0.699\}$.

**36.** $2^{2x} = 5$
$\log 2^{2x} = \log 5$
$2x \log 2 = \log 5$
$x = \dfrac{\log 5}{2 \log 2} \approx 1.161$

The solution set is $\left\{\dfrac{\log 5}{2 \log 2}\right\} \approx \{1.161\}$. If we had taken the natural logarithm of both sides, the solution set would be $\left\{\dfrac{\ln 5}{2 \ln 2}\right\} \approx \{1.161\}$.

**38.** $\left(\dfrac{1}{3}\right)^{2x} = 4$
$\log\left(\dfrac{1}{3}\right)^{2x} = \log 4$
$2x \log\left(\dfrac{1}{3}\right) = \log 4$
$x = \dfrac{\log 4}{2 \log\left(\frac{1}{3}\right)} \approx -0.631$

The solution set is $\left\{\dfrac{\log 4}{2 \log\left(\frac{1}{3}\right)}\right\} \approx \{-0.631\}$. If we had taken the natural logarithm of both sides, the solution set would be $\left\{\dfrac{\ln 4}{2 \ln\left(\frac{1}{3}\right)}\right\} \approx \{-0.631\}$.

**40.** $3 \cdot 4^x - 5 = 10$
$3 \cdot 4^x = 15$
$4^x = 5$
$\log 4^x = \log 5$
$x \log 4 = \log 5$
$x = \dfrac{\log 5}{\log 4} \approx 1.161$

The solution set is $\left\{\dfrac{\log 5}{\log 4}\right\} \approx \{1.161\}$. If we had taken the natural logarithm of both sides, the solution set would be $\left\{\dfrac{\ln 5}{\ln 4}\right\} \approx \{1.161\}$.

42. $\dfrac{1}{2}e^x = 4$
$e^x = 8$
$\ln e^x = \ln 8$
$x = \ln 8 \approx 2.079$
The solution set is $\{\ln 8\} \approx \{2.079\}$.

44. $0.4^x = 2^{x-3}$
$\log 0.4^x = \log 2^{x-3}$
$x \log 0.4 = (x-3) \log 2$
$x \log 0.4 = x \log 2 - 3 \log 2$
$x \log 0.4 - x \log 2 = -3 \log 2$
$x(\log 0.4 - \log 2) = -3 \log 2$
$x = \dfrac{-3 \log 2}{\log 0.4 - \log 2} \approx 1.292$

The solution set is $\left\{\dfrac{-3 \log 2}{\log 0.4 - \log 2}\right\} \approx \{1.292\}$. If we had taken the natural logarithm of both sides, the solution set would be $\left\{\dfrac{-3 \ln 2}{\ln 0.4 - \ln 2}\right\} \approx \{1.292\}$.

46. $\log_6 x + \log_6 (x+5) = 2$
$\log_6 [x(x+5)] = 2$
$\log_6 (x^2 + 5x) = 2$
$x^2 + 5x = 6^2$
$x^2 + 5x = 36$
$x^2 + 5x - 36 = 0$
$(x+9)(x-4) = 0$
$x+9 = 0 \quad \text{or} \quad x-4 = 0$
$x = -9 \quad \text{or} \quad x = 4$
The apparent solution $x = -9$ is extraneous because it causes the argument of a logarithm to be negative. The solution set is $\{4\}$.

48. $3^{2x} = 4$
$\log 3^{2x} = \log 4$
$2x \log 3 = \log 4$
$x = \dfrac{\log 4}{2 \log 3} \approx 0.631$

The solution set is $\left\{\dfrac{\log 4}{2 \log 3}\right\} \approx \{0.631\}$. If we had taken the natural logarithm of both sides, the solution set would be $\left\{\dfrac{\ln 4}{2 \ln 3}\right\} \approx \{0.631\}$.

50. $5 \log_4 x = \log_4 32$
$\log_4 x^5 = \log_4 32$
$x^5 = 32$
$x = \sqrt[5]{32}$
$x = 2$
The solution set is $\{2\}$.

52. $-4e^x = -16$
$e^x = 4$
$\ln e^x = \ln 4$
$x = \ln 4 \approx 1.386$
The solution set is $\{\ln 4\} \approx \{1.386\}$.

54. $9^x = 27^{x-4}$
$(3^2)^x = (3^3)^{x-4}$
$3^{2x} = 3^{3(x-4)}$
$2x = 3(x-4)$
$2x = 3x - 12$
$-x = -12$
$x = 12$
The solution set is $\{12\}$.

56. $\log_7 x = \log_7 8$
$x = 8$
The solution set is $\{8\}$.

58. $\log_3 (x-5) + \log_3 (x+1) = \log_3 7$
$\log_3 [(x-5)(x+1)] = \log_3 7$
$\log_3 (x^2 - 4x - 5) = \log_3 7$
$x^2 - 4x - 5 = 7$
$x^2 - 4x - 12 = 0$
$(x+2)(x-6) = 0$
$x+2 = 0 \quad \text{or} \quad x-6 = 0$
$x = -2 \quad \text{or} \quad x = 6$
The apparent solution $x = -2$ is extraneous because it causes the argument of a logarithm to be negative. The solution set is $\{6\}$.

60. a. Note that 9.65 billion = 9,650 million. We need to determine the year when $P = 9,650$ million. So we solve the equation

Chapter 9: Exponential and Logarithmic Functions

$$9{,}650 = 6{,}448(1.0126)^{t-2005}$$
$$\frac{9{,}650}{6{,}448} = (1.0126)^{t-2005}$$
$$\log\frac{9{,}650}{6{,}448} = \log(1.01026)^{t-2005}$$
$$\log\frac{9{,}650}{6{,}448} = (t-2005)\log(1.0126)$$
$$\frac{\log\frac{9{,}650}{6{,}448}}{\log 1.0126} = t - 2005$$
$$\frac{\log\frac{9{,}650}{6{,}448}}{\log 1.0126} + 2005 = t$$
$$2037.200 \approx t$$

Thus, according to the model, the world population will reach 9.65 billion in about the year 2037.

b. Note that 11.55 billion = 11,550 million. We need to determine the year when $P = 11{,}550$ million. So we solve the equation
$$11{,}550 = 6{,}448(1.0126)^{t-2005}$$
$$\frac{11{,}550}{6{,}448} = (1.0126)^{t-2005}$$
$$\log\frac{11{,}550}{6{,}448} = \log(1.01026)^{t-2005}$$
$$\log\frac{11{,}550}{6{,}448} = (t-2005)\log(1.0126)$$
$$\frac{\log\frac{11{,}550}{6{,}448}}{\log 1.0126} = t - 2005$$
$$\frac{\log\frac{11{,}550}{6{,}448}}{\log 1.0126} + 2005 = t$$
$$2051.554 \approx t$$

Thus, according to the model, the world population will reach 11.55 billion around the year 2052.

62. We first write the model with the parameters $P = 8000$, $r = 0.04$, and $n = 4$ to obtain
$$A = 8000\left(1 + \frac{0.04}{4}\right)^{4t} \text{ or } A = 8000(1.01)^{4t}.$$

a. We need to determine the time until $A = \$10{,}000$, so we solve the equation

$$10{,}000 = 8000(1.01)^{4t}$$
$$1.25 = (1.01)^{4t}$$
$$\log 1.25 = \log(1.01)^{4t}$$
$$\log 1.25 = 4t \log 1.01$$
$$\frac{\log 1.25}{4 \log 1.01} = t$$
$$5.606 \approx t$$

Thus, after approximately 5.6 years (5 years, 7 months), the account will be worth $10,000.

b. We need to determine the time until $A = \$24{,}000$, so we solve the equation
$$24{,}000 = 8000(1.01)^{4t}$$
$$3 = (1.01)^{4t}$$
$$\log 3 = \log(1.01)^{4t}$$
$$\log 3 = 4t \log 1.01$$
$$\frac{\log 3}{4 \log 1.01} = t$$
$$27.602 \approx t$$

Thus, after approximately 27.6 years (27 years, 7 months), the account will be worth $24,000.

64. a. We need to determine the time until $V = \$10{,}000$. So we solve the equation
$$10{,}000 = 19{,}282(0.84)^t$$
$$\frac{10{,}000}{19{,}282} = (0.84)^t$$
$$\log\frac{10{,}000}{19{,}282} = \log(0.84)^t$$
$$\log\frac{10{,}000}{19{,}282} = t \log 0.84$$
$$\frac{\log\frac{10{,}000}{19{,}282}}{\log 0.84} = t$$
$$3.766 \approx t$$

According to the model, the car will be worth $10,000 after about 3.766 years.

b. We need to determine the time until $V = \$5000$. So we solve the equation

$$5000 = 19{,}282(0.84)^t$$
$$\frac{5000}{19{,}282} = (0.84)^t$$
$$\log\frac{5000}{19{,}282} = \log(0.84)^t$$
$$\log\frac{5000}{19{,}282} = t\log 0.84$$
$$\frac{\log\frac{5000}{19{,}282}}{\log 0.84} = t$$
$$7.741 \approx t$$

According to the model, the car will be worth $5000 in about 7.741 years.

c. We need to determine the time until $V = \$1000$. So we solve the equation
$$1000 = 19{,}282(0.84)^t$$
$$\frac{1000}{19{,}282} = (0.84)^t$$
$$\log\frac{1000}{19{,}282} = \log(0.84)^t$$
$$\log\frac{1000}{19{,}282} = t\log 0.84$$
$$\frac{\log\frac{1000}{19{,}282}}{\log 0.84} = t$$
$$16.972 \approx t$$

According to the model, the car will be worth $1000 in about 16.972 years.

**66. a.** We need to determine the time until $A = 90$ grams. So we solve the equation
$$90 = 100\left(\frac{1}{2}\right)^{t/19.255}$$
$$0.9 = \left(\frac{1}{2}\right)^{t/19.255}$$
$$\log 0.9 = \log\left(\frac{1}{2}\right)^{t/19.255}$$
$$\log 0.9 = \frac{t}{19.255}\log\left(\frac{1}{2}\right)$$
$$\frac{19.255 \cdot \log 0.9}{\log(1/2)} = t$$
$$2.927 \approx t$$

Thus, 90 grams of carbon-10 will be left after approximately 2.927 seconds.

b. We need to determine the time until $A = 25$ grams. So we solve the equation
$$25 = 100\left(\frac{1}{2}\right)^{t/19.255}$$
$$0.25 = \left(\frac{1}{2}\right)^{t/19.255}$$
$$\log 0.25 = \log\left(\frac{1}{2}\right)^{t/19.255}$$
$$\log 0.25 = \frac{t}{19.255}\log\left(\frac{1}{2}\right)$$
$$\frac{19.255 \cdot \log 0.25}{\log(1/2)} = t$$
$$38.51 = t$$

Thus, 25 grams of carbon-10 will be left after 38.51 seconds.

c. We need to determine the time until $A = 10$ grams. So we solve the equation
$$10 = 100\left(\frac{1}{2}\right)^{t/19.255}$$
$$0.1 = \left(\frac{1}{2}\right)^{t/19.255}$$
$$\log 0.1 = \log\left(\frac{1}{2}\right)^{t/19.255}$$
$$\log 0.1 = \frac{t}{19.255}\log\left(\frac{1}{2}\right)$$
$$\frac{19.255 \cdot \log 0.1}{\log(1/2)} = t$$
$$63.964 \approx t$$

Thus, 10 grams of carbon-10 will be left after approximately 63.964 seconds.

**68. a.** We need to determine the time until $u = 120°F$. So we solve the equation
$$120 = 70 + 100e^{-0.045t}$$
$$50 = 100e^{-0.045t}$$
$$0.5 = e^{-0.045t}$$
$$\ln(0.5) = \ln e^{-0.045t}$$
$$\ln(0.5) = -0.045t$$
$$\frac{\ln(0.5)}{-0.045} = t$$
$$15.403 \approx t$$

According to the model, the temperature of the coffee will be $120°F$ after about 15.403 minutes.

b. We need to determine the time until $u = 100°F$. So we solve the equation

## Chapter 9: Exponential and Logarithmic Functions

$100 = 70 + 100e^{-0.045t}$
$30 = 100e^{-0.045t}$
$0.3 = e^{-0.045t}$
$\ln(0.3) = \ln e^{-0.045t}$
$\ln(0.3) = -0.045t$
$\dfrac{\ln(0.3)}{-0.045} = t$
$26.755 \approx t$

According to the model, the temperature of the coffee will be $100°F$ after about 26.755 minutes.

**70. a.** We need to determine the time at which $L = 10$ words. So we solve the equation
$10 = 50(1 - e^{-0.0223t})$
$0.2 = 1 - e^{-0.0223t}$
$e^{-0.0223t} = 0.8$
$\ln e^{-0.0223t} = \ln 0.8$
$-0.0223t = \ln 0.8$
$t = \dfrac{\ln 0.8}{-0.0223}$
$t \approx 10.006$

According to the model, the student must study about 10 minutes in order to learn 10 words.

**b.** We need to determine the time at which $L = 40$ words. So we solve the equation
$40 = 50(1 - e^{-0.0223t})$
$0.8 = 1 - e^{-0.0223t}$
$e^{-0.0223t} = 0.2$
$\ln e^{-0.0223t} = \ln 0.2$
$-0.0223t = \ln 0.2$
$t = \dfrac{\ln 0.2}{-0.0223}$
$t \approx 72.172$

According to the model, the student must study about 72 minutes (or 1.2 hours) in order to learn 40 words.

**72.** Bank B offers the better deal. It has a slightly higher compounding factor as can be seen below:

Bank A: $A = P\left(1 + \dfrac{0.04}{365}\right)^{365t}$
$= P\left[\left(1 + \dfrac{0.04}{365}\right)^{365}\right]^{t}$
$\approx P(1.040808493)^{t}$

Bank B: $A = P\left(1 + \dfrac{0.041}{4}\right)^{365t}$
$= P\left[\left(1 + \dfrac{0.041}{4}\right)^{4}\right]^{t}$
$\approx P(1.041634694)^{t}$

**74.** $f(x) = 5x + 2$

**a.** $f(3) = 5(3) + 2 = 15 + 2 = 17$

**b.** $f(-2) = 5(-2) + 2 = -10 + 2 = -8$

**c.** $f(0) = 5(0) + 2 = 0 + 2 = 2$

**76.** $f(x) = \dfrac{x+3}{x-2}$

**a.** $f(3) = \dfrac{3+3}{3-2} = \dfrac{6}{1} = 6$

**b.** $f(-2) = \dfrac{-2+3}{-2-2} = \dfrac{1}{-4} = -\dfrac{1}{4}$

**c.** $f(0) = \dfrac{0+3}{0-2} = \dfrac{3}{-2} = -\dfrac{3}{2}$

**78.** $f(x) = \sqrt{x+5}$

a. $f(3) = \sqrt{3+5} = \sqrt{8} = \sqrt{4 \cdot 2} = 2\sqrt{2}$

b. $f(-2) = \sqrt{-2+5} = \sqrt{3}$

c. $f(0) = \sqrt{0+5} = \sqrt{5}$

80. Let $Y_1 = e^x$ and $Y_2 = -3x+2$.

 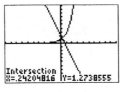

The solution set is approximately $\{0.24\}$.

82. Let $Y_1 = e^x$ and $Y_2 = x+2$.

 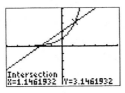

84. Let $Y_1 = e^x + \ln x$ and $Y_2 = 2$.

 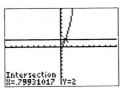

The solution set is approximately $\{0.80\}$.

86. Let $Y_1 = \ln x$ and $Y_2 = x^2 + 1$.

 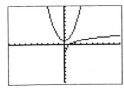

Since there are no points of intersection, the equation has no solution. The solution set is { } or $\varnothing$.

## Chapter 9 Review

1. $f(x) = 3x+5$; $g(x) = 2x-1$

   a. $g(5) = 2(5) - 1 = 9$
   $f(9) = 3(9) + 5 = 32$
   $(f \circ g)(5) = f(g(5)) = f(9) = 32$

   b. $f(-3) = 3(-3) + 5 = -4$
   $g(-4) = 2(-4) - 1 = -9$
   $(g \circ f)(-3) = g(f(-3)) = g(-4) = -9$

   c. $f(-2) = 3(-2) + 5 = -1$
   $f(-1) = 3(-1) + 5 = 2$
   $(f \circ f)(-2) = f(f(-2)) = f(-1) = 2$

   d. $g(4) = 2(4) - 1 = 7$
   $g(7) = 2(7) - 1 = 13$
   $(g \circ g)(4) = g(g(4)) = g(7) = 13$

2. $f(x) = x-3$; $g(x) = 5x+2$

   a. $g(5) = 5(5) + 2 = 27$
   $f(27) = 27 - 3 = 24$
   $(f \circ g)(5) = f(g(5)) = f(27) = 24$

   b. $f(-3) = -3 - 3 = -6$
   $g(-6) = 5(-6) + 2 = -28$
   $(g \circ f)(-3) = g(f(-3)) = g(-6) = -28$

   c. $f(-2) = -2 - 3 = -5$
   $f(-5) = -5 - 3 = -8$
   $(f \circ f)(-2) = f(f(-2)) = f(-5) = -8$

   d. $g(4) = 5(4) + 2 = 22$
   $g(22) = 5(22) + 2 = 112$
   $(g \circ g)(4) = g(g(4)) = g(22) = 112$

3. $f(x) = 2x^2 + 1$; $g(x) = x+5$

   a. $g(5) = 5 + 5 = 10$
   $f(10) = 2(10)^2 + 1 = 201$
   $(f \circ g)(5) = f(g(5)) = f(10) = 201$

## Chapter 9: Exponential and Logarithmic Functions

b. $f(-3) = 2(-3)^2 + 1 = 19$
$g(19) = 19 + 5 = 24$
$(g \circ f)(-3) = g(f(-3)) = g(19) = 24$

c. $f(-2) = 2(-2)^2 + 1 = 9$
$f(9) = 2(9)^2 + 1 = 163$
$(f \circ f)(-2) = f(f(-2)) = f(9) = 163$

d. $g(4) = 4 + 5 = 9$
$g(9) = 9 + 5 = 14$
$(g \circ g)(4) = g(g(4)) = g(9) = 14$

4. $f(x) = x - 3$; $g(x) = x^2 + 1$

   a. $g(5) = (5)^2 + 1 = 26$
   $f(26) = 26 - 3 = 23$
   $(f \circ g)(5) = f(g(5)) = f(26) = 23$

   b. $f(-3) = -3 - 3 = -6$
   $g(-6) = (-6)^2 + 1 = 37$
   $(g \circ f)(-3) = g(f(-3)) = g(-6) = 37$

   c. $f(-2) = -2 - 3 = -5$
   $f(-5) = -5 - 3 = -8$
   $(f \circ f)(-2) = f(f(-2)) = f(-5) = -8$

   d. $g(4) = (4)^2 + 1 = 17$
   $g(17) = (17)^2 + 1 = 290$
   $(g \circ g)(4) = g(g(4)) = g(17) = 290$

5. $f(x) = x + 1$; $g(x) = 5x$

   a. $(f \circ g)(x) = f(g(x)) = f(5x) = 5x + 1$

   b. $(g \circ f)(x) = g(f(x))$
   $= g(x+1)$
   $= 5(x+1)$
   $= 5x + 5$

   c. $(f \circ f)(x) = f(f(x))$
   $= f(x+1)$
   $= (x+1) + 1$
   $= x + 2$

   d. $(g \circ g)(x) = g(g(x)) = g(5x) = 5(5x) = 25x$

6. $f(x) = 2x - 3$; $g(x) = x + 6$

   a. $(f \circ g)(x) = f(g(x))$
   $= f(x+6)$
   $= 2(x+6) - 3$
   $= 2x + 12 - 3$
   $= 2x + 9$

   b. $(g \circ f)(x) = g(f(x))$
   $= g(2x - 3)$
   $= (2x - 3) + 6$
   $= 2x + 3$

   c. $(f \circ f)(x) = f(f(x))$
   $= f(2x - 3)$
   $= 2(2x - 3) - 3$
   $= 4x - 6 - 3$
   $= 4x - 9$

   d. $(g \circ g)(x) = g(g(x))$
   $= g(x + 6)$
   $= (x + 6) + 6$
   $= x + 12$

7. $f(x) = x^2 + 1$; $g(x) = 2x + 1$

   a. $(f \circ g)(x) = f(g(x))$
   $= f(2x + 1)$
   $= (2x + 1)^2 + 1$
   $= 4x^2 + 4x + 1 + 1$
   $= 4x^2 + 4x + 2$

   b. $(g \circ f)(x) = g(f(x))$
   $= g(x^2 + 1)$
   $= 2(x^2 + 1) + 1$
   $= 2x^2 + 2 + 1$
   $= 2x^2 + 3$

   c. $(f \circ f)(x) = f(f(x))$
   $= f(x^2 + 1)$
   $= (x^2 + 1)^2 + 1$
   $= x^4 + 2x^2 + 1 + 1$
   $= x^4 + 2x^2 + 2$

d. $(g \circ g)(x) = g(g(x))$
$= g(2x+1)$
$= 2(2x+1)+1$
$= 4x+2+1$
$= 4x+3$

8. $f(x) = \dfrac{2}{x+1}$; $g(x) = \dfrac{1}{x}$

   a. $(f \circ g)(x) = f(g(x))$
   $= f\left(\dfrac{1}{x}\right)$
   $= \dfrac{2}{\dfrac{1}{x}+1}$
   $= \dfrac{2}{\dfrac{1+x}{x}}$
   $= \dfrac{2x}{x+1}$ where $x \neq -1, 0$.

   b. $(g \circ f)(x) = g(f(x))$
   $= g\left(\dfrac{2}{x+1}\right)$
   $= \dfrac{1}{\dfrac{2}{x+1}}$
   $= 1 \cdot \dfrac{x+1}{2}$
   $= \dfrac{x+1}{2}$ where $x \neq -1$.

   c. $(f \circ f)(x) = f(f(x))$
   $= f\left(\dfrac{2}{x+1}\right)$
   $= \dfrac{2}{\dfrac{2}{x+1}+1}$
   $= \dfrac{2}{\dfrac{2+x+1}{x+1}}$
   $= 2 \cdot \dfrac{x+1}{x+3}$
   $= \dfrac{2(x+1)}{x+3}$ where $x \neq -1, -3$.

   d. $(g \circ g)(x) = g(g(x))$
   $= g\left(\dfrac{1}{x}\right)$
   $= \dfrac{1}{\dfrac{1}{x}}$
   $= 1 \cdot \dfrac{x}{1}$
   $= x$ where $x \neq 0$.

9. The function is not one-to-one. There is an element in the range (8) that corresponds to more than one element in the domain ($-5$ and $-1$).

10. The function is one-to-one. Each element in the range corresponds to exactly one element in the domain.

11. The graph passes the horizontal line test, so the graph is that of a one-to-one function.

12. The graph fails the horizontal line test. Therefore, the function is not one-to-one.

13. Inverse:

    | Height (inches) | Age |
    |---|---|
    | 69 | 24 |
    | 71 | 59 |
    | 72 | 29 |
    | 73 | 81 |
    | 74 | 37 |

14. Inverse:

    | Quantity Demanded | Price ($) |
    |---|---|
    | 112 | 300 |
    | 129 | 200 |
    | 144 | 170 |
    | 161 | 150 |
    | 176 | 130 |

15. To obtain the inverse, we switch the $x$- and $y$-coordinates.
    Inverse: $\{(3,-5),(1,-3),(-3,1),(9,2)\}$

16. To obtain the inverse, we switch the $x$- and $y$-coordinates.
    Inverse: $\{(1,-20),(4,-15),(3,5),(2,25)\}$

17. To plot the inverse, switch the $x$- and $y$-coordinates in each point and connect the corresponding points. The graphs of the function (shaded) and the line $y = x$ (dashed) are included for reference.

## Chapter 9: Exponential and Logarithmic Functions

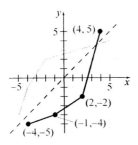

18. To plot the inverse, switch the $x$- and $y$-coordinates in each point and connect the corresponding points. The graphs of the function (shaded) and the line $y = x$ (dashed) are included for reference.

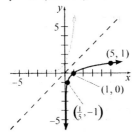

19. $f(x) = 5x$
$$y = 5x$$
$$x = 5y$$
$$\frac{x}{5} = y$$
$$f^{-1}(x) = \frac{x}{5}$$

Check: $f(f^{-1}(x)) = 5\left(\frac{x}{5}\right) = x$

$$f^{-1}(f(x)) = \frac{5x}{5} = x$$

20. $H(x) = 2x + 7$
$$y = 2x + 7$$
$$x = 2y + 7$$
$$x - 7 = 2y$$
$$\frac{x-7}{2} = y$$
$$H^{-1}(x) = \frac{x-7}{2}$$

Check:
$$H(H^{-1}(x)) = 2\left(\frac{x-7}{2}\right) + 7 = x - 7 + 7 = x$$
$$H^{-1}(H(x)) = \frac{(2x+7)-7}{2} = \frac{2x}{2} = x$$

21. $P(x) = \dfrac{4}{x+2}$
$$y = \frac{4}{x+2}$$
$$x = \frac{4}{y+2}$$
$$x(y+2) = 4$$
$$y + 2 = \frac{4}{x}$$
$$y = \frac{4}{x} - 2$$
$$P^{-1}(x) = \frac{4}{x} - 2$$

Check:
$$P(P^{-1}(x)) = \frac{4}{\left(\frac{4}{x} - 2\right) + 2} = \frac{4}{\frac{4}{x}} = 4 \cdot \frac{x}{4} = x$$
$$P^{-1}(P(x)) = \frac{4}{\frac{4}{x+2}} - 2$$
$$= 4 \cdot \frac{x+2}{4} - 2 = (x+2) - 2 = x$$

22. $g(x) = 2x^3 - 1$
$$y = 2x^3 - 1$$
$$x = 2y^3 - 1$$
$$x + 1 = 2y^3$$
$$\frac{x+1}{2} = y^3$$
$$\sqrt[3]{\frac{x+1}{2}} = y$$
$$g^{-1}(x) = \sqrt[3]{\frac{x+1}{2}}$$

Check:
$$g(g^{-1}(x)) = 2\left(\sqrt[3]{\frac{x+1}{2}}\right)^3 - 1$$
$$= 2\left(\frac{x+1}{2}\right) - 1 = (x+1) - 1 = x$$
$$g^{-1}(g(x)) = \sqrt[3]{\frac{(2x^3 - 1) + 1}{2}} = \sqrt[3]{\frac{2x^3}{2}} = \sqrt[3]{x^3} = x$$

23. a. $7^{1.7} \approx 27.332$
    b. $7^{1.73} \approx 28.975$
    c. $7^{1.732} \approx 29.088$
    d. $7^{1.7321} \approx 29.093$
    e. $7^{\sqrt{3}} \approx 29.091$

24. a. $10^{3.1} \approx 1258.925$
    b. $10^{3.14} \approx 1380.384$
    c. $10^{3.142} \approx 1386.756$
    d. $10^{3.1416} \approx 1385.479$
    e. $10^{\pi} \approx 1385.456$

25. a. $e^{0.5} \approx 1.649$
    b. $e^{-1} \approx 0.368$
    c. $e^{1.5} \approx 4.482$
    d. $e^{-0.8} \approx 0.449$
    e. $e^{\sqrt{\pi}} \approx 5.885$

26. Locate some points on the graph of $f(x) = 9^x$.

| $x$ | $f(x) = 9^x$ | $(x, f(x))$ |
|---|---|---|
| $-2$ | $f(-2) = 9^{-2} = \dfrac{1}{9^2} = \dfrac{1}{81}$ | $\left(-2, \dfrac{1}{81}\right)$ |
| $-1$ | $f(-1) = 9^{-1} = \dfrac{1}{9^1} = \dfrac{1}{9}$ | $\left(-1, \dfrac{1}{9}\right)$ |
| $0$ | $f(0) = 9^0 = 1$ | $(0, 1)$ |
| $1$ | $f(1) = 9^1 = 9$ | $(1, 9)$ |

Plot the points and connect them with a smooth curve.

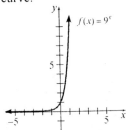

The domain of $f$ is all real numbers or, using interval notation, $(-\infty, \infty)$. The range of $f$ is $\{y \mid y > 0\}$ or, using interval notation, $(0, \infty)$.

27. Locate some points on the graph of $g(x) = \left(\dfrac{1}{9}\right)^x$.

| $x$ | $g(x) = \left(\dfrac{1}{9}\right)^x$ | $(x, g(x))$ |
|---|---|---|
| $-1$ | $g(-1) = \left(\dfrac{1}{9}\right)^{-1} = 9^1 = 9$ | $(-1, 9)$ |
| $0$ | $g(0) = \left(\dfrac{1}{9}\right)^0 = 1$ | $(0, 1)$ |
| $1$ | $g(1) = \left(\dfrac{1}{9}\right)^1 = \dfrac{1}{9}$ | $\left(1, \dfrac{1}{9}\right)$ |
| $2$ | $g(1) = \left(\dfrac{1}{9}\right)^2 = \dfrac{1}{81}$ | $\left(2, \dfrac{1}{81}\right)$ |

Plot the points and connect them with a smooth curve.

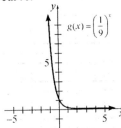

The domain of $g$ is all real numbers or, using interval notation, $(-\infty, \infty)$. The range of $g$ is $\{y \mid y > 0\}$ or, using interval notation, $(0, \infty)$.

28. Locate some points on the graph of $H(x) = 4^{x-2}$.

| $x$ | $H(x) = 4^{x-2}$ | $(x, H(x))$ |
|---|---|---|
| $0$ | $H(0) = 4^{0-2} = 4^{-2} = \dfrac{1}{4^2} = \dfrac{1}{16}$ | $\left(0, \dfrac{1}{16}\right)$ |
| $1$ | $H(1) = 4^{1-2} = 4^{-1} = \dfrac{1}{4^1} = \dfrac{1}{4}$ | $\left(1, \dfrac{1}{4}\right)$ |
| $2$ | $H(2) = 4^{2-2} = 4^0 = 1$ | $(2, 1)$ |
| $3$ | $H(3) = 4^{3-2} = 4^1 = 4$ | $(3, 4)$ |

Plot the points and connect them with a smooth curve.

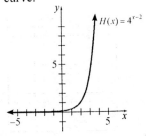

The domain of $H$ is all real numbers or, using interval notation, $(-\infty, \infty)$. The range of $H$ is $\{y \mid y > 0\}$ or, using interval notation, $(0, \infty)$.

29. Locate some points on the graph of $h(x) = 4^x - 2$.

| $x$ | $h(x) = 4^x - 2$ | $(x, h(x))$ |
|---|---|---|
| $-2$ | $h(-2) = 4^{-2} - 2 = \dfrac{1}{16} - 2 = -\dfrac{31}{16}$ | $\left(-2, -\dfrac{31}{16}\right)$ |
| $-1$ | $h(-1) = 4^{-1} - 2 = \dfrac{1}{4} - 2 = -\dfrac{7}{4}$ | $\left(-1, -\dfrac{7}{4}\right)$ |
| $0$ | $h(0) = 4^0 - 2 = 1 - 2 = -1$ | $(0, -1)$ |
| $1$ | $h(1) = 4^1 - 2 = 4 - 2 = 2$ | $(1, 2)$ |

Plot the points and connect them with a smooth curve.

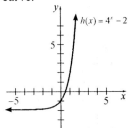

The domain of $h$ is all real numbers or, using interval notation, $(-\infty, \infty)$. The range of $h$ is $\{y \mid y > -2\}$ or, using interval notation, $(-2, \infty)$.

30. The number $e$ is defined as the number that the expression $\left(1 + \dfrac{1}{n}\right)^n$ approaches as $n$ becomes unbounded in the positive direction.

31. $2^x = 64$
    $2^x = 2^6$
    $x = 6$
    The solution set is $\{6\}$.

32. $25^{x-2} = 125$
    $(5^2)^{x-2} = 5^3$
    $5^{2(x-2)} = 5^3$
    $5^{2x-4} = 5^3$
    $2x - 4 = 3$
    $2x = 7$
    $x = \dfrac{7}{2}$
    The solution set is $\left\{\dfrac{7}{2}\right\}$.

33. $27^x \cdot 3^{x^2} = 9^2$
    $(3^3)^x \cdot 3^{x^2} = (3^2)^2$
    $3^{3x} \cdot 3^{x^2} = 3^4$
    $3^{x^2 + 3x} = 3^4$
    $x^2 + 3x = 4$
    $x^2 + 3x - 4 = 0$
    $(x+4)(x-1) = 0$
    $x + 4 = 0$ or $x - 1 = 0$
    $x = -4$ or $x = 1$
    The solution set is $\{-4, 1\}$.

34. $\left(\dfrac{1}{4}\right)^x = 16$
    $(4^{-1})^x = 4^2$
    $4^{-x} = 4^2$
    $-x = 2$
    $x = -2$
    The solution set is $\{-2\}$.

35. $(e^2)^{x-1} = e^x \cdot e^7$
    $e^{2(x-1)} = e^{x+7}$
    $e^{2x-2} = e^{x+7}$
    $2x - 2 = x + 7$
    $x - 2 = 7$
    $x = 9$
    The solution set is $\{9\}$.

36. $(2^x)^x = 512$
    $2^{x^2} = 2^9$
    $x^2 = 9$
    $x = \pm\sqrt{9}$
    $x = \pm 3$
    The solution set is $\{-3, 3\}$.

37. a. We use the compound interest formula with $P = \$2500$, $r = 0.045$, $t = 25$, and $n = 1$, so that $A = 2500\left(1 + \dfrac{0.045}{1}\right)^{1(25)} \approx \$7513.59$.

    b. We use the compound interest formula with $P = \$2500$, $r = 0.045$, $t = 25$, and $n = 4$, so that $A = 2500\left(1 + \dfrac{0.045}{4}\right)^{4(25)} \approx \$7652.33$.

c. We use the compound interest formula with $P = \$2500$, $r = 0.045$, $t = 25$, and $n = 12$, so that $A = 2500\left(1 + \dfrac{0.045}{12}\right)^{12(25)} \approx \$7684.36$.

d. We use the compound interest formula with $P = \$2500$, $r = 0.045$, $t = 25$, and $n = 365$, so that $A = 2500\left(1 + \dfrac{0.045}{365}\right)^{365(25)} \approx \$7700.01$.

38. a. $A(1) = 100\left(\dfrac{1}{2}\right)^{1/3.5} \approx 82.034$.

    After 1 day, approximately 82.034 grams of radon gas will be left in the sample.

    b. $A(3.5) = 100\left(\dfrac{1}{2}\right)^{3.5/3.5} = 50$.

    After 3.5 days, 50 grams of radon gas will be left in the sample.

    c. $A(7) = 100\left(\dfrac{1}{2}\right)^{7/3.5} = 25$.

    After 7 days, 25 grams of radon gas will be left in the sample.

    d. $A(30) = 100\left(\dfrac{1}{2}\right)^{30/3.5} \approx 0.263$.

    After 30 days, approximately 0.263 gram of radon gas will be left in the sample.

39. a. $P(2006) = 1.998(1.052)^{2006-2000} \approx 2.708$.

    According to the model, the population of Nevada in 2006 will be approximately 2.708 million people.

    b. $P(2010) = 1.998(1.052)^{2010-2000} \approx 3.317$.

    According to the model, the population of Nevada in 2010 will be approximately 3.317 million people.

40. a. $u(15) = 72 + 278e^{-0.0835(15)} \approx 151.449$.

    According to the model, the temperature of the cake after 15 minutes will be approximately $151.449°F$.

    b. $u(30) = 72 + 278e^{-0.0835(30)} \approx 94.706$.

    According to the model, the temperature of the cake after 30 minutes will be approximately $94.706°F$.

41. If $3^4 = 81$, then $\log_3 81 = 4$.

42. If $4^{-3} = \dfrac{1}{64}$, then $\log_4\left(\dfrac{1}{64}\right) = -3$.

43. If $b^3 = 5$, then $\log_b 5 = 3$.

44. If $10^{3.74} = x$, then $\log x = 3.74$.

45. If $\log_8 2 = \dfrac{1}{3}$, then $8^{1/3} = 2$.

46. If $\log_5 18 = r$, then $5^r = 18$.

47. If $\ln(x+3) = 2$, then $e^2 = x+3$.

48. If $\log x = -4$, then $10^{-4} = x$.

49. Let $y = \log_8 128$. Then,
$$8^y = 128$$
$$(2^3)^y = 2^7$$
$$2^{3y} = 2^7$$
$$3y = 7$$
$$y = \dfrac{7}{3}$$
Thus, $\log_8 128 = \dfrac{7}{3}$.

50. Let $y = \log_6 1$. Then,
$$6^y = 1$$
$$6^y = 6^0$$
$$y = 0$$
Thus, $\log_6 1 = 0$.

51. Let $y = \log \dfrac{1}{100}$. Then,
$$10^y = \dfrac{1}{100}$$
$$10^y = \dfrac{1}{10^2}$$
$$10^y = 10^{-2}$$
$$y = -2$$
Thus, $\log \dfrac{1}{100} = -2$.

## Chapter 9: Exponential and Logarithmic Functions

**52.** Let $y = \log_9 27$. Then
$$9^y = 27$$
$$(3^2)^y = 3^3$$
$$3^{2y} = 3^3$$
$$2y = 3$$
$$y = \frac{3}{2}$$
Thus, $\log_9 27 = \frac{3}{2}$.

**53.** The domain of $f(x) = \log_2(x+5)$ is the set of all real numbers $x$ such that
$$x + 5 > 0$$
$$x > -5$$
Thus, the domain of $f(x) = \log_2(x+5)$ is $\{x \mid x > -5\}$, or using interval notation, $(-5, \infty)$.

**54.** The domain of $g(x) = \log_8(7-3x)$ is the set of all real numbers $x$ such that
$$7 - 3x > 0$$
$$-3x > -7$$
$$x < \frac{-7}{-3}$$
$$x < \frac{7}{3}$$
Thus, the domain of $g(x) = \log_8(7-3x)$ is $\left\{x \mid x < \frac{7}{3}\right\}$, or using interval notation, $\left(-\infty, \frac{7}{3}\right)$.

**55.** The domain of $h(x) = \ln(3x)$ is the set of all real numbers $x$ such that
$$3x > 0$$
$$x > 0$$
Thus, the domain of $h(x) = \ln(3x)$ is $\{x \mid x > 0\}$, or using interval notation, $(0, \infty)$.

**56.** The domain of $F(x) = \log_{1/3}(4x+10)$ is the set of all real numbers $x$ such that
$$4x + 10 > 0$$
$$4x > -10$$
$$x > \frac{-10}{4}$$
$$x > -\frac{5}{2}$$
Thus, the domain of $F(x) = \log_{1/3}(4x+10)$ is $\left\{x \mid x > -\frac{5}{2}\right\}$, or using interval notation, $\left(-\frac{5}{2}, \infty\right)$.

**57.** Rewrite $y = f(x) = \log_{5/2} x$ as $x = \left(\frac{5}{2}\right)^y$. Locate some points on the graph of $x = \left(\frac{5}{2}\right)^y$.

| $y$ | $x = \left(\frac{5}{2}\right)^y$ | $(x, y)$ |
|---|---|---|
| $-2$ | $x = \left(\frac{5}{2}\right)^{-2} = \left(\frac{2}{5}\right)^2 = \frac{4}{25}$ | $\left(\frac{4}{25}, -2\right)$ |
| $-1$ | $x = \left(\frac{5}{2}\right)^{-1} = \left(\frac{2}{5}\right)^1 = \frac{2}{5}$ | $\left(\frac{2}{5}, -1\right)$ |
| $0$ | $x = \left(\frac{5}{2}\right)^0 = 1$ | $(1, 0)$ |
| $1$ | $x = \left(\frac{5}{2}\right)^1 = \frac{5}{2}$ | $\left(\frac{5}{2}, 1\right)$ |
| $2$ | $x = \left(\frac{5}{2}\right)^2 = \frac{25}{4}$ | $\left(\frac{25}{4}, 2\right)$ |

Plot the points and connect them with a smooth curve.

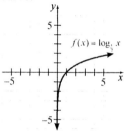

**58.** Rewrite $y = g(x) = \log_{2/5} x$ as $x = \left(\frac{2}{5}\right)^y$. Locate some points on the graph of $x = \left(\frac{2}{5}\right)^y$.

| $y$ | $x = \left(\frac{2}{5}\right)^y$ | $(x, y)$ |
|---|---|---|
| $-2$ | $x = \left(\frac{2}{5}\right)^{-2} = \left(\frac{5}{2}\right)^2 = \frac{25}{4}$ | $\left(\frac{25}{4}, -2\right)$ |
| $-1$ | $x = \left(\frac{2}{5}\right)^{-1} = \left(\frac{5}{2}\right)^1 = \frac{5}{2}$ | $\left(\frac{5}{2}, -1\right)$ |
| $0$ | $x = \left(\frac{2}{5}\right)^0 = 1$ | $(1, 0)$ |
| $1$ | $x = \left(\frac{2}{5}\right)^1 = \frac{2}{5}$ | $\left(\frac{2}{5}, 1\right)$ |
| $2$ | $x = \left(\frac{2}{5}\right)^2 = \frac{4}{25}$ | $\left(\frac{4}{25}, 2\right)$ |

Plot the points and connect them with a smooth curve.

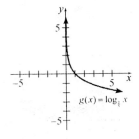

59. $\ln 24 \approx 3.178$

60. $\ln \dfrac{5}{6} \approx -0.182$

61. $\log 257 \approx 2.410$

62. $\log 0.124 \approx -0.907$

63. $\log_7(4x-19) = 2$
    $4x - 19 = 7^2$
    $4x - 19 = 49$
    $4x = 68$
    $x = 17$
    The solution set is $\{17\}$.

64. $\log_{1/3}(x^2 + 8x) = -2$
    $x^2 + 8x = \left(\dfrac{1}{3}\right)^{-2}$
    $x^2 + 8x = 3^2$
    $x^2 + 8x = 9$
    $x^2 + 8x - 9 = 0$
    $(x+9)(x-1) = 0$
    $x + 9 = 0$ or $x - 1 = 0$
    $x = -9$ or $x = 1$
    The solution set is $\{-9, 1\}$.

65. $\log_a \dfrac{4}{9} = -2$
    $a^{-2} = \dfrac{4}{9}$
    $a^2 = \dfrac{9}{4}$
    $a = \pm\sqrt{\dfrac{9}{4}} = \pm\dfrac{3}{2}$
    Since the base of a logarithm must always be positive, we know that $a = -\dfrac{3}{2}$ is extraneous.
    The solution set is $\left\{\dfrac{3}{2}\right\}$.

66. $\ln e^{5x} = 30$
    $5x = 30$
    $x = 6$
    The solution set is $\{6\}$.

67. $\log(6 - 7x) = 3$
    $6 - 7x = 10^3$
    $6 - 7x = 1000$
    $-7x = 994$
    $x = -142$
    The solution set is $\{-142\}$.

68. $\log_b 75 = 2$
    $b^2 = 75$
    $b = \pm\sqrt{75}$
    $b = \pm 5\sqrt{3}$
    Since the base of a logarithm must always be positive, we know that $b = -5\sqrt{3}$ is extraneous.
    The solution set is $\{5\sqrt{3}\}$.

69. We evaluate $L(x) = 10 \log \dfrac{x}{10^{-12}}$ at $x = 10^{-4}$.
    $L(10^{-4}) = 10 \log \dfrac{10^{-4}}{10^{-12}}$
    $= 10 \log 10^{-4-(-12)}$
    $= 10 \log 10^8$
    $= 10(8)$
    $= 80$
    The loudness of the vacuum cleaner is 80 decibels.

70. We solve $M(x) = \log\left(\dfrac{x}{10^{-3}}\right)$ for $M(x) = 8$.
    $M(x) = 8$
    $\log\left(\dfrac{x}{10^{-3}}\right) = 8$
    $\dfrac{x}{10^{-3}} = 10^8$
    $x = 10^8 \cdot 10^{-3}$
    $= 10^{8+(-3)}$
    $= 10^5$
    $= 100{,}000$
    The seismographic reading of the Great New Madrid Earthquake 100 kilometers from the epicenter would have been 100,000 millimeters.

*Chapter 9: Exponential and Logarithmic Functions*

71. $\log_4 4^{21} = 21$

72. $7^{\log_7 9.34} = 9.34$

73. $\log_5 5 = 1$

74. $\log_9 1 = 0$

75. $\log_4 12 - \log_4 3 = \log_4 \dfrac{12}{3} = \log_4 4 = 1$

76. $12^{\log_{12} 2 + \log_{12} 8} = 12^{\log_{12}(2 \cdot 8)} = 12^{\log_{12} 16} = 16$

77. $\log_7 \left(\dfrac{xy}{z}\right) = \log_7(xy) - \log_7 z$
    $= \log_7 x + \log_7 y - \log_7 z$

78. $\log_3 \left(\dfrac{81}{x^2}\right) = \log_3 81 - \log_3 x^2$
    $= \log_3 3^4 - \log_3 x^2$
    $= 4 - 2\log_3 x$

79. $\log 1000 r^4 = \log 1000 + \log r^4$
    $= \log 10^3 + \log r^4$
    $= 3 + 4\log r$

80. $\ln \sqrt{\dfrac{x-1}{x}} = \ln \left(\dfrac{x-1}{x}\right)^{\frac{1}{2}}$
    $= \dfrac{1}{2} \ln \left(\dfrac{x-1}{x}\right)$
    $= \dfrac{1}{2} \left[\ln(x-1) - \ln x\right]$
    $= \dfrac{1}{2} \ln(x-1) - \dfrac{1}{2} \ln x$

81. $4\log_3 x + 2\log_3 y = \log_3 x^4 + \log_3 y^2$
    $= \log_3 (x^4 y^2)$

82. $\dfrac{1}{4} \ln x + \ln 7 - 2\ln 3 = \ln x^{\frac{1}{4}} + \ln 7 - \ln 3^2$
    $= \ln \sqrt[4]{x} + \ln 7 - \ln 9$
    $= \ln \left(7\sqrt[4]{x}\right) - \ln 9$
    $= \ln \left(\dfrac{7\sqrt[4]{x}}{9}\right)$

83. $\log_2 3 - \log_2 6 = \log_2 \dfrac{3}{6} = \log_2 \dfrac{1}{2} = \log_2 2^{-1} = -1$

84. $\log_6 (x^2 - 7x + 12) - \log_6 (x - 3)$
    $= \log_6 \left[(x-4)(x-3)\right] - \log_6 (x-3)$
    $= \log_6 (x-4) + \log_6 (x-3) - \log_6 (x-3)$
    $= \log_6 (x-4)$

85. Using common logarithms:
    $\log_6 50 = \dfrac{\log 50}{\log 6} \approx 2.183$

86. Using common logarithms:
    $\log_\pi 2 = \dfrac{\log 2}{\log \pi} \approx 0.606$

87. Using natural logarithms:
    $\log_{2/3} 6 = \dfrac{\ln 6}{\ln \left(\dfrac{2}{3}\right)} \approx -4.419$

88. Using natural logarithms:
    $\log_{\sqrt{5}} 20 = \dfrac{\ln 20}{\ln \sqrt{5}} \approx 3.723$

89. $3\log_4 x = \log_4 1000$
    $\log_4 x^3 = \log_4 1000$
    $x^3 = 1000$
    $x = \sqrt[3]{1000}$
    $x = 10$
    The solution set is $\{10\}$.

90. $\log_3 x + \log_3 (x+6) = 3$
    $\log_3 [x(x+6)] = 3$
    $\log_3 (x^2 + 6x) = 3$
    $x^2 + 6x = 3^3$
    $x^2 + 6x = 27$
    $x^2 + 6x - 27 = 0$
    $(x+9)(x-3) = 0$
    $x + 9 = 0$ or $x - 3 = 0$
    $x = -9$ or $x = 3$
    The apparent solution $x = -9$ is extraneous because it causes the argument of a logarithm to be negative. The solution set is $\{3\}$.

**91.** $\ln(x+2) - \ln x = \ln(x+1)$
$$\ln\left(\frac{x+2}{x}\right) = \ln(x+1)$$
$$\frac{x+2}{x} = x+1$$
$$x(x+1) = x+2$$
$$x^2 + x = x+2$$
$$x^2 = 2$$
$$x = \pm\sqrt{2} \approx \pm 1.414$$

The apparent solution $x = -\sqrt{2} \approx -1.414$ is extraneous because it causes the argument of a logarithm to be negative. The solution set is $\{\sqrt{2}\} \approx \{1.414\}$.

**92.** $\frac{1}{3}\log_{12} x = 2\log_{12} 2$
$$\log_{12} x^{\frac{1}{3}} = \log_{12} 2^2$$
$$\log_{12} \sqrt[3]{x} = \log_{12} 4$$
$$\sqrt[3]{x} = 4$$
$$x = 4^3 = 64$$

The solution set is $\{64\}$.

**93.** $2^x = 15$
$$\log 2^x = \log 15$$
$$x \log 2 = \log 15$$
$$x = \frac{\log 15}{\log 2} \approx 3.907$$

The solution set is $\left\{\frac{\log 15}{\log 2}\right\} \approx \{3.907\}$. If we had taken the natural logarithm of both sides, the solution set would be $\left\{\frac{\ln 15}{\ln 2}\right\} \approx \{3.907\}$.

**94.** $10^{3x} = 27$
$$\log 10^{3x} = \log 27$$
$$3x = \log 27$$
$$x = \frac{\log 27}{3} \approx 0.477$$

The solution set is $\left\{\frac{\log 27}{3}\right\} \approx \{0.477\}$.

**95.** $\frac{1}{3}e^{7x} = 13$
$$e^{7x} = 39$$
$$\ln e^{7x} = \ln 39$$
$$7x = \ln 39$$
$$x = \frac{\ln 39}{7} \approx 0.523$$

The solution set is $\left\{\frac{\ln 39}{7}\right\} \approx \{0.523\}$.

**96.** $3^x = 2^{x+1}$
$$\log 3^x = \log 2^{x+1}$$
$$x \log 3 = (x+1) \log 2$$
$$x \log 3 = x \log 2 + \log 2$$
$$x \log 3 - x \log 2 = \log 2$$
$$x(\log 3 - \log 2) = \log 2$$
$$x = \frac{\log 2}{\log 3 - \log 2} \approx 1.710$$

The solution set is $\left\{\frac{\log 2}{\log 3 - \log 2}\right\} \approx \{1.710\}$. If we had taken the natural logarithm of both sides, the solution set would be $\left\{\frac{\ln 2}{\ln 3 - \ln 2}\right\} \approx \{1.710\}$.

**97. a.** We need to determine the time until $A = 75$ grams. So we solve the equation
$$75 = 100\left(\frac{1}{2}\right)^{t/3.5}$$
$$0.75 = \left(\frac{1}{2}\right)^{t/3.5}$$
$$\log 0.75 = \log\left(\frac{1}{2}\right)^{t/3.5}$$
$$\log 0.75 = \frac{t}{3.5}\log\left(\frac{1}{2}\right)$$
$$\frac{3.5 \log 0.75}{\log\left(\frac{1}{2}\right)} = t$$

So, $t = \frac{3.5 \log 0.75}{\log\left(\frac{1}{2}\right)} \approx 1.453$ days.

Thus, 75 grams of radon gas will be left after approximately 1.453 days.

**b.** We need to determine the time until $A = 1$ gram. So we solve the equation

$$1 = 100\left(\frac{1}{2}\right)^{t/3.5}$$

$$0.01 = \left(\frac{1}{2}\right)^{t/3.5}$$

$$\log 0.01 = \log\left(\frac{1}{2}\right)^{t/3.5}$$

$$\log 0.01 = \frac{t}{3.5}\log\left(\frac{1}{2}\right)$$

$$\frac{3.5\log 0.01}{\log\left(\frac{1}{2}\right)} = t$$

So, $t = \dfrac{3.5\log 0.01}{\log\left(\frac{1}{2}\right)} \approx 23.253$ days.

Thus, 1 gram of radon gas will be left after approximately 23.253 days.

**98. a.** We need to determine the year when $P = 3.0$ million people. So we solve the equation

$$3.0 = 1.998(1.052)^{t-2000}$$

$$\frac{3.0}{1.998} = (1.052)^{t-2000}$$

$$\log\frac{3.0}{1.998} = \log(1.052)^{t-2000}$$

$$\log\frac{3.0}{1.998} = (t-2000)\log 1.052$$

$$\frac{\log\frac{3.0}{1.998}}{\log 1.052} = t - 2000$$

$$\frac{\log\frac{3.0}{1.998}}{\log 1.052} + 2000 = t$$

$$2008.018 \approx t$$

According to the model, the population of Nevada will be 3.0 million people in about 2008.

**b.** We need to determine the year when $P = 4.5$ million people. So we solve the equation

$$4.5 = 1.998(1.052)^{t-2000}$$

$$\frac{4.5}{1.998} = (1.052)^{t-2000}$$

$$\log\frac{4.5}{1.998} = \log(1.052)^{t-2000}$$

$$\log\frac{4.5}{1.998} = (t-2000)\log 1.052$$

$$\frac{\log\frac{4.5}{1.998}}{\log 1.052} = t - 2000$$

$$\frac{\log\frac{4.5}{1.998}}{\log 1.052} + 2000 = t$$

$$2016.017 \approx t$$

According to the model, the population of Nevada will be 4.5 million people in about 2016.

## Chapter 9 Test

1. The function is not one-to-one. There is an element in the range (4) that corresponds to more than one element in the domain (1 and $-1$).

2. $f(x) = 4x - 3$

   $y = 4x - 3$

   $x = 4y - 3$

   $x + 3 = 4y$

   $\dfrac{x+3}{4} = y$

   $f^{-1}(x) = \dfrac{x+3}{4}$

   Check:

   $f(f^{-1}(x)) = 4\left(\dfrac{x+3}{4}\right) - 3 = x + 3 - 3 = x$

   $f^{-1}(f(x)) = \dfrac{(4x-3)+3}{4} = \dfrac{4x}{4} = x$

3. **a.** $3.1^{3.1} \approx 33.360$

   **b.** $3.14^{3.14} \approx 36.338$

   **c.** $3.142^{3.142} \approx 36.494$

   **d.** $3.1416^{3.1416} \approx 36.463$

   **e.** $\pi^\pi \approx 36.462$

# Chapter 9 Test

4. If $4^x = 19$, then $\log_4 19 = x$.

5. If $\log_b x = y$, then $b^y = x$.

6. a. Let $y = \log_3\left(\dfrac{1}{27}\right)$. Then,
   $$3^y = \dfrac{1}{27}$$
   $$3^y = \dfrac{1}{3^3}$$
   $$3^y = 3^{-3}$$
   $$y = -3$$
   Thus, $\log_3\left(\dfrac{1}{27}\right) = -3$.

   b. Let $y = \log 10{,}000$. Then,
   $$10^y = 10{,}000$$
   $$10^y = 10^4$$
   $$y = 4$$
   Thus, $\log 10{,}000 = 4$.

7. The domain of $f(x) = \log_5(7 - 4x)$ is the set of all real numbers $x$ such that
   $$7 - 4x > 0$$
   $$-4x > -7$$
   $$x < \dfrac{-7}{-4}$$
   $$x < \dfrac{7}{4}$$
   Thus, the domain of $f(x) = \log_5(7 - 4x)$ is $\left\{x \mid x < \dfrac{7}{4}\right\}$ or, using interval notation, $\left(-\infty, \dfrac{7}{4}\right)$.

8. Locate some points on the graph of $f(x) = 6^x$.

   | $x$ | $f(x) = 6^x$ | $(x, f(x))$ |
   |---|---|---|
   | $-2$ | $f(-2) = 6^{-2} = \dfrac{1}{6^2} = \dfrac{1}{36}$ | $\left(-2, \dfrac{1}{36}\right)$ |
   | $-1$ | $f(-1) = 6^{-1} = \dfrac{1}{6^1} = \dfrac{1}{6}$ | $\left(-1, \dfrac{1}{6}\right)$ |
   | $0$ | $f(0) = 6^0 = 1$ | $(0, 1)$ |
   | $1$ | $f(1) = 6^1 = 6$ | $(1, 6)$ |

   Plot the points and connect them with a smooth curve.

   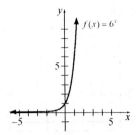

   The domain of $f$ is all real numbers or, using interval notation, $(-\infty, \infty)$. The range of $f$ is $\{y \mid y > 0\}$ or, using interval notation, $(0, \infty)$.

9. Rewrite $y = g(x) = \log_{1/9} x$ as $x = \left(\dfrac{1}{9}\right)^y$. Locate some points on the graph of $x = \left(\dfrac{1}{9}\right)^y$.

   | $y$ | $x = \left(\dfrac{1}{9}\right)^y$ | $(x, y)$ |
   |---|---|---|
   | $-1$ | $x = \left(\dfrac{1}{9}\right)^{-1} = \left(\dfrac{9}{1}\right)^1 = 9$ | $(9, -1)$ |
   | $0$ | $x = \left(\dfrac{1}{9}\right)^0 = 1$ | $(1, 0)$ |
   | $1$ | $x = \left(\dfrac{1}{9}\right)^1 = \dfrac{1}{9}$ | $\left(\dfrac{1}{9}, 1\right)$ |
   | $2$ | $x = \left(\dfrac{1}{9}\right)^2 = \dfrac{1}{81}$ | $\left(\dfrac{1}{81}, 2\right)$ |

   Plot the points and connect them with a smooth curve.

   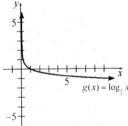

   The domain of $g$ is $\{x \mid x > 0\}$ or, using interval notation, $(0, \infty)$. The range of $g$ is all real numbers or, using interval notation, $(-\infty, \infty)$.

## Chapter 9: Exponential and Logarithmic Functions

**10. a.** $\log_7 7^{10} = 10$

**b.** $3^{\log_3 15} = 15$

**11.** $\log_4 \dfrac{\sqrt{x}}{y^3} = \log_4 \sqrt{x} - \log_4 y^3$
$= \log_4 x^{\frac{1}{2}} - \log_4 y^3$
$= \dfrac{1}{2}\log_4 x - 3\log_4 y$

**12.** $4\log M + 3\log N = \log M^4 + \log N^3$
$= \log(M^4 N^3)$

**13.** Using common logarithms:
$\log_{3/4} 10 = \dfrac{\log 10}{\log \dfrac{3}{4}} \approx -8.004$

**14.** $4^{x+1} = 2^{3x+1}$
$(2^2)^{x+1} = 2^{3x+1}$
$2^{2x+2} = 2^{3x+1}$
$2x+2 = 3x+1$
$-x+2 = 1$
$-x = -1$
$x = 1$
The solution set is $\{1\}$.

**15.** $5^{x^2} \cdot 125 = 25^{2x}$
$5^{x^2} \cdot 5^3 = (5^2)^{2x}$
$5^{x^2+3} = 5^{4x}$
$x^2 + 3 = 4x$
$x^2 - 4x + 3 = 0$
$(x-1)(x-3) = 0$
$x-1 = 0$ or $x-3 = 0$
$x = 1$ or $x = 3$
The solution set is $\{1, 3\}$.

**16.** $\log_a 64 = 3$
$a^3 = 64$
$a = \sqrt[3]{64} = 4$
The solution set is $\{4\}$.

**17.** $\log_2(x^2 - 33) = 8$
$2^8 = x^2 - 33$
$256 = x^2 - 33$
$289 = x^2$
$x = \pm\sqrt{289} = \pm 17$
The solution set is $\{-17, 17\}$.

**18.** $2\log_7(x-3) = \log_7 3 + \log_7 12$
$\log_7(x-3)^2 = \log_7(3 \cdot 12)$
$\log_7(x-3)^2 = \log_7 36$
$(x-3)^2 = 36$
$x - 3 = \pm 6$
$x = 3 \pm 6$
$x = 9$ or $x = -3$
The apparent solution $x = -3$ is extraneous because it causes the argument of a logarithm to be negative. The solution set is $\{9\}$.

**19.** $3^{x-1} = 17$
$\log 3^{x-1} = \log 17$
$(x-1)\log 3 = \log 17$
$x \log 3 - \log 3 = \log 17$
$x \log 3 = \log 17 + \log 3$
$x = \dfrac{\log 17 + \log 3}{\log 3} \approx 3.579$

The solution set is $\left\{\dfrac{\log 17 + \log 3}{\log 3}\right\} \approx \{3.579\}$. If we had taken the natural logarithm of both sides, the solution set would be $\left\{\dfrac{\ln 17 + \ln 3}{\ln 3}\right\} \approx \{3.579\}$.

**20.** $\log(x-2) + \log(x+2) = 2$
$\log[(x-2)(x+2)] = 2$
$\log(x^2 - 4) = 2$
$10^2 = x^2 - 4$
$x^2 - 4 = 100$
$x^2 = 104$
$x = \pm\sqrt{104}$
$x = \pm 2\sqrt{26} \approx \pm 10.198$
The apparent solution $x = -2\sqrt{26} \approx -10.198$ is extraneous because it causes the argument of a logarithm to be negative. The solution set is $\{2\sqrt{26}\} \approx \{10.198\}$.

**21. a.** Evaluate $P(t) = 31.9(1.008)^{t-2002}$ at $t = 2010$.

$P(2010) = 31.9(1.008)^{2010-2002} \approx 34.000$.

According to the model, the population of Canada in 2010 will be about 34 million people.

**b.** We need to determine the year when $P = 50$ million people. So we solve the equation

$$50 = 31.9(1.008)^{t-2002}$$

$$\frac{50}{31.9} = (1.008)^{t-2002}$$

$$\log\frac{50}{31.9} = \log(1.008)^{t-2002}$$

$$\log\frac{50}{31.9} = (t-2002)\log 1.008$$

$$\frac{\log\frac{50}{31.9}}{\log 1.008} = t - 2002$$

$$\frac{\log\frac{50}{31.9}}{\log 1.008} + 2002 = t$$

$$2058.402 \approx t$$

According to the model, the population of Canada will be 50 million people in about 2058.

**22.** We evaluate $L(x) = 10\log\frac{x}{10^{-12}}$ at $x = 10^{-11}$.

$$L(10^{-11}) = 10\log\frac{10^{-11}}{10^{-12}}$$
$$= 10\log 10^{-11-(-12)}$$
$$= 10\log 10^{1}$$
$$= 10(1)$$
$$= 10$$

The loudness of rustling leaves is 10 decibels.

## Cumulative Review Chapters R – 9

**1.** $3(5-2x)+8 = 4(x-7)+1$
$15-6x+8 = 4x-28+1$
$-6x+23 = 4x-27$
$-10x = -50$
$x = 5$
The solution set is $\{5\}$.

**2.** $5-3|x-2| \geq -7$
$-3|x-2| \geq -12$
$|x-2| \leq 4$
$-4 \leq x-2 \leq 4$
$-2 \leq x \leq 6$
The solution set is $\{x \mid -2 \leq x \leq 6\}$ or, using interval notation, $[-2, 6]$.

**3.** We need to find all values of $x$ that cause the denominator $2x^2 - x - 21$ to equal 0.

$2x^2 - x - 21 = 0$
$(2x-7)(x+3) = 0$
$2x - 7 = 0$ or $x + 3 = 0$
$x = \frac{7}{2}$ $\qquad x = -3$

Thus, the domain of $f(x) = \frac{9-x^2}{2x^2-x-21}$ is

$\left\{x \mid x \neq -3 \text{ and } x \neq \frac{7}{2}\right\}$.

**4.** $4x + 3y = 6$
$3y = -4x + 6$
$y = \frac{-4x+6}{3}$
$y = -\frac{4}{3}x + 2$

The slope is $-\frac{4}{3}$ and the $y$-intercept is 2. Begin at the point $(0, 2)$ and move to the right 3 units and down 4 units to find the point $(3, -2)$.

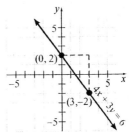

5. The slope of the line we seek is
$$m = \frac{y_2 - y_1}{x_2 - x_1} = \frac{-4-17}{5-(-10)} = \frac{-21}{15} = -\frac{7}{5}.$$
Thus, the equation of the line we seek is:
$$y - y_1 = m(x - x_1)$$
$$y - (-4) = -\frac{7}{5}(x - 5)$$
$$y + 4 = -\frac{7}{5}x + 7$$
$$y = -\frac{7}{5}x + 3 \text{ or } 7x + 5y = 15$$

6. $\begin{cases} x - 2y + z = 4 & (1) \\ y - z = -2 & (2) \\ 2z = -6 & (3) \end{cases}$

Solve equation (3) for $z$:
$$2z = -6$$
$$z = -3$$
Substituting $-3$ for $z$ into equation (2), we obtain
$$y - (-3) = -2$$
$$y + 3 = -2$$
$$y = -5$$
Substituting $-3$ for $z$ and $-5$ for $y$ into equation (1), we obtain
$$x - 2(-5) + (-3) = 4$$
$$x + 10 - 3 = 4$$
$$x + 7 = 4$$
$$x = -3$$
The solution is the ordered triple $(-3, -5, -3)$.

7. $\begin{vmatrix} 4 & 0 & -2 \\ 2 & -1 & 1 \\ 1 & 3 & 1 \end{vmatrix} = 4\begin{vmatrix} -1 & 1 \\ 3 & 1 \end{vmatrix} - 0\begin{vmatrix} 2 & 1 \\ 1 & 1 \end{vmatrix} + (-2)\begin{vmatrix} 2 & -1 \\ 1 & 3 \end{vmatrix}$
$$= 4[-1(1) - 3(1)] - 0[2(1) - 1(1)]$$
$$\qquad + (-2)[2(3) - 1(-1)]$$
$$= 4(-1 - 3) - 0(2 - 1) - 2(6 + 1)$$
$$= 4(-4) - 0(1) - 2(7)$$
$$= -16 - 0 - 14$$
$$= -30$$

8. $\begin{cases} x + 2y \geq 8 \\ 2x - y < 1 \end{cases}$

First, graph the inequality $x + 2y \geq 8$. To do so, replace the inequality symbol with an equal sign to obtain $x + 2y = 8$. Because the inequality is not strict, graph $x + 2y = 8$ $\left(y = -\frac{1}{2}x + 4\right)$ using a solid line.

Test Point: $(0,0)$: $(0) + 2(0) \not\geq 8$

Therefore, the half-plane not containing $(0,0)$ is the solution set of $x + 2y \geq 8$.

Second, graph the inequality $2x - y < 1$. To do so, replace the inequality symbol with an equal sign to obtain $2x - y = 1$. Because the inequality is strict, graph $2x - y = 1$ $(y = 2x - 1)$ using a dashed line.

Test Point: $(0,0)$: $2(0) - (0) < 1$

Therefore, the half-plane containing $(0,0)$ is the solution set of $2x - y < 1$.

The overlapping shaded region (that is, the shaded region in the graph below) is the solution to the system of linear inequalities.

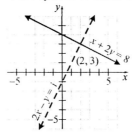

9. $(m^2 - 5m + 13) - (6 - 2m - 3m^2)$
$$= m^2 - 5m + 13 - 6 + 2m + 3m^2$$
$$= 4m^2 - 3m + 7$$

10. $(2n + 3)(n^2 - 4n + 6)$
$$= 2n(n^2) + 2n(-4n) + 2n(6) + 3(n^2)$$
$$\qquad + 3(-4n) + 3(6)$$
$$= 2n^3 - 8n^2 + 12n + 3n^2 - 12n + 18$$
$$= 2n^3 - 5n^2 + 18$$

11. $16a^2 + 8ab + b^2 = (4a)^2 + 2(4a)(b) + b^2$
$$= (4a + b)^2$$

12. $a \cdot c = 6(7) = 42$.

The factors of 42 that add to $-17$ (the linear coefficient) are $-3$ and $-14$. Thus,
$$6y^2 - 17y + 7 = 6y^2 - 3y - 14y + 7$$
$$= 3y(2y-1) - 7(2y-1)$$
$$= (3y-7)(2y-1)$$

13.
$$\frac{2x^2 - 9x - 5}{x^2 - 3x - 10} \cdot \frac{3x^2 + 2x - 8}{2x^2 - 13x - 7}$$
$$= \frac{(2x+1)(x-5)}{(x+2)(x-5)} \cdot \frac{(3x-4)(x+2)}{(2x+1)(x-7)}$$
$$= \frac{\cancel{(2x+1)}\cancel{(x-5)}}{\cancel{(x+2)}\cancel{(x-5)}} \cdot \frac{(3x-4)\cancel{(x+2)}}{\cancel{(2x+1)}(x-7)}$$
$$= \frac{3x-4}{x-7}$$

14.
$$\frac{4}{p^2 - 6p + 5} + \frac{2}{p^2 - 3p - 10}$$
$$= \frac{4}{(p-5)(p-1)} + \frac{2}{(p-5)(p+2)}$$
$$= \frac{4(p+2)}{(p-5)(p-1)(p+2)} + \frac{2(p-1)}{(p-5)(p-1)(p+2)}$$
$$= \frac{4(p+2) + 2(p-1)}{(p-5)(p-1)(p+2)}$$
$$= \frac{4p + 8 + 2p - 2}{(p-5)(p-1)(p+2)}$$
$$= \frac{6p + 6}{(p-5)(p-1)(p+2)} = \frac{6(p+1)}{(p-5)(p-1)(p+2)}$$

15.
$$\frac{2}{x-5} = \frac{x-2}{x+1} + \frac{6x-12}{x^2 - 4x - 5}$$
$$\frac{2}{x-5} = \frac{x-2}{x+1} + \frac{6x-12}{(x-5)(x+1)}$$
$$(x-5)(x+1)\left(\frac{2}{x-5}\right)$$
$$= (x-5)(x+1)\left(\frac{x-2}{x+1} + \frac{6x-12}{(x-5)(x+1)}\right)$$
$$2(x+1) = (x-5)(x-2) + (6x-12)$$
$$2x + 2 = x^2 - 2x - 5x + 10 + 6x - 12$$
$$2x + 2 = x^2 - x - 2$$
$$0 = x^2 - 3x - 4$$
$$0 = (x+1)(x-4)$$
$$x + 1 = 0 \quad \text{or} \quad x - 4 = 0$$
$$x = -1 \quad \text{or} \quad x = 4$$

Since $x = -1$ is not in the domain of the variable, it is an extraneous solution. Thus, the solution set is $\{4\}$.

16. $\sqrt{150} + 4\sqrt{6} - \sqrt{24} = \sqrt{25 \cdot 6} + 4\sqrt{6} - \sqrt{4 \cdot 6}$
$$= 5\sqrt{6} + 4\sqrt{6} - 2\sqrt{6}$$
$$= 7\sqrt{6}$$

17. $\frac{1+\sqrt{5}}{3-\sqrt{5}} = \frac{1+\sqrt{5}}{3-\sqrt{5}} \cdot \frac{3+\sqrt{5}}{3+\sqrt{5}}$
$$= \frac{3 + \sqrt{5} + 3\sqrt{5} + 5}{9 + 3\sqrt{5} - 3\sqrt{5} - 5} = \frac{8 + 4\sqrt{5}}{4} = 2 + \sqrt{5}$$

18. $\sqrt{x-8} + \sqrt{x} = 4$
$$\sqrt{x-8} = 4 - \sqrt{x}$$
$$\left(\sqrt{x-8}\right)^2 = \left(4 - \sqrt{x}\right)^2$$
$$x - 8 = 16 - 8\sqrt{x} + x$$
$$8\sqrt{x} = 24$$
$$\sqrt{x} = 3$$
$$\left(\sqrt{x}\right)^2 = (3)^2$$
$$x = 9$$

Check:
$$\sqrt{9-8} + \sqrt{9} \stackrel{?}{=} 4$$
$$\sqrt{1} + \sqrt{9} \stackrel{?}{=} 4$$
$$1 + 3 \stackrel{?}{=} 4$$
$$4 = 4 \checkmark$$

The solution set is $\{9\}$.

19. $3x^2 = 4x + 6$
$$3x^2 - 4x - 6 = 0$$

Because this equation does not easily factor, solve by using the quadratic formula. For this equation, $a = 3$, $b = -4$, and $c = -6$.
$$x = \frac{-(-4) \pm \sqrt{(-4)^2 - 4(3)(-6)}}{2(3)}$$
$$= \frac{4 \pm \sqrt{88}}{6} = \frac{4 \pm 2\sqrt{22}}{6} = \frac{2 \pm \sqrt{22}}{3}$$

The solution set is $\left\{\frac{2-\sqrt{22}}{3}, \frac{2+\sqrt{22}}{3}\right\}$.

20. $2a - 7\sqrt{a} + 6 = 0$
$$2\left(\sqrt{a}\right)^2 - 7\left(\sqrt{a}\right) + 6 = 0$$

Let $u = \sqrt{a}$.
$$2u^2 - 7u + 6 = 0$$
$$(2u - 3)(u - 2) = 0$$

$2u - 3 = 0$ or $u - 2 = 0$
$u = \dfrac{3}{2}$ or $u = 2$
$\sqrt{a} = \dfrac{3}{2}$ or $\sqrt{a} = 2$
$a = \left(\dfrac{3}{2}\right)^2$ or $a = 2^2$
$a = \dfrac{9}{4}$ or $a = 4$

Check:
$a = \dfrac{9}{4}$: $2 \cdot \dfrac{9}{4} - 7\sqrt{\dfrac{9}{4}} + 6 \stackrel{?}{=} 0$
$\dfrac{9}{2} - 7 \cdot \dfrac{3}{2} + 6 \stackrel{?}{=} 0$
$\dfrac{9}{2} - \dfrac{21}{2} + 6 \stackrel{?}{=} 0$
$0 = 0$ ✓

$a = 4$: $2 \cdot 4 - 7\sqrt{4} + 6 \stackrel{?}{=} 0$
$8 - 7 \cdot 2 + 6 \stackrel{?}{=} 0$
$8 - 14 + 6 \stackrel{?}{=} 0$
$0 = 0$ ✓

Both check; the solution set is $\left\{\dfrac{9}{4}, 4\right\}$.

21. $f(x) = -x^2 + 6x - 4$
$a = -1, b = 6, c = -4$
The graph opens down because the coefficient on $x^2$ is negative.

vertex:
$x = -\dfrac{b}{2a} = -\dfrac{6}{2(-1)} = 3$
$y = f(3) = -(3)^2 + 6(3) - 4 = 5$
The vertex is $(3, 5)$. The axis of symmetry is $x = 3$.

y-intercept:
$f(0) = -(0)^2 + 6(0) - 4 = -4$

x-intercepts:
$b^2 - 4ac = 6^2 - 4(-1)(-4) = 20$
There are two distinct x-intercepts. We find these by solving

$f(x) = 0$
$-x^2 + 6x - 4 = 0$
$x = \dfrac{-6 \pm \sqrt{20}}{2(-1)} = \dfrac{-6 \pm 2\sqrt{5}}{-2} = 3 \pm \sqrt{5}$
$x \approx 0.76$ or $x \approx 5.24$

Graph:
The y-intercept point, $(0, -4)$, is three units to the left of the axis of symmetry. Therefore, if we move three units to the right of the axis of symmetry, we obtain the point $(6, -4)$ which must also be on the graph.

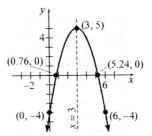

22. $3x^2 + 2x - 8 < 0$
Solve: $3x^2 + 2x - 8 = 0$
$(3x - 4)(x + 2) = 0$
$3x - 4 = 0$ or $x + 2 = 0$
$x = \dfrac{4}{3}$ or $x = -2$

Determine where each factor is positive and negative and where the product of these factors is positive and negative.

| Interval | $(-\infty, -2)$ | $-2$ | $\left(-2, \dfrac{4}{3}\right)$ | $\dfrac{4}{3}$ | $\left(\dfrac{4}{3}, \infty\right)$ |
|---|---|---|---|---|---|
| $3x - 4$ | $----$ | $-$ | $----$ | $0$ | $++++$ |
| $x + 2$ | $----$ | $0$ | $++++$ | $+$ | $++++$ |
| $(3x-4)(x+2)$ | $++++$ | $0$ | $----$ | $0$ | $++++$ |

The inequality is strict, so $-2$ and $\dfrac{4}{3}$ are not part of the solution. Now, $(3x - 4)(x + 2)$ is less than zero where the product is negative. The solution is $\left\{x \mid -2 < x < \dfrac{4}{3}\right\}$ or, using interval notation, $\left(-2, \dfrac{4}{3}\right)$.

23. Locate some points on the graph of $g(x) = 3^x - 4$.

| $x$ | $g(x) = 3^x - 4$ | $(x, g(x))$ |
|---|---|---|
| $-2$ | $g(-2) = 3^{-2} - 4 = \dfrac{1}{9} - 4 = -\dfrac{35}{9}$ | $\left(-2, -\dfrac{35}{9}\right)$ |
| $-1$ | $g(-1) = 3^{-1} - 4 = \dfrac{1}{3} - 4 = -\dfrac{11}{3}$ | $\left(-1, -\dfrac{11}{3}\right)$ |
| $0$ | $g(0) = 3^0 - 4 = 1 - 4 = -3$ | $(0, -3)$ |
| $1$ | $g(1) = 3^1 - 4 = 3 - 4 = -1$ | $(1, -1)$ |
| $2$ | $g(2) = 3^2 - 4 = 9 - 4 = 5$ | $(2, 5)$ |

Plot the points and connect them with a smooth curve.

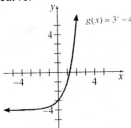

24. Let $y = \log_9\left(\dfrac{1}{27}\right)$. Then,
$$9^y = \dfrac{1}{27}$$
$$(3^2)^y = \dfrac{1}{3^3}$$
$$3^{2y} = 3^{-3}$$
$$2y = -3$$
$$y = -\dfrac{3}{2}$$

Thus, $\log_9\left(\dfrac{1}{27}\right) = -\dfrac{3}{2}$.

# Chapter 10

## 10.1 Preparing for Distance and Midpoint Formulas

1. a. $\sqrt{64} = 8$

   b. $\sqrt{24} = \sqrt{4 \cdot 6} = 2\sqrt{6}$

2. $c^2 = a^2 + b^2$
   $c^2 = 6^2 + 8^2$
   $c^2 = 36 + 64$
   $c^2 = 100$
   $c = \sqrt{100} = 10$
   The length of the hypotenuse is 10.

## 10.1 Exercises

2. $M = \left(\dfrac{x_1 + x_2}{2}, \dfrac{y_1 + y_2}{2}\right)$

4. TRUE.

6. Answers may vary. One possibility follows: The distance formula can be used to verify that the distances from the each endpoint to the midpoint of the segment are equal. In other words, the distance formula can be use to verify that $d(P_1, M) = d(P_2, M)$

8. $d(P_1, P_2) = \sqrt{(x_2 - x_1)^2 + (y_2 - y_1)^2}$
   $= \sqrt{(-2 - 0)^2 + (6 - 0)^2}$
   $= \sqrt{(-2)^2 + 6^2}$
   $= \sqrt{4 + 36}$
   $= \sqrt{40}$
   $= 2\sqrt{10} \approx 6.32$

10. $d(P_1, P_2) = \sqrt{(x_2 - x_1)^2 + (y_2 - y_1)^2}$
    $= \sqrt{(3 - (-3))^2 + (5 - 1)^2}$
    $= \sqrt{6^2 + 4^2}$
    $= \sqrt{36 + 16}$
    $= \sqrt{52}$
    $= 2\sqrt{13} \approx 7.21$

12. $d(P_1, P_2) = \sqrt{(x_2 - x_1)^2 + (y_2 - y_1)^2}$
    $= \sqrt{(4 - 1)^2 + (7 - 3)^2}$
    $= \sqrt{3^2 + 4^2}$
    $= \sqrt{9 + 16}$
    $= \sqrt{25}$
    $= 5$

14. $d(P_1, P_2) = \sqrt{(x_2 - x_1)^2 + (y_2 - y_1)^2}$
    $= \sqrt{(14 - (-10))^2 + (4 - (-3))^2}$
    $= \sqrt{24^2 + 7^2}$
    $= \sqrt{576 + 49}$
    $= \sqrt{625}$
    $= 25$

16. $d(P_1, P_2) = \sqrt{(x_2 - x_1)^2 + (y_2 - y_1)^2}$
    $= \sqrt{(-1 - (-1))^2 + (0 - 2)^2}$
    $= \sqrt{0^2 + (-2)^2}$
    $= \sqrt{0 + 4}$
    $= \sqrt{4}$
    $= 2$

18. $d(P_1, P_2) = \sqrt{(x_2 - x_1)^2 + (y_2 - y_1)^2}$
    $= \sqrt{(-1 - 5)^2 + (-4 - 0)^2}$
    $= \sqrt{(-6)^2 + (-4)^2}$
    $= \sqrt{36 + 16}$
    $= \sqrt{52}$
    $= 2\sqrt{13} \approx 7.21$

20. $d(P_1, P_2) = \sqrt{(x_2 - x_1)^2 + (y_2 - y_1)^2}$
    $= \sqrt{(3\sqrt{6} - \sqrt{6})^2 + (10\sqrt{2} - (-2\sqrt{2}))^2}$
    $= \sqrt{(2\sqrt{6})^2 + (12\sqrt{2})^2}$
    $= \sqrt{4(6) + 144(2)}$
    $= \sqrt{24 + 288}$
    $= \sqrt{312}$
    $= 2\sqrt{78} \approx 17.66$

22. $d(P_1, P_2) = \sqrt{(x_2 - x_1)^2 + (y_2 - y_1)^2}$
$= \sqrt{(0.3 - (-1.7))^2 + (2.6 - 1.3)^2}$
$= \sqrt{2^2 + 1.3^2}$
$= \sqrt{4 + 1.69}$
$= \sqrt{5.69} \approx 2.39$

24. $M = \left(\dfrac{x_1 + x_2}{2}, \dfrac{y_1 + y_2}{2}\right)$
$= \left(\dfrac{1+5}{2}, \dfrac{3+7}{2}\right) = \left(\dfrac{6}{2}, \dfrac{10}{2}\right) = (3, 5)$

26. $M = \left(\dfrac{x_1 + x_2}{2}, \dfrac{y_1 + y_2}{2}\right)$
$= \left(\dfrac{-10+14}{2}, \dfrac{-3+7}{2}\right) = \left(\dfrac{4}{2}, \dfrac{4}{2}\right) = (2, 2)$

28. $M = \left(\dfrac{x_1 + x_2}{2}, \dfrac{y_1 + y_2}{2}\right)$
$= \left(\dfrac{-1+3}{2}, \dfrac{2+9}{2}\right) = \left(\dfrac{2}{2}, \dfrac{11}{2}\right) = \left(1, \dfrac{11}{2}\right)$

30. $M = \left(\dfrac{x_1 + x_2}{2}, \dfrac{y_1 + y_2}{2}\right)$
$= \left(\dfrac{5+(-1)}{2}, \dfrac{0+(-4)}{2}\right) = \left(\dfrac{4}{2}, \dfrac{-4}{2}\right) = (2, -2)$

32. $M = \left(\dfrac{x_1 + x_2}{2}, \dfrac{y_1 + y_2}{2}\right)$
$= \left(\dfrac{\sqrt{6} + 3\sqrt{6}}{2}, \dfrac{-2\sqrt{2} + 10\sqrt{2}}{2}\right)$
$= \left(\dfrac{4\sqrt{6}}{2}, \dfrac{8\sqrt{2}}{2}\right)$
$= (2\sqrt{6}, 4\sqrt{2})$

34. $M = \left(\dfrac{x_1 + x_2}{2}, \dfrac{y_1 + y_2}{2}\right)$
$= \left(\dfrac{-1.7 + 0.3}{2}, \dfrac{1.3 + 2.6}{2}\right)$
$= \left(\dfrac{-1.4}{2}, \dfrac{3.9}{2}\right)$
$= (-0.7, 1.95)$

36. a.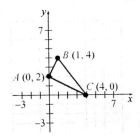

b. $d(A, B) = \sqrt{(1-0)^2 + (4-2)^2}$
$= \sqrt{1^2 + 2^2}$
$= \sqrt{1 + 4}$
$= \sqrt{5} \approx 2.24$

$d(B, C) = \sqrt{(4-1)^2 + (0-4)^2}$
$= \sqrt{3^2 + (-4)^2}$
$= \sqrt{9 + 16}$
$= \sqrt{25}$
$= 5$

$d(A, C) = \sqrt{(4-0)^2 + (0-2)^2}$
$= \sqrt{4^2 + (-2)^2}$
$= \sqrt{16 + 4}$
$= \sqrt{20}$
$= 2\sqrt{5} \approx 4.47$

c. To verify that triangle $ABC$ is a right triangle, we show that
$[d(A,B)]^2 + [d(A,C)]^2 \stackrel{?}{=} [d(A,C)]^2$
$(\sqrt{5})^2 + (2\sqrt{5})^2 \stackrel{?}{=} 5^2$
$5 + 4 \cdot 5 \stackrel{?}{=} 25$
$5 + 20 \stackrel{?}{=} 25$
$25 = 25 \leftarrow$ True
Therefore, triangle $ABC$ is a right triangle.

d. The length of the "base" of the triangle is $d(A, C) = 2\sqrt{5}$ and the length of the "height" of the triangle is $d(A, B) = \sqrt{5}$. Thus, the area of triangle $ABC$ is
Area $= \dfrac{1}{2} \cdot$ base $\cdot$ height $= \dfrac{1}{2} \cdot 2\sqrt{5} \cdot \sqrt{5} = 5$ square units.

## Chapter 10: Conics

**38. a.**

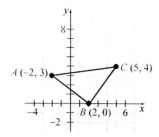

**b.** 
$$d(A,B) = \sqrt{(2-(-2))^2 + (0-3)^2}$$
$$= \sqrt{4^2 + (-3)^2}$$
$$= \sqrt{16+9}$$
$$= \sqrt{25}$$
$$= 5$$

$$d(B,C) = \sqrt{(5-2)^2 + (4-0)^2}$$
$$= \sqrt{3^2 + 4^2}$$
$$= \sqrt{9+16}$$
$$= \sqrt{25}$$
$$= 5$$

$$d(A,C) = \sqrt{(5-(-2))^2 + (4-3)^2}$$
$$= \sqrt{7^2 + 1^2}$$
$$= \sqrt{49+1}$$
$$= \sqrt{50} = 5\sqrt{2} \approx 7.07$$

**c.** To verify that triangle $ABC$ is a right triangle, we show that
$$[d(A,B)]^2 + [d(B,C)]^2 \stackrel{?}{=} [d(A,C)]^2$$
$$5^2 + 5^2 \stackrel{?}{=} (5\sqrt{2})^2$$
$$25+25 \stackrel{?}{=} 25 \cdot 2$$
$$50 = 50 \leftarrow \text{True}$$
Therefore, triangle $ABC$ is a right triangle.

**d.** The length of the "base" of the triangle is $d(B,C) = 5$ and the length of the "height" of the triangle is $d(A,B) = 5$. Thus, the area of triangle $ABC$ is
$$\text{Area} = \frac{1}{2} \cdot \text{base} \cdot \text{height} = \frac{1}{2} \cdot 5 \cdot 5 = \frac{25}{2} \text{ square units.}$$

**40.** We want to find $y$ such that the distance between $P_1 = (0, 3)$ and $P_2 = (4, y)$ is 5.

$$d(P_1, P_2) = \sqrt{(x_2-x_1)^2 + (y_2-y_1)^2}$$
$$5 = \sqrt{(4-0)^2 + (y-3)^2}$$
$$5 = \sqrt{4^2 + (y-3)^2}$$
$$5 = \sqrt{16 + (y-3)^2}$$
$$25 = 16 + (y-3)^2$$
$$9 = (y-3)^2$$
$$y - 3 = \pm\sqrt{9}$$
$$y - 3 = \pm 3$$
$$y = 3 \pm 3$$
$$y = 0 \text{ or } y = 6$$

Thus, the distance between the points $(0,3)$ and $(4,0)$ is 5, and the distance between $(0,3)$ and $(4,6)$ is 5.

**42.** We want to find $x$ such that the distance between $P_1 = (-4, 2)$ and $P_2 = (x, -3)$ is 13.

$$d(P_1, P_2) = \sqrt{(x_2-x_1)^2 + (y_2-y_1)^2}$$
$$13 = \sqrt{(x-(-4))^2 + (-3-2)^2}$$
$$13 = \sqrt{(x+4)^2 + (-5)^2}$$
$$13 = \sqrt{(x+4)^2 + 25}$$
$$169 = (x+4)^2 + 25$$
$$144 = (x+4)^2$$
$$x + 4 = \pm\sqrt{144}$$
$$x + 4 = \pm 12$$
$$x = -4 \pm 12$$
$$x = -16 \text{ or } x = 8$$

Thus, the distance between the points $(-4, 2)$ and $(-16, -3)$ is 13, and the distance between $(-4, 2)$ and $(8, -3)$ is 13.

Madison and State Street to U.S. Cellular Field is approximately 35.13 blocks.

**c.** 
$$d = \sqrt{(-10-(-3))^2 + (36-(-35))^2}$$
$$= \sqrt{(-7)^2 + 71^2}$$
$$= \sqrt{49 + 5041}$$
$$= \sqrt{5090} \approx 71.34$$

The distance "as the crow flies" from Wrigley Field to U.S. Cellular Field is approximately 71.34 blocks.

Section 10.1 Distance and Midpoint Formulas

44. a. Home plate: (0, 0); first base: (90, 0); second base: (90, 90); third base: (0, 90)

    b. The distance from the center fielder to second base is
    $$d = \sqrt{(310-90)^2 + (260-90)^2}$$
    $$= \sqrt{220^2 + 170^2}$$
    $$= \sqrt{48,400 + 28,900}$$
    $$= \sqrt{77,300}$$
    $$= 10\sqrt{773} \approx 278.03 \text{ feet}$$

    c. The distance from the shortstop to second base is
    $$d = \sqrt{(60-90)^2 + (100-90)^2}$$
    $$= \sqrt{(-30)^2 + 10^2}$$
    $$= \sqrt{900 + 100}$$
    $$= \sqrt{1000}$$
    $$= 10\sqrt{10} \approx 31.62 \text{ feet}$$

46. a. The distance between the point $(1, 2)$ and $P = (x, y) = (x, x^2 - 4)$ is
    $$d = \sqrt{(x-1)^2 + (y-2)^2}$$
    $$= \sqrt{(x-1)^2 + [(x^2-4)-2]^2}$$
    $$= \sqrt{(x-1)^2 + (x^2-6)^2}$$
    $$= \sqrt{x^2 - 2x + 1 + x^4 - 12x^2 + 36}$$
    $$= \sqrt{x^4 - 11x^2 - 2x + 37}$$

    b. $x = 0$: $d = \sqrt{0^4 - 11(0)^2 - 2(0) + 37} = \sqrt{37}$

    c. $x = 3$: $d = \sqrt{3^4 - 11(3)^2 - 2(3) + 37}$
    $$= \sqrt{81 - 99 - 6 + 37}$$
    $$= \sqrt{13}$$

48. $8^2 = 8 \cdot 8 = 64$; $\sqrt{64} = 8$

50. $(-3)^3 = (-3) \cdot (-3) \cdot (-3) = -27$; $\sqrt[3]{-27} = -3$

**10.2 Preparing for Circles**

1. Start: $x^2 - 8x$

   Add: $\left(\frac{1}{2} \cdot (-8)\right)^2 = 16$

   Result: $x^2 - 8x + 16$

   Factored Form: $(x-4)^2$

**10.2 Exercises**

2. radius

4. TRUE.

6. No, circles are not functions because they fail the vertical line test.

8. Yes, $3x^2 - 12x + 3y^2 - 15 = 0$ is the equation of a circle. Putting the equation in standard form, we obtain:
   $$\frac{3x^2 - 12x + 3y^2 - 15}{3} = \frac{0}{3}$$
   $$x^2 - 4x + y^2 - 5 = 0$$
   $$(x^2 - 4x) + y^2 = 5$$
   $$(x^2 - 4x + 4) + y^2 = 5 + 4$$
   $$(x - 2)^2 + y^2 = 9$$
   $$(x - 2)^2 + (y - 0)^2 = 3^2$$
   The center is $(2, 0)$, and the radius is 3.

10. The center of the circle is $(1, -1)$. The radius of the circle is the distance from the center point to the point $(1, 2)$ on the circle. Thus,
    $$r = \sqrt{(1-1)^2 + (2-(-1))^2} = 3.$$
    The equation of the circle is
    $$(x - h)^2 + (y - k)^2 = r^2$$
    $$(x - 1)^2 + (y - (-1))^2 = 3^2$$
    $$(x - 1)^2 + (y + 1)^2 = 9$$

12. The center of the circle will be the midpoint of the line segment with endpoints $(1, -1)$ and $(1, 9)$.

ISM – 609

## Chapter 10: Conics

Thus, $(h, k) = \left( \dfrac{1+1}{2}, \dfrac{9+(-1)}{2} \right) = (1, 4)$. The radius of the circle will be the distance from the center point $(1, 4)$ to a point on the circle, say $(1, 9)$. Thus, $r = \sqrt{(1-1)^2 + (9-4)^2} = 5$.

The equation of the circle is
$$(x-h)^2 + (y-k)^2 = r^2$$
$$(x-1)^2 + (y-4)^2 = 5^2$$
$$(x-1)^2 + (y-4)^2 = 25$$

**14.** $(x-h)^2 + (y-k)^2 = r^2$
$(x-0)^2 + (y-0)^2 = 5^2$
$x^2 + y^2 = 25$

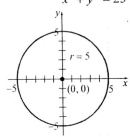

**16.** $(x-h)^2 + (y-k)^2 = r^2$
$(x-3)^2 + (y-1)^2 = 4^2$
$(x-3)^2 + (y-1)^2 = 16$

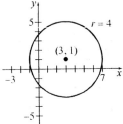

**18.** $(x-h)^2 + (y-k)^2 = r^2$
$(x-1)^2 + (y-(-4))^2 = 3^2$
$(x-1)^2 + (y+4)^2 = 9$

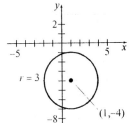

**20.** $(x-h)^2 + (y-k)^2 = r^2$
$(x-1)^2 + (y-0)^2 = 2^2$
$(x-1)^2 + y^2 = 4$

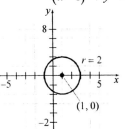

**22.** $(x-h)^2 + (y-k)^2 = r^2$
$(x-(-4))^2 + (y-4)^2 = 4^2$
$(x+4)^2 + (y-4)^2 = 16$

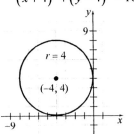

**24.** $(x-h)^2 + (y-k)^2 = r^2$
$(x-5)^2 + (y-2)^2 = \left(\sqrt{7}\right)^2$
$(x-5)^2 + (y-2)^2 = 7$

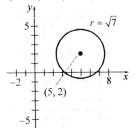

**26.** $x^2 + y^2 = 144$
$(x-0)^2 + (y-0)^2 = 12^2$

The center is $(h, k) = (0, 0)$, and the radius is $r = 12$.

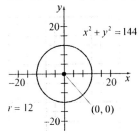

28. $(x-2)^2+(y-3)^2=9$
    $(x-2)^2+(y-3)^2=3^2$
    The center is $(h,k)=(2,3)$, and the radius is $r=3$.

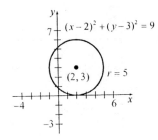

30. $(x-5)^2+(y+2)^2=49$
    $(x-5)^2+(y-(-2))^2=7^2$
    The center is $(h,k)=(5,-2)$, and the radius is $r=7$.

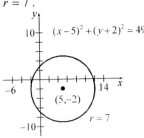

32. $(x-6)^2+y^2=36$
    $(x-6)^2+(y-0)^2=6^2$
    The center is $(h,k)=(6,0)$, and the radius is $r=6$.

34. $(x-2)^2+(y+2)^2=\dfrac{1}{4}$
    $(x-2)^2+(y-(-2))^2=\left(\dfrac{1}{2}\right)^2$
    The center is $(h,k)=(2,-2)$, and the radius is $r=\dfrac{1}{2}$.

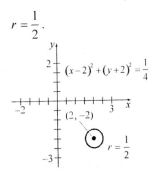

36. $x^2+y^2+2x-8y+8=0$
    $(x^2+2x)+(y^2-8y)=-8$
    $(x^2+2x+1)+(y^2-8y+16)=-8+1+16$
    $(x+1)^2+(y-4)^2=9$
    $(x-(-1))^2+(y-4)^2=3^2$
    The center is $(h,k)=(-1,4)$; nd the radius is $r=3$.

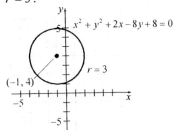

38. $x^2+y^2+4x-12y+36=0$
    $(x^2+4x)+(y^2-12y)=-36$
    $(x^2+4x+4)+(y^2-12y+36)=-36+4+36$
    $(x+2)^2+(y-6)^2=4$
    $(x-(-2))^2+(y-6)^2=2^2$
    The center is $(h,k)=(-2,6)$; the radius is $r=2$.

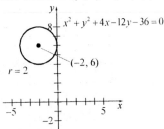

40. $2x^2+2y^2-28x+20y+20=0$
    $\dfrac{2x^2+2y^2-28x+20y+20}{2}=\dfrac{0}{2}$
    $x^2+y^2-14x+10y+10=0$
    $(x^2-14x)+(y^2+10y)=-10$
    $(x^2-14x+49)+(y^2+10y+25)=-10+49+25$
    $(x-7)^2+(y+5)^2=64$
    $(x-7)^2+(y-(-5))^2=8^2$
    The center is $(h,k)=(7,-5)$; the radius is $r=8$.

## Chapter 10: Conics

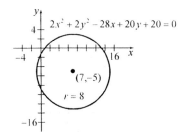

**42.** The radius of the circle will be the distance from the center point $(0,3)$ to the point on the circle $(3,7)$. Thus, $r = \sqrt{(3-0)^2 + (7-3)^2} = \sqrt{25} = 5$.
The equation of the circle is
$(x-h)^2 + (y-k)^2 = r^2$
$(x-0)^2 + (y-3)^2 = 5^2$
$x^2 + (y-3)^2 = 25$

**44.** Since the center of the circle is $(2,-3)$ and since the circle is tangent to the $x$-axis, the circle must contain the point $(2,0)$. The radius of the circle will be the distance from the center point $(2,-3)$ to the point on the circle $(2,0)$. Thus,
$r = \sqrt{(2-2)^2 + (0-(-3))^2} = 3$.
The equation of the circle is
$(x-h)^2 + (y-k)^2 = r^2$
$(x-2)^2 + (y-(-3))^2 = 3^2$
$(x-2)^2 + (y+3)^2 = 9$

**46.** The center of the circle will be the midpoint of the diameter with endpoints $(-5,-3)$ and $(7,2)$.
Thus, $(h,k) = \left(\dfrac{-5+7}{2}, \dfrac{-3+2}{2}\right) = \left(1, -\dfrac{1}{2}\right)$. The radius of the circle will be the distance from the center point $\left(1, -\dfrac{1}{2}\right)$ to one of the endpoints of the diameter, say $(7,2)$. Thus,
$r = \sqrt{(7-1)^2 + \left(2-\left(-\dfrac{1}{2}\right)\right)^2} = \sqrt{\dfrac{169}{4}} = \dfrac{13}{2}$.
The equation of the circle is
$(x-h)^2 + (y-k)^2 = r^2$
$(x-1)^2 + \left(y-\left(-\dfrac{1}{2}\right)\right)^2 = \left(\dfrac{13}{2}\right)^2$
$(x-1)^2 + \left(y+\dfrac{1}{2}\right)^2 = \dfrac{169}{4}$

**48.** The radius of the circle $(x-1)^2 + (y-4)^2 = 49$ is $r = \sqrt{49} = 7$. Thus, the area of the circle is
$A = \pi r^2 = \pi (7)^2 = 49\pi$ square units.
The circumference of the circle is
$C = 2\pi r = 2\pi(7) = 14\pi$ units.

**50.** To find the area of the shaded region, we must subtract the area of the square from the area of the circle. Note that the circle has radius $r = \sqrt{25} = 5$. Thus, the area of the circle is
$A_{\text{circle}} = \pi(5)^2 = 25\pi$ square units.

To find the area of the square we must first find the length of the square's sides. To do so, we recognize that the $x$- and $y$-coordinates of the corner point in the first quadrant where the square touches the circle must be equal. That is, $x = y$.

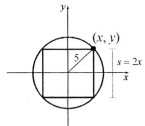

Thus, $x^2 + y^2 = 25$
$x^2 + x^2 = 25$
$2x^2 = 25$
$x^2 = \dfrac{25}{2}$
$x = \sqrt{\dfrac{25}{2}} = \dfrac{5\sqrt{2}}{2}$

This means the length of each side of the square is $s = 2x = 2\left(\dfrac{5\sqrt{2}}{2}\right) = 5\sqrt{2}$, and the area of the square is $A_{\text{square}} = s^2 = (5\sqrt{2})^2 = 50$ square units.

Finally, the area of the shaded region is:
$A_{\text{shaded region}} = A_{\text{circle}} - A_{\text{square}}$
$= 25\pi - 50 \approx 28.54$ square units

**52.** The circle shown lies completely in the first quadrant. This means the center point will be of the form $(h,k)$ with $h > r > 0$ and $k > r > 0$.

**a.** $(x-2)^2 + (y+3)^2 = 1$ has center $(2,-3)$, which does not lie in the first quadrant.

Thus, this equation could not have the graph shown.

b. $(x-3)^2 + (y-4)^2 = 4$ has center $(3,4)$ and radius $r = 2$, which meets the proper conditions. Thus, this equation could have the graph shown.

c. $(x+3)^2 + (y+4)^2 = 9$ has center $(-3,-4)$, which does not lie in the first quadrant. Thus, this equation could not have the graph shown.

d. $(x-5)^2 + (y-5)^2 = 25$ has center $(5,5)$ and radius $r = 5$. This means the graph of this circle would be tangent to both the $x$- and $y$-axis. Thus, this equation could not have the graph shown.

e. $x^2 + y^2 + 8x + 10y + 32 = 0$
$(x^2 + 8x) + (y^2 + 10y) = -32$
$(x^2 + 8x + 16) + (y^2 + 10y + 25) = -32 + 16 + 25$
$(x+4)^2 + (y+5)^2 = 9$

This circle has center $(-4,-5)$, which does not lie in the first quadrant. Thus, this equation could not have the graph shown

f. $x^2 + y^2 - 4x - 6y - 3 = 0$
$(x^2 - 4x) + (y^2 - 6y) = 3$
$(x^2 - 4x + 4) + (y^2 - 6y + 9) = 3 + 4 + 9$
$(x-2)^2 + (y-3)^2 = 16$

This circle has center $(2,3)$ and the radius is $r = 4$. Since $r > h$ and $r > k$, the circle does not lie entirely within the first quadrant. Thus, this equation could not have the graph shown.

54. $2x + 5y = 20$
To use point plotting, let $x = -5, 0,$ and $5$.
$x = -5$: $2(-5) + 5y = 20$
$-10 + 5y = 20$
$5y = 30$
$y = 6$
$x = 0$: $2(0) + 5y = 20$
$5y = 20$
$y = 4$

$x = 5$: $2(5) + 5y = 20$
$10 + 5y = 20$
$5y = 10$
$y = 2$

Thus, the points $(-5, 6)$, $(0, 4)$, and $(5, 2)$ are on the graph.

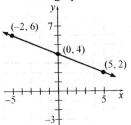

Using properties of linear functions, we put the equation in slope-intercept form:
$2x + 5y = 20$
$5y = -2x + 20$
$y = \dfrac{-2x + 20}{5} = -\dfrac{2}{5}x + 4$

The slope is $-\dfrac{2}{5}$ and the $y$-intercept is 4. Begin at $(0, 4)$ and move to the right 5 units and down 2 units to find point $(5, 2)$.

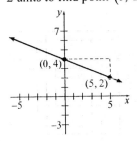

56. $F(x) = -3x^2 + 12x - 12$
To use point plotting, let $x = 0, 1, 2, 3,$ and $4$.
$F(0) = -3(0)^2 + 12(0) - 12 = 0 + 0 - 12 = -12$
$F(1) = -3(1)^2 + 12(1) - 12 = -3 + 12 - 12 = -3$
$F(2) = -3(2)^2 + 12(2) - 12 = -12 + 24 - 12 = 0$
$F(3) = -3(3)^2 + 12(3) - 12 = -27 + 36 - 12 = -3$
$F(4) = -3(4)^2 + 12(4) - 12 = -48 + 48 - 12 = -12$

Thus, the points $(0,-12)$, $(1,-3)$, $(2, 0)$, $(3,-3)$, and $(4,-12)$ are on the graph.

## Chapter 10: Conics

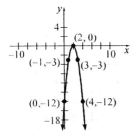

Using properties of quadratic functions, for $F(x) = -3x^2 + 12x - 12$, $a = -3$, $b = 12$, and $c = -12$. The graph opens down because the coefficient on $x^2$ is negative.

vertex:
$$x = -\frac{b}{2a} = -\frac{12}{2(-3)} = 2$$
$$F(2) = -3(2)^2 + 12(2) - 12 = 0$$

The vertex is $(2, 0)$ and the axis of symmetry is $x = 2$.

$y$-intercept:
$$F(0) = -3(0)^2 + 12(0) - 12 = -12$$

$x$-intercepts:
$$b^2 - 4ac = 12^2 - 4(-3)(-12) = 0$$

There one $x$-intercept. We find these by solving
$$F(x) = 0$$
$$-3x^2 + 12x - 12 = 0$$
$$-3(x-2)(x-2) = 0$$
$$x - 2 = 0 \text{ or } x - 2 = 0$$
$$x = 2 \text{ or } \quad x = 2$$

Graph:
The $y$-intercept point, $(0, -12)$, is two units to the left of the axis of symmetry. Therefore, if we move two units to the right of the axis of symmetry, we obtain the point $(4, -12)$ which must also be on the graph.

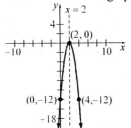

**58.** Answers will vary.

**60.** $x^2 + y^2 = 144$
$$y^2 = 144 - x^2$$
$$y = \pm\sqrt{144 - x^2}$$
Let $Y_1 = \sqrt{144 - x^2}$ and $Y_2 = -\sqrt{144 - x^2}$.

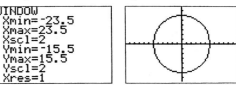

The graph here agrees with that in Problem 26.

**62.** $(x-2)^2 + (y-3)^2 = 9$
$$(y-3)^2 = 9 - (x-2)^2$$
$$y - 3 = \pm\sqrt{9 - (x-2)^2}$$
$$y = 3 \pm \sqrt{9 - (x-2)^2}$$
Let $Y_1 = 3 + \sqrt{9 - (x-2)^2}$ and
$Y_2 = 3 - \sqrt{9 - (x-2)^2}$.

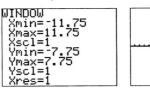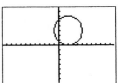

The graph here agrees with that in Problem 28.

**64.** $(x-5)^2 + (y+2)^2 = 49$
$$(y+2)^2 = 49 - (x-5)^2$$
$$y + 2 = \pm\sqrt{49 - (x-5)^2}$$
$$y = -2 \pm \sqrt{49 - (x-5)^2}$$
Let $Y_1 = -2 + \sqrt{49 - (x-5)^2}$ and
$Y_2 = -2 - \sqrt{49 - (x-5)^2}$.

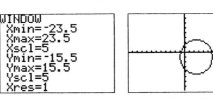

The graph here agrees with that in Problem 30.

Section 10.2 Circles

66. $(x-6)^2 + y^2 = 36$

$$y^2 = 36 - (x-6)^2$$
$$y = \pm\sqrt{36-(x-6)^2}$$

Let $Y_1 = \sqrt{36-(x-6)^2}$ and $Y_2 = -\sqrt{36-(x-6)^2}$.

 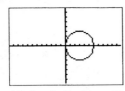

The graph here agrees with that in Problem 32.

68. $(x-2)^2 + (y+2)^2 = \dfrac{1}{4}$

$$(y+2)^2 = \frac{1}{4} - (x-2)^2$$
$$y+2 = \pm\sqrt{\frac{1}{4} - (x-2)^2}$$
$$y = -2 \pm \sqrt{\frac{1}{4} - (x-2)^2}$$

Let $Y_1 = -2 + \sqrt{\dfrac{1}{4} - (x-2)^2}$ and $Y_2 = -2 - \sqrt{\dfrac{1}{4} - (x-2)^2}$.

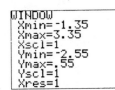 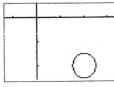

The graph here agrees with that in Problem 34.

## 10.3 Preparing for Parabolas

1. For the function $f(x) = -3(x+4)^2 - 5$, we see that $a = -3$, $h = -4$, and $k = -5$. Thus, the vertex is $(h, k) = (-4, -5)$, and the axis of symmetry is $x = -4$. The parabola opens down since $a = -3 < 0$.

2. For the function $f(x) = 2x^2 - 8x + 1$, we see that $a = 2$, $b = -8$, and $c = 1$. The x-coordinate of the vertex is $x = -\dfrac{b}{2a} = -\dfrac{(-8)}{2(2)} = 2$. The y-coordinate of the vertex is

   $f\left(-\dfrac{b}{2a}\right) = f(2)$
   $= 2(2)^2 - 8(2) + 1$
   $= 8 - 16 + 1$
   $= -7$

   Thus, the vertex is $(2, -7)$ and the axis of symmetry is the line $x = 2$. The parabola opens up because $a = 2 > 0$.

3. Start: $x^2 - 12x$

   Add: $\left(\dfrac{1}{2} \cdot (-12)\right)^2 = 36$

   Result: $x^2 - 12x + 36$

   Factored Form: $(x-6)^2$

4. $(x-3)^2 = 25$
   $x - 3 = \pm\sqrt{25}$
   $x - 3 = \pm 5$
   $x = 3 \pm 5$
   $x = 3 - 5$ or $x = 3 + 5$
   $x = -2$ or $x = 8$

   The solution set is $\{-2, 8\}$.

## 10.3 Exercises

2. parabola

4. TRUE.

6. TRUE.

8. $(y-k)^2 = 4a(x-h)$
   $(y-k)^2 = -4a(x-h)$
   $(x-h)^2 = 4a(y-k)$
   $(x-h)^2 = -4a(y-k)$

10. Answers may vary. One possibility follows: The discussion in this section takes a geometric approach while the discussion ins Sections 8.4 and 8.5 relied more on algebra.

12. h. The parabola opens downward and has vertex $(-1, 2)$ and focus $(-1, 0)$, so the equation is of the form $(x-h)^2 = -4a(y-k)$, with

## Chapter 10: Conics

$a = 2 - 0 = 2$, $h = -1$, and $k = 2$:
$$(x-(-1))^2 = -4 \cdot 2(y-2)$$
$$(x+1)^2 = -8(y-2)$$

14. f. The parabola opens to the left and has vertex $(-1, 2)$ and focus $(-3, 2)$, so the equation is of the form $(y-k)^2 = -4a(x-h)$, with $a = -1-(-3) = 2$, $h = -1$, and $k = 2$:
$$(y-2)^2 = -4 \cdot 2(x-(-1))$$
$$(y-2)^2 = -8(x+1)$$

16. d. The parabola opens to downwars and has vertex $(0, 0)$ and focus $(0, -2)$, so the equation is of the form $x^2 = -4ay$ with $a = 2$:
$$x^2 = -4(2)y$$
$$x^2 = -8y$$

18. g. The parabola opens upward and has vertex $(-1, 2)$ and focus $(-1, 4)$, so the equation is of the form $(x-h)^2 = 4a(y-k)$, with $a = 4 - 2 = 2$, $h = -1$, and $k = 2$:
$$(x-(-1))^2 = 4 \cdot 2(y-2)$$
$$(x+1)^2 = 8(y-2)$$

20. The distance from the vertex $(0, 0)$ to the focus $(0, 5)$ is $a = 5$. Because the focus lies on the positive $y$-axis, we know that the parabola will open upward and the axis of symmetry is the $y$-axis. This means the equation of the parabola is of the form $x^2 = 4ay$ with $a = 5$:
$$x^2 = 4(5)y$$
$$x^2 = 20y$$
The directrix is the line $y = -5$. To help graph the parabola, we plot the two points on the graph above and below the focus. Let $y = 5$:
$$x^2 = 20(5) = 100$$
$$x = \pm 10$$
The points $(-10, 5)$ and $(10, 5)$ are on the graph.

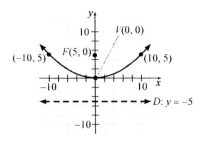

22. The distance from the vertex $(0, 0)$ to the focus $(-8, 0)$ is $a = 8$. Because the focus lies on the negative $x$-axis, we know that the parabola will open to the left and the axis of symmetry is the $x$-axis. This means the equation of the parabola is of the form $y^2 = -4ax$ with $a = 8$:
$$y^2 = -4(8)x$$
$$y^2 = -32x$$
The directrix is the line $x = 8$. To help graph the parabola, we plot the two points on the graph above and below the focus. Let $x = -8$:
$$y^2 = -32(-8) = 256$$
$$y = \pm 16$$
The points $(-8, -16)$ and $(-8, 16)$ are on the graph.

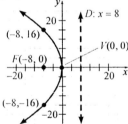

23. The vertex is at the origin and the axis of

24. The vertex is at the origin and the axis of symmetry is the $x$-axis, so the parabola either opens to the left or right. Because the graph contains the point $(2, 2)$, which is in quadrant I, the parabola must open to the right. Therefore, the equation of the parabola is of the form $y^2 = 4ax$. Now $y = 2$ when $x = 2$, so
$$y^2 = 4ax$$
$$2^2 = 4a(2)$$
$$4 = 8a$$
$$a = \frac{4}{8} = \frac{1}{2}$$

Section 10.3 Parabolas

The equation of the parabola is
$$y^2 = 4\left(\frac{1}{2}\right)x$$
$$y^2 = 2x$$

With $a = \frac{1}{2}$, we know that the focus is $\left(\frac{1}{2}, 0\right)$ and the directrix is the line $x = -\frac{1}{2}$. To help graph the parabola, we plot the two points above and below the focus. Let $x = \frac{1}{2}$:

$$y^2 = 2\left(\frac{1}{2}\right)$$
$$y^2 = 1$$
$$y = \pm 1$$

The points $\left(\frac{1}{2}, -1\right)$ and $\left(\frac{1}{2}, 1\right)$ are on the graph.

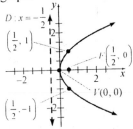

26. The vertex is at the origin and the directrix is the line $x = -4$, so the focus must be $(4, 0)$. Accordingly, the parabola opens to the right with $a = 4$. This means the equation of the parabola is of the form $y^2 = 4ax$ with $a = 4$:
$$y^2 = 4(4)x$$
$$y^2 = 16x$$
To help graph the parabola, we plot the two points on the graph above and below the focus. Let $x = 4$:
$$y^2 = 16(4) = 64$$
$$y = \pm 8$$
The points $(4, -8)$ and $(4, 8)$ are on the graph.

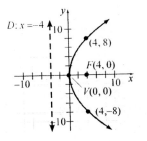

28. Notice that the directrix is the horizontal line $y = 2$ and that the focus $(0, -2)$ is on the $y$-axis. Thus, the axis of symmetry is the $y$-axis, the vertex is at the origin, and the parabola opens downward. Now, the distance from the vertex $(0, 0)$ to the focus $(0, -2)$ is $a = 2$, so the equation of the parabola is of the form $x^2 = -4ay$ with $a = 2$:
$$x^2 = -4(2)y$$
$$x^2 = -8y$$
To help graph the parabola, we plot the two points on the graph to the left and right of the focus. Let $y = -2$:
$$x^2 = -8(-2) = 16$$
$$x = \pm 4$$
The points $(-4, -2)$ and $(4, -2)$ are on the graph.

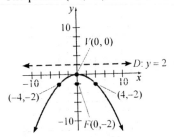

30. Notice that $x^2 = 28y$ is of the form $x^2 = 4ay$, where $4a = 28$, so that $a = 7$. Now, the graph of an equation of the form $x^2 = 4ay$ will be a parabola that opens upward with the vertex at the origin, focus at $(0, a)$, and directrix of $y = -a$. Thus, the graph of $x^2 = 28y$ is a parabola that opens upward with vertex $(0, 0)$, focus $(0, 7)$, and directrix $y = -7$. To help graph the parabola, we plot the two points on the graph to the left and to the right of the focus. Let $y = 7$:
$$x^2 = 28(7)$$
$$x^2 = 196$$
$$x = \pm 14$$

The points $(-14, 7)$ and $(14, 7)$ are on the graph.

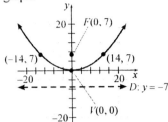

32. Notice that $y^2 = 10x$ is of the form $y^2 = 4ax$, where $4a = 10$, so that $a = \frac{10}{4} = \frac{5}{2}$. Now, the graph of an equation of the form $y^2 = 4ax$ will be a parabola that opens to the right with the vertex at the origin, focus at $(a, 0)$, and directrix of $x = -a$. Thus, the graph of $y^2 = 10x$ is a parabola that opens to the right with vertex $(0,0)$, focus $\left(\frac{5}{2}, 0\right)$, and directrix $x = -\frac{5}{2}$. To help graph the parabola, we plot the two points on the graph above and below the focus. Let $x = \frac{5}{2}$:

$$y^2 = 10\left(\frac{5}{2}\right)$$
$$y^2 = 25$$
$$y = \pm 5$$

The points $\left(\frac{5}{2}, -5\right)$ and $\left(\frac{5}{2}, 5\right)$ are on the graph.

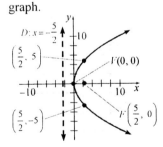

33

34. Notice that $x^2 = -16y$ is of the form $x^2 = -4ay$, where $-4a = -16$, so that $a = 4$. Now, the graph of an equation of the form $x^2 = -4ay$ will be a parabola that opens downward with the vertex at the origin, focus at $(0, -a)$, and directrix of $y = a$. Thus, the graph of $x^2 = -16y$ is a parabola that opens downward with vertex $(0,0)$, focus $(0,-4)$, and directrix $y = 4$. To help graph the parabola, we plot the two points on the graph to the left and to the right of the focus. Let $y = -4$:

$$x^2 = -16(-4)$$
$$x^2 = 64$$
$$x = \pm 8$$

The points $(-8, -4)$ and $(8, -4)$ are on the graph.

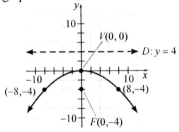

36. Notice that $(x+4)^2 = -4(y-1)$ is of the form $(x-h)^2 = -4a(y-k)$, where $-4a = -4$, so that $a = 1$ and $(h,k) = (-4, 1)$. Now, the graph of an equation of the form $(x-h)^2 = -4a(y-k)$ will be a parabola that opens downward with vertex at $(h,k)$, focus at $(h, k-a)$, and directrix of $y = k+a$. Note that $k-a = 1-1 = 0$ and $k+a = 1+1 = 2$. Thus, the graph of $(x+4)^2 = -4(y-1)$ is a parabola that opens downward with vertex $(-4, 1)$, focus $(-4, 0)$, and directrix $y = 2$. To help graph the parabola, we plot the two points to the left and to the right of the focus. Let $y = 0$:

$$(x+4)^2 = -4(0-1)$$
$$(x+4)^2 = -4(-1)$$
$$(x+4)^2 = 4$$
$$x+4 = \pm 2$$
$$x = -4 \pm 2$$
$$x = -6 \text{ or } x = -2$$

The points $(-6, 0)$ and $(-2, 0)$ are on the

graph.

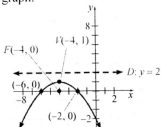

38. Notice that $(y-2)^2 = 12(x+5)$ is of the form $(y-k)^2 = 4a(x-h)$, where $(h,k) = (-5,2)$ and $4a = 12$, so that $a = 3$. Now, the graph of an equation of the form $(y-k)^2 = 4a(x-h)$ will be a parabola that opens to the right with the vertex at $(h,k)$, focus at $(h+a,k)$, and directrix of $x = h-a$. Note that $h+a = -5+3 = -2$ and $h-a = -5-3 = -8$. Thus, the graph of $(y-2)^2 = 12(x+5)$ is a parabola that opens to the right with vertex $(-5, 2)$, focus $(-2, 2)$, and directrix $x = -8$. To help graph the parabola, we plot the two points on the graph above and below the focus. Let $x = -2$:

$$(y-2)^2 = 12(-2+5)$$
$$(y-2)^2 = 12(3)$$
$$(y-2)^2 = 36$$
$$y-2 = \pm 6$$
$$y = 2 \pm 6$$
$$y = -4 \text{ or } y = 8$$

The points $(-2,-4)$ and $(-2, 8)$ are on the graph.

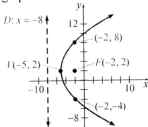

40. Notice that $(x-6)^2 = 2(y-2)$ is of the form $(x-h)^2 = 4a(y-k)$, where $4a = 2$, so that $a = \frac{2}{4} = \frac{1}{2}$ and $(h,k) = (6, 2)$. Now, the graph of an equation of the form $(x-h)^2 = 4a(y-k)$ will be a parabola that opens upward with vertex at $(h,k)$, focus at $(h,k+a)$, and directrix of $y = k-a$. Note that $k+a = 2+\frac{1}{2} = \frac{5}{2}$ and $k-a = 2-\frac{1}{2} = \frac{3}{2}$. Thus, the graph of $(x-6)^2 = 2(y-2)$ is a parabola that opens upward with vertex $(6, 2)$, focus $\left(6, \frac{5}{2}\right)$, and directrix $y = \frac{3}{2}$. To help graph the parabola, we plot the two points to the left and to the right of the focus. Let $y = \frac{5}{2}$:

$$(x-6)^2 = 2\left(\frac{5}{2}-2\right)$$
$$(x-6)^2 = 2\left(\frac{1}{2}\right)$$
$$(x-6)^2 = 1$$
$$x-6 = \pm 1$$
$$x = 6 \pm 1$$
$$x = 5 \text{ or } x = 7$$

The points $\left(5, \frac{5}{2}\right)$ and $\left(7, \frac{5}{2}\right)$ are on the graph.

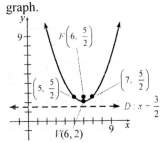

42. We complete the square in $x$ to write the equation in standard form:
$$x^2 + 2x - 8y + 25 = 0$$
$$x^2 + 2x = 8y - 25$$
$$x^2 + 2x + 1 = 8y - 25 + 1$$
$$x^2 + 2x + 1 = 8y - 24$$
$$(x+1)^2 = 8(y-3)$$

Notice that $(x+1)^2 = 8(y-3)$ is of the form $(x-h)^2 = 4a(y-k)$, where $4a = 8$, so that $a = 2$ and $(h,k) = (-1, 3)$. Now, the graph of an equation of the form $(x-h)^2 = 4a(y-k)$ will be a parabola that opens upward with vertex at $(h,k)$, focus at $(h,k+a)$, and directrix of $y = k-a$. Note that $k+a = 3+2 = 5$ and $k-a = 3-2 = 1$. Thus, the graph of

## Chapter 10: Conics

$(x+1)^2 = 8(y-3)$ is a parabola that opens upward with vertex $(-1, 3)$, focus $(-1, 5)$, and directrix $y = 1$. To help graph the parabola, we plot the two points to the left and to the right of the focus. Let $y = 5$:
$$(x+1)^2 = 8(5-3)$$
$$(x+1)^2 = 8(2)$$
$$(x+1)^2 = 16$$
$$x+1 = \pm 4$$
$$x = -1 \pm 4$$
$$x = -5 \text{ or } x = 3$$
The points $(-5, 5)$ and $(3, 5)$ are on the graph.

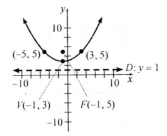

44. We complete the square in $y$ to write the equation in standard form:
$$y^2 - 8y + 16x - 16 = 0$$
$$y^2 - 8y = -16x + 16$$
$$y^2 - 8y + 16 = -16x + 16 + 16$$
$$y^2 - 8y + 16 = -16x + 32$$
$$(y-4)^2 = -16(x-2)$$
Notice that $(y-4)^2 = -16(x-2)$ is of the form $(y-k)^2 = -4a(x-h)$, where $(h,k) = (2,4)$ and $-4a = -16$, so that $a = 4$. Now, the graph of an equation of the form $(y-k)^2 = -4a(x-h)$ will be a parabola that opens to the left with the vertex at $(h,k)$, focus at $(h-a,k)$, and directrix of $x = h + a$. Note that $h - a = 2 - 4 = -2$ and $h + a = 2 + 4 = 6$. Thus, the graph of $(y-4)^2 = -16(x-2)$ is a parabola that opens to the left with vertex $(2,4)$, focus $(-2,4)$, and directrix $x = 6$. To help graph the parabola, we plot the two points on the graph above and below the focus. Let $x = -2$:
$$(y-4)^2 = -16(-2-2)$$
$$(y-4)^2 = -16(-4)$$
$$(y-4)^2 = 64$$
$$y - 4 = \pm 8$$
$$y = 4 \pm 8$$

$y = -4$ or $y = 12$
The points $(-2,-4)$ and $(-2, 12)$ are on the graph.

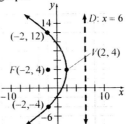

46. We complete the square in $x$ to write the equation in standard form:
$$x^2 - 4x + 10y + 4 = 0$$
$$x^2 - 4x = -10y - 4$$
$$x^2 - 4x + 4 = -10y - 4 + 4$$
$$x^2 - 4x + 4 = -10y$$
$$(x-2)^2 = -10y$$
Notice that $(x-2)^2 = -10y$ is of the form $(x-h)^2 = -4a(y-k)$, where $(h,k) = (2,0)$ and $-4a = -10$, so that $a = \dfrac{-10}{-4} = \dfrac{5}{2}$. Now, the graph of an equation of the form $(x-h)^2 = -4a(y-k)$ will be a parabola that opens downward with the vertex at $(h,k)$, focus at $(h, k-a)$, and directrix of $y = k + a$. Note that $k - a = 0 - \dfrac{5}{2} = -\dfrac{5}{2}$ and $k + a = 0 + \dfrac{5}{2} = \dfrac{5}{2}$.
Thus, the graph of $(x-2)^2 = -10y$ is a parabola that opens downward with vertex $(2,0)$, focus $\left(0, -\dfrac{5}{2}\right)$, and directrix $y = \dfrac{5}{2}$. To help graph the parabola, we plot the two points on the graph to the left and right of the focus. Let $y = -\dfrac{5}{2}$:
$$(x-2)^2 = -10\left(-\dfrac{5}{2}\right)$$
$$(x-2)^2 = 25$$
$$x - 2 = \pm 5$$
$$x = 2 \pm 5$$
$$x = -3 \text{ or } x = 7$$
The points $\left(-3, -\dfrac{5}{2}\right)$ and $\left(7, -\dfrac{5}{2}\right)$ are on the

graph.

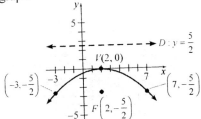

48. The parabola opens to the left and has vertex $(0, 0)$, so the equation must have the form $y^2 = -4ax$. Because the point $(-6, 2)$ is on the parabola, we let $x = -6$ and $y = 2$ to determine $a$:
$$2^2 = -4a(-6)$$
$$4 = 24a$$
$$a = \frac{4}{24} = \frac{1}{6}$$

Thus, the equation of the parabola is
$$y^2 = -4\left(\frac{1}{6}\right)x$$
$$y^2 = -\frac{2}{3}x$$

50. To solve this problem, we draw a parabola on a Cartesian plane so that the vertex is the origin and the focus is on the positive $y$-axis. Since the focus is 1 inch from the vertex, it must be located at the point $(0, 1)$. Let $y$ represent the depth of the headlight. Since the width of the parabola is 5 inches, we represent two points on the graph of the parabola: $(-2.5, y)$ and $(2.5, y)$.

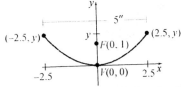

The equation of the parabola has the form $x^2 = 4ay$. Since the focus point is 1 unit from the vertex, we know that $a = 1$. Thus, the equation of the parabola is
$$x^2 = 4(1)y$$
$$x^2 = 4y$$

Finally, to find the depth of the headlight, we find $y$ when $x = 2.5$:
$$2.5^2 = 4y$$
$$6.25 = 4y$$
$$1.5625 = y$$

The depth of the headlight is 1.5625 inches (or $1\frac{9}{16}$ inches).

52. To solve this problem, we draw a parabola on a Cartesian plane so that the vertex is the origin and the focus is on the positive $y$-axis. The distance between the towers is 400 feet, and the height of the towers is 80 feet. Therefore, we know two points on the graph of the parabola: $(-200, 80)$ and $(200, 80)$.

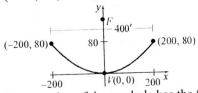

The equation of the parabola has the form $x^2 = 4ay$. Since $(200, 80)$ is a point on the graph, we have
$$200^2 = 4a(80)$$
$$40{,}000 = 320a$$
$$a = \frac{40{,}000}{320} = 125$$

Then the equation of the parabola is
$$x^2 = 4(125)y$$
$$x^2 = 500y$$

To find the height of the cable at a point 100 feet from the center of the bridge, let $x = 100$:
$$100^2 = 500y$$
$$10{,}000 = 500y$$
$$20 = y$$

At a point 100 feet from the center of the bridge, the height of the cable is 20 feet.

54. To solve this problem, we draw a parabola on a Cartesian plane so that the vertex is on the positive $y$-axis and the base of the bridge is along the $x$-axis. The height of the arch at a distance 30 feet from the center is 15 feet, so the point $(30, 15)$ is on the graph of parabola. The span of the bridge is 120 feet, so we know two other points on the graph: $(-60, 0)$ and $(60, 0)$.

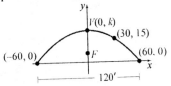

Because the parabola opens downward, its equation has the form $(x - h)^2 = -4a(y - k)$. Because the vertex lies on the $y$-axis, we know $h = 0$, so the

equation simplifies further to $x^2 = -4a(y-k)$.
Substituting (30, 15) into this equation, we obtain
$$30^2 = -4a(15-k)$$
$$900 = -60a + 4ak$$
Substituting (60, 0) into the equation, we obtain
$$60^2 = -4a(0-k)$$
$$3600 = 4ak$$
We use these result to form a system of equations:
$$\begin{cases} 900 = -60a + 4ak \\ 3600 = 4ak \end{cases}$$
To solve the system, we substitute 3600 for $4ak$ (from the second equation) into the first equation:
$$900 = -60a + 3600$$
$$-2700 = -60a$$
$$45 = a$$
Back-substituting 45 for $a$ into the second equation we obtain: $3600 = 4(45)k$
$$3600 = 180k$$
$$20 = k$$
Thus, the vertex of the parabola is (0, 20), and the height of the arch at its center is 20 feet.

56. The parabola opens downward, so the equation must have the form $(x-h)^2 = -4a(y-k)$. Substituting the vertex $(h,k) = (1,3)$ into this equation, we obtain $(x-1)^2 = -4a(y-3)$. Because the point (5, 1) is on the parabola, we let $x = 5$ and $y = 1$ to determine $a$:
$$(5-1)^2 = -4a(1-3)$$
$$4^2 = -4a(-2)$$
$$16 = 8a$$
$$2 = a$$
Thus, the equation of the parabola is
$$(x-1)^2 = -4 \cdot 2(y-3)$$
$$(x-1)^2 = -8(y-3)$$

58. The parabola opens to the right, so the equation must have the form $(y-k)^2 = 4a(x-h)$. Substituting the vertex $(h,k) = (-1,-1)$ into this equation, we obtain $(y+1)^2 = 4a(x+1)$. Because the point (3,1) is on the parabola, we let $x = 3$ and $y = 1$ to determine $a$:

$$(1+1)^2 = 4a(3+1)$$
$$2^2 = 4a(4)$$
$$4 = 16a$$
$$a = \frac{4}{16} = \frac{1}{4}$$
Thus, the equation of the parabola is
$$(y+1)^2 = 4 \cdot \frac{1}{4}(x+1)$$
$$(y+1)^2 = x+1$$

60. $(y-3)^2 = 12(x+1)$

    a. Let $x = 2$ and $y = 9$:
    $$(9-3)^2 \stackrel{?}{=} 12(2+1)$$
    $$6^2 \stackrel{?}{=} 12(3)$$
    $$36 = 36 \leftarrow \text{True}$$
    Thus, the point (2, 9) in on the parabola.

    b. Notice that $(y-3)^2 = 12(x+1)$ is of the form $(y-k)^2 = 4a(x-h)$, where $4a = 12$, so that $a = 3$. Thus, the focus of the parabola is $F(-1+3, 3) = (2, 3)$, and the directrix is $D: x = -1 - 3 = -4$. Now, the distance from the point $P(2, 9)$ to the focus $F(2, 3)$ is
    $$d(F,P) = \sqrt{(2-2)^2 + (9-3)^2} = \sqrt{36} = 6$$, and the distance from the point $P(2, 9)$ to the directrix $D: x = -4$ is $d(P,D) = 2-(-4) = 6$. Thus, $d(F,P) = d(P,D) = 6$.

62. For $y = (x+3)^2 = x^2 + 6x + 9$, $a = 1$, $b = 6$, and $c = 9$. The graph opens up because the coefficient on $x^2$ is positive.

    vertex:
    $$x = -\frac{b}{2a} = -\frac{6}{2(1)} = -3$$
    $$y = (-3)^2 + 6(-3) + 9 = 0$$
    The vertex is $(-3, 0)$ and the axis of symmetry is $x = -3$.

    y-intercept:
    $$x = 0: y = (0)^2 + 6(0) + 9 = 9$$

    x-intercepts:
    $$b^2 - 4ac = (6)^2 - 4(1)(9) = 0$$

There is one $x$-intercepts. We find it by solving
$$x^2 + 6x + 9 = 0$$
$$(x+3)^2 = 0$$
$$x + 3 = 0$$
$$x = -3$$

Graph:

The $y$-intercept point, $(0, 9)$, is three units to the right of the axis of symmetry. Therefore, if we move three units to the left of the axis of symmetry, we obtain the point $(-6, 9)$ which must also be on the graph.

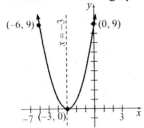

64. First write the equation in the form
$y = a(x-h)^2 + k$:
$$4(y+2) = (x-2)^2$$
$$y + 2 = \frac{1}{4}(x-2)^2$$
$$y = \frac{1}{4}(x-2)^2 - 2$$

Begin with the graph of $y = x^2$, then shift the graph 2 units to the right to obtain the graph of $y = (x-2)^2$. Multiply the $y$-coordinate by $\frac{1}{4}$ to obtain the graph of $y = \frac{1}{4}(x-2)^2$. Lastly, shift the graph down 2 units to obtain the graph of $y = \frac{1}{4}(x-2)^2 - 2$.

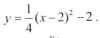
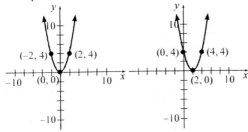

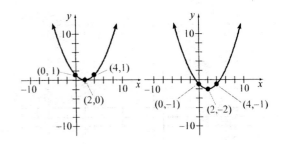

66. Answers will vary.

68. $x^2 = -8y$
$$-\frac{1}{8}x^2 = y$$

Let $Y_1 = -\frac{1}{8}x^2$.

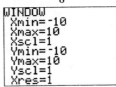
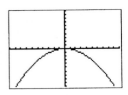

70. $y^2 = 10x$
$$y = \pm\sqrt{10x}$$

Let $Y_1 = -\sqrt{10x}$ and $Y_2 = \sqrt{10x}$.

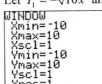
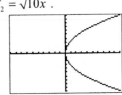

72. $(x+4)^2 = -4(y-1)$
$$-\frac{1}{4}(x+4)^2 = y - 1$$
$$-\frac{1}{4}(x+4)^2 + 1 = y$$

Let $Y_1 = -\frac{1}{4}(x+4)^2 + 1$.

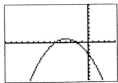

## Chapter 10: Conics

74. $(y-2)^2 = 12(x+5)$

    $y - 2 = \pm\sqrt{12(x+5)}$

    $y = 2 \pm \sqrt{12(x+5)}$

    Let $Y_1 = 2 - \sqrt{12(x+5)}$ and $Y_2 = 2 + \sqrt{12(x+5)}$.

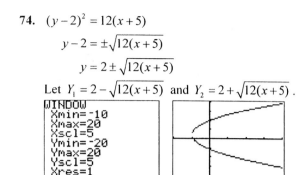

76. $x^2 + 2x - 8y + 25 = 0$

    $x^2 + 2x + 25 = 8y$

    $\dfrac{1}{8}(x^2 + 2x + 25) = y$

    Let $Y_1 = \dfrac{1}{8}(x^2 + 2x + 25)$.

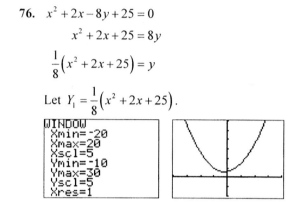

78. $y^2 - 8y + 16x - 16 = 0$

    $y^2 - 8y = -16x + 16$

    $y^2 - 8y + 16 = -16x + 16 + 16$

    $(y-4)^2 = -16x + 32$

    $y - 4 = \pm\sqrt{-16x + 32}$

    $y = 4 \pm \sqrt{-16x + 32}$

    Let $Y_1 = 4 - \sqrt{-16x+32}$ and $Y_2 = 4 + \sqrt{-16x+32}$.

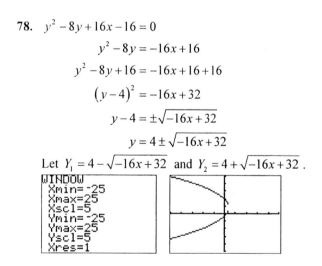

## 10.4 Preparing for Ellipses

1. Start: $x^2 + 10x$

   Add: $\left(\dfrac{1}{2} \cdot 10\right)^2 = 25$

   Result: $x^2 + 10x + 25$

   Factored Form: $(x+5)^2$

2. Begin with the graph of $y = x^2$, then shift the graph 2 units to the left to obtain the graph of $y = (x+2)^2$. Shift this graph down 1 unit to obtain the graph of $f(x) = (x+2)^2 - 1$.

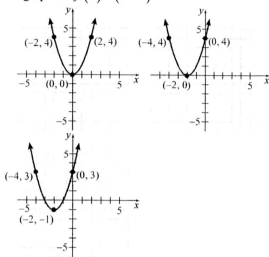

## 10.4 Exercises

2. major axis

4. FALSE. The equation of an ellipse centered at the origin with vertex $(a, 0)$ and focus $(c, 0)$ is $\dfrac{x^2}{a^2} + \dfrac{y^2}{b^2} = 1$.

6. FALSE. The center of $\dfrac{(x-3)^2}{25} + \dfrac{(y+1)^2}{16} = 1$ is $(3, -1)$.

8. Answers will vary. One possibility follows:

   The ellipse given by the equation $\dfrac{x^2}{25} + \dfrac{y^2}{9} = 1$, or $\dfrac{x^2}{5^2} + \dfrac{y^2}{3^2} = 1$, has the x-axis as its major axis. Its vertices will be $(\pm 5, 0)$. Its y-intercepts will be $(0, \pm 3)$. Its foci will be $\left(\pm\sqrt{25-9}, 0\right) = (\pm 4, 0)$.

   The ellipse given by the equation $\dfrac{x^2}{9} + \dfrac{y^2}{25} = 1$, or $\dfrac{x^2}{3^2} + \dfrac{y^2}{5^2} = 1$, has the y-axis as its major axis. Its

vertices will be $(0,\pm 5)$. Its $x$-intercepts will be $(\pm 3,0)$. Its foci will be $\left(0,\pm\sqrt{25-9}\right)=(0,\pm 4)$.
Both ellipses are centered at (0, 0) and have the exact same shape and size.

10. b. The center of the ellipse is the origin, the major axis is the $x$-axis, and the vertices are $(\pm 4,0)$, and the $y$-intercepts are $(0,\pm 1)$. Thus, the equation is of the form $\dfrac{x^2}{a^2}+\dfrac{y^2}{b^2}=1$, with $a=3$ and $b=1$. Thus, the equation is
$$\dfrac{x^2}{3^2}+\dfrac{y^2}{1^2}=1$$
$$\dfrac{x^2}{9}+y^2=1$$

12. a. The center of the ellipse is the origin, the major axis is the $y$-axis, and the vertices are $(0,\pm 3)$, and the $x$-intercepts are $(\pm 1,0)$. Thus, the equation is of the form $\dfrac{x^2}{b^2}+\dfrac{y^2}{a^2}=1$, with $a=3$ and $b=1$.
Thus, the equation is
$$\dfrac{x^2}{1^2}+\dfrac{y^2}{3^2}=1$$
$$x^2+\dfrac{y^2}{9}=1$$

14. $\dfrac{x^2}{25}+\dfrac{y^2}{4}=1$

The larger number, 25, is in the denominator of the $x^2$-term. This means that the major axis is the $x$-axis and that the equation of the ellipse is of the form $\dfrac{x^2}{a^2}+\dfrac{y^2}{b^2}=1$, so that $a^2=25$ and $b^2=4$. The center of the ellipse is $(0,0)$. Because $b^2=a^2-c^2$, or $c^2=a^2-b^2$, we have that $c^2=25-4=21$, so that $c=\pm\sqrt{21}$. Since the major axis is the $x$-axis, the foci are $\left(-\sqrt{21},0\right)$ and $\left(\sqrt{21},0\right)$. To find the $x$-intercepts (vertices), let $y=0$; to find the $y$-intercepts, let $x=0$:

$x$-intercepts:  $y$-intercepts:
$\dfrac{x^2}{25}+\dfrac{0^2}{4}=1$  $\dfrac{0^2}{25}+\dfrac{y^2}{4}=1$
$\dfrac{x^2}{25}=1$  $\dfrac{y^2}{4}=1$
$x^2=25$  $y^2=4$
$x=\pm 5$  $y=\pm 2$

The intercepts are $(-5,0)$, $(5,0)$, $(0,-2)$, and $(0,2)$.

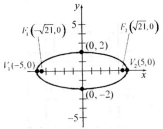

16. $\dfrac{x^2}{16}+\dfrac{y^2}{36}=1$

The larger number, 36, is in the denominator of the $y^2$-term. This means that the major axis is the $y$-axis and that the equation of the ellipse is of the form $\dfrac{x^2}{b^2}+\dfrac{y^2}{a^2}=1$, so that $a^2=36$ and $b^2=16$. The center of the ellipse is $(0,0)$. Because $b^2=a^2-c^2$, or $c^2=a^2-b^2$, we have that $c^2=36-16=20$, so that $c=\pm\sqrt{20}=\pm 2\sqrt{5}$. Since the major axis is the $y$-axis, the foci are $\left(0,-2\sqrt{5}\right)$ and $\left(0,2\sqrt{5}\right)$. To find the $x$-intercepts, let $y=0$; to find the $y$-intercepts (vertices), let $x=0$:

$x$-intercepts:  $y$-intercepts:
$\dfrac{x^2}{16}+\dfrac{0^2}{36}=1$  $\dfrac{0^2}{16}+\dfrac{y^2}{36}=1$
$\dfrac{x^2}{16}=1$  $\dfrac{y^2}{36}=1$
$x^2=16$  $y^2=36$
$x=\pm 4$  $y=\pm 6$

The intercepts are $(-4,0)$, $(4,0)$, $(0,-6)$, and $(0,6)$.

## Chapter 10: Conics

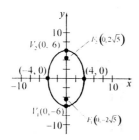

18. $\dfrac{x^2}{121}+\dfrac{y^2}{100}=1$

The larger number, 121, is in the denominator of the $x^2$-term. This means that the major axis is the $x$-axis and that the equation of the ellipse is of the form $\dfrac{x^2}{a^2}+\dfrac{y^2}{b^2}=1$, so that $a^2=121$ and $b^2=100$. The center of the ellipse is $(0,0)$. Because $b^2=a^2-c^2$, or $c^2=a^2-b^2$, we have that $c^2=121-100=21$, so that $c=\pm\sqrt{21}$. Since the major axis is the $x$-axis, the foci are $\left(-\sqrt{21},0\right)$ and $\left(\sqrt{21},0\right)$. To find the $x$-intercepts (vertices), let $y=0$; to find the $y$-intercepts, let $x=0$:

$x$-intercepts:  
$\dfrac{x^2}{121}+\dfrac{0^2}{100}=1$  
$\dfrac{x^2}{121}=1$  
$x^2=121$  
$x=\pm 11$

$y$-intercepts:  
$\dfrac{0^2}{121}+\dfrac{y^2}{100}=1$  
$\dfrac{y^2}{100}=1$  
$y^2=100$  
$y=\pm 10$

The intercepts are $(-11,0)$, $(11,0)$, $(0,-10)$, and $(0,10)$.

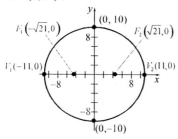

20. $\dfrac{x^2}{64}+y^2=1$

$\dfrac{x^2}{64}+\dfrac{y^2}{1}=1$

The larger number, 64, is in the denominator of the $x^2$-term. This means that the major axis is the $x$-axis and that the equation of the ellipse is of the form $\dfrac{x^2}{a^2}+\dfrac{y^2}{b^2}=1$, so that $a^2=64$ and $b^2=1$. The center of the ellipse is $(0,0)$. Because $b^2=a^2-c^2$, or $c^2=a^2-b^2$, we have that $c^2=64-1=63$, so that $c=\pm\sqrt{63}=\pm 3\sqrt{7}$. Since the major axis is the $x$-axis, the foci are $\left(-3\sqrt{7},0\right)$ and $\left(3\sqrt{7},0\right)$. To find the $x$-intercepts (vertices), let $y=0$; to find the $y$-intercepts, let $x=0$:

$x$-intercepts:  
$\dfrac{x^2}{64}+0^2=1$  
$\dfrac{x^2}{64}=1$  
$x^2=64$  
$x=\pm 8$

$y$-intercepts:  
$\dfrac{0^2}{64}+y^2=1$  
$y^2=1$  
$y=\pm 1$

The intercepts are $(-8,0)$, $(8,0)$, $(0,-1)$, and $(0,1)$.

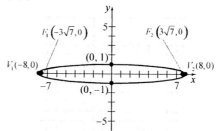

22. $9x^2+y^2=81$

$\dfrac{9x^2+y^2}{81}=\dfrac{81}{81}$

$\dfrac{x^2}{9}+\dfrac{y^2}{81}=1$

The larger number, 81, is in the denominator of the $y^2$-term. This means that the major axis is the $y$-axis and that the equation of the ellipse is of the form $\dfrac{x^2}{b^2}+\dfrac{y^2}{a^2}=1$, so that $a^2=81$ and $b^2=9$. The center of the ellipse is $(0,0)$. Because $b^2=a^2-c^2$, or $c^2=a^2-b^2$, we have that $c^2=81-9=72$, so that $c=\pm\sqrt{72}=\pm 6\sqrt{2}$. Since the major axis is the $y$-axis, the foci are $\left(0,-6\sqrt{2}\right)$ and $\left(0,6\sqrt{2}\right)$. To find the $y$-intercepts (vertices), let $x=0$:

$x$-intercepts:  $y$-intercepts:

$$9x^2 + 0^2 = 81 \qquad 9(0)^2 + y^2 = 81$$
$$9x^2 = 81 \qquad y^2 = 81$$
$$x^2 = 9 \qquad y = \pm 9$$
$$x = \pm 3$$

The intercepts are $(-3, 0)$, $(3, 0)$, $(0, -9)$, and $(0, 9)$.

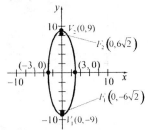

24. The given focus $(2, 0)$ and the given vertex $(5, 0)$ lie on the $x$-axis. Thus, the equation of the ellipse is of the form $\dfrac{x^2}{a^2} + \dfrac{y^2}{b^2} = 1$. The distance from the center of the ellipse to the vertex is $a = 5$ units. The distance from the center of the ellipse to the focus is $c = 2$ units. Because $b^2 = a^2 - c^2$, we have that $b^2 = 5^2 - 2^2 = 25 - 4 = 21$. Thus, the equation of the ellipse is $\dfrac{x^2}{25} + \dfrac{y^2}{21} = 1$.

To help graph the ellipse, find the $y$-intercepts:

Let $x = 0$: $\dfrac{0^2}{25} + \dfrac{y^2}{21} = 1$
$$\dfrac{y^2}{21} = 1$$
$$y^2 = 21$$
$$y = \pm\sqrt{21}$$

The $y$-intercepts are $(0, -\sqrt{21})$ and $(0, \sqrt{21})$.

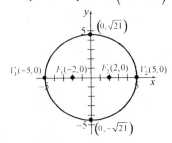

26. The given focus $(0, -1)$ and the given vertex $(0, 5)$ lie on the $y$-axis. Thus, the equation of the ellipse is of the form $\dfrac{x^2}{b^2} + \dfrac{y^2}{a^2} = 1$. The distance from the center of the ellipse to the vertex is $a = 5$ units. The distance from the center of the ellipse to the focus is $c = 1$ units. Because $b^2 = a^2 - c^2$, we have that $b^2 = 5^2 - 1^2 = 25 - 1 = 24$. Thus, the equation of the ellipse is $\dfrac{x^2}{24} + \dfrac{y^2}{25} = 1$.

To help graph the ellipse, find the $x$-intercepts:

Let $y = 0$: $\dfrac{x^2}{24} + \dfrac{0^2}{25} = 1$
$$\dfrac{x^2}{24} = 1$$
$$x^2 = 24$$
$$x = \pm\sqrt{24} = \pm 2\sqrt{6}$$

The $x$-intercepts are $(-2\sqrt{6}, 0)$ and $(2\sqrt{6}, 0)$.

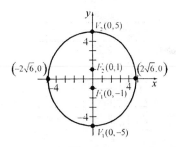

28. The given foci $(0, \pm 2)$ and the given vertices $(0, \pm 7)$ lie on the $y$-axis. The center of the ellipse is the midpoint between the two vertices (or foci). Thus, the center of the ellipse is $(0, 0)$, and its equation is of the form $\dfrac{x^2}{b^2} + \dfrac{y^2}{a^2} = 1$. The distance from the center of the ellipse to the vertex is $a = 7$ units. The distance from the center of the ellipse to the focus is $c = 2$ units. Because $b^2 = a^2 - c^2$, we have that $b^2 = 7^2 - 2^2 = 49 - 4 = 45$. Thus, the equation of the ellipse is $\dfrac{x^2}{45} + \dfrac{y^2}{49} = 1$.

To help graph the ellipse, find the $x$-intercepts:

## Chapter 10: Conics

Let $y = 0$: $\dfrac{x^2}{45} + \dfrac{0^2}{49} = 1$

$\dfrac{x^2}{45} = 1$

$x^2 = 45$

$x = \pm\sqrt{45} = \pm 3\sqrt{5}$

The x-intercepts are $\left(-3\sqrt{5}, 0\right)$ and $\left(3\sqrt{5}, 0\right)$.

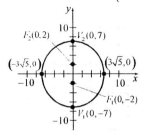

30. The given foci $(\pm 6, 0)$ lie on the x-axis. The center of the ellipse is the midpoint between the two foci. Thus, center of the ellipse is $(0, 0)$ and its equation is of the form $\dfrac{x^2}{a^2} + \dfrac{y^2}{b^2} = 1$. Now, the length of the major axis is 20, so the vertices must be $(\pm 10, 0)$, and the distance from the center of the ellipse to each vertex is $a = 10$ units. The distance from the center of the ellipse to each focus is $c = 6$ units. Because $b^2 = a^2 - c^2$, we have that $b^2 = 10^2 - 6^2 = 100 - 36 = 64$. Thus, the equation of the ellipse is $\dfrac{x^2}{100} + \dfrac{y^2}{64} = 1$.

To help graph the ellipse, we find the y-intercepts:

Let $y = 0$: $\dfrac{0^2}{100} + \dfrac{y^2}{64} = 1$

$\dfrac{y^2}{64} = 1$

$y^2 = 64$

$y = \pm 8$

The x-intercepts are $(0, -8)$ and $(0, 8)$.

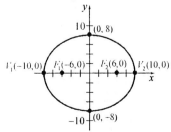

32. $\dfrac{(x-1)^2}{36} + \dfrac{(y+4)^2}{100} = 1$

The center of the ellipse is $(h, k) = (1, -4)$. Because the larger number, 100, is in the denominator of the $y^2$-term, the major axis is parallel to the y-axis. Because $a^2 = 100$ and $b^2 = 36$, we have that $c^2 = a^2 - b^2 = 100 - 36 = 64$. The vertices are $a = 10$ units below and above the center at $V_1(1, -14)$ and $V_2(1, 6)$. The foci are $c = \sqrt{64} = 8$ units below and above the center at $F_1(1, -12)$ and $F_2(1, 4)$. We plot the points $b = 6$ units to the left and right of the center point at $(-5, -4)$ and $(7, -4)$.

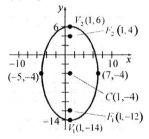

34. $\dfrac{(x+5)^2}{64} + \dfrac{(y+1)^2}{16} = 1$

The center of the ellipse is $(h, k) = (-5, -1)$. Because the larger number, 64, is in the denominator of the $x^2$-term, the major axis is parallel to the x-axis. Because $a^2 = 64$ and $b^2 = 16$, we have that $c^2 = a^2 - b^2 = 64 - 16 = 48$. The vertices are $a = 8$ units to the left and right of the center at $V_1(-13, -1)$ and $V_2(3, -1)$. The foci are $c = \sqrt{48} = 4\sqrt{3}$ units to the left and right of the center at $F_1\left(-5 - 4\sqrt{3}, -1\right)$ and $F_2\left(-5 + 4\sqrt{3}, -1\right)$. We plot the points $b = 4$ units above and below the center point at $(-5, 3)$ and $(-5, -5)$.

Section 10.4 Ellipses

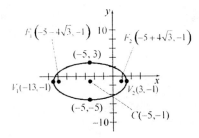

36. $\dfrac{(x+8)^2}{81}+(y-3)^2=1$

$\dfrac{(x+8)^2}{81}+\dfrac{(y-3)^2}{1}=1$

The center of the ellipse is $(h,k)=(-8,3)$. Because the larger number, 81, is in the denominator of the $x^2$-term, the major axis is parallel to the $x$-axis. Because $a^2=81$ and $b^2=1$, we have that $c^2=a^2-b^2=81-1=80$. The vertices are $a=9$ units to the left and right of the center at $V_1(-17,3)$ and $V_2(1,3)$. The foci are $c=\sqrt{80}=4\sqrt{5}$ units to the left and right of the center at $F_1\left(-8-4\sqrt{5},3\right)$ and $F_2\left(-8+4\sqrt{5},3\right)$. We plot the points $b=1$ units below and above the center point at $(-8,2)$ and $(-8,4)$.

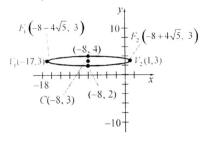

38. $9(x-3)^2+(y-4)^2=81$

$\dfrac{9(x-3)^2+(y-4)^2}{81}=\dfrac{81}{81}$

$\dfrac{(x-3)^2}{9}+\dfrac{(y-4)^2}{81}=1$

The center of the ellipse is $(h,k)=(3,4)$. Because the larger number, 81, is in the denominator of the $y^2$-term, the major axis is parallel to the $y$-axis. Because $a^2=81$ and $b^2=9$, we have that $c^2=a^2-b^2=81-9=72$. The vertices are $a=9$ units below and above the center at $V_1(3,-5)$ and $V_2(3,13)$. The foci are $c=\sqrt{72}=6\sqrt{2}$ units below and above the

center at $F_1\left(3,4-6\sqrt{2}\right)$ and $F_2\left(3,4+6\sqrt{2}\right)$. We plot the points $b=3$ units to the left and right of the center point at $(0,4)$ and $(6,4)$.

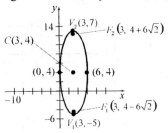

40. $16x^2+9y^2-128x+54y-239=0$

$16x^2-128x+9y^2+54y=239$

$16(x^2-8x)+9(y^2+6y)=239$

$16(x^2-8x+16)+9(y^2+6y+9)$
$\qquad =239+16(16)+9(9)$

$16(x-4)^2+9(y+3)^2=576$

$\dfrac{16(x-4)^2+9(y+3)^2}{576}=\dfrac{576}{576}$

$\dfrac{(x-4)^2}{36}+\dfrac{(y+3)^2}{64}=1$

The center of the ellipse is $(h,k)=(4,-3)$. Because the larger number, 64, is in the denominator of the $y^2$-term, the major axis is parallel to the $y$-axis. Because $a^2=64$ and $b^2=36$, we have that $c^2=a^2-b^2=64-36=28$. The vertices are $a=8$ units below and above the center at $V_1(4,-11)$ and $V_2(4,5)$. The foci are $c=\sqrt{28}=2\sqrt{7}$ units below and above the center at $F_1\left(4,-3-2\sqrt{7}\right)$ and $F_2\left(4,-3+2\sqrt{7}\right)$. We plot the points $b=6$ units to the left and right of the center point at $(-2,-3)$ and $(10,-3)$.

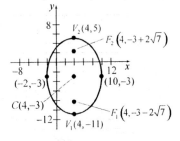

41. a. To solve this problem, we draw the ellipse on a Cartesian plane so that the $x$-axis coincides with the water and the $y$-axis

passes through the center of the arch. Thus, the origin is the center of the ellipse. Now, the "center" of the arch is 10 meters above the water, so the point $(0,10)$ is on the ellipse. The river is 30 meters wide, so the two points $(-15,0)$ and $(15,0)$ are the two vertices of the ellipse and the major axis is along the x-axis.

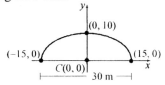

The equation of the ellipse must have the form $\dfrac{x^2}{a^2}+\dfrac{y^2}{b^2}=1$. The distance from the center of the ellipse to each vertex is $a=15$. Also, the height of the semi-ellipse is 10 feet, so $b=10$. Thus, the equation of the ellipse is

$$\frac{x^2}{15^2}+\frac{y^2}{10^2}=1$$
$$\frac{x^2}{225}+\frac{y^2}{100}=1$$

**b.** To determine if the barge can fit through the opening of the bridge, we center it beneath the arch so that the points $(-9,7)$ and $(9,7)$ represent the top corners of the barge. Now, we determine the height of the arch above the water at points 9 meters to the left and right of the center by substituting $x=9$ into the equation of the ellipse and solving for y:

$$\frac{9^2}{225}+\frac{y^2}{100}=1$$
$$\frac{9}{25}+\frac{y^2}{100}=1$$
$$\frac{y^2}{100}=\frac{16}{25}$$
$$y^2=64$$
$$y=8$$

At points 9 meters from the center of the river, the arch is 8 meters above the surface of the water. Thus, the barge can fit through the opening with about 1 meter of clearance above the top corners.

**c.** No. From part (b), the barge only has 1 meter of clearance when the water level is at it normal stage. Thus, if the water level increases by 1.1 meters, then the barge will not be able to pass under the bridge.

**42. a.** To solve this problem, we draw the ellipse on a Cartesian plane so that the x-axis coincides with the water and the y-axis passes through the center of the arch. Thus, the origin is the center of the ellipse. Now, the "center" of the arch is 15 meters above the water, so the point $(0,15)$ is on the ellipse. The river is 45.6 meters wide, so the two points $(-22.8)$ and $(22.8,0)$ are the two vertices of the ellipse and the major axis is along the x-axis.

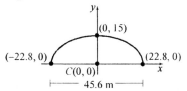

The equation of the ellipse must have the form $\dfrac{x^2}{a^2}+\dfrac{y^2}{b^2}=1$. The distance from the center of the ellipse to each vertex is $a=22.8$. Also, the height of the semi-ellipse is 15 feet, so $b=15$. Thus, the equation of the ellipse is

$$\frac{x^2}{22.8^2}+\frac{y^2}{15^2}=1$$
$$\frac{x^2}{519.84}+\frac{y^2}{225}=1$$

**b.** To determine if the barge can fit through the opening of the bridge, we center it beneath the arch so that the points $(-10,12)$ and $(10,12)$ represent the top corners of the barge. Now, we determine the height of the arch above the water at points 10 meters to the left and right of the center by substituting $x=10$ into the equation of the ellipse and solving for $y$:

$$\frac{10^2}{519.84} + \frac{y^2}{225} = 1$$

$$\frac{100}{519.84} + \frac{y^2}{225} = 1$$

$$\frac{y^2}{225} = \frac{419.84}{519.84}$$

$$y^2 = \frac{94,464}{519.84}$$

$$y = \sqrt{\frac{94,464}{519.84}} \approx 13.48$$

At points 10 meters from the center of the river, the arch is approximately 13.48 meters above the surface of the water. Thus, the barge can fit through the opening with about 1.48 meters of clearance above the top corners.

c.  No. From part (b), the barge only has 1.48 meters of clearance when the water level is at it normal stage. Thus, if the water level increases by 1.5 meters, then the barge will not be able to pass under the bridge.

44. To find the amphelion of Mars, we recognize that Aphelion = 2(Mean distance) − Perihilion . Thus, the amphelion of Mars is
$2 \cdot 142 - 128.5 = 155.5$ million miles.

To find the equation of the elliptical orbit of Mars, we draw the ellipse on a Cartesian plane so that the center of the ellipse is at the origin and the major axis is along the $x$-axis. The equation of the ellipse is of the form $\frac{x^2}{a^2} + \frac{y^2}{b^2} = 1$. Now, the mean distance of Mars from the Sun is 142 million miles, so the distance from the center of the orbit to each vertex is $a = 142$ million miles. Since the aphelion of Mars is 155.5 million miles, then distance from the center of the orbit to each focus is $c = 155.5 - 142 = 13.5$ million miles.

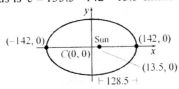

Now, because $b^2 = a^2 - c^2$, we have that
$b^2 = 142^2 - 13.5^2 = 20,164 - 182.25 = 19,981.75$.
Thus, the equation that describes the orbit of Mars is

$$\frac{x^2}{20,164} + \frac{y^2}{19,981.75} = 1.$$

46. The mean distance of Pluto from the Sun is $4551 + 897.5 = 5448.5$ million miles.

To find the aphelion of Pluto, we recognize that Aphelion = 2(Mean distance) − Perihelion . Thus, the perihelion of Pluto is
$2 \cdot 5448.5 - 4551 = 6346$ million miles.

To find the equation of the elliptical orbit of Pluto, we draw the ellipse on a Cartesian plane so that the center of the ellipse is at the origin and the major axis is along the $x$-axis. The equation of the ellipse is of the form $\frac{x^2}{a^2} + \frac{y^2}{b^2} = 1$. Now, we found above that the mean distance of Pluto from the Sun is 5448.5 million miles, so the distance from the center of the orbit to each vertex is $a = 5448.5$ million miles. We are given the distance from the center of the orbit to the Sun (that is, to each focus) is $c = 897.5$ million miles.

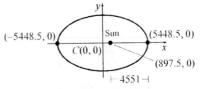

Now, because $b^2 = a^2 - c^2$, we have that
$b^2 = 5448.5^2 - 897.5^2$
$= 29,686,152.25 - 805,506.25$
$= 28,880,646$

Thus, the equation that describes the orbit of Pluto is

$$\frac{x^2}{29,686,152.25} + \frac{y^2}{28,880,646} = 1.$$

48. The major axis of the ellipse is the line $x = 2$, which is parallel to the $y$-axis, so the equation of the ellipse is of the form $\frac{(x-h)^2}{b^2} + \frac{(y-k)^2}{a^2} = 1$.

The center is $(h, k) = (2, 1)$. The vertices are $(2, -2)$ and $(2, 4)$, so $a = 4 - 1 = 3$.

The points $(0, 1)$ and $(4, 1)$ are the points on the ellipse that are left and right of the center, so

## Chapter 10: Conics

$b = 4 - 2 = 2$. Thus, the equation is
$$\frac{(x-2)^2}{2^2} + \frac{(y-1)^2}{3^2} = 1$$
$$\frac{(x-2)^2}{4} + \frac{(y-1)^2}{9} = 1$$

**50.** The major axis of the ellipse is $x$-axis, so the equation of the ellipse is of the form $\frac{(x-h)^2}{a^2} + \frac{(y-k)^2}{b^2} = 1$. The center is $(h,k) = (1,0)$. The vertices are $(-3, 0)$ and $(5, 0)$, so $a = 5 - 1 = 4$. The points $(1,-1)$ and $(1, 1)$ are the points on the ellipse that are below and above the center, so $b = 1 - 0 = 1$. Thus, the equation is
$$\frac{(x-1)^2}{4^2} + \frac{(y-0)^2}{1^2} = 1$$
$$\frac{(x-1)^2}{16} + y^2 = 1$$

**52.** Answer may vary. Possibilities follow:
Upon examining the graphs of the ellipses we have constructed in this section and calculating the eccentricity for them, we make the following generalizations:

**a.** The closer the eccentricity, $e$, is to 0, the more circular the ellipse becomes.

**b.** When the eccentricity is close to 0.5, the ellipse looks "egg shaped", as we would expect an ellipse to look.

**c.** As the eccentricity gets closer to 1, the more elongated the ellipse becomes.

**54.** $f(x) = \frac{x-1}{x^2+4}$

$f(5) = \frac{5-1}{5^2+4} = \frac{4}{29} \approx 0.13793$

$f(10) = \frac{10-1}{10^2+4} = \frac{9}{104} \approx 0.08654$

$f(100) = \frac{100-1}{100^2+4} = \frac{99}{10,004} \approx 0.00990$

$f(1000) = \frac{1000-1}{1000^2+4} = \frac{999}{1,000,004} \approx 0.00100$

| $x$ | 5 | 10 | 100 | 100 |
|---|---|---|---|---|
| $f(x)$ | 0.13793 | 0.08654 | 0.00990 | 0.00100 |

**56.** $f(x) = \frac{3x^2 - x + 1}{x^2 + 1}$

$f(5) = \frac{3(5)^2 - 5 + 1}{5^2 + 1} = \frac{71}{26} \approx 2.73077$

$f(10) = \frac{3(10)^2 - 10 + 1}{10^2 + 1} = \frac{291}{101} \approx 2.88119$

$f(100) = \frac{3(100)^2 - 100 + 1}{100^2 + 1} = \frac{29,901}{10,001} \approx 2.98980$

$f(1000) = \frac{3(1000)^2 - 1000 + 1}{1000^2 + 1}$
$= \frac{2,998,999}{1,000,001} \approx 2.99900$

| $x$ | 5 | 10 | 100 | 100 |
|---|---|---|---|---|
| $f(x)$ | 2.73077 | 2.88119 | 2.98980 | 2.99900 |

**58.** $f(x) = \frac{x^2 - 3x + 5}{x + 2}$; $g(x) = x - 5$

$f(5) = \frac{5^2 - 3(5) + 5}{5 + 2} = \frac{15}{7} \approx 2.14286$

$f(10) = \frac{10^2 - 3(10) + 5}{10 + 2} = \frac{75}{12} = 6.25$

$f(100) = \frac{100^2 - 3(100) + 5}{100 + 2} = \frac{9705}{102} \approx 95.14706$

$f(1000) = \frac{1000^2 - 3(1000) + 5}{1000 + 2}$
$= \frac{997,005}{1002} \approx 995.01497$

$g(5) = 5 - 5 = 0$
$g(10) = 10 - 5 = 5$
$g(100) = 100 - 5 = 95$
$g(1000) = 1000 - 5 = 995$

| $x$ | 5 | 10 | 100 | 100 |
|---|---|---|---|---|
| $f(x)$ | 2.14286 | 6.25 | 95.14706 | 995.01497 |
| $g(x)$ | 0 | 5 | 95 | 995 |

**60.** For Problem 57:
$$\begin{array}{r} x+2 \phantom{xxxx} \\ x+1 \overline{\smash{)}x^2 + 3x + 1} \\ \underline{x^2 + x \phantom{xxx}} \\ 2x + 1 \\ \underline{2x + 2} \\ -1 \end{array}$$

Thus, $f(x) = \frac{x^2 + 3x + 1}{x + 1} = x + 2 + \frac{-1}{x + 1}$

ISM − 632

**Section 10.4** Ellipses

For Problem 58:
$$x+2\overline{)x^2-3x+5}$$
quotient $x-5$, with steps $x^2+2x$; $-5x+5$; $-5x-10$; remainder $15$.

Thus, $f(x) = \dfrac{x^2-3x+5}{x+2} = x-5+\dfrac{15}{x+2}$

The function values for $f$ and $g$ get closer to each other as $x$ gets larger.

In both problems, the function $g$ represents the linear function that the rational function $f$ will approach as $x$ increases without bound.

**62.** $\dfrac{x^2}{25}+\dfrac{y^2}{4}=1$

$\dfrac{y^2}{4}=1-\dfrac{x^2}{25}$

$y^2 = 4\left(1-\dfrac{x^2}{25}\right)$

$y = \pm\sqrt{4\left(1-\dfrac{x^2}{25}\right)} = \pm 2\sqrt{1-\dfrac{x^2}{25}}$

Let $Y_1 = 2\sqrt{1-\dfrac{x^2}{25}}$ and $Y_2 = -2\sqrt{1-\dfrac{x^2}{25}}$.

WINDOW
Xmin=-9.4
Xmax=9.4
Xscl=1
Ymin=-6.2
Ymax=6.2
Yscl=1
Xres=1

**64.** $9x^2+y^2=81$

$y^2 = 81-9x^2$

$y = \pm\sqrt{81-9x^2}$

Let $Y_1 = \sqrt{81-9x^2}$ and $Y_2 = -\sqrt{81-9x^2}$.

WINDOW
Xmin=-14.1
Xmax=14.1
Xscl=2
Ymin=-9.3
Ymax=9.3
Yscl=2
Xres=1

**66.** $\dfrac{(x-1)^2}{36}+\dfrac{(y+4)^2}{100}=1$

$\dfrac{(y+4)^2}{100} = 1-\dfrac{(x-1)^2}{36}$

$(y+4)^2 = 100\left(1-\dfrac{(x-1)^2}{36}\right)$

$y+4 = \pm\sqrt{100\left(1-\dfrac{(x-1)^2}{36}\right)}$

$y+4 = \pm 10\sqrt{1-\dfrac{(x-1)^2}{36}}$

$y = -4 \pm 5\sqrt{1-\dfrac{(x-3)^2}{9}}$

Let $Y_1 = -4+5\sqrt{1-\dfrac{(x-3)^2}{9}}$ and

$Y_2 = -4-5\sqrt{1-\dfrac{(x-3)^2}{9}}$.

WINDOW
Xmin=-14.1
Xmax=14.1
Xscl=2
Ymin=-9.3
Ymax=9.3
Yscl=2
Xres=1

**68.** $\dfrac{(x+5)^2}{64}+\dfrac{(y+1)^2}{16}=1$

$\dfrac{(y+1)^2}{16} = 1-\dfrac{(x+5)^2}{64}$

$(y+1)^2 = 16\left(1-\dfrac{(x+5)^2}{64}\right)$

$y+1 = \pm\sqrt{16\left(1-\dfrac{(x+5)^2}{64}\right)}$

$y+1 = \pm 4\sqrt{1-\dfrac{(x+5)^2}{64}}$

$y = -1 \pm 4\sqrt{1-\dfrac{(x+5)^2}{64}}$

Let $Y_1 = -1+4\sqrt{1-\dfrac{(x+5)^2}{64}}$ and

$Y_2 = -1-4\sqrt{1-\dfrac{(x+5)^2}{64}}$.

WINDOW
Xmin=-14.4
Xmax=4.4
Xscl=1
Ymin=-6.2
Ymax=6.2
Yscl=1
Xres=1

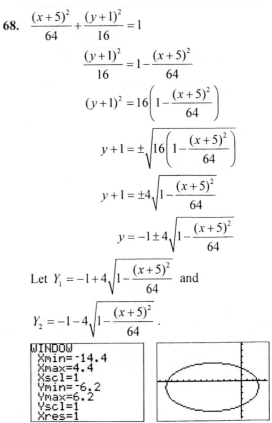

ISM – 633

## 10.5 Preparing for Hyperbolas

1. Start: $x^2 - 5x$

   Add: $\left(\frac{1}{2} \cdot (-5)\right)^2 = \frac{25}{4}$

   Result: $x^2 - 5x + \frac{25}{4}$

   Factored Form: $\left(x - \frac{5}{2}\right)^2$

2. $y^2 = 64$

   $y = \pm\sqrt{64}$

   $y = \pm 8$

   The solution set is $\{-8, 8\}$.

## Section 10.5

2. transverse axis

4. TRUE

6. FALSE

8. Answers may vary. One possibility follows: The standard-form equations for hyperbolas and ellipses are very similar. The difference is that the standard-form equations for hyperbolas will involve subtraction between the $x^2$ and $y^2$ terms while the standard-form equations of an ellipse will involve addition of the $x^2$ and $y^2$ terms.

10. c. The transverse axis is the $y$-axis and the hyperbola opens up and down. Also, the distance between the vertices $(0, \pm 2)$ and the center $(0,0)$ is $a = 2$ units. This means that the equation of the hyperbola is of the form

    $$\frac{y^2}{a^2} - \frac{x^2}{b^2} = 1$$
    $$\frac{y^2}{2^2} - \frac{x^2}{b^2} = 1$$
    $$\frac{y^2}{4} - \frac{x^2}{b^2} = 1$$

    Of the list of provided equations, only equation (c): $\frac{y^2}{4} - x^2 = 1$ is of this form.

12. d. The transverse axis is the $y$-axis and the hyperbola opens up and down. Also, the distance between the vertices $(0, \pm 1)$ and the center $(0,0)$ is $a = 1$ unit. This means that the equation of the hyperbola is of the form

    $$\frac{y^2}{a^2} - \frac{x^2}{b^2} = 1$$
    $$\frac{y^2}{1^2} - \frac{x^2}{b^2} = 1$$
    $$y^2 - \frac{x^2}{b^2} = 1$$

    Of the list of provided equations, only equation (d): $y^2 - \frac{x^2}{4} = 1$ is of this form.

14. $\frac{x^2}{9} - \frac{y^2}{16} = 1$

    Notice the equation is of the form $\frac{x^2}{a^2} - \frac{y^2}{b^2} = 1$.

    Because the $x^2$-term is first, the transverse axis is the $x$-axis and the hyperbola opens left and right. The center of the hyperbola is the origin. We have that $a^2 = 9$ and $b^2 = 16$. Because $c^2 = a^2 + b^2$, we have that $c^2 = 9 + 16 = 25$, so that $c = \sqrt{25} = 5$. The vertices are $(\pm a, 0) = (\pm 3, 0)$, and the foci are $(\pm c, 0) = (\pm 5, 0)$. Since $a = 3$ and $b = 4$, the equations of the asymptotes are $y = \frac{4}{3}x$ and $y = -\frac{4}{3}x$. To help graph the hyperbola, we form the rectangle using the points $(\pm a, 0) = (\pm 3, 0)$ and $(0, \pm b) = (0, \pm 4)$. The diagonals are the asymptotes.

    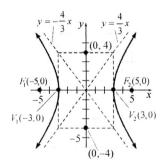

**Section 10.5** Hyperbolas

16. $\dfrac{y^2}{81} - \dfrac{x^2}{9} = 1$

Notice the equation is of the form $\dfrac{y^2}{a^2} - \dfrac{x^2}{b^2} = 1$.

Because the $y^2$-term is first, the transverse axis is the $y$-axis and the hyperbola opens up and down. The center of the hyperbola is the origin. We have that $a^2 = 81$ and $b^2 = 9$. Because $c^2 = a^2 + b^2$, we have that $c^2 = 81 + 9 = 90$, so that $c = \sqrt{90} = 3\sqrt{10}$. The vertices are $(0, \pm a) = (0, \pm 9)$, and the foci are $(0, \pm c) = \left(0, \pm 3\sqrt{10}\right)$. Since $a = 81$ and $b = 9$, the equations of the asymptotes are $y = \dfrac{9}{3}x = 3x$ and $y = -\dfrac{9}{3}x = -3x$. To help graph the hyperbola, we form the rectangle using the points $(0, \pm a) = (0, \pm 9)$ and $(\pm b, 0) = (\pm 3, 0)$. The diagonals are the asymptotes.

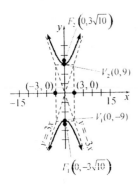

18. $x^2 - 9y^2 = 36$

$\dfrac{x^2 - 9y^2}{36} = \dfrac{36}{36}$

$\dfrac{x^2}{36} - \dfrac{y^2}{4} = 1$

Notice the equation is of the form $\dfrac{x^2}{a^2} - \dfrac{y^2}{b^2} = 1$.

Because the $x^2$-term is first, the transverse axis is the $x$-axis and the hyperbola opens left and right. The center of the hyperbola is the origin. We have that $a^2 = 36$ and $b^2 = 4$. Because $c^2 = a^2 + b^2$, we have that $c^2 = 36 + 4 = 40$, so that $c = \sqrt{40} = 2\sqrt{10}$. The vertices are $(\pm a, 0) = (\pm 6, 0)$, and the foci are $(\pm c, 0) = \left(\pm 2\sqrt{10}, 0\right)$. Since $a = 6$ and $b = 2$, the equations of the asymptotes are

$y = \dfrac{2}{6}x = \dfrac{1}{3}x$ and $y = -\dfrac{2}{6}x = -\dfrac{1}{3}x$. To help graph the hyperbola, we form the rectangle using the points $(\pm a, 0) = (\pm 6, 0)$ and $(0, \pm b) = (0, \pm 2)$. The diagonals are the asymptotes.

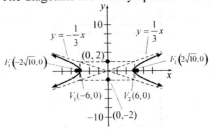

20. $4y^2 - 9x^2 = 36$

$\dfrac{4y^2 - 9x^2}{36} = \dfrac{36}{36}$

$\dfrac{y^2}{9} - \dfrac{x^2}{4} = 1$

Notice the equation is of the form $\dfrac{y^2}{a^2} - \dfrac{x^2}{b^2} = 1$.

Because the $y^2$-term is first, the transverse axis is the $y$-axis and the hyperbola opens up and down. The center of the hyperbola is the origin. We have that $a^2 = 9$ and $b^2 = 4$. Because $c^2 = a^2 + b^2$, we have that $c^2 = 9 + 4 = 13$, so that $c = \sqrt{13}$. The vertices are $(0, \pm a) = (0, \pm 3)$, and the foci are $(0, \pm c) = \left(0, \pm\sqrt{13}\right)$. Since $a = 3$ and $b = 2$, the equations of the asymptotes are $y = \dfrac{3}{2}x$ and $y = -\dfrac{3}{2}x$. To help graph the hyperbola, we form the rectangle using the points $(0, \pm a) = (0, \pm 3)$ and $(\pm b, 0) = (\pm 2, 0)$. The diagonals are the asymptotes.

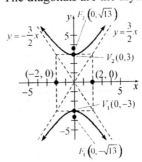

22. The given center $(0,0)$, focus $(-4,0)$, and vertex $(-1,0)$ all lie on the $x$-axis. Thus, the transverse axis is the $x$-axis and the hyperbola

Chapter 10: Conics

opens left and right. This means that the equation of the hyperbola is of the form $\frac{x^2}{a^2} - \frac{y^2}{b^2} = 1$. Now, the distance between the given vertex $(-1, 0)$ and the center $(0, 0)$ is $a = 1$ unit. Likewise, the distance between the given focus $(-4, 0)$ and the center is $c = 4$ units. Because $b^2 = c^2 - a^2$, we have that $b^2 = 4^2 - 1^2 = 16 - 1 = 15$. Thus, the equation of the hyperbola is $\frac{x^2}{1} - \frac{y^2}{15} = 1$ or $x^2 - \frac{y^2}{15} = 1$.

To help graph the hyperbola, we first find the vertex and focus that were not given. Since the center is at $(0, 0)$ and since one vertex is at $(-1, 0)$, the other vertex must be at $(1, 0)$. Similarly, one focus is at $(-4, 0)$, so the other focus is at $(4, 0)$. Next, plot the points on the graph above and below the foci. Let $x = \pm 4$:

$$(\pm 4)^2 - \frac{y^2}{15} = 1$$
$$16 - \frac{y^2}{15} = 1$$
$$-\frac{y^2}{15} = -15$$
$$y^2 = 225$$
$$y = \pm 15$$

The points above and below the foci are $(4, 15)$, $(4, -15)$, $(-4, 15)$, and $(-4, -15)$.

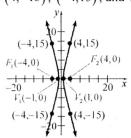

24. The given vertices $(0, \pm 6)$ and focus $(0, 8)$ all lie on the $y$-axis. Thus, the transverse axis is the $y$-axis and the hyperbola opens up and down. The center of the hyperbola is the midpoint between the two vertices. Therefore, the center is $(0, 0)$ and the equation of the hyperbola is of the form $\frac{y^2}{a^2} - \frac{x^2}{b^2} = 1$. Now, the distance from the center $(0, 0)$ to each vertex is $a = 6$ units.

Likewise, the distance from the center to the given focus $(0, 8)$ is $c = 8$ units. Also, $b^2 = c^2 - a^2$, so $b^2 = 8^2 - 6^2 = 64 - 36 = 28$. Thus, the equation of the hyperbola is $\frac{y^2}{36} - \frac{x^2}{28} = 1$.

To help graph the hyperbola, we first find the focus that was not given. Since the center is at $(0, 0)$ and since one focus is at $(0, 8)$, the other focus must be at $(0, -8)$. Next, plot the points on the graph to the left and right of the foci. Let $y = \pm 8$:

$$\frac{(\pm 8)^2}{36} - \frac{x^2}{28} = 1$$
$$\frac{16}{9} - \frac{x^2}{28} = 1$$
$$-\frac{x^2}{28} = -\frac{7}{9}$$
$$x^2 = \frac{196}{9}$$
$$x = \pm \frac{14}{3}$$

The points to the left and right of the foci are $\left(\frac{14}{3}, 8\right)$, $\left(-\frac{14}{3}, 8\right)$, $\left(\frac{14}{3}, -8\right)$, and $\left(-\frac{14}{3}, -8\right)$.

26. The given foci $(\pm 5, 0)$ and vertex $(-3, 0)$ all lie on the $x$-axis. Thus, the transverse axis is the $x$-axis and the hyperbola opens left and right. The center of the hyperbola is the midpoint between the two vertices. Therefore, the center is $(0, 0)$ and the equation of the hyperbola is of the form $\frac{x^2}{a^2} - \frac{y^2}{b^2} = 1$. Now, the distance between the given vertex $(-3, 0)$ and the center $(0, 0)$ is $a = 3$ units. Likewise, the distance between the given foci and the center is $c = 5$

units. Also, $b^2 = c^2 - a^2$, so $b^2 = 5^2 - 3^2 = 25 - 9 = 16$. Thus, the equation of the hyperbola is $\dfrac{x^2}{9} - \dfrac{y^2}{16} = 1$.

To help graph the hyperbola, we first find the vertex that was not given. Since the center is at $(0,0)$ and since one vertex is at $(-3,0)$, the other vertex must be at $(3,0)$. Next, plot the points on the graph above and below the foci. Let $x = \pm 5$:

$$\dfrac{(\pm 5)^2}{9} - \dfrac{y^2}{16} = 1$$

$$\dfrac{25}{9} - \dfrac{y^2}{16} = 1$$

$$-\dfrac{y^2}{16} = -\dfrac{16}{9}$$

$$y^2 = \dfrac{256}{9}$$

$$y = \pm \dfrac{16}{3}$$

The points above and below the foci are $\left(5, \dfrac{16}{3}\right)$, $\left(5, -\dfrac{16}{3}\right)$, $\left(-5, \dfrac{16}{3}\right)$, and $\left(-5, -\dfrac{16}{3}\right)$.

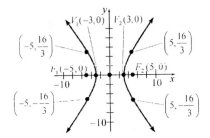

28. The given vertices $(0, \pm 4)$ both lie on the $y$-axis. Thus, the transverse axis is the $y$-axis and the hyperbola opens up and down. The center of the hyperbola is the midpoint between the two vertices. Therefore, the center is $(0,0)$ and the equation of the hyperbola is of the form $\dfrac{y^2}{a^2} - \dfrac{x^2}{b^2} = 1$. Now, the distance from the center $(0,0)$ to each vertex is $a = 4$ units. We are given that an asymptote of the hyperbola is $y = 2x$. Now the equation of the asymptote is of the form is $y = \dfrac{a}{b}x$ and $a = 4$. So,

$$\dfrac{a}{b} = 2$$

$$\dfrac{4}{b} = 2$$

$$4 = 2b$$

$$2 = b$$

Thus, the equation of the hyperbola is

$$\dfrac{y^2}{4^2} - \dfrac{x^2}{2^2} = 1$$

$$\dfrac{y^2}{16} - \dfrac{x^2}{4} = 1$$

To help graph the hyperbola, we first find the foci. Since $c^2 = a^2 + b^2$, we have that $c^2 = 16 + 4 = 20$, so $c = \sqrt{20} = 2\sqrt{5}$. Since the center is $(0,0)$, the foci are $\left(0, \pm 2\sqrt{5}\right)$. We form the rectangle using the points $(0, \pm a) = (0, \pm 4)$ and $(\pm b, 0) = (\pm 2, 0)$. The diagonals are the two asymptotes $y = 2x$ and $y = -2x$.

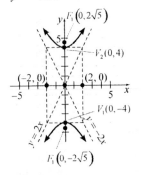

30. The given foci $(\pm 9, 0)$ both lie on the $x$-axis. Thus, the transverse axis is the $x$-axis and the hyperbola opens left and right. The center of the hyperbola is the midpoint between the two vertices. Therefore, the center is $(0,0)$ and the equation of the hyperbola is of the form $\dfrac{x^2}{a^2} - \dfrac{y^2}{b^2} = 1$. We are given that an asymptote of the hyperbola is $y = -3x$. Now the equation of the asymptote is of the form $y = \dfrac{b}{a}x$. So,

$\dfrac{b}{a} = -3$ which means $b = -3a$. Because each focus is 9 units from the center, we have $c = 9$ and $c^2 = 81$. So,

## Chapter 10: Conics

$$a^2 + b^2 = c^2$$
$$a^2 + (-3a)^2 = 81$$
$$a^2 + 9a^2 = 81$$
$$10a^2 = 81$$
$$a^2 = 8.1$$
$$a = \sqrt{8.1}$$

Since $b = -3a$, we have that $b^2 = 9a^2 = 9(8.1) = 72.9$.
Thus, the equation of the hyperbola is
$$\frac{x^2}{8.1} - \frac{y^2}{72.9} = 1.$$
To help graph the hyperbola, we find the vertices. Since $a = \sqrt{8.1}$ and the center is $(0,0)$, the vertices are $(\pm\sqrt{8.1}, 0)$. We form the rectangle using the points $(\pm a, 0) = (\pm\sqrt{8.1}, 0)$ and $(0, \pm b) = (0, \pm\sqrt{72.9})$. The diagonals are the two asymptotes $y = -3x$ and $y = 3x$.

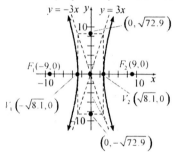

32. The graph of the hyperbola opens up and down and the center is $(0,0)$, so the equation is of the form $\frac{y^2}{a^2} - \frac{x^2}{b^2} = 1$. From the graph, we observe that the asymptotes are $y = -x$ and $y = x$. Now the equations of the asymptotes are of the form $y = -\frac{a}{b}x$ and $y = \frac{a}{b}x$. So, $\frac{a}{b} = 1$ which means $a = b$. We also observe that the vertices are $(0, \pm 1)$, so the distance from the center to each vertex is $a = 1$. Thus, $a = b = 1$ and $a^2 = b^2 = 1$. Thus, the equation of the hyperbola is $\frac{y^2}{1} - \frac{x^2}{1} = 1$ or $y^2 - x^2 = 1$.

34. The graph of the hyperbola opens left and right and the center is $(0,0)$, so the equation is of the form $\frac{x^2}{a^2} - \frac{y^2}{b^2} = 1$. From the graph, we observe that the asymptotes are $y = -2x$ and $y = 2x$. Now the equations of the asymptotes are of the form $y = -\frac{b}{a}x$ and $y = \frac{b}{a}x$. So, $\frac{b}{a} = 2$ which means $b = 2a$. We also observe that the vertices are $(\pm 2, 0)$, so the distance from the center to each vertex is $a = 2$ and $b = 2(2) = 4$. Thus, $a^2 = 4$ and $b^2 = 16$, and the equation of the hyperbola is $\frac{x^2}{4} - \frac{y^2}{16} = 1$.

36. Answers may vary. One possibility follows: When the eccentricity, $e$, of a hyperbola is close to 1, the hyperbola will be narrower. As $e$ gets larger the hyperbola will be wider.

38. $\begin{cases} 3x + 4y = 3 & (1) \\ -6x + 2y = -\dfrac{7}{2} & (2) \end{cases}$

We use elimination to solve the system. Multiply both sides of equation (2) by $-2$, and add the result to equation (1).
$$3x + 4y = 3$$
$$12x - 4y = 7$$
$$\overline{15x = 10}$$
$$x = \frac{10}{15} = \frac{2}{3}$$

Substituting $\frac{2}{3}$ for $x$ into equation (1), we obtain
$$3\left(\frac{2}{3}\right) + 4y = 3$$
$$2 + 4y = 3$$
$$4y = 1$$
$$y = \frac{1}{4}$$

The solution is the ordered pair $\left(\frac{2}{3}, \frac{1}{4}\right)$.

40. $\begin{cases} -2x + y = 8 & (1) \\ x - \dfrac{1}{2}y = -4 & (2) \end{cases}$

We use elimination to solve the system. Multiply both sides of equation (2) by 2, and add the result to equation (1).

$-2x + y = 8$
$\underline{2x - y = -8}$
$0 = 0$

The system is dependent. The solution is $\{(x, y) | -2x + y = 8\}$.

42. Answers will vary.

44. $\dfrac{x^2}{9} - \dfrac{y^2}{16} = 1$

$-\dfrac{y^2}{16} = 1 - \dfrac{x^2}{9}$

$-16\left(-\dfrac{y^2}{16}\right) = -16\left(1 - \dfrac{x^2}{9}\right)$

$y^2 = 16\left(\dfrac{x^2}{9} - 1\right)$

$y = \pm\sqrt{16\left(\dfrac{x^2}{9} - 1\right)}$

$y = \pm 4\sqrt{\dfrac{x^2}{9} - 1}$

Let $Y_1 = 4\sqrt{\dfrac{x^2}{9} - 1}$ and $Y_2 = -4\sqrt{\dfrac{x^2}{9} - 1}$.

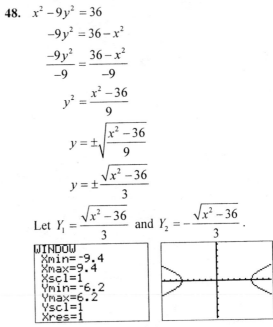

46. $\dfrac{y^2}{81} - \dfrac{x^2}{9} = 1$

$\dfrac{y^2}{81} = \dfrac{x^2}{9} + 1$

$y^2 = 81\left(\dfrac{x^2}{9} + 1\right)$

$y = \pm\sqrt{81\left(\dfrac{x^2}{9} + 1\right)}$

$y = \pm 9\sqrt{\dfrac{x^2}{9} + 1}$

Let $Y_1 = 9\sqrt{\dfrac{x^2}{9} + 1}$ and $Y_2 = -9\sqrt{\dfrac{x^2}{9} + 1}$.

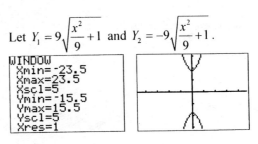

48. $x^2 - 9y^2 = 36$

$-9y^2 = 36 - x^2$

$\dfrac{-9y^2}{-9} = \dfrac{36 - x^2}{-9}$

$y^2 = \dfrac{x^2 - 36}{9}$

$y = \pm\sqrt{\dfrac{x^2 - 36}{9}}$

$y = \pm\dfrac{\sqrt{x^2 - 36}}{3}$

Let $Y_1 = \dfrac{\sqrt{x^2 - 36}}{3}$ and $Y_2 = -\dfrac{\sqrt{x^2 - 36}}{3}$.

50. $4y^2 - 9x^2 = 36$

$4y^2 = 9x^2 + 36$

$y^2 = \dfrac{9x^2 + 36}{4}$

$y = \pm\sqrt{\dfrac{9x^2 + 36}{4}}$

$y = \pm\dfrac{\sqrt{9x^2 + 36}}{2}$

Let $Y_1 = \dfrac{\sqrt{9x^2 + 36}}{2}$ and $Y_2 = -\dfrac{\sqrt{9x^2 + 36}}{2}$.

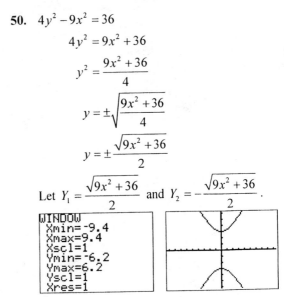

## Chapter 10: Conics

**Putting the Concepts Together (Sections 10.1 – 10.5)**

1.  $d(P_1, P_2) = \sqrt{(x_2 - x_1)^2 + (y_2 - y_1)^2}$
    $= \sqrt{(3-(-6))^2 + (-2-4)^2}$
    $= \sqrt{(9)^2 + (-6)^2}$
    $= \sqrt{81 + 36}$
    $= \sqrt{117}$
    $= 3\sqrt{13}$

2.  $M = \left(\dfrac{x_1 + x_2}{2}, \dfrac{y_1 + y_2}{2}\right)$
    $= \left(\dfrac{-3+5}{2}, \dfrac{1+(-7)}{2}\right) = \left(\dfrac{2}{2}, \dfrac{-6}{2}\right) = (1, -3)$

3.  $(x+2)^2 + (y-8)^2 = 36$
    $(x-(-2))^2 + (y-8)^2 = 6^2$
    The center is $(h, k) = (-2, 8)$, and the radius is $r = 6$.

    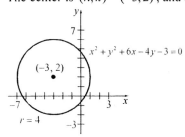

4.  $x^2 + y^2 + 6x - 4y - 3 = 0$
    $(x^2 + 6x) + (y^2 - 4y) = 3$
    $(x^2 + 6x + 9) + (y^2 - 4y + 4) = 3 + 9 + 4$
    $(x+3)^2 + (y-2)^2 = 16$
    $(x-(-3))^2 + (y-2)^2 = 4^2$
    The center is $(h, k) = (-3, 2)$, and the radius is $r = 4$.

5.  The radius of the circle will be the distance from the center point $(0, 0)$ to the point on the circle $(-5, 12)$. Thus, $r = \sqrt{(-5-0)^2 + (12-0)^2} = 13$.

    The equation of the circle is
    $(x-h)^2 + (y-k)^2 = r^2$
    $(x-0)^2 + (y-0)^2 = 13^2$
    $x^2 + y^2 = 169$

6.  The center of the circle will be the midpoint of the diameter with endpoints $(-1, 5)$ and $(5, -3)$.
    Thus, $(h, k) = \left(\dfrac{-1+5}{2}, \dfrac{5+(-3)}{2}\right) = (2, 1)$. The radius of the circle will be the distance from the center point $(2, 1)$ to one of the endpoints of the diameter, say $(-1, 5)$. Thus,
    $r = \sqrt{(-1-2)^2 + (5-1)^2} = \sqrt{25} = 5$.
    The equation of the circle is
    $(x-h)^2 + (y-k)^2 = r^2$
    $(x-2)^2 + (y-1)^2 = 5^2$
    $(x-2)^2 + (y-1)^2 = 25$

7.  Notice that $(x+2)^2 = -4(y-4)$ is of the form $(x-h)^2 = -4a(y-k)$, where $(h, k) = (-2, 4)$ and $-4a = -4$, so that $a = 1$. Now, the graph of an equation of the form $(x-h)^2 = -4a(y-k)$ will be a parabola that opens downward with the vertex at $(h, k)$, focus at $(h, k-a)$, and directrix of $y = k + a$. Note that $k - a = 4 - 1 = 3$ and $k + a = 4 + 1 = 5$. Thus, the graph of $(x+2)^2 = -4(y-4)$ is a parabola that opens downward with vertex $(-2, 4)$, focus $(-2, 3)$, and directrix $y = 5$. To help graph the parabola, we plot the two points on the graph to the left and right of the focus. Let $y = 3$:
    $(x+2)^2 = -4(3-4)$
    $(x+2)^2 = 4$
    $x+2 = \pm 2$
    $x = -2 \pm 2$
    $x = -4$ or $x = 0$
    The points $(-4, 3)$ and $(0, 3)$ are on the graph.

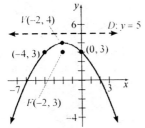

8. We complete the square in $y$ to write the equation in standard form:
$$y^2 + 2y - 8x + 25 = 0$$
$$y^2 + 2y = 8x - 25$$
$$y^2 + 2y + 1 = 8x - 25 + 1$$
$$y^2 + 2y + 1 = 8x - 24$$
$$(y+1)^2 = 8(x-3)$$

Notice that $(y+1)^2 = 8(x-3)$ is of the form $(y-k)^2 = 4a(x-h)$, where $(h,k) = (3,-1)$ and $4a = 8$, so that $a = 2$. Now, the graph of an equation of the form $(y-k)^2 = 4a(x-h)$ will be a parabola that opens to the right with the vertex at $(h,k)$, focus at $(h+a,k)$, and directrix of $x = h - a$. Note that $h + a = 3 + 2 = 5$ and $h - a = 3 - 2 = 1$. Thus, the graph of $(y+1)^2 = 8(x-3)$ is a parabola that opens to the right with vertex $(3,-1)$, focus $(5,-1)$, and directrix $x = 1$. To help graph the parabola, we plot the two points on the graph below and above the focus. Let $x = 5$:
$$(y+1)^2 = 8(5-3)$$
$$(y+1)^2 = 16$$
$$y + 1 = \pm 4$$
$$y = -1 \pm 4$$
$$y = -5 \text{ or } y = 3$$

The points $(5,-5)$ and $(5,3)$ are on the graph.

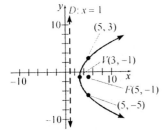

9. The distance from the vertex $(h,k) = (-1,-2)$ to the focus $(-1,-5)$ is $a = 3$. The vertex and focus both lie on the vertical line $x = -1$, which is the axis of symmetry (parallel to the $y$-axis).

Because the focus is below the vertex, we know that the parabola will open downward. This means the equation of the parabola is of the form $(x-h)^2 = -4a(y-k)$ with $a = 3$, $h = -1$, and $k = -2$:
$$(x-(-1))^2 = -4 \cdot 3(y-(-2))$$
$$(x+1)^2 = -12(y+2)$$

10. The vertex is $(-3,3)$ and the axis of symmetry is parallel to the $x$-axis, so the axis of symmetry is the line $y = 3$ and the parabola either opens to the left or right. Because the graph contains the point $(-1,7)$, which is to the right of the vertex, the parabola must open to the right. Therefore, the equation of the parabola is of the form $(y-k)^2 = 4a(x-h)$, with $h = -3$ and $k = 3$. Now $y = 7$ when $x = -1$, so
$$(7-3)^2 = 4a(-1-(-3))$$
$$(4)^2 = 4a(2)$$
$$16 = 8a$$
$$a = 2$$
The equation of the parabola is
$$(y-3)^2 = 4 \cdot 2(x-(-3))$$
$$(y-3)^2 = 8(x+3)$$

11. $x^2 + 9y^2 = 81$
$$\frac{x^2 + 9y^2}{81} = \frac{81}{81}$$
$$\frac{x^2}{81} + \frac{y^2}{9} = 1$$

The larger number, 81, is in the denominator of the $x^2$-term. This means that the major axis is the $x$-axis and that the equation of the ellipse is of the form $\frac{x^2}{a^2} + \frac{y^2}{b^2} = 1$, so that $a^2 = 81$ and $b^2 = 9$. The center of the ellipse is $(0,0)$. Because $c^2 = a^2 - b^2$, we have that $c^2 = 81 - 9 = 72$, so that $c = \pm\sqrt{72} = \pm 6\sqrt{2}$. Since the major axis is the $x$-axis, the foci are

## Chapter 10: Conics

$(-6\sqrt{2}, 0)$ and $(6\sqrt{2}, 0)$. To find the $x$-intercepts (vertices), let $y = 0$; to find the $y$-intercepts, let $x = 0$:

$x$-intercepts:
$x^2 + 9(0)^2 = 81$
$x^2 = 81$
$x = \pm 9$

$y$-intercepts:
$0^2 + 9y^2 = 81$
$9y^2 = 81$
$y^2 = 9$
$y = \pm 3$

The intercepts are $(-9, 0)$, $(9, 0)$, $(0, -3)$, and $(0, 3)$.

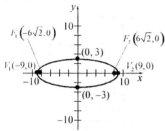

12. $\dfrac{(x+1)^2}{36} + \dfrac{(y-2)^2}{49} = 1$

The center of the ellipse is $(h, k) = (-1, 2)$. Because the larger number, 49, is the denominator of the $y^2$-term, the major axis is parallel to the $y$-axis. Because $a^2 = 49$ and $b^2 = 36$, we have that $c^2 = a^2 - b^2 = 49 - 36 = 13$. The vertices are $a = 7$ units below and above the center at $V_1(-1, -5)$ and $V_2(-1, 9)$. The foci are $c = \sqrt{13}$ units below and above the center at $F_1(-1, 2 - \sqrt{13})$ and $F_2(-1, 2 + \sqrt{13})$. We plot the points $b = 6$ units to the left and right of the center point at $(-7, 2)$ and $(5, 2)$.

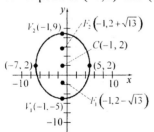

13. The given foci $(0, \pm 6)$ and the given vertices $(0, \pm 9)$ lie on the $y$-axis. The center of the ellipse is the midpoint between the two vertices (or foci). Thus, the center of the ellipse is $(0, 0)$ and its equation is of the form $\dfrac{x^2}{b^2} + \dfrac{y^2}{a^2} = 1$. The distance from the center of the ellipse to the vertex is $a = 9$ units. The distance from the center of the ellipse to the focus is $c = 6$ units. Because $b^2 = a^2 - c^2$, we have that $b^2 = 9^2 - 6^2 = 81 - 36 = 45$. Thus, the equation of the ellipse is $\dfrac{x^2}{45} + \dfrac{y^2}{81} = 1$.

14. The center is $(h, k) = (3, -4)$. The center, focus, and vertex given all lie on the horizontal line $y = -4$. Therefore, the major axis is parallel to the $x$-axis, and the equation of the ellipse is of the form $\dfrac{(x-h)^2}{a^2} + \dfrac{(y-k)^2}{b^2} = 1$. Now, $a = 4$ is the distance from the center $(3, -4)$ to a vertex $(7, -4)$, and $c = 3$ is the distance from the center to a focus $(6, -4)$. Also, $b^2 = a^2 - c^2$, so $b^2 = 4^2 - 3^2 = 16 - 9 = 7$. Thus, the equation of the ellipse is

$\dfrac{(x-3)^2}{16} + \dfrac{(y-(-4))^2}{7} = 1$

$\dfrac{(x-3)^2}{16} + \dfrac{(y+4)^2}{7} = 1$

15. $\dfrac{y^2}{81} - \dfrac{x^2}{9} = 1$

Notice the equation is of the form $\dfrac{y^2}{a^2} - \dfrac{x^2}{b^2} = 1$. Because the $y^2$-term is first, the transverse is along the $y$-axis and the hyperbola opens up and down. The center of the hyperbola is $(0, 0)$. We have that $a^2 = 81$ and $b^2 = 9$. Because $c^2 = a^2 + b^2$, we have that $c^2 = 81 + 9 = 90$, so that $c = \sqrt{90} = 3\sqrt{10}$. The vertices are $(0, \pm a) = (0, \pm 9)$, and the foci are $(0, \pm c) = (0, \pm 3\sqrt{10})$. Since $a = 9$ and $b = 3$, the equations of the asymptotes are $y = \dfrac{9}{3}x = 3x$ and $y = -\dfrac{9}{3}x = -3x$. To help graph the hyperbola, we form the rectangle using the

**Putting the Concepts Together (Sections 10.1–10.5)**

points $(0, \pm a) = (0, \pm 9)$ and $(\pm b, 0) = (\pm 3, 0)$. The diagonals are the asymptotes.

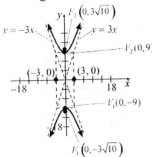

16. $25x^2 - y^2 = 25$

$$\frac{25x^2 - y^2}{25} = \frac{25}{25}$$

$$\frac{x^2}{1} - \frac{y^2}{25} = 1$$

Notice the equation is of the form $\frac{x^2}{a^2} - \frac{y^2}{b^2} = 1$.

Because the $x^2$-term is first, the transverse is along the x-axis and the hyperbola opens left and right. The center of the hyperbola is $(0, 0)$. We have that $a^2 = 1$ and $b^2 = 25$. Because $c^2 = a^2 + b^2$, we have that $c^2 = 1 + 25 = 26$, so that $c = \sqrt{26}$. The vertices are $(\pm a, 0) = (\pm 1, 0)$, and the foci are $(\pm c, 0) = (\pm \sqrt{26}, 0)$. Since $a = 1$ and $b = 5$, the equations of the asymptotes are $y = \frac{5}{1}x = 5x$ and $y = -\frac{5}{1}x = -5x$. To help graph the hyperbola, we form the rectangle using the points $(\pm a, 0) = (\pm 1, 0)$ and $(0, \pm b) = (0, \pm 5)$. The diagonals are the asymptotes.

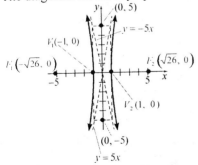

17. The given center $(0, 0)$, focus $(0, -5)$, and vertex $(0, -2)$ all lie on the y-axis. Thus, the transverse axis is the y-axis and the hyperbola

opens up and down. This means that the equation of the hyperbola is $\frac{y^2}{a^2} - \frac{x^2}{b^2} = 1$. Now, the distance between the given vertex $(0, -2)$ and the center $(0, 0)$ is $a = 2$ units. Likewise, the distance between the given focus $(0, -5)$ and the center is $c = 5$ units. Because $b^2 = c^2 - a^2$, we have that $b^2 = 5^2 - 2^2 = 25 - 4 = 21$. Thus, the equation of the hyperbola is $\frac{y^2}{4} - \frac{x^2}{21} = 1$.

18. The light bulb should be located at the focus of the flood light, so we need to find where the focus of the flood light is. To solve this problem, we draw a parabola on a Cartesian plane so that the vertex is the origin and the focus is on the positive y-axis. The width of the parabola is 36 inches, and the depth is 12 inches. Therefore, we know two points on the graph of the parabola: $(-18, 12)$ and $(18, 12)$.

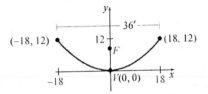

The equation of the parabola has the form $x^2 = 4ay$. Since $(18, 12)$ is a point on the graph, we have

$18^2 = 4a(12)$

$324 = 48a$

$a = \frac{324}{48} = 6.75$

The light bulb should be located 6.75 inches above the vertex, along its axis of symmetry.

**10.6 Preparing for Nonlinear Systems of Equations**

1. $\begin{cases} y = 2x - 5 & (1) \\ 2x - 3y = 7 & (2) \end{cases}$

Substituting $2x - 5$ for y in equation (2), we obtain

ISM – 643

## Chapter 10: Conics

$2x - 3(2x - 5) = 7$
$2x - 6x + 15 = 7$
$-4x + 15 = 7$
$-4x = -8$
$x = 2$

Substituting 2 for $x$ in equation (1), we obtain
$y = 2(2) - 5 = 4 - 5 = -1$.

The solution is the ordered pair $(2, -1)$.

$6y = 15$
$y = \dfrac{15}{6}$
$y = \dfrac{5}{2}$

Substituting $\dfrac{5}{2}$ for $y$ in equation (1), we obtain

$2x - 4\left(\dfrac{5}{2}\right) = -11$
$2x - 10 = -11$
$2x = -1$
$x = -\dfrac{1}{2}$

The solution is the ordered pair $\left(-\dfrac{1}{2}, \dfrac{5}{2}\right)$.

3. $\begin{cases} 3x - 5y = 4 & (1) \\ -6x + 10y = -8 & (2) \end{cases}$

Multiply both sides of equation (1) by 2, and add the result to equation (2).
$6x - 10y = 8$
$-6x + 10y = -8$
―――――――――
$0 = 0$

The system is dependent. The solution is $\{(x, y) \mid 3x - 5y = 4\}$.

## 10.6 Exercises

2. FALSE. See Exercise 15.

4. Answers will vary.

6. $\begin{cases} y = x^3 + 2 \\ y = x + 2 \end{cases}$

First, graph each equation in the system.

2. $\begin{cases} 2x - 4y = -11 & (1) \\ -x + 5y = 13 & (2) \end{cases}$

Multiply both sides of equation (2) by 2, and add the result to equation (1).
$2x - 4y = -11$
$-2x + 10y = 26$
―――――――――

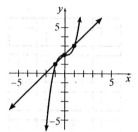

The system apparently has three solutions. Now substitute $x^3 + 2$ for $y$ into the second equation:
$x^3 + 2 = x + 2$
$x^3 - x = 0$
$x(x + 1)(x - 1) = 0$
$x = 0$ or $x = -1$ or $x = 1$

Substitute these $x$-values into the first equation to find the corresponding $y$-values:
$x = 0$: $y = 0^3 + 2 = 2$
$x = -1$: $y = (-1)^3 + 2 = -1 + 2 = 1$
$x = 1$: $y = 1^3 + 2 = 1 + 2 = 3$

All three pairs check, so the solutions are $(-1, 1)$, $(0, 2)$ and $(1, 3)$.

8. $\begin{cases} y = \sqrt{100 - x^2} \\ x + y = 14 \end{cases}$

First, graph each equation in the system.

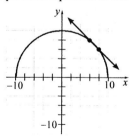

The system apparently has two solutions. Now substitute $\sqrt{100 - x^2}$ for $y$ into the second equation:
$x + \sqrt{100 - x^2} = 14$
$\sqrt{100 - x^2} = 14 - x$

$$\left(\sqrt{100-x^2}\right)^2 = (14-x)^2$$
$$100 - x^2 = 196 - 28x + x^2$$
$$-2x^2 + 28x - 96 = 0$$
$$x^2 - 14x + 48 = 0$$
$$(x-6)(x-8) = 0$$
$$x = 6 \text{ or } x = 8$$

Substitute these $x$-values into the first equation to find the corresponding $y$-values:

$x = 6$: $y = \sqrt{100-6^2} = \sqrt{100-36} = \sqrt{64} = 8$

$x = 8$: $y = \sqrt{100-8^2} = \sqrt{100-64} = \sqrt{36} = 6$

Both pairs check, so the solutions are $(6, 8)$ and $(8, 6)$.

10. $\begin{cases} x^2 + y^2 = 16 \\ y = x^2 - 4 \end{cases}$

First, graph each equation in the system.

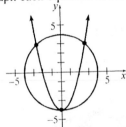

The system apparently has three solutions. Now solve the first equation for $x^2$: $x^2 = 16 - y^2$.

Substitute the result for $x^2$ into the second equation:
$$y = (16 - y^2) - 4$$
$$y = 12 - y^2$$
$$y^2 + y - 12 = 0$$
$$(y+4)(y-3) = 0$$
$$y = -4 \text{ or } y = 3$$

Substitute these $y$-values into the first equation to find the corresponding $x$-values:

$y = -4$: $x^2 + (-4)^2 = 16$
$$x^2 + 16 = 16$$
$$x^2 = 0$$
$$x = 0$$

$y = 3$: $x^2 + 3^2 = 16$
$$x^2 + 9 = 16$$
$$x^2 = 7$$
$$x = \pm\sqrt{7}$$

All three pairs check, so the solutions are $(0, -4)$, $\left(-\sqrt{7}, 3\right)$, and $\left(\sqrt{7}, 3\right)$.

12. $\begin{cases} xy = 1 \\ x^2 - y = 0 \end{cases}$

First, graph each equation in the system.

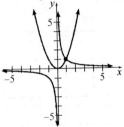

The system apparently has one solution. Now solve the second equation for $y$: $y = x^2$.
Substitute the result into the first equation.
$$x(x^2) = 1$$
$$x^3 = 1$$
$$x = \sqrt[3]{1} = 1$$

Substitute 1 for $x$ into the equation $y = x^2$ to find the corresponding $y$-value.

$x = 1$: $y = 1^2 = 1$

The ordered pair checks, so the solution is $(1, 1)$.

14. $\begin{cases} x^2 + y^2 = 8 \\ x^2 + y^2 + 4y = 0 \end{cases}$

First, graph each equation in the system.

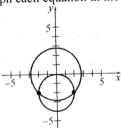

The system apparently has two solutions. Now multiply the first equation by $-1$ and add the result to the second equation:

$$\begin{array}{r} -x^2 - y^2 \phantom{+4y} = -8 \\ x^2 + y^2 + 4y = 0 \\ \hline 4y = -8 \\ y = -2 \end{array}$$

Substitute this $y$-value into the first equation to find the corresponding $x$-values:

## Chapter 10: Conics

$y = -2:\ x^2 + (-2)^2 = 8$
$\qquad\qquad x^2 + 4 = 8$
$\qquad\qquad x^2 = 4$
$\qquad\qquad x = \pm\sqrt{4} = \pm 2$

Both ordered pairs check, so the solutions are $(-2,-2)$ and $(2,-2)$.

**16.** $\begin{cases} 4x^2 + 16y^2 = 16 \\ 2x^2 - 2y^2 = 8 \end{cases}$

First, graph each equation in the system.

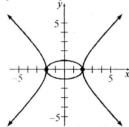

The system apparently has two solutions. Now multiply the second equation by $-2$ and add the result to the first equation:

$4x^2 + 16y^2 = 16$
$\underline{-4x^2 + 4y^2 = -16}$
$\qquad\ \ 20y^2 = 0$
$\qquad\ \ \ \ y^2 = 0$
$\qquad\ \ \ \ y = 0$

Substitute 0 for $y$ into the first equation to find the $x$ values.

$4x^2 + 16(0)^2 = 16$
$\qquad\ \ 4x^2 = 16$
$\qquad\ \ \ x^2 = 4$
$\qquad\ \ \ x = \pm 2$

Both ordered pairs check, so the solutions are $(-2,0)$ and $(2,0)$.

**18.** $\begin{cases} 2x^2 + y^2 = 18 \\ x^2 - y^2 = 9 \end{cases}$

First, graph each equation in the system.

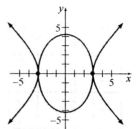

The system apparently has three solutions. Now multiply the second equation by $-1$ and add the result to the first equation:

$2x^2 + y^2 = 18$
$\underline{\ x^2 - y^2 = 9\ }$
$3x^2\qquad = 27$
$\quad x^2 = 9$
$\quad x = \pm 3$

Substitute the result for $x^2$ into the first equation:

$2(9) + y^2 = 18$
$\qquad\ y^2 = 0$
$\qquad\ y = 0$

Both ordered pairs check, so the solutions are $(-3, 0)$ and $(3, 0)$.

**20.** $\begin{cases} 2x^2 - 5x + y = 12 \\ 14x - 2y = -16 \end{cases}$

First, graph each equation in the system.

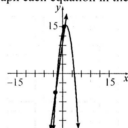

The system apparently has two solutions. Now divide the second equation by 2 and add the result to the first equation:

$2x^2 - 5x + y = 12$
$\underline{\ \ \ \ 7x - y = -8\ }$
$2x^2 + 2x\ \ \ = 4$
$\quad x^2 + x = 2$
$\quad x^2 + x - 2 = 0$
$(x+2)(x-1) = 0$
$x = -2\ \text{or}\ x = 1$

Substitute these results into the second equation to find the corresponding $y$-values.

ISM – 646

## Section 10.6 Nonlinear Systems of Equations

$x = -2$: $14(-2) - 2y = -16$
$-28 - 2y = -16$
$-2y = 12$
$y = -6$

$x = 1$: $14(1) - 2y = -16$
$14 - 2y = -16$
$-2y = -30$
$y = 15$

Both pairs check, so the solutions are $(-2, -6)$ and $(1, 15)$.

22. $\begin{cases} y = x^2 + 4x + 5 \\ x - y = 9 \end{cases}$

First, graph each equation in the system.

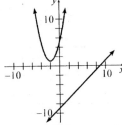

The system apparently has no solution. Now solve the second equation for $y$: $y = x - 9$. Substitute the result into the first equation.
$x - 9 = x^2 + 4x + 5$
$0 = x^2 + 3x + 13$
The discriminant of this resulting quadratic equation is $D = 3^2 - 4(1)(13) = 9 - 52 = -43$, which means the equation has no real solutions. Therefore, the system is inconsistent, and the solution set is $\varnothing$.

24. $\begin{cases} x^2 + y^2 = 25 \\ x^2 - y^2 = 25 \end{cases}$

First, graph each equation in the system.

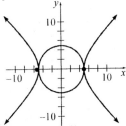

The system apparently has two solutions. Now add the two equations:

$x^2 + y^2 = 25$
$\underline{x^2 - y^2 = 25}$
$2x^2 \phantom{+y^2} = 50$
$x^2 = 25$
$x = \pm 5$

Substitute 25 for $x^2$ into the first equation to find the $y$-values.
$25 + y^2 = 25$
$y^2 = 0$
$y = 0$

Both pairs check, so the solutions are $(-5, 0)$ and $(5, 0)$.

26. $\begin{cases} (x+4)^2 + (y-2)^2 = 100 \\ 8x + y = 18 \end{cases}$

First, graph each equation in the system.

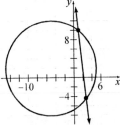

The system apparently has two solutions. Now solve the second equation for $y$: $y = -8x + 18$.
Substitute the result into the first equation:
$(x+5)^2 + [(-8x+18) - 2]^2 = 100$
$(x+5)^2 + (-8x+16)^2 = 100$
$x^2 + 10x + 25 + 64x^2 - 256x + 256 = 100$
$65x^2 - 246x + 181 = 0$
$(65x - 181)(x - 1) = 0$
$65x - 181 = 0$ or $x - 1 = 0$
$x = \dfrac{181}{65}$ or $x = 1$

Substitute these $x$-values into the equation $y = -8x + 18$ to find the corresponding $y$-values:

$x = \dfrac{181}{65}$: $y = -8\left(\dfrac{181}{65}\right) + 18 = -\dfrac{278}{65}$

$x = 1$: $y = -8(1) + 18 = 10$

Both pairs check, so the solutions are $(1, 10)$ and $\left(\dfrac{181}{65}, -\dfrac{278}{65}\right)$.

## Chapter 10: Conics

**28.** $\begin{cases} (x+2)^2 + (y-1)^2 = 4 \\ y^2 - 2y - x = 5 \end{cases}$

First, graph each equation in the system.

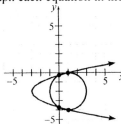

The system apparently has four solutions. Now expand the first equation and simplify:
$$(x+2)^2 + (y-1)^2 = 4$$
$$x^2 + 4x + 4 + y^2 - 2y + 1 = 4$$
$$x^2 + 4x + y^2 - 2y = -1$$

Multiply the second equation by $-1$ and add the result to the expanded form of the first equation:
$$x^2 + 4x + y^2 - 2y = -1$$
$$\underline{\phantom{xx} x - y^2 + 2y = -5}$$
$$x^2 + 5x \phantom{xxxxxxx} = -6$$
$$x^2 + 5x + 6 = 0$$
$$(x+3)(x+2) = 0$$
$$x = -3 \text{ or } x = -2$$

Substitute these $x$-values into the first equation to find the corresponding $y$-values.

$x = -3$: $(-3+2)^2 + (y-1)^2 = 4$
$$1 + (y-1)^2 = 4$$
$$(y-1)^2 = 3$$
$$y - 1 = \pm\sqrt{3}$$
$$y = 1 \pm \sqrt{3}$$

$x = -2$: $(-2+2)^2 + (y-1)^2 = 4$
$$0 + (y-1)^2 = 4$$
$$(y-1)^2 = 4$$
$$y - 1 = \pm 2$$
$$y = 1 \pm 2$$
$$y = -1 \text{ or } y = 3$$

All four pairs check, so the solutions are $\left(-3, 1-\sqrt{3}\right)$, $\left(-3, 1+\sqrt{3}\right)$, $(-2, -1)$, and $(-2, 3)$.

**30.** $\begin{cases} 9x^2 + 4y^2 = 36 \\ x^2 + (y-7)^2 = 4 \end{cases}$

First, graph each equation in the system.

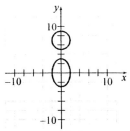

The system apparently has no solution. Now expand the second equation and solve it for $x^2$:
$$x^2 + (y-7)^2 = 4$$
$$x^2 + y^2 - 14y + 49 = 4$$
$$x^2 = -y^2 + 14y - 45$$

Substitute the result for $x^2$ into the first equation:
$$9(-y^2 + 14y - 45) + 4y^2 = 36$$
$$-9y^2 + 126y - 405 + 4y^2 = 36$$
$$-5y^2 + 126y - 441 = 0$$
$$5y^2 - 126y + 441 = 0$$
$$y = \frac{-(-126) \pm \sqrt{(-126)^2 - 4(5)(441)}}{2(5)}$$
$$= \frac{126 \pm \sqrt{7056}}{10} = \frac{126 \pm 84}{10}$$
$$y = \frac{126 - 84}{10} = \frac{42}{10} = 4.2 \text{ or } y = \frac{126 + 84}{10} = \frac{210}{10} = 21$$

Substitute these $y$-values into the first equation to find the corresponding $x$-values.

$y = 4.2$: $9x^2 + 4(4.2)^2 = 36$
$$9x^2 + 70.6 = 36$$
$$9x^2 = -34.56$$
$$x^2 = -3.84$$
$$x = \pm\sqrt{-3.84} \text{ which is not real.}$$

$y = 21$: $9x^2 + 4(21)^2 = 36$
$$9x^2 + 1764 = 36$$
$$9x^2 = -1728$$
$$x^2 = -192$$
$$x = \pm\sqrt{-192} \text{ which is not real.}$$

Since neither of the $y$-values result in real $x$-values, the system is inconsistent, and the solution set is $\varnothing$.

**32.** $\begin{cases} y - 2x = 1 \\ 2x^2 + y^2 = 1 \end{cases}$

First, graph each equation in the system.

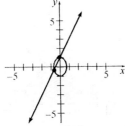

The system apparently has two solutions. Now solve the first equation for $y$: $y = 2x + 1$.
Substitute the result into the first equation:
$$2x^2 + (2x+1)^2 = 1$$
$$2x^2 + 4x^2 + 4x + 1 = 1$$
$$6x^2 + 4x = 0$$
$$2x(3x + 2) = 0$$
$$x = 0 \text{ or } x = -\frac{2}{3}$$

Substitute these $x$-values into the equation $y = 2x + 1$ to find the corresponding $y$-values.

$x = 0$: $y = 2(0) + 1 = 1$

$x = -\frac{2}{3}$: $y = 2\left(-\frac{2}{3}\right) + 1 = -\frac{4}{3} + 1 = -\frac{1}{3}$

Both pairs check, so the solutions are $\left(-\frac{2}{3}, -\frac{1}{3}\right)$ and $(0, 1)$.

34. $\begin{cases} 4x^2 + 3y^2 = 4 \\ 6y^2 - 2x^2 = 3 \end{cases}$

First, graph each equation in the system.

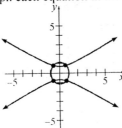

The system apparently has four solutions. Now multiply the second equation by 2 and add the result to the first equation:
$$4x^2 + 3y^2 = 4$$
$$-4x^2 + 12y^2 = 6$$

$$15y^2 = 10$$
$$y^2 = \frac{10}{15} = \frac{2}{3}$$
$$y = \pm\sqrt{\frac{2}{3}} = \pm\frac{\sqrt{6}}{3}$$

Substitute $\frac{2}{3}$ for $y^2$ into the first equation to find the $x$-values:
$$4x^2 + 3\left(\frac{2}{3}\right) = 4$$
$$4x^2 + 2 = 4$$
$$4x^2 = 2$$
$$x^2 = \frac{2}{4} = \frac{1}{2}$$
$$x = \pm\sqrt{\frac{1}{2}} = \pm\frac{\sqrt{2}}{2}$$

All four pairs check, so the solutions are
$\left(-\frac{\sqrt{2}}{2}, -\frac{\sqrt{6}}{3}\right)$, $\left(-\frac{\sqrt{2}}{2}, \frac{\sqrt{6}}{3}\right)$, $\left(\frac{\sqrt{2}}{2}, -\frac{\sqrt{6}}{3}\right)$, and $\left(\frac{\sqrt{2}}{2}, \frac{\sqrt{6}}{3}\right)$.

36. $\begin{cases} x^2 + y^2 = 65 \\ y = -x^2 + 9 \end{cases}$

First, graph each equation in the system.

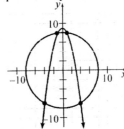

The system apparently has four solutions. Now Solve the second equation for $x^2$: $x^2 = -y + 9$

Substitute this result for $x^2$ into the first equation:
$$(-y + 9) + y^2 = 65$$
$$y^2 - y - 56 = 0$$
$$(y + 7)(y - 8) = 0$$
$$y = -7 \text{ or } y = 8$$

Substitute these $y$-values into the equation $x^2 = -y + 9$ to find the corresponding $x$-values:

$y = -7$: $x^2 = -(-7) + 9 = 7 + 9 = 16$
$$x = \pm\sqrt{16} = \pm 4$$

## Chapter 10: Conics

$y = 8$: $x^2 = -8 + 9 = 1$
$x = \pm\sqrt{1} = \pm 1$

All four pairs check, so the solutions are
$(-4,-7)$, $(4,-7)$, $(-1, 8)$, and $(1, 8)$.

38. Let $x$ and $y$ represent the two numbers.
$$\begin{cases} x + y = 8 \\ x^2 + y^2 = 160 \end{cases}$$
Solve the first equation for $x$: $x = -y + 8$.
Substitute the result into the second equation:
$$(-y+8)^2 + y^2 = 160$$
$$(y^2 - 16y + 64) + y^2 = 160$$
$$2y^2 - 16y - 96 = 0$$
$$y^2 - 8y - 48 = 0$$
$$(y+4)(y-12) = 0$$
$$y = -4 \text{ or } y = 12$$
Substitute these $y$-values into the equation $x = -y + 8$ to find the corresponding $x$-values.
$y = -4$: $x = -(-4) + 8 = 12$
$y = 12$: $x = -12 + 8 = -4$
Note that the two outcomes ultimately result in the same two numbers. The two numbers are $-4$ and $12$.

40. Let $x$ represent the length and $y$ represent the width of the rectangle.
$$\begin{cases} 2x + 2y = 64 \\ xy = 240 \end{cases}$$
Solve the first equation for $y$: $y = 32 - x$.
Substitute the result into the second equation:
$$x(32 - x) = 240$$
$$32x - x^2 = 240$$
$$x^2 - 32x + 240 = 0$$
$$(x-12)(x-20) = 0$$
$$x = 12 \text{ or } x = 20$$
Substitute these $x$-values into the equation $y = 32 - x$ to find the corresponding $y$-values:
$x = 12$: $y = 32 - 12 = 20$
$x = 20$: $y = 32 - 20 = 12$
Note that the two outcomes result in the same overall dimensions. Assuming the length is the longer of the two sides, the length is 20 meters and the width is 12 meters.

42. Filling in the "missing" sides in the figure, we obtain the following.

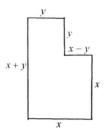

We obtain the following equation from the perimeter of the region:
$$x + x + y + y + (x+y) + (x-y) = 132$$
$$4x + 2y = 132$$
$$2x + y = 66$$
From the area, we get the equation $x^2 + y^2 = 900$.
Thus, we have the following system of equations:
$$\begin{cases} 2x + y = 66 \\ x^2 + y^2 = 900 \end{cases}$$
Solve the first equation for $y$: $y = -2x + 66$.
Substitute the result into the second equation:
$$x^2 + (-2x + 66)^2 = 900$$
$$x^2 + 4x^2 - 264x + 4356 = 900$$
$$5x^2 - 264x + 3456 = 0$$
$$(5x - 144)(x - 24) = 0$$
$$x = \frac{144}{5} = 28.8 \text{ or } x = 24$$
Substitute these $x$-values into the equation $y = -2x + 66$ to find the corresponding $y$-values:
$x = 28.2$: $y = -2(28.8) + 66 = 8.4$
$x = 24$: $y = -2(24) + 66 = 18$
The lengths are either $x = 28.8$ yards and $y = 8.4$ yards or $x = 24$ yards and $y = 18$ yards.

44. $$\begin{cases} x^3 - 2x^2 + y^2 + 3y - 4 = 0 \\ x - 2 + \dfrac{y^2 - y}{x^2} = 0 \end{cases}$$

Multiply the second equation by $-x^2$:
$$-x^2\left(x - 2 + \frac{y^2 - y}{x^2}\right) = -x^2(0)$$
$$-x^3 + 2x^2 - (y^2 - y) = 0$$
$$-x^3 + 2x^2 - y^2 + y = 0$$
Add the result to the first equation:

$$x^3 - 2x^2 + y^2 + 3y - 4 = 0$$
$$\underline{-x^3 + 2x^2 - y^2 + y \phantom{aaaa} = 0}$$
$$4y - 4 = 0$$
$$4y = 4$$
$$y = 1$$

Substitute 1 for $y$ into the equation $-x^3 + 2x^2 - y^2 + y = 0$ to find the corresponding $x$-values:
$$-x^3 + 2x^2 - (1)^2 + 1 = 0$$
$$-x^3 + 2x^2 - 1 + 1 = 0$$
$$-x^3 + 2x^2 = 0$$
$$-x^2(x - 2) = 0$$
$$x = 0 \text{ or } x = 2$$

The apparent solution $x = 0$ is extraneous because the denominator of a fraction cannot be 0. Thus, solution is $(2, 1)$.

46. $\begin{cases} \ln x = 5 \ln y \\ \log_2 x = 3 + 2 \log_2 y \end{cases}$

Write the first equation in exponential form:
$$\ln x = 5 \ln y$$
$$\ln x = \ln y^5$$
$$x = y^5$$

Write the second equation in exponential form:
$$\log_2 x = 3 + 2 \log_2 y$$
$$\log_2 x = \log_2 8 + \log_2 y^2$$
$$\log_2 x = \log_2 (8y^2)$$
$$x = 8y^2$$

Thus, we have that
$$y^5 = 8y^2$$
$$y^5 - 8y^2 = 0$$
$$y^2(y^3 - 8) = 0$$
$$y^2 = 0 \text{ or } y^3 - 8 = 0$$
$$y = 0 \phantom{aaaa} y^3 = 8$$
$$y = 2$$

The apparent $y$-value 0 is extraneous because the argument of a logarithm must be positive. Thus, the only possible $y$-value is 2.
Substitute 2 for $y$ into the equation $x = y^5$ to find the corresponding $x$-value: $x = 2^5 = 32$. The solution is $(32, 2)$.

48. The graph of a polynomial of degree 3 can intersect a circle a most 6 times. The graph of a polynomial of degree 4 can intersect a circle a most 8 times. The graph of a polynomial of degree $n$ can intersect a circle a most $2n$ times.

Explanations may vary. One possibility follows: The graph of a polynomial of degree $n$ will change directions (i.e., from upward to downward, or vice versa) at most $n$ times. Each time it changes directions, it has the possibility of intersecting the circle in 2 locations. Thus, the number of times the graph of a polynomial of degree $n$ can intersect a circle is at most $2n$.

50. a. $f(2) = 3(2) + 4 = 6 + 4 = 10$

   b. $g(2) = 2^2 = 4$

52. a. $f(4) = 3(4) + 4 = 12 + 4 = 16$

   b. $g(4) = 2^4 = 16$

54. a. $f(2) - f(1) = 10 - 7 = 3$

   b. $f(3) - f(2) = 13 - 10 = 3$

   c. $f(4) - f(3) = 16 - 13 = 3$

   d. $f(5) - f(4) = 19 - 16 = 3$

   e. $\dfrac{g(2)}{g(1)} = \dfrac{4}{2} = 2$

   f. $\dfrac{g(3)}{g(2)} = \dfrac{8}{4} = 2$

   g. $\dfrac{g(4)}{g(3)} = \dfrac{16}{8} = 2$

   h. $\dfrac{g(5)}{g(4)} = \dfrac{32}{16} = 2$

   i. In general, $f(n+1) - f(n) = 3$ for integer $n \geq 1$. In general, $\dfrac{g(n+1)}{g(n)} = 2$ for integer $n \geq 1$.

## Chapter 10: Conics

**56.** $\begin{cases} y = x^2 + 4x + 5 \\ x - y = 9 \end{cases}$ $(y = x - 9)$

Let $Y_1 = x^2 + 4x + 5$ and $Y_2 = x - 9$.

 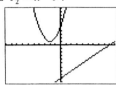

Because the graphs do not intersect, the system has no solution, $\varnothing$.

**58.** $\begin{cases} x^2 + y^2 = 25 & \left(y = \pm\sqrt{25 - x^2}\right) \\ x^2 - y^2 = 25 & \left(y = \pm\sqrt{x^2 - 25}\right) \end{cases}$

Let $Y_1 = \sqrt{25 - x^2}$, $Y_2 = -\sqrt{25 - x^2}$,
$Y_3 = \sqrt{x^2 - 25}$, and $Y_4 = -\sqrt{x^2 - 25}$.

 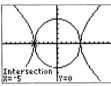

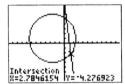

The solutions are approximately $(2.785, -4.277)$ and $(1, 10)$.

**62.** $\begin{cases} (x+2)^2 + (y-1)^2 = 4 & \left(y = 1 \pm \sqrt{4 - (x+2)^2}\right) \\ y^2 - 2y - x = 5 & \left(y = 1 \pm \sqrt{x + 6}\right) \end{cases}$

Let $Y_1 = 1 + \sqrt{4 - (x+2)^2}$, $Y_2 = 1 - \sqrt{4 - (x+2)^2}$,
$Y_3 = 1 + \sqrt{x + 6}$, and $Y_4 = 1 - \sqrt{x + 6}$.

 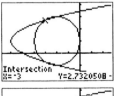

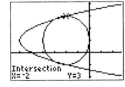 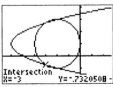

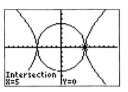

The solutions are $(-5, 0)$ and $(5, 0)$.

**60.** $\begin{cases} (x+5)^2 + (y-2)^2 = 100 & \left(y = 2 \pm \sqrt{100 - (x+5)^2}\right) \\ 8x + y = 18 & (y = -8x + 18) \end{cases}$

Let $Y_1 = 2 + \sqrt{100 - (x+5)^2}$,
$Y_2 = 2 - \sqrt{100 - (x+5)^2}$, and $Y_3 = -8x + 18$.

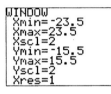 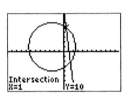

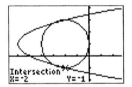

The solutions are approximately $(-3, 2.732)$, $(-3, -0.732)$, $(-2, -1)$, and $(-2, 3)$.

**64.** $\begin{cases} 9x^2 + 4y^2 = 36 & \left(y = \pm\sqrt{9 - \dfrac{9}{4}x^2}\right) \\ x^2 + (y-7)^2 = 4 & \left(y = 7 \pm \sqrt{4 - x^2}\right) \end{cases}$

Let $Y_1 = \sqrt{9 - \dfrac{9}{4}x^2}$, $Y_2 = -\sqrt{9 - \dfrac{9}{4}x^2}$,
$Y_3 = 7 + \sqrt{4 - x^2}$, and $Y_4 = 7 - \sqrt{4 - x^2}$.

 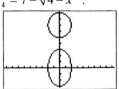

Because the graphs do not intersect, the system has no solution, $\varnothing$.

# Chapter 10 Review

1. $d(P_1, P_2) = \sqrt{(x_2 - x_1)^2 + (y_2 - y_1)^2}$
   $= \sqrt{(-4 - 0)^2 + (-3 - 0)^2}$
   $= \sqrt{(-4)^2 + (-3)^2}$
   $= \sqrt{16 + 9}$
   $= \sqrt{25}$
   $= 5$

2. $d(P_1, P_2) = \sqrt{(x_2 - x_1)^2 + (y_2 - y_1)^2}$
   $= \sqrt{(5 - (-3))^2 + (-4 - 2)^2}$
   $= \sqrt{8^2 + (-6)^2}$
   $= \sqrt{64 + 36}$
   $= \sqrt{100}$
   $= 10$

3. $d(P_1, P_2) = \sqrt{(x_2 - x_1)^2 + (y_2 - y_1)^2}$
   $= \sqrt{(5 - (-1))^2 + (3 - 1)^2}$
   $= \sqrt{6^2 + 2^2}$
   $= \sqrt{36 + 4}$
   $= \sqrt{40}$
   $= 2\sqrt{10} \approx 6.32$

4. $d(P_1, P_2) = \sqrt{(x_2 - x_1)^2 + (y_2 - y_1)^2}$
   $= \sqrt{(6 - 6)^2 + (-1 - (-7))^2}$
   $= \sqrt{0^2 + 6^2}$
   $= \sqrt{0 + 36}$
   $= \sqrt{36}$
   $= 6$

5. $d(P_1, P_2) = \sqrt{(x_2 - x_1)^2 + (y_2 - y_1)^2}$
   $= \sqrt{(4\sqrt{7} - \sqrt{7})^2 + (5\sqrt{3} - (-\sqrt{3}))^2}$
   $= \sqrt{(3\sqrt{7})^2 + (6\sqrt{3})^2}$
   $= \sqrt{9(7) + 36(3)}$
   $= \sqrt{63 + 108}$
   $= \sqrt{171}$
   $= 3\sqrt{19} \approx 13.08$

6. $d(P_1, P_2) = \sqrt{(x_2 - x_1)^2 + (y_2 - y_1)^2}$
   $= \sqrt{(1.3 - (-0.2))^2 + (3.7 - 1.7)^2}$
   $= \sqrt{1.5^2 + 2^2}$
   $= \sqrt{2.25 + 4}$
   $= \sqrt{6.25}$
   $= 2.5$

7. $M = \left(\dfrac{x_1 + x_2}{2}, \dfrac{y_1 + y_2}{2}\right)$
   $= \left(\dfrac{-1 + (-3)}{2}, \dfrac{6 + 4}{2}\right) = \left(\dfrac{-4}{2}, \dfrac{10}{2}\right) = (-2, 5)$

8. $M = \left(\dfrac{x_1 + x_2}{2}, \dfrac{y_1 + y_2}{2}\right)$
   $= \left(\dfrac{7 + 5}{2}, \dfrac{0 + (-4)}{2}\right) = \left(\dfrac{12}{2}, \dfrac{-4}{2}\right) = (6, -2)$

9. $M = \left(\dfrac{x_1 + x_2}{2}, \dfrac{y_1 + y_2}{2}\right)$
   $= \left(\dfrac{-\sqrt{3} + (-7\sqrt{3})}{2}, \dfrac{2\sqrt{6} + (-8\sqrt{6})}{2}\right)$
   $= \left(\dfrac{-8\sqrt{3}}{2}, \dfrac{-6\sqrt{6}}{2}\right)$
   $= (-4\sqrt{3}, -3\sqrt{6})$

10. $M = \left(\dfrac{x_1 + x_2}{2}, \dfrac{y_1 + y_2}{2}\right)$
    $= \left(\dfrac{5 + 0}{2}, \dfrac{-2 + 3}{2}\right) = \left(\dfrac{5}{2}, \dfrac{1}{2}\right)$

11. $M = \left(\dfrac{x_1 + x_2}{2}, \dfrac{y_1 + y_2}{2}\right)$
    $= \left(\dfrac{\frac{1}{4} + \frac{5}{4}}{2}, \dfrac{\frac{2}{3} + \frac{1}{3}}{2}\right) = \left(\dfrac{\frac{3}{2}}{2}, \dfrac{1}{2}\right) = \left(\dfrac{3}{4}, \dfrac{1}{2}\right)$

12. a.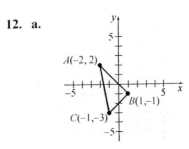

## Chapter 10: Conics

**b.** $d(A,B) = \sqrt{(1-(-2))^2 + (-1-2)^2}$
$= \sqrt{3^2 + (-3)^2}$
$= \sqrt{9+9}$
$= \sqrt{18}$
$= 3\sqrt{2} \approx 4.24$

$d(B,C) = \sqrt{(-1-1)^2 + (-3-(-1))^2}$
$= \sqrt{(-2)^2 + (-2)^2}$
$= \sqrt{4+4}$
$= \sqrt{8}$
$= 2\sqrt{2} \approx 2.83$

$d(A,C) = \sqrt{(-1-(-2))^2 + (-3-2)^2}$
$= \sqrt{1^2 + (-5)^2}$
$= \sqrt{1+25}$
$= \sqrt{26} \approx 5.10$

**c.** To determine if triangle $ABC$ is a right triangle, we check to see if
$[d(A,B)]^2 + [d(B,C)]^2 \stackrel{?}{=} [d(A,C)]^2$
$(3\sqrt{2})^2 + (2\sqrt{2})^2 \stackrel{?}{=} (\sqrt{26})^2$
$9 \cdot 2 + 4 \cdot 2 \stackrel{?}{=} 26$
$18 + 8 \stackrel{?}{=} 26$
$26 = 26 \leftarrow$ True
Therefore, triangle $ABC$ is a right triangle.

**d.** The length of the "base" of the triangle is $d(B,C) = 2\sqrt{2}$ and the length of the "height" of the triangle is $d(A,B) = 3\sqrt{2}$. Thus, the area of triangle $ABC$ is
Area $= \dfrac{1}{2} \cdot$ base $\cdot$ height $= \dfrac{1}{2} \cdot 2\sqrt{2} \cdot 3\sqrt{2} = 6$
square units.

**13.** The center of the circle will be the midpoint of the line segment with endpoints $(-6,1)$ and $(2,1)$. Thus, $(h,k) = \left(\dfrac{-6+2}{2}, \dfrac{1+1}{2}\right) = (-2,1)$.

The radius of the circle will be the distance from the center point $(-2,1)$ to a point on the circle, say $(2,1)$. Thus, $r = \sqrt{(2-(-2))^2 + (1-1)^2} = 4$.

The equation of the circle is
$(x-h)^2 + (y-k)^2 = r^2$
$(x-(-2))^2 + (y-1)^2 = 4^2$
$(x+2)^2 + (y-1)^2 = 16$

**14.** The center of the circle will be the midpoint of the line segment with endpoints $(2,3)$ and $(8,3)$. Thus, $(h,k) = \left(\dfrac{2+8}{2}, \dfrac{3+3}{2}\right) = (5,3)$.

The radius of the circle will be the distance from the center point $(5,3)$ to a point on the circle, say $(8,3)$. Thus, $r = \sqrt{(8-5)^2 + (3-3)^2} = 3$.

The equation of the circle is
$(x-h)^2 + (y-k)^2 = r^2$
$(x-5)^2 + (y-3)^2 = 3^2$
$(x-5)^2 + (y-3)^2 = 9$

**15.** $(x-h)^2 + (y-k)^2 = r^2$
$(x-0)^2 + (y-0)^2 = 4^2$
$x^2 + y^2 = 16$

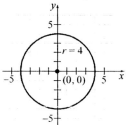

**16.** $(x-h)^2 + (y-k)^2 = r^2$
$(x-(-3))^2 + (y-1)^2 = 3^2$
$(x+3)^2 + (y-1)^2 = 9$

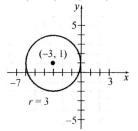

17. $(x-h)^2 + (y-k)^2 = r^2$
$(x-5)^2 + (y-(-2))^2 = 1^2$
$(x-5)^2 + (y+2)^2 = 1$

18. $(x-h)^2 + (y-k)^2 = r^2$
$(x-4)^2 + (y-0)^2 = (\sqrt{7})^2$
$(x-4)^2 + y^2 = 7$

19. The radius of the circle will be the distance from the center point $(2,-1)$ to the point on the circle $(5,3)$. Thus, $r = \sqrt{(5-2)^2 + (3-(-1))^2} = 5$. The equation of the circle is
$(x-h)^2 + (y-k)^2 = r^2$
$(x-2)^2 + (y-(-1))^2 = 5^2$
$(x-2)^2 + (y+1)^2 = 25$

20. The center of the circle will be the midpoint of the diameter with endpoints $(-3,-1)$ and $(1,7)$. Thus, $(h,k) = \left(\dfrac{-3+1}{2}, \dfrac{-1+7}{2}\right) = (-1,3)$. The radius of the circle will be the distance from the center point $(-1,3)$ to one of the endpoints of the diameter, say $(1,7)$. Thus,
$r = \sqrt{(1-(-1))^2 + (7-3)^2} = \sqrt{20} = 2\sqrt{5}$.

The equation of the circle is
$(x-h)^2 + (y-k)^2 = r^2$
$(x-(-1))^2 + (y-3)^2 = (2\sqrt{5})^2$
$(x+1)^2 + (y-3)^2 = 20$

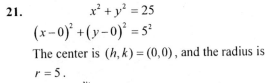

21. $x^2 + y^2 = 25$
$(x-0)^2 + (y-0)^2 = 5^2$
The center is $(h,k) = (0,0)$, and the radius is $r = 5$.

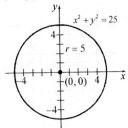

22. $(x-1)^2 + (y-2)^2 = 4$
$(x-1)^2 + (y-2)^2 = 2^2$
The center is $(h,k) = (1,2)$, and the radius is $r = 2$.

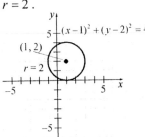

23. $x^2 + (y-4)^2 = 16$
$(x-0)^2 + (y-4)^2 = 4^2$
The center is $(h,k) = (0,4)$, and the radius is $r = 4$.

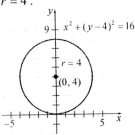

24. $(x+1)^2 + (y+6)^2 = 49$
$(x-(-1))^2 + (y-(-6))^2 = 7^2$
The center is $(h,k) = (-1,-6)$, and the radius is

## Chapter 10: Conics

$r = 7$.

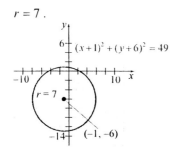

$r = 6$.

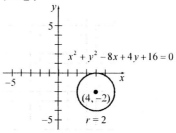

**25.** $(x+2)^2 + \left(y - \dfrac{3}{2}\right)^2 = \dfrac{1}{4}$

$(x-(-2))^2 + \left(y - \dfrac{3}{2}\right)^2 = \left(\dfrac{1}{2}\right)^2$

The center is $(h,k) = \left(-2, \dfrac{3}{2}\right)$, and the radius is $r = \dfrac{1}{2}$.

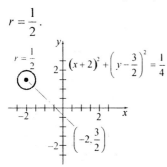

**28.** $x^2 + y^2 - 8x + 4y + 16 = 0$

$(x^2 - 8x) + (y^2 + 4y) = -16$

$(x^2 - 8x + 16) + (y^2 + 4y + 4) = -16 + 16 + 4$

$(x-4)^2 + (y+2)^2 = 4$

$(x-4)^2 + (y-(-2))^2 = 2^2$

The center is $(h,k) = (4,-2)$, and the radius is $r = 2$.

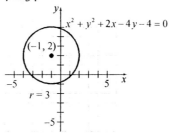

**26.** $(x+3)^2 + (y+3)^2 = 4$

$(x-(-3))^2 + (y-(-3))^2 = 2^2$

The center is $(h,k) = (-3,-3)$, and the radius is $r = 2$.

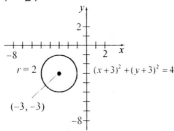

**29.** $x^2 + y^2 + 2x - 4y - 4 = 0$

$(x^2 + 2x) + (y^2 - 4y) = 4$

$(x^2 + 2x + 1) + (y^2 - 4y + 4) = 4 + 1 + 4$

$(x+1)^2 + (y-2)^2 = 9$

$(x-(-1))^2 + (y-2)^2 = 3^2$

The center is $(h,k) = (-1, 2)$, and the radius is $r = 3$.

**27.** $x^2 + y^2 + 6x + 10y - 2 = 0$

$(x^2 + 6x) + (y^2 + 10y) = 2$

$(x^2 + 6x + 9) + (y^2 + 10y + 25) = 2 + 9 + 25$

$(x+3)^2 + (y+5)^2 = 36$

$(x-(-3))^2 + (y-(-5))^2 = 6^2$

The center is $(h,k) = (-3,-5)$, and the radius is

30. $$x^2 + y^2 - 10x - 2y + 17 = 0$$
$$(x^2 - 10x) + (y^2 - 2y) = -17$$
$$(x^2 - 10x + 25) + (y^2 - 2y + 1) = -17 + 25 + 1$$
$$(x-5)^2 + (y-1)^2 = 9$$
$$(x-5)^2 + (y-1)^2 = 3^2$$

The center is $(h,k) = (5,1)$, and the radius is $r = 3$.

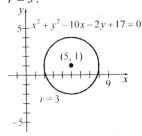

31. The distance from the vertex $(0,0)$ to the focus $(0,-3)$ is $a = 3$. Because the focus lies on the negative $y$-axis, we know that the parabola will open downward and the axis of symmetry is the $y$-axis. This means the equation of the parabola is of the form $x^2 = -4ay$ with $a = 3$:
$$x^2 = -4(3)y$$
$$x^2 = -12y$$

The directrix is the line $y = 3$. To help graph the parabola, we plot the two points on the graph to the left and right of the focus. Let $y = -3$:
$$x^2 = -12(-3) = 36$$
$$x = \pm 6$$

The points $(-6,-3)$ and $(6,-3)$ are on the graph.

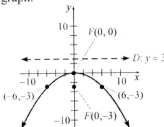

32. Notice that the directrix is the vertical line $x = 4$ and that the focus $(-4, 0)$ is on the $x$-axis. Thus, the axis of symmetry is the $x$-axis, the vertex is at the origin, and the parabola opens to the left. Now, the distance from the vertex $(0,0)$ to the focus $(-4, 0)$ is $a = 4$, so the equation of the parabola is of the form $y^2 = -4ax$ with $a = 4$:
$$y^2 = -4(4)x$$
$$y^2 = -16x$$

To help graph the parabola, we plot the two points on the graph above and below the focus. Let $x = -4$:
$$y^2 = -16(-4) = 64$$
$$y = \pm 8$$

The points $(-4,-8)$ and $(-4, 8)$ are on the graph.

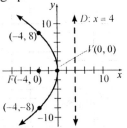

33. The vertex is at the origin and the axis of symmetry is the $x$-axis, so the parabola either opens to the left or to the right. Because the graph contains the point $(8,-2)$, which is in quadrant IV, the parabola must open to the right. Therefore, the equation of the parabola is of the form $y^2 = 4ax$. Now $y = -2$ when $x = 8$, so
$$y^2 = 4ax$$
$$(-2)^2 = 4a(8)$$
$$4 = 32a$$
$$a = \frac{4}{32} = \frac{1}{8}$$

The equation of the parabola is
$$y^2 = 4\left(\frac{1}{8}\right)x$$
$$y^2 = \frac{1}{2}x$$

With $a = \frac{1}{8}$, we know that the focus is $\left(\frac{1}{8}, 0\right)$ and the directrix is the line $x = -\frac{1}{8}$. To help graph the parabola, we plot the two points on the graph above and below the focus. Let $x = \frac{1}{8}$:

ISM – 657

## Chapter 10: Conics

$$y^2 = \frac{1}{2}\left(\frac{1}{8}\right) = \frac{1}{16}$$
$$y = \pm\frac{1}{4}$$

The points $\left(\frac{1}{8}, -\frac{1}{4}\right)$ and $\left(\frac{1}{8}, \frac{1}{4}\right)$ are on the graph.

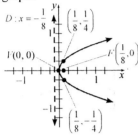

34. The vertex is at the origin and the directrix is the line $y = -2$, so the focus must be $(0, 2)$. Accordingly, the parabola opens upward with $a = 2$. This means the equation of the parabola is of the form $x^2 = 4ay$ with $a = 2$:
$$x^2 = 4(2)y$$
$$x^2 = 8y$$
To help graph the parabola, we plot the two points on the graph to the left and right of the focus. Let $y = 2$:
$$x^2 = 8(2) = 16$$
$$x = \pm 4$$
The points $(-4, 2)$ and $(4, 2)$ are on the graph.

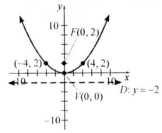

35. Notice that $x^2 = 2y$ is of the form $x^2 = 4ay$, where $4a = 2$, so that $a = \frac{2}{4} = \frac{1}{2}$. Now, the graph of an equation of the form $x^2 = 4ay$ will be a parabola that opens upward with the vertex at the origin, focus at $(0, a)$, and directrix of $y = -a$. Thus, the graph of $x^2 = 2y$ is a parabola that opens upward with vertex $(0, 0)$,

focus $\left(0, \frac{1}{2}\right)$, and directrix $y = -\frac{1}{2}$. To help graph the parabola, we plot the two points on the graph to the left and to the right of the focus. Let $y = \frac{1}{2}$:
$$x^2 = 2\left(\frac{1}{2}\right)$$
$$x^2 = 1$$
$$x = \pm 1$$
The points $\left(-1, \frac{1}{2}\right)$ and $\left(1, \frac{1}{2}\right)$ are on the graph.

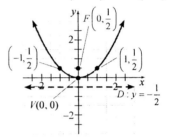

36. Notice that $y^2 = 16x$ is of the form $y^2 = 4ax$, where $4a = 16$, so that $a = 4$. Now, the graph of an equation of the form $y^2 = 4ax$ will be a parabola that opens to the right with the vertex at the origin, focus at $(a, 0)$, and directrix of $x = -a$. Thus, the graph of $y^2 = 16x$ is a parabola that opens to the right with vertex $(0, 0)$, focus $(4, 0)$, and directrix $x = -4$. To help graph the parabola, we plot the two points on the graph above and below the focus. Let $x = 4$:
$$y^2 = 16(4)$$
$$y^2 = 64$$
$$y = \pm 8$$
The points $(4, -8)$ and $(4, 8)$ are on the graph.

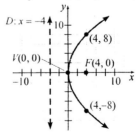

37. Notice that $(x+1)^2 = 8(y-3)$ is of the form $(x-h)^2 = 4a(y-k)$, where $4a = 8$, so that

$a = 2$ and $(h,k) = (-1,3)$. Now, the graph of an equation of the form $(x-h)^2 = 4a(y-k)$ will be a parabola that opens upward with vertex at $(h,k)$, focus at $(h, k+a)$, and directrix of $y = k - a$. Note that $k + a = 3 + 2 = 5$ and $k - a = 3 - 2 = 1$. Thus, the graph of $(x+1)^2 = 8(y-3)$ is a parabola that opens upward with vertex $(-1,3)$, focus $(-1,5)$, and directrix $y = 1$. To help graph the parabola, we plot the two points on the left and on the right of the focus. Let $y = 5$:
$$(x+1)^2 = 8(5-3)$$
$$(x+1)^2 = 16$$
$$x + 1 = \pm 4$$
$$x = -1 \pm 4$$
$$x = -5 \text{ or } x = 3$$
The points $(-5, 5)$ and $(3, 5)$ are on the graph.

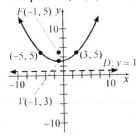

**38.** Notice that $(y-4)^2 = -2(x+3)$ is of the form $(y-k)^2 = -4a(x-h)$, where $(h,k) = (-3,4)$ and $-4a = -2$, so that $a = \dfrac{-2}{-4} = \dfrac{1}{2}$. Now, the graph of an equation of the form $(y-k)^2 = -4a(x-h)$ will be a parabola that opens to the left with the vertex at $(h,k)$, focus at $(h-a, k)$, and directrix of $x = h + a$. Note that $h - a = -3 - \dfrac{1}{2} = -\dfrac{7}{2}$ and $h + a = -3 + \dfrac{1}{2} = -\dfrac{5}{2}$. Thus, the graph of $(y-4)^2 = -2(x+3)$ is a parabola that opens to the left with vertex $(-3, 4)$, focus $\left(-\dfrac{7}{2}, 4\right)$, and directrix $x = -\dfrac{5}{2}$. To help graph the parabola, we plot the two points on the graph above and

below the focus. Let $x = -\dfrac{7}{2}$:
$$(y-4)^2 = -2\left(-\dfrac{7}{2} + 3\right)$$
$$(y-4)^2 = -2\left(-\dfrac{1}{2}\right)$$
$$(y-4)^2 = 1$$
$$y - 4 = \pm 1$$
$$y = 4 \pm 1$$
$$y = 3 \text{ or } y = 5$$
The points $\left(-\dfrac{7}{2}, 3\right)$ and $\left(-\dfrac{7}{2}, 5\right)$ are on the graph.

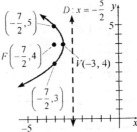

**39.** We complete the square in $x$ to write the equation in standard form:
$$x^2 - 10x + 3y + 19 = 0$$
$$x^2 - 10x = -3y - 19$$
$$x^2 - 10x + 25 = -3y - 19 + 25$$
$$x^2 - 10x + 25 = -3y + 6$$
$$(x-5)^2 = -3(y-2)$$
Notice that $(x-5)^2 = -3(y-2)$ is of the form $(x-h)^2 = -4a(y-k)$, where $(h,k) = (5,2)$ and $-4a = -3$, so that $a = \dfrac{3}{4}$. Now, the graph of an equation of the form $(x-h)^2 = -4a(y-k)$ will be a parabola that opens downward with the vertex at $(h,k)$, focus at $(h, k-a)$, and directrix of $y = k + a$. Note that $k - a = 2 - \dfrac{3}{4} = \dfrac{5}{4}$ and $k + a = 2 + \dfrac{3}{4} = \dfrac{11}{4}$. Thus, the graph of $(x-5)^2 = -3(y-2)$ is a parabola that opens downward with vertex $(5,2)$, focus $\left(5, \dfrac{5}{4}\right)$, and directrix $y = \dfrac{11}{4}$. To help graph the parabola,

## Chapter 10: Conics

we plot the two points on the graph to the left and right of the focus. Let $y = \dfrac{5}{4}$:

$$(x-5)^2 = -3\left(\dfrac{5}{4} - 2\right)$$

$$(x-5)^2 = -3\left(-\dfrac{3}{4}\right)$$

$$(x-5)^2 = \dfrac{9}{4}$$

$$x - 5 = \pm\dfrac{3}{2}$$

$$x = 5 \pm \dfrac{3}{2}$$

$$x = \dfrac{7}{2} \text{ or } x = \dfrac{13}{2}$$

The points $\left(\dfrac{7}{2}, \dfrac{5}{4}\right)$ and $\left(\dfrac{13}{2}, \dfrac{5}{4}\right)$ are on the graph.

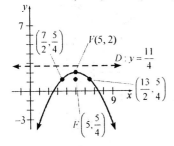

**40.** The receiver should be located at the focus of the dish, so we need to find where the focus of the dish is. To solve this problem, we draw a parabola on a Cartesian plane so that the vertex is the origin and the focus is on the positive $y$-axis. The width of the dish is 300 feet, and the depth is 44 feet. Therefore, we know two points on the graph of the parabola: $(-150, 44)$ and $(150, 44)$.

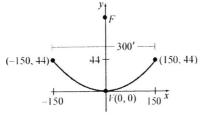

The equation of the parabola has the form $x^2 = 4ay$. Since $(150, 44)$ is a point on the graph, we have

$$150^2 = 4a(44)$$
$$22{,}500 = 176a$$
$$a = \dfrac{22{,}500}{176} \approx 127.84$$

The receiver should be located approximately 127.84 feet above the center of the dish, along its axis of symmetry.

**41.** $\dfrac{x^2}{9} + y^2 = 1$

$$\dfrac{x^2}{9} + \dfrac{y^2}{1} = 1$$

The larger number, 9, is in the denominator of the $x^2$-term. This means that the major axis is the $x$-axis and that the equation of the ellipse is of the form $\dfrac{x^2}{a^2} + \dfrac{y^2}{b^2} = 1$, so that $a^2 = 9$ and $b^2 = 1$. The center of the ellipse is $(0,0)$. Because $b^2 = a^2 - c^2$, or $c^2 = a^2 - b^2$, we have that $c^2 = 9 - 1 = 8$, so that $c = \pm\sqrt{8} = \pm 2\sqrt{2}$. Since the major axis is the $x$-axis, the foci are $\left(-2\sqrt{2}, 0\right)$ and $\left(2\sqrt{2}, 0\right)$. To find the $x$-intercepts (vertices), let $y = 0$; to find the $y$-intercepts, let $x = 0$:

$x$-intercepts:
$$\dfrac{x^2}{9} + 0^2 = 1$$
$$\dfrac{x^2}{9} = 1$$
$$x^2 = 9$$
$$x = \pm 3$$

$y$-intercepts:
$$\dfrac{0^2}{9} + y^2 = 1$$
$$y^2 = 1$$
$$y = \pm 1$$

The intercepts are $(-3, 0)$, $(3, 0)$, $(0, -1)$, and $(0, 1)$.

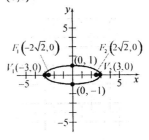

**42.** $9x^2 + 4y^2 = 36$

$$\dfrac{9x^2 + 4y^2}{36} = \dfrac{36}{36}$$

$$\dfrac{x^2}{4} + \dfrac{y^2}{9} = 1$$

The larger number, 9, is in the denominator of the $y^2$-term. This means that the major axis is the $y$-axis and that the equation of the ellipse is of the form $\dfrac{x^2}{b^2} + \dfrac{y^2}{a^2} = 1$, so that $a^2 = 9$ and

$b^2 = 4$. The center of the ellipse is $(0,0)$. Because $b^2 = a^2 - c^2$, or $c^2 = a^2 - b^2$, we have that $c^2 = 9 - 4 = 5$, so that $c = \pm\sqrt{5}$. Since the major axis is the $y$-axis, the foci are $\left(0, -\sqrt{5}\right)$ and $\left(0, \sqrt{5}\right)$. To find the $x$-intercepts, let $y = 0$; to find the $y$-intercepts (vertices), let $x = 0$:

$x$-intercepts:
$9x^2 + 4(0)^2 = 36$
$9x^2 = 36$
$x^2 = 4$
$x = \pm 2$

$y$-intercepts:
$9(0)^2 + 4y^2 = 36$
$4y^2 = 36$
$y^2 = 9$
$y = \pm 3$

The intercepts are $(-2,0)$, $(2,0)$, $(0,-3)$, and $(0,3)$.

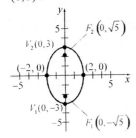

43. The given focus $(0,3)$ and the given vertex $(0,5)$ lie on the $y$-axis. Thus, the equation of the ellipse is of the form $\dfrac{x^2}{b^2} + \dfrac{y^2}{a^2} = 1$. The distance from the center of the ellipse to the vertex is $a = 5$ units. The distance from the center of the ellipse to the focus is $c = 3$ units. Because $b^2 = a^2 - c^2$, we have that $b^2 = 5^2 - 3^2 = 25 - 9 = 16$. Thus, the equation of the ellipse is $\dfrac{x^2}{16} + \dfrac{y^2}{25} = 1$.

To help graph the ellipse, find the $x$-intercepts:
Let $y = 0$: $\dfrac{x^2}{16} + \dfrac{0^2}{25} = 1$
$\dfrac{x^2}{16} = 1$
$x^2 = 16$
$x = \pm 4$

The $x$-intercepts are $(-4,0)$ and $(4,0)$.

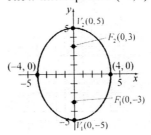

44. The given focus $(-2,0)$ and the given vertex $(-6,0)$ lie on the $x$-axis. Thus, the equation of the ellipse is of the form $\dfrac{x^2}{a^2} + \dfrac{y^2}{b^2} = 1$. The distance from the center of the ellipse to the vertex is $a = 6$ units. The distance from the center of the ellipse to the focus is $c = 2$ units. Because $b^2 = a^2 - c^2$, we have that $b^2 = 6^2 - 2^2 = 36 - 4 = 32$. Thus, the equation of the ellipse is $\dfrac{x^2}{36} + \dfrac{y^2}{32} = 1$.

To help graph the ellipse, find the $y$-intercepts:

Let $x = 0$: $\dfrac{0^2}{36} + \dfrac{y^2}{32} = 1$
$\dfrac{y^2}{32} = 1$
$y^2 = 32$
$y = \pm\sqrt{32} = \pm 4\sqrt{2}$

The $y$-intercepts are $\left(0, -4\sqrt{2}\right)$ and $\left(0, 4\sqrt{2}\right)$.

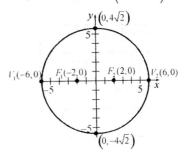

## Chapter 10: Conics

**45.** The given foci $(\pm 8, 0)$ and the given vertices $(\pm 10, 0)$ lie on the $x$-axis. The center of the ellipse is the midpoint between the two vertices (or foci). Thus, the center of the ellipse is $(0,0)$ and its equation is of the form $\dfrac{x^2}{a^2} + \dfrac{y^2}{b^2} = 1$. The distance from the center of the ellipse to the vertex is $a = 10$ units. The distance from the center of the ellipse to the focus is $c = 8$ units. Because $b^2 = a^2 - c^2$, we have that $b^2 = 10^2 - 8^2 = 100 - 64 = 36$. Thus, the equation of the ellipse is $\dfrac{x^2}{100} + \dfrac{y^2}{36} = 1$.

To help graph the ellipse, find the $y$-intercepts:

Let $x = 0$: $\dfrac{0^2}{100} + \dfrac{y^2}{36} = 1$

$\dfrac{y^2}{36} = 1$

$y^2 = 36$

$y = \pm 6$

The $y$-intercepts are $(0, -6)$ and $(0, 6)$.

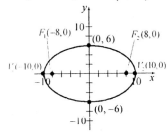

**46.** $\dfrac{(x-1)^2}{49} + \dfrac{(y+2)^2}{25} = 1$

The center of the ellipse is $(h, k) = (1, -2)$. Because the larger number, 49, is the denominator of the $x^2$-term, the major axis is parallel to the $x$-axis. Because $a^2 = 49$ and $b^2 = 25$, we have that $c^2 = a^2 - b^2 = 49 - 25 = 24$. The vertices are $a = 7$ units to the left and right of the center at $V_1(-6, -2)$ and $V_2(8, -2)$. The foci are $c = \sqrt{24} = 2\sqrt{6}$ units to the left and right of the center at $F_1\left(1 - 2\sqrt{6}, -2\right)$ and $F_2\left(1 + 2\sqrt{6}, -2\right)$.

We plot the points $b = 5$ units above and below the center point at $(1, 3)$ and $(1, -7)$.

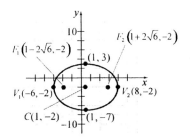

**47.** $25(x+3)^2 + 9(y-4)^2 = 225$

$\dfrac{25(x+3)^2 + 9(y-4)^2}{225} = \dfrac{225}{225}$

$\dfrac{(x+3)^2}{9} + \dfrac{(y-4)^2}{25} = 1$

The center of the ellipse is $(h, k) = (-3, 4)$. Because the larger number, 25, is the denominator of the $y^2$-term, the major axis is parallel to the $y$-axis. Because $a^2 = 25$ and $b^2 = 9$, we have that $c^2 = a^2 - b^2 = 25 - 9 = 16$. The vertices are $a = 5$ units below and above the center at $V_1(-3, -1)$ and $V_2(-3, 9)$. The foci are $c = \sqrt{16} = 4$ units below and above the center at $F_1(-3, 0)$ and $F_2(-3, 8)$. We plot the points $b = 3$ units to the left and right of the center point at $(-6, 4)$ and $(0, 4)$.

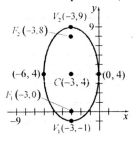

**48. a.** To solve this problem, we draw the ellipse on a Cartesian plane so that the $x$-axis coincides with the water and the $y$-axis passes through the center of the arch. Thus, the origin is the center of the ellipse. Now, the "center" of the arch is 16 feet above the water, so the point $(0, 16)$ is on the ellipse. The river is 60 feet wide, so the two points $(-30, 0)$ and $(30, 0)$ are the two vertices of the ellipse and the major axis is along the $x$-axis.

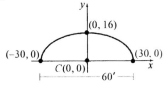

ISM - 662

The equation of the ellipse must have the form $\dfrac{x^2}{a^2}+\dfrac{y^2}{b^2}=1$. The distance from the center of the ellipse to each vertex is $a=30$. Also, the height of the semi-ellipse is 16 feet, so $b=16$. Thus, the equation of the ellipse is

$$\dfrac{x^2}{30^2}+\dfrac{y^2}{16^2}=1$$

$$\dfrac{x^2}{900}+\dfrac{y^2}{256}=1$$

b. To determine if the barge can fit through the opening of the bridge, we center it beneath the arch so that the points $(-12.5,12)$ and $(12.5,12)$ represent the top corners of the barge. Now, we determine the height of the arch above the water at points 12.5 feet to the left and right of the center by substituting $x=12.5$ into the equation of the ellipse and solving for $y$:

$$\dfrac{12.5^2}{900}+\dfrac{y^2}{256}=1$$

$$\dfrac{156.25}{900}+\dfrac{y^2}{256}=1$$

$$\dfrac{y^2}{256}=\dfrac{119}{144}$$

$$y^2=\dfrac{1904}{9}$$

$$y=\sqrt{\dfrac{1904}{9}}\approx 14.54$$

At points 12.5 feet from the center of the river, the arch is approximately 14.54 feet above the surface of the water. Thus, the barge can fit through the opening with about 2.54 feet of clearance above the top corners.

49. $\dfrac{x^2}{4}-\dfrac{y^2}{9}=1$

Notice the equation is of the form $\dfrac{x^2}{a^2}-\dfrac{y^2}{b^2}=1$.

Because the $x^2$-term is first, the transverse axis is the $x$-axis and the hyperbola opens left and right. The center of the hyperbola is the origin. We have that $a^2=4$ and $b^2=9$. Because $c^2=a^2+b^2$, we have that $c^2=4+9=13$, so that $c=\sqrt{13}$. The vertices are $(\pm a,0)=(\pm 2,0)$, and the foci are $(\pm c,0)=(\pm\sqrt{13},0)$. To help graph the hyperbola,

we plot the points on the graph above and below the foci. Let $x=\pm\sqrt{13}$:

$$\dfrac{(\pm\sqrt{13})^2}{4}-\dfrac{y^2}{9}=1$$

$$\dfrac{13}{4}-\dfrac{y^2}{9}=1$$

$$-\dfrac{y^2}{9}=-\dfrac{9}{4}$$

$$y^2=\dfrac{81}{4}$$

$$y=\pm\dfrac{9}{2}$$

The points above and below the foci are $\left(\sqrt{13},\dfrac{9}{2}\right)$, $\left(\sqrt{13},-\dfrac{9}{2}\right)$, $\left(-\sqrt{13},\dfrac{9}{2}\right)$, and $\left(-\sqrt{13},-\dfrac{9}{2}\right)$.

50. $\dfrac{y^2}{25}-\dfrac{x^2}{49}=1$

Notice the equation is of the form $\dfrac{y^2}{a^2}-\dfrac{x^2}{b^2}=1$.

Because the $y^2$-term is first, the transverse axis is the $y$-axis and the hyperbola opens up and down. The center of the hyperbola is the origin. We have that $a^2=25$ and $b^2=49$. Because $c^2=a^2+b^2$, we have that $c^2=25+49=74$, so that $c=\sqrt{74}$. The vertices are $(0,\pm a)=(0,\pm 5)$, and the foci are $(0,\pm c)=(0,\pm\sqrt{74})$. To help graph the hyperbola, we plot the points on the graph to the left and right of the foci. Let $y=\pm\sqrt{74}$:

$$\dfrac{(\pm\sqrt{74})^2}{25}-\dfrac{x^2}{49}=1$$

$$\dfrac{74}{25}-\dfrac{x^2}{49}=1$$

$$-\dfrac{x^2}{49}=-\dfrac{49}{25}$$

## Chapter 10: Conics

$$x^2 = \frac{2401}{25}$$
$$x = \pm\frac{49}{5}$$

The points to the left and right of the foci are $\left(\frac{49}{5}, \sqrt{74}\right)$, $\left(-\frac{49}{5}, \sqrt{74}\right)$, $\left(\frac{49}{5}, -\sqrt{74}\right)$, and $\left(-\frac{49}{5}, -\sqrt{74}\right)$.

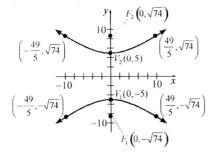

**51.** $16y^2 - 25x^2 = 400$
$$\frac{16y^2 - 25x^2}{400} = \frac{400}{400}$$
$$\frac{y^2}{25} - \frac{x^2}{16} = 1$$

Notice the equation is of the form $\frac{y^2}{a^2} - \frac{x^2}{b^2} = 1$.

Because the $y^2$-term is first, the transverse axis is the $y$-axis and the hyperbola opens up and down. The center of the hyperbola is the origin. We have that $a^2 = 25$ and $b^2 = 16$. Because $c^2 = a^2 + b^2$, we have that $c^2 = 25 + 16 = 41$, so that $c = \sqrt{41}$. The vertices are $(0, \pm a) = (0, \pm 5)$, and the foci are $(0, \pm c) = \left(0, \pm\sqrt{41}\right)$. To help graph the hyperbola, we plot the points on the graph to the left and right of the foci. Let $y = \pm\sqrt{41}$:

$$\frac{\left(\pm\sqrt{41}\right)^2}{25} - \frac{x^2}{16} = 1$$
$$\frac{41}{25} - \frac{x^2}{16} = 1$$
$$-\frac{x^2}{16} = -\frac{16}{25}$$
$$x^2 = \frac{256}{25}$$
$$x = \pm\frac{16}{5}$$

The points to the left and right of the foci are $\left(\frac{16}{5}, \sqrt{41}\right)$, $\left(\frac{16}{5}, -\sqrt{41}\right)$, $\left(-\frac{16}{5}, \sqrt{41}\right)$, and $\left(-\frac{16}{5}, -\sqrt{41}\right)$.

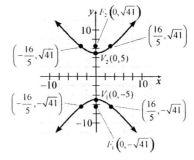

**52.** $\frac{x^2}{36} - \frac{y^2}{36} = 1$

Notice the equation is of the form $\frac{x^2}{a^2} - \frac{y^2}{b^2} = 1$.

Because the $x^2$-term is first, the transverse is along the $x$-axis and the hyperbola opens left and right. The center of the hyperbola is $(0,0)$. We have that $a^2 = 36$ and $b^2 = 36$. Because $c^2 = a^2 + b^2$, we have that $c^2 = 36 + 36 = 72$, so that $c = \sqrt{72} = 6\sqrt{2}$. The vertices are $(\pm a, 0) = (\pm 6, 0)$, and the foci are $(\pm c, 0) = \left(\pm 6\sqrt{2}, 0\right)$. Since $a = 6$ and $b = 6$, the equations of the asymptotes are $y = \frac{6}{6}x = x$ and $y = -\frac{6}{6}x = -x$. To help graph the hyperbola, we form the rectangle using the points $(\pm a, 0) = (\pm 6, 0)$ and $(0, \pm b) = (0, \pm 6)$. The diagonals are the asymptotes.

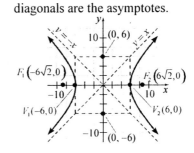

ISM – 664

## Chapter 10 Review

53. $\dfrac{y^2}{25} - \dfrac{x^2}{4} = 1$

Notice the equation is of the form $\dfrac{y^2}{a^2} - \dfrac{x^2}{b^2} = 1$.

Because the $y^2$-term is first, the transverse is along the $y$-axis and the hyperbola opens up and down. The center of the hyperbola is $(0,0)$. We have that $a^2 = 25$ and $b^2 = 4$. Because $c^2 = a^2 + b^2$, we have that $c^2 = 25 + 4 = 29$, so that $c = \sqrt{29}$. The vertices are $(0, \pm a) = (0, \pm 5)$, and the foci are $(0, \pm c) = \left(0, \pm\sqrt{29}\right)$. Since $a = 5$ and $b = 2$, the equations of the asymptotes are $y = \dfrac{5}{2}x$ and $y = -\dfrac{5}{2}x$. To help graph the hyperbola, we form the rectangle using the points $(0, \pm a) = (0, \pm 5)$ and $(\pm b, 0) = (\pm 2, 0)$. The diagonals are the asymptotes.

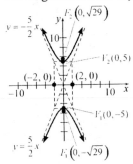

54. The given center $(0,0)$, focus $(-4,0)$, and vertex $(-3,0)$ all lie on the $x$-axis. Thus, the transverse axis is the $x$-axis and the hyperbola opens left and right. This means that the equation of the hyperbola is of the form $\dfrac{x^2}{a^2} - \dfrac{y^2}{b^2} = 1$. Now, the distance between the given vertex $(-3,0)$ and the center $(0,0)$ is $a = 3$ units. Likewise, the distance between the given focus $(-4,0)$ and the center is $c = 4$ units. Because $b^2 = c^2 - a^2$, we have that $b^2 = 4^2 - 3^2 = 16 - 9 = 7$. Thus, the equation of the hyperbola is $\dfrac{x^2}{9} - \dfrac{y^2}{7} = 1$.

To help graph the hyperbola, we first find the vertex and focus that were not given. Since the center is at $(0,0)$ and since one vertex is at $(-3,0)$, the other vertex must be at $(3,0)$.

Similarly, one the focus is at $(-4,0)$, so the other focus is at $(4,0)$. Next, plot the points on the graph above and below the foci. Let $x = \pm 4$:

$$\dfrac{(\pm 4)^2}{9} - \dfrac{y^2}{7} = 1$$

$$\dfrac{16}{9} - \dfrac{y^2}{7} = 1$$

$$-\dfrac{y^2}{7} = -\dfrac{7}{9}$$

$$y^2 = \dfrac{49}{9}$$

$$y = \pm \dfrac{7}{3}$$

The points above and below the foci are $\left(4, \dfrac{7}{3}\right)$, $\left(4, -\dfrac{7}{3}\right)$, $\left(-4, \dfrac{7}{3}\right)$, and $\left(-4, -\dfrac{7}{3}\right)$.

55. The given vertices $(0, \pm 3)$ and focus $(0,5)$ all lie on the $y$-axis. Thus, the transverse axis is the $y$-axis and the hyperbola opens up and down. The center of the hyperbola is the midpoint between the two vertices. Therefore, the center is $(0,0)$ and the equation of the hyperbola is of the form $\dfrac{y^2}{a^2} - \dfrac{x^2}{b^2} = 1$. Now, the distance from the center $(0,0)$ to each vertex is $a = 3$ units. Likewise, the distance from the center to the given focus $(0,5)$ is $c = 5$ units. Also, $b^2 = c^2 - a^2$, so $b^2 = 5^2 - 3^2 = 25 - 9 = 16$. Thus, the equation of the hyperbola is

$$\dfrac{y^2}{9} - \dfrac{x^2}{16} = 1.$$

To help graph the hyperbola, we first find the focus that was not given. Since the center is at $(0,0)$ and since one focus is at $(0,5)$, the other focus must be at $(0,-5)$. Next, plot the points on the graph to the left and right of the foci. Let $y = \pm 5$:

ISM - 665

## Chapter 10: Conics

$$\frac{(\pm 5)^2}{9} - \frac{x^2}{16} = 1$$

$$\frac{25}{9} - \frac{x^2}{16} = 1$$

$$-\frac{x^2}{16} = -\frac{16}{9}$$

$$x^2 = \frac{256}{9}$$

$$x = \pm \frac{16}{3}$$

The points to the left and right of the foci are $\left(\frac{16}{3}, 5\right)$, $\left(-\frac{16}{3}, 5\right)$, $\left(\frac{16}{3}, -5\right)$, and $\left(-\frac{16}{3}, -5\right)$.

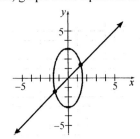

**56.** The given vertices $(0, \pm 4)$ both lie on the $y$-axis. Thus, the transverse axis is the $y$-axis and the hyperbola opens up and down. The center of the hyperbola is the midpoint between the two vertices. Therefore, the center is $(0, 0)$ and the equation of the hyperbola is of the form $\frac{y^2}{a^2} - \frac{x^2}{b^2} = 1$. Now, the distance from the center $(0,0)$ to each vertex is $a = 4$ units. We are given that an asymptote of the hyperbola is $y = \frac{4}{3}x$. Now the equation of the asymptote is of the form is $y = \frac{a}{b}x$ and $a = 4$. So,

$$\frac{a}{b} = \frac{4}{3}$$

$$\frac{4}{b} = \frac{4}{3}$$

$$4b = 12$$

$$b = 3$$

Thus, the equation of the hyperbola is

$$\frac{y^2}{4^2} - \frac{x^2}{3^2} = 1.$$

$$\frac{y^2}{16} - \frac{x^2}{9} = 1$$

To help graph the hyperbola, we first find the foci. Since $c^2 = a^2 + b^2$, we have that $c^2 = 16 + 9 = 25$, so $c = 5$. Since the center is $(0, 0)$, the foci are $(0, \pm 5)$. We form the rectangle using the points $(0, \pm a) = (0, \pm 4)$ and $(\pm b, 0) = (\pm 3, 0)$. The diagonals are the two asymptotes $y = \frac{4}{3}x$ and $y = -\frac{4}{3}x$.

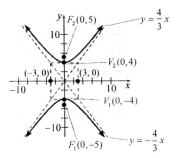

**57.** $\begin{cases} 4x^2 + y^2 = 10 \\ y = x \end{cases}$

First, graph each equation in the system.

The system apparently has two solutions. Now substitute $x$ for $y$ into the first equation:

$$4x^2 + x^2 = 10$$
$$5x^2 = 10$$
$$x^2 = 2$$
$$x = \pm\sqrt{2}$$

Substitute these $x$-values into the second equation to find the corresponding $y$-values:

$x = \sqrt{2}$:  $y = x = \sqrt{2}$
$x = -\sqrt{2}$:  $y = x = -\sqrt{2}$

Both pairs check, so the solutions are $\left(\sqrt{2}, \sqrt{2}\right)$ and $\left(-\sqrt{2}, -\sqrt{2}\right)$.

**58.** $\begin{cases} y = 2x^2 + 1 \\ y = x + 2 \end{cases}$

First, graph each equation in the system.

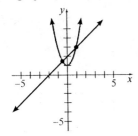

The system apparently has two solutions. Now substitute $2x^2 + 1$ for $y$ into the second equation:
$$2x^2 + 1 = x + 2$$
$$2x^2 - x - 1 = 0$$
$$(2x+1)(x-1) = 0$$
$$x = -\frac{1}{2} \text{ or } x = 1$$

Substitute these $x$-values into the first equation to find the corresponding $y$-values:

$x = -\frac{1}{2}$: $y = 2\left(-\frac{1}{2}\right)^2 + 1 = 2\left(\frac{1}{4}\right) + 1 = \frac{1}{2} + 1 = \frac{3}{2}$

$x = 1$: $y = 2(1)^2 + 1 = 2(1) + 1 = 2 + 1 = 3$

Both pairs check, so the solutions are $\left(-\frac{1}{2}, \frac{3}{2}\right)$ and $(1, 3)$.

**59.** $\begin{cases} 6x - y = 5 \\ xy = 1 \end{cases}$

First, graph each equation in the system.

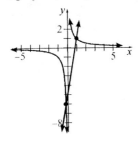

The system apparently has two solutions. Now solve the first equation for $y$: $y = 6x - 5$.
Substitute the result into the second equation.
$$x(6x - 5) = 1$$
$$6x^2 - 5x - 1 = 0$$
$$(6x + 1)(x - 1) = 0$$
$$x = -\frac{1}{6} \text{ or } x = 1$$

Substitute these $x$-values into the equation $y = 6x - 5$ to find the corresponding $y$-values.

$x = -\frac{1}{6}$: $y = 6\left(-\frac{1}{6}\right) - 5 = -1 - 5 = -6$

$x = 1$: $y = 6(1) - 5 = 6 - 5 = 1$

Both pairs check, so the solutions are $\left(-\frac{1}{6}, -6\right)$ and $(1, 1)$.

**60.** $\begin{cases} x^2 + y^2 = 26 \\ x^2 - 2y^2 = 23 \end{cases}$

First, graph each equation in the system.

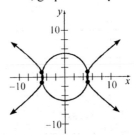

The system apparently has four solutions. Now solve the first equation for $x^2$: $x^2 = 26 - y^2$.
Substitute the result into the second equation.
$$(26 - y^2) - 2y^2 = 23$$
$$-3y^2 = -3$$
$$y^2 = 1$$
$$y = \pm 1$$

Substitute 1 for $y^2$ into the first equation.
$$x^2 + 1 = 26$$
$$x^2 = 25$$
$$x = \pm 5$$

All four pairs check, so the solutions are $(-5, -1)$, $(-5, 1)$, $(5, -1)$, and $(5, 1)$.

## Chapter 10: Conics

**61.** $\begin{cases} 4x - y^2 = 0 \\ 2x^2 + y^2 = 16 \end{cases}$

First, graph each equation in the system.

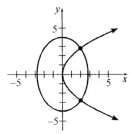

The system apparently has two solutions. Now add the two equations:

$4x - y^2 = 0$
$2x^2 \phantom{+0} + y^2 = 16$
$\overline{2x^2 \phantom{+y^2} + 4x = 16}$
$2x^2 + 4x - 16 = 0$
$x^2 + 2x - 8 = 0$
$(x+4)(x-2) = 0$
$x = -4 \text{ or } x = 2$

Substitute these $x$-values into the first equation to find the corresponding $y$-values:

$x = -4:\ 4(-4) - y^2 = 0$
$\phantom{x = -4:\ }-16 - y^2 = 0$
$\phantom{x = -4:\ }-16 = y^2$
$\phantom{x = -4:\ }y = \pm\sqrt{-16}$ (not real)

$x = 2:\ 4(2) - y^2 = 0$
$\phantom{x = 2:\ }8 - y^2 = 0$
$\phantom{x = 2:\ }8 = y^2$
$\phantom{x = 2:\ }y = \pm\sqrt{8} = \pm 2\sqrt{2}$

Both real-number pairs check, so the solutions are $\left(2, -2\sqrt{2}\right)$ and $\left(2, 2\sqrt{2}\right)$.

**62.** $\begin{cases} x^2 - y = -2 \\ x^2 + y = 4 \end{cases}$

First, graph each equation in the system.

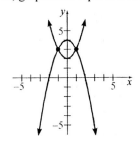

The system apparently has two solutions. Now multiply the first equation by $-1$ and add the result to the second equation:

$-x^2 + y = 2$
$x^2 + y = 4$
$\overline{\phantom{-x^2 + }2y = 6}$
$\phantom{-x^2 + }y = 3$

Substitute this $y$-value into the second equation to find the corresponding $x$-values.

$x^2 + 3 = 4$
$x^2 = 1$
$x = \pm 1$

Both pairs check, so the solutions are $(-1, 3)$ and $(1, 3)$.

**63.** $\begin{cases} 4x^2 - 2y^2 = 2 \\ -x^2 + y^2 = 2 \end{cases}$

First, graph each equation in the system.

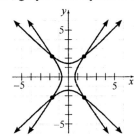

The system apparently has four solutions. Now multiply the second equation by 2 and add the result to the first equation:

$4x^2 - 2y^2 = 2$
$-2x^2 + 2y^2 = 4$
$\overline{2x^2 \phantom{+ 2y^2} = 6}$
$x^2 = 3$
$x = \pm\sqrt{3}$

Substitute 3 for $x^2$ into the second equation:

$-3 + y^2 = 2$
$y^2 = 5$
$y = \pm\sqrt{5}$

All four pairs check, so the solutions are $\left(-\sqrt{3}, -\sqrt{5}\right)$, $\left(-\sqrt{3}, \sqrt{5}\right)$, $\left(\sqrt{3}, -\sqrt{5}\right)$, and $\left(\sqrt{3}, \sqrt{5}\right)$.

64. $\begin{cases} x^2 + y^2 = 8x \\ y^2 = 3x \end{cases}$

First, graph each equation in the system.

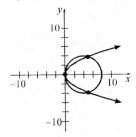

The system apparently has three solutions. Now multiply the second equation by $-1$ and add the result to the first equation:

$x^2 + y^2 = 8x$
$\underline{\phantom{x^2}\ -y^2 = -3x}$
$x^2 \phantom{+y^2} = 5x$
$x^2 - 5x = 0$
$x(x-5) = 0$
$x = 0$ or $x = 5$

Substitute these $x$-values into the second equation to find the corresponding $y$-values:

$x = 0$: $y^2 = 3(0)$
$\phantom{x = 0:\ } y^2 = 0$
$\phantom{x = 0:\ } y = 0$

$x = 5$: $y^2 = 3(5)$
$\phantom{x = 5:\ } y^2 = 15$
$\phantom{x = 5:\ } y = \pm\sqrt{15}$

All three pairs check, so the solutions are $(0,0)$, $\left(5, -\sqrt{15}\right)$, and $\left(5, \sqrt{15}\right)$.

65. $\begin{cases} y = x+2 \\ y = x^2 \end{cases}$

First, graph each equation in the system.

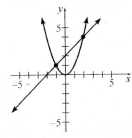

The system apparently has two solutions. Now Substitute $x^2$ for $y$ into the first equation:

$x^2 = x + 2$
$x^2 - x - 2 = 0$
$(x-2)(x+1) = 0$
$x = 2$ or $x = -1$

Substitute these results into the first equation to find the corresponding $y$ values.

$x = 2$: $y = 2 + 2 = 4$
$x = -1$: $y = -1 + 2 = 1$

Both pairs check, so the solutions are $(2, 4)$ and $(-1, 1)$.

66. $\begin{cases} x^2 + 2y = 9 \\ 5x - 2y = 5 \end{cases}$

First, graph each equation in the system.

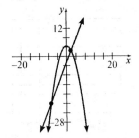

The system apparently has two solutions. Now add the two equations:

$x^2 \phantom{+5x} + 2y = 9$
$\underline{\phantom{x^2+}\ 5x - 2y = 5}$
$x^2 \phantom{+2y} + 5x = 14$
$x^2 + 5x - 14 = 0$
$(x+7)(x-2) = 0$
$x = -7$ or $x = 2$

Substitute these results into the first equation to find the corresponding $y$ values.

$x = -7$: $(-7)^2 + 2y = 9$
$\phantom{x = -7:\ } 49 + 2y = 9$
$\phantom{x = -7:\ } 2y = -40$
$\phantom{x = -7:\ } y = -20$

$x = 2$: $(2)^2 + 2y = 9$
$\phantom{x = 2:\ } 4 + 2y = 9$
$\phantom{x = 2:\ } 2y = 5$
$\phantom{x = 2:\ } y = \dfrac{5}{2}$

Both pairs check, so the solutions are $(-7, -20)$ and $\left(2, \dfrac{5}{2}\right)$.

## Chapter 10: Conics

**67.** $\begin{cases} x^2 + y^2 = 36 \\ x - y = -6 \end{cases}$

First, graph each equation in the system.

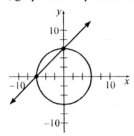

The system apparently has two solutions. Now solve the second equation for $y$: $y = x + 6$.
Substitute the result into the first equation:
$$x^2 + (x+6)^2 = 36$$
$$x^2 + (x^2 + 12x + 36) = 36$$
$$2x^2 + 12x = 0$$
$$2x(x+6) = 0$$
$$2x = 0 \text{ or } x + 6 = 0$$
$$x = 0 \text{ or } \quad x = -6$$

Substitute these $x$ values into the equation $y = x + 6$ to find the corresponding $y$ values:
$x = 0$: $y = 0 + 6 = 6$
$x = -6$: $y = -6 + 6 = 0$

Both pairs check, so the solutions are $(0, 6)$ and $(-6, 0)$.

**68.** $\begin{cases} y = 2x - 4 \\ y^2 = 4x \end{cases}$

First, graph each equation in the system.

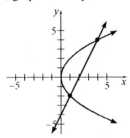

The system apparently has two solutions. Now substitute $2x - 4$ for $y$ into the second equation:
$$(2x-4)^2 = 4x$$
$$4x^2 - 16x + 16 = 4x$$
$$4x^2 - 20x + 16 = 0$$
$$x^2 - 5x + 4 = 0$$
$$(x-4)(x-1) = 0$$
$$x = 4 \text{ or } x = 1$$

Substitute these results into the first equation to find the corresponding $y$ values.
$x = 4$: $y = 2(4) - 4 = 4$
$x = 1$: $y = 2(1) - 4 = -2$

Both pairs check, so the solutions are $(4, 4)$ and $(1, -2)$.

**69.** $\begin{cases} x^2 + y^2 = 9 \\ x + y = 7 \end{cases}$

First, graph each equation in the system.

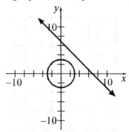

The system apparently has no solution. Now solve the second equation for $y$: $y = 7 - x$.
Substitute the result into the first equation:
$$x^2 + (7-x)^2 = 9$$
$$x^2 + (49 - 14x + x^2) = 9$$
$$2x^2 - 14x + 49 = 9$$
$$2x^2 - 14x + 40 = 0$$
$$x = \frac{-(-14) \pm \sqrt{(-14)^2 - 4(2)(40)}}{2(2)}$$
$$= \frac{14 \pm \sqrt{-124}}{4}$$
$$= \frac{14 \pm 2\sqrt{31}\,i}{4}$$
$$= \frac{7}{2} \pm \frac{\sqrt{31}}{2}i$$

Since the solutions are not real, the system is inconsistent. The solution set is $\varnothing$.

**70.** $\begin{cases} 2x^2 + 3y^2 = 14 \\ x^2 - y^2 = -3 \end{cases}$

First, graph each equation in the system.

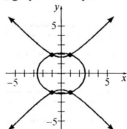

The system apparently has four solutions. Now multiply the second equation by 3 and add the result to the first equation.

$$3x^2 - 3y^2 = -9$$
$$\underline{2x^2 + 3y^2 = 14}$$
$$5x^2 \phantom{+3y^2} = 5$$
$$x^2 = 1$$
$$x = \pm 1$$

Substitute 1 for $x^2$ into the second equation to find the $y$ values.
$$1 - y^2 = -3$$
$$-y^2 = -4$$
$$y^2 = 4$$
$$y = \pm 2$$

All four pairs check, so the solutions are $(-1,-2)$, $(-1, 2)$, $(1,-2)$, and $(1, 2)$.

71. $\begin{cases} x^2 + y^2 = 16 \\ x^2 + 4y = 16 \end{cases}$

First, graph each equation in the system.

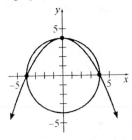

The system apparently has three solutions. Now multiply the second equation by $-1$ and add the result to the first equation:
$$x^2 + y^2 \phantom{-4y} = 16$$
$$\underline{-x^2 \phantom{+y^2} - 4y = -16}$$
$$y^2 - 4y = 0$$
$$y(y-4) = 0$$
$$y = 0 \text{ or } y = 4$$

Substitute these $y$-values into the first equation to find the corresponding $x$-values:
$$y = 0: \quad x^2 + 0^2 = 16$$
$$x^2 = 16$$
$$x = \pm 4$$
$$y = 4: \quad x^2 + 4^2 = 16$$
$$x^2 + 16 = 16$$
$$x^2 = 0$$
$$x = 0$$

All three pairs check, so the solutions are $(-4, 0)$, $(4, 0)$, and $(0, 4)$.

72. $\begin{cases} x = 4 - y^2 \\ x = 2y + 4 \end{cases}$

First, graph each equation in the system.

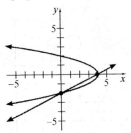

The system apparently has two solutions. Now substitute $2y + 4$ for $x$ into the first equation:
$$2y + 4 = 4 - y^2$$
$$y^2 + 2y = 0$$
$$y(y+2) = 0$$
$$y = 0 \text{ or } y = -2$$

Substitute these $y$-values into the second equation to find the corresponding $x$-values:
$$y = 0: \quad x = 2(0) + 4 = 4$$
$$y = -2: \quad x = 2(-2) + 4 = 0$$

Both pairs check, so the solutions are $(4, 0)$ and $(0, -2)$.

73. Let $x$ represent the larger of the two numbers and $y$ represent the smaller of the two numbers.
$$\begin{cases} x + y = 12 \\ x^2 - y^2 = 24 \end{cases}$$

Solve the first equation for $y$: $y = 12 - x$.
Substitute the result into the second equation:
$$x^2 - (12 - x)^2 = 24$$
$$x^2 - (144 - 24x + x^2) = 24$$
$$x^2 - 144 + 24x - x^2 = 24$$
$$24x = 168$$
$$x = 7$$

Substitute 7 for $x$ into the first equation to find the corresponding $y$-value:
$$7 + y = 12$$
$$y = 5$$

The two numbers are 7 and 5.

## Chapter 10: Conics

**74.** Let $x$ represent the length and $y$ represent the width of the rectangle.
$$\begin{cases} 2x+2y = 34 \\ xy = 60 \end{cases}$$
Solve the first equation for $y$: $y = 17-x$.
Substitute the result into the second equation:
$$x(17-x) = 60$$
$$17x - x^2 = 60$$
$$x^2 - 17x + 60 = 0$$
$$(x-12)(x-5) = 0$$
$$x = 12 \text{ or } x = 5$$
Substitute these $x$-values into the second equation to find the corresponding $y$-values:
$x = 12$: $12y = 60$
$y = 5$
$x = 5$: $5y = 60$
$y = 12$
Note that the two outcomes result in the same overall dimensions. Assuming the length is the longer of the two sides, the length is 12 centimeters and the width is 5 centimeters.

**75.** Let $x$ represent the length and $y$ represent the width of the rectangle.

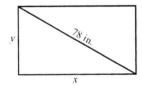

$$\begin{cases} xy = 2160 \\ x^2 + y^2 = 78^2 \end{cases}$$
Solve the first equation for $y$: $y = \dfrac{2160}{x}$.
Substitute the result into the second equation:
$$x^2 + \left(\dfrac{2160}{x}\right)^2 = 78^2$$
$$x^2 + \dfrac{4,665,600}{x^2} = 6084$$
$$x^4 + 4,665,600 = 6084x^2$$
$$x^4 - 6084x^2 + 4,665,600 = 0$$
$$(x^2 - 5184)(x^2 - 900) = 0$$
$$(x-72)(x+72)(x-30)(x+30) = 0$$
$x = 72$ or $x = -72$ or $x = 30$ or $x = -30$
Since the length cannot be negative, we discard $x = -72$ and $x = -30$, so we are left with the possible answers $x = 72$ and $x = 30$. Substitute these $x$-values into the first equation to find the corresponding $y$-values:

$x = 72$: $72y = 2160$
$y = 30$
$x = 30$: $30y = 2160$
$y = 72$
Note that the two outcomes result in the same overall dimensions. Assuming the length is the longer of the two sides, the length is 72 inches and the width is 30 inches.

**76.** Let $x$ and $y$ represent the lengths of the two legs of the right triangle.

$$\begin{cases} x+y+15 = 36 \\ x^2 + y^2 = 15^2 \end{cases}$$
Solve the first equation for $y$: $y = 21-x$.
Substitute the result into the second equation:
$$x^2 + (21-x)^2 = 15^2$$
$$x^2 + 441 - 42x + x^2 = 225$$
$$2x^2 - 42x + 216 = 0$$
$$x^2 - 21x + 108 = 0$$
$$(x-12)(x-9) = 0$$
$$x = 12 \text{ or } x = 9$$
Substitute these $x$-values into the first equation to find the corresponding $y$-values:
$x = 12$: $12 + y + 15 = 36$
$y + 27 = 36$
$y = 9$
$x = 9$: $9 + y + 15 = 36$
$y + 24 = 36$
$y = 12$
Note that the two outcomes result in the same overall dimensions. The lengths of the two legs are 12 inches and 9 inches.

### Chapter 10 Test

**1.** $d(P_1, P_2) = \sqrt{(x_2 - x_1)^2 + (y_2 - y_1)^2}$
$= \sqrt{(3-(-1))^2 + (-5-3)^2}$
$= \sqrt{4^2 + (-8)^2}$
$= \sqrt{16 + 64}$
$= \sqrt{80}$
$= 4\sqrt{5}$

2. $M = \left(\dfrac{x_1 + x_2}{2}, \dfrac{y_1 + y_2}{2}\right)$

$= \left(\dfrac{-7+5}{2}, \dfrac{6+(-2)}{2}\right) = \left(\dfrac{-2}{2}, \dfrac{4}{2}\right) = (-1, 2)$

3. $(x-4)^2 + (y+1)^2 = 9$

$(x-4)^2 + (y-(-1))^2 = 3^2$

The center is $(h, k) = (4, -1)$, and the radius is $r = 3$.

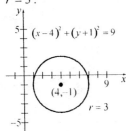

4. $x^2 + y^2 + 10x - 4y + 13 = 0$

$(x^2 + 10x) + (y^2 - 4y) = -13$

$(x^2 + 10x + 25) + (y^2 - 4y + 4) = -13 + 25 + 4$

$(x+5)^2 + (y-2)^2 = 16$

$(x-(-5))^2 + (y-2)^2 = 4^2$

The center is $(h, k) = (-5, 2)$, and the radius is $r = 4$.

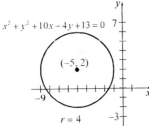

5. $(x-h)^2 + (y-k)^2 = r^2$

$(x-(-3))^2 + (y-7)^2 = 6^2$

$(x+3)^2 + (y-7)^2 = 36$

6. The radius of the circle will be the distance from the center point $(-5, 8)$ to the point on the circle $(3, 2)$. Thus, $r = \sqrt{(3-(-5))^2 + (2-8)^2} = 10$.

The equation of the circle is

$(x-h)^2 + (y-k)^2 = r^2$

$(x-(-5))^2 + (y-8)^2 = 10^2$

$(x+5)^2 + (y-8)^2 = 100$

7. Notice that $(y+2)^2 = 4(x-1)$ is of the form $(y-k)^2 = 4a(x-h)$, where $(h, k) = (1, -2)$ and $4a = 4$, so that $a = 1$. Now, the graph of an equation of the form $(y-k)^2 = 4a(x-h)$ will be a parabola that opens to the right with the vertex at $(h, k)$, focus at $(h+a, k)$, and directrix of $x = h - a$. Note that $h + a = 1 + 1 = 2$ and $h - a = 1 - 1 = 0$. Thus, the graph of $(y+2)^2 = 4(x-1)$ is a parabola that opens to the right with vertex $(1, -2)$, focus $(2, -2)$, and directrix $x = 0$. To help graph the parabola, we plot the two points on the graph above and below the focus. Let $x = 2$:

$(y+2)^2 = 4(2-1)$

$(y+2)^2 = 4$

$y + 2 = \pm 2$

$y = -2 \pm 2$

$y = -4$ or $y = 0$

The points $(2, -4)$ and $(2, 0)$ are on the graph.

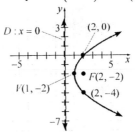

8. We complete the square in $x$ to write the equation in standard form:

$x^2 - 4x + 3y - 8 = 0$

$x^2 - 4x = -3y + 8$

$x^2 - 4x + 4 = -3y + 8 + 4$

$x^2 - 4x + 4 = -3y + 12$

$(x-2)^2 = -3(y-4)$

Notice that $(x-2)^2 = -3(y-4)$ is of the form $(x-h)^2 = -4a(y-k)$, where $(h, k) = (2, 4)$ and $-4a = -3$, so that $a = \dfrac{3}{4}$. Now, the graph of an equation of the form $(x-h)^2 = -4a(y-k)$ will be a parabola that opens downward with the vertex at $(h, k)$, focus at $(h, k-a)$, and directrix of $y = k + a$. Note that $k - a = 4 - \dfrac{3}{4} = \dfrac{13}{4}$ and $k + a = 4 + \dfrac{3}{4} = \dfrac{19}{4}$. Thus, the graph of

$(x-2)^2 = -3(y-4)$ is a parabola that opens downward with vertex $(2,4)$, focus $\left(2, \frac{13}{4}\right)$, and directrix $y = \frac{19}{4}$. To help graph the parabola, we plot the two points on the graph to the left and right of the focus. Let $y = \frac{13}{4}$:

$$(x-2)^2 = -3\left(\frac{13}{4} - 4\right)$$
$$(x-2)^2 = -3\left(-\frac{3}{4}\right)$$
$$(x-2)^2 = \frac{9}{4}$$
$$x - 2 = \pm\frac{3}{2}$$
$$x = 2 \pm \frac{3}{2}$$
$$x = \frac{1}{2} \text{ or } x = \frac{7}{2}$$

The points $\left(\frac{1}{2}, \frac{13}{4}\right)$ and $\left(\frac{7}{2}, \frac{13}{4}\right)$ are on the graph.

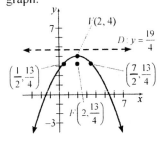

9. The distance from the vertex $(0,0)$ to the focus $(0,-4)$ is $a = 4$. Because the focus lies on the negative $y$-axis, we know that the parabola will open downward and the axis of symmetry is the $y$-axis. This means the equation of the parabola is of the form $x^2 = -4ay$ with $a = 4$:
$$x^2 = -4(4)y$$
$$x^2 = -16y$$

10. Notice that the directrix $x = -1$ is a vertical line and that the focus is $(3,4)$. Therefore, the axis of symmetry must be the line $y = 4$. Because the focus is to the right of the directrix, the parabola must open to the right. Since the vertex must be the point on the axis of symmetry that is midway between the focus and the directrix, the vertex is $(1,4)$. Now, the distance from the vertex $(1,4)$ to the focus $(3,4)$ is $a = 2$, so the equation of the parabola is of the form $(y-k)^2 = 4a(x-h)$ with $a = 2$, $h = 1$, and $k = 4$:
$$(y-4)^2 = 4 \cdot 2(x-1)$$
$$(y-4)^2 = 8(x-1)$$

11. $9x^2 + 25y^2 = 225$
$$\frac{9x^2 + 25y^2}{225} = \frac{225}{225}$$
$$\frac{x^2}{25} + \frac{y^2}{9} = 1$$

The larger number, 25, is in the denominator of the $x^2$-term. This means that the major axis is the $x$-axis and that the equation of the ellipse is of the form $\frac{x^2}{a^2} + \frac{y^2}{b^2} = 1$, so that $a^2 = 25$ and $b^2 = 9$. The center of the ellipse is $(0,0)$. Because $c^2 = a^2 - b^2$, we have that $c^2 = 25 - 9 = 16$, so that $c = \pm\sqrt{16} = \pm 4$. Since the major axis is the $x$-axis, the foci are $(-4, 0)$ and $(4, 0)$. To find the $x$-intercepts (vertices), let $y = 0$; to find the $y$-intercepts, let $x = 0$:

$x$-intercepts:
$9x^2 + 25(0)^2 = 225$
$9x^2 = 225$
$x^2 = 25$
$x = \pm 5$

$y$-intercepts:
$9(0)^2 + 25y^2 = 225$
$25y^2 = 225$
$y^2 = 9$
$y = \pm 3$

The intercepts are $(-5, 0)$, $(5, 0)$, $(0, -3)$, and $(0, 3)$.

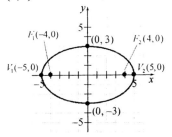

12. $\dfrac{(x-2)^2}{9}+\dfrac{(y+4)^2}{16}=1$

    The center of the ellipse is $(h,k)=(2,-4)$. Because the larger number, 16, is the denominator of the $y^2$-term, the major axis is parallel to the $y$-axis. Because $a^2=16$ and $b^2=9$, we have that $c^2=a^2-b^2=16-9=7$. The vertices are $a=4$ units below and above the center at $V_1(2,-8)$ and $V_2(2,0)$. The foci are $c=\sqrt{7}$ units below and above the center at $F_1\left(2,-4-\sqrt{7}\right)$ and $F_2\left(2,-4+\sqrt{7}\right)$. We plot the points $b=3$ units to the left and right of the center point at $(-1,-4)$ and $(5,-4)$.

    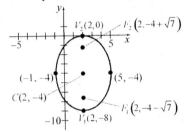

13. The given focus $(0,-4)$ and the given vertex $(0,-5)$ lie on the $y$-axis. Thus, the equation of the ellipse is of the form $\dfrac{x^2}{b^2}+\dfrac{y^2}{a^2}=1$. The distance from the center of the ellipse to the vertex is $a=5$ units. The distance from the center of the ellipse to the focus is $c=4$ units. Because $b^2=a^2-c^2$, we have that $b^2=5^2-4^2=25-16=9$. Thus, the equation of the ellipse is $\dfrac{x^2}{9}+\dfrac{y^2}{25}=1$.

14. The given vertices and focus all lie on the vertical line $x=-1$. Therefore, the major axis is parallel to the $y$-axis, and the equation of the ellipse is of the form $\dfrac{(x-h)^2}{b^2}+\dfrac{(y-k)^2}{a^2}=1$.

    Now, the center will be the midpoint between the two vertices $(-1,7)$ and $(-1,-3)$, which is $(h,k)=(-1,2)$. Now, $a=5$ is the distance from the center $(-1,2)$ to a vertex $(-1,7)$, and $c=3$ is the distance from the center to a focus $(-1,-1)$. Also, $b^2=a^2-c^2=5^2-3^2=25-9=16$. Thus, the equation of the ellipse is

    $\dfrac{(x-(-1))^2}{16}+\dfrac{(y-2)^2}{25}=1$

    $\dfrac{(x+1)^2}{16}+\dfrac{(y-2)^2}{25}=1$

15. $x^2-\dfrac{y^2}{4}=1$

    $\dfrac{x^2}{1}-\dfrac{y^2}{4}=1$

    Notice the equation is of the form $\dfrac{x^2}{a^2}-\dfrac{y^2}{b^2}=1$.

    Because the $x^2$-term is first, the transverse is along the $x$-axis and the hyperbola opens left and right. The center of the hyperbola is $(0,0)$. We have that $a^2=1$ and $b^2=4$. Because $c^2=a^2+b^2$, we have that $c^2=1+4=5$, so that $c=\sqrt{5}$. The vertices are $(\pm a,0)=(\pm 1,0)$, and the foci are $(\pm c,0)=\left(\pm\sqrt{5},0\right)$. Since $a=1$ and $b=2$, the equations of the asymptotes are $y=\dfrac{2}{1}x=2x$ and $y=-\dfrac{2}{1}x=-2x$. To help graph the hyperbola, we form the rectangle using the points $(\pm a,0)=(\pm 1,0)$ and $(0,\pm b)=(0,\pm 2)$. The diagonals are the asymptotes.

    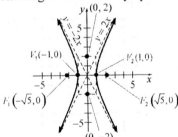

16. $16y^2-25x^2=1600$

    $\dfrac{16y^2-25x^2}{1600}=\dfrac{1600}{1600}$

    $\dfrac{y^2}{100}-\dfrac{x^2}{64}=1$

    Notice the equation is of the form $\dfrac{y^2}{a^2}-\dfrac{x^2}{b^2}=1$.

    Because the $y^2$-term is first, the transverse is along the $y$-axis and the hyperbola opens up and down. The center of the hyperbola is $(0,0)$. We have that $a^2=100$ and $b^2=64$. Because $c^2=a^2+b^2$, we have that $c^2=100+64=164$,

## Chapter 10: Conics

so that $c = \sqrt{164} = 2\sqrt{41}$. The vertices are $(0, \pm a) = (0, \pm 10)$, and the foci are $(0, \pm c) = \left(0, \pm 2\sqrt{41}\right)$. Since $a = 10$ and $b = 8$, the equations of the asymptotes are $y = \frac{10}{8}x = \frac{5}{4}x$ and $y = -\frac{10}{8}x = -\frac{5}{4}x$. To help graph the hyperbola, we form the rectangle using the points $(0, \pm a) = (0, \pm 10)$ and $(\pm b, 0) = (\pm 8, 0)$. The diagonals are the asymptotes.

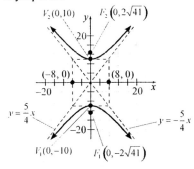

17. The given foci $(\pm 8, 0)$ and vertex $(-3, 0)$ all lie on the $x$-axis. Thus, the transverse axis is the $x$-axis and the hyperbola opens left and right. The center of the hyperbola is the midpoint between the two foci. Therefore, the center is $(0, 0)$ and the equation of the hyperbola is of the form $\frac{x^2}{a^2} - \frac{y^2}{b^2} = 1$. Now, the distance between the given vertex $(-3, 0)$ and the center $(0, 0)$ is $a = 3$ units. Likewise, the distance between the given foci and the center is $c = 8$ units. Also, $b^2 = c^2 - a^2$, so $b^2 = 8^2 - 3^2 = 64 - 9 = 55$. Thus, the equation of the hyperbola is $\frac{x^2}{9} - \frac{y^2}{55} = 1$.

18. $\begin{cases} x^2 + y^2 = 17 \\ x + y = -3 \end{cases}$

    First, graph each equation in the system.

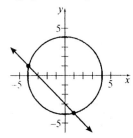

The system apparently has two solutions. Now solve the second equation for $y$: $y = -x - 3$. Substitute the result into the first equation:
$$x^2 + (-x-3)^2 = 17$$
$$x^2 + (x^2 + 6x + 9) = 17$$
$$2x^2 + 6x - 8 = 0$$
$$x^2 + 3x - 4 = 0$$
$$(x+4)(x-1) = 0$$
$$x = -4 \text{ or } x = 1$$

Substitute these results into the first equation to find the corresponding $y$-values.
$x = -4$: $-4 + y = -3$
$\qquad y = 1$
$x = 1$: $1 + y = -3$
$\qquad y = -4$

Both pairs check, so the solutions are $(-4, 1)$ and $(1, -4)$.

19. $\begin{cases} x^2 + y^2 = 9 \\ 4x^2 - y^2 = 16 \end{cases}$

    First, graph each equation in the system.

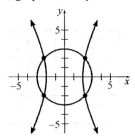

The system apparently has four solutions. Now add the two equations:
$$x^2 + y^2 = 9$$
$$4x^2 - y^2 = 16$$
$$\overline{5x^2 \qquad = 25}$$
$$x^2 = 5$$
$$x = \pm\sqrt{5}$$

Substitute 5 for $x^2$ into the first equation:
$$5 + y^2 = 9$$
$$y^2 = 4$$
$$y = \pm\sqrt{4} = \pm 2$$

All four pairs check, so the solutions are $\left(-\sqrt{5}, -2\right), \left(-\sqrt{5}, 2\right), \left(\sqrt{5}, -2\right)$, and $\left(\sqrt{5}, 2\right)$.

20. a. To solve this problem, we draw the ellipse on a Cartesian plane so that the x-axis coincides with the water and the y-axis passes through the center of the arch. Thus, the origin is the center of the ellipse. Now, the "center" of the arch is 10 feet above the water, so the point $(0,10)$ is on the ellipse. The creek is 30 feet wide, so the two points $(-15,0)$ and $(15,0)$ are the two vertices of the ellipse and the major axis is along the x-axis.

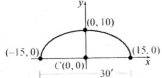

The equation of the ellipse must have the form $\dfrac{x^2}{a^2} + \dfrac{y^2}{b^2} = 1$. The distance from the center of the ellipse to each vertex is $a = 15$. Also, the height of the semi-ellipse is 10 feet, so $b = 10$. Thus, the equation of the ellipse is

$$\frac{x^2}{15^2} + \frac{y^2}{10^2} = 1$$
$$\frac{x^2}{225} + \frac{y^2}{100} = 1$$

b. Substitute $x = 12$ into the equation, and solve for $y$:

$$\frac{12^2}{225} + \frac{y^2}{100} = 1$$
$$\frac{144}{225} + \frac{y^2}{100} = 1$$
$$\frac{y^2}{100} = \frac{9}{25}$$
$$y^2 = 36$$
$$y = 6$$

Thus, the height of the arch at a distance 12 feet from the center of the creek is 6 feet.

# Chapter 11

## 11.1 Preparing for Sequences

1. $f(x) = x^2 - 4$

   a. $f(3) = (3)^2 - 4 = 9 - 4 = 5$

   b. $f(-7) = (-7)^2 - 4 = 49 - 4 = 45$

2. $g(x) = 2x - 3$
   $g(1) = 2(1) - 3 = 2 - 3 = -1$
   $g(2) = 2(2) - 3 = 4 - 3 = 1$
   $g(3) = 2(3) - 3 = 6 - 3 = 3$
   Therefore,
   $g(1) + g(2) + g(3) = -1 + 1 + 3 = 3$

3. In the function $f(n) = n^2 - 4$, the independent variable is $n$ (the input variable).

## 11.1 Exercises

2. infinite; finite

4. True; a sequence is a function whose domain is the set of natural numbers (positive integers).

6. False; since the domain of a sequence is defined to be the set of positive integers, 0 is not in the domain.

8. The graph of the sequence $\{3n+1\}$ consists of a set of disconnected points while the graph of the function $y = 3x + 1$ consists of a continuous set of points.

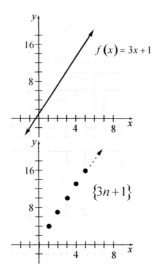

10. A sequence alternates when the signs of the terms alternate between positive and negative.

12. $\{n - 4\}$
    $a_1 = 1 - 4 = -3$
    $a_2 = 2 - 4 = -2$
    $a_3 = 3 - 4 = -1$
    $a_4 = 4 - 4 = 0$
    $a_5 = 5 - 4 = 1$
    The first five terms of the sequence are $-3$, $-2$, $-1$, 0, and 1.

14. $\left\{\dfrac{n+4}{n}\right\}$
    $a_1 = \dfrac{1+4}{1} = 5$
    $a_2 = \dfrac{2+4}{2} = 3$
    $a_3 = \dfrac{3+4}{3} = \dfrac{7}{3}$
    $a_4 = \dfrac{4+4}{4} = 2$
    $a_5 = \dfrac{5+4}{5} = \dfrac{9}{5}$
    The first five terms of the sequence are 5, 3, $\dfrac{7}{3}$, 2, and $\dfrac{9}{5}$.

**Section 11.1** Sequences

16. $\{(-1)^{n+1} \cdot n\}$

    $a_1 = (-1)^{1+1} \cdot 1 = 1$
    $a_2 = (-1)^{2+1} \cdot 2 = -2$
    $a_3 = (-1)^{3+1} \cdot 3 = 3$
    $a_4 = (-1)^{4+1} \cdot 4 = -4$
    $a_5 = (-1)^{5+1} \cdot 5 = 5$

    The first five terms of the sequence are 1, $-2$, 3, $-4$, and 5.

18. $\{3^n - 1\}$

    $a_1 = 3^1 - 1 = 3 - 1 = 2$
    $a_2 = 3^2 - 1 = 9 - 1 = 8$
    $a_3 = 3^3 - 1 = 27 - 1 = 26$
    $a_4 = 3^4 - 1 = 81 - 1 = 80$
    $a_5 = 3^5 - 1 = 243 - 1 = 242$

    The first five terms of the sequence are 2, 8, 26, 80, and 242.

20. $\left\{\dfrac{3n}{3^n}\right\}$

    $a_1 = \dfrac{3 \cdot 1}{3^1} = \dfrac{3}{3} = 1$
    $a_2 = \dfrac{3 \cdot 2}{3^2} = \dfrac{6}{9} = \dfrac{2}{3}$
    $a_3 = \dfrac{3 \cdot 3}{3^3} = \dfrac{9}{27} = \dfrac{1}{3}$
    $a_4 = \dfrac{3 \cdot 4}{3^4} = \dfrac{12}{81} = \dfrac{4}{27}$
    $a_5 = \dfrac{3 \cdot 5}{3^5} = \dfrac{15}{243} = \dfrac{5}{81}$

    The first five terms of the sequence are 1, $\dfrac{2}{3}$, $\dfrac{1}{3}$, $\dfrac{4}{27}$, and $\dfrac{5}{81}$.

22. $\left\{\dfrac{n^2}{2}\right\}$

    $a_1 = \dfrac{1^2}{2} = \dfrac{1}{2}$, $a_2 = \dfrac{2^2}{2} = \dfrac{4}{2} = 2$, $a_3 = \dfrac{3^2}{2} = \dfrac{9}{2}$,
    $a_4 = \dfrac{4^2}{2} = \dfrac{16}{2} = 8$, $a_5 = \dfrac{5^2}{2} = \dfrac{25}{2}$

    The first five terms of the sequence are $\dfrac{1}{2}$, 2, $\dfrac{9}{2}$, 8, and $\dfrac{25}{2}$.

24. The terms are all multiples of 5 with the first term being $5 \cdot 1$, the second term being $5 \cdot 2$, and so on. A formula for the $n$th term is given by $a_n = 5n$.

26. Notice that the first term is $\dfrac{1}{2}$, the second term is $\dfrac{2}{2}$, the third term is $\dfrac{3}{2}$, and so on. Each term is a fraction with the denominator equaling 2. The numerator is equal to the term number. A formula for the $n$th term is given by $a_n = \dfrac{n}{2}$.

28. When $n = 1$, we have that $a_1 = 1^3 - 1$; when $n = 2$, we have that $a_2 = 2^3 - 1$; when $n = 3$, we have that $a_3 = 3^3 - 1$. Notice that each term is equal to 1 less than the cube of the term number. Therefore, a formula for the $n$th term is given by $a_n = n^3 - 1$.

30. Notice that the terms alternate signs with the first term being positive. Ignoring the signs, also notice that each term is a fraction with the numerator equal 1 and the denominator being a power of 2. The power is 1 less than the term number. Therefore, a formula for the $n$th term is given by $a_n = \left(-\dfrac{1}{2}\right)^{n-1}$.

32. $\displaystyle\sum_{i=1}^{5}(3i+2) = (3 \cdot 1 + 2) + (3 \cdot 2 + 2) + (3 \cdot 3 + 2)$
    $\qquad + (3 \cdot 4 + 2) + (3 \cdot 5 + 2)$
    $= 5 + 8 + 11 + 14 + 17$
    $= 55$

34. $\displaystyle\sum_{i=1}^{4}\dfrac{i^3}{2} = \dfrac{1^3}{2} + \dfrac{2^3}{2} + \dfrac{3^3}{2} + \dfrac{4^3}{2}$
    $= \dfrac{1}{2} + \dfrac{8}{2} + \dfrac{27}{2} + \dfrac{64}{2}$
    $= 50$

## Chapter 11: Sequences, Series, and the Binomial Expansion

36. $\sum_{k=1}^{4} 3^k = 3^1 + 3^2 + 3^3 + 3^4$
    $= 3 + 9 + 27 + 81$
    $= 120$

38. $\sum_{k=1}^{8} \left[(-1)^k \cdot k\right]$
    $= \left[(-1)^1 \cdot 1\right] + \left[(-1)^2 \cdot 2\right] + \left[(-1)^3 \cdot 3\right]$
    $+ \left[(-1)^4 \cdot 4\right] + \left[(-1)^5 \cdot 5\right] + \left[(-1)^6 \cdot 6\right]$
    $+ \left[(-1)^7 \cdot 7\right] + \left[(-1)^8 \cdot 8\right]$
    $= -1 + 2 + (-3) + 4 + (-5) + 6 + (-7) + 8$
    $= 4$

40. $\sum_{j=1}^{8} 2 = 2 + 2 + 2 + 2 + 2 + 2 + 2 + 2 = 16$

42. $\sum_{k=5}^{10}(k+4) = (5+4) + (6+4) + (7+4) + (8+4)$
    $+ (9+4) + (10+4)$
    $= 9 + 10 + 11 + 12 + 13 + 14$
    $= 69$

44. The sum $1 + 3 + 5 + \ldots + 17$ consists of 9 consecutive odd numbers from 1 to 17. The odd numbers can be expressed as $2k - 1$ where $k$ goes from 1 to 9.
    $1 + 3 + 5 + \ldots + 17 = \sum_{k=1}^{9}(2k-1)$

46. To see the pattern, note that the first term can be written as $1 = \dfrac{1}{2^0}$. Thus, each term is a fraction whose numerator is 1 and whose denominator is a power of 2. The power is 1 less than the term number, so there are 16 terms.
    $1 + \dfrac{1}{2} + \dfrac{1}{4} + \ldots + \dfrac{1}{2^{15}} = \sum_{k=1}^{16} \dfrac{1}{2^{k-1}}$

48. From the final term, we can see that the sequence is alternating because of the $(-1)$ to a power. Ignoring the sign, each term is a power of the fraction $\tfrac{2}{3}$. Since the first term is $\left(\tfrac{2}{3}\right)^1$ and the last term is $\left(\tfrac{2}{3}\right)^{15}$, there are 15 terms in the sequence. Note that the exponent on $(-1)$ is 1 more than the term number.
    $\dfrac{2}{3} - \dfrac{4}{9} + \dfrac{8}{27} + \ldots + (-1)^{15+1}\left(\dfrac{2}{3}\right)^{15} = \sum_{i=1}^{15}(-1)^{i+1}\left(\dfrac{2}{3}\right)^i$

50. To see the pattern, note that the first term can be written as $3 \cdot \left(\dfrac{1}{2}\right)^0$. Thus, each term is the product of 3 and a power of $\tfrac{1}{2}$. The power is always 1 less than the term number so there are 12 terms in the sequence.
    $3 + 3 \cdot \dfrac{1}{2} + 3 \cdot \dfrac{1}{4} + \ldots + 3 \cdot \left(\dfrac{1}{2}\right)^{11} = \sum_{k=1}^{12}\left[3 \cdot \left(\dfrac{1}{2}\right)^{k-1}\right]$

52. $a_n = 5{,}000\left(1 + \dfrac{0.08}{12}\right)^n$

    a. After 1 month, we have $n = 1$.
    $a_1 = 5{,}000\left(1 + \dfrac{0.08}{12}\right)^1 = 5033.33$
    After 1 month, the account will have a value of \$5033.33.

    b. After 1 year, we have $n = 12$ months.
    $a_{12} = 5{,}000\left(1 + \dfrac{0.08}{12}\right)^{12} = 5415.00$
    After 1 year, the account will have a value of \$5415.00.

    c. After 10 years, we have $n = 120$ months.
    $a_{120} = 5{,}000\left(1 + \dfrac{0.08}{12}\right)^{120} = 11{,}098.20$
    After 10 years, the account will have a value of \$11,098.20.

54. Population: $p_n = 6448(1.0126)^n$ where $n$ is the number of years after 2005.

    a. In 2010, we have $n = 5$.
    $p_{10} = 6448(1.0126)^5 \approx 6865$
    The world population in 2010 is predicted to be about 6,865 million.

    b. In 2050, we have $n = 45$.
    $p_{50} = 6448(1.0126)^{45} \approx 11{,}327$

The world population in 2050 is predicted to be about 11,327 million.

56. The diagonal sums form the Fibonacci sequence.

| Diagonal | Sum |
|---|---|
| 1 | 1 |
| 2 | 1 |
| 3 | $1+1=2$ |
| 4 | $1+2=3$ |
| 5 | $1+3+1=5$ |
| 6 | $1+4+3=8$ |
| 7 | $1+5+6+1=13$ |

58. $b_1 = 20$
$b_2 = 3b_1 = 3(20) = 60$
$b_3 = 3b_2 = 3(60) = 180$
$b_4 = 3b_3 = 3(180) = 540$
$b_5 = 3b_4 = 3(540) = 1620$
The first five terms of the sequence are 20, 60, 180, 540, and 1620.

60. $c_1 = 1,000$
$c_2 = 1.01c_1 + 100 = 1.01(1000) + 100 = 1110$
$c_3 = 1.01c_2 + 100 = 1.01(1110) + 100 = 1221.10$
$c_4 = 1.01c_3 + 100$
$\phantom{c_4}= 1.01(1221.10) + 100 = 1333.311$
$c_5 = 1.01c_4 + 100 = 1.01(1333.311) + 100$
$\phantom{c_5}= 1446.64411$
The first five terms of the sequence are 1000, 1110, 1221.10, 1333.311, and 1446.64411.

64. a. The function is in slope-intercept form so the slope is $m = 2$.

66. $f(x) = 4x - 6$
$f(2) - f(1) = 2 - (-2) = 4$
$f(3) - f(2) = 6 - 2 = 4$
$f(4) - f(3) = 10 - 6 = 4$

$f(x) = 2x - 10$
$f(2) - f(1) = -6 - (-8) = 2$
$f(3) - f(2) = -4 - (-6) = 2$
$f(4) - f(3) = -2 - (-4) = 2$

$f(x) = -5x + 8$
$f(2) - f(1) = -2 - 3 = -5$
$f(3) - f(2) = -7 - (-2) = -5$
$f(4) - f(3) = -12 - (-7) = -5$

The difference in the value of the function for consecutive values of the independent variable is equal to the slope of the function.

68. $\{n - 4\}$

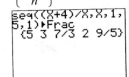

The first five terms of the sequence are $-3$, $-2$, $-1$, 0, and 1.

70. $\left\{\dfrac{n+4}{n}\right\}$

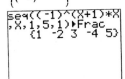

The first five terms of the sequence are 5, 3, $\dfrac{7}{3}$, 2, and $\dfrac{9}{5}$.

72. $\{(-1)^{n+1} \cdot n\}$

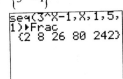

The first five terms of the sequence are 1, $-2$, 3, $-4$, and 5.

74. $\{3^n - 1\}$

The first five terms of the sequence are 2, 8, 26, 80, and 242.

## Chapter 11: Sequences, Series, and the Binomial Expansion

76. $\sum_{i=1}^{5}(3i+2)$

```
sum(seq(3X+2,X,1
,5,1)
            55
```

The sum is 55.

78. $\sum_{i=1}^{4}\frac{i^3}{2}$

```
sum(seq(X^3/2,X,
1,4,1)
            50
```

The sum is 50.

80. $\sum_{k=1}^{4}3^k$

```
sum(seq(3^X,X,1,
4,1)
           120
```

The sum is 120.

81. $\sum_{k=1}^{5}\left[(-1)^{k+1}\cdot 2k\right]$

```
sum(seq((-1)^(X+
1)*2X,X,1,5,1)
             6
```

The sum is 6.

82. $\sum_{k=1}^{8}\left[(-1)^k \cdot k\right]$

```
sum(seq((-1)^X*X
,X,1,8,1)
             4
```

The sum is 4.

### 11.2 Preparing for Arithmetic Sequences

1. $y=-3x+1$ is in the (slope-intercept) form $y=mx+b$. Therefore, the slope is $m=-3$.

2. $g(x)=5x+2$
$g(3)=5(3)+2$
$\phantom{g(3)}=15+2$
$\phantom{g(3)}=17$

3. $\begin{cases} x-3y=-17 \quad (1) \\ 2x+y=1 \quad (2) \end{cases}$

Multiply both sides of equation (1) by $-2$.
$\begin{cases} -2x+6y=34 \quad (1) \\ 2x+y=1 \quad (2) \end{cases}$

Add equation (1) and equation (2).
$\begin{array}{r} \begin{cases} -2x+6y=34 \quad (1) \\ 2x+y=1 \quad (2) \end{cases} \\ \hline 7y=35 \\ y=5 \end{array}$

Substitute this result into the original equation (1) and solve for $x$.
$x-3y=-17$
$x-3(5)=-17$
$x-15=-17$
$x=-2$

The solution is the ordered pair $(-2,5)$.

### 11.2 Exercises

2. $a_n = a+(n-1)d$

4. False; in an arithmetic sequence, the difference between consecutive terms is the common difference.

6. Answers may vary. To get from one term to the next, we add the common difference. We begin with the first term and do not have to add the common difference for that term. Therefore, to get to the $n$th term requires us to add the common difference $n-1$ times to the first term. That is, $a_n = a+(n-1)d$.

8. $\{n-1\}$
$a_n - a_{n-1} = [n-1]-[(n-1)-1]$
$\phantom{a_n - a_{n-1}} = n-1-(n-2)$
$\phantom{a_n - a_{n-1}} = n-1-n+2=1$
$a_1 = 1-1 = 0$

This is an arithmetic sequence whose first term

is $a = 0$ and whose common difference is $d = 1$.
The first four terms are 0, 1, 2, and 3.

10. $\{10n+1\}$

$$a_n - a_{n-1} = [10n+1] - [10(n-1)+1]$$
$$= 10n+1 - (10n-10+1)$$
$$= 10n+1 - (10n-9)$$
$$= 10n+1 - 10n + 9 = 10$$
$$a_1 = 10(1) + 1 = 11$$

This is an arithmetic sequence whose first term is $a = 11$ and whose common difference is $d = 10$.
The first four terms are 11, 21, 31, and 41.

12. $\{5 - 2n\}$

$$a_n - a_{n-1} = [5 - 2n] - [5 - 2(n-1)]$$
$$= 5 - 2n - (5 - 2n + 2)$$
$$= 5 - 2n - (7 - 2n)$$
$$= 5 - 2n - 7 + 2n$$
$$= -2$$
$$a_1 = 5 - 2(1) = 3$$

This is an arithmetic sequence whose first term is $a = 3$ and whose common difference is $d = -2$.
The first four terms are 3, 1, $-1$, and $-3$.

14. $\left\{\dfrac{1}{4}n + \dfrac{3}{4}\right\}$

$$a_n - a_{n-1} = \left[\dfrac{1}{4}n + \dfrac{3}{4}\right] - \left[\dfrac{1}{4}(n-1) + \dfrac{3}{4}\right]$$
$$= \dfrac{1}{4}n + \dfrac{3}{4} - \left(\dfrac{1}{4}n - \dfrac{1}{4} + \dfrac{3}{4}\right)$$
$$= \dfrac{1}{4}n + \dfrac{3}{4} - \dfrac{1}{4}n - \dfrac{2}{4}$$
$$= \dfrac{1}{4}$$
$$a_1 = \dfrac{1}{4}(1) + \dfrac{3}{4} = \dfrac{4}{4} = 1$$

This is an arithmetic sequence whose first term is $a = 1$ and whose common difference is $d = \dfrac{1}{4}$.
The first four terms are 1, $\dfrac{5}{4}$, $\dfrac{6}{4} = \dfrac{3}{2}$, and $\dfrac{7}{4}$.

16. $a = 8$, $d = 3$

$$a_n = a + (n-1)d$$
$$= 8 + (n-1)3$$
$$= 8 + 3n - 3$$
$$= 3n + 5$$

To find the fifth term we let $n = 5$.
$$a_5 = 3(5) + 5 = 20$$
The fifth term of the sequence is 20.

18. $a = 12$, $d = -3$

$$a_n = a + (n-1)d = 12 + (n-1)(-3)$$
$$= 12 - 3n + 3$$
$$= -3n + 15$$

To find the fifth term we let $n = 5$.
$$a_5 = -3(5) + 15 = -15 + 15 = 0$$
The fifth term of the sequence is 0.

20. $a = -3$, $d = \dfrac{1}{2}$

$$a_n = a + (n-1)d = -3 + (n-1)\dfrac{1}{2}$$
$$= -3 + \dfrac{1}{2}n - \dfrac{1}{2}$$
$$= \dfrac{1}{2}n - \dfrac{7}{2}$$

To find the fifth term we let $n = 5$.
$$a_5 = \dfrac{1}{2}(5) - \dfrac{7}{2} = -\dfrac{2}{2} = -1$$
The fifth term is $-1$.

22. $a = -\dfrac{4}{3}$, $d = -\dfrac{2}{3}$

$$a_n = a + (n-1)d = -\dfrac{4}{3} + (n-1)\left(-\dfrac{2}{3}\right)$$
$$= -\dfrac{4}{3} - \dfrac{2}{3}n + \dfrac{2}{3}$$
$$= -\dfrac{2}{3}n - \dfrac{2}{3}$$

To find the fifth term, we let $n = 5$.
$$a_5 = -\dfrac{2}{3}(5) - \dfrac{2}{3} = -\dfrac{10}{3} - \dfrac{2}{3} = -\dfrac{12}{3} = -4$$
The fifth term is $-4$.

## Chapter 11: Sequences, Series, and the Binomial Expansion

24. $-5, -1, 3, 7, \ldots$
    Notice that the difference between consecutive terms is $d = 4$. Since the first term is $a = -5$, the $n$th term can be written as
    $$a_n = -5 + (n-1)(4)$$
    $$= -5 + 4n - 4$$
    $$= 4n - 9$$
    To find the twentieth term, we let $n = 20$.
    $$a_{20} = 4(20) - 9 = 80 - 9 = 71$$
    The twentieth term of the sequence is 71.

26. $20, 14, 8, 2, \ldots$
    Notice that the difference between consecutive terms is $d = -6$. Since the first term is $a = 20$, the $n$th term can be written as
    $$a_n = 20 + (n-1)(-6)$$
    $$= 20 - 6n + 6$$
    $$= -6n + 26$$
    To find the twentieth term, we let $n = 20$.
    $$a_{20} = -6(20) + 26 = -120 + 26 = -94$$
    The twentieth term of the sequence is $-94$.

28. $10, \dfrac{19}{2}, 9, \dfrac{17}{2}, \ldots$
    Notice that the difference between consecutive terms is $d = -\dfrac{1}{2}$. Since the first term is $a = 10$, the $n$th term can be written as
    $$a_n = 10 + (n-1)\left(-\dfrac{1}{2}\right)$$
    $$= 10 - \dfrac{1}{2}n + \dfrac{1}{2}$$
    $$= -\dfrac{1}{2}n + \dfrac{21}{2}$$
    To find the twentieth term, we let $n = 20$.
    $$a_{20} = -\dfrac{1}{2}(20) + \dfrac{21}{2} = -\dfrac{20}{2} + \dfrac{21}{2} = \dfrac{1}{2}$$
    The twentieth term of the sequence is $\dfrac{1}{2}$.

30. We know that the $n$th term of an arithmetic sequence is given by $a_n = a + (n-1)d$ where $a$ is the first term and $d$ is the common difference. Since $a_5 = 7$ and $a_9 = 19$, we have
    $$\begin{cases} a_5 = a + (5-1)d \\ a_9 = a + (9-1)d \end{cases} \text{ or } \begin{cases} 7 = a + 4d \quad (1) \\ 19 = a + 8d \quad (2) \end{cases}$$
    This is a system of linear equations in two variables, $a$ and $d$. We can solve the system by elimination. If we subtract equation (2) from equation (1), we obtain
    $$-12 = -4d$$
    $$3 = d$$
    Let $d = 3$ in equation (1) to find $a$.
    $$7 = a + 4(3)$$
    $$7 = a + 12$$
    $$-5 = a$$
    The first term is $a = -5$ and the common difference is $d = 3$. Thus, a formula for the $n$th term is
    $$a_n = -5 + (n-1)(3)$$
    $$= -5 + 3n - 3$$
    $$= 3n - 8$$

32. We know that the $n$th term of an arithmetic sequence is given by $a_n = a + (n-1)d$ where $a$ is the first term and $d$ is the common difference. Since $a_2 = -9$ and $a_8 = 15$, we have
    $$\begin{cases} a_2 = a + (2-1)d \\ a_8 = a + (8-1)d \end{cases} \text{ or } \begin{cases} -9 = a + d \quad (1) \\ 15 = a + 7d \quad (2) \end{cases}$$
    This is a system of linear equations in two variables, $a$ and $d$. We can solve the system by elimination. If we subtract equation (2) from equation (1), we obtain
    $$-24 = -6d$$
    $$4 = d$$
    Let $d = 4$ in equation (1) to find $a$.
    $$-9 = a + 4$$
    $$-13 = a$$
    The first term is $a = -13$ and the common difference is $d = 4$. Thus, a formula for the $n$th term is
    $$a_n = -13 + (n-1)(4)$$
    $$= -13 + 4n - 4$$
    $$= 4n - 17$$

34. We know that the $n$th term of an arithmetic sequence is given by $a_n = a + (n-1)d$ where $a$ is the first term and $d$ is the common difference. Since $a_6 = -8$ and $a_{12} = -38$, we have
    $$\begin{cases} a_6 = a + (6-1)d \\ a_{12} = a + (12-1)d \end{cases} \text{ or } \begin{cases} -8 = a + 5d \quad (1) \\ -38 = a + 11d \quad (2) \end{cases}$$
    This is a system of linear equations in two variables, $a$ and $d$. We can solve the system by elimination. If we subtract equation (2) from equation (1), we obtain

$30 = -6d$

$-5 = d$

Let $d = -5$ in equation (1) to find $a$.

$-8 = a + 5(-5)$

$-8 = a - 25$

$17 = a$

The first term is $a = 17$ and the common difference is $d = -5$. Thus, a formula for the $n$th term is

$a_n = 17 + (n-1)(-5)$

$= 17 - 5n + 5$

$= -5n + 22$

**36.** We know that the $n$th term of an arithmetic sequence is given by $a_n = a + (n-1)d$ where $a$ is the first term and $d$ is the common difference. Since $a_5 = 5$ and $a_{13} = 7$, we have

$\begin{cases} a_5 = a + (5-1)d \\ a_{13} = a + (13-1)d \end{cases}$ or $\begin{cases} 5 = a + 4d \quad (1) \\ 7 = a + 12d \quad (2) \end{cases}$

This is a system of linear equations in two variables, $a$ and $d$. We can solve the system by elimination. If we subtract equation (2) from equation (1), we obtain

$-2 = -8d$

$\dfrac{1}{4} = d$

Let $d = \dfrac{1}{4}$ in equation (1) to find $a$.

$5 = a + 4\left(\dfrac{1}{4}\right)$

$5 = a + 1$

$4 = a$

The first term is $a = 4$ and the common difference is $d = \dfrac{1}{4}$. Thus, a formula for the $n$th term is

$a_n = 4 + (n-1)\left(\dfrac{1}{4}\right)$

$= 4 + \dfrac{1}{4}n - \dfrac{1}{4}$

$= \dfrac{1}{4}n + \dfrac{15}{4}$

**38.** We know that the first term is $a = 1$ and the common difference is $d = 8 - 1 = 7$. The sum of the first $n = 40$ terms of this arithmetic sequence

is given by

$S_{40} = \dfrac{40}{2}[2a + (40-1)d]$

$= 20[2(1) + 39(7)]$

$= 20(275)$

$= 5500$

**40.** We know that the first term is $a = -9$ and the common difference is $d = -5 - (-9) = 4$. The sum of the first $n = 75$ terms of this arithmetic sequence is given by

$S_{75} = \dfrac{75}{2}[2a + (75-1)d]$

$= \dfrac{75}{2}[2(-9) + 74(4)]$

$= \dfrac{75}{2}(278)$

$= 10,425$

**42.** We know that the first term is $a = 12$ and the common difference is $d = 4 - 12 = -8$. The sum of the first $n = 50$ terms of this arithmetic sequence is given by

$S_{50} = \dfrac{50}{2}[2a + (50-1)d]$

$= 25[2(12) + 49(-8)]$

$= 25(24 - 392)$

$= 25(-368)$

$= -9200$

**44.** The first term of the sequence is given by $a_1 = 2(1) - 13 = -11$ and the 80th term is given by $a_{80} = 2(80) - 13 = 147$. We can now use the formula $S_n = \dfrac{n}{2}[a + a_n]$ to find the sum.

$S_{80} = \dfrac{80}{2}(-11 + 147) = 40(136) = 5440$.

The sum of the first 80 terms is 5440.

**46.** The first term of the sequence is given by $a_1 = -6(1) + 25 = 19$ and the 35th term is given by $a_{35} = -6(35) + 25 = -185$. We can now use the formula $S_n = \dfrac{n}{2}[a + a_n]$ to find the sum.

$$S_{35} = \frac{35}{2}[19+(-185)]$$
$$= \frac{35}{2}(-166)$$
$$= -2905$$
The sum of the first 35 terms is $-2905$.

48. The first term of the sequence is given by
$$a_1 = 7 - \frac{3}{2}(1) = \frac{11}{2}$$ and the 28th term is given by
$$a_{28} = 7 - \frac{3}{2}(28) = 7 - 42 = -35.$$ We can now use the formula $S_n = \frac{n}{2}[a+a_n]$ to find the sum.
$$S_{28} = \frac{28}{2}\left[\frac{11}{2}+(-35)\right]$$
$$= 14\left(-\frac{59}{2}\right)$$
$$= -413$$
The sum of the first 28 terms is $-413$.

50. To be an arithmetic sequence, the difference between successive terms must be constant. Therefore, we start by finding the differences between the terms.
$$d_1 = (3x+2)-(2x)$$
$$= 3x+2-2x$$
$$= x+2$$
$$d_2 = (5x+3)-(3x+2)$$
$$= 5x+3-3x-2$$
$$= 2x+1$$
Now we set the two differences equal to each other and solve the resulting equation for $x$.
$$d_1 = d_2$$
$$x+2 = 2x+1$$
$$2 = x+1$$
$$1 = x$$

52. Since the first row has 46 bricks, we know that $a = 46$. Each row has two fewer bricks so we know that $d = -2$. We don't know how many rows are in the pile, but we do know that the last row has $a_n = 2$ bricks. We can use this to determine the number of rows in the pile.

$$a_n = a+(n-1)d$$
$$2 = 46+(n-1)(-2)$$
$$-44 = (n-1)(-2)$$
$$22 = n-1$$
$$23 = n$$
So, there are 23 rows in the pile. To find the total number of bricks, we use the summation formula
$$S_n = \frac{n}{2}[a+a_n]$$
$$= \frac{23}{2}(46+2) = \frac{23}{2}(48)$$
$$= 552 \text{ bricks}$$

54. The number of red triangles in each row forms an arithmetic sequence with $a = 1$, $d = 1$, and $n = 20$. The number of blue triangles in each row forms an arithmetic sequence with $a = 1$, $d = 1$, and $n = 19$.
$$S_{20} = \frac{20}{2}(2(1)+(20-1)(1)) = 10(2+19) = 210$$
$$S_{19} = \frac{19}{2}(2(1)+(19-1)(1)) = \frac{19}{2}(2+18) = 190$$
The mosaic will require 210 red triangles and 190 blue triangles.

56. From the terms listed, we see that the first term is $a = -9$, the last term is $a_n = 219$, and the common difference is $d = -5-(-9) = 4$. We can determine the number of terms in the sequence by using the formula for the $n$th term.
$$a_n = a+(n-1)d$$
$$219 = -9+(n-1)(4)$$
$$228 = (n-1)(4)$$
$$57 = n-1$$
$$58 = n$$
There are 58 terms in the sequence.

58. From the terms listed, we see that the first term is $a = 99$, the last term is $a_n = -339$, and the common difference is $d = 93-99 = -6$. We can determine the number of terms in the sequence by using the formula for the $n$th term.

## Section 11.2 Arithmetic Sequences

$a_n = a + (n-1)d$
$-339 = 99 + (n-1)(-6)$
$-438 = (n-1)(-6)$
$73 = n - 1$
$74 = n$
There are 74 terms in the sequence.

60. Since the number of seats in each row increases by a constant number, they form an arithmetic sequence with $a = 14$, $d = 4$, and $S_n = 2{,}030$. The number of rows can be found by using the summation formula

$S_n = \dfrac{n}{2}[2a + (n-1)d]$

$2030 = \dfrac{n}{2}[2(14) + (n-1)(4)]$

$4060 = n(28 + 4n - 4)$
$4060 = n(24 + 4n)$
$0 = 4n^2 + 24n - 4060$
$0 = n^2 + 6n - 1015$
$0 = (n-29)(n+35)$
$n = 29$ or $\cancel{n = -35}$

The section contains 29 rows.

62. a. $f(x) = 4^x$ is in the form $y = a^x$ so the base is 4.

    b. $f(1) = 4^1 = 4$
       $f(2) = 4^2 = 16$
       $f(3) = 4^3 = 64$
       $f(4) = 4^4 = 256$

64. $f(x) = 3^x$

$\dfrac{f(2)}{f(1)} = \dfrac{9}{3} = 3$

$\dfrac{f(3)}{f(2)} = \dfrac{27}{9} = 3$

$\dfrac{f(4)}{f(3)} = \dfrac{81}{27} = 3$

$f(x) = 4^x$

$\dfrac{f(2)}{f(1)} = \dfrac{16}{4} = 4$

$\dfrac{f(3)}{f(2)} = \dfrac{64}{16} = 4$

$\dfrac{f(4)}{f(3)} = \dfrac{256}{64} = 4$

$f(x) = 10\left(\dfrac{1}{2}\right)^x$

$\dfrac{f(2)}{f(1)} = \dfrac{5/2}{5} = \dfrac{1}{2}$; $\dfrac{f(3)}{f(2)} = \dfrac{5/4}{5/2} = \dfrac{1}{2}$;

$\dfrac{f(4)}{f(3)} = \dfrac{5/8}{5/4} = \dfrac{1}{2}$

The ratio equals the base.

66. $\{2.67n - 1.23\}$; $n = 25$

```
sum(seq(2.67X-1.
23,X,1,25,1)
                837
```

The sum of the first 25 terms of this arithmetic sequence is 837.

68. $a = -11.8$; $d = -8.2 - (-11.8) = 3.6$

$a_n = a + (n-1)d$
$= -11.8 + (n-1)(3.6)$
$= -11.8 + 3.6n - 3.6$
$= 3.6n - 15.4$

$\{3.6n - 15.4\}$; $n = 30$

```
sum(seq(3.6X-15.
4,X,1,30,1)
                1212
```

The sum of the first 30 terms of this arithmetic sequence is 1212.

## 11.3 Preparing for Geometric Sequences and Series

1. $g(x) = 4^x$
   $g(1) = 4^1 = 4$
   $g(2) = 4^2 = 16$
   $g(3) = 4^3 = 64$

## Chapter 11: Sequences, Series, and the Binomial Expansion

2. $\dfrac{x^4}{x^3} = x^{4-3} = x$

### 11.3 Exercises

2. $\dfrac{a}{1-r}$

4. True

6. A geometric series has a finite sum if $-1 < r < 1$.

8. $\{(-2)^n\}$

$\dfrac{a_n}{a_{n-1}} = \dfrac{(-2)^n}{(-2)^{n-1}} = (-2)^{n-(n-1)} = -2$

$a_1 = (-2)^1 = -2$

$a_2 = (-2)^2 = 4$

$a_3 = (-2)^3 = -8$

$a_4 = (-2)^4 = 16$

The common ratio is $r = -2$ and the first four terms of the sequence are $-2$, $4$, $-8$, and $16$.

10. $\left\{\dfrac{2^n}{3}\right\}$

$\dfrac{a_n}{a_{n-1}} = \dfrac{\frac{2^n}{3}}{\frac{2^{n-1}}{3}} = \dfrac{2^n}{3} \cdot \dfrac{3}{2^{n-1}} = 2^{n-(n-1)} = 2$

$a_1 = \dfrac{2^1}{3} = \dfrac{2}{3}$; $a_2 = \dfrac{2^2}{3} = \dfrac{4}{3}$;

$a_3 = \dfrac{2^3}{3} = \dfrac{8}{3}$; $a_4 = \dfrac{2^4}{3} = \dfrac{16}{3}$

The common ratio is $r = 2$ and the first four terms of the sequence are $\dfrac{2}{3}, \dfrac{4}{3}, \dfrac{8}{3}$, and $\dfrac{16}{3}$.

12. $\left\{-10 \cdot \left(\dfrac{1}{2}\right)^n\right\}$

$\dfrac{a_n}{a_{n-1}} = \dfrac{-10 \cdot \left(\frac{1}{2}\right)^n}{-10 \cdot \left(\frac{1}{2}\right)^{n-1}} = \left(\dfrac{1}{2}\right)^{n-(n-1)} = \dfrac{1}{2}$

$a_1 = -10 \cdot \left(\dfrac{1}{2}\right)^1 = -10 \cdot \dfrac{1}{2} = -5$

$a_2 = -10 \cdot \left(\dfrac{1}{2}\right)^2 = -10 \cdot \dfrac{1}{4} = -\dfrac{5}{2}$

$a_3 = -10 \cdot \left(\dfrac{1}{2}\right)^3 = -10 \cdot \dfrac{1}{8} = -\dfrac{5}{4}$

$a_4 = -10 \cdot \left(\dfrac{1}{2}\right)^3 = -10 \cdot \dfrac{1}{16} = -\dfrac{5}{8}$

The common ratio is $r = \dfrac{1}{2}$ and the first four terms of the sequence are $-5, -\dfrac{5}{2}, -\dfrac{5}{4}$, and $-\dfrac{5}{8}$.

14. $\left\{\dfrac{3^{-n}}{2^{n-1}}\right\}$

$\dfrac{a_n}{a_{n-1}} = \dfrac{\frac{3^{-n}}{2^{n-1}}}{\frac{3^{-(n-1)}}{2^{(n-1)-1}}} = \dfrac{3^{-n}}{2^{n-1}} \cdot \dfrac{2^{n-2}}{3^{-n+1}}$

$= 3^{-n-(-n+1)} \cdot 2^{n-2-(n-1)}$

$= 3^{-1} \cdot 2^{-1} = \dfrac{1}{6}$

$a_1 = \dfrac{3^{-1}}{2^{1-1}} = \dfrac{1}{3 \cdot 1} = \dfrac{1}{3}$

$a_2 = \dfrac{3^{-2}}{2^{2-1}} = \dfrac{1}{9 \cdot 2} = \dfrac{1}{18}$

$a_3 = \dfrac{3^{-3}}{2^{3-1}} = \dfrac{1}{27 \cdot 4} = \dfrac{1}{108}$

$a_4 = \dfrac{3^{-4}}{2^{4-1}} = \dfrac{1}{81 \cdot 8} = \dfrac{1}{648}$

The common ratio is $r = \dfrac{1}{6}$ and the first four terms of the sequence are $\dfrac{1}{3}, \dfrac{1}{18}, \dfrac{1}{108}$, and $\dfrac{1}{648}$.

16. $\{8 - 3n\}$

$a_n - a_{n-1} = [8 - 3n] - [8 - 3(n-1)]$
$= [8 - 3n] - [8 - 3n + 3]$
$= 8 - 3n - 11 + 3n$
$= -3$

Since the difference between consecutive terms is constant, the sequence is arithmetic. The common difference is $d = -3$.

Section 11.3   Geometric Sequences and Series

18. $\{n^2 - 2\}$

$$a_n - a_{n-1} = [n^2 - 2] - [(n-1)^2 - 2]$$
$$= [n^2 - 2] - [n^2 - 2n + 1 - 2]$$
$$= n^2 - 2 - n^2 + 2n + 1$$
$$= 2n - 1$$

Since the difference between consecutive terms is not constant, the sequence is not arithmetic.

$$\frac{a_n}{a_{n-1}} = \frac{n^2 - 2}{(n-1)^2 - 2} = \frac{n^2 - 2}{n^2 - 2n - 1}$$

Since the ratio of consecutive terms is not constant, the sequence is not geometric.

20. $\left\{\dfrac{2}{3^n}\right\}$

$$a_n - a_{n-1} = \frac{2}{3^n} - \frac{2}{3^{n-1}} = \frac{2}{3^n} - \frac{6}{3^n} = -\frac{4}{3^n}$$

Since the difference between consecutive terms is not constant, the sequence is not arithmetic.

$$\frac{a_n}{a_{n-1}} = \frac{\frac{2}{3^n}}{\frac{2}{3^{n-1}}} = \frac{2}{3^n} \cdot \frac{3^{n-1}}{2} = 3^{n-1-(n)} = 3^{-1} = \frac{1}{3}$$

Since the ratio of consecutive terms is constant, the sequence is geometric. The common ratio is $r = \dfrac{1}{3}$.

22. $a_2 - a_1 = 20 - 100 = -80$
$a_3 - a_2 = 4 - 20 = -16$

The difference between consecutive terms is not constant, so the sequence is not arithmetic.

$$\frac{a_2}{a_1} = \frac{20}{100} = \frac{1}{5}$$
$$\frac{a_3}{a_2} = \frac{4}{20} = \frac{1}{5}$$
$$\frac{a_4}{a_3} = \frac{\frac{4}{5}}{4} = \frac{4}{5} \cdot \frac{1}{4} = \frac{1}{5}$$

The ratio of consecutive terms is constant so the sequence is geometric. The common ratio is $r = \dfrac{1}{5}$.

24. $a_2 - a_1 = 12 - 15 = -3$
$a_3 - a_2 = 9 - 12 = -3$
$a_4 - a_3 = 6 - 9 = -3$

The difference between consecutive terms is constant so the sequence is arithmetic. The common difference is $d = -3$.

26. $a_2 - a_1 = -2 - 5 = -7$
$a_3 - a_2 = 3 - (-2) = 5$

The difference between consecutive terms is not constant so the sequence is not arithmetic.

$$\frac{a_2}{a_1} = \frac{-2}{5} = -\frac{2}{5}$$
$$\frac{a_3}{a_2} = \frac{3}{-2} = -\frac{3}{2}$$

The ratio of consecutive terms is not constant so the sequence is not geometric.

28. a. $a_n = a \cdot r^{n-1} = 2 \cdot 3^{n-1}$

b. $a_8 = 2 \cdot 3^{8-1} = 2 \cdot 3^7 = 2 \cdot 2187 = 4374$

30. a. $a_n = a \cdot r^{n-1} = 30 \cdot \left(\dfrac{1}{3}\right)^{n-1}$

b. $a_8 = 30 \cdot \left(\dfrac{1}{3}\right)^{8-1} = 30 \cdot \left(\dfrac{1}{3}\right)^7 = \dfrac{30}{2187} = \dfrac{10}{729}$

32. a. $a_n = a \cdot r^{n-1} = 1 \cdot (-4)^{n-1} = (-4)^{n-1}$

b. $a_8 = (-4)^{8-1} = (-4)^7 = -16{,}384$

34. a. $a_n = a \cdot r^{n-1} = 500 \cdot (1.04)^{n-1}$

b. $a_8 = 500 \cdot (1.04)^{8-1} = 500 \cdot (1.04)^7 \approx 657.97$

36. $r = \dfrac{3}{1} = 3$; $a = 1$

$a_n = a \cdot r^{n-1}$
$a_{12} = 1 \cdot 3^{12-1} = 3^{11} = 177{,}147$

38. $r = \dfrac{-20}{10} = -2$; $a = 10$

$a_n = a \cdot r^{n-1}$
$a_8 = 10 \cdot (-2)^{8-1} = 10 \cdot (-2)^7 = 10(-128) = -1280$

ISM - 689

## Chapter 11: Sequences, Series, and the Binomial Expansion

40. $r = \dfrac{0.04}{0.4} = \dfrac{1}{10}$ ; $a = 0.4$

    $a_n = a \cdot r^{n-1}$

    $a_{10} = 0.4 \cdot \left(\dfrac{1}{10}\right)^{10-1} = 0.4 \cdot \left(\dfrac{1}{10}\right)^9$

    $= 0.0000000004$

42. $3 + 9 + 27 + \ldots + 3^{10}$

    This is a geometric sequence with $a = 3$ and common ratio $r = 3$. We wish to find the sum of the first 10 terms, $S_{10}$.

    $S_n = a \cdot \dfrac{1 - r^n}{1 - r}$

    $S_{10} = 3 \cdot \dfrac{1 - 3^{10}}{1 - 3} = 3 \cdot \dfrac{-59{,}048}{-2} = 88{,}572$

    $3 + 9 + 27 + \ldots + 3^{10} = 88{,}572$

44. $10 + 5 + \dfrac{5}{2} + \ldots + 10 \cdot \left(\dfrac{1}{2}\right)^{12-1}$

    This is a geometric sequence with $a = 10$ and common ratio $r = \dfrac{1}{2}$. We wish to find the sum of the first 12 terms, $S_{12}$.

    $S_n = a \cdot \dfrac{1 - r^n}{1 - r}$

    $S_{12} = 10 \cdot \dfrac{1 - \left(\tfrac{1}{2}\right)^{12}}{1 - \tfrac{1}{2}} = 10 \cdot \dfrac{\tfrac{4095}{4096}}{\tfrac{1}{2}} = 20 \cdot \dfrac{4095}{4096}$

    $= \dfrac{20{,}475}{1{,}024} \approx 19.99511719$

    $10 + 5 + \dfrac{5}{2} + \ldots + 10 \cdot \left(\dfrac{1}{2}\right)^{12-1} \approx 19.99511719$

46. $\sum_{n=1}^{12}\left[5 \cdot 2^n\right] = \sum_{n=1}^{12}\left[5 \cdot 2 \cdot 2^{n-1}\right] = \sum_{n=1}^{12}\left[10 \cdot 2^{n-1}\right]$

    Here we want to find the sum of the first 12 terms of a geometric sequence with $a = 10$ and common ratio $r = 2$.

    $S_n = a \cdot \dfrac{1 - r^n}{1 - r}$

    $S_{12} = 10 \cdot \dfrac{1 - 2^{12}}{1 - 2} = 10 \cdot \dfrac{-4095}{-1} = 40{,}950$

    $\sum_{n=1}^{12}\left[5 \cdot 2^n\right] = 40{,}950$

48. $\sum_{n=1}^{14}\left[10 \cdot \left(\dfrac{1}{2}\right)^{n-1}\right]$

    Here we want to find the sum of the first 14 terms of a geometric sequence with $a = 10$ and common ratio $r = \dfrac{1}{2}$.

    $S_n = a \cdot \dfrac{1 - r^n}{1 - r}$

    $S_{14} = 10 \cdot \dfrac{1 - \left(\tfrac{1}{2}\right)^{14}}{1 - \tfrac{1}{2}} = 10 \cdot \dfrac{\tfrac{16383}{16384}}{\tfrac{1}{2}} = 20 \cdot \dfrac{16383}{16384}$

    $= \dfrac{81{,}915}{4{,}096} \approx 19.9987793$

    $\sum_{n=1}^{14}\left[10 \cdot \left(\dfrac{1}{2}\right)^{n-1}\right] \approx 19.9987793$

50. This is an infinite geometric series with $a = 1$ and common ratio $r = \dfrac{1/3}{1} = \dfrac{1}{3}$. Since the common ratio, $r$, is between $-1$ and $1$, we can use the formula for the sum of an infinite geometric series.

    $1 + \dfrac{1}{3} + \dfrac{1}{9} + \ldots = \dfrac{1}{1 - \tfrac{1}{3}} = \dfrac{1}{\tfrac{2}{3}} = \dfrac{3}{2}$

52. This is an infinite geometric series with $a = 20$ and common ratio $r = \dfrac{5}{20} = \dfrac{1}{4}$. Since the common ratio, $r$, is between $-1$ and $1$, we can use the formula for the sum of an infinite geometric series.

    $20 + 5 + \dfrac{5}{4} + \ldots = \dfrac{20}{1 - \tfrac{1}{4}} = \dfrac{20}{\tfrac{3}{4}} = 20 \cdot \dfrac{4}{3} = \dfrac{80}{3}$

54. This is an infinite geometric series with $a = 12$ and common ratio $r = \dfrac{-3}{12} = -\dfrac{1}{4}$. Since the common ratio, $r$, is between $-1$ and $1$, we can use the formula for the sum of an infinite geometric series.

    $12 - 3 + \dfrac{3}{4} - \dfrac{3}{16} + \ldots = \dfrac{12}{1 - \left(-\tfrac{1}{4}\right)} = \dfrac{12}{\tfrac{5}{4}} = 12 \cdot \dfrac{4}{5} = \dfrac{48}{5}$

56. This is an infinite geometric series with $a = 10 \cdot \left(\frac{1}{3}\right)^1 = \frac{10}{3}$ and common ratio $r = \frac{1}{3}$.

 Since the common ratio, $r$, is between $-1$ and $1$, we can use the formula for the sum of an infinite geometric series.
 $$\sum_{n=1}^{\infty}\left(10 \cdot \left(\frac{1}{3}\right)^n\right) = \frac{\frac{10}{3}}{1-\frac{1}{3}} = \frac{\frac{10}{3}}{\frac{2}{3}} = \frac{10}{3} \cdot \frac{3}{2} = 5$$

58. This is an infinite geometric series with $a = 100$ and common ratio $r = -\frac{1}{2}$. Since the common ratio, $r$, is between $-1$ and $1$, we can use the formula for the sum of an infinite geometric series.
 $$\sum_{n=1}^{\infty}\left(100 \cdot \left(-\frac{1}{2}\right)^{n-1}\right) = \frac{100}{1-\left(-\frac{1}{2}\right)} = \frac{100}{\frac{3}{2}}$$
 $$= 100 \cdot \frac{2}{3}$$
 $$= \frac{200}{3}$$

60. $0.\overline{3} = 0.3 + 0.03 + 0.003 + \ldots = \sum_{n=1}^{\infty}\left[0.3 \cdot \left(\frac{1}{10}\right)^{n-1}\right]$

 This is an infinite geometric series with $a = 0.3$ and common ratio $r = \frac{1}{10}$. Since the common ratio, $r$, is between $-1$ and $1$, we can use the formula for the sum of an infinite geometric series.
 $$0.\overline{3} = \frac{0.3}{1-\frac{1}{10}} = \frac{0.3}{0.9} = \frac{1}{3}$$

62. $0.\overline{45} = 0.45 + 0.0045 + 0.000045 + \ldots$
 $$= \sum_{n=1}^{\infty}\left[0.45 \cdot \left(\frac{1}{100}\right)^{n-1}\right]$$

 This is an infinite geometric series with $a = 0.45$ and common ratio $r = \frac{1}{100}$. Since the common ratio, $r$, is between $-1$ and $1$, we can use the formula for the sum of an infinite geometric series.
 $$0.\overline{45} = \frac{0.45}{1-\frac{1}{100}} = \frac{0.45}{0.99} = \frac{5}{11}$$

64. To be a geometric sequence, the ratio of consecutive terms must be the same. Therefore, we need to solve the equation
 $$\frac{x}{x-1} = \frac{x+2}{x}$$
 $$x(x-1) \cdot \frac{x}{x-1} = x(x-1) \cdot \frac{x+2}{x}$$
 $$x^2 = (x-1)(x+2)$$
 $$x^2 = x^2 + x - 2$$
 $$0 = x - 2$$
 $$x = 2$$

 b. Your salary at the beginning of the tenth year is the value of the tenth term in the sequence.
 $$a_{10} = a \cdot r^{10-1} = 40,000 \cdot (1.05)^9 \approx 62,053$$
 Your salary at the beginning of the tenth year will be about $62,053.

 c. Your cumulative earnings after completing your tenth year is the sum of the first ten terms of the sequence.
 $$S_{10} = a \cdot \frac{1-r^{10}}{1-r}$$
 $$= 40,000 \cdot \frac{1-(1.05)^{10}}{1-1.05}$$
 $$\approx 40,000 \cdot (12.577893)$$
 $$\approx 503,116$$
 Your cumulative earnings after finishing your tenth year will be about $503,116.

66. Your annual salaries will form a geometric sequence with $a = 45,000$ and common ratio $r = 1.04$.

 a. Your salary at the beginning of the second year is the value of the second term in the sequence.
 $$a_2 = a \cdot r^{2-1} = 45,000 \cdot (1.04)^1 = 46,800$$
 Your salary at the beginning of the second year will be $46,800.

 b. Your salary at the beginning of the tenth year is the value of the tenth term in the sequence.
 $$a_{10} = a \cdot r^{10-1} = 45,000 \cdot (1.04)^9 \approx 64,049$$
 Your salary at the beginning of the tenth year will be about $64,049.

## Chapter 11: Sequences, Series, and the Binomial Expansion

c. Your cumulative earnings after completing your tenth year is the sum of the first ten terms of the sequence.
$$S_{10} = a \cdot \frac{1-r^{10}}{1-r}$$
$$= 45{,}000 \cdot \frac{1-(1.04)^{10}}{1-1.04}$$
$$\approx 45{,}000 \cdot (12.006107)$$
$$\approx 540{,}275$$
Your cumulative earnings after finishing your tenth year will be about \$540,275.

68. The value of the car at the beginning of the year forms a geometric sequence with $a = 16{,}000$ and common ratio $r = 0.9$. The value of the car after you have owned it for four years will be the fifth term of the sequence (the beginning of the fifth year of ownership).
$$a_5 = 16{,}000 \cdot (0.9)^{5-1}$$
$$= 16{,}000 \cdot (0.9)^4 = 10{,}497.60$$
After four years of ownership, the car will be worth \$10,497.60.

70. The heights of the ball after each bounce form a geometric sequence with $a = 24$ (that is, $30(0.8) = 24$) and common ratio $r = 0.8$.

   a. $a_4 = 24 \cdot (0.8)^{4-1} = 24(0.8)^3 = 12.288$
   After the fourth bounce, the ball will reach a height of 12.288 feet.

   b. Using the result from part (a),
   $a_5 = 0.8 a_4 = (0.8)(12.288) = 9.8304$
   After the fifth bounce, the ball will reach a height of 9.8304 feet.

   c. Here we want to determine the term number, $n$. Since units are in feet, we need to use $\frac{1}{2}$ foot instead of 6 inches.

$$a_n = \tfrac{1}{2}$$
$$a \cdot r^{n-1} = \tfrac{1}{2}$$
$$24(0.8)^{n-1} = \tfrac{1}{2}$$
$$(0.8)^{n-1} = \tfrac{1}{48}$$
$$\ln\left[(0.8)^{n-1}\right] = \ln\left(\tfrac{1}{48}\right)$$
$$(n-1)(\ln 0.8) = \ln\left(\tfrac{1}{48}\right)$$
$$n-1 = \frac{\ln\left(\tfrac{1}{48}\right)}{(\ln 0.8)}$$
$$n = \frac{\ln\left(\tfrac{1}{48}\right)}{(\ln 0.8)} + 1 \approx 18.35$$
The ball will bounce to a height of less than 6 inches after the 19th bounce.

   d. For each bounce, the ball travels a distance equal to twice the height achieved (since it goes up and then back to the ground). We can find the sum of the infinite series for the heights and double that amount. In addition, we must add the original 30 feet traveled prior to the first bounce.
   $$S_\infty = \frac{a}{1-r} = \frac{24}{1-0.8} = \frac{24}{0.2} = 120$$
   Therefore, the ball travels a total distance of $2(120) + 30 = 240 + 30 = 270$ feet.

72. To compare the options, we need to determine the total amount that would be earned under each plan.

Option A:
Under this plan, your client would receive \$1,575,000 for each of the five years. Thus, the total amount earned over the 5-year contract would be
$5(\$1{,}575{,}000) = \$7{,}875{,}000$

## Section 11.3 Geometric Sequences and Series

Option B:
The annual salaries under this plan form a geometric sequence with $a = 1,500,000$ and $r = 1.04$. Thus, the total amount earned over the 5-year contract would be
$$S_5 = 1,500,000 \cdot \frac{1-1.04^5}{1-1.04} = \$8,124,483.84$$

Option C:
The annual salaries under this plan form an arithmetic sequence with $a = 1,500,000$ and $d = 80,000$. Thus, the total amount earned over the 5-year contract would be
$$S_5 = \frac{5}{2}(2(1,500,000) + 4(80,000))$$
$$= 2.5(3,000,000 + 320,000)$$
$$= 2.5(3,320,000)$$
$$= \$8,300,000$$

Option C is better for earning the most money over the 5-year contract. Option A provides the least amount over the 5 years.

74. The multiplier in this case is the geometric series with first term $a = 1$ and common ratio $r = 0.96$.
$$1 + 0.96 + (0.96)^2 + \ldots = \frac{1}{1-0.96} = 25$$
The multiplier is 25.

76. $\frac{1+i}{1+r} = \frac{1+.03}{1+.1} = \frac{1.03}{1.1} \approx 0.93636$

The price is an infinite geometric series with $a = 3$ and a common ratio of $\frac{1.03}{1.1}$. The maximum price is given by
$$\text{Price} = \frac{3}{1 - \frac{1.03}{1.1}} \approx 47.14$$
The maximum price you should pay for the stock

78. $A = P \cdot \frac{(1+i)^n - 1}{i}$

In this case we have $P = 400$, $n = 4 \cdot 12 = 48$, and $i = \frac{0.04}{12}$.

$$A = 400 \cdot \left[ \frac{\left(1 + \frac{0.04}{12}\right)^{48} - 1}{\frac{0.04}{12}} \right]$$
$$= 400 \cdot [51.959601]$$
$$= \$20,783.84$$
After 4 years, Jolene will have $20,783.84 for a down payment.

80. $A = P \cdot \frac{(1+i)^n - 1}{i}$

In this case we have $P = 1,500$, $n = 15 \cdot 2 = 30$, and $i = \frac{0.1}{2} = 0.05$.

$$A = 1500 \cdot \left[ \frac{(1+0.05)^{30} - 1}{0.05} \right]$$
$$= 1500 \cdot [66.43885]$$
$$= \$99,658.27$$
After 15 years, Raymont's IRA will be worth $99,658.27.

82. $A = P \cdot \frac{(1+i)^n - 1}{i}$

In this case we have $A = 2,000,000$, $n = 25 \cdot 4 = 100$, and $i = \frac{0.12}{4} = 0.03$.

$$2,000,000 = P \cdot \left[ \frac{(1+0.03)^{100} - 1}{0.03} \right]$$
$$2,000,000 = P \cdot [607.28773]$$
$$P = \$3293.33$$
Sophia will need to contribute $3293.33, or about $3293, each quarter in order to achieve her goal.

84. $0.85\overline{9} = 0.85 + 0.009 + 0.0009 + \ldots$
After the 0.85, the rest of the sum forms an infinite geometric series with $a = 0.009$ and common ratio $r = 0.1$. We can find the sum of this part as $S_\infty = \frac{a}{1-r} = \frac{0.009}{1-0.1} = \frac{0.009}{0.9} = 0.01$

Thus, $0.85\overline{9} = 0.85 + 0.01 = 0.86 = \frac{86}{100} = \frac{43}{50}$

86. A sequence of constant terms is both arithmetic (common difference of $d = 0$) and geometric (common ratio of $r = 1$) at the same time.

## Chapter 11: Sequences, Series, and the Binomial Expansion

88. $\dfrac{1}{3} = 0.333333... = 0.\overline{3}$

    $\dfrac{2}{3} = 0.666666... = 0.\overline{6}$

90. $0.999999...$ can be written as
    $0.9 + 0.09 + 0.009 + 0.0009 + ...$
    This is an infinite geometric series with $a = 0.9$ and $r = 0.1$. Using the formula for the sum of an infinite series, we get
    $$S_\infty = \dfrac{a}{1-r} = \dfrac{0.9}{1-0.1} = \dfrac{0.9}{0.9} = 1$$
    Therefore, $0.999999... = 1$.

92. This is a geometric series with $a = 3$ and common ratio $r = 1.6$. Based on the form of the last term, it appears that there are $n = 20$ terms in the series.

    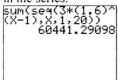

    The sum is approximately 60,441.29098.

93. $\displaystyle\sum_{n=1}^{20}\left[1.2(1.05)^n\right]$

    This is a geometric series with $a = 1.26$ (that is, $1.2(1.05)^1$), common ratio $r = 1.05$, and $n = 20$ terms.

    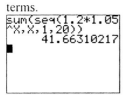

    The sum is approximately 41.66310217.

94. $\displaystyle\sum_{n=1}^{25}\left[1.3(0.55)^n\right]$

    This is a geometric series with $a = 0.715$ (that is, $1.3(0.55)^1$), common ratio $r = 0.55$, and $n = 25$ terms.

    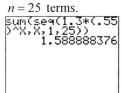

    The sum is approximately 1.588888376.

### Putting the Concepts Together (Sections 11.1 – 11.3)

1. $\dfrac{3/16}{3/4} = \dfrac{3}{16} \cdot \dfrac{4}{3} = \dfrac{1}{4}$

   $\dfrac{3/64}{3/16} = \dfrac{3}{64} \cdot \dfrac{16}{3} = \dfrac{1}{4}$

   $\dfrac{3/256}{3/64} = \dfrac{3}{256} \cdot \dfrac{64}{3} = \dfrac{1}{4}$

   The ratio of consecutive terms is a constant so the sequence is geometric with $a = \dfrac{3}{4}$ and common ratio $r = \dfrac{1}{4}$.

2. $a_n - a_{n-1} = \left[2(n+3)\right] - \left[2(n-1+3)\right]$
   $= 2(n+3) - 2(n+2)$
   $= 2n + 6 - 2n - 4$
   $= 2$

   The difference between consecutive terms is a constant so the sequence is arithmetic with $a = 2(1+3) = 8$ and common difference $d = 2$.

3. $a_n - a_{n-1} = \left[\dfrac{7n+2}{9}\right] - \left[\dfrac{7(n-1)+2}{9}\right]$
   $= \left(\dfrac{7n+2}{9}\right) - \left(\dfrac{7n-7+2}{9}\right)$
   $= \dfrac{7n+2-7n+5}{9}$
   $= \dfrac{7}{9}$

   The difference between consecutive terms is a constant so the sequence is arithmetic with $a = \dfrac{7(1)+2}{9} = 1$ and common difference $d = \dfrac{7}{9}$.

4. $-4 - 1 = -5$
   $9 - (-4) = 13$
   $\dfrac{-4}{1} = -4$
   $\dfrac{9}{-4} = -\dfrac{9}{4}$

   The sequence is neither arithmetic nor geometric. The difference between consecutive terms is not constant so the sequence is not arithmetic. Likewise, the ratio of consecutive terms is not constant so the sequence is not geometric.

## Putting the Concepts Together (Sections 11.1 – 11.3)

5. $\dfrac{a_n}{a_{n-1}} = \dfrac{3 \cdot 2^{n+1}}{3 \cdot 2^{(n-1)+1}} = \dfrac{3 \cdot 2^n \cdot 2}{3 \cdot 2^n} = 2$

   The ratio of consecutive terms is a constant so the sequence is geometric with $a = 3 \cdot 2^{1+1} = 12$ and common ratio $r = 2$.

6. $a_n - a_{n-1} = \left[n^2 - 5\right] - \left[(n-1)^2 - 5\right]$
   $= n^2 - 5 - \left(n^2 - 2n + 1 - 5\right)$
   $= n^2 - 5 - n^2 + 2n + 4$
   $= 2n - 1$

   $\dfrac{a_n}{a_{n-1}} = \dfrac{n^2 - 5}{(n-1)^2 - 5} = \dfrac{n^2 - 5}{n^2 - 2n - 4}$

   The sequence is neither arithmetic nor geometric. The difference between consecutive terms is not constant so the sequence is not arithmetic. Likewise, the ratio of consecutive terms is not constant so the sequence is not geometric.

7. $\sum\limits_{k=1}^{6} [3k + 4]$
   $= (3 \cdot 1 + 4) + (3 \cdot 2 + 4) + (3 \cdot 3 + 4)$
   $+ (3 \cdot 4 + 4) + (3 \cdot 5 + 4) + (3 \cdot 6 + 4)$
   $= 3 + 4 + 6 + 4 + 9 + 4 + 12 + 4 + 15 + 4 + 18 + 4$
   $= 87$

8. Each term is written in the form $\dfrac{1}{2(6+i)}$, where $i$ is the term number, and there are $n = 12$ terms.

   $\dfrac{1}{2(6+1)} + \dfrac{1}{2(6+2)} + \ldots + \dfrac{1}{2(6+12)} = \sum\limits_{i=1}^{12} \dfrac{1}{2(6+i)}$

9. $a_n = a + (n-1)d$
   $= 25 + (n-1)(-2)$
   $= 25 - 2n + 2$
   $= 27 - 2n$
   $a_1 = 27 - 2(1) = 25$
   $a_2 = 27 - 2(2) = 23$
   $a_3 = 27 - 2(3) = 21$
   $a_4 = 27 - 2(4) = 19$
   $a_5 = 27 - 2(5) = 17$

   The first five terms of the sequence are 25, 23, 21, 19, and 17.

10. $a_n = a + (n-1)d$
    $a_4 = a + (4-1)(11) = 9$
    $a + 33 = 9$
    $a = -24$
    $a_n = -24 + (n-1)(11)$
    $= -24 + 11n - 11$
    $= 11n - 35$
    $a_1 = 11(1) - 35 = -24$
    $a_2 = 11(2) - 35 = -13$
    $a_3 = 11(3) - 35 = -2$
    $a_4 = 11(4) - 35 = 9$
    $a_5 = 11(5) - 35 = 20$

    The first five terms of the sequence are $-24$, $-13$, $-2$, $9$, and $20$.

11. $a_n = a \cdot r^{n-1}$
    $a_4 = a \cdot \left(\dfrac{1}{5}\right)^{4-1} = \dfrac{9}{25}$
    $a \cdot \dfrac{1}{125} = \dfrac{9}{25}$ $\rightarrow$ $a = 125 \cdot \dfrac{9}{25} = 45$

    $a_n = 45 \cdot \left(\dfrac{1}{5}\right)^{n-1}$

    $a_1 = 45 \cdot \left(\dfrac{1}{5}\right)^{1-1} = 45$
    $a_2 = 45 \cdot \left(\dfrac{1}{5}\right)^{2-1} = 9$
    $a_3 = 45 \cdot \left(\dfrac{1}{5}\right)^{3-1} = \dfrac{9}{5}$
    $a_4 = 45 \cdot \left(\dfrac{1}{5}\right)^{4-1} = \dfrac{9}{25}$
    $a_5 = 45 \cdot \left(\dfrac{1}{5}\right)^{5-1} = \dfrac{9}{125}$

    The first five terms of the sequence are $45$, $9$, $\dfrac{9}{5}$, $\dfrac{9}{25}$, and $\dfrac{9}{125}$.

12. $a_n = a \cdot r^{n-1}$
    $= 150 \cdot (1.04)^{n-1}$
    $a_1 = 150 \cdot (1.04)^{1-1} = 150$
    $a_2 = 150 \cdot (1.04)^{2-1} = 156$
    $a_3 = 150 \cdot (1.04)^{3-1} = 162.24$
    $a_4 = 150 \cdot (1.04)^{4-1} = 168.7296$
    $a_5 = 150 \cdot (1.04)^{5-1} = 175.478784$

## Chapter 11: Sequences, Series, and the Binomial Expansion

The first five terms of the sequence are 150, 156, 162.24, 168.7296, and 175.478784.

13. The terms form a geometric sequence with first term $a=2$, common ratio $r=3$, and $n=11$ terms.
$$S_{11} = 2 \cdot \frac{1-3^{11}}{1-3} = 177{,}146$$

14. The terms form an arithmetic sequence with first term $a=2$, common difference $d=5$, and $n=20$ terms.
$$a_{20} = 2 + (20-1) \cdot 5 = 97$$
$$S_{20} = \frac{20}{2}[2+97] = 990$$

15. This is an infinite geometric series with $a=1000$ and $r=\frac{1}{10}$.
$$S_\infty = \frac{a}{1-r} = \frac{1000}{1-\frac{1}{10}} = \frac{1000}{\frac{9}{10}} = \frac{10{,}000}{9} \text{ or } 1111\frac{1}{9}$$

16. The number of people who can be seated forms an arithmetic sequence with first term $a=4$ and common difference $d=2$. The number of tables required is the same as the term number, $n$. To find the number of tables required to seat a party of 24 people, we solve the following for $n$:
$$a_n = a + (n-1)d$$
$$24 = 4 + (n-1)(2)$$
$$24 = 4 + 2n - 2$$
$$24 = 2n + 2$$
$$22 = 2n$$
$$11 = n$$
A party of 24 people would require 11 tables.

### 11.4 Preparing for the Binomial Theorem

1. $(x-5)^2 = x^2 - 2 \cdot x \cdot 5 + 5^2$
   $= x^2 - 10x + 25$

2. $(2x+3)^2 = (2x+3)(2x+3)$
   $= \left((2x)^2 + 2 \cdot 2x \cdot 3 + 3^2\right)$
   $= 4x^2 + 12x + 9$

### 11.4 Exercises

2. 1; 1; 3,628,800

4. False; $\binom{n}{j} = \frac{n!}{j!(n-j)!}$

6. 
```
      1
     1 1
    1 2 1
   1 3 3 1
```

7. Beginning with $n$, the exponents of $x$ decrease by 1 in subsequent terms until the exponent is 0.

8. Beginning with 0, the exponents of $a$ increase by 1 in subsequent terms until the exponent is $n$.

9. The degree of each monomial (sum of the exponents) in the expansion of $(x+a)^n$ is equal to $n$.

10. In general, the $k$th term of the expansion of $(x+a)^n$ is given by the expression
$$k\text{th term} = \binom{n}{k-1} a^{k-1} x^{n-k+1}$$

12. $(x-1)^4 = (x+(-1))^4$
$= \binom{4}{0}x^4 + \binom{4}{1}(-1)^1 \cdot x^{4-1} + \binom{4}{2}(-1)^2 \cdot x^{4-2} + \binom{4}{3}(-1)^3 \cdot x^{4-3} + \binom{4}{4}(-1)^4$
$= x^4 + 4(-1)x^3 + 6(1)x^2 + 4(-1)x + 1$
$= x^4 - 4x^3 + 6x^2 - 4x + 1$

14. $(x+5)^5 = \binom{5}{0}x^5 + \binom{5}{1}5^1 \cdot x^{5-1} + \binom{5}{2}5^2 \cdot x^{5-2} + \binom{5}{3}5^3 \cdot x^{5-3} + \binom{5}{4}5^4 \cdot x^{5-4} + \binom{5}{5}5^5$
$= x^5 + 5 \cdot 5x^4 + 10 \cdot 25x^3 + 10 \cdot 125x^2 + 5 \cdot 625x + 3125$
$= x^5 + 25x^4 + 250x^3 + 1250x^2 + 3125x + 3125$

16. $(2q+3)^4 = ((2q)+3)^4$
$= \binom{4}{0}(2q)^4 + \binom{4}{1}3^1 \cdot (2q)^{4-1} + \binom{4}{2}3^2 \cdot (2q)^{4-2} + \binom{4}{3}3^3 \cdot (2q)^{4-3} + \binom{4}{4}3^4$
$= 16q^4 + 4 \cdot 3 \cdot 8q^3 + 6 \cdot 9 \cdot 4q^2 + 4 \cdot 27 \cdot 2q + 81$
$= 16q^4 + 96q^3 + 216q^2 + 216q + 81$

18. $(3w-4)^4 = \binom{4}{0}(3w)^4 + \binom{4}{1}(-4)^1 \cdot (3w)^{4-1} + \binom{4}{2}(-4)^2 \cdot (3w)^{4-2} + \binom{4}{3}(-4)^3 \cdot (3w)^{4-3} + \binom{4}{4}(-4)^4$
$= 81w^4 + 4 \cdot (-4) \cdot 27w^3 + 6 \cdot 16 \cdot 9w^2 + 4 \cdot (-64) \cdot 3w + 256$
$= 81w^4 - 432w^3 + 864w^2 - 768w + 256$

20. $(y^2-3)^4 = \binom{4}{0}(y^2)^4 + \binom{4}{1}(-3)^1 \cdot (y^2)^{4-1} + \binom{4}{2}(-3)^2 \cdot (y^2)^{4-2} + \binom{4}{3}(-3)^3 \cdot (y^2)^{4-3} + \binom{4}{4}(-3)^4$
$= y^8 + 4 \cdot (-3)y^6 + 6 \cdot 9y^4 + 4 \cdot (-27)y^2 + 81$
$= y^8 - 12y^6 + 54y^4 - 108y^2 + 81$

22. $(3b^2+2)^5 = ((3b^2)+2)^5$
$= \binom{5}{0}(3b^2)^5 + \binom{5}{1}2^1 \cdot (3b^2)^{5-1} + \binom{5}{2}2^2 \cdot (3b^2)^{5-2} + \binom{5}{3}2^3 \cdot (3b^2)^{5-3} + \binom{5}{4}2^4 \cdot (3b^2)^{5-4} + \binom{5}{5}2^5$
$= 243b^{10} + 5 \cdot 2 \cdot 81b^8 + 10 \cdot 4 \cdot 27b^6 + 10 \cdot 8 \cdot 9b^4 + 5 \cdot 16 \cdot 3b^2 + 32$
$= 243b^{10} + 810b^8 + 1080b^6 + 720b^4 + 240b^2 + 32$

24. $(p-3)^6$
$= (p+(-3))^6$
$= \binom{6}{0}p^6 + \binom{6}{1}(-3)^1 \cdot p^{6-1} + \binom{6}{2}(-3)^2 \cdot p^{6-2} + \binom{6}{3}(-3)^3 \cdot p^{6-3} + \binom{6}{4}(-3)^4 \cdot p^{6-4} + \binom{6}{5}(-3)^5 \cdot p^{6-5} + \binom{6}{6}(-3)^6$
$= p^6 + 6 \cdot (-3)p^5 + 15 \cdot 9p^4 + 20 \cdot (-27)p^3 + 15 \cdot 81p^2 + 6 \cdot (-243)p + 729$
$= p^6 - 18p^5 + 135p^4 - 540p^3 + 1215p^2 - 1458p + 729$

## Chapter 11: Sequences, Series, and the Binomial Expansion

**26.** $(3x^2 + y^3)^4 = ((3x^2) + (y^3))^4$

$= \binom{4}{0}(3x^2)^4 + \binom{4}{1}(3x^2)^{4-1} \cdot (y^3)^1 + \binom{4}{2}(3x^2)^{4-2} \cdot (y^3)^2 + \binom{4}{3}(3x^2)^{4-3} \cdot (y^3)^3 + \binom{4}{4}(y^3)^4$

$= 81x^8 + 4 \cdot 27x^6 \cdot (y^3) + 6 \cdot 9x^4 \cdot (y^6) + 4 \cdot 3x^2 \cdot (y^9) + y^{12}$

$= 81x^8 + 108x^6 y^3 + 54x^4 y^6 + 12x^2 y^9 + y^{12}$

**28.** $(1.001)^5 = (1 + 0.001)^5$

$= \binom{5}{0}1^5 + \binom{5}{1}1^{5-1}(0.001)^1 + \binom{5}{2}1^{5-2}(0.001)^2 + \binom{5}{3}1^{5-3}(0.001)^3 + \binom{5}{4}1^{5-4}(0.001)^4 + \binom{5}{5}(0.001)^5$

$= 1 + 5(0.001) + 10(0.000001) + 10(0.000000001) + 5(0.000000000001) + (0.000000000000001)$

$= 1 + 0.005 + 0.00001 + 0.00000001 + 0.000000000005 + 0.000000000000001$

$\approx 1.00501$

**30.** $(0.997)^5 = (1 - 0.003)^5$

$= (1 + (-0.003))^5$

$= \binom{5}{0}1^5 + \binom{5}{1}1^{5-1}(-0.003)^1 + \binom{5}{2}1^{5-2}(-0.003)^2 + \binom{5}{3}1^{5-3}(-0.003)^3 + \binom{5}{4}1^{5-4}(-0.003)^4 + \binom{5}{5}(-0.003)^5$

$= 1 + 5(-0.003) + 10(0.000009) + 10(-0.000000027) + 5(0.000000000081) + (-0.000000000000243)$

$= 1 - 0.015 + 0.00009 - 0.00000027 + 0.000000000405 - 0.000000000000243$

$= 1.000090000405 - 0.015000270000243$

$\approx 0.98509$

**32.** The fourth term of the expansion of $(x-1)^{10} = (x + (-1))^{10}$ is

$\binom{10}{3}(-1)^3 x^{10-3} = 120 \cdot (-1) x^7 = -120x^7$

**34.** The seventh term of the expansion of $(3p + 1)^9 = ((3p) + 1)^9$ is

$\binom{9}{6}(1)^6 \cdot (3p)^{9-6} = 84 \cdot 27 p^3 = 2268 p^3$

**36.** $\binom{n}{j} = \dfrac{n!}{j!(n-j)!} = \dfrac{n!}{(n-j)!j!} = \dfrac{n!}{(n-j)!(n-(n-j))!} = \binom{n}{n-j}$

**38.** $g(x) = x^5 + 3$

$g(z+1) = (z+1)^5 + 3$

$= \left[\binom{5}{0}z^5 + \binom{5}{1}1^1 \cdot z^{5-1} + \binom{5}{2}1^2 \cdot z^{5-2} + \binom{5}{3}1^3 \cdot z^{5-3} + \binom{5}{4}1^4 \cdot z^{5-4} + \binom{5}{5}1^5\right] + 3$

$= \left[z^5 + 5z^4 + 10z^3 + 10z^2 + 5z + 1\right] + 3$

$= z^5 + 5z^4 + 10z^3 + 10z^2 + 5z + 4$

**40.** $h(x) = 2x^5 + 5x^4$

$h(a+3) = 2(a+3)^5 + 5(a+3)^4$

$= 2\left[\binom{5}{0}a^5 + \binom{5}{1}3^1 \cdot a^{5-1} + \binom{5}{2}3^2 \cdot a^{5-2} + \binom{5}{3}3^3 \cdot a^{5-3} + \binom{5}{4}3^4 \cdot a^{5-4} + \binom{5}{5}3^5\right]$

$+ 5\left[\binom{4}{0}a^4 + \binom{4}{1}3^1 \cdot a^{4-1} + \binom{4}{2}3^2 \cdot a^{4-2} + \binom{4}{3}3^3 \cdot a^{4-3} + \binom{4}{4}3^4\right]$

$= 2\left[a^5 + 15a^4 + 90a^3 + 270a^2 + 405a + 243\right] + 5\left[a^4 + 12a^3 + 54a^2 + 108a + 81\right]$

$= 2a^5 + 30a^4 + 180a^3 + 540a^2 + 810a + 486 + 5a^4 + 60a^3 + 270a^2 + 540a + 405$

$= 2a^5 + 35a^4 + 240a^3 + 810a^2 + 1350a + 891$

## Chapter 11 Review

**1.** $a_n = -3n + 2$

$a_1 = -3(1) + 2 = -1$
$a_2 = -3(2) + 2 = -4$
$a_3 = -3(3) + 2 = -7$
$a_4 = -3(4) + 2 = -10$
$a_5 = -3(5) + 2 = -13$

The first five terms of the sequence are $-1$, $-4$, $-7$, $-10$, and $-13$.

**2.** $a_n = \dfrac{n-2}{n+4}$

$a_1 = \dfrac{1-2}{1+4} = \dfrac{-1}{5} = -\dfrac{1}{5}$; $a_2 = \dfrac{2-2}{2+4} = \dfrac{0}{6} = 0$;

$a_3 = \dfrac{3-2}{3+4} = \dfrac{1}{7}$; $a_4 = \dfrac{4-2}{4+4} = \dfrac{2}{8} = \dfrac{1}{4}$;

$a_5 = \dfrac{5-2}{5+4} = \dfrac{3}{9} = \dfrac{1}{3}$

The first five terms of the sequence are $-\dfrac{1}{5}$, $0$, $\dfrac{1}{7}$, $\dfrac{1}{4}$, and $\dfrac{1}{3}$.

**3.** $a_n = 5^n + 1$

$a_1 = 5^1 + 1 = 5 + 1 = 6$
$a_2 = 5^2 + 1 = 25 + 1 = 26$
$a_3 = 5^3 + 1 = 125 + 1 = 126$
$a_4 = 5^4 + 1 = 625 + 1 = 626$
$a_5 = 5^5 + 1 = 3125 + 1 = 3126$

The first five terms of the sequence are 6, 26, 126, 626, and 3126.

**4.** $a_n = (-1)^{n-1} \cdot 3n$

$a_1 = (-1)^{1-1} \cdot 3(1) = (-1)^0 \cdot 3 = 3$
$a_2 = (-1)^{2-1} \cdot 3(2) = (-1)^1 \cdot 6 = -6$
$a_3 = (-1)^{3-1} \cdot 3(3) = (-1)^2 \cdot 9 = 9$
$a_4 = (-1)^{4-1} \cdot 3(4) = (-1)^3 \cdot 12 = -12$
$a_5 = (-1)^{5-1} \cdot 3(5) = (-1)^4 \cdot 15 = 15$

The first five terms of the sequence are 3, $-6$, 9, $-12$, and 15.

**5.** $a_n = \dfrac{n^2}{n+1}$

$a_1 = \dfrac{1^2}{1+1} = \dfrac{1}{2}$, $a_2 = \dfrac{2^2}{2+1} = \dfrac{4}{3}$,

$a_3 = \dfrac{3^2}{3+1} = \dfrac{9}{4}$, $a_4 = \dfrac{4^2}{4+1} = \dfrac{16}{5}$,

$a_5 = \dfrac{5^2}{5+1} = \dfrac{25}{6}$

The first five terms of the sequence are $\dfrac{1}{2}$, $\dfrac{4}{3}$, $\dfrac{9}{4}$, $\dfrac{16}{5}$, and $\dfrac{25}{6}$.

**6.** $a_n = \dfrac{\pi^n}{n}$

$a_1 = \dfrac{\pi^1}{1} = \pi$; $a_2 = \dfrac{\pi^2}{2}$;

$a_3 = \dfrac{\pi^3}{3}$; $a_4 = \dfrac{\pi^4}{4}$;

$a_5 = \dfrac{\pi^5}{5}$

Chapter 11: Sequences, Series, and the Binomial Theorem

The first five terms of the sequence are $\pi$, $\dfrac{\pi^2}{2}$, $\dfrac{\pi^3}{3}$, $\dfrac{\pi^4}{4}$, and $\dfrac{\pi^5}{5}$.

7. The terms can be expressed as multiples of $-3$.
$-3 = -3 \cdot 1$
$-6 = -3 \cdot 2$
$-9 = -3 \cdot 3$
$-12 = -3 \cdot 4$
$-15 = -3 \cdot 5$
The $n$th term of the sequence is given by $a_n = -3n$.

8. The terms are rational numbers with a denominator of 3. The numerators are consecutive integers beginning with 1. The $n$th term of the sequence is given by $a_n = \dfrac{n}{3}$.

9. The terms can be expressed as the product of 5 and a power of 2.
$5 = 5 \cdot 2^0$
$10 = 5 \cdot 2^1$
$20 = 5 \cdot 2^2$
$40 = 5 \cdot 2^3$
$80 = 5 \cdot 2^4$
The exponent is one less than the term number. Therefore, the $n$th term of the sequence is given by $a_n = 5 \cdot 2^{n-1}$.

10. Rewrite as $-\dfrac{1}{2}, \dfrac{2}{2}, -\dfrac{3}{2}, \dfrac{4}{2}, \ldots$
The terms are rational numbers with alternating signs. The denominator is always 2, and the numerators are consecutive integers beginning with 1. The $n$th term of the sequence is given by
$$a_n = (-1)^n \cdot \dfrac{n}{2}.$$

11. The terms can be expressed as 5 more than the square of the term number.
$6 = 1^2 + 5$
$9 = 2^2 + 5$
$14 = 3^2 + 5$
$21 = 4^2 + 5$
$30 = 5^2 + 5$
The $n$th term of the sequence is given by $a_n = n^2 + 5$.

12. Rewrite the terms as $\dfrac{0}{2}, \dfrac{1}{3}, \dfrac{2}{4}, \dfrac{3}{5}, \ldots$
The terms are rational numbers whose numerators are consecutive integers beginning with 0 and whose denominators are consecutive integers beginning with 2. The $n$th term of the sequence is given by $a_n = \dfrac{n-1}{n+1}$.

13. $\displaystyle\sum_{k=1}^{5}(5k-2)$
$= (5(1)-2) + (5(2)-2) + (5(3)-2)$
$\quad + (5(4)-2) + (5(5)-2)$
$= 3 + 8 + 13 + 18 + 23$
$= 65$

14. $\sum_{k=1}^{6}\left(\dfrac{k+2}{2}\right)$
$=\dfrac{1+2}{2}+\dfrac{2+2}{2}+\dfrac{3+2}{2}+\dfrac{4+2}{2}+\dfrac{5+2}{2}+\dfrac{6+2}{2}$
$=\dfrac{3}{2}+\dfrac{4}{2}+\dfrac{5}{2}+\dfrac{6}{2}+\dfrac{7}{2}+\dfrac{8}{2}$
$=\dfrac{33}{2}$

15. $\sum_{i=1}^{5}(-2i)$
$=(-2\cdot1)+(-2\cdot2)+(-2\cdot3)+(-2\cdot4)+(-2\cdot5)$
$=-2-4-6-8-10$
$=-30$

16. $\sum_{i=1}^{4}\dfrac{i^2-1}{3}=\dfrac{1^2-1}{3}+\dfrac{2^2-1}{3}+\dfrac{3^2-1}{3}+\dfrac{4^2-1}{3}$
$=\dfrac{0}{3}+\dfrac{3}{3}+\dfrac{8}{3}+\dfrac{15}{3}$
$=\dfrac{26}{3}$

17. Each term takes on the form $4+3i$ and there are $n=15$ terms.
$(4+3\cdot1)+(4+3\cdot2)+\ldots+(4+3\cdot15)=\sum_{i=1}^{15}(4+3i)$

18. Each term takes on the form $\dfrac{1}{3^i}$ and there are $n=8$ terms.
$\dfrac{1}{3^1}+\dfrac{1}{3^2}+\ldots+\dfrac{1}{3^8}=\sum_{i=1}^{8}\dfrac{1}{3^i}$

19. Each term takes on the form $\dfrac{i^3+1}{i+1}$ and there are $n=10$ terms.
$\dfrac{1^3+1}{1+1}+\dfrac{2^3+1}{2+1}+\ldots+\dfrac{10^3+1}{10+1}=\sum_{i=1}^{10}\dfrac{i^3+1}{i+1}$

20. This is an alternating series. Each term takes on the form $(-1)^{i-1}\cdot i^2$ and there are $n=7$ terms.
$(-1)^{1-1}\cdot 1^2+(-1)^{2-1}\cdot 2^2+\ldots+(-1)^{7-1}\cdot 7^2$
$=\sum_{i=1}^{7}\left[(-1)^{i-1}\cdot i^2\right]$

21. $10-4=6$
$16-10=6$
$22-16=6$
The sequence is arithmetic with a common difference of $d=6$.

22. $\dfrac{1}{2}-(-1)=\dfrac{3}{2}$
$2-\dfrac{1}{2}=\dfrac{3}{2}$
$\dfrac{7}{2}-2=\dfrac{3}{2}$
The sequence is arithmetic with a common difference of $d=\dfrac{3}{2}$.

23. $-5-(-2)=-3$
$-9-(-5)=-4$
$-14-(-9)=-5$
The differences between consecutive terms is not constant. Therefore, the sequence is not arithmetic.

24. $3-(-1)=4$
$-5-3=-8$
$7-(-5)=12$
The difference between consecutive terms is not constant. Therefore, the sequence is not arithmetic.

25. $a_n-a_{n-1}=[4n+7]-[4(n-1)+7]$
$=4n+7-(4n-4+7)$
$=4n+7-4n+4-7$
$=4$
The sequence is arithmetic with a common difference of $d=4$.

26. $a_n-a_{n-1}=\left[\dfrac{n+1}{2n}\right]-\left[\dfrac{(n-1)+1}{2(n-1)}\right]$
$=\dfrac{n+1}{2n}-\dfrac{n}{2(n-1)}$
$=\dfrac{(n+1)(n-1)}{2n(n-1)}-\dfrac{n(n)}{2n(n-1)}$
$=\dfrac{n^2-1-n^2}{2n(n-1)}$
$=\dfrac{-1}{2n(n-1)}$
The difference between consecutive terms is not constant. Therefore, the sequence is not arithmetic.

Chapter 11: Sequences, Series, and the Binomial Theorem

27. $a_n = a + (n-1)d$
$= 3 + (n-1)(8)$
$= 3 + 8n - 8$
$= 8n - 5$
$a_{25} = 8(25) - 5 = 200 - 5 = 195$

28. $a_n = a + (n-1)d$
$= -4 + (n-1)(-3)$
$= -4 - 3n + 3$
$= -3n - 1$
$a_{25} = -3(25) - 1 = -75 - 1 = -76$

29. $d = \frac{20}{3} - 7 = -\frac{1}{3}$; $a = 7$
$a_n = a + (n-1)d$
$= 7 + (n-1)\left(-\frac{1}{3}\right)$
$= 7 - \frac{1}{3}n + \frac{1}{3}$
$= -\frac{1}{3}n + \frac{22}{3}$
$a_{25} = -\frac{1}{3}(25) + \frac{22}{3} = -\frac{3}{3} = -1$

30. $d = 17 - 11 = 6$; $a = 11$
$a_n = a + (n-1)d$
$= 11 + (n-1)(6)$
$= 11 + 6n - 6$
$= 6n + 5$
$a_{25} = 6(25) + 5 = 150 + 5 = 155$

31. Since $a_3 = 7$ and $a_8 = 25$, we have
$\begin{cases} 7 = a + (3-1)d \\ 25 = a + (8-1)d \end{cases}$ or $\begin{cases} 7 = a + 2d \\ 25 = a + 7d \end{cases}$
Subtract the second equation from the first equation to obtain
$-18 = -5d$
$\frac{18}{5} = d$
Let $d = \frac{18}{5}$ in the first equation to find $a$.

$7 = a + 2\left(\frac{18}{5}\right)$
$7 = a + \frac{36}{5}$
$-\frac{1}{5} = a$
The $n$th term of the sequence is given by
$a_n = a + (n-1)d$
$= -\frac{1}{5} + (n-1)\left(\frac{18}{5}\right)$
$= -\frac{1}{5} + \frac{18}{5}n - \frac{18}{5}$
$= \frac{18}{5}n - \frac{19}{5}$
$a_{25} = \frac{18}{5}(25) - \frac{19}{5} = \frac{450}{5} - \frac{19}{5} = \frac{431}{5}$

32. Since $a_4 = -20$ and $a_7 = -32$, we have
$\begin{cases} -20 = a + (4-1)d \\ -32 = a + (7-1)d \end{cases}$ or $\begin{cases} -20 = a + 3d \\ -32 = a + 6d \end{cases}$
Subtract the second equation from the first equation to obtain
$12 = -3d$
$-4 = d$
Let $d = -4$ in the first equation to find $a$.
$-20 = a + 3(-4)$
$-20 = a - 12$
$-8 = a$
The $n$th term of the sequence is given by
$a_n = a + (n-1)d$
$= -8 + (n-1)(-4)$
$= -8 - 4n + 4$
$= -4n - 4$
$a_{25} = -4(25) - 4 = -100 - 4 = -104$

33. We know the first term is $a = -1$ and the common difference is $d = 9 - (-1) = 10$.
$S_n = \frac{n}{2}[2a + (n-1)d]$
$S_{30} = \frac{30}{2}[2(-1) + (30-1)(10)]$
$= 15[-2 + 29(10)]$
$= 15(-2 + 290)$
$= 15(288)$
$= 4320$

**34.** We know the first term is $a = 5$ and the common difference is $d = 2 - 5 = -3$.
$$S_n = \frac{n}{2}\left[2a + (n-1)d\right]$$
$$S_{40} = \frac{40}{2}\left[2(5) + (40-1)(-3)\right]$$
$$= 20\left[10 + 39(-3)\right]$$
$$= 20(10 - 117)$$
$$= 20(-107)$$
$$= -2140$$

**35.** $a_n = -2n - 7$
$$a = a_1 = -2(1) - 7 = -9$$
$$a_{60} = -2(60) - 7 = -127$$
$$S_n = \frac{n}{2}\left[a + a_n\right]$$
$$S_{60} = \frac{60}{2}\left[-9 + (-127)\right]$$
$$= 30(-136)$$
$$= -4080$$

**36.** $a_n = \frac{1}{4}n + 3$
$$a = a_1 = \frac{1}{4}(1) + 3 = \frac{13}{4}$$
$$a_{50} = \frac{1}{4}(50) + 3 = \frac{62}{4}$$
$$S_n = \frac{n}{2}\left[a + a_n\right]$$
$$S_{50} = \frac{50}{2}\left[\frac{13}{4} + \frac{62}{4}\right]$$
$$= 25\left(\frac{75}{4}\right)$$
$$= \frac{1875}{4} \quad \text{or} \quad 468.75$$

**37.** We could list terms of the sequence formed by the years when the Brood X cicada returns, or we could use the general formula for the $n$th term. We have $a = 2004$ and $d = 17$. We wish to know the first time when $a_n \geq 2101$.
$$a_n \geq 2101$$
$$a + (n-1)d \geq 2101$$
$$2004 + (n-1)(17) \geq 2101$$
$$2004 + 17n - 17 \geq 2101$$
$$17n + 1987 \geq 2101$$
$$17n \geq 114$$
$$n \geq \frac{114}{17} \approx 6.71$$
The Brood X cicada will first appear in the 22$^{\text{nd}}$ century when $n = 7$.
$$a_7 = 2004 + (7-1)(17) = 2106$$
The Brood X cicada will first appear in the 22$^{\text{nd}}$ century in the year 2106.

**38.** Since the distance to each line is the same as the distance back to the goal line, we can consider the sequence of distances run away from the goal line and double our result. The sequence of distances is 10, 20, 30, 40, and 50. This is an arithmetic sequence with first term $a = 10$ and fifth term $a_5 = 50$.
$$S_5 = \frac{5}{2}\left[10 + 50\right] = \frac{5}{2}(60) = 150$$
Doubling this result gives a total distance of 300 yards run by each player during wind sprints.

**39.** $\dfrac{2}{1/3} = 2 \cdot \dfrac{3}{1} = 6$
$$\frac{12}{2} = 6$$
$$\frac{72}{12} = 6$$
The ratio of consecutive terms is constant so the sequence is geometric with common ratio $r = 6$.

**40.** $\dfrac{3}{-1} = -3$
$$\frac{-9}{3} = -3$$
$$\frac{27}{-9} = -3$$
The ratio of consecutive terms is constant so the sequence is geometric with common ratio $r = -3$.

## Chapter 11: Sequences, Series, and the Binomial Theorem

41. $\frac{1}{1}=1$

    $\frac{2}{1}=2$

    $\frac{6}{2}=3$

    The ratio of consecutive terms is not constant so the sequence is not geometric.

42. $\frac{4}{6}=\frac{2}{3}$

    $\frac{8/3}{4}=\frac{8}{3}\cdot\frac{1}{4}=\frac{2}{3}$

    $\frac{16/9}{8/3}=\frac{16}{9}\cdot\frac{3}{8}=\frac{2}{3}$

    The ratio of consecutive terms is constant so the sequence is geometric with common ratio $r=\frac{2}{3}$.

43. $\frac{a_n}{a_{n-1}}=\frac{5\cdot(-2)^n}{5\cdot(-2)^{n-1}}=\frac{(-2)^{n-1}\cdot(-2)}{(-2)^{n-1}}=-2$

    The ratio of consecutive terms is constant so the sequence is geometric with common ratio $r=-2$.

44. $\frac{a_n}{a_{n-1}}=\frac{3n-14}{3(n-1)-14}=\frac{3n-14}{3n-17}$

    The ratio of consecutive terms is not constant so the sequence is not geometric.

45. $a_n=a\cdot r^{n-1}$

    $a_n=4\cdot 3^{n-1}$

    $a_{10}=4\cdot 3^{10-1}=4\cdot 3^9=4(19,683)=78,732$

46. $a_n=a\cdot r^{n-1}$

    $a_n=8\cdot\left(\frac{1}{4}\right)^{n-1}$

    $a_{10}=8\cdot\left(\frac{1}{4}\right)^{10-1}$

    $=8\cdot\left(\frac{1}{4}\right)^9$

    $=8\cdot\left(\frac{1}{262,144}\right)$

    $=\frac{1}{32,768}$

47. $a_n=a\cdot r^{n-1}$

    $a_n=5\cdot(-2)^{n-1}$

    $a_{10}=5\cdot(-2)^{10-1}=5\cdot(-2)^9=5(-512)=-2560$

48. $a_n=a\cdot r^{n-1}$

    $a_n=1000\cdot(1.08)^{n-1}$

    $a_{10}=1000\cdot(1.08)^{10-1}$

    $=1000\cdot(1.08)^9$

    $\approx 1000(1.999005)$

    $=1999.005$

49. This is a geometric series with first term $a=2$, common ratio $r=\frac{4}{2}=2$, and $n=15$ terms.

    $S_n=a\cdot\frac{1-r^n}{1-r}$

    $S_{15}=2\cdot\frac{1-2^{15}}{1-2}$

    $=2\cdot\frac{-32,767}{-1}$

    $=65,534$

50. This is a geometric series with first term $a=40$, common ratio $r=\frac{1}{8}$, and $n=13$ terms.

    $S_n=a\cdot\frac{1-r^n}{1-r}$

    $S_{13}=40\cdot\frac{1-\left(\frac{1}{8}\right)^{13}}{1-\frac{1}{8}}$

    $\approx 45.71428571$

51. This is a geometric series with $n=12$ terms. The first term is $a=\frac{3}{4}\cdot 2^{1-1}=\frac{3}{4}$ and the common ratio is $r=2$.

    $S_n=a\cdot\frac{1-r^n}{1-r}$

    $S_{12}=\frac{3}{4}\cdot\frac{1-2^{12}}{1-2}$

    $=\frac{3}{4}\cdot\frac{-4095}{-1}$

    $=\frac{12,285}{4}$ or $3071.25$

**52.** This is a geometric series with $n = 16$ terms. The first term is $a = -4(3^1) = -12$ and the common ratio is $r = 3$.
$$S_n = a \cdot \frac{1-r^n}{1-r}$$
$$S_{16} = -12 \cdot \frac{1-3^{16}}{1-3}$$
$$= -12(21,523,360)$$
$$= -258,280,320$$

**53.** This is an infinite geometric series with first term $a = 20 \cdot \left(\frac{1}{4}\right)^1 = 5$ and common ratio $r = \frac{1}{4}$.
Since the ratio is between $-1$ and $1$, we can use the formula for the sum of an infinite geometric series.
$$S_\infty = \frac{a}{1-r} = \frac{5}{1-\frac{1}{4}} = \frac{5}{3/4} = 5 \cdot \frac{4}{3} = \frac{20}{3}$$

**54.** This is an infinite geometric series with first term $a = 50 \cdot \left(-\frac{1}{2}\right)^{1-1} = 50$ and common ratio $r = -\frac{1}{2}$. Since the ratio is between $-1$ and $1$, we can use the formula for the sum of an infinite geometric series.
$$S_\infty = \frac{a}{1-r} = \frac{50}{1-\left(-\frac{1}{2}\right)} = \frac{50}{3/2} = 50 \cdot \frac{2}{3} = \frac{100}{3}$$

**55.** This is an infinite geometric series with first term $a = 1$ and common ratio $r = \frac{1/5}{1} = \frac{1}{5}$. The ratio is between $-1$ and $1$, so we can use the formula for the sum of an infinite geometric series.
$$S_\infty = \frac{a}{1-r} = \frac{1}{1-\frac{1}{5}} = \frac{1}{4/5} = \frac{5}{4}$$

**56.** This is an infinite geometric series with first term $a = 0.8$ and common ratio $r = \frac{0.08}{0.8} = 0.1$. Since the ratio is between $-1$ and $1$, we can use the formula for the sum of an infinite geometric series.
$$S_\infty = \frac{a}{1-r} = \frac{0.8}{1-0.1} = \frac{0.8}{0.9} = \frac{8}{9}$$

**57.** Here we consider a geometric sequence with first term $a = 200$ and common ratio $r = \frac{1}{2}$. Each term in the sequence represents the amount after each half-life, or the amount remaining after every 12 years. Since $\frac{72}{12} = 6$, we are interested in the 7th term of the sequence (that is, after 6 common ratios).
$$a_n = a \cdot r^{n-1}$$
$$a_7 = 200 \cdot \left(\frac{1}{2}\right)^{7-1}$$
$$= 200 \cdot \left(\frac{1}{2}\right)^6$$
$$= \frac{200}{64}$$
$$= 3.125$$
After 72 years, there will be 3.125 grams of the Tritium remaining.

**58.** The number of emails in each cycle forms a geometric sequence with first term $a = 5$ and common ratio $r = 5$. To find the total e-mails sent after 15 minutes, we need to find the sum of the first 15 terms of this sequence.
$$S_n = a \cdot \frac{1-r^n}{1-r}$$
$$S_{15} = 5 \cdot \frac{1-5^{15}}{1-5}$$
$$\approx 3.815 \times 10^{10}$$
After 15 minutes a total of about 38.15 billion e-mails will have been sent.

**59.** $A = P \cdot \left[\frac{(1+i)^n - 1}{i}\right]$
Note that the total contribution each quarter is the sum of Scott's contribution and the matching contribution from his employer.
In this case we have $P = 900 + 450 = 1350$, $n = 25 \cdot 4 = 100$, and $i = \frac{0.07}{4} = 0.0175$.
$$A = 1350 \cdot \left[\frac{(1+0.0175)^{100} - 1}{0.0175}\right]$$
$$= 1350 \cdot [266.7517679]$$
$$= \$360,114.89$$
After 25 years, Scott's 403(b) will be worth $\$360,114.89$.

## Chapter 11: Sequences, Series, and the Binomial Theorem

60. <u>Lump sum option:</u>
    For this option, we use the compound interest formula.
    $A = P(1+i)^n$
    In this case we have $P = \$28,000,000$,
    $n = 26 \cdot 1 = 26$, and $i = \dfrac{0.065}{1} = 0.065$.
    $A = 28,000,000 \cdot (1+0.065)^{26} = \$143,961,987.40$

    <u>Annuity option:</u>
    For this option, we use the annuity formula.
    $A = P \cdot \left[ \dfrac{(1+i)^n - 1}{i} \right]$
    In this case we have $P = 2,000,000$,
    $n = 26 \cdot 1 = 26$, and $i = \dfrac{0.065}{1} = 0.065$.
    $A = 2,000,000 \cdot \left[ \dfrac{(1+0.065)^{26} - 1}{0.065} \right]$
    $= 2,000,000 \cdot [63.71537769]$
    $= \$127,430,755.40$
    The lump sum option would yield more money after 26 years.

61. $A = P \cdot \left[ \dfrac{(1+i)^n - 1}{i} \right]$
    In this case we have $A = 2,500,000$,
    $n = 40 \cdot 12 = 480$, and $i = \dfrac{0.09}{12} = 0.0075$.
    $2,500,000 = P \cdot \left[ \dfrac{(1+0.0075)^{480} - 1}{0.0075} \right]$
    $2,500,000 = P \cdot [4681.320273]$
    $P = \$534.04$
    Sheri would need to contribute $534.04, or about $534, each month to reach her goal.

62. $A = P \cdot \left[ \dfrac{(1+i)^n - 1}{i} \right]$
    In this case we have $P = 400$, $n = 10 \cdot 12 = 120$,
    and $i = \dfrac{0.0525}{12} = 0.004375$.
    $A = 400 \cdot \left[ \dfrac{(1+0.004375)^{120} - 1}{0.004375} \right]$
    $= 400 \cdot [157.3769632]$
    $= \$62,950.79$

    $\dfrac{62,950.79}{340} \approx 185.15$
    When Samantha turns 18, the plan will be worth $62,950.79 and will cover about 185 credit hours.

63. $5! = 5 \cdot 4 \cdot 3 \cdot 2 \cdot 1 = 120$

64. $\dfrac{11!}{7!} = \dfrac{11 \cdot 10 \cdot 9 \cdot 8 \cdot 7!}{7!} = 11 \cdot 10 \cdot 9 \cdot 8 = 7920$

65. $\dfrac{10!}{6!} = \dfrac{10 \cdot 9 \cdot 8 \cdot 7 \cdot 6!}{6!} = 10 \cdot 9 \cdot 8 \cdot 7 = 5040$

66. $\dfrac{13!}{6!7!} = \dfrac{13 \cdot 12 \cdot 11 \cdot 10 \cdot 9 \cdot 8 \cdot 7!}{6 \cdot 5 \cdot 4 \cdot 3 \cdot 2 \cdot 1 \cdot 7!}$
    $= \dfrac{13 \cdot 12 \cdot 11 \cdot 10 \cdot 9 \cdot 8}{6 \cdot 5 \cdot 4 \cdot 3 \cdot 2 \cdot 1}$
    $= 13 \cdot 11 \cdot 3 \cdot 4$
    $= 1716$

67. $\binom{7}{3} = \dfrac{7!}{3!4!}$
    $= \dfrac{7 \cdot 6 \cdot 5 \cdot 4!}{3 \cdot 2 \cdot 1 \cdot 4!}$
    $= \dfrac{7 \cdot 6 \cdot 5}{3 \cdot 2 \cdot 1}$
    $= 35$

68. $\binom{10}{5} = \dfrac{10!}{5!5!}$
    $= \dfrac{10 \cdot 9 \cdot 8 \cdot 7 \cdot 6 \cdot 5!}{5! \cdot 5 \cdot 4 \cdot 3 \cdot 2 \cdot 1}$
    $= \dfrac{10 \cdot 9 \cdot 8 \cdot 7 \cdot 6}{5 \cdot 4 \cdot 3 \cdot 2 \cdot 1}$
    $= 252$

## Chapter 11 Review

69. $\binom{8}{8} = \dfrac{8!}{8!\,0!} = 1$

70. $\binom{6}{0} = \dfrac{6!}{0!\,6!} = 1$

71. $(z+1)^4 = \binom{4}{0}z^4 + \binom{4}{1}z^3 + \binom{4}{2}z^2 + \binom{4}{3}z + \binom{4}{4}$
$= 1 \cdot z^4 + 4 \cdot z^3 + 6 \cdot z^2 + 4 \cdot z + 1$
$= z^4 + 4z^3 + 6z^2 + 4z + 1$

72. $(y-3)^5 = \binom{5}{0}y^5 + \binom{5}{1}y^4 \cdot (-3) + \binom{5}{2}y^3 \cdot (-3)^2 + \binom{5}{3}y^2 \cdot (-3)^3 + \binom{5}{4}y \cdot (-3)^4 + \binom{5}{5}\cdot(-3)^5$
$= 1 \cdot y^5 + 5 \cdot y^4 \cdot (-3) + 10 \cdot y^3 \cdot 9 + 10 \cdot y^2 \cdot (-27) + 5 \cdot y \cdot 81 + (-243)$
$= y^5 - 15y^4 + 90y^3 - 270y^2 + 405y - 243$

73. $(3y+4)^6$
$= \binom{6}{0}(3y)^6 + \binom{6}{1}(3y)^5(4) + \binom{6}{2}(3y)^4(4)^2 + \binom{6}{3}(3y)^3(4)^3 + \binom{6}{4}(3y)^2(4)^4 + \binom{6}{5}(3y)(4)^5 + \binom{6}{6}(4)^6$
$= 1 \cdot 729y^6 + 6 \cdot 243y^5 \cdot 4 + 15 \cdot 81y^4 \cdot 16 + 20 \cdot 27y^3 \cdot 64 + 15 \cdot 9y^2 \cdot 256 + 6 \cdot 3y \cdot 1024 + 4096$
$= 729y^6 + 5832y^5 + 19{,}440y^4 + 34{,}560y^3 + 34{,}560y^2 + 18{,}432y + 4096$

74. $(2x^2-3)^4 = \binom{4}{0}(2x^2)^4 + \binom{4}{1}(2x^2)^3(-3) + \binom{4}{2}(2x^2)^2(-3)^2 + \binom{4}{3}(2x^2)(-3)^3 + \binom{4}{4}(-3)^4$
$= 16x^8 + 4 \cdot 8x^6 \cdot (-3) + 6 \cdot 4x^4 \cdot 9 + 4 \cdot 2x^2 \cdot (-27) + 81$
$= 16x^8 - 96x^6 + 216x^4 - 216x^2 + 81$

75. $(3p-2q)^4 - \binom{4}{0}(3p)^4 + \binom{4}{1}(3p)^3(-2q) + \binom{4}{2}(3p)^2(-2q)^2 + \binom{4}{3}(3p)(-2q)^3 + \binom{4}{4}(-2q)^4$
$= 81p^4 + 4 \cdot 27p^3 \cdot (-2q) + 6 \cdot 9p^2 \cdot (4q^2) + 4 \cdot 3p \cdot (-8q^3) + 16q^4$
$= 81p^4 - 216p^3q + 216p^2q^2 - 96pq^3 + 16q^4$

76. $(a^3+3b)^5 = \binom{5}{0}(a^3)^5 + \binom{5}{1}(a^3)^4(3b) + \binom{5}{2}(a^3)^3(3b)^2 + \binom{5}{3}(a^3)^2(3b)^3 + \binom{5}{4}(a^3)(3b)^4 + \binom{5}{5}(3b)^5$
$= a^{15} + 5a^{12} \cdot 3b + 10a^9 \cdot 9b^2 + 10a^6 \cdot 27b^3 + 5a^3 \cdot 81b^4 + 243b^5$
$= a^{15} + 15a^{12}b + 90a^9b^2 + 270a^6b^3 + 405a^3b^4 + 243b^5$

77. The fourth term of the expansion of $(x-2)^8 = (x+(-2))^8$ is
$\binom{8}{3}(-2)^3 x^5 = 56x^5 \cdot (-8) = -448x^5$

78. The seventh term of the expansion of $(2x+1)^{11} = ((2x)+1)^{11}$ is
$\binom{11}{6}1^6(2x)^5 = 462 \cdot 32x^5 = 14{,}784x^5$

## Chapter 11: Sequences, Series, and the Binomial Theorem

**Chapter 11 Test**

1. $-7-(-15) = 8$
   $1-(-7) = 8$
   $9-1 = 8$
   The difference between consecutive terms is a constant so the sequence is arithmetic with $a = -15$ and common difference $d = 8$.

2. $\dfrac{a_n}{a_{n-1}} = \dfrac{(-4)^n}{(-4)^{n-1}} = \dfrac{-4 \cdot (-4)^{n-1}}{(-4)^{n-1}} = -4$

   The ratio of consecutive terms is a constant so the sequence is geometric with $a = (-4)^1 = -4$ and common ratio $r = -4$.

3. $a_n = \dfrac{4}{n!}$; $a_{n-1} = \dfrac{4}{(n-1)!}$

   $a_n - a_{n-1} = \dfrac{4}{n!} - \dfrac{4}{(n-1)!} = \dfrac{4}{n!} - \dfrac{4n}{n(n-1)!}$
   $= \dfrac{4-4n}{n!}$

   $\dfrac{a_n}{a_{n-1}} = \dfrac{\frac{4}{n!}}{\frac{4}{(n-1)!}} = \dfrac{4}{n!} \cdot \dfrac{(n-1)!}{4} = \dfrac{1}{n}$

   The sequence is neither arithmetic nor geometric. The difference between consecutive terms is not constant so the sequence is not arithmetic. Likewise, the ratio of consecutive terms is not constant so the sequence is not geometric.

4. $a_n - a_{n-1} = \left[\dfrac{2n-3}{5}\right] - \left[\dfrac{2(n-1)-3}{5}\right]$
   $= \dfrac{2n-3-(2n-2-3)}{5}$
   $= \dfrac{2n-3-2n+5}{5} = \dfrac{2}{5}$

   The difference between consecutive terms is constant so the sequence is arithmetic with $a = \dfrac{2(1)-3}{5} = -\dfrac{1}{5}$ and common difference $d = \dfrac{2}{5}$.

5. $2-(-3) = 5$
   $0-2 = 2$
   $\dfrac{2}{-3} = -\dfrac{2}{3}$
   $\dfrac{0}{2} = 0$

   The sequence is neither arithmetic nor geometric. The difference between consecutive terms is not constant so the sequence is not arithmetic. Likewise, the ratio of consecutive terms is not constant so the sequence is not geometric.

6. $\dfrac{a_n}{a_{n-1}} = \dfrac{7 \cdot 3^n}{7 \cdot 3^{n-1}} = \dfrac{7 \cdot 3 \cdot 3^{n-1}}{7 \cdot 3^{n-1}} = 3$

   The ratio of consecutive terms is a constant so the sequence is geometric with $a = 7 \cdot 3^1 = 21$ and common ratio $r = 3$.

7. $\sum_{i=1}^{5}\left[\dfrac{3}{i^2}+2\right]$

   $= \left(\dfrac{3}{1^2}+2\right) + \left(\dfrac{3}{2^2}+2\right) + \left(\dfrac{3}{3^2}+2\right) + \left(\dfrac{3}{4^2}+2\right) + \left(\dfrac{3}{5^2}+2\right)$

   $= 3+2+\dfrac{3}{4}+2+\dfrac{1}{3}+2+\dfrac{3}{16}+2+\dfrac{3}{25}+2$

   $= 13 + \dfrac{900+400+225+144}{1200} = 13 + \dfrac{1669}{1200}$

   $= 14\dfrac{469}{1200}$ or $\dfrac{17269}{1200}$

8. Note that $\dfrac{2}{3} = \dfrac{4}{6}$, $\dfrac{3}{4} = \dfrac{6}{8}$, and $\dfrac{5}{6} = \dfrac{10}{12}$.

   Each term can be written as $\dfrac{i+2}{i+4}$, where $i$ is the term number, and there are $n = 8$ terms.

   $\dfrac{3}{5} + \dfrac{2}{3} + \dfrac{5}{7} + \dfrac{3}{4} + \ldots + \dfrac{5}{6} = \sum_{i=1}^{8}\dfrac{i+2}{i+4}$

9. $a_n = a + (n-1)d$
   $= 6 + (n-1)(10)$
   $= 6 + 10n - 10$
   $= 10n - 4$
   $a_1 = 10 \cdot 1 - 4 = 6$
   $a_2 = 10 \cdot 2 - 4 = 16$
   $a_3 = 10 \cdot 3 - 4 = 26$
   $a_4 = 10 \cdot 4 - 4 = 36$
   $a_5 = 10 \cdot 5 - 4 = 46$

   The first five terms of the sequence are 6, 16, 26, 36, and 46.

10. $a_n = a + (n-1)d$
$= 0 + (n-1)(-4)$
$= -4n + 4$
$= 4 - 4n$
$a_1 = 4 - 4 \cdot 1 = 0$
$a_2 = 4 - 4 \cdot 2 = -4$
$a_3 = 4 - 4 \cdot 3 = -8$
$a_4 = 4 - 4 \cdot 4 = -12$
$a_5 = 4 - 4 \cdot 5 = -16$
The first five terms of the sequence are $0$, $-4$, $-8$, $-12$, and $-16$.

11. $a_n = a \cdot r^{n-1}$
$= 10 \cdot 2^{n-1}$
$a_1 = 10 \cdot 2^{1-1} = 10$
$a_2 = 10 \cdot 2^{2-1} = 20$
$a_3 = 10 \cdot 2^{3-1} = 40$
$a_4 = 10 \cdot 2^{4-1} = 80$
$a_5 = 10 \cdot 2^{5-1} = 160$
The first five terms of the sequence are $10$, $20$, $40$, $80$, and $160$.

12. $a_n = a \cdot r^{n-1}$
$a_3 = a \cdot (-3)^{3-1} = 9$
$a(-3)^2 = 9$
$9a = 9$
$a = 1$
$a_n = 1 \cdot (-3)^{n-1}$
$= (-3)^{n-1}$
$a_1 = (-3)^{1-1} = 1$
$a_2 = (-3)^{2-1} = -3$
$a_3 = (-3)^{3-1} = 9$
$a_4 = (-3)^{4-1} = -27$
$a_5 = (-3)^{5-1} = 81$
The first five terms of the sequence are $1$, $-3$, $9$, $-27$, and $81$.

13. The terms form an arithmetic sequence with $n = 20$ terms, common difference $d = 4$, and first term $a = -2$.
$S_n = \dfrac{n}{2}[2a + (n-1)d]$
$S_{20} = \dfrac{20}{2}[2(-2) + (20-1)(4)]$
$= 10(72)$
$= 720$

14. The terms form a geometric sequence with $n = 12$ terms, common ratio $r = -3$, and first term $a = \dfrac{1}{9}$.
$S_n = a \cdot \dfrac{1 - r^n}{1 - r}$
$S_{12} = \dfrac{1}{9} \cdot \dfrac{1 - (-3)^{12}}{1 - (-3)}$
$= \dfrac{1}{9} \cdot \dfrac{1 - 3^{12}}{1 + 3}$
$= -\dfrac{132,860}{9}$

15. The terms form an infinite geometric series with first term $a = 216$ and common ratio $r = \dfrac{1}{3}$.
$S_\infty = \dfrac{a}{1-r} = \dfrac{216}{1 - \left(\dfrac{1}{3}\right)} = \dfrac{216}{2/3} = 216 \cdot \dfrac{3}{2} = 324$

16. $\dfrac{15!}{8!7!} = \dfrac{15 \cdot 14 \cdot 13 \cdot 12 \cdot 11 \cdot 10 \cdot 9 \cdot 8!}{8! \cdot 7 \cdot 6 \cdot 5 \cdot 4 \cdot 3 \cdot 2 \cdot 1}$
$= \dfrac{15 \cdot 14 \cdot 13 \cdot 12 \cdot 11 \cdot 10 \cdot 9}{7 \cdot 6 \cdot 5 \cdot 4 \cdot 3 \cdot 2 \cdot 1}$
$= 6435$

17. $\binom{12}{5} = \dfrac{12!}{5!7!} = \dfrac{12 \cdot 11 \cdot 10 \cdot 9 \cdot 8 \cdot 7!}{5 \cdot 4 \cdot 3 \cdot 2 \cdot 1 \cdot 7!}$
$= \dfrac{12 \cdot 11 \cdot 10 \cdot 9 \cdot 8}{5 \cdot 4 \cdot 3 \cdot 2 \cdot 1}$
$= 792$

## Chapter 11: Sequences, Series, and the Binomial Theorem

18. $(5m-2)^4 = \binom{4}{0}(5m)^4 + \binom{4}{1}(5m)^3(-2) + \binom{4}{2}(5m)^2(-2)^2 + \binom{4}{3}(5m)(-2)^3 + \binom{4}{4}(-2)^4$
$= 625m^4 + 4 \cdot 125m^3 \cdot (-2) + 6 \cdot 25m^2 \cdot 4 + 4 \cdot 5m \cdot (-8) + 16$
$= 625m^4 - 1000m^3 + 600m^2 - 160m + 16$

19. The average tuition values form a geometric sequence with first term $a = 4694$ and common ratio $r = 1.14$. To determine the average tuition and fees for the 2023-2024 school year, we need the 21$^{st}$ term of the sequence.
$a_{21} = a \cdot r^{21-1}$
$= 4694 \cdot 1.14^{20}$
$\approx 64,512$

If the percent increase continues, the average tuition and fees for in-state students would be about \$64,512 during the 2023-2024 academic year.

20. $1639 - 1631 = 8$
$1761 - 1639 = 122$
$1769 - 1761 = 8$
$1874 - 1769 = 105$
$1882 - 1874 = 8$
$2004 - 1882 = 122$
Following this pattern, the next three Venus transits should occur after intervals of 8, 105, and 8 years.
$2004 + 8 = 2012$
$2012 + 105 = 2117$
$2117 + 8 = 2125$
The next three Venus transits should occur in 2012, 2117, and 2125.

**Cumulative Review Chapters R-11**

1. $\frac{1}{2}(x+2) = \frac{5}{4}(x-3y)$
$4 \cdot \frac{1}{2}(x+2) = 4 \cdot \frac{5}{4}(x-3y)$
$2(x+2) = 5(x-3y)$
$2x + 4 = 5x - 15y$
$15y = 3x - 4$
$y = \frac{3x-4}{15}$ or $y = \frac{1}{5}x - \frac{4}{15}$

2. $f(x) = x^2 - x + 7$
$f(2) = (2)^2 - (2) + 7 = 4 - 2 + 7 = 9$
$f(-3) = (-3)^2 - (-3) + 7 = 9 + 3 + 7 = 19$

## Cumulative Review Chapters R-11

3. $\frac{1}{2}x - 2 = \frac{1}{3}(x+1) + 3$
   
   $6\left(\frac{1}{2}x - 2\right) = 6\left(\frac{1}{3}(x+1) + 3\right)$
   
   $3x - 12 = 2(x+1) + 18$
   
   $3x - 12 = 2x + 2 + 18$
   
   $3x - 2x = 20 + 12$
   
   $x = 32$
   
   The solution set is $\{32\}$.

4. $5x^2 - 3x = 2$
   
   $5x^2 - 3x - 2 = 0$
   
   $(5x+2)(x-1) = 0$
   
   $5x + 2 = 0$ or $x - 1 = 0$
   
   $5x = -2 \qquad x = 1$
   
   $x = -\frac{2}{5}$
   
   The solution set is $\left\{-\frac{2}{5}, 1\right\}$.

5. $3x^2 + 7x - 2 = 0$
   
   $a = 3, b = 7, c = -2$
   
   $x = \frac{-b \pm \sqrt{b^2 - 4ac}}{2a}$
   
   $= \frac{-7 \pm \sqrt{7^2 - 4(3)(-2)}}{2(3)}$
   
   $= \frac{-7 \pm \sqrt{49 + 24}}{6}$
   
   $= \frac{-7 \pm \sqrt{73}}{6}$
   
   The solution set is $\left\{\frac{-7 - \sqrt{73}}{6}, \frac{-7 + \sqrt{73}}{6}\right\}$.

6. $\sqrt{2x+1} - 3 = 8$
   
   $\sqrt{2x+1} = 11$
   
   $\left(\sqrt{2x+1}\right)^2 = 11^2$
   
   $2x + 1 = 121$
   
   $2x = 120$
   
   $x = 60$
   
   The solution set is $\{60\}$.

7. $4^{x+1} = 8^{2x-3}$
   
   $\left(2^2\right)^{x+1} = \left(2^3\right)^{2x-3}$
   
   $2^{2x+2} = 2^{6x-9}$
   
   Therefore, we get
   
   $2x + 2 = 6x - 9$
   
   $-4x = -11$
   
   $x = \frac{11}{4}$
   
   The solution set is $\left\{\frac{11}{4}\right\}$.

8. $x^2(2x+1) + 40 = (x^2 - 8)(x - 5)$
   
   $2x^3 + x^2 + 40 = x^3 - 8x - 5x^2 + 40$
   
   $x^3 + 6x^2 + 8x = 0$
   
   $x(x^2 + 6x + 8) = 0$
   
   $x(x+4)(x+2) = 0$
   
   $x = 0$ or $x + 4 = 0$ or $x + 2 = 0$
   
   $\qquad\qquad x = -4 \qquad x = -2$
   
   The solution set is $\{-4, -2, 0\}$.

9. $\frac{2}{3}x + 1 > \frac{1}{4}x - \frac{3}{2}$
   
   $12\left(\frac{2}{3}x + 1\right) > 12\left(\frac{1}{4}x - \frac{3}{2}\right)$
   
   $8x + 12 > 3x - 18$
   
   $5x > -30$
   
   $\frac{5x}{5} > \frac{-30}{5}$
   
   $x > -6$
   
   Interval: $(-6, \infty)$

10. $3x^2 - 2x \leq 3 - 10x$
    
    $3x^2 + 8x - 3 \leq 0$
    
    $(3x - 1)(x + 3) \leq 0$
    
    $3x - 1 = 0 \qquad x + 3 = 0$
    
    $3x = 1 \qquad\quad x = -3$
    
    $x = \frac{1}{3}$

## Chapter 11: Sequences, Series, and the Binomial Theorem

| Interval | $(-\infty,-3)$ | $-3$ | $\left(-3,\frac{1}{3}\right)$ | $\frac{1}{3}$ | $\left(\frac{1}{3},\infty\right)$ |
|---|---|---|---|---|---|
| $(3x-1)$ | $---$ | $-$ | $---$ | $0$ | $+++$ |
| $(x+3)$ | $---$ | $0$ | $+++$ | $+$ | $+++$ |
| $(3x-1)(x+3)$ | $+++$ | $0$ | $---$ | $0$ | $+++$ |

The inequality is non-strict, so $\frac{1}{3}$ and $-3$ are part of the solution. Now, $(3x-1)(x+3)$ is less than zero where the product is negative. The solution is $\left\{x \mid -3 \le x \le \frac{1}{3}\right\}$ or, using interval notation, $\left[-3,\frac{1}{3}\right]$.

11. $2x^2 - 5x - 18$
    $ac = 2(-18) = -36$
    We are looking for two factors of $-36$ whose sum is $-5$. Since the product is negative, the factors will have opposite signs. The sum is negative so the factor with the largest absolute value will be negative.

    | factor 1 | factor 2 | sum |
    |---|---|---|
    | 1 | $-36$ | $-35$ |
    | 2 | $-18$ | $-16$ |
    | 3 | $-12$ | $-9$ |
    | 4 | $-9$ | $-5$ ← okay |

    $2x^2 - 5x - 18 = 2x^2 + 4x - 9x - 18$
    $= 2x(x+2) - 9(x+2)$
    $= (x+2)(2x-9)$

12. $6x^3 - 3x^2 + 4x - 2 = 3x^2(2x-1) + 2(2x-1)$
    $= (2x-1)(3x^2+2)$

13. $(5x-3)(4x^2-2x+1)$
    $= 20x^3 - 10x^2 + 5x - 12x^2 + 6x - 3$
    $= 20x^3 - 22x^2 + 11x - 3$

14. $\dfrac{x}{x+4} - \dfrac{3}{x-1} = \dfrac{x(x-1)}{(x+4)(x-1)} - \dfrac{3(x+4)}{(x+4)(x-1)}$
    $= \dfrac{x(x-1) - 3(x+4)}{(x+4)(x-1)}$
    $= \dfrac{x^2 - x - 3x - 12}{(x+4)(x-1)}$
    $= \dfrac{x^2 - 4x - 12}{(x+4)(x-1)}$
    $= \dfrac{(x-6)(x+2)}{(x+4)(x-1)}$

15. $\dfrac{3-i}{2+i} = \dfrac{(3-i)(2-i)}{(2+i)(2-i)}$
    $= \dfrac{6 - 2i - 3i + i^2}{4 - i^2}$
    $= \dfrac{6 - 5i - 1}{4 - (-1)}$
    $= \dfrac{5 - 5i}{5}$
    $= 1 - i$

16. Because of the two radicals, we need
    $x - 15 \ge 0$ and $2x - 5 \ge 0$
    $x \ge 15$ $\quad\quad\quad 2x \ge 5$
    $\quad\quad\quad\quad\quad\quad x \ge \dfrac{5}{2}$

    Therefore, the domain is $\{x \mid x \ge 15\}$ or $[15, \infty)$.

17. Begin by finding the slope of the line connecting the two points.
    $m = \dfrac{y_2 - y_1}{x_2 - x_1} = \dfrac{4 - (-3)}{1 - 2} = \dfrac{7}{-1} = -7$
    $y - y_1 = m(x - x_1)$
    $y - (-3) = -7(x - 2)$
    $y + 3 = -7x + 14$
    $y = -7x + 11$

18. $2x + 3y = 5$
    $x - 2y = 6$
    We can solve this system by using elimination. Multiply the second equation by $-2$ and add the two equations.

$2x+3y=5$
$-2x+4y=-12$
$\overline{\phantom{-2x+}7y=-7}$
$\phantom{-2x+7}y=-1$

Let $y=-1$ in the second equation to find $x$.
$x-2(-1)=6$
$x+2=6$
$x=4$

The ordered pair $(4,-1)$ is the solution to the system.

19. $f(x)=2x^2-8x-3$

The graph of $f$ will be a parabola. The parabola will open up because $a=2>0$.
The x-coordinate of the vertex is
$x=-\dfrac{b}{2a}=-\dfrac{(-8)}{2(2)}=2$

The y-coordinate of the vertex is
$f(2)=2(2)^2-8(2)-3=-11$

The vertex is $(2,-11)$ and the axis of symmetry is $x=2$.
$f(0)=2(0)^2-8(0)-3=-3$
The y-intercept is $-3$.

To find the x-intercepts we solve the equation $2x^2-8x-3=0$.

$x=\dfrac{-(-8)\pm\sqrt{(-8)^2-4(2)(-3)}}{2(2)}$

$=\dfrac{8\pm\sqrt{88}}{4}$

$=\dfrac{8\pm2\sqrt{22}}{4}$

$=\dfrac{4\pm\sqrt{22}}{2}$

The graph will have x-intercepts of $x\approx-0.35$ and $x\approx 4.35$.

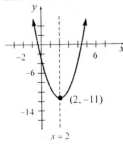

20. $(x-h)^2+(y-k)^2=r^2$
$(x-4)^2+(y-(-3))^2=6^2$
$(x-4)^2+(y+3)^2=36$

To sketch the graph of the circle, we can plot four additional points that are $r=6$ units above, below, left, and right of the center. These points are $(4,3),(4,-9),(-2,-3)$, and $(10,-3)$.

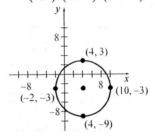

21. Begin by writing the equation in standard form.
$\dfrac{4x^2}{64}+\dfrac{y^2}{64}=\dfrac{64}{64}$

$\dfrac{x^2}{16}+\dfrac{y^2}{64}=1$

The larger number, 64, is in the denominator of the $y^2$-term. This means that the major axis is the y-axis and the equation of the ellipse is of the form $\dfrac{x^2}{b^2}+\dfrac{y^2}{a^2}=1$ so that $a^2=64$ and $b^2=16$.

The center of the ellipse is the origin, $(0,0)$.
$c^2=a^2-b^2=64-16=48$, so that $c=\pm 4\sqrt{3}$.
The foci are $(0,-4\sqrt{3})$ and $(0,4\sqrt{3})$.

$\dfrac{x^2}{16}+\dfrac{0^2}{64}=1 \qquad \dfrac{0^2}{16}+\dfrac{y^2}{64}=1$

$\dfrac{x^2}{16}=1 \qquad \dfrac{y^2}{64}=1$

$x^2=16 \qquad y^2=64$

$x=\pm 4 \qquad y=\pm 8$

The intercepts are $(-4,0),(4,0),(0,-8)$, and $(0,8)$.

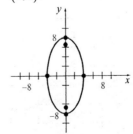

## Chapter 11: Sequences, Series, and the Binomial Theorem

22. $S_n = \dfrac{n}{2}\left[2a+(n-1)d\right]$

    $S_{20} = \dfrac{20}{2}\left[2(-47)+(20-1)(12)\right]$

    $= 10(134)$

    $= 1340$

23. For this infinite geometric series we have $a = 2$ and $r = \dfrac{3/2}{2} = \dfrac{3}{4}$.

    $S_\infty = \dfrac{a}{1-r} = \dfrac{2}{1-\frac{3}{4}} = \dfrac{2}{\frac{1}{4}} = 8$

24. 

    | Machine | # of lots | time (hrs) | rate |
    |---|---|---|---|
    | Robomower | 1 | 5 | $\dfrac{1}{5}$ |
    | Mowbot | 1 | 6 | $\dfrac{1}{6}$ |
    | Together | 1 | $t$ | $\dfrac{1}{t}$ |

    We can't add the times, but we can add the rates.

    $\dfrac{1}{5} + \dfrac{1}{6} = \dfrac{1}{t}$

    $\dfrac{11}{30} = \dfrac{1}{t}$

    $t = \dfrac{30}{11} \approx 2.73$

    It would take about 2.73 hours to cut the lot if both machines worked together.

25. Let $x$ = metric tons of pure aluminum.

    We can solve the problem by writing an equation for the total amount of manganese.

    $\text{Mn}_{tot} = \text{Mn}_{init} + \text{Mn}_{added}$

    $(\%)(\text{wt.})_{tot} = (\%)(\text{wt.})_{init} + (\%)(\text{wt.})_{added}$

    $(1.2)(100+x) = (2.5)(100) + (0)(x)$

    $120 + 1.2x = 250$

    $1.2x = 130$

    $x = \dfrac{130}{1.2}$

    $x = \dfrac{325}{3}$ or $108\dfrac{1}{3}$

    $108\dfrac{1}{3}$ metric tons of pure aluminum must be added.

# Appendix A

**Preparing for the Library of Functions**

1. $y = x^2$

| $x$ | $y = x^2$ | $(x, y)$ |
|---|---|---|
| $-3$ | $y = (-3)^2 = 9$ | $(-3, 9)$ |
| $-2$ | $y = (-2)^2 = 4$ | $(-2, 4)$ |
| $0$ | $y = (0)^2 = 0$ | $(0, 0)$ |
| $2$ | $y = (2)^2 = 4$ | $(2, 4)$ |
| $3$ | $y = (3)^2 = 9$ | $(3, 9)$ |

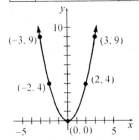

2. $y = x^3$

| $x$ | $y = x^3$ | $(x, y)$ |
|---|---|---|
| $-2$ | $y = (-2)^3 = -8$ | $(-2, -8)$ |
| $-1$ | $y = (-1)^3 = -1$ | $(-1, -1)$ |
| $0$ | $y = (0)^3 = 0$ | $(0, 0)$ |
| $1$ | $y = (1)^3 = 1$ | $(1, 1)$ |
| $2$ | $y = (2)^3 = 8$ | $(2, 8)$ |

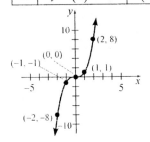

3. $x = y^2$

| $y$ | $x = y^2$ | $(x, y)$ |
|---|---|---|
| $-2$ | $x = (-2)^2 = 4$ | $(4, -2)$ |
| $-1$ | $x = (-1)^2 = 1$ | $(1, -1)$ |
| $0$ | $x = (0)^2 = 0$ | $(0, 0)$ |
| $1$ | $x = (1)^2 = 1$ | $(1, 1)$ |
| $2$ | $x = (2)^2 = 4$ | $(4, 2)$ |

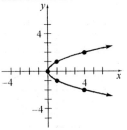

**Appendix A Exercises**

1. constant function

2. square root function

3. False; the domain of the square function is all real numbers, but the range is the set of nonnegative real numbers.

4. True

5. Square function, (c).

6. Constant function, (a)

7. Square root function, (e)

8. Absolute value function, (g)

9. Linear function, (b)

10. Cube function, (d)

11. Reciprocal function, (f)

12. Cube root function, (h)

*Appendix A*

**13.** $f(x) = x^2$

| $x$ | $y = f(x) = x^2$ | $(x, y)$ |
|---|---|---|
| $-3$ | $y = (-3)^2 = 9$ | $(-3, 9)$ |
| $-2$ | $y = (-2)^2 = 4$ | $(-2, 4)$ |
| $0$ | $y = (0)^2 = 0$ | $(0, 0)$ |
| $2$ | $y = (2)^2 = 4$ | $(2, 4)$ |
| $3$ | $y = (3)^2 = 9$ | $(3, 9)$ |

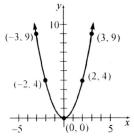

**14.** $f(x) = x^3$

| $x$ | $y = f(x) = x^3$ | $(x, y)$ |
|---|---|---|
| $-2$ | $y = (-2)^3 = -8$ | $(-2, -8)$ |
| $-1$ | $y = (-1)^3 = -1$ | $(-1, -1)$ |
| $0$ | $y = (0)^3 = 0$ | $(0, 0)$ |
| $1$ | $y = (1)^3 = 1$ | $(1, 1)$ |
| $2$ | $y = (2)^3 = 8$ | $(2, 8)$ |

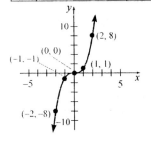

**15.** $f(x) = \sqrt{x}$

| $x$ | $y = f(x) = \sqrt{x}$ | $(x, y)$ |
|---|---|---|
| $0$ | $y = \sqrt{0} = 0$ | $(0, 0)$ |
| $1$ | $y = \sqrt{1} = 1$ | $(1, 1)$ |
| $4$ | $y = \sqrt{4} = 2$ | $(4, 2)$ |
| $9$ | $y = \sqrt{9} = 3$ | $(9, 3)$ |

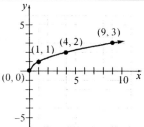

**16.** $f(x) = \sqrt[3]{x}$

| $x$ | $y = f(x) = \sqrt[3]{x}$ | $(x, y)$ |
|---|---|---|
| $-8$ | $y = \sqrt[3]{-8} = -2$ | $(-8, -2)$ |
| $-1$ | $y = \sqrt[3]{-1} = -1$ | $(-1, -1)$ |
| $0$ | $y = \sqrt[3]{0} = 0$ | $(0, 0)$ |
| $1$ | $y = \sqrt[3]{1} = 1$ | $(1, 1)$ |
| $8$ | $y = \sqrt[3]{8} = 2$ | $(8, 2)$ |

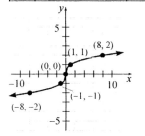

17. $f(x) = \dfrac{1}{x}$

| $x$ | $y = f(x) = \dfrac{1}{x}$ | $(x, y)$ |
|---|---|---|
| $-5$ | $y = \dfrac{1}{-5} = -\dfrac{1}{5}$ | $\left(-5, -\dfrac{1}{5}\right)$ |
| $-1$ | $y = \dfrac{1}{-1} = -1$ | $(-1, -1)$ |
| $-\dfrac{1}{2}$ | $y = \dfrac{1}{-1/2} = -2$ | $\left(-\dfrac{1}{2}, -2\right)$ |
| $\dfrac{1}{2}$ | $y = \dfrac{1}{1/2} = 2$ | $\left(\dfrac{1}{2}, 2\right)$ |
| $1$ | $y = \dfrac{1}{1} = 1$ | $(1, 1)$ |
| $5$ | $y = \dfrac{1}{5}$ | $\left(5, \dfrac{1}{5}\right)$ |

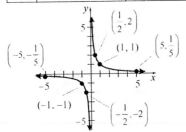

18. $f(x) = |x|$

| $x$ | $y = f(x) = |x|$ | $(x, y)$ |
|---|---|---|
| $-5$ | $y = |-5| = 5$ | $(-5, 5)$ |
| $-3$ | $y = |-3| = 3$ | $(-3, 3)$ |
| $0$ | $y = |0| = 0$ | $(0, 0)$ |
| $3$ | $y = |3| = 3$ | $(3, 3)$ |
| $5$ | $y = |5| = 5$ | $(5, 5)$ |

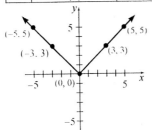